Student's Solutions Manual
to accompany Jon Rogawski's

Multivariable CALCULUS

GREGORY P. DRESDEN
Washington and Lee University

With Chapter 11 contributed by
BRIAN BRADIE
Christopher Newport University

W. H. Freeman and Company
New York

ISBN-13: 978-0-7167-9880-4
ISBN-10: 0-7167-9880-8

Printed in the United States of America

Third printing

W. H. Freeman and Company, 41 Madison Avenue, New York, NY 10010
Houndmills, Basingstoke RG21 6XS, England
www.whfreeman.com

CONTENTS

11 | INFINITE SERIES

11.1 Sequences (ET Section 10.1)

Preliminary Questions

1. What is a_4 for the sequence $a_n = n^2 - n$?

SOLUTION Substituting $n = 4$ in the expression for a_n gives

$$a_4 = 4^2 - 4 = 12.$$

2. Which of the following sequences converge to zero?

(a) $\dfrac{n^2}{n^2+1}$ **(b)** 2^n **(c)** $\left(\dfrac{-1}{2}\right)^n$

SOLUTION

(a) This sequence does not converge to zero:

$$\lim_{n\to\infty} \frac{n^2}{n^2+1} = \lim_{x\to\infty} \frac{x^2}{x^2+1} = \lim_{x\to\infty} \frac{1}{1+\frac{1}{x^2}} = \frac{1}{1+0} = 1.$$

(b) This sequence does not converge to zero: this is a geometric sequence with $r = 2 > 1$; hence, the sequence diverges to ∞.

(c) Recall that if $|a_n|$ converges to 0, then a_n must also converge to zero. Here,

$$\left|\left(-\frac{1}{2}\right)^n\right| = \left(\frac{1}{2}\right)^n,$$

which is a geometric sequence with $0 < r < 1$; hence, $(\frac{1}{2})^n$ converges to zero. It therefore follows that $(-\frac{1}{2})^n$ converges to zero.

3. Let a_n be the nth decimal approximation to $\sqrt{2}$. That is, $a_1 = 1$, $a_2 = 1.4$, $a_3 = 1.41$, etc. What is $\lim_{n\to\infty} a_n$?

SOLUTION $\lim_{n\to\infty} a_n = \sqrt{2}$.

4. Which sequence is defined recursively?

(a) $a_n = \sqrt{2 + n^{-1}}$ **(b)** $b_n = \sqrt{4 + b_{n-1}}$

SOLUTION

(a) a_n can be computed directly, since it depends on n only and not on preceding terms. Therefore a_n is defined explicitly and not recursively.

(b) b_n is computed in terms of the preceding term b_{n-1}, hence the sequence $\{b_n\}$ is defined recursively.

5. Theorem 5 says that every convergent sequence is bounded. Which of the following statements follow from Theorem 5 and which are false? If false, give a counterexample.

(a) If $\{a_n\}$ is bounded, then it converges.

(b) If $\{a_n\}$ is not bounded, then it diverges.

(c) If $\{a_n\}$ diverges, then it is not bounded.

SOLUTION

(a) This statement is false. The sequence $a_n = \cos \pi n$ is bounded since $-1 \leq \cos \pi n \leq 1$ for all n, but it does not converge: since $a_n = \cos n\pi = (-1)^n$, the terms assume the two values 1 and -1 alternately, hence they do not approach one value.

(b) By Theorem 5, a converging sequence must be bounded. Therefore, if a sequence is not bounded, it certainly does not converge.

(c) The statement is false. The sequence $a_n = (-1)^n$ is bounded, but it does not approach one limit.

Exercises

1. Match the sequence with the general term:

$a_1, a_2, a_3, a_4, \ldots$	**General term**
(a) $\frac{1}{2}, \frac{2}{3}, \frac{3}{4}, \frac{4}{5}, \ldots$	(i) $\cos \pi n$
(b) $-1, 1, -1, 1, \ldots$	(ii) $\frac{n!}{2^n}$
(c) $1, -1, 1, -1, \ldots$	(iii) $(-1)^{n+1}$
(d) $\frac{1}{2}, \frac{2}{4}, \frac{6}{8}, \frac{24}{16} \ldots$	(iv) $\frac{n}{n+1}$

SOLUTION

(a) The numerator of each term is the same as the index of the term, and the denominator is one more than the numerator; hence $a_n = \frac{n}{n+1}$, $n = 1, 2, 3, \ldots$.

(b) The terms of this sequence are alternating between -1 and 1 so that the positive terms are in the even places. Since $\cos \pi n = 1$ for even n and $\cos \pi n = -1$ for odd n, we have $a_n = \cos \pi n$, $n = 1, 2, \ldots$.

(c) The terms a_n are 1 for odd n and -1 for even n. Hence, $a_n = (-1)^{n+1}$, $n = 1, 2, \ldots$

(d) The numerator of each term is $n!$, and the denominator is 2^n; hence, $a_n = \frac{n!}{2^n}$, $n = 1, 2, 3, \ldots$.

In Exercises 3–10, calculate the first four terms of the following sequences, starting with $n = 1$.

3. $c_n = \dfrac{2^n}{n!}$

SOLUTION Setting $n = 1, 2, 3, 4$ in the formula for c_n gives

$$c_1 = \frac{2^1}{1!} = \frac{2}{1} = 2, \qquad c_2 = \frac{2^2}{2!} = \frac{4}{2} = 2,$$

$$c_3 = \frac{2^3}{3!} = \frac{8}{6} = \frac{4}{3}, \qquad c_4 = \frac{2^4}{4!} = \frac{16}{24} = \frac{2}{3}.$$

5. $a_1 = 3, \quad a_{n+1} = 1 + a_n^2$

SOLUTION For $n = 1, 2, 3$ we have:

$$a_2 = a_{1+1} = 1 + a_1^2 = 1 + 3^2 = 10;$$

$$a_3 = a_{2+1} = 1 + a_2^2 = 1 + 10^2 = 101;$$

$$a_4 = a_{3+1} = 1 + a_3^2 = 1 + 101^2 = 10{,}202.$$

The first four terms of $\{a_n\}$ are 3, 10, 101, 10,202.

7. $c_n = 1 + \dfrac{1}{2} + \dfrac{1}{3} + \cdots + \dfrac{1}{n}$

SOLUTION

$$c_1 = 1;$$

$$c_2 = 1 + \frac{1}{2} = \frac{3}{2};$$

$$c_3 = 1 + \frac{1}{2} + \frac{1}{3} = \frac{3}{2} + \frac{1}{3} = \frac{11}{6};$$

$$c_4 = 1 + \frac{1}{2} + \frac{1}{3} + \frac{1}{4} = \frac{11}{6} + \frac{1}{4} = \frac{25}{12}.$$

9. $b_1 = 2, \quad b_2 = 5, \quad b_n = b_{n-1} + 2b_{n-2}$

SOLUTION We need to find b_3 and b_4. Setting $n = 3$ and $n = 4$ and using the given values for b_1 and b_2 we obtain:

$$b_3 = b_{3-1} + 2b_{3-2} = b_2 + 2b_1 = 5 + 2 \cdot 2 = 9;$$

$$b_4 = b_{4-1} + 2b_{4-2} = b_3 + 2b_2 = 9 + 2 \cdot 5 = 19.$$

The first four terms of the sequence $\{b_n\}$ are 2, 5, 9, 19.

11. Find a formula for the nth term of the following sequence:

(a) $\frac{1}{1}, \frac{-1}{8}, \frac{1}{27}, \dots$

(b) $\frac{2}{6}, \frac{3}{7}, \frac{4}{8}, \dots$

SOLUTION

(a) The denominators are the third powers of the positive integers starting with $n = 1$. Also, the sign of the terms is alternating with the sign of the first term being positive. Thus,

$$a_1 = \frac{1}{1^3} = \frac{(-1)^{1+1}}{1^3}; \quad a_2 = -\frac{1}{2^3} = \frac{(-1)^{2+1}}{2^3}; \quad a_3 = \frac{1}{3^3} = \frac{(-1)^{3+1}}{3^3}.$$

This rule leads to the following formula for the nth term:

$$a_n = \frac{(-1)^{n+1}}{n^3}.$$

(b) Assuming a starting index of $n = 1$, we see that each numerator is one more than the index and the denominator is four more than the numerator. Thus, the general term a_n is

$$a_n = \frac{n+1}{n+5}.$$

In Exercises 13–26, use Theorem 1 to determine the limit of the sequence or state that the sequence diverges.

13. $a_n = 4$

SOLUTION We have $a_n = f(n)$ where $f(x) = 4$; thus,

$$\lim_{n\to\infty} a_n = \lim_{x\to\infty} f(x) = \lim_{x\to\infty} 4 = 4.$$

15. $a_n = 5 - \frac{9}{n^2}$

SOLUTION We have $a_n = f(n)$ where $f(x) = 5 - \frac{9}{x^2}$; thus,

$$\lim_{n\to\infty}\left(5 - \frac{9}{n^2}\right) = \lim_{x\to\infty}\left(5 - \frac{9}{x^2}\right) = 5 - 0 = 5.$$

17. $c_n = -2^{-n}$

SOLUTION We have $c_n = f(n)$ where $f(x) = -2^{-x}$; thus,

$$\lim_{n\to\infty}\left(-2^{-n}\right) = \lim_{x\to\infty} -2^{-x} = \lim_{x\to\infty} -\frac{1}{2^x} = 0.$$

19. $z_n = \left(\frac{1}{3}\right)^n$

SOLUTION We have $z_n = f(n)$ where $f(x) = \left(\frac{1}{3}\right)^x$; thus,

$$\lim_{n\to\infty}\left(\frac{1}{3}\right)^n = \lim_{x\to\infty}\left(\frac{1}{3}\right)^x = 0.$$

21. $a_n = \frac{(-1)^n n^2 + n}{4n^2 + 1}$

SOLUTION We examine the terms with n odd and n even separately. With n odd, the terms take the form

$$\frac{-n^2 + n}{4n^2 + 1} \quad \text{and} \quad \lim_{n\to\infty}\frac{-n^2 + n}{4n^2 + 1} = \lim_{x\to\infty}\frac{-x^2 + x}{4x^2 + 1} = -\frac{1}{4}.$$

On the other hand, with n even, the terms take the form

$$\frac{n^2 + n}{4n^2 + 1} \quad \text{and} \quad \lim_{n\to\infty}\frac{n^2 + n}{4n^2 + 1} = \lim_{x\to\infty}\frac{x^2 + x}{4x^2 + 1} = \frac{1}{4}.$$

Because these two limits are different, the general term of the given sequence does not approach one value. Hence, the sequence diverges.

23. $a_n = \dfrac{n}{\sqrt{n^3+1}}$

SOLUTION We have $a_n = f(n)$ where $f(x) = \dfrac{x}{\sqrt{x^3+1}}$; thus,

$$\lim_{n\to\infty} \frac{n}{\sqrt{n^3+1}} = \lim_{x\to\infty} \frac{x}{\sqrt{x^3+1}} = \lim_{x\to\infty} \frac{\frac{x}{x^{3/2}}}{\frac{\sqrt{x^3+1}}{x^{3/2}}} = \lim_{x\to\infty} \frac{\frac{1}{\sqrt{x}}}{\sqrt{1+\frac{1}{x^3}}} = \frac{0}{\sqrt{1+0}} = \frac{0}{1} = 0.$$

25. $a_n = \cos \pi n$

SOLUTION The terms in the odd places are -1 and the terms in the even places are 1. Therefore, the general term of the sequence does not approach one limit. Hence, the sequence diverges.

27. Let $a_n = \dfrac{n}{n+1}$. Find a number M such that:
(a) $|a_n - 1| \le 0.001$ for $n \ge M$.
(b) $|a_n - 1| \le 0.00001$ for $n \ge M$.
Then use the limit definition to prove that $\lim_{n\to\infty} a_n = 1$.

SOLUTION
(a) We have

$$|a_n - 1| = \left|\frac{n}{n+1} - 1\right| = \left|\frac{n-(n+1)}{n+1}\right| = \left|\frac{-1}{n+1}\right| = \frac{1}{n+1}.$$

Therefore $|a_n - 1| \le 0.001$ provided $\frac{1}{n+1} \le 0.001$, that is, $n \ge 999$. It follows that we can take $M = 999$.

(b) By part (a), $|a_n - 1| \le 0.00001$ provided $\frac{1}{n+1} \le 0.00001$, that is, $n \ge 99999$. It follows that we can take $M = 99999$.

We now prove formally that $\lim_{n\to\infty} a_n = 1$. Using part (a), we know that

$$|a_n - 1| = \frac{1}{n+1} < \epsilon,$$

provided $n > \frac{1}{\epsilon} - 1$. Thus, Let $\epsilon > 0$ and take $M = \frac{1}{\epsilon} - 1$. Then, for $n > M$, we have

$$|a_n - 1| = \frac{1}{n+1} < \frac{1}{M+1} = \epsilon.$$

29. Use the limit definition to prove that $\lim_{n\to\infty} n^{-2} = 0$.

SOLUTION We see that

$$|n^{-2} - 0| = \left|\frac{1}{n^2}\right| = \frac{1}{n^2} < \epsilon$$

provided

$$n > \frac{1}{\sqrt{\epsilon}}.$$

Thus, let $\epsilon > 0$ and take $M = \frac{1}{\sqrt{\epsilon}}$. Then, for $n > M$, we have

$$|n^{-2} - 0| = \left|\frac{1}{n^2}\right| = \frac{1}{n^2} < \frac{1}{M^2} = \epsilon.$$

31. Find $\lim_{n\to\infty} 2^{1/n}$.

SOLUTION Because 2^x is a continuous function,

$$\lim_{n\to\infty} 2^{1/n} = \lim_{x\to\infty} 2^{1/x} = 2^{\lim_{x\to\infty}(1/x)} = 2^0 = 1.$$

33. Find $\lim_{n\to\infty} n^{1/n}$.

SOLUTION Let $a_n = n^{1/n}$. Take the natural logarithm of both sides of this expression to obtain

$$\ln a_n = \ln n^{1/n} = \frac{\ln n}{n}.$$

Thus,

$$\lim_{n\to\infty} (\ln a_n) = \lim_{n\to\infty} \frac{\ln n}{n} = \lim_{x\to\infty} \frac{\ln x}{x} = \lim_{x\to\infty} \frac{1}{x} = 0.$$

Because $f(x) = e^x$ is a continuous function, it follows that

$$\lim_{n\to\infty} a_n = \lim_{n\to\infty} e^{\ln a_n} = e^{\lim_{n\to\infty}(\ln a_n)} = e^0 = 1.$$

That is,

$$\lim_{n\to\infty} n^{1/n} = 1.$$

35. Find $\lim_{n\to\infty} \left(1 + \frac{1}{n}\right)^n$.

SOLUTION Let $a_n = \left(1 + \frac{1}{n}\right)^n$. Taking the natural logarithm of both sides of this expression yields

$$\ln a_n = \ln\left(1 + \frac{1}{n}\right)^n = n\ln\left(1 + \frac{1}{n}\right) = \frac{\ln\left(1 + \frac{1}{n}\right)}{\frac{1}{n}}.$$

Thus,

$$\lim_{n\to\infty} (\ln a_n) = \lim_{x\to\infty} \frac{\ln\left(1 + \frac{1}{x}\right)}{\frac{1}{x}} = \lim_{x\to\infty} \frac{\frac{d}{dx}\left(\ln\left(1 + \frac{1}{x}\right)\right)}{\frac{d}{dx}\left(\frac{1}{x}\right)} = \lim_{x\to\infty} \frac{\frac{1}{1+\frac{1}{x}} \cdot \left(-\frac{1}{x^2}\right)}{-\frac{1}{x^2}} = \lim_{x\to\infty} \frac{1}{1 + \frac{1}{x}} = \frac{1}{1+0} = 1.$$

Because $f(x) = e^x$ is a continuous function, it follows that

$$\lim_{n\to\infty} a_n = \lim_{n\to\infty} e^{\ln a_n} = e^{\lim_{n\to\infty}(\ln a_n)} = e^1 = e.$$

37. Use the Squeeze Theorem to find $\lim_{n\to\infty} a_n$, where
$a_n = \frac{1}{\sqrt{n^4 + n^8}}$ by proving that

$$\frac{1}{\sqrt{2n^4}} \le a_n \le \frac{1}{\sqrt{2n^2}}$$

SOLUTION For all $n > 1$ we have $n^4 < n^8$, so the quotient $\frac{1}{\sqrt{n^4+n^8}}$ is smaller than $\frac{1}{\sqrt{n^4+n^4}}$ and larger than $\frac{1}{\sqrt{n^8+n^8}}$. That is,

$$a_n < \frac{1}{\sqrt{n^4 + n^4}} = \frac{1}{\sqrt{n^4 \cdot 2}} = \frac{1}{\sqrt{2n^2}}; \text{ and}$$

$$a_n > \frac{1}{\sqrt{n^8 + n^8}} = \frac{1}{\sqrt{2n^8}} = \frac{1}{\sqrt{2n^4}}.$$

Now, since $\lim_{n\to\infty} \frac{1}{\sqrt{2n^4}} = \lim_{n\to\infty} \frac{1}{\sqrt{2n^2}} = 0$, the Squeeze Theorem for Sequences implies that $\lim_{n\to\infty} a_n = 0$.

39. Evaluate $\lim_{n\to\infty} n \sin\frac{1}{n}$.

SOLUTION We have $a_n = f(n)$ where $f(x) = x\sin\frac{1}{x}$. Thus,

$$\lim_{n\to\infty} n\sin\frac{1}{n} = \lim_{x\to\infty} x\sin\frac{1}{x} = \lim_{x\to\infty} \frac{\sin\frac{1}{x}}{\frac{1}{x}}.$$

The limit $\lim_{x\to 0} \frac{\sin x}{x} = 1$ implies that $\lim_{x\to\infty} \frac{\sin\frac{1}{x}}{\frac{1}{x}} = 1$. Hence,

$$\lim_{n\to\infty} n\sin\frac{1}{n} = 1.$$

41. Which statement is equivalent to the assertion $\lim_{n\to\infty} a_n = L$? Explain.

(a) For every $\epsilon > 0$, the interval $(L-\epsilon, L+\epsilon)$ contains at least one element of the sequence $\{a_n\}$.

(b) For every $\epsilon > 0$, the interval $(L-\epsilon, L+\epsilon)$ contains all but at most finitely many elements of the sequence $\{a_n\}$.

SOLUTION Statement (b) is equivalent to Definition 1 of the limit, since the assertion "$|a_n - L| < \epsilon$ for all $n > M$" means that $L-\epsilon < a_n < L+\epsilon$ for all $n > M$; that is, the interval $(L-\epsilon, L+\epsilon)$ contains all the elements a_n except (maybe) the finite number of elements $a_1, a_2, \ldots, a_M$.

Statement (a) is not equivalent to the assertion $\lim_{n\to\infty} a_n = L$. We show this, by considering the following sequence:

$$a_n = \begin{cases} \frac{1}{n} & \text{for odd } n \\ 1 + \frac{1}{n} & \text{for even } n \end{cases}$$

Clearly for every $\epsilon > 0$, the interval $(-\epsilon, \epsilon) = (L-\epsilon, L+\epsilon)$ for $L = 0$ contains at least one element of $\{a_n\}$, but the sequence diverges (rather than converges to $L = 0$). Since the terms in the odd places converge to 0 and the terms in the even places converge to 1. Hence, a_n does not approach one limit.

In Exercises 43–63, determine the limit of the sequence or show that the sequence diverges by using the appropriate Limit Laws or theorems.

43. $a_n = \dfrac{3n^2 + n + 2}{2n^2 - 3}$

SOLUTION

$$\lim_{n\to\infty} \frac{3n^2 + n + 2}{2n^2 - 3} = \lim_{x\to\infty} \frac{3x^2 + x + 2}{2x^2 - 3} = \frac{3}{2}.$$

45. $a_n = 3 + \left(-\frac{1}{2}\right)^n$

SOLUTION By the Limit Laws for Sequences we have:

$$\lim_{n\to\infty}\left(3 + \left(-\frac{1}{2}\right)^n\right) = \lim_{n\to\infty} 3 + \lim_{n\to\infty}\left(-\frac{1}{2}\right)^n = 3 + \lim_{n\to\infty}\left(-\frac{1}{2}\right)^n.$$

Now,

$$-\left(\frac{1}{2}\right)^n \le \left(-\frac{1}{2}\right)^n \le \left(\frac{1}{2}\right)^n.$$

Because

$$\lim_{n\to\infty}\left(\frac{1}{2}\right)^n = 0,$$

by the Limit Laws for Sequences,

$$\lim_{n\to\infty} -\left(\frac{1}{2}\right)^n = -\lim_{n\to\infty}\left(\frac{1}{2}\right)^n = 0.$$

Thus, we have

$$\lim_{n\to\infty}\left(-\frac{1}{2}\right)^n = 0,$$

and

$$\lim_{n\to\infty}\left(3 + \left(-\frac{1}{2}\right)^n\right) = 3 + 0 = 3.$$

47. $b_n = \tan^{-1}\left(1 - \frac{2}{n}\right)$

SOLUTION Because $f(x) = \tan^{-1} x$ is a continuous function, it follows that

$$\lim_{n\to\infty} a_n = \lim_{x\to\infty} \tan^{-1}\left(1 - \frac{2}{x}\right) = \tan^{-1}\left(\lim_{x\to\infty}\left(1 - \frac{2}{x}\right)\right) = \tan^{-1} 1 = \frac{\pi}{4}.$$

49. $c_n = \ln\left(\frac{2n+1}{3n+4}\right)$

SOLUTION Because $f(x) = \ln x$ is a continuous function, it follows that

$$\lim_{n\to\infty} c_n = \lim_{x\to\infty} \ln\left(\frac{2x+1}{3x+4}\right) = \ln\left(\lim_{x\to\infty}\frac{2x+1}{3x+4}\right) = \ln\frac{2}{3}.$$

51. $y_n = \frac{e^n + 3^n}{5^n}$

SOLUTION We rewrite the general term of the sequence as follows:

$$\frac{e^n + 3^n}{5^n} = \frac{e^n}{5^n} + \frac{3^n}{5^n} = \left(\frac{e}{5}\right)^n + \left(\frac{3}{5}\right)^n.$$

Because $0 < \frac{e}{5} < 1$ and $0 < \frac{3}{5} < 1$, the geometric sequences $\left(\frac{e}{5}\right)^n$ and $\left(\frac{3}{5}\right)^n$ converge to 0; hence,

$$\lim_{n\to\infty}\frac{e^n + 3^n}{5^n} = \lim_{n\to\infty}\left(\left(\frac{e}{5}\right)^n + \left(\frac{3}{5}\right)^n\right) = \lim_{n\to\infty}\left(\frac{e}{5}\right)^n + \lim_{n\to\infty}\left(\frac{3}{5}\right)^n = 0 + 0 = 0.$$

53. $a_n = \frac{e^n}{2^{n^2}}$

SOLUTION Using the natural logarithm, we find

$$\ln a_n = \ln\frac{e^n}{2^{n^2}} = \ln e^n - \ln 2^{n^2} = n - n^2\ln 2.$$

Thus,

$$\lim_{n\to\infty}\ln a_n = \lim_{n\to\infty} n\,(1 - n\ln 2) = \lim_{x\to\infty} x\,(1 - x\ln 2) = -\infty.$$

Because $f(x) = e^x$ is a continuous function, it follows that

$$\lim_{n\to\infty} a_n = \lim_{n\to\infty} e^{\ln a_n} = e^{\lim_{n\to\infty}(\ln a_n)} = 0.$$

55. $b_n = \frac{n^3 + 2e^{-n}}{3n^3 + 4e^{-n}}$

SOLUTION We rewrite the general term of the sequence as follows:

$$a_n = \frac{n^3 + 2e^{-n}}{3n^3 + 4e^{-n}} = \frac{\frac{n^3}{n^3} + \frac{2e^{-n}}{n^3}}{\frac{3n^3}{n^3} + \frac{4e^{-n}}{n^3}} = \frac{1 + \frac{2}{n^3e^n}}{3 + \frac{4}{n^3e^n}}.$$

Thus,

$$\lim_{n\to\infty} a_n = \lim_{x\to\infty}\frac{1 + \frac{2}{x^3e^x}}{3 + \frac{4}{x^3e^x}} = \frac{\lim_{x\to\infty}\left(1 + \frac{2}{x^3e^x}\right)}{\lim_{x\to\infty}\left(3 + \frac{4}{x^3e^x}\right)} = \frac{\lim_{x\to\infty} 1 + 2\lim_{x\to\infty}\frac{1}{x^3e^x}}{\lim_{x\to\infty} 3 + 4\lim_{x\to\infty}\frac{1}{x^3e^x}} = \frac{1 + 2\cdot 0}{3 + 4\cdot 0} = \frac{1}{3}.$$

57. $a_n = \frac{3 - 4^n}{2 + 7\cdot 3^n}$

SOLUTION Divide the numerator and denominator by 3^n to obtain

$$a_n = \frac{3 - 4^n}{2 + 7\cdot 3^n} = \frac{\frac{3}{3^n} - \frac{4^n}{3^n}}{\frac{2}{3^n} + \frac{7\cdot 3^n}{3^n}} = \frac{\frac{3}{3^n} - \left(\frac{4}{3}\right)^n}{\frac{2}{3^n} + 7}.$$

We examine the limits of the numerator and the denominator:

$$\lim_{n\to\infty}\left(\frac{3}{3^n} - \left(\frac{4}{3}\right)^n\right) = 3\lim_{n\to\infty}\left(\frac{1}{3}\right)^n - 3\lim_{n\to\infty}\left(\frac{4}{3}\right)^n = 3\cdot 0 - \infty = -\infty,$$

whereas

$$\lim_{n\to\infty}\left(\frac{2}{3^n} + 7\right) = \lim_{n\to\infty}\frac{2}{3^n} + \lim_{n\to\infty} 7 = 2\lim_{n\to\infty}\left(\frac{1}{3}\right)^n + \lim_{n\to\infty} 7 = 2\cdot 0 + 7 = 7.$$

Thus, $\lim_{n\to\infty} a_n = -\infty$; that is, the sequence diverges.

59. $a_n = \dfrac{(-4{,}000)^n}{n!}$

SOLUTION Let $a_n = \frac{(-4000)^n}{n!}$ and note

$$-\left(\frac{4000^n}{n!}\right) \le a_n \le \frac{4000^n}{n!}.$$

For $n > 4001$,

$$\left(\frac{4000}{1} \cdot \frac{4000}{2} \cdot \cdots \cdot \frac{4000}{4000}\right)\left(\frac{4000}{4001} \cdot \frac{4000}{4002} \cdot \cdots \cdot \frac{4000}{n-1}\right) \cdot \frac{4000}{n} = \frac{4000^{4000}}{4000!}\left(\frac{4000}{4001} \cdot \frac{4000}{4002} \cdot \cdots \cdot \frac{4000}{n-1}\right) \cdot \frac{4000}{n}.$$

Each one of the factors in the brackets is less than 1, so we have

$$0 < \frac{4000^n}{n!} < \frac{4000^{4000}}{4000!} \cdot \frac{4000}{n} = \frac{4000^{4001}}{4000!} \cdot \frac{1}{n}.$$

Since $\frac{4000^{4001}}{4000!} \cdot \frac{1}{n}$ tends to zero, the Squeeze Theorem guarantees that

$$\lim_{n\to\infty} A_n = \lim_{n\to\infty} \frac{(-4000)^n}{n!} = 0.$$

61. $a_n = \cos \dfrac{\pi}{n}$

SOLUTION By the Theorem on Sequences Defined by a Function, we have:

$$\lim_{n\to\infty} \cos \frac{\pi}{n} = \lim_{x\to\infty} \cos \frac{\pi}{x}.$$

The limit $\lim_{x\to\infty} \frac{\pi}{x} = 0$ and the continuity of $\cos u$ at $u = 0$ imply:

$$\lim_{x\to\infty} \cos \frac{\pi}{x} = \cos\left(\lim_{x\to\infty} \frac{\pi}{x}\right) = \cos 0 = 1.$$

Thus,

$$\lim_{n\to\infty} \cos \frac{\pi}{n} = 1.$$

63. $a_n = \sqrt[n]{n}$

SOLUTION Let $a_n = n^{1/n}$. Taking the natural logarithm of both sides of this expression yields

$$\ln a_n = \ln n^{1/n} = \frac{1}{n} \ln n = \frac{\ln n}{n}.$$

Thus,

$$\lim_{n\to\infty} (\ln a_n) = \lim_{n\to\infty} \frac{\ln n}{n} = \lim_{x\to\infty} \frac{\ln x}{x} = \lim_{x\to\infty} \frac{1/x}{1} = 0.$$

Because $f(x) = e^x$ is a continuous function, it follows that

$$\lim_{n\to\infty} a_n = \lim_{n\to\infty} e^{\ln a_n} = e^{\lim_{n\to\infty}(\ln a_n)} = e^0 = 1.$$

65. Show that $a_n = \dfrac{3n^2}{n^2+2}$ is strictly increasing. Find an upper bound.

SOLUTION Let $f(x) = \frac{3x^2}{x^2+2}$. Then

$$f'(x) = \frac{6x(x^2+2) - 3x^2 \cdot 2x}{(x^2+2)^2} = \frac{12x}{(x^2+2)^2}.$$

$f'(x) > 0$ for $x > 0$, hence f is strictly increasing on this interval. It follows that $a_n = f(n)$ is also strictly increasing. We now show that $M = 3$ is an upper bound for a_n, by writing:

$$a_n = \frac{3n^2}{n^2+2} \le \frac{3n^2+6}{n^2+2} = \frac{3(n^2+2)}{n^2+2} = 3.$$

That is, $a_n \le 3$ for all n.

67. Use the limit definition to prove that the limit does not change if a finite number of terms are added or removed from a convergent sequence.

SOLUTION Suppose that $\{a_n\}$ is a sequence such that $\lim_{n\to\infty} a_n = L$. For every $\epsilon > 0$, there is a number M such that $|a_n - L| < \epsilon$ for all $n > M$. That is, the inequality $|a_n - L| < \epsilon$ holds for all the terms of $\{a_n\}$ except possibly a finite number of terms. If we add a finite number of terms, these terms may not satisfy the inequality $|a_n - L| < \epsilon$, but there are still only a finite number of terms that do not satisfy this inequality. By removing terms from the sequence, the number of terms in the new sequence that do not satisfy $|a_n - L| < \epsilon$ are no more than in the original sequence. Hence the new sequence also converges to L.

69. Let $\{a_n\}$ be a sequence such that $\lim\limits_{n\to\infty} |a_n|$ exists and is nonzero. Show that $\lim\limits_{n\to\infty} a_n$ exists if and only if there exists an integer M such that the sign of a_n does not change for $n > M$.

SOLUTION Let $\{a_n\}$ be a sequence such that $\lim\limits_{n\to\infty} |a_n|$ exists and is nonzero. Suppose $\lim\limits_{n\to\infty} a_n$ exists and let $L = \lim\limits_{n\to\infty} a_n$. Note that L cannot be zero for then $\lim\limits_{n\to\infty} |a_n|$ would also be zero. Now, choose $\epsilon < |L|$. Then there exists an integer M such that $|a_n - L| < \epsilon$, or $L - \epsilon < a_n < L + \epsilon$, for all $n > M$. If $L < 0$, then $-2L < a_n < 0$, whereas if $L > 0$, then $0 < a_n < 2L$; that is, a_n does not change for $n > M$.

Now suppose that there exists an integer M such that a_n does not change for $n > M$. If $a_n > 0$ for $n > M$, then $a_n = |a_n|$ for $n > M$ and

$$\lim_{n\to\infty} a_n = \lim_{n\infty} |a_n|.$$

On the other hand, if $a_n < 0$ for $n > M$, then $a_n = -|a_n|$ for $n > M$ and

$$\lim_{n\to\infty} a_n = \lim_{n\infty} -|a_n| = -\lim_{n\to\infty} |a_n|.$$

In either case, $\lim\limits_{n\to\infty} a_n$ exists. Thus, $\lim\limits_{n\to\infty} a_n$ exists if and only if there exists an integer M such that the sign of a_n does not change for $n > M$.

71. Show, by giving an example, that there exist *divergent* sequences $\{a_n\}$ and $\{b_n\}$ such that $\{a_n + b_n\}$ converges.

SOLUTION Let $a_n = 2^n$ and $b_n = -2^n$. Then $\{a_n\}$ and $\{b_n\}$ are divergent geometric sequences. However, since $a_n + b_n = 2^n - 2^n = 0$, the sequence $\{a_n + b_n\}$ is the constant sequence with all the terms equal zero, so it converges to zero.

73. Use the limit definition to prove that if $\{a_n\}$ is a convergent sequence of integers with limit L, then there exists a number M such that $a_n = L$ for all $n \ge M$.

SOLUTION Suppose $\{a_n\}$ converges to L, and let $\epsilon = \frac{1}{2}$. Then, there exists a number M such that

$$|a_n - L| < \frac{1}{2}$$

for all $n \ge M$. In other words, for all $n \ge M$,

$$L - \frac{1}{2} < a_n < L + \frac{1}{2}.$$

However, we are given that $\{a_n\}$ is a sequence of integers. Thus, it must be that $a_n = L$ for all $n \ge M$.

75. Prove that the following sequence is bounded and increasing. Then find its limit:

$$a_1 = \sqrt{5}, \quad a_2 = \sqrt{5 + \sqrt{5}}, \quad a_3 = \sqrt{5 + \sqrt{5 + \sqrt{5}}}, \ldots$$

SOLUTION Notice that this sequence is defined recursively by the formula:

$$a_{n+1} = \sqrt{5 + a_n}.$$

First, let's show that the sequence is bounded. All the terms in the sequence are positive, so 0 is a lower bound. Now, $a_1 = \sqrt{5} < 3$. If we suppose that $a_n < 3$ for some n, it then follows that

$$a_{n+1} = \sqrt{5 + a_n} < \sqrt{5 + 3} = \sqrt{8} < \sqrt{9} = 3.$$

Thus, by mathematical induction, $a_n < 3$ for all n, and 3 is an upper bound. Next, let's show that the sequence is increasing. Observe that $a_1 = \sqrt{5} \approx 2.236$ and $a_2 = \sqrt{5 + \sqrt{5}} \approx 2.690$. Thus, $a_2 > a_1$. If we suppose that $a_n > a_{n-1}$ for some n, it then follows that

$$a_{n+1} = \sqrt{5 + a_n} > \sqrt{5 + a_{n-1}} = a_n;$$

hence, by mathematical induction, $a_{n+1} > a_n$ for all n.

Now, since $\{a_n\}$ is increasing with upper bound, this sequence is convergent. Let $\lim_{n\to\infty} a_n = L$. Then, by Exercise 68, $\lim_{n\to\infty} a_{n+1} = L$ as well. It follows that

$$L = \lim_{n\to\infty} a_{n+1} = \lim_{n\to\infty} \sqrt{5+a_n} = \sqrt{\lim_{n\to\infty}(5+a_n)} = \sqrt{5+\lim_{n\to\infty} a_n} = \sqrt{5+L}.$$

That is,

$$L^2 = 5 + L$$

$$L^2 - L - 5 = 0 \Rightarrow L_{1,2} = \frac{1 \pm \sqrt{1+20}}{2}.$$

Since $a_n \geq 0$ for all n, the appropriate solution is:

$$\lim_{n\to\infty} a_n = \frac{1+\sqrt{21}}{2}.$$

77. Find the limit of the sequence

$$c_n = \frac{1}{\sqrt{n^2+1}} + \frac{1}{\sqrt{n^2+2}} + \cdots + \frac{1}{\sqrt{n^2+n}}$$

Hint: Show that

$$\frac{n}{\sqrt{n^2+n}} \leq c_n \leq \frac{n}{\sqrt{n^2+1}}$$

SOLUTION Since each of the n terms in the sum defining c_n is not smaller than $\frac{1}{\sqrt{n^2+n}}$ and not larger than $\frac{1}{\sqrt{n^2+1}}$ we obtain the following inequalities:

$$c_n \geq \underbrace{\frac{1}{\sqrt{n^2+n}} + \cdots + \frac{1}{\sqrt{n^2+n}}}_{n \text{ terms}} = n \cdot \frac{1}{\sqrt{n^2+n}} = \frac{n}{\sqrt{n^2+n}};$$

$$c_n \leq \underbrace{\frac{1}{\sqrt{n^2+1}} + \cdots + \frac{1}{\sqrt{n^2+1}}}_{n \text{ terms}} = n \cdot \frac{1}{\sqrt{n^2+1}} = \frac{n}{\sqrt{n^2+1}}.$$

Thus,

$$\frac{n}{\sqrt{n^2+n}} \leq c_n \leq \frac{n}{\sqrt{n^2+1}}.$$

We now compute the limits of the two sequences:

$$\lim_{n\to\infty} \frac{n}{\sqrt{n^2+1}} = \lim_{n\to\infty} \frac{\frac{n}{n}}{\frac{\sqrt{n^2+1}}{n}} = \lim_{n\to\infty} \frac{1}{\frac{\sqrt{n^2+1}}{\sqrt{n^2}}} = \lim_{n\to\infty} \frac{1}{\sqrt{1+\frac{1}{n^2}}} = 1;$$

$$\lim_{n\to\infty} \frac{n}{\sqrt{n^2+n}} = \lim_{n\to\infty} \frac{\frac{n}{n}}{\frac{\sqrt{n^2+n}}{n}} = \lim_{n\to\infty} \frac{1}{\frac{\sqrt{n^2+n}}{\sqrt{n^2}}} = \lim_{n\to\infty} \frac{1}{\sqrt{1+\frac{1}{n}}} = 1.$$

By the Squeeze Theorem we conclude that:

$$\lim_{n\to\infty} c_n = 1.$$

Further Insights and Challenges

79. Let $b_n = \frac{\sqrt[n]{n!}}{n}$.

(a) Show that $\ln b_n = \frac{\ln(n!) - n\ln n}{n} = \frac{1}{n}\sum_{k=1}^{n} \ln\frac{k}{n}$.

(b) Show that $\ln b_n$ converges to $\int_0^1 \ln x\,dx$ and conclude that $b_n \to e^{-1}$.

SOLUTION

(a) Let $b_n = \frac{(n!)^{1/n}}{n}$. Then

$$\ln b_n = \ln (n!)^{1/n} - \ln n = \frac{1}{n}\ln (n!) - \ln n = \frac{\ln (n!) - n\ln n}{n} = \frac{1}{n}\left[\ln (n!) - \ln n^n\right] = \frac{1}{n}\ln \frac{n!}{n^n}$$
$$= \frac{1}{n}\ln\left(\frac{1}{n}\cdot\frac{2}{n}\cdot\frac{3}{n}\cdot\dots\cdot\frac{n}{n}\right) = \frac{1}{n}\left(\ln\frac{1}{n} + \ln\frac{2}{n} + \ln\frac{3}{n} + \dots + \ln\frac{n}{n}\right) = \frac{1}{n}\sum_{k=1}^{n}\ln\frac{k}{n}.$$

(b) By part (a) we have,

$$\lim_{n\to\infty}(\ln b_n) = \lim_{n\to\infty}\frac{1}{n}\sum_{k=1}^{n}\ln\frac{k}{n}.$$

Notice that $\frac{1}{n}\sum_{k=1}^{n}\ln\frac{k}{n}$ is the nth right-endpoint approximation of the integral of $\ln x$ over the interval $[0, 1]$. Hence,

$$\lim_{n\to\infty}\frac{1}{n}\sum_{k=1}^{n}\ln\frac{k}{n} = \int_0^1 \ln x\,dx.$$

We compute the improper integral using integration by parts, with $u = \ln x$ and $v' = 1$. Then $u' = \frac{1}{x}$, $v = x$ and

$$\int_0^1 \ln x\,dx = x\ln x\Big|_0^1 - \int_0^1 \frac{1}{x}x\,dx = 1\cdot\ln 1 - \lim_{x\to 0+}(x\ln x) - \int_0^1 dx$$
$$= 0 - \lim_{x\to 0+}(x\ln x) - x\Big|_0^1 = -1 - \lim_{x\to 0+}(x\ln x).$$

We compute the remaining limit using L'Hôpital's Rule. This gives:

$$\lim_{x\to 0+}(x\cdot\ln x) = \lim_{x\to 0+}\frac{\ln x}{\frac{1}{x}} = \lim_{x\to 0+}\frac{\frac{1}{x}}{-\frac{1}{x^2}} = \lim_{x\to 0+}(-x) = 0.$$

Thus,

$$\lim_{n\to\infty}\ln b_n = \int_0^1 \ln x\,dx = -1,$$

and

$$\lim_{n\to\infty} b_n = e^{-1}.$$

81. Let $c_n = \frac{1}{n} + \frac{1}{n+1} + \frac{1}{n+2} + \dots + \frac{1}{2n}$.

(a) Calculate c_1, c_2, c_3, c_4.

(b) Use a comparison of rectangles with the area under $y = x^{-1}$ over the interval $[n, 2n]$ to prove that

$$\int_n^{2n}\frac{dx}{x} + \frac{1}{2n} \le c_n \le \int_n^{2n}\frac{dx}{x} + \frac{1}{n}$$

(c) Use the Squeeze Theorem to determine $\lim_{n\to\infty} c_n$.

SOLUTION

(a)

$$c_1 = 1 + \frac{1}{2} = \frac{3}{2};$$
$$c_2 = \frac{1}{2} + \frac{1}{3} + \frac{1}{4} = \frac{13}{12};$$
$$c_3 = \frac{1}{3} + \frac{1}{4} + \frac{1}{5} + \frac{1}{6} = \frac{19}{20};$$
$$c_4 = \frac{1}{4} + \frac{1}{5} + \frac{1}{6} + \frac{1}{7} + \frac{1}{8} = \frac{743}{840};$$

(b) We consider the left endpoint approximation to the integral of $y = \frac{1}{x}$ over the interval $[n, 2n]$. Since the function $y = \frac{1}{x}$ is decreasing, the left endpoint approximation is greater than $\int_n^{2n}\frac{dx}{x}$; that is,

$$\int_n^{2n}\frac{dx}{x} \le \frac{1}{n}\cdot 1 + \frac{1}{n+1}\cdot 1 + \frac{1}{n+2}\cdot 1 + \dots + \frac{1}{2n-1}\cdot 1.$$

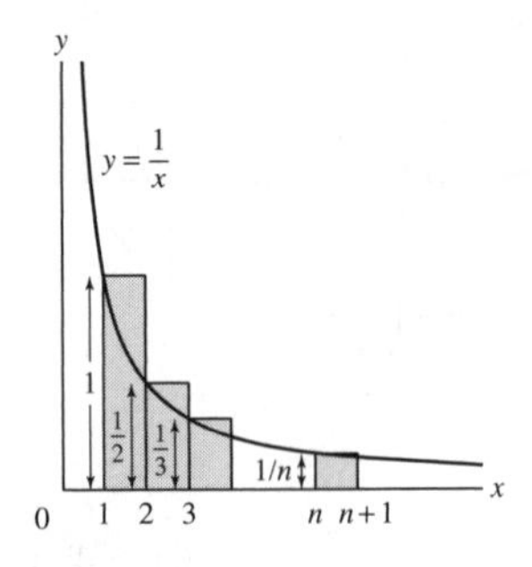

We express the right hand-side of the inequality in terms of c_n, obtaining:

$$\int_n^{2n} \frac{dx}{x} \le c_n - \frac{1}{2n}.$$

We now consider the right endpoint approximation to the integral $\int_n^{2n} \frac{dx}{x}$; that is,

$$\frac{1}{n+1} \cdot 1 + \frac{1}{n+2} \cdot 1 + \cdots + \frac{1}{2n} \cdot 1 \le \int_n^{2n} \frac{dx}{x}.$$

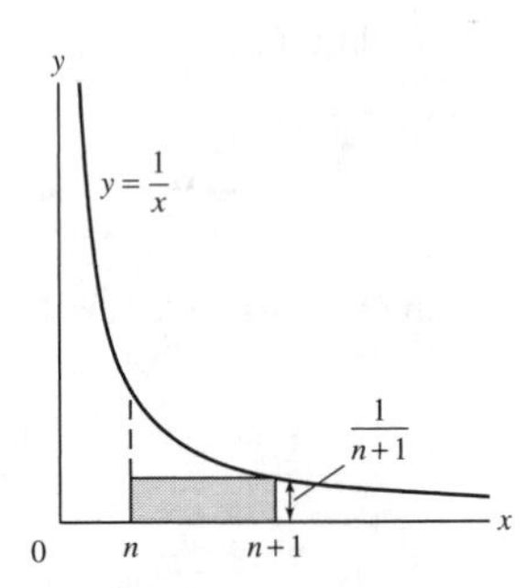

We express the left hand-side of the inequality in terms of c_n, obtaining:

$$c_n - \frac{1}{n} \le \int_n^{2n} \frac{dx}{x}.$$

Thus,

$$\int_n^{2n} \frac{dx}{x} + \frac{1}{2n} \le c_n \le \int_n^{2n} \frac{dx}{x} + \frac{1}{n}.$$

(c) With

$$\int_n^{2n} \frac{dx}{x} = \ln x\big|_n^{2n} = \ln 2n - \ln n = \ln \frac{2n}{n} = \ln 2,$$

the result from part (b) becomes

$$\ln 2 + \frac{1}{2n} \le c_n \le \ln 2 + \frac{1}{n}.$$

Because

$$\lim_{n\to\infty} \frac{1}{2n} = \lim_{n\to\infty} \frac{1}{n} = 0,$$

it follows from the Squeeze Theorem that

$$\lim_{n\to\infty} c_n = \ln 2.$$

11.2 Summing an Infinite Series (ET Section 10.2)

Preliminary Questions

1. What role do partial sums play in defining the sum of an infinite series?

SOLUTION The sum of an infinite series is defined as the limit of the sequence of partial sums. If the limit of this sequence does not exist, the series is said to diverge.

2. What is the sum of the following infinite series?

$$\frac{1}{4}+\frac{1}{8}+\frac{1}{16}+\frac{1}{32}+\frac{1}{64}+\cdots$$

SOLUTION This is a geometric series with $c=\frac{1}{4}$ and $r=\frac{1}{2}$. The sum of the series is therefore

$$\frac{\frac{1}{4}}{1-\frac{1}{2}}=\frac{\frac{1}{4}}{\frac{1}{2}}=\frac{1}{2}.$$

3. What happens if you apply the formula for the sum of a geometric series to the following series? Is the formula valid?

$$1+3+3^2+3^3+3^4+\cdots$$

SOLUTION This is a geometric series with $c=1$ and $r=3$. Applying the formula for the sum of a geometric series then gives

$$\sum_{n=0}^{\infty}3^n=\frac{1}{1-3}=-\frac{1}{2}.$$

Clearly, this is not valid: a series with all positive terms cannot have a negative sum. The formula is not valid in this case because a geometric series with $r=3$ diverges.

4. Arvind asserts that $\sum_{n=1}^{\infty}\frac{1}{n^2}=0$ because $\frac{1}{n^2}$ tends to zero. Is this valid reasoning?

SOLUTION Arvind's reasoning is not valid. Though the terms in the series do tend to zero, the general term in the sequence of partial sums,

$$S_n=1+\frac{1}{2^2}+\frac{1}{3^2}+\cdots+\frac{1}{n^2},$$

is clearly larger than 1. The sum of the series therefore cannot be zero.

5. Colleen claims that $\sum_{n=1}^{\infty}\frac{1}{\sqrt{n}}$ converges because $\lim_{n\to\infty}\frac{1}{\sqrt{n}}=0$. Is this valid reasoning?

SOLUTION Colleen's reasoning is not valid. Although the general term of a convergent series must tend to zero, a series whose general term tends to zero need not converge. In the case of $\sum_{n=1}^{\infty}\frac{1}{\sqrt{n}}$, the series diverges even though its general term tends to zero.

6. Find an N such that $S_N>25$ for the series $\sum_{n=1}^{\infty}2$.

SOLUTION The Nth partial sum of the series is:

$$S_N=\sum_{n=1}^{N}2=\underbrace{2+\cdots+2}_{N}=2N.$$

7. Does there exist an N such that $S_N>25$ for the series $\sum_{n=1}^{\infty}2^{-n}$? Explain.

SOLUTION The series $\sum_{n=1}^{\infty}2^{-n}$ is a convergent geometric series with the common ratio $r=\frac{1}{2}$. The sum of the series is:

$$S=\frac{\frac{1}{2}}{1-\frac{1}{2}}=1.$$

Notice that the sequence of partial sums $\{S_N\}$ is increasing and converges to 1; therefore $S_N\le 1$ for all N. Thus, there does not exist an N such that $S_N>25$.

8. Give an example of a divergent infinite series whose general term tends to zero.

SOLUTION Consider the series $\sum_{n=1}^{\infty} \frac{1}{n^{\frac{9}{10}}}$. The general term tends to zero, since $\lim_{n \to \infty} \frac{1}{n^{\frac{9}{10}}} = 0$. However, the Nth partial sum satisfies the following inequality:

$$S_N = \frac{1}{1^{\frac{9}{10}}} + \frac{1}{2^{\frac{9}{10}}} + \cdots + \frac{1}{N^{\frac{9}{10}}} \geq \frac{N}{N^{\frac{9}{10}}} = N^{1-\frac{9}{10}} = N^{\frac{1}{10}}.$$

That is, $S_N \geq N^{\frac{1}{10}}$ for all N. Since $\lim_{N \to \infty} N^{\frac{1}{10}} = \infty$, the sequence of partial sums S_n diverges; hence, the series $\sum_{n=1}^{\infty} \frac{1}{n^{\frac{9}{10}}}$ diverges.

Exercises

1. Find a formula for the general term a_n (not the partial sum) of the infinite series.

(a) $\frac{1}{3} + \frac{1}{9} + \frac{1}{27} + \frac{1}{81} + \cdots$

(b) $\frac{1}{1} + \frac{5}{2} + \frac{25}{4} + \frac{125}{8} + \cdots$

(c) $\frac{1}{1} - \frac{2^2}{2 \cdot 1} + \frac{3^3}{3 \cdot 2 \cdot 1} - \frac{4^4}{4 \cdot 3 \cdot 2 \cdot 1} + \cdots$

(d) $\frac{2}{1^2+1} + \frac{1}{2^2+1} + \frac{2}{3^2+1} + \frac{1}{4^2+1} + \cdots$

SOLUTION

(a) The denominators of the terms are powers of 3, starting with the first power. Hence, the general term is:

$$a_n = \frac{1}{3^n}.$$

(b) The numerators are powers of 5, and the denominators are the same powers of 2. The first term is $a_1 = 1$ so,

$$a_n = \left(\frac{5}{2}\right)^{n-1}.$$

(c) The general term of this series is,

$$a_n = (-1)^{n+1} \frac{n^n}{n!}.$$

(d) Notice that the numerators of a_n equal 2 for odd values of n and 1 for even values of n. Thus,

$$a_n = \begin{cases} \frac{2}{n^2+1} & \text{odd } n \\ \frac{1}{n^2+1} & \text{even } n \end{cases}$$

The formula can also be rewritten as follows:

$$a_n = \frac{1 + \frac{(-1)^{n+1}+1}{2}}{n^2+1}.$$

In Exercises 3–6, compute the partial sums S_2, S_4, and S_6.

3. $1 + \frac{1}{2^2} + \frac{1}{3^2} + \frac{1}{4^2} + \cdots$

SOLUTION

$$S_2 = 1 + \frac{1}{2^2} = \frac{5}{4};$$

$$S_4 = 1 + \frac{1}{2^2} + \frac{1}{3^2} + \frac{1}{4^2} = \frac{205}{144};$$

$$S_6 = 1 + \frac{1}{2^2} + \frac{1}{3^2} + \frac{1}{4^2} + \frac{1}{5^2} + \frac{1}{6^2} = \frac{5369}{3600}.$$

5. $\frac{1}{1\cdot 2}+\frac{1}{2\cdot 3}+\frac{1}{3\cdot 4}+\cdots$

SOLUTION

$$S_2=\frac{1}{1\cdot 2}+\frac{1}{2\cdot 3}=\frac{1}{2}+\frac{1}{6}=\frac{4}{6}=\frac{2}{3};$$
$$S_4=S_2+a_3+a_4=\frac{2}{3}+\frac{1}{3\cdot 4}+\frac{1}{4\cdot 5}=\frac{2}{3}+\frac{1}{12}+\frac{1}{20}=\frac{4}{5};$$
$$S_6=S_4+a_5+a_6=\frac{4}{5}+\frac{1}{5\cdot 6}+\frac{1}{6\cdot 7}=\frac{4}{5}+\frac{1}{30}+\frac{1}{42}=\frac{6}{7}.$$

7. Compute S_5, S_{10}, and S_{15} for the series

$$S=\frac{1}{2\cdot 3\cdot 4}-\frac{1}{4\cdot 5\cdot 6}+\frac{1}{6\cdot 7\cdot 8}-\frac{1}{8\cdot 9\cdot 10}+\cdots$$

This series S is known to converge to $\frac{\pi-3}{4}$. Do your calculations support this conclusion?

SOLUTION The formula for the general term in the series is

$$a_n=\frac{(-1)^{n+1}}{2n(2n+1)(2n+2)}.$$

Thus,

$$S_5=\frac{1}{2(3)(4)}-\frac{1}{4(5)(6)}+\cdots-\frac{1}{8(9)(10)}+\frac{1}{10(11)(12)}=0.035678;$$
$$S_{10}=\frac{1}{2(3)(4)}-\frac{1}{4(5)(6)}+\cdots+\frac{1}{18(19)(20)}-\frac{1}{20(21)(22)}=0.035352;$$
$$S_{15}=\frac{1}{2(3)(4)}-\frac{1}{4(5)(6)}+\cdots-\frac{1}{28(29)(30)}+\frac{1}{30(31)(32)}=0.035413.$$

Using a calculator we find

$$\frac{\pi-3}{4}=0.035398,$$

which is consistent with our partial sum calculations.

9. Calculate S_3, S_4, and S_5 and then find the sum of the telescoping series

$$S=\sum_{n=1}^{\infty}\left(\frac{1}{n+1}-\frac{1}{n+2}\right)$$

SOLUTION

$$S_3=\left(\frac{1}{2}-\frac{1}{3}\right)+\left(\frac{1}{3}-\frac{1}{4}\right)+\left(\frac{1}{4}-\frac{1}{5}\right)=\frac{1}{2}-\frac{1}{5}=\frac{3}{10};$$
$$S_4=S_3+\left(\frac{1}{5}-\frac{1}{6}\right)=\frac{1}{2}-\frac{1}{6}=\frac{1}{3};$$
$$S_5=S_4+\left(\frac{1}{6}-\frac{1}{7}\right)=\frac{1}{2}-\frac{1}{7}=\frac{5}{14}.$$

The general term in the sequence of partial sums is

$$S_N=\left(\frac{1}{2}-\frac{1}{3}\right)+\left(\frac{1}{3}-\frac{1}{4}\right)+\left(\frac{1}{4}-\frac{1}{5}\right)+\cdots+\left(\frac{1}{N+1}-\frac{1}{N+2}\right)=\frac{1}{2}-\frac{1}{N+2};$$

thus,

$$S=\lim_{N\to\infty}S_N=\lim_{N\to\infty}\left(\frac{1}{2}-\frac{1}{N+2}\right)=\frac{1}{2}.$$

The sum of the telescoping series is therefore $\frac{1}{2}$.

11. Write $\sum_{n=3}^{\infty}\frac{1}{n(n-1)}$ as a telescoping series and find its sum.

SOLUTION By partial fraction decomposition

$$\frac{1}{n(n-1)} = \frac{1}{n-1} - \frac{1}{n},$$

so

$$\sum_{n=3}^{\infty} \frac{1}{n(n-1)} = \sum_{n=3}^{\infty} \left(\frac{1}{n-1} - \frac{1}{n} \right).$$

The general term in the sequence of partial sums for this series is

$$S_N = \left(\frac{1}{2} - \frac{1}{3} \right) + \left(\frac{1}{3} - \frac{1}{4} \right) + \left(\frac{1}{4} - \frac{1}{5} \right) + \cdots + \left(\frac{1}{N-1} - \frac{1}{N} \right) = \frac{1}{2} - \frac{1}{N};$$

thus,

$$S = \lim_{N\to\infty} S_N = \lim_{N\to\infty} \left(\frac{1}{2} - \frac{1}{N} \right) = \frac{1}{2}.$$

In Exercises 13–16, use Theorem 2 to prove that the following series diverge.

13. $\sum_{n=1}^{\infty} (-1)^n n^2$

SOLUTION The general term $a_n = (-1)^n n^2$ does not tend to zero. In fact, because $\lim_{n\to\infty} n^2 = \infty$, $\lim_{n\to\infty} a_n$ does not exist. By Theorem 2, we conclude that the given series diverges.

15. $\sum_{n=0}^{\infty} (\sqrt{4n^2+1} - n)$

SOLUTION The general term of the series satisfies

$$\sqrt{4n^2+1} - n > \sqrt{4n^2} - n = n$$

Thus the general term tends to infinity. The series diverges by Theorem 2.

17. Which of these series converge?

(a) $\sum_{n=1}^{\infty} \left(\frac{1}{\sqrt{n}} - \frac{1}{\sqrt{n+1}} \right)$

(b) $\sum_{n=1}^{\infty} (\ln n - \ln(n+1))$

SOLUTION

(a) This series converges. The general term in the sequence of partial sums is

$$S_N = \left(1 - \frac{1}{\sqrt{2}} \right) + \left(\frac{1}{\sqrt{2}} - \frac{1}{\sqrt{3}} \right) + \left(\frac{1}{\sqrt{3}} - \frac{1}{\sqrt{4}} \right) + \cdots + \left(\frac{1}{\sqrt{N}} - \frac{1}{\sqrt{N+1}} \right) = 1 - \frac{1}{\sqrt{N+1}}.$$

Because

$$\lim_{N\to\infty} S_N = \lim_{N\to\infty} \left(1 - \frac{1}{\sqrt{N+1}} \right) = 1,$$

the series converges to 1.

(b) This series diverges. The general term in the sequence of partial sums is

$$S_N = (\ln 1 - \ln 2) + (\ln 2 - \ln 3) + (\ln 3 - \ln 4) + \cdots + (\ln N - \ln(N+1)) = -\ln(N+1).$$

Because

$$\lim_{N\to\infty} S_N = \lim_{N\to\infty} -\ln(N+1) = -\infty,$$

we conclude the given series diverges.

In Exercises 18–31, use the formula for the sum of a geometric series to find the sum or state that the series diverges.

19. $\frac{1}{3^3} + \frac{1}{3^4} + \frac{1}{3^5} + \cdots$

SOLUTION This is a geometric series with $c = \frac{1}{27}$ and $r = \frac{1}{3}$. Thus,

$$\sum_{n=3}^{\infty} \left(\frac{1}{3}\right)^n = \frac{\frac{1}{27}}{1 - \frac{1}{3}} = \frac{\frac{1}{27}}{\frac{2}{3}} = \frac{1}{18}.$$

21. $\sum_{n=3}^{\infty} \frac{3^n}{11^n}$

SOLUTION This is a geometric series with $c = \left(\frac{3}{11}\right)^3 = \frac{27}{1331}$ and $r = \frac{3}{11}$. Thus,

$$\sum_{n=3}^{\infty} \frac{3^n}{11^n} = \frac{\frac{27}{1331}}{1 - \frac{3}{11}} = \frac{\frac{27}{1331}}{\frac{8}{11}} = \frac{27}{968}.$$

23. $1 + \frac{2}{7} + \frac{2^2}{7^2} + \frac{2^3}{7^3} + \cdots$

SOLUTION This is a geometric series with $c = 1$ and $r = \frac{2}{7}$. Thus,

$$\sum_{n=0}^{\infty} \left(\frac{2}{7}\right)^n = \frac{1}{1 - \frac{2}{7}} = \frac{1}{\frac{5}{7}} = \frac{7}{5}.$$

25. $\sum_{n=2}^{\infty} e^{3-2n}$

SOLUTION Rewrite the series as

$$\sum_{n=2}^{\infty} e^3 e^{-2n} = \sum_{n=2}^{\infty} e^3 \left(\frac{1}{e^2}\right)^n$$

to recognize it as a geometric series with $c = e^3 \left(\frac{1}{e^2}\right)^2 = \frac{1}{e}$ and $r = \frac{1}{e^2}$. Thus,

$$\sum_{n=2}^{\infty} e^{3-2n} = \frac{\frac{1}{e}}{1 - \frac{1}{e^2}} = \frac{e}{e^2 - 1}.$$

27. $\sum_{n=0}^{\infty} \frac{93^n + 4^{n-2}}{5^n}$

SOLUTION Rewrite the series as

$$\sum_{n=0}^{\infty} \frac{93^n + 4^{n-2}}{5^n} = \sum_{n=0}^{\infty} \left(\frac{93^n}{5^n} + \frac{4^{n-2}}{5^n}\right) = \sum_{n=0}^{\infty} \left(\frac{93}{5}\right)^n + \sum_{n=0}^{\infty} \frac{1}{16} \cdot \left(\frac{4}{5}\right)^n,$$

which is a sum of two geometric series. The first series has $c = \left(\frac{93}{5}\right)^0 = 1$ and $r = \frac{93}{5}$; the second has $c = \frac{1}{16}\left(\frac{4}{5}\right)^0 = \frac{1}{16}$ and $r = \frac{4}{5}$. Because the first series has $r > 1$, that series diverges. Consequently, the original series also diverges.

29. $\frac{2^3}{7} + \frac{2^4}{7^2} + \frac{2^5}{7^3} + \frac{2^6}{7^4} + \cdots$

SOLUTION This is a geometric series with $c = \frac{8}{7}$ and $r = \frac{2}{7}$. Thus,

$$\sum_{n=0}^{\infty} \frac{8}{7} \cdot \left(\frac{2}{7}\right)^n = \frac{\frac{8}{7}}{1 - \frac{2}{7}} = \frac{\frac{8}{7}}{\frac{5}{7}} = \frac{8}{5}.$$

31. $\frac{64}{49} + \frac{8}{7} + 1 + \frac{7}{8} + \frac{49}{64} + \frac{343}{512} + \cdots$

SOLUTION This is a geometric series with $c = \frac{64}{49}$ and $r = \frac{7}{8}$. Thus,

$$\sum_{n=0}^{\infty} \frac{64}{49} \cdot \left(\frac{7}{8}\right)^n = \frac{\frac{64}{49}}{1 - \frac{7}{8}} = \frac{\frac{64}{49}}{\frac{1}{8}} = \frac{512}{49}.$$

33. Which of the following series are divergent?

(a) $\displaystyle\sum_{n=0}^{\infty} \frac{2^n}{5^n}$

(b) $\displaystyle\sum_{n=3}^{\infty} 1.5^n$

(c) $\displaystyle\sum_{n=0}^{\infty} \frac{5^n}{2^n}$

(d) $\displaystyle\sum_{n=0}^{\infty} (0.4)^n$

SOLUTION

(a) The series $\displaystyle\sum_{n=0}^{\infty} \frac{2^n}{5^n} = \sum_{n=0}^{\infty} \left(\frac{2}{5}\right)^n$ is a geometric series with common ratio $r = \frac{2}{5}$. Since $|r| < 1$, the series converges.

(b) The series $\displaystyle\sum_{n=3}^{\infty} 1.5^n$ is a geometric series with common ratio $r = 1.5$. Since $r > 1$, the series diverges.

(c) The series $\displaystyle\sum_{n=0}^{\infty} \frac{5^n}{2^n} = \sum_{n=0}^{\infty} \left(\frac{5}{2}\right)^n$ is a geometric series with common ratio $r = \frac{5}{2}$. Since $r > 1$, the series diverges.

(d) The series $\displaystyle\sum_{n=0}^{\infty} 0.4^n$ is a geometric series with common ratio $r = 0.4$. Since $|r| < 1$, the series converges.

35. Let $S = \displaystyle\sum_{n=1}^{\infty} a_n$ be an infinite series such that $S_N = 5 - \frac{2}{N^2}$.

(a) What are the values of $\displaystyle\sum_{n=1}^{10} a_n$ and $\displaystyle\sum_{n=4}^{16} a_n$?

(b) What is the value of a_3?

(c) Find a general formula for a_n.

(d) Find the sum $\displaystyle\sum_{n=1}^{\infty} a_n$.

SOLUTION

(a)

$$\sum_{n=1}^{10} a_n = S_{10} = 5 - \frac{2}{10^2} = \frac{249}{50};$$

$$\sum_{n=4}^{16} a_n = (a_1 + \cdots + a_{16}) - (a_1 + a_2 + a_3) = S_{16} - S_3 = \left(5 - \frac{2}{16^2}\right) - \left(5 - \frac{2}{3^2}\right) = \frac{2}{9} - \frac{2}{256} = \frac{494}{2304}.$$

(b)

$$a_3 = (a_1 + a_2 + a_3) - (a_1 + a_2) = S_3 - S_2 = \left(5 - \frac{2}{3^2}\right) - \left(5 - \frac{2}{2^2}\right) = \frac{1}{2} - \frac{2}{9} = \frac{5}{18}.$$

(c) Since $a_n = S_n - S_{n-1}$, we have:

$$a_n = S_n - S_{n-1} = \left(5 - \frac{2}{n^2}\right) - \left(5 - \frac{2}{(n-1)^2}\right) = \frac{2}{(n-1)^2} - \frac{2}{n^2}$$

$$= \frac{2\left(n^2 - (n-1)^2\right)}{(n(n-1))^2} = \frac{2\left(n^2 - n^2 + 2n - 1\right)}{(n(n-1))^2} = \frac{2(2n-1)}{n^2(n-1)^2}.$$

(d) The sum $\displaystyle\sum_{n=1}^{\infty} a_n$ is the limit of the sequence of partial sums $\{S_N\}$. Hence:

$$\sum_{n=1}^{\infty} a_n = \lim_{N\to\infty} S_N = \lim_{N\to\infty} \left(5 - \frac{2}{N^2}\right) = 5.$$

37. Use the method of Example 5 to show that $\sum_{k=1}^{\infty} \frac{1}{k^{1/3}}$ diverges.

SOLUTION Each term in the Nth partial sum is greater than or equal to $\frac{1}{N^{\frac{1}{3}}}$, hence:

$$S_N = \frac{1}{1^{1/3}} + \frac{1}{2^{1/3}} + \frac{3}{3^{1/3}} + \cdots + \frac{1}{N^{1/3}} \geq \frac{1}{N^{1/3}} + \frac{1}{N^{1/3}} + \frac{1}{N^{1/3}} + \cdots + \frac{1}{N^{1/3}} = N \cdot \frac{1}{N^{1/3}} = N^{2/3}.$$

Since $\lim_{N\to\infty} N^{2/3} = \infty$, it follows that

$$\lim_{N\to\infty} S_N = \infty.$$

Thus, the series $\sum_{k=1}^{\infty} \frac{1}{k^{1/3}}$ diverges.

39. A ball dropped from a height of 10 ft begins to bounce. Each time it strikes the ground, it returns to two-thirds of its previous height. What is the total distance traveled by the ball if it bounces infinitely many times?

SOLUTION The distance traveled by the ball is shown in the accompanying figure:

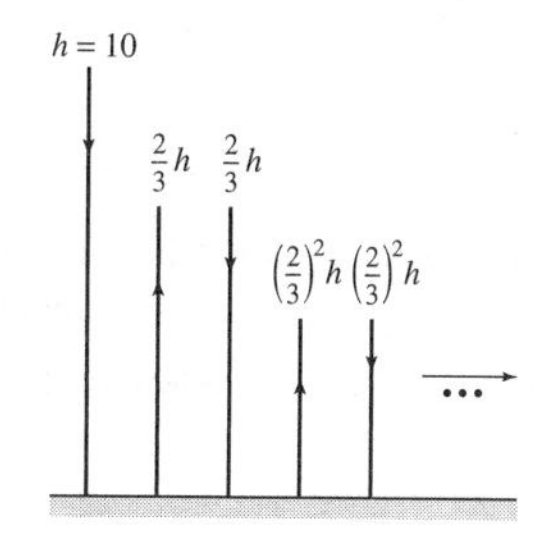

The total distance d traveled by the ball is given by the following infinite sum:

$$d = h + 2 \cdot \frac{2}{3}h + 2 \cdot \left(\frac{2}{3}\right)^2 h + 2 \cdot \left(\frac{2}{3}\right)^3 h + \cdots = h + 2h\left(\frac{2}{3} + \left(\frac{2}{3}\right)^2 + \left(\frac{2}{3}\right)^3 + \cdots\right) = h + 2h\sum_{n=1}^{\infty}\left(\frac{2}{3}\right)^n.$$

We use the formula for the sum of a geometric series to compute the sum of the resulting series:

$$d = h + 2h \cdot \frac{\left(\frac{2}{3}\right)^1}{1 - \frac{2}{3}} = h + 2h(2) = 5h.$$

With $h = 10$ feet, it follows that the total distance traveled by the ball is 50 feet.

41. Find the sum of $\frac{1}{1\cdot 3} + \frac{1}{3\cdot 5} + \frac{1}{5\cdot 7} + \cdots$.

SOLUTION We may write this sum as

$$\sum_{n=1}^{\infty} \frac{1}{(2n-1)(2n+1)} = \sum_{n=1}^{\infty} \frac{1}{2}\left(\frac{1}{2n-1} - \frac{1}{2n+1}\right).$$

The general term in the sequence of partial sums is

$$S_N = \frac{1}{2}\left(\frac{1}{1} - \frac{1}{3}\right) + \frac{1}{2}\left(\frac{1}{3} - \frac{1}{5}\right) + \frac{1}{2}\left(\frac{1}{5} - \frac{1}{7}\right) + \cdots + \frac{1}{2}\left(\frac{1}{2N-1} - \frac{1}{2N+1}\right) = \frac{1}{2}\left(1 - \frac{1}{2N+1}\right);$$

thus,

$$\lim_{N\to\infty} S_N = \lim_{N\to\infty} \frac{1}{2}\left(1 - \frac{1}{2N+1}\right) = \frac{1}{2},$$

and

$$\sum_{n=1}^{\infty} \frac{1}{(2n-1)(2n+1)} = \frac{1}{2}.$$

43. Let $\{b_n\}$ be a sequence and let $a_n = b_n - b_{n-1}$. Show that $\sum_{n=1}^{\infty} a_n$ converges if and only if $\lim_{n\to\infty} b_n$ exists.

SOLUTION Let $a_n = b_n - b_{n-1}$. The general term in the sequence of partial sums for the series $\sum_{n=1}^{\infty} a_n$ is then

$$S_N = (b_1 - b_0) + (b_2 - b_1) + (b_3 - b_2) + \cdots + (b_N - b_{N-1}) = b_N - b_0.$$

Now, if $\lim_{N\to\infty} b_N$ exists, then so does $\lim_{N\to\infty} S_N$ and $\sum_{n=1}^{\infty} a_n$ converges. On the other hand, if $\sum_{n=1}^{\infty} a_n$ converges, then $\lim_{N\to\infty} S_N$ exists, which implies that $\lim_{N\to\infty} b_N$ also exists. Thus, $\sum_{n=1}^{\infty} a_n$ converges if and only if $\lim_{n\to\infty} b_n$ exists.

45. Find the total length of the infinite zigzag path in Figure 4 (each zag occurs at an angle of $\frac{\pi}{4}$).

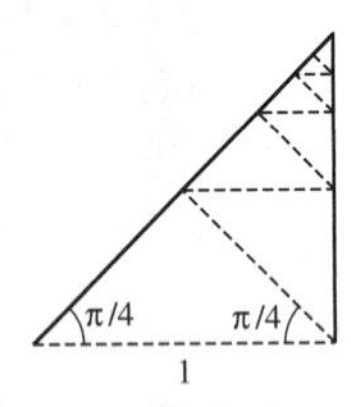

FIGURE 4

SOLUTION Because the angle at the lower left in Figure 4 has measure $\frac{\pi}{4}$ and each zag in the path occurs at an angle of $\frac{\pi}{4}$, every triangle in the figure is an isosceles right triangle. Accordingly, the length of each new segment in the path is $\frac{1}{\sqrt{2}}$ times the length of the previous segment. Since the first segment has length 1, the total length of the path is

$$\sum_{n=0}^{\infty} \left(\frac{1}{\sqrt{2}}\right)^n = \frac{1}{1 - \frac{1}{\sqrt{2}}} = \frac{\sqrt{2}}{\sqrt{2} - 1} = 2 + \sqrt{2}.$$

47. Show that if a is a positive integer, then

$$\sum_{n=1}^{\infty} \frac{1}{n(n+a)} = \frac{1}{a}\left(1 + \frac{1}{2} + \cdots + \frac{1}{a}\right)$$

SOLUTION By partial fraction decomposition

$$\frac{1}{n\,(n+a)} = \frac{A}{n} + \frac{B}{n+a};$$

clearing the denominators gives

$$1 = A(n+a) + Bn.$$

Setting $n = 0$ then yields $A = \frac{1}{a}$, while setting $n = -a$ yields $B = -\frac{1}{a}$. Thus,

$$\frac{1}{n\,(n+a)} = \frac{\frac{1}{a}}{n} - \frac{\frac{1}{a}}{n+a} = \frac{1}{a}\left(\frac{1}{n} - \frac{1}{n+a}\right),$$

and

$$\sum_{n=1}^{\infty} \frac{1}{n(n+a)} = \sum_{n=1}^{\infty} \frac{1}{a}\left(\frac{1}{n} - \frac{1}{n+a}\right).$$

For $N > a$, the Nth partial sum is

$$S_N = \frac{1}{a}\left(1 + \frac{1}{2} + \frac{1}{3} + \cdots + \frac{1}{a}\right) - \frac{1}{a}\left(\frac{1}{N+1} + \frac{1}{N+2} + \frac{1}{N+3} + \cdots + \frac{1}{N+a}\right).$$

Thus,

$$\sum_{n=1}^{\infty} \frac{1}{n(n+a)} = \lim_{N\to\infty} S_N = \frac{1}{a}\left(1 + \frac{1}{2} + \frac{1}{3} + \cdots + \frac{1}{a}\right).$$

Further Insights and Challenges

49. Professor George Andrews of Pennsylvania State University observed that geometric sums can be used to calculate the derivative of $f(x) = x^N$ in a new way. By Eq. (2),

$$1 + r + r^2 + \cdots + r^{N-1} = \frac{1 - r^N}{1 - r} \qquad \boxed{7}$$

Assume that $a \neq 0$ and let $x = ra$. Show that

$$f'(a) = \lim_{x \to a} \frac{x^N - a^N}{x - a} = a^{N-1} \lim_{r \to 1} \frac{r^N - 1}{r - 1}$$

Then use Eq. (7) to evaluate the limit on the right.

SOLUTION According to the definition of derivative of $f(x)$ at $x = a$

$$f'(a) = \lim_{x \to a} \frac{x^N - a^N}{x - a}.$$

Now, let $x = ra$. Then $x \to a$ if and only if $r \to 1$, and

$$f'(a) = \lim_{x \to a} \frac{x^N - a^N}{x - a} = \lim_{r \to 1} \frac{(ra)^N - a^N}{ra - a} = \lim_{r \to 1} \frac{a^N\left(r^N - 1\right)}{a(r - 1)} = a^{N-1} \lim_{r \to 1} \frac{r^N - 1}{r - 1}.$$

By Eq. (7),

$$\frac{1 - r^N}{1 - r} = \frac{r^N - 1}{r - 1} = 1 + r + r^2 + \cdots + r^{N-1},$$

so

$$\lim_{r \to 1} \frac{r^N - 1}{r - 1} = \lim_{r \to 1} \left(1 + r + r^2 + \cdots + r^{N-1}\right) = 1 + 1 + 1^2 + \cdots + 1^{N-1} = N.$$

Therefore, $f'(a) = a^{N-1} \cdot N = Na^{N-1}$

51. Cantor's Disappearing Table (following Larry Knop of Hamilton College) Take a table of length L (Figure 6). At stage 1, remove the section of length $L/4$ centered at the midpoint. Two sections remain, each with a length less than $L/2$. At stage 2, remove sections of length $L/4^2$ from each of these two sections (this stage removes $L/8$ of the table). Now four sections remain, each of length less than $L/4$. At stage 3, remove the four central sections of length $L/4^3$, etc.

(a) Show that at the Nth stage, each remaining section has length less than $L/2^N$ and that the total amount of table removed is

$$L\left(\frac{1}{4} + \frac{1}{8} + \frac{1}{16} + \cdots + \frac{1}{2^{N+1}}\right)$$

(b) Show that in the limit as $N \to \infty$, precisely one-half of the table remains.

This result is curious, because there are no nonzero intervals of table left (at each stage, the remaining sections have a length less than $L/2^N$). So the table has "disappeared." However, we can place any object longer than $L/4$ on the table and it will not fall through since it will not fit through any of the removed sections.

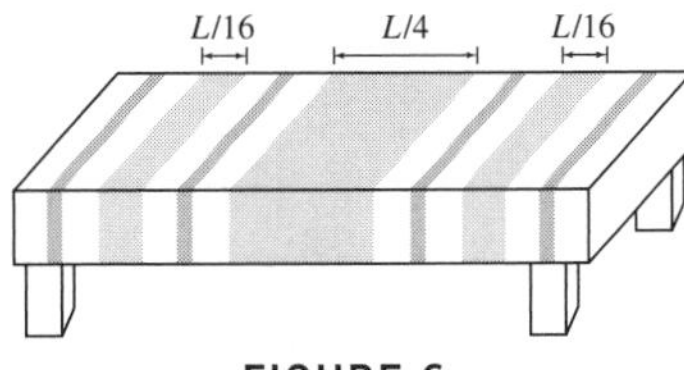

FIGURE 6

SOLUTION

(a) After the Nth stage, the total amount of table that has been removed is

$$\frac{L}{4} + \frac{2L}{4^2} + \frac{4L}{4^3} + \cdots + \frac{2^{N-1}L}{4^N} = L\left(\frac{1}{4} + \frac{1}{8} + \frac{1}{16} + \cdots + \frac{2^{N-1}}{2^{2N}}\right) = L\left(\frac{1}{4} + \frac{1}{8} + \frac{1}{16} + \cdots + \frac{1}{2^{N+1}}\right)$$

At the first stage ($N = 1$), there are two remaining sections each of length

$$\frac{L - \frac{L}{4}}{2} = \frac{3L}{8} < \frac{L}{2}.$$

Suppose that at the Kth stage, each of the 2^K remaining sections has length less than $\frac{L}{2^K}$. The $(K+1)$st stage is obtained by removing the section of length $\frac{L}{4^{K+1}}$ centered at the midpoint of each segment in the Kth stage. Let a_k and a_{K+1}, respectively, denote the length of each segment in the Kth and $(K+1)$st stage. Then,

$$a_{K+1} = \frac{a_K - \frac{L}{4^{K+1}}}{2} < \frac{\frac{L}{2^K} - \frac{L}{4^{K+1}}}{2} = \frac{L}{2^K}\left(\frac{1 - \frac{1}{2^{K+2}}}{2}\right) < \frac{L}{2^K} \cdot \frac{1}{2} = \frac{L}{2^{K+1}}.$$

Thus, by mathematical induction, each remaining section at the Nth stage has length less than $\frac{L}{2^N}$.

(b) From part (a), we know that after N stages, the amount of the table that has been removed is

$$L\left(\frac{1}{4} + \frac{1}{8} + \frac{1}{16} + \cdots + \frac{1}{2^{N+1}}\right) = \sum_{n=1}^{N} \frac{1}{2^{n+1}}.$$

As $N \to \infty$, the amount of the table that has been removed becomes a geometric series whose sum is

$$L \sum_{n=1}^{\infty} \frac{1}{2}\left(\frac{1}{2}\right)^n = L\frac{\frac{1}{4}}{1 - \frac{1}{2}} = \frac{1}{2}L.$$

Thus, the amount of table that remains is $L - \frac{1}{2}L = \frac{1}{2}L$.

11.3 Convergence of Series with Positive Terms (ET Section 10.3)

Preliminary Questions

1. Let $S = \sum_{n=1}^{\infty} a_n$. If the partial sums S_N are increasing, then (choose correct conclusion)

(a) $\{a_n\}$ is an increasing sequence.

(b) $\{a_n\}$ is a positive sequence.

SOLUTION The correct response is **(b)**. Recall that $S_N = a_1 + a_2 + a_3 + \cdots + a_N$; thus, $S_N - S_{N-1} = a_N$. If S_N is increasing, then $S_N - S_{N-1} \ge 0$. It then follows that $a_N \ge 0$; that is, $\{a_n\}$ is a positive sequence.

2. What are the hypotheses of the Integral Test?

SOLUTION The hypotheses for the Integral Test are: A function $f(x)$ such that $a_n = f(n)$ must be positive, decreasing, and continuous for $x \ge 1$.

3. Which test would you use to determine whether $\sum_{n=1}^{\infty} n^{-3.2}$ converges?

SOLUTION Because $n^{-3.2} = \frac{1}{n^{3.2}}$, we see that the indicated series is a p-series with $p = 3.2 > 1$. Therefore, the series converges.

4. Which test would you use to determine whether $\sum_{n=1}^{\infty} \frac{1}{2^n + \sqrt{n}}$ converges?

SOLUTION Because

$$\frac{1}{2^n + \sqrt{n}} < \frac{1}{2^n} = \left(\frac{1}{2}\right)^n,$$

and

$$\sum_{n=1}^{\infty} \left(\frac{1}{2}\right)^n$$

is a convergent geometric series, the comparison test would be an appropriate choice to establish that the given series converges.

5. Ralph hopes to investigate the convergence of $\sum_{n=1}^{\infty} \frac{e^{-n}}{n}$ by comparing it with $\sum_{n=1}^{\infty} \frac{1}{n}$. Is Ralph on the right track?

SOLUTION No, Ralph is not on the right track. For $n \geq 1$,

$$\frac{e^{-n}}{n} < \frac{1}{n};$$

however, $\sum_{n=1}^{\infty} \frac{1}{n}$ is a divergent series. The Comparison Test therefore does not allow us to draw a conclusion about the convergence or divergence of the series $\sum_{n=1}^{\infty} \frac{e^{-n}}{n}$.

Exercises

In Exercises 1–14, use the Integral Test to determine whether the infinite series is convergent.

1. $\sum_{n=1}^{\infty} \frac{1}{n^4}$

SOLUTION Let $f(x) = \frac{1}{x^4}$. This function is continuous, positive and decreasing on the interval $x \geq 1$, so the Integral Test applies. Moreover,

$$\int_1^{\infty} \frac{dx}{x^4} = \lim_{R\to\infty} \int_1^R x^{-4}\,dx = -\frac{1}{3} \lim_{R\to\infty} \left(\frac{1}{R^3} - 1 \right) = \frac{1}{3}.$$

The integral converges; hence, the series $\sum_{n=1}^{\infty} \frac{1}{n^4}$ also converges.

3. $\sum_{n=1}^{\infty} n^{-1/3}$

SOLUTION Let $f(x) = x^{-\frac{1}{3}} = \frac{1}{\sqrt[3]{x}}$. This function is continuous, positive and decreasing on the interval $x \geq 1$, so the Integral Test applies. Moreover,

$$\int_1^{\infty} x^{-1/3}\,dx = \lim_{R\to\infty} \int_1^R x^{-1/3}\,dx = \frac{3}{2} \lim_{R\to\infty} \left(R^{2/3} - 1 \right) = \infty.$$

The integral diverges; hence, the series $\sum_{n=1}^{\infty} n^{-1/3}$ also diverges.

5. $\sum_{n=25}^{\infty} \frac{n^2}{(n^3+9)^{5/2}}$

SOLUTION Let $f(x) = \frac{x^2}{(x^3+9)^{5/2}}$. This function is positive and continuous for $x \geq 25$. Moreover, because

$$f'(x) = \frac{2x(x^3+9)^{5/2} - x^2 \cdot \frac{5}{2}(x^3+9)^{3/2} \cdot 3x^2}{(x^3+9)^5} = \frac{x(36 - 11x^3)}{2(x^3+9)^{7/2}},$$

we see that $f'(x) < 0$ for $x \geq 25$, so f is decreasing on the interval $x \geq 25$. The Integral Test therefore applies. To evaluate the improper integral, we use the substitution $u = x^3 + 9$, $du = 3x^2 dx$. We then find

$$\int_{25}^{\infty} \frac{x^2}{(x^3+9)^{5/2}}\,dx = \lim_{R\to\infty} \int_{25}^{R} \frac{x^2}{(x^3+9)^{5/2}}\,dx = \frac{1}{3} \lim_{R\to\infty} \int_{15634}^{R^3+9} \frac{du}{u^{5/2}}$$

$$= -\frac{2}{9} \lim_{R\to\infty} \left(\frac{1}{(R^3+9)^{3/2}} - \frac{1}{15634^{3/2}} \right) = \frac{2}{9 \cdot 15634^{3/2}}.$$

The integral converges; hence, the series $\sum_{n=25}^{\infty} \frac{n^2}{(n^3+9)^{5/2}}$ also converges.

7. $\sum_{n=1}^{\infty} \frac{1}{n^2+1}$

SOLUTION Let $f(x) = \frac{1}{x^2+1}$. This function is positive, decreasing and continuous on the interval $x \geq 1$, hence the Integral Test applies. Moreover,

$$\int_1^{\infty} \frac{dx}{x^2+1} = \lim_{R\to\infty} \int_1^R \frac{dx}{x^2+1} = \lim_{R\to\infty} \left(\tan^{-1} R - \frac{\pi}{4}\right) = \frac{\pi}{2} - \frac{\pi}{4} = \frac{\pi}{4}.$$

The integral converges; hence, the series $\sum_{n=1}^{\infty} \frac{1}{n^2+1}$ also converges.

9. $\sum_{n=1}^{\infty} ne^{-n^2}$

SOLUTION Let $f(x) = xe^{-x^2}$. This function is continuous and positive on the interval $x \geq 1$. Moreover, because

$$f'(x) = 1 \cdot e^{-x^2} + x \cdot e^{-x^2} \cdot (-2x) = e^{-x^2}\left(1 - 2x^2\right),$$

we see that $f'(x) < 0$ for $x \geq 1$, so f is decreasing on this interval. To compute the improper integral we make the substitution $u = x^2$, $du = 2x\,dx$. Then, we find

$$\int_1^{\infty} xe^{-x^2}\,dx = \lim_{R\to\infty} \int_1^R xe^{-x^2}\,dx = \frac{1}{2}\int_1^{R^2} e^{-u}\,du = -\frac{1}{2}\lim_{R\to\infty}\left(e^{-R^2} - e^{-1}\right) = \frac{1}{2e}.$$

The integral converges; hence, the series $\sum_{n=1}^{\infty} ne^{-n^2}$ also converges.

11. $\sum_{n=1}^{\infty} \frac{1}{2^{\ln n}}$

SOLUTION Note that

$$2^{\ln n} = (e^{\ln 2})^{\ln n} = (e^{\ln n})^{\ln 2} = n^{\ln 2}.$$

Thus,

$$\sum_{n=1}^{\infty} \frac{1}{2^{\ln n}} = \sum_{n=1}^{\infty} \frac{1}{n^{\ln 2}}.$$

Now, let $f(x) = \frac{1}{x^{\ln 2}}$. This function is positive, continuous and decreasing on the interval $x \geq 1$; therefore, the Integral Test applies. Moreover,

$$\int_1^{\infty} \frac{dx}{x^{\ln 2}} = \lim_{R\to\infty} \int_1^R \frac{dx}{x^{\ln 2}} = \frac{1}{1 - \ln 2}\lim_{R\to\infty} (R^{1-\ln 2} - 1) = \infty,$$

because $1 - \ln 2 > 0$. The integral diverges; hence, the series $\sum_{n=1}^{\infty} \frac{1}{2^{\ln n}}$ also diverges.

13. $\sum_{n=1}^{\infty} \frac{\ln n}{n^2}$

SOLUTION Let $f(x) = \frac{\ln x}{x^2}$. Because

$$f'(x) = \frac{\frac{1}{x}\cdot x^2 - 2x\ln x}{x^4} = \frac{x(1 - 2\ln x)}{x^4} = \frac{1 - 2\ln x}{x^3},$$

we see that $f'(x) < 0$ for $x > \sqrt{e} \approx 1.65$. We conclude that f is decreasing on the interval $x \geq 2$. Since f is also positive and continuous on this interval, the Integral Test can be applied. By Integration by Parts, we find

$$\int \frac{\ln x}{x^2}\,dx = -\frac{\ln x}{x} + \int x^{-2}\,dx = -\frac{\ln x}{x} - \frac{1}{x} + C;$$

therefore,

$$\int_2^{\infty} \frac{\ln x}{x^2}\,dx = \lim_{R\to\infty} \int_2^R \frac{\ln x}{x^2}\,dx = \lim_{R\to\infty}\left(\frac{1}{2} + \frac{\ln 2}{2} - \frac{1}{R} - \frac{\ln R}{R}\right) = \frac{1 + \ln 2}{2} - \lim_{R\to\infty}\frac{\ln R}{R}.$$

We compute the resulting limit using L'Hôpital's Rule:

$$\lim_{R\to\infty} \frac{\ln R}{R} = \lim_{R\to\infty} \frac{1/R}{1} = 0.$$

Hence,

$$\int_2^\infty \frac{\ln x}{x^2}\,dx = \frac{1+\ln 2}{2}.$$

The integral converges; therefore, the series $\sum_{n=2}^{\infty} \frac{\ln n}{n^2}$ also converges. Since the convergence of the series is not affected by adding the finite sum $\sum_{n=1}^{1} \frac{\ln n}{n^2}$, the series $\sum_{n=1}^{\infty} \frac{\ln n}{n^2}$ also converges.

15. Use the Comparison Test to show that $\sum_{n=1}^{\infty} \frac{1}{n^3+8n}$ converges. *Hint:* Compare with $\sum_{n=1}^{\infty} n^{-3}$.

SOLUTION We compare the series with the p-series $\sum_{n=1}^{\infty} n^{-3}$. For $n \ge 1$,

$$\frac{1}{n^3+8n} \le \frac{1}{n^3}.$$

Since $\sum_{n=1}^{\infty} \frac{1}{n^3}$ converges (it is a p-series with $p = 3 > 1$), the series $\sum_{n=1}^{\infty} \frac{1}{n^3+8n}$ also converges by the Comparison Test.

17. Let $S = \sum_{n=1}^{\infty} \frac{1}{n+\sqrt{n}}$. Verify that for $n \ge 1$

$$\frac{1}{n+\sqrt{n}} \le \frac{1}{n}, \qquad \frac{1}{n+\sqrt{n}} \le \frac{1}{\sqrt{n}}$$

Can either inequality be used to show that S diverges? Show that $\frac{1}{n+\sqrt{n}} \ge \frac{1}{2n}$ and conclude that S diverges.

SOLUTION For $n \ge 1$, $n + \sqrt{n} \ge n$ and $n + \sqrt{n} \ge \sqrt{n}$. Taking the reciprocal of each of these inequalities yields

$$\frac{1}{n+\sqrt{n}} \le \frac{1}{n} \quad \text{and} \quad \frac{1}{n+\sqrt{n}} \le \frac{1}{\sqrt{n}}.$$

These inequalities indicate that the series $\sum_{n=1}^{\infty} \frac{1}{n+\sqrt{n}}$ is smaller than both $\sum_{n=1}^{\infty} \frac{1}{n}$ and $\sum_{n=1}^{\infty} \frac{1}{\sqrt{n}}$; however, $\sum_{n=1}^{\infty} \frac{1}{n}$ and $\sum_{n=1}^{\infty} \frac{1}{\sqrt{n}}$ both diverge so neither inequality allows us to show that S diverges.

On the other hand, for $n \ge 1$, $n \ge \sqrt{n}$, so $2n \ge n + \sqrt{n}$ and

$$\frac{1}{n+\sqrt{n}} \ge \frac{1}{2n}.$$

The series $\sum_{n=1}^{\infty} \frac{1}{2n} = 2\sum_{n=1}^{\infty} \frac{1}{n}$ diverges, since the harmonic series diverges. The Comparison Test then lets us conclude that the larger series $\sum_{n=1}^{\infty} \frac{1}{n+\sqrt{n}}$ also diverges.

In Exercises 19–31, use the Comparison Test to determine whether the infinite series is convergent.

19. $\sum_{n=1}^{\infty} \frac{1}{n2^n}$

SOLUTION We compare with the geometric series $\sum_{n=1}^{\infty} \left(\frac{1}{2}\right)^n$. For $n \ge 1$,

$$\frac{1}{n2^n} \le \frac{1}{2^n} = \left(\frac{1}{2}\right)^n.$$

Since $\sum_{n=1}^{\infty}\left(\frac{1}{2}\right)^n$ converges (it is a geometric series with $r = \frac{1}{2}$), we conclude by the Comparison Test that $\sum_{n=1}^{\infty}\frac{1}{n2^n}$ also converges.

21. $\sum_{k=1}^{\infty}\frac{k^{1/3}}{k^2+k}$

SOLUTION For $k \geq 1$,

$$\frac{k^{1/3}}{k^2+k} \leq \frac{k^{1/3}}{k^2} = \frac{1}{k^{5/3}}.$$

The series $\sum_{k=1}^{\infty}\frac{1}{k^{5/3}}$ is a p-series with $p = \frac{5}{3} > 1$, so it converges. By the Comparison Test we can therefore conclude that the series $\sum_{k=1}^{\infty}\frac{k^{1/3}}{k^2+k}$ also converges.

23. $\sum_{n=1}^{\infty}\frac{1}{\sqrt{n^3+1}}$

SOLUTION For $n \geq 1$,

$$\frac{1}{\sqrt{n^3+1}} \leq \frac{1}{\sqrt{n^3}} = \frac{1}{n^{3/2}}.$$

The series $\sum_{n=1}^{\infty}\frac{1}{n^{\frac{3}{2}}}$ converges, since it is a p-series with $p = \frac{3}{2} > 1$. By the Comparison Test we can therefore conclude that the series $\sum_{n=1}^{\infty}\frac{1}{\sqrt{n^3+1}}$ also converges.

25. $\sum_{k=1}^{\infty}\frac{\sin^2 k}{k^2}$

SOLUTION For $k \geq 1$, $0 \leq \sin^2 k \leq 1$, so

$$0 \leq \frac{\sin^2 k}{k^2} \leq \frac{1}{k^2}.$$

The series $\sum_{k=1}^{\infty}\frac{1}{k^2}$ is a p-series with $p = 2 > 1$, so it converges. By the Comparison Test we can therefore conclude that the series $\sum_{k=1}^{\infty}\frac{\sin^2 k}{k^2}$ also converges.

27. $\sum_{m=1}^{\infty}\frac{4}{m!+4^m}$

SOLUTION For $m \geq 1$,

$$\frac{4}{m!+4^m} \leq \frac{4}{4^m} = \left(\frac{1}{4}\right)^{m-1}.$$

The series $\sum_{m=1}^{\infty}\left(\frac{1}{4}\right)^{m-1}$ is a geometric series with $r = \frac{1}{4}$, so it converges. By the Comparison Test we can therefore conclude that the series $\sum_{m=1}^{\infty}\frac{4}{m!+4^m}$ also converges.

29. $\sum_{k=1}^{\infty}2^{-k^2}$

SOLUTION For $k \geq 1$, $k^2 \geq k$ and

$$\frac{1}{2^{k^2}} \leq \frac{1}{2^k} = \left(\frac{1}{2}\right)^k.$$

The series $\sum_{k=1}^{\infty}\left(\frac{1}{2}\right)^k$ is a geometric series with $r = \frac{1}{2}$, so it converges. By the Comparison Test we can therefore conclude that the series $\sum_{k=1}^{\infty}\frac{1}{2^{k^2}} = \sum_{k=1}^{\infty}2^{-k^2}$ also converges.

31. $\sum_{n=1}^{\infty}\frac{\ln n}{n^3 + 3\ln n}$

SOLUTION For $n \geq 1$, $0 \leq \ln n \leq n$ and

$$0 \leq \frac{\ln n}{n^3 + 3\ln n} \leq \frac{n}{n^3} = \frac{1}{n^2}.$$

The series $\sum_{n=1}^{\infty}\frac{1}{n^2}$ is a p-series with $p = 2 > 1$, so it converges. By the Comparison Test we can therefore conclude that the series $\sum_{n=1}^{\infty}\frac{\ln n}{n^3 + 3\ln n}$ also converges.

33. Does $\sum_{n=1}^{\infty}\frac{n}{\sqrt{n^2 + c}}$ converge for any c?

SOLUTION Because

$$\lim_{n\to\infty}\frac{n}{\sqrt{n^2 + c}} = \lim_{n\to\infty}\frac{n}{n} = 1 \neq 0,$$

it follows from the Divergence Test that the series $\sum_{n=1}^{\infty}\frac{n}{\sqrt{n^2 + c}}$ diverges for all values of c.

In Exercises 34–42, use the Limit Comparison Test to prove convergence or divergence of the infinite series.

35. $\sum_{n=2}^{\infty}\frac{1}{n^2 - \sqrt{n}}$

SOLUTION Let $a_n = \frac{1}{n^2 - \sqrt{n}}$. For large n, $\frac{1}{n^2 - \sqrt{n}} \approx \frac{1}{n^2}$, so we apply the Limit Comparison Test with $b_n = \frac{1}{n^2}$. We find

$$L = \lim_{n\to\infty}\frac{a_n}{b_n} = \lim_{n\to\infty}\frac{\frac{1}{n^2 - \sqrt{n}}}{\frac{1}{n^2}} = \lim_{n\to\infty}\frac{n^2}{n^2 - \sqrt{n}} = 1.$$

The series $\sum_{n=1}^{\infty}\frac{1}{n^2}$ is a p-series with $p = 2 > 1$, so it converges; hence, the series $\sum_{n=2}^{\infty}\frac{1}{n^2}$ also converges. Because L exists, by the Limit Comparison Test we can conclude that the series $\sum_{n=2}^{\infty}\frac{1}{n^2 - \sqrt{n}}$ converges.

37. $\sum_{n=2}^{\infty}\frac{n^3}{\sqrt{n^7 - 2n^2 + 1}}$

SOLUTION Let a_n be the general term of our series. Observe that

$$a_n = \frac{n^3}{\sqrt{n^7 - 2n^2 + 1}} = \frac{n^{-3} \cdot n^3}{n^{-3} \cdot \sqrt{n^7 - 2n^2 + 1}} = \frac{1}{\sqrt{n - 2n^{-4} + n^{-6}}}$$

This suggests that apply the Limit Comparison Test, comparing our series with

$$\sum_{n=2}^{\infty} b_n = \sum_{n=2}^{\infty}\frac{1}{n^{1/2}}$$

The ratio of the terms is

$$\frac{a_n}{b_n} = \frac{1}{\sqrt{n - 2n^{-4} + n^{-6}}} \cdot \frac{\sqrt{n}}{1} = \frac{1}{\sqrt{1 - 2n^{-5} + n^{-7}}}$$

Hence

$$\lim_{n\to\infty} \frac{a_n}{b_n} = \lim_{n\to\infty} \frac{1}{\sqrt{1-2n^{-5}+n^{-7}}} = 1$$

The p-series $\sum_{n=2}^{\infty} \frac{1}{n^{1/2}}$ diverges since $p = 1/2 < 1$. Therefore, our original series diverges.

39. $\sum_{n=1}^{\infty} \frac{e^n+n}{e^{2n}-n^2}$

SOLUTION Let

$$a_n = \frac{e^n+n}{e^{2n}-n^2} = \frac{e^n+n}{(e^n-n)(e^n+n)} = \frac{1}{e^n-n}.$$

For large n,

$$\frac{1}{e^n-n} \approx \frac{1}{e^n} = e^{-n},$$

so we apply the Limit Comparison Test with $b_n = e^{-n}$. We find

$$L = \lim_{n\to\infty} \frac{a_n}{b_n} = \lim_{n\to\infty} \frac{\frac{1}{e^n-n}}{e^{-n}} = \lim_{n\to\infty} \frac{e^n}{e^n-n} = 1.$$

The series $\sum_{n=1}^{\infty} e^{-n} = \sum_{n=1}^{\infty} \left(\frac{1}{e}\right)^n$ is a geometric series with $r = \frac{1}{e} < 1$, so it converges. Because L exists, by the Limit Comparison Test we can conclude that the series $\sum_{n=1}^{\infty} \frac{e^n+n}{e^{2n}-n^2}$ also converges.

41. $\sum_{n=1}^{\infty} \left(1 - \cos\frac{1}{n}\right)$ *Hint:* Compare with $\sum_{n=1}^{\infty} n^{-2}$.

SOLUTION Let $a_n = 1 - \cos\frac{1}{n}$, and apply the Limit Comparison Test with $b_n = \frac{1}{n^2}$. We find

$$L = \lim_{n\to\infty} \frac{a_n}{b_n} = \lim_{n\to\infty} \frac{1-\cos\frac{1}{n}}{\frac{1}{n^2}} = \lim_{x\to\infty} \frac{1-\cos\frac{1}{x}}{\frac{1}{x^2}} = \lim_{x\to\infty} \frac{-\frac{1}{x^2}\sin\frac{1}{x}}{-\frac{2}{x^3}} = \frac{1}{2}\lim_{x\to\infty} \frac{\sin\frac{1}{x}}{\frac{1}{x}}.$$

As $x \to \infty$, $u = \frac{1}{x} \to 0$, so

$$L = \frac{1}{2}\lim_{x\to\infty} \frac{\sin\frac{1}{x}}{\frac{1}{x}} = \frac{1}{2}\lim_{u\to 0} \frac{\sin u}{u} = \frac{1}{2}.$$

The series $\sum_{n=1}^{\infty} \frac{1}{n^2}$ is a p-series with $p = 2 > 1$, so it converges. Because L exists, by the Limit Comparison Test we can conclude that the series $\sum_{n=1}^{\infty} \left(1 - \cos\frac{1}{n}\right)$ also converges.

43. Show that if $a_n \geq 0$ and $\lim_{n\to\infty} n^2 a_n$ exists, then $\sum_{n=1}^{\infty} a_n$ converges. *Hint:* Show that if M is larger than $\lim_{n\to\infty} n^2 a_n$, then $a_n \leq M/n^2$ for n sufficiently large.

SOLUTION Let $\lim_{n\to\infty} n^2 a_n = L$. Then there exists an integer N such that, for all $n \geq N$,

$$|n^2 a_n - L| < \frac{1}{2}.$$

Thus,

$$0 \leq n^2 a_n < L + \frac{1}{2} \quad \text{or} \quad 0 \leq a_n < \frac{L+\frac{1}{2}}{n^2}.$$

The series $\sum_{n=1}^{\infty} \frac{1}{n^2}$ is a p-series with $p = 2 > 1$, so it converges; hence, the series $\sum_{n=N}^{\infty} \frac{L + \frac{1}{2}}{n^2}$ also converges. By the Comparison Test we can therefore conclude that the series $\sum_{n=N}^{\infty} a_n$ converges. This series continues to converge if we add the finite sum $\sum_{n=1}^{N-1} a_n$; hence, the series $\sum_{n=1}^{\infty} a_n$ converges.

45. Show that $\sum_{n=2}^{\infty} (\ln n)^{-2}$ diverges. *Hint:* Show that for x sufficiently large, $\ln x < x^{1/2}$.

SOLUTION Using L'Hôpital's Rule,

$$\lim_{x \to \infty} \frac{x^{1/2}}{\ln x} = \lim_{x \to \infty} \frac{\frac{1}{2} x^{-1/2}}{\frac{1}{x}} = \lim_{x \to \infty} \frac{1}{2} x^{1/2} = \infty;$$

thus, there exists an integer N such that

$$\frac{n^{1/2}}{\ln n} > 1 \quad \text{or} \quad n^{1/2} > \ln n$$

for all $n \geq N$. Hence,

$$\frac{1}{\ln n} > \frac{1}{n^{1/2}} \quad \text{and} \quad \frac{1}{(\ln n)^2} > \frac{1}{n}$$

for all $n \geq N$. The harmonic series diverges, so the series $\sum_{n=N}^{\infty} \frac{1}{n}$ also diverges. By the Comparison Test we can therefore conclude that the series $\sum_{n=N}^{\infty} \frac{1}{(\ln n)^2}$ diverges. It follows that the series $\sum_{n=2}^{\infty} (\ln n)^{-2}$ also diverges.

47. For which a does $\sum_{n=2}^{\infty} \frac{1}{n^a \ln n}$ converge?

SOLUTION First consider the case $a > 1$. For $n \geq 3$, $\ln n > 1$ and

$$\frac{1}{n^a \ln n} < \frac{1}{n^a}.$$

The series $\sum_{n=1}^{\infty} \frac{1}{n^a}$ is a p-series with $p = a > 1$, so it converges; hence, $\sum_{n=3}^{\infty} \frac{1}{n^a}$ also converges. By the Comparison Test we can therefore conclude that the series $\sum_{n=3}^{\infty} \frac{1}{n^a \ln n}$ converges, which implies the series $\sum_{n=2}^{\infty} \frac{1}{n^a \ln n}$ also converges.

For $a \leq 1$, $n^a \leq n$ so

$$\frac{1}{n^a \ln n} \geq \frac{1}{n \ln n}$$

for $n \geq 2$. Let $f(x) = \frac{1}{x \ln x}$. For $x \geq 2$, this function is continuous, positive and decreasing, so the Integral Test applies. Using the substitution $u = \ln x$, $du = \frac{1}{x}\, dx$, we find

$$\int_2^{\infty} \frac{dx}{x \ln x} = \lim_{R \to \infty} \int_2^R \frac{dx}{x \ln x} = \lim_{R \to \infty} \int_{\ln 2}^{\ln R} \frac{du}{u} = \lim_{R \to \infty} (\ln(\ln R) - \ln(\ln 2)) = \infty.$$

The integral diverges; hence, the series $\sum_{n=2}^{\infty} \frac{1}{n \ln n}$ also diverges. By the Comparison Test we can therefore conclude that the series $\sum_{n=2}^{\infty} \frac{1}{n^a \ln n}$ diverges.

To summarize,

$$\sum_{n=2}^{\infty} \frac{1}{n^a \ln n} \text{ converges for } a > 1 \text{ and diverges for } a \leq 1.$$

In Exercises 49–74, determine convergence or divergence using any method covered so far.

49. $\sum_{n=4}^{\infty} \frac{1}{n^2-9}$

SOLUTION Apply the Limit Comparison Test with $a_n = \frac{1}{n^2-9}$ and $b_n = \frac{1}{n^2}$:

$$L = \lim_{n\to\infty} \frac{a_n}{b_n} = \lim_{n\to\infty} \frac{\frac{1}{n^2-9}}{\frac{1}{n^2}} = \lim_{n\to\infty} \frac{n^2}{n^2-9} = 1.$$

Since the p-series $\sum_{n=1}^{\infty} \frac{1}{n^2}$ converges, the series $\sum_{n=4}^{\infty} \frac{1}{n^2}$ also converges. Because L exists, by the Limit Comparison Test we can conclude that the series $\sum_{n=4}^{\infty} \frac{1}{n^2-9}$ converges.

51. $\sum_{n=1}^{\infty} \frac{\sqrt{n}}{4n+9}$

SOLUTION Apply the Limit Comparison Test with $a_n = \frac{\sqrt{n}}{4n+9}$ and $b_n = \frac{1}{\sqrt{n}}$:

$$L = \lim_{n\to\infty} \frac{a_n}{b_n} = \lim_{n\to\infty} \frac{\frac{\sqrt{n}}{4n+9}}{\frac{1}{\sqrt{n}}} = \lim_{n\to\infty} \frac{n}{4n+9} = \frac{1}{4}.$$

The series $\sum_{n=1}^{\infty} \frac{1}{\sqrt{n}}$ is a divergent p-series. Because $L > 0$, by the Limit Comparison Test we can conclude that the series $\sum_{n=1}^{\infty} \frac{\sqrt{n}}{4n+9}$ also diverges.

53. $\sum_{n=1}^{\infty} \frac{1}{3^{n^2}}$

SOLUTION Because $n^2 \ge n$ for $n \ge 1$, $3^{n^2} \ge 3^n$ and

$$\frac{1}{3^{n^2}} \le \frac{1}{3^n} = \left(\frac{1}{3}\right)^n.$$

The series $\sum_{n=1}^{\infty} \left(\frac{1}{3}\right)^n$ is a geometric series with $r = \frac{1}{3}$, so it converges. By the Comparison Test we can therefore conclude that the series $\sum_{n=1}^{\infty} \frac{1}{3^{n^2}}$ also converges.

55. $\sum_{n=2}^{\infty} \frac{1}{n^{3/2} \ln n}$

SOLUTION For $n \ge 3$, $\ln n > 1$, so $n^{3/2} \ln n > n^{3/2}$ and

$$\frac{1}{n^{3/2} \ln n} < \frac{1}{n^{3/2}}.$$

The series $\sum_{n=1}^{\infty} \frac{1}{n^{3/2}}$ is a convergent p-series, so the series $\sum_{n=3}^{\infty} \frac{1}{n^{3/2}}$ also converges. By the Comparison Test we can therefore conclude that the series $\sum_{n=3}^{\infty} \frac{1}{n^{3/2} \ln n}$ converges. Hence, the series $\sum_{n=2}^{\infty} \frac{1}{n^{3/2} \ln n}$ also converges.

57. $\sum_{n=2}^{\infty} \frac{1}{n^{1/2} \ln n}$

SOLUTION By L'Hôpital's Rule

$$\lim_{x\to\infty} \frac{x^{1/4}}{\ln x} = \lim_{x\to\infty} \frac{\frac{1}{4}x^{-3/4}}{\frac{1}{x}} = \lim_{x\to\infty} \frac{1}{4}x^{1/4} = \infty;$$

so there exists an integer N such that for all $n \geq N$

$$\frac{n^{1/4}}{\ln n} \geq 1 \quad \text{or} \quad n^{1/4} \geq \ln n.$$

Therefore, for $n \geq N$,

$$\frac{1}{n^{1/2}\ln n} \geq \frac{1}{n^{3/4}}.$$

The series $\sum_{n=1}^{\infty} \frac{1}{n^{3/4}}$ is a divergent p-series, so $\sum_{n=N}^{\infty} \frac{1}{n^{3/4}}$ also diverges. By the Comparison Test we can therefore conclude that $\sum_{n=N}^{\infty} \frac{1}{n^{1/2}\ln n}$ diverges. Hence, the series $\sum_{n=2}^{\infty} \frac{1}{n^{1/2}\ln n}$ also diverges.

59. $\sum_{n=2}^{\infty} \frac{n}{e^{n^2}}$

SOLUTION Let $f(x) = xe^{-x^2}$. This function is continuous and positive for $x \geq 2$. Moreover, as

$$f'(x) = x(-2xe^{-x^2}) + e^{-x^2} = (1 - 2x^2)e^{-x^2},$$

we see that $f'(x) < 0$ for $x \geq 2$, so f is decreasing on this interval. The Integral Test therefore applies. Now,

$$\int_2^{\infty} xe^{-x^2}\,dx = \lim_{R\to\infty} \int_2^R xe^{-x^2}\,dx = -\frac{1}{2}\lim_{R\to\infty}\left(e^{-R^2} - e^{-4}\right) = \frac{1}{2}e^{-4}.$$

The integral converges; hence, the series $\sum_{n=2}^{\infty} \frac{n}{e^{n^2}}$ also converges.

61. $\sum_{n=1}^{\infty} \frac{2^n}{3^n - n}$

SOLUTION Apply the Limit Comparison Test with $a_n = \frac{2^n}{3^n - n}$ and $b_n = \frac{2^n}{3^n}$:

$$L = \lim_{n\to\infty} \frac{a_n}{b_n} = \lim_{n\to\infty} \frac{\frac{2^n}{3^n - n}}{\frac{2^n}{3^n}} = \lim_{n\to\infty} \frac{1}{1 - \frac{n}{3^n}}.$$

Now,

$$\lim_{n\to\infty} \frac{n}{3^n} = \lim_{x\to\infty} \frac{x}{3^x} = \lim_{x\to\infty} \frac{1}{3^x \ln 3} = 0,$$

so

$$L = \lim_{n\to\infty} \frac{a_n}{b_n} = \frac{1}{1 - 0} = 1.$$

The series $\sum_{n=1}^{\infty} \left(\frac{2}{3}\right)^n$ is a convergent geometric series. Because L exists, by the Limit Comparison Test we can conclude that the series $\sum_{n=1}^{\infty} \frac{2^n}{3^n - n}$ also converges.

63. $\sum_{n=1}^{\infty} \frac{\tan^{-1} n}{n^2}$

SOLUTION Apply the Limit Comparison Test with $a_n = \frac{\tan^{-1} n}{n^2}$ and $b_n = \frac{1}{n^2}$:

$$L = \lim_{n\to\infty} \frac{a_n}{b_n} = \lim_{n\to\infty} \frac{\frac{\tan^{-1} n}{n^2}}{\frac{1}{n^2}} = \lim_{n\to\infty} \tan^{-1} n = \frac{\pi}{4}.$$

The series $\sum_{n=1}^{\infty} \frac{1}{n^2}$ is a convergent p-series. Because L exists, by the Limit Comparison Test we can conclude that the series $\sum_{n=1}^{\infty} \frac{\tan^{-1} n}{n^2}$ also converges.

65. $\displaystyle\sum_{n=1}^{\infty} \frac{\ln n}{n^3}$

SOLUTION Apply the Limit Comparison Test with $a_n = \dfrac{\ln n}{n^3}$ and $b_n = \dfrac{1}{n^2}$:

$$L = \lim_{n\to\infty} \frac{a_n}{b_n} = \lim_{n\to\infty} \frac{\frac{\ln n}{n^3}}{\frac{1}{n^2}} = \lim_{n\to\infty} \frac{\ln n}{n} = 0.$$

The series $\displaystyle\sum_{n=1}^{\infty} \frac{1}{n^2}$ is a convergent p-series. Because L exists, by the Limit Comparison Test we can conclude that the series $\displaystyle\sum_{n=1}^{\infty} \frac{\ln n}{n^3}$ also converges.

67. $\displaystyle\sum_{n=1}^{\infty} \frac{2+(-1)^n}{n^{3/2}}$

SOLUTION For $n \geq 1$

$$0 < \frac{2+(-1)^n}{n^{3/2}} \leq \frac{2+1}{n^{3/2}} = \frac{3}{n^{3/2}}.$$

The series $\displaystyle\sum_{n=1}^{\infty} \frac{1}{n^{3/2}}$ is a convergent p-series; hence, the series $\displaystyle\sum_{n=1}^{\infty} \frac{3}{n^{3/2}}$ also converges. By the Comparison Test we can therefore conclude that the series $\displaystyle\sum_{n=1}^{\infty} \frac{2+(-1)^n}{n^{3/2}}$ converges.

69. $\displaystyle\sum_{n=1}^{\infty} \frac{2n+1}{4^n}$

SOLUTION For $n \geq 3$, $2n+1 < 2^n$, so

$$\frac{2n+1}{4^n} < \frac{2^n}{4^n} = \left(\frac{1}{2}\right)^n.$$

The series $\displaystyle\sum_{n=1}^{\infty} \left(\frac{1}{2}\right)^n$ is a convergent geometric series, so $\displaystyle\sum_{n=3}^{\infty} \left(\frac{1}{2}\right)^n$ also converges. By the Comparison Test we can therefore conclude that the series $\displaystyle\sum_{n=3}^{\infty} \frac{2n+1}{4^n}$ converges. Finally, the series $\displaystyle\sum_{n=1}^{\infty} \frac{2n+1}{4^n}$ converges.

71. $\displaystyle\sum_{n=1}^{\infty} \frac{n^2-n}{n^5+n}$

SOLUTION First rewrite $a_n = \dfrac{n^2-n}{n^5+n} = \dfrac{n\,(n-1)}{n\,(n^4+1)} = \dfrac{n-1}{n^4+1}$ and observe

$$\frac{n-1}{n^4+1} < \frac{n}{n^4} = \frac{1}{n^3}$$

for $n \geq 1$. The series $\displaystyle\sum_{n=1}^{\infty} \frac{1}{n^3}$ is a convergent p-series, so by the Comparison Test we can conclude that the series $\displaystyle\sum_{n=1}^{\infty} \frac{n^2-n}{n^5+n}$ also converges.

73. $\displaystyle\sum_{n=2}^{\infty} \frac{1}{n^{1.2}\ln n}$

SOLUTION For $n \geq 3$, $\ln n > 1$, so

$$\frac{1}{n^{1.2}\ln n} < \frac{1}{n^{1.2}}.$$

The series $\sum_{n=1}^{\infty} \frac{1}{n^{1.2}}$ is a convergent p-series, so the series $\sum_{n=3}^{\infty} \frac{1}{n^{1.2}}$ also converges. By the Comparison Test we can therefore conclude that $\sum_{n=3}^{\infty} \frac{1}{n^{1.2} \ln n}$ converges. Finally, $\sum_{n=2}^{\infty} \frac{1}{n^{1.2} \ln n}$ also converges.

Approximating Infinite Sums *In Exercises 75–77, let $a_n = f(n)$, where $f(x)$ is a continuous, decreasing function such that*

$$\int_1^{\infty} f(x)\, dx$$

converges.

75. Show that

$$\int_1^{\infty} f(x)\, dx \le \sum_{n=1}^{\infty} a_n \le a_1 + \int_1^{\infty} f(x)\, dx \qquad \boxed{4}$$

SOLUTION From the proof of the Integral Test, we know that

$$a_2 + a_3 + a_4 + \cdots + a_N \le \int_1^N f(x)\, dx \le \int_1^{\infty} f(x)\, dx;$$

that is,

$$S_N - a_1 \le \int_1^{\infty} f(x)\, dx \quad \text{or} \quad S_N \le a_1 + \int_1^{\infty} f(x)\, dx.$$

Also from the proof of the Integral test, we know that

$$\int_1^N f(x)\, dx \le a_1 + a_2 + a_3 + \cdots + a_{N-1} = S_N - a_N \le S_N.$$

Thus,

$$\int_1^N f(x)\, dx \le S_N \le a_1 + \int_1^{\infty} f(x)\, dx.$$

Taking the limit as $N \to \infty$ yields Eq. (4), as desired.

77. Let $S = \sum_{n=1}^{\infty} a_n$. Arguing as in Exercise 75, show that

$$\sum_{n=1}^{M} a_n + \int_{M+1}^{\infty} f(x)\, dx \le S \le \sum_{n=1}^{M+1} a_n + \int_{M+1}^{\infty} f(x)\, dx \qquad \boxed{5}$$

Conclude that

$$0 \le S - \left(\sum_{n=1}^{M} a_n + \int_{M+1}^{\infty} f(x)\, dx \right) \le a_{M+1} \qquad \boxed{6}$$

This yields a method for approximating S with an error of at most a_{M+1}.

SOLUTION Following the proof of the Integral Test and the argument in Exercise 75, but starting with $n = M + 1$ rather than $n = 1$, we obtain

$$\int_{M+1}^{\infty} f(x)\, dx \le \sum_{n=M+1}^{\infty} a_n \le a_{M+1} + \int_{M+1}^{\infty} f(x)\, dx.$$

Adding $\sum_{n=1}^{M} a_n$ to each part of this inequality yields

$$\sum_{n=1}^{M} a_n + \int_{M+1}^{\infty} f(x)\, dx \le \sum_{n=1}^{\infty} a_n = S \le \sum_{n=1}^{M+1} a_n + \int_{M+1}^{\infty} f(x)\, dx.$$

Subtracting $\sum_{n=1}^{M} a_n + \int_{M+1}^{\infty} f(x)\,dx$ from each part of this last inequality then gives us

$$0 \le S - \left(\sum_{n=1}^{M} a_n + \int_{M+1}^{\infty} f(x)\,dx\right) \le a_{M+1}.$$

79. CAS Apply Eq. (5) with $M = 40{,}000$ to show that

$$1.644934066 \le \sum_{n=1}^{\infty} \frac{1}{n^2} \le 1.644934068$$

Is this consistent with Euler's result, according to which this infinite series has sum $\pi^2/6$?

SOLUTION Using Eq. (5) with $f(x) = \frac{1}{x^2}$, $a_n = \frac{1}{n^2}$ and $M = 40000$, we find

$$S_{40000} + \int_{40001}^{\infty} \frac{dx}{x^2} \le \sum_{n=1}^{\infty} \frac{1}{n^2} \le S_{40001} + \int_{40001}^{\infty} \frac{dx}{x^2}.$$

Now,

$$S_{40000} = 1.6449090672;$$

$$S_{40001} = S_{40000} + \frac{1}{40001} = 1.6449090678;$$

and

$$\int_{40001}^{\infty} \frac{dx}{x^2} = \lim_{R\to\infty} \int_{40001}^{R} \frac{dx}{x^2} = -\lim_{R\to\infty}\left(\frac{1}{R} - \frac{1}{40001}\right) = \frac{1}{40001} = 0.0000249994.$$

Thus,

$$1.6449090672 + 0.0000249994 \le \sum_{n=1}^{\infty} \frac{1}{n^2} \le 1.6449090678 + 0.0000249994,$$

or

$$1.6449340665 \le \sum_{n=1}^{\infty} \frac{1}{n^2} \le 1.6449340672.$$

Since $\frac{\pi^2}{6} \approx 1.6449340668$, our approximation is consistent with Euler's result.

81. CAS Using a CAS and Eq. (6), determine the value of $\sum_{n=1}^{\infty} n^{-5}$ to within an error less than 10^{-4}.

SOLUTION Using Eq. (6) with $f(x) = x^{-5}$ and $a_n = n^{-5}$, we have

$$0 \le \sum_{n=1}^{\infty} n^{-5} - \left(\sum_{n=1}^{M+1} n^{-5} + \int_{M+1}^{\infty} x^{-5}\,dx\right) \le (M+1)^{-5}.$$

To guarantee an error less than 10^{-4}, we need $(M+1)^{-5} \le 10^{-4}$. This yields $M \ge 10^{4/5} - 1 \approx 5.3$, so we choose $M = 6$. Now,

$$\sum_{n=1}^{7} n^{-5} = 1.0368498887,$$

and

$$\int_{7}^{\infty} x^{-5}\,dx = \lim_{R\to\infty} \int_{7}^{R} x^{-5}\,dx = -\frac{1}{4}\lim_{R\to\infty}\left(R^{-4} - 7^{-4}\right) = \frac{1}{4\cdot 7^4} = 0.0001041233.$$

Thus,

$$\sum_{n=1}^{\infty} n^{-5} \approx \sum_{n=1}^{7} n^{-5} + \int_{7}^{\infty} x^{-5}\,dx = 1.0368498887 + 0.0001041233 = 1.0369540120.$$

83. Let p_n denote the nth prime number ($p_1 = 2$, $p_2 = 3$, etc.). It is known that there is a constant C such that $p_n \le Cn \ln n$. Prove the divergence of

$$\sum_{n=1}^{\infty} \frac{1}{p_n} = \frac{1}{2} + \frac{1}{3} + \frac{1}{5} + \frac{1}{7} + \frac{1}{11} + \cdots$$

SOLUTION Since $p_n \le Cn \ln n$ for $n \ge 2$,

$$\frac{1}{p_n} \ge \frac{1}{C} \cdot \frac{1}{n \ln n}.$$

Now, let $f(x) = \dfrac{1}{x \ln x}$. For $x \ge 2$, this function is continuous, positive and decreasing, so the Integral Test applies. Using the substitution $u = \ln x$, $du = \frac{1}{x}\,dx$, we find

$$\int_2^{\infty} \frac{dx}{x \ln x} = \lim_{R\to\infty} \int_2^R \frac{dx}{x \ln x} = \lim_{R\to\infty} \int_{\ln 2}^{\ln R} \frac{du}{u} = \lim_{R\to\infty} (\ln(\ln R) - \ln(\ln 2)) = \infty.$$

The integral diverges; hence, the series $\sum_{n=2}^{\infty} \frac{1}{n \ln n}$ also diverges. By the Comparison Test we can therefore conclude that the series $\sum_{n=2}^{\infty} \frac{1}{p_n}$ diverges; hence, $\sum_{n=1}^{\infty} \frac{1}{p_n}$ also diverges.

Further Insights and Challenges

85. Use the Integral Test to prove again that the geometric series $\sum_{n=1}^{\infty} r^n$ converges if $0 < r < 1$ and diverges if $r > 1$.

SOLUTION Let $f(x) = r^x$ on the interval $x \ge 1$. For $0 < r < 1$ this function is decreasing, continuous and positive, so the Integral Test applies.

$$\int_1^{\infty} r^x\,dx = \lim_{R\to\infty} \int_1^R r^x\,dx = \frac{1}{\ln r} \lim_{R\to\infty} (r^R - r) = -\frac{r}{\ln r}.$$

The integral converges; hence, the series $\sum_{n=1}^{\infty} r^n$ also converges. Now, if $r > 1$, $\lim_{n\to\infty} r^n = \infty$, and the series $\sum_{n=1}^{\infty} r^n$ diverges by the Divergence Test.

87. Kummer's Acceleration Method Suppose we wish to approximate $S = \sum_{n=1}^{\infty} \frac{1}{n^2}$. There is a similar telescoping series whose value can be computed exactly (see Example 1 in Section 11.2):

$$\sum_{n=1}^{\infty} \frac{1}{n(n+1)} = 1$$

(a) Verify that

$$S = \sum_{n=1}^{\infty} \frac{1}{n(n+1)} + \sum_{n=1}^{\infty} \left(\frac{1}{n^2} - \frac{1}{n(n+1)}\right)$$

Thus for M large,

$$S \approx 1 + \sum_{n=1}^{M} \frac{1}{n^2(n+1)} \qquad \boxed{7}$$

(b) Explain what has been gained. Why is (7) a better approximation to S than $\sum_{n=1}^{M} \frac{1}{n^2}$?

(c) CAS Compute

$$\sum_{n=1}^{1{,}000} \frac{1}{n^2}, \qquad 1 + \sum_{n=1}^{100} \frac{1}{n^2(n+1)}$$

Which is a better approximation to S, whose exact value is $\pi^2/6$?

SOLUTION

(a) Because the series $\sum_{n=1}^{\infty} \frac{1}{n^2}$ and $\sum_{n=1}^{\infty} \frac{1}{n(n+1)}$ both converge,

$$\sum_{n=1}^{\infty} \frac{1}{n(n+1)} + \sum_{n=1}^{\infty} \left(\frac{1}{n^2} - \frac{1}{n(n+1)} \right) = \sum_{n=1}^{\infty} \frac{1}{n(n+1)} + \sum_{n=1}^{\infty} \frac{1}{n^2} - \sum_{n=1}^{\infty} \frac{1}{n(n+1)} = \sum_{n=1}^{\infty} \frac{1}{n^2} = S.$$

Now,

$$\frac{1}{n^2} - \frac{1}{n(n+1)} = \frac{n+1}{n^2(n+1)} - \frac{n}{n^2(n+1)} = \frac{1}{n^2(n+1)},$$

so, for M large,

$$S \approx 1 + \sum_{n=1}^{M} \frac{1}{n^2(n+1)}.$$

(b) The series $\sum_{n=1}^{\infty} \frac{1}{n^2(n+1)}$ converges more rapidly than $\sum_{n=1}^{\infty} \frac{1}{n^2}$ since the degree of n in the denominator is larger.

(c) Using a computer algebra system, we find

$$\sum_{n=1}^{1000} \frac{1}{n^2} = 1.6439345667 \quad \text{and} \quad 1 + \sum_{n=1}^{100} \frac{1}{n^2(n+1)} = 1.6448848903.$$

The second sum is more accurate because it is closer to the exact solution $\frac{\pi^2}{6} \approx 1.6449340668$.

11.4 Absolute and Conditional Convergence (ET Section 10.4)

Preliminary Questions

1. Suppose that $S = \sum_{n=0}^{\infty} a_n$ is conditionally convergent. Which of the following statements are correct?

(a) $\sum_{n=0}^{\infty} |a_n|$ may or may not converge.

(b) S may or may not converge.

(c) $\sum_{n=0}^{\infty} |a_n|$ diverges.

SOLUTION By definition, because $\sum_{n=1}^{\infty} a_n$ is conditionally convergent, we know that $\sum_{n=1}^{\infty} a_n$ converges but $\sum_{n=1}^{\infty} |a_n|$ diverges. Thus:

(a) This statement is incorrect: $\sum_{n=0}^{\infty} |a_n|$ must diverge.

(b) This statement is incorrect: S must converge.

(c) This statement is correct.

2. Which of the following statements is equivalent to Theorem 1?

(a) If $\sum_{n=0}^{\infty} |a_n|$ diverges, then $\sum_{n=0}^{\infty} a_n$ also diverges.

(b) If $\sum_{n=0}^{\infty} a_n$ diverges, then $\sum_{n=0}^{\infty} |a_n|$ also diverges.

(c) If $\sum_{n=0}^{\infty} a_n$ converges, then $\sum_{n=0}^{\infty} |a_n|$ also converges.

SOLUTION The correct answer is **(b)**: If $\sum_{n=0}^{\infty} a_n$ diverges, then $\sum_{n=0}^{\infty} |a_n|$ also diverges. Take $a_n = (-1)^n \frac{1}{n}$ to see that statements **(a)** and **(c)** are not true in general.

3. Lathika argues that $\sum_{n=1}^{\infty}(-1)^n\sqrt{n}$ is an alternating series and therefore converges. Is Lathika right?

SOLUTION No. Although $\sum_{n=1}^{\infty}(-1)^n\sqrt{n}$ is an alternating series, the terms $a_n = \sqrt{n}$ do not form a decreasing sequence that tends to zero. In fact, $a_n = \sqrt{n}$ is an increasing sequence that tends to ∞, so $\sum_{n=1}^{\infty}(-1)^n\sqrt{n}$ diverges by the Divergence Test.

4. Give an example of a series such that $\sum a_n$ converges but $\sum |a_n|$ diverges.

SOLUTION The series $\sum \frac{(-1)^n}{\sqrt[3]{n}}$ converges by the Leibniz Test, but the positive series $\sum \frac{1}{\sqrt[3]{n}}$ is a divergent p-series.

Exercises

1. Show that $\sum_{n=0}^{\infty}\frac{(-1)^n}{2^n}$ converges absolutely.

SOLUTION The positive series $\sum_{n=0}^{\infty}\frac{1}{2^n}$ is a geometric series with $r = \frac{1}{2}$. Thus, the positive series converges, and the given series converges absolutely.

In Exercises 3–12, determine whether the series converges absolutely, conditionally, or not at all.

3. $\sum_{n=1}^{\infty}\frac{(-1)^n}{\sqrt{n}}$

SOLUTION Let $a_n = \frac{1}{\sqrt{n}}$. Then a_n forms a decreasing sequence that tends to zero; hence, the series $\sum_{n=1}^{\infty}\frac{(-1)^n}{\sqrt{n}}$ converges by the Leibniz Test. However, the positive series $\sum_{n=1}^{\infty}\frac{1}{\sqrt{n}}$ is a divergent p-series, so the original series converges conditionally.

5. $\sum_{n=1}^{\infty}\frac{(-1)^{n-1}}{(1.1)^n}$

SOLUTION The positive series $\sum_{n=1}^{\infty}\left(\frac{1}{1.1}\right)^n$ is a convergent geometric series; thus, the original series converges absolutely.

7. $\sum_{n=2}^{\infty}\frac{(-1)^{n+1}}{n\ln n}$

SOLUTION Let $a_n = \frac{1}{n\ln n}$. Then a_n forms a decreasing sequence (note that n and $\ln n$ are both increasing functions of n) that tends to zero; hence, the series $\sum_{n=2}^{\infty}\frac{(-1)^{n+1}}{n\ln n}$ converges by the Leibniz Test. However, the positive series $\sum_{n=2}^{\infty}\frac{1}{n\ln n}$ diverges, so the original series converges conditionally.

9. $\sum_{n=1}^{\infty}\frac{\sin n\pi}{\sqrt{n}}$

SOLUTION $\sin n\pi = 0$ for all n, so the general term of the series is zero. Consequently, the series, as well as the positive series, converge to zero; that is, the series converges absolutely.

11. $\sum_{n=1}^{\infty}\frac{\cos\frac{1}{n}}{n^2}$

SOLUTION The positive series is $\sum_{n=1}^{\infty}\frac{|\cos\frac{1}{n}|}{n^2}$. Because

$$\frac{|\cos\frac{1}{n}|}{n^2} \le \frac{1}{n^2}$$

for all n, the Comparison Test and the convergence of the p-series $\sum_{n=1}^{\infty} \frac{1}{n^2}$ imply that the series $\sum_{n=1}^{\infty} \frac{|\cos \frac{1}{n}|}{n^2}$ converges. Hence, the original series converges absolutely.

13. Let $S = \sum_{n=1}^{\infty} (-1)^{n+1} \frac{1}{n^3}$.

(a) Calculate S_n for $1 \le n \le 10$.

(b) Use Eq. (2) to show that $0.9 \le S \le 0.902$.

SOLUTION

(a)

$$S_1 = 1 \qquad S_6 = S_5 - \frac{1}{6^3} = 0.899782407$$

$$S_2 = 1 - \frac{1}{2^3} = \frac{7}{8} = 0.875 \qquad S_7 = S_6 + \frac{1}{7^3} = 0.902697859$$

$$S_3 = S_2 + \frac{1}{3^3} = 0.912037037 \qquad S_8 = S_7 - \frac{1}{8^3} = 0.900744734$$

$$S_4 = S_3 - \frac{1}{4^3} = 0.896412037 \qquad S_9 = S_8 + \frac{1}{9^3} = 0.902116476$$

$$S_5 = S_4 + \frac{1}{5^3} = 0.904412037 \qquad S_{10} = S_9 - \frac{1}{10^3} = 0.901116476$$

(b) By Eq. (2),

$$|S_{10} - S| \le a_{11} = \frac{1}{11^3},$$

so

$$S_{10} - \frac{1}{11^3} \le S \le S_{10} + \frac{1}{11^3},$$

or

$$0.900365161 \le S \le 0.901867791.$$

15. Approximate $\sum_{n=1}^{\infty} \frac{(-1)^{n+1}}{n^4}$ to three decimal places.

SOLUTION Let $S = \sum_{n=1}^{\infty} \frac{(-1)^{n+1}}{n^4}$, so that $a_n = \frac{1}{n^4}$. By Eq. (2),

$$|S_N - S| \le a_{N+1} = \frac{1}{(N+1)^4}.$$

To guarantee accuracy to three decimal places, we must choose N so that

$$\frac{1}{(N+1)^4} < 5 \times 10^{-4} \quad \text{or} \quad N > \sqrt[4]{2000} - 1 \approx 5.7.$$

The smallest value that satisfies the required inequality is then $N = 6$. Thus,

$$S \approx S_6 = 1 - \frac{1}{2^4} + \frac{1}{3^4} - \frac{1}{4^4} + \frac{1}{5^4} - \frac{1}{6^4} = 0.946767824.$$

In Exercises 17–18, use Eq. (2) to approximate the value of the series to within an error of at most 10^{-5}.

17. $\sum_{n=1}^{\infty} \frac{(-1)^{n+1}}{n(n+2)(n+3)}$

SOLUTION Let $S = \sum_{n=1}^{\infty} \frac{(-1)^{n+1}}{n\,(n+2)\,(n+3)}$, so that $a_n = \frac{1}{n\,(n+2)\,(n+3)}$. By Eq. (2),

$$|S_N - S| \le a_{N+1} = \frac{1}{(N+1)(N+3)(N+4)}.$$

We must choose N so that

$$\frac{1}{(N+1)(N+3)(N+4)} \leq 10^{-5} \quad \text{or} \quad (N+1)(N+3)(N+4) \geq 10^5.$$

For $N = 43$, the product on the left hand side is 95,128, while for $N = 44$ the product is 101,520; hence, the smallest value of N which satisfies the required inequality is $N = 44$. Thus,

$$S \approx S_{44} = \sum_{n=1}^{44} \frac{(-1)^{n+1}}{n(n+2)(n+3)} = 0.0656746.$$

In Exercises 19–26, determine convergence or divergence by any method.

19. $\displaystyle\sum_{n=1}^{\infty} \frac{1}{3^n + 5^n}$

SOLUTION For $n \geq 1$

$$\frac{1}{3^n + 5^n} \leq \frac{1}{3^n} = \left(\frac{1}{3}\right)^n.$$

The series $\displaystyle\sum_{n=1}^{\infty} \left(\frac{1}{3}\right)^n$ is a convergent geometric series, so the Comparison Test implies that the series $\displaystyle\sum_{n=1}^{\infty} \frac{1}{3^n + 5^n}$ also converges.

21. $\displaystyle\sum_{n=1}^{\infty} \frac{(-1)^n}{\sqrt{n^2+1}}$

SOLUTION This is an alternating series with $a_n = \dfrac{1}{\sqrt{n^2+1}}$. Because a_n is a decreasing sequence that converges to zero, the series $\displaystyle\sum_{n=1}^{\infty} \frac{(-1)^n}{\sqrt{n^2+1}}$ converges by the Leibniz Test.

23. $\displaystyle\sum_{n=1}^{\infty} \frac{3^n + (-1)^n 2^n}{5^n}$

SOLUTION The series $\displaystyle\sum_{n=1}^{\infty} \frac{3^n}{5^n} = \sum_{n=1}^{\infty} \left(\frac{3}{5}\right)^n$ is a convergent geometric series, as is the series $\displaystyle\sum_{n=1}^{\infty} \frac{(-1)^n 2^n}{5^n} = \sum_{n=1}^{\infty} \left(-\frac{2}{5}\right)^n$. Hence,

$$\sum_{n=1}^{\infty} \frac{3^n + (-1)^n 2^n}{5^n} = \sum_{n=1}^{\infty} \left(\frac{3}{5}\right)^n + \sum_{n=1}^{\infty} \left(-\frac{2}{5}\right)^n$$

also converges.

25. $\displaystyle\sum_{n=1}^{\infty} (-1)^n n e^{-n}$

SOLUTION This is an alternating series with $a_n = ne^{-n}$. Consider the function $f(x) = xe^{-x}$. Using L'Hôpital's Rule,

$$\lim_{x\to\infty} \frac{x}{e^x} = \lim_{x\to\infty} \frac{1}{e^x} = 0.$$

Moreover,

$$f'(x) = \frac{e^x - xe^x}{e^{2x}} = \frac{1-x}{e^x},$$

so $f'(x) < 0$ and f is decreasing for $x > 1$. Therefore, $\{a_n\}$ is a decreasing sequence which converges to zero, and $\displaystyle\sum_{n=1}^{\infty} (-1)^n n e^{-n}$ converges by the Leibniz Test.

27. Show that

$$S = \frac{1}{2} - \frac{1}{2} + \frac{1}{3} - \frac{1}{3} + \frac{1}{4} - \frac{1}{4}$$

converges by computing the partial sums. Does it converge absolutely?

SOLUTION The sequence of partial sums is

$$S_1 = \frac{1}{2}$$
$$S_2 = S_1 - \frac{1}{2} = 0$$
$$S_3 = S_2 + \frac{1}{3} = \frac{1}{3}$$
$$S_4 = S_3 - \frac{1}{3} = 0$$

and, in general,

$$S_N = \begin{cases} \frac{1}{N}, & \text{for odd } N \\ 0, & \text{for even } N \end{cases}$$

Thus, $\lim_{N\to\infty} S_N = 0$, and the series converges to 0. The positive series is

$$\frac{1}{2} + \frac{1}{2} + \frac{1}{3} + \frac{1}{3} + \frac{1}{4} + \frac{1}{4} + \cdots = 2\sum_{n=2}^{\infty} \frac{1}{n};$$

which diverges. Therefore, the original series converges conditionally, not absolutely.

29. Determine whether the following series converges conditionally:

$$1 - \frac{1}{3} + \frac{1}{2} - \frac{1}{5} + \frac{1}{3} - \frac{1}{7} + \frac{1}{4} - \frac{1}{9} + \frac{1}{5} - \frac{1}{11} + \cdots$$

SOLUTION Although this is an alternating series, the sequence of terms $\{a_n\}$ is not decreasing, so we cannot apply the Leibniz Test. However, we may express the series as

$$\sum_{n=1}^{\infty} \left(\frac{1}{n} - \frac{1}{2n+1} \right) = \sum_{n=1}^{\infty} \frac{n+1}{n(2n+1)}.$$

Using the Limit Comparison Test and comparing with the harmonic series, we find

$$L = \lim_{n\to\infty} \frac{\frac{n+1}{n(2n+1)}}{\frac{1}{n}} = \lim_{n\to\infty} \frac{n+1}{2n+1} = \frac{1}{2}.$$

Because $L > 0$, we conclude that the series

$$1 - \frac{1}{3} + \frac{1}{2} - \frac{1}{5} + \frac{1}{3} - \frac{1}{7} + \frac{1}{4} - \frac{1}{9} + \frac{1}{5} - \frac{1}{11} + \cdots$$

diverges.

Further Insights and Challenges

31. Prove the following variant of the Leibniz Test: If $\{a_n\}$ is a positive, decreasing sequence with $\lim_{n\to\infty} a_n = 0$, then the series

$$a_1 + a_2 - 2a_3 + a_4 + a_5 - 2a_6 + \cdots$$

converges. *Hint:* Show that S_{3N} is increasing and bounded by $a_1 + a_2$, and continue as in the proof of the Leibniz Test.

SOLUTION Following the hint, we first examine the sequence $\{S_{3N}\}$. Now,

$$S_{3N+3} = S_{3(N+1)} = S_{3N} + a_{3N+1} + a_{3N+2} - 2a_{3N+3} = S_{3N} + (a_{3N+1} - a_{3N+3}) + (a_{3N+2} - a_{3N+3}) \ge S_{3N}$$

because $\{a_n\}$ is a decreasing sequence. Moreover,

$$S_{3N} = a_1 + a_2 - \sum_{k=1}^{N-1} (2a_{3k} - a_{3k+1} - a_{3k+2}) - 2a_{3N}$$
$$= a_1 + a_2 - \sum_{k=1}^{N-1} \left[(a_{3k} - a_{3k+1}) + (a_{3k} - a_{3k+2}) - 2a_{3N} \right] \le a_1 + a_2$$

again because $\{a_n\}$ is a decreasing sequence. Thus, $\{S_{3N}\}$ is an increasing sequence with an upper bound; hence, $\{S_{3N}\}$ converges. Next,

$$S_{3N+1} = S_{3N} + a_{3N+1} \quad \text{and} \quad S_{3N+2} = S_{3N} + a_{3N+1} + a_{3N+2}.$$

Given that $\lim_{n\to\infty} a_n = 0$, it follows that

$$\lim_{N\to\infty} S_{3N+1} = \lim_{N\to\infty} S_{3N+2} = \lim_{N\to\infty} S_{3N}.$$

Having just established that $\lim_{N\to\infty} S_{3N}$ exists, it follows that the sequences $\{S_{3N+1}\}$ and $\{S_{3N+2}\}$ converge to the same limit. Finally, we can conclude that the sequence of partial sums $\{S_N\}$ converges, so the given series converges.

33. Prove the conditional convergence of

$$R = 1 + \frac{1}{2} + \frac{1}{3} - \frac{3}{4} + \frac{1}{5} + \frac{1}{6} + \frac{1}{7} - \frac{3}{8} + \cdots$$

SOLUTION Using Exercise 31 as a template, we first examine the sequence $\{R_{4N}\}$. Now,

$$R_{4N+4} = R_{4(N+1)} = R_{4N} + \frac{1}{4N+1} + \frac{1}{4N+2} + \frac{1}{4N+3} - \frac{3}{4N+4}$$
$$= R_N + \left(\frac{1}{4N+1} - \frac{1}{4N+4}\right) + \left(\frac{1}{4N+2} - \frac{1}{4N+4}\right) + \left(\frac{1}{4N+3} - \frac{1}{4N+4}\right) \geq R_{4N}.$$

Moreover,

$$R_{4N} = 1 + \frac{1}{2} + \frac{1}{3} - \sum_{k=1}^{N-1}\left(\frac{3}{4k} - \frac{1}{4k+1} - \frac{1}{4k+2} - \frac{1}{4k+3}\right) - \frac{3}{4N} \leq 1 + \frac{1}{2} + \frac{1}{3}.$$

Thus, $\{R_{4N}\}$ is an increasing sequence with an upper bound; hence, $\{R_{4N}\}$ converges. Next,

$$R_{4N+1} = R_{4N} + \frac{1}{4N+1};$$
$$R_{4N+2} = R_{4N} + \frac{1}{4N+1} + \frac{1}{4N+2}; \text{ and}$$
$$R_{4N+3} = R_{4N} + \frac{1}{4N+1} + \frac{1}{4N+2} + \frac{1}{4N+3},$$

so

$$\lim_{n\to\infty} R_{4N+1} = \lim_{N\to\infty} R_{4N+2} = \lim_{N\to\infty} R_{4N+3} = \lim_{N\to\infty} R_{4N}.$$

Having just established that $\lim_{N\to\infty} R_{4N}$ exists, it follows that the sequences $\{R_{4N+1}\}$, $\{R_{4N+2}\}$ and $\{R_{4N+3}\}$ converge to the same limit. Finally, we can conclude that the sequence of partial sums $\{R_N\}$ converges, so the series R converges.

Now, consider the positive series

$$R^+ = 1 + \frac{1}{2} + \frac{1}{3} + \frac{3}{4} + \frac{1}{5} + \frac{1}{6} + \frac{1}{7} + \frac{3}{8} + \cdots$$

Because the terms in this series are greater than or equal to the corresponding terms in the divergent harmonic series, it follows from the Comparison Test that R^+ diverges. Thus, by definition, R converges conditionally.

35. **Assumptions Matter** Show by counterexample that the Leibniz Test does not remain true if $\{a_n\}$ tends to zero but we drop the assumption that the sequence a_n is decreasing. *Hint:* Consider

$$R = \frac{1}{2} - \frac{1}{4} + \frac{1}{3} - \frac{1}{8} + \frac{1}{4} - \frac{1}{16} + \cdots + \left(\frac{1}{n} - \frac{1}{2^n}\right) + \cdots$$

SOLUTION Let

$$R = \frac{1}{2} - \frac{1}{4} + \frac{1}{3} - \frac{1}{8} + \frac{1}{4} - \frac{1}{16} + \cdots + \left(\frac{1}{n+1} - \frac{1}{2^{n+1}}\right) + \cdots$$

This is an alternating series with

$$a_n = \begin{cases} \dfrac{1}{k+1}, & n = 2k-1 \\[2ex] \dfrac{1}{2^{k+1}}, & n = 2k \end{cases}$$

Note that $a_n \to 0$ as $n \to \infty$, but the sequence $\{a_n\}$ is not decreasing. We will now establish that R diverges.
For sake of contradiction, suppose that R converges. The geometric series

$$\sum_{n=1}^{\infty} \frac{1}{2^{n+1}}$$

converges, so the sum of R and this geometric series must also converge; however,

$$R + \sum_{n=1}^{\infty} \frac{1}{2^{n+1}} = \sum_{n=2}^{\infty} \frac{1}{n},$$

which diverges because the harmonic series diverges. Thus, the series R must diverge.

37. We say that $\{b_n\}$ is a rearrangement of $\{a_n\}$ if $\{b_n\}$ has the same terms as $\{a_n\}$ but occurring in a different order. Show that if $\{b_n\}$ is a rearrangement of $\{a_n\}$ and $S = \sum_{n=1}^{\infty} a_n$ converges absolutely, then $T = \sum_{n=1}^{\infty} b_n$ also converges absolutely. (This result does not hold if S is only conditionally convergent.) *Hint:* Prove that the partial sums $\sum_{n=1}^{N} |b_n|$ are bounded. It can be shown further that $S = T$.

SOLUTION Suppose the series $S = \sum_{n=1}^{\infty} a_n$ converges absolutely and denote the corresponding positive series by

$$S^+ = \sum_{n=1}^{\infty} |a_n|.$$

Further, let $T_N = \sum_{n=1}^{N} |b_n|$ denote the Nth partial sum of the series $\sum_{n=1}^{\infty} |b_n|$. Because $\{b_n\}$ is a rearrangement of $\{a_n\}$, we know that

$$0 \le T_N \le \sum_{n=1}^{\infty} |a_n| = S^+;$$

that is, the sequence $\{T_N\}$ is bounded. Moreover,

$$T_{N+1} = \sum_{n=1}^{N+1} |b_n| = T_N + |b_{N+1}| \ge T_N;$$

that is, $\{T_N\}$ is increasing. It follows that $\{T_N\}$ converges, so the series $\sum_{n=1}^{\infty} |b_n|$ converges, which means the series $\sum_{n=1}^{\infty} b_n$ converges absolutely.

11.5 The Ratio and Root Tests (ET Section 10.5)

Preliminary Questions

1. In the Ratio Test, is ρ equal to $\lim_{n\to\infty} \left|\frac{a_{n+1}}{a_n}\right|$ or $\lim_{n\to\infty} \left|\frac{a_n}{a_{n+1}}\right|$?

SOLUTION In the Ratio Test ρ is the limit $\lim_{n\to\infty} \left|\frac{a_{n+1}}{a_n}\right|$.

2. Is the Ratio Test conclusive for $\sum_{n=1}^{\infty} \frac{1}{2^n}$? Is it conclusive for $\sum_{n=1}^{\infty} \frac{1}{n}$?

SOLUTION The general term of $\sum_{n=1}^{\infty} \frac{1}{2^n}$ is $a_n = \frac{1}{2^n}$; thus,

$$\left|\frac{a_{n+1}}{a_n}\right| = \frac{1}{2^{n+1}} \cdot \frac{2^n}{1} = \frac{1}{2},$$

and

$$\rho = \lim_{n\to\infty} \left|\frac{a_{n+1}}{a_n}\right| = \frac{1}{2} < 1.$$

Consequently, the Ratio Test guarantees that the series $\sum_{n=1}^{\infty} \frac{1}{2^n}$ converges.

The general term of $\sum_{n=1}^{\infty} \frac{1}{n}$ is $a_n = \frac{1}{n}$; thus,

$$\left|\frac{a_{n+1}}{a_n}\right| = \frac{1}{n+1} \cdot \frac{n}{1} = \frac{n}{n+1},$$

and

$$\rho = \lim_{n\to\infty} \left|\frac{a_{n+1}}{a_n}\right| = \lim_{n\to\infty} \frac{n}{n+1} = 1.$$

The Ratio Test is therefore inconclusive for the series $\sum_{n=1}^{\infty} \frac{1}{n}$.

3. Can the Ratio Test be used to show convergence if the series is only conditionally convergent?

SOLUTION No. The Ratio Test can only establish absolute convergence and divergence, not conditional convergence.

Exercises

In Exercises 1–18, apply the Ratio Test to determine convergence or divergence, or state that the Ratio Test is inconclusive.

1. $\sum_{n=1}^{\infty} \frac{1}{5^n}$

SOLUTION With $a_n = \frac{1}{5^n}$,

$$\left|\frac{a_{n+1}}{a_n}\right| = \frac{1}{5^{n+1}} \cdot \frac{5^n}{1} = \frac{1}{5} \quad \text{and} \quad \rho = \lim_{n\to\infty} \left|\frac{a_{n+1}}{a_n}\right| = \frac{1}{5} < 1.$$

Therefore, the series $\sum_{n=1}^{\infty} \frac{1}{5^n}$ converges by the Ratio Test.

3. $\sum_{n=1}^{\infty} \frac{(-1)^{n-1}}{n^n}$

SOLUTION With $a_n = \frac{(-1)^{n-1}}{n^n}$,

$$\left|\frac{a_{n+1}}{a_n}\right| = \frac{1}{(n+1)^{n+1}} \cdot \frac{n^n}{1} = \frac{1}{n+1}\left(\frac{n}{n+1}\right)^n = \frac{1}{n+1}\left(1+\frac{1}{n}\right)^{-n},$$

and

$$\rho = \lim_{n\to\infty} \left|\frac{a_{n+1}}{a_n}\right| = 0 \cdot \frac{1}{e} = 0 < 1.$$

Therefore, the series $\sum_{n=1}^{\infty} \frac{(-1)^{n-1}}{n^n}$ converges by the Ratio Test.

5. $\sum_{n=1}^{\infty} \frac{n}{n^2+1}$

SOLUTION With $a_n = \frac{n}{n^2+1}$,

$$\left|\frac{a_{n+1}}{a_n}\right| = \frac{n+1}{(n+1)^2+1} \cdot \frac{n^2+1}{n} = \frac{n+1}{n} \cdot \frac{n^2+1}{n^2+2n+2},$$

and

$$\rho = \lim_{n\to\infty} \left|\frac{a_{n+1}}{a_n}\right| = 1 \cdot 1 = 1.$$

Therefore, for the series $\sum_{n=1}^{\infty} \frac{n}{n^2+1}$, the Ratio Test is inconclusive.

We can show that this series diverges by using the Limit Comparison Test and comparing with the divergent harmonic series.

7. $\displaystyle\sum_{n=1}^{\infty} \frac{2^n}{n^{100}}$

SOLUTION With $a_n = \frac{2^n}{n^{100}}$,

$$\left|\frac{a_{n+1}}{a_n}\right| = \frac{2^{n+1}}{(n+1)^{100}} \cdot \frac{n^{100}}{2^n} = 2\left(\frac{n}{n+1}\right)^{100} \quad \text{and} \quad \rho = \lim_{n\to\infty}\left|\frac{a_{n+1}}{a_n}\right| = 2 \cdot 1^{100} = 2 > 1.$$

Therefore, the series $\displaystyle\sum_{n=1}^{\infty} \frac{2^n}{n^{100}}$ diverges by the Ratio Test.

9. $\displaystyle\sum_{n=1}^{\infty} \frac{10^n}{2^{n^2}}$

SOLUTION With $a_n = \frac{10^n}{2^{n^2}}$,

$$\left|\frac{a_{n+1}}{a_n}\right| = \frac{10^{n+1}}{2^{(n+1)^2}} \cdot \frac{2^{n^2}}{10^n} = 10 \cdot \frac{1}{2^{2n+1}} \quad \text{and} \quad \rho = \lim_{n\to\infty}\left|\frac{a_{n+1}}{a_n}\right| = 10 \cdot 0 = 0 < 1.$$

Therefore, the series $\displaystyle\sum_{n=1}^{\infty} \frac{10^n}{2^{n^2}}$ converges by the Ratio Test.

11. $\displaystyle\sum_{n=1}^{\infty} \frac{e^n}{n^n}$

SOLUTION With $a_n = \frac{e^n}{n^n}$,

$$\left|\frac{a_{n+1}}{a_n}\right| = \frac{e^{n+1}}{(n+1)^{n+1}} \cdot \frac{n^n}{e^n} = \frac{e}{n+1}\left(\frac{n}{n+1}\right)^n = \frac{e}{n+1}\left(1+\frac{1}{n}\right)^{-n},$$

and

$$\rho = \lim_{n\to\infty}\left|\frac{a_{n+1}}{a_n}\right| = 0 \cdot \frac{1}{e} = 0 < 1.$$

Therefore, the series $\displaystyle\sum_{n=1}^{\infty} \frac{e^n}{n^n}$ converges by the Ratio Test.

13. $\displaystyle\sum_{n=0}^{\infty} (-1)^n \frac{n!}{4^n}$

SOLUTION With $a_n = (-1)^n \frac{n!}{4^n}$,

$$\left|\frac{a_{n+1}}{a_n}\right| = \frac{(n+1)!}{4^{n+1}} \cdot \frac{4^n}{n!} = \frac{n+1}{4} \quad \text{and} \quad \rho = \lim_{n\to\infty}\left|\frac{a_{n+1}}{a_n}\right| = \infty > 1.$$

Therefore, the series $\displaystyle\sum_{n=0}^{\infty} (-1)^n \frac{n!}{4^n}$ diverges by the Ratio Test.

15. $\displaystyle\sum_{n=2}^{\infty} \frac{1}{n \ln n}$

SOLUTION With $a_n = \frac{1}{n \ln n}$,

$$\left|\frac{a_{n+1}}{a_n}\right| = \frac{1}{(n+1)\ln(n+1)} \cdot \frac{n \ln n}{1} = \frac{n}{n+1} \frac{\ln n}{\ln(n+1)},$$

and

$$\rho = \lim_{n\to\infty}\left|\frac{a_{n+1}}{a_n}\right| = 1 \cdot \lim_{n\to\infty} \frac{\ln n}{\ln(n+1)}.$$

Now,

$$\lim_{n\to\infty} \frac{\ln n}{\ln(n+1)} = \lim_{x\to\infty} \frac{\ln x}{\ln(x+1)} = \lim_{x\to\infty} \frac{1/(x+1)}{1/x} = \lim_{x\to\infty} \frac{x}{x+1} = 1.$$

Thus, $\rho = 1$, and the Ratio Test is inconclusive for the series $\sum_{n=2}^{\infty} \frac{1}{n \ln n}$.

Using the Integral Test, we can show that the series $\sum_{n=2}^{\infty} \frac{1}{n \ln n}$ diverges.

17. $\sum_{n=1}^{\infty} \frac{n^2}{(2n+1)!}$

SOLUTION With $a_n = \frac{n^2}{(2n+1)!}$,

$$\left|\frac{a_{n+1}}{a_n}\right| = \frac{(n+1)^2}{(2n+3)!} \cdot \frac{(2n+1)!}{n^2} = \left(\frac{n+1}{n}\right)^2 \frac{1}{(2n+3)(2n+2)},$$

and

$$\rho = \lim_{n\to\infty} \left|\frac{a_{n+1}}{a_n}\right| = 1^2 \cdot 0 = 0 < 1.$$

Therefore, the series $\sum_{n=1}^{\infty} \frac{n^2}{(2n+1)!}$ converges by the Ratio Test.

19. Show that $\sum_{n=1}^{\infty} n^k 3^{-n}$ converges for all exponents k.

SOLUTION With $a_n = n^k 3^{-n}$,

$$\left|\frac{a_{n+1}}{a_n}\right| = \frac{(n+1)^k 3^{-(n+1)}}{n^k 3^{-n}} = \frac{1}{3}\left(1 + \frac{1}{n}\right)^k,$$

and, for all k,

$$\rho = \lim_{n\to\infty} \left|\frac{a_{n+1}}{a_n}\right| = \frac{1}{3} \cdot 1 = \frac{1}{3} < 1.$$

Therefore, the series $\sum_{n=1}^{\infty} n^k 3^{-n}$ converges for all exponents k by the Ratio Test.

21. Show that $\sum_{n=1}^{\infty} 2^n x^n$ converges if $|x| < \frac{1}{2}$.

SOLUTION With $a_n = 2^n x^n$,

$$\left|\frac{a_{n+1}}{a_n}\right| = \frac{2^{n+1}|x|^{n+1}}{2^n|x|^n} = 2|x| \quad \text{and} \quad \rho = \lim_{n\to\infty} \left|\frac{a_{n+1}}{a_n}\right| = 2|x|.$$

Therefore, $\rho < 1$ and the series $\sum_{n=1}^{\infty} 2^n x^n$ converges by the Ratio Test provided $|x| < \frac{1}{2}$.

23. Show that $\sum_{n=1}^{\infty} \frac{r^n}{n}$ converges if $|r| < 1$.

SOLUTION With $a_n = \frac{r^n}{n}$,

$$\left|\frac{a_{n+1}}{a_n}\right| = \frac{|r|^{n+1}}{n+1} \cdot \frac{n}{|r|^n} = |r|\frac{n}{n+1} \quad \text{and} \quad \rho = \lim_{n\to\infty} \left|\frac{a_{n+1}}{a_n}\right| = 1 \cdot |r| = |r|.$$

Therefore, by the Ratio Test, the series $\sum_{n=1}^{\infty} \frac{r^n}{n}$ converges provided $|r| < 1$.

25. Show that $\sum_{n=1}^{\infty} \frac{n!}{n^n}$ converges. *Hint:* Use $\lim_{n\to\infty} \left(1 + \frac{1}{n}\right)^n = e$.

SOLUTION With $a_n = \frac{n!}{n^n}$,

$$\left|\frac{a_{n+1}}{a_n}\right| = \frac{(n+1)!}{(n+1)^{n+1}} \cdot \frac{n^n}{n!} = \left(\frac{n}{n+1}\right)^n = \left(1 + \frac{1}{n}\right)^{-n},$$

and

$$\rho = \lim_{n\to\infty}\left|\frac{a_{n+1}}{a_n}\right| = \frac{1}{e} < 1.$$

Therefore, the series $\sum_{n=1}^{\infty} \frac{n!}{n^n}$ converges by the Ratio Test.

In Exercises 26–31, assume that $|a_{n+1}/a_n|$ converges to $\rho = \frac{1}{3}$. What can you say about the convergence of the given series?

27. $\sum_{n=1}^{\infty} n^3 a_n$

SOLUTION Let $b_n = n^3 a_n$. Then

$$\rho = \lim_{n\to\infty}\left|\frac{b_{n+1}}{b_n}\right| = \lim_{n\to\infty}\left(\frac{n+1}{n}\right)^3\left|\frac{a_{n+1}}{a_n}\right| = 1^3 \cdot \frac{1}{3} = \frac{1}{3} < 1.$$

Therefore, the series $\sum_{n=1}^{\infty} n^3 a_n$ converges by the Ratio Test.

29. $\sum_{n=1}^{\infty} 3^n a_n$

SOLUTION Let $b_n = 3^n a_n$. Then

$$\rho = \lim_{n\to\infty}\left|\frac{b_{n+1}}{b_n}\right| = \lim_{n\to\infty}\frac{3^{n+1}}{3^n}\left|\frac{a_{n+1}}{a_n}\right| = 3 \cdot \frac{1}{3} = 1.$$

Therefore, the Ratio Test is inconclusive for the series $\sum_{n=1}^{\infty} 3^n a_n$.

31. $\sum_{n=1}^{\infty} a_n^2$

SOLUTION Let $b_n = a_n^2$. Then

$$\rho = \lim_{n\to\infty}\left|\frac{b_{n+1}}{b_n}\right| = \lim_{n\to\infty}\left|\frac{a_{n+1}}{a_n}\right|^2 = \left(\frac{1}{3}\right)^2 = \frac{1}{9} < 1.$$

Therefore, the series $\sum_{n=1}^{\infty} a_n^2$ converges by the Ratio Test.

33. Is the Ratio Test conclusive for the p-series $\sum_{n=1}^{\infty} \frac{1}{n^p}$?

SOLUTION With $a_n = \frac{1}{n^p}$,

$$\left|\frac{a_{n+1}}{a_n}\right| = \frac{1}{(n+1)^p} \cdot \frac{n^p}{1} = \left(\frac{n}{n+1}\right)^p \quad \text{and} \quad \rho = \lim_{n\to\infty}\left|\frac{a_{n+1}}{a_n}\right| = 1^p = 1.$$

Therefore, the Ratio Test is inconclusive for the p-series $\sum_{n=1}^{\infty} \frac{1}{n^p}$.

In Exercises 34–39, use the Root Test to determine convergence or divergence (or state that the test is inconclusive).

35. $\sum_{n=1}^{\infty} \frac{1}{n^n}$

SOLUTION With $a_n = \frac{1}{n^n}$,

$$\sqrt[n]{a_n} = \sqrt[n]{\frac{1}{n^n}} = \frac{1}{n} \quad \text{and} \quad \lim_{n\to\infty} \sqrt[n]{a_n} = 0 < 1.$$

Therefore, the series $\sum_{n=0}^{\infty} \frac{1}{n^n}$ converges by the Root Test.

37. $\sum_{k=0}^{\infty} \left(\frac{k}{3k+1}\right)^k$

SOLUTION With $a_k = \left(\frac{k}{3k+1}\right)^k$,

$$\sqrt[k]{a_k} = \sqrt[k]{\left(\frac{k}{3k+1}\right)^k} = \frac{k}{3k+1} \quad \text{and} \quad \lim_{k\to\infty} \sqrt[k]{a_k} = \frac{1}{3} < 1.$$

Therefore, the series $\sum_{k=0}^{\infty} \left(\frac{k}{3k+1}\right)^k$ converges by the Root Test.

39. $\sum_{n=4}^{\infty} \left(1 + \frac{1}{n}\right)^{-n^2}$

SOLUTION With $a_k = \left(1 + \frac{1}{n}\right)^{-n^2}$,

$$\sqrt[n]{a_n} = \sqrt[n]{\left(1 + \frac{1}{n}\right)^{-n^2}} = \left(1 + \frac{1}{n}\right)^{-n} \quad \text{and} \quad \lim_{n\to\infty} \sqrt[n]{a_n} = e^{-1} < 1.$$

Therefore, the series $\sum_{k=0}^{\infty} \left(1 + \frac{1}{n}\right)^{-n^2}$ converges by the Root Test.

In Exercises 41–52, determine convergence or divergence using any method covered in the text so far.

41. $\sum_{n=1}^{\infty} \frac{2^n + 4^n}{7^n}$

SOLUTION Because the series

$$\sum_{n=1}^{\infty} \frac{2^n}{7^n} = \sum_{n=1}^{\infty} \left(\frac{2}{7}\right)^n \quad \text{and} \quad \sum_{n=1}^{\infty} \frac{4^n}{7^n} = \sum_{n=1}^{\infty} \left(\frac{4}{7}\right)^n$$

are both convergent geometric series, it follows that

$$\sum_{n=1}^{\infty} \frac{2^n + 4^n}{7^n} = \sum_{n=1}^{\infty} \left(\frac{2}{7}\right)^n + \sum_{n=1}^{\infty} \left(\frac{4}{7}\right)^n$$

also converges.

43. $\sum_{n=1}^{\infty} \frac{n^3}{5^n}$

SOLUTION The presence of the exponential term suggests applying the Ratio Test. With $a_n = \frac{n^3}{5^n}$,

$$\left|\frac{a_{n+1}}{a_n}\right| = \frac{(n+1)^3}{5^{n+1}} \cdot \frac{5^n}{n^3} = \frac{1}{5}\left(1 + \frac{1}{n}\right)^3 \quad \text{and} \quad \rho = \lim_{n\to\infty} \left|\frac{a_{n+1}}{a_n}\right| = \frac{1}{5} \cdot 1^3 = \frac{1}{5} < 1.$$

Therefore, the series $\sum_{n=1}^{\infty} \frac{n^3}{5^n}$ converges by the Ratio Test.

45. $\sum_{n=2}^{\infty} \frac{1}{\sqrt{n^3 - n^2}}$

SOLUTION This series is similar to a p-series; because

$$\frac{1}{\sqrt{n^3-n^2}} \approx \frac{1}{\sqrt{n^3}} = \frac{1}{n^{3/2}}$$

for large n, we will apply the Limit Comparison Test comparing with the p-series with $p = \frac{3}{2}$. Now,

$$L = \lim_{n\to\infty} \frac{\frac{1}{\sqrt{n^3-n^2}}}{\frac{1}{n^{3/2}}} = \lim_{n\to\infty} \sqrt{\frac{n^3}{n^3-n^2}} = 1.$$

The p-series with $p = \frac{3}{2}$ converges and L exists; therefore, the series $\sum_{n=2}^{\infty} \frac{1}{\sqrt{n^3-n^2}}$ also converges.

47. $\sum_{n=1}^{\infty} \frac{n^2+4n}{3n^4+9}$

SOLUTION This series is similar to a p-series; because

$$\frac{n^2+4n}{3n^4+9} \approx \frac{n^2}{\sqrt{3n^4}} = \frac{1}{3n^2}$$

for large n, we will apply the Limit Comparison Test comparing with the p-series with $p = 2$. Now,

$$L = \lim_{n\to\infty} \frac{\frac{n^2+4n}{3n^4+9}}{\frac{1}{n^2}} = \lim_{n\to\infty} \frac{n^4+4n^3}{3n^4+9} = \frac{1}{3}.$$

The p-series with $p = 2$ converges and L exists; therefore, the series $\sum_{n=1}^{\infty} \frac{n^2+4n}{3n^4+9}$ also converges.

49. $\sum_{n=1}^{\infty} \sin \frac{1}{n^2}$

SOLUTION Here, we will apply the Limit Comparison Test, comparing with the p-series with $p = 2$. Now,

$$L = \lim_{n\to\infty} \frac{\sin \frac{1}{n^2}}{\frac{1}{n^2}} = \lim_{u\to 0} \frac{\sin u}{u} = 1,$$

where $u = \frac{1}{n^2}$. The p-series with $p = 2$ converges and L exists; therefore, the series $\sum_{n=1}^{\infty} \sin \frac{1}{n^2}$ also converges.

51. $\sum_{n=1}^{\infty} \left(\frac{n}{n+12}\right)^n$

SOLUTION Because the general term has the form of a function of n raised to the nth power, we might be tempted to use the Root Test; however, the Root Test is inconclusive for this series. Instead, note

$$\lim_{n\to\infty} a_n = \lim_{n\to\infty} \left(1+\frac{12}{n}\right)^{-n} = \lim_{n\to\infty} \left[\left(1+\frac{12}{n}\right)^{n/12}\right]^{-12} = e^{-12} \neq 0.$$

Therefore, the series diverges by the Divergence Test.

Further Insights and Challenges

53. **Proof of the Root Test** Let $S = \sum_{n=0}^{\infty} a_n$ be a positive series and assume that $L = \lim_{n\to\infty} \sqrt[n]{a_n}$ exists.

(a) Show that S converges if $L < 1$. *Hint:* Choose R with $\rho < R < 1$ and show that $a_n \le R^n$ for n sufficiently large. Then compare with the geometric series $\sum R^n$.

(b) Show that S diverges if $L > 1$.

SOLUTION Suppose $\lim_{n\to\infty} \sqrt[n]{a_n} = L$ exists.

(a) If $L < 1$, let $\epsilon = \dfrac{1-L}{2}$. By the definition of a limit, there is a positive integer N such that

$$-\epsilon \leq \sqrt[n]{a_n} - L \leq \epsilon$$

for $n \geq N$. From this, we conclude that

$$0 \leq \sqrt[n]{a_n} \leq L + \epsilon$$

for $n \geq N$. Now, let $R = L + \epsilon$. Then

$$R = L + \frac{1-L}{2} = \frac{L+1}{2} < \frac{1+1}{2} = 1,$$

and

$$0 \leq \sqrt[n]{a_n} \leq R \quad \text{or} \quad 0 \leq a_n \leq R^n$$

for $n \geq N$. Because $0 \leq R < 1$, the series $\sum_{n=N}^{\infty} R^n$ is a convergent geometric series, so the series $\sum_{n=N}^{\infty} a_n$ converges by the Comparison Test. Therefore, the series $\sum_{n=0}^{\infty} a_n$ also converges.

(b) If $L > 1$, let $\epsilon = \dfrac{L-1}{2}$. By the definition of a limit, there is a positive integer N such that

$$-\epsilon \leq \sqrt[n]{a_n} - L \leq \epsilon$$

for $n \geq N$. From this, we conclude that

$$L - \epsilon \leq \sqrt[n]{a_n}$$

for $n \geq N$. Now, let $R = L - \epsilon$. Then

$$R = L - \frac{L-1}{2} = \frac{L+1}{2} > \frac{1+1}{2} = 1,$$

and

$$R \leq \sqrt[n]{a_n} \quad \text{or} \quad R^n \leq a_n$$

for $n \geq N$. Because $R > 1$, the series $\sum_{n=N}^{\infty} R^n$ is a divergent geometric series, so the series $\sum_{n=N}^{\infty} a_n$ diverges by the Comparison Test. Therefore, the series $\sum_{n=0}^{\infty} a_n$ also diverges.

55. Let $S = \displaystyle\sum_{n=1}^{\infty} \frac{c^n n!}{n^n}$, where c is a constant.

(a) Prove that S converges absolutely if $|c| < e$ and diverges if $|c| > e$.

(b) It is known that $\displaystyle\lim_{n\to\infty} \frac{e^n n!}{n^{n+1/2}} = \sqrt{2\pi}$. Verify this numerically.

(c) Use the Limit Comparison Test to prove that S diverges for $c = e$.

SOLUTION

(a) With $a_n = \frac{c^n n!}{n^n}$,

$$\left|\frac{a_{n+1}}{a_n}\right| = \frac{|c|^{n+1}(n+1)!}{(n+1)^{n+1}} \cdot \frac{n^n}{|c|^n n!} = |c|\left(\frac{n}{n+1}\right)^n = |c|\left(1 + \frac{1}{n}\right)^{-n},$$

and

$$\rho = \lim_{n\to\infty}\left|\frac{a_{n+1}}{a_n}\right| = |c|e^{-1}.$$

Thus, by the Ratio Test, the series $\displaystyle\sum_{n=1}^{\infty} \frac{c^n n!}{n^n}$ converges when $|c|e^{-1} < 1$, or when $|c| < e$. The series diverges when $|c| > e$.

(b) The table below lists the value of $\frac{e^n n!}{n^{n+1/2}}$ for several increasing values of n. Since $\sqrt{2\pi} = 2.506628275$, the numerical evidence verifies that

$$\lim_{n\to\infty} \frac{e^n n!}{n^{n+1/2}} = \sqrt{2\pi}.$$

n	100	1000	10000	100000
$\frac{e^n n!}{n^{n+1/2}}$	2.508717995	2.506837169	2.506649163	2.506630363

(c) With $c = e$, the series S becomes $\sum_{n=1}^{\infty} \frac{e^n n!}{n^n}$. Using the result from part (b),

$$L = \lim_{n\to\infty} \frac{\frac{e^n n!}{n^n}}{\sqrt{n}} = \lim_{n\to\infty} \frac{e^n n!}{n^{n+1/2}} = \sqrt{2\pi}.$$

Because the series $\sum_{n=1}^{\infty} \sqrt{n}$ diverges by the Divergence Test and $L > 0$, we conclude that $\sum_{n=1}^{\infty} \frac{e^n n!}{n^n}$ diverges by the Limit Comparison Test.

11.6 Power Series (ET Section 10.6)

Preliminary Questions

1. Suppose that $\sum a_n x^n$ converges for $x = 5$. Must it also converge for $x = 4$? What about $x = -3$?

SOLUTION The power series $\sum a_n x^n$ is centered at $x = 0$. Because the series converges for $x = 5$, the radius of convergence must be at least 5 and the series converges absolutely at least for the interval $|x| < 5$. Both $x = 4$ and $x = -3$ are inside this interval, so the series converges for $x = 4$ and for $x = -3$.

2. Suppose that $\sum a_n (x-6)^n$ converges for $x = 10$. At which of the points (a)–(d) must it also converge?

(a) $x = 8$ **(b)** $x = 12$ **(c)** $x = 2$ **(d)** $x = 0$

SOLUTION The given power series is centered at $x = 6$. Because the series converges for $x = 10$, the radius of convergence must be at least $|10 - 6| = 4$ and the series converges absolutely at least for the interval $|x - 6| < 4$, or $2 < x < 10$.

(a) $x = 8$ is inside the interval $2 < x < 10$, so the series converges for $x = 8$.

(b) $x = 12$ is not inside the interval $2 < x < 10$, so the series may or may not converge for $x = 12$.

(c) $x = 2$ is an endpoint of the interval $2 < x < 10$, so the series may or may not converge for $x = 2$.

(d) $x = 0$ is not inside the interval $2 < x < 10$, so the series may or may not converge for $x = 0$.

3. Suppose that $F(x)$ is a power series with radius of convergence $R = 12$. What is the radius of convergence of $F(3x)$?

SOLUTION If the power series $F(x)$ has radius of convergence $R = 12$, then the power series $F(3x)$ has radius of convergence $R = \frac{12}{3} = 4$.

4. The power series $F(x) = \sum_{n=1}^{\infty} nx^n$ has radius of convergence $R = 1$. What is the power series expansion of $F'(x)$ and what is its radius of convergence?

SOLUTION We obtain the power series expansion for $F'(x)$ by differentiating the power series expansion for $F(x)$ term-by-term. Thus,

$$F'(x) = \sum_{n=1}^{\infty} n^2 x^{n-1}.$$

The radius of convergence for this series is $R = 1$, the same as the radius of convergence for the series expansion for $F(x)$.

Exercises

1. Use the Ratio Test to determine the radius of convergence of $\sum_{n=0}^{\infty} \frac{x^n}{2^n}$.

SOLUTION With $a_n = \frac{x^n}{2^n}$,

$$\left|\frac{a_{n+1}}{a_n}\right| = \frac{|x|^{n+1}}{2^{n+1}} \cdot \frac{2^n}{|x|^n} = \frac{|x|}{2} \quad \text{and} \quad \rho = \lim_{n\to\infty} \left|\frac{a_{n+1}}{a_n}\right| = \frac{|x|}{2}.$$

By the Ratio Test, the series converges when $\rho = \frac{|x|}{2} < 1$, or $|x| < 2$, and diverges when $\rho = \frac{|x|}{2} > 1$, or $|x| > 2$. The radius of convergence is therefore $R = 2$.

3. Show that the following three power series have the same radius of convergence. Then show that (a) diverges at both endpoints, (b) converges at one endpoint but diverges at the other, and (c) converges at both endpoints.

(a) $\sum_{n=1}^{\infty} \frac{x^n}{3^n}$ **(b)** $\sum_{n=1}^{\infty} \frac{x^n}{n3^n}$ **(c)** $\sum_{n=1}^{\infty} \frac{x^n}{n^2 3^n}$

SOLUTION

(a) With $a_n = \frac{1}{3^n}$,

$$\left|\frac{a_{n+1}}{a_n}\right| = \frac{1}{3^{n+1}} \cdot \frac{3^n}{1} = \frac{1}{3} \quad \text{and} \quad r = \lim_{n\to\infty} \left|\frac{a_{n+1}}{a_n}\right| = \frac{1}{3}.$$

The radius of convergence is therefore $R = r^{-1} = 3$. For the endpoint $x = 3$, the series becomes

$$\sum_{n=1}^{\infty} \frac{3^n}{3^n} = \sum_{n=1}^{\infty} 1,$$

which diverges by the Divergence Test. For the endpoint $x = -3$, the series becomes

$$\sum_{n=1}^{\infty} \frac{(-3)^n}{3^n} = \sum_{n=1}^{\infty} (-1)^n,$$

which also diverges by the Divergence Test.

(b) With $a_n = \frac{1}{n3^n}$,

$$\left|\frac{a_{n+1}}{a_n}\right| = \frac{1}{(n+1)3^{n+1}} \cdot \frac{n3^n}{1} = \frac{1}{3}\left(\frac{n}{n+1}\right) \quad \text{and} \quad r = \lim_{n\to\infty} \left|\frac{a_{n+1}}{a_n}\right| = \frac{1}{3} \cdot 1 = \frac{1}{3}.$$

The radius of convergence is therefore $R = r^{-1} = 3$. For the endpoint $x = 3$, the series becomes

$$\sum_{n=1}^{\infty} \frac{3^n}{n3^n} = \sum_{n=1}^{\infty} \frac{1}{n},$$

which is the divergent harmonic series. For the endpoint $x = -3$, the series becomes

$$\sum_{n=1}^{\infty} \frac{(-3)^n}{n3^n} = \sum_{n=1}^{\infty} \frac{(-1)^n}{n},$$

which converges by the Leibniz Test.

(c) With $a_n = \frac{1}{n^2 3^n}$,

$$\left|\frac{a_{n+1}}{a_n}\right| = \frac{1}{(n+1)^2 3^{n+1}} \cdot \frac{n^2 3^n}{1} = \frac{1}{3}\left(\frac{n}{n+1}\right)^2 \quad \text{and} \quad r = \lim_{n\to\infty} \left|\frac{a_{n+1}}{a_n}\right| = \frac{1}{3} \cdot 1^2 = \frac{1}{3}.$$

The radius of convergence is therefore $R = r^{-1} = 3$. For the endpoint $x = 3$, the series becomes

$$\sum_{n=1}^{\infty} \frac{3^n}{n^2 3^n} = \sum_{n=1}^{\infty} \frac{1}{n^2},$$

which is a convergent p-series. For the endpoint $x = -3$, the series becomes

$$\sum_{n=1}^{\infty} \frac{(-3)^n}{n^2 3^n} = \sum_{n=1}^{\infty} \frac{(-1)^n}{n^2},$$

which converges by the Leibniz Test.

5. Show that $\sum_{n=0}^{\infty} n^n x^n$ diverges for all $x \neq 0$.

SOLUTION With $a_n = n^n$,

$$\left|\frac{a_{n+1}}{a_n}\right| = \frac{(n+1)^{n+1}}{n^n} = \left(1 + \frac{1}{n}\right)^n (n+1) \quad \text{and} \quad r = \lim_{n\to\infty}\left|\frac{a_{n+1}}{a_n}\right| = \infty.$$

The radius of convergence is therefore $R = r^{-1} = 0$. In other words, the power series converges only for $x = 0$.

In Exercises 7–26, find the values of x for which the following power series converge.

7. $\sum_{n=1}^{\infty} nx^n$

SOLUTION With $a_n = n$,

$$\left|\frac{a_{n+1}}{a_n}\right| = \frac{n+1}{n} \quad \text{and} \quad r = \lim_{n\to\infty}\left|\frac{a_{n+1}}{a_n}\right| = 1.$$

The radius of convergence is therefore $R = r^{-1} = 1$, and the series converges absolutely on the interval $|x| < 1$, or $-1 < x < 1$. For the endpoint $x = 1$, the series becomes $\sum_{n=1}^{\infty} n$, which diverges by the Divergence Test. For the endpoint $x = -1$, the series becomes $\sum_{n=1}^{\infty} (-1)^n n$, which also diverges by the Divergence Test. Thus, the series $\sum_{n=1}^{\infty} nx^n$ converges for $-1 < x < 1$ and diverges elsewhere.

9. $\sum_{n=1}^{\infty} \frac{2^n x^n}{n}$

SOLUTION With $a_n = \frac{2^n}{n}$,

$$\left|\frac{a_{n+1}}{a_n}\right| = \frac{2^{n+1}}{n+1} \cdot \frac{n}{2^n} = 2\frac{n}{n+1} \quad \text{and} \quad r = \lim_{n\to\infty}\left|\frac{a_{n+1}}{a_n}\right| = 2 \cdot 1 = 2.$$

The radius of convergence is therefore $R = r^{-1} = \frac{1}{2}$, and the series converges absolutely on the interval $|x| < \frac{1}{2}$, or $-\frac{1}{2} < x < \frac{1}{2}$. For the endpoint $x = \frac{1}{2}$, the series becomes $\sum_{n=1}^{\infty} \frac{1}{n}$, which is the divergent harmonic series. For the endpoint $x = -\frac{1}{2}$, the series becomes $\sum_{n=1}^{\infty} \frac{(-1)^n}{n}$, which converges by the Leibniz Test. Thus, the series $\sum_{n=1}^{\infty} \frac{2^n x^n}{n}$ converges for $-\frac{1}{2} \le x < \frac{1}{2}$ and diverges elsewhere.

11. $\sum_{n=2}^{\infty} \frac{x^n}{\ln n}$

SOLUTION With $a_n = \frac{1}{\ln n}$,

$$\left|\frac{a_{n+1}}{a_n}\right| = \frac{1}{\ln(n+1)} \cdot \frac{\ln n}{1} = \frac{\ln n}{\ln(n+1)} \quad \text{and} \quad r = \lim_{n\to\infty}\left|\frac{a_{n+1}}{a_n}\right| = 1.$$

The radius of convergence is therefore $R = r^{-1} = 1$, and the series converges absolutely on the interval $|x| < 1$, or $-1 < x < 1$. For the endpoint $x = 1$, the series becomes $\sum_{n=1}^{\infty} \frac{1}{\ln n}$. Because $\frac{1}{\ln n} > \frac{1}{n}$ and $\sum_{n=1}^{\infty} \frac{1}{n}$ is the divergent harmonic series, the endpoint series diverges by the Comparison Test. For the endpoint $x = -1$, the series becomes $\sum_{n=1}^{\infty} \frac{(-1)^n}{\ln n}$, which converges by the Leibniz Test. Thus, the series $\sum_{n=2}^{\infty} \frac{x^n}{\ln n}$ converges for $-1 \le x < 1$ and diverges elsewhere.

13. $\sum_{n=1}^{\infty} \frac{x^n}{(n!)^2}$

SOLUTION With $a_n = \frac{1}{(n!)^2}$,

$$\left|\frac{a_{n+1}}{a_n}\right| = \frac{1}{((n+1)!)^2} \cdot \frac{(n!)^2}{1} = \left(\frac{1}{n+1}\right)^2 \quad \text{and} \quad r = \lim_{n\to\infty}\left|\frac{a_{n+1}}{a_n}\right| = 0.$$

The radius of convergence is therefore $R = r^{-1} = \infty$, and the series converges absolutely for all x.

15. $\sum_{n=1}^{\infty} (-1)^n n^4 (x+4)^n$

SOLUTION With $a_n = (-1)^n n^4$,

$$\left|\frac{a_{n+1}}{a_n}\right| = \frac{(n+1)^4}{n^4} = \left(1 + \frac{1}{n}\right)^4 \quad \text{and} \quad r = \lim_{n\to\infty}\left|\frac{a_{n+1}}{a_n}\right| = 1^4 = 1.$$

The radius of convergence is therefore $R = r^{-1} = 1$, and the series converges absolutely on the interval $|x+4| < 1$, or $-5 < x < -3$. For the endpoint $x = -3$, the series becomes $\sum_{n=1}^{\infty} (-1)^n n^4$, which diverges by the Divergence Test. For the endpoint $x = -5$, the series becomes $\sum_{n=1}^{\infty} n^4$, which also diverges by the Divergence Test. Thus, the series $\sum_{n=1}^{\infty} (-1)^n n^4 (x+4)^n$ converges for $-5 < x < -3$ and diverges elsewhere.

17. $\sum_{n=0}^{\infty} \frac{n}{2^n} x^n$

SOLUTION With $a_n = \frac{n}{2^n}$,

$$\left|\frac{a_{n+1}}{a_n}\right| = \frac{n+1}{2^{n+1}} \cdot \frac{2^n}{n} = \frac{1}{2}\left(1 + \frac{1}{n}\right) \quad \text{and} \quad r = \lim_{n\to\infty}\left|\frac{a_{n+1}}{a_n}\right| = \frac{1}{2} \cdot 1 = \frac{1}{2}.$$

The radius of convergence is therefore $R = r^{-1} = 2$, and the series converges absolutely on the interval $|x| < 2$, or $-2 < x < 2$. For the endpoint $x = 2$, the series becomes $\sum_{n=1}^{\infty} n$, which diverges by the Divergence Test. For the endpoint $x = -2$, the series becomes $\sum_{n=1}^{\infty} (-1)^n n$, which also diverges by the Divergence Test. Thus, the series $\sum_{n=0}^{\infty} \frac{n}{2^n} x^n$ converges for $-2 < x < 2$ and diverges elsewhere.

19. $\sum_{n=1}^{\infty} \frac{(x-4)^n}{n^4}$

SOLUTION With $a_n = \frac{1}{n^4}$,

$$\left|\frac{a_{n+1}}{a_n}\right| = \frac{1}{(n+1)^4} \cdot \frac{n^4}{1} = \left(\frac{n}{n+1}\right)^4 \quad \text{and} \quad r = \lim_{n\to\infty}\left|\frac{a_{n+1}}{a_n}\right| = 1^4 = 1.$$

The radius of convergence is therefore $R = r^{-1} = 1$, and the series converges absolutely on the interval $|x-4| < 1$, or $3 < x < 5$. For the endpoint $x = 5$, the series becomes $\sum_{n=1}^{\infty} \frac{1}{n^4}$, which is a convergent p-series. For the endpoint $x = 3$, the series becomes $\sum_{n=1}^{\infty} \frac{(-1)^n}{n^4}$, which converges by the Leibniz Test. Thus, the series $\sum_{n=1}^{\infty} \frac{(x-4)^n}{n^4}$ converges for $3 \le x \le 5$ and diverges elsewhere.

21. $\sum_{n=10}^{\infty} n!\,(x+5)^n$

SOLUTION With $a_n = n!$,

$$\left|\frac{a_{n+1}}{a_n}\right| = \frac{(n+1)!}{n!} = n+1 \quad \text{and} \quad r = \lim_{n\to\infty}\left|\frac{a_{n+1}}{a_n}\right| = \infty.$$

The radius of convergence is therefore $R = r^{-1} = 0$, and the series converges absolutely only for $x = -5$.

23. $\sum_{n=12}^{\infty} e^n (x-2)^n$

SOLUTION With $a_n = e^n$,

$$\left|\frac{a_{n+1}}{a_n}\right| = \frac{e^{n+1}}{e^n} = e \quad \text{and} \quad r = \lim_{n\to\infty}\left|\frac{a_{n+1}}{a_n}\right| = e.$$

The radius of convergence is therefore $R = r^{-1} = e^{-1}$, and the series converges absolutely on the interval $|x-2| < e^{-1}$, or $2 - e^{-1} < x < 2 + e^{-1}$. For the endpoint $x = 2 + e^{-1}$, the series becomes $\sum_{n=1}^{\infty} 1$, which diverges by the Divergence Test. For the endpoint $x = 2 - e^{-1}$, the series becomes $\sum_{n=1}^{\infty} (-1)^n$, which also diverges by the Divergence Test. Thus, the series $\sum_{n=12}^{\infty} e^n (x-2)^n$ converges for $2 - e^{-1} < x < 2 + e^{-1}$ and diverges elsewhere.

25. $\displaystyle\sum_{n=1}^{\infty} \frac{x^n}{n - 4\ln n}$

SOLUTION With $a_n = \frac{1}{n-4\ln n}$,

$$\left|\frac{a_{n+1}}{a_n}\right| = \frac{1}{n+1-4\ln(n+1)} \cdot \frac{n-4\ln n}{1} = \frac{1 - 4\frac{\ln n}{n}}{1 + \frac{1}{n} - 4\frac{\ln(n+1)}{n}},$$

and

$$r = \lim_{n\to\infty}\left|\frac{a_{n+1}}{a_n}\right| = 1.$$

The radius of convergence is therefore $R = r^{-1} = 1$, and the series converges absolutely on the interval $|x| < 1$, or $-1 < x < 1$. For the endpoint $x = 1$, the series becomes $\sum_{n=1}^{\infty} \frac{1}{n - 4\ln n}$. Because $\frac{1}{n-4\ln n} > \frac{1}{n}$ and $\sum_{n=1}^{\infty} \frac{1}{n}$ is the divergent harmonic series, the endpoint series diverges by the Comparison Test. For the endpoint $x = -1$, the series becomes $\sum_{n=1}^{\infty} \frac{(-1)^n}{n - 4\ln n}$, which converges by the Leibniz Test. Thus, the series $\sum_{n=1}^{\infty} \frac{x^n}{n - 4\ln n}$ converges for $-1 \le x < 1$ and diverges elsewhere.

In Exercises 27–34, use Eq. (1) to expand the function in a power series with center $c = 0$ and determine the set of x for which the expansion is valid.

27. $f(x) = \dfrac{1}{1-3x}$

SOLUTION Substituting $3x$ for x in Eq. (1), we obtain

$$\frac{1}{1-3x} = \sum_{n=0}^{\infty} (3x)^n = \sum_{n=0}^{\infty} 3^n x^n.$$

This series is valid for $|3x| < 1$, or $|x| < \frac{1}{3}$.

29. $f(x) = \dfrac{1}{3-x}$

SOLUTION First write

$$\frac{1}{3-x} = \frac{1}{3} \cdot \frac{1}{1 - \frac{x}{3}}.$$

Substituting $\frac{x}{3}$ for x in Eq. (1), we obtain

$$\frac{1}{1-\frac{x}{3}} = \sum_{n=0}^{\infty} \left(\frac{x}{3}\right)^n = \sum_{n=0}^{\infty} \frac{x^n}{3^n};$$

Thus,

$$\frac{1}{3-x} = \frac{1}{3}\sum_{n=0}^{\infty} \frac{x^n}{3^n} = \sum_{n=0}^{\infty} \frac{x^n}{3^{n+1}}.$$

This series is valid for $|x/3| < 1$, or $|x| < 3$.

31. $f(x) = \dfrac{1}{1+x^9}$

SOLUTION Substituting $-x^9$ for x in Eq. (1), we obtain

$$\frac{1}{1+x^9} = \sum_{n=0}^{\infty}(-x^9)^n = \sum_{n=0}^{\infty}(-1)^n x^{9n}.$$

This series is valid for $|-x^9| < 1$, or $|x| < 1$.

33. $f(x) = \dfrac{1}{1+3x^7}$

SOLUTION Substituting $-3x^7$ for x in Eq. (1), we obtain

$$\frac{1}{1+3x^7} = \sum_{n=0}^{\infty}(-3x^7)^n = \sum_{n=0}^{\infty}(-3)^n x^{7n}.$$

This series is valid for $|-3x^7| < 1$, or $|x| < \frac{1}{\sqrt[7]{3}}$.

35. Use the equalities

$$\frac{1}{1-x} = \frac{1}{-3-(x-4)} = \frac{-\frac{1}{3}}{1+\left(\frac{x-4}{3}\right)}$$

to show that for $|x-4| < 3$

$$\frac{1}{1-x} = \sum_{n=0}^{\infty}(-1)^{n+1}\frac{(x-4)^n}{3^{n+1}}$$

SOLUTION Substituting $-\frac{x-4}{3}$ for x in Eq. (1), we obtain

$$\frac{1}{1+\left(\frac{x-4}{3}\right)} = \sum_{n=0}^{\infty}\left(-\frac{x-4}{3}\right)^n = \sum_{n=0}^{\infty}(-1)^n\frac{(x-4)^n}{3^n}.$$

Thus,

$$\frac{1}{1-x} = -\frac{1}{3}\sum_{n=0}^{\infty}(-1)^n\frac{(x-4)^n}{3^n} = \sum_{n=0}^{\infty}(-1)^{n+1}\frac{(x-4)^n}{3^{n+1}}.$$

This series is valid for $|-\frac{x-4}{3}| < 1$, or $|x-4| < 3$.

37. Use the method of Exercise 35 to expand $\dfrac{1}{4-x}$ in a power series with center $c = 5$. Determine the set of x for which the expansion is valid.

SOLUTION First write

$$\frac{1}{4-x} = \frac{1}{-1-(x-5)} = -\frac{1}{1+(x-5)}.$$

Substituting $-(x-5)$ for x in Eq. (1), we obtain

$$\frac{1}{1+(x-5)} = \sum_{n=0}^{\infty}(-(x-5))^n = \sum_{n=0}^{\infty}(-1)^n(x-5)^n.$$

Thus,

$$\frac{1}{4-x} = -\sum_{n=0}^{\infty}(-1)^n(x-5)^n = \sum_{n=0}^{\infty}(-1)^{n+1}(x-5)^n.$$

This series is valid for $|-(x-5)| < 1$, or $|x-5| < 1$.

39. Give an example of a power series that converges for x in $[2, 6)$.

SOLUTION The power series must be centered at $c = \frac{6+2}{2} = 4$, with radius of convergence $R = 2$. Consider the following series:

$$\sum_{n=1}^{\infty} \frac{(x-4)^n}{n2^n}.$$

With $a_n = \frac{1}{n2^n}$,

$$r = \lim_{n\to\infty} \frac{n2^n}{(n+1)2^{n+1}} = \frac{1}{2} \lim_{n\to\infty} \frac{n}{n+1} = \frac{1}{2}.$$

The radius of convergence is therefore $R = r^{-1} = 2$, and the series converges absolutely for $|x-4| < 2$, or $2 < x < 6$. For the endpoint $x = 6$, the series becomes $\sum_{n=1}^{\infty} \frac{(6-4)^n}{n \cdot 2^n} = \sum_{n=1}^{\infty} \frac{1}{n}$, which is the divergent harmonic series. For the endpoint $x = 2$, the series becomes $\sum_{n=1}^{\infty} \frac{(2-4)^n}{n \cdot 2^n} = \sum_{n=1}^{\infty} \frac{(-1)^n}{n}$, which converges by the Leibniz Test. Therefore, the series converges for $2 \le x < 6$, as desired.

41. Use Exercise 40 to prove that

$$\ln \frac{3}{2} = \frac{1}{2} - \frac{1}{2 \cdot 2^2} + \frac{1}{3 \cdot 2^3} - \frac{1}{4 \cdot 2^4} + \cdots$$

Use your knowledge of alternating series to find an N such that the partial sum S_N approximates $\ln \frac{3}{2}$ to within an error of at most 10^{-3}. Confirm this using a calculator to compute both S_N and $\ln \frac{3}{2}$.

SOLUTION In the previous exercise we found that

$$\ln(1+x) = \sum_{n=0}^{\infty} (-1)^n \frac{x^{n+1}}{n+1}.$$

Setting $x = \frac{1}{2}$ yields:

$$\ln \frac{3}{2} = \sum_{n=1}^{\infty} (-1)^{n-1} \frac{\left(\frac{1}{2}\right)^n}{n} = \sum_{n=1}^{\infty} \frac{(-1)^{n-1}}{n2^n} = \frac{1}{2} - \frac{1}{2 \cdot 2^2} + \frac{1}{3 \cdot 2^3} - \frac{1}{4 \cdot 2^4} + \cdots$$

Note that the series for $\ln \frac{3}{2}$ is an alternating series with $a_n = \frac{1}{n2^n}$. The error in approximating $\ln \frac{3}{2}$ by the partial sum S_N is therefore bounded by

$$\left| \ln \frac{3}{2} - S_N \right| < a_{N+1} = \frac{1}{(N+1)2^{N+1}}.$$

To obtain an error of at most 10^{-3}, we must find an N such that

$$\frac{1}{(N+1)2^{N+1}} < 10^{-3} \quad \text{or} \quad (N+1)2^{N+1} > 1000.$$

For $N = 6$, $(N+1)2^{N+1} = 7 \cdot 2^7 = 896 < 1000$, but for $N = 7$, $(N+1)2^{N+1} = 8 \cdot 2^8 = 2048 > 1000$; hence, the smallest value for N is $N = 7$. The corresponding approximation is

$$S_7 = \frac{1}{2} - \frac{1}{2 \cdot 2^2} + \frac{1}{3 \cdot 2^3} - \frac{1}{4 \cdot 2^4} + \frac{1}{5 \cdot 2^5} - \frac{1}{6 \cdot 2^6} + \frac{1}{7 \cdot 2^7} = 0.405803571.$$

Now, $\ln \frac{3}{2} = 0.405465108$, so

$$\left| \ln \frac{3}{2} - S_7 \right| = 3.385 \times 10^{-4} < 10^{-3}.$$

43. Show that for $|x| < 1$

$$\frac{1+2x}{1+x+x^2} = 1 + x - 2x^2 + x^3 + x^4 - 2x^5 + x^6 + x^7 - 2x^8 + \cdots$$

Hint: Use the hint from Exercise 42.

SOLUTION The terms in the series on the right-hand side are either of the form x^n or $-2x^n$ for some n. Because

$$\lim_{n\to\infty} \sqrt[n]{2} = \lim_{n\to\infty} \sqrt[n]{1} = 1,$$

it follows that

$$\lim_{n\to\infty} \sqrt[n]{|a_n|} = |x|.$$

Hence, by the Root Test, the series converges absolutely for $|x| < 1$.

By Exercise 37 of Section 11.4, any rearrangement of the terms of an absolutely convergent series yields another absolutely convergent series with the same sum as the original series. If we let S denote the sum of the series, then

$$S = \left(1 + x^3 + x^6 + \cdots\right) + \left(x + x^4 + x^7 + \cdots\right) - 2\left(x^2 + x^5 + x^8 + \cdots\right)$$

$$= \frac{1}{1-x^3} + \frac{x}{1-x^3} - \frac{2x^2}{1-x^3} = \frac{1+x-2x^2}{1-x^3} = \frac{(1-x)(2x+1)}{(1-x)(1+x+x^2)} = \frac{2x+1}{1+x+x^2}.$$

45. Use the power series for $y = e^x$ to show that

$$\frac{1}{e} = \frac{1}{2!} - \frac{1}{3!} + \frac{1}{4!} - \cdots$$

Use your knowledge of alternating series to find an N such that the partial sum S_N approximates e^{-1} to within an error of at most 10^{-3}. Confirm this using a calculator to compute both S_N and e^{-1}.

SOLUTION Recall that the series for e^x is

$$\sum_{n=0}^{\infty} \frac{x^n}{n!} = 1 + x + \frac{x^2}{2!} + \frac{x^3}{3!} + \frac{x^4}{4!} + \cdots.$$

Setting $x = -1$ yields

$$e^{-1} = 1 - 1 + \frac{1}{2!} - \frac{1}{3!} + \frac{1}{4!} - + \cdots = \frac{1}{2!} - \frac{1}{3!} + \frac{1}{4!} - + \cdots.$$

This is an alternating series with $a_n = \frac{1}{(n+1)!}$. The error in approximating e^{-1} with the partial sum S_N is therefore bounded by

$$|S_N - e^{-1}| \le a_{N+1} = \frac{1}{(N+2)!}.$$

To make the error at most 10^{-3}, we must choose N such that

$$\frac{1}{(N+2)!} \le 10^{-3} \quad \text{or} \quad (N+2)! \ge 1000.$$

For $N = 4$, $(N+2)! = 6! = 720 < 1000$, but for $N = 5$, $(N+2)! = 7! = 5040$; hence, $N = 5$ is the smallest value that satisfies the error bound. The corresponding approximation is

$$S_5 = \frac{1}{2!} - \frac{1}{3!} + \frac{1}{4!} - \frac{1}{5!} + \frac{1}{6!} = 0.368055555$$

Now, $e^{-1} = 0.367879441$, so

$$|S_5 - e^{-1}| = 1.761 \times 10^{-4} < 10^{-3}.$$

47. Find a power series $P(x)$ satisfying the differential equation:

$$y'' - xy' + y = 0 \qquad \boxed{10}$$

with initial condition $y(0) = 1$, $y'(0) = 0$. What is the radius of convergence of the power series?

SOLUTION Let $P(x) = \sum_{n=0}^{\infty} a_n x^n$. Then

$$P'(x) = \sum_{n=1}^{\infty} n a_n x^{n-1} \quad \text{and} \quad P''(x) = \sum_{n=2}^{\infty} n(n-1) a_n x^{n-2}.$$

Note that $P(0) = a_0$ and $P'(0) = a_1$; in order to satisfy the initial conditions $P(0) = 1$, $P'(0) = 0$, we must have $a_0 = 1$ and $a_1 = 0$. Now,

$$\begin{aligned} P''(x) - xP'(x) + P(x) &= \sum_{n=2}^{\infty} n(n-1)a_n x^{n-2} - \sum_{n=1}^{\infty} na_n x^n + \sum_{n=0}^{\infty} a_n x^n \\ &= \sum_{n=0}^{\infty} (n+2)(n+1)a_{n+2}x^n - \sum_{n=1}^{\infty} na_n x^n + \sum_{n=0}^{\infty} a_n x^n \\ &= 2a_2 + a_0 + \sum_{n=1}^{\infty} \left[(n+2)(n+1)a_{n+2} - na_n + a_n\right] x^n. \end{aligned}$$

In order for this series to be equal to zero, the coefficient of x^n must be equal to zero for each n; thus, $2a_2 + a_0 = 0$ and $(n+2)(n+1)a_{n+2} - (n-1)a_n = 0$, or

$$a_2 = -\frac{1}{2}a_0 \quad \text{and} \quad a_{n+2} = \frac{n-1}{(n+2)(n+1)}a_n.$$

Starting from $a_1 = 0$, we calculate

$$\begin{aligned} a_3 &= \frac{1-1}{(3)(2)}a_1 = 0; \\ a_5 &= \frac{2}{(5)(4)}a_3 = 0; \\ a_7 &= \frac{4}{(7)(6)}a_5 = 0; \end{aligned}$$

and, in general, all of the odd coefficients are zero. As for the even coefficients, we have $a_0 = 1$, $a_2 = -\frac{1}{2}$,

$$\begin{aligned} a_4 &= \frac{1}{(4)(3)}a_2 = -\frac{1}{4!}; \\ a_6 &= \frac{3}{(6)(5)}a_4 = -\frac{3}{6!}; \\ a_8 &= \frac{5}{(8)(7)}a_6 = -\frac{15}{8!} \end{aligned}$$

and so on. Thus,

$$P(x) = 1 - \frac{1}{2}x^2 - \frac{1}{4!}x^4 - \frac{3}{6!}x^6 - \frac{15}{8!}x^8 - \cdots$$

To determine the radius of convergence, treat this as a series in the variable x^2, and observe that

$$r = \lim_{k\to\infty} \left|\frac{a_{2k+2}}{a_{2k}}\right| = \lim_{k\to\infty} \frac{2k-1}{(2k+2)(2k+1)} = 0.$$

Thus, the radius of convergence is $R = r^{-1} = \infty$.

49. Prove that $J_2(x) = \sum_{k=0}^{\infty} \frac{(-1)^k}{2^{2k+2}\, k!\,(k+3)!}x^{2k+2}$ is a solution of the Bessel differential equation of order two:

$$x^2y'' + xy' + (x^2 - 4)y = 0$$

SOLUTION Let $J_2(x) = \sum_{k=0}^{\infty} \frac{(-1)^k}{2^{2k+2}\, k!\,(k+2)!}x^{2k+2}$. Then

$$\begin{aligned} J_2'(x) &= \sum_{k=0}^{\infty} \frac{(-1)^k(k+1)}{2^{2k+1}\, k!\,(k+2)!}x^{2k+1} \\ J_2''(x) &= \sum_{k=0}^{\infty} \frac{(-1)^k(k+1)(2k+1)}{2^{2k+1}\, k!\,(k+2)!}x^{2k} \end{aligned}$$

and

$$x^2J_2''(x) + xJ_2'(x) + (x^2-4)J_2(x) = \sum_{k=0}^{\infty} \frac{(-1)^k(k+1)(2k+1)}{2^{2k+1}\, k!\,(k+2)!}x^{2k+2} + \sum_{k=0}^{\infty} \frac{(-1)^k(k+1)}{2^{2k+1}\, k!\,(k+2)!}x^{2k+2}$$

$$
\begin{aligned}
&- \sum_{k=0}^{\infty} \frac{(-1)^k}{2^{2k+2}\, k!\,(k+2)!} x^{2k+4} - \sum_{k=0}^{\infty} \frac{(-1)^k}{2^{2k}\, k!\,(k+2)!} x^{2k+2} \\
&= \sum_{k=0}^{\infty} \frac{(-1)^k k(k+2)}{2^{2k} k!(k+2)!} x^{2k+2} + \sum_{k=1}^{\infty} \frac{(-1)^{k-1}}{2^{2k}\,(k-1)!\,(k+1)!} x^{2k+2} \\
&= \sum_{k=1}^{\infty} \frac{(-1)^k}{2^{2k}\,(k-1)!(k+1)!} x^{2k+2} - \sum_{k=1}^{\infty} \frac{(-1)^k}{2^{2k}\,(k-1)!(k+1)!} x^{2k+2} = 0.
\end{aligned}
$$

51. Let $C(x) = 1 - \frac{x^2}{2!} + \frac{x^4}{4!} - \frac{x^6}{6!} + \cdots$.

(a) Show that $C(x)$ has an infinite radius of convergence.

(b) Prove that $C(x)$ and $f(x) = \cos x$ are both solutions of $y'' = -y$ with initial conditions $y(0) = 1$, $y'(0) = 0$. This initial value problem has a unique solution, so it follows that $C(x) = \cos x$ for all x.

SOLUTION

(a) Consider the series

$$C(x) = 1 - \frac{x^2}{2!} + \frac{x^4}{4!} - \frac{x^6}{6!} + \cdots = \sum_{n=0}^{\infty} (-1)^n \frac{x^{2n}}{(2n)!}.$$

With $a_n = (-1)^n \frac{x^{2n}}{(2n)!}$,

$$\left| \frac{a_{n+1}}{a_n} \right| = \frac{|x|^{2n+2}}{(2n+2)!} \cdot \frac{(2n)!}{|x|^{2n}} = \frac{|x|^2}{(2n+2)(2n+1)},$$

and

$$r = \lim_{n\to\infty} \left| \frac{a_{n+1}}{a_n} \right| = 0.$$

The radius of convergence for $C(x)$ is therefore $R = r^{-1} = \infty$.

(b) Differentiating the series defining $C(x)$ term-by-term, we find

$$C'(x) = \sum_{n=1}^{\infty} (-1)^n (2n) \frac{x^{2n-1}}{(2n)!} = \sum_{n=1}^{\infty} (-1)^n \frac{x^{2n-1}}{(2n-1)!}$$

and

$$
\begin{aligned}
C''(x) &= \sum_{n=1}^{\infty} (-1)^n (2n-1) \frac{x^{2n-2}}{(2n-1)!} = \sum_{n=1}^{\infty} (-1)^n \frac{x^{2n-2}}{(2n-2)!} \\
&= \sum_{n=0}^{\infty} (-1)^{n+1} \frac{x^{2n}}{(2n)!} = -\sum_{n=0}^{\infty} (-1)^n \frac{x^{2n}}{(2n)!} = -C(x).
\end{aligned}
$$

Moreover, $C(0) = 1$ and $C'(0) = 0$.

53. Find all values of x such that the following series converges:

$$F(x) = 1 + 3x + x^2 + 27x^3 + x^4 + 243x^5 + \cdots$$

SOLUTION Observe that $F(x)$ can be written as the sum of two geometric series:

$$F(x) = \left(1 + x^2 + x^4 + \cdots\right) + \left(3x + 27x^3 + 243x^5 + \cdots\right) = \sum_{n=0}^{\infty} (x^2)^n + \sum_{n=0}^{\infty} 3x(9x^2)^n$$

The first geometric series converges for $|x^2| < 1$, or $|x| < 1$; the second geometric series converges for $|9x^2| < 1$, or $|x| < \frac{1}{3}$. Since both geometric series must converge for $F(x)$ to converge, we find that $F(x)$ converges for $|x| < \frac{1}{3}$, the intersection of the intervals of convergence for the two geometric series.

55. Why is it impossible to expand $f(x) = |x|$ as a power series that converges in an interval around $x = 0$? Explain this using Theorem 3.

SOLUTION Suppose that there exists a $c > 0$ such that f can be represented by a power series on the interval $(-c, c)$; that is,

$$|x| = \sum_{n=0}^{\infty} a_n x^n$$

for $|x| < c$. Then it follows by Theorem 3 that $|x|$ is differentiable on $(-c, c)$. This contradicts the well known property that $f(x) = |x|$ is not differentiable at the point $x = 0$.

Further Insights and Challenges

57. Suppose that the coefficients of $F(x) = \sum_{n=0}^{\infty} a_n x^n$ are *periodic*, that is, for some whole number $M > 0$, we have $a_{M+n} = a_n$. Prove that $F(x)$ converges absolutely for $|x| < 1$ and that

$$F(x) = \frac{a_0 + a_1 x + \cdots + a_{M-1} x^{M-1}}{1 - x^M}$$

Hint: Use the hint for Exercise 42.

SOLUTION Suppose the coefficients of $F(x)$ are periodic, with $a_{M+n} = a_n$ for some whole number M and all n. The $F(x)$ can be written as the sum of M geometric series:

$$F(x) = a_0\left(1 + x^M + x^{2M} + \cdots\right) + a_1\left(x + x^{M+1} + x^{2M+1} + \cdots\right) +$$
$$= a_2\left(x^2 + x^{M+2} + x^{2M+2} + \cdots\right) + \cdots + a_{M-1}\left(x^{M-1} + x^{2M-1} + x^{3M-1} + \cdots\right)$$
$$= \frac{a_0}{1 - x^M} + \frac{a_1 x}{1 - x^M} + \frac{a_2 x^2}{1 - x^M} + \cdots + \frac{a_{M-1} x^{M-1}}{1 - x^M} = \frac{a_0 + a_1 x + a_2 x^2 + \cdots + a_{M-1} x^{M-1}}{1 - x^M}.$$

As each geometric series converges absolutely for $|x| < 1$, it follows that $F(x)$ also converges absolutely for $|x| < 1$.

11.7 Taylor Series (ET Section 10.7)

Preliminary Questions

1. Determine $f(0)$ and $f'''(0)$ for a function $f(x)$ with Maclaurin series

$$T(x) = 3 + 2x + 12x^2 + 5x^3 + \cdots$$

SOLUTION The Maclaurin series for a function f has the form

$$f(0) + \frac{f'(0)}{1!}x + \frac{f''(0)}{2!}x^2 + \frac{f'''(0)}{3!}x^3 + \cdots$$

Matching this general expression with the given series, we find $f(0) = 3$ and $\frac{f'''(0)}{3!} = 5$. From this latter equation, it follows that $f'''(0) = 30$.

2. Determine $f(-2)$ and $f^{(4)}(-2)$ for a function with Taylor series

$$T(x) = 3(x + 2) + (x + 2)^2 - 4(x + 2)^3 + 2(x + 2)^4 + \cdots$$

SOLUTION The Taylor series for a function f centered at $x = -2$ has the form

$$f(-2) + \frac{f'(-2)}{1!}(x + 2) + \frac{f''(-2)}{2!}(x + 2)^2 + \frac{f'''(-2)}{3!}(x + 2)^3 + \frac{f^{(4)}(-2)}{4!}(x + 2)^4 + \cdots$$

Matching this general expression with the given series, we find $f(-2) = 0$ and $\frac{f^{(4)}(-2)}{4!} = 2$. From this latter equation, it follows that $f^{(4)}(-2) = 48$.

3. What is the easiest way to find the Maclaurin series for the function $f(x) = \sin(x^2)$?

SOLUTION The easiest way to find the Maclaurin series for $\sin\left(x^2\right)$ is to substitute x^2 for x in the Maclaurin series for $\sin x$.

4. What is the Taylor series for $f(x)$ centered at $c = 3$ if $f(3) = 4$ and $f'(x)$ has a Taylor expansion

$$f'(x) = \sum_{n=1}^{\infty} \frac{(x-3)^n}{n}$$

SOLUTION Integrating the series for $f'(x)$ term-by-term gives

$$f(x) = C + \sum_{n=1}^{\infty} \frac{(x-3)^{n+1}}{n(n+1)}.$$

Substituting $x = 3$ then yields

$$f(3) = C = 4;$$

so

$$f(x) = 4 + \sum_{n=1}^{\infty} \frac{(x-3)^{n+1}}{n(n+1)}.$$

5. Let $T(x)$ be the Maclaurin series of $f(x)$. Which of the following guarantees that $f(2) = T(2)$?

(a) $T(x)$ converges for $x = 2$.

(b) The remainder $R_k(2)$ approaches a limit as $k \to \infty$.

(c) The remainder $R_k(2)$ approaches zero as $k \to \infty$.

SOLUTION The correct response is **(c)**: $f(2) = T(2)$ if and only if the remainder $R_k(2)$ approaches zero as $k \to \infty$.

Exercises

1. Write out the first four terms of the Maclaurin series of $f(x)$ if

$$f(0) = 2, \quad f'(0) = 3, \quad f''(0) = 4, \quad f'''(0) = 12$$

SOLUTION The first four terms of the Maclaurin series of $f(x)$ are

$$f(0) + f'(0)x + \frac{f''(0)}{2!}x^2 + \frac{f'''(0)}{3!}x^3 = 2 + 3x + \frac{4}{2}x^2 + \frac{12}{6}x^3 = 2 + 3x + 2x^2 + 2x^3.$$

In Exercises 3–20, find the Maclaurin series.

3. $f(x) = \dfrac{1}{1-2x}$

SOLUTION Substituting $2x$ for x in the Maclaurin series for $\frac{1}{1-x}$ gives

$$\frac{1}{1-2x} = \sum_{n=0}^{\infty} (2x)^n = \sum_{n=0}^{\infty} 2^n x^n.$$

This series is valid for $|2x| < 1$, or $|x| < \frac{1}{2}$.

5. $f(x) = \cos 3x$

SOLUTION Substituting $3x$ for x in the Maclaurin series for $\cos x$ gives

$$\cos 3x = \sum_{n=0}^{\infty} (-1)^n \frac{(3x)^{2n}}{(2n)!} = \sum_{n=0}^{\infty} (-1)^n \frac{9^n x^{2n}}{(2n)!}.$$

This series is valid for all x.

7. $f(x) = \sin(x^2)$

SOLUTION Substituting x^2 for x in the Maclaurin series for $\sin x$ gives

$$\sin x^2 = \sum_{n=0}^{\infty} (-1)^n \frac{(x^2)^{2n+1}}{(2n+1)!} = \sum_{n=0}^{\infty} (-1)^n \frac{x^{4n+2}}{(2n+1)!}.$$

This series is valid for all x.

9. $f(x) = \ln(1 - x^2)$

SOLUTION Substituting $-x^2$ for x in the Maclaurin series for $\ln(1+x)$ gives

$$\ln(1-x^2) = \sum_{n=1}^{\infty} \frac{(-1)^{n-1}(-x^2)^n}{n} = \sum_{n=1}^{\infty} \frac{(-1)^{2n-1}x^{2n}}{n} = -\sum_{n=1}^{\infty} \frac{x^{2n}}{n}.$$

This series is valid for $|x| < 1$.

11. $f(x) = \tan^{-1}(x^2)$

SOLUTION Substituting x^2 for x in the Maclaurin series for $\tan^{-1} x$ gives

$$\tan^{-1}(x^2) = \sum_{n=0}^{\infty}(-1)^n \frac{(x^2)^{2n+1}}{2n+1} = \sum_{n=0}^{\infty}(-1)^n \frac{x^{4n+2}}{2n+1}.$$

This series is valid for $|x| \le 1$.

13. $f(x) = e^{x-2}$

SOLUTION $e^{x-2} = e^{-2}e^x$; thus,

$$e^{x-2} = e^{-2}\sum_{n=0}^{\infty} \frac{x^n}{n!} = \sum_{n=0}^{\infty} \frac{x^n}{e^2 n!}.$$

This series is valid for all x.

15. $f(x) = \ln(1-5x)$

SOLUTION Substituting $-5x$ for x in the Maclaurin series for $\ln(1+x)$ gives

$$\ln(1-5x) = \sum_{n=1}^{\infty} \frac{(-1)^{n-1}(-5x)^n}{n} = \sum_{n=1}^{\infty} \frac{(-1)^{2n-1}5^n x^n}{n} = -\sum_{n=1}^{\infty} \frac{5^n x^n}{n}.$$

This series is valid for $|5x| < 1$, or $|x| < \frac{1}{5}$, and for $x = -\frac{1}{5}$.

17. $f(x) = \sinh x$

SOLUTION Recall that

$$\sinh x = \frac{1}{2}(e^x - e^{-x}).$$

Therefore,

$$\sinh x = \frac{1}{2}\left(\sum_{n=0}^{\infty} \frac{x^n}{n!} - \sum_{n=0}^{\infty} \frac{(-x)^n}{n!}\right) = \sum_{n=0}^{\infty} \frac{x^n}{2(n!)}\left(1-(-1)^n\right).$$

Now,

$$1-(-1)^n = \begin{cases} 0, & n \text{ even} \\ 2, & n \text{ odd} \end{cases}$$

so

$$\sinh x = \sum_{k=0}^{\infty} 2\frac{x^{2k+1}}{2(2k+1)!} = \sum_{k=0}^{\infty} \frac{x^{2k+1}}{(2k+1)!}.$$

This series is valid for all x.

19. $f(x) = \dfrac{1-\cos(x^2)}{x}$

SOLUTION Substituting x^2 for x in the Maclaurin series for $\cos x$ gives

$$\cos x^2 = \sum_{n=0}^{\infty}(-1)^n \frac{(x^2)^{2n}}{(2n)!} = \sum_{n=0}^{\infty}(-1)^n \frac{x^{4n}}{(2n)!} = 1 + \sum_{n=1}^{\infty}(-1)^n \frac{x^{4n}}{(2n)!}.$$

Thus,

$$1-\cos x^2 = 1 - \left(1 + \sum_{n=1}^{\infty}(-1)^n \frac{x^{4n}}{(2n)!}\right) = \sum_{n=1}^{\infty}(-1)^{n+1} \frac{x^{4n}}{(2n)!},$$

and

$$\frac{1-\cos(x^2)}{x} = \frac{1}{x}\sum_{n=1}^{\infty}(-1)^{n+1}\frac{x^{4n}}{(2n)!} = \sum_{n=1}^{\infty}(-1)^{n+1}\frac{x^{4n-1}}{(2n)!}.$$

21. Use multiplication to find the first four terms in the Maclaurin series for $f(x) = e^x \sin x$.

SOLUTION Multiply the fifth-order Taylor Polynomials for e^x and $\sin x$:

$$\left(1 + x + \frac{x^2}{2} + \frac{x^3}{6} + \frac{x^4}{24} + \frac{x^5}{120}\right)\left(x - \frac{x^3}{6} + \frac{x^5}{120}\right)$$

$$= x + x^2 - \frac{x^3}{6} + \frac{x^3}{2} - \frac{x^4}{6} + \frac{x^4}{6} + \frac{x^5}{120} - \frac{x^5}{12} + \frac{x^5}{24} + \text{higher-order terms}$$

$$= x + x^2 + \frac{x^3}{3} - \frac{x^5}{30} + \text{higher-order terms.}$$

The first four terms in the Maclaurin series for $f(x) = e^x \sin x$ are therefore

$$x + x^2 + \frac{x^3}{3} - \frac{x^5}{30}.$$

23. Find the first four terms of the Maclaurin series for $f(x) = e^x \ln(1-x)$.

SOLUTION Multiply the fourth order Taylor Polynomials for e^x and $\ln(1-x)$:

$$\left(1 + x + \frac{x^2}{2} + \frac{x^3}{6} + \frac{x^4}{24}\right)\left(-x - \frac{x^2}{2} - \frac{x^3}{3} - \frac{x^4}{4}\right)$$

$$= -x - \frac{x^2}{2} - x^2 - \frac{x^3}{3} - \frac{x^3}{2} - \frac{x^3}{2} - \frac{x^4}{4} - \frac{x^4}{3} - \frac{x^4}{4} - \frac{x^4}{6} + \text{higher-order terms}$$

$$= -x - \frac{3x^2}{2} - \frac{4x^3}{3} - x^4 + \text{higher-order terms.}$$

The first four terms of the Maclaurin series for $f(x) = e^x \ln(1-x)$ are therefore

$$-x - \frac{3x^2}{2} - \frac{4x^3}{3} - x^4.$$

25. Write out the first five terms of the binomial series for $f(x) = (1+x)^{-3/2}$.

SOLUTION The first five generalized binomial coefficients for $a = -\frac{3}{2}$ are

$$1, \quad -\frac{3}{2}, \quad \frac{-\frac{3}{2}(-\frac{5}{2})}{2!} = \frac{15}{8}, \quad \frac{-\frac{3}{2}(-\frac{5}{2})(-\frac{7}{2})}{3!} = -\frac{35}{16}, \quad \frac{-\frac{3}{2}(-\frac{5}{2})(-\frac{7}{2})(-\frac{9}{2})}{4!} = \frac{315}{128}.$$

Therefore, the first five terms in the binomial series for $f(x) = (1+x)^{-3/2}$ are

$$1 - \frac{3}{2}x + \frac{15}{8}x^2 - \frac{35}{16}x^3 + \frac{315}{128}x^4.$$

27. Find the first four terms of the Maclaurin for $f(x) = e^{(e^x)}$.

SOLUTION With $f(x) = e^{(e^x)}$, we find

$$f'(x) = e^{(e^x)} \cdot e^x$$

$$f''(x) = e^{(e^x)} \cdot e^x + e^{(e^x)} \cdot e^{2x} = e^{(e^x)}\left(e^{2x} + e^x\right)$$

$$f'''(x) = e^{(e^x)}\left(2e^{2x} + e^x\right) + e^{(e^x)}\left(e^{2x} + e^x\right)e^x$$

$$= e^{(e^x)}\left(e^{3x} + 3e^{2x} + e^x\right)$$

and

$$f(0) = e, \quad f'(0) = e, \quad f''(0) = 2e, \quad f'''(0) = 5e.$$

Therefore, the first four terms of the Maclaurin for $f(x) = e^{(e^x)}$ are

$$e + ex + ex^2 + \frac{5e}{6}x^3.$$

29. Find the Taylor series for $\sin x$ at $c = \frac{\pi}{2}$.

SOLUTION Because

$$\sin x = \cos\left(x - \frac{\pi}{2}\right),$$

we obtain

$$\sin x = \sum_{n=0}^{\infty} \frac{(-1)^n}{(2n)!}\left(x - \frac{\pi}{2}\right)^{2n},$$

by substituting $x - \frac{\pi}{2}$ for x in the Maclaurin series for $\cos x$.

In Exercises 31–40, find the Taylor series centered at c.

31. $f(x) = \frac{1}{x}, \quad c = 1$

SOLUTION Write

$$\frac{1}{x} = \frac{1}{1 + (x - 1)},$$

and then substitute $-(x-1)$ for x in the Maclaurin series for $\frac{1}{1-x}$ to obtain

$$\frac{1}{x} = \sum_{n=0}^{\infty} [-(x-1)]^n = \sum_{n=0}^{\infty} (-1)^n (x-1)^n.$$

This series is valid for $|x - 1| < 1$.

33. $f(x) = \frac{1}{1-x}, \quad c = 5$

SOLUTION Write

$$\frac{1}{1-x} = \frac{1}{-4 - (x-5)} = -\frac{1}{4} \cdot \frac{1}{1 + \frac{x-5}{4}}.$$

Substituting $-\frac{x-5}{4}$ for x in the Maclaurin series for $\frac{1}{1-x}$ yields

$$\frac{1}{1 + \frac{x-5}{4}} = \sum_{n=0}^{\infty}\left(-\frac{x-5}{4}\right)^n = \sum_{n=0}^{\infty}(-1)^n \frac{(x-5)^n}{4^n}.$$

Thus,

$$\frac{1}{1-x} = -\frac{1}{4}\sum_{n=0}^{\infty}(-1)^n \frac{(x-5)^n}{4^n} = \sum_{n=0}^{\infty}(-1)^{n+1}\frac{(x-5)^n}{4^{n+1}}.$$

This series is valid for $\left|\frac{x-5}{4}\right| < 1$, or $|x - 5| < 4$.

35. $f(x) = x^4 + 3x - 1, \quad c = 2$

SOLUTION To determine the Taylor series with center $c = 2$, we compute

$$f'(x) = 4x^3 + 3, \quad f''(x) = 12x^2, \quad f'''(x) = 24x,$$

and $f^{(4)}(x) = 24$. All derivatives of order five and higher are zero. Now,

$$f(2) = 21, \quad f'(2) = 35, \quad f''(2) = 48, \quad f'''(2) = 48,$$

and $f^{(4)}(2) = 24$. Therefore, the Taylor series is

$$21 + 35(x-2) + \frac{48}{2}(x-2)^2 + \frac{48}{6}(x-2)^3 + \frac{24}{24}(x-2)^4,$$

or

$$21 + 35(x-2) + 24(x-2)^2 + 8(x-2)^3 + (x-2)^4.$$

37. $f(x) = e^{3x}, \quad c = -1$

SOLUTION Write

$$e^{3x} = e^{3(x+1)-3} = e^{-3}e^{3(x+1)}.$$

Now, substitute $3(x+1)$ for x in the Maclaurin series for e^x to obtain

$$e^{3(x+1)} = \sum_{n=0}^{\infty} \frac{(3(x+1))^n}{n!} = \sum_{n=0}^{\infty} \frac{3^n}{n!}(x+1)^n.$$

Thus,

$$e^{3x} = e^{-3} \sum_{n=0}^{\infty} \frac{3^n}{n!}(x+1)^n = \sum_{n=0}^{\infty} \frac{3^n e^{-3}}{n!}(x+1)^n,$$

This series is valid for all x.

39. $f(x) = \frac{1}{1-x^2}, \quad c = 3$

SOLUTION By partial fraction decomposition

$$\frac{1}{1-x^2} = \frac{\frac{1}{2}}{1-x} + \frac{\frac{1}{2}}{1+x},$$

so

$$\frac{1}{1-x^2} = \frac{\frac{1}{2}}{-2-(x-3)} + \frac{\frac{1}{2}}{4+(x-3)} = -\frac{1}{4} \cdot \frac{1}{1+\frac{x-3}{2}} + \frac{1}{8} \cdot \frac{1}{1+\frac{x-3}{4}}.$$

Substituting $-\frac{x-3}{2}$ for x in the Maclaurin series for $\frac{1}{1-x}$ gives

$$\frac{1}{1+\frac{x-3}{2}} = \sum_{n=0}^{\infty} \left(-\frac{x-3}{2}\right)^n = \sum_{n=0}^{\infty} \frac{(-1)^n}{2^n}(x-3)^n,$$

while substituting $-\frac{x-3}{4}$ for x in the same series gives

$$\frac{1}{1+\frac{x-3}{4}} = \sum_{n=0}^{\infty} \left(-\frac{x-3}{4}\right)^n = \sum_{n=0}^{\infty} \frac{(-1)^n}{4^n}(x-3)^n.$$

Thus,

$$\frac{1}{1-x^2} = -\frac{1}{4} \sum_{n=0}^{\infty} \frac{(-1)^n}{2^n}(x-3)^n + \frac{1}{8} \sum_{n=0}^{\infty} \frac{(-1)^n}{4^n}(x-3)^n = \sum_{n=0}^{\infty} \frac{(-1)^{n+1}}{2^{n+2}}(x-3)^n + \sum_{n=0}^{\infty} \frac{(-1)^n}{2^{2n+3}}(x-3)^n$$

$$= \sum_{n=0}^{\infty} \left(\frac{(-1)^{n+1}}{2^{n+2}} + \frac{(-1)^n}{2^{2n+3}} \right)(x-3)^n = \sum_{n=0}^{\infty} \frac{(-1)^{n+1}(2^{n+1}-1)}{2^{2n+3}}(x-3)^n.$$

This series is valid for $|x-3| < 2$.

41. Find the Maclaurin series for $f(x) = \frac{1}{\sqrt{1-9x^2}}$ (see Example 10).

SOLUTION From Example 10, we know that for $|x| < 1$,

$$\frac{1}{\sqrt{1-x^2}} = \sum_{n=0}^{\infty} \frac{1 \cdot 3 \cdot 5 \cdots (2n-1)}{2 \cdot 4 \cdot 6 \cdots (2n)} x^{2n},$$

so

$$\frac{1}{\sqrt{1-9x^2}} = \frac{1}{\sqrt{1-(3x)^2}} = \sum_{n=0}^{\infty} \frac{1 \cdot 3 \cdot 5 \cdots (2n-1)}{2 \cdot 4 \cdot 6 \cdots (2n)} (3x)^{2n} = \sum_{n=0}^{\infty} \frac{1 \cdot 3 \cdot 5 \cdots (2n-1)}{2 \cdot 4 \cdot 6 \cdots (2n)} 9^n x^{2n}.$$

This series is valid for $9x^2 < 1$, or $|x| < \frac{1}{3}$.

43. Use the first five terms of the Maclaurin series in Exercise 42 to approximate $\sin^{-1} \frac{1}{2}$. Compare the result with the calculator value.

SOLUTION From Exercise 42 we know that for $|x| < 1$,

$$\sin^{-1} x = x + \sum_{n=1}^{\infty} \frac{1 \cdot 3 \cdot 5 \cdots (2n-1)}{2 \cdot 4 \cdot 6 \cdots (2n)} \frac{x^{2n+1}}{2n+1}.$$

The first five terms of the series are:

$$x + \frac{1}{2}\frac{x^3}{3} + \frac{1 \cdot 3}{2 \cdot 4}\frac{x^5}{5} + \frac{1 \cdot 3 \cdot 5}{2 \cdot 4 \cdot 6}\frac{x^7}{7} + \frac{1 \cdot 3 \cdot 5 \cdot 7}{2 \cdot 4 \cdot 6 \cdot 8}\frac{x^9}{9} = x + \frac{x^3}{6} + \frac{3x^5}{40} + \frac{5x^7}{112} + \frac{35x^9}{1152}$$

Setting $x = \frac{1}{2}$, we obtain the following approximation:

$$\sin^{-1}\frac{1}{2} \approx \frac{1}{2} + \frac{\left(\frac{1}{2}\right)^3}{6} + \frac{3 \cdot \left(\frac{1}{2}\right)^5}{40} + \frac{5 \cdot \left(\frac{1}{2}\right)^7}{112} + \frac{35 \cdot \left(\frac{1}{2}\right)^9}{1152} \approx 0.52358519539.$$

The calculator value is $\sin^{-1}\frac{1}{2} \approx 0.5235988775$.

45. Use the Maclaurin series for $\ln(1+x)$ and $\ln(1-x)$ to show that

$$\frac{1}{2}\ln\left(\frac{1+x}{1-x}\right) = x + \frac{x^3}{3} + \frac{x^5}{5} + \cdots$$

What can you conclude by comparing this result with that of Exercise 44?

SOLUTION Using the Maclaurin series for $\ln(1+x)$ and $\ln(1-x)$, we have for $|x| < 1$

$$\ln(1+x) - \ln(1-x) = \sum_{n=1}^{\infty} \frac{(-1)^{n-1}}{n} x^n - \sum_{n=1}^{\infty} \frac{(-1)^{n-1}}{n} (-x)^n$$

$$= \sum_{n=1}^{\infty} \frac{(-1)^{n-1}}{n} x^n + \sum_{n=1}^{\infty} \frac{x^n}{n} = \sum_{n=1}^{\infty} \frac{1 + (-1)^{n-1}}{n} x^n.$$

Since $1 + (-1)^{n-1} = 0$ for even n and $1 + (-1)^{n-1} = 2$ for odd n,

$$\ln(1+x) - \ln(1-x) = \sum_{k=0}^{\infty} \frac{2}{2k+1} x^{2k+1}.$$

Thus,

$$\frac{1}{2}\ln\left(\frac{1+x}{1-x}\right) = \frac{1}{2}\left(\ln(1+x) - \ln(1-x)\right) = \frac{1}{2}\sum_{k=0}^{\infty} \frac{2}{2k+1} x^{2k+1} = \sum_{k=0}^{\infty} \frac{x^{2k+1}}{2k+1}.$$

Observe that this is the same series we found in Exercise 44; therefore,

$$\frac{1}{2}\ln\left(\frac{1+x}{1-x}\right) = \tanh^{-1} x.$$

47. Use the Maclaurin expansion for e^{-t^2} to express $\int_0^x e^{-t^2}\,dt$ as an alternating power series in t.

(a) How many terms of the infinite series are needed to approximate the integral for $x = 1$ to within an error of at most 0.001?

(b) CAS Carry out the computation and check your answer using a computer algebra system.

SOLUTION Substituting $-t^2$ for t in the Maclaurin series for e^t yields

$$e^{-t^2} = \sum_{n=0}^{\infty} \frac{(-t^2)^n}{n!} = \sum_{n=0}^{\infty} (-1)^n \frac{t^{2n}}{n!};$$

thus,

$$\int_0^x e^{-t^2}\,dt = \sum_{n=0}^{\infty} (-1)^n \frac{x^{2n+1}}{n!(2n+1)}.$$

(a) For $x = 1$,

$$\int_0^1 e^{-t^2}\,dt = \sum_{n=0}^{\infty}(-1)^n \frac{1}{n!(2n+1)}.$$

This is an alternating series with $a_n = \frac{1}{n!(2n+1)}$; therefore, the error incurred by using S_N to approximate the value of the definite integral is bounded by

$$\left|\int_0^1 e^{-t^2}\,dt - S_N\right| \le a_{N+1} = \frac{1}{(N+1)!(2N+3)}.$$

To guarantee the error is at most 0.001, we must choose N so that

$$\frac{1}{(N+1)!(2N+3)} < 0.001 \quad \text{or} \quad (N+1)!(2N+3) > 1000.$$

For $N = 3$, $(N+1)!(2N+3) = 4! \cdot 9 = 216 < 1000$ and for $N = 4$, $(N+1)!(2N+3) = 5! \cdot 11 = 1320 > 1000$; thus, the smallest acceptable value for N is $N = 4$. The corresponding approximation is

$$S_4 = \sum_{n=0}^{4} \frac{(-1)^n}{n!(2n+1)} = 1 - \frac{1}{3} + \frac{1}{2! \cdot 5} - \frac{1}{3! \cdot 7} + \frac{1}{4! \cdot 9} = 0.747486772.$$

(b) Using a computer algebra system, we find

$$\int_0^1 e^{-t^2}\,dt = 0.746824133;$$

therefore

$$\left|\int_0^1 e^{-t^2}\,dt - S_4\right| = 6.626 \times 10^{-4} < 10^{-3}.$$

In Exercises 49–52, express the definite integral as an infinite series and find its value to within an error of at most 10^{-4}.

49. $\displaystyle\int_0^1 \cos(x^2)\,dx$

SOLUTION Substituting x^2 for x in the Maclaurin series for $\cos x$ yields

$$\cos(x^2) = \sum_{n=0}^{\infty}(-1)^n \frac{(x^2)^{2n}}{(2n)!} = \sum_{n=0}^{\infty}(-1)^n \frac{x^{4n}}{(2n)!};$$

therefore,

$$\int_0^1 \cos(x^2)\,dx = \sum_{n=0}^{\infty}(-1)^n \left.\frac{x^{4n+1}}{(2n)!(4n+1)}\right|_0^1 = \sum_{n=0}^{\infty} \frac{(-1)^n}{(2n)!(4n+1)}.$$

This is an alternating series with $a_n = \frac{1}{(2n)!(4n+1)}$; therefore, the error incurred by using S_N to approximate the value of the definite integral is bounded by

$$\left|\int_0^1 \cos(x^2)\,dx - S_N\right| \le a_{N+1} = \frac{1}{(2N+2)!(4N+5)}.$$

To guarantee the error is at most 0.0001, we must choose N so that

$$\frac{1}{(2N+2)!(4N+5)} < 0.0001 \quad \text{or} \quad (2N+2)!(4N+5) > 10000.$$

For $N = 2$, $(2N+2)!(4N+5) = 6! \cdot 13 = 9360 < 10000$ and for $N = 3$, $(2N+2)!(4N+5) = 8! \cdot 17 = 685440 > 10000$; thus, the smallest acceptable value for N is $N = 3$. The corresponding approximation is

$$S_3 = \sum_{n=0}^{3} \frac{(-1)^n}{(2n)!(4n+1)} = 1 - \frac{1}{5 \cdot 2!} + \frac{1}{9 \cdot 4!} - \frac{1}{13 \cdot 6!} = 0.904522792.$$

51. $\displaystyle\int_0^2 e^{-x^3}\,dx$

SOLUTION Substituting $-x^3$ for x in the Maclaurin series for e^x yields

$$e^{-x^3} = \sum_{n=0}^{\infty} \frac{(-x^3)^n}{n!} = \sum_{n=0}^{\infty} (-1)^n \frac{x^{3n}}{n!};$$

therefore,

$$\int_0^1 e^{-x^3}\,dx = \sum_{n=0}^{\infty} (-1)^n \left. \frac{x^{3n+1}}{n!(3n+1)} \right|_0^1 = \sum_{n=0}^{\infty} \frac{(-1)^n}{n!(3n+1)}.$$

This is an alternating series with $a_n = \frac{1}{n!(3n+1)}$; therefore, the error incurred by using S_N to approximate the value of the definite integral is bounded by

$$\left| \int_0^1 e^{-x^3}\,dx - S_N \right| \le a_{N+1} = \frac{1}{(N+1)!(3N+4)}.$$

To guarantee the error is at most 0.0001, we must choose N so that

$$\frac{1}{(N+1)!(3N+4)} < 0.0001 \quad \text{or} \quad (N+1)!(3N+4) > 10000.$$

For $N = 4$, $(N+1)!(3N+4) = 5! \cdot 16 = 1920 < 10000$ and for $N = 5$, $(N+1)!(3N+4) = 6! \cdot 19 = 13680 > 10000$; thus, the smallest acceptable value for N is $N = 5$. The corresponding approximation is

$$S_5 = \sum_{n=0}^{5} \frac{(-1)^n}{n!(3n+1)} = 0.807446200.$$

In Exercises 53–56, express the integral as an infinite series.

53. $\displaystyle\int_0^x \frac{1-\cos(t)}{t}\,dt$, for all x

SOLUTION The Maclaurin series for $\cos t$ is

$$\cos t = \sum_{n=0}^{\infty} (-1)^n \frac{t^{2n}}{(2n)!} = 1 + \sum_{n=1}^{\infty} (-1)^n \frac{t^{2n}}{(2n)!},$$

so

$$1 - \cos t = -\sum_{n=1}^{\infty} (-1)^n \frac{t^{2n}}{(2n)!} = \sum_{n=1}^{\infty} (-1)^{n+1} \frac{t^{2n}}{(2n)!},$$

and

$$\frac{1-\cos t}{t} = \frac{1}{t} \sum_{n=1}^{\infty} (-1)^{n+1} \frac{t^{2n}}{(2n)!} = \sum_{n=1}^{\infty} (-1)^{n+1} \frac{t^{2n-1}}{(2n)!}.$$

Thus,

$$\int_0^x \frac{1-\cos(t)}{t}\,dt = \sum_{n=1}^{\infty} (-1)^{n+1} \left. \frac{t^{2n}}{(2n)!2n} \right|_0^x = \sum_{n=1}^{\infty} (-1)^{n+1} \frac{x^{2n}}{(2n)!2n}.$$

55. $\displaystyle\int_0^x \ln(1+t^2)\,dt$, for $|x| < 1$

SOLUTION Substituting t^2 for t in the Maclaurin series for $\ln(1+t)$ yields

$$\ln(1+t^2) = \sum_{n=1}^{\infty} (-1)^{n-1} \frac{(t^2)^n}{n} = \sum_{n=1}^{\infty} (-1)^n \frac{t^{2n}}{n}.$$

Thus,

$$\int_0^x \ln(1+t^2)\,dt = \sum_{n=1}^{\infty} (-1)^n \left. \frac{t^{2n+1}}{n(2n+1)} \right|_0^x = \sum_{n=1}^{\infty} (-1)^n \frac{x^{2n+1}}{n(2n+1)}.$$

57. Which function has Maclaurin series $\sum_{n=0}^{\infty}(-1)^n 2^n x^n$?

SOLUTION We recognize that

$$\sum_{n=0}^{\infty}(-1)^n 2^n x^n = \sum_{n=0}^{\infty}(-2x)^n$$

is the Maclaurin series for $\frac{1}{1-x}$ with x replaced by $-2x$. Therefore,

$$\sum_{n=0}^{\infty}(-1)^n 2^n x^n = \frac{1}{1-(-2x)} = \frac{1}{1+2x}.$$

In Exercises 59–62, find the first four terms of the Taylor series.

59. $f(x) = \sin(x^2)\cos(x^2)$

SOLUTION Substituting x^2 for x in the Maclaurin series for $\sin x$ and $\cos x$, we find

$$\begin{aligned}\sin(x^2)\cos(x^2) &= \left(x^2 - \frac{x^6}{6} + \frac{x^{10}}{120} - \frac{x^{14}}{5040} + \cdots\right)\left(1 - \frac{x^4}{2} + \frac{x^8}{24} - \frac{x^{12}}{720} + \cdots\right)\\ &= x^2 - \frac{x^6}{6} - \frac{x^6}{2} + \frac{x^{10}}{24} + \frac{x^{10}}{12} + \frac{x^{10}}{120} - \frac{x^{14}}{720} - \frac{x^{14}}{144} - \frac{x^{14}}{240} - \frac{x^{14}}{5040} + \cdots\\ &= x^2 - \frac{2}{3}x^6 + \frac{2}{15}x^{10} - \frac{4}{315}x^{14} + \cdots.\end{aligned}$$

61. $f(x) = e^{\sin x}$

SOLUTION Substituting $\sin x$ for x in the Maclaurin series for e^x and then using the Maclaurin series for $\sin x$, we find

$$\begin{aligned}e^{\sin x} &= 1 + \sin x + \frac{\sin^2 x}{2} + \frac{\sin^3 x}{6} + \frac{\sin^4 x}{24} + \cdots\\ &= 1 + \left(x - \frac{x^3}{6} + \cdots\right) + \frac{1}{2}\left(x - \frac{x^3}{6} + \cdots\right)^2 + \frac{1}{6}(x - \cdots)^3 + \frac{1}{24}(x - \cdots)^4\\ &= 1 + x + \frac{1}{2}x^2 - \frac{1}{6}x^3 + \frac{1}{6}x^3 - \frac{1}{6}x^4 + \frac{1}{24}x^4 + \cdots\\ &= 1 + x + \frac{1}{2}x^2 - \frac{1}{8}x^4 + \cdots.\end{aligned}$$

In Exercises 63–66, find the functions with the following Maclaurin series (refer to Table 1).

63. $1 + x^3 + \frac{x^6}{2!} + \frac{x^9}{3!} + \frac{x^{12}}{4!} + \cdots$

SOLUTION We recognize

$$1 + x^3 + \frac{x^6}{2!} + \frac{x^9}{3!} + \frac{x^{12}}{4!} + \cdots = \sum_{n=0}^{\infty}\frac{x^{3n}}{n!} = \sum_{n=0}^{\infty}\frac{(x^3)^n}{n!}$$

as the Maclaurin series for e^x with x replaced by x^3. Therefore,

$$1 + x^3 + \frac{x^6}{2!} + \frac{x^9}{3!} + \frac{x^{12}}{4!} + \cdots = e^{x^3}.$$

65. $1 - \frac{5^3x^3}{3!} + \frac{5^5x^5}{5!} - \frac{5^7x^7}{7!} + \cdots$

SOLUTION Note

$$1 - \frac{5^3x^3}{3!} + \frac{5^5x^5}{5!} - \frac{5^7x^7}{7!} + \cdots = 1 - 5x + \left(5x - \frac{5^3x^3}{3!} + \frac{5^5x^5}{5!} - \frac{5^7x^7}{7!} + \cdots\right)$$

$$= 1 - 5x + \sum_{n=0}^{\infty} (-1)^n \frac{(5x)^{2n+1}}{(2n+1)!}.$$

The series is the Maclaurin series for $\sin x$ with x replaced by $5x$, so

$$1 - \frac{5^3 x^3}{3!} + \frac{5^5 x^5}{5!} - \frac{5^7 x^7}{7!} + \cdots = 1 - 5x + \sin(5x).$$

67. When a voltage V is applied to a series circuit consisting of a resistor R and an inductor L, the current at time t is

$$I(t) = \left(\frac{V}{R}\right)\left(1 - e^{-Rt/L}\right)$$

Expand $I(t)$ in a Maclaurin series. Show that $I(t) \approx Vt/L$ if R is small.

SOLUTION Substituting $-\frac{Rt}{L}$ for t in the Maclaurin series for e^t gives

$$e^{-Rt/L} = \sum_{n=0}^{\infty} \frac{\left(-\frac{Rt}{L}\right)^n}{n!} = \sum_{n=0}^{\infty} \frac{(-1)^n}{n!}\left(\frac{R}{L}\right)^n t^n = 1 + \sum_{n=1}^{\infty} \frac{(-1)^n}{n!}\left(\frac{R}{L}\right)^n t^n$$

Thus,

$$1 - e^{-Rt/L} = 1 - \left(1 + \sum_{n=1}^{\infty} \frac{(-1)^n}{n!}\left(\frac{R}{L}\right)^n t^n\right) = \sum_{n=1}^{\infty} \frac{(-1)^{n+1}}{n!}\left(\frac{Rt}{L}\right)^n,$$

and

$$I(t) = \frac{V}{R}\sum_{n=1}^{\infty} \frac{(-1)^{n+1}}{n!}\left(\frac{Rt}{L}\right)^n = \frac{Vt}{L} + \frac{V}{R}\sum_{n=2}^{\infty} \frac{(-1)^{n+1}}{n!}\left(\frac{Rt}{L}\right)^n.$$

If Rt/L is small, then the terms in the series are even smaller, and we find

$$V(t) \approx \frac{Vt}{L}.$$

69. Find the Maclaurin series for $f(x) = \cos(\sqrt{x})$ and use it to determine $f^{(5)}(0)$.

SOLUTION Substituting $\sqrt{x}$ for x in the Maclaurin series for $\cos x$

$$\cos(\sqrt{x}) = \sum_{n=0}^{\infty} (-1)^n \frac{(\sqrt{x})^{2n}}{(2n)!} = \sum_{n=0}^{\infty} (-1)^n \frac{x^n}{(2n)!}.$$

The coefficient of x^5 in this series is

$$\frac{(-1)^5}{10!} = -\frac{1}{10!} = \frac{f^{(5)}(0)}{5!},$$

so

$$f^{(5)}(0) = -\frac{5!}{10!} = -\frac{1}{6 \cdot 7 \cdot 8 \cdot 9 \cdot 10} = -\frac{1}{30240}.$$

71. Use the binomial series to find $f^{(8)}(0)$ for $f(x) = \sqrt{1 - x^2}$.

SOLUTION We obtain the Maclaurin series for $f(x) = \sqrt{1 - x^2}$ by substituting $-x^2$ for x in the binomial series with $a = \frac{1}{2}$. This gives

$$\sqrt{1 - x^2} = \sum_{n=0}^{\infty} \binom{\frac{1}{2}}{n} \left(-x^2\right)^n = \sum_{n=0}^{\infty} (-1)^n \binom{\frac{1}{2}}{n} x^{2n}.$$

The coefficient of x^8 is

$$(-1)^4 \binom{\frac{1}{2}}{4} = \frac{\frac{1}{2}\left(\frac{1}{2} - 1\right)\left(\frac{1}{2} - 2\right)\left(\frac{1}{2} - 3\right)}{4!} = -\frac{15}{16 \cdot 4!} = \frac{f^{(8)}(0)}{8!},$$

so

$$f^{(8)}(0) = \frac{-15 \cdot 8!}{16 \cdot 4!} = -1575.$$

73. Does the Taylor series for $f(x) = (1+x)^{3/4}$ converge to $f(x)$ at $x = 2$? Give numerical evidence to support your answer.

SOLUTION The Taylor series for $f(x) = (1+x)^{3/4}$ converges to $f(x)$ for $|x| < 1$; because $x = 2$ is not contained on this interval, the series does not converge to $f(x)$ at $x = 2$. The graph below displays

$$S_N = \sum_{n=0}^{N} \binom{\frac{3}{4}}{n} 2^n$$

for $0 \le N \le 14$. The divergent nature of the sequence of partial sums is clear.

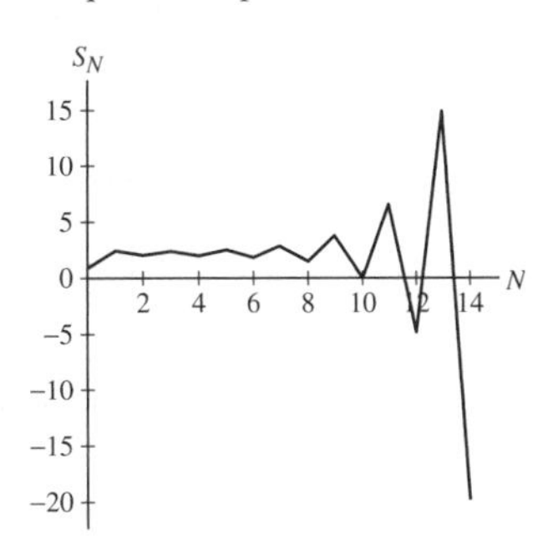

75. Explain the steps required to verify that the Maclaurin series for $f(x) = \tan^{-1} x$ converges to $f(x)$ at $x = 0.5$.

SOLUTION Recall that the Maclaurin series for $\tan^{-1} x$ can be obtained by term-by-term integration of the Maclaurin series for $\frac{1}{1+x^2}$. Now, we know that the geometric series

$$\sum_{n=0}^{\infty} (-x^2)^n$$

converges to $\frac{1}{1+x^2}$ for $|x| < 1$. It then follows from Theorem 3 of Section 11.6 that the Maclaurin series for $f(x) = \tan^{-1} x$ converges to $f(x)$ for $|x| < 1$. Because $x = 0.5$ is inside this interval, the Maclaurin series for $f(x) = \tan^{-1} x$ converges to $f(x)$ at $x = 0.5$.

77. How many terms of the Maclaurin series of $f(x) = \ln(1+x)$ are needed to compute $\ln 1.2$ to within an error of at most 0.0001? Make the computation and compare the result with the calculator value.

SOLUTION Substitute $x = 0.2$ into the Maclaurin series for $\ln(1+x)$ to obtain:

$$\ln 1.2 = \sum_{n=1}^{\infty} (-1)^{n-1} \frac{(0.2)^n}{n} = \sum_{n=1}^{\infty} (-1)^{n-1} \frac{1}{5^n n}.$$

This is an alternating series with $a_n = \dfrac{1}{n \cdot 5^n}$. Using the error bound for alternating series

$$|\ln 1.2 - S_N| \le a_{N+1} = \frac{1}{(N+1)5^{N+1}},$$

so we must choose N so that

$$\frac{1}{(N+1)5^{N+1}} < 0.0001 \quad \text{or} \quad (N+1)5^{N+1} > 10000.$$

For $N = 3$, $(N+1)5^{N+1} = 4 \cdot 5^4 = 2500 < 10,000$, and for $N = 4$, $(N+1)5^{N+1} = 5 \cdot 5^5 = 15,625 > 10,000$; thus, the smallest acceptable value for N is $N = 4$. The corresponding approximation is:

$$S_4 = \sum_{n=1}^{4} \frac{(-1)^{n-1}}{5^n \cdot n} = \frac{1}{5} - \frac{1}{5^2 \cdot 2} + \frac{1}{5^3 \cdot 3} - \frac{1}{5^4 \cdot 4} = 0.182266666.$$

Now, $\ln 1.2 = 0.182321556$, so

$$|\ln 1.2 - S_4| = 5.489 \times 10^{-5} < 0.0001.$$

In Exercises 78–79, let

$$f(x) = \frac{1}{(1-x)(1-2x)}$$

79. Find the Taylor series for $f(x)$ at $c = 2$. *Hint:* Rewrite the identity of Exercise 78 as

$$f(x) = \frac{2}{-3-2(x-2)} - \frac{1}{-1-(x-2)}$$

SOLUTION Using the given identity,

$$f(x) = \frac{2}{-3-2(x-2)} - \frac{1}{-1-(x-2)} = -\frac{2}{3}\frac{1}{1+\frac{2}{3}(x-2)} + \frac{1}{1+(x-2)}.$$

Substituting $-\frac{2}{3}(x-2)$ for x in the Maclaurin series for $\frac{1}{1-x}$ yields

$$\frac{1}{1+\frac{2}{3}(x-2)} = \sum_{n=0}^{\infty}(-1)^n\left(\frac{2}{3}\right)^n(x-2)^n,$$

and substituting $-(x-2)$ for x in the same Maclaurin series yields

$$\frac{1}{1+(x-2)} = \sum_{n=0}^{\infty}(-1)^n(x-2)^n.$$

The first series is valid for $\left|-\frac{2}{3}(x-2)\right| < 1$, or $|x-2| < \frac{3}{2}$, and the second series is valid for $|x-2| < 1$; therefore, the two series together are valid for $|x-2| < 1$. Finally, for $|x-2| < 1$,

$$f(x) = -\frac{2}{3}\sum_{n=0}^{\infty}(-1)^n\left(\frac{2}{3}\right)^n(x-2)^n + \sum_{n=0}^{\infty}(-1)^n(x-2)^n = \sum_{n=0}^{\infty}(-1)^n\left[1-\left(\frac{2}{3}\right)^{n+1}\right](x-2)^n.$$

81. Use Example 11 and the approximation $\sin x \approx x$ to show that the period T of a pendulum released at an angle θ has the following second-order approximation:

$$T \approx 2\pi\sqrt{\frac{L}{g}}\left(1+\frac{\theta^2}{16}\right)$$

SOLUTION The period T of a pendulum of length L released from an angle θ is

$$T = 4\sqrt{\frac{L}{g}}E(k),$$

where $g \approx 9.8\ \text{m/s}^2$ is the acceleration due to gravity, $E(k)$ is the elliptic function of the first kind and $k = \sin\frac{\theta}{2}$. From Example 11, we know that

$$E(k) = \frac{\pi}{2}\sum_{n=0}^{\infty}\left(\frac{1\cdot 3\cdot 5\cdots(2n-1)}{2\cdot 4\cdot 6\cdots(2n)}\right)^2 k^{2n}.$$

Using the approximation $\sin x \approx x$, we have

$$k = \sin\frac{\theta}{2} \approx \frac{\theta}{2};$$

moreover, using the first two terms of the series for $E(k)$, we find

$$E(k) \approx \frac{\pi}{2}\left[1+\left(\frac{1}{2}\right)^2\left(\frac{\theta}{2}\right)^2\right] = \frac{\pi}{2}\left(1+\frac{\theta^2}{16}\right).$$

Therefore,

$$T = 4\sqrt{\frac{L}{g}}E(k) \approx 2\pi\sqrt{\frac{L}{g}}\left(1+\frac{\theta^2}{16}\right).$$

Further Insights and Challenges

In Exercises 83–84, we investigate the convergence of the binomial series

$$T_a(x) = \sum_{n=0}^{\infty}\binom{a}{n}x^n$$

83. Prove that $T_a(x)$ has radius of convergence $R = 1$ if a is not a whole number. What is the radius of convergence if a is a whole number?

SOLUTION Suppose that a is not a whole number. Then

$$\binom{a}{n} = \frac{a\,(a-1)\cdots(a-n+1)}{n!}$$

is never zero. Moreover,

$$\left|\frac{\binom{a}{n+1}}{\binom{a}{n}}\right| = \left|\frac{a(a-1)\cdots(a-n+1)(a-n)}{(n+1)!} \cdot \frac{n!}{a(a-1)\cdots(a-n+1)}\right| = \left|\frac{a-n}{n+1}\right|,$$

so, by the formula for the radius of convergence

$$r = \lim_{n\to\infty}\left|\frac{a-n}{n+1}\right| = 1.$$

The radius of convergence of $T_a(x)$ is therefore $R = r^{-1} = 1$.

If a is a whole number, then $\binom{a}{n} = 0$ for all $n > a$. The infinite series then reduces to a polynomial of degree a, so it converges for all x (i.e. $R = \infty$).

85. The function $G(k) = \int_0^{\pi/2} \sqrt{1 - k^2 \sin^2 t}\, dt$ is called an **elliptic function of the second kind**. Prove that for $|k| < 1$

$$G(k) = \frac{\pi}{2} - \frac{\pi}{2}\sum_{n=1}^{\infty}\left(\frac{1\cdot 3\cdots(2n-1)}{2\cdots 4\cdot(2n)}\right)^2 \frac{k^{2n}}{2n-1}$$

SOLUTION For $|k| < 1$, $|k^2 \sin^2 t| < 1$ for all t. Substituting $-k^2\sin^2 t$ for t in the binomial series for $a = \frac{1}{2}$, we find

$$\begin{aligned}
\sqrt{1-k^2\sin^2 t} &= 1 + \sum_{n=1}^{\infty}\binom{\frac{1}{2}}{n}\left(-k^2\sin^2 t\right)^n \\
&= 1 + \sum_{n=1}^{\infty}(-1)^n \frac{\frac{1}{2}\left(\frac{1}{2}-1\right)\left(\frac{1}{2}-2\right)\cdots\left(\frac{1}{2}-n+1\right)}{n!}k^{2n}\sin^{2n} t \\
&= 1 + \sum_{n=1}^{\infty}(-1)^n \frac{1(1-2)(1-4)\cdots(1-2(n-1))}{2^n n!}k^{2n}\sin^{2n} t \\
&= 1 + \sum_{n=1}^{\infty}(-1)^n(-1)^{n-1} \frac{(2-1)(4-1)\cdots(2n-3)}{2^n n!}k^{2n}\sin^{2n} t \\
&= 1 - \sum_{n=1}^{\infty}\frac{1\cdot 3\cdot 5\cdots(2n-3)}{2\cdot 4\cdot 6\cdots(2n)}k^{2n}\sin^{2n} t.
\end{aligned}$$

Integrating from 0 to $\frac{\pi}{2}$ term-by-term, we obtain

$$G(k) = \frac{\pi}{2} - \sum_{n=1}^{\infty}\frac{1\cdot 3\cdot 5\cdots(2n-3)}{2\cdot 4\cdot 6\cdots(2n)}k^{2n}\int_0^{\pi/2}\sin^{2n} t\,dt.$$

Finally, using the formula

$$\int_0^{\pi/2}\sin^{2n} t\,dt = \frac{1\cdot 3\cdot 5\cdots(2n-1)}{2\cdot 4\cdot 6\cdots(2n)}\frac{\pi}{2},$$

we arrive at

$$G(k) = \frac{\pi}{2} - \frac{\pi}{2}\sum_{n=1}^{\infty}\left(\frac{1\cdot 3\cdot 5\cdots(2n-3)}{2\cdot 4\cdot 6\cdots(2n)}\right)^2(2n-1)k^{2n} = \frac{\pi}{2} - \frac{\pi}{2}\sum_{n=1}^{\infty}\left(\frac{1\cdot 3\cdot 5\cdots(2n-1)}{2\cdot 4\cdot 6\cdots(2n)}\right)^2\frac{k^{2n}}{2n-1}.$$

87. Use Exercise 85 to prove that if $a < b$ and a/b is near 1 (a nearly circular ellipse), then

$$L \approx \frac{\pi}{2}\left(3b + \frac{a^2}{b}\right)$$

Hint: Use the first two terms of the series for $G(k)$.

SOLUTION From the previous exercise, we know that

$$L = 4bG(k), \quad \text{where} \quad k = \sqrt{1 - \frac{a^2}{b^2}}.$$

Following the hint and using only the first two terms of the series expansion for $G(k)$ from Exercise 85, we find

$$L \approx 4b\left(\frac{\pi}{2} - \frac{\pi}{2}\left(\frac{1}{2}\right)^2 k^2\right) = \frac{\pi}{2}\left(4b - b\left(1 - \frac{a^2}{b^2}\right)\right) = \frac{\pi}{2}\left(3b + \frac{a^2}{b}\right).$$

CHAPTER REVIEW EXERCISES

1. Let $a_n = \dfrac{n-3}{n!}$ and $b_n = a_{n+3}$. Calculate the first three terms in the sequence:

(a) a_n^2

(b) b_n

(c) $a_n b_n$

(d) $2a_{n+1} - 3a_n$

SOLUTION

(a)

$$a_1^2 = \left(\frac{1-3}{1!}\right)^2 = (-2)^2 = 4;$$

$$a_2^2 = \left(\frac{2-3}{2!}\right)^2 = \left(-\frac{1}{2}\right)^2 = \frac{1}{4};$$

$$a_3^2 = \left(\frac{3-3}{3!}\right)^2 = 0.$$

(b)

$$b_1 = a_4 = \frac{4-3}{4!} = \frac{1}{24};$$

$$b_2 = a_5 = \frac{5-3}{5!} = \frac{1}{60};$$

$$b_3 = a_6 = \frac{6-3}{6!} = \frac{1}{240}.$$

(c) Using the formula for a_n and the values in (b) we obtain:

$$a_1b_1 = \frac{1-3}{1!} \cdot \frac{1}{24} = -\frac{1}{12};$$

$$a_2b_2 = \frac{2-3}{2!} \cdot \frac{1}{60} = -\frac{1}{120};$$

$$a_3b_3 = \frac{3-3}{3!} \cdot \frac{1}{240} = 0.$$

(d)

$$2a_2 - 3a_1 = 2\left(-\frac{1}{2}\right) - 3(-2) = 5;$$

$$2a_3 - 3a_2 = 2 \cdot 0 - 3\left(-\frac{1}{2}\right) = \frac{3}{2};$$

$$2a_4 - 3a_3 = 2 \cdot \frac{1}{24} - 3 \cdot 0 = \frac{1}{12}.$$

In Exercises 3–8, compute the limit (or state that it does not exist) assuming that $\lim_{n\to\infty} a_n = 2$.

3. $\lim_{n\to\infty} (5a_n - 2a_n^2)$

SOLUTION

$$\lim_{n\to\infty} \left(5a_n - 2a_n^2\right) = 5\lim_{n\to\infty} a_n - 2\lim_{n\to\infty} a_n^2 = 5\lim_{n\to\infty} a_n - 2\left(\lim_{n\to\infty} a_n\right)^2 = 5\cdot 2 - 2\cdot 2^2 = 2.$$

5. $\lim_{n\to\infty} e^{a_n}$

SOLUTION The function $f(x) = e^x$ is continuous, hence:

$$\lim_{n\to\infty} e^{a_n} = e^{\lim_{n\to\infty} a_n} = e^2.$$

7. $\lim_{n\to\infty} (-1)^n a_n$

SOLUTION Because $\lim_{n\to\infty} a_n \neq 0$, it follows that $\lim_{n\to\infty} (-1)^n a_n$ does not exist.

In Exercises 9–22, determine the limit of the sequence or show that the sequence diverges.

9. $a_n = \sqrt{n+5} - \sqrt{n+2}$

SOLUTION First rewrite a_n as follows:

$$a_n = \frac{\left(\sqrt{n+5} - \sqrt{n+2}\right)\left(\sqrt{n+5} + \sqrt{n+2}\right)}{\sqrt{n+5} + \sqrt{n+2}} = \frac{(n+5)-(n+2)}{\sqrt{n+5} + \sqrt{n+2}} = \frac{3}{\sqrt{n+5} + \sqrt{n+2}}.$$

Thus,

$$\lim_{n\to\infty} a_n = \lim_{n\to\infty} \frac{3}{\sqrt{n+5} + \sqrt{n+2}} = 0.$$

11. $a_n = 2^{1/n^2}$

SOLUTION The function $f(x) = 2^x$ is continuous, so

$$\lim_{n\to\infty} a_n = \lim_{n\to\infty} 2^{1/n^2} = 2^{\lim_{n\to\infty}(1/n^2)} = 2^0 = 1.$$

13. $b_m = 1 + (-1)^m$

SOLUTION Because $1 + (-1)^m$ is equal to 0 for m odd and is equal to 2 for m even, the sequence $\{b_m\}$ does not approach one limit; hence this sequence diverges.

15. $b_n = \tan^{-1}\left(\frac{n+2}{n+5}\right)$

SOLUTION The function $\tan^{-1} x$ is continuous, so

$$\lim_{n\to\infty} b_n = \lim_{n\to\infty} \tan^{-1}\left(\frac{n+2}{n+5}\right) = \tan^{-1}\left(\lim_{n\to\infty} \frac{n+2}{n+5}\right) = \tan^{-1} 1 = \frac{\pi}{4}.$$

17. $b_n = \sqrt{n^2+n} - \sqrt{n^2+1}$

SOLUTION Rewrite b_n as

$$b_n = \frac{\left(\sqrt{n^2+n} - \sqrt{n^2+1}\right)\left(\sqrt{n^2+n} + \sqrt{n^2+1}\right)}{\sqrt{n^2+n} + \sqrt{n^2+1}} = \frac{\left(n^2+n\right) - \left(n^2+1\right)}{\sqrt{n^2+n} + \sqrt{n^2+1}} = \frac{n-1}{\sqrt{n^2+n} + \sqrt{n^2+1}}.$$

Then

$$\lim_{n\to\infty} b_n = \lim_{n\to\infty} \frac{\frac{n}{n} - \frac{1}{n}}{\sqrt{\frac{n^2}{n^2} + \frac{n}{n^2}} + \sqrt{\frac{n^2}{n^2} + \frac{1}{n^2}}} = \lim_{n\to\infty} \frac{1 - \frac{1}{n}}{\sqrt{1 + \frac{1}{n}} + \sqrt{1 + \frac{1}{n^2}}} = \frac{1-0}{\sqrt{1+0} + \sqrt{1+0}} = \frac{1}{2}.$$

19. $a_n = \frac{100^n}{n!} - \frac{3 + \pi^n}{5^n}$

SOLUTION For $n > 100$,

$$0 \le \frac{100^n}{n!} = \left(\frac{100}{1}\cdot\frac{100}{2}\cdots\frac{100}{100}\right)\frac{100}{101}\cdot\frac{100}{102}\cdot\frac{100}{n} < \frac{100^{100}}{99!n};$$

therefore,

$$\lim_{n\to\infty}\frac{100^n}{n!} = 0$$

by the Squeeze Theorem. Moreover,

$$\lim_{n\to\infty}\left(\frac{3+\pi^n}{5^n}\right) = \lim_{n\to\infty}\frac{3}{5^n} + \lim_{n\to\infty}\left(\frac{\pi}{5}\right)^n = 0+0=0.$$

Thus,

$$\lim_{n\to\infty} a_n = 0+0=0.$$

21. $c_n = \left(1+\frac{3}{n}\right)^n$

SOLUTION Write

$$c_n = \left(1+\frac{1}{n/3}\right)^n = \left[\left(1+\frac{1}{n/3}\right)^{n/3}\right]^3.$$

Then, because x^3 is a continuous function,

$$\lim_{n\to\infty} c_n = \left[\lim_{n\to\infty}\left(1+\frac{1}{n/3}\right)^{n/3}\right]^3 = e^3.$$

23. Use the Squeeze Theorem to show that $\lim_{n\to\infty}\frac{\arctan(n^2)}{\sqrt{n}} = 0.$

SOLUTION For all x,

$$-\frac{\pi}{2} < \arctan x < \frac{\pi}{2},$$

so

$$-\frac{\pi/2}{\sqrt{n}} < \frac{\arctan(n^2)}{\sqrt{n}} < \frac{\pi/2}{\sqrt{n}},$$

for all n. Because

$$\lim_{n\to\infty}\left(-\frac{\pi/2}{\sqrt{n}}\right) = \lim_{n\to\infty}\frac{\pi/2}{\sqrt{n}} = 0,$$

it follows by the Squeeze Theorem that

$$\lim_{n\to\infty}\frac{\arctan(n^2)}{\sqrt{n}} = 0.$$

25. Given $a_n = \frac{1}{2}3^n - \frac{1}{3}2^n$,

(a) Calculate $\lim_{n\to\infty} a_n$.

(b) Calculate $\lim_{n\to\infty}\frac{a_{n+1}}{a_n}$.

SOLUTION

(a) Because

$$\frac{1}{2}3^n - \frac{1}{3}2^n \ge \frac{1}{2}3^n - \frac{1}{3}3^n = \frac{3^n}{6}$$

and

$$\lim_{n\to\infty}\frac{3^n}{6} = \infty,$$

we conclude that

$$\lim_{n\to\infty} a_n = \infty.$$

(b) $$\lim_{n\to\infty} \frac{a_{n+1}}{a_n} = \lim_{n\to\infty} \frac{\frac{1}{2}3^{n+1} - \frac{1}{3}2^{n+1}}{\frac{1}{2}3^n - \frac{1}{3}2^n} = \lim_{n\to\infty} \frac{3^{n+2} - 2^{n+2}}{3^{n+1} - 2^{n+1}} = \lim_{n\to\infty} \frac{3 - 2\left(\frac{2}{3}\right)^{n+1}}{1 - \left(\frac{2}{3}\right)^{n+1}} = \frac{3-0}{1-0} = 3.$$

27. Calculate the partial sums S_4 and S_7 of the series $\sum_{n=1}^{\infty} \frac{n-2}{n^2+2n}$.

SOLUTION

$$S_4 = -\frac{1}{3} + 0 + \frac{1}{15} + \frac{2}{24} = -\frac{11}{60} = -0.183333;$$

$$S_7 = -\frac{1}{3} + 0 + \frac{1}{15} + \frac{2}{24} + \frac{3}{35} + \frac{4}{48} + \frac{5}{63} = \frac{287}{4410} = 0.065079.$$

29. Find the sum $\frac{4}{9} + \frac{8}{27} + \frac{16}{81} + \frac{32}{243} + \cdots$.

SOLUTION This is a geometric series with common ratio $r = \frac{2}{3}$. Therefore,

$$\frac{4}{9} + \frac{8}{27} + \frac{16}{81} + \frac{32}{243} + \cdots = \frac{\frac{4}{9}}{1-\frac{2}{3}} = \frac{4}{3}.$$

31. Find the sum $\sum_{n=0}^{\infty} \frac{2^{n+1}}{3^n}$.

SOLUTION Note

$$\sum_{n=0}^{\infty} \frac{2^{n+1}}{3^n} = 2\sum_{n=0}^{\infty} \frac{2^n}{3^n} = 2\sum_{n=0}^{\infty} \left(\frac{2}{3}\right)^n;$$

therefore,

$$\sum_{n=0}^{\infty} \frac{2^{n+1}}{3^n} = 2\frac{1}{1-\frac{2}{3}} = 2 \cdot 3 = 6.$$

33. Give an example of divergent series $\sum_{n=1}^{\infty} a_n$, $\sum_{n=1}^{\infty} b_n$ such that $\sum_{n=1}^{\infty} (a_n + b_n) = 1$.

SOLUTION Let $a_n = \left(\frac{1}{2}\right)^n + 1$, $b_n = -1$. The corresponding series diverge by the Divergence Test; however,

$$\sum_{n=1}^{\infty} (a_n + b_n) = \sum_{n=1}^{\infty} \left(\frac{1}{2}\right)^n = \frac{\frac{1}{2}}{1-\frac{1}{2}} = 1.$$

In Exercises 35–38, use the Integral Test to determine if the infinite series converges.

35. $\sum_{n=1}^{\infty} \frac{n^2}{n^3+1}$

SOLUTION Let $f(x) = \frac{x^2}{x^3+1}$. This function is continuous and positive for $x \geq 1$. Because

$$f'(x) = \frac{(x^3+1)(2x) - x^2(3x^2)}{(x^3+1)^2} = \frac{x(2-x^3)}{(x^3+1)^2},$$

we see that $f'(x) < 0$ and f is decreasing on the interval $x \geq 2$. Therefore, the Integral Test applies on the interval $x \geq 2$. Now,

$$\int_2^{\infty} \frac{x^2}{x^3+1}\,dx = \lim_{R\to\infty} \int_2^R \frac{x^2}{x^3+1}\,dx = \frac{1}{3}\lim_{R\to\infty} \left(\ln(R^3+1) - \ln 9\right) = \infty.$$

The integral diverges; hence, the series $\sum_{n=2}^{\infty} \frac{n^2}{n^3+1}$ diverges, as does the series $\sum_{n=1}^{\infty} \frac{n^2}{n^3+1}$.

37. $\sum_{n=1}^{\infty} \frac{n^3}{e^{n^4}}$

SOLUTION Let $f(x) = x^3 e^{-x^4}$. This function is continuous and positive for $x \geq 1$. Because

$$f'(x) = x^3 \left(-4x^3 e^{-x^4}\right) + 3x^2 e^{-x^4} = x^2 e^{-x^4} \left(3 - 4x^4\right),$$

we see that $f'(x) < 0$ and f is decreasing on the interval $x \geq 1$. Therefore, the Integral Test applies on the interval $x \geq 1$. Now,

$$\int_1^{\infty} x^3 e^{-x^4}\, dx = \lim_{R \to \infty} \int_1^R x^3 e^{-x^4}\, dx = -\frac{1}{4} \lim_{R \to \infty} \left(e^{-R^4} - e^{-1}\right) = \frac{1}{4e}.$$

The integral converges; hence, the series $\sum_{n=1}^{\infty} \frac{n^3}{e^{n^4}}$ also converges.

In Exercises 39–46, use the Comparison or Limit Comparison Test to determine whether the infinite series converges.

39. $\sum_{n=1}^{\infty} \frac{1}{(n+1)^2}$

SOLUTION For all $n \geq 1$,

$$0 < \frac{1}{n+1} < \frac{1}{n} \quad \text{so} \quad \frac{1}{(n+1)^2} < \frac{1}{n^2}.$$

The series $\sum_{n=1}^{\infty} \frac{1}{n^2}$ is a convergent p-series, so the series $\sum_{n=1}^{\infty} \frac{1}{(n+1)^2}$ converges by the Comparison Test.

41. $\sum_{n=2}^{\infty} \frac{n^2+1}{n^{3.5}-2}$

SOLUTION Apply the Limit Comparison Test with $a_n = \frac{n^2+1}{n^{3.5}-2}$ and $b_n = \frac{1}{n^{1.5}}$. Now,

$$L = \lim_{n \to \infty} \frac{\frac{n^2+1}{n^{3.5}-2}}{\frac{1}{n^{1.5}}} = \lim_{n \to \infty} \frac{n^{3.5} + n^{1.5}}{n^{3.5} - 2} = 1.$$

Because L exists and $\sum_{n=1}^{\infty} \frac{1}{n^{1.5}}$ is a convergent p-series, we conclude by the Limit Comparison Test that the series $\sum_{n=2}^{\infty} \frac{n^2+1}{n^{3.5}-2}$ also converges.

43. $\sum_{n=2}^{\infty} \frac{\ln n}{1.5^n}$

SOLUTION Apply the Limit Comparison Test with $a_n = \frac{\ln n}{1.5^n}$ and $b_n = \frac{1}{1.25^n}$. Then,

$$L = \lim_{n \to \infty} \frac{a_n}{b_n} = \lim_{n \to \infty} \frac{\ln n}{1.5^n} \cdot \frac{1.25^n}{1} = \lim_{n \to \infty} \frac{\ln n}{(6/5)^n} = 0,$$

because the logarithm grows more slowly than the exponential function. The series $\sum_{n=2}^{\infty} \frac{1}{1.25^n}$ is a convergent geometric series; because L exists, we may therefore conclude by the Limit Comparison Test that the series $\sum_{n=2}^{\infty} \frac{\ln n}{1.5^n}$ also converges.

45. $\sum_{n=1}^{\infty} \frac{1}{3^n - 2^n}$

SOLUTION Apply the Limit Comparison Test with $a_n = \frac{1}{3^n-2^n}$ and $b_n = \frac{1}{3^n}$. Then,

$$L = \lim_{n \to \infty} \frac{a_n}{b_n} = \lim_{n \to \infty} \frac{3^n}{3^n - 2^n} = \lim_{n \to \infty} \frac{1}{1 - \left(\frac{2}{3}\right)^n} = 1.$$

The series $\sum_{n=1}^{\infty} \frac{1}{3^n}$ is a convergent geometric series; because L exists, we may therefore conclude by the Limit Comparison Test that the series $\sum_{n=1}^{\infty} \frac{1}{3^n - 2^n}$ also converges.

47. Show that $\sum_{n=2}^{\infty} \left(1 - \sqrt{1 - \frac{1}{n}}\right)$ diverges. *Hint:* Show that

$$1 - \sqrt{1 - \frac{1}{n}} \geq \frac{1}{2n}$$

SOLUTION

$$1 - \sqrt{1 - \frac{1}{n}} = 1 - \sqrt{\frac{n-1}{n}} = \frac{\sqrt{n} - \sqrt{n-1}}{\sqrt{n}} = \frac{n - (n-1)}{\sqrt{n}(\sqrt{n} + \sqrt{n-1})} = \frac{1}{n + \sqrt{n^2 - n}}$$

$$\geq \frac{1}{n + \sqrt{n^2}} = \frac{1}{2n}.$$

The series $\sum_{n=2}^{\infty} \frac{1}{2n}$ diverges, so the series $\sum_{n=2}^{\infty} \left(1 - \sqrt{1 - \frac{1}{n}}\right)$ also diverges by the Comparison Test.

49. Let $S = \sum_{n=1}^{\infty} \frac{n}{(n^2+1)^2}$.

(a) Show that S converges.

(b) CAS Use Eq. (5) in Exercise 77 of Section 11.3 with $M = 99$ to approximate S. What is the maximum size of the error?

SOLUTION

(a) For $n \geq 1$,

$$\frac{n}{(n^2+1)^2} < \frac{n}{(n^2)^2} = \frac{1}{n^3}.$$

The series $\sum_{n=1}^{\infty} \frac{1}{n^3}$ is a convergent p-series, so the series $\sum_{n=1}^{\infty} \frac{n}{(n^2+1)^2}$ also converges by the Comparison Test.

(b) With $a_n = \frac{n}{(n^2+1)^2}$, $f(x) = \frac{x}{(x^2+1)^2}$ and $M = 99$, Eq. (5) in Exercise 77 of Section 11.3 becomes

$$\sum_{n=1}^{99} \frac{n}{(n^2+1)^2} + \int_{100}^{\infty} \frac{x}{(x^2+1)^2}\,dx \leq S \leq \sum_{n=1}^{100} \frac{n}{(n^2+1)^2} + \int_{100}^{\infty} \frac{x}{(x^2+1)^2}\,dx,$$

or

$$0 \leq S - \left(\sum_{n=1}^{99} \frac{n}{(n^2+1)^2} + \int_{100}^{\infty} \frac{x}{(x^2+1)^2}\,dx\right) \leq \frac{100}{(100^2+1)^2}.$$

Now,

$$\sum_{n=1}^{99} \frac{n}{(n^2+1)^2} = 0.397066274; \text{ and}$$

$$\int_{100}^{\infty} \frac{x}{(x^2+1)^2}\,dx = \lim_{R\to\infty} \int_{100}^{R} \frac{x}{(x^2+1)^2}\,dx = \frac{1}{2} \lim_{R\to\infty} \left(-\frac{1}{R^2+1} + \frac{1}{100^2+1}\right)$$

$$= \frac{1}{20002} = 0.000049995;$$

thus,

$$S \approx 0.397066274 + 0.000049995 = 0.397116269.$$

The bound on the error in this approximation is

$$\frac{100}{(100^2+1)^2} = 9.998 \times 10^{-7}.$$

In Exercises 50–53, determine whether the series converges absolutely. If not, determine whether it converges conditionally.

51. $\sum_{n=1}^{\infty} \frac{(-1)^n}{n^{1.1} \ln(n+1)}$

SOLUTION Consider the corresponding positive series $\sum_{n=1}^{\infty} \frac{1}{n^{1.1} \ln(n+1)}$. Because

$$\frac{1}{n^{1.1} \ln(n+1)} < \frac{1}{n^{1.1}}$$

and $\sum_{n=1}^{\infty} \frac{1}{n^{1.1}}$ is a convergent p-series, we can conclude by the Comparison Test that $\sum_{n=1}^{\infty} \frac{(-1)^n}{n^{1.1} \ln(n+1)}$ also converges. Thus, $\sum_{n=1}^{\infty} \frac{(-1)^n}{n^{1.1} \ln(n+1)}$ converges absolutely.

53. $\sum_{n=1}^{\infty} \frac{\cos\left(\frac{\pi}{4} + 2\pi n\right)}{\sqrt{n}}$

SOLUTION $\cos\left(\frac{\pi}{4} + 2\pi n\right) = \cos\frac{\pi}{4} = \frac{\sqrt{2}}{2}$, so

$$\sum_{n=1}^{\infty} \frac{\cos\left(\frac{\pi}{4} + 2\pi n\right)}{\sqrt{n}} = \frac{\sqrt{2}}{2} \sum_{n=1}^{\infty} \frac{1}{\sqrt{n}}.$$

This is a divergent p-series, so the series $\sum_{n=1}^{\infty} \frac{\cos\left(\frac{\pi}{4} + 2\pi n\right)}{\sqrt{n}}$ diverges.

55. How many terms of the series are needed to calculate Catalan's constant $K = \sum_{k=0}^{\infty} \frac{(-1)^k}{(2k+1)^2}$ to three decimal places? Carry out the calculation.

SOLUTION Using the error bound for an alternating series, we have

$$|S_N - K| \leq \frac{1}{(2(N+1)+1)^2} = \frac{1}{(2N+3)^2}.$$

For accuracy to three decimal places, we must choose N so that

$$\frac{1}{(2N+3)^2} < 5 \times 10^{-3} \quad \text{or} \quad (2N+3)^2 > 2000.$$

Solving for N yields

$$N > \frac{1}{2}\left(\sqrt{2000} - 3\right) \approx 20.9.$$

Thus,

$$K \approx \sum_{k=0}^{21} \frac{(-1)^k}{(2k+1)^2} = 0.915707728.$$

57. Let $\sum_{n=1}^{\infty} a_n$ be an absolutely convergent series. Determine whether the following series are convergent or divergent:

(a) $\sum_{n=1}^{\infty} \left(a_n + \frac{1}{n^2}\right)$

(b) $\sum_{n=1}^{\infty} (-1)^n a_n$

(c) $\sum_{n=1}^{\infty} \frac{1}{1 + a_n^2}$

(d) $\sum_{n=1}^{\infty} \frac{|a_n|}{n}$

SOLUTION Because $\sum_{n=1}^{\infty} a_n$ converges absolutely, we know that $\sum_{n=1}^{\infty} a_n$ converges and that $\sum_{n=1}^{\infty} |a_n|$ converges.

(a) Because we know that $\sum_{n=1}^{\infty} a_n$ converges and the series $\sum_{n=1}^{\infty} \frac{1}{n^2}$ is a convergent p-series, the sum of these two series, $\sum_{n=1}^{\infty} \left(a_n + \frac{1}{n^2}\right)$ also converges.

(b) We have,

$$\sum_{n=1}^{\infty} \left|(-1)^n a_n\right| = \sum_{n=1}^{\infty} |a_n|$$

Because $\sum_{n=1}^{\infty} |a_n|$ converges, it follows that $\sum_{n=1}^{\infty} (-1)^n a_n$ converges absolutely, which implies that $\sum_{n=1}^{\infty} (-1)^n a_n$ converges.

(c) Because $\sum_{n=1}^{\infty} a_n$ converges, $\lim_{n\to\infty} a_n = 0$. Therefore,

$$\lim_{n\to\infty} \frac{1}{1+a_n^2} = \frac{1}{1+0^2} = 1 \neq 0,$$

and the series $\sum_{n=1}^{\infty} \frac{1}{1+a_n^2}$ diverges by the Divergence Test.

(d) $\frac{|a_n|}{n} \le |a_n|$ and the series $\sum_{n=1}^{\infty} |a_n|$ converges, so the series $\sum_{n=1}^{\infty} \frac{|a_n|}{n}$ also converges by the Comparison Test.

In Exercises 58–65, apply the Ratio Test to determine convergence or divergence, or state that the Ratio Test is inconclusive.

59. $\sum_{n=1}^{\infty} \frac{\sqrt{n+1}}{n^8}$

SOLUTION With $a_n = \frac{\sqrt{n+1}}{n^8}$,

$$\left|\frac{a_{n+1}}{a_n}\right| = \frac{\sqrt{n+2}}{(n+1)^8} \cdot \frac{n^8}{\sqrt{n+1}} = \sqrt{\frac{n+2}{n+1}} \left(\frac{n}{n+1}\right)^8,$$

and

$$\rho = \lim_{n\to\infty} \left|\frac{a_{n+1}}{a_n}\right| = 1 \cdot 1^8 = 1.$$

Because $\rho = 1$, the Ratio Test is inconclusive.

61. $\sum_{n=1}^{\infty} \frac{n^4}{n!}$

SOLUTION With $a_n = \frac{n^4}{n!}$,

$$\left|\frac{a_{n+1}}{a_n}\right| = \frac{(n+1)^4}{(n+1)!} \cdot \frac{n!}{n^4} = \frac{(n+1)^3}{n^4} \quad \text{and} \quad \rho = \lim_{n\to\infty} \frac{a_{n+1}}{a_n} = 0.$$

Because $\rho < 1$, the series converges by the Ratio Test.

63. $\sum_{n=4}^{\infty} \frac{\ln n}{n^{3/2}}$

SOLUTION With $a_n = \frac{\ln n}{n^{3/2}}$,

$$\left|\frac{a_{n+1}}{a_n}\right| = \frac{\ln(n+1)}{(n+1)^{3/2}} \cdot \frac{n^{3/2}}{\ln n} = \left(\frac{n}{n+1}\right)^{3/2} \frac{\ln(n+1)}{\ln n},$$

and

$$\rho = \lim_{n\to\infty} \left|\frac{a_{n+1}}{a_n}\right| = 1^{3/2} \cdot 1 = 1.$$

Because $\rho = 1$, the Ratio Test is inconclusive.

65. $\sum_{n=1}^{\infty} \left(\frac{n}{4}\right)^n \frac{1}{n!}$

SOLUTION With $a_n = \left(\frac{n}{4}\right)^n \frac{1}{n!}$,

$$\left|\frac{a_{n+1}}{a_n}\right| = \left(\frac{n+1}{4}\right)^{n+1} \frac{1}{(n+1)!} \cdot \left(\frac{4}{n}\right)^n n! = \frac{1}{4}\left(\frac{n+1}{n}\right)^n = \frac{1}{4}\left(1+\frac{1}{n}\right)^n,$$

and

$$\rho = \lim_{n\to\infty} \left|\frac{a_{n+1}}{a_n}\right| = \frac{1}{4}e.$$

Because $\rho = \frac{e}{4} < 1$, the series converges by the Ratio Test.

In Exercises 66–69, apply the Root Test to determine convergence or divergence, or state that the Root Test is inconclusive.

67. $\sum_{n=1}^{\infty} \left(\frac{2}{n}\right)^n$

SOLUTION With $a_n = \left(\frac{2}{n}\right)^n$,

$$L = \lim_{n\to\infty} \sqrt[n]{\left(\frac{2}{n}\right)^n} = \lim_{n\to\infty} \frac{2}{n} = 0.$$

Because $L < 1$, the series converges by the Root Test.

69. $\sum_{n=1}^{\infty} \left(\cos\frac{1}{n}\right)^{n^3}$

SOLUTION With $a_n = \left(\cos\frac{1}{n}\right)^{n^3}$,

$$L = \lim_{n\to\infty} \sqrt[n]{a_n} = \lim_{n\to\infty} \sqrt[n]{\cos\left(\frac{1}{n}\right)^{n^3}} = \lim_{n\to\infty} \cos\left(\frac{1}{n}\right)^{n^2} = \lim_{x\to\infty} \cos\left(\frac{1}{x}\right)^{x^2}.$$

Now,

$$\ln L = \lim_{x\to\infty} x^2 \ln\cos\left(\frac{1}{x}\right) = \lim_{x\to\infty} \frac{\ln\cos\left(\frac{1}{x}\right)}{\frac{1}{x^2}} = \lim_{x\to\infty} \frac{\frac{1}{\cos\left(\frac{1}{x}\right)}\left(-\sin\left(\frac{1}{x}\right)\right)\left(-\frac{1}{x^2}\right)}{-\frac{2}{x^3}}$$

$$= -\frac{1}{2}\lim_{x\to\infty} \frac{1}{\cos\left(\frac{1}{x}\right)} \cdot \lim_{x\to\infty} \frac{\sin\left(\frac{1}{x}\right)}{\frac{1}{x}} = -\frac{1}{2}\cdot 1 \cdot 1 = -\frac{1}{2}.$$

Therefore, $L = e^{-1/2}$. Because $L < 1$, the series converges by the Root Test.

In Exercises 71–84, determine convergence or divergence using any method covered in the text.

71. $\sum_{n=1}^{\infty} \left(\frac{2}{3}\right)^n$

SOLUTION This is a geometric series with ratio $r = \frac{2}{3} < 1$; hence, the series converges.

73. $\sum_{n=1}^{\infty} e^{-0.02n}$

SOLUTION This is a geometric series with common ratio $r = \frac{1}{e^{0.02}} \approx 0.98 < 1$; hence, the series converges.

75. $\sum_{n=1}^{\infty} \frac{(-1)^{n-1}}{\sqrt{n}+\sqrt{n+1}}$

SOLUTION In this alternating series, $a_n = \frac{1}{\sqrt{n}+\sqrt{n+1}}$. The sequence $\{a_n\}$ is decreasing, and

$$\lim_{n\to\infty} a_n = 0;$$

therefore the series converges by the Leibniz Test.

77. $\displaystyle\sum_{n=10}^{\infty} \frac{(-1)^n}{\log n}$

SOLUTION The sequence $a_n = \frac{1}{\log n}$ is decreasing for $n \geq 10$ and

$$\lim_{n\to\infty} a_n = 0;$$

therefore, the series converges by the Leibniz Test.

79. $\displaystyle\sum_{n=1}^{\infty} \frac{e^n}{n!}$

SOLUTION With $a_n = \frac{e^n}{n!}$,

$$\left|\frac{a_{n+1}}{a_n}\right| = \frac{e^{n+1}}{(n+1)!} \cdot \frac{n!}{e^n} = \frac{e}{n+1} \quad \text{and} \quad \rho = \lim_{n\to\infty} \left|\frac{a_{n+1}}{a_n}\right| = 0.$$

Because $\rho < 1$, the series converges by the Ratio Test.

81. $\displaystyle\sum_{n=1}^{\infty} \frac{1}{n - 100.1}$

SOLUTION For $n \geq 101$, the sequence $\frac{1}{n-100.1}$ is positive and

$$\frac{1}{n - 100.1} > \frac{1}{n}.$$

Now, $\displaystyle\sum_{n=101}^{\infty} \frac{1}{n}$ diverges, so $\displaystyle\sum_{n=101}^{\infty} \frac{1}{n - 100.1}$ diverges by the Comparison Test. Since adding the finite sum $\displaystyle\sum_{n=1}^{100} \frac{1}{n - 100.1}$ does not affect the convergence of the series, it follows that $\displaystyle\sum_{n=1}^{\infty} \frac{1}{n - 100.1}$ diverges as well.

83. $\displaystyle\sum_{n=1}^{\infty} \sin^2 \frac{\pi}{n}$

SOLUTION For all $x > 0$, $\sin x < x$. Therefore, $\sin^2 x < x^2$, and for $x = \frac{\pi}{n}$,

$$\sin^2 \frac{\pi}{n} < \frac{\pi^2}{n^2} = \pi^2 \cdot \frac{1}{n^2}.$$

The series $\displaystyle\sum_{n=1}^{\infty} \frac{1}{n^2}$ is a convergent p-series, so the series $\displaystyle\sum_{n=1}^{\infty} \sin^2 \frac{\pi}{n}$ also converges by the Comparison Test.

In Exercises 85–90, find the values of x for which the power series converges.

85. $\displaystyle\sum_{n=0}^{\infty} \frac{2^n x^n}{n!}$

SOLUTION With $a_n = \frac{2^n}{n!}$,

$$\left|\frac{a_{n+1}}{a_n}\right| = \frac{2^{n+1}}{(n+1)!} \cdot \frac{n!}{2^n} = \frac{2}{n+1} \quad \text{and} \quad r = \lim_{n\to\infty} \left|\frac{a_{n+1}}{a_n}\right| = 0.$$

The radius of convergence is therefore $R = r^{-1} = \infty$, and the series converges for all x.

87. $\displaystyle\sum_{n=0}^{\infty} \frac{n^6 (x-3)^n}{n^8 + 1}$

SOLUTION With $a_n = \frac{n^6}{n^8+1}$,

$$r = \lim_{n\to\infty} \left|\frac{a_{n+1}}{a_n}\right| = \lim_{n\to\infty} \left(\frac{n+1}{n}\right)^6 \frac{n^8+1}{(n+1)^8+1} = 1.$$

The radius of convergence is therefore $R = r^{-1} = 1$, and the series converges absolutely for $|x - 3| < 1$, or $2 < x < 4$. For the endpoint $x = 4$, the series becomes $\displaystyle\sum_{n=0}^{\infty} \frac{n^6}{n^8 + 1}$, which converges by the Comparison Test comparing with the convergent p-series $\sum_{n=1}^{\infty} \frac{1}{n^2}$. For the endpoint $x = 2$, the series becomes $\sum_{n=0}^{\infty} \frac{n^6(-1)^n}{n^8+1}$, which converges by the Leibniz Test. The series $\sum_{n=0}^{\infty} \frac{n^6(x-3)^n}{n^8+1}$ therefore converges for $2 \le x \le 4$.

89. $\sum_{n=0}^{\infty} (nx)^n$

SOLUTION With $a_n = n^n$,

$$\left|\frac{a_{n+1}}{a_n}\right| = \frac{(n+1)^{n+1}}{n^n} = (n+1)\left(\frac{n+1}{n}\right)^n = (n+1)\left(1+\frac{1}{n}\right)^n,$$

and

$$r = \lim_{n\to\infty} \left|\frac{a_{n+1}}{a_n}\right| = \infty.$$

The radius of convergence is therefore $R = r^{-1} = 0$, and the series converges only for $x = 0$.

91. Expand the function $f(x) = \frac{2}{4-3x}$ as a power series centered at $c = 0$. Determine the values of x for which the series converges.

SOLUTION Write

$$\frac{2}{4-3x} = \frac{1}{2}\frac{1}{1-\frac{3}{4}x}.$$

Substituting $\frac{3}{4}x$ for x in the Maclaurin series for $\frac{1}{1-x}$, we obtain

$$\frac{1}{1-\frac{3}{4}x} = \sum_{n=0}^{\infty} \left(\frac{3}{4}\right)^n x^n.$$

This series converges for $\left|\frac{3}{4}x\right| < 1$, or $|x| < \frac{4}{3}$. Hence, for $|x| < \frac{4}{3}$,

$$\frac{2}{4-3x} = \frac{1}{2}\sum_{n=0}^{\infty} \left(\frac{3}{4}\right)^n x^n.$$

93. Let $F(x) = \sum_{k=0}^{\infty} \frac{x^{2k}}{2^k \cdot k!}$.

(a) Show that $F(x)$ has infinite radius of convergence.

(b) Show that $y = F(x)$ is a solution to the differential equation

$$y'' = xy' + y$$

satisfying $y(0) = 1$, $y'(0) = 0$.

(c) CAS Plot the partial sums S_N for $N = 1, 3, 5, 7$ on the same set of axes.

SOLUTION

(a) With $a_k = \frac{x^{2k}}{2^k \cdot k!}$,

$$\left|\frac{a_{k+1}}{a_k}\right| = \frac{|x|^{2k+2}}{2^{k+1}\cdot(k+1)!} \cdot \frac{2^k \cdot k!}{|x|^{2k}} = \frac{x^2}{2(k+1)},$$

and

$$\rho = \lim_{n\to\infty} \left|\frac{a_{k+1}}{a_k}\right| = x^2 \cdot 0 = 0.$$

Because $\rho < 1$ for all x, we conclude that the series converges for all x; that is, $R = \infty$.

(b) Let

$$y = F(x) = \sum_{k=0}^{\infty} \frac{x^{2k}}{2^k \cdot k!}.$$

Then

$$y' = \sum_{k=1}^{\infty} \frac{2kx^{2k-1}}{2^k k!} = \sum_{k=1}^{\infty} \frac{x^{2k-1}}{2^{k-1}(k-1)!},$$

$$y'' = \sum_{k=1}^{\infty} \frac{(2k-1)x^{2k-2}}{2^{k-1}(k-1)!},$$

and

$$xy' + y = x\sum_{k=1}^{\infty} \frac{x^{2k-1}}{2^{k-1}(k-1)!} + \sum_{k=0}^{\infty} \frac{x^{2k}}{2^k k!} = \sum_{k=1}^{\infty} \frac{x^{2k}}{2^{k-1}(k-1)!} + 1 + \sum_{k=1}^{\infty} \frac{x^{2k}}{2^k k!}$$

$$= 1 + \sum_{k=1}^{\infty} \frac{(2k+1)x^{2k}}{2^k k!} = \sum_{k=0}^{\infty} \frac{(2k+1)x^{2k}}{2^k k!} = \sum_{k=1}^{\infty} \frac{(2k-1)x^{2k-2}}{2^{k-1}(k-1)!} = y''.$$

Moreover,

$$y(0) = 1 + \sum_{k=1}^{\infty} \frac{0^{2k}}{2^k k!} = 1 \quad \text{and} \quad y'(0) = \sum_{k=1}^{\infty} \frac{0^{2k-1}}{2^{k-1}(k-1)!} = 0.$$

Thus, $\sum_{k=0}^{\infty} \frac{x^{2k}}{2^k k!}$ is the solution to the equation $y'' = xy' + y$ satisfying $y(0) = 1$, $y'(0) = 0$.

(c) The partial sums S_1, S_3, S_5 and S_7 are plotted in the figure below.

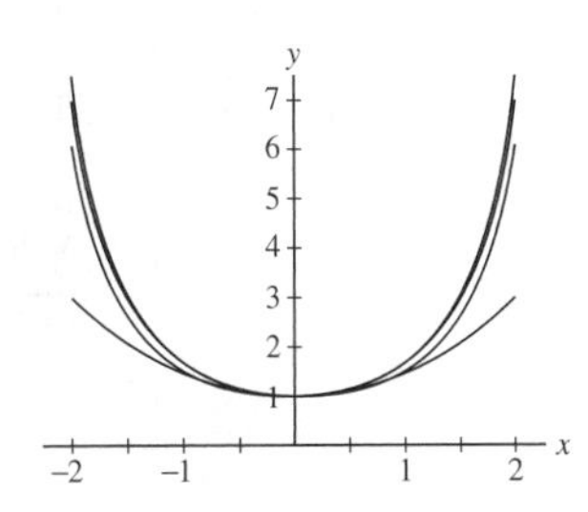

95. Use the Maclaurin series for $f(x) = \cos x$ to calculate the limit $\lim_{x\to 0} \frac{1 - \frac{x^2}{2} - \cos x}{x^4}$.

SOLUTION Using the Maclaurin series for $\cos x$, we find

$$\cos x = \sum_{n=0}^{\infty} (-1)^n \frac{x^{2n}}{(2n)!} = 1 - \frac{x^2}{2} + \frac{x^4}{24} + \sum_{n=3}^{\infty} (-1)^n \frac{x^{2n}}{(2n)!}.$$

Thus,

$$1 - \frac{x^2}{x} - \cos x = -\frac{x^4}{24} - \sum_{n=3}^{\infty} (-1)^n \frac{x^{2n}}{(2n)!},$$

and

$$\frac{1 - \frac{x^2}{2} - \cos x}{x^4} = -\frac{1}{24} - \sum_{n=3}^{\infty} (-1)^n \frac{x^{2n-4}}{(2n)!}.$$

Finally,

$$\lim_{x\to 0} \frac{1 - \frac{x^2}{2} - \cos x}{x^4} = \lim_{x\to 0} \left(-\frac{1}{24} - \sum_{n=3}^{\infty} (-1)^n \frac{x^{2n-4}}{(2n)!} \right) = -\frac{1}{24} + 0 = -\frac{1}{24}.$$

In Exercises 96–103, find the Taylor series centered at c.

97. $f(x) = e^{2x}$, $c = -1$

SOLUTION Write:

$$e^{2x} = e^{2(x+1)-2} = e^{-2}e^{2(x+1)}.$$

Substituting $2(x+1)$ for x in the Maclaurin series for e^x yields

$$e^{2(x+1)} = \sum_{n=0}^{\infty} \frac{(2(x+1))^n}{n!} = \sum_{n=0}^{\infty} \frac{2^n}{n!}(x+1)^n;$$

hence,

$$e^{2x} = e^{-2} \sum_{n=0}^{\infty} \frac{2^n (x+1)^n}{n!}.$$

99. $f(x) = \ln \frac{x}{2}, \quad c = 2$

SOLUTION Write

$$\ln \frac{x}{2} = \ln\left(\frac{(x-2)+2}{2}\right) = \ln\left(1 + \frac{x-2}{2}\right).$$

Substituting $\frac{x-2}{2}$ for x in the Maclaurin series for $\ln(1+x)$ yields

$$\ln \frac{x}{2} = \sum_{n=1}^{\infty} \frac{(-1)^{n+1}\left(\frac{x-2}{2}\right)^n}{n} = \sum_{n=1}^{\infty} \frac{(-1)^{n+1}(x-2)^n}{n \cdot 2^n}.$$

This series is valid for $|x-2| < 2$.

101. $f(x) = \sqrt{x} \arctan \sqrt{x}, \quad c = 0$

SOLUTION Substituting $\sqrt{x}$ for x in the Maclaurin series for $\arctan x$ yields

$$\arctan \sqrt{x} = \sum_{n=0}^{\infty} \frac{(-1)^n \sqrt{x}^{2n+1}}{2n+1} = \sum_{n=0}^{\infty} \frac{(-1)^n x^{n+\frac{1}{2}}}{2n+1}.$$

Thus,

$$\sqrt{x} \arctan \sqrt{x} = \sqrt{x} \sum_{n=0}^{\infty} \frac{(-1)^n x^{n+\frac{1}{2}}}{2n+1} = \sum_{n=0}^{\infty} \frac{(-1)^n x^{n+1}}{2n+1}.$$

103. $f(x) = e^{x-1}, \quad c = -1$

SOLUTION Write

$$e^{x-1} = e^{x+1-1-1} = e^{-2} e^{x+1}.$$

Substituting $x+1$ for x in the Maclaurin series for e^x yields

$$e^{x+1} = \sum_{n=0}^{\infty} \frac{(x+1)^n}{n!};$$

hence,

$$e^{x-1} = e^{-2} \sum_{n=0}^{\infty} \frac{(x+1)^n}{n!} = \sum_{n=0}^{\infty} \frac{(x+1)^n}{n! e^2}.$$

105. Use the Maclaurin series of $\sin x$ and $\sqrt{1+x}$ to calculate $f^{(4)}(0)$, where $f(x) = (\sin x)\sqrt{1+x}$.

SOLUTION Recall that the coefficient of x^4 in the Maclaurin series for $f(x) = (\sin x)\sqrt{1+x}$ is $\frac{f^{(4)}(0)}{4!}$. Now, we can obtain the Maclaurin series for $f(x)$ by multiplying the Maclaurin series for $\sqrt{1+x}$:

$$\sqrt{1+x} = 1 + \frac{1}{2}x - \frac{1}{8}x^2 + \frac{1}{16}x^3 - \frac{5}{128}x^4 + \cdots$$

by the Maclaurin series for $\sin x$:

$$\sin x = x - \frac{x^3}{3!} + \cdots$$

The term involving x^4 in this product is

$$x\left(\frac{1}{16}x^3\right) - \frac{x^3}{3!}\left(\frac{1}{2}x\right) = x^4\left(\frac{1}{16} - \frac{1}{12}\right) = -\frac{1}{48}x^4.$$

Therefore,

$$\frac{f^{(4)}(0)}{4!} = -\frac{1}{48} \quad \text{and} \quad f^{(4)}(0) = -\frac{1}{48} \cdot 24 = -\frac{1}{2}.$$

107. Find the Maclaurin series of the function $F(x) = \int_0^x \frac{e^t - 1}{t}\,dt$.

SOLUTION Subtracting 1 from the Maclaurin series for e^t yields

$$e^t - 1 = \sum_{n=0}^{\infty} \frac{t^n}{n!} - 1 = 1 + \sum_{n=1}^{\infty} \frac{t^n}{n!} - 1 = \sum_{n=1}^{\infty} \frac{t^n}{n!}.$$

Thus,

$$\frac{e^t - 1}{t} = \frac{1}{t}\sum_{n=1}^{\infty} \frac{t^n}{n!} = \sum_{n=1}^{\infty} \frac{t^{n-1}}{n!}.$$

Finally, integrating term-by-term yields

$$\int_0^x \frac{e^t - 1}{t}\,dt = \int_0^x \sum_{n=1}^{\infty} \frac{t^{n-1}}{n!}\,dt = \sum_{n=1}^{\infty} \int_0^x \frac{t^{n-1}}{n!}\,dt = \sum_{n=1}^{\infty} \frac{x^n}{n!\,n}.$$

12 PARAMETRIC EQUATIONS, POLAR COORDINATES, AND CONIC SECTIONS

12.1 Parametric Equations (ET Section 11.1)

Preliminary Questions

1. Describe the shape of the curve $x = 3\cos t$, $y = 3\sin t$.

SOLUTION For all t,

$$x^2 + y^2 = (3\cos t)^2 + (3\sin t)^2 = 9(\cos^2 t + \sin^2 t) = 9 \cdot 1 = 9,$$

therefore the curve is on the circle $x^2 + y^2 = 9$. Also, each point on the circle $x^2 + y^2 = 9$ can be represented in the form $(3\cos t, 3\sin t)$ for some value of t. We conclude that the curve $x = 3\cos t$, $y = 3\sin t$ is the circle of radius 3 centered at the origin.

2. How does $x = 4 + 3\cos t$, $y = 5 + 3\sin t$ differ from the curve in the previous question?

SOLUTION In this case we have

$$(x-4)^2 + (y-5)^2 = (3\cos t)^2 + (3\sin t)^2 = 9(\cos^2 t + \sin^2 t) = 9 \cdot 1 = 9$$

Therefore, the given equations parametrize the circle of radius 3 centered at the point $(4, 5)$.

3. What is the maximum height of a particle whose path has parametric equations $x = t^9$, $y = 4 - t^2$?

SOLUTION The particle's height is $y = 4 - t^2$. To find the maximum height we set the derivative equal to zero and solve:

$$\frac{dy}{dt} = \frac{d}{dt}(4 - t^2) = -2t = 0 \quad \text{or} \quad t = 0$$

The maximum height is $y(0) = 4 - 0^2 = 4$.

4. Can the parametric curve $(t, \sin t)$ be represented as a graph $y = f(x)$? What about $(\sin t, t)$?

SOLUTION In the parametric curve $(t, \sin t)$ we have $x = t$ and $y = \sin t$, therefore, $y = \sin x$. That is, the curve can be represented as a graph of a function. In the parametric curve $(\sin t, t)$ we have $x = \sin t$, $y = t$, therefore $x = \sin y$. This equation does not define y as a function of x, therefore the parametric curve $(\sin t, t)$ cannot be represented as a graph of a function $y = f(x)$.

5. Match the derivatives with a verbal description:

(a) $\dfrac{dx}{dt}$ **(b)** $\dfrac{dy}{dt}$ **(c)** $\dfrac{dy}{dx}$

(i) Slope of the tangent line to the curve

(ii) Vertical rate of change with respect to time

(iii) Horizontal rate of change with respect to time

SOLUTION

(a) The derivative $\dfrac{dx}{dt}$ is the horizontal rate of change with respect to time.

(b) The derivative $\dfrac{dy}{dt}$ is the vertical rate of change with respect to time.

(c) The derivative $\dfrac{dy}{dx}$ is the slope of the tangent line to the curve.

Hence, (a) $\leftrightarrow$ (iii), (b) $\leftrightarrow$ (ii), (c) $\leftrightarrow$ (i)

Exercises

1. Find the coordinates at times $t = 0, 2, 4$ of a particle following the path $x = 1 + t^3$, $y = 9 - 3t^2$.

SOLUTION Substituting $t = 0$, $t = 2$, and $t = 4$ into $x = 1 + t^3$, $y = 9 - 3t^2$ gives the coordinates of the particle at these times respectively. That is,

$$\begin{aligned}
(t = 0) &\quad x = 1 + 0^3 = 1,\ y = 9 - 3 \cdot 0^2 = 9 &&\Rightarrow (1, 9)\\
(t = 2) &\quad x = 1 + 2^3 = 9,\ y = 9 - 3 \cdot 2^2 = -3 &&\Rightarrow (9, -3)\\
(t = 4) &\quad x = 1 + 4^3 = 65,\ y = 9 - 3 \cdot 4^2 = -39 &&\Rightarrow (65, -39).
\end{aligned}$$

3. Show that the path traced by the bullet in Example 2 is a parabola by eliminating the parameter.

SOLUTION The path traced by the bullet is given by the following parametric equations:

$$x = 200t, \quad y = 400t - 16t^2$$

We eliminate the parameter. Since $x = 200t$, we have $t = \frac{x}{200}$. Substituting into the equation for y we obtain:

$$y = 400t - 16t^2 = 400 \cdot \frac{x}{200} - 16\left(\frac{x}{200}\right)^2 = 2x - \frac{x^2}{2500}$$

The equation $y = -\frac{x^2}{2500} + 2x$ is the equation of a parabola.

5. Graph the parametric curves. Include arrows indicating the direction of motion.

(a) (t, t), $\quad -\infty < t < \infty$

(b) $(\sin t, \sin t)$, $\quad 0 \le t \le 2\pi$

(c) (e^t, e^t), $\quad -\infty < t < \infty$

(d) (t^3, t^3), $\quad -1 \le t \le 1$

SOLUTION

(a) For the trajectory $c(t) = (t, t)$, $-\infty < t < \infty$ we have $y = x$. Also the two coordinates tend to ∞ and $-\infty$ as $t \to \infty$ and $t \to -\infty$ respectively. The graph is shown next:

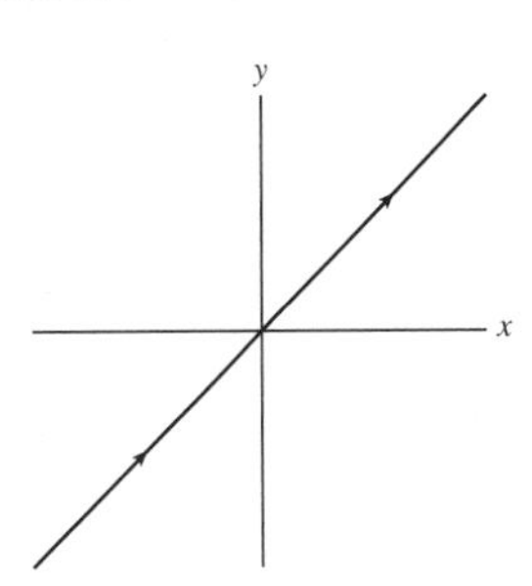

(b) For the curve $c(t) = (\sin t, \sin t)$, $0 \le t \le 2\pi$, we have $y = x$. $\sin t$ is increasing for $0 < t \le \frac{\pi}{2}$, decreasing for $\frac{\pi}{2} \le t \le \frac{3\pi}{2}$ and increasing again for $\frac{3\pi}{2} \le t < 2\pi$. Hence the particle moves from $c(0) = (0, 0)$ to $c(\frac{\pi}{2}) = (1, 1)$, then moves back to $c(\frac{3\pi}{2}) = (-1, -1)$ and then returns to $c(2\pi) = (0, 0)$. We obtain the following trajectory:

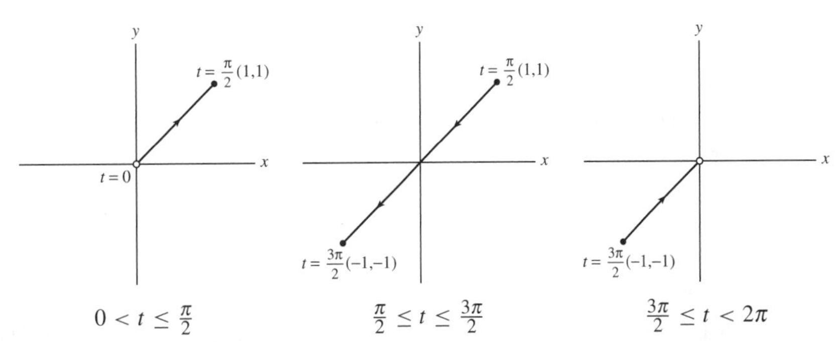

These three parts of the trajectory are shown together in the next figure:

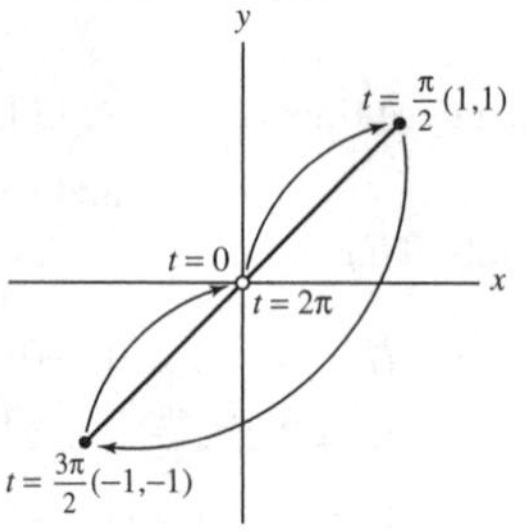

(c) For the trajectory $c(t) = (e^t, e^t)$, $-\infty < t < \infty$, we have $y = x$. However since $\lim_{t\to-\infty} e^t = 0$ and $\lim_{t\to\infty} e^t = \infty$, the trajectory is the part of the line $y = x$, $0 < x$.

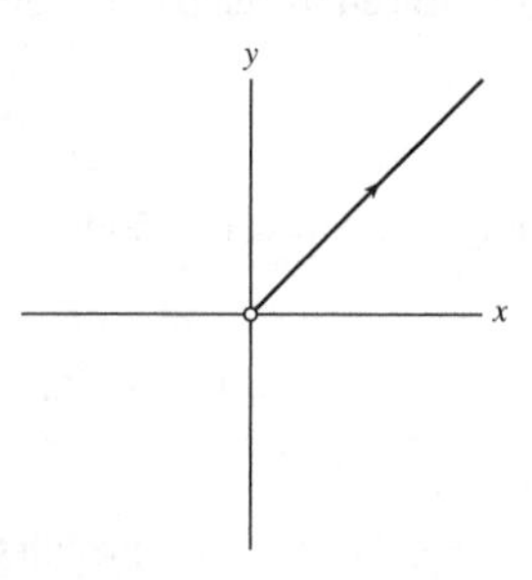

(d) For the trajectory $c(t) = (t^3, t^3)$, $-1 \le t \le 1$, we have again $y = x$. Since the function t^3 is increasing the particle moves in one direction starting at $((-1)^3, (-1)^3) = (-1, -1)$ and ending at $(1^3, 1^3) = (1, 1)$. The trajectory is shown next:

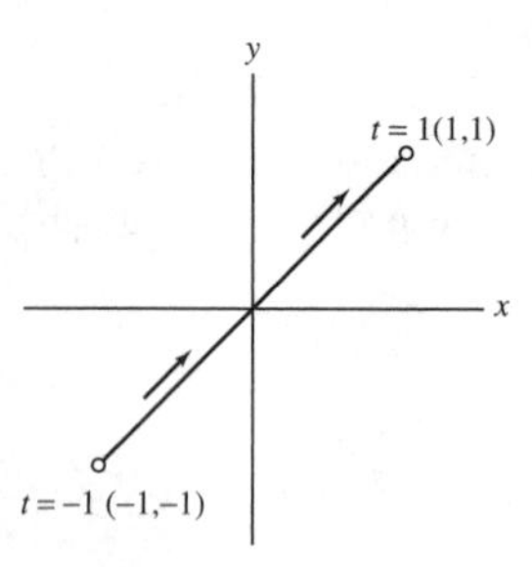

In Exercises 7–14, express in the form $y = f(x)$ by eliminating the parameter.

7. $x = t + 3, \quad y = 4t$

SOLUTION We eliminate the parameter. Since $x = t + 3$, we have $t = x - 3$. Substituting into $y = 4t$ we obtain

$$y = 4t = 4(x - 3) \Rightarrow y = 4x - 12$$

9. $x = t, \quad y = \tan^{-1}(t^3 + e^t)$

SOLUTION Replacing t by x in the equation for y we obtain $y = \tan^{-1}(x^3 + e^x)$.

11. $x = e^{-2t}, \quad y = 6e^{4t}$

SOLUTION We eliminate the parameter. Since $x = e^{-2t}$, we have $-2t = \ln x$ or $t = -\frac{1}{2}\ln x$. Substituting in $y = 6e^{4t}$ we get

$$y = 6e^{4t} = 6e^{4\cdot(-\frac{1}{2}\ln x)} = 6e^{-2\ln x} = 6e^{\ln x^{-2}} = 6x^{-2} \Rightarrow y = \frac{6}{x^2}, \quad x > 0.$$

13. $x = \ln t, \quad y = 2 - t$

SOLUTION Since $x = \ln t$ we have $t = e^x$. Substituting in $y = 2 - t$ we obtain $y = 2 - e^x$.

In Exercises 15–18, graph the curve and draw an arrow specifying the direction corresponding to motion.

15. $x = \frac{1}{2}t, \quad y = 2t^2$

SOLUTION Let $c(t) = (x(t), y(t)) = (\frac{1}{2}t, 2t^2)$. Then $c(-t) = (-x(t), y(t))$ so the curve is symmetric with respect to the y-axis. Also, the function $\frac{1}{2}t$ is increasing. Hence there is only one direction of motion on the curve. The corresponding function is the parabola $y = 2 \cdot (2x)^2 = 8x^2$. We obtain the following trajectory:

y

$t = 0$ x

17. $x = \pi t, \quad y = \sin t$

SOLUTION We find the function by eliminating t. Since $x = \pi t$, we have $t = \frac{x}{\pi}$. Substituting $t = \frac{x}{\pi}$ into $y = \sin t$ we get $y = \sin \frac{x}{\pi}$. We obtain the following curve:

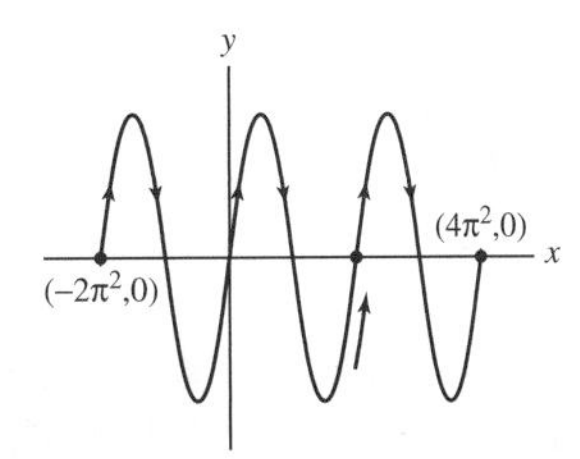

19. Match the parametrizations (a)–(d) below with their plots in Figure 14 and draw an arrow indicating the direction of motion.

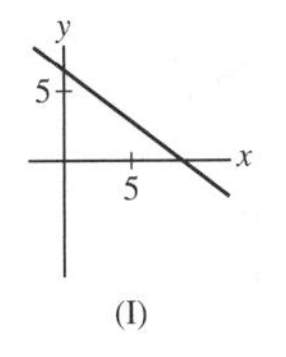

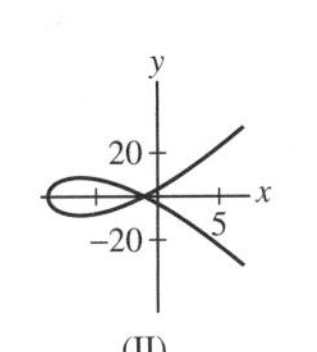

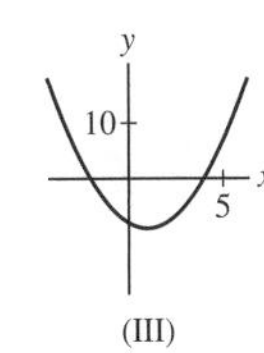

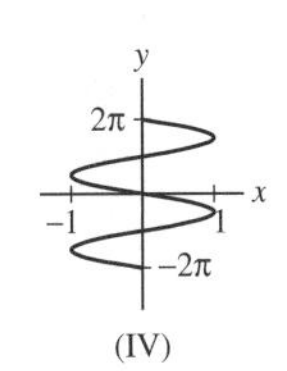

FIGURE 14

(a) $c(t) = (\sin t, -t)$

(b) $c(t) = (t^2 - 9, -t^3 - 8)$

(c) $c(t) = (1 - t, t^2 - 9)$

(d) $c(t) = (4t + 2, 5 - 3t)$

SOLUTION

(a) In the curve $c(t) = (\sin t, -t)$ the x-coordinate is varying between -1 and 1 so this curve corresponds to plot IV. As t increases, the y-coordinate $y = -t$ is decreasing so the direction of motion is downward.

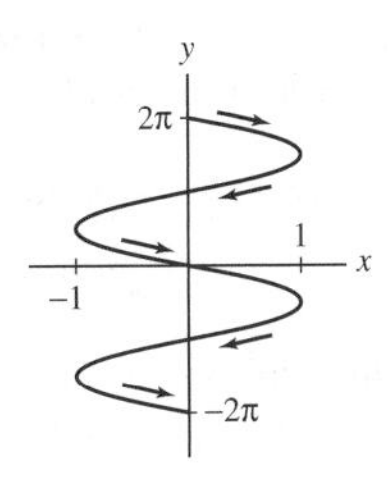

(IV) $c(t) = (\sin t, -t)$

(b) The curve $c(t) = (t^2 - 9, -t^3 - 8)$ intersects the x-axis where $y = -t^3 - 8 = 0$, or $t = -2$. The x-intercept is $(-5, 0)$. The y-intercepts are obtained where $x = t^2 - 9 = 0$, or $t = \pm 3$. The y-intercepts are $(0, -35)$ and $(0, 19)$. As t increases from $-\infty$ to 0, x and y decrease, and as t increases from 0 to ∞, x increases and y decreases. We obtain the following trajectory:

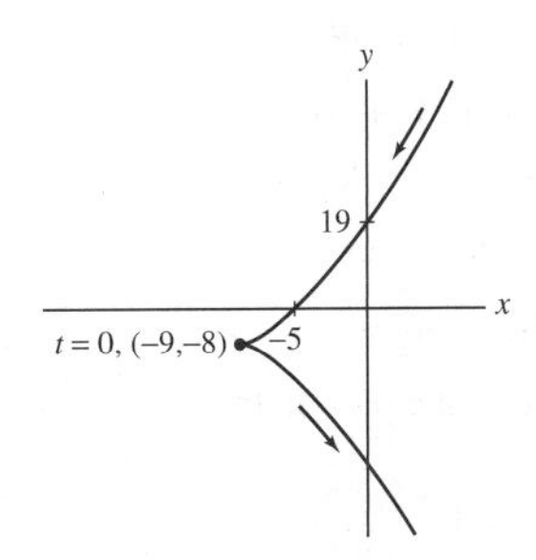

(II)

(c) The curve $c(t) = (1 - t, t^2 - 9)$ intersects the y-axis where $x = 1 - t = 0$, or $t = 1$. The y-intercept is $(0, -8)$. The x-intercepts are obtained where $t^2 - 9 = 0$ or $t = \pm 3$. These are the points $(-2, 0)$ and $(4, 0)$. Setting $t = 1 - x$ we get

$$y = t^2 - 9 = (1 - x)^2 - 9 = x^2 - 2x - 8.$$

As t increases the x coordinate decreases and we obtain the following trajectory:

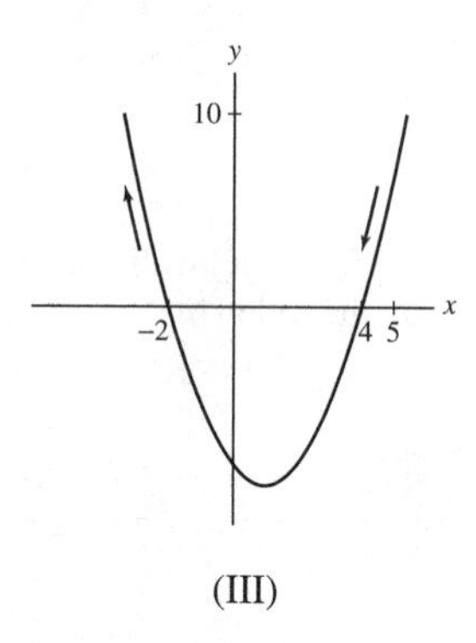

(III)

(d) The curve $c(t) = (4t + 2, 5 - 3t)$ is a straight line, since eliminating t in $x = 4t + 2$ and substituting in $y = 5 - 3t$ gives $y = 5 - 3 \cdot \frac{x-2}{4} = -\frac{3}{4}x + \frac{13}{2}$ which is the equation of a line. As t increases, the x coordinate $x = 4t + 2$ increases and the y-coordinate $y = 5 - 3t$ decreases. We obtain the following trajectory:

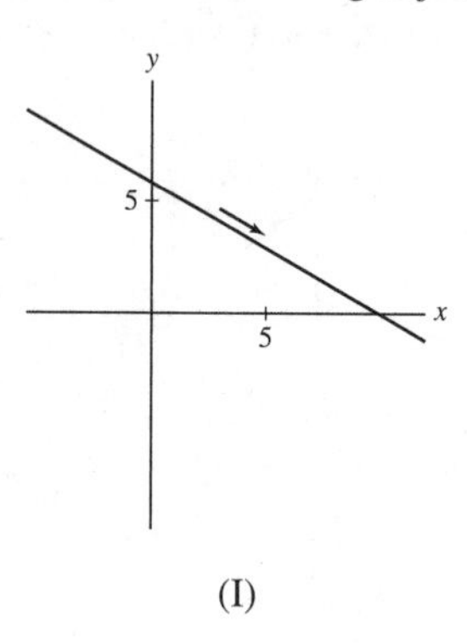

(I)

21. Find an interval of t-values such that $c(t) = (\cos t, \sin t)$ traces the lower half of the unit circle.

SOLUTION For $t = \pi$, we have $c(\pi) = (-1, 0)$. As t increases from π to 2π, the x-coordinate of $c(t)$ increases from -1 to 1, and the y-coordinate decreases from 0 to -1 (at $t = 3\pi/2$) and then returns to 0. Thus, for t in $[\pi, 2\pi]$, the equation traces the lower part of the circle.

In Exercises 23–34, find parametric equations for the given curve.

23. $y = 9 - 4x$

SOLUTION This is a line through $P = (0, 9)$ with slope $m = -4$. Using the parametric representation of a line, as given in Example 3, we obtain $c(t) = (t, 9 - 4t)$.

25. $4x - y^2 = 5$

SOLUTION We define the parameter $t = y$. Then, $x = \dfrac{5 + y^2}{4} = \dfrac{5 + t^2}{4}$, giving us the parametrization $c(t) = \left(\dfrac{5 + t^2}{4}, t\right)$.

27. $(x + 9)^2 + (y - 4)^2 = 49$

SOLUTION This is a circle of radius 7 centered at $(-9, 4)$. Using the parametric representation of a circle we get $c(t) = (-9 + 7\cos t, 4 + 7\sin t)$.

29. Line through $(2, 5)$ perpendicular to $y = 3x$

SOLUTION The line perpendicular to $y = 3x$ has slope $m = -\frac{1}{3}$. We use the parametric representation of a line given in Example 3 to obtain the parametrization $c(t) = (2 + t, 5 - \frac{1}{3}t)$.

31. $\left(\frac{x}{4}\right)^2 + \left(\frac{y}{9}\right)^2 = 1$

SOLUTION This is an ellipse with $a = 4$ and $b = 9$. Hence using the parametric representation given in Example 4 we get $c(t) = (4\cos t, 9\sin t)$.

33. The parabola $y = x^2$ translated so that its minimum occurs at $(2, 3)$

SOLUTION The equation of the translated parabola is

$$y - 3 = (x - 2)^2 \Rightarrow y = 3 + (x - 2)^2.$$

We let $t = x - 2$, hence $x = t + 2$ and $y = 3 + t^2$. We obtain the representation $c(t) = (t + 2, t^2 + 3)$.

35. Describe the parametrized curve $c(t) = (\sin^2 t, \cos^2 t)$ for $0 \le t \le \pi$.

SOLUTION The graphs of $x = \sin^2 t$ and $y = \cos^2 t$ for $0 \le t \le \pi$ are shown in the following figures:

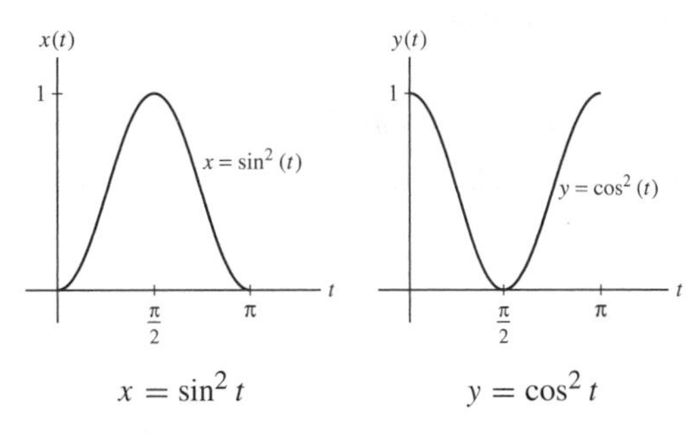

Now $\sin^2 t + \cos^2 t = 1$, hence $x + y = 1$ and so $y = 1 - x$ (for $0 \le x \le 1$). The path starts at the point $c(0) = (0, 1)$. As t increases from $t = 0$ to $t = \frac{\pi}{2}$, the x coordinate increases and the y coordinate decreases. As t increases from $t = \frac{\pi}{2}$ to $t = \pi$, the x coordinate decreases and the y coordinate increases, while returning from $c(\frac{\pi}{2}) = (1, 0)$ back to $c(\pi) = (0, 1)$. We obtain the following curve:

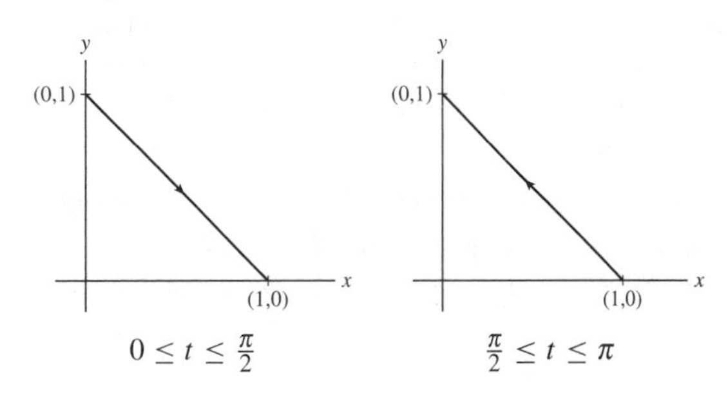

37. Find a parametrization $c(t)$ of the line $y = 3x - 4$ such that $c(0) = (2, 2)$.

SOLUTION Let $x(t) = t + a$ and $y(t) = 3x - 4 = 3(t + a) - 4$. We want $x(0) = 2$, thus we must use $a = 2$. Our line is $c(t) = (x(t), y(t)) = (t + 2, 3(t + 2) - 4) = (t + 2, 3t + 2)$.

39. Show that $x = \cosh t$, $y = \sinh t$ parametrizes the hyperbola $x^2 - y^2 = 1$. Calculate $\frac{dy}{dx}$ as a function of t.

Generalize to obtain a parametrization of $\left(\frac{x}{a}\right)^2 - \left(\frac{y}{b}\right)^2 = 1$.

SOLUTION We check that $x = \cosh t$, $y = \sinh t$ satisfy the equation of the hyperbola

$$x^2 - y^2 = (\cosh t)^2 - (\sinh t)^2 = 1.$$

We now find $\frac{dy}{dx}$ using the formula for the slope of the tangent line:

$$\frac{dy}{dx} = \frac{\frac{dy}{dt}}{\frac{dx}{dt}} = \frac{\frac{d}{dt}(\sinh t)}{\frac{d}{dt}(\cosh t)} = \frac{\cosh t}{\sinh t} = \coth t.$$

We can generalize to obtain the following parametrization of $(\frac{x}{a})^2 - (\frac{y}{b})^2 = 1$:

$$x = a\cosh t, \quad y = b\sinh t.$$

Next, we check that $x = a\cosh t$ and $y = b\sin t$ satisfy the equation $(\frac{x}{a})^2 - (\frac{y}{b})^2 = 1$:

$$\left(\frac{x}{a}\right)^2 - \left(\frac{y}{b}\right)^2 = \left(\frac{a\cosh t}{a}\right)^2 - \left(\frac{b\sinh t}{b}\right)^2 = (\cosh t)^2 - (\sinh t)^2 = 1.$$

41. Which of (I) or (II) is the graph of $x(t)$ for the parametric curve in Figure 16(A)? Which represents $y(t)$?

(A) (I) (II)

FIGURE 16

SOLUTION As indicated by Figure 16(A), the y-coordinate is decreasing and then increasing, so plot I is the graph of y. Figure 16(A) also shows that the x-coordinate is increasing, decreasing and then increasing, so plot II is the graph for x.

43. Sketch $c(t) = (t^2, \sin t)$ for $-2\pi \le t \le 2\pi$.

SOLUTION The graphs of $x(t) = t^2$ and $y = \sin t$ on the interval $-2\pi \le t \le 2\pi$ are shown in the following figures:

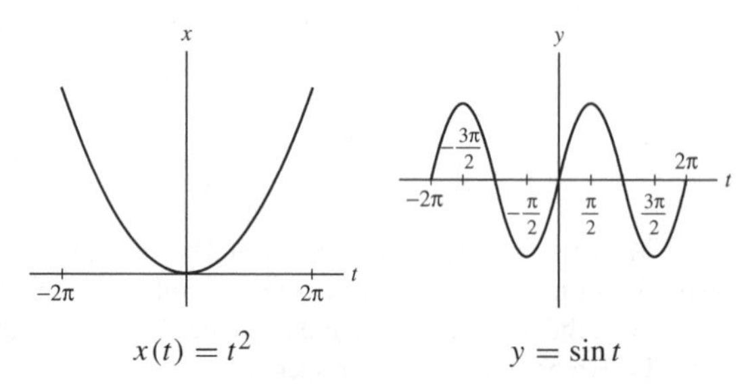

$x(t) = t^2$ $\qquad$ $y = \sin t$

Since $c(-t) = (x(-t), y(-t)) = (x(t), -y(t))$, $c(t)$ and $c(-t)$ are symmetric with respect to the x-axis, it suffices to graph the curve for $t \ge 0$. We have $c(0) = (0, 0)$. The x-coordinate increases as t increases so the curve is directed to the right. The y-coordinate is positive and increasing for $0 < t < \frac{\pi}{2}$, positive and decreasing for $\frac{\pi}{2} < t < \pi$, negative and decreasing for $\pi < t < \frac{3\pi}{2}$, and negative and increasing for $\frac{3\pi}{2} < t < 2\pi$.

Therefore, starting at the origin, the curve first goes up then at $c(\frac{\pi}{2}) = (\frac{\pi^2}{4}, 1)$ it turns down, crosses the x-axis at $c(\pi) = (\pi^2, 0)$, turns up again at $c(\frac{3\pi}{2}) = (\frac{9\pi^2}{4}, -1)$ ending at $c(2\pi) = (4\pi^2, 0)$. The part of the path for $t \le 0$ is obtained by reflecting across the x-axis. We obtain the following path:

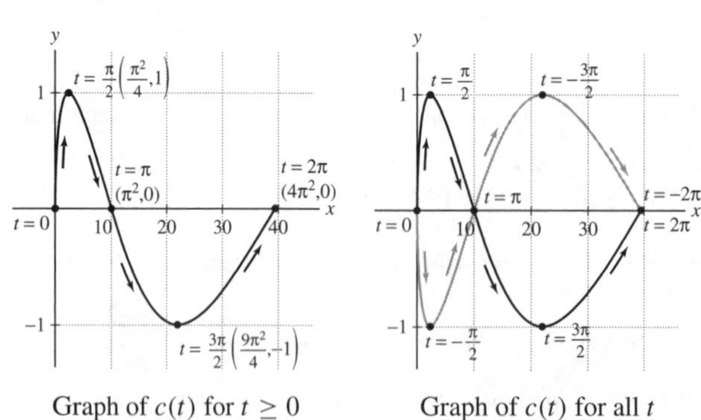

Graph of $c(t)$ for $t \ge 0$ $\qquad$ Graph of $c(t)$ for all t

In Exercises 45–48, use Eq. (7) to find $\dfrac{dy}{dx}$ *at the given point.*

45. $(t^3, t^2 - 1)$, $\quad t = -4$

SOLUTION By Eq. (7) we have

$$\frac{dy}{dx} = \frac{y'(t)}{x'(t)} = \frac{(t^2-1)'}{(t^3)'} = \frac{2t}{3t^2} = \frac{2}{3t}$$

Substituting $t = -4$ we get

$$\frac{dy}{dx} = \left.\frac{2}{3t}\right|_{t=-4} = \frac{2}{3\cdot(-4)} = -\frac{1}{6}.$$

47. $(s^{-1} - 3s, s^3)$, $\quad s = -1$

SOLUTION Using Eq. (7) we get

$$\frac{dy}{dx} = \frac{y'(s)}{x'(s)} = \frac{(s^3)'}{(s^{-1} - 3s)'} = \frac{3s^2}{-s^{-2} - 3} = \frac{3s^4}{-1 - 3s^2}$$

Substituting $s = -1$ we obtain

$$\frac{dy}{dx} = \left.\frac{3s^4}{-1 - 3s^2}\right|_{s=-1} = \frac{3 \cdot (-1)^4}{-1 - 3 \cdot (-1)^2} = -\frac{3}{4}.$$

In Exercises 49–52, find an equation $y = f(x)$ for the parametric curve and compute $\frac{dy}{dx}$ in two ways: using Eq. (7) and by differentiating $f(x)$.

49. $c(t) = (2t + 1, 1 - 9t)$

SOLUTION Since $x = 2t + 1$, we have $t = \frac{x-1}{2}$. Substituting in $y = 1 - 9t$ we have

$$y = 1 - 9\left(\frac{x-1}{2}\right) = -\frac{9}{2}x + \frac{11}{2}$$

Differentiating $y = -\frac{9}{2}x + \frac{11}{2}$ gives $\frac{dy}{dx} = -\frac{9}{2}$. We now find $\frac{dy}{dx}$ using Eq. (7):

$$\frac{dy}{dx} = \frac{y'(t)}{x'(t)} = \frac{(1 - 9t)'}{(2t + 1)'} = -\frac{9}{2}$$

51. $x = s^3, \quad y = s^6 + s^{-3}$

SOLUTION We find y as a function of x:

$$y = s^6 + s^{-3} = \left(s^3\right)^2 + \left(s^3\right)^{-1} = x^2 + x^{-1}.$$

We now differentiate $y = x^2 + x^{-1}$. This gives

$$\frac{dy}{dx} = 2x - x^{-2}.$$

Alternatively, we can use Eq. (7) to obtain the following derivative:

$$\frac{dy}{dx} = \frac{y'(s)}{x'(s)} = \frac{\left(s^6 + s^{-3}\right)'}{\left(s^3\right)'} = \frac{6s^5 - 3s^{-4}}{3s^2} = 2s^3 - s^{-6}.$$

Hence, since $x = s^3$,

$$\frac{dy}{dx} = 2x - x^{-2}.$$

In Exercises 53–56, let $c(t) = (t^2 - 9, t^2 - 8t)$ (see Figure 17).

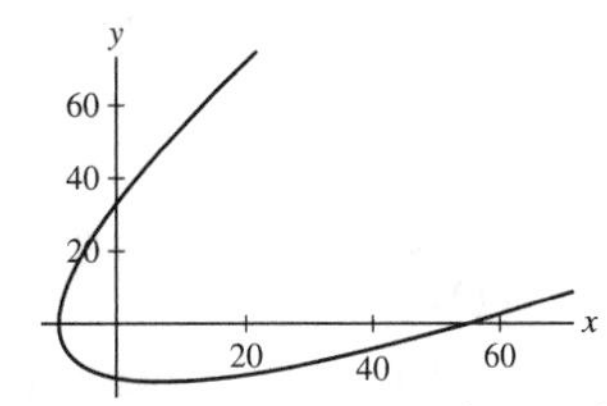

FIGURE 17 Plot of $c(t) = (t^2 - 9, t^2 - 8t)$.

53. Draw an arrow indicating the direction of motion and determine the interval of t-values corresponding to the portion of the curve in each of the four quadrants.

SOLUTION We plot the functions $x(t) = t^2 - 9$ and $y(t) = t^2 - 8t$:

$x = t^2 - 9$ $\qquad\qquad$ $y = t^2 - 8t$

We note carefully where each of these graphs are positive or negative, increasing or decreasing. In particular, $x(t)$ is decreasing for $t < 0$, increasing for $t > 0$, positive for $|t| > 3$, and negative for $|t| < 3$. Likewise, $y(t)$ is decreasing for $t < 4$, increasing for $t > 4$, positive for $t > 8$ or $t < 0$, and negative for $0 < t < 8$. We now draw arrows on the path following the decreasing/increasing behavior of the coordinates as indicated above. We obtain:

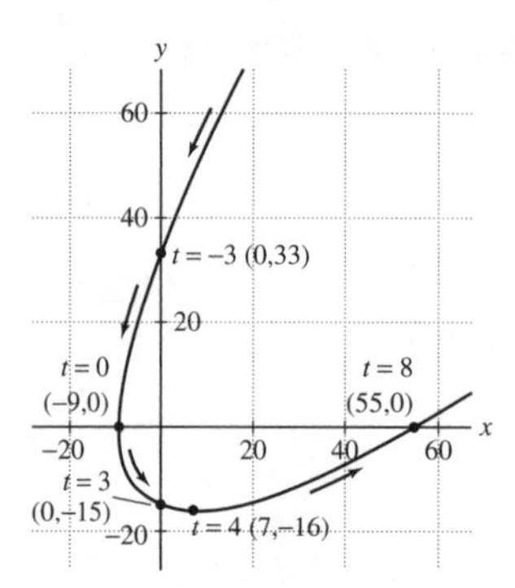

This plot also shows that:

- The graph is in the first quadrant for $t < -3$ or $t > 8$.
- The graph is in the second quadrant for $-3 < t < 0$.
- The graph is in the third quadrant for $0 < t < 3$.
- The graph is in the fourth quadrant for $3 < t < 8$.

55. Find the points where the tangent has slope $\frac{1}{2}$.

SOLUTION The slope of the tangent at t is

$$\frac{dy}{dx} = \frac{\left(t^2 - 8t\right)'}{\left(t^2 - 9\right)'} = \frac{2t - 8}{2t} = 1 - \frac{4}{t}$$

The point where the tangent has slope $\frac{1}{2}$ corresponds to the value of t that satisfies

$$\frac{dy}{dx} = 1 - \frac{4}{t} = \frac{1}{2} \Rightarrow \frac{4}{t} = \frac{1}{2} \Rightarrow t = 8.$$

We substitute $t = 8$ in $x(t) = t^2 - 9$ and $y(t) = t^2 - 8t$ to obtain the following point:

$$\begin{aligned} x(8) &= 8^2 - 9 = 55 \\ y(8) &= 8^2 - 8 \cdot 8 = 0 \end{aligned} \quad \Rightarrow \quad (55, 0)$$

57. Find the equation of the ellipse represented parametrically by $x = 4\cos t$, $y = 7\sin t$. Calculate the slope of the tangent line at the point $(2\sqrt{2}, 7\sqrt{2}/2)$.

SOLUTION In Example 4 it is shown that the ellipse

$$\left(\frac{x}{a}\right)^2 + \left(\frac{y}{b}\right)^2 = 1$$

has the parametrization $x = a\cos t$, $y = b\sin t$. Therefore for the ellipse represented by the parametric equations $x = 4\cos t$, $y = 7\sin t$, we have $a = 4$, $b = 7$. The equation of the ellipse is therefore

$$\left(\frac{x}{4}\right)^2 + \left(\frac{y}{7}\right)^2 = 1.$$

The slope of the tangent line at t is:

$$m = \frac{dy}{dx} = \frac{y'(t)}{x'(t)} = \frac{(7\sin t)'}{(4\cos t)'} = \frac{7\cos t}{-4\sin t} \tag{1}$$

we compute the value of t corresponding to the point $(2\sqrt{2}, \frac{7\sqrt{2}}{2})$, by solving the following equations:

$$x(t) = 2\sqrt{2} = 4\cos t$$

$$y(t) = \frac{7\sqrt{2}}{2} = 7\sin t$$

The first equation implies that $\cos t = \frac{\sqrt{2}}{2}$ and the second equation implies that $\sin t = \frac{\sqrt{2}}{2}$. Setting these values in (1) gives the following slope:

$$m = \frac{7 \cdot \frac{\sqrt{2}}{2}}{-4 \cdot \frac{\sqrt{2}}{2}} = -\frac{7}{4}$$

In Exercises 58–60, refer to the Bézier curve defined by Eqs. (8) and (9).

59. Find and plot the Bézier curve $c(t)$ passing through the control points

$$P_0 = (3, 2), \quad P_1 = (0, 2), \quad P_2 = (5, 4), \quad P_3 = (2, 4)$$

SOLUTION Setting $a_0 = 3$, $a_1 = 0$, $a_2 = 5$, $a_3 = 2$, and $b_0 = 2$, $b_1 = 2$, $b_2 = 4$, $b_3 = 4$ into Eq. (8)–(9) and simplifying gives

$$\begin{aligned} x(t) &= 3(1-t)^3 + 0 + 15t^2(1-t) + 2t^3 \\ &= 3(1 - 3t + 3t^2 - t^3) + 15t^2 - 15t^3 + 2t^3 = 3 - 9t + 24t^2 - 16t^3 \\ y(t) &= 2(1-t)^3 + 6t(1-t)^2 + 12t^2(1-t) + 4t^3 \\ &= 2(1 - 3t + 3t^2 - t^3) + 6t(1 - 2t + t^2) + 12t^2 - 12t^3 + 4t^3 \\ &= 2 - 6t + 6t^2 - 2t^3 + 6t - 12t^2 + 6t^3 + 12t^2 - 12t^3 + 4t^3 = 2 + 6t^2 - 4t^3 \end{aligned}$$

We obtain the following equation

$$c(t) = (3 - 9t + 24t^2 - 16t^3, 2 + 6t^2 - 4t^3), \quad 0 \le t \le 1.$$

The graph of the Bézier curve is shown in the following figure:

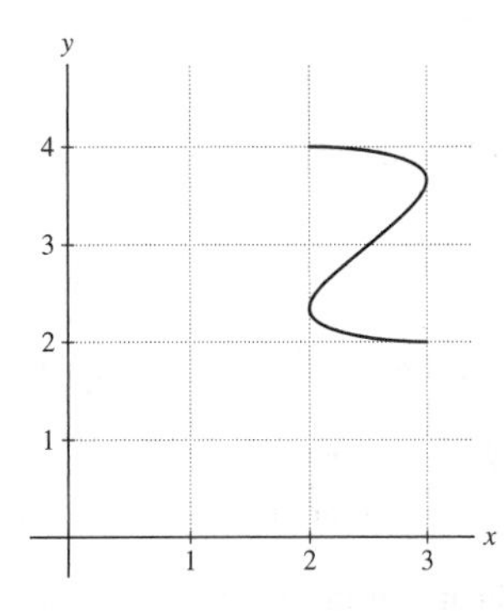

61. A bullet fired from a gun follows the trajectory

$$x = at, \quad y = bt - 16t^2 \quad (a, b > 0)$$

Show that the bullet leaves the gun at an angle $\theta = \tan^{-1}\left(\frac{b}{a}\right)$ and lands at a distance $\frac{ab}{16}$ from the origin.

SOLUTION The height of the bullet equals the value of the y-coordinate. When the bullet leaves the gun, $y(t) = t(b - 16t) = 0$. The solutions to this equation are $t = 0$ and $t = \frac{b}{16}$, with $t = 0$ corresponding to the moment the bullet leaves the gun. We find the slope m of the tangent line at $t = 0$:

$$\frac{dy}{dx} = \frac{y'(t)}{x'(t)} = \frac{b - 32t}{a} \Rightarrow m = \left.\frac{b - 32t}{a}\right|_{t=0} = \frac{b}{a}$$

It follows that $\tan\theta = \frac{b}{a}$ or $\theta = \tan^{-1}\left(\frac{b}{a}\right)$. The bullet lands at $t = \frac{b}{16}$. We find the distance of the bullet from the origin at this time, by substituting $t = \frac{b}{16}$ in $x(t) = at$. This gives

$$x\left(\frac{b}{16}\right) = \frac{ab}{16}$$

63. CAS Plot the astroid $x = \cos^3\theta$, $y = \sin^3\theta$ and find the equation of the tangent line at $\theta = \frac{\pi}{3}$.

SOLUTION The graph of the astroid $x = \cos^3\theta$, $y = \sin^3\theta$ is shown in the following figure:

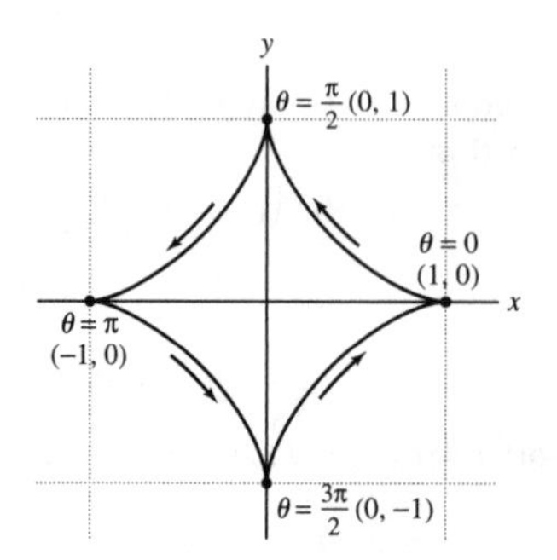

The slope of the tangent line at $\theta = \frac{\pi}{3}$ is

$$m = \left.\frac{dy}{dx}\right|_{\theta=\pi/3} = \left.\frac{(\sin^3\theta)'}{(\cos^3\theta)'}\right|_{\theta=\pi/3} = \left.\frac{3\sin^2\theta\cos\theta}{3\cos^2\theta(-\sin\theta)}\right|_{\theta=\pi/3} = -\left.\frac{\sin\theta}{\cos\theta}\right|_{\theta=\pi/3} = -\tan\theta\Big|_{\pi/3} = -\sqrt{3}$$

We find the point of tangency:

$$\left(x\left(\frac{\pi}{3}\right), y\left(\frac{\pi}{3}\right)\right) = \left(\cos^3\frac{\pi}{3}, \sin^3\frac{\pi}{3}\right) = \left(\frac{1}{8}, \frac{3\sqrt{3}}{8}\right)$$

The equation of the tangent line at $\theta = \frac{\pi}{3}$ is, thus,

$$y - \frac{3\sqrt{3}}{8} = -\sqrt{3}\left(x - \frac{1}{8}\right) \Rightarrow y = -\sqrt{3}x + \frac{\sqrt{3}}{2}$$

65. Find the points with horizontal tangent line on the cycloid with parametric equation (5).

SOLUTION The parametric equations of the cycloid are

$$x = t - \sin t, \quad y = 1 - \cos t$$

We find the slope of the tangent line at t:

$$\frac{dy}{dx} = \frac{(1-\cos t)'}{(t-\sin t)'} = \frac{\sin t}{1-\cos t}$$

The tangent line is horizontal where it has slope zero. That is,

$$\frac{dy}{dx} = \frac{\sin t}{1-\cos t} = 0 \quad \Rightarrow \quad \begin{array}{c}\sin t = 0\\ \cos t \neq 1\end{array} \quad \Rightarrow \quad t = (2k-1)\pi, \quad k = 0, \pm 1, \pm 2, \dots$$

We find the coordinates of the points with horizontal tangent line, by substituting $t = (2k-1)\pi$ in $x(t)$ and $y(t)$. This gives

$$x = (2k-1)\pi - \sin(2k-1)\pi = (2k-1)\pi$$

$$y = 1 - \cos((2k-1)\pi) = 1 - (-1) = 2$$

The required points are

$$((2k-1)\pi, 2), \quad k = 0, \pm 1, \pm 2, \dots$$

67. A *curtate cycloid* (Figure 19) is the curve traced by a point at a distance h from the center of a circle of radius R rolling along the x-axis where $h < R$. Show that this curve has parametric equations $x = Rt - h \sin t$, $y = R - h \cos t$.

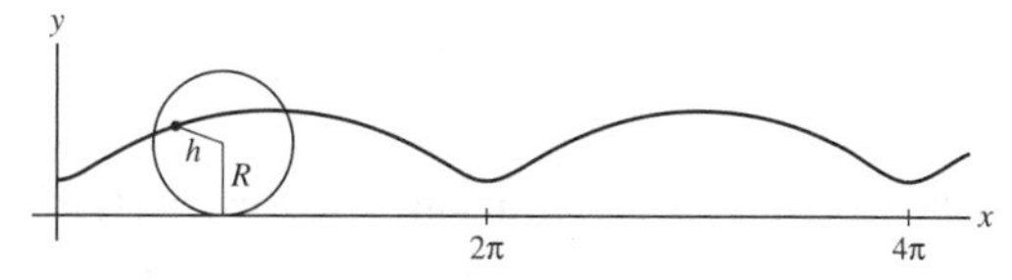

FIGURE 19 Curtate cycloid.

SOLUTION Let P be a point at a distance h from the center C of the circle. Assume that at $t = 0$, the line of CP is passing through the origin. When the circle rolls a distance Rt along the x-axis, the length of the arc $\widehat{SQ}$ (see figure) is also Rt and the angle $\angle SCQ$ has radian measure t. We compute the coordinates x and y of P.

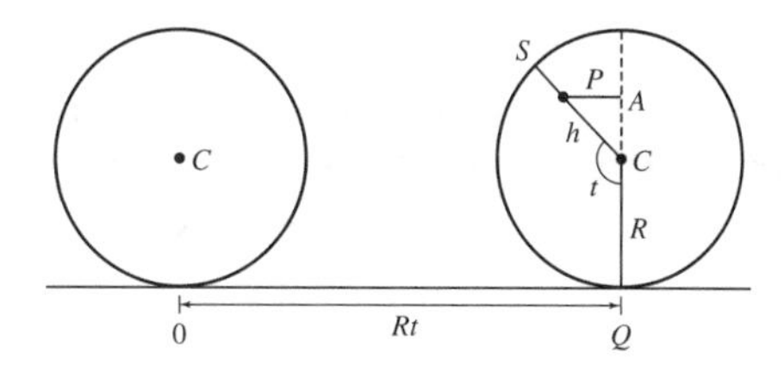

$$x = Rt - \overline{PA} = Rt - h\sin(\pi - t) = Rt - h\sin t$$
$$y = R + \overline{AC} = R + h\cos(\pi - t) = R - h\cos t$$

We obtain the following parametrization:

$$x = Rt - h\sin t, \quad y = R - h\cos t.$$

69. Show that the line of slope t through $(-1, 0)$ intersects the unit circle in the point with coordinates

$$x = \frac{1 - t^2}{t^2 + 1}, \qquad y = \frac{2t}{t^2 + 1}$$

10

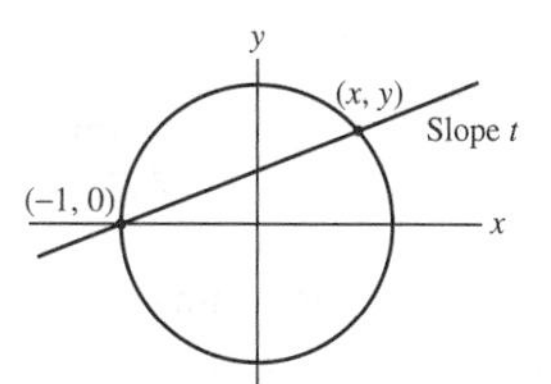

FIGURE 20 Unit circle.

Conclude that these equations parametrize the unit circle with the point $(-1, 0)$ excluded (Figure 20). Show further that $t = \dfrac{y}{x + 1}$.

SOLUTION The equation of the line of slope t through $(-1, 0)$ is $y = t(x + 1)$. The equation of the unit circle is $x^2 + y^2 = 1$. Hence, the line intersects the unit circle at the points (x, y) that satisfy the equations:

$$y = t(x + 1) \qquad (1)$$
$$x^2 + y^2 = 1 \qquad (2)$$

Substituting y from equation (1) into equation (2) and solving for x we obtain

$$x^2 + t^2(x + 1)^2 = 1$$
$$x^2 + t^2x^2 + 2t^2x + t^2 = 1$$
$$(1 + t^2)x^2 + 2t^2x + (t^2 - 1) = 0$$

This gives

$$x_{1,2} = \frac{-2t^2 \pm \sqrt{4t^4 - 4(t^2 + 1)(t^2 - 1)}}{2(1 + t^2)} = \frac{-2t^2 \pm 2}{2(1 + t^2)} = \frac{\pm 1 - t^2}{1 + t^2}$$

So $x_1 = -1$ and $x_2 = \frac{1-t^2}{t^2+1}$. The solution $x = -1$ corresponds to the point $(-1, 0)$. We are interested in the second point of intersection that is varying as t varies. Hence the appropriate solution is

$$x = \frac{1-t^2}{t^2+1}$$

We find the y-coordinate by substituting x in equation (1). This gives

$$y = t(x+1) = t\left(\frac{1-t^2}{t^2+1} + 1\right) = t \cdot \frac{1-t^2+t^2+1}{t^2+1} = \frac{2t}{t^2+1}$$

We conclude that the line and the unit circle intersect, besides at $(-1, 0)$, at the point with the following coordinates:

$$x = \frac{1-t^2}{t^2+1}, \quad y = \frac{2t}{t^2+1} \tag{3}$$

Since these points determine all the points on the unit circle except for $(-1, 0)$ and no other points, the equations in (3) parametrize the unit circle with the point $(-1, 0)$ excluded.

We show that $t = \frac{y}{x+1}$. Using (3) we have

$$\frac{y}{x+1} = \frac{\frac{2t}{t^2+1}}{\frac{1-t^2}{t^2+1} + 1} = \frac{\frac{2t}{t^2+1}}{\frac{1-t^2+t^2+1}{t^2+1}} = \frac{\frac{2t}{t^2+1}}{\frac{2}{t^2+1}} = \frac{2t}{2} = t.$$

71. Use the results of Exercise 70 to show that the asymptote of the folium is the line $x + y = -a$. Hint: show that $\lim_{t\to-1}(x+y) = -a$.

SOLUTION We must show that as $x \to \infty$ or $x \to -\infty$ the graph of the folium is getting arbitrarily close to the line $x + y = -a$, and the derivative $\frac{dy}{dx}$ is approaching the slope -1 of the line.

In Exercise 70 we showed that $x \to \infty$ when $t \to (-1^-)$ and $x \to -\infty$ when $t \to (-1^+)$. We first show that the graph is approaching the line $x + y = -a$ as $x \to \infty$ or $x \to -\infty$, by showing that $\lim_{t\to-1-} x + y = \lim_{t\to-1+} x + y = -a$.

For $x(t) = \frac{3at}{1+t^3}$, $y(t) = \frac{3at^2}{1+t^3}$, $a > 0$, calculated in Exercise 70, we obtain using L'Hôpital's Rule:

$$\lim_{t\to-1-}(x+y) = \lim_{t\to-1-} \frac{3at+3at^2}{1+t^3} = \lim_{t\to-1-} \frac{3a+6at}{3t^2} = \frac{3a-6a}{3} = -a$$

$$\lim_{t\to-1+}(x+y) = \lim_{t\to-1+} \frac{3at+3at^2}{1+t^3} = \lim_{t\to-1+} \frac{3a+6at}{3t^2} = \frac{3a-6a}{3} = -a$$

We now show that $\frac{dy}{dx}$ is approaching -1 as $t \to -1-$ and as $t \to -1+$. We use $\frac{dy}{dx} = \frac{6at-3at^4}{3a-6at^3}$ computed in Exercise 70 to obtain

$$\lim_{t\to-1-} \frac{dy}{dx} = \lim_{t\to-1-} \frac{6at-3at^4}{3a-6at^3} = \frac{-9a}{9a} = -1$$

$$\lim_{t\to-1+} \frac{dy}{dx} = \lim_{t\to-1+} \frac{6at-3at^4}{3a-6at^3} = \frac{-9a}{9a} = -1$$

We conclude that the line $x + y = -a$ is an asymptote of the folium as $x \to \infty$ and as $x \to -\infty$.

73. Verify that the **tractrix** curve ($\ell > 0$)

$$c(t) = \left(t - \ell \tanh \frac{t}{\ell}, \ell \operatorname{sech} \frac{t}{\ell}\right)$$

has the following property: For all t, the segment from $c(t)$ to $(0, t)$ is tangent to the curve and has length ℓ (Figure 22).

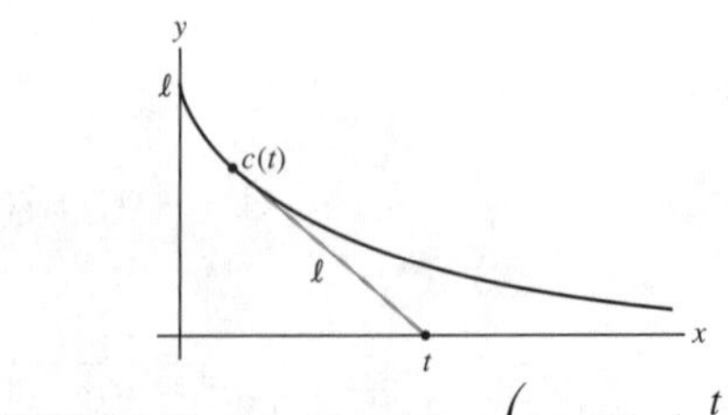

FIGURE 22 The tractrix $c(t) = \left(t - \ell \tanh \frac{t}{\ell}, \ell \operatorname{sech} \frac{t}{\ell}\right)$.

SOLUTION Let $P = c(t)$ and $Q = (t, 0)$.

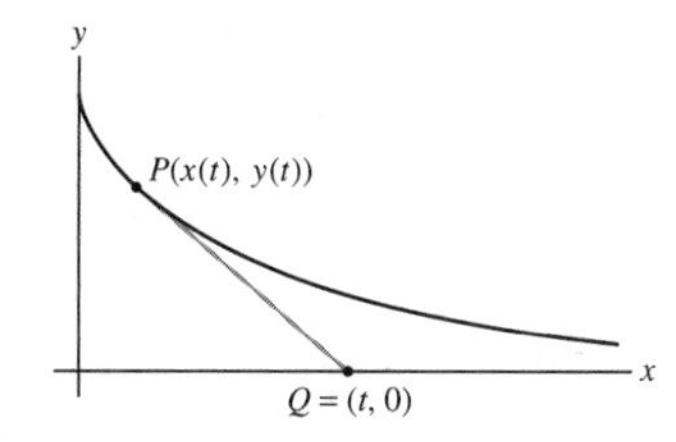

The slope of the segment $\overline{PQ}$ is

$$m_1 = \frac{y(t) - 0}{x(t) - t} = \frac{\ell \operatorname{sech}\left(\frac{t}{\ell}\right)}{-\ell \tanh\left(\frac{t}{\ell}\right)} = -\frac{1}{\sinh\left(\frac{t}{\ell}\right)}$$

We compute the slope of the tangent line at P:

$$m_2 = \frac{dy}{dx} = \frac{y'(t)}{x'(t)} = \frac{\left(\ell \operatorname{sech}\left(\frac{t}{\ell}\right)\right)'}{\left(t - \ell \tanh\left(\frac{t}{\ell}\right)\right)'} = \frac{\ell \cdot \frac{1}{\ell}\left(-\operatorname{sech}\left(\frac{t}{\ell}\right)\tanh\left(\frac{t}{\ell}\right)\right)}{1 - \ell \cdot \frac{1}{\ell}\operatorname{sech}^2\left(\frac{t}{\ell}\right)}$$

$$= -\frac{-\operatorname{sech}\left(\frac{t}{\ell}\right)\tanh\left(\frac{t}{\ell}\right)}{1 - \operatorname{sech}^2\left(\frac{t}{\ell}\right)} = \frac{-\operatorname{sech}\left(\frac{t}{\ell}\right)\tanh\left(\frac{t}{\ell}\right)}{-\tanh^2\left(\frac{t}{\ell}\right)} = \frac{-\operatorname{sech}\left(\frac{t}{\ell}\right)}{\tanh\left(\frac{t}{\ell}\right)} = -\frac{1}{\sinh\left(\frac{t}{\ell}\right)}$$

Since $m_1 = m_2$, we conclude that the segment from $c(t)$ to $(t, 0)$ is tangent to the curve.

We now show that $|\overline{PQ}| = \ell$:

$$|\overline{PQ}| = \sqrt{(x(t) - t)^2 + (y(t) - 0)^2} = \sqrt{\left(-\ell \tanh \frac{t}{\ell}\right)^2 + \left(\ell \operatorname{sech}\left(\frac{t}{\ell}\right)\right)^2}$$

$$= \sqrt{\ell^2\left(\tanh^2\left(\frac{t}{\ell}\right) + \operatorname{sech}^2\left(\frac{t}{\ell}\right)\right)} = \ell\sqrt{\operatorname{sech}^2\left(\frac{t}{\ell}\right)\sinh^2\left(\frac{t}{\ell}\right) + \operatorname{sech}^2\left(\frac{t}{\ell}\right)}$$

$$= \ell \operatorname{sech}\left(\frac{t}{\ell}\right)\sqrt{\sin h^2\left(\frac{t}{\ell}\right) + 1} = \ell \operatorname{sech}\left(\frac{t}{\ell}\right)\cosh\left(\frac{t}{\ell}\right) = \ell \cdot 1 = \ell$$

75. Let A and B be the points where the ray of angle θ intersects the two concentric circles of radii $r < R$ centered at the origin (Figure 23). Let P be the point of intersection of the horizontal line through A and the vertical line through B. Express the coordinates of P as a function of θ and describe the curve traced by P for $0 \le \theta \le 2\pi$.

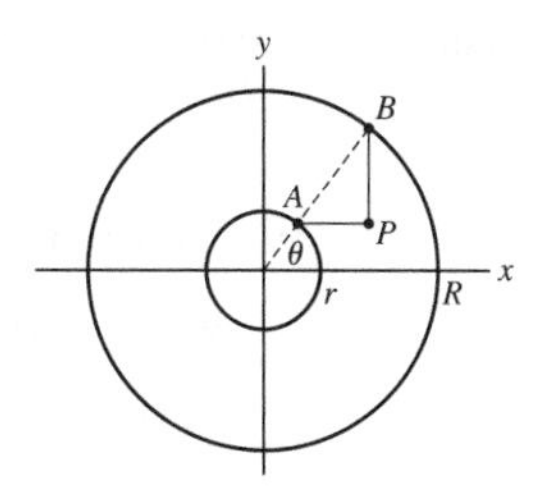

FIGURE 23

SOLUTION We use the parametric representation of a circle to determine the coordinates of the points A and B. That is,

$$A = (r\cos\theta, r\sin\theta), \quad B = (R\cos\theta, R\sin\theta)$$

The coordinates of P are therefore

$$P = (R\cos\theta, r\sin\theta)$$

In order to identify the curve traced by P, we notice that the x and y coordinates of P satisfy $\frac{x}{R} = \cos\theta$ and $\frac{y}{r} = \sin\theta$. Hence

$$\left(\frac{x}{R}\right)^2 + \left(\frac{y}{r}\right)^2 = \cos^2\theta + \sin^2\theta = 1.$$

The equation

$$\left(\frac{x}{R}\right)^2 + \left(\frac{y}{r}\right)^2 = 1$$

is the equation of ellipse. Hence, the coordinates of P, $(R\cos\theta, r\sin\theta)$ describe an ellipse for $0 \le \theta \le 2\pi$.

Further Insights and Challenges

77. Derive the formula for the slope of the tangent line to a parametric curve $c(t) = (x(t), y(t))$ using a different method than that presented in the text. Assume that $x'(t_0)$ and $y'(t_0)$ exist and that $x'(t_0) \neq 0$. Show that

$$\lim_{h\to 0} \frac{y(t_0+h) - y(t_0)}{x(t_0+h) - x(t_0)} = \frac{y'(t_0)}{x'(t_0)}$$

Then explain why this limit is equal to the slope dy/dx. Draw a diagram showing that the ratio in the limit is the slope of a secant line.

SOLUTION Since $y'(t_0)$ and $x'(t_0)$ exist, we have the following limits:

$$\lim_{h\to 0} \frac{y(t_0+h) - y(t_0)}{h} = y'(t_0), \quad \lim_{h\to 0} \frac{x(t_0+h) - x(t_0)}{h} = x'(t_0) \tag{1}$$

We use Basic Limit Laws, the limits in (1) and the given data $x'(t_0) \neq 0$, to write

$$\lim_{h\to 0} \frac{y(t_0+h) - y(t_0)}{x(t_0+h) - x(t_0)} = \lim_{h\to 0} \frac{\frac{y(t_0+h)-y(t_0)}{h}}{\frac{x(t_0+h)-x(t_0)}{h}} = \frac{\lim_{h\to 0} \frac{y(t_0+h)-y(t_0)}{h}}{\lim_{h\to 0} \frac{x(t_0+h)-x(t_0)}{h}} = \frac{y'(t_0)}{x'(t_0)}$$

Notice that the quotient $\frac{y(t_0+h) - y(t_0)}{x(t_0+h) - x(t_0)}$ is the slope of the secant line determined by the points $P = (x(t_0), y(t_0))$ and $Q = (x(t_0+h), y(t_0+h))$. Hence, the limit of the quotient as $h \to 0$ is the slope of the tangent line at P, that is the derivative $\frac{dy}{dx}$.

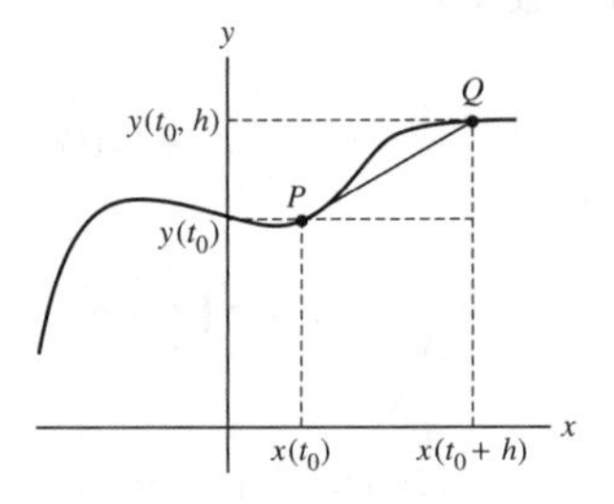

In Exercises 79–82, use Eq. (11) to find $\frac{d^2y}{dx^2}$.

79. $x = t^3 + t^2, \quad y = 7t^2 - 4, \quad t = 2$

SOLUTION We find the first and second derivatives of $x(t)$ and $y(t)$:

$$\begin{aligned}
x'(t) &= 3t^2 + 2t \Rightarrow x'(2) = 3 \cdot 2^2 + 2 \cdot 2 = 16\\
x''(t) &= 6t + 2 \quad \Rightarrow x''(2) = 6 \cdot 2 + 2 = 14\\
y'(t) &= 14t \quad \Rightarrow y'(2) = 14 \cdot 2 = 28\\
y''(t) &= 14 \quad \Rightarrow y''(2) = 14
\end{aligned}$$

Using Eq. (11) we get

$$\left.\frac{d^2y}{dx^2}\right|_{t=2} = \left.\frac{x'(t)y''(t) - y'(t)x''(t)}{x'(t)^3}\right|_{t=2} = \frac{16 \cdot 14 - 28 \cdot 14}{16^3} = \frac{-21}{512}$$

81. $x = 8t + 9, \quad y = 1 - 4t, \quad t = -3$

SOLUTION We compute the first and second derivatives of $x(t)$ and $y(t)$:

$$\begin{aligned}
x'(t) &= 8 \quad \Rightarrow x'(-3) = 8\\
x''(t) &= 0 \quad \Rightarrow x''(-3) = 0\\
y'(t) &= -4 \Rightarrow y'(-3) = -4\\
y''(t) &= 0 \quad \Rightarrow y''(-3) = 0
\end{aligned}$$

Using Eq. (11) we get

$$\left.\frac{d^2y}{dx^2}\right|_{t=-3} = \frac{x'(-3)y''(-3) - y'(-3)x''(-3)}{x'(-3)^3} = \frac{8 \cdot 0 - (-4) \cdot 0}{8^3} = 0$$

83. Use Eq. (11) to find the t-intervals on which $c(t) = (t^2, t^3 - 4t)$ is concave up.

SOLUTION The curve is concave up where $\frac{d^2y}{dx^2} > 0$. Thus,

$$\frac{x'(t)y''(t) - y'(t)x''(t)}{x'(t)^3} > 0 \tag{1}$$

We compute the first and second derivatives:

$$x'(t) = 2t, \qquad x''(t) = 2$$
$$y'(t) = 3t^2 - 4, \quad y''(t) = 6t$$

Substituting in (1) and solving for t gives

$$\frac{12t^2 - (6t^2 - 8)}{8t^3} = \frac{6t^2 + 8}{8t^3}$$

Since $6t^2 + 8 > 0$ for all t, the quotient is positive if $8t^3 > 0$. We conclude that the curve is concave up for $t > 0$.

85. Area under a Parametrized Curve Let $c(t) = (x(t), y(t))$ be a parametrized curve such that $x'(t) > 0$ and $y(t) > 0$ (Figure 25). Show that the area A under $c(t)$ for $t_0 \le t \le t_1$ is

$$A = \int_{t_0}^{t_1} y(t)x'(t)\,dt \tag{12}$$

Hint: $x(t)$ is increasing and therefore has an inverse, say, $t = g(x)$. Observe that $c(t)$ is the graph of the function $y(g(x))$ and apply the Change of Variables formula to $A = \int_{x(t_0)}^{x(t_1)} y(g(x))\,dx$.

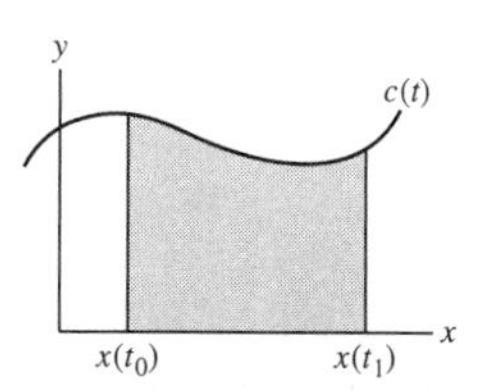

FIGURE 25

SOLUTION Let $x_0 = x(t_0)$ and $x_1 = x(t_1)$. We are given that $x'(t) > 0$, hence $x = x(t)$ is an increasing function of t, so it has an inverse function $t = g(x)$. The area A is given by $\int_{x_0}^{x_1} y(g(x))\,dx$. Recall that y is a function of t and $t = g(x)$, so the height y at any point x is given by $y = y(g(x))$. We find the new limits of integration. Since $x_0 = x(t_0)$ and $x_1 = x(t_1)$, the limits for t are t_0 and t_1, respectively. Also since $x'(t) = \frac{dx}{dt}$, we have $dx = x'(t)dt$. Performing this substitution gives

$$A = \int_{x_0}^{x_1} y(g(x))\,dx = \int_{t_0}^{t_1} y(g(x))x'(t)\,dt.$$

Since $g(x) = t$, we have $A = \int_{t_0}^{t_1} y(t)x'(t)\,dt$.

87. What does Eq. (12) say if $c(t) = (t, f(t))$?

SOLUTION In the parametrization $x(t) = t$, $y(t) = f(t)$ we have $x'(t) = 1$, $t_0 = x(t_0)$, $t_1 = x(t_1)$. Hence Eq. (12) becomes

$$A = \int_{t_0}^{t_1} y(t)x'(t)\,dt = \int_{x(t_0)}^{x(t_1)} f(t)\,dt$$

We see that in this parametrization Eq. (12) is the familiar formula for the area under the graph of a positive function.

89. Use Eq. (12) to show that the area under one arch of the cycloid $c(t)$ (Figure 26) generated by a circle of radius R is equal to three times the area of the circle. Recall that

$$c(t) = (Rt - R\sin t, R - R\cos t)$$

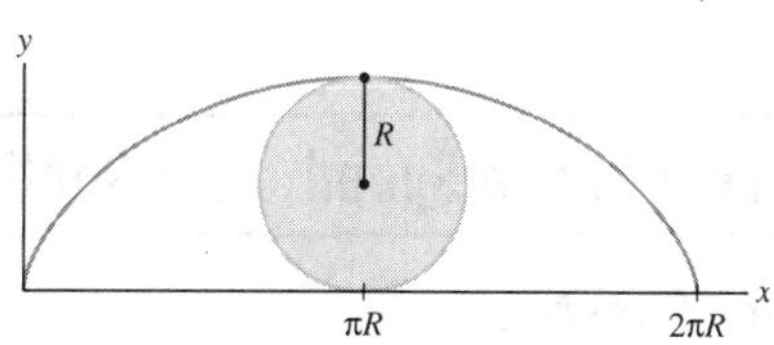

FIGURE 26 The area of the generating circle is one-third the area of one arch of the cycloid.

SOLUTION This reduces to

$$\int_0^{2\pi} (R - R\cos t)(Rt - R\sin t)'\,dt = \int_0^{2\pi} R^2(1-\cos t)^2\,dt = 3\pi R^2.$$

In Exercises 90–91, refer to Figure 27.

91. Show that the parametrization of the ellipse by the angle θ is

$$x = \frac{ab\cos\theta}{\sqrt{a^2\sin^2\theta + b^2\cos^2\theta}}$$
$$y = \frac{ab\sin\theta}{\sqrt{a^2\sin^2\theta + b^2\cos^2\theta}}$$

SOLUTION We consider the ellipse

$$\frac{x^2}{a^2} + \frac{y^2}{b^2} = 1.$$

For the angle θ we have $\tan\theta = \frac{y}{x}$, hence,

$$y = x\tan\theta \tag{1}$$

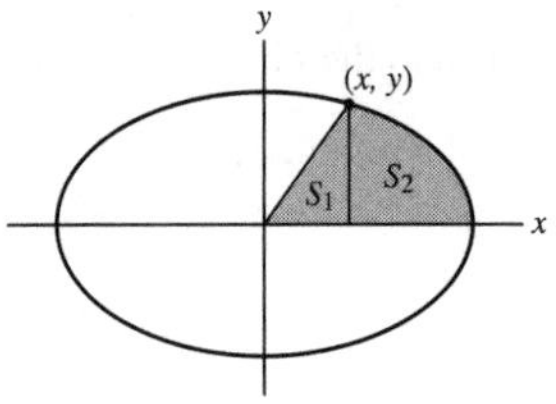

$$\frac{x^2}{a^2} + \frac{y^2}{b^2} = 1$$

Substituting in the equation of the ellipse and solving for x we obtain

$$\frac{x^2}{a^2} + \frac{x^2\tan^2\theta}{b^2} = 1$$
$$b^2x^2 + a^2x^2\tan^2\theta = a^2b^2$$
$$(a^2\tan^2\theta + b^2)x^2 = a^2b^2$$
$$x^2 = \frac{a^2b^2}{a^2\tan^2\theta + b^2} = \frac{a^2b^2\cos^2\theta}{a^2\sin^2\theta + b^2\cos^2\theta}$$

We now take the square root. Since the sign of the x-coordinate is the same as the sign of $\cos\theta$, we take the positive root, obtaining

$$x = \frac{ab\cos\theta}{\sqrt{a^2\sin^2\theta + b^2\cos^2\theta}} \tag{2}$$

Hence by (1), the y-coordinate is

$$y = x\tan\theta = \frac{ab\cos\theta\tan\theta}{\sqrt{a^2\sin^2\theta + b^2\cos^2\theta}} = \frac{ab\sin\theta}{\sqrt{a^2\sin^2\theta + b^2\cos^2\theta}} \tag{3}$$

Equalities (2) and (3) give the following parametrization for the ellipse:

$$c_1(\theta) = \left(\frac{ab\cos\theta}{\sqrt{a^2\sin^2\theta + b^2\cos^2\theta}}, \frac{ab\sin\theta}{\sqrt{a^2\sin^2\theta + b^2\cos^2\theta}}\right)$$

12.2 Arc Length and Speed (ET Section 11.2)

Preliminary Questions

1. What is the definition of arc length?

SOLUTION A curve can be approximated by a polygonal path obtained by connecting points

$$p_0 = c(t_0),\ p_1 = c(t_1),\ \ldots,\ p_N = c(t_N)$$

on the path with segments. One gets an approximation by summing the lengths of the segments. The definition of arc length is the limit of that approximation when increasing the number of points so that the lengths of the segments approach zero. In doing so, we obtain the following theorem for the arc length:

$$S = \int_a^b \sqrt{x'(t)^2 + y'(t)^2}\, dt,$$

which is the length of the curve $c(t) = (x(t), y(t))$ for $a \le t \le b$.

2. What is the interpretation of $\sqrt{x'(t)^2 + y'(t)^2}$ for a particle following the trajectory $(x(t), y(t))$?

SOLUTION The expression $\sqrt{x'(t)^2 + y'(t)^2}$ denotes the speed at time t of a particle following the trajectory $(x(t), y(t))$.

3. A particle travels along a path from $(0, 0)$ to $(3, 4)$. What is the displacement? Can the distance traveled be determined from the information given?

SOLUTION The net displacement is the distance between the initial point $(0, 0)$ and the endpoint $(3, 4)$. That is

$$\sqrt{(3-0)^2 + (4-0)^2} = \sqrt{25} = 5.$$

The distance traveled can be determined only if the trajectory $c(t) = (x(t), y(t))$ of the particle is known.

4. A particle traverses the parabola $y = x^2$ with constant speed 3 cm/s. What is the distance traveled during the first minute? *Hint:* No computation is necessary.

SOLUTION Since the speed is constant, the distance traveled is the following product: $L = st = 3 \cdot 60 = 180$ cm.

Exercises

1. Use Eq. (4) to calculate the length of the semicircle

$$x = 3\sin t, \quad y = 3\cos t, \qquad 0 \le t \le \pi$$

SOLUTION We substitute $x' = 3\cos t$, $y' = -3\sin t$, $a = 0$, and $b = \pi$ in the equation and compute the resulting integral. We obtain the following length:

$$S = \int_0^{\pi} \sqrt{x'(t)^2 + y'(t)^2}\, dt = \int_0^{\pi} \sqrt{9\cos^2 t + 9\sin^2 t}\, dt = \int_0^{\pi} \sqrt{9}\, dt = 3\pi.$$

In Exercises 3–12, use Eq. (4) to find the length of the path over the given interval.

3. $(3t + 1, 9 - 4t), \quad 0 \le t \le 2$

SOLUTION Since $x = 3t + 1$ and $y = 9 - 4t$ we have $x' = 3$ and $y' = -4$. Hence, the length of the path is

$$S = \int_0^2 \sqrt{3^2 + (-4)^2}\, dt = 5\int_0^2 dt = 10.$$

5. $(2t^2, 3t^2 - 1), \quad 0 \le t \le 4$

SOLUTION Since $x = 2t^2$ and $y = 3t^2 - 1$, we have $x' = 4t$ and $y' = 6t$. By the formula for the arc length we get

$$S = \int_0^4 \sqrt{x'(t)^2 + y'(t)^2}\, dt = \int_0^4 \sqrt{16t^2 + 36t^2}\, dt = \sqrt{52}\int_0^4 t\, dt = \sqrt{52} \cdot \frac{t^2}{2}\bigg|_0^4 = 16\sqrt{13}$$

7. $(3t^2, 4t^3), \quad 1 \le t \le 4$

SOLUTION We have $x = 3t^2$ and $y = 4t^3$. Hence $x' = 6t$ and $y' = 12t^2$. By the formula for the arc length we get

$$S = \int_1^4 \sqrt{x'(t)^2 + y'(t)^2}\, dt = \int_1^4 \sqrt{36t^2 + 144t^4}\, dt = 6\int_1^4 \sqrt{1 + 4t^2}t\, dt.$$

Using the substitution $u = 1 + 4t^2$, $du = 8t\, dt$ we obtain

$$S = \frac{6}{8}\int_5^{65} \sqrt{u}\, du = \frac{3}{4} \cdot \frac{2}{3}u^{3/2}\bigg|_5^{65} = \frac{1}{2}(65^{3/2} - 5^{3/2}) \approx 256.43$$

9. $(\sin 3t, \cos 3t), \quad 0 \le t \le \pi$

SOLUTION We have $x = \sin 3t$, $y = \cos 3t$, hence $x' = 3\cos 3t$ and $y' = -3\sin 3t$. By the formula for the arc length we obtain:

$$S = \int_0^{\pi} \sqrt{x'(t)^2 + y'(t)^2}\, dt = \int_0^{\pi} \sqrt{9\cos^2 3t + 9\sin^2 3t}\, dt = \int_0^{\pi} \sqrt{9}\, dt = 3\pi$$

11. $(\sin\theta - \theta\cos\theta, \cos\theta + \theta\sin\theta), \quad 0 \le \theta \le 2$

SOLUTION We have $x = \sin\theta - \theta\cos\theta$ and $y = \cos\theta + \theta\sin\theta$. Hence, $x' = \cos\theta - (\cos\theta - \theta\sin\theta) = \theta\sin\theta$ and $y' = -\sin\theta + \sin\theta + \theta\cos\theta = \theta\cos\theta$. Using the formula for the arc length we obtain:

$$S = \int_0^2 \sqrt{x'(\theta)^2 + y'(\theta)^2}\, d\theta = \int_0^2 \sqrt{(\theta\sin\theta)^2 + (\theta\cos\theta)^2}\, d\theta$$

$$= \int_0^2 \sqrt{\theta^2(\sin^2\theta + \cos^2\theta)}\, d\theta = \int_0^2 \theta\, d\theta = \left.\frac{\theta^2}{2}\right|_0^2 = 2$$

13. Show that one arch of a cycloid generated by a circle of radius R has length $8R$.

SOLUTION Recall from earlier that the cycloid generated by a circle of radius R has parametric equations $x = Rt - R\sin t$, $y = R - R\cos t$. Hence, $x' = R - R\cos t$, $y' = R\sin t$. Using the identity $\sin^2\frac{t}{2} = \frac{1-\cos t}{2}$, we get

$$x'(t)^2 + y'(t)^2 = R^2(1-\cos t)^2 + R^2\sin^2 t = R^2(1 - 2\cos t + \cos^2 t + \sin^2 t)$$

$$= R^2(1 - 2\cos t + 1) = 2R^2(1-\cos t) = 4R^2\sin^2\frac{t}{2}$$

One arch of the cycloid is traced as t varies from 0 to 2π. Hence, using the formula for the arc length we obtain:

$$S = \int_0^{2\pi} \sqrt{x'(t)^2 + y'(t)^2}\, dt = \int_0^{2\pi} \sqrt{4R^2\sin^2\frac{t}{2}}\, dt = 2R\int_0^{2\pi} \sin\frac{t}{2}\, dt = 4R\int_0^{\pi} \sin u\, du$$

$$= -4R\cos u\Big|_0^{\pi} = -4R(\cos\pi - \cos 0) = 8R$$

15. Find the length of the spiral $c(t) = (t\cos t, t\sin t)$ for $0 \le t \le 2\pi$ to three decimal places (Figure 8). *Hint:* Use the formula

$$\int \sqrt{1+t^2}\, dt = \frac{1}{2}t\sqrt{1+t^2} + \frac{1}{2}\ln\left(t + \sqrt{1+t^2}\right)$$

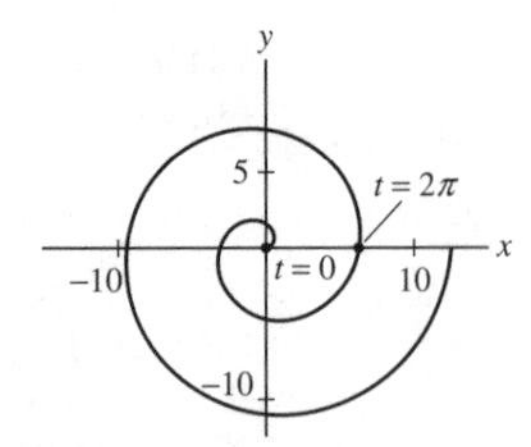

FIGURE 8 The spiral $c(t) = (t\cos t, t\sin t)$.

SOLUTION We use the formula for the arc length:

$$S = \int_0^{2\pi} \sqrt{x'(t)^2 + y'(t)^2}\, dt \tag{1}$$

Differentiating $x = t\cos t$ and $y = t\sin t$ yields

$$x'(t) = \frac{d}{dt}(t\cos t) = \cos t - t\sin t$$

$$y'(t) = \frac{d}{dt}(t\sin t) = \sin t + t\cos t$$

Thus,

$$\sqrt{x'(t)^2 + y'(t)^2} = \sqrt{(\cos t - t\sin t)^2 + (\sin t + t\cos t)^2}$$

$$= \sqrt{\cos^2 t - 2t\cos t\sin t + t^2\sin^2 t + \sin^2 t + 2t\sin t\cos t + t^2\cos^2 t}$$

$$= \sqrt{(\cos^2 t + \sin^2 t)(1+t^2)} = \sqrt{1+t^2}$$

We substitute into (1) and use the integral given in the hint to obtain the following arc length:

$$S = \int_0^{2\pi} \sqrt{1+t^2}\,dt = \frac{1}{2}t\sqrt{1+t^2} + \frac{1}{2}\ln\left(t+\sqrt{1+t^2}\right)\Big|_0^{2\pi}$$

$$= \frac{1}{2}\cdot 2\pi\sqrt{1+(2\pi)^2} + \frac{1}{2}\ln\left(2\pi+\sqrt{1+(2\pi)^2}\right) - \left(0+\frac{1}{2}\ln 1\right)$$

$$= \pi\sqrt{1+4\pi^2} + \frac{1}{2}\ln\left(2\pi+\sqrt{1+4\pi^2}\right) \approx 21.256$$

In Exercises 16–19, determine the speed $s(t)$ of a particle with a given trajectory at time t_0 (in units of meters and seconds).

17. $(3\sin 5t, 8\cos 5t)$, $t = \frac{\pi}{4}$

SOLUTION We have $x = 3\sin 5t$, $y = 8\cos 5t$, hence $x' = 15\cos 5t$, $y' = -40\sin 5t$. Thus, the speed of the particle at time t is

$$\frac{ds}{dt} = \sqrt{x'(t)^2 + y'(t)^2} = \sqrt{225\cos^2 5t + 1600\sin^2 5t}$$

$$= \sqrt{225(\cos^2 5t + \sin^2 5t) + 1375\sin^2 5t} = 5\sqrt{9+55\sin^2 5t}$$

Thus,

$$\frac{ds}{dt} = 5\sqrt{9+55\sin^2 5t}.$$

The speed at time $t = \frac{\pi}{4}$ is thus

$$\left.\frac{ds}{dt}\right|_{t=\pi/4} = 5\sqrt{9+55\sin^2\left(5\cdot\frac{\pi}{4}\right)} \cong 30.21 \text{ m/s}$$

19. $(\ln(t^2+1), t^3)$, $t = 1$

SOLUTION We have $x = \ln(t^2+1)$, $y = t^3$, so $x' = \frac{2t}{t^2+1}$ and $y' = 3t^2$. The speed of the particle at time t is thus

$$\frac{ds}{dt} = \sqrt{x'(t)^2 + y'(t)^2} = \sqrt{\frac{4t^2}{(t^2+1)^2} + 9t^4} = t\sqrt{\frac{4}{(t^2+1)^2} + 9t^2}.$$

The speed at time $t = 1$ is

$$\left.\frac{ds}{dt}\right|_{t=1} = \sqrt{\frac{4}{2^2} + 9} = \sqrt{10} \approx 3.16 \text{ m/s}.$$

21. Find the minimum speed of a particle with trajectory $c(t) = (t^3 - 4t, t^2 + 1)$ for $t \geq 0$. *Hint:* It is easier to find the minimum of the square of the speed.

SOLUTION We first find the speed of the particle. We have $x(t) = t^3 - 4t$, $y(t) = t^2 + 1$, hence $x'(t) = 3t^2 - 4$ and $y'(t) = 2t$. The speed is thus

$$\frac{ds}{dt} = \sqrt{(3t^2-4)^2 + (2t)^2} = \sqrt{9t^4 - 24t^2 + 16 + 4t^2} = \sqrt{9t^4 - 20t^2 + 16}.$$

The square root function is an increasing function, hence the minimum speed occurs at the value of t where the function $f(t) = 9t^4 - 20t^2 + 16$ has minimum value. Since $\lim_{t\to\infty} f(t) = \infty$, f has a minimum value on the interval $0 \leq t < \infty$, and it occurs at a critical point or at the endpoint $t = 0$. We find the critical point of f on $t \geq 0$:

$$f'(t) = 36t^3 - 40t = 4t(9t^2 - 10) = 0 \Rightarrow t = 0, t = \sqrt{\frac{10}{9}}.$$

We compute the values of f at these points:

$$f(0) = 9\cdot 0^4 - 20\cdot 0^2 + 16 = 16$$

$$f\left(\sqrt{\frac{10}{9}}\right) = 9\left(\sqrt{\frac{10}{9}}\right)^4 - 20\left(\sqrt{\frac{10}{9}}\right)^2 + 16 = \frac{44}{9} \approx 4.89$$

We conclude that the minimum value of f on $t \geq 0$ is 4.89. The minimum speed is therefore

$$\left(\frac{ds}{dt}\right)_{\min} \approx \sqrt{4.89} \approx 2.21.$$

23. Find the speed of the cycloid $c(t) = (4t - 4\sin t, 4 - 4\cos t)$ at points where the tangent line is horizontal.

SOLUTION We first find the points where the tangent line is horizontal. The slope of the tangent line is the following quotient:

$$\frac{dy}{dx} = \frac{dy/dt}{dx/dt} = \frac{4\sin t}{4 - 4\cos t} = \frac{\sin t}{1 - \cos t}.$$

To find the points where the tangent line is horizontal we solve the following equation for $t \geq 0$:

$$\frac{dy}{dx} = 0, \quad \frac{\sin t}{1 - \cos t} = 0 \Rightarrow \sin t = 0 \quad \text{and} \quad \cos t \neq 1.$$

Now, $\sin t = 0$ and $t \geq 0$ at the points $t = \pi k$, $k = 0, 1, 2, \ldots$. Since $\cos \pi k = (-1)^k$, the points where $\cos t \neq 1$ are $t = \pi k$ for k odd. The points where the tangent line is horizontal are, therefore:

$$t = \pi(2k - 1), \quad k = 1, 2, 3, \ldots$$

The speed at time t is given by the following expression:

$$\begin{aligned}\frac{ds}{dt} &= \sqrt{x'(t)^2 + y'(t)^2} = \sqrt{(4 - 4\cos t)^2 + (4\sin t)^2} \\ &= \sqrt{16 - 32\cos t + 16\cos^2 t + 16\sin^2 t} = \sqrt{16 - 32\cos t + 16} \\ &= \sqrt{32(1 - \cos t)} = \sqrt{32 \cdot 2\sin^2 \frac{t}{2}} = 8\left|\sin \frac{t}{2}\right|\end{aligned}$$

That is, the speed of the cycloid at time t is

$$\frac{ds}{dt} = 8\left|\sin \frac{t}{2}\right|.$$

We now substitute

$$t = \pi(2k - 1), \quad k = 1, 2, 3, \ldots$$

to obtain

$$\frac{ds}{dt} = 8\left|\sin \frac{\pi(2k-1)}{2}\right| = 8|(-1)^{k+1}| = 8$$

CAS *In Exercises 25–28, plot the curve and use the Midpoint Rule with $N = 10$, 20, 30, and 50 to approximate its length.*

25. $c(t) = (\cos t, e^{\sin t})$ for $0 \leq t \leq 2\pi$

SOLUTION The curve of $c(t) = (\cos t, e^{\sin t})$ for $0 \leq t \leq 2\pi$ is shown in the figure below:

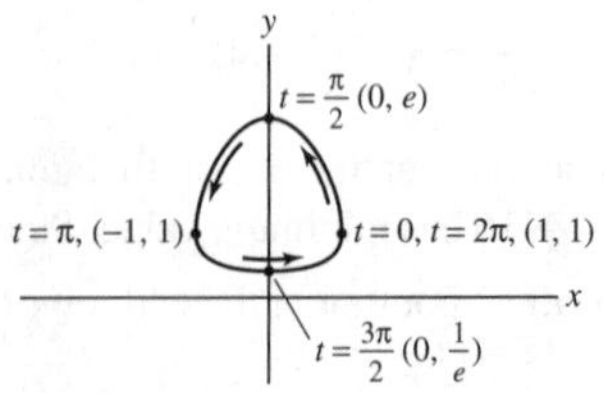

y
$t = \frac{\pi}{2}$ (0, e)
$t = \pi$, (−1, 1)
$t = 0, t = 2\pi$, (1, 1)
x
$t = \frac{3\pi}{2}$ $(0, \frac{1}{e})$

$c(t) = (\cos t, e^{\sin t})$, $0 \leq t \leq 2\pi$.

The length of the curve is given by the following integral:

$$S = \int_0^{2\pi} \sqrt{x'(t)^2 + y'(t)^2}\,dt = \int_0^{2\pi} \sqrt{(-\sin t)^2 + (\cos t\, e^{\sin t})^2}\,dt.$$

That is, $S = \int_0^{2\pi} \sqrt{\sin^2 t + \cos^2 t\, e^{2\sin t}}\, dt$. We approximate the integral using the Mid-Point Rule with $N = 10, 20, 30, 50$. For $f(t) = \sqrt{\sin^2 t + \cos^2 t\, e^{2\sin t}}$ we obtain

$$(N = 10):\quad \Delta x = \frac{2\pi}{10} = \frac{\pi}{5},\ c_i = \left(i - \frac{1}{2}\right)\cdot\frac{\pi}{5}$$

$$M_{10} = \frac{\pi}{5}\sum_{i=1}^{10} f(c_i) = 6.903734$$

$$(N = 20):\quad \Delta x = \frac{2\pi}{20} = \frac{\pi}{10},\ c_i = \left(i - \frac{1}{2}\right)\cdot\frac{\pi}{10}$$

$$M_{20} = \frac{\pi}{10}\sum_{i=1}^{20} f(c_i) = 6.915035$$

$$(N = 30):\quad \Delta x = \frac{2\pi}{30} = \frac{\pi}{15},\ c_i = \left(i - \frac{1}{2}\right)\cdot\frac{\pi}{15}$$

$$M_{30} = \frac{\pi}{15}\sum_{i=1}^{30} f(c_i) = 6.914949$$

$$(N = 50):\quad \Delta x = \frac{2\pi}{50} = \frac{\pi}{25},\ c_i = \left(i - \frac{1}{2}\right)\cdot\frac{\pi}{25}$$

$$M_{50} = \frac{\pi}{25}\sum_{i=1}^{50} f(c_i) = 6.914951$$

27. The ellipse $\left(\frac{x}{5}\right)^2 + \left(\frac{y}{3}\right)^2 = 1$

SOLUTION We use the parametrization given in Example 4, section 12.1, that is, $c(t) = (5\cos t, 3\sin t)$, $0 \le t \le 2\pi$. The curve is shown in the figure below:

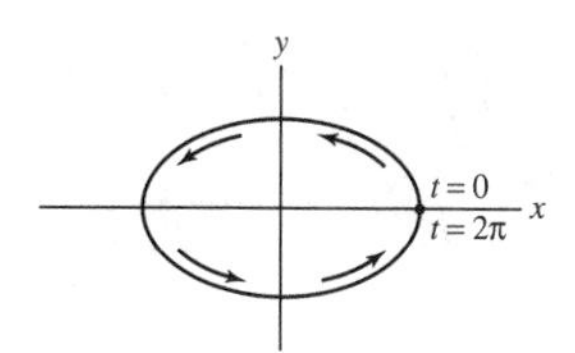

$c(t) = (5\cos t, 3\sin t),\ 0 \le t \le 2\pi.$

The length of the curve is given by the following integral:

$$S = \int_0^{2\pi} \sqrt{x'(t)^2 + y'(t)^2}\, dt = \int_0^{2\pi} \sqrt{(-5\sin t)^2 + (3\cos t)^2}\, dt$$

$$= \int_0^{2\pi} \sqrt{25\sin^2 t + 9\cos^2 t}\, dt = \int_0^{2\pi} \sqrt{9(\sin^2 t + \cos^2 t) + 16\sin^2 t}\, dt = \int_0^{2\pi} \sqrt{9 + 16\sin^2 t}\, dt.$$

That is,

$$S = \int_0^{2\pi} \sqrt{9 + 16\sin^2 t}\, dt.$$

We approximate the integral using the Mid-Point Rule with $N = 10, 20, 30, 50$, for $f(t) = \sqrt{9 + 16\sin^2 t}$. We obtain

$$(N = 10):\quad \Delta x = \frac{2\pi}{10} = \frac{\pi}{5},\ c_i = \left(i - \frac{1}{2}\right)\cdot\frac{\pi}{5}$$

$$M_{10} = \frac{\pi}{5}\sum_{i=1}^{10} f(c_i) = 25.528309$$

$$(N = 20):\quad \Delta x = \frac{2\pi}{20} = \frac{\pi}{10},\ c_i = \left(i - \frac{1}{2}\right)\cdot\frac{\pi}{10}$$

$$M_{20} = \frac{\pi}{10}\sum_{i=1}^{20} f(c_i) = 25.526999$$

$$(N = 30): \quad \Delta x = \frac{2\pi}{30} = \frac{\pi}{15}, c_i = \left(i - \frac{1}{2}\right) \cdot \frac{\pi}{15}$$

$$M_{30} = \frac{\pi}{15} \sum_{i=1}^{30} f(c_i) = 25.526999$$

$$(N = 50): \quad \Delta x = \frac{2\pi}{50} = \frac{\pi}{25}, c_i = \left(i - \frac{1}{2}\right) \cdot \frac{\pi}{25}$$

$$M_{50} = \frac{\pi}{25} \sum_{i=1}^{50} f(c_i) = 25.526999$$

29. Let $a > b$ and set $k = \sqrt{1 - \frac{b^2}{a^2}}$. Use a parametric representation to show that the ellipse $\left(\frac{x}{a}\right)^2 + \left(\frac{y}{b}\right)^2 = 1$ has length $L = 4aG\left(\frac{\pi}{2}, k\right)$, where

$$G(\theta, k) = \int_0^\theta \sqrt{1 - k^2 \sin^2 t}\, dt$$

is the *elliptic integral of the second kind.*

SOLUTION Since the ellipse is symmetric with respect to the x and y axis, its length L is four times the length of the part of the ellipse which is in the first quadrant. This part is represented by the following parametrization: $x(t) = a \sin t$, $y(t) = b \cos t$, $0 \le t \le \frac{\pi}{2}$. Using the formula for the arc length we get:

$$L = 4 \int_0^{\pi/2} \sqrt{x'(t)^2 + y'(t)^2}\, dt = 4 \int_0^{\pi/2} \sqrt{(a \cos t)^2 + (-b \sin t)^2}\, dt$$

$$= 4 \int_0^{\pi/2} \sqrt{a^2 \cos^2 t + b^2 \sin^2 t}\, dt$$

We rewrite the integrand as follows:

$$L = 4 \int_0^{\pi/2} \sqrt{a^2 \cos^2 t + a^2 \sin^2 t + (b^2 - a^2) \sin^2 t}\, dt$$

$$= 4 \int_0^{\pi/2} \sqrt{a^2 (\cos^2 t + \sin^2 t) + (b^2 - a^2) \sin^2 t}\, dt$$

$$= 4 \int_0^{\pi/2} \sqrt{a^2 + (b^2 - a^2) \sin^2 t}\, dt = 4a \int_0^{\pi/2} \sqrt{\frac{a^2}{a^2} + \frac{b^2 - a^2}{a^2} \sin^2 t}\, dt$$

$$= 4a \int_0^{\pi/2} \sqrt{1 - \left(1 - \frac{b^2}{a^2}\right) \sin^2 t}\, dt = 4a \int_0^{\pi/2} \sqrt{1 - k^2 \sin^2 t}\, dt = 4aG\left(\frac{\pi}{2}, k\right)$$

where $k = \sqrt{1 - \frac{b^2}{a^2}}$.

In Exercises 30–33, use Eq. (5) to compute the surface area of the given surface.

31. The surface generated by revolving the astroid with parametrization $c(t) = (\cos^3 t, \sin^3 t)$ about the x-axis for $0 \le t \le \frac{\pi}{2}$

SOLUTION We have $x(t) = \cos^3 t$, $y(t) = \sin^3 t$, $x'(t) = -3 \cos^2 t \sin t$, $y'(t) = 3 \sin^2 t \cos t$. Hence,

$$x'(t)^2 + y'(t)^2 = 9 \cos^4 t \sin^2 t + 9 \sin^4 t \cos^2 t = 9 \cos^2 t \sin^2 t (\cos^2 t + \sin^2 t) = 9 \cos^2 t \sin^2 t$$

Using the formula for the surface area we get

$$S = 2\pi \int_0^{\pi/2} y(t) \sqrt{x'(t)^2 + y'(t)^2}\, dt = 2\pi \int_0^{\pi/2} \sin^3 t \cdot 3 \cos t \sin t\, dt = 6\pi \int_0^{\pi/2} \sin^4 t \cos t\, dt$$

We compute the integral using the substitution $u = \sin t$ $du = \cos t\, dt$. We obtain

$$S = 6\pi \int_0^1 u^4\, du = 6\pi \frac{u^5}{5} \bigg|_0^1 = \frac{6\pi}{5}.$$

33. The surface generated by revolving one arch of the cycloid $c(t) = (t - \sin t, 1 - \cos t)$ about the x-axis

SOLUTION One arch of the cycloid is traced as t varies from 0 to 2π. Since $x(t) = t - \sin t$ and $y(t) = 1 - \cos t$, we have $x'(t) = 1 - \cos t$ and $y'(t) = \sin t$. Hence, using the identity $1 - \cos t = 2\sin^2 \frac{t}{2}$, we get

$$x'(t)^2 + y'(t)^2 = (1 - \cos t)^2 + \sin^2 t = 1 - 2\cos t + \cos^2 t + \sin^2 t = 2 - 2\cos t = 4\sin^2 \frac{t}{2}$$

By the formula for the surface area we obtain:

$$S = 2\pi \int_0^{2\pi} y(t)\sqrt{x'(t)^2 + y'(t)^2}\,dt = 2\pi \int_0^{2\pi} (1 - \cos t) \cdot 2\sin \frac{t}{2}\,dt$$

$$= 2\pi \int_0^{2\pi} 2\sin^2 \frac{t}{2} \cdot 2\sin \frac{t}{2}\,dt = 8\pi \int_0^{2\pi} \sin^3 \frac{t}{2}\,dt = 16\pi \int_0^{\pi} \sin^3 u\,du$$

We use a reduction formula to compute this integral, obtaining

$$S = 16\pi \left[\frac{1}{3}\cos^3 u - \cos u\right]\Big|_0^{\pi} = 16\pi \left[\frac{4}{3}\right] = \frac{64\pi}{3}$$

Further Insights and Challenges

35. CAS Let $a \geq b > 0$ and set $k = \dfrac{2\sqrt{ab}}{a - b}$. Show that the **trochoid**

$$x = at - b\sin t, \qquad y = a - b\cos t, \qquad 0 \leq t \leq T$$

has length $2(a - b)G\left(\frac{T}{2}, k\right)$ with $G(\theta, k)$ as in Exercise 29.

SOLUTION We have $x'(t) = a - b\cos t$, $y'(t) = b\sin t$. Hence,

$$x'(t)^2 + y'(t)^2 = (a - b\cos t)^2 + (b\sin t)^2 = a^2 - 2ab\cos t + b^2\cos^2 t + b^2\sin^2 t$$
$$= a^2 + b^2 - 2ab\cos t$$

The length of the trochoid for $0 \leq t \leq T$ is

$$L = \int_0^T \sqrt{a^2 + b^2 - 2ab\cos t}\,dt$$

We rewrite the integrand as follows to bring it to the required form. We use the identity $1 - \cos t = 2\sin^2 \frac{t}{2}$ to obtain

$$L = \int_0^T \sqrt{(a - b)^2 + 2ab - 2ab\cos t}\,dt = \int_0^T \sqrt{(a - b)^2 + 2ab(1 - \cos t)}\,dt$$

$$= \int_0^T \sqrt{(a - b)^2 + 4ab\sin^2 \frac{t}{2}}\,dt = \int_0^T \sqrt{(a - b)^2 \left(1 + \frac{4ab}{(a - b)^2}\sin^2 \frac{t}{2}\right)}\,dt$$

$$= (a - b)\int_0^T \sqrt{1 + k^2\sin^2 \frac{t}{2}}\,dt$$

(where $k = \frac{2\sqrt{ab}}{a-b}$).

Substituting $u = \frac{t}{2}$, $du = \frac{1}{2}\,dt$, we get

$$L = 2(a - b)\int_0^{T/2} \sqrt{1 + k^2\sin^2 u}\,du = 2(a - b)E(T/2, k)$$

37. The acceleration due to gravity on the surface of the earth is $g = \dfrac{Gm_e}{R_e^2} = 9.8\ \mathrm{m/s^2}$, where $R_e = 6{,}378$ km. Use Exercise 36(b) to show that a satellite orbiting at the earth's surface would have period $T_e = 2\pi\sqrt{R_e/g} \approx 84.5$ min. Then estimate the distance R_m from the moon to the center of the earth. Assume that the period of the moon (sidereal month) is $T_m \approx 27.43$ days.

SOLUTION By part (b) of Exercise 36, it follows that

$$\frac{R_e^3}{T_e^2} = \frac{Gm_e}{4\pi^2} \Rightarrow T_e^2 = \frac{4\pi^2 R_e^3}{Gm_e} = \frac{4\pi^2 R_e}{\frac{Gm_e}{R_e^2}} = \frac{4\pi^2 R_e}{g}$$

Hence,

$$T_e = 2\pi\sqrt{\frac{R_e}{g}} = 2\pi\sqrt{\frac{6378 \cdot 10^3}{9.8}} \approx 5068.8 \text{ s} \approx 84.5 \text{ min}.$$

In part (b) of Exercise 36 we showed that $\frac{R^3}{T^2}$ is the same for all orbits. It follows that this quotient is the same for the satellite orbiting at the earth's surface and for the moon orbiting around the earth. Thus,

$$\frac{R_m^3}{T_m^2} = \frac{R_e^3}{T_e^2} \Rightarrow R_m = R_e\left(\frac{T_m}{T_e}\right)^{2/3}.$$

Setting $T_m = 27.43 \cdot 1440 = 39499.2$ minutes, $T_e = 84.5$ minutes, and $R_e = 6378$ km we get

$$R_m = 6378\left(\frac{39499.2}{84.5}\right)^{2/3} \approx 384154 \text{ km}.$$

12.3 Polar Coordinates (ET Section 11.3)

Preliminary Questions

1. If P and Q have the same radial coordinate, then (choose the correct answer):

(a) P and Q lie on the same circle with the center at the origin.

(b) P and Q lie on the same ray based at the origin.

SOLUTION Two points with the same radial coordinate are equidistant from the origin, therefore they lie on the same circle centered at the origin. The angular coordinate defines a ray based at the origin. Therefore, if the two points have the same angular coordinate, they lie on the same ray based at the origin.

2. Give two polar coordinate representations for the point $(x, y) = (0, 1)$, one with negative r and one with positive r.

SOLUTION The point $(0, 1)$ is on the y-axis, distant one unit from the origin, hence the polar representation with positive r is $(r, \theta) = \left(1, \frac{\pi}{2}\right)$. The point $(r, \theta) = \left(-1, \frac{\pi}{2}\right)$ is the reflection of $(r, \theta) = \left(1, \frac{\pi}{2}\right)$ through the origin, hence we must add π to return to the original point.

We obtain the following polar representation of $(0, 1)$ with negative r:

$$(r, \theta) = \left(-1, \frac{\pi}{2} + \pi\right) = \left(-1, \frac{3\pi}{2}\right).$$

3. Does a point (r, θ) have more than one representation in rectangular coordinates?

SOLUTION The rectangular coordinates are determined uniquely by the relations $x = r\cos\theta$, $y = r\sin\theta$. Therefore a point (r, θ) has exactly one representation in rectangular coordinates.

4. Describe the curves with polar equations

(a) $r = 2$ **(b)** $r^2 = 2$ **(c)** $r\cos\theta = 2$

SOLUTION

(a) Converting to rectangular coordinates we get

$$\sqrt{x^2 + y^2} = 2 \quad \text{or} \quad x^2 + y^2 = 2^2.$$

This is the equation of the circle of radius 2 centered at the origin.

(b) We convert to rectangular coordinates, obtaining $x^2 + y^2 = 2$. This is the equation of the circle of radius $\sqrt{2}$, centered at the origin.

(c) We convert to rectangular coordinates. Since $x = r\cos\theta$ we obtain the following equation: $x = 2$. This is the equation of the vertical line through the point $(2, 0)$.

5. If $f(-\theta) = f(\theta)$, then the curve $r = f(\theta)$ is symmetric with respect to the (choose the correct answer):

(a) x-axis **(b)** y-axis **(c)** origin

SOLUTION The equality $f(-\theta) = f(\theta)$ for all θ implies that whenever a point (r, θ) is on the curve, also the point $(r, -\theta)$ is on the curve. Since the point $(r, -\theta)$ is the reflection of (r, θ) with respect to the x-axis, we conclude that the curve is symmetric with respect to the x-axis.

Exercises

1. Find polar coordinates for each of the seven points plotted in Figure 17.

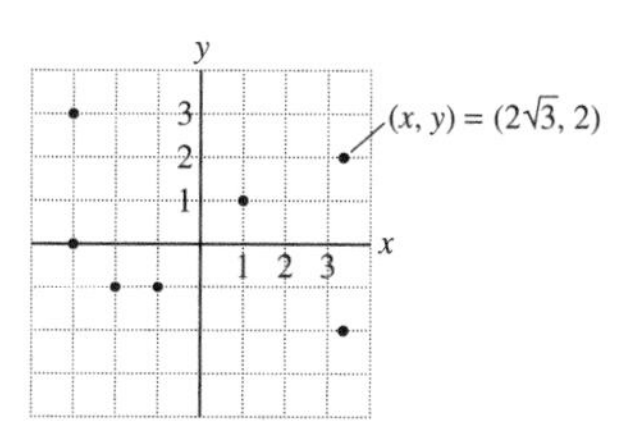

FIGURE 17

SOLUTION We mark the points as shown in the figure.

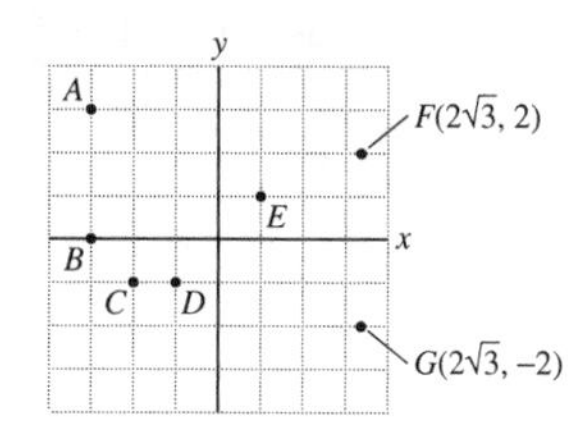

Using the data given in the figure for the x and y coordinates and the quadrants in which the point are located, we obtain:

(A): $\begin{array}{l} r = \sqrt{(-3)^2 + 3^2} = \sqrt{18} \\ \theta = \pi - \frac{\pi}{4} = \frac{3\pi}{4} \end{array} \Rightarrow (r, \theta) = \left(3\sqrt{2}, \frac{3\pi}{4}\right)$

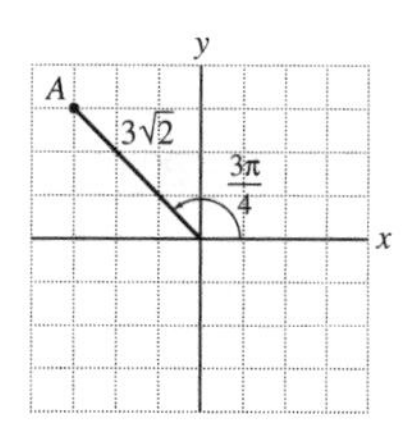

(B): $\begin{array}{l} r = 3 \\ \theta = \pi \end{array} \Rightarrow (r, \theta) = (3, \pi)$

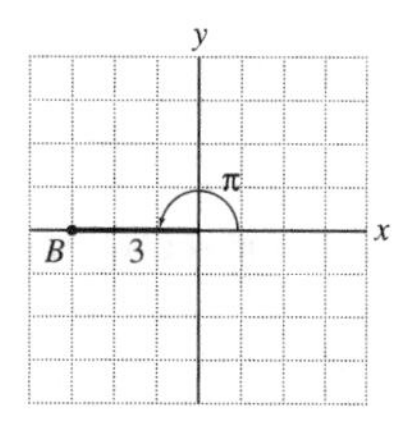

(C): $\begin{array}{l} r = \sqrt{2^2 + 1^2} = \sqrt{5} \approx 2.2 \\ \theta = \tan^{-1}\left(\frac{-1}{-2}\right) = \tan^{-1}\left(\frac{1}{2}\right) = \pi + 0.46 \approx 3.6 \end{array} \Rightarrow (r, \theta) \approx \left(\sqrt{5}, 3.6\right)$

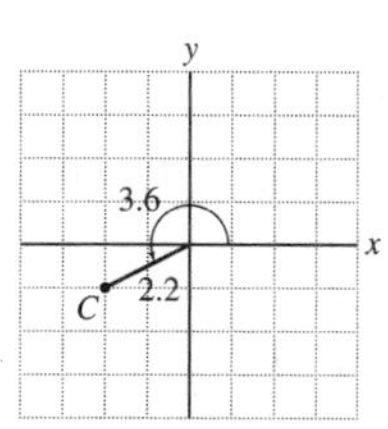

(D): $\begin{array}{l} r = \sqrt{1^2 + 1^2} = \sqrt{2} \approx 1.4 \\ \theta = \pi + \frac{\pi}{4} = \frac{5\pi}{4} \end{array} \Rightarrow (r, \theta) \approx \left(\sqrt{2}, \frac{5\pi}{4}\right)$

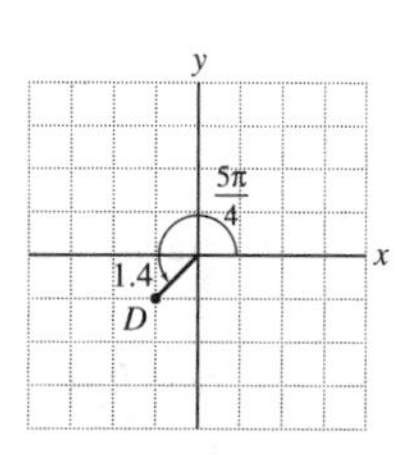

(E): $r = \sqrt{1^2 + 1^2} = \sqrt{2} \approx 1.4$, $\theta = \tan^{-1}\left(\frac{1}{1}\right) = \frac{\pi}{4}$ $\Rightarrow (r, \theta) \approx \left(\sqrt{2}, \frac{\pi}{4}\right)$

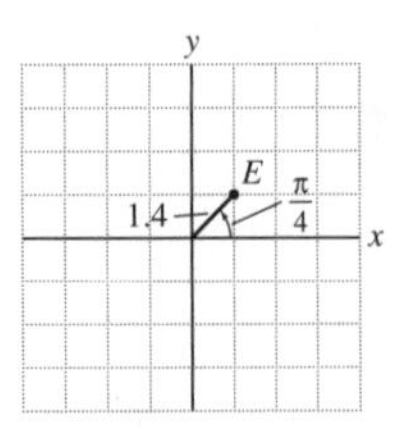

(F): $r = \sqrt{\left(2\sqrt{3}\right)^2 + 2^2} = \sqrt{16} = 4$, $\theta = \tan^{-1}\left(\frac{2}{2\sqrt{3}}\right) = \tan^{-1}\left(\frac{1}{\sqrt{3}}\right) = \frac{\pi}{6}$ $\Rightarrow (r, \theta) = \left(4, \frac{\pi}{6}\right)$

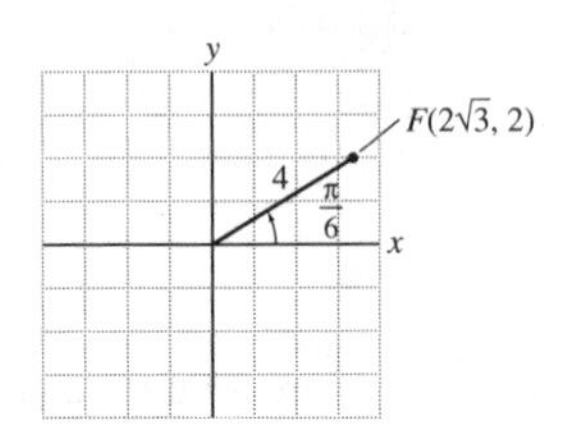

(G): G is the reflection of F about the x axis, hence the two points have equal radial coordinates, and the angular coordinate of G is obtained from the angular coordinate of F: $\theta = 2\pi - \frac{\pi}{6} = \frac{11\pi}{6}$. Hence, the polar coordinates of G are $\left(4, \frac{11\pi}{6}\right)$.

3. Convert from rectangular to polar coordinates:

(a) $(1, 0)$ **(b)** $(3, \sqrt{3})$

(c) $(-2, 2)$ **(d)** $(-1, \sqrt{3})$

SOLUTION

(a) The point $(1, 0)$ is on the positive x axis distanced one unit from the origin. Hence, $r = 1$ and $\theta = 0$. Thus, $(r, \theta) = (1, 0)$.

(b) The point $\left(3, \sqrt{3}\right)$ is in the first quadrant so $\theta = \tan^{-1}\left(\frac{\sqrt{3}}{3}\right) = \frac{\pi}{6}$. Also, $r = \sqrt{3^2 + \left(\sqrt{3}\right)^2} = \sqrt{12}$. Hence, $(r, \theta) = \left(\sqrt{12}, \frac{\pi}{6}\right)$.

(c) The point $(-2, 2)$ is in the second quadrant. Hence,

$$\theta = \tan^{-1}\left(\frac{2}{-2}\right) = \tan^{-1}(-1) = \pi - \frac{\pi}{4} = \frac{3\pi}{4}.$$

Also, $r = \sqrt{(-2)^2 + 2^2} = \sqrt{8}$. Hence, $(r, \theta) = \left(\sqrt{8}, \frac{3\pi}{4}\right)$.

(d) The point $\left(-1, \sqrt{3}\right)$ is in the second quadrant, hence,

$$\theta = \tan^{-1}\left(\frac{\sqrt{3}}{-1}\right) = \tan^{-1}\left(-\sqrt{3}\right) = \pi - \frac{\pi}{3} = \frac{2\pi}{3}.$$

Also, $r = \sqrt{(-1)^2 + \left(\sqrt{3}\right)^2} = \sqrt{4} = 2$. Hence, $(r, \theta) = \left(2, \frac{2\pi}{3}\right)$.

5. Convert from polar to rectangular coordinates:

(a) $(3, \frac{\pi}{6})$ **(b)** $(6, \frac{3\pi}{4})$ **(c)** $(5, -\frac{\pi}{2})$

SOLUTION

(a) Since $r = 3$ and $\theta = \frac{\pi}{6}$, we have:

$$\begin{aligned} x &= r\cos\theta = 3\cos\frac{\pi}{6} = 3 \cdot \frac{\sqrt{3}}{2} \approx 2.6 \\ y &= r\sin\theta = 3\sin\frac{\pi}{6} = 3 \cdot \frac{1}{2} = 1.5 \end{aligned} \quad \Rightarrow \quad (x, y) \approx (2.6, 1.5).$$

(b) For $\left(6, \frac{3\pi}{4}\right)$ we have $r = 6$ and $\theta = \frac{3\pi}{4}$. Hence,

$$\begin{aligned} x &= r\cos\theta = 6\cos\frac{3\pi}{4} \approx -4.24 \\ y &= r\sin\theta = 6\sin\frac{3\pi}{4} \approx 4.24 \end{aligned} \quad \Rightarrow \quad (x, y) \approx (-4.24, 4.24).$$

(c) Since $r = 5$ and $\theta = -\frac{\pi}{2}$ we have

$$\begin{aligned} x &= r\cos\theta = 5\cos\left(-\frac{\pi}{2}\right) = 5 \cdot 0 = 0 \\ y &= r\sin\theta = 5\sin\left(-\frac{\pi}{2}\right) = 5 \cdot (-1) = -5 \end{aligned} \quad \Rightarrow \quad (x, y) = (0, -5)$$

7. Which of the following are possible polar coordinates for the point P with rectangular coordinates $(0, -2)$?

(a) $\left(2, \frac{\pi}{2}\right)$ **(b)** $\left(2, \frac{7\pi}{2}\right)$

(c) $\left(-2, -\frac{3\pi}{2}\right)$ **(d)** $\left(-2, \frac{7\pi}{2}\right)$

(e) $\left(-2, -\frac{\pi}{2}\right)$ **(f)** $\left(2, -\frac{7\pi}{2}\right)$

SOLUTION The point P has distance 2 from the origin and the angle between $\overline{OP}$ and the positive x-axis in the positive direction is $\frac{3\pi}{2}$. Hence, $(r, \theta) = \left(2, \frac{3\pi}{2}\right)$ is one choice for the polar coordinates for P.

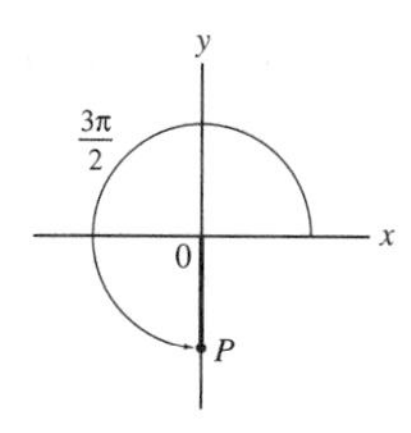

The polar coordinates $(2, \theta)$ are possible for P if $\theta - \frac{3\pi}{2}$ is a multiple of 2π. The polar coordinate $(-2, \theta)$ are possible for P if $\theta - \frac{3\pi}{2}$ is an odd multiple of π. These considerations lead to the following conclusions:

(a) $\left(2, \frac{\pi}{2}\right)$ $\frac{\pi}{2} - \frac{3\pi}{2} = -\pi \Rightarrow \left(2, \frac{\pi}{2}\right)$ does not represent P.

(b) $\left(2, \frac{7\pi}{2}\right)$ $\frac{7\pi}{2} - \frac{3\pi}{2} = 2\pi \Rightarrow \left(2, \frac{7\pi}{2}\right)$ represents P.

(c) $\left(-2, -\frac{3\pi}{2}\right)$ $-\frac{3\pi}{2} - \frac{3\pi}{2} = -3\pi \Rightarrow \left(-2, -\frac{3\pi}{2}\right)$ represents P.

(d) $\left(-2, \frac{7\pi}{2}\right)$ $\frac{7\pi}{2} - \frac{3\pi}{2} = 2\pi \Rightarrow \left(-2, \frac{7\pi}{2}\right)$ does not represent P.

(e) $\left(-2, -\frac{\pi}{2}\right)$ $-\frac{\pi}{2} - \frac{3\pi}{2} = -2\pi \Rightarrow \left(-2, -\frac{\pi}{2}\right)$ does not represent P.

(f) $\left(2, -\frac{7\pi}{2}\right)$ $-\frac{7\pi}{2} - \frac{3\pi}{2} = -5\pi \Rightarrow \left(2, -\frac{7\pi}{2}\right)$ does not represent P.

9. Find the equation in polar coordinates of the line through the origin with slope $\frac{1}{2}$.

SOLUTION A line of slope $m = \frac{1}{2}$ makes an angle $\theta_0 = \tan^{-1}\frac{1}{2} \approx 0.46$ with the positive x-axis. The equation of the line is $\theta \approx 0.46$, while r is arbitrary.

11. Which of the two equations, $r = 2\sec\theta$ and $r = 2\csc\theta$, defines a horizontal line?

SOLUTION The equation $r = 2\csc\theta$ is the polar equation of a horizontal line, as it can be written as $r = 2/\sin\theta$, so $r\sin\theta = 2$, which becomes $y = 2$. On the other hand, the equation $r = 2\sec\theta$ is the polar equation of a vertical line, as it can be written as $r = 2/\cos\theta$, so $r\cos\theta = 2$, which becomes $x = 2$.

In Exercises 12–17, convert to an equation in rectangular coordinates.

13. $r = \sin\theta$

SOLUTION Multiplying by r and substituting $y = r\sin\theta$ and $r^2 = x^2 + y^2$ gives

$$\begin{aligned} r^2 &= r\sin\theta \\ x^2 + y^2 &= y \end{aligned}$$

We move the y and then complete the square to obtain

$$x^2 + y^2 - y = 0$$

$$x^2 + \left(y - \frac{1}{2}\right)^2 = \left(\frac{1}{2}\right)^2$$

Thus, $r = \sin\theta$ is the equation of a circle of radius $\frac{1}{2}$ and center $\left(0, \frac{1}{2}\right)$.

15. $r = 2\csc\theta$

SOLUTION We multiply the equation by $\sin\theta$ and substitute $y = r\sin\theta$. We get

$$r\sin\theta = 2$$

$$y = 2$$

Thus, $r = 2\csc\theta$ is the equation of the line $y = 2$.

17. $r = \dfrac{1}{2 - \cos\theta}$

SOLUTION We multiply the equation by $2 - \cos\theta$. Then we substitute $x = r\cos\theta$ and $r = \sqrt{x^2 + y^2}$, to obtain

$$r\,(2 - \cos\theta) = 1$$

$$2r - r\cos\theta = 1$$

$$2\sqrt{x^2 + y^2} - x = 1$$

Moving the x, then squaring and simplifying, we obtain

$$2\sqrt{x^2 + y^2} = x + 1$$

$$4\left(x^2 + y^2\right) = x^2 + 2x + 1$$

$$3x^2 - 2x + 4y^2 = 1$$

We complete the square:

$$3\left(x^2 - \frac{2}{3}x\right) + 4y^2 = 1$$

$$3\left(x - \frac{1}{3}\right)^2 + 4y^2 = \frac{4}{3}$$

$$\frac{\left(x - \frac{1}{3}\right)^2}{\frac{4}{9}} + \frac{y^2}{\frac{1}{3}} = 1$$

This is the equation of the ellipse shown in the figure:

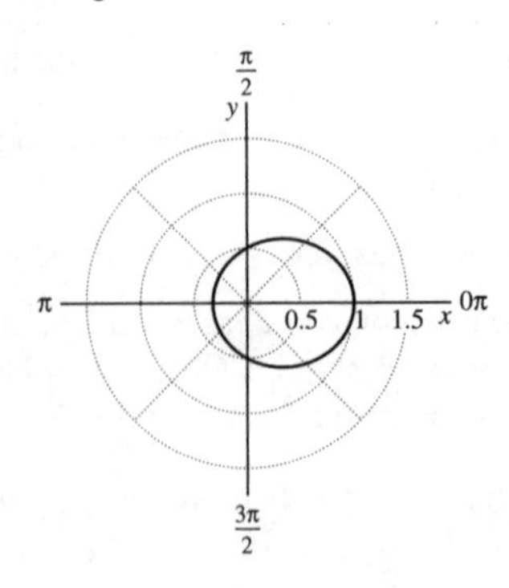

In Exercises 18–21, convert to an equation in polar coordinates.

19. $x = 5$

SOLUTION Substituting $x = r\cos\theta$ gives the polar equation $r\cos\theta = 5$ or $r = 5\sec\theta$.

21. $xy = 1$

SOLUTION We substitute $x = r\cos\theta$, $y = r\sin\theta$ to obtain

$$(r\cos\theta)(r\sin\theta) = 1$$
$$r^2\cos\theta\sin\theta = 1$$

Using the identity $\cos\theta\sin\theta = \frac{1}{2}\sin 2\theta$ yields

$$r^2 \cdot \frac{\sin 2\theta}{2} = 1 \Rightarrow r^2 = 2\csc 2\theta.$$

23. Find the values of θ in the plot of $r = 4\cos\theta$ corresponding to points A, B, C, D in Figure 19. Then indicate the portion of the graph traced out as θ varies in the following intervals:

(a) $0 \le \theta \le \frac{\pi}{2}$ **(b)** $\frac{\pi}{2} \le \theta \le \pi$ **(c)** $\pi \le \theta \le \frac{3\pi}{2}$

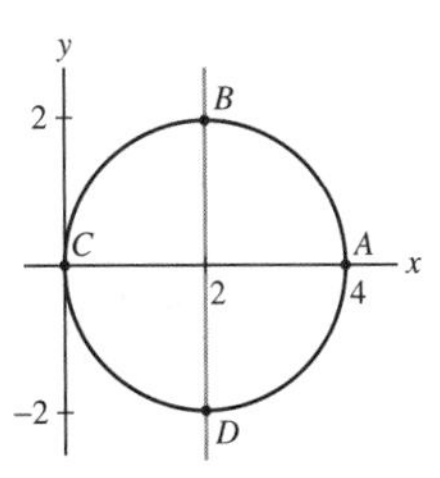

FIGURE 19 Plot of $r = 4\cos\theta$.

SOLUTION The point A is on the x-axis hence $\theta = 0$. The point B is in the first quadrant with $x = y = 2$ hence $\theta = \tan^{-1}\left(\frac{2}{2}\right) = \tan^{-1}(1) = \frac{\pi}{4}$. The point C is at the origin. Thus,

$$r = 0 \Rightarrow 4\cos\theta = 0 \Rightarrow \theta = \frac{\pi}{2}, \frac{3\pi}{2}.$$

The point D is in the fourth quadrant with $x = 2$, $y = -2$, hence

$$\theta = \tan^{-1}\left(\frac{-2}{2}\right) = \tan^{-1}(-1) = 2\pi - \frac{\pi}{4} = \frac{7\pi}{4}.$$

$0 \le \theta \le \frac{\pi}{2}$ represents the first quadrant, hence the points (r, θ) where $r = 4\cos\theta$ and $0 \le \theta \le \frac{\pi}{2}$ are the points on the circle which are in the first quadrant, as shown below:

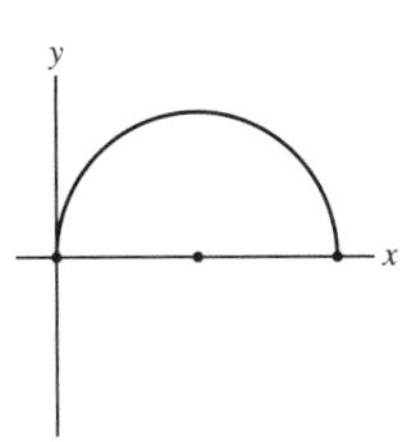

If we insist that $r \ge 0$, then since $\frac{\pi}{2} \le \theta \le \pi$ represents the second quadrant and $\pi \le \theta \le \frac{3\pi}{2}$ represents the third quadrant, and since the circle $r = 4\cos\theta$ has no points in the left xy -plane, then there are no points for (b) and (c). However, if we allow $r < 0$ then (b) represents the semi-circle

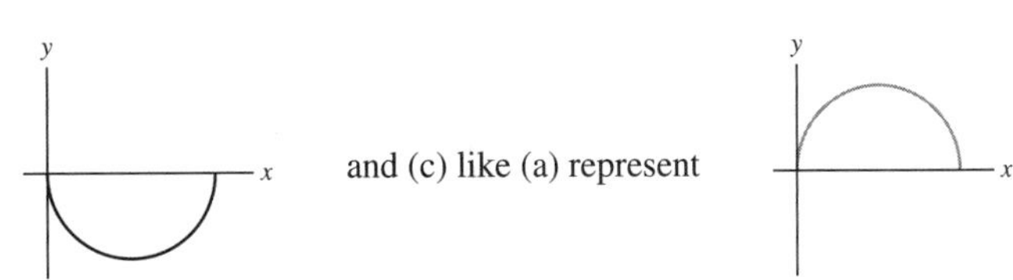

25. Match each equation in rectangular coordinates with its equation in polar coordinates.

(a) $x^2 + y^2 = 2$ **(i)** $r^2(1 + 2\sin^2\theta) = 4$
(b) $x^2 + (y-1)^2 = 1$ **(ii)** $r(\cos\theta + \sin\theta) = 4$
(c) $x^2 - y^2 = 4$ **(iii)** $r = 2\sin\theta$
(d) $x + y = 4$ **(iv)** $r = 2$

SOLUTION

(a) Since $x^2 + y^2 = r^2$, we have $r^2 = 4$ or $r = 2$.

(b) Using Example 7, the equation of the circle $x^2 + (y-1)^2 = 1$ has polar equation $r = 2\sin\theta$.
(c) Setting $x = r\cos\theta$, $y = r\sin\theta$ in $x^2 - y^2 = 4$ gives

$$x^2 - y^2 = r^2\cos^2\theta - r^2\sin^2\theta = r^2\left(\cos^2\theta - \sin^2\theta\right) = 4.$$

We now use the identity $\cos^2\theta - \sin^2\theta = 1 - 2\sin^2\theta$ to obtain the following equation:

$$r^2\left(1 - 2\sin^2\theta\right) = 4.$$

(d) Setting $x = r\cos\theta$ and $y = r\sin\theta$ in $x + y = 4$ we get:

$$x + y = 4$$
$$r\cos\theta + r\sin\theta = 4$$

so

$$r(\cos\theta + \sin\theta) = 4$$

27. Show that $r = \sin\theta + \cos\theta$ is the equation of the circle of radius $1/\sqrt{2}$ whose center in rectangular coordinates is $\left(\frac{1}{2}, \frac{1}{2}\right)$. Then find the values of θ between 0 and π such that $(\theta, r(\theta))$ yields the points A, B, C, and D in Figure 20.

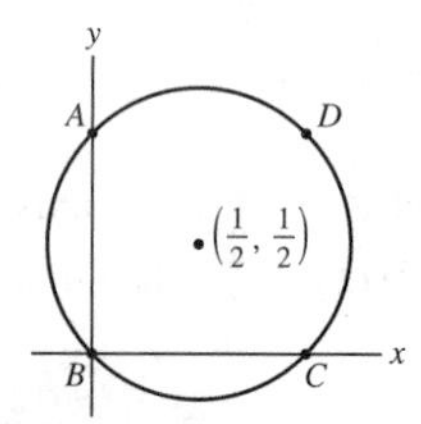

FIGURE 20 Plot of $r = \sin\theta + \cos\theta$.

SOLUTION We show that the rectangular equation of $r = \sin\theta + \cos\theta$ is

$$\left(x - \frac{1}{2}\right)^2 + \left(y - \frac{1}{2}\right)^2 = \frac{1}{2}.$$

We multiply the polar equation by r and substitute $r^2 = x^2 + y^2$, $r\sin\theta = y$, $r\cos\theta = x$. This gives

$$r = \sin\theta + \cos\theta$$
$$r^2 = r\sin\theta + r\cos\theta$$
$$x^2 + y^2 = y + x$$

Transferring sides and completing the square yields

$$x^2 - x + y^2 - y = 0$$
$$\left(x - \frac{1}{2}\right)^2 + \left(y - \frac{1}{2}\right)^2 = \frac{1}{4} + \frac{1}{4} = \frac{1}{2}$$

The point A corresponds to $\theta = \frac{\pi}{2}$. Hence, $r = \sin\frac{\pi}{2} + \cos\frac{\pi}{2} = 1 + 0 = 1$. That is, for the point A we have $(\theta, r) = \left(\frac{\pi}{2}, 1\right)$. The point B corresponds to $r = 0$, that is, $\sin\theta + \cos\theta = 0$. Solving for $0 \le \theta \le \pi$ we get $\sin\theta = -\cos\theta$ or $\tan\theta = -1$, hence $\theta = \frac{3\pi}{4}$. That is, $(\theta, r) = \left(\frac{3\pi}{4}, 0\right)$. The point C corresponds to $\theta = 0$. Hence, $r = \sin 0 + \cos 0 = 1$. That is, $(\theta, r) = (0, 1)$. The point D is on the line $y = x$, hence $\theta = \frac{\pi}{4}$. The corresponding value of r is

$$r = \sin\frac{\pi}{4} + \cos\frac{\pi}{4} = 2\frac{\sqrt{2}}{2} = \sqrt{2}.$$

Thus, $(\theta, r) = \left(\frac{\pi}{4}, \sqrt{2}\right)$.

29. Sketch the graph of $r = 3\cos\theta - 1$ (see Example 8).

SOLUTION We first choose some values of θ between 0 and π and mark the corresponding points on the graph. Then we use symmetry (due to $\cos(2\pi - \theta) = \cos\theta$) to plot the other half of the graph by reflecting the first half through the x-axis. Since $r = 3\cos\theta - 1$ is periodic, the entire curve is obtained as θ varies from 0 to 2π. We start with the values $\theta = 0, \frac{\pi}{6}, \frac{\pi}{3}, \frac{\pi}{2}, \frac{2\pi}{3}, \frac{5\pi}{6}, \pi$, and compute the corresponding values of r:

$$r = 3\cos 0 - 1 = 3 - 1 = 2 \Rightarrow A = (2, 0)$$

$$r = 3\cos\frac{\pi}{6} - 1 = \frac{3\sqrt{3}}{2} - 1 \approx 1.6 \Rightarrow B = \left(1.6, \frac{\pi}{6}\right)$$

$$r = 3\cos\frac{\pi}{3} - 1 = \frac{3}{2} - 1 = 0.5 \Rightarrow C = \left(0.5, \frac{\pi}{3}\right)$$

$$r = 3\cos\frac{\pi}{2} - 1 = 3 \cdot 0 - 1 = -1 \Rightarrow D = \left(-1, \frac{\pi}{2}\right)$$

$$r = 3\cos\frac{2\pi}{3} - 1 = -2.5 \Rightarrow E = \left(-2.5, \frac{2\pi}{3}\right)$$

$$r = 3\cos\frac{5\pi}{6} - 1 = -3.6 \Rightarrow F = \left(-3.6, \frac{5\pi}{6}\right)$$

$$r = 3\cos\pi - 1 = -4 \Rightarrow G = (-4, \pi)$$

The graph begins at the point $(r, \theta) = (2, 0)$ and moves toward the other points in this order, as θ varies from 0 to π. Since r is negative for $\frac{\pi}{2} \le \theta \le \pi$, the curve continues into the fourth quadrant, rather than into the second quadrant. We obtain the following graph:

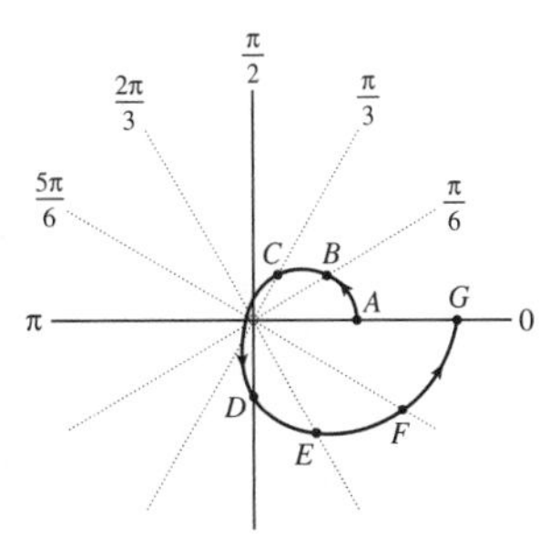

Now we have half the curve and we use symmetry to plot the rest. Reflecting the first half through the x axis we obtain the whole curve:

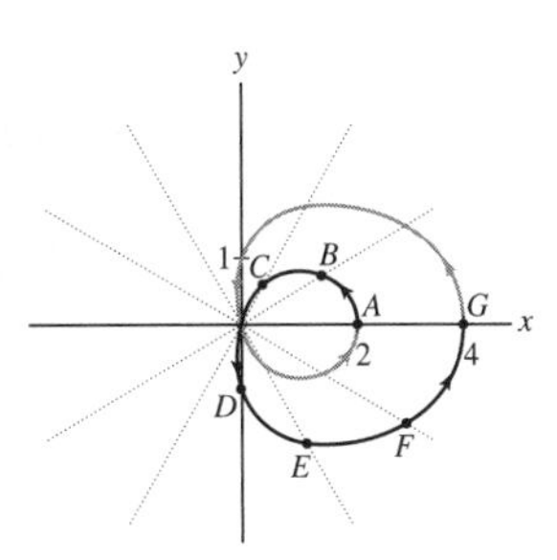

31. Figure 21 displays the graphs of $r = \sin 2\theta$ in rectangular coordinates and in polar coordinates, where it is a "rose with four petals." Identify (a) the points in (B) corresponding to the points labeled A–I in (A), and (b) the parts of the curve in (B) corresponding to the angle intervals $\left[0, \frac{\pi}{2}\right]$, $\left[\frac{\pi}{2}, \pi\right]$, $\left[\pi, \frac{3\pi}{2}\right]$, and $\left[\frac{3\pi}{2}, 2\pi\right]$.

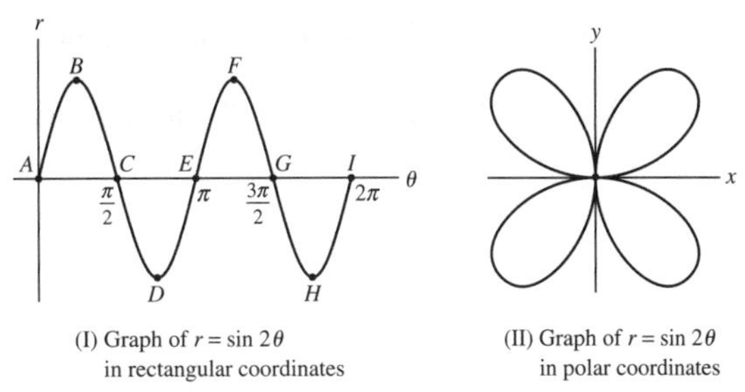

FIGURE 21 Rose with four petals.

SOLUTION

(a) The graph (I) gives the following polar coordinates of the labeled points:

$$A:\quad \theta = 0,\quad r = 0$$

$$B:\quad \theta = \frac{\pi}{4},\quad r = \sin\frac{2\pi}{4} = 1$$

$$C:\quad \theta = \frac{\pi}{2},\quad r = 0$$

$$D:\quad \theta = \frac{3\pi}{4},\quad r = \sin\frac{2\cdot 3\pi}{4} = -1$$

$$E:\quad \theta = \pi,\quad r = 0$$

$$F:\quad \theta = \frac{5\pi}{4},\quad r = 1$$

$$G:\quad \theta = \frac{3\pi}{2},\quad r = 0$$

$$H:\quad \theta = \frac{7\pi}{4},\quad r = -1$$

$$I:\quad \theta = 2\pi,\quad r = 0.$$

Since the maximal value of $|r|$ is 1, the points with $r = 1$ or $r = -1$ are the furthest points from the origin. The corresponding quadrant is determined by the value of θ and the sign of r. If $r_0 < 0$, the point (r_0, θ_0) is on the ray $\theta = -\theta_0$. These considerations lead to the following identification of the points in the xy plane. Notice that A, C, G, E, and I are the same point.

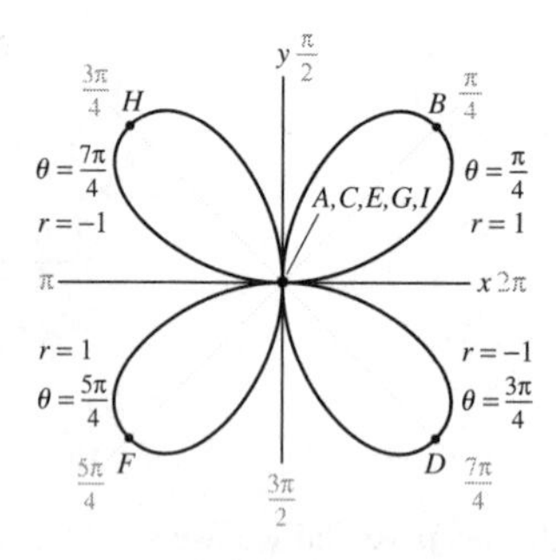

(b) We use the graph (I) to find the sign of $r = \sin 2\theta$: $0 \le \theta \le \frac{\pi}{2} \Rightarrow r \ge 0 \Rightarrow (r, \theta)$ is in the first quadrant. $\frac{\pi}{2} \le \theta \le \pi \Rightarrow r \le 0 \Rightarrow (r, \theta)$ is in the fourth quadrant. $\pi \le \theta \le \frac{3\pi}{2} \Rightarrow r \ge 0 \Rightarrow (r, \theta)$ is in the third quadrant. $\frac{3\pi}{2} \le \theta \le 2\pi \Rightarrow r \le 0 \Rightarrow (r, \theta)$ is in the second quadrant. That is,

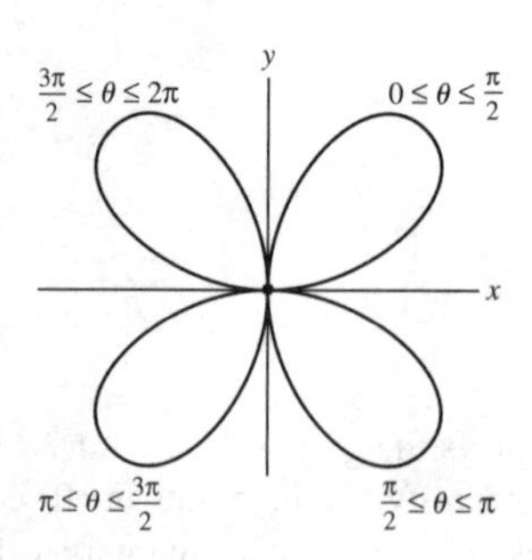

33. CAS Plot the **cissoid** $r = 2\sin\theta\tan\theta$ and show that its equation in rectangular coordinates is $y^2 = \dfrac{x^3}{2-x}$.

SOLUTION Using a CAS we obtain the following curve of the cissoid:

We substitute $\sin\theta = \frac{y}{r}$ and $\tan\theta = \frac{y}{x}$ in $r = 2\sin\theta\tan\theta$ to obtain

$$r = 2\frac{y}{r}\cdot\frac{y}{x}.$$

Multiplying by rx, setting $r^2 = x^2 + y^2$ and simplifying, yields

$$\begin{aligned} r^2x &= 2y^2 \\ (x^2+y^2)x &= 2y^2 \\ x^3 + y^2x &= 2y^2 \\ y^2(2-x) &= x^3 \end{aligned}$$

so

$$y^2 = \frac{x^3}{2-x}$$

35. Show that $r = a\cos\theta + b\sin\theta$ is the equation of a circle passing through the origin. Express the radius and center (in rectangular coordinates) in terms of a and b.

SOLUTION We multiply the equation by r and then make the substitution $x = r\cos\theta$, $y = r\sin\theta$, and $r^2 = x^2 + y^2$. This gives

$$\begin{aligned} r^2 &= ar\cos\theta + br\sin\theta \\ x^2 + y^2 &= ax + by \end{aligned}$$

Transferring sides and completing the square yields

$$x^2 - ax + y^2 - by = 0$$

$$\left(x^2 - 2\cdot\frac{a}{2}x + \left(\frac{a}{2}\right)^2\right) + \left(y^2 - 2\cdot\frac{b}{2}y + \left(\frac{b}{2}\right)^2\right) = \left(\frac{a}{2}\right)^2 + \left(\frac{b}{2}\right)^2$$

$$\left(x - \frac{a}{2}\right)^2 + \left(y - \frac{b}{2}\right)^2 = \frac{a^2+b^2}{4}$$

This is the equation of the circle with radius $\frac{\sqrt{a^2+b^2}}{2}$ centered at the point $\left(\frac{a}{2}, \frac{b}{2}\right)$. By plugging in $x = 0$ and $y = 0$ it is clear that the circle passes through the origin.

37. Use the identity $\cos 2\theta = \cos^2\theta - \sin^2\theta$ to find a polar equation of the hyperbola $x^2 - y^2 = 1$.

SOLUTION We substitute $x = r\cos\theta$, $y = r\sin\theta$ in $x^2 - y^2 = 1$ to obtain

$$\begin{aligned} r^2\cos^2\theta - r^2\sin^2\theta &= 1 \\ r^2(\cos^2\theta - \sin^2\theta) &= 1 \end{aligned}$$

Using the identity $\cos 2\theta = \cos^2\theta - \sin^2\theta$ we obtain the following equation of the hyperbola:

$$r^2\cos 2\theta = 1 \quad \text{or} \quad r^2 = \sec 2\theta.$$

39. Show that $\cos 3\theta = \cos^3\theta - 3\cos\theta\sin^2\theta$ and use this identity to find an equation in rectangular coordinates for the curve $r = \cos 3\theta$.

SOLUTION We use the identities $\cos(\alpha + \beta) = \cos\alpha\cos\beta - \sin\alpha\sin\beta$, $\cos 2\alpha = \cos^2\alpha - \sin^2\alpha$, and $\sin 2\alpha = 2\sin\alpha\cos\alpha$ to write

$$\begin{aligned}\cos 3\theta &= \cos(2\theta + \theta) = \cos 2\theta\cos\theta - \sin 2\theta\sin\theta \\ &= (\cos^2\theta - \sin^2\theta)\cos\theta - 2\sin\theta\cos\theta\sin\theta \\ &= \cos^3\theta - \sin^2\theta\cos\theta - 2\sin^2\theta\cos\theta \\ &= \cos^3\theta - 3\sin^2\theta\cos\theta\end{aligned}$$

Using this identity we may rewrite the equation $r = \cos 3\theta$ as follows:

$$r = \cos^3\theta - 3\sin^2\theta\cos\theta \tag{1}$$

Since $x = r\cos\theta$ and $y = r\sin\theta$, we have $\cos\theta = \frac{x}{r}$ and $\sin\theta = \frac{y}{r}$. Substituting into (1) gives:

$$\begin{aligned}r &= \left(\frac{x}{r}\right)^3 - 3\left(\frac{y}{r}\right)^2\left(\frac{x}{r}\right) \\ r &= \frac{x^3}{r^3} - \frac{3y^2x}{r^3}\end{aligned}$$

We now multiply by r^3 and make the substitution $r^2 = x^2 + y^2$ to obtain the following equation for the curve:

$$\begin{aligned}r^4 &= x^3 - 3y^2x \\ (x^2 + y^2)^2 &= x^3 - 3y^2x\end{aligned}$$

In Exercises 41–45, find an equation in polar coordinates of the line $\mathcal{L}$ with given description.

41. The point on $\mathcal{L}$ closest to the origin has polar coordinates $\left(2, \frac{\pi}{9}\right)$.

SOLUTION In Example 5, it is shown that the polar equation of the line where (r, α) is the point on the line closest to the origin is $r = d\sec(\theta - \alpha)$. Setting $(d, \alpha) = \left(2, \frac{\pi}{9}\right)$ we obtain the following equation of the line:

$$r = 2\sec\left(\theta - \frac{\pi}{9}\right).$$

43. $\mathcal{L}$ is tangent to the circle $r = 2\sqrt{10}$ at the point with rectangular coordinates $(-2, -6)$.

SOLUTION

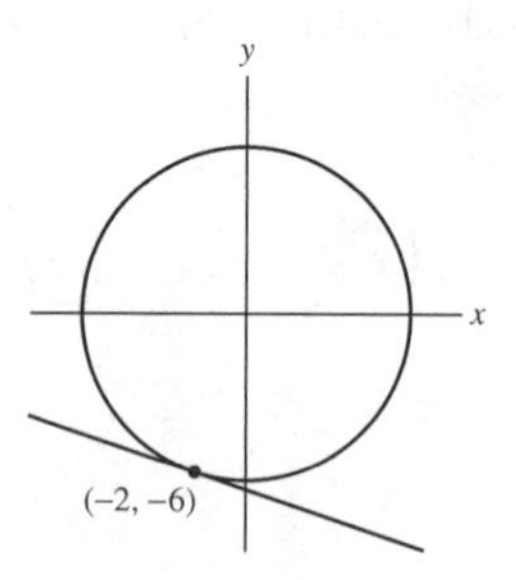

Since $\mathcal{L}$ is tangent to the circle at the point $(-2, -6)$, this is the point on $\mathcal{L}$ closest to the center of the circle which is at the origin. Therefore, we may use the polar coordinates (d, α) of this point in the equation of the line:

$$r = d\sec(\theta - \alpha) \tag{1}$$

We thus must convert the coordinates $(-2, -6)$ to polar coordinates. This point is in the third quadrant so $\pi < \alpha < \frac{3\pi}{2}$. We get

$$\begin{aligned}d &= \sqrt{(-2)^2 + (-6)^2} = \sqrt{40} = 2\sqrt{10} \\ \alpha &= \tan^{-1}\left(\frac{-6}{-2}\right) = \tan^{-1}3 \approx \pi + 1.25 \approx 4.39\end{aligned}$$

Substituting in (1) yields the following equation of the line:

$$r = 2\sqrt{10}\sec(\theta - 4.39).$$

45. $y = 4x - 9$.

SOLUTION Substituting $y = r\sin\theta$ and $x = r\cos\theta$ in $y = 4x - 9$, gives

$$r\sin\theta = 4r\cos\theta - 9$$

$$4r\cos\theta - r\sin\theta = 9$$

$$r(4\cos\theta - \sin\theta) = 9$$

so

$$r = \frac{9}{4\cos\theta - \sin\theta}$$

47. Distance Formula Use the Law of Cosines (Figure 23) to show that the distance d between two points with polar coordinates (r, θ) and (r_0, θ_0) is

$$d^2 = r^2 + r_0^2 - 2rr_0\cos(\theta - \theta_0) \qquad \boxed{2}$$

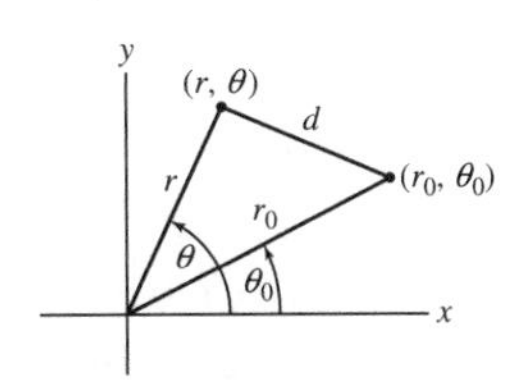

FIGURE 23

SOLUTION Note that the angle between the line segments r and r_0 has measurement $\theta - \theta_0$. Thus, by the Law of Cosines,

$$d^2 = r^2 + r_0^2 - 2rr_0\cos(\theta - \theta_0)$$

49. Show that the cardiod $r = 1 + \sin\theta$ has equation

$$x^2 + y^2 = (x^2 + y^2 - x)^2$$

SOLUTION We write the equation of the cardioid in the form $1 = r\sin\theta$ and multiply by r. This gives $r = r^2 - r\sin\theta$. We now make the substitution $r = \sqrt{x^2 + y^2}$ and $r\sin\theta = y$ to obtain

$$\sqrt{x^2 + y^2} = x^2 + y^2 - y.$$

Finally, we square both sides to obtain

$$x^2 + y^2 = \left(x^2 + y^2 - y\right)^2.$$

51. The Derivative in Polar Coordinates A polar curve $r = f(\theta)$ has parametric equations (since $x = r\cos\theta$ and $y = r\sin\theta$):

$$x = f(\theta)\cos\theta, \quad y = f(\theta)\sin\theta$$

Apply Theorem 1 of Section 12.1 to prove the formula

$$\frac{dy}{dx} = \frac{f(\theta)\cos\theta + f'(\theta)\sin\theta}{-f(\theta)\sin\theta + f'(\theta)\cos\theta} \qquad \boxed{3}$$

where $f'(\theta) = df/d\theta$.

SOLUTION By the formula for the derivative we have

$$\frac{dy}{dx} = \frac{y'(\theta)}{x'(\theta)} \qquad (1)$$

We differentiate the functions $x = f(\theta)\cos\theta$ and $y = f(\theta)\sin\theta$ using the Product Rule for differentiation. This gives

$$y'(\theta) = f'(\theta)\sin\theta + f(\theta)\cos\theta$$

$$x'(\theta) = f'(\theta)\cos\theta - f(\theta)\sin\theta$$

Substituting in (1) gives

$$\frac{dy}{dx} = \frac{f'(\theta)\sin\theta + f(\theta)\cos\theta}{f'(\theta)\cos\theta - f(\theta)\sin\theta} = \frac{f(\theta)\cos\theta + f'(\theta)\sin\theta}{-f(\theta)\sin\theta + f'(\theta)\cos\theta}.$$

53. Find the equation in rectangular coordinates of the tangent line to $r = 4\cos 3\theta$ at $\theta = \frac{\pi}{6}$.

SOLUTION We have $f(\theta) = 4\cos 3\theta$. By Eq. (3),

$$m = \frac{4\cos 3\theta\cos\theta - 12\sin 3\theta\sin\theta}{-4\cos 3\theta\sin\theta - 12\sin 3\theta\cos\theta}.$$

Setting $\theta = \frac{\pi}{6}$ yields

$$m = \frac{4\cos\left(\frac{\pi}{2}\right)\cos\left(\frac{\pi}{6}\right) - 12\sin\left(\frac{\pi}{2}\right)\sin\left(\frac{\pi}{6}\right)}{-4\cos\left(\frac{\pi}{2}\right)\sin\left(\frac{\pi}{6}\right) - 12\sin\left(\frac{\pi}{2}\right)\cos\left(\frac{\pi}{6}\right)} = \frac{-12\sin\frac{\pi}{6}}{-12\cos\frac{\pi}{6}} = \tan\frac{\pi}{6} = \frac{1}{\sqrt{3}}.$$

We identify the point of tangency. For $\theta = \frac{\pi}{6}$ we have $r = 4\cos\frac{3\pi}{6} = 4\cos\frac{\pi}{2} = 0$. The point of tangency is the origin. The tangent line is the line through the origin with slope $\frac{1}{\sqrt{3}}$. This is the line $y = \frac{x}{\sqrt{3}}$.

Further Insights and Challenges

55. Let c be a fixed constant. Explain the relationship between the graphs of:

(a) $y = f(x + c)$ and $y = f(x)$ (rectangular)

(b) $r = f(\theta + c)$ and $r = f(\theta)$ (polar)

(c) $y = f(x) + c$ and $y = f(x)$ (rectangular)

(d) $r = f(\theta) + c$ and $r = f(\theta)$ (polar)

SOLUTION

(a) For $c > 0$, $y = f(x + c)$ shifts the graph of $y = f(x)$ by c units to the left. If $c < 0$, the result is a shift to the right. It is a horizontal translation.

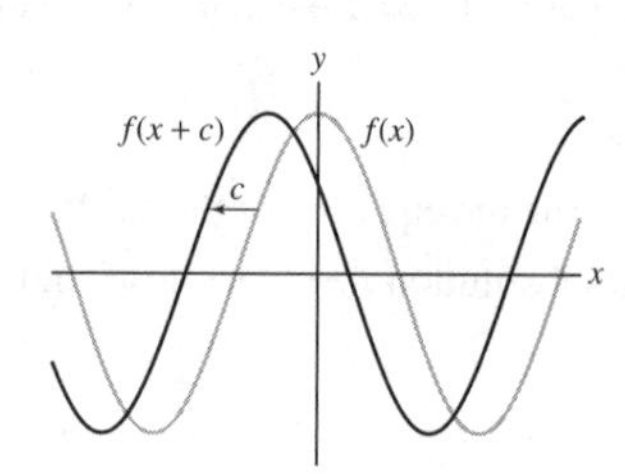

(b) As in part (a), the graph of $r = f(\theta + c)$ is a shift of the graph of $r = f(\theta)$ by c units in θ. Thus, the graph in polar coordinates is rotated by angle c as shown in the following figure:

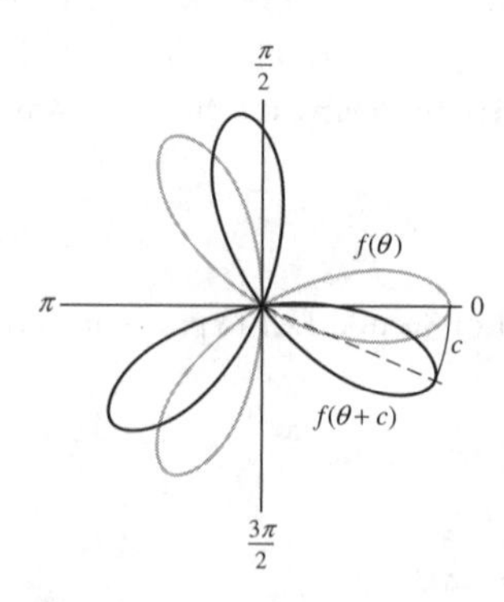

(c) $y = f(x) + c$ shifts the graph vertically upward by c units if $c > 0$, and downward by $(-c)$ units if $c < 0$. It is a vertical translation.

(d) The graph of $r = f(\theta) + c$ is a shift of the graph of $r = f(\theta)$ by c units in r. In the corresponding graph, in polar coordinates, each point with $f(\theta) > 0$ moves on the ray connecting it to the origin c units away from the origin if $c > 0$ and $(-c)$ units toward the origin if $c < 0$, and vice-versa for $f(\theta) < 0$.

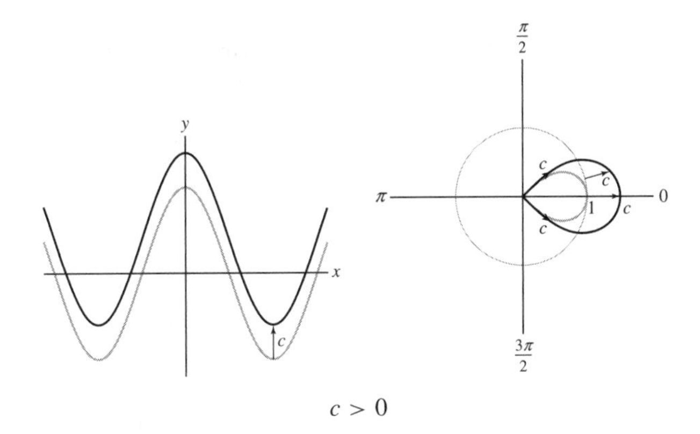

$c > 0$

57. GU Use a graphing utility to convince yourself that graphs of the polar equations $r = f_1(\theta) = 2\cos\theta - 1$ and $r = f_2(\theta) = 2\cos\theta + 1$ have the same graph. Then explain why. *Hint:* Show that the points $(f_1(\theta+\pi), \theta+\pi)$ and $(f_2(\theta), \theta)$ coincide.

SOLUTION The graphs of $r = 2\cos\theta - 1$ and $r = 2\cos\theta + 1$ in the xy-plane coincide as shown in the graph obtained using a CAS.

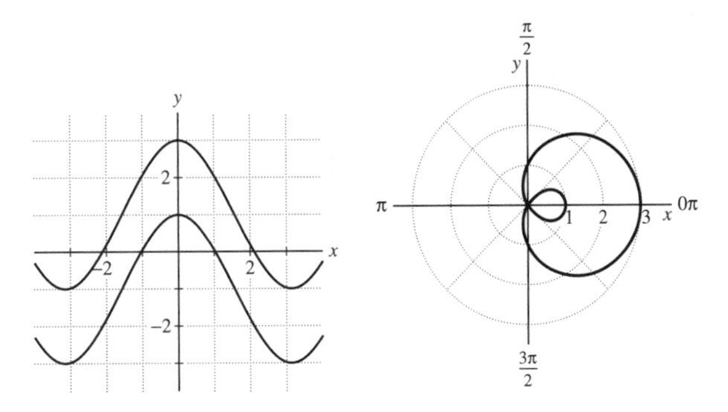

Recall that (r, θ) and $(-r, \theta+\pi)$ represent the same point. Replacing θ by $\theta+\pi$ and r by $(-r)$ in $r = 2\cos\theta - 1$ we obtain

$$-r = 2\cos(\theta+\pi) - 1$$
$$-r = -2\cos\theta - 1$$
$$r = 2\cos\theta + 1$$

Thus, the two equations define the same graph. (One could also convert both equations to rectangular coordinates and note that they come out identical.)

12.4 Area and Arc Length in Polar Coordinates (ET Section 11.4)

Preliminary Questions

1. True or False: The area under the curve with polar equation $r = f(\theta)$ is equal to the integral of $f(\theta)$.

SOLUTION The statement is false. Consider the circle $r = 1$. Its area is π, yet the integral of 1 from 0 to 2π is 2π. Thus. we see that the integral does not give the area under the curve.

2. Polar coordinates are best suited to finding the area (choose one):

(a) Under a curve between $x = a$ and $x = b$.

(b) Bounded by a curve and two rays through the origin.

SOLUTION Polar coordinates are best suited to finding the area bounded by a curve and two rays through the origin. The formula for the area in polar coordinates gives the area of this region.

3. True or False: The formula for area in polar coordinates is valid only if $f(\theta) \geq 0$.

SOLUTION The statement is false. The formula for the area

$$\frac{1}{2}\int_{\alpha}^{\beta} f(\theta)^2\, d\theta$$

always gives the actual (positive) area, even if $f(\theta)$ takes on negative values.

4. The horizontal line $y = 1$ has polar equation $r = \csc\theta$. Which area is represented by the integral $\frac{1}{2}\int_{\pi/6}^{\pi/2} \csc^2\theta\, d\theta$ (Figure 13)?

(a) $\square ABCD$ **(b)** $\triangle ABC$ **(c)** $\triangle ACD$

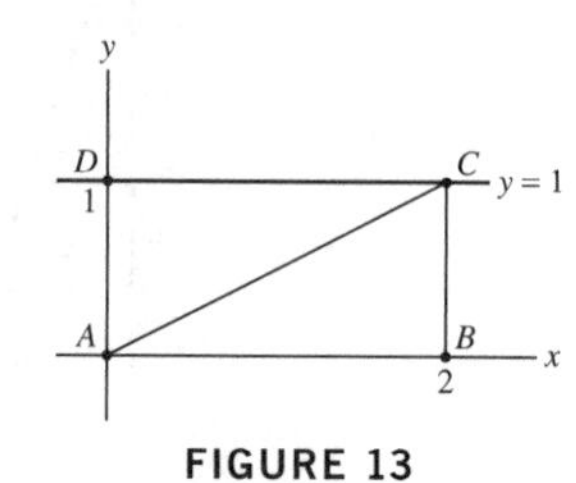

FIGURE 13

SOLUTION This integral represents an area taken from $\theta = \pi/6$ to $\theta = \pi/2$, which can only be the triangle $\triangle ACD$, as seen in part (c).

Exercises

1. Sketch the area bounded by the circle $r = 5$ and the rays $\theta = \frac{\pi}{2}$ and $\theta = \pi$, and compute its area as an integral in polar coordinates.

SOLUTION The region bounded by the circle $r = 5$ and the rays $\theta = \frac{\pi}{2}$ and $\theta = \pi$ is the shaded region in the figure. The area of the region is given by the following integral:

$$\frac{1}{2}\int_{\pi/2}^{\pi} r^2\, d\theta = \frac{1}{2}\int_{\pi/2}^{\pi} 5^2\, d\theta = \frac{25}{2}\left(\pi - \frac{\pi}{2}\right) = \frac{25\pi}{4}$$

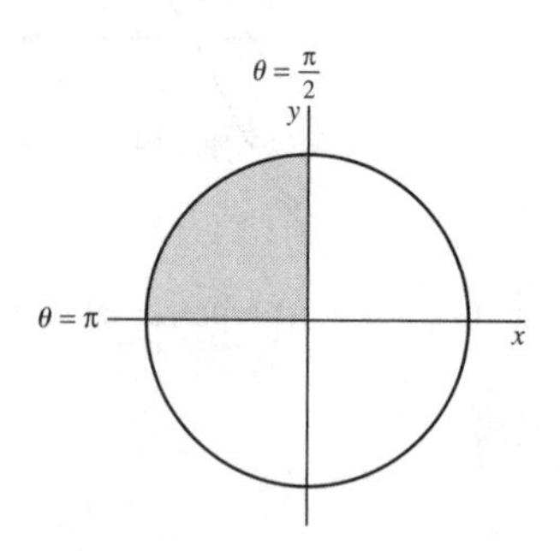

3. Calculate the area of the circle $r = 4\sin\theta$ as an integral in polar coordinates (see Figure 4). Be careful to choose the correct limits of integration.

SOLUTION The equation $r = 4\sin\theta$ defines a circle of radius 2 tangent to the x-axis at the origin as shown in the figure:

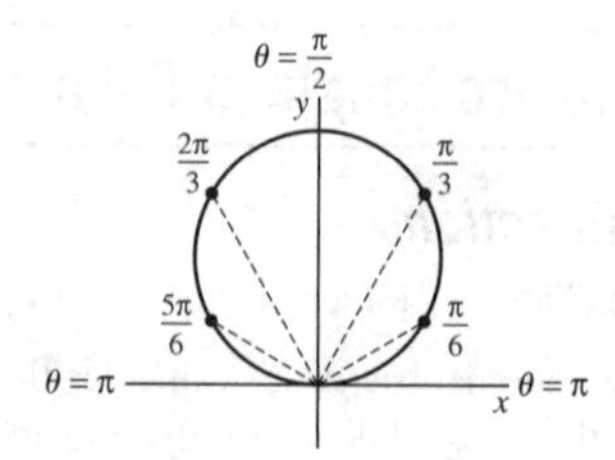

The circle is traced as θ varies from 0 to π. We use the area in polar coordinates and the identity

$$\sin^2\theta = \frac{1}{2}(1 - \cos 2\theta)$$

to obtain the following area:

$$A = \frac{1}{2}\int_0^{\pi} r^2\, d\theta = \frac{1}{2}\int_0^{\pi} (4\sin\theta)^2\, d\theta = 8\int_0^{\pi} \sin^2\theta\, d\theta = 4\int_0^{\pi} (1 - \cos 2\theta)\, d\theta = 4\left[\theta - \frac{\sin 2\theta}{2}\right]_0^{\pi}$$

$$= 4\left(\left(\pi - \frac{\sin 2\pi}{2}\right) - 0\right) = 4\pi.$$

5. Find the total area enclosed by the cardioid $r = 1 - \cos\theta$ (Figure 15).

FIGURE 15 The cardioid $r = 1 - \cos\theta$.

SOLUTION We graph $r = 1 - \cos\theta$ in r and θ (cartesian, not polar, this time):

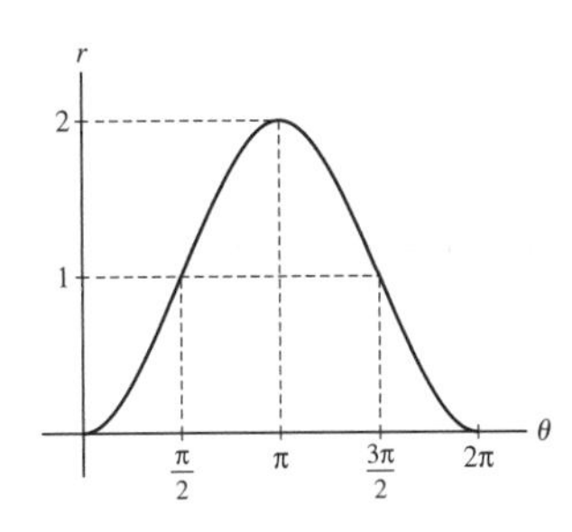

We see that as θ varies from 0 to π, the radius r increases from 0 to 2, so we get the upper half of the cardioid (the lower half is obtained as θ varies from π to 2π and consequently r decreases from 2 to 0). Since the cardioid is symmetric with respect to the x-axis we may compute the upper area and double the result. Using

$$\cos^2\theta = \frac{\cos 2\theta + 1}{2}$$

we get

$$\begin{aligned} A &= 2 \cdot \frac{1}{2}\int_0^{\pi} r^2\,d\theta = \int_0^{\pi} (1-\cos\theta)^2\,d\theta = \int_0^{\pi} \left(1 - 2\cos\theta + \cos^2\theta\right)d\theta \\ &= \int_0^{\pi} \left(1 - 2\cos\theta + \frac{\cos 2\theta + 1}{2}\right)d\theta = \int_0^{\pi}\left(\frac{3}{2} - 2\cos\theta + \frac{1}{2}\cos 2\theta\right)d\theta \\ &= \frac{3}{2}\theta - 2\sin\theta + \frac{1}{4}\sin 2\theta\Big|_0^{\pi} = \frac{3\pi}{2} \end{aligned}$$

The total area enclosed by the cardioid is $A = \frac{3\pi}{2}$.

7. Find the area of one leaf of the "four-petaled rose" $r = \sin 2\theta$ (Figure 16).

SOLUTION We consider the graph of $r = \sin 2\theta$ in cartesian and in polar coordinates:

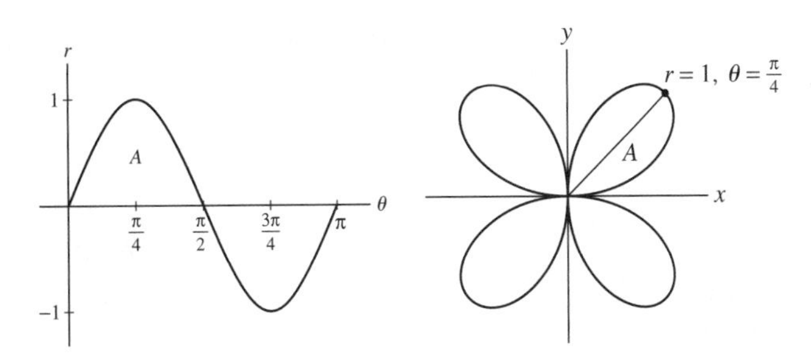

We see that as θ varies from 0 to $\frac{\pi}{4}$ the radius r is increasing from 0 to 1, and when θ varies from $\frac{\pi}{4}$ to $\frac{\pi}{2}$, r is decreasing back to zero. Hence, the leaf in the first quadrant is traced as θ varies from 0 to $\frac{\pi}{2}$. The area of the leaf (the four leaves have equal areas) is thus

$$A = \frac{1}{2}\int_0^{\pi/2} r^2\,d\theta = \frac{1}{2}\int_0^{\pi/2} \sin^2 2\theta\,d\theta.$$

Using the identity

$$\sin^2 2\theta = \frac{1 - \cos 4\theta}{2}$$

we get

$$A = \frac{1}{2}\int_0^{\pi/2}\left(\frac{1}{2} - \frac{\cos 4\theta}{2}\right)d\theta = \frac{1}{2}\left(\frac{\theta}{2} - \frac{\sin 4\theta}{8}\right)\Big|_0^{\pi/2} = \frac{1}{2}\left(\left(\frac{\pi}{4} - \frac{\sin 2\pi}{8}\right) - 0\right) = \frac{\pi}{8}$$

The area of one leaf is $A = \frac{\pi}{8} \approx 0.39$.

9. Find the area enclosed by one loop of the lemniscate with equation $r^2 = \cos 2\theta$ (Figure 17). Choose your limits of integration carefully.

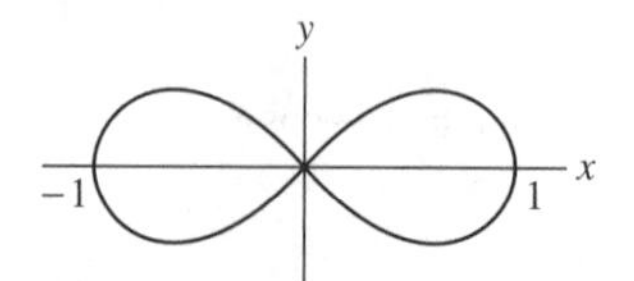

FIGURE 17 The lemniscate $r^2 = \cos 2\theta$.

SOLUTION We sketch the graph of $r^2 = \cos 2\theta$ in the $\left(r^2, \theta\right)$ plane; for $-\frac{\pi}{4} \le \theta \le \frac{\pi}{4}$:

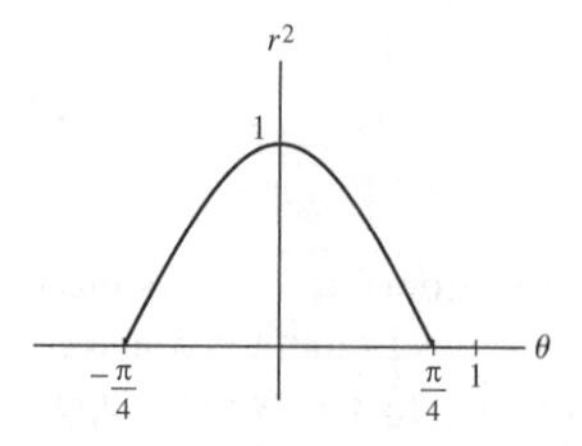

We see that as θ varies from $-\frac{\pi}{4}$ to 0, r^2 increases from 0 to 1, hence r also increases from 0 to 1. Then, as θ varies from 0 to $\frac{\pi}{4}$, r^2, so r decreases from 1 to 0. This gives the right-hand loop of the lemniscate.

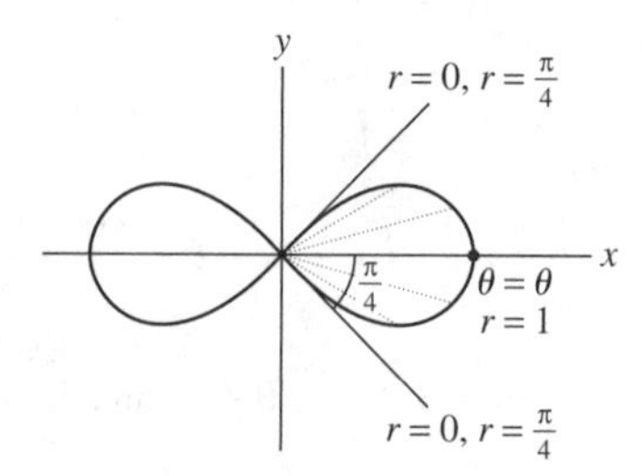

We compute the area enclosed by the right-hand loop, by the following integral:

$$\frac{1}{2}\int_{-\pi/4}^{\pi/4} r^2\,d\theta = \frac{1}{2}\int_{-\pi/4}^{\pi/4}\cos 2\theta\,d\theta = \frac{1}{2}\frac{\sin 2\theta}{2}\Big|_{-\pi/4}^{\pi/4} = \frac{1}{4}\left(\sin\frac{\pi}{2} - \sin\left(-\frac{\pi}{2}\right)\right) = \frac{1}{2}$$

Since the lemniscate is symmetric with respect to the y-axis, the total area A enclosed by the lemniscate is twice the area enclosed by the right-hand loop. Thus,

$$A = 2 \cdot \frac{1}{2} = 1.$$

11. Find the area enclosed by the cardioid $r = a(1 + \cos\theta)$, where $a > 0$.

SOLUTION The graph of $r = a\,(1 + \cos\theta)$ in the $r\theta$-plane for $0 \le \theta \le 2\pi$ and the cardioid in the xy-plane are shown in the following figures:

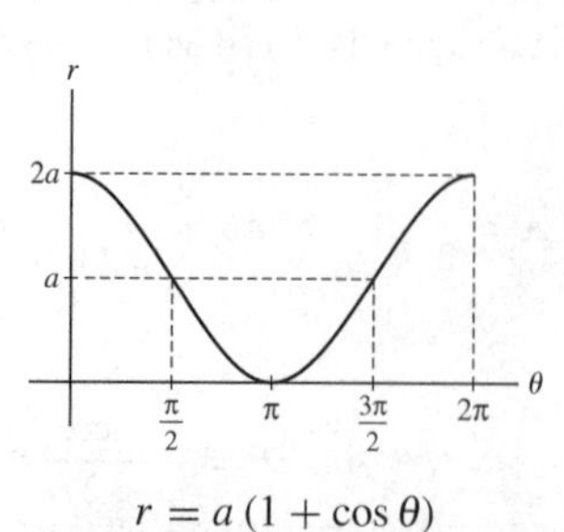

$r = a\,(1 + \cos\theta)$

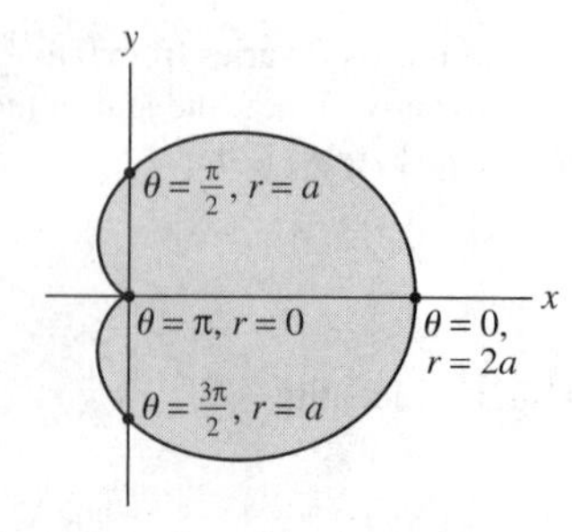

The cardioid $r = a\,(1 + \cos\theta)$, $a > 0$

As θ varies from 0 to π the radius r decreases from $2a$ to 0, and this gives the upper part of the cardioid.

The lower part is traced as θ varies from π to 2π and consequently r increases from 0 back to $2a$. We compute the area enclosed by the upper part of the cardioid and the x-axis, using the following integral (we use the identity $\cos^2\theta = \frac{1}{2} + \frac{1}{2}\cos 2\theta$):

$$\frac{1}{2}\int_0^{\pi} r^2\,d\theta = \frac{1}{2}\int_0^{\pi} a^2(1+\cos\theta)^2\,d\theta = \frac{a^2}{2}\int_0^{\pi}\left(1+2\cos\theta+\cos^2\theta\right)d\theta$$

$$= \frac{a^2}{2}\int_0^{\pi}\left(1+2\cos\theta+\frac{1}{2}+\frac{1}{2}\cos 2\theta\right)d\theta = \frac{a^2}{2}\int_0^{\pi}\left(\frac{3}{2}+2\cos\theta+\frac{1}{2}\cos 2\theta\right)d\theta$$

$$= \frac{a^2}{2}\left[\frac{3\theta}{2}+2\sin\theta+\frac{1}{4}\sin 2\theta\right]\Big|_0^{\pi} = \frac{a^2}{2}\left[\frac{3\pi}{2}+2\sin\pi+\frac{1}{4}\sin 2\pi - 0\right] = \frac{3\pi a^2}{4}$$

Using symmetry, the total area A enclosed by the cardioid is

$$A = 2\cdot\frac{3\pi a^2}{4} = \frac{3\pi a^2}{2}$$

13. Find the area of region A in Figure 18.

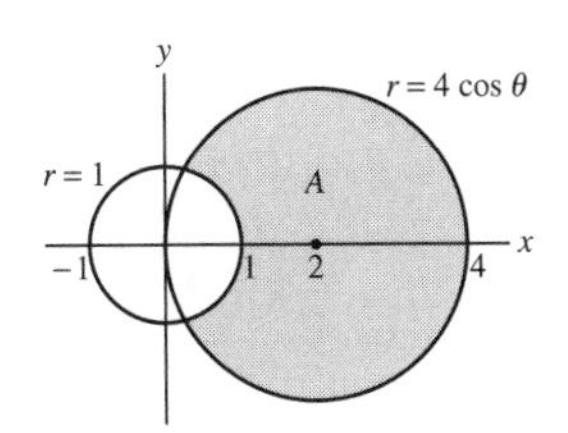

FIGURE 18

SOLUTION We first find the values of θ at the points of intersection of the two circles, by solving the following equation for $-\frac{\pi}{2} \le x \le \frac{\pi}{2}$:

$$4\cos\theta = 1 \Rightarrow \cos\theta = \frac{1}{4} \Rightarrow \theta_1 = \cos^{-1}\left(\frac{1}{4}\right)$$

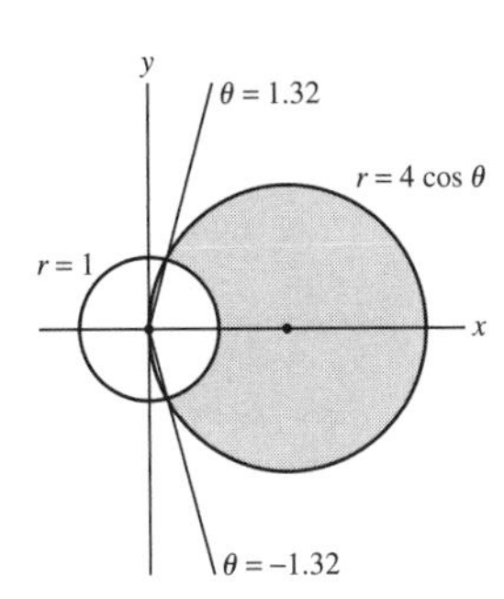

We now compute the area using the formula for the area between two curves:

$$A = \frac{1}{2}\int_{-\theta_1}^{\theta_1}\left((4\cos\theta)^2 - 1^2\right)d\theta = \frac{1}{2}\int_{-\theta_1}^{\theta_1}\left(16\cos^2\theta - 1\right)d\theta$$

Using the identity $\cos^2\theta = \frac{\cos 2\theta+1}{2}$ we get

$$A = \frac{1}{2}\int_{-\theta_1}^{\theta_1}\left(\frac{16(\cos 2\theta+1)}{2} - 1\right)d\theta = \frac{1}{2}\int_{-\theta_1}^{\theta_1}(8\cos 2\theta + 7)\,d\theta = \frac{1}{2}(4\sin 2\theta + 7\theta)\Big|_{-\theta_1}^{\theta_1}$$

$$= 4\sin 2\theta_1 + 7\theta_1 = 8\sin\theta_1\cos\theta_1 + 7\theta_1 = 8\sqrt{1-\cos^2\theta_1}\cos\theta_1 + 7\theta_1$$

Using the fact that $\cos\theta_1 = \frac{1}{4}$ we get

$$A = \frac{\sqrt{15}}{2} + 7\cos^{-1}\left(\frac{1}{4}\right) \approx 11.163$$

15. Find the area of the inner loop of the limaçon with polar equation $r = 2\cos\theta - 1$ (Figure 20).

SOLUTION We consider the graph of $r = 2\cos\theta - 1$ in cartesian and in polar, for $-\frac{\pi}{2} \le x \le \frac{\pi}{2}$:

$r = 2\cos\theta - 1$

As θ varies from $-\frac{\pi}{3}$ to 0, r increases from 0 to 1. As θ varies from 0 to $\frac{\pi}{3}$, r decreases from 1 back to 0. Hence, the inner loop of the limaçon is traced as θ varies from $-\frac{\pi}{3}$ to $\frac{\pi}{3}$. The area of the shaded region is thus

$$A = \frac{1}{2}\int_{-\pi/3}^{\pi/3} r^2\,d\theta = \frac{1}{2}\int_{-\pi/3}^{\pi/3} (2\cos\theta - 1)^2\,d\theta = \frac{1}{2}\int_{-\pi/3}^{\pi/3} \left(4\cos^2\theta - 4\cos\theta + 1\right)d\theta$$

$$= \frac{1}{2}\int_{-\pi/3}^{\pi/3} (2(\cos 2\theta + 1) - 4\cos\theta + 1)\,d\theta = \frac{1}{2}\int_{-\pi/3}^{\pi/3} (2\cos 2\theta - 4\cos\theta + 3)\,d\theta$$

$$= \frac{1}{2}(\sin 2\theta - 4\sin\theta + 3\theta)\Big|_{-\pi/3}^{\pi/3} = \frac{1}{2}\left(\left(\sin\frac{2\pi}{3} - 4\sin\frac{\pi}{3} + \pi\right) - \left(\sin\left(-\frac{2\pi}{3}\right) - 4\sin\left(-\frac{\pi}{3}\right) - \pi\right)\right)$$

$$= \frac{\sqrt{3}}{2} - \frac{4\sqrt{3}}{2} + \pi = \pi - \frac{3\sqrt{3}}{2} \approx 0.54$$

17. Find the area of the part of the circle $r = \sin\theta + \cos\theta$ in the fourth quadrant (see Exercise 27 in Section 12.3).

SOLUTION The value of θ corresponding to the point B is the solution of $r = \sin\theta + \cos\theta = 0$ for $-\pi \le \theta \le \pi$.

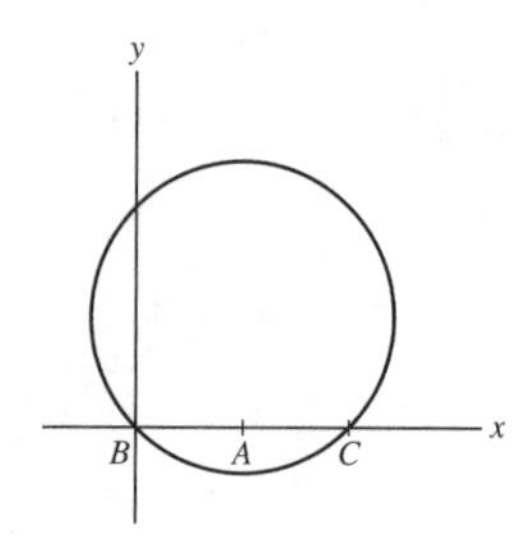

That is,

$$\sin\theta + \cos\theta = 0 \Rightarrow \sin\theta = -\cos\theta \Rightarrow \tan\theta = -1 \Rightarrow \theta = -\frac{\pi}{4}$$

At the point C, we have $\theta = 0$. The part of the circle in the fourth quadrant is traced if θ varies between $-\frac{\pi}{4}$ and 0. This leads to the following area:

$$A = \frac{1}{2}\int_{-\pi/4}^{0} r^2\,d\theta = \frac{1}{2}\int_{-\pi/4}^{0} (\sin\theta + \cos\theta)^2\,d\theta = \frac{1}{2}\int_{-\pi/4}^{0} \left(\sin^2\theta + 2\sin\theta\cos\theta + \cos^2\theta\right)d\theta$$

Using the identities $\sin^2\theta + \cos^2\theta = 1$ and $2\sin\theta\cos\theta = \sin 2\theta$ we get:

$$A = \frac{1}{2}\int_{-\pi/4}^{0} (1 + \sin 2\theta)\,d\theta = \frac{1}{2}\left(\theta - \frac{\cos 2\theta}{2}\right)\Big|_{-\pi/4}^{0}$$

$$= \frac{1}{2}\left(\left(0 - \frac{1}{2}\right) - \left(-\frac{\pi}{4} - \frac{\cos\left(\frac{-\pi}{2}\right)}{2}\right)\right) = \frac{1}{2}\left(\frac{\pi}{4} - \frac{1}{2}\right) = \frac{\pi}{8} - \frac{1}{4} \approx 0.14.$$

19. Find the area between the two curves in Figure 22(A).

SOLUTION We compute the area A between the two curves as the difference between the area A_1 of the region enclosed in the outer curve $r = 2 + \cos 2\theta$ and the area A_2 of the region enclosed in the inner curve $r = \sin 2\theta$. That is,

$$A = A_1 - A_2.$$

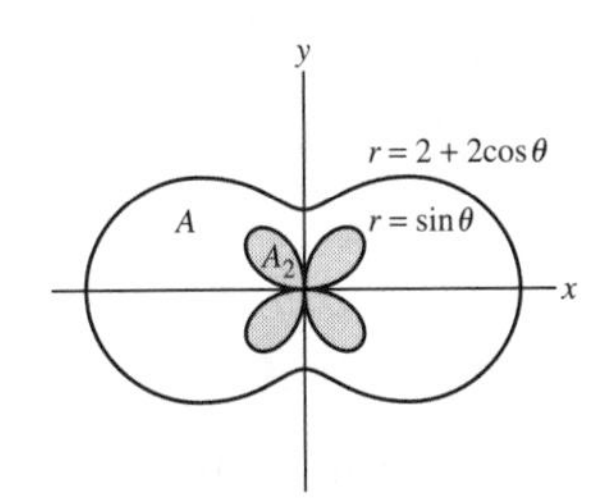

In Exercise 8 we showed that $A_2 = \frac{\pi}{2}$, hence,

$$A = A_1 - \frac{\pi}{2} \qquad (1)$$

We compute the area A_1.

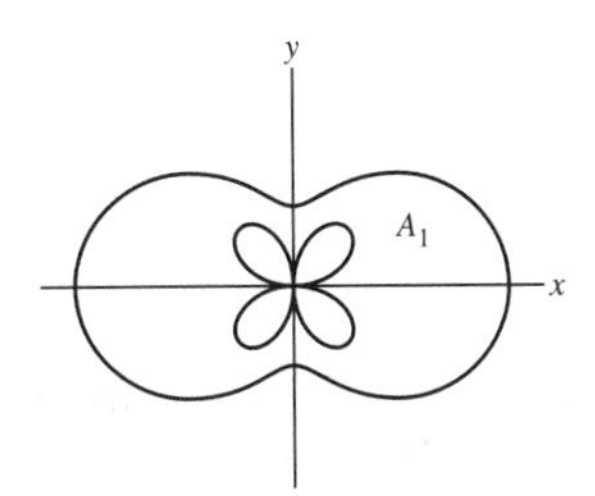

Using symmetry, the area is four times the area enclosed in the first quadrant. That is,

$$A_1 = 4 \cdot \frac{1}{2} \int_0^{\pi/2} r^2 \, d\theta = 2 \int_0^{\pi/2} (2 + \cos 2\theta)^2 \, d\theta = 2 \int_0^{\pi/2} \left(4 + 4\cos 2\theta + \cos^2 2\theta\right) d\theta$$

Using the identity $\cos^2 2\theta = \frac{1}{2}\cos 4\theta + \frac{1}{2}$ we get

$$A_1 = 2 \int_0^{\pi/2} \left(4 + 4\cos 2\theta + \frac{1}{2}\cos 4\theta + \frac{1}{2}\right) d\theta = 2 \int_0^{\pi/2} \left(\frac{9}{2} + \frac{1}{2}\cos 4\theta + 4\cos 2\theta\right) d\theta$$

$$= 2\left(\frac{9\theta}{2} + \frac{\sin 4\theta}{8} + 2\sin 2\theta\right)\Big|_0^{\pi/2} = 2\left(\left(\frac{9\pi}{4} + \frac{\sin 2\pi}{8} + 2\sin \pi\right) - 0\right) = \frac{9\pi}{2} \qquad (2)$$

Combining (1) and (2) we obtain

$$A = \frac{9\pi}{2} - \frac{\pi}{2} = 4\pi.$$

21. Find the area inside both curves in Figure 23.

SOLUTION The area we need to find is the area of the shaded region in the figure.

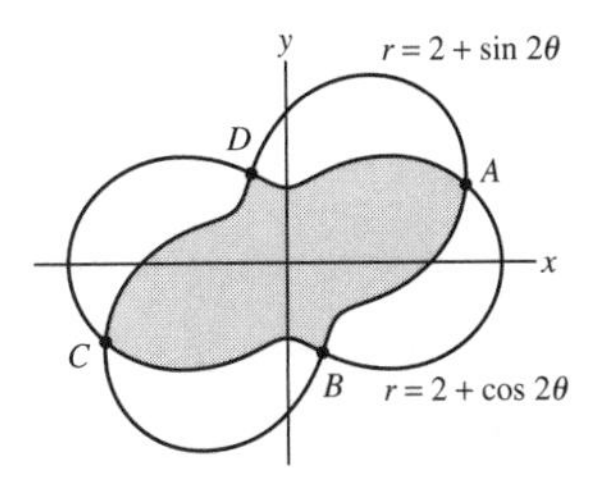

We first find the values of θ at the points of intersection A, B, C, and D of the two curves, by solving the following equation for $-\pi \le \theta \le \pi$:

$$2 + \cos 2\theta = 2 + \sin 2\theta$$
$$\cos 2\theta = \sin 2\theta$$
$$\tan 2\theta = 1 \Rightarrow 2\theta = \frac{\pi}{4} + \pi k \Rightarrow \theta = \frac{\pi}{8} + \frac{\pi k}{2}$$

The solutions for $-\pi \le \theta \le \pi$ are

$$A: \quad \theta = \frac{\pi}{8}.$$

$$B: \quad \theta = -\frac{3\pi}{8}.$$

$$C: \quad \theta = -\frac{7\pi}{8}.$$

$$D: \quad \theta = \frac{5\pi}{8}.$$

Using symmetry, we compute the shaded area in the figure below and multiply it by 4:

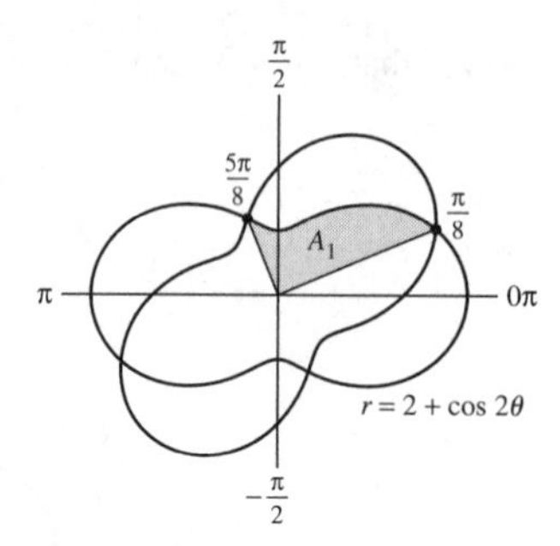

$$A = 4 \cdot A_1 = 4 \cdot \frac{1}{2} \cdot \int_{\pi/8}^{5\pi/8} (2+\cos 2\theta)^2 \, d\theta = 2\int_{\pi/8}^{5\pi/8} \left(4 + 4\cos 2\theta + \cos^2 2\theta\right) d\theta$$

$$= 2\int_{\pi/8}^{5\pi/8} \left(4 + 4\cos 2\theta + \frac{1+\cos 4\theta}{2}\right) d\theta = \int_{\pi/8}^{5\pi/8} (9 + 8\cos 2\theta + \cos 4\theta) \, d\theta$$

$$= 9\theta + 4\sin 2\theta + \frac{\sin 4\theta}{4}\Big|_{\pi/8}^{5\pi/8} = 9\left(\frac{5\pi}{8} - \frac{\pi}{8}\right) + 4\left(\sin\frac{5\pi}{4} - \sin\frac{\pi}{4}\right) + \frac{1}{4}\left(\sin\frac{5\pi}{2} - \sin\frac{\pi}{2}\right) = \frac{9\pi}{2} - 4\sqrt{2}$$

23. Figure 24 suggests that the circle $r = \sin\theta$ lies inside the spiral $r = \theta$. Which inequality from Chapter 2 assures us that this is the case? Find the area between the curves $r = \theta$ and $r = \sin\theta$ in the first quadrant.

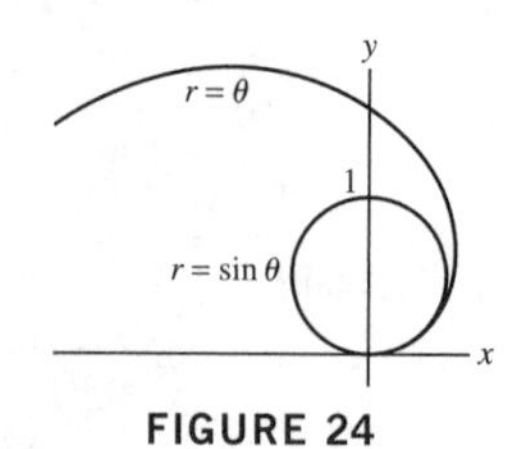

FIGURE 24

SOLUTION The inequality $|\sin\theta| \le |\theta|$ assures us that for all θ, the corresponding point on the spiral has radial coordinate with absolute value greater than that of the corresponding point on the circle. Hence, the circle lies inside the spiral.

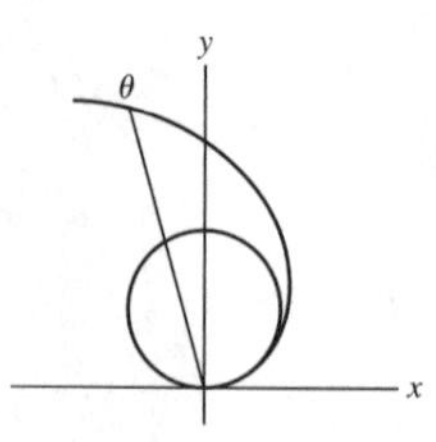

The area between the circle and the spiral in the first quadrant is traced as θ varies from 0 to $\frac{\pi}{2}$.

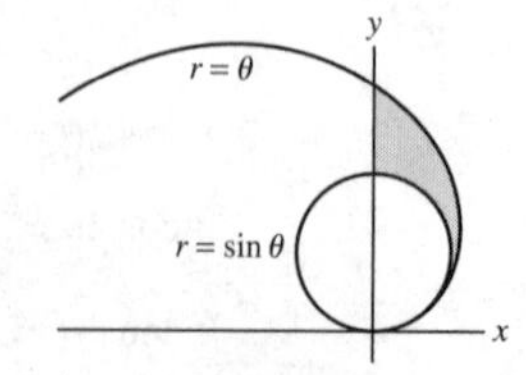

Using the formula for the area between two curves and the identity $\sin^2\theta = \frac{1}{2} - \frac{1}{2}\cos 2\theta$ we obtain

$$A = \frac{1}{2}\int_0^{\pi/2} \left(\theta^2 - \sin^2\theta\right) d\theta = \frac{1}{2}\int_0^{\pi/2} \left(\theta^2 - \frac{1}{2} + \frac{1}{2}\cos 2\theta\right) d\theta$$

$$= \frac{1}{2}\left(\frac{\theta^3}{3} - \frac{\theta}{2} + \frac{1}{4}\sin 2\theta\right)\Bigg|_0^{\pi/2} = \frac{1}{2}\left(\frac{\pi^3}{24} - \frac{\pi}{4} + \frac{\sin\pi}{4} - 0\right) = \frac{\pi^3}{48} - \frac{\pi}{8} \approx 0.25$$

25. Find the length of the spiral $r = \theta$ for $0 \le \theta \le A$.

SOLUTION We use the formula for the arc length. In this case $f(\theta) = \theta$, $f'(\theta) = 1$. Using integration formulas we get:

$$S = \int_0^A \sqrt{\theta^2 + 1^2}\, d\theta = \int_0^A \sqrt{\theta^2 + 1}\, d\theta = \frac{\theta}{2}\sqrt{\theta^2 + 1} + \frac{1}{2}\ln\left|\theta + \sqrt{\theta^2 + 1}\right|\Bigg|_0^A$$

$$= \frac{A}{2}\sqrt{A^2 + 1} + \frac{1}{2}\ln\left|A + \sqrt{A^2 + 1}\right|$$

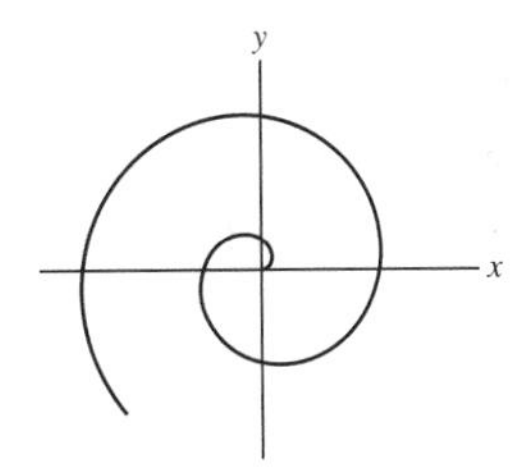

The spiral $r = \theta$

27. Sketch the segment $r = \sec\theta$ for $0 \le \theta \le A$. Then compute its length in two ways: as an integral in polar coordinates and using trigonometry.

SOLUTION The line $r = \sec\theta$ has the rectangular equation $x = 1$. The segment AB for $0 \le \theta \le A$ is shown in the figure.

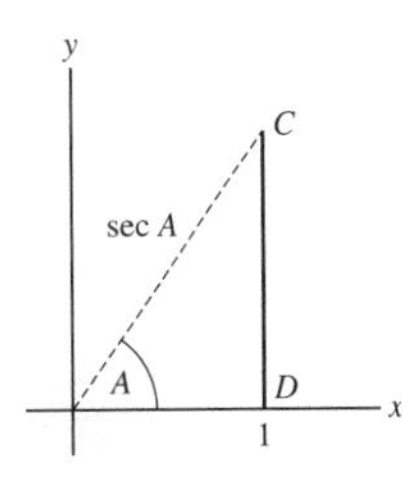

Using trigonometry, the length of the segment $\overline{AB}$ is

$$L = \overline{AB} = \overline{0B}\tan A = 1 \cdot \tan A = \tan A$$

Alternatively, we use the integral in polar coordinates with $f(\theta) = \sec(\theta)$ and $f'(\theta) = \tan\theta\sec\theta$. This gives

$$L = \int_0^A \sqrt{(\sec\theta)^2 + (\tan\theta\sec\theta)^2}\, d\theta = \int_0^A \sqrt{1 + \tan^2\theta}\sec\theta\, d\theta = \int_0^A \sec^2\theta\, d\theta = \tan\theta\Big|_0^A = \tan A.$$

The two answers agree, as expected.

29. Find the length of the cardioid $r = 1 + \cos\theta$. *Hint:* Use the identity $1 + \cos\theta = 2\cos^2\left(\frac{\theta}{2}\right)$ to evaluate the arc length integral.

SOLUTION In the equation of the cardioid, $f(\theta) = 1 + \cos\theta$. Using the formula for arc length in polar coordinates we have:

$$L = \int_\alpha^\beta \sqrt{f(\theta)^2 + f'(\theta)^2}\, d\theta \tag{1}$$

We compute the integrand:

$$\sqrt{f(\theta)^2 + f'(\theta)^2} = \sqrt{(1 + \cos\theta)^2 + (-\sin\theta)^2} = \sqrt{1 + 2\cos\theta + \cos^2\theta + \sin^2\theta} = \sqrt{2(1 + \cos\theta)}$$

We identify the interval of θ. Since $-1 \le \cos\theta \le 1$, every $0 \le \theta \le 2\pi$ corresponds to nonnegative value of r. Hence, θ varies from 0 to 2π. By (1) we obtain

$$L = \int_0^{2\pi} \sqrt{2(1 + \cos\theta)}\, d\theta$$

Since $1 + \cos\theta = 2\cos^2(\theta/2)$, and by the symmetry of the graph,

$$L = 2\int_0^{\pi} \sqrt{2(1+\cos\theta)}\,d\theta = 2\int_0^{\pi} 2\cos(\theta/2)\,d\theta = 8\sin\frac{\theta}{2}\Big|_0^{\pi} = 8$$

31. Find the length of the *equiangular spiral* $r = e^{\theta}$ for $0 \le \theta \le 2\pi$.

SOLUTION Since $f(\theta) = e^{\theta}$, by the formula for the arc length we have:

$$L = \int_0^{2\pi} \sqrt{f'(\theta)^2 + f(\theta)}\,d\theta + \int_0^{2\pi} \sqrt{\left(e^{\theta}\right)^2 + \left(e^{\theta}\right)^2}\,d\theta = \int_0^{2\pi} \sqrt{2e^{2\theta}}\,d\theta$$

$$= \sqrt{2}\int_0^{2\pi} e^{\theta}\,d\theta = \sqrt{2}e^{\theta}\Big|_0^{2\pi} = \sqrt{2}\left(e^{2\pi} - e^{0}\right) = \sqrt{2}\left(e^{2\pi} - 1\right) \approx 755.9$$

In Exercises 33–36, express the length of the curve as an integral but do not evaluate it.

33. $r = e^{a\theta}, \quad 0 \le \theta \le \pi$

SOLUTION We use the formula for the arc length in polar coordinates. For this curve $f(\theta) = e^{a\theta}$, $f'(\theta) = ae^{a\theta}$, hence,

$$L = \int_0^{\pi} \sqrt{f(\theta)^2 + f'(\theta)^2}\,d\theta = \int_0^{\pi} \sqrt{e^{2a\theta} + a^2e^{2a\theta}}\,d\theta = \int_0^{\pi} \sqrt{1+a^2}e^{a\theta}\,d\theta$$

35. $r = (2 - \cos\theta)^{-1}, \quad 0 \le \theta \le 2\pi$

SOLUTION We have $f(\theta) = (2-\cos\theta)^{-1}$, $f'(\theta) = -(2-\cos\theta)^{-2}\sin\theta$, hence,

$$\sqrt{f^2(\theta) + f'(\theta)^2} = \sqrt{(2-\cos\theta)^{-2} + (2-\cos\theta)^{-4}\sin^2\theta} = \sqrt{(2-\cos\theta)^{-4}\left((2-\cos\theta)^2 + \sin^2\theta\right)}$$

$$= (2-\cos\theta)^{-2}\sqrt{4 - 4\cos\theta + \cos^2\theta + \sin^2\theta} = (2-\cos\theta)^{-2}\sqrt{5-4\cos\theta}$$

Using the integral for the arc length we get

$$L = \int_0^{2\pi} \sqrt{5-4\cos\theta}(2-\cos\theta)^{-2}\,d\theta.$$

Further Insights and Challenges

37. Suppose that the polar coordinates of a moving particle at time t are $(r(t), \theta(t))$. Prove that the particle's speed is equal to $\sqrt{(dr/dt)^2 + r^2(d\theta/dt)^2}$.

SOLUTION The speed of the particle in rectangular coordinates is:

$$\frac{ds}{dt} = \sqrt{x'(t)^2 + y'(t)^2} \tag{1}$$

We need to express the speed in polar coordinates. The x and y coordinates of the moving particles as functions of t are

$$x(t) = r(t)\cos\theta(t), \qquad y(t) = r(t)\sin\theta(t)$$

We differentiate $x(t)$ and $y(t)$, using the Product Rule for differentiation. We obtain (omitting the independent variable t)

$$x' = r'\cos\theta - r(\sin\theta)\,\theta'$$

$$y' = r'\sin\theta - r(\cos\theta)\,\theta'$$

Hence,

$$x'^2 + y'^2 = \left(r'\cos\theta - r\theta'\sin\theta\right)^2 + \left(r'\sin\theta + r\theta'\cos\theta\right)^2$$

$$= r'^2\cos^2\theta - 2r'r\theta'\cos\theta\sin\theta + r^2\theta'^2\sin^2\theta + r'^2\sin^2\theta + 2r'r\theta'\sin^2\theta\cos\theta + r^2\theta'^2\cos^2\theta$$

$$= r'^2\left(\cos^2\theta + \sin^2\theta\right) + r^2\theta'^2\left(\sin^2\theta + \cos^2\theta\right) = r'^2 + r^2\theta'^2 \tag{2}$$

Substituting (2) into (1) we get

$$\frac{ds}{dt} = \sqrt{r'^2 + r^2\theta'^2} = \sqrt{\left(\frac{dr}{dt}\right)^2 + r^2\left(\frac{d\theta}{dt}\right)^2}$$

12.5 Conic Sections (ET Section 11.5)

Preliminary Questions

1. Which of the following equations defines an ellipse? Which does not define a conic section?

(a) $4x^2 - 9y^2 = 12$ **(b)** $-4x + 9y^2 = 0$

(c) $4y^2 + 9x^2 = 12$ **(d)** $4x^3 + 9y^3 = 12$

SOLUTION

(a) This is the equation of the hyperbola $\left(\frac{x}{\sqrt{3}}\right)^2 - \left(\frac{y}{\frac{2}{\sqrt{3}}}\right)^2 = 1$, which is a conic section.

(b) The equation $-4x + 9y^2 = 0$ can be rewritten as $x = \frac{9}{4}y^2$, which defines a parabola. This is a conic section.

(c) The equation $4y^2 + 9x^2 = 12$ can be rewritten in the form $\left(\frac{y}{\sqrt{3}}\right)^2 + \left(\frac{x}{\frac{2}{\sqrt{3}}}\right)^2 = 1$, hence it is the equation of an ellipse, which is a conic section.

(d) This is not the equation of a conic section, since it is not an equation of degree two in x and y.

2. For which conic sections do the vertices lie between the foci?

SOLUTION If the vertices lie between the foci, the conic section is a hyperbola.

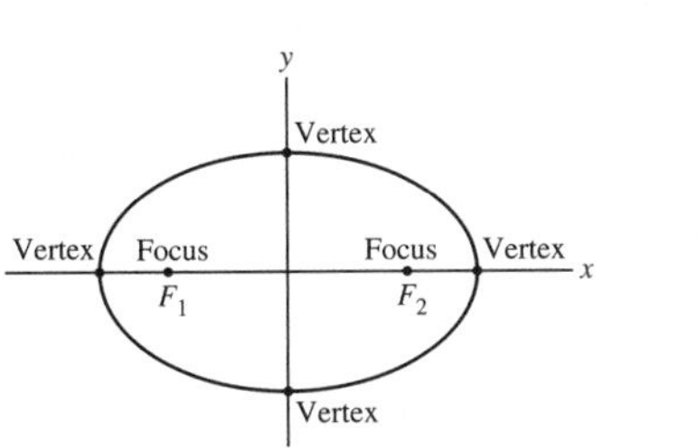

ellipse: foci between vertices

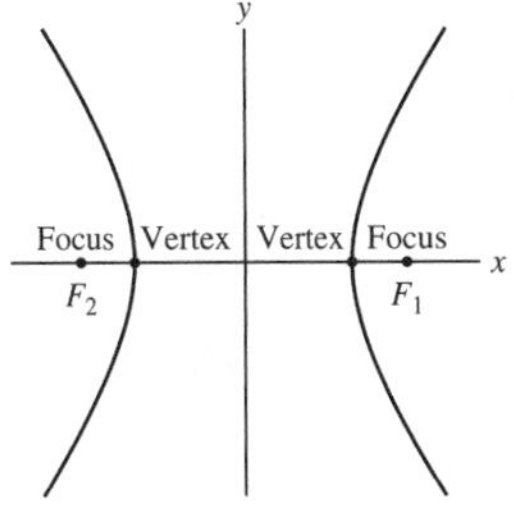

hyperbola: vertices between foci

3. What are the foci of $\left(\frac{x}{a}\right)^2 + \left(\frac{y}{b}\right)^2 = 1$ if $a < b$?

SOLUTION If $a < b$ the foci of the ellipse $\left(\frac{x}{a}\right)^2 + \left(\frac{y}{b}\right)^2 = 1$ are at the points $(0, c)$ and $(0, -c)$ on the y-axis, where $c = \sqrt{b^2 - a^2}$.

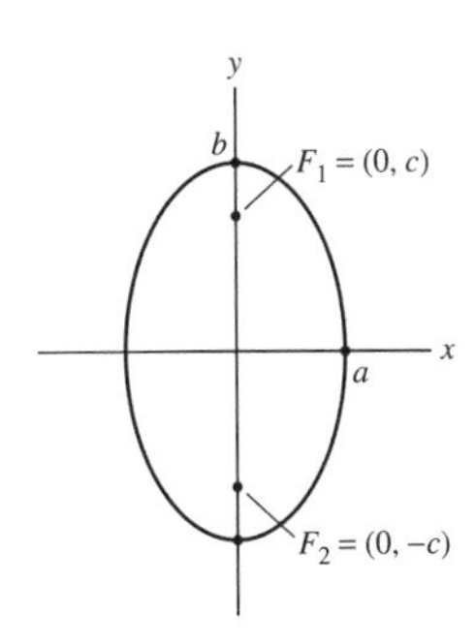

$\left(\frac{x}{a}\right)^2 + \left(\frac{y}{b}\right)^2 = 1$; $a < b$

4. For a hyperbola in standard position, the set of points equidistant from the foci is the y-axis. Use the definition $PF_1 - PF_2 = \pm K$ to explain why the hyperbola does not intersect the y-axis.

SOLUTION The points on the hyperbola are the point such that the difference of the distances to the two foci is $\pm k$, for some constant $k > 0$. For the points on the y-axis, this difference is zero, hence they are not on the hyperbola. Therefore, the hyperbola does not intersect the y-axis.

5. What is the geometric interpretation of the quantity $\frac{b}{a}$ in the equation of a hyperbola in standard position?

SOLUTION The vertices, i.e., the points where the focal axis intersects the hyperbola, are at the points $(a, 0)$ and $(-a, 0)$. The values $\pm\frac{b}{a}$ are the slopes of the two asymptotes of the hyperbola.

Hyperbola in standard position

Exercises

In Exercises 1–8, find the vertices and foci of the conic section.

1. $\left(\frac{x}{9}\right)^2 + \left(\frac{y}{4}\right)^2 = 1$

SOLUTION This is an ellipse in standard position with $a = 9$ and $b = 4$. Hence, $c = \sqrt{9^2 - 4^2} = \sqrt{65} \approx 8.06$. The foci are at $F_1 = (-8.06, 0)$ and $F_2 = (8.06, 0)$, and the vertices are $(9, 0)$, $(-9, 0)$, $(0, 4)$, $(0, -4)$.

3. $\frac{x^2}{9} + \frac{y^2}{4} = 1$

SOLUTION Writing the equation in the from $\left(\frac{x}{3}\right)^2 + \left(\frac{y}{2}\right)^2 = 1$ we get an ellipse with $a = 3$ and $b = 2$. Hence $c = \sqrt{3^2 - 2^2} = \sqrt{5} \approx 2.24$. The foci are at $F_1 = (-2.24, 0)$ and $F_2 = (2.24, 0)$ and the vertices are $(3, 0)$, $(-3, 0)$, $(0, 2)$, $(0, -2)$.

5. $\left(\frac{x}{4}\right)^2 - \left(\frac{y}{9}\right)^2 = 1$

SOLUTION For this hyperbola $a = 4$, $b = 9$ and $c = \sqrt{4^2 + 9^2} \approx 9.85$. The foci are at $F_1 = (9.85, 0)$ and $F_2 = (-9.85, 0)$ and the vertices are $A = (4, 0)$ and $A' = (-4, 0)$.

7. $\left(\frac{x-3}{7}\right)^2 - \left(\frac{y+1}{4}\right)^2 = 1$

SOLUTION We first consider the hyperbola $\left(\frac{x}{7}\right)^2 - \left(\frac{y}{4}\right)^2 = 1$. For this hyperbola, $a = 7$, $b = 4$ and $c = \sqrt{7^2 + 4^2} \approx 8.06$. Hence, the foci are at $(8.06, 0)$ and $(-8.06, 0)$ and the vertices are at $(7, 0)$ and $(-7, 0)$. Since the given hyperbola is obtained by translating the center of the hyperbola $\left(\frac{x}{7}\right)^2 - \left(\frac{y}{4}\right)^2 = 1$ to the point $(3, -1)$, the foci are at $F_1 = (8.06 + 3, 0 - 1) = (11.06, -1)$ and $F_2 = (-8.06 + 3, 0 - 1) = (-5.06, -1)$ and the vertices are $A = (7 + 3, 0 - 1) = (10, -1)$ and $A' = (-7 + 3, 0 - 1) = (-4, -1)$.

In Exercises 9–12, consider the ellipse

$$\left(\frac{x-12}{5}\right)^2 + \left(\frac{y-9}{7}\right)^2 = 1$$

Find the equation of the translated ellipse.

9. Translated so that its center is at the origin

SOLUTION Recall that the equation

$$\frac{(x-h)^2}{a^2} + \frac{(y-k)^2}{b^2} = 1$$

describes an ellipse with center (h, k). Thus, for our ellipse to be located at the origin, it must have equation

$$\frac{x^2}{5^2} + \frac{y^2}{7^2} = 1$$

11. Translated three units down

SOLUTION Recall that the equation

$$\frac{(x-h)^2}{a^2} + \frac{(y-k)^2}{b^2} = 1$$

describes an ellipse with center (h, k). Thus, for our ellipse to be moved 3 units down, it must have equation

$$\frac{(x-12)^2}{5^2} + \frac{(y-6)^2}{7^2} = 1$$

In Exercises 13–16, find the equation of the ellipse with the given properties.

13. Vertices at $(\pm 9, 0)$ and $(0, \pm 16)$

SOLUTION The equation is $\left(\frac{x}{a}\right)^2 + \left(\frac{y}{b}\right)^2 = 1$ with $a = 9$ and $b = 16$. That is,

$$\left(\frac{x}{9}\right)^2 + \left(\frac{y}{16}\right)^2 = 1$$

15. Foci $(0, \pm 6)$ and two vertices at $(\pm 4, 0)$

SOLUTION The equation of the ellipse is $\left(\frac{x}{a}\right)^2 + \left(\frac{y}{b}\right)^2 = 1$. The foci are $(0, \pm c)$ on the y-axis with $c = 6$, and two vertices are at $(\pm a, 0)$ with $a = 4$. We use the relation $c = \sqrt{b^2 - a^2}$ to find b:

$$b = \sqrt{a^2 + c^2} = \sqrt{4^2 + 6^2} \approx 7.2$$

Therefore the equation is

$$\left(\frac{x}{4}\right)^2 + \left(\frac{y}{7.2}\right)^2 = 1.$$

In Exercises 17–22, find the equation of the hyperbola with the given properties.

17. Vertices $(\pm 3, 0)$ and foci at $(\pm 5, 0)$

SOLUTION The equation is $\left(\frac{x}{a}\right)^2 - \left(\frac{y}{b}\right)^2 = 1$. The vertices are $(\pm a, 0)$ with $a = 3$ and the foci $(\pm c, 0)$ with $c = 5$. We use the relation $c = \sqrt{a^2 + b^2}$ to find b:

$$b = \sqrt{c^2 - a^2} = \sqrt{5^2 - 3^2} = \sqrt{16} = 4$$

Therefore, the equation of the hyperbola is

$$\left(\frac{x}{3}\right)^2 - \left(\frac{y}{4}\right)^2 = 1.$$

19. Vertices $(\pm 4, 0)$ and asymptotes $y = \pm 3x$

SOLUTION The equation is $\left(\frac{x}{a}\right)^2 - \left(\frac{y}{b}\right)^2 = 1$. The vertices are at $(\pm a, 0)$ with $a = 4$ and the asymptotes are $y = \pm \frac{b}{a}x$ with $\frac{b}{a} = 3$. Hence $b = 3a = 3 \cdot 4 = 12$, and the equation of the hyperbola is

$$\left(\frac{x}{4}\right)^2 - \left(\frac{y}{12}\right)^2 = 1$$

21. Vertices $(0, -5)$, $(0, 4)$ and foci $(0, -8)$, $(0, 7)$

SOLUTION The center of the parabola is at $\frac{-5+4}{2} = -0.5$. Thus, the equation has the form $\left(\frac{y+0.5}{b}\right)^2 - \left(\frac{x}{a}\right)^2 = 1$. Since $b = 4.5$ and $c = 7.5$, we quickly find that $a = 6$, giving us the equation

$$\left(\frac{y+0.5}{4.5}\right)^2 - \left(\frac{x}{6}\right)^2 = 1.$$

In Exercises 23–30, find the equation of the parabola with the given properties.

23. Vertex $(0, 0)$, focus $(2, 0)$

SOLUTION Since the focus is on the x-axis rather than on the y-axis, the equation is $x = \frac{y^2}{4c}$. Since $c = 2$ we get $x = \frac{y^2}{8}$.

25. Vertex $(0, 0)$, directrix $y = -5$

SOLUTION The equation is $y = \frac{1}{4c}x^2$. The directrix is $y = -c$ with $c = 5$, hence $y = \frac{1}{20}x^2$.

27. Focus $(0, 4)$, directrix $y = -4$

SOLUTION The focus is $(0, c)$ with $c = 4$ and the directrix is $y = -c$ with $c = 4$, hence the equation of the parabola is

$$y = \frac{1}{4c}x^2 = \frac{x^2}{16}.$$

29. Focus $(2, 0)$, directrix $x = -2$

SOLUTION The focus is on the x-axis rather than on the y-axis and the directrix is a vertical line rather than horizontal as in the parabola in standard position. Therefore, the equation of the parabola is obtained by interchanging x and y in $y = \frac{1}{4c}x^2$. Also, by the given information $c = 2$. Hence, $x = \frac{1}{4c}y^2 = \frac{1}{4\cdot 2}y^2$ or $x = \frac{y^2}{8}$.

In Exercises 31–40, find the vertices, foci, axes, center (if an ellipse or a hyperbola) and asymptotes (if a hyperbola) of the conic section.

31. $x^2 + 4y^2 = 16$

SOLUTION We first divide the equation by 16 to convert it to the equation in standard form:

$$\frac{x^2}{16} + \frac{4y^2}{16} = 1 \Rightarrow \frac{x^2}{16} + \frac{y^2}{4} = 1 \Rightarrow \left(\frac{x}{4}\right)^2 + \left(\frac{y}{2}\right)^2 = 1$$

For this ellipse, $a = 4$ and $b = 2$ hence $c = \sqrt{4^2 - 2^2} = \sqrt{12} \approx 3.5$. Since $a > b$ we have:

- The vertices are at $(\pm 4, 0)$, $(0, \pm 2)$.
- The foci are $F_1 = (-3.5, 0)$ and $F_2 = (3.5, 0)$.
- The focal axis is the x-axis and the conjugate axis is the y-axis.
- The ellipse is centered at the origin.

33. $4x^2 + y^2 = 16$

SOLUTION We divide the equation by 16 to rewrite it in the standard form:

$$\frac{4x^2}{16} + \frac{y^2}{16} = 1 \Rightarrow \frac{x^2}{4} + \frac{y^2}{16} = 1 \Rightarrow \left(\frac{x}{2}\right)^2 + \left(\frac{y}{4}\right)^2 = 1$$

This is the equation of an ellipse with $a = 2$, $b = 4$. Since $a < b$ the focal axis is the y-axis. Also, $c = \sqrt{4^2 - 2^2} = \sqrt{12} \approx 3.5$. We get:

- The vertices are at $(\pm 2, 0)$, $(0, \pm 4)$.
- The foci are $(0, \pm 3.5)$.
- The focal axis is the y-axis and the conjugate axis is the x-axis.
- The center is at the origin.

35. $4x^2 - 3y^2 + 8x + 30y = 215$

SOLUTION Since there is no cross term, we complete the square of the terms involving x and y separately:

$$4x^2 - 3y^2 + 8x + 30y = 4\left(x^2 + 2x\right) - 3\left(y^2 - 10y\right) = 4(x+1)^2 - 4 - 3(y-5)^2 + 75 = 215$$

Hence,

$$4(x+1)^2 - 3(y-5)^2 = 144$$

$$\frac{4(x+1)^2}{144} - \frac{3(y-5)^2}{144} = 1$$

$$\left(\frac{x+1}{6}\right)^2 - \left(\frac{y-5}{\sqrt{48}}\right)^2 = 1$$

This is the equation of the hyperbola obtained by translating the hyperbola $\left(\frac{x}{6}\right)^2 - \left(\frac{y}{\sqrt{48}}\right)^2 = 1$ one unit to the left and five units upwards. Since $a = 6$, $b = \sqrt{48}$, we have $c = \sqrt{36 + 48} = \sqrt{84} \sim 9.2$. We obtain the following table:

	Standard position	Translated hyperbola
vertices	$(6, 0), (-6, 0)$	$(5, 5), (-7, 5)$
foci	$(\pm 9.2, 0)$	$(8.2, 5), (-10.2, 5)$
focal axis	The x-axis	$y = 5$
conjugate axis	The y-axis	$x = -1$
center	The origin	$(-1, 5)$
asymptotes	$y = \pm 1.15x$	$y = -1.15x + 3.85$ $y = 1.15x + 6.15$

37. $y = 4(x-4)^2$

SOLUTION By Exercise 36, the parabola $y = 4x^2$ has the vertex at the origin, the focus at $\left(0, \frac{1}{16}\right)$ and its axis is the y-axis. Our parabola is a translation of the standard parabola four units to the right. Hence its vertex is at $(4, 0)$, the focus is at $\left(4, \frac{1}{16}\right)$ and its axis is the vertical line $x = 4$.

39. $4x^2 + 25y^2 - 8x - 10y = 20$

SOLUTION Since there are no cross terms this conic section is obtained by translating a conic section in standard position. To identify the conic section we complete the square of the terms involving x and y separately:

$$
\begin{aligned}
4x^2 + 25y^2 - 8x - 10y &= 4\left(x^2 - 2x\right) + 25\left(y^2 - \frac{2}{5}y\right) \\
&= 4(x-1)^2 - 4 + 25\left(y - \frac{1}{5}\right)^2 - 1 \\
&= 4(x-1)^2 + 25\left(y - \frac{1}{5}\right)^2 - 5 = 20
\end{aligned}
$$

Hence,

$$
\begin{aligned}
l4(x-1)^2 + 25\left(y - \frac{1}{5}\right)^2 &= 25 \\
\frac{4}{25}(x-1)^2 + \left(y - \frac{1}{5}\right)^2 &= 1 \\
\left(\frac{x-1}{\frac{5}{2}}\right)^2 + \left(y - \frac{1}{5}\right)^2 &= 1
\end{aligned}
$$

This is the equation of the ellipse obtained by translating the ellipse in standard position $\left(\frac{x}{\frac{5}{2}}\right)^2 + y^2 = 1$ one unit to the right and $\frac{1}{5}$ unit upward. Since $a = \frac{5}{2}$, $b = 1$ we have $c = \sqrt{\left(\frac{5}{2}\right)^2 - 1} \approx 2.3$, so we obtain the following table:

	Standard position	Translated ellipse
Vertices	$\left(\pm\frac{5}{2}, 0\right)$, $(0, \pm 1)$	$\left(1 \pm \frac{5}{2}, \frac{1}{5}\right)$, $\left(1, \frac{1}{5} \pm 1\right)$
Foci	$(-2.3, 0)$, $(2.3, 0)$	$\left(-1.3, \frac{1}{5}\right)$, $\left(3.3, \frac{1}{5}\right)$
Focal axis	The x-axis	$y = \frac{1}{5}$
Conjugate axis	The y-axis	$x = 1$
Center	The origin	$\left(1, \frac{1}{5}\right)$

In Exercises 41–44, use the Discriminant Test to determine the type of the conic section defined by the equation. You may assume that the equation is nondegenerate. Plot the curve if you have a computer algebra system.

41. $4x^2 + 5xy + 7y^2 = 24$

SOLUTION Here, $D = 25 - 4 \cdot 4 \cdot 7 = -87$, so the conic section is an ellipse.

43. $2x^2 - 8xy - 3y^2 - 4 = 0$

SOLUTION Here, $D = 64 - 4 \cdot 2 \cdot (-3) = 88$, giving us a hyperbola.

45. Show that $\frac{b}{a} = \sqrt{1 - e^2}$ for a standard ellipse of eccentricity e.

SOLUTION By the definition of eccentricity:

$$e = \frac{c}{a} \tag{1}$$

For the ellipse in standard position, $c = \sqrt{a^2 - b^2}$. Substituting into (1) and simplifying yields

$$e = \frac{\sqrt{a^2 - b^2}}{a} = \sqrt{\frac{a^2 - b^2}{a^2}} = \sqrt{1 - \left(\frac{b}{a}\right)^2}$$

We square the two sides and solve for $\frac{b}{a}$:

$$e^2 = 1 - \left(\frac{b}{a}\right)^2 \Rightarrow \left(\frac{b}{a}\right)^2 = 1 - e^2 \Rightarrow \frac{b}{a} = \sqrt{1 - e^2}$$

47. Explain why the dots in Figure 22 lie on a parabola. Where are the focus and directrix located?

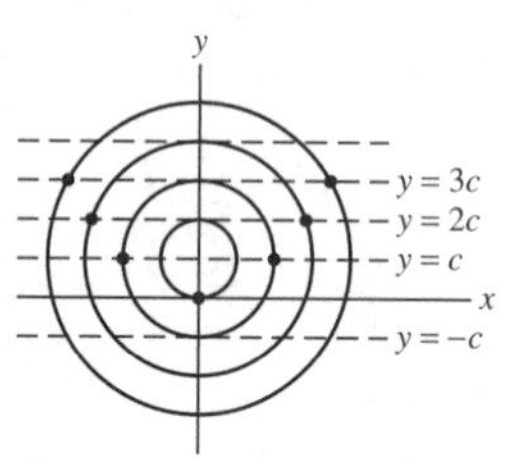

FIGURE 22

SOLUTION All the circles are centered at $(0, c)$ and the kth circle has radius kc. Hence the indicated point P_k on the kth circle has a distance kc from the point $F = (0, c)$. The point P_k also has distance kc from the line $y = -c$. That is, the indicated point on each circle is equidistant from the point $F = (0, c)$ and the line $y = -c$, hence it lies on the parabola with focus at $F = (0, c)$ and directrix $y = -c$.

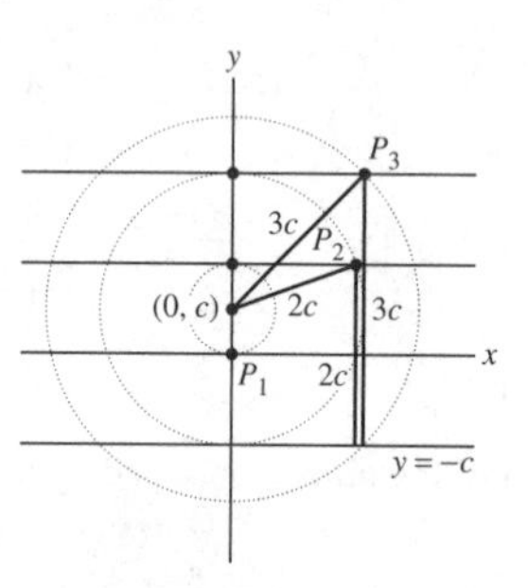

49. Kepler's First Law states that the orbits of the planets around the sun are ellipses with the sun at one focus. The orbit of Pluto has an eccentricity of approximately $e = 0.25$ and the **perihelion** (closest distance to the sun) of Pluto's orbit is approximately 2.7 billion miles. Find the **aphelion** (farthest distance from the sun).

SOLUTION We define an xy-coordinate system so that the orbit is an ellipse in standard position, as shown in the figure.

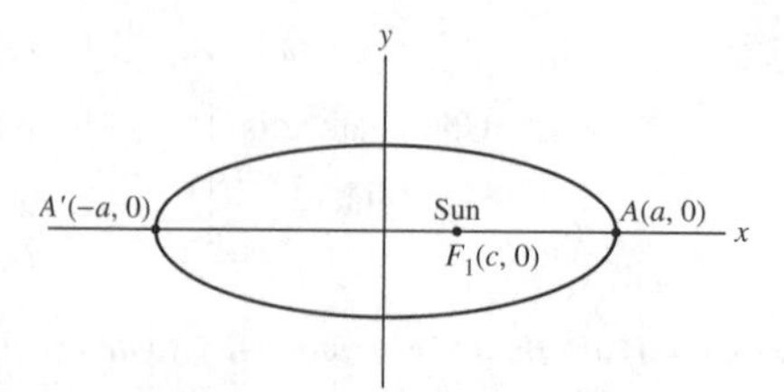

The aphelion is the length of $\overline{A'F_1}$, that is $a + c$. By the given data, we have

$$0.25 = e = \frac{c}{a} \Rightarrow c = 0.25a$$

$$a - c = 2.7 \Rightarrow c = a - 2.7$$

Equating the two expressions for c we get

$$0.25a = a - 2.7$$

$$0.75a = 2.7 \Rightarrow a = \frac{2.7}{0.75} = 3.6, \; c = 3.6 - 2.7 = 0.9$$

The aphelion is thus

$$\overline{A'F_0} = a + c = 3.6 + 0.9 = 4.5 \text{ billion miles.}$$

In Exercises 51–54, find the polar equation of the conic with given eccentricity and directrix.

51. $e = \frac{1}{2}, \quad x = 3$

SOLUTION Substituting $e = \frac{1}{2}$ and $d = 3$ in the polar equation of a conic section we obtain

$$r = \frac{ed}{1 + e\cos\theta} = \frac{\frac{1}{2} \cdot 3}{1 + \frac{1}{2}\cos\theta} = \frac{3}{2 + \cos\theta} \Rightarrow r = \frac{3}{2 + \cos\theta}$$

53. $e = 1, \quad x = 4$

SOLUTION We substitute $e = 1$ and $d = 4$ in the polar equation of a conic section to obtain

$$r = \frac{ed}{1 + e\cos\theta} = \frac{1 \cdot 4}{1 + 1 \cdot \cos\theta} = \frac{4}{1 + \cos\theta} \Rightarrow r = \frac{4}{1 + \cos\theta}$$

In Exercises 55–58, identify the type of conic, the eccentricity, and the equation of the directrix.

55. $r = \dfrac{8}{1 + 4\cos\theta}$

SOLUTION Matching with the polar equation $r = \frac{ed}{1+e\cos\theta}$ we get $ed = 8$ and $e = 4$ yielding $d = 2$. Since $e > 1$, the conic section is a hyperbola, having eccentricity $e = 4$ and directrix $x = 2$ (referring to the focus-directrix definition (11)).

57. $r = \dfrac{8}{4 + 3\cos\theta}$

SOLUTION We first rewrite the equation in the form $r = \frac{ed}{1+e\cos\theta}$, obtaining

$$r = \frac{2}{1 + \frac{3}{4}\cos\theta}$$

Hence, $ed = 2$ and $e = \frac{3}{4}$ yielding $d = \frac{8}{3}$. Since $e < 1$, the conic section is an ellipse, having eccentricity $e = \frac{3}{4}$ and directrix $x = \frac{8}{3}$.

59. Show that $r = f_1(\theta)$ and $r = f_2(\theta)$ define the same curves in polar coordinates if $f_1(\theta) = -f_2(\theta + \pi)$, and use this to show that the following define the same conic section:

$$r = \frac{de}{1 - e\cos\theta}, \qquad r = \frac{-de}{1 + e\cos\theta}$$

SOLUTION The curve $r = f_2(\theta)$ can be parametrized using θ as a parameter:

$$x = f_2(\theta)\cos\theta \qquad (1)$$
$$y = f_2(\theta)\sin\theta$$

Using the identities $\cos(\theta + \pi) = -\cos\theta$ and $\sin(\theta + \pi) = -\sin\theta$, we obtain the following parametrization for $r = f_1(\theta)$:

$$\begin{aligned} x &= f_1(\theta)\cos\theta = -f_2(\theta + \pi)\cos\theta \\ &= f_2(\theta + \pi)(-\cos\theta) = f_2(\theta + \pi)\cos(\theta + \pi) \\ y &= f_1(\theta)\sin\theta = -f_2(\theta + \pi)\sin\theta \\ &= f_2(\theta + \pi)(-\sin\theta) = f_2(\theta + \pi)\sin(\theta + \pi) \end{aligned}$$

Using $t = \theta + \pi$ as the parameter we get

$$x = f_2(t)\cos t \qquad (2)$$

$$y = f_2(t)\sin t$$

The parametrizations (1) and (2) define the same curve. We now consider the polar equations:

$$r = \frac{de}{1 - e\cos\theta} = f_1(\theta), \quad r = \frac{-de}{1 + e\cos\theta} = f_2(\theta) \tag{3}$$

The following equality holds:

$$-f_2(\theta + \pi) = -\frac{-de}{1 + e\cos(\theta + \pi)} = \frac{de}{1 - e\cos\theta} = f_1(\theta)$$

We conclude that the polar equations in (3) define the same conic section.

61. Find the equation of the ellipse $r = \dfrac{4}{2 + \cos\theta}$ in rectangular coordinates.

SOLUTION We use the polar equation of the ellipse with focus at the origin and directrix at $x = d$:

$$r = \frac{ed}{1 + e\cos\theta}$$

Dividing the numerator and denominator of the given equation by 2 yields

$$r = \frac{2}{1 + \frac{1}{2}\cos\theta}.$$

We identify the values $e = \frac{1}{2}$ and $ed = 2$. Hence, $d = \frac{2}{e} = 4$. The focus-directrix of the ellipse is $\overline{PO} = eD$. That is, $\sqrt{x^2 + y^2} = \frac{1}{2}|4 - x|$.

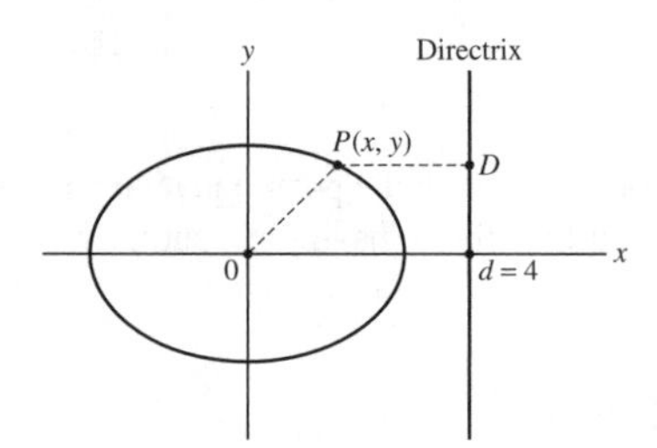

We square the two sides, simplify, and complete the square to obtain

$$4\left(x^2 + y^2\right) = 16 - 8x + x^2$$

$$3x^2 + 8x + 4y^2 = 16$$

$$3\left(x^2 + \frac{8}{3}x\right) + 4y^2 = 16$$

$$3\left(x + \frac{4}{3}\right)^2 + 4y^2 = \frac{64}{3}$$

$$\frac{\left(x + \frac{4}{3}\right)^2}{\frac{64}{9}} + \frac{y^2}{\frac{16}{3}} = 1 \Rightarrow \left(\frac{x + \frac{4}{3}}{\frac{8}{3}}\right)^2 + \left(\frac{y}{\frac{4}{\sqrt{3}}}\right)^2 = 1.$$

63. Let $e > 1$. Show that the vertices of the hyperbola $r = \dfrac{de}{1 + e\cos\theta}$ have x-coordinates $\dfrac{ed}{e+1}$ and $\dfrac{ed}{e-1}$.

SOLUTION Since the focus is at the origin and the hyperbola is to the right (see figure), the two vertices have positive x coordinates. The corresponding values of θ at the vertices are $\theta = 0$ and $\theta = \pi$. Hence, since $e > 1$ we obtain

$$x_A = |r(0)| = \left|\frac{de}{1 + e\cos 0}\right| = \frac{de}{1 + e}$$

$$x_{A'} = |r(\pi)| = \left|\frac{de}{1 + e\cos\pi}\right| = \left|\frac{de}{1 - e}\right| = \frac{de}{e - 1}$$

Further Insights and Challenges

65. Verify Theorem 4 in the case $0 < e < 1$. *Hint:* Repeat the proof of Theorem 4, but set $c = d/(e^{-2} - 1)$.

SOLUTION We follow closely the proof of Theorem 4 in the book, which covered the case $e > 1$. This time, for $0 < e < 1$, we prove that $PF = ePD$ defines an ellipse. We choose our coordinate axes so that the focus F lies on the x-axis with coordinates $F = (c, 0)$ and so that the directrix is vertical, lying to the right of F at a distance d from F. As suggested by the hint, we set $c = \frac{d}{e^{-2}-1}$, but since we are working towards an ellipse, we will also need to let $b = \sqrt{a^2 - c^2}$ as opposed to the $\sqrt{c^2 - a^2}$ from the original proof of Theorem 4. Here's the complete list of definitions:

$$c = \frac{d}{e^{-2} - 1}, \quad a = \frac{c}{e}, \quad b = \sqrt{a^2 - c^2}$$

The directrix is the line

$$x = c + d = c + c(e^{-2} - 1) = ce^{-2} = \frac{a}{e}$$

Now, the equation

$$PF = e \cdot PD$$

for the points $P = (x, y)$, $F = (c, 0)$, and $D = (a/e, y)$ becomes

$$\sqrt{(x - c)^2 + y^2} = e \cdot \sqrt{(x - (a/e))^2}$$

Returning to the proof of Theorem 4, we see that this is the same equation that appears in the middle of the proof of the Theorem. As seen there, this equation can be transformed into

$$\frac{x^2}{a^2} - \frac{y^2}{a^2(e^2 - 1)} = 1$$

and this is equivalent to

$$\frac{x^2}{a^2} + \frac{y^2}{a^2(1 - e^2)} = 1$$

Since $a^2(1 - e^2) = a^2 - a^2e^2 = a^2 - c^2 = b^2$, then we obtain the equation of the ellipse

$$\frac{x^2}{a^2} + \frac{y^2}{b^2} = 1$$

Reflective Property of the Ellipse *In Exercises 67–69, we prove that the focal radii at a point on an ellipse make equal angles with the tangent line. Let $P = (x_0, y_0)$ be a point on the ellipse $\left(\frac{x}{a}\right)^2 + \left(\frac{y}{b}\right)^2 = 1$ $(a > b)$ with foci $F_1 = (-c, 0)$ and $F_2 = (c, 0)$, and eccentricity e (Figure 24).*

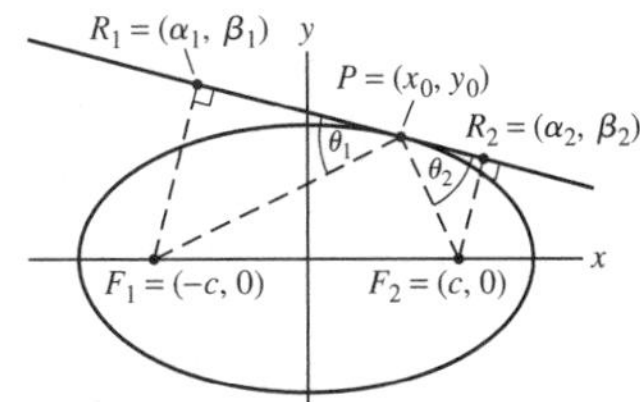

FIGURE 24 The ellipse $\left(\frac{x}{a}\right)^2 + \left(\frac{y}{b}\right)^2 = 1$.

67. Show that $PF_1 = a + x_0e$ and $PF_2 = a - x_0e$. *Hints:*

(a) Show that $PF_1{}^2 - PF_2{}^2 = 4x_0c$.

(b) Divide the previous relation by $PF_1 + PF_2 = 2a$, and conclude that $PF_1 - PF_2 = 2x_0e$.

SOLUTION Using the distance formula we have

$$\overline{PF_1}^2 = (x_0 + c)^2 + y^2; \quad \overline{PF_2}^2 = (x_0 - c)^2 + y^2$$

Hence,

$$\overline{PF_1}^2 - \overline{PF_2}^2 = (x_0 + c)^2 + y^2 - (x_0 - c)^2 - y^2$$

$$= (x_0 + c)^2 - (x_0 - c)^2$$
$$= {x_0}^2 + 2x_0c + c^2 - {x_0}^2 + 2x_0c - c^2 = 4x_0c$$

That is, $\overline{PF_1} + \overline{PF_2} = 4x_0c$. We use the identity $u^2 - v^2 = (u - v)(u + v)$ to write this as

$$\left(\overline{PF_1} - \overline{PF_2}\right)\left(\overline{PF_1} + \overline{PF_2}\right) = 4x_0c \qquad (1)$$

Since P lies on the ellipse $\left(\frac{x}{a}\right)^2 + \left(\frac{y}{b}\right)^2 = 1$ we have

$$\overline{PF_1} + \overline{PF_2} = 2a \qquad (2)$$

Substituting in (1) gives

$$\left(\overline{PF_1} - \overline{PF_2}\right) \cdot 2a = 4x_0c$$

We divide by a and use the eccentricity $e = \frac{c}{a}$ to obtain

$$\overline{PF_1} - \overline{PF_2} = 2x_0e$$

69. Define R_1 and R_2 as in the figure, so that $\overline{F_1R_1}$ and $\overline{F_2R_2}$ are perpendicular to the tangent line.

(a) Show that $\dfrac{\alpha_1 + c}{\beta_1} = \dfrac{\alpha_2 - c}{\beta_2} = \dfrac{A}{B}$.

(b) Use (a) and the distance formula to show that

$$\frac{F_1R_1}{F_2R_2} = \frac{\beta_1}{\beta_2}$$

(c) Solve for β_1 and β_2:

$$\beta_1 = \frac{B(1 + Ac)}{A^2 + B^2}, \qquad \beta_2 = \frac{B(1 - Ac)}{A^2 + B^2}$$

(d) Show that $\dfrac{F_1R_1}{F_2R_2} = \dfrac{PF_1}{PF_2}$. Conclude that $\theta_1 = \theta_2$.

SOLUTION

(a) Since $R_1 = (\alpha_1, \beta_1)$ and $R_2 = (\alpha_2, \beta_2)$ lie on the tangent line at P, that is on the line $Ax + By = 1$, we have

$$A\alpha_1 + B\beta_1 = 1 \quad \text{and} \quad A\alpha_2 + B\beta_2 = 1$$

The slope of the line R_1F_1 is $\frac{\beta_1}{\alpha_1 + c}$ and it is perpendicular to the tangent line having slope $-\frac{A}{B}$. Similarly, the slope of the line R_2F_2 is $\frac{\beta_2}{\alpha_2 - c}$ and it is also perpendicular to the tangent line. Hence,

$$\frac{\alpha_1 + c}{\beta_1} = \frac{A}{B} \quad \text{and} \quad \frac{\alpha_2 - c}{\beta_2} = \frac{A}{B}.$$

(b) Using the distance formula, we have

$$\overline{R_1F_1}^2 = (\alpha_1 + c)^2 + \beta_1^2$$

Thus,

$$\overline{R_1F_1}^2 = \beta_1^2\left(\left(\frac{\alpha_1 + c}{\beta_1}\right)^2 + 1\right) \qquad (1)$$

By part (a), $\frac{\alpha_1 + c}{\beta_1} = \frac{A}{B}$. Substituting in (1) gives

$$\overline{R_1F_1}^2 = \beta_1^2\left(\frac{A^2}{B^2} + 1\right) = \beta_1^2\left(1 + B^{-2}A^2\right) \qquad (2)$$

Likewise,

$$\overline{R_2F_2}^2 = (\alpha_2 - c)^2 + {\beta_2}^2 = {\beta_2}^2\left(\left(\frac{\alpha_2 - c}{\beta_2}\right)^2 + 1\right)$$

but since $\frac{\alpha_2 - c}{\beta_2} = \frac{A}{B}$, we get that

$$\overline{R_2F_2}^2 = \beta_2^2\left(\frac{A^2}{B^2}+1\right). \tag{3}$$

Dividing, we find that

$$\frac{\overline{R_1F_1}^2}{\overline{R_2F_2}^2} = \frac{\beta_1^2}{\beta_2^2} \quad \text{so} \quad \frac{\overline{R_1F_1}}{\overline{R_2F_2}} = \frac{\beta_1}{\beta_2},$$

as desired.

(c), (d) In part (a) we show that

$$\begin{cases} A\alpha_1 + B\beta_1 = 1 \\ \dfrac{\beta_1}{\alpha_1 + c} = \dfrac{B}{A} \end{cases}$$

Solving for β_1 gives $\beta_1 = \frac{B(1+Ac)}{A^2+B^2}$. Substituting in (2) we obtain

$$\overline{R_1F_1}^2 = \frac{B^2(1+Ac)^2}{\left(A^2+B^2\right)^2}\left(1+\frac{A^2}{B^2}\right) = \frac{B^2(1+Ac)^2(A^2+B^2)}{\left(A^2+B^2\right)^2B^2} = \frac{(1+Ac)^2}{A^2+B^2}$$

Similarly solving the equations in part (a) for β_2 and using equation (3) yields

$$\begin{cases} A\alpha_2 + B\beta_2 = 1 \\ \dfrac{\beta_2}{\alpha_2 - c} = \dfrac{B}{A} \end{cases} \quad \Rightarrow \quad \beta_2 = \frac{B\,(1-Ac)}{A^2+B^2} \tag{4}$$

Using the distance formula for $\overline{R_2F_2}$ we have

$$\overline{R_2F_2}^2 = (\alpha_2 - c)^2 + \beta_2^2 = \beta_2^2\left(\left(\frac{A_2-c}{\beta_2}\right)^2+1\right)$$

Substituting $\frac{A_2-c}{\beta_2} = \frac{A}{B}$ from part (a) and β_2 in (3) we get

$$\overline{R_2F_2}^2 = \frac{B^2(1-Ac)^2}{\left(A^2+B^2\right)^2}\left(\frac{A^2}{B^2}+1\right) = \frac{B^2(1-Ac)^2\left(A^2+B^2\right)}{\left(A^2+B^2\right)^2B^2} = \frac{(1-Ac)^2}{A^2+B^2}$$

Using the expression for $\overline{R_1F_1}$ and $\overline{R_2F_2}$ obtained above, we get

$$\frac{\overline{R_1F_1}}{\overline{R_2F_2}} = \frac{\frac{1+Ac}{\sqrt{A^2+B^2}}}{\frac{1-Ac}{\sqrt{A^2+B^2}}} = \frac{1+Ac}{1-Ac}$$

Substituting $c = ea$ and $A = \frac{x_0}{a^2}$ we obtain

$$\frac{\overline{R_1F_1}}{\overline{R_2F_2}} = \frac{1+\frac{x_0ea}{a^2}}{1-\frac{x_0ea}{a^2}} = \frac{1+\frac{x_0e}{a}}{1-\frac{x_0e}{a}} = \frac{a+x_0e}{a-x_0e}$$

Now by Exercise 67, we have $\overline{PF_1} = a + x_0e$ and $\overline{PF_2} = a - x_0e$, where $P = (x_0, y_0)$. By Exercise 69, $\frac{\overline{R_1F_1}}{\overline{R_2F_2}} = \frac{a+ex_0}{a-ex_0}$. Using these results we derive the following equality:

$$\frac{\overline{PF_1}}{\overline{PF_2}} = \frac{\overline{R_1F_1}}{\overline{R_2F_2}} \Rightarrow \frac{\overline{R_1F_1}}{\overline{PF_1}} = \frac{\overline{R_2F_2}}{\overline{PF_2}}$$

By $\frac{\overline{R_1F_1}}{\overline{PF_1}} = \sin\theta_1$ and $\frac{\overline{R_2F_2}}{\overline{PF_2}} = \sin\theta_2$ we get $\sin\theta_1 = \sin\theta_2$, which implies that $\theta_1 = \theta_2$ since the two angles are acute.

71. Show that $y = \dfrac{x^2}{4c}$ is the equation of a parabola with directrix $y = -c$, focus $(0, c)$, and the vertex at the origin, as stated in Theorem 3.

SOLUTION The points $P = (x, y)$ on the parabola are equidistant from $F = (0, c)$ and the line $y = -c$.

That is, by the distance formula, we have

$$\overline{PF} = \overline{PD}$$
$$\sqrt{x^2 + (y-c)^2} = |y + c|$$

Squaring and simplifying yields

$$x^2 + (y-c)^2 = (y+c)^2$$
$$x^2 + y^2 - 2yc + c^2 = y^2 + 2yc + c^2$$
$$x^2 - 2yc = 2yc$$
$$x^2 = 4yc \Rightarrow y = \frac{x^2}{4c}$$

Thus, we showed that the points that are equidistant from the focus $F = (0, c)$ and the directrix $y = -c$ satisfy the equation $y = \frac{x^2}{4c}$.

73. If we rewrite the general equation of degree two (13) in terms of variables x' and y' that are related to x and y by equations (14) and (15), we obtain a new equation of degree two in x' and y' of the same form but with different coefficients:

$$A'x^2 + B'xy + C'y^2 + D'x + E'y + F' = 0$$

(a) Show that $B' = B\cos 2\theta + (C - A)\sin 2\theta$.

(b) Show that if $B \neq 0$, then we obtain $B' = 0$ for

$$\theta = \frac{1}{2}\cot^{-1}\frac{A-C}{B}$$

This proves that it is always possible to eliminate the cross term Bxy by rotating the axes through a suitable angle.

SOLUTION

(a) If we plug in $x = x'\cos\theta - y'\sin\theta$ and $y = x'\sin\theta + y'\cos\theta$ into the equation $Ax^2 + Bxy + Cy^2 + Dx + Ey + F = 0$, we will get a very ugly mess. Fortunately, we only care about the $x'y'$ term, so we really only need to look at the $Ax^2 + Bxy + Cy^2$ part of the formula. In fact, we only need to pull out those terms which have an $x'y'$ in them. From the Ax^2 term, after replacing x with $x = x'\cos\theta - y'\sin\theta$, we will get an $x'y'$ term of $-2Ax'y'\cos\theta\sin\theta$. Likewise, from the Bxy term we will get $Bx'y'\cos^2\theta - \sin^2\theta$, and from the Cy^2 term we get $2Cx'y'\cos\theta\sin\theta$. Adding these together, we see that the (new) B', the coefficient of the (new) $x'y'$ term, will be

$$-2A\cos\theta\sin\theta + B\cos^2\theta - \sin^2\theta + 2Cx'y'\cos\theta\sin\theta$$

which simplifies to $B\cos 2\theta + (C - A)\sin 2\theta$, as desired.

(b) Setting $B' = 0$, we get $0 = B\cos 2\theta + (C - A)\sin 2\theta$, so $B\cos 2\theta = (A - C)\sin 2\theta$, so $\cot 2\theta = \frac{A-C}{B}$, giving us $2\theta = \cot^{-1}\frac{A-C}{B}$, and thus $\theta = \frac{1}{2}\cot^{-1}\frac{A-C}{B}$.

CHAPTER REVIEW EXERCISES

1. Which of the following curves pass through the point $(1, 4)$?

(a) $c(t) = (t^2, t + 3)$

(b) $c(t) = (t^2, t - 3)$

(c) $c(t) = (t^2, 3 - t)$

(d) $c(t) = (t - 3, t^2)$

SOLUTION To check whether it passes through the point $(1, 4)$, we solve the equations $c(t) = (1, 4)$ for the given curves.

(a) Comparing the second coordinate of the curve and the point yields:

$$t + 3 = 4$$
$$t = 1$$

We substitute $t = 1$ in the first coordinate, to obtain

$$t^2 = 1^2 = 1$$

Hence the curve passes through $(1, 4)$.

(b) Comparing the second coordinate of the curve and the point yields:

$$t - 3 = 4$$
$$t = 7$$

We substitute $t = 7$ in the first coordinate to obtain

$$t^2 = 7^2 = 49 \neq 1$$

Hence the curve does not pass through $(1, 4)$.

(c) Comparing the second coordinate of the curve and the point yields

$$3 - t = 4$$
$$t = -1$$

We substitute $t = -1$ in the first coordinate, to obtain

$$t^2 = (-1)^2 = 1$$

Hence the curve passes through $(1, 4)$.

(d) Comparing the first coordinate of the curve and the point yields

$$t - 3 = 1$$
$$t = 4$$

We substitute $t = 4$ in the second coordinate, to obtain:

$$t^2 = 4^2 = 16 \neq 4$$

Hence the curve does not pass through $(1, 4)$.

3. Find parametric equations for the circle of radius 2 with center $(1, 1)$. Use the equations to find the points of intersection of the circle with the x- and the y-axes.

SOLUTION Using the standard technique for parametric equations of curves, we obtain

$$c(t) = (1 + 2\cos t, 1 + 2\sin t)$$

We compare the x coordinate of $c(t)$ to 0:

$$1 + 2\cos t = 0$$
$$\cos t = -\frac{1}{2}$$
$$t = \pm\frac{2\pi}{3}$$

Substituting in the y coordinate yields

$$1 + 2\sin\left(\pm\frac{2\pi}{3}\right) = 1 \pm 2\frac{\sqrt{3}}{2} = 1 \pm \sqrt{3}$$

Hence, the intersection points with the y-axis are $(0, 1 \pm \sqrt{3})$. We compare the y coordinate of $c(t)$ to 0:

$$1 + 2\sin t = 0$$
$$\sin t = -\frac{1}{2}$$
$$t = -\frac{\pi}{6} \quad \text{or} \quad \frac{7}{6}\pi$$

Substituting in the x coordinates yields

$$1+2\cos\left(-\frac{\pi}{6}\right)=1+2\frac{\sqrt{3}}{2}=1+\sqrt{3}$$

$$1+2\cos\left(\frac{7}{6}\pi\right)=1-2\cos\left(\frac{\pi}{6}\right)=1-2\frac{\sqrt{3}}{2}=1-\sqrt{3}$$

Hence, the intersection points with the x-axis are $(1\pm\sqrt{3},0)$.

5. Find a parametrization $c(\theta)$ of the unit circle such that $c(0)=(-1,0)$.

SOLUTION The unit circle has the parametrization

$$c(t)=(\cos t,\sin t)$$

This parametrization does not satisfy $c(0)=(-1,0)$. We replace the parameter t by a parameter θ so that $t=\theta+\alpha$, to obtain another parametrization for the circle:

$$c^*(\theta)=(\cos(\theta+\alpha),\sin(\theta+\alpha)) \tag{1}$$

We need that $c^*(0)=(1,0)$, that is,

$$c^*(0)=(\cos\alpha,\sin\alpha)=(-1,0)$$

Hence

$$\begin{aligned}\cos\alpha&=-1\\ \sin\alpha&=0\end{aligned}\quad\Rightarrow\quad\alpha=\pi$$

Substituting in (1) we obtain the following parametrization:

$$c^*(\theta)=(\cos(\theta+\pi),\sin(\theta+\pi))$$

7. Find a path $c(t)$ that traces the line $y=2x+1$ from $(1,3)$ to $(3,7)$ for $0\le t\le 1$.

SOLUTION Solution 1: By one of the examples in section 12.1, the line through $P=(1,3)$ with slope 2 has the parametrization

$$c(t)=(1+t,3+2t)$$

But this parametrization does not satisfy $c(1)=(3,7)$. We replace the parameter t by a parameter s so that $t=\alpha s+\beta$. We get

$$c^*(s)=\big(1+\alpha s+\beta,3+2(\alpha s+\beta)\big)=(\alpha s+\beta+1,2\alpha s+2\beta+3)$$

We need that $c^*(0)=(1,3)$ and $c^*(1)=(3,7)$. Hence,

$$\begin{aligned}c^*(0)&=(1+\beta,3+2\beta)=(1,3)\\ c^*(1)&=(\alpha+\beta+1,2\alpha+2\beta+3)=(3,7)\end{aligned}$$

We obtain the equations

$$\begin{aligned}1+\beta&=1\\ 3+2\beta&=3\\ \alpha+\beta+1&=3\\ 2\alpha+2\beta+3&=7\end{aligned}\quad\Rightarrow\quad\beta=0,\alpha=2$$

Substituting in (1) gives

$$c^*(s)=(2s+1,4s+3)$$

Solution 2: The segment from $(1,3)$ to $(3,7)$ has the following vector parametrization:

$$(1-t)\langle1,3\rangle+t\langle3,7\rangle=\langle1-t+3t,3(1-t)+7t\rangle=\langle1+2t,3+4t\rangle$$

The parametrization is thus

$$c(t)=(1+2t,3+4t)$$

In Exercises 9–12, express the parametric curve in the form $y = f(x)$.

9. $c(t) = (4t - 3, 10 - t)$

SOLUTION We use the given equation to express t in terms of x.

$$x = 4t - 3$$
$$4t = x + 3$$
$$t = \frac{x+3}{4}$$

Substituting in the equation of y yields

$$y = 10 - t = 10 - \frac{x+3}{4} = -\frac{x}{4} + \frac{37}{4}$$

That is,

$$y = -\frac{x}{4} + \frac{37}{4}$$

11. $c(t) = \left(3 - \frac{2}{t}, t^3 + \frac{1}{t}\right)$

SOLUTION We use the given equation to express t in terms of x:

$$x = 3 - \frac{2}{t}$$
$$\frac{2}{t} = 3 - x$$
$$t = \frac{2}{3-x}$$

Substituting in the equation of y yields

$$y = \left(\frac{2}{3-x}\right)^3 + \frac{1}{2/(3-x)} = \frac{8}{(3-x)^3} + \frac{3-x}{2}$$

13. Find all points visited twice by the path $c(t) = (t^2, \sin t)$. Plot $c(t)$ with a graphing utility.

SOLUTION For every point, if the curve passes through it twice, then its x coordinate and y coordinate are obtained twice by the functions $x(t)$, $y(t)$. We first calculate the x coordinate of these points. Since $x(t) = t^2$, every x coordinate is obtained twice—once for t and once for $-t$. We now check for which of the above x coordinates, the y coordinates are equal as well. We substitute the above condition in the formula for the y coordinates. We have $\sin t = \sin(-t)$. Since $\sin t = -\sin(-t)$ for all $t \in \mathbb{R}$, we can add it to the former equation. We obtain $2 \sin t = 0$ or $t = \pi k$ where $k \in \mathbb{Z}$. We substitute the t values in the parametric equation to find the desired points. We obtain

$$(\pi^2 k^2, 0) \quad \text{where } k \in \mathbb{Z}$$

The path $c(t)$ is shown in the following figure.

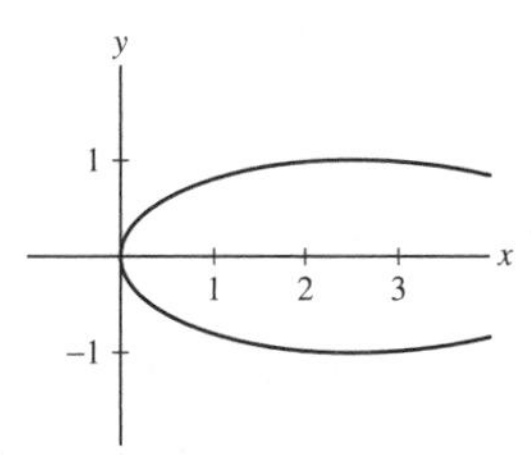

In Exercises 14–17, calculate $\frac{dy}{dx}$ at the point indicated.

15. $c(\theta) = (\tan^2 \theta, \cos \theta), \quad \theta = \frac{\pi}{4}$

SOLUTION The parametric equations are $x = \tan^2\theta$, $y = \cos\theta$. We use the theorem on the slope of the tangent line to find $\frac{dy}{dx}$:

$$\frac{dy}{dx} = \frac{\frac{dy}{d\theta}}{\frac{dx}{d\theta}} = \frac{-\sin\theta}{2\tan\theta\sec^2\theta} = -\frac{\cos^3\theta}{2}$$

We now substitute $\theta = \frac{\pi}{4}$ to obtain

$$\left.\frac{dy}{dx}\right|_{\theta=\pi/4} = -\frac{\cos^3\frac{\pi}{4}}{2} = -\frac{1}{4\sqrt{2}}$$

17. $c(t) = (\ln t, 3t^2 - t)$, $\quad P = (0, 2)$

SOLUTION The parametric equations are $x = \ln t$, $y = 3t^2 - t$. We use the theorem for the slope of the tangent line to find $\frac{dy}{dx}$:

$$\frac{dy}{dx} = \frac{\frac{dy}{dt}}{\frac{dx}{dt}} = \frac{6t-1}{\frac{1}{t}} = 6t^2 - t \tag{1}$$

We now must identify the value of t corresponding to the point $P = (0, 2)$ on the curve. We solve the following equations:

$$\begin{aligned} \ln t &= 0 \\ 3t^2 - t &= 2 \end{aligned} \quad \Rightarrow \quad t = 1$$

Substituting $t = 1$ in (1) we obtain

$$\left.\frac{dy}{dx}\right|_P = 6 \cdot 1^2 - 1 = 5$$

19. Find the points on $(t + \sin t, t - 2\sin t)$ where the tangent is vertical or horizontal.

SOLUTION We use the theorem for the slope of the tangent line to find $\frac{dy}{dx}$:

$$\frac{dy}{dx} = \frac{\frac{dy}{dt}}{\frac{dx}{dt}} = \frac{1-2\cos t}{1+\cos t}$$

We find the values of t for which the denominator is zero. We ignore the numerator, since when $1 + \cos t = 0$, $1 - 2\cos t = 3 \neq 0$.

$$\begin{aligned} 1 + \cos t &= 0 \\ \cos t &= -1 \\ t &= \pi + 2\pi k \quad \text{where } k \in \mathbb{Z} \end{aligned}$$

We now find the values of t for which the numerator is 0:

$$\begin{aligned} 1 - 2\cos t &= 0 \\ 1 &= 2\cos t \\ \frac{1}{2} &= \cos t \\ t &= \pm\frac{\pi}{3} + 2\pi k \quad \text{where } k \in \mathbb{Z} \end{aligned}$$

Note that the denominator is not zero at these points. Thus, we have vertical tangents at $t = \pi + 2\pi k$ and horizontal tangents at $t = \pm\pi/3 + 2\pi k$.

21. Find the speed at $t = \frac{\pi}{4}$ of a particle whose position at time t seconds is $c(t) = (\sin 4t, \cos 3t)$.

SOLUTION We use the parametric definition to find the speed. We obtain

$$\frac{ds}{dt} = \sqrt{((\sin 4t)')^2 + ((\cos 3t)')^2} = \sqrt{(4\cos 4t)^2 + (-3\sin 3t)^2} = \sqrt{16\cos^2 4t + 9\sin^2 3t}$$

At time $t = \frac{\pi}{4}$ the speed is

$$\left.\frac{ds}{dt}\right|_{t=\pi/4} = \sqrt{16\cos^2\pi + 9\sin^2\frac{3\pi}{4}} = \sqrt{16 + 9\cdot\frac{1}{2}} = \sqrt{20.5} \approx 4.53$$

23. Find the length of $(3e^t - 3, 4e^t + 7)$ for $0 \le t \le 1$.

SOLUTION We use the formula for arc length, to obtain

$$s = \int_0^1 \sqrt{((3e^t - 3)')^2 + ((4e^t + 7)')^2}\,dt = \int_0^1 \sqrt{(3e^t)^2 + (4e^t)^2}\,dt$$

$$= \int_0^1 \sqrt{9e^{2t} + 16e^{2t}}\,dt = \int_0^1 \sqrt{25e^{2t}}\,dt = \int_0^1 5e^t\,dt = 5e^t\Big|_0^1 = 5(e-1)$$

In Exercises 24–25, let $c(t) = (e^{-t}\cos t, e^{-t}\sin t)$.

25. Find the first positive value of t_0 such that the tangent line to $c(t_0)$ is vertical and calculate the speed at $t = t_0$.

SOLUTION The curve has a vertical tangent where $\lim_{t\to t_0} \left|\frac{dy}{dx}\right| = \infty$. We first find $\frac{dy}{dx}$ using the theorem for the slope of a tangent line:

$$\frac{dy}{dx} = \frac{\frac{dy}{dt}}{\frac{dx}{dt}} = \frac{(e^{-t}\sin t)'}{(e^{-t}\cos t)'} = \frac{-e^{-t}\sin t + e^{-t}\cos t}{-e^{-t}\cos t - e^{-t}\sin t}$$

$$= -\frac{\cos t - \sin t}{\cos t + \sin t} = \frac{\sin t - \cos t}{\sin t + \cos t}$$

We now search for t_0 such that $\lim_{t\to t_0} \left|\frac{dy}{dx}\right| = \infty$. In our case, this happens when the denominator is 0, but the numerator is not, thus:

$$\sin t_0 + \cos t_0 = 0$$

$$\cos t_0 = -\sin t_0$$

$$\cos -t_0 = \sin -t_0$$

$$-t_0 = \frac{\pi}{4} - \pi$$

$$t_0 = \frac{3}{4}\pi$$

We now use the formula for the speed, to find the speed at t_0.

$$\frac{ds}{dt} = \sqrt{((e^{-t}\sin t)')^2 + ((e^{-t}\cos t)')^2}$$

$$= \sqrt{(-e^{-t}\cos t - e^{-t}\sin t)^2 + (-e^{-t}\sin t + e^{-t}\cos t)^2}$$

$$= \sqrt{e^{-2t}(\cos t + \sin t)^2 + e^{-2t}(\cos t - \sin t)^2}$$

$$= e^{-t}\sqrt{\cos^2 t + 2\sin t\cos t + \sin^2 t + \cos^2 t - 2\sin t\cos t + \sin^2 t} = e^{-t}\sqrt{2}$$

Next we substitute $t = \frac{3}{4}\pi$, to obtain

$$e^{-t_0}\sqrt{2} = e^{-3\pi/4}\sqrt{2}$$

27. Convert the points $(x, y) = (1, -3), (3, -1)$ from rectangular to polar coordinates.

SOLUTION We convert the given points from cartesian coordinates to polar coordinates. For the first point we have

$$r = \sqrt{x^2 + y^2} = \sqrt{1^2 + (-3)^2} = \sqrt{10}$$

$$\theta = \arctan\frac{y}{x} = \arctan -3 = 5.034$$

For the second point we have

$$r = \sqrt{x^2 + y^2} = \sqrt{3^2 + (-1)^2}$$

$$= \sqrt{10}\theta = \arctan\frac{y}{x} = \arctan\frac{-1}{3} = -0.321,\ 5.961$$

29. Write $(x+y)^2 = xy + 6$ as an equation in polar coordinates.

SOLUTION We use the formula for converting from cartesian coordinates to polar coordinates to substitute r and θ for x and y:

$$\begin{aligned}
(x+y)^2 &= xy+6\\
x^2+2xy+y^2 &= xy+6\\
x^2+y^2 &= -xy+6\\
r^2 &= -(r\cos\theta)(r\sin\theta)+6\\
r^2 &= -r^2\cos\theta\sin\theta+6\\
r^2(1+\sin\theta\cos\theta) &= 6\\
r^2 &= \frac{6}{1+\sin\theta\cos\theta}\\
r^2 &= \frac{6}{1+\frac{\sin 2\theta}{2}}\\
r^2 &= \frac{12}{2+\sin 2\theta}
\end{aligned}$$

31. Show that $r = \dfrac{4}{7\cos\theta - \sin\theta}$ is the polar equation of a line.

SOLUTION We use the formula for converting from polar coordinates to cartesian coordinates to substitute x and y for r and θ:

$$\begin{aligned}
r &= \frac{4}{7\cos\theta-\sin\theta}\\
1 &= \frac{4}{7r\cos\theta - r\sin\theta}\\
1 &= \frac{4}{7x-y}\\
7x-y &= 4\\
y &= 7x-4
\end{aligned}$$

We obtained a linear function. Since the original equation in polar coordinates represents the same curve, it represents a straight line as well.

33. Calculate the area of the circle $r = 3\sin\theta$ bounded by the rays $\theta = \frac{\pi}{3}$ and $\theta = \frac{2\pi}{3}$.

SOLUTION We use the formula for area in polar coordinates to obtain

$$A = \frac{1}{2}\int_{\pi/3}^{2\pi/3}(3\sin\theta)^2\,d\theta = \frac{9}{2}\int_{\pi/3}^{2\pi/3}\sin^2\theta\,d\theta = \frac{9}{4}\int_{\pi/3}^{2\pi/3}(1-\cos 2\theta)\,d\theta = \frac{9}{4}\left(\theta - \frac{\sin 2\theta}{2}\bigg|_{\pi/3}^{2\pi/3}\right)$$

$$= \frac{9}{4}\left(\frac{\pi}{3} - \frac{1}{2}\left(\sin\frac{4\pi}{3} - \sin\frac{2\pi}{3}\right)\right) = \frac{9}{4}\left(\frac{\pi}{3} - \frac{1}{2}\left(-\frac{\sqrt{3}}{2} - \frac{\sqrt{3}}{2}\right)\right) = \frac{9}{4}\left(\frac{\pi}{3} + \frac{\sqrt{3}}{2}\right)$$

35. The equation $r = \sin(n\theta)$, where $n \geq 2$ is even, is a "rose" of $2n$ petals (Figure 1). Compute the total area of the flower and show that it does not depend on n.

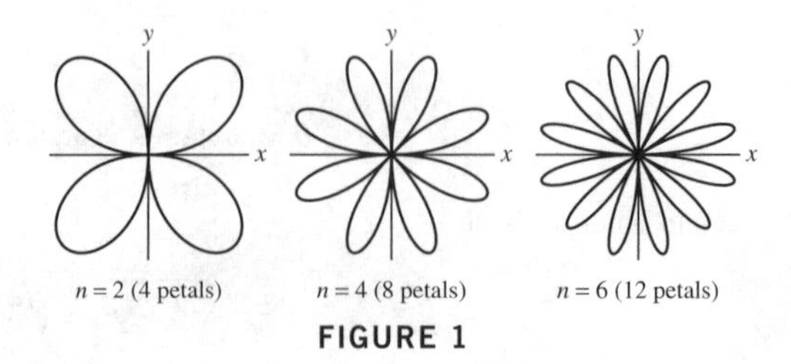

FIGURE 1

SOLUTION We calculate the total area of the flower, that is, the area between the rays $\theta = 0$ and $\theta = 2\pi$, using the formula for area in polar coordinates:

$$A = \frac{1}{2}\int_0^{2\pi} \sin^2 2n\theta\, d\theta = \frac{1}{4}\int_0^{2\pi} (1 - \cos 4n\theta)\, d\theta = \frac{1}{4}\left(\theta - \frac{\sin 4n\theta}{4n}\Big|_0^{2\pi}\right)$$

$$= \frac{\pi}{2} - \frac{1}{16n}(\sin 8n\pi - \sin 0) = \frac{\pi}{2}$$

Since the area is $\frac{\pi}{2}$ for every $n \in \mathbb{Z}$, the area is independent of n.

37. Find the shaded area in Figure 3.

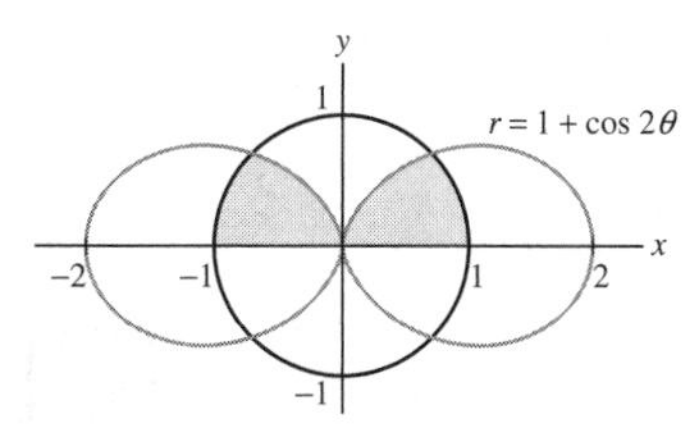

FIGURE 3

SOLUTION We first find the points of intersection between the unit circle and the function.

$$1 = 1 + \cos 2\theta$$

$$\cos 2\theta = 0$$

$$2\theta = \frac{\pi}{2} + \pi n$$

$$\theta = \frac{\pi}{4} + \frac{\pi}{2}n$$

We now find the area of the shaded figure in the first quadrant. This has two parts. The first, from 0 to $\pi/4$, is just an octant of the unit circle, and thus has area $\pi/8$. The second, from $\pi/4$ to $\pi/2$, is found as follows:

$$A = \frac{1}{2}\int_{\pi/4}^{\pi/2} (1 + \cos 2\theta)^2\, d\theta = \frac{1}{2}\int_{\pi/4}^{\pi/2} 1 + 2\cos 2\theta + \cos^2 2\theta\, d\theta = \frac{1}{2}\int_{\pi/4}^{\pi/2} \frac{3}{2} + 2\cos 2\theta + \frac{1}{2}\cos 4\theta\, d\theta$$

$$= \frac{1}{2}\left(\frac{3\theta}{2} + \sin 2\theta + \frac{1}{8}\sin 4\theta\right)\Big|_{\pi/4}^{\pi/2} = \frac{1}{2}\left(\frac{3\pi}{8} - 1\right)$$

The total area in the first quadrant is thus $\frac{5\pi}{16} - \frac{1}{2}$; multiply by 2 to get the total area of $\frac{5\pi}{8} - 1$.

39. *CAS* Figure 5 shows the graph of $r = e^{0.5\theta}\sin\theta$ for $0 \le \theta \le 2\pi$. Use a CAS to approximate the difference in length between the outer and inner loops.

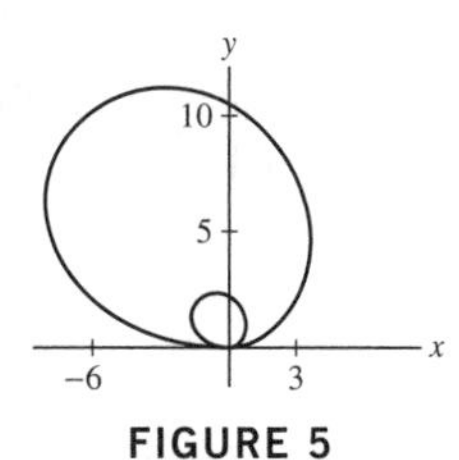

FIGURE 5

SOLUTION We note that the inner loop is the curve for $\theta \in [0, \pi]$, and the outer loop is the curve for $\theta \in [\pi, 2\pi]$. We express the length of these loops using the formula for the arc length. The length of the inner loop is

$$s_1 = \int_0^{\pi} \sqrt{(e^{0.5\theta}\sin\theta)^2 + ((e^{0.5\theta}\sin\theta)')^2}\,d\theta = \int_0^{\pi} \sqrt{e^{\theta}\sin^2\theta + \left(\frac{e^{0.5\theta}\sin\theta}{2} + e^{0.5\theta}\cos\theta\right)^2}\, d\theta$$

and the length of the outer loop is

$$s_2 = \int_{\pi}^{2\pi} \sqrt{e^{\theta}\sin^2\theta + \left(\frac{e^{0.5\theta}\sin\theta}{2} + e^{0.5\theta}\cos\theta\right)^2}\, d\theta$$

We now use the CAS to calculate the arc length of each of the loops. We obtain that the length of the inner loop is 7.5087 and the length of the outer loop is 36.121, hence the outer one is 4.81 times longer than the inner one.

In Exercises 40–43, identify the conic section. Find the vertices and foci.

41. $x^2 - 2y^2 = 4$

SOLUTION We divide the equation by 4 to obtain

$$\left(\frac{x}{2}\right)^2 - \left(\frac{y}{\sqrt{2}}\right)^2 = 1$$

This is a hyperbola in standard position, its foci are $\left(\pm\sqrt{2^2 + \sqrt{2}^2}, 0\right) = (\pm\sqrt{6}, 0)$, and its vertices are $(\pm 2, 0)$.

43. $(y - 3)^2 = 2x^2 - 1$

SOLUTION We simplify the equation:

$$(y - 3)^2 = 2x^2 - 1$$
$$2x^2 - (y - 3)^2 = 1$$
$$\left(\frac{x}{\frac{1}{\sqrt{2}}}\right)^2 - (y - 3)^2 = 1$$

This is a hyperbola shifted 3 units on the y-axis. Therefore, its foci are $\left(\pm\sqrt{\left(\frac{1}{\sqrt{2}}\right)^2 + 1}, 3\right) = \left(\pm\sqrt{\frac{3}{2}}, 3\right)$ and its vertices are $\left(\pm\frac{1}{\sqrt{2}}, 3\right)$.

45. Find the equation of a standard hyperbola with vertices at $(\pm 8, 0)$ and asymptotes $y = \pm\frac{3}{4}x$.

SOLUTION Since the asymptotes of the hyperbola are $y = \pm\frac{3}{4}x$, and the equation of the asymptotes for a general hyperbola in standard position is $y = \pm\frac{b}{a}x$, we conclude that $\frac{b}{a} = \frac{3}{4}$. We are given that the vertices are $(\pm 8, 0)$, thus $a = 8$. We substitute and solve for b:

$$\frac{b}{a} = \frac{3}{4}$$
$$\frac{b}{8} = \frac{3}{4}$$
$$b = 6$$

Next we use a and b to construct the equation of the hyperbola:

$$\left(\frac{x}{8}\right)^2 - \left(\frac{y}{6}\right)^2 = 1.$$

47. Find the equation of a standard ellipse with foci at $(\pm 8, 0)$ and eccentricity $\frac{1}{8}$.

SOLUTION If the foci are on the x-axis, then $a > b$, and $c = \sqrt{a^2 - b^2}$. We are given that $e = \frac{1}{8}$, and $c = 8$. Substituting and solving for a and b yields

$$e = \frac{c}{a}$$
$$c = \sqrt{a^2 - b^2}$$
$$\frac{1}{8} = \frac{8}{a}$$
$$64 = a$$
$$8 = \sqrt{64^2 - b^2}$$
$$64 = 64^2 - b^2$$
$$b^2 = 64 \cdot 63$$
$$b = 8\sqrt{63}$$

We use a and b to construct the equation of the ellipse:

$$\left(\frac{x}{64}\right)^2 + \left(\frac{y}{8\sqrt{63}}\right)^2 = 1.$$

49. Show that the "conic section" with equation $x^2 - 4x + y^2 + 5 = 0$ has no points.

SOLUTION We complete the squares in the given equation:

$$x^2 - 4x + 4y^2 + 5 = 0$$
$$x^2 - 4x + 4 - 4 + 4y^2 + 5 = 0$$
$$(x - 2)^2 + 4y^2 = -1$$

Since $(x - 2)^2 \geq 0$ and $y^2 \geq 0$, there is no point satisfying the equation, hence it cannot represent a conic section.

51. The orbit of Jupiter is an ellipse with the sun at a focus. Find the eccentricity of the orbit if the perihelion (closest distance to the sun) equals 740×10^6 km and the aphelion (farthest distance to the sun) equals 816×10^6 km.

SOLUTION For the sake of simplicity, we treat all numbers in units of 10^6 km. By Kepler's First Law we conclude that the sun is at one of the foci of the ellipse. Therefore, the closest and farthest points to the sun are vertices. Moreover, they are the vertices on the x-axis, hence we conclude that the distance between the two vertices is

$$2a = 740 + 816 = 1556$$

Since the distance between each focus and the vertex that is closest to it is the same distance, and since $a = 778$, we conclude that the distance between the foci is

$$c = a - 740 = 38$$

We substitute this in the formula for the eccentricity to obtain:

$$e = \frac{c}{a} = 0.0488.$$

13 | VECTOR GEOMETRY

13.1 Vectors in the Plane (ET Section 12.1)

Preliminary Questions

1. Answer true or false. Every nonzero vector is:

(a) Equivalent to a vector based at the origin.

(b) Equivalent to a unit vector based at the origin.

(c) Parallel to a vector based at the origin.

(d) Parallel to a unit vector based at the origin.

SOLUTION

(a) This statement is true. Translating the vector so that it is based on the origin, we get an equivalent vector based at the origin.

(b) Equivalent vectors have equal lengths, hence vectors that are not unit vectors, are not equivalent to a unit vector.

(c) This statement is true. A vector based at the origin such that the line through this vector is parallel to the line through the given vector, is parallel to the given vector.

(d) Since parallel vectors do not necessarily have equal lengths, the statement is true by the same reasoning as in (c).

2. What is the length of $-3\mathbf{a}$ if $\|\mathbf{a}\| = 5$?

SOLUTION Using properties of the length we get

$$\|-3\mathbf{a}\| = |-3|\|\mathbf{a}\| = 3\|\mathbf{a}\| = 3 \cdot 5 = 15$$

3. Suppose that $\mathbf{v}$ has components $\langle 3, 1\rangle$. How, if at all, do the components change if you translate $\mathbf{v}$ horizontally two units to the left?

SOLUTION Translating $\mathbf{v} = \langle 3, 1\rangle$ yields an equivalent vector, hence the components are not changed.

4. What are the components of the zero vector based at $P = \langle 3, 5\rangle$?

SOLUTION The components of the zero vector are always $\langle 0, 0\rangle$, no matter where it is based.

5. Are the following true or false?

(a) The vectors $\mathbf{v}$ and $-2\mathbf{v}$ are parallel.

(b) The vectors $\mathbf{v}$ and $-2\mathbf{v}$ point in the same direction.

SOLUTION

(a) The lines through $\mathbf{v}$ and $-2\mathbf{v}$ are parallel, therefore these vectors are parallel.

(b) The vector $-2\mathbf{v}$ is a scalar multiple of $\mathbf{v}$, where the scalar is negative. Therefore $-2\mathbf{v}$ points in the opposite direction as $\mathbf{v}$.

6. Explain the commutativity of vector addition in terms of the Parallelogram Law.

SOLUTION To determine the vector $\mathbf{v} + \mathbf{w}$, we translate $\mathbf{w}$ to the equivalent vector $\mathbf{w}'$ whose tail coincides with the head of $\mathbf{v}$. The vector $\mathbf{v} + \mathbf{w}$ is the vector pointing from the tail of $\mathbf{v}$ to the head of $\mathbf{w}'$.

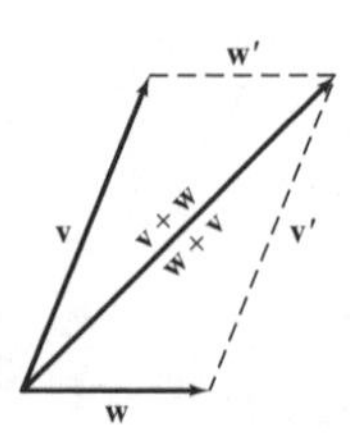

To determine the vector $\mathbf{w} + \mathbf{v}$, we translate $\mathbf{v}$ to the equivalent vector $\mathbf{v}'$ whose tail coincides with the head of $\mathbf{w}$. Then $\mathbf{w} + \mathbf{v}$ is the vector pointing from the tail of $\mathbf{w}$ to the head of $\mathbf{v}'$. In either case, the resulting vector is the vector with the tail at the basepoint of $\mathbf{v}$ and $\mathbf{w}$, and head at the opposite vertex of the parallelogram. Therefore $\mathbf{v} + \mathbf{w} = \mathbf{w} + \mathbf{v}$.

Exercises

1. Sketch the vectors $\mathbf{v}_1, \mathbf{v}_2, \mathbf{v}_3, \mathbf{v}_4$ with tail P and head Q, and compute their lengths. Are any two of these vectors equivalent?

	$\mathbf{v}_1$	$\mathbf{v}_2$	$\mathbf{v}_3$	$\mathbf{v}_4$
P	$(2, 4)$	$(-1, 3)$	$(-1, 3)$	$(4, 1)$
Q	$(4, 4)$	$(1, 3)$	$(2, 4)$	$(6, 3)$

SOLUTION Using the definitions we obtain the following answers:

$$\mathbf{v}_1 = \overrightarrow{PQ} = \langle 4-2, 4-4\rangle = \langle 2, 0\rangle \qquad \mathbf{v}_2 = \langle 1-(-1), 3-3\rangle = \langle 2, 0\rangle$$

$$\|\mathbf{v}_1\| = \sqrt{2^2+0^2} = 2 \qquad \|\mathbf{v}_2\| = \sqrt{2^2+0^2} = 2$$

$$\mathbf{v}_3 = \langle 2-(-1), 4-3\rangle = \langle 3, 1\rangle \qquad \mathbf{v}_4 = \langle 6-4, 3-1\rangle = \langle 2, 2\rangle$$

$$\|\mathbf{v}_3\| = \sqrt{3^2+1^2} = \sqrt{10} \qquad \|\mathbf{v}_4\| = \sqrt{2^2+2^2} = \sqrt{8} = 2\sqrt{2}$$

$\mathbf{v}_1$ and $\mathbf{v}_2$ are parallel and have the same length, hence they are equivalent.

3. What is the terminal point of the vector $\mathbf{a} = \langle 1, 3\rangle$ based at $P = (2, 2)$? Sketch $\mathbf{a}$ and the vector $\mathbf{a}_0$ based at the origin equivalent to $\mathbf{a}$.

SOLUTION The terminal point Q of the vector $\mathbf{a}$ is located 1 unit to the right and 3 units up from $P = (2, 2)$. Therefore, $Q = (2+1, 2+3) = (3, 5)$. The vector $\mathbf{a}_0$ equivalent to $\mathbf{a}$ based at the origin is shown in the figure, along with the vector $\mathbf{a}$.

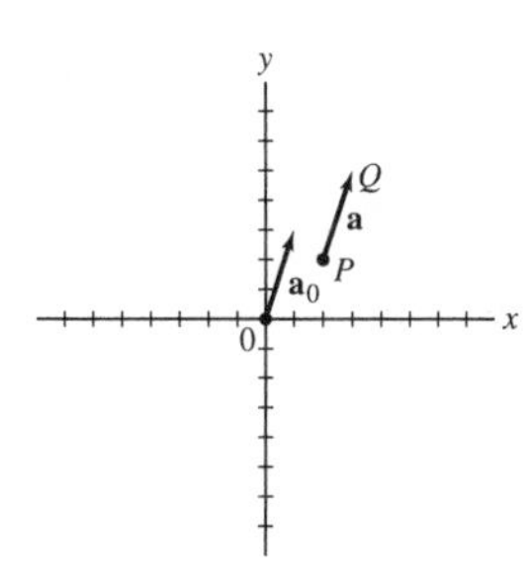

In Exercises 5–8, find the components of $\overrightarrow{PQ}$.

5. $P = (3, 2), \quad Q = (2, 7)$

SOLUTION Using the definition of the components of a vector we have $\overrightarrow{PQ} = \langle 2-3, 7-2\rangle = \langle -1, 5\rangle$.

7. $P = (3, 5), \quad Q = (1, -4)$

SOLUTION By the definition of the components of a vector, we obtain $\overrightarrow{PQ} = \langle 1-3, -4-5\rangle = \langle -2, -9\rangle$.

In Exercises 9–14, calculate.

9. $\langle 2, 1\rangle + \langle 3, 4\rangle$

SOLUTION Using vector algebra we have $\langle 2, 1\rangle + \langle 3, 4\rangle = \langle 2+3, 1+4\rangle = \langle 5, 5\rangle$.

11. $5\,\langle 6, 2\rangle$

SOLUTION $5\langle 6, 2\rangle = \langle 5\cdot 6, 5\cdot 2\rangle = \langle 30, 10\rangle$

13. $\left\langle -\frac{1}{2}, \frac{5}{3}\right\rangle + \left\langle 3, \frac{10}{3}\right\rangle$

SOLUTION The vector sum is $\left\langle -\frac{1}{2}, \frac{5}{3}\right\rangle + \left\langle 3, \frac{10}{3}\right\rangle = \left\langle -\frac{1}{2}+3, \frac{5}{3}+\frac{10}{3}\right\rangle = \left\langle \frac{5}{2}, 5\right\rangle$.

15. Which of the vectors (A)–(C) in Figure 21 is equivalent to $\mathbf{v} - \mathbf{w}$?

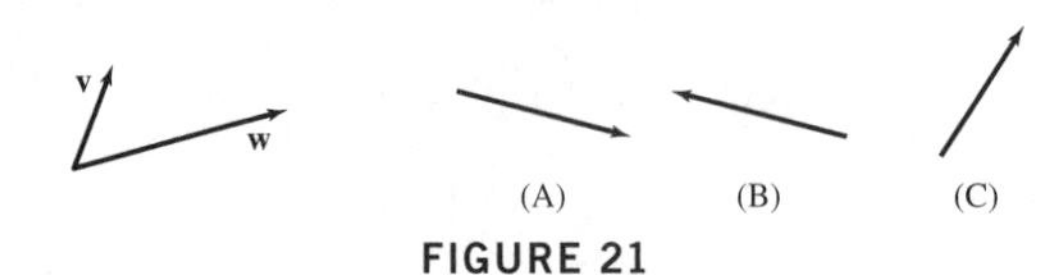

FIGURE 21

SOLUTION The vector $-\mathbf{w}$ has the same length as $\mathbf{w}$ but points in the opposite direction. The sum $\mathbf{v} + (-\mathbf{w})$, which is the difference $\mathbf{v} - \mathbf{w}$, is obtained by the parallelogram law. This vector is the vector shown in (b).

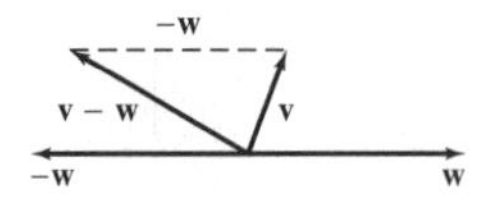

17. Sketch $2\mathbf{v}$, $-\mathbf{w}$, $\mathbf{v}+\mathbf{w}$, and $2\mathbf{v}-\mathbf{w}$ for the vectors in Figure 23.

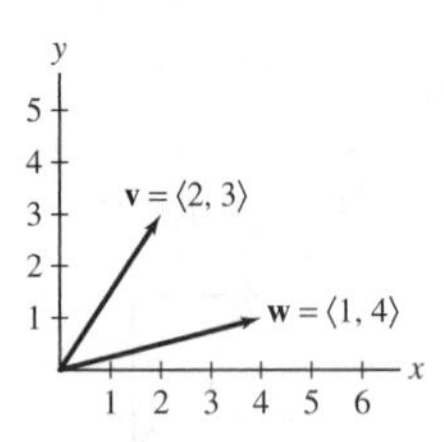

FIGURE 23

SOLUTION The scalar multiple $2\mathbf{v}$ points in the same direction as $\mathbf{v}$ and its length is twice the length of $\mathbf{v}$. It is the vector $2\mathbf{v} = \langle 4, 6\rangle$.

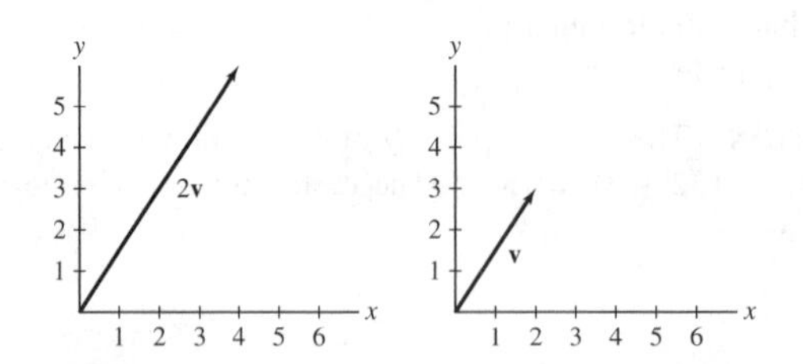

$-\mathbf{w}$ has the same length as $\mathbf{w}$ but points to the opposite direction. It is the vector $-\mathbf{w} = \langle -4, -1\rangle$.

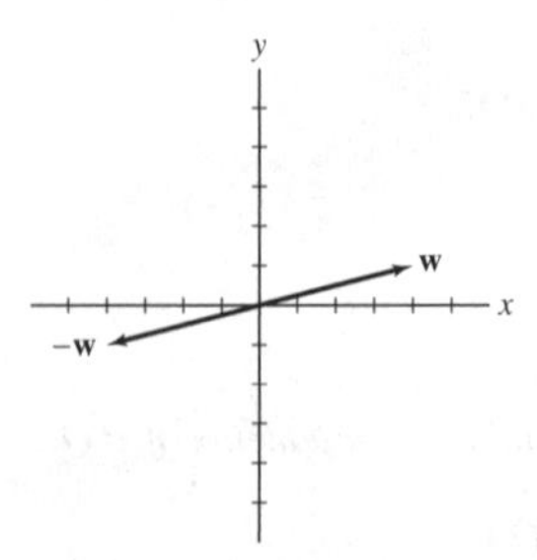

The vector sum $\mathbf{v}+\mathbf{w}$ is the vector:

$$\mathbf{v}+\mathbf{w} = \langle 2, 3\rangle + \langle 4, 1\rangle = \langle 6, 4\rangle.$$

This vector is shown in the following figure:

The vector $2\mathbf{v} - \mathbf{w}$ is

$$2\mathbf{v} - \mathbf{w} = 2\langle 2, 3\rangle - \langle 4, 1\rangle = \langle 4, 6\rangle - \langle 4, 1\rangle = \langle 0, 5\rangle$$

It is shown next:

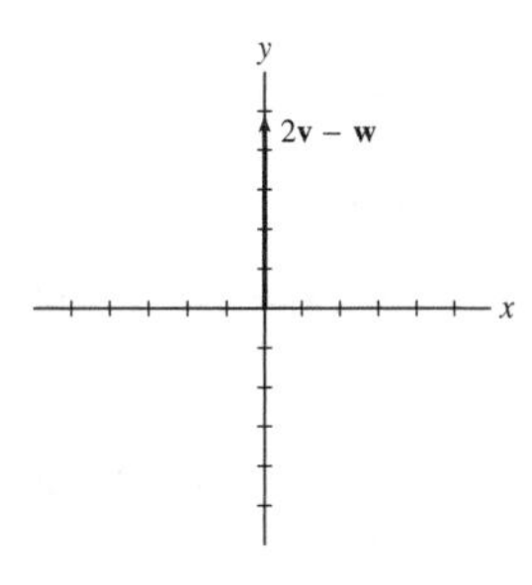

19. Sketch $\mathbf{v} = \langle 0, 2\rangle$, $\mathbf{w} = \langle -2, 4\rangle$, $3\mathbf{v} + \mathbf{w}$, $2\mathbf{v} - 2\mathbf{w}$.

SOLUTION We compute the vectors and then sketch them:

$$3\mathbf{v} + \mathbf{w} = 3\langle 0, 2\rangle + \langle -2, 4\rangle = \langle 0, 6\rangle + \langle -2, 4\rangle = \langle -2, 10\rangle$$

$$2\mathbf{v} - 2\mathbf{w} = 2\langle 0, 2\rangle - 2\langle -2, 4\rangle = \langle 0, 4\rangle - \langle -4, 8\rangle = \langle 4, -4\rangle$$

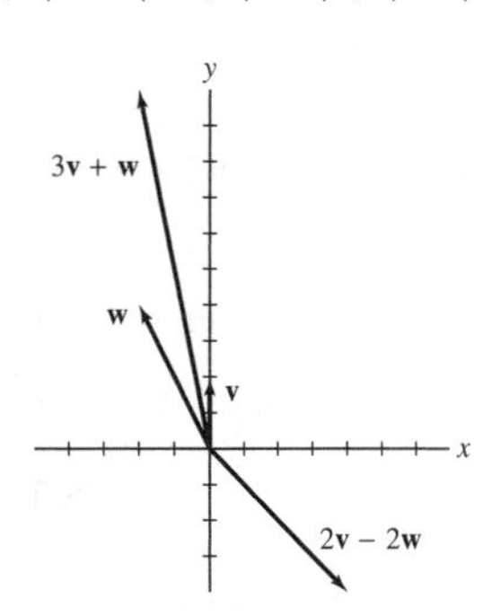

21. Sketch the vector $\mathbf{v}$ such that $\mathbf{v} + \mathbf{v}_1 + \mathbf{v}_2 = \mathbf{0}$ for $\mathbf{v}_1$ and $\mathbf{v}_2$ in Figure 24(A).

SOLUTION Since $\mathbf{v} + \mathbf{v}_1 + \mathbf{v}_2 = 0$, we have that $\mathbf{v} = -\mathbf{v}_1 - \mathbf{v}_2$, and since $\mathbf{v}_1 = \langle 1, 3\rangle$ and $\mathbf{v}_2 = \langle -3, 1\rangle$, then $\mathbf{v} = -\mathbf{v}_1 - \mathbf{v}_2 = \langle 2, -4\rangle$, as seen in this picture.

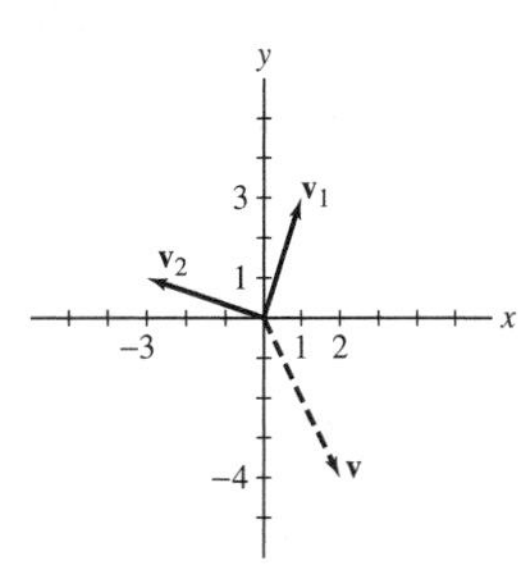

23. Let $\mathbf{v} = \overrightarrow{PQ}$, where $P = (-2, 5)$, $Q = (1, -2)$. Which of the vectors with the following given tails and heads are equivalent to $\mathbf{v}$?

(a) $(-3, 3)$, $(0, 4)$

(b) $(0, 0)$, $(3, -7)$

(c) $(-1, 2)$, $(2, -5)$

(d) $(4, -5)$, $(1, 4)$

SOLUTION Two vectors are equivalent if they have the same components. We thus compute the vectors and check whether this condition is satisfied.

$$\mathbf{v} = \overrightarrow{PQ} = \langle 1-(-2), -2-5\rangle = \langle 3, -7\rangle$$

(a) $\langle 0-(-3), 4-3\rangle = \langle 3, 1\rangle$

(b) $\langle 3-0, -7-0\rangle = \langle 3, -7\rangle$

(c) $\langle 2-(-1), -5-2\rangle = \langle 3, -7\rangle$

(d) $\langle 1-4, 4-(-5)\rangle = \langle -3, 9\rangle$

We see that the vectors in (b) and (c) are equivalent to **v**.

In Exercises 25–28, sketch the vectors $\overrightarrow{AB}$ and $\overrightarrow{PQ}$, and determine whether they are equivalent.

25. $A = (1, 1), \quad B = (3, 7), \quad P = (4, -1), \quad Q = (6, 5)$

SOLUTION We compute the vectors and check whether they have the same components:

$$\begin{array}{l} \overrightarrow{AB} = \langle 3-1, 7-1\rangle = \langle 2, 6\rangle \\ \overrightarrow{PQ} = \langle 6-4, 5-(-1)\rangle = \langle 2, 6\rangle \end{array} \quad \Rightarrow \quad \text{The vectors are equivalent.}$$

27. $A = (-3, 2), \quad B = (0, 0), \quad P = (0, 0), \quad Q = (3, -2)$

SOLUTION We compute the vectors $\overrightarrow{AB}$ and $\overrightarrow{PQ}$:

$$\begin{array}{l} \overrightarrow{AB} = \langle 0-(-3), 0-2\rangle = \langle 3, -2\rangle \\ \overrightarrow{PQ} = \langle 3-0, -2-0\rangle = \langle 3, -2\rangle \end{array} \quad \Rightarrow \quad \text{The vectors are equivalent.}$$

In Exercises 29–32, are $\overrightarrow{AB}$ and $\overrightarrow{PQ}$ parallel (and if so, do they point in the same direction)?

29. $A = (1, 1), \quad B = (3, 4), \quad P = (1, 1), \quad Q = (7, 10)$

SOLUTION We compute the vectors $\overrightarrow{AB}$ and $\overrightarrow{PQ}$:

$$\overrightarrow{AB} = \langle 3-1, 4-1\rangle = \langle 2, 3\rangle$$
$$\overrightarrow{PQ} = \langle 7-1, 10-1\rangle = \langle 6, 9\rangle$$

Since $\overrightarrow{AB} = \frac{1}{3}\langle 6, 9\rangle$, the vectors are parallel and point in the same direction.

31. $A = (2, 2), \quad B = (-6, 3), \quad P = (9, 5), \quad Q = (17, 4)$

SOLUTION We compute the vectors $\overrightarrow{AB}$ and $\overrightarrow{PQ}$:

$$\overrightarrow{AB} = \langle -6-2, 3-2\rangle = \langle -8, 1\rangle$$
$$\overrightarrow{PQ} = \langle 17-9, 4-5\rangle = \langle 8, -1\rangle$$

Since $\overrightarrow{AB} = -\overrightarrow{PQ}$, the vectors are parallel and point in opposite directions.

In Exercises 33–36, let $R = (-2, 7)$. Calculate the following.

33. The length of $\overrightarrow{OR}$

SOLUTION Since $\overrightarrow{OR} = \langle -2, 7\rangle$, the length of the vector is $\|\overrightarrow{OR}\| = \sqrt{(-2)^2 + 7^2} = \sqrt{53}$.

35. The point P such that $\overrightarrow{PR}$ has components $\langle -2, 7\rangle$

SOLUTION Denoting $P = (x_0, y_0)$ we have:

$$\overrightarrow{PR} = \langle -2-x_0, 7-y_0\rangle = \langle -2, 7\rangle$$

Equating corresponding components yields:

$$\begin{array}{r} -2-x_0 = -2 \\ 7-y_0 = 7 \end{array} \quad \Rightarrow \quad x_0 = 0, \ y_0 = 0 \quad \Rightarrow \quad P = (0, 0)$$

In Exercises 37–44, find the given vector.

37. Unit vector $\mathbf{e_v}$ where $\mathbf{v} = \langle 3, 4\rangle$

SOLUTION The unit vector $\mathbf{e_v}$ is the following vector:

$$\mathbf{e_v} = \frac{1}{\|\mathbf{v}\|}\mathbf{v}$$

We find the length of $\mathbf{v} = \langle 3, 4\rangle$:

$$\|\mathbf{v}\| = \sqrt{3^2 + 4^2} = \sqrt{25} = 5$$

Thus

$$\mathbf{e_v} = \frac{1}{5}\langle 3, 4\rangle = \left\langle \frac{3}{5}, \frac{4}{5}\right\rangle.$$

39. Unit vector in the direction of $\mathbf{u} = \langle -1, -1\rangle$

SOLUTION The unit vector in the direction of $\mathbf{u} = \langle -1, -1\rangle$ is the vector:

$$\mathbf{e_u} = \frac{1}{\|\mathbf{u}\|}\mathbf{u}$$

We compute the length of $\mathbf{u}$:

$$\|\mathbf{u}\| = \sqrt{(-1)^2 + (-1)^2} = \sqrt{2}$$

Hence

$$\mathbf{e_u} = \frac{1}{\sqrt{2}}\langle -1, -1\rangle = \left\langle -\frac{1}{\sqrt{2}}, -\frac{1}{\sqrt{2}}\right\rangle.$$

41. Unit vector in the direction opposite to $\mathbf{v} = \langle -2, 4\rangle$

SOLUTION We first compute the unit vector $\mathbf{e_v}$ in the direction of $\mathbf{v}$ and then multiply by -1 to obtain a unit vector in the opposite direction. This gives:

$$\mathbf{e_v} = \frac{1}{\|\mathbf{v}\|}\mathbf{v} = \frac{1}{\sqrt{(-2)^2 + 4^2}}\langle -2, 4\rangle = \frac{1}{\sqrt{20}}\langle -2, 4\rangle = \left\langle -\frac{2}{2\sqrt{5}}, \frac{4}{2\sqrt{5}}\right\rangle = \left\langle -\frac{1}{\sqrt{5}}, \frac{2}{\sqrt{5}}\right\rangle$$

The desired vector is thus

$$-\mathbf{e_v} = -\left\langle -\frac{1}{\sqrt{5}}, \frac{2}{\sqrt{5}}\right\rangle = \left\langle \frac{1}{\sqrt{5}}, -\frac{2}{\sqrt{5}}\right\rangle.$$

43. Vector of length 2 making an angle of 30° with the x-axis

SOLUTION The desired vector is

$$2\langle \cos 30^\circ, \sin 30^\circ\rangle = 2\left\langle \frac{\sqrt{3}}{2}, \frac{1}{2}\right\rangle = \left\langle \sqrt{3}, 1\right\rangle.$$

45. Find all scalars λ such that $\lambda\,\langle 2, 3\rangle$ has length one.

SOLUTION We have:

$$\|\lambda\,\langle 2, 3\rangle\| = |\lambda|\|\langle 2, 3\rangle\| = |\lambda|\sqrt{2^2 + 3^2} = |\lambda|\sqrt{13}$$

The scalar λ must satisfy

$$\begin{aligned} |\lambda|\sqrt{13} &= 1 \\ |\lambda| &= \frac{1}{\sqrt{13}} \end{aligned} \quad\Rightarrow\quad \lambda_1 = \frac{1}{\sqrt{13}},\ \lambda_2 = -\frac{1}{\sqrt{13}}$$

47. What are the coordinates a and b in the parallelogram in Figure 25(B)?

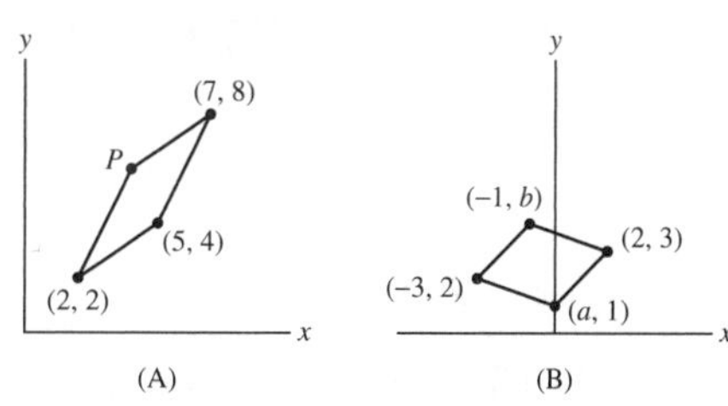

FIGURE 25

SOLUTION We denote the points in the figure by A, B, C and D.

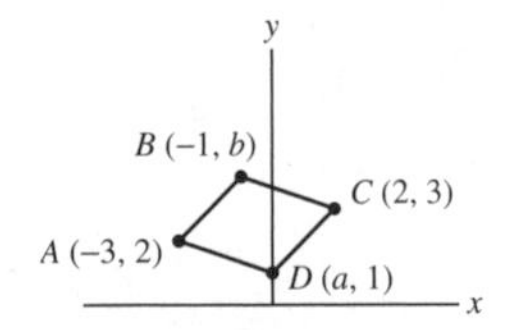

We compute the following vectors:

$$\overrightarrow{AB} = \langle -1 - (-3), b - 2 \rangle = \langle 2, b - 2 \rangle$$
$$\overrightarrow{DC} = \langle 2 - a, 3 - 1 \rangle = \langle 2 - a, 2 \rangle$$

Since $\overrightarrow{AB} = \overrightarrow{PC}$, the two vectors have the same components. That is,

$$\begin{aligned} 2 &= 2 - a \\ b - 2 &= 2 \end{aligned} \quad \Rightarrow \quad \begin{aligned} a &= 0 \\ b &= 4 \end{aligned}$$

49. Find the components and length of the following vectors:

(a) $4\mathbf{i} + 3\mathbf{j}$ **(b)** $2\mathbf{i} - 3\mathbf{j}$ **(c)** $\mathbf{i} + \mathbf{j}$ **(d)** $\mathbf{i} - 3\mathbf{j}$

SOLUTION

(a) Since $\mathbf{i} = \langle 1, 0 \rangle$ and $\mathbf{j} = \langle 0, 1 \rangle$, using vector algebra we have:

$$4\mathbf{i} + 3\mathbf{j} = 4\langle 1, 0 \rangle + 3\langle 0, 1 \rangle = \langle 4, 0 \rangle + \langle 0, 3 \rangle = \langle 4 + 0, 0 + 3 \rangle = \langle 4, 3 \rangle$$

The length of the vector is:

$$\|4\mathbf{i} + 3\mathbf{j}\| = \sqrt{4^2 + 3^2} = 5$$

(b) We use vector algebra and the definition of the standard basis vector to compute the components of the vector $2\mathbf{i} - 3\mathbf{j}$:

$$2\mathbf{i} - 3\mathbf{j} = 2\langle 1, 0 \rangle - 3\langle 0, 1 \rangle = \langle 2, 0 \rangle - \langle 0, 3 \rangle = \langle 2 - 0, 0 - 3 \rangle = \langle 2, -3 \rangle$$

The length of this vector is:

$$\|2\mathbf{i} - 3\mathbf{j}\| = \sqrt{2^2 + (-3)^2} = \sqrt{13}$$

(c) We find the components of the vector $\mathbf{i} + \mathbf{j}$:

$$\mathbf{i} + \mathbf{j} = \langle 1, 0 \rangle + \langle 0, 1 \rangle = \langle 1 + 0, 0 + 1 \rangle = \langle 1, 1 \rangle$$

The length of this vector is:

$$\|\mathbf{i} + \mathbf{j}\| = \sqrt{1^2 + 1^2} = \sqrt{2}$$

(d) We find the components of the vector $\mathbf{i} - 3\mathbf{j}$, using vector algebra:

$$\mathbf{i} - 3\mathbf{j} = \langle 1, 0 \rangle - 3\langle 0, 1 \rangle = \langle 1, 0 \rangle - \langle 0, 3 \rangle = \langle 1 - 0, 0 - 3 \rangle = \langle 1, -3 \rangle$$

The length of this vector is

$$\|\mathbf{i} - 3\mathbf{j}\| = \sqrt{1^2 + (-3)^2} = \sqrt{10}$$

In Exercises 50–53, calculate the linear combination.

51. $6(\mathbf{i} + 9\mathbf{j}) + 2(\mathbf{i} - 4\mathbf{j})$

SOLUTION Using the definition of vector operations we get

$$6(\mathbf{i} + 9\mathbf{j}) + 2(\mathbf{i} - 4\mathbf{j}) = (6\mathbf{i} + 54\mathbf{j}) + (2\mathbf{i} - 8\mathbf{j}) = 8\mathbf{i} + 46\mathbf{j}.$$

53. $3(\mathbf{i} + \mathbf{j}) + 7(3\mathbf{i} - \mathbf{j})$

SOLUTION Using the basic vector operations gives

$$3(\mathbf{i} + \mathbf{j}) + 7(3\mathbf{i} - \mathbf{j}) = (3\mathbf{i} + 3\mathbf{j}) + (21\mathbf{i} - 7\mathbf{j}) = (3 + 21)\mathbf{i} + (3 - 7)\mathbf{j} = 24\mathbf{i} - 4\mathbf{j}$$

55. Express $\mathbf{u} = \langle 3, -1 \rangle$ as a linear combination $\mathbf{u} = r\mathbf{v} + s\mathbf{w}$, where $\mathbf{v} = \langle 2, 1 \rangle$ and $\mathbf{w} = \langle 1, 3 \rangle$. Sketch the vectors $\mathbf{u}$, $\mathbf{v}$, $\mathbf{w}$ and the parallelogram formed by $r\mathbf{v}$ and $s\mathbf{w}$.

SOLUTION We have

$$\mathbf{u} = \langle 3, -1 \rangle = r\mathbf{v} + s\mathbf{w} = r\langle 2, 1 \rangle + s\langle 1, 3 \rangle$$

which becomes the two equations

$$3 = 2r + s$$
$$-1 = r + 3s$$

Solving the second equation for r gives $r = -1 - 3s$, and substituting that into the first equation gives $3 = 2(-1 - 3s) + s = -2 - 6s + s$, so $5 = -5s$, so $s = -1$, and thus $r = 2$. In other words,

$$\mathbf{u} = \langle 3, -1 \rangle = 2\langle 2, 1 \rangle - 1\langle 1, 3 \rangle$$

as seen in this sketch:

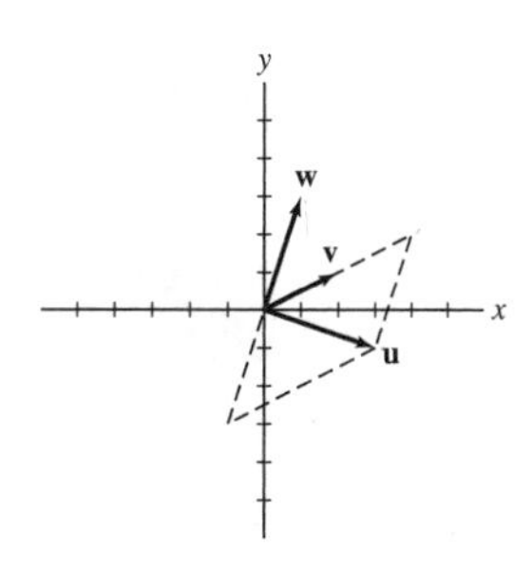

57. Sketch the parallelogram spanned by $\mathbf{v} = \langle 1, 4 \rangle$ and $\mathbf{w} = \langle 5, 2 \rangle$. Add the vector $\mathbf{u} = \langle 2, 3 \rangle$ to the sketch and express $\mathbf{u}$ as a linear combination of $\mathbf{v}$ and $\mathbf{w}$.

SOLUTION We have

$$\mathbf{u} = \langle 2, 3 \rangle = r\mathbf{v} + s\mathbf{w} = r\langle 1, 4 \rangle + s\langle 5, 2 \rangle$$

which becomes the two equations

$$2 = r + 5s$$
$$3 = 4r + 2s$$

Solving the first equation for r gives

$$r = 2 - 5s$$

and substituting that into the first equation gives

$$3 = 4(2 - 5s) + 2s = 8 - 18s$$

So $18s = 5$, so $s = 5/18$, and thus $r = 11/18$. In other words,

$$\mathbf{u} = \langle 2, 3 \rangle = \frac{11}{18}\langle 1, 4 \rangle + \frac{5}{18}\langle 5, 2 \rangle$$

as seen in this picture:

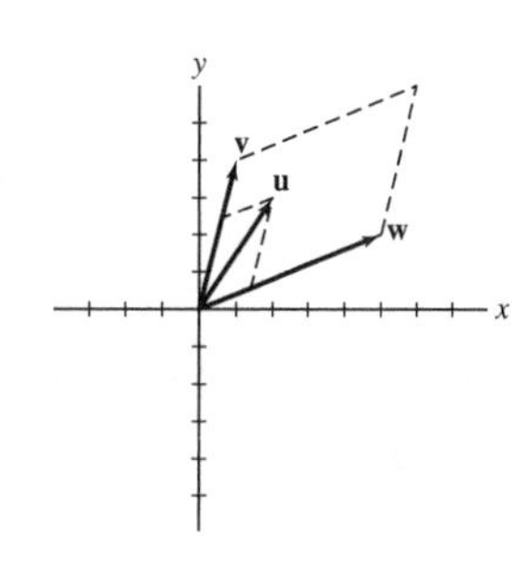

59. Determine the magnitude of the forces $\mathbf{F}_1$ and $\mathbf{F}_2$ in Figure 28, assuming that there is no net force on the object.

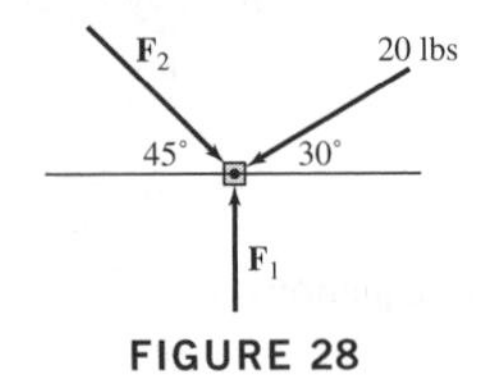

FIGURE 28

SOLUTION We denote $\|\mathbf{F}_1\| = f_1$ and $\|\mathbf{F}_2\| = f_2$.

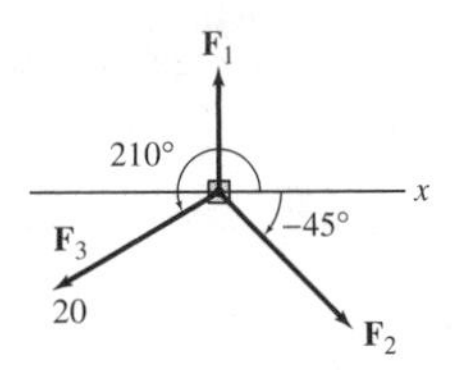

Since there is no net force on the object, we have

$$\mathbf{F}_1 + \mathbf{F}_2 + \mathbf{F}_3 = 0 \tag{1}$$

We find the forces:

$$\mathbf{F}_1 = f_1\langle 0, 1\rangle = \langle 0, f_1\rangle$$

$$\mathbf{F}_2 = f_2\langle\cos(-45^\circ), \sin(-45^\circ)\rangle = f_2\left\langle\frac{\sqrt{2}}{2}, -\frac{\sqrt{2}}{2}\right\rangle = \langle 0.707 f_2, -0.707 f_2\rangle$$

$$\mathbf{F}_3 = 20\langle\cos 210^\circ, \sin 210^\circ\rangle = \langle -17.32, -10\rangle$$

We substitute the forces in (1):

$$\langle 0, f_1\rangle + \langle 0.707 f_2, -0.707 f_2\rangle + \langle -17.32, -10\rangle = \langle 0, 0\rangle$$

$$\langle 0.707 f_2 - 17.32, f_1 - 0.707 f_2 - 10\rangle = \langle 0, 0\rangle$$

Equating corresponding components we obtain

$$0.707 f_2 - 17.32 = 0$$

$$f_1 - 0.707 f_2 - 10 = 0$$

The first equation gives $f_2 = 24.5$. Substituting in the second equation and solving for f_1 gives

$$f_1 - 0.707 \cdot 24.5 - 10 = 0 \Rightarrow f_1 = 27.32$$

The magnitude of the forces $\mathbf{F}_1$ and $\mathbf{F}_2$ are $f_1 = 27.32$ lb and $f_2 = 24.5$ lb respectively.

Further Insights and Challenges

In Exercises 61–63, refer to Figure 30, which shows a robotic arm consisting of two segments of lengths L_1 and L_2.

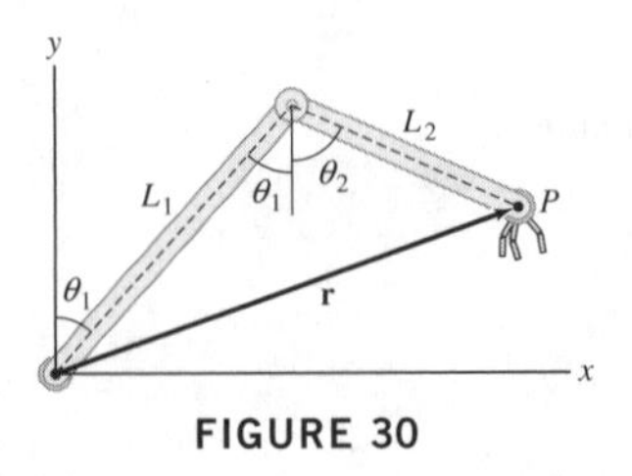

FIGURE 30

61. Find the components of the vector $\mathbf{r} = \overrightarrow{OP}$ in terms of θ_1 and θ_2.

SOLUTION We denote by A the point in the figure.

By the parallelogram law we have

$$\mathbf{r} = \overrightarrow{OA} + \overrightarrow{AP} \tag{1}$$

We find the vectors $\overrightarrow{OA}$ and $\overrightarrow{AP}$:

- The vector $\overrightarrow{OA}$ has length L_1 and it makes an angle of $90^\circ - \theta_1$ with the x-axis.
- The vector $\overrightarrow{AP}$ has length L_2 and it makes an angle of $-(90^\circ - \theta_2) = \theta_2 - 90^\circ$ with the x-axis.

Hence,

$$\overrightarrow{OA} = L_1 \left\langle \cos(90^\circ - \theta_1), \sin(90^\circ - \theta_1) \right\rangle = L_1 \langle \sin\theta_1, \cos\theta_1 \rangle = \langle L_1 \sin\theta_1, L_1 \cos\theta_1 \rangle$$

$$\overrightarrow{AP} = L_2 \left\langle \cos(\theta_2 - 90^\circ), \sin(\theta_2 - 90^\circ) \right\rangle = L_2 \langle \sin\theta_2, -\cos\theta_2 \rangle = \langle L_2 \sin\theta_2, -L_2 \cos\theta_2 \rangle$$

Substituting into (1) we obtain

$$\mathbf{r} = \langle L_1 \sin\theta_1, L_1 \cos\theta_1 \rangle + \langle L_2 \sin\theta_2 - L_2 \cos\theta_2 \rangle$$

$$\mathbf{r} = \langle L_1 \sin\theta_1 + L_2 \sin\theta_2, L_1 \cos\theta_1 - L_2 \cos\theta_2 \rangle$$

Thus, the x component of $\mathbf{r}$ is $L_1 \sin\theta_1 + L_2 \sin\theta_2$ and the y component is $L_1 \cos\theta_1 - L_2 \cos\theta_2$.

63. Let $L_1 = 5$ and $L_2 = 3$. Show that the set of points reachable by the robotic arm with $\theta_1 = \theta_2$ is an ellipse.

SOLUTION Substituting $L_1 = 5$, $L_2 = 3$, and $\theta_1 = \theta_2 = \theta$ in the formula for $\mathbf{r}$ obtained in Exercise 61 we get

$$\begin{aligned} \mathbf{r} &= \langle L_1 \sin\theta_1 + L_2 \sin\theta_2, L_1 \cos\theta_1 - L_2 \cos\theta_2 \rangle \\ &= \langle 5\sin\theta + 3\sin\theta, 5\cos\theta - 3\cos\theta \rangle = \langle 8\sin\theta, 2\cos\theta \rangle \end{aligned}$$

Thus, the x and y components of $\mathbf{r}$ are

$$x = 8\sin\theta, \; y = 2\cos\theta$$

so $\frac{x}{8} = \sin\theta$, $\frac{y}{2} = \cos\theta$. Using the identity $\sin^2\theta + \cos^2\theta = 1$ we get

$$\left(\frac{x}{8}\right)^2 + \left(\frac{y}{2}\right)^2 = 1,$$

which is the formula of an ellipse.

65. Use vectors to prove that the segments joining the midpoints of opposite sides of a quadrilateral bisect each other (Figure 32). *Hint:* Show that the midpoints of these segments are the terminal points of $\frac{1}{4}(2\mathbf{u} + \mathbf{v} + \mathbf{z})$ and $\frac{1}{4}(2\mathbf{v} + \mathbf{w} + \mathbf{u})$.

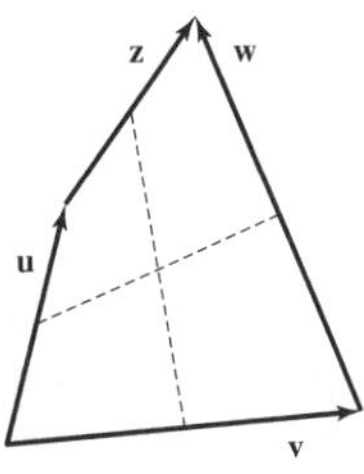

FIGURE 32

SOLUTION We denote by A, B, C, D the corresponding points in the figure and by E, F, G, H the midpoints of the sides $\overline{AB}$, $\overline{BC}$, $\overline{CD}$ and $\overline{AD}$, respectively. Also, O is the midpoint of $\overline{FH}$ and O' is the midpoint of $\overline{EG}$.

We must show that O and O' are the same point. Using the Parallelogram Law we have

$$\overrightarrow{AO} = \overrightarrow{AH} + \overrightarrow{HO} = \frac{1}{2}\mathbf{v} + \frac{1}{2}\overrightarrow{HF}$$

$$\overrightarrow{HF} = \overrightarrow{HD} + \overrightarrow{DC} + \overrightarrow{CF} = \frac{1}{2}\mathbf{v} + \mathbf{w} - \frac{1}{2}\mathbf{z}$$

Hence,

$$\overrightarrow{AO} = \frac{1}{2}\mathbf{v} + \frac{1}{2}\left(\frac{1}{2}\mathbf{v} + \mathbf{w} - \frac{1}{2}\mathbf{z}\right) = \frac{3}{4}\mathbf{v} + \frac{1}{2}\mathbf{w} - \frac{1}{4}\mathbf{z} \tag{1}$$

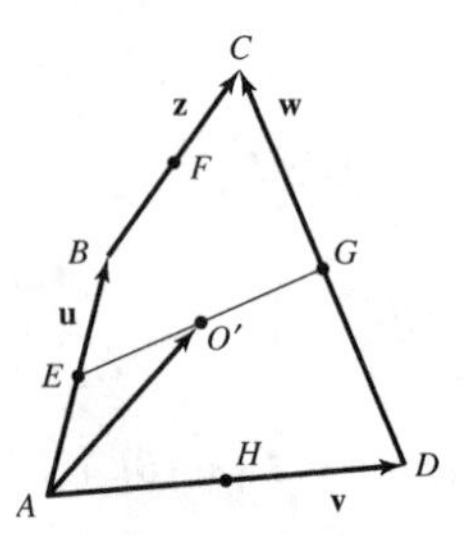

Similarly,

$$\overrightarrow{AO'} = \overrightarrow{AD} + \overrightarrow{DG} + \overrightarrow{GO'} = \mathbf{v} + \frac{1}{2}\mathbf{w} + \frac{1}{2}\overrightarrow{GE}$$

$$\overrightarrow{GE} = \overrightarrow{GD} + \overrightarrow{DA} + \overrightarrow{AE} = -\frac{1}{2}\mathbf{w} - \mathbf{v} + \frac{1}{2}\mathbf{u}$$

Hence,

$$\overrightarrow{AO'} = \mathbf{v} + \frac{1}{2}\mathbf{w} + \frac{1}{2}\left(-\frac{1}{2}\mathbf{w} - \mathbf{v} + \frac{1}{2}\mathbf{u}\right) = \frac{1}{2}\mathbf{v} + \frac{1}{4}\mathbf{w} + \frac{1}{4}\mathbf{u} \tag{2}$$

To show that $\overrightarrow{AO} = \overrightarrow{AO'}$ we must express $\mathbf{u}$ in terms of $\mathbf{v}$, $\mathbf{w}$ and $\mathbf{z}$. We have

$$\mathbf{v} + \mathbf{w} - \mathbf{z} - \mathbf{u} = 0 \Rightarrow \mathbf{u} = \mathbf{v} + \mathbf{w} - \mathbf{z}$$

Substituting into (2) we get

$$\overrightarrow{AO'} = \frac{1}{2}\mathbf{v} + \frac{1}{4}\mathbf{w} + \frac{1}{4}(\mathbf{v} + \mathbf{w} - \mathbf{z}) = \frac{1}{2}\mathbf{v} + \frac{1}{4}\mathbf{w} + \frac{1}{4}\mathbf{v} + \frac{1}{4}\mathbf{w} - \frac{1}{4}\mathbf{z} = \frac{3}{4}\mathbf{v} + \frac{1}{2}\mathbf{w} - \frac{1}{4}\mathbf{z} \tag{3}$$

By (1) and (3) we conclude that $\overrightarrow{AO} = \overrightarrow{AO'}$. It means that the points O and O' are the same point, in other words, the segment $\overline{FH}$ and $\overline{EG}$ bisect each other.

13.2 Vectors in Three Dimensions (ET Section 12.2)

Preliminary Questions

1. What is the terminal point of the vector $\mathbf{v} = \langle 3, 2, 1\rangle$ based at the point $P = (1, 1, 1)$?

SOLUTION We denote the terminal point by $Q = (a, b, c)$. Then by the definition of components of a vector, we have

$$\langle 3, 2, 1\rangle = \langle a - 1, b - 1, c - 1\rangle$$

Equivalent vectors have equal components respectively, thus,

$$3 = a - 1 \qquad a = 4$$

$$2 = b - 1 \quad \Rightarrow \quad b = 3$$
$$1 = c - 1 \qquad\qquad c = 2$$

The terminal point of **v** is thus $Q = (4, 3, 2)$.

2. What are the components of the vector $\mathbf{v} = \langle 3, 2, 1 \rangle$ based at the point $P = (1, 1, 1)$?

SOLUTION The component of $\mathbf{v} = \langle 3, 2, 1 \rangle$ are $\langle 3, 2, 1 \rangle$ regardless of the base point. The component of **v** and the base point $P = (1, 1, 1)$ determine the head $Q = (a, b, c)$ of the vector, as found in the previous exercise.

3. If $\mathbf{v} = -3\mathbf{w}$, then (choose the correct answer):

(a) **v** and **w** are parallel.

(b) **v** and **w** point in the same direction.

SOLUTION The vectors **v** and **w** lie on parallel lines, hence these vectors are parallel. Since **v** is a scalar multiple of **w** by a negative scalar, **v** and **w** point in opposite directions. Thus, (a) is correct and (b) is not.

4. Which of the following is a direction vector for the line through $P = (3, 2, 1)$ and $Q = (1, 1, 1)$?

(a) $\langle 3, 2, 1 \rangle$ **(b)** $\langle 1, 1, 1 \rangle$ **(c)** $\langle 2, 1, 0 \rangle$

SOLUTION Any vector that is parallel to the vector $\overrightarrow{PQ}$ is a direction vector for the line through P and Q. We compute the vector $\overrightarrow{PQ}$:

$$\overrightarrow{PQ} = \langle 1 - 3, 1 - 2, 1 - 1 \rangle = \langle -2, -1, 0 \rangle.$$

The vectors $\langle 3, 2, 1 \rangle$ and $\langle 1, 1, 1 \rangle$ are not constant multiples of $\overrightarrow{PQ}$, hence they are not parallel to $\overrightarrow{PQ}$. However $\langle 2, 1, 0 \rangle = -1\langle -2, -1, 0 \rangle = -\overrightarrow{PQ}$, hence the vector $\langle 2, 1, 0 \rangle$ is parallel to $\overrightarrow{PQ}$. Therefore, the vector $\langle 2, 1, 0 \rangle$ is a direction vector for the line through P and Q.

5. How many different direction vectors does a line have?

SOLUTION All the vectors that are parallel to a line are also direction vectors for that line. Therefore, there are infinitely many direction vectors for a line.

6. True or false: If **v** is a direction vector for a line $\mathcal{L}$, then $-\mathbf{v}$ is also a direction vector for $\mathcal{L}$.

SOLUTION True. Every vector that is parallel to **v** is a direction vector for the line L. Since $-\mathbf{v}$ is parallel to **v**, it is also a direction vector for L.

Exercises

1. Sketch the vector $\mathbf{v} = \langle 1, 3, 2 \rangle$ and compute its length.

SOLUTION The vector $\mathbf{v} = \langle 1, 3, 2 \rangle$ is shown in the following figure:

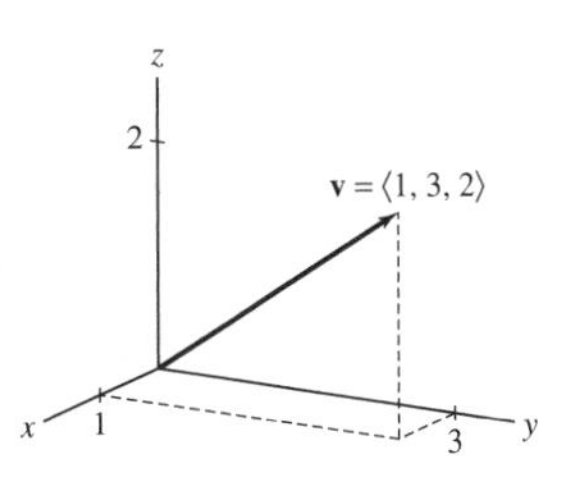

The length of **v** is

$$\|\mathbf{v}\| = \sqrt{1^2 + 3^2 + 2^2} = \sqrt{14}$$

3. Sketch the vector $\mathbf{v} = \langle 1, 1, 0 \rangle$ based at $P = (0, 1, 1)$. Describe this vector in the form $\overrightarrow{PQ}$ for some point Q and sketch the vector $\mathbf{v}_0$ based at the origin equivalent to **v**.

SOLUTION The vector $\mathbf{v} = \langle 1, 1, 0 \rangle$ based at $P = (0, 1, 1)$ is shown in the figure:

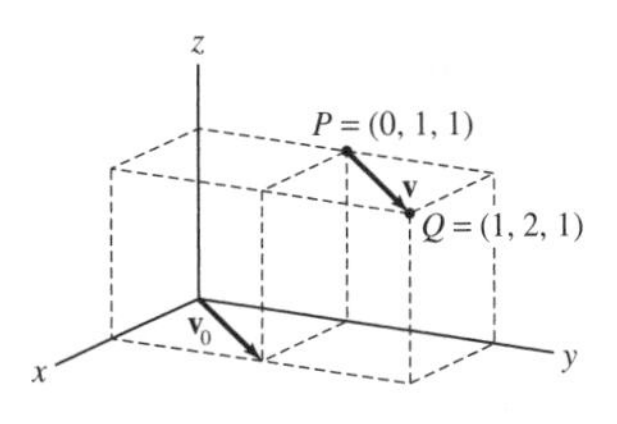

The head Q of the vector $\mathbf{v} = \overrightarrow{PQ}$ is at the point $Q = (0 + 1, 1 + 1, 1 + 0) = (1, 2, 1)$.

The vector $\mathbf{v}_0$ based at the origin and equivalent to $\mathbf{v}$ is

$$v_0 = \langle 1, 1, 0 \rangle = \overrightarrow{OS}, \text{ where } S = (1, 1, 0).$$

In Exercises 5–8, find the components of the vector $\overrightarrow{PQ}$.

5. $P = (1, 0, 1), \quad Q = (2, 1, 0)$

SOLUTION By the definition of the vector components we have

$$\overrightarrow{PQ} = \langle 2 - 1, 1 - 0, 0 - 1 \rangle = \langle 1, 1, -1 \rangle$$

7. $P = (4, 6, 0), \quad Q = \left(-\frac{1}{2}, \frac{9}{2}, 1\right)$

SOLUTION Using the definition of vector components we have

$$\overrightarrow{PQ} = \left\langle -\frac{1}{2} - 4, \frac{9}{2} - 6, 1 - 0 \right\rangle = \left\langle -\frac{9}{2}, -\frac{3}{2}, 1 \right\rangle$$

In Exercises 9–12, let $R = (1, 4, 3)$.

9. Calculate the length of $\overrightarrow{OR}$.

SOLUTION The length of $\overrightarrow{OR}$ is the distance from $R = (1, 4, 3)$ to the origin. That is,

$$\|\overrightarrow{OR}\| = \sqrt{(1-0)^2 + (4-0)^2 + (3-0)^2} = \sqrt{26} \approx 5.1.$$

11. Find the point P such that $\mathbf{w} = \overrightarrow{PR}$ has components $\langle 3, -2, 3 \rangle$ and sketch $\mathbf{w}$.

SOLUTION Denoting $P = (x_0, y_0, z_0)$ we get

$$\overrightarrow{PR} = \langle 1 - x_0, 4 - y_0, 3 - z_0 \rangle = \langle 3, -2, 3 \rangle$$

Equating corresponding components gives

$$\begin{aligned} 1 - x_0 &= 3 \\ 4 - y_0 &= -2 \quad \Rightarrow \quad x_0 = -2, \; y_0 = 6, \; z_0 = 0 \\ 3 - z_0 &= 3 \end{aligned}$$

The point P is, thus, $P = (-2, 6, 0)$.

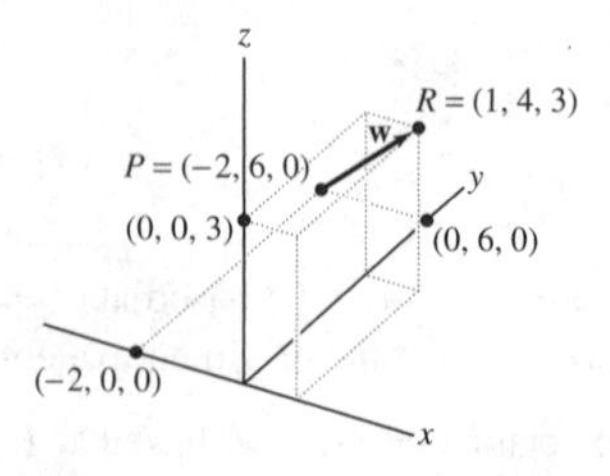

13. Let $\mathbf{v} = \langle 4, 8, 12 \rangle$. Which of the following vectors is parallel to $\mathbf{v}$? Which point in the same direction?

(a) $\langle 2, 4, 6 \rangle$

(b) $\langle -1, -2, 3 \rangle$

(c) $\langle -7, -14, -21 \rangle$

(d) $\langle 6, 10, 14 \rangle$

SOLUTION A vector is parallel to $\mathbf{v}$ if it is a scalar multiple of $\mathbf{v}$. It points in the same direction if the multiplying scalar is positive. Using these properties we obtain the following answer:

(a) $\langle 2, 4, 6 \rangle = \frac{1}{2}\mathbf{v} \Rightarrow$ The vectors are parallel and point in the same direction.

(b) $\langle -1, -2, 3 \rangle$ is not a scalar multiple of $\mathbf{v}$, hence these vectors are not parallel.

(c) $\langle -7, -14, -21\rangle = -\frac{7}{4}\mathbf{v} \Rightarrow$ The vectors are parallel but point in opposite directions.

(d) $\langle 6, 10, 14\rangle$ is not a constant multiple of $\mathbf{v}$, hence these vectors are not parallel.

In Exercises 14–17, determine whether $\overrightarrow{AB}$ is equivalent to $\overrightarrow{PQ}$.

15. $A = (1, 4, 1)$, $B = (-2, 2, 0)$, $P = (2, 5, 7)$, $Q = (-3, 2, 1)$

SOLUTION We compute the two vectors:

$$\overrightarrow{AB} = \langle -2-1, 2-4, 0-1\rangle = \langle -3, -2, -1\rangle$$
$$\overrightarrow{PQ} = \langle -3-2, 2-5, 1-7\rangle = \langle -5, -3, -6\rangle$$

The components of $\overrightarrow{AB}$ and $\overrightarrow{PQ}$ are not equal, hence they are not a translate of each other, that is, the vectors are not equivalent.

17. $A = (1, 1, 0)$, $B = (3, 3, 5)$, $P = (2, -9, 7)$, Q= (4,-7,13)

SOLUTION The vectors $\overrightarrow{AB}$ and $\overrightarrow{PQ}$ are the following vectors:

$$\overrightarrow{AB} = \langle 3-1, 3-1, 5-0\rangle = \langle 2, 2, 5\rangle$$
$$\overrightarrow{PQ} = \langle 4-2, -7-(-9), 13-7\rangle = \langle 2, 2, 6\rangle$$

The z-coordinates of the vectors are not equal, hence the vectors are not equivalent.

In Exercises 18–21, calculate the linear combinations.

19. $-2\langle 8, 11, 3\rangle + 4\langle 2, 1, 1\rangle$

SOLUTION Using the operations of vector addition and scalar multiplication we have

$$-2\langle 8, 11, 3\rangle + 4\langle 2, 1, 1\rangle = \langle -16, -22, -6\rangle + \langle 8, 4, 4\rangle = \langle -8, -18, -2\rangle$$

21. $-\langle 4, 3, 8\rangle + \langle 8, 3, 3\rangle$

SOLUTION Using the operations on vectors we have

$$-\langle 4, 3, 8\rangle + \langle 8, 3, 3\rangle = \langle -4, -3, -8\rangle + \langle 8, 3, 3\rangle = \langle 4, 0, -5\rangle$$

In Exercises 22–25, find the given vector.

23. $\mathbf{e_w}$, where $\mathbf{w} = \langle 4, -2, -1\rangle$

SOLUTION We first find the length of $\mathbf{w}$:

$$\|\mathbf{w}\| = \sqrt{4^2 + (-2)^2 + 1^2} = \sqrt{21}$$

Hence,

$$\mathbf{e_w} = \frac{1}{\|\mathbf{w}\|}\mathbf{w} = \left\langle \frac{4}{\sqrt{21}}, \frac{-2}{\sqrt{21}}, \frac{-1}{\sqrt{21}}\right\rangle$$

25. Unit vector in the direction opposite to $\mathbf{v} = \langle -4, 4, 2\rangle$

SOLUTION A unit vector in the direction opposite to $\mathbf{v} = \langle -4, 4, 2\rangle$ is the following vector:

$$-\mathbf{e_v} = -\frac{1}{\|\mathbf{v}\|}\mathbf{v}$$

We compute the length of $\mathbf{v}$:

$$\|\mathbf{v}\| = \sqrt{(-4)^2 + 4^2 + 2^2} = 6$$

The desired vector is, thus,

$$-\mathbf{e_v} = -\frac{1}{6}\langle -4, 4, 2\rangle = \left\langle \frac{-4}{-6}, \frac{4}{-6}, \frac{2}{-6}\right\rangle = \left\langle \frac{2}{3}, -\frac{2}{3}, -\frac{1}{3}\right\rangle$$

In Exercises 27–34, find a vector parametrization for the line with the given description.

27. Passes through $P = (1, 2, -8)$, direction vector $\mathbf{v} = \langle 2, 1, 3\rangle$

SOLUTION The vector parametrization for the line is

$$\mathbf{r}(t) = \overrightarrow{OP} + t\mathbf{v}$$

Inserting the given data we get

$$\mathbf{r}(t) = \langle 1, 2, -8\rangle + t\langle 2, 1, 3\rangle = \langle 1 + 2t, 2 + t, -8 + 3t\rangle$$

29. Passes through $P = (4, 0, 8)$, direction vector $\mathbf{v} = 7\mathbf{i} + 4\mathbf{k}$

SOLUTION Since $\mathbf{v} = 7\mathbf{i} + 4\mathbf{k} = \langle 7, 0, 4\rangle$ we obtain the following parametrization:

$$\mathbf{r}(t) = \overrightarrow{OP} + t\mathbf{v} = \langle 4, 0, 8\rangle + t\langle 7, 0, 4\rangle = \langle 4 + 7t, 0, 8 + 4t\rangle$$

31. Passes through $(1, 1, 1)$ and $(3, -5, 2)$

SOLUTION We use the equation of the line through two points P and Q:

$$\mathbf{r}(t) = (1 - t)\overrightarrow{OP} + t\overrightarrow{OQ}$$

Since $\overrightarrow{OP} = \langle 1, 1, 1\rangle$ and $\overrightarrow{OQ} = \langle 3, -5, 2\rangle$ we obtain

$$\mathbf{r}(t) = (1 - t)\langle 1, 1, 1\rangle + t\langle 3, -5, 2\rangle = \langle 1 - t, 1 - t, 1 - t\rangle + \langle 3t, -5t, 2t\rangle = \langle 1 + 2t, 1 - 6t, 1 + t\rangle$$

33. Passes through O and $(4, 1, 1)$

SOLUTION By the equation of the line through two points we get

$$\mathbf{r}(t) = (1 - t)\langle 0, 0, 0\rangle + t\langle 4, 1, 1\rangle = \langle 0, 0, 0\rangle + \langle 4t, t, t\rangle = \langle 4t, t, t\rangle$$

In Exercises 35–40, find parametric equations of the line with the given description.

35. Perpendicular to the xy-plane passing through the origin

SOLUTION A direction vector for the line is a vector parallel to the z-axis, for instance, we may choose $\mathbf{v} = \langle 0, 0, 1\rangle$. The line passes through the origin $(0, 0, 0)$, hence we obtain the following parametrization:

$$\mathbf{r}(t) = \langle 0, 0, 0\rangle + t\langle 0, 0, 1\rangle = \langle 0, 0, t\rangle$$

or $x = 0$, $y = 0$, $z = t$.

37. Perpendicular to the yz-plane passing through the point $(0, 0, 2)$

SOLUTION The direction vector is parallel to the x-axis. We may choose $\mathbf{v} = \langle 1, 0, 0\rangle$. Also $\overrightarrow{OP_0} = \langle 0, 0, 2\rangle$ so we obtain

$$\mathbf{r}(t) = \overrightarrow{OP_0} + t\mathbf{v} = \langle 0, 0, 2\rangle + t\langle 1, 0, 0\rangle = \langle t, 0, 2\rangle$$

or $x = t$, $y = 0$, $z = 2$.

39. Passes through $(2, 3, 1)$ in the direction $\mathbf{v} = \langle 2, 1, -2\rangle$

SOLUTION Using the equation of a line we get

$$\mathbf{r}(t) = \overrightarrow{OP_0} + t\mathbf{v} = \langle 2, 3, 1\rangle + t\langle 2, 1, -2\rangle = \langle 2 + 2t, 3 + t, 1 - 2t\rangle$$

Which yields the parametric equations $x = 2 + 2t$, $y = 3 + t$, $z = 1 - 2t$.

In Exercises 41–44, let $P = (2, 1, -1)$ and $Q = (4, 7, 7)$. Find the coordinates of each of the following.

41. The midpoint of $\overline{PQ}$

SOLUTION We first parametrize the line through $P = (2, 1, -1)$ and $Q = (4, 7, 7)$:

$$\mathbf{r}(t) = (1 - t)\langle 2, 1, -1\rangle + t\langle 4, 7, 7\rangle = \langle 2 + 2t, 1 + 6t, -1 + 8t\rangle$$

The midpoint of $\overline{PQ}$ occurs at $t = \frac{1}{2}$, that is,

$$\text{midpoint} = \mathbf{r}\left(\frac{1}{2}\right) = \left\langle 2 + 2\cdot\frac{1}{2}, 1 + 6\cdot\frac{1}{2}, -1 + 8\cdot\frac{1}{2}\right\rangle = \langle 3, 4, 3\rangle$$

The midpoint of $\overline{PQ}$ is the terminal point of the vector $\mathbf{r}(t)$, that is, $(3, 4, 3)$. (One could also use the midpoint formula to arrive at the same solution.)

43. The point R such that Q is the midpoint of $\overline{PR}$

SOLUTION We denote $R = (x_0, y_0, z_0)$. By the formula for the midpoint of a segment we have

$$\langle 4, 7, 7 \rangle = \left\langle \frac{2 + x_0}{2}, \frac{1 + y_0}{2}, \frac{-1 + z_0}{2} \right\rangle$$

Equating corresponding components we get

$$\begin{aligned} 4 &= \frac{2 + x_0}{2} \\ 7 &= \frac{1 + y_0}{2} \quad \Rightarrow \quad x_0 = 6, \ y_0 = 13, \ z_0 = 15 \quad \Rightarrow \quad R = (6, 13, 15) \\ 7 &= \frac{-1 + z_0}{2} \end{aligned}$$

45. Show that $\mathbf{r}_1(t)$ and $\mathbf{r}_2(t)$ define the same line, where

$$\mathbf{r}_1(t) = \langle 3, -1, 4 \rangle + t\,\langle 8, 12, -6 \rangle$$
$$\mathbf{r}_2(t) = \langle 11, 11, -2 \rangle + t\,\langle 4, 6, -3 \rangle$$

Hint: Show that $\mathbf{r}_2$ passes through $(3, -1, 4)$ and that the direction vectors for $\mathbf{r}_1$ and $\mathbf{r}_2$ are parallel.

SOLUTION We observe first that the direction vectors of $\mathbf{r}_1(t)$ and $\mathbf{r}_2(t)$ are multiples of each other:

$$\langle 8, 12, -6 \rangle = 2\,\langle 4, 6, -3 \rangle$$

Therefore $\mathbf{r}_1(t)$ and $\mathbf{r}_2(t)$ are parallel. To show they coincide, it suffices to prove that they share a point in common, so we verify that $\mathbf{r}_1(0) = \langle 3, -1, 4 \rangle$ lies on $\mathbf{r}_2(t)$ by solving for t:

$$\begin{aligned} \langle 3, -1, 4 \rangle &= \langle 11, 11, -2 \rangle + t\,\langle 4, 6, -3 \rangle \\ \langle 3, -1, 4 \rangle - \langle 11, 11, -2 \rangle &= t\,\langle 4, 6, -3 \rangle \\ \langle -8, -12, 6 \rangle &= t\,\langle 4, 6, -3 \rangle \end{aligned}$$

This equation is satisfied for $t = -2$, so $\mathbf{r}_1$ and $\mathbf{r}_2$ coincide.

47. Find two different vector parametrizations of the line through $P = (5, 5, 2)$ with direction vector $\mathbf{v} = \langle 0, -2, 1 \rangle$.

SOLUTION Two different parameterizations are

$$\mathbf{r}_1(t) = \langle 5, 5, 2 \rangle + t\,\langle 0, -2, 1 \rangle$$
$$\mathbf{r}_2(t) = \langle 5, 5, 2 \rangle + t\,\langle 0, -20, 10 \rangle$$

49. Show that the lines $\mathbf{r}_1(t) = \langle -1, 2, 2 \rangle + t\,\langle 4, -2, 1 \rangle$ and $\mathbf{r}_2(t) = \langle 0, 1, 1 \rangle + t\,\langle 2, 0, 1 \rangle$ do not intersect.

SOLUTION The two lines intersect if there exist parameter values t_1 and t_2 such that

$$\begin{aligned} \langle -1, 2, 2 \rangle + t_1\langle 4, -2, 1 \rangle &= \langle 0, 1, 1 \rangle + t_2\langle 2, 0, 1 \rangle \\ \langle -1 + 4t_1, 2 - 2t_1, 2 + t_1 \rangle &= \langle 2t_2, 1, 1 + t_2 \rangle \end{aligned}$$

Equating corresponding components yields

$$\begin{aligned} -1 + 4t_1 &= 2t_2 \\ 2 - 2t_1 &= 1 \\ 2 + t_1 &= 1 + t_2 \end{aligned}$$

The second equation implies $t_1 = \frac{1}{2}$. Substituting into the first and third equations we get

$$\begin{aligned} -1 + 4 \cdot \frac{1}{2} = 2t_2 \quad &\Rightarrow \quad t_2 = \frac{1}{2} \\ 2 + \frac{1}{2} = 1 + t_2 \quad &\Rightarrow \quad t_2 = \frac{3}{2} \end{aligned}$$

We conclude that the equations do not have solutions, which means that the two lines do not intersect.

51. Determine if the lines $\mathbf{r}_1(t) = \langle 0, 1, 1 \rangle + t\,\langle 1, 1, 2 \rangle$ and $\mathbf{r}_2(s) = \langle 2, 0, 3 \rangle + s\,\langle 1, 4, 4 \rangle$ intersect and, if so, find the point of intersection.

SOLUTION The lines intersect if there exist parameter values t and s such that

$$\langle 0, 1, 1\rangle + t\langle 1, 1, 2\rangle = \langle 2, 0, 3\rangle + s\langle 1, 4, 4\rangle$$

$$\langle t, 1+t, 1+2t\rangle = \langle 2+s, 4s, 3+4s\rangle \tag{1}$$

Equating corresponding components we get

$$t = 2 + s$$
$$1 + t = 4s$$
$$1 + 2t = 3 + 4s$$

Substituting t from the first equation into the second equation we get

$$\begin{aligned} 1 + 2 + s &= 4s \\ 3s &= 3 \end{aligned} \quad \Rightarrow \quad s = 1,\ t = 2 + s = 3$$

We now check whether $s = 1$, $t = 3$ satisfy the third equation:

$$1 + 2 \cdot 3 = 3 + 4 \cdot 1$$
$$7 = 7$$

We conclude that $s = 1$, $t = 3$ is the solution of (1), hence the two lines intersect. To find the point of intersection we substitute $s = 1$ in the right-hand side of (1) to obtain

$$\langle 2 + 1, 4 \cdot 1, 3 + 4 \cdot 1\rangle = \langle 3, 4, 7\rangle$$

The point of intersection is the terminal point of this vector, that is, $(3, 4, 7)$.

53. Find the components of the vector $\mathbf{v}$ whose tail and head are the midpoints of segments $\overline{AC}$ and $\overline{BC}$ in Figure 18.

SOLUTION We denote by P and Q the midpoints of the segments $\overline{AC}$ and $\overline{BC}$ respectively. Thus,

$$\mathbf{v} = \overrightarrow{PQ} \tag{1}$$

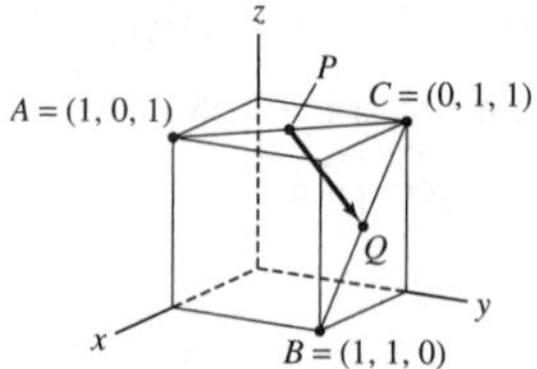

We use the formula for the midpoint of a segment to find the coordinates of the points P and Q. This gives

$$P = \left(\frac{1+0}{2}, \frac{0+1}{2}, \frac{1+1}{2}\right) = \left(\frac{1}{2}, \frac{1}{2}, 1\right)$$

$$Q = \left(\frac{1+0}{2}, \frac{1+1}{2}, \frac{0+1}{2}\right) = \left(\frac{1}{2}, 1, \frac{1}{2}\right)$$

Substituting in (1) yields the following vector:

$$\mathbf{v} = \overrightarrow{PQ} = \left\langle \frac{1}{2} - \frac{1}{2}, 1 - \frac{1}{2}, \frac{1}{2} - 1 \right\rangle = \left\langle 0, \frac{1}{2}, -\frac{1}{2} \right\rangle.$$

Further Insights and Challenges

In Exercises 55–59, we consider the equations of a line in ***symmetric form****.*

$$\frac{x - x_0}{a} = \frac{y - y_0}{b} = \frac{z - z_0}{c} \tag{11}$$

55. Let $\mathcal{L}$ be the line through $P_0 = (x_0, y_0, c_0)$ with direction vector $\mathbf{v} = \langle a, b, c\rangle$. Show that $\mathcal{L}$ is defined by the symmetric equations (11). *Hint:* Use the vector parametrization to show that every point on $\mathcal{L}$ satisfies (11).

SOLUTION $\mathcal{L}$ is given by vector parametrization

$$\mathbf{r}(t) = \langle x_0, y_0, z_0 \rangle + t \langle a, b, c \rangle$$

which gives us the equations

$$x = x_0 + at$$
$$y = y_0 + bt$$
$$z = z_0 + ct.$$

Solving for t gives

$$t = \frac{x - x_0}{a}$$
$$t = \frac{y - y_0}{b}$$
$$t = \frac{z - z_0}{c}$$

Setting each equation equal to the other gives Equation (11).

57. Find a vector parametrization for the line with symmetric equations

$$\frac{x-5}{9} = \frac{y+3}{7} = z - 10$$

SOLUTION Using $(x_0, y_0, z_0) = (5, -3, 10)$ and $\langle a, b, c \rangle = \langle 9, 7, 1 \rangle$ gives

$$\mathbf{r}(t) = \langle 5, -3, 10 \rangle + t \langle 9, 7, 1 \rangle$$

59. Show that the line in the plane through (x_0, y_0) of slope m has symmetric equations

$$\frac{x - x_0}{1} = \frac{y - y_0}{m}$$

SOLUTION The line through (x_0, y_0) of slope m has equation $y - y_0 = m(x - x_0)$, which becomes $x - x_0 = \frac{1}{m}(y - y_0)$, which becomes

$$\frac{x - x_0}{1} = \frac{y - y_0}{m}$$

61. A median of a tetrahedron is a segment joining a vertex to the centroid of the opposite face. Show that the medians of the tetrahedron with vertices at the terminal points of vectors **u**, **v**, **w**, **z** intersect at the terminal point of $\frac{1}{4}(\mathbf{u} + \mathbf{v} + \mathbf{w} + \mathbf{z})$ [Figure 19(B)].

SOLUTION We denote by M_1, M_2, M_3, and M_4 the centroids of the faces ABC, DBC, ADC, and ADB respectively.

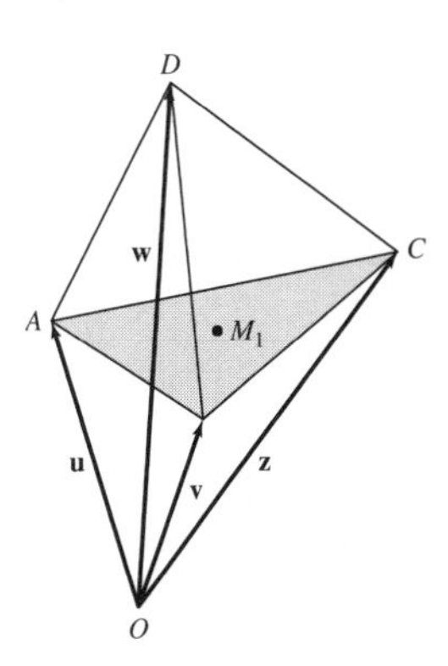

By Exercise 60 the following holds:

$$\overrightarrow{OM_1} = \frac{1}{3}(\mathbf{u} + \mathbf{v} + \mathbf{z})$$
$$\overrightarrow{OM_2} = \frac{1}{3}(\mathbf{w} + \mathbf{v} + \mathbf{z})$$

$$\overrightarrow{OM_3} = \frac{1}{3}(\mathbf{u} + \mathbf{w} + \mathbf{z})$$

$$\overrightarrow{OM_4} = \frac{1}{3}(\mathbf{u} + \mathbf{v} + \mathbf{w})$$

We rewrite the linear combination $\frac{1}{4}(\mathbf{u} + \mathbf{v} + \mathbf{w} + \mathbf{z})$ as follows:

$$\frac{1}{4}(\mathbf{u} + \mathbf{v} + \mathbf{w} + \mathbf{z}) = \frac{3}{4} \cdot \frac{1}{3}(\mathbf{u} + \mathbf{v} + \mathbf{z}) + \frac{1}{4}\mathbf{w} = \frac{3}{4}\overrightarrow{OM_1} + \frac{1}{4}\mathbf{w} \tag{1}$$

$$= \frac{3}{4} \cdot \frac{1}{3}(\mathbf{w} + \mathbf{v} + \mathbf{z}) + \frac{1}{4}\mathbf{u} = \frac{3}{4}\overrightarrow{OM_2} + \frac{1}{4}\mathbf{u} \tag{2}$$

$$= \frac{3}{4} \cdot \frac{1}{3}(\mathbf{u} + \mathbf{w} + \mathbf{z}) + \frac{1}{4}\mathbf{v} = \frac{3}{4}\overrightarrow{OM_3} + \frac{1}{4}\mathbf{v} \tag{3}$$

$$= \frac{3}{4} \cdot \frac{1}{3}(\mathbf{u} + \mathbf{v} + \mathbf{w}) + \frac{1}{4}\mathbf{z} = \frac{3}{4}\overrightarrow{OM_4} + \frac{1}{4}\mathbf{z} \tag{4}$$

We denote by S the terminal point of the vector

$$\frac{1}{4}(\mathbf{u} + \mathbf{v} + \mathbf{w} + \mathbf{z}).$$

By the first equation above, S lies on the segment joining the vertex D (at the tip of the vector $\mathbf{w}$) to the centroid M_1 of the opposite face (at the tip of the vector $\overrightarrow{OM_1}$), because we have S in the form $t \cdot \overrightarrow{OM_1} + (1 - t) \cdot \mathbf{w}$ for $t = \frac{3}{4}$. Likewise, by the second equation, S lies on the segment joining the vertex A with the centroid M_2 of the opposite face. A similar conclusion follows from the third and fourth equations. We conclude that S lies on the four medians of the tetrahedron, hence S is the intersection point of the medians.

13.3 Dot Product and the Angle Between Two Vectors (ET Section 12.3)

Preliminary Questions

1. Is the dot product of two vectors a scalar or a vector?

SOLUTION The dot product of two vectors is the sum of products of scalars, hence it is a scalar.

2. What can you say about the angle between **a** and **b** if $\mathbf{a} \cdot \mathbf{b} < 0$?

SOLUTION Since the cosine of the angle between **a** and **b** satisfies $\cos\theta = \frac{\mathbf{a}\cdot\mathbf{b}}{\|\mathbf{a}\|\|\mathbf{b}\|}$, also $\cos\theta < 0$. By definition $0 \le \theta \le \pi$, but since $\cos\theta < 0$ then θ is in $[\pi/2, \pi]$. In other words, the angle between **a** and **b** is obtuse.

3. Suppose that **v** is orthogonal to both **u** and **w**. Which property of dot products allows us to conclude that **v** is orthogonal to $\mathbf{u} + \mathbf{w}$?

SOLUTION One property is that two vectors are orthogonal if and only if the dot product of the two vectors is zero. The second property is the Distributive Law. Since **v** is orthogonal to **u** and **w**, we have $\mathbf{v} \cdot \mathbf{u} = 0$ and $\mathbf{v} \cdot \mathbf{w} = 0$. Therefore,

$$\mathbf{v} \cdot (\mathbf{u} + \mathbf{w}) = \mathbf{v} \cdot \mathbf{u} + \mathbf{v} \cdot \mathbf{w} = 0 + 0 = 0$$

We conclude that **v** is orthogonal to $\mathbf{u} + \mathbf{w}$.

4. What is $\text{proj}_{\mathbf{v}}(\mathbf{v})$?

SOLUTION The projection of **v** along itself is **v**, since

$$\text{proj}_{\mathbf{v}}(\mathbf{v}) = \left(\frac{\mathbf{v} \cdot \mathbf{v}}{\mathbf{v} \cdot \mathbf{v}}\right)\mathbf{v} = \mathbf{v}$$

5. What is the difference, if any, between the projection of **u** along **v** and the projection along the unit vector $\mathbf{e_v}$?

SOLUTION The projection of **u** along **v** is the vector

$$\text{proj}_{\mathbf{v}}(\mathbf{u}) = (\mathbf{u} \cdot \mathbf{e_v})\mathbf{e_v} \tag{1}$$

The projection of **u** along $\mathbf{e_v}$ is the vector

$$\text{proj}_{\mathbf{e_v}}(\mathbf{u}) = \left(\mathbf{u} \cdot \mathbf{e_{e_v}}\right)\mathbf{e_{e_v}} \tag{2}$$

Since $\mathbf{e_{e_v}} = \mathbf{e_v}$, the vectors in (1) and (2) are identical. That is, the projection of **u** along **v** is the projection of **u** along $\mathbf{e_v}$.

6. Suppose that $\text{proj}_{\mathbf{v}}(\mathbf{u}) = \mathbf{v}$. Determine:

(a) $\text{proj}_{2\mathbf{v}}(\mathbf{u})$

(b) $\text{proj}_{\mathbf{v}}(2\mathbf{u})$

SOLUTION

(a) The projection of **u** along 2**v** is the following vector:

$$\text{proj}_{2\mathbf{v}}(\mathbf{u}) = \left(\frac{\mathbf{u}\cdot 2\mathbf{v}}{(2\mathbf{v})\cdot(2\mathbf{v})}\right)2\mathbf{v} = \left(\frac{2\mathbf{u}\cdot\mathbf{v}}{4\mathbf{v}\cdot\mathbf{v}}\right)2\mathbf{v} = \left(\frac{4\mathbf{u}\cdot\mathbf{v}}{4\mathbf{v}\cdot\mathbf{v}}\right)\mathbf{v} = \left(\frac{\mathbf{u}\cdot\mathbf{v}}{\mathbf{v}\cdot\mathbf{v}}\right)\mathbf{v} = \text{proj}_{\mathbf{v}}(\mathbf{u}) = \mathbf{v}$$

(b) The projection of 2**u** along **v** is the following vector:

$$\text{proj}_{\mathbf{v}}(2\mathbf{u}) = \left(\frac{2\mathbf{u}\cdot\mathbf{v}}{\mathbf{v}\cdot\mathbf{v}}\right)\mathbf{v} = 2\left(\frac{\mathbf{u}\cdot\mathbf{v}}{\mathbf{v}\cdot\mathbf{v}}\right)\mathbf{v} = 2\text{proj}_{\mathbf{v}}(\mathbf{u}) = 2\cdot\mathbf{v} = 2\mathbf{v}$$

7. Let $\mathbf{u}_{||}$ be the projection of **u** along **v**. Which of the following is the projection **u** along the vector 2**v**?

(a) $\frac{1}{2}\mathbf{u}_{||}$ **(b)** $\mathbf{u}_{||}$ **(c)** $2\mathbf{u}_{||}$

SOLUTION Since $u_{\|}$ is the projection of **u** along **v**, we have,

$$u_{\|} = \left(\frac{\mathbf{u}\cdot\mathbf{v}}{\mathbf{v}\cdot\mathbf{v}}\right)\mathbf{v}$$

The projection of **u** along the vector 2**v** is

$$\text{proj}_{2\mathbf{v}}(\mathbf{u}) = \left(\frac{\mathbf{u}\cdot 2\mathbf{v}}{2\mathbf{v}\cdot 2\mathbf{v}}\right)2\mathbf{v} = \left(\frac{2\mathbf{u}\cdot\mathbf{v}}{4\mathbf{v}\cdot\mathbf{v}}\right)2\mathbf{v} = \left(\frac{4\mathbf{u}\cdot\mathbf{v}}{4\mathbf{v}\cdot\mathbf{v}}\right)\mathbf{v} = \left(\frac{\mathbf{u}\cdot\mathbf{v}}{\mathbf{v}\cdot\mathbf{v}}\right)\mathbf{v} = u_{\|}$$

That is, $u_{\|}$ is the projection of **u** along 2**v**. Notice that the projection of **u** along **v** is the projection of **u** along the unit vector $\mathbf{e}_{\mathbf{v}}$, hence it depends on the direction of **v** rather than on the length of **v**. Therefore, the projection of **u** along **v** and along 2**v** is the same vector.

8. Let θ be the angle between **u** and **v**. Which of the following is equal to the $\cos\theta$?

(a) $\mathbf{u}\cdot\mathbf{v}$ **(b)** $\mathbf{u}\cdot\mathbf{e}_{\mathbf{v}}$ **(c)** $\mathbf{e}_{\mathbf{u}}\cdot\mathbf{e}_{\mathbf{v}}$

SOLUTION By the Theorems on the Dot Product and the Angle Between Vectors, we have

$$\cos\theta = \frac{\mathbf{u}\cdot\mathbf{v}}{\|\mathbf{u}\|\|\mathbf{v}\|} = \frac{\mathbf{u}}{\|\mathbf{u}\|}\cdot\frac{\mathbf{v}}{\|\mathbf{v}\|} = \mathbf{e}_{\mathbf{u}}\cdot\mathbf{e}_{\mathbf{v}}$$

The correct answer is (c).

Exercises

In Exercises 1–12, compute the dot product.

1. $\langle 1, 2, 1\rangle \cdot \langle 4, 3, 5\rangle$

SOLUTION Using the definition of the dot product we obtain

$$\langle 1, 2, 1\rangle \cdot \langle 4, 3, 5\rangle = 1\cdot 4 + 2\cdot 3 + 1\cdot 5 = 15$$

3. $\langle 0, 1, 0\rangle \cdot \langle 7, 41, -3\rangle$

SOLUTION The dot product is

$$\langle 0, 1, 0\rangle \cdot \langle 7, 41, -3\rangle = 0\cdot 7 + 1\cdot 41 + 0\cdot(-3) = 41$$

5. $\langle 3, 1\rangle \cdot \langle 4, -7\rangle$

SOLUTION The dot product of the two vectors is the following scalar:

$$\langle 3, 1\rangle \cdot \langle 4, -7\rangle = 3\cdot 4 + 1\cdot(-7) = 5$$

7. $\mathbf{i}\cdot\mathbf{k}$

SOLUTION The vectors **i** and **k** are orthogonal, hence $\mathbf{i}\cdot\mathbf{k} = 0$.

9. $(\mathbf{i}+\mathbf{j})\cdot(\mathbf{i}+\mathbf{k})$

SOLUTION By the distributive law and the orthogonality of **i**, **j**, and **k**,

$$(\mathbf{i}+\mathbf{j})\cdot(\mathbf{i}+\mathbf{k}) = \mathbf{i}\cdot(\mathbf{i}+\mathbf{k}) + \mathbf{j}(\mathbf{i}+\mathbf{k}) = \mathbf{i}\cdot\mathbf{i} + \mathbf{i}\cdot\mathbf{k} + \mathbf{j}\cdot\mathbf{i} + \mathbf{j}\cdot\mathbf{k} = \|\mathbf{i}\|^2 + 0 + 0 + 0 = 1$$

11. $(\mathbf{i}+\mathbf{j}+\mathbf{k})\cdot(3\mathbf{i}+2\mathbf{j}-5\mathbf{k})$

SOLUTION We use properties of the dot product to obtain

$$(\mathbf{i}+\mathbf{j}+\mathbf{k})\cdot(3\mathbf{i}+2\mathbf{j}-5\mathbf{k}) = 3\mathbf{i}\cdot\mathbf{i}+2\mathbf{i}\cdot\mathbf{j}-5\mathbf{i}\cdot\mathbf{k}+3\mathbf{j}\cdot\mathbf{i}+2\mathbf{j}\cdot\mathbf{j}-5\mathbf{j}\cdot\mathbf{k}+3\mathbf{k}\cdot\mathbf{i}+2\mathbf{k}\cdot\mathbf{j}-5\mathbf{k}\cdot\mathbf{k}$$
$$= 3\|\mathbf{i}\|^2+2\|\mathbf{j}\|^2-5\|\mathbf{k}\|^2 = 3\cdot 1+2\cdot 1-5\cdot 1 = 0$$

In Exercises 13–18, determine whether the two vectors are orthogonal and if not, whether the angle between them is acute or obtuse.

13. $\langle 1, 1, 1\rangle, \quad \langle 1, -2, -2\rangle$

SOLUTION We compute the dot product of the two vectors:

$$\langle 1, 1, 1\rangle \cdot \langle 1, -2, -2\rangle = 1\cdot 1 + 1\cdot(-2) + 1\cdot(-2) = -3$$

Since the dot product is negative, the angle between the vectors is obtuse.

15. $\langle 0, 2, 4\rangle, \quad \langle 3, 1, 0\rangle$

SOLUTION We find the dot product of the two vectors:

$$\langle 0, 2, 4\rangle \cdot \langle 3, 1, 0\rangle = 0\cdot 3 + 2\cdot 1 + 4\cdot 0 = 2$$

The dot product is positive, hence the angle between the vectors is acute.

17. $\langle 4, 3\rangle, \quad \langle 2, -4\rangle$

SOLUTION We find the dot product:

$$\langle 4, 3\rangle \cdot \langle 2, -4\rangle = 4\cdot 2 + 3\cdot(-4) = -4$$

The dot product is negative, hence the angle between the vectors is obtuse.

In Exercises 19–22, find the cosine of the angle between the vectors.

19. $\langle 0, 3, 1\rangle, \quad \langle 4, 0, 0\rangle$

SOLUTION Since $\langle 0, 3, 1\rangle \cdot \langle 4, 0, 0\rangle = 0\cdot 4 + 3\cdot 0 + 1\cdot 0 = 0$, the vectors are orthogonal, that is, the angle between them is $\theta = 90^\circ$ and $\cos\theta = 0$.

21. $\mathbf{i}+\mathbf{j}, \quad \mathbf{j}+2\mathbf{k}$

SOLUTION We use the formula for the cosine of the angle between two vectors. Let $\mathbf{v} = \mathbf{i}+\mathbf{j}$ and $\mathbf{w} = \mathbf{j}+2\mathbf{k}$. We compute the following values:

$$\|\mathbf{v}\| = \|\mathbf{i}+\mathbf{j}\| = \sqrt{1^2+1^2} = \sqrt{2}$$
$$\|\mathbf{w}\| = \|\mathbf{j}+2\mathbf{k}\| = \sqrt{1^2+2^2} = \sqrt{5}$$
$$\mathbf{v}\cdot\mathbf{w} = (\mathbf{i}+\mathbf{j})\cdot(\mathbf{j}+2\mathbf{k}) = \mathbf{i}\cdot\mathbf{j}+2\mathbf{i}\cdot\mathbf{k}+\mathbf{j}\cdot\mathbf{j}+2\mathbf{j}\cdot\mathbf{k} = \|\mathbf{j}\|^2 = 1$$

Hence,

$$\cos\theta = \frac{\mathbf{v}\cdot\mathbf{w}}{\|\mathbf{v}\|\|\mathbf{w}\|} = \frac{1}{\sqrt{2}\sqrt{5}} = \frac{1}{\sqrt{10}}.$$

In Exercises 23–28, find the angle between the vectors.

23. $\langle 1, 2\rangle, \quad \langle 3, 5\rangle$

SOLUTION Letting $\mathbf{v} = \langle 1, 2\rangle$ and $\mathbf{w} = \langle 3, 5\rangle$, we have

$$\|\mathbf{v}\| = \sqrt{1^2+2^2} = \sqrt{5}$$
$$\|\mathbf{w}\| = \sqrt{3^2+5^2} = \sqrt{34}$$
$$\mathbf{v}\cdot\mathbf{w} = \langle 1, 2\rangle \cdot \langle 3, 5\rangle = 1\cdot 3 + 2\cdot 5 = 13$$

Using the formula for the angle between two vectors we obtain

$$\cos\theta = \frac{\mathbf{v}\cdot\mathbf{w}}{\|\mathbf{v}\|\|\mathbf{w}\|} = \frac{13}{\sqrt{5}\sqrt{34}} = 0.997 \quad\Rightarrow\quad \theta = \cos^{-1} 0.997 \approx 4.4^\circ$$

25. $\langle 1, 1, 1\rangle, \quad \langle 1, -1, 1\rangle$

SOLUTION Letting $\mathbf{v} = \langle 1, 1, 1\rangle$ and $\mathbf{w} = \langle 1, -1, 1\rangle$, we have

$$\|\mathbf{v}\| = \sqrt{1^2 + 1^2 + 1^2} = \sqrt{3}$$
$$\|\mathbf{w}\| = \sqrt{1^2 + (-1)^2 + 1^2} = \sqrt{3}$$
$$\mathbf{v}\cdot\mathbf{w} = \langle 1, 1, 1\rangle \cdot \langle 1, -1, 1\rangle = 1\cdot 1 + 1\cdot(-1) + 1\cdot 1 = 1$$

Using the formula for the cosine of the angle between $\mathbf{v}$ and $\mathbf{w}$ we get

$$\cos\theta = \frac{\mathbf{v}\cdot\mathbf{w}}{\|\mathbf{v}\|\|\mathbf{w}\|} = \frac{1}{\sqrt{3}\cdot\sqrt{3}} = \frac{1}{3} \quad\Rightarrow\quad \theta \approx 70.53^\circ.$$

27. $\langle \pi, e, 3\rangle$, $\left\langle \cos 1, \tan 1, e^2\right\rangle$

SOLUTION We denote $\mathbf{v} = \langle \pi, e, 3\rangle$ and $\mathbf{w} = \langle \cos 1, \tan 1, e^2\rangle$. Note that $\cos 1$ and $\tan 1$ must be calculated in radians! Hence,

$$\|\mathbf{v}\| = \sqrt{\pi^2 + e^2 + 9} \approx 5.124$$
$$\|\mathbf{w}\| = \sqrt{\cos^2 1 + \tan^2 1 + e^4} \approx 7.57$$
$$\mathbf{v}\cdot\mathbf{w} = \pi\cos 1 + e\tan 1 + 3e^2 \approx 28.1$$

Using the formula for the cosine of the angle we have

$$\cos\theta = \frac{\mathbf{v}\cdot\mathbf{w}}{\|\mathbf{v}\|\|\mathbf{w}\|} = \frac{28.1}{5.124\cdot 7.57} = 0.724 \quad\Rightarrow\quad \theta = \cos^{-1} 0.724 \approx 43.57^\circ$$

29. Find all values of b for which the vectors are orthogonal.

(a) $\langle b, 3, 2\rangle$, $\langle 1, b, 1\rangle$

(b) $\langle 4, -2, 7\rangle$, $\left\langle b^2, b, 0\right\rangle$

SOLUTION

(a) The vectors are orthogonal if and only if the scalar product is zero. That is,

$$\langle b, 3, 2\rangle \cdot \langle 1, b, 1\rangle = 0$$
$$b\cdot 1 + 3\cdot b + 2\cdot 1 = 0$$
$$4b + 2 = 0 \quad\Rightarrow\quad b = -\frac{1}{2}$$

(b) We set the scalar product of the two vectors equal to zero and solve for b. This gives

$$\langle 4, -2, 7\rangle \cdot \langle b^2, b, 0\rangle = 0$$
$$4b^2 - 2b + 7\cdot 0 = 0$$
$$2b(2b - 1) = 0 \quad\Rightarrow\quad b = 0 \text{ or } b = \frac{1}{2}$$

31. Find two vectors (which are not multiples of each other) that are both orthogonal to $\langle 2, 0, -3\rangle$.

SOLUTION We denote by $\langle a, b, c\rangle$, a vector orthogonal to $\langle 2, 0, -3\rangle$. Hence,

$$\langle a, b, c\rangle \cdot \langle 2, 0, -3\rangle = 0$$
$$2a + 0 - 3c = 0$$
$$2a - 3c = 0 \quad\Rightarrow\quad a = \frac{3}{2}c$$

Thus, the vectors orthogonal to $\langle 2, 0, -3\rangle$ are of the form

$$\left\langle \frac{3}{2}c, b, c\right\rangle.$$

We may find two such vectors by setting $c = 0, b = 1$ and $c = 2, b = 2$. We obtain

$$\mathbf{v}_1 = \langle 0, 1, 0\rangle, \quad \mathbf{v}_2 = \langle 3, 2, 2\rangle.$$

In Exercises 33–36, assume that $\mathbf{u}\cdot\mathbf{v} = 2$, $\|\mathbf{u}\| = 1$, *and* $\|\mathbf{v}\| = 3$ *and evaluate the expression.*

33. $\mathbf{u} \cdot (4\mathbf{v})$

SOLUTION Using properties of the dot product we get

$$\mathbf{u} \cdot (4\mathbf{v}) = 4(\mathbf{u} \cdot \mathbf{v}) = 4 \cdot 2 = 8.$$

35. $2\mathbf{u} \cdot (3\mathbf{u} - \mathbf{v})$

SOLUTION By properties of the dot product we obtain

$$\begin{aligned} 2\mathbf{u} \cdot (3\mathbf{u} - \mathbf{v}) &= (2\mathbf{u}) \cdot (3\mathbf{u}) - (2\mathbf{u}) \cdot \mathbf{v} = 6(\mathbf{u} \cdot \mathbf{u}) - 2(\mathbf{u} \cdot \mathbf{v}) \\ &= 6\|\mathbf{u}\|^2 - 2(\mathbf{u} \cdot \mathbf{v}) = 6 \cdot 1^2 - 2 \cdot 2 = 2 \end{aligned}$$

In Exercises 37–40, simplify the expression.

37. $(\mathbf{v} - \mathbf{w}) \cdot \mathbf{v} + \mathbf{v} \cdot \mathbf{w}$

SOLUTION By properties of the dot product we obtain

$$(\mathbf{v} - \mathbf{w}) \cdot \mathbf{v} + \mathbf{v} \cdot \mathbf{w} = \mathbf{v} \cdot \mathbf{v} - \mathbf{w} \cdot \mathbf{v} + \mathbf{v} \cdot \mathbf{w} = \|\mathbf{v}\|^2 - \mathbf{v} \cdot \mathbf{w} + \mathbf{v} \cdot \mathbf{w} = \|\mathbf{v}\|^2$$

39. $(\mathbf{v} + \mathbf{w}) \cdot \mathbf{v} - (\mathbf{v} + \mathbf{w}) \cdot \mathbf{w}$

SOLUTION We use properties of the dot product to write

$$\begin{aligned} (\mathbf{v} + \mathbf{w}) \cdot \mathbf{v} - (\mathbf{v} + \mathbf{w}) \cdot \mathbf{w} &= \mathbf{v} \cdot \mathbf{v} + \mathbf{w} \cdot \mathbf{v} - \mathbf{v} \cdot \mathbf{w} - \mathbf{w} \cdot \mathbf{w} \\ &= \|\mathbf{v}\|^2 + \mathbf{w} \cdot \mathbf{v} - \mathbf{w} \cdot \mathbf{v} - \|\mathbf{w}\|^2 = \|\mathbf{v}\|^2 - \|\mathbf{w}\|^2 \end{aligned}$$

In Exercises 41–48, find the projection $\text{proj}_{\mathbf{v}}(\mathbf{u})$.

41. $\mathbf{u} = \langle -1, 2, 0 \rangle, \quad \mathbf{v} = \langle 2, 0, 1 \rangle$

SOLUTION The projection of $\mathbf{u}$ along $\mathbf{v}$ is the following vector:

$$\text{proj}_{\mathbf{v}}(\mathbf{u}) = \left(\frac{\mathbf{u} \cdot \mathbf{v}}{\mathbf{v} \cdot \mathbf{v}}\right) \mathbf{v}$$

We compute the values in this expression:

$$\begin{aligned} \mathbf{u} \cdot \mathbf{v} &= \langle -1, 2, 0 \rangle \cdot \langle 2, 0, 1 \rangle = -1 \cdot 2 + 2 \cdot 0 + 0 \cdot 1 = -2 \\ \mathbf{v} \cdot \mathbf{v} &= \|\mathbf{v}\|^2 = 2^2 + 0^2 + 1^2 = 5 \end{aligned}$$

Hence,

$$\text{proj}_{\mathbf{v}}(\mathbf{u}) = -\frac{2}{5}\langle 2, 0, 1 \rangle = \left\langle -\frac{4}{5}, 0, -\frac{2}{5} \right\rangle.$$

43. $\mathbf{u} = \langle 1, 1, 1 \rangle, \quad \mathbf{v} = \langle 1, 1, 0 \rangle$

SOLUTION We first compute the following dot products:

$$\begin{aligned} \mathbf{u} \cdot \mathbf{v} &= \langle 1, 1, 1 \rangle \cdot \langle 1, 1, 0 \rangle = 1 \cdot 1 + 1 \cdot 1 + 1 \cdot 0 = 2 \\ \mathbf{v} \cdot \mathbf{v} &= \|\mathbf{v}\|^2 = 1^2 + 1^2 + 0^2 = 2 \end{aligned}$$

The projection of $\mathbf{u}$ along $\mathbf{v}$ is the following vector:

$$\text{proj}_{\mathbf{v}}(\mathbf{u}) = \left(\frac{\mathbf{u} \cdot \mathbf{v}}{\mathbf{v} \cdot \mathbf{v}}\right) \mathbf{v} = \frac{2}{2}\mathbf{v} = \mathbf{v} = \langle 1, 1, 0 \rangle$$

45. $\mathbf{u} = 5\mathbf{i} + 7\mathbf{j} - 4\mathbf{k}, \quad \mathbf{v} = \mathbf{k}$

SOLUTION The projection of $\mathbf{u}$ along $\mathbf{v}$ is the following vector:

$$\text{proj}_{\mathbf{v}}(\mathbf{u}) = \left(\frac{\mathbf{u} \cdot \mathbf{v}}{\mathbf{v} \cdot \mathbf{v}}\right) \mathbf{v}$$

We compute the dot products:

$$\mathbf{u} \cdot \mathbf{v} = (5\mathbf{i} + 7\mathbf{j} - 4\mathbf{k}) \cdot \mathbf{k} = -4\mathbf{k} \cdot \mathbf{k} = -4$$

$$\mathbf{v} \cdot \mathbf{v} = \|\mathbf{v}\|^2 = \|\mathbf{k}\|^2 = 1$$

Hence,

$$\text{proj}_{\mathbf{v}}(\mathbf{u}) = \frac{-4}{1}\mathbf{k} = -4\mathbf{k}$$

47. $\mathbf{u} = \langle a, b, c \rangle, \quad \mathbf{v} = \mathbf{i}$

SOLUTION The component of $\mathbf{u}$ along $\mathbf{v}$ is a, since

$$\mathbf{u} \cdot \mathbf{e_v} = (a\mathbf{i} + b\mathbf{j} + c\mathbf{k}) \cdot \mathbf{i} = a$$

Therefore, the projection of $\mathbf{u}$ along $\mathbf{v}$ is the vector

$$\text{proj}_{\mathbf{v}}(\mathbf{u}) = (\mathbf{u} \cdot \mathbf{e_v})\mathbf{e_v} = a\mathbf{i}$$

In Exercises 49–54, find the decomposition $\mathbf{a} = \mathbf{a}_{\|} + \mathbf{a}_{\perp}$ *with respect to* $\mathbf{b}$.

49. $\mathbf{a} = \langle -2, 1, 1 \rangle, \quad \mathbf{b} = \langle 1, 0, 0 \rangle$

SOLUTION We compute $\mathbf{a}_{\|}$ in two steps:

Step 1. Compute $\mathbf{a} \cdot \mathbf{b}$ and $\mathbf{b} \cdot \mathbf{b}$

$$\mathbf{a} \cdot \mathbf{b} = \langle -2, 1, 1 \rangle \cdot \langle 1, 0, 0 \rangle = -2 \cdot 1 + 1 \cdot 0 + 1 \cdot 0 = -2$$

$$\mathbf{b} \cdot \mathbf{b} = \|\mathbf{b}\|^2 = 1^2 + 0^2 + 0^2 = 1$$

Step 2. We find $\mathbf{a}_{\|}$:

$$\mathbf{a}_{\|} = \left(\frac{\mathbf{a} \cdot \mathbf{b}}{\mathbf{b} \cdot \mathbf{b}}\right)\mathbf{b} = \frac{-2}{1}\langle 1, 0, 0 \rangle = \langle -2, 0, 0 \rangle$$

The orthogonal part is the difference:

$$\mathbf{a}_{\perp} = \mathbf{a} - \mathbf{a}_{\|} = \langle -2, 1, 1 \rangle - \langle -2, 0, 0, \rangle = \langle 0, 1, 1 \rangle$$

Thus,

$$\mathbf{a} = \langle -2, 1, 1 \rangle = \mathbf{a}_{\|} + \mathbf{a}_{\perp} = \langle -2, 0, 0, \rangle + \langle 0, 1, 1 \rangle$$

51. $\mathbf{a} = \langle 1, 0 \rangle, \quad \mathbf{b} = \langle 1, 1 \rangle$

SOLUTION

Step 1. We compute $\mathbf{a} \cdot \mathbf{b}$ and $\mathbf{b} \cdot \mathbf{b}$

$$\mathbf{a} \cdot \mathbf{b} = \langle 1, 0 \rangle \cdot \langle 1, 1 \rangle = 1 \cdot 1 + 0 \cdot 1 = 1$$

$$\mathbf{b} \cdot \mathbf{b} = \|\mathbf{b}\|^2 = 1^2 + 1^2 = 2$$

Step 2. We find the projection of $\mathbf{a}$ along $\mathbf{b}$:

$$\mathbf{a}_{\|} = \left(\frac{\mathbf{a} \cdot \mathbf{b}}{\mathbf{b} \cdot \mathbf{b}}\right)\mathbf{b} = \frac{1}{2}\langle 1, 1 \rangle = \left\langle \frac{1}{2}, \frac{1}{2} \right\rangle$$

Step 3. We find the orthogonal part as the difference:

$$\mathbf{a}_{\perp} = \mathbf{a} - \mathbf{a}_{\|} = \langle 1, 0 \rangle - \left\langle \frac{1}{2}, \frac{1}{2} \right\rangle = \left\langle \frac{1}{2}, -\frac{1}{2} \right\rangle$$

Hence,

$$\mathbf{a} = \mathbf{a}_{\|} + \mathbf{a}_{\perp} = \left\langle \frac{1}{2}, \frac{1}{2} \right\rangle + \left\langle \frac{1}{2}, -\frac{1}{2} \right\rangle.$$

53. $\mathbf{a} = \langle 4, 2 \rangle, \quad \mathbf{b} = \langle -1, 2 \rangle$

SOLUTION Since $\mathbf{a} \cdot \mathbf{b} = 4 \cdot (-1) + 2 \cdot 2 = 0$, the vectors $\mathbf{a}$ and $\mathbf{b}$ are orthogonal, hence the projection of $\mathbf{a}$ along $\mathbf{b}$ is the zero vector. Thus we have

$$\mathbf{a}_{\|} = \mathbf{0}, \quad \mathbf{a}_{\perp} = \mathbf{a} = \langle 4, 2 \rangle.$$

55. Let $\mathbf{e}_\theta = \langle \cos\theta, \sin\theta \rangle$. Show that $\mathbf{e}_\theta$ is a unit vector making an angle θ with the x-axis. Show that $\mathbf{e}_\theta \cdot \mathbf{e}_\psi = \cos(\theta - \psi)$ for any two angles θ and ψ.

SOLUTION First, $\mathbf{e}_\theta$ is a unit vector since by a trigonometric identity we have

$$\|\mathbf{e}_\theta\| = \sqrt{\cos^2\theta + \sin^2\theta} = \sqrt{1} = 1$$

The cosine of the angle α between $\mathbf{e}_\theta$ and the vector $\mathbf{i}$ in the direction of the positive x-axis is

$$\cos\alpha = \frac{\mathbf{e}_\theta \cdot \mathbf{i}}{\|\mathbf{e}_\theta\| \cdot \|\mathbf{i}\|} = \mathbf{e}_\theta \cdot \mathbf{i} = ((\cos\theta)\mathbf{i} + (\sin\theta)\mathbf{j}) \cdot \mathbf{i} = \cos\theta$$

The solution of $\cos\alpha = \cos\theta$ for angles between 0° and 180° is $\alpha = \theta$. That is, the vector $\mathbf{e}_\theta$ makes an angle θ with the x-axis. We now use the trigonometric identity

$$\cos\theta\cos\psi + \sin\theta\sin\psi = \cos(\theta - \psi)$$

to obtain the following equality:

$$\mathbf{e}_\theta \cdot \mathbf{e}_\psi = \langle \cos\theta, \sin\theta \rangle \cdot \langle \cos\psi, \sin\psi \rangle = \cos\theta\cos\psi + \sin\theta\sin\psi = \cos(\theta - \psi)$$

57. Determine $\|\mathbf{v} + \mathbf{w}\|$ if $\mathbf{v}$ and $\mathbf{w}$ are unit vectors separated by an angle of 30°.

SOLUTION We use the relation of the dot product with length and properties of the dot product to write

$$\begin{aligned}\|\mathbf{v} + \mathbf{w}\|^2 &= (\mathbf{v} + \mathbf{w}) \cdot (\mathbf{v} + \mathbf{w}) = \mathbf{v} \cdot \mathbf{v} + \mathbf{v} \cdot \mathbf{w} + \mathbf{w} \cdot \mathbf{v} + \mathbf{w} \cdot \mathbf{w} \\ &= \|\mathbf{v}\|^2 + 2\mathbf{v} \cdot \mathbf{w} + \|\mathbf{w}\|^2 = 1^2 + 2\mathbf{v} \cdot \mathbf{w} + 1^2 = 2 + 2\mathbf{v} \cdot \mathbf{w}\end{aligned}$$

We now find $\mathbf{v} \cdot \mathbf{w}$:

$$\mathbf{v} \cdot \mathbf{w} = \|\mathbf{v}\|\|\mathbf{w}\| \cos 30^\circ = 1 \cdot 1 \cos 30^\circ = \frac{\sqrt{3}}{2}$$

Hence,

$$\|\mathbf{v} + \mathbf{w}\|^2 = 2 + 2 \cdot \frac{\sqrt{3}}{2} = 2 + \sqrt{3} \quad \Rightarrow \quad \|\mathbf{v} + \mathbf{w}\| = \sqrt{2 + \sqrt{3}} \approx 1.93$$

59. Suppose that $\|\mathbf{v}\| = 2$ and $\|\mathbf{w}\| = 3$, and the angle between $\mathbf{v}$ and $\mathbf{w}$ is 120°. Determine:

(a) $\mathbf{v} \cdot \mathbf{w}$ **(b)** $\|2\mathbf{v} + \mathbf{w}\|$ **(c)** $\|2\mathbf{v} - 3\mathbf{w}\|$

SOLUTION

(a) We use the relation between the dot product and the angle between two vectors to write

$$\mathbf{v} \cdot \mathbf{w} = \|\mathbf{v}\|\|\mathbf{w}\| \cos\theta = 2 \cdot 3\cos 120^\circ = 6 \cdot \left(-\frac{1}{2}\right) = -3$$

(b) By the relation of the dot product with length and by properties of the dot product we have

$$\begin{aligned}\|2\mathbf{v} + \mathbf{w}\|^2 &= (2\mathbf{v} + \mathbf{w}) \cdot (2\mathbf{v} + \mathbf{w}) = 4\mathbf{v} \cdot \mathbf{v} + 2\mathbf{v} \cdot \mathbf{w} + 2\mathbf{w} \cdot \mathbf{v} + \mathbf{w} \cdot \mathbf{w} \\ &= 4\|\mathbf{v}\|^2 + 4\mathbf{v} \cdot \mathbf{w} + \|\mathbf{w}\|^2\end{aligned}$$

We now substitute $\mathbf{v} \cdot \mathbf{w} = -3$ from part (a) and the given information, obtaining

$$\|2\mathbf{v} + \mathbf{w}\|^2 = 4 \cdot 2^2 + 4(-3) + 3^2 = 13 \quad \Rightarrow \quad \|2\mathbf{v} + \mathbf{w}\| = \sqrt{13} \approx 3.61$$

(c) We express the length in terms of a dot product and use properties of the dot product. This gives

$$\begin{aligned}\|2\mathbf{v} - 3\mathbf{w}\|^2 &= (2\mathbf{v} - 3\mathbf{w}) \cdot (2\mathbf{v} - 3\mathbf{w}) = 4\mathbf{v} \cdot \mathbf{v} - 6\mathbf{v} \cdot \mathbf{w} - 6\mathbf{w} \cdot \mathbf{v} + 9\mathbf{w} \cdot \mathbf{w} \\ &= 4\|\mathbf{v}\|^2 - 12\mathbf{v} \cdot \mathbf{w} + 9\|\mathbf{w}\|^2\end{aligned}$$

Substituting $\mathbf{v} \cdot \mathbf{w} = -3$ from part (a) and the given values yields

$$\|2\mathbf{v} - 3\mathbf{w}\|^2 = 4 \cdot 2^2 - 12(-3) + 9 \cdot 3^2 = 133 \quad \Rightarrow \quad \|2\mathbf{v} - 3\mathbf{w}\| = \sqrt{133} \approx 11.53$$

61. Find all three angles in the triangle in Figure 11.

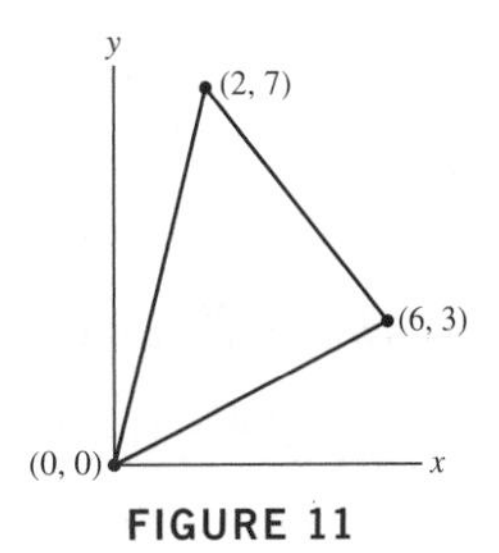

FIGURE 11

SOLUTION We denote by **u**, **v** and **w** the vectors and by θ_1, θ_2, and θ_3 the angles shown in the figure. We compute the vectors:

$$\mathbf{u} = \langle 2, 7 \rangle$$
$$\mathbf{v} = \langle 6, 3 \rangle$$
$$\mathbf{w} = \langle 6-2, 3-7 \rangle = \langle 4, -4 \rangle$$

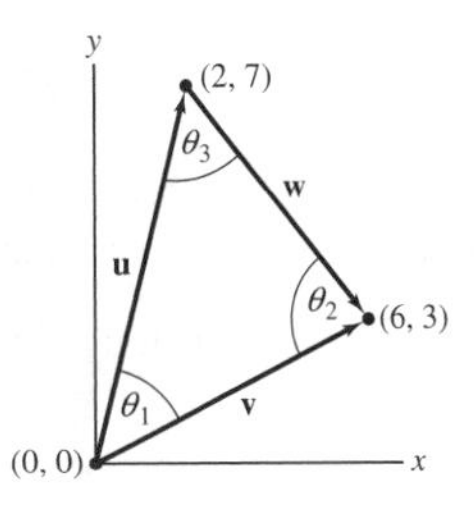

Since the angles are acute the cosines are positive, so we have

$$\cos\theta_1 = \frac{|\mathbf{u}\cdot\mathbf{v}|}{\|\mathbf{u}\|\,\|\mathbf{v}\|},$$
$$\cos\theta_2 = \frac{|\mathbf{v}\cdot\mathbf{w}|}{\|\mathbf{v}\|\,\|\mathbf{w}\|},$$
$$\cos\theta_3 = 180 - (\theta_1 + \theta_2) \tag{1}$$

We compute the lengths and the dot products in (1):

$$\mathbf{u}\cdot\mathbf{v} = \langle 2, 7\rangle \cdot \langle 6, 3\rangle = 2\cdot 6 + 7\cdot 3 = 33$$
$$\mathbf{v}\cdot\mathbf{w} = \langle 6, 3\rangle \cdot \langle 4, -4\rangle = 6\cdot 4 + 3\cdot(-4) = 12$$
$$\|\mathbf{u}\| = \sqrt{2^2+7^2} = \sqrt{53}$$
$$\|\mathbf{v}\| = \sqrt{6^2+3^2} = \sqrt{45}$$
$$\|\mathbf{w}\| = \sqrt{4^2+(-4)^2} = \sqrt{32}$$

Substituting in (1) and solving for acute angles yields

$$\cos\theta_1 = \frac{33}{\sqrt{53}\sqrt{45}} \approx 0.676 \quad\Rightarrow\quad \theta_1 \approx 47.47^\circ$$
$$\cos\theta_2 = \frac{12}{\sqrt{45}\sqrt{32}} \approx 0.316 \quad\Rightarrow\quad \theta_2 \approx 71.58^\circ$$

The sum of the angles in a triangle is 180°, hence

$$\theta_3 = 180^\circ - (47.47 + 71.58) \approx 60.95^\circ.$$

In Exercises 62–65, refer to Figure 12.

63. Find the angle between $\overline{AB}$ and $\overline{AD}$.

SOLUTION The cosine of the angle β between the vectors $\overrightarrow{AB}$ and $\overrightarrow{AD}$ is

$$\cos\beta = \frac{\overrightarrow{AB}\cdot\overrightarrow{AD}}{\|\overrightarrow{AB}\|\,\|\overrightarrow{AD}\|} \tag{1}$$

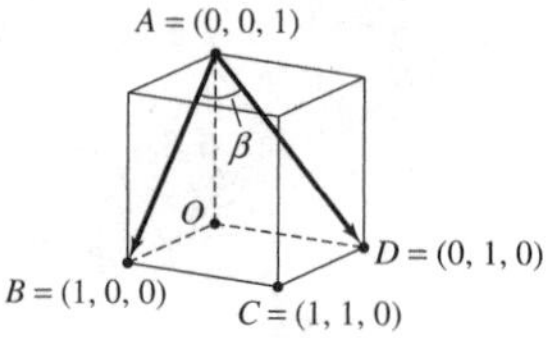

We compute the vectors $\overrightarrow{AB}$ and $\overrightarrow{AD}$ and then calculate their dot product and lengths. This gives

$$\overrightarrow{AB} = \langle 1-0, 0-0, 0-1 \rangle = \langle 1, 0, -1 \rangle$$

$$\overrightarrow{AD} = \langle 0-0, 1-0, 0-1 \rangle = \langle 0, 1, -1 \rangle$$

$$\overrightarrow{AB} \cdot \overrightarrow{AD} = \langle 1, 0, -1 \rangle \cdot \langle 0, 1, -1 \rangle = 1 \cdot 0 + 0 \cdot 1 + (-1) \cdot (-1) = 1$$

$$\|\overrightarrow{AB}\| = \sqrt{1^2 + 0^2 + (-1)^2} = \sqrt{2}$$

$$\|\overrightarrow{AD}\| = \sqrt{0^2 + 1^2 + (-1)^2} = \sqrt{2}$$

Substituting in (1) and solving for $0 \le \beta \le 90^\circ$ gives

$$\cos \beta = \frac{1}{\sqrt{2} \cdot \sqrt{2}} = \frac{1}{2} \quad \Rightarrow \quad \beta = 60^\circ.$$

It's interesting to note that we could have done this problem in a much simpler way. The triangle ABD is equilateral since each side is the diagonal of a unit square. Hence, all interior angles of the triangle are 60 degrees!

65. Calculate the projection of $\overrightarrow{AD}$ along $\overrightarrow{AB}$.

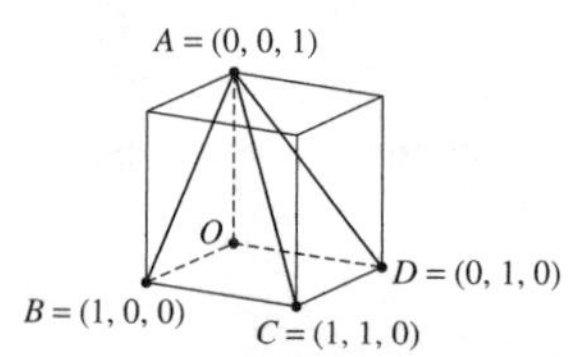

FIGURE 12 Unit cube in $\mathbf{R}^3$.

SOLUTION The projection of $\overrightarrow{AD}$ along $\overrightarrow{AB}$ is the following vector:

$$\text{proj}_{\overrightarrow{AB}}(\overrightarrow{AD}) = \left(\frac{\overrightarrow{AD} \cdot \overrightarrow{AB}}{\overrightarrow{AB} \cdot \overrightarrow{AB}} \right) \overrightarrow{AB} \tag{1}$$

We compute the vectors $\overrightarrow{AB}$ and $\overrightarrow{AD}$ and then calculate the dot product appearing in (1). We obtain

$$\overrightarrow{AB} = \langle 1-0, 0-0, 0-1 \rangle = \langle 1, 0, -1 \rangle$$

$$\overrightarrow{AD} = \langle 0-0, 1-0, 0-1 \rangle = \langle 0, 1, -1 \rangle$$

$$\overrightarrow{AB} \cdot \overrightarrow{AD} = \langle 1, 0, -1 \rangle \cdot \langle 0, 1, -1 \rangle = 1 \cdot 0 + 0 \cdot 1 + (-1) \cdot (-1) = 1$$

$$\overrightarrow{AB} \cdot \overrightarrow{AB} = \|\overrightarrow{AB}\|^2 = 1^2 + 0^2 + (-1)^2 = 2$$

Substituting in (1) gives

$$\text{proj}_{\overrightarrow{AB}}(\overrightarrow{AD}) = \frac{1}{2} \langle 1, 0, -1 \rangle = \left\langle \frac{1}{2}, 0, -\frac{1}{2} \right\rangle.$$

67. Let $\mathbf{v}$, $\mathbf{w}$, and $\mathbf{a}$ be nonzero vectors such that $\mathbf{v} \cdot \mathbf{a} = \mathbf{w} \cdot \mathbf{a}$. Is it true that $\mathbf{v} = \mathbf{w}$? Either prove this or give a counterexample.

SOLUTION The equality $\mathbf{v} \cdot \mathbf{a} = \mathbf{w} \cdot \mathbf{a}$ is equivalent to the following equality:

$$\mathbf{v} \cdot \mathbf{a} = \mathbf{w} \cdot \mathbf{a}$$

$$\mathbf{v} \cdot \mathbf{a} - \mathbf{w} \cdot \mathbf{a} = 0$$

$$(\mathbf{v} - \mathbf{w}) \cdot \mathbf{a} = 0$$

That is, $\mathbf{v} - \mathbf{w}$ is orthogonal to $\mathbf{a}$ rather than $\mathbf{v} = \mathbf{w}$. Consider the following counterexample:

$$\mathbf{a} = \langle 1, 0, 1 \rangle; \quad \mathbf{v} = \langle 3, 1, 1 \rangle; \quad \mathbf{w} = \langle 4, 1, 0 \rangle$$

Obviously, $\mathbf{v} \neq \mathbf{w}$, but $\mathbf{v} \cdot \mathbf{a} = \mathbf{w} \cdot \mathbf{a}$ since

$$\mathbf{v} \cdot \mathbf{a} = \langle 3, 1, 1 \rangle \cdot \langle 1, 0, 1 \rangle = 3 \cdot 1 + 1 \cdot 0 + 1 \cdot 1 = 4$$

$$\mathbf{w} \cdot \mathbf{a} = \langle 4, 1, 0 \rangle \cdot \langle 1, 0, 1 \rangle = 4 \cdot 1 + 1 \cdot 0 + 0 \cdot 1 = 4$$

69. Calculate the force (in Newtons) required to push a 40-kg wagon up a 10° incline (Figure 13). One N is equal to 1 kg-m/s^2, and the force due to gravity on the wagon is $\mathbf{F} = 40g$ N, where $g = 9.8$.

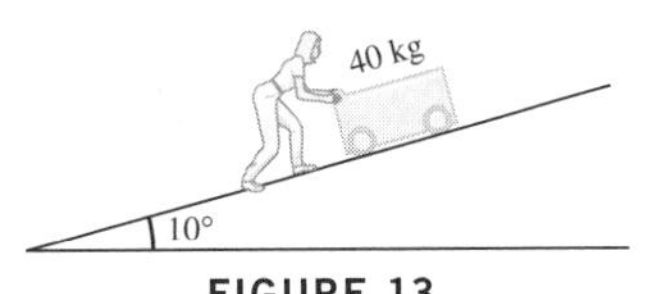

FIGURE 13

SOLUTION The magnitude of the force required to push the wagon equals the component of the force $\mathbf{F}_g$ due to gravity along the ramp. Resolving $\mathbf{F}_g$ into a sum $\mathbf{F}_g = \mathbf{F}_{\|} + \mathbf{F}_{\perp}$, where $\mathbf{F}_{\|}$ is the force along the ramp and $\mathbf{F}_{\perp}$ is the force orthogonal to the ramp, we need to find the magnitude of $\mathbf{F}_{\|}$. The angle between $\mathbf{F}_g$ and the ramp is $90° - 10° = 80°$. Hence,

$$\mathbf{F}_{\|} = \|\mathbf{F}_g\| \cos 80° = 40 \cdot 9.8 \cdot \cos 80° \approx 68.07N.$$

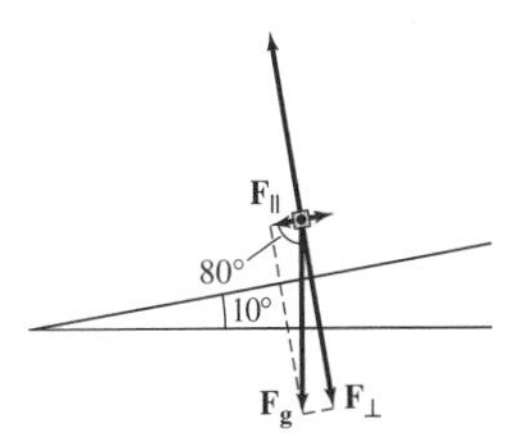

Therefore the minimum force required to push the wagon is 68.07 N. (Actually, this is the force required to keep the wagon from sliding down the hill; any slight amount greater than this force will serve to push it up the hill.)

71. A 10-kg mass hangs from two ropes (of negligible weight) as in Figure 15. Rope 1 exerts a force of magnitude $\mathbf{F}_1$ acting in the direction $\overrightarrow{PA}$ and rope 2 exerts a force $\mathbf{F}_2$ in the direction $\overrightarrow{PB}$. The sum of these forces balances the force of gravity $\mathbf{F}_g = 10g$ N acting downward at P (where $g = 9.8$ m/s^2). Determine $\mathbf{F}_1$ and $\mathbf{F}_2$.

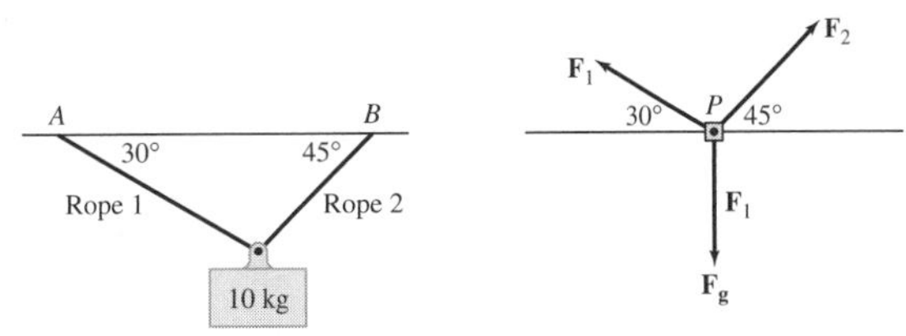

FIGURE 15

SOLUTION Since the sum of $\mathbf{F}_1$ and $\mathbf{F}_2$ balance the force of gravity, we have

$$\mathbf{F}_1 + \mathbf{F}_2 + \mathbf{F}_g = 0 \tag{1}$$

We resolve $\mathbf{F}_1$, $\mathbf{F}_2$, and $\mathbf{F}_g$ into a sum of a force along the ground and a force orthogonal to the ground.

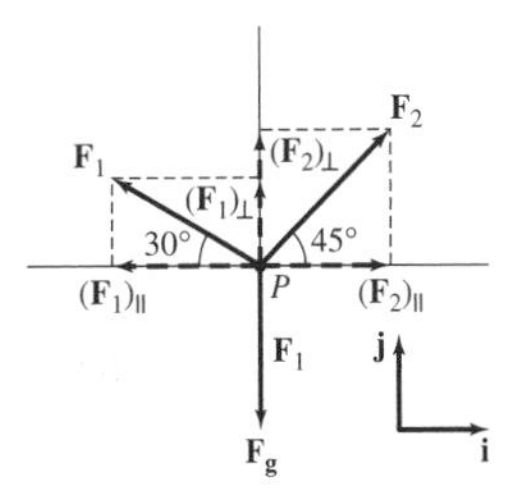

This gives

$$(\mathbf{F}_1)_{\|} = -(\|\mathbf{F}_1\| \cos 30°)\mathbf{i}, \qquad (\mathbf{F}_1)_{\perp} = (\|\mathbf{F}_1\| \cos 60°)\mathbf{j}$$

$$(\mathbf{F}_2)_{\|} = (\|\mathbf{F}_2\| \cos 45°)\mathbf{i}, \qquad (\mathbf{F}_2)_{\perp} = (\|\mathbf{F}_2\| \cos 45°)\mathbf{j}$$

$$(\mathbf{F}_g)_{\|} = \mathbf{0}, \qquad (\mathbf{F}_g)_{\perp} = -10g\mathbf{j} \approx -98\mathbf{j}$$

Substituting these forces in (1) gives

$$\left(-\frac{\|\mathbf{F}_1\|\sqrt{3}}{2}\mathbf{i}+\frac{\|\mathbf{F}_1\|}{2}\mathbf{j}\right)+\left(\frac{\|\mathbf{F}_2\|\sqrt{2}}{2}\mathbf{i}+\frac{\|\mathbf{F}_2\|\sqrt{2}}{2}\mathbf{j}\right)-98\mathbf{j}=0$$

$$\frac{1}{2}\left(\sqrt{2}\|\mathbf{F}_2\|-\sqrt{3}\|\mathbf{F}_1\|\right)\mathbf{i}+\frac{1}{2}\left(\|\mathbf{F}_1\|+\sqrt{2}\|\mathbf{F}_2\|-98\right)\mathbf{j}=0$$

We now equate each component to zero, to obtain

$$\begin{aligned}\frac{\sqrt{2}\|\mathbf{F}_2\|-\sqrt{3}\|\mathbf{F}_1\|}{2}&=0\\ \frac{\|\mathbf{F}_1\|+\sqrt{2}\|\mathbf{F}_2\|}{2}-98&=0\end{aligned}\quad\Rightarrow\quad\begin{aligned}\|\mathbf{F}_1\|&\approx 71.7\\ \|\mathbf{F}_2\|&\approx 87.8\end{aligned}$$

We conclude that

$$\mathbf{F}_1=(-71.7\cos 30^\circ)\mathbf{i}+(71.7\cos 60^\circ)\mathbf{j}=-62.1\mathbf{i}+35.9\mathbf{j}$$

$$\mathbf{F}_2=(87.8\cos 45^\circ)\mathbf{i}+(87.8\cos 45^\circ)\mathbf{j}=62.1\mathbf{i}+62.1\mathbf{j}$$

(As in the statement of the problems, all units are in Newtons.)

73. Let P and Q be antipodal (opposite) points on a sphere of radius r centered at the origin and let R be a third point on the sphere (Figure 17). Prove that $\overline{PR}$ and $\overline{QR}$ are orthogonal.

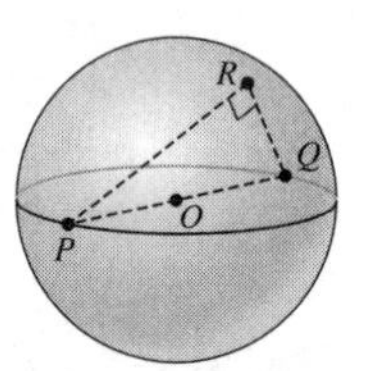

FIGURE 17

SOLUTION We denote the vectors $\overrightarrow{OP}$ and $\overrightarrow{OR}$ by

$$\mathbf{v}=\overrightarrow{OP},\quad \mathbf{w}=\overrightarrow{OR}$$

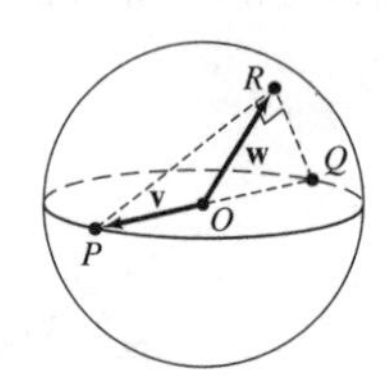

Thus,

$$\overrightarrow{PR}=\overrightarrow{PO}+\overrightarrow{OR}=-\mathbf{v}+\mathbf{w}$$

$$\overrightarrow{RQ}=\overrightarrow{RO}+\overrightarrow{OQ}=-\mathbf{w}-\mathbf{v}$$

We now show that $\overrightarrow{PR}\cdot\overrightarrow{RQ}=0$:

$$\begin{aligned}\overrightarrow{PR}\cdot\overrightarrow{RQ}&=(-\mathbf{v}+\mathbf{w})\cdot(-\mathbf{w}-\mathbf{v})=(\mathbf{v}-\mathbf{w})\cdot(\mathbf{v}+\mathbf{w})\\&=\mathbf{v}\cdot\mathbf{v}+\mathbf{v}\cdot\mathbf{w}-\mathbf{w}\cdot\mathbf{v}-\mathbf{w}\cdot\mathbf{w}=\|\mathbf{v}\|^2-\|\mathbf{w}\|^2\end{aligned}$$

The lengths $\|\mathbf{v}\|$ and $\|\mathbf{w}\|$ equal the radius r of the sphere, hence,

$$\overrightarrow{PR}\cdot\overrightarrow{RQ}=\|\mathbf{v}\|^2-\|\mathbf{w}\|^2=r^2-r^2=0$$

The dot product of $\overrightarrow{PR}$ and $\overrightarrow{RQ}$ is zero, therefore the two vectors are orthogonal.

75. Use Exercise 74 to show that $\mathbf{v}$ and $\mathbf{w}$ are orthogonal if and only if $\|\mathbf{v}-\mathbf{w}\|=\|\mathbf{v}+\mathbf{w}\|$.

SOLUTION In Exercise 74 we showed that

$$\|\mathbf{v}+\mathbf{w}\|^2-\|\mathbf{v}-\mathbf{w}\|^2=4\mathbf{v}\cdot\mathbf{w}$$

The vectors $\mathbf{v}\cdot\mathbf{w}$ are orthogonal if and only if $\mathbf{v}\cdot\mathbf{w}=0$. That is, if and only if

$$\|\mathbf{v}+\mathbf{w}\|^2-\|\mathbf{v}-\mathbf{w}\|^2=0$$

or

$$\|\mathbf{v}+\mathbf{w}\| = \|\mathbf{v}-\mathbf{w}\|.$$

77. Verify the Distributive Law:

$$\mathbf{u}\cdot(\mathbf{v}+\mathbf{w}) = \mathbf{u}\cdot\mathbf{v}+\mathbf{u}\cdot\mathbf{w}$$

SOLUTION We denote the components of the vectors **u**, **v**, and **w** by

$$\mathbf{u} = \langle a_1, a_2, a_3\rangle; \quad \mathbf{v} = \langle b_1, b_2, b_3\rangle; \quad \mathbf{w} = \langle c_1, c_2, c_3\rangle$$

We compute the left-hand side:

$$\begin{aligned}\mathbf{u}\cdot(\mathbf{v}+\mathbf{w}) &= \langle a_1, a_2, a_3\rangle\,(\langle b_1, b_2, b_3\rangle + \langle c_1, c_2, c_3\rangle)\\ &= \langle a_1, a_2, a_3\rangle\cdot\langle b_1+c_1, b_2+c_2, b_3+c_3\rangle\\ &= \langle a_1(b_1+c_1), a_2(b_2+c_2), a_3(b_3+c_3)\rangle\end{aligned}$$

Using the distributive law for scalars and the definitions of vector sum and the dot product we get

$$\begin{aligned}\mathbf{u}\cdot(\mathbf{v}+\mathbf{w}) &= \langle a_1b_1+a_1c_1, a_2b_2+a_2c_2, a_3b_3+a_3c_3\rangle\\ &= \langle a_1b_1, a_2b_2, a_3b_3\rangle + \langle a_1c_1, a_2c_2, a_3c_3\rangle\\ &= \langle a_1, a_2, a_3\rangle\cdot\langle b_1, b_2, b_3\rangle + \langle a_1, a_2, a_3\rangle\cdot\langle c_1, c_2, c_3\rangle\\ &= \mathbf{u}\cdot\mathbf{v}+\mathbf{u}\cdot\mathbf{w}\end{aligned}$$

Further Insights and Challenges

79. Prove the Law of Cosines: $c^2 = a^2 + b^2 - 2ab\cos\theta$ by referring to Figure 18. *Hint:* Consider the right triangle $\triangle PQR$.

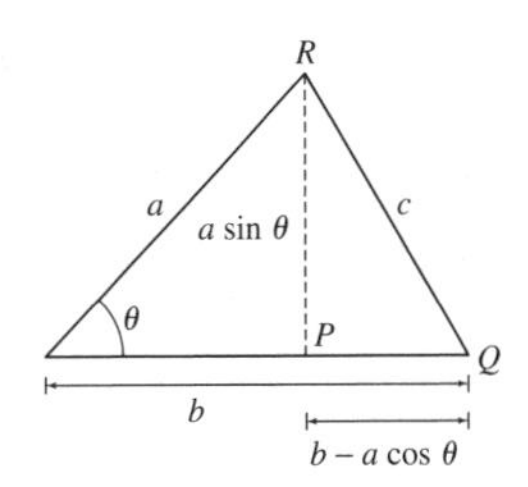

FIGURE 18

SOLUTION We denote the vertices of the triangle by S, Q, and R. Since $\overrightarrow{RQ} = \overrightarrow{RS} + \overrightarrow{SQ}$, we have

$$\begin{aligned}c^2 = \|\overrightarrow{RQ}\|^2 &= \overrightarrow{RQ}\cdot\overrightarrow{RQ} = \left(\overrightarrow{RS}+\overrightarrow{SQ}\right)\cdot\left(\overrightarrow{RS}+\overrightarrow{SQ}\right)\\ &= \overrightarrow{RS}\cdot\overrightarrow{RS} + \overrightarrow{RS}\cdot\overrightarrow{SQ} + \overrightarrow{SQ}\cdot\overrightarrow{RS} + \overrightarrow{SQ}\cdot\overrightarrow{SQ}\\ &= \|\overrightarrow{RS}\|^2 + 2\overrightarrow{RS}\cdot\overrightarrow{SQ} + \|\overrightarrow{SQ}\|^2\\ c^2 &= a^2 + 2\overrightarrow{RS}\cdot\overrightarrow{SQ} + b^2 \end{aligned} \qquad (1)$$

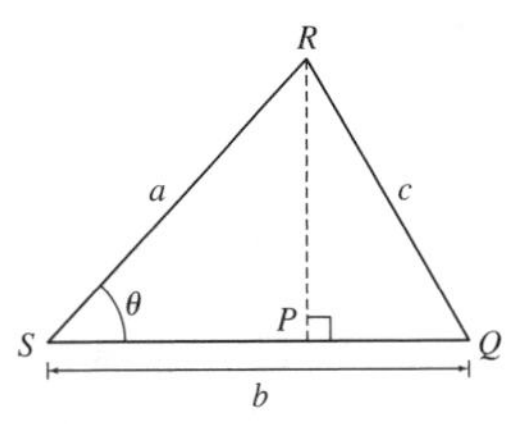

We find the dot product $\overrightarrow{RS}\cdot\overrightarrow{SQ}$. The angle between the vectors $\overrightarrow{RS}$ and $\overrightarrow{SQ}$ is θ, hence,

$$\overrightarrow{SR}\cdot\overrightarrow{SQ} = \|\overrightarrow{SR}\|\cdot\|\overrightarrow{SQ}\|\cos\theta = ab\cos\theta.$$

Therefore,

$$\overrightarrow{RS}\cdot\overrightarrow{SQ} = -\overrightarrow{SR}\cdot\overrightarrow{SQ} = -ab\cos\theta \qquad (2)$$

Substituting (2) in (1) yields

$$c^2 = a^2 - 2ab\cos\theta + b^2 = a^2 + b^2 - 2ab\cos\theta.$$

(Note that we did not need to use the point P.)

81. Use (7) to prove the Triangle Inequality

$$\|\mathbf{v} + \mathbf{w}\| \le \|\mathbf{v}\| + \|\mathbf{w}\|$$

Hint: First use the Triangle Inequality for numbers to prove

$$|(\mathbf{v} + \mathbf{w}) \cdot (\mathbf{v} + \mathbf{w})| \le |(\mathbf{v} + \mathbf{w}) \cdot \mathbf{v}| + |(\mathbf{v} + \mathbf{w}) \cdot \mathbf{w}|$$

SOLUTION Using the relation between the length and dot product we have

$$\begin{aligned}\|\mathbf{v} + \mathbf{w}\|^2 &= (\mathbf{v} + \mathbf{w}) \cdot (\mathbf{v} + \mathbf{w}) = \mathbf{v} \cdot \mathbf{v} + \mathbf{v} \cdot \mathbf{w} + \mathbf{w} \cdot \mathbf{v} + \mathbf{w} \cdot \mathbf{w} \\ &= \|\mathbf{v}\|^2 + 2\mathbf{v} \cdot \mathbf{w} + \|\mathbf{w}\|^2 \qquad (1)\end{aligned}$$

Obviously, $\mathbf{v} \cdot \mathbf{w} \le |\mathbf{v} \cdot \mathbf{w}|$. Also, by the Cauchy–Schwarz inequality $|\mathbf{v} \cdot \mathbf{w}| \le \|\mathbf{v}\|\|\mathbf{w}\|$. Therefore, $\mathbf{v} \cdot \mathbf{w} \le \|\mathbf{v}\|\|\mathbf{w}\|$, and combining this with (1) we get

$$\|\mathbf{v}+\mathbf{w}\|^2 = \|\mathbf{v}\|^2 + 2\mathbf{v} \cdot \mathbf{w} + \|\mathbf{w}\|^2 \le \|\mathbf{v}\|^2 + 2\|\mathbf{v}\|\|\mathbf{w}\| + \|\mathbf{w}\|^2 = (\|\mathbf{v}\| + \|\mathbf{w}\|)^2$$

That is,

$$\|\mathbf{v}+\mathbf{w}\|^2 \le (\|\mathbf{v}\| + \|\mathbf{w}\|)^2$$

Taking the square roots of both sides and recalling that the length is nonnegative, we get

$$\|\mathbf{v}+\mathbf{w}\| \le \|\mathbf{v}\| + \|\mathbf{w}\|$$

83. Let $\mathbf{v} = \langle x, y \rangle$ and

$$\mathbf{v}_\theta = \langle x\cos\theta + y\sin\theta, -x\sin\theta + y\cos\theta \rangle$$

Prove that the angle between $\mathbf{v}$ and $\mathbf{v}_\theta$ is θ.

SOLUTION The dot product of the vectors $\mathbf{v}$ and $\mathbf{v}_\theta$ is

$$\begin{aligned}\mathbf{v} \cdot \mathbf{v}_\theta &= \langle x, y \rangle \cdot \langle x\cos\theta + y\sin\theta, -x\sin\theta + y\cos\theta \rangle \\ &= x(x\cos\theta + y\sin\theta) + y(-x\sin\theta + y\cos\theta) \\ &= x^2\cos\theta + xy\sin\theta - xy\sin\theta + y^2\cos\theta \\ &= (x^2 + y^2)\cos\theta\end{aligned}$$

That is,

$$\mathbf{v} \cdot \mathbf{v}_\theta = (x^2 + y^2)\cos\theta \qquad (1)$$

On the other hand, if α denotes the angle between $\mathbf{v}$ and $\mathbf{v}_\theta$, we have

$$\mathbf{v} \cdot \mathbf{v}_\theta = \|\mathbf{v}\|\|\mathbf{v}_\theta\|\cos\alpha \qquad (2)$$

We compute the lengths. Using the identities $\cos^2\theta + \sin^2\theta = 1$ and $2\sin\theta\cos\theta = \sin 2\theta$, we obtain

$$\begin{aligned}\|\mathbf{v}\| &= \sqrt{\langle x, y \rangle} = \sqrt{x^2 + y^2} \\ \|\mathbf{v}_\theta\| &= \sqrt{(x\cos\theta + y\sin\theta)^2 + (-x\sin\theta + y\cos\theta)^2} \\ &= \sqrt{x^2\cos^2\theta + xy\sin 2\theta + y^2\sin^2\theta + x^2\sin^2\theta - xy\sin 2\theta + y^2\cos^2\theta} \\ &= \sqrt{x^2(\cos^2\theta + \sin^2\theta) + y^2(\sin^2\theta + \cos^2\theta)} = \sqrt{x^2 \cdot 1 + y^2 \cdot 1} = \sqrt{x^2 + y^2}\end{aligned}$$

Substituting the lengths in (2) yields

$$\mathbf{v} \cdot \mathbf{v}_\theta = \sqrt{x^2 + y^2} \cdot \sqrt{x^2 + y^2}\cos\alpha = (x^2 + y^2)\cos\alpha \qquad (3)$$

We now equate (1) and (3) to obtain

$$(x^2 + y^2)\cos\theta = (x^2 + y^2)\cos\alpha$$
$$\cos\theta = \cos\alpha$$

The solution for angles between 0° and 180° is $\alpha = 0$. That is, the angle between $\mathbf{v}$ and $\mathbf{v}_\theta$ is θ.

85. Find the direction cosines of $\mathbf{v} = \langle 3, 6, -2 \rangle$.

SOLUTION Let α, β, γ denote the angles between $\mathbf{v}$ and the unit vectors $\mathbf{i}$, $\mathbf{j}$, $\mathbf{k}$ respectively. We need to compute $\cos\alpha$, $\cos\beta$, and $\cos\gamma$. Using the formula for the angle between two vectors and the lengths $\|\mathbf{i}\| = \|\mathbf{j}\| = \|\mathbf{k}\| = 1$, $\|\mathbf{v}\| = \sqrt{3^2 + 6^2 + (-2)^2} = 7$ we get

$$\cos\alpha = \frac{\mathbf{v}\cdot\mathbf{i}}{\|\mathbf{v}\|\,\|\mathbf{i}\|} = \frac{(3\mathbf{i} + 6\mathbf{j} - 2\mathbf{k})\cdot\mathbf{i}}{7} = \frac{3}{7}$$
$$\cos\beta = \frac{\mathbf{v}\cdot\mathbf{j}}{\|\mathbf{v}\|\,\|\mathbf{j}\|} = \frac{(3\mathbf{i} + 6\mathbf{j} - 2\mathbf{k})\cdot\mathbf{j}}{7} = \frac{6}{7}$$
$$\cos\gamma = \frac{\mathbf{v}\cdot\mathbf{k}}{\|\mathbf{v}\|\,\|\mathbf{k}\|} = \frac{(3\mathbf{i} + 6\mathbf{j} - 2\mathbf{k})\cdot\mathbf{k}}{7} = -\frac{2}{7}$$

87. Sketch the plane consisting of all points $X = (x, y, z)$ equidistant from the points $P = (0, 1, 0)$ and $Q = (0, 0, 1)$. Use Eq. (8) to show that X lies on this plane if and only if $y = z$.

SOLUTION As seen in the solution to Problem 86, the point $X = (x, y, z)$ lies on the plane iff Eq. (8) holds. Using this equation with $X = (x, y, z)$, $P = (0, 1, 0)$, and $Q = (0, 0, 1)$ gives

$$\langle x, y, z \rangle \cdot \langle 0, -1, 1 \rangle = \frac{1}{2}(1^2 - 1^2) = 0$$

This gives us $0x - 1y + 1z = 0$, which gives us $y = z$, as desired.

13.4 The Cross Product (ET Section 12.4)

Preliminary Questions

1. What is the $(1, 3)$ minor of the matrix $\begin{vmatrix} 3 & 4 & 2 \\ -5 & -1 & 1 \\ 4 & 0 & 3 \end{vmatrix}$?

SOLUTION The $(1, 3)$ minor is obtained by crossing out the first row and third column of the matrix. That is,

$$\begin{vmatrix} 3 & 4 & 2 \\ -5 & -1 & 1 \\ 4 & 0 & 3 \end{vmatrix} \Rightarrow \begin{vmatrix} -5 & -1 \\ 4 & 0 \end{vmatrix}$$

2. The angle between two unit vectors $\mathbf{e}$ and $\mathbf{f}$ is $\frac{\pi}{6}$. What is the length of $\mathbf{e} \times \mathbf{f}$?

SOLUTION We use the Formula for the Length of the Cross Product:

$$\|\mathbf{e} \times \mathbf{f}\| = \|\mathbf{e}\|\,\|\mathbf{f}\| \sin\theta$$

Since $\mathbf{e}$ and $\mathbf{f}$ are unit vectors, $\|\mathbf{e}\| = \|\mathbf{f}\| = 1$. Also $\theta = \frac{\pi}{6}$, therefore,

$$\|\mathbf{e} \times \mathbf{f}\| = 1 \cdot 1 \cdot \sin\frac{\pi}{6} = \frac{1}{2}$$

The length of $\mathbf{e} \times \mathbf{f}$ is $\frac{1}{2}$.

3. What is $\mathbf{u} \times \mathbf{w}$, assuming that $\mathbf{w} \times \mathbf{u} = \langle 2, 2, 1 \rangle$?

SOLUTION By anti-commutativity of the cross product, we have

$$\mathbf{u} \times \mathbf{w} = -\mathbf{w} \times \mathbf{u} = -\langle 2, 2, 1 \rangle = \langle -2, -2, -1 \rangle$$

4. Find the cross product without using the formula:

(a) $\langle 4, 8, 2 \rangle \times \langle 4, 8, 2 \rangle$

(b) $\langle 4, 8, 2 \rangle \times \langle 2, 4, 1 \rangle$

SOLUTION By properties of the cross product, the cross product of parallel vectors is the zero vector. In particular, the cross product of a vector with itself is the zero vector. Since $\langle 4, 8, 2\rangle = 2\langle 2, 4, 1\rangle$, the vectors $\langle 4, 8, 2\rangle$ and $\langle 2, 4, 1\rangle$ are parallel. We conclude that

$$\langle 4, 8, 2\rangle \times \langle 4, 8, 2\rangle = \mathbf{0} \quad \text{and} \quad \langle 4, 8, 2\rangle \times \langle 2, 4, 1\rangle = \mathbf{0}.$$

5. What are $\mathbf{i} \times \mathbf{j}$ and $\mathbf{i} \times \mathbf{k}$?

SOLUTION The cross product $\mathbf{i} \times \mathbf{j}$ and $\mathbf{i} \times \mathbf{k}$ are determined by the right-hand rule. We can also use the following figure to determine these cross-products:

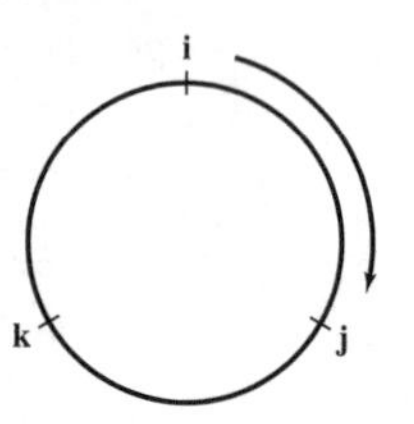

We get

$$\mathbf{i} \times \mathbf{j} = \mathbf{k} \text{ and } \mathbf{i} \times \mathbf{k} = -\mathbf{j}$$

6. When is the cross product $\mathbf{v} \times \mathbf{w}$ equal to zero?

SOLUTION The cross product $\mathbf{v} \times \mathbf{w}$ is equal to zero if one of the vectors $\mathbf{v}$ or $\mathbf{w}$ (or both) is the zero vector, or if $\mathbf{v}$ and $\mathbf{w}$ are parallel vectors.

Exercises

In Exercises 1–4, calculate the 2×2-determinant.

1. $\begin{vmatrix} 1 & 2 \\ 4 & 3 \end{vmatrix}$

SOLUTION Using the definition of 2×2 determinant we get

$$\begin{vmatrix} 1 & 2 \\ 4 & 3 \end{vmatrix} = 1 \cdot 3 - 2 \cdot 4 = -5$$

3. $\begin{vmatrix} -6 & 9 \\ 1 & 1 \end{vmatrix}$

SOLUTION We evaluate the determinant to obtain

$$\begin{vmatrix} -6 & 9 \\ 1 & 1 \end{vmatrix} = -6 \cdot 1 - 9 \cdot 1 = -15$$

In Exercises 5–10, calculate the 3×3-determinant.

5. $\begin{vmatrix} 1 & 2 & 1 \\ 4 & -3 & 0 \\ 1 & 0 & 1 \end{vmatrix}$

SOLUTION Using the definition of 3×3 determinant we obtain

$$\begin{aligned} \begin{vmatrix} 1 & 2 & 1 \\ 4 & -3 & 0 \\ 1 & 0 & 1 \end{vmatrix} &= 1 \begin{vmatrix} -3 & 0 \\ 0 & 1 \end{vmatrix} - 2 \begin{vmatrix} 4 & 0 \\ 1 & 1 \end{vmatrix} + 1 \begin{vmatrix} 4 & -3 \\ 1 & 0 \end{vmatrix} \\ &= 1 \cdot (-3 \cdot 1 - 0 \cdot 0) - 2 \cdot (4 \cdot 1 - 0 \cdot 1) + 1 \cdot (4 \cdot 0 - (-3) \cdot 1) \\ &= -3 - 8 + 3 = -8 \end{aligned}$$

7. $\begin{vmatrix} 1 & 0 & 1 \\ 1 & 3 & 1 \\ -2 & 0 & 3 \end{vmatrix}$

SOLUTION We use the definition to write

$$\begin{vmatrix} 1 & 0 & 1 \\ 1 & 3 & 1 \\ -2 & 0 & 3 \end{vmatrix} = 1\begin{vmatrix} 3 & 1 \\ 0 & 3 \end{vmatrix} - 0\begin{vmatrix} 1 & 1 \\ -2 & 3 \end{vmatrix} + 1\begin{vmatrix} 1 & 3 \\ -2 & 0 \end{vmatrix}$$
$$= (3 \cdot 3 - 1 \cdot 0) - 0 + 1 \cdot (1 \cdot 0 - 3(-2)) = 9 + 6 = 15$$

9. $\begin{vmatrix} 1 & 0 & 0 \\ 0 & 1 & 0 \\ 0 & 0 & 1 \end{vmatrix}$

SOLUTION Using the definition of 3×3 determinant yields

$$\begin{vmatrix} 1 & 0 & 0 \\ 0 & 1 & 0 \\ 0 & 0 & 1 \end{vmatrix} = 1\begin{vmatrix} 1 & 0 \\ 0 & 1 \end{vmatrix} - 0\begin{vmatrix} 0 & 0 \\ 0 & 1 \end{vmatrix} + 0\begin{vmatrix} 0 & 1 \\ 0 & 0 \end{vmatrix}$$
$$= 1(1 \cdot 1 - 0 \cdot 0) - 0 + 0 = 1$$

In Exercises 11–16, calculate $\mathbf{v} \times \mathbf{w}$.

11. $\mathbf{v} = \langle 1, 2, 1 \rangle, \quad \mathbf{w} = \langle 3, 1, 1 \rangle$

SOLUTION Using the definition of the cross product we get

$$\mathbf{v} \times \mathbf{w} = \begin{vmatrix} \mathbf{i} & \mathbf{j} & \mathbf{k} \\ 1 & 2 & 1 \\ 3 & 1 & 1 \end{vmatrix} = \begin{vmatrix} 2 & 1 \\ 1 & 1 \end{vmatrix}\mathbf{i} - \begin{vmatrix} 1 & 1 \\ 3 & 1 \end{vmatrix}\mathbf{j} + \begin{vmatrix} 1 & 2 \\ 3 & 1 \end{vmatrix}\mathbf{k}$$
$$= (2 - 1)\mathbf{i} - (1 - 3)\mathbf{j} + (1 - 6)\mathbf{k} = \mathbf{i} + 2\mathbf{j} - 5\mathbf{k}$$

13. $\mathbf{v} = \left\langle \frac{2}{3}, 1, \frac{1}{2} \right\rangle, \quad \mathbf{w} = \langle 4, -6, 3 \rangle$

SOLUTION We have

$$\mathbf{v} \times \mathbf{w} = \begin{vmatrix} \mathbf{i} & \mathbf{j} & \mathbf{k} \\ \frac{2}{3} & 1 & \frac{1}{2} \\ 4 & -6 & 3 \end{vmatrix} = \begin{vmatrix} 1 & \frac{1}{2} \\ -6 & 3 \end{vmatrix}\mathbf{i} - \begin{vmatrix} \frac{2}{3} & \frac{1}{2} \\ 4 & 3 \end{vmatrix}\mathbf{j} + \begin{vmatrix} \frac{2}{3} & 1 \\ 4 & -6 \end{vmatrix}\mathbf{k}$$
$$= (3 + 3)\,\mathbf{i} - (2 - 2)\,\mathbf{j} + (-4 - 4)\,\mathbf{k} = 6\mathbf{i} - 8\mathbf{k}$$

15. $\mathbf{v} = \left\langle \frac{1}{3}, 1, \frac{1}{3} \right\rangle, \quad \mathbf{w} = \langle -1, -1, 2 \rangle$

SOLUTION The cross product is the following vector:

$$\mathbf{v} \times \mathbf{w} = \begin{vmatrix} \mathbf{i} & \mathbf{j} & \mathbf{k} \\ \frac{1}{3} & 1 & \frac{1}{3} \\ -1 & -1 & 2 \end{vmatrix} = \begin{vmatrix} 1 & \frac{1}{3} \\ -1 & 2 \end{vmatrix}\mathbf{i} - \begin{vmatrix} \frac{1}{3} & \frac{1}{3} \\ -1 & 2 \end{vmatrix}\mathbf{j} + \begin{vmatrix} \frac{1}{3} & 1 \\ -1 & -1 \end{vmatrix}\mathbf{k}$$
$$= \left(2 + \frac{1}{3}\right)\mathbf{i} - \left(\frac{2}{3} + \frac{1}{3}\right)\mathbf{j} + \left(-\frac{1}{3} + 1\right)\mathbf{k} = \frac{7}{3}\mathbf{i} - \mathbf{j} + \frac{2}{3}\mathbf{k}$$

In Exercises 17–20, calculate the cross product.

17. $(\mathbf{i} + \mathbf{j}) \times \mathbf{k}$

SOLUTION We use basic properties of the cross product to obtain

$$(\mathbf{i} + \mathbf{j}) \times \mathbf{k} = \mathbf{i} \times \mathbf{k} + \mathbf{j} \times \mathbf{k} = -\mathbf{j} + \mathbf{i}$$

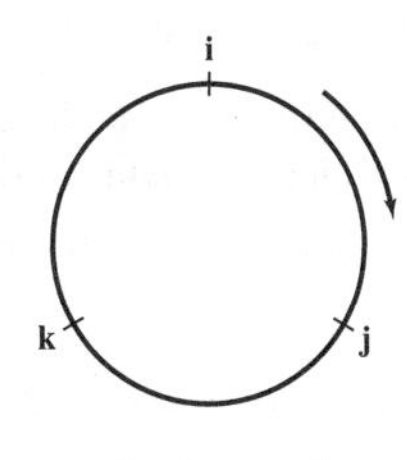

$$\mathbf{i} \times \mathbf{k} = -\mathbf{j}$$
$$\mathbf{j} \times \mathbf{k} = \mathbf{i}$$

19. $(\mathbf{i} + 2\mathbf{k}) \times (\mathbf{j} - \mathbf{k})$

SOLUTION Using the distributive law we obtain

$$\begin{aligned}(\mathbf{i} + 2\mathbf{k}) \times (\mathbf{j} - \mathbf{k}) &= (\mathbf{i} + 2\mathbf{k}) \times \mathbf{j} + (\mathbf{i} + 2\mathbf{k}) \times (-\mathbf{k}) = \mathbf{i} \times \mathbf{j} + 2\mathbf{k} \times \mathbf{j} - \mathbf{i} \times \mathbf{k} - 2\mathbf{k} \times \mathbf{k} \\ &= \mathbf{k} - 2\mathbf{i} + \mathbf{j} - 0 = -2\mathbf{i} + \mathbf{j} + \mathbf{k}\end{aligned}$$

In Exercises 21–26, calculate the cross product assuming that $\mathbf{u} \times \mathbf{v} = \langle 1, 1, 0 \rangle$, $\mathbf{u} \times \mathbf{w} = \langle 0, 3, 1 \rangle$, *and* $\mathbf{v} \times \mathbf{w} = \langle 2, -1, 1 \rangle$.

21. $\mathbf{v} \times \mathbf{u}$

SOLUTION Using the properties of the cross product we obtain

$$\mathbf{v} \times \mathbf{u} = -\mathbf{u} \times \mathbf{v} = \langle -1, -1, 0 \rangle$$

23. $(3\mathbf{u} + 4\mathbf{w}) \times \mathbf{w}$

SOLUTION Using the properties of the cross product we obtain

$$(3\mathbf{u} + 4\mathbf{w}) \times \mathbf{w} = 3\mathbf{u} \times \mathbf{w} + 4\mathbf{w} \times \mathbf{w} = \langle 0, 9, 3 \rangle$$

25. $\mathbf{w} \times (\mathbf{u} + \mathbf{v})$

SOLUTION Using the properties of the cross product we obtain

$$\mathbf{w} \times (\mathbf{u} + \mathbf{v}) = \mathbf{w} \times \mathbf{u} + \mathbf{w} \times \mathbf{v} = -\mathbf{u} \times \mathbf{w} - \mathbf{v} \times \mathbf{w} = \langle -2, -2, -2 \rangle .$$

27. Let $\mathbf{v} = \langle a, b, c \rangle$. Calculate $\mathbf{v} \times \mathbf{i}$, $\mathbf{v} \times \mathbf{j}$, and $\mathbf{v} \times \mathbf{k}$.

SOLUTION We write $\mathbf{v} = a\mathbf{i} + b\mathbf{j} + c\mathbf{k}$ and use the distributive law:

$$\begin{aligned}\mathbf{v} \times \mathbf{i} &= (a\mathbf{i} + b\mathbf{j} + c\mathbf{k}) \times \mathbf{i} = a\mathbf{i} \times \mathbf{i} + b\mathbf{j} \times \mathbf{i} + c\mathbf{k} \times \mathbf{i} = a \cdot \mathbf{0} - b\mathbf{k} + c\mathbf{j} = -b\mathbf{k} + c\mathbf{j} = \langle 0, c, -b \rangle \\ \mathbf{v} \times \mathbf{j} &= (a\mathbf{i} + b\mathbf{j} + c\mathbf{k}) \times \mathbf{j} = a\mathbf{i} \times \mathbf{j} + b\mathbf{j} \times \mathbf{j} + c\mathbf{k} \times \mathbf{j} = a\mathbf{k} + b\mathbf{0} - c\mathbf{i} = a\mathbf{k} - c\mathbf{i} = \langle -c, 0, a \rangle \\ \mathbf{v} \times \mathbf{k} &= (a\mathbf{i} + b\mathbf{j} + c\mathbf{k}) \times \mathbf{k} = a\mathbf{i} \times \mathbf{k} + b\mathbf{j} \times \mathbf{k} + c\mathbf{k} \times \mathbf{k} = -a\mathbf{j} + b\mathbf{i} + c\mathbf{0} = -a\mathbf{j} + b\mathbf{i} = \langle b, -a, 0 \rangle\end{aligned}$$

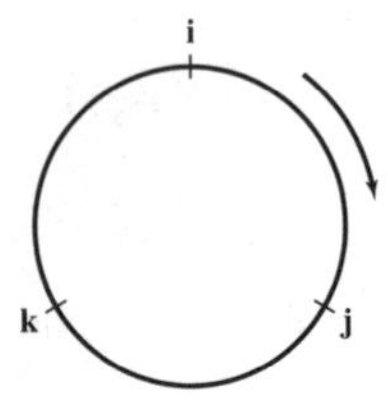

In Exercises 29–30, refer to Figure 15.

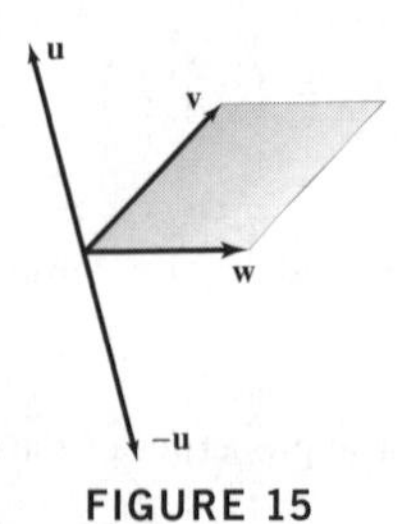

FIGURE 15

29. Which of $\mathbf{u}$ and $-\mathbf{u}$ is equal to $\mathbf{v} \times \mathbf{w}$?

SOLUTION The direction of $\mathbf{v} \times \mathbf{w}$ is determined by the right-hand rule, that is, our thumb points in the direction of $\mathbf{v} \times \mathbf{w}$ when the fingers of our right hand curl from $\mathbf{v}$ to $\mathbf{w}$. Therefore $\mathbf{v} \times \mathbf{w}$ equals $-\mathbf{u}$ rather than $\mathbf{u}$.

31. Let $\mathbf{v} = \langle 3, 0, 0 \rangle$ and $\mathbf{w} = \langle 0, 1, -1 \rangle$. Determine $\mathbf{u} = \mathbf{v} \times \mathbf{w}$ using the geometric properties of the cross product rather than the formula.

SOLUTION The cross product $\mathbf{u} = \mathbf{v} \times \mathbf{w}$ is orthogonal to $\mathbf{v}$.

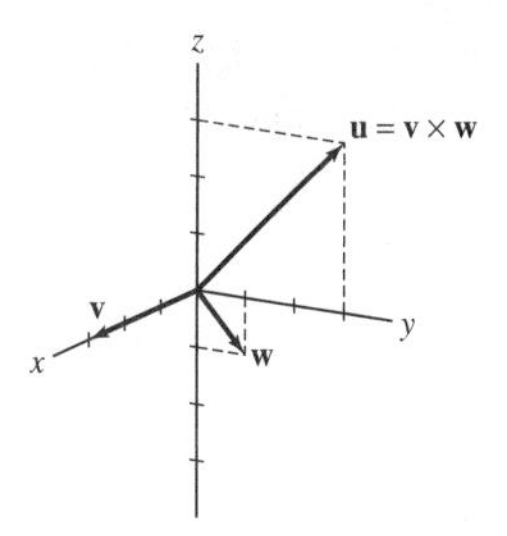

Since $\mathbf{v}$ lies along the x-axis, $\mathbf{u}$ lies in the yz-plane, therefore $\mathbf{u} = \langle 0, b, c\rangle$. $\mathbf{u}$ is also orthogonal to $\mathbf{w}$, so $\mathbf{u} \cdot \mathbf{w} = 0$. This gives $\mathbf{u} \cdot \mathbf{w} = \langle 0, b, c\rangle \cdot \langle 0, 1, -1\rangle = b - c = 0 \Rightarrow b = c$. Thus, $\mathbf{u} = \langle 0, b, b\rangle$. By the right-hand rule, $\mathbf{u}$ points to the positive z-direction so $b > 0$. We compute the length of $\mathbf{u}$. Since $\mathbf{v} \cdot \mathbf{w} = \langle 3, 0, 0\rangle \cdot \langle 0, 1, -1\rangle = 0$, $\mathbf{v}$ and $\mathbf{w}$ are orthogonal. Hence,

$$\|\mathbf{v} \times \mathbf{w}\| = \|\mathbf{v}\|\|\mathbf{w}\| \sin\frac{\pi}{2} = \|\mathbf{v}\|\|\mathbf{w}\| = 3 \cdot \sqrt{2}.$$

Also since $b > 0$, we have

$$\|\mathbf{u}\| = \|\langle 0, b, b\rangle\| = \sqrt{2b^2} = b\sqrt{2}$$

Equating the lengths gives

$$b\sqrt{2} = 3\sqrt{2} \quad \Rightarrow \quad b = 3.$$

We conclude that $\mathbf{u} = \mathbf{v} \times \mathbf{w} = \langle 0, 3, 3\rangle$.

33. Show that if $\mathbf{v}$ and $\mathbf{w}$ lie in the yz-plane, then $\mathbf{v} \times \mathbf{w}$ is a multiple of $\mathbf{i}$.

SOLUTION $\mathbf{v} \times \mathbf{w}$ is orthogonal to $\mathbf{v}$ and $\mathbf{w}$. Since $\mathbf{v}$ and $\mathbf{w}$ lie in the yz-plane, $\mathbf{v} \times \mathbf{w}$ must lie along the x axis which is perpendicular to yz-plane. That is, $\mathbf{v} \times \mathbf{w}$ is a scalar multiple of the unit vector $\mathbf{i}$.

35. Let $\mathbf{e}$ and $\mathbf{e}'$ be unit vectors in $\mathbf{R}^3$ such that $\mathbf{e} \perp \mathbf{e}'$. Reason geometrically to find the cross product $\mathbf{e} \times (\mathbf{e}' \times \mathbf{e})$.

SOLUTION Let $\mathbf{u} = \mathbf{e} \times (\mathbf{e}' \times \mathbf{e})$ and $\mathbf{v} = \mathbf{e}' \times \mathbf{e}$. The vector $\mathbf{v}$ is orthogonal to $\mathbf{e}'$ and $\mathbf{e}$, hence $\mathbf{v}$ is orthogonal to the plane π defined by $\mathbf{e}'$ and $\mathbf{e}$. Now $\mathbf{u}$ is orthogonal to $\mathbf{v}$, hence $\mathbf{u}$ lies in the plane π orthogonal to $\mathbf{v}$. $\mathbf{u}$ is orthogonal to $\mathbf{e}$, which is in this plane, hence $\mathbf{u}$ is a multiple of $\mathbf{e}'$:

$$\mathbf{u} = \lambda \mathbf{e}' \qquad (1)$$

The right-hand rule implies that $\mathbf{u}$ is in the direction of $\mathbf{e}'$, hence $\lambda > 0$. To find λ, we compute the length of $\mathbf{u}$:

$$\|\mathbf{v}\| = \|\mathbf{e}' \times \mathbf{e}\| = \|\mathbf{e}'\|\|\mathbf{e}\| \sin\frac{\pi}{2} = 1 \cdot 1 \cdot 1 = 1$$

$$\|\mathbf{u}\| = \|\mathbf{e} \times \mathbf{v}\| = \|\mathbf{e}\|\|\mathbf{v}\| \sin\frac{\pi}{2} = 1 \cdot 1 \cdot 1 = 1 \qquad (2)$$

Combining (1), (2), and $\lambda > 0$ we conclude that

$$\mathbf{u} = \mathbf{e} \times (\mathbf{e}' \times \mathbf{e}) = \mathbf{e}'.$$

37. An electron moving with velocity $\mathbf{v}$ in the plane experiences a force $\mathbf{F} = q(\mathbf{v} \times \mathbf{B})$, where q is the charge on the electron and $\mathbf{B}$ is a uniform magnetic field pointing directly out of the page. Which of the two vectors $\mathbf{F}_1$ or $\mathbf{F}_2$ in Figure 16 represents the force on the electron? Remember that q is negative.

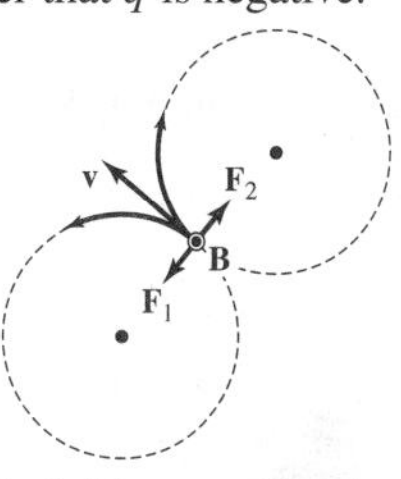

FIGURE 16 The magnetic field vector $\mathbf{B}$ points directly out of the page.

SOLUTION Since the magnetic field **B** points directly out of the page (toward us), the right-hand rule implies that the cross product $\mathbf{v} \times \mathbf{B}$ is in the direction of $\mathbf{F}_2$ (see figure).

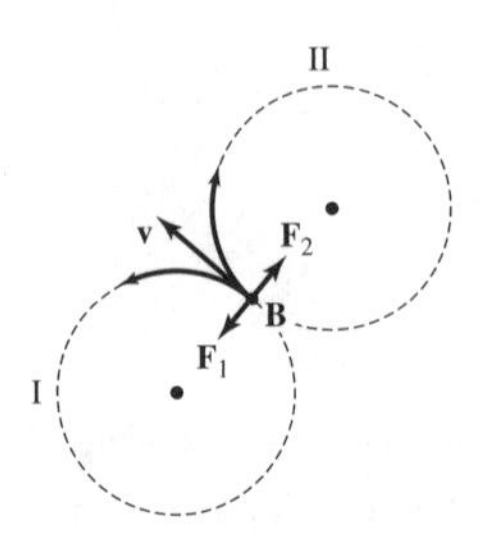

Since $\mathbf{F} = q(\mathbf{v} \times \mathbf{B})$ and $q < 0$, the force **F** on the electron is represented by the opposite vector $\mathbf{F}_1$.

39. Verify identity (10) for vectors $\mathbf{v} = \langle 3, -2, 2 \rangle$ and $\mathbf{w} = \langle 4, -1, 2 \rangle$.

SOLUTION We compute the cross product $\mathbf{v} \times \mathbf{w}$:

$$\mathbf{v} \times \mathbf{w} = \begin{vmatrix} \mathbf{i} & \mathbf{j} & \mathbf{k} \\ 3 & -2 & 2 \\ 4 & -1 & 2 \end{vmatrix} = \begin{vmatrix} -2 & 2 \\ -1 & 2 \end{vmatrix}\mathbf{i} - \begin{vmatrix} 3 & 2 \\ 4 & 2 \end{vmatrix}\mathbf{j} + \begin{vmatrix} 3 & -2 \\ 4 & -1 \end{vmatrix}\mathbf{k}$$

$$= (-4+2)\mathbf{i} - (6-8)\mathbf{j} + (-3+8)\mathbf{k} = -2\mathbf{i} + 2\mathbf{j} + 5\mathbf{k} = \langle -2, 2, 5 \rangle$$

We now find the dot product $\mathbf{v} \cdot \mathbf{w}$:

$$\mathbf{v} \cdot \mathbf{w} = \langle 3, -2, 2 \rangle \cdot \langle 4, -1, 2 \rangle = 3 \cdot 4 + (-2) \cdot (-1) + 2 \cdot 2 = 18$$

Finally we compute the squares of the lengths of **v**, **w** and $\mathbf{v} \times \mathbf{w}$:

$$\|\mathbf{v}\|^2 = 3^2 + (-2)^2 + 2^2 = 17$$
$$\|\mathbf{w}\|^2 = 4^2 + (-1)^2 + 2^2 = 21$$
$$\|\mathbf{v} \times \mathbf{w}\|^2 = (-2)^2 + 2^2 + 5^2 = 33$$

We now verify the equality:

$$\|\mathbf{v}\|^2\|\mathbf{w}\|^2 - (\mathbf{v} \cdot \mathbf{w})^2 = 17 \cdot 21 - 18^2 = 33 = \|\mathbf{v} \times \mathbf{w}\|^2$$

41. Find the area of the parallelogram spanned by **v** and **w** in Figure 17.

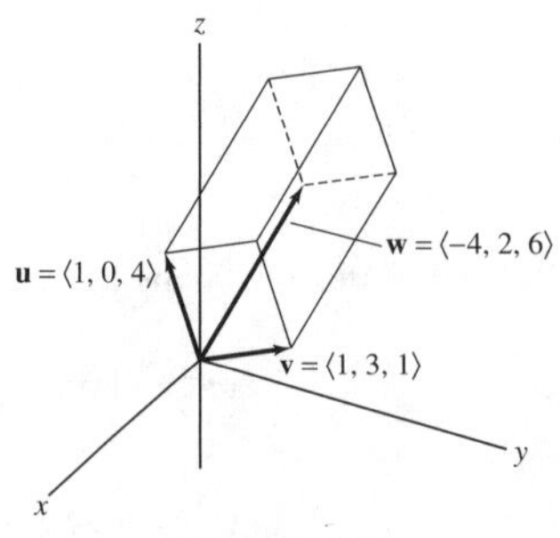

FIGURE 17

SOLUTION The area of the parallelogram equals the length of the cross product of the two vectors $\mathbf{v} = \langle 1, 3, 1 \rangle$ and $\mathbf{w} = \langle -4, 2, 6 \rangle$. We calculate the cross product as follows:

$$\mathbf{v} \times \mathbf{w} = \begin{vmatrix} \mathbf{i} & \mathbf{j} & \mathbf{k} \\ 1 & 3 & 1 \\ -4 & 2 & 6 \end{vmatrix} = (18-2)\mathbf{i} - (6+4)\mathbf{j} + (2+12)\mathbf{k} = 16\mathbf{i} - 10\mathbf{j} + 14\mathbf{k}$$

The length of this vector $16\mathbf{i} - 10\mathbf{j} + 14\mathbf{k}$ is $\sqrt{16^2 + 10^2 + 14^2} = 2\sqrt{138}$. Thus, the area of the parallelogram is $2\sqrt{138}$.

43. Sketch and compute the volume of the parallelepiped spanned by

$$\mathbf{u} = \langle 1, 0, 0 \rangle, \qquad \mathbf{v} = \langle 0, 2, 0 \rangle, \qquad \mathbf{w} = \langle 1, 1, 2 \rangle$$

SOLUTION Using $\mathbf{u} = \langle 1, 0, 0 \rangle$, $\mathbf{v} = \langle 0, 2, 0 \rangle$, and $\mathbf{w} = \langle 1, 1, 2 \rangle$, the volume is given by the following scalar triple product:

$$\mathbf{u} \cdot (\mathbf{v} \times \mathbf{w}) = \begin{vmatrix} 1 & 0 & 0 \\ 0 & 2 & 0 \\ 1 & 1 & 2 \end{vmatrix} = 1(4 - 0) - 0 + 0 = 4.$$

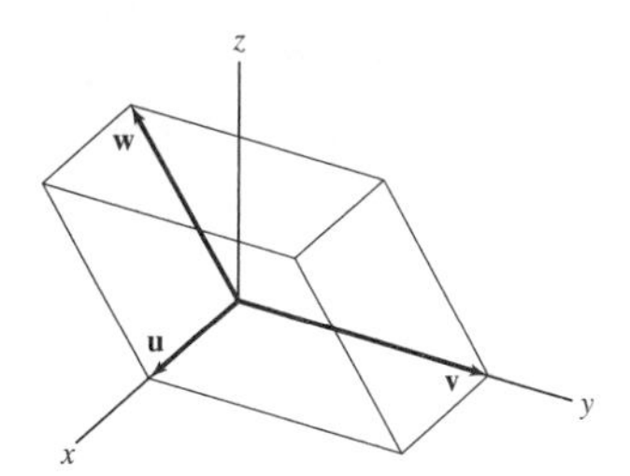

45. Calculate the area of the parallelogram spanned by $\mathbf{u} = \langle 1, 0, 3 \rangle$ and $\mathbf{v} = \langle 2, 1, 1 \rangle$.

SOLUTION The area of the parallelogram is the length of the vector $\mathbf{u} \times \mathbf{v}$. We first compute this vector:

$$\mathbf{u} \times \mathbf{v} = \begin{vmatrix} \mathbf{i} & \mathbf{j} & \mathbf{k} \\ 1 & 0 & 3 \\ 2 & 1 & 1 \end{vmatrix} = \begin{vmatrix} 0 & 3 \\ 1 & 1 \end{vmatrix} \mathbf{i} - \begin{vmatrix} 1 & 3 \\ 2 & 1 \end{vmatrix} \mathbf{j} + \begin{vmatrix} 1 & 0 \\ 2 & 1 \end{vmatrix} \mathbf{k} = -3\mathbf{i} - (1 - 6)\mathbf{j} + \mathbf{k} = -3\mathbf{i} + 5\mathbf{j} + \mathbf{k}$$

The area A is the length

$$A = \|\mathbf{u} \times \mathbf{v}\| = \sqrt{(-3)^2 + 5^2 + 1^2} = \sqrt{35} \approx 5.92.$$

47. Sketch the triangle with vertices O, $P = (3, 3, 0)$, and $Q = (0, 3, 3)$, and compute its area using cross products.

SOLUTION The triangle OPQ is shown in the following figure.

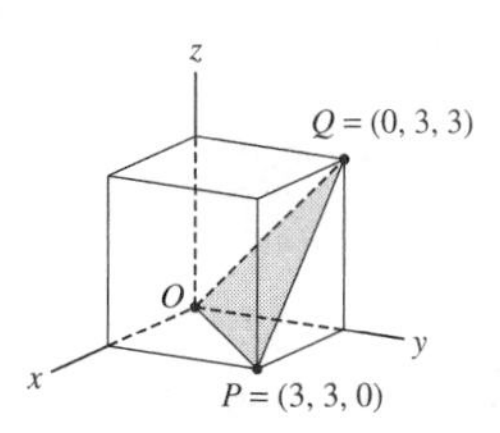

The area S of the triangle is half of the area of the parallelogram determined by the vectors $\overrightarrow{OP} = \langle 3, 3, 0 \rangle$ and $\overrightarrow{OQ} = \langle 0, 3, 3 \rangle$. Thus,

$$S = \frac{1}{2} \|\overrightarrow{OP} \times \overrightarrow{OQ}\| \tag{1}$$

We compute the cross product:

$$\overrightarrow{OP} \times \overrightarrow{OQ} = \begin{vmatrix} \mathbf{i} & \mathbf{j} & \mathbf{k} \\ 3 & 3 & 0 \\ 0 & 3 & 3 \end{vmatrix} = \begin{vmatrix} 3 & 0 \\ 3 & 3 \end{vmatrix} \mathbf{i} - \begin{vmatrix} 3 & 0 \\ 0 & 3 \end{vmatrix} \mathbf{j} + \begin{vmatrix} 3 & 3 \\ 0 & 3 \end{vmatrix} \mathbf{k}$$

$$= 9\mathbf{i} - 9\mathbf{j} + 9\mathbf{k} = 9\langle 1, -1, 1 \rangle$$

Substituting into (1) gives

$$S = \frac{1}{2} \|9\langle 1, -1, 1 \rangle\| = \frac{9}{2} \|\langle 1, -1, 1 \rangle\| = \frac{9}{2} \sqrt{1^2 + (-1)^2 + 1^2} = \frac{9\sqrt{3}}{2} \approx 7.8$$

The area of the triangle is $S = \frac{9\sqrt{3}}{2} \approx 7.8$.

In Exercises 49–51, verify the identity using the formula for the cross product.

49. $\mathbf{v} \times \mathbf{w} = -\mathbf{w} \times \mathbf{v}$

SOLUTION Let $\mathbf{v} = \langle a, b, c \rangle$ and $\mathbf{w} = \langle d, e, f \rangle$. By the definition of the cross product we have

$$\mathbf{v} \times \mathbf{w} = \begin{vmatrix} \mathbf{i} & \mathbf{j} & \mathbf{k} \\ a & b & c \\ d & e & f \end{vmatrix} = \begin{vmatrix} b & c \\ e & f \end{vmatrix} \mathbf{i} - \begin{vmatrix} a & c \\ d & f \end{vmatrix} \mathbf{j} + \begin{vmatrix} a & b \\ d & e \end{vmatrix} \mathbf{k} = (bf - ec)\mathbf{i} - (af - dc)\mathbf{j} + (ae - db)\mathbf{k}$$

We also have

$$-\mathbf{w} \times \mathbf{v} = \begin{vmatrix} \mathbf{i} & \mathbf{j} & \mathbf{k} \\ -d & -e & -f \\ a & b & c \end{vmatrix} = (-ec + bf)\mathbf{i} - (-dc + af)\mathbf{j} + (-db + ea)\mathbf{k}$$

Thus, $\mathbf{v} \times \mathbf{w} = -\mathbf{w} \times \mathbf{v}$, as desired.

51. $(\mathbf{u} + \mathbf{v}) \times \mathbf{w} = \mathbf{u} \times \mathbf{w} + \mathbf{v} \times \mathbf{w}$

SOLUTION We let $\mathbf{u} = \langle a_1, a_2, a_3 \rangle$, $\mathbf{v} = \langle b_1, b_2, b_3 \rangle$ and $\mathbf{w} = \langle c_1, c_2, c_3 \rangle$. Computing the left-hand side gives

$$(\mathbf{u} + \mathbf{v}) \times \mathbf{w} = \langle a_1 + b_1, a_2 + b_2, a_3 + b_3 \rangle \times \langle c_1, c_2, c_3 \rangle = \begin{vmatrix} \mathbf{i} & \mathbf{j} & \mathbf{k} \\ a_1 + b_1 & a_2 + b_2 & a_3 + b_3 \\ c_1 & c_2 & c_3 \end{vmatrix}$$

$$= \begin{vmatrix} a_2 + b_2 & a_3 + b_3 \\ c_2 & c_3 \end{vmatrix} \mathbf{i} - \begin{vmatrix} a_1 + b_1 & a_3 + b_3 \\ c_1 & c_3 \end{vmatrix} \mathbf{j} + \begin{vmatrix} a_1 + b_1 & a_2 + b_2 \\ c_1 & c_2 \end{vmatrix} \mathbf{k}$$

$$= (c_3(a_2 + b_2) - c_2(a_3 + b_3))\,\mathbf{i} - (c_3\,(a_1 + b_1) - c_1\,(a_3 + b_3))\,\mathbf{j} + (c_2(a_1 + b_1) - c_1(a_2 + b_2))\,\mathbf{k}$$

We now compute the right-hand-side of the equality:

$$\mathbf{u} \times \mathbf{w} + \mathbf{v} \times \mathbf{w} = \begin{vmatrix} \mathbf{i} & \mathbf{j} & \mathbf{k} \\ a_1 & a_2 & a_3 \\ c_1 & c_2 & c_3 \end{vmatrix} + \begin{vmatrix} \mathbf{i} & \mathbf{j} & \mathbf{k} \\ b_1 & b_2 & b_3 \\ c_1 & c_2 & c_3 \end{vmatrix}$$

$$= \begin{vmatrix} a_2 & a_3 \\ c_2 & c_3 \end{vmatrix} \mathbf{i} - \begin{vmatrix} a_1 & a_3 \\ c_1 & c_3 \end{vmatrix} \mathbf{j} + \begin{vmatrix} a_1 & a_2 \\ c_1 & c_2 \end{vmatrix} \mathbf{k} + \begin{vmatrix} b_2 & b_3 \\ c_2 & c_3 \end{vmatrix} \mathbf{i} - \begin{vmatrix} b_1 & b_3 \\ c_1 & c_3 \end{vmatrix} \mathbf{j} + \begin{vmatrix} b_1 & b_2 \\ c_1 & c_2 \end{vmatrix} \mathbf{k}$$

$$= (a_2c_3 - a_3c_2)\mathbf{i} - (a_1c_3 - a_3c_1)\mathbf{j} + (a_1c_2 - a_2c_1)\mathbf{k}$$
$$+ (b_2c_3 - b_3c_2)\mathbf{i} - (b_1c_3 - b_3c_1)\mathbf{j} + (b_1c_2 - b_2c_1)\mathbf{k}$$

$$= (a_2c_3 - a_3c_2 + b_2c_3 - b_3c_2)\mathbf{i} - (a_1c_3 - a_3c_1 + b_1c_3 - b_3c_1)\mathbf{j} + (a_1c_2 - a_2c_1 + b_1c_2 - b_2c_1)\mathbf{k}$$

$$= (c_3(a_2 + b_2) - c_2(a_3 + b_3))\,\mathbf{i} - (c_3(a_1 + b_1) - c_1(a_3 + b_3))\,\mathbf{j} + (c_2(a_1 + b_1) - c_1(a_2 + b_2))\,\mathbf{k}$$

The results are the same. Hence,

$$(\mathbf{u} + \mathbf{v}) \times \mathbf{w} = \mathbf{u} \times \mathbf{w} + \mathbf{v} \times \mathbf{w}.$$

53. Show that $(\mathbf{i} \times \mathbf{j}) \times \mathbf{j} \neq \mathbf{i} \times (\mathbf{j} \times \mathbf{j})$. Conclude that the Associative Law does not hold for cross products.

SOLUTION Using the cross products of the unit vectors **i**, **j**, and **k**, we obtain

$$(\mathbf{i} \times \mathbf{j}) \times \mathbf{j} = \mathbf{k} \times \mathbf{j} = -\mathbf{i}$$
$$\mathbf{i} \times (\mathbf{j} \times \mathbf{j}) = \mathbf{i} \times \mathbf{0} = \mathbf{0}$$

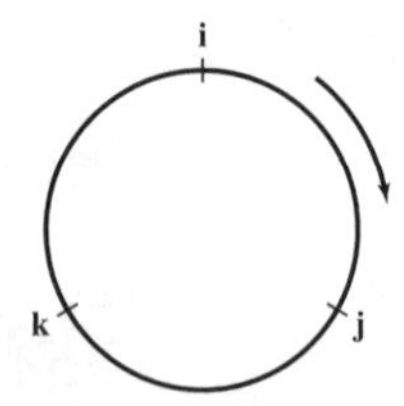

Since $(\mathbf{i} \times \mathbf{j}) \times \mathbf{j} \neq \mathbf{i} \times (\mathbf{j} \times \mathbf{j})$ the associative law does not hold for cross products.

55. Use Theorem 1(ii) to show that $\mathbf{v} \times \mathbf{w} = \mathbf{0}$ if and only if $\mathbf{w} = \lambda \mathbf{v}$ for some scalar λ or $\mathbf{v} = \mathbf{0}$.

SOLUTION $\mathbf{v} \times \mathbf{w} = \mathbf{0}$ if and only if $\|\mathbf{v} \times \mathbf{w}\| = 0$, that is, using Theorem 2 (b), if and only if

$$\|\mathbf{v}\|\,\|\mathbf{w}\| \sin\theta = 0$$

where θ is the angle between **v** and **w**. This equality holds only if at least one of the vectors **v** or **w** is the zero vector or $\sin\theta = 0$. The solutions of $\sin\theta = 0$ for angles between 0 and 180° are $\theta = 0$ and $\theta = 180°$, that is, **v** and **w** are parallel vectors. To summarize, we conclude that $\mathbf{v} \times \mathbf{w} = \mathbf{0}$ if and only if $\mathbf{v} = \mathbf{0}$ or $\mathbf{w} = \mathbf{0}$ or $\mathbf{w} = \lambda\mathbf{v}$. This can be written as

$$\mathbf{v} = \mathbf{0} \quad \text{or} \quad \mathbf{w} = \lambda\mathbf{v}.$$

57. Formulate and prove analogs of the result in Exercise 56 for the **i** and **j** components of $\mathbf{v} \times \mathbf{w}$.

SOLUTION The analogs for the $\mathbf{i}$ and $\mathbf{j}$ components of $\mathbf{v} \times \mathbf{w}$ are the following statements:

(a) The area of the parallelogram spanned by the projections $\mathbf{v}'$ and $\mathbf{w}'$ of vectors $\mathbf{v}$, $\mathbf{w}$ onto the xz-plane is equal to the absolute value of the $\mathbf{j}$-component of $\mathbf{v} \times \mathbf{w}$.

(b) The area of the parallelogram spanned by the projections $\mathbf{v}'$ and $\mathbf{w}'$ of vectors $\mathbf{v}$, $\mathbf{w}$ onto the yz-plane is equal to the absolute value of the $\mathbf{i}$-component of $\mathbf{v} \times \mathbf{w}$.

(c) If $\mathbf{v} = \langle a_1, a_2, a_3 \rangle$ and $\mathbf{w} = \langle b_1, b_2, b_3 \rangle$ then $\mathbf{v}' = \langle a_1, 0, a_3 \rangle$ and $\mathbf{w}' = \langle b_1, 0, b_3 \rangle$. The area S of the parallelogram spanned by $\mathbf{v}'$ and $\mathbf{w}'$ is

$$S = \|\mathbf{v}' \times \mathbf{w}'\| \tag{1}$$

We compute the cross product:

$$\mathbf{v}' \times \mathbf{w}' = \begin{vmatrix} \mathbf{i} & \mathbf{j} & \mathbf{k} \\ a_1 & 0 & a_3 \\ b_1 & 0 & b_3 \end{vmatrix} = \begin{vmatrix} 0 & a_3 \\ 0 & b_3 \end{vmatrix} \mathbf{i} - \begin{vmatrix} a_1 & a_3 \\ b_1 & b_3 \end{vmatrix} \mathbf{j} + \begin{vmatrix} a_1 & 0 \\ b_1 & 0 \end{vmatrix} \mathbf{k}$$
$$= 0\mathbf{i} - (a_1b_3 - a_3b_1)\mathbf{j} + 0\mathbf{k} = -(a_1b_3 - a_3b_1)\mathbf{j}$$

Combining with (1) we get

$$S = \| -(a_1b_3 - a_3b_1)\mathbf{j}\| = |a_1b_3 - a_3b_1| \tag{2}$$

We now find the cross product $\mathbf{v} \times \mathbf{w}$:

$$\mathbf{v} \times \mathbf{w} = \begin{vmatrix} \mathbf{i} & \mathbf{j} & \mathbf{k} \\ a_1 & a_2 & a_3 \\ b_1 & b_2 & b_3 \end{vmatrix} = \begin{vmatrix} a_2 & a_3 \\ b_2 & b_3 \end{vmatrix} \mathbf{i} - \begin{vmatrix} a_1 & a_3 \\ b_1 & b_3 \end{vmatrix} \mathbf{j} + \begin{vmatrix} a_1 & a_2 \\ b_1 & b_2 \end{vmatrix} \mathbf{k}$$

The $\mathbf{j}$-component of the cross product is

$$-\begin{vmatrix} a_1 & a_3 \\ b_1 & b_3 \end{vmatrix} = -(a_1b_3 - a_3b_1) \tag{3}$$

By (2) and (3) we obtain the desired result.

Proof of (b). In this case, $\mathbf{v}' = \langle 0, a_2, a_3 \rangle$ and $\mathbf{w}' = \langle 0, b_2, b_3 \rangle$. The area S of the parallelogram spanned by $\mathbf{v}'$ and $\mathbf{w}'$ is

$$S = \|\mathbf{v}' \times \mathbf{w}'\| \tag{4}$$

We compute the cross product:

$$\mathbf{v}' \times \mathbf{w}' = \begin{vmatrix} \mathbf{i} & \mathbf{j} & \mathbf{k} \\ 0 & a_2 & a_3 \\ 0 & b_2 & b_3 \end{vmatrix} = \begin{vmatrix} a_2 & a_3 \\ b_2 & b_3 \end{vmatrix} \mathbf{i} - \begin{vmatrix} 0 & a_3 \\ 0 & b_3 \end{vmatrix} \mathbf{j} + \begin{vmatrix} 0 & a_2 \\ 0 & b_2 \end{vmatrix} \mathbf{k} = (a_2b_3 - a_3b_2)\mathbf{i}$$

Hence, by (4) we get

$$S = \| (a_2b_3 - a_3b_2)\,\mathbf{i}\| = |a_2b_3 - a_3b_2|\|\mathbf{i}\| = |a_2b_3 - a_3b_2| \tag{5}$$

We now identify the $\mathbf{i}$ component of $\mathbf{v} \times \mathbf{w}$ as seen in the proof of part (a):

$$\begin{vmatrix} a_2 & a_3 \\ b_2 & b_3 \end{vmatrix} = a_2b_3 - a_3b_2 \tag{6}$$

By (5) and (6) we obtain the desired result.

59. Show that three points P, Q, R are collinear (lie on a line) if and only if $\overrightarrow{PQ} \times \overrightarrow{PR} = \mathbf{0}$.

SOLUTION The points P, Q, and R lie on one line if and only if the vectors $\overrightarrow{PQ}$ and $\overrightarrow{PR}$ are parallel. By basic properties of the cross product this is equivalent to $\overrightarrow{PQ} \times \overrightarrow{PR} = \mathbf{0}$.

61. Find a vector $\mathbf{X}$ such that $\langle 1, 1, 1 \rangle \times \mathbf{X} = \langle 1, -1, 0 \rangle$.

SOLUTION Let $\mathbf{X} = \langle a, b, c \rangle$. We compute the cross product:

$$\langle 1,1,1\rangle \times \langle a,b,c\rangle = \begin{vmatrix} \mathbf{i} & \mathbf{j} & \mathbf{k} \\ 1 & 1 & 1 \\ a & b & c \end{vmatrix} = \begin{vmatrix} 1 & 1 \\ b & c \end{vmatrix}\mathbf{i} - \begin{vmatrix} 1 & 1 \\ a & c \end{vmatrix}\mathbf{j} + \begin{vmatrix} 1 & 1 \\ a & b \end{vmatrix}\mathbf{k}$$

$$= (c-b)\mathbf{i} - (c-a)\mathbf{j} + (b-a)\mathbf{k} = \langle c-b, a-c, b-a \rangle$$

The equation for $\mathbf{X}$ is, thus,

$$\langle c-b, a-c, b-a\rangle = \langle 1, -1, 0\rangle$$

Equating corresponding components we get

$$c - b = 1$$
$$a - c = -1$$
$$b - a = 0$$

The third equation implies $a = b$. Substituting in the first and second equations gives

$$\begin{matrix} c - a = 1 \\ a - c = -1 \end{matrix} \quad \Rightarrow \quad c = a + 1$$

The solution is thus, $b = a$, $c = a + 1$. The corresponding solutions $\mathbf{X}$ are

$$\mathbf{X} = \langle a, b, c\rangle = \langle a, a, a+1\rangle$$

One possible solution is obtained for $a = 1$, that is, $\mathbf{X} = \langle 1, 1, 2\rangle$.

63. Suppose that vectors $\mathbf{u}$, $\mathbf{v}$, and $\mathbf{w}$ are mutually orthogonal, that is, $\mathbf{u} \perp \mathbf{v}$, $\mathbf{u} \perp \mathbf{w}$, and $\mathbf{v} \perp \mathbf{w}$. Prove that $(\mathbf{u} \times \mathbf{v}) \times \mathbf{w} = \mathbf{0}$ and $\mathbf{u} \times (\mathbf{v} \times \mathbf{w}) = \mathbf{0}$.

SOLUTION The cross product $\mathbf{u} \times \mathbf{v}$ is orthogonal to $\mathbf{u}$ and $\mathbf{v}$, hence it is parallel to $\mathbf{w}$. The cross product of parallel vectors is the zero vector, hence $(\mathbf{u} \times \mathbf{v}) \times \mathbf{w} = \mathbf{0}$. Similarly, the cross product $\mathbf{v} \times \mathbf{w}$ is orthogonal to $\mathbf{v}$ and $\mathbf{w}$, hence it is parallel to $\mathbf{u}$. Since the cross product of parallel vectors is the zero vector, we conclude that $\mathbf{u} \times (\mathbf{v} \times \mathbf{w}) = \mathbf{0}$.

*Exercises 64–67 deal with torque: When a force $\mathbf{F}$ acts on an object with position vector $\mathbf{r}$, the **torque** about the origin O is the vector quantity $\tau = \mathbf{r} \times \mathbf{F}$. If several forces $\mathbf{F}_j$ act at positions $\mathbf{r}_j$, then the net torque is the sum $\tau = \sum \mathbf{r}_j \times \mathbf{F}_j$. By Newton's Laws, τ (units: N-m or lb-ft) is equal to the rate of change of angular momentum. Torque is a measure of how much the force causes the object to rotate.*

65. Calculate the net torque about O at P, assuming that a 30-kg mass is attached at P [Figure 20(B)]. The force $\mathbf{F}_g$ due to gravity on a mass m has magnitude $9.8m$ m/s^2 in the downward direction.

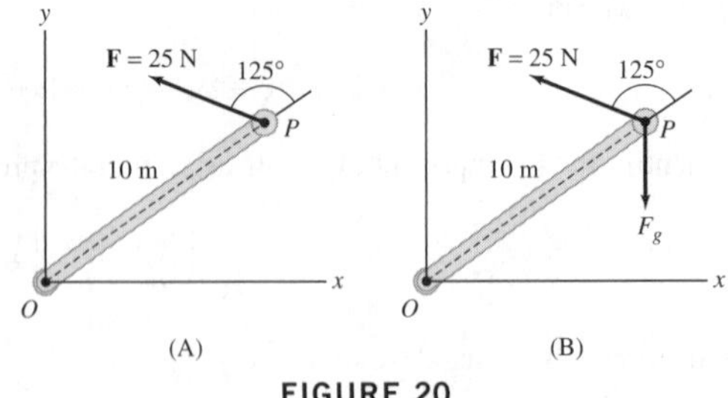

FIGURE 20

SOLUTION We denote by τ_1 and τ_2 the torques due to the forces $\mathbf{F}$ and $\mathbf{F}_g$ respectively. Let θ denote the angle between the arm and the x-axis, and $\mathbf{r} = \overrightarrow{OP}$ the position vector. The net torque about O at P is

$$\tau = \mathbf{r} \times \mathbf{F} + \mathbf{r} \times \mathbf{F}_g = \tau_1 + \tau_2 \tag{1}$$

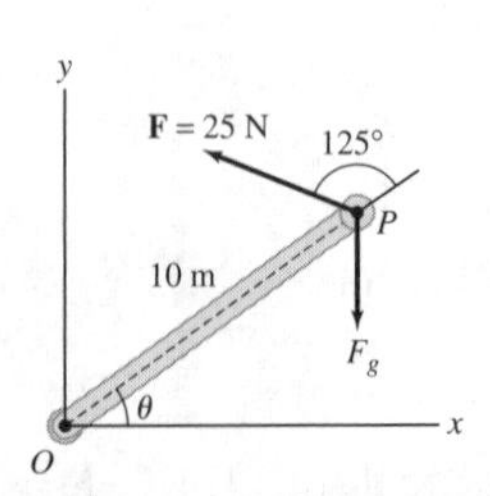

In Exercise 64 we found that

$$\tau_1 = 204.79\mathbf{k} \tag{2}$$

We compute the torque τ_2:

$$\begin{aligned}\tau_2 = \mathbf{r} \times \mathbf{F}_g &= 10\,(\cos\theta\mathbf{i} + \sin\theta\mathbf{j}) \times 9.8 \cdot 30\,(-\mathbf{j}) \\ &= 2940\,(\cos\theta\mathbf{i} + \sin\theta\mathbf{j}) \times (-\mathbf{j}) \qquad (3) \\ &= 2940\cos\theta\,(-\mathbf{k}) = -2940\cos\theta\mathbf{k}\end{aligned}$$

Combining (1), (2), and (3) we obtain

$$\tau = 204.79\mathbf{k} - 2940\cos\theta\mathbf{k} = (204.79 - 2940\cos\theta)\mathbf{k}$$

67. Continuing with Exercise 66, suppose that $L_1 = 5$ ft, $L_2 = 3$ ft, and $m_1 = 30$ lb, $m_2 = 20$ lb, $m_3 = 50$ lb. If the angles θ_1, θ_2 are equal (say, to θ), what is the maximum allowable value of θ if we assume that the robotic arm can sustain a maximum torque of 400 ft-lb?

SOLUTION Setting the given values $L_1 = 5$, $L_2 = 3$, $m_1 = 30$, $m_2 = 20$, $m_3 = 50$, and $\theta_1 = \theta_2 = \theta$ in the formula for τ obtained in Exercise 66 we get

$$\tau = -\,(5\,(15 + 20 + 50)\sin\theta + 3\,(10 + 50)\sin\theta)\,\mathbf{k} = -605\sin\theta\mathbf{k}$$

Thus,

$$\|\tau\| = 605\sin\theta$$

Since the maximum torque sustained by the robotic arm is 400 ft-lbs, we have

$$605\sin\theta \le 400$$

$$\sin\theta \le \frac{400}{605} \approx 0.661$$

The solution for acute angles is

$$\theta \le 41.39^\circ$$

The maximum allowable value of θ is $\theta = 41.39^\circ$.

Further Insights and Challenges

69. Use the diagonal rule to calculate $\begin{vmatrix} 2 & 4 & 3 \\ 0 & 1 & -7 \\ -1 & 5 & 3 \end{vmatrix}$.

SOLUTION We form the following matrix:

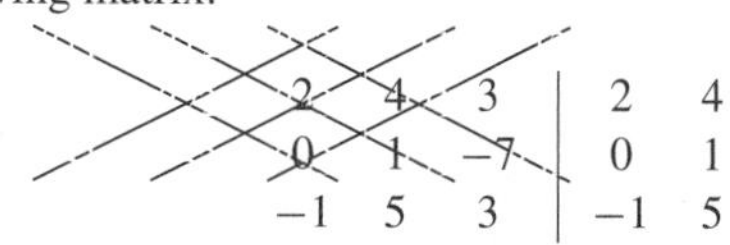

We now form the diagonals which slant from left to right and the diagonals which slant from right to left and assign corresponding sign to each diagonal:

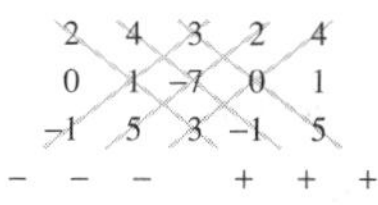

We add the products for the diagonals with a positive sign and subtract the products for the diagonals with a negative sign. This gives

$$\begin{vmatrix} 2 & 4 & 3 \\ 0 & 1 & -7 \\ -1 & 5 & 3 \end{vmatrix} = 2 \cdot 1 \cdot 3 + 4 \cdot (-7) \cdot (-1) + 3 \cdot 0 \cdot 5 - 3 \cdot 1 \cdot (-1) - 2 \cdot (-7) \cdot 5 - 4 \cdot 0 \cdot 3 = 107$$

71. Use Eq. (10) to prove the Cauchy–Schwarz inequality:

$$|\mathbf{v} \cdot \mathbf{w}| \le \|\mathbf{v}\|\ \|\mathbf{w}\|$$

Show that equality holds if and only if $\mathbf{w}$ is a multiple of $\mathbf{v}$ or at least one of $\mathbf{v}$ and $\mathbf{w}$ is zero.

SOLUTION Transferring sides in Eq. (10) we get

$$(\mathbf{v}\cdot\mathbf{w})^2 = \|\mathbf{v}\|^2\|\mathbf{w}\|^2 - \|\mathbf{v}\times\mathbf{w}\|^2 \tag{1}$$

Since $\|\mathbf{v}\times\mathbf{w}\|^2 \ge 0$, we have

$$(\mathbf{v}\cdot\mathbf{w})^2 \le \|\mathbf{v}\|^2\|\mathbf{w}\|^2$$

Taking the square root of both sides gives

$$|\mathbf{v}\cdot\mathbf{w}| \le \|\mathbf{v}\|\|\mathbf{w}\|$$

Equality $|\mathbf{v}\cdot\mathbf{w}| = \|\mathbf{v}\|\|\mathbf{w}\|$ holds if and only if $(\mathbf{v}\cdot\mathbf{w})^2 = \|\mathbf{v}\|^2\|\mathbf{w}\|^2$, that is by (1), if and only if $\|\mathbf{v}\times\mathbf{w}\| = 0$, or $\mathbf{v}\times\mathbf{w} = \mathbf{0}$. This is equivalent to $\mathbf{w} = \lambda\mathbf{v}$ for some scalar λ, or $\mathbf{v} = \mathbf{0}$ (Theorem 3 (c)).

73. Suppose that **u**, **v**, **w** are nonzero and

$$(\mathbf{u}\times\mathbf{v})\times\mathbf{w} = \mathbf{u}\times(\mathbf{v}\times\mathbf{w}) = \mathbf{0}$$

Show that **u**, **v**, and **w** are either mutually parallel or mutually perpendicular. *Hint:* Use Exercise 72.

SOLUTION First notice that since $\mathbf{u}\times(\mathbf{v}\times\mathbf{w}) = \mathbf{0}$, also,

$$(\mathbf{v}\times\mathbf{w})\times\mathbf{u} = -\mathbf{u}\times(\mathbf{v}\times\mathbf{w}) = \mathbf{0}$$

Using Exercise 72 for nonzero vectors, we obtain

$$(\mathbf{u}\times\mathbf{v})\times\mathbf{w} = \mathbf{0} \Rightarrow \mathbf{u}\parallel\mathbf{v} \text{ or } (\mathbf{w}\perp\mathbf{u} \text{ and } \mathbf{w}\perp\mathbf{v}) \tag{1}$$

$$(\mathbf{v}\times\mathbf{w})\times\mathbf{u} = \mathbf{0} \Rightarrow \mathbf{w}\parallel\mathbf{v} \text{ or } (\mathbf{u}\perp\mathbf{v} \text{ and } \mathbf{u}\perp\mathbf{w}) \tag{2}$$

We consider each of the two possibilities in (1):

Case 1: $\mathbf{u}\parallel\mathbf{v}$ In this case, **u** is not orthogonal to **v**. Hence, by (2), $\mathbf{w}\parallel\mathbf{v}$ must hold. Thus, the vectors **u**, **v** and **w** are parallel.

Case 2: $\mathbf{w}\perp\mathbf{u}$ and $\mathbf{w}\perp\mathbf{v}$ In this case, **w** and **v** are not parallel. Hence, by (2), $\mathbf{u}\perp\mathbf{v}$ and $\mathbf{u}\perp\mathbf{w}$ must hold. Thus, the vectors **u**, **v** and **w** are mutually perpendicular.

Conclusion: If $(\mathbf{u}\times\mathbf{v})\times\mathbf{w} = \mathbf{0}$ and $\mathbf{u}\times(\mathbf{v}\times\mathbf{w}) = \mathbf{0}$ for nonzero vectors **u**, **v**, **w** then these vectors are parallel or mutually perpendicular.

75. Use Exercise 74 to prove the identity

$$(\mathbf{a}\times\mathbf{b})\times\mathbf{c} - \mathbf{a}\times(\mathbf{b}\times\mathbf{c}) = (\mathbf{a}\cdot\mathbf{b})\mathbf{c} - (\mathbf{b}\cdot\mathbf{c})\mathbf{a}$$

SOLUTION We have

$$\begin{aligned}(\mathbf{a}\times\mathbf{b})\times\mathbf{c} - \mathbf{a}\times(\mathbf{b}\times\mathbf{c}) &= -\mathbf{c}\times(\mathbf{a}\times\mathbf{b}) - \mathbf{a}\times(\mathbf{b}\times\mathbf{c})\\ &= -[(\mathbf{c}\cdot\mathbf{b})\mathbf{a} - (\mathbf{c}\cdot\mathbf{a})\mathbf{b}] - [(\mathbf{a}\cdot\mathbf{c})\mathbf{b} - (\mathbf{a}\cdot\mathbf{b})\mathbf{c}]\\ &= -(\mathbf{c}\cdot\mathbf{b})\mathbf{a} + (\mathbf{a}\cdot\mathbf{b})\mathbf{c} = (\mathbf{a}\cdot\mathbf{b})\mathbf{c} - (\mathbf{b}\cdot\mathbf{c})\mathbf{a}\end{aligned}$$

as desired.

77. Show that if **a**, **b** are nonzero vectors such that $\mathbf{a}\perp\mathbf{b}$, then there exists a vector **x** such that

$$\mathbf{x}\times\mathbf{a} = \mathbf{b} \tag{13}$$

Hint: Show that if **x** is orthogonal to **b** and is not a multiple of **a**, then $\mathbf{x}\times\mathbf{a}$ is a multiple of **b**.

SOLUTION We define the following vectors:

$$\mathbf{x} = \frac{\mathbf{b}\times\mathbf{a}}{\|\mathbf{a}\|^2}, \qquad \mathbf{c} = \mathbf{x}\times\mathbf{a} \tag{1}$$

We show that $\mathbf{c} = \mathbf{b}$. Since **x** is orthogonal to **a** and **b**, **x** is orthogonal to the plane of **a** and **b**. But **c** is orthogonal to **x**, hence **c** is contained in the plane of **a** and **b**, that is, **a**, **b** and **c** are in the same plane. Now the vectors **a**, **b** and **c** are in one plane, and the vectors **c** and **b** are orthogonal to **a**.

$$\text{It follows that } \mathbf{c} \text{ and } \mathbf{b} \text{ are parallel.} \tag{2}$$

We now show that $\|\mathbf{c}\| = \|\mathbf{b}\|$. We use the cross-product identity to obtain

$$\|\mathbf{c}\|^2 = \|\mathbf{x}\times\mathbf{a}\|^2 = \|\mathbf{x}\|^2\|\mathbf{a}\|^2 - (\mathbf{x}\cdot\mathbf{a})^2$$

$\mathbf{x}$ is orthogonal to $\mathbf{a}$, hence $\mathbf{x} \cdot \mathbf{a} = 0$, and we obtain

$$\|\mathbf{c}\|^2 = \|\mathbf{x}\|^2 \|\mathbf{a}\|^2 = \left\| \frac{\mathbf{b} \times \mathbf{a}}{\|\mathbf{a}\|^2} \right\|^2 \|\mathbf{a}\|^2 = \frac{1}{\|\mathbf{a}\|^4} \|\mathbf{b} \times \mathbf{a}\|^2 \|\mathbf{a}\|^2 = \frac{1}{\|\mathbf{a}\|^2} \|\mathbf{b} \times \mathbf{a}\|^2$$

By the given data, $\mathbf{a}$ and $\mathbf{b}$ are orthogonal vectors, so,

$$\|\mathbf{c}\|^2 = \frac{1}{\|\mathbf{a}\|^2} \left(\|\mathbf{b}\|^2 \|\mathbf{a}\|^2 \right) = \|\mathbf{b}\|^2 \Rightarrow \|\mathbf{c}\| = \|\mathbf{b}\| \tag{3}$$

By (2) and (3) it follows that $\mathbf{c} = \mathbf{b}$ or $\mathbf{c} = -\mathbf{b}$. We thus proved that the vector $\mathbf{x} = \dfrac{\mathbf{b} \times \mathbf{a}}{\|\mathbf{a}\|^2}$ satisfies $\mathbf{x} \times \mathbf{a} = \mathbf{b}$ or $\mathbf{x} \times \mathbf{a} = -\mathbf{b}$. If $\mathbf{x} \times \mathbf{a} = -\mathbf{b}$, then $(-\mathbf{x}) \times \mathbf{a} = \mathbf{b}$. Hence, there exists a vector $\mathbf{x}$ such that $\mathbf{x} \times \mathbf{a} = \mathbf{b}$.

79. Consider the tetrahedron spanned by vectors $\mathbf{a}$, $\mathbf{b}$, and $\mathbf{c}$ as in Figure 24(A). Let A, B, C be the faces containing the origin O and let D be the fourth face opposite O. For each face F, let $\mathbf{v}_F$ be the vector normal to the face, pointing outside the tetrahedron, of magnitude equal to twice the area of F. Prove the relations

$$\mathbf{v}_A + \mathbf{v}_B + \mathbf{v}_C = \mathbf{a} \times \mathbf{b} + \mathbf{b} \times \mathbf{c} + \mathbf{c} \times \mathbf{a}$$

$$\mathbf{v}_A + \mathbf{v}_B + \mathbf{v}_C + \mathbf{v}_D = 0$$

Hint: Show that $\mathbf{v}_D = (\mathbf{c} - \mathbf{b}) \times (\mathbf{b} - \mathbf{a})$.

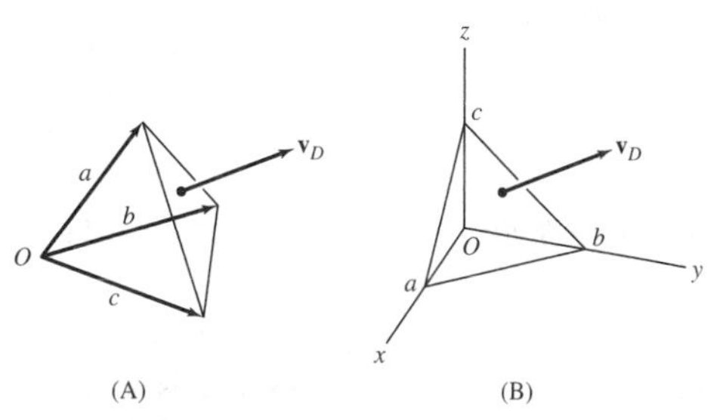

FIGURE 24 The vector $\mathbf{v}_D$ is perpendicular to the face.

SOLUTION We first show that $\mathbf{v}_D = (\mathbf{c} - \mathbf{a}) \times (\mathbf{b} - \mathbf{a})$.

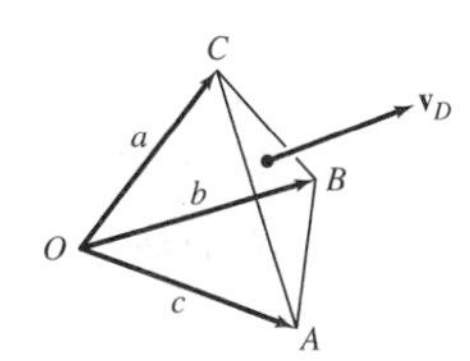

Since $\mathbf{v}_D$ is normal to the face D, it is orthogonal to the vectors $\mathbf{c} - \mathbf{a}$ and $\mathbf{b} - \mathbf{a}$, hence it is parallel to the cross product of these two vectors. In other words, there exists a scalar $\lambda > 0$ such that (using the right-hand rule)

$$\mathbf{v}_D = \lambda\, (\mathbf{c} - \mathbf{a}) \times (\mathbf{b} - \mathbf{a})$$

The area of the face D is half of the area of the parallelogram spanned by $mathbfc - \mathbf{a}$ and $\mathbf{b} - \mathbf{a}$. The area of the parallelogram is $\|(\mathbf{c} - \mathbf{a}) \times (\mathbf{b} - \mathbf{a})\|$. Hence,

$$\|\mathbf{v}_D\| = \|\,(\mathbf{c} - \mathbf{a}) \times (\mathbf{b} - \mathbf{a})\,\|$$

Combining the above equations we have

$$\mathbf{v}_D = (\mathbf{c} - \mathbf{a}) \times (\mathbf{b} - \mathbf{a})$$

Now, since $\mathbf{v}_A$ is normal to the face A, it is orthogonal to the vectors $\mathbf{a}$ and $\mathbf{c}$, therefore it is parallel to $\mathbf{c} \times \mathbf{a}$. The area of the face A is $\frac{1}{2}\|\mathbf{c} \times \mathbf{a}\|$, which is also half the length of $\mathbf{v}_A$. Hence, using the right-hand rule we get

$$\mathbf{v}_A = \mathbf{c} \times \mathbf{a}$$

Similarly, we have

$$\mathbf{v}_B = \mathbf{a} \times \mathbf{b}, \quad \mathbf{v}_C = \mathbf{b} \times \mathbf{c}$$

Combining gives us

$$\mathbf{v}_A + \mathbf{v}_B + \mathbf{v}_C = (\mathbf{c} \times \mathbf{a} + \mathbf{a} \times \mathbf{b} + \mathbf{b} \times \mathbf{c}) = (\mathbf{a} \times \mathbf{b} + \mathbf{b} \times \mathbf{c} + \mathbf{c} \times \mathbf{a})$$

We evaluate $\mathbf{v}_D$ using the distributive law:

$$\mathbf{v}_D = (\mathbf{c}-\mathbf{a}) \times (\mathbf{b}-\mathbf{a}) = (\mathbf{c}\times\mathbf{b} - \mathbf{a}\times\mathbf{b} - \mathbf{c}\times\mathbf{a} + \mathbf{a}\times\mathbf{a}) = -\mathbf{b}\times\mathbf{c} - \mathbf{a}\times\mathbf{b} - \mathbf{c}\times\mathbf{a}$$

Hence,

$$\mathbf{v}_A + \mathbf{v}_B + \mathbf{v}_C + \mathbf{v}_D = (\mathbf{a}\times\mathbf{b} + \mathbf{b}\times\mathbf{c} + \mathbf{c}\times\mathbf{a} - \mathbf{b}\times\mathbf{c} - \mathbf{a}\times\mathbf{b} - \mathbf{c}\times\mathbf{a}) = \mathbf{0}$$

13.5 Planes in Three-Space (ET Section 12.5)

Preliminary Questions

1. What is the equation of the plane parallel to $3x + 4 - z = 5$ passing through the origin?

SOLUTION The two planes are parallel, therefore the vector $\mathbf{n} = \langle 3, 4, -1\rangle$ that is normal to the given plane is also normal to the plane we need to find. This plane is passing through the origin, hence we may substitute $\langle x_0, y_0, z_0\rangle = \langle 0, 0, 0\rangle$ in the vector form of the equation of the plane. This gives

$$\mathbf{n}\cdot\langle x, y, z\rangle = \mathbf{n}\cdot\langle x_0, y_0, z_0\rangle$$

$$\langle 3, 4, -1\rangle\cdot\langle x, y, z\rangle = \langle 3, 4, -1\rangle\cdot\langle 0, 0, 0\rangle = 0$$

or in scalar form

$$3x + 4y - z = 0$$

2. The vector $\mathbf{k}$ is normal to which of the following planes?

(a) $x = 1$ **(b)** $y = 1$ **(c)** $z = 1$

SOLUTION The planes $x = 1$, $y = 1$, and $z = 1$ are orthogonal to the x, y, and z-axes respectively. Since the plane $z = 1$ is orthogonal to the z-axis, the vector $\mathbf{k}$ is normal to this plane.

3. Which of the following planes are not parallel to the plane $x + y + z = 1$?

(a) $2x + 2y + 2z = 1$ **(b)** $x + y + z = 3$

(c) $x - y + z = 0$

SOLUTION The two planes are parallel if vectors that are normal to the planes are parallel. The vector $\mathbf{n} = \langle 1, 1, 1\rangle$ is normal to the plane $x + y + z = 1$. We identify the following normals:

- $\mathbf{v} = \langle 2, 2, 2\rangle$ is normal to plane (a)
- $\mathbf{u} = \langle 1, 1, 1\rangle$ is normal to plane (b)
- $\mathbf{w} = \langle 1, -1, 1\rangle$ is normal to plane (c)

The vectors $\mathbf{v}$ and $\mathbf{u}$ are parallel to $\mathbf{n}$, whereas $\mathbf{w}$ is not. (These vectors are not constant multiples of each other). Therefore, only plane (c) is not parallel to the plane $x + y + z = 1$.

4. To which coordinate plane is the plane $y = 1$ parallel?

SOLUTION The plane $y = 1$ is parallel to the xz-plane.

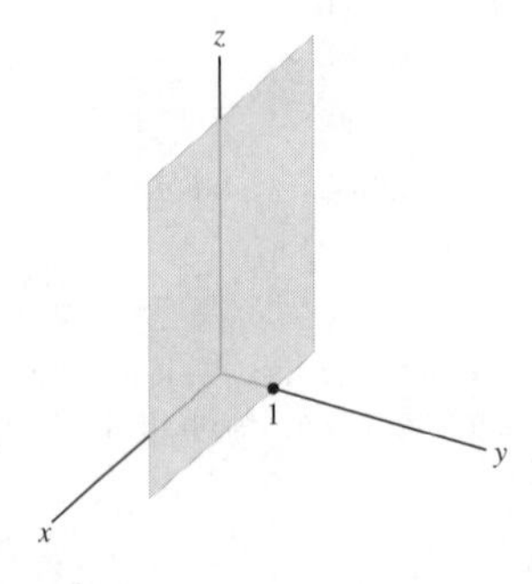

5. Which of the following planes contains the z-axis?

(a) $z = 1$ **(b)** $x + y = 1$ **(c)** $x + y = 0$

SOLUTION The points on the z-axis are the points with zero x and y coordinates. A plane contains the z-axis if and only if the points $(0, 0, c)$ satisfy the equation of the plane for all values of c.

(a) Plane (a) does not contain the z-axis, rather it is orthogonal to this axis. Only the point $(0, 0, 1)$ is on the plane.

(b) $x = 0$ and $y = 0$ do not satisfy the equation of the plane, since $0 + 0 \neq 1$. Therefore the plane does not contain the z-axis.

(c) The plane $x + y = 0$ contains the z-axis since $x = 0$ and $y = 0$ satisfy the equation of the plane.

6. Suppose that a plane $\mathcal{P}$ with normal vector $\mathbf{n}$ and a line $\mathcal{L}$ with direction vector $\mathbf{v}$ both pass through the origin and that $\mathbf{n} \cdot \mathbf{v} = 0$. Which of the following statements is correct?

(a) $\mathcal{L}$ is contained in the $\mathcal{P}$.

(b) $\mathcal{L}$ is orthogonal to $\mathcal{P}$.

SOLUTION The direction vector of the line $\mathcal{L}$ is orthogonal to the vector $\mathbf{n}$ that is normal to the plane. Therefore, $\mathcal{L}$ is either parallel or contained in the plane. Since the origin is common to $\mathcal{L}$ and $\mathcal{P}$, the line is contained in the plane. That is, statement (a) is correct.

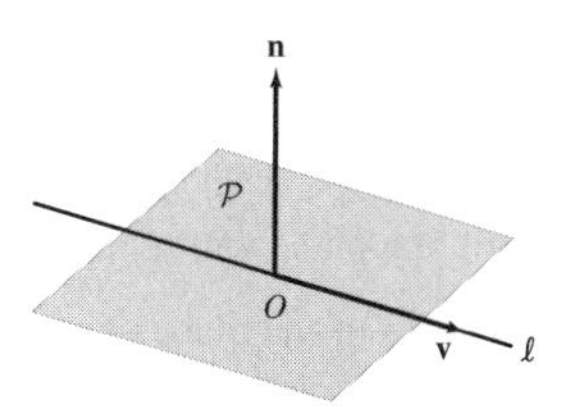

Exercises

In Exercises 1–8, write an equation of the plane with normal vector **n** *passing through the given point in vector form and the scalar forms (3) and (4).*

1. $\mathbf{n} = \langle 1, 3, 2 \rangle$, $(4, -1, 1)$

SOLUTION The vector equation is

$$\langle 1, 3, 2 \rangle \cdot \langle x, y, z \rangle = \langle 1, 3, 2 \rangle \cdot \langle 4, -1, 1 \rangle = 4 - 3 + 2 = 3$$

To obtain the scalar forms we compute the dot product on the left-hand side of the previous equation:

$$x + 3y + 2z = 3$$

or in the other scalar form:

$$(x - 4) + 3(y + 1) + 2(z - 1) + 4 - 3 + 2 = 3$$
$$(x - 4) + 3(y + 1) + 2(z - 1) = 0$$

3. $\mathbf{n} = \langle -1, 2, 1 \rangle$, $(4, 1, 5)$

SOLUTION The vector form is

$$\langle -1, 2, 1 \rangle \cdot \langle x, y, z \rangle = \langle -1, 2, 1 \rangle \cdot \langle 4, 1, 5 \rangle = -4 + 2 + 5 = 3$$

To obtain the scalar form we compute the dot product above:

$$-x + 2y + z = 3$$

or in the other scalar form:

$$-(x - 4) + 2(y - 1) + (z - 5) = 3 + 4 - 2 - 5 = 0$$
$$-(x - 4) + 2(y - 1) + (z - 5) = 0$$

5. $\mathbf{n} = \mathbf{i}$, $(3, 1, -9)$

SOLUTION We find the vector form of the equation of the plane. We write the vector $\mathbf{n} = \mathbf{i}$ as $\mathbf{n} = \langle 1, 0, 0 \rangle$ and obtain

$$\langle 1, 0, 0 \rangle \cdot \langle x, y, z \rangle = \langle 1, 0, 0 \rangle \cdot \langle 3, 1, -9 \rangle = 3 + 0 + 0 = 3$$

Computing the dot product above gives the scalar form:

$$x + 0 + 0 = 3$$
$$x = 3$$

Or in the other scalar form:

$$(x - 3) + 0 \cdot (y - 1) + 0 \cdot (z + 9) = 3 - 3 = 0$$

7. $\mathbf{n} = \mathbf{k}$, $(6, 7, 2)$

SOLUTION We write the normal $\mathbf{n} = \mathbf{k}$ in the form $\mathbf{n} = \langle 0, 0, 1\rangle$ and obtain the following vector form of the equation of the plane:

$$\langle 0, 0, 1\rangle \cdot \langle x, y, z\rangle = \langle 0, 0, 1\rangle \cdot \langle 6, 7, 2\rangle = 0 + 0 + 2 = 2$$

We compute the dot product to obtain the scalar form:

$$0x + 0y + 1z = 2$$
$$z = 2$$

or in the other scalar form:

$$0(x - 6) + 0(y - 7) + 1(z - 2) = 0$$

9. Find an equation of any plane with normal $\mathbf{n} = \langle 2, 1, 1\rangle$ *not* passing through the point $P = (0, 1, 1)$.

SOLUTION The equations of the planes with normal $\mathbf{n} = \langle 2, 1, 1\rangle$ are

$$2x + y + z = d \qquad (1)$$

The planes that are not passing through the point $P = (0, 1, 1)$ are the planes for which

$$2 \cdot 0 + 1 + 1 \neq d \quad \Rightarrow \quad d \neq 2$$

For instance we choose $d = 1$, to obtain the equation of one of these planes:

$$2x + y + z = 1.$$

In Exercises 11–14, find an equation of the plane passing through the three points given.

11. $P = (2, -1, 4)$, $Q = (1, 1, 1)$, $R = (3, 1, -2)$

SOLUTION We go through the steps below:

Step 1. Find the normal vector $\mathbf{n}$. The vectors $\mathbf{a} = \overrightarrow{PQ}$ and $\mathbf{b} = \overrightarrow{PR}$ lie on the plane, hence the cross product $\mathbf{n} = \mathbf{a} \times \mathbf{b}$ is normal to the plane. We compute the cross product:

$$\mathbf{a} = \overrightarrow{PQ} = \langle 1 - 2, 1 - (-1), 1 - 4\rangle = \langle -1, 2, -3\rangle$$
$$\mathbf{b} = \overrightarrow{PR} = \langle 3 - 2, 1 - (-1), -2 - 4\rangle = \langle 1, 2, -6\rangle$$
$$\mathbf{n} = \mathbf{a} \times \mathbf{b} = \begin{vmatrix} \mathbf{i} & \mathbf{j} & \mathbf{k} \\ -1 & 2 & -3 \\ 1 & 2 & -6 \end{vmatrix} = \begin{vmatrix} 2 & -3 \\ 2 & -6 \end{vmatrix}\mathbf{i} - \begin{vmatrix} -1 & -3 \\ 1 & -6 \end{vmatrix}\mathbf{j} + \begin{vmatrix} -1 & 2 \\ 1 & 2 \end{vmatrix}\mathbf{k}$$
$$= -6\mathbf{i} - 9\mathbf{j} - 4\mathbf{k} = \langle -6, -9, -4\rangle$$

Step 2. Choose a point on the plane. We choose any one of the three points on the plane, for instance $Q = (1, 1, 1)$. Using the vector form of the equation of the plane we get

$$\mathbf{n} \cdot \langle x, y, z\rangle = \mathbf{n} \cdot \langle x_0, y_0, z_0\rangle$$
$$\langle -6, -9, -4\rangle \cdot \langle x, y, z\rangle = \langle -6, -9, -4\rangle \cdot \langle 1, 1, 1\rangle$$

Computing the dot products we obtain the following equation:

$$-6x - 9y - 4z = -6 - 9 - 4 = -19$$
$$6x + 9y + 4z = 19$$

13. $P = (1, 0, 0)$, $Q = (0, 1, 1)$, $R = (2, 0, 1)$

SOLUTION We use the vector form of the equation of the plane:

$$\mathbf{n} \cdot \langle x, y, z\rangle = d \qquad (1)$$

To find the normal vector to the plane, $\mathbf{n}$, we first compute the vectors $\overrightarrow{PQ}$ and $\overrightarrow{PR}$ that lie in the plane, and then find the cross product of these vectors. This gives

$$\overrightarrow{PQ} = \langle 0, 1, 1\rangle - \langle 1, 0, 0\rangle = \langle -1, 1, 1\rangle$$

$$\overrightarrow{PR} = \langle 2, 0, 1\rangle - \langle 1, 0, 0\rangle = \langle 1, 0, 1\rangle$$

$$\mathbf{n} = \overrightarrow{PQ} \times \overrightarrow{PR} = \begin{vmatrix} \mathbf{i} & \mathbf{j} & \mathbf{k} \\ -1 & 1 & 1 \\ 1 & 0 & 1 \end{vmatrix} = \begin{vmatrix} 1 & 1 \\ 0 & 1 \end{vmatrix}\mathbf{i} - \begin{vmatrix} -1 & 1 \\ 1 & 1 \end{vmatrix}\mathbf{j} + \begin{vmatrix} -1 & 1 \\ 1 & 0 \end{vmatrix}\mathbf{k}$$

$$= \mathbf{i} + 2\mathbf{j} - \mathbf{k} = \langle 1, 2, -1\rangle \qquad (2)$$

We now choose any one of the three points in the plane, say $P = (1, 0, 0)$, and compute d:

$$d = \mathbf{n} \cdot \overrightarrow{OP} = \langle 1, 2, -1\rangle \cdot \langle 1, 0, 0\rangle = 1 \cdot 1 + 2 \cdot 0 + (-1) \cdot 0 = 1 \qquad (3)$$

Finally we substitute (2) and (3) into (1) to obtain the following equation of the plane:

$$\langle 1, 2, -1\rangle \cdot \langle x, y, z\rangle = 1$$

$$x + 2y - z = 1$$

In Exercises 15–20, find a vector normal to the plane with the given equation.

15. $9x - 4y - 11z = 2$

SOLUTION Using the scalar form of the equation of the plane, a vector normal to the plane is the coefficients vector:

$$\mathbf{n} = \langle 9, -4, -11\rangle$$

17. $3x + 25y + 9z = 45$

SOLUTION We write the equation in the vector form

$$\langle 3, 25, 9\rangle \cdot \langle x, y, z\rangle = 45$$

The vector $\mathbf{n} = \langle 3, 25, 9\rangle$ is normal to the plane.

19. $x - z = 0$

SOLUTION We write the equation in vector form $\langle 1, 0, -1\rangle \cdot \langle x, y, z\rangle = 0$ and identify $\mathbf{n} = \langle 1, 0, -1\rangle$ as a vector normal to the plane.

In Exercises 21–26, find the equation of the plane with the given description.

21. Passes through O and is parallel to $4x - 9y + z = 3$

SOLUTION The vector $\mathbf{n} = \langle 4, -9, 1\rangle$ is normal to the plane $4x - 9y + z = 3$, and so is also normal to the parallel plane. Setting $\mathbf{n} = \langle 4, -9, 1\rangle$ and $(x_0, y_0, z_0) = (0, 0, 0)$ in the vector equation of the plane yields

$$\langle 4, -9, 1\rangle \cdot \langle x, y, z\rangle = \langle 4, -9, 1\rangle \cdot \langle 0, 0, 0\rangle = 0$$

$$4x - 9y + z = 0$$

23. Passes through $(4, 1, 9)$ and is parallel to $x = 3$

SOLUTION The vector form of the plane $x = 3$ is

$$\langle 1, 0, 0\rangle \cdot \langle x, y, z\rangle = 3$$

Hence, $\mathbf{n} = \langle 1, 0, 0\rangle$ is normal to this plane. This vector is also normal to the parallel plane. Setting $(x_0, y_0, z_0) = (4, 1, 9)$ and $\mathbf{n} = \langle 1, 0, 0\rangle$ in the vector equation of the plane yields

$$\langle 1, 0, 0\rangle \cdot \langle x, y, z\rangle = \langle 1, 0, 0\rangle \cdot \langle 4, 1, 9\rangle = 4 + 0 + 0 = 4$$

or

$$x + 0 + 0 = 4 \quad \Rightarrow \quad x = 4$$

25. Contains the lines $\mathbf{r}_1(t) = \langle t, 2t, 3t\rangle$ and $\mathbf{r}_2(t) = \langle 3t, t, 8t\rangle$

SOLUTION Since the plane contains the lines $\ell_1(t) = \langle t, 2t, 3t\rangle$ and $\ell_2(t) = \langle 3t, t, 8t\rangle$, the direction vectors $\mathbf{v}_1 = \langle 1, 2, 3\rangle$ and $\mathbf{v}_2 = \langle 3, 1, 8\rangle$ of the lines lie in the plane. Therefore the cross product $\mathbf{n} = \mathbf{v}_1 \times \mathbf{v}_2$ is normal to the plane. We compute the cross product:

$$\mathbf{n} = \langle 1, 2, 3\rangle \times \langle 3, 1, 8\rangle = \begin{vmatrix} \mathbf{i} & \mathbf{j} & \mathbf{k} \\ 1 & 2 & 3 \\ 3 & 1 & 8 \end{vmatrix} = \begin{vmatrix} 2 & 3 \\ 1 & 8 \end{vmatrix}\mathbf{i} - \begin{vmatrix} 1 & 3 \\ 3 & 8 \end{vmatrix}\mathbf{j} + \begin{vmatrix} 1 & 2 \\ 3 & 1 \end{vmatrix}\mathbf{k}$$

$$= 13\mathbf{i} + \mathbf{j} - 5\mathbf{k} = \langle 13, 1, -5 \rangle$$

We now must choose a point on the plane. Since the line $\ell_1(t) = \langle t, 2t, 3t \rangle$ is contained in the plane, all of its points are on the plane. We choose the point corresponding to $t = 1$, that is,

$$\langle x_0, y_0, z_0 \rangle = \langle 1, 2 \cdot 1, 3 \cdot 1 \rangle = \langle 1, 2, 3 \rangle$$

We now use the vector equation of the plane to determine the equation of the desired plane:

$$\begin{aligned} \mathbf{n} \cdot \langle x, y, z \rangle &= \mathbf{n} \cdot \langle x_0, y_0, z_0 \rangle \\ \langle 13, 1, -5 \rangle \cdot \langle x, y, z \rangle &= \langle 13, 1, -5 \rangle \cdot \langle 1, 2, 3 \rangle \\ 13x + y - 5z &= 13 + 2 - 15 = 0 \\ 13x + y - 5z &= 0 \end{aligned}$$

27. Are the planes $\frac{1}{2}x + 2x - y = 5$ and $3x + 12x - 6y = 1$ parallel?

SOLUTION The planes $2\frac{1}{2}x - y = 5$ and $15x - 6y = 1$, are parallel if and only if the vectors $\mathbf{n}_1 = \left\langle 2\frac{1}{2}, -1, 0 \right\rangle$ and $\mathbf{n}_2 = \langle 15, -6, 0 \rangle$ normal to the planes are parallel. Since $\mathbf{n}_2 = 6\mathbf{n}_1$ the planes are parallel.

29. Find an equation of the plane $\mathcal{P}$ in Figure 9.

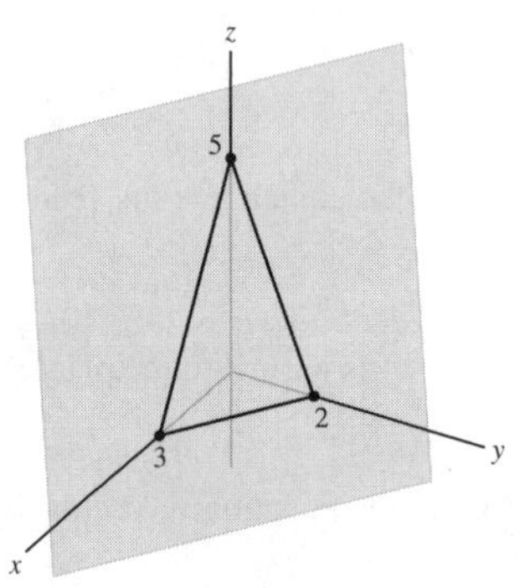

FIGURE 9

SOLUTION We must find the equation of the plane passing though the points $P = (3, 0, 0)$, $Q = (0, 2, 0)$, and $R = (0, 0, 5)$.

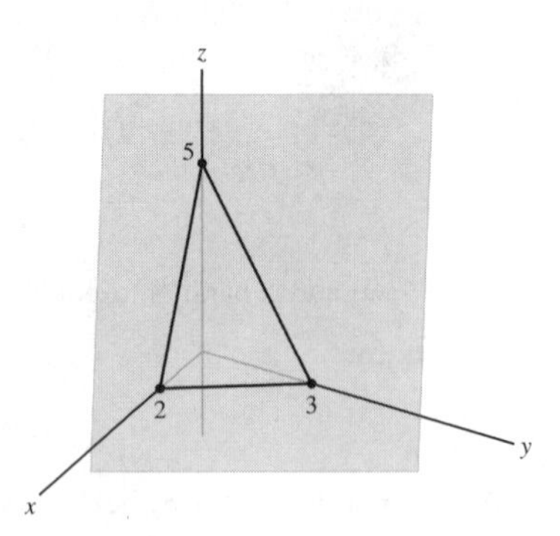

We use the following steps:

Step 1. Find a normal vector $\mathbf{n}$. The vectors $\mathbf{a} = \overrightarrow{PQ}$ and $\mathbf{b} = \overrightarrow{PR}$ lie in the plane, hence the cross product $\mathbf{n} = \mathbf{a} \times \mathbf{b}$ is normal to the plane. We compute the cross product:

$$\mathbf{a} = \overrightarrow{PQ} = \langle 0 - 3, 2 - 0, 0 - 0 \rangle = \langle -3, 2, 0 \rangle$$

$$\mathbf{b} = \overrightarrow{PR} = \langle 0 - 3, 0 - 0, 5 - 0 \rangle = \langle -3, 0, 5 \rangle$$

$$\mathbf{n} = \mathbf{a} \times \mathbf{b} = \begin{vmatrix} \mathbf{i} & \mathbf{j} & \mathbf{k} \\ -3 & 2 & 0 \\ -3 & 0 & 5 \end{vmatrix} = \begin{vmatrix} 2 & 0 \\ 0 & 5 \end{vmatrix} \mathbf{i} - \begin{vmatrix} -3 & 0 \\ -3 & 5 \end{vmatrix} \mathbf{j} + \begin{vmatrix} -3 & 2 \\ -3 & 0 \end{vmatrix} \mathbf{k}$$

$$= 10\mathbf{i} + 15\mathbf{j} + 6\mathbf{k} = \langle 10, 15, 6 \rangle$$

Step 2. Choose a point on the plane. We choose one of the points on the plane, say $P = (3, 0, 0)$. Substituting $\mathbf{n} = \langle 10, 15, 6 \rangle$ and $(x_0, y_0, z_0) = (3, 0, 0)$ in the vector form of the equation of the plane gives

$$\mathbf{n} \cdot \langle x, y, z \rangle = \mathbf{n} \cdot \langle x_0, y_0, z_0 \rangle$$

$$\langle 10, 15, 6 \rangle \cdot \langle x, y, z \rangle = \langle 10, 15, 6 \rangle \cdot \langle 3, 0, 0 \rangle$$

Computing the dot products we get the following scalar form of the equation of the plane:

$$10x + 15y + 6z = 10 \cdot 3 + 0 + 0 = 30$$

$$10x + 15y + 6z = 30$$

In Exercises 30–33, find the intersection of the line and plane.

31. $2x + y = 3, \quad \mathbf{r}(t) = \langle 2, -1, -1 \rangle + t \langle 1, 2, -4 \rangle$

SOLUTION The parametric equations of the line are

$$x = 2 + t, \quad y = -1 + 2t, \quad z = -1 - 4t \tag{1}$$

We substitute the parametric equations in the equation of the plane and solve for t, to find the value of t for which (x, y, z) lies on the plane. We obtain

$$2x + y = 3$$

$$2(2 + t) + (-1 + 2t) = 3$$

$$4 + 2t - 1 + 2t = 3$$

$$4t = 0 \quad \Rightarrow \quad t = 0$$

We find the coordinates of the point P of intersection by substituting $t = 0$ in the parametric equations (1). We obtain

$$x = 2 + 0 = 2, \quad y = -1 + 2 \cdot 0 = -1, \quad z = -1 - 4 \cdot 0 = -1$$

That is,

$$P = (2, -1, -1).$$

33. $x - z = 6, \quad \mathbf{r}(t) = \langle 1, 0, -1 \rangle + t \langle 4, 9, 2 \rangle$

SOLUTION The parametric equations of the line are

$$x = 1 + 4t, \quad y = 9t, \quad z = -1 + 2t \tag{1}$$

We substitute the parametric equations in the equation of the plane and solve for t:

$$x - z = 6$$

$$1 + 4t - (-1 + 2t) = 6$$

$$1 + 4t + 1 - 2t = 6$$

$$2t = 4 \quad \Rightarrow \quad t = 2$$

The value of the parameter at the point P of intersection is $t = 2$. We find the coordinates of P by substituting $t = 2$ in (1). This gives

$$x = 1 + 4 \cdot 2 = 9, \quad y = 9 \cdot 2 = 18, \quad z = -1 + 2 \cdot 2 = 3$$

That is,

$$P = (9, 18, 3).$$

35. The plane $3x - 4y + 2z = 8$ intersects the xy-plane in a line. What is the equation of this line?

SOLUTION We need to find the trace of the plane $3x - 4y + 2z = 8$ in the xy-plane. The xy-plane has equation $z = 0$, so the intersection of the plane with the xy-plane must satisfy both $z = 0$ and $3x - 4y + 2z = 8$.

This is the set of all points $(x, y, 0)$ such that $3x - 4y + 2 \cdot 0 = 8$, hence the trace of the plane in the xy-plane is the line $3x - 4y = 8$ in the xy-plane.

In Exercises 36–41, find the trace of the plane in the given coordinate plane.

37. $3x - 9y + 4z = 5, \quad xz$

SOLUTION The trace of the plane in the xz coordinate plane is obtained by substituting $y = 0$ in the equation of the plane $3x - 9y + 4z = 5$. This gives the line $3x + 4z = 5$ in the xz-plane.

39. $3x + 4z = -2, \quad xz$

SOLUTION The xz-plane has equation $y = 0$, hence the intersection of the plane $3x + 4z = -2$ with the xz-plane is the set of all points $(x, 0, z)$ such that $3x + 4z = -2$. This is a line in the xz-plane.

41. $-x + y = 4, \quad yz$

SOLUTION The trace of the plane $-x + y = 4$ on the yz-plane is the set of all points that satisfy both the equation of the plane and the equation $x = 0$ of the yz-plane. That is, the set of all points $(0, y, z)$ such that $-0 + y = 4$, or $y = 4$. This is a vertical line in the yz-plane.

43. Give equations for two distinct planes whose trace in the xy-plane has equation $4x + 3y = 8$.

SOLUTION The xy-plane has the equation $z = 0$, hence the trace of a plane $ax + by + cz = 0$ in the xy-plane is obtained by substituting $z = 0$ in the equation of the plane. Therefore, the following two planes have trace $4x + 3y = 8$ in the xy-plane:

$$4x + 3y + z = 8; \quad 4x + 3y - 5z = 8$$

45. Find all planes in $\mathbf{R}^3$ whose intersection with the xy-plane is the line $\mathbf{r}(t) = t\,\langle 2, 1, 0\rangle$.

SOLUTION The intersection of the plane $ax + by + cz = d$ with the xy-plane is obtained by substituting $z = 0$ in the equation of the plane. This gives the line $ax + by = d$, in the xy-plane. We find the equation of the line $\mathbf{l}\,(t) = t\langle 2, 1, 0\rangle$. On this line we have

$$\begin{matrix} x = 2t \\ y = t \end{matrix} \quad \Rightarrow \quad y = \frac{1}{2}x \Rightarrow x - 2y = 0$$

We thus must have $d = 0$ and $\frac{b}{a} = -2$, $a \neq 0$. That is, $d = 0$, $b = -2a$, $a \neq 0$. Notice that c can have any value. Hence, the planes are

$$ax - 2ay + cz = 0, a \neq 0$$

47. Let $\mathcal{P}$ be the plane $\mathbf{n} \cdot \langle x, y, z\rangle = d$, where $\mathbf{n} \neq \mathbf{0}$, and let $\mathcal{P}_1$ be the parallel plane $\mathbf{n} \cdot \langle x, y, z\rangle = d_1$ (Figure 10).

(a) Show that the line through $\mathbf{n}$ intersects $\mathcal{P}$ at the terminal point of the vector $\left(\frac{d}{\|\mathbf{n}\|}\right)\mathbf{e_n}$.

(b) Show that the distance between $\mathcal{P}$ and $\mathcal{P}_1$ is $\dfrac{|d - d_1|}{\|\mathbf{n}\|}$.

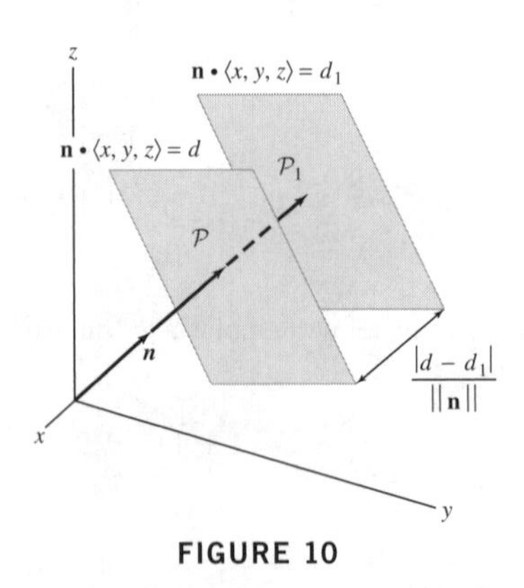

FIGURE 10

SOLUTION We place the vector $\mathbf{n}$ so that the ray through $\mathbf{n}$ passes through the origin.

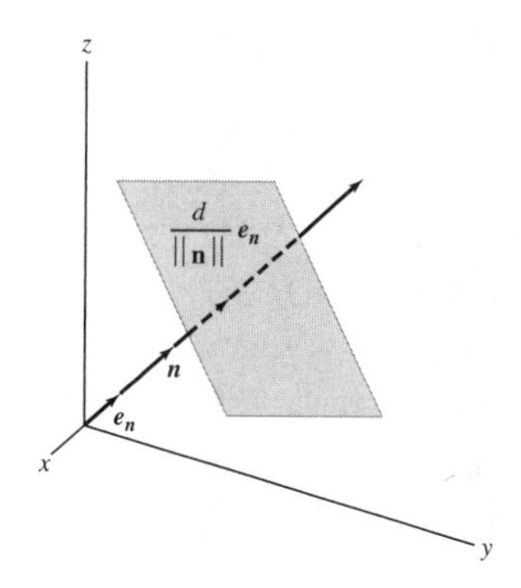

Let $\mathbf{v}$ be the vector $\mathbf{v} = \frac{d}{\|\mathbf{n}\|}\mathbf{e_n}$. Then $\mathbf{v}$ is parallel to $\mathbf{n}$ and the two vectors are on the same ray. Now we must show that the terminal point of $\mathbf{v}$ lies on the plane $\mathbf{n} \cdot \langle x, y, z\rangle = d$, that is, we must show that

$$\mathbf{n} \cdot \frac{d}{\|\mathbf{n}\|}\mathbf{e_n} = d \tag{1}$$

Since $\frac{\mathbf{n}}{\|\mathbf{n}\|} = \mathbf{e_n}$ we have

$$\mathbf{n} \cdot \frac{d}{\|\mathbf{n}\|}\mathbf{e_n} = d\frac{\mathbf{n}}{\|\mathbf{n}\|} \cdot \mathbf{e_n} = d \cdot \mathbf{e_n} \cdot \mathbf{e_n} = d\|\mathbf{e_n}\|^2 = d \cdot 1 = d$$

We now show that the distance between the plane $\mathbf{n} \cdot \langle x, y, z\rangle = d$ and the parallel plane $\mathbf{n} \cdot \langle x, y, z\rangle = d_1$ is $\frac{|d-d_1|}{\|\mathbf{n}\|}$. The two planes are parallel, hence they have the same normal n. By the previous result, the ray through $\mathbf{n}$ intersects the planes at the terminal points of the vectors $\mathbf{v} = \frac{d}{\|\mathbf{n}\|}\mathbf{e_n}$ and $\mathbf{v}_1 = \frac{d_1}{\|\mathbf{n}\|}\mathbf{e_n}$ respectively.

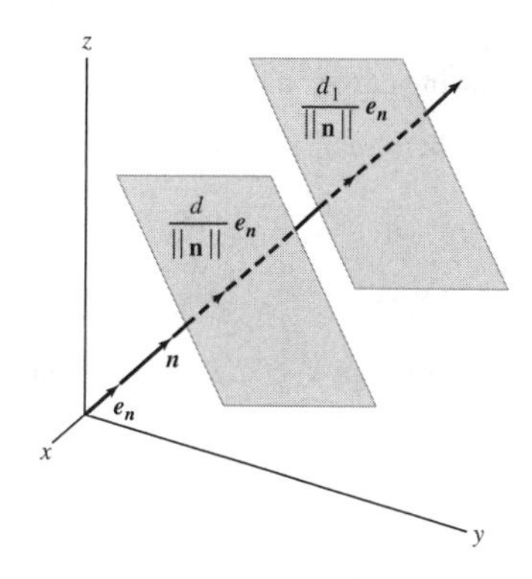

Therefore, the distance between the planes is the length of the difference vector $\mathbf{v} - \mathbf{v}_1$. We find it here:

$$\|\mathbf{v} - \mathbf{v}_1\| = \left\|\frac{d}{\|\mathbf{n}\|}\mathbf{e_n} - \frac{d_1}{\|\mathbf{n}\|}\mathbf{e_n}\right\| = \left\|\frac{d-d_1}{\|\mathbf{n}\|}\mathbf{e_n}\right\| = \frac{|d-d_1|}{\|\mathbf{n}\|}\|\mathbf{e_n}\| = \frac{|d-d_1|}{\|\mathbf{n}\|} \cdot 1 = \frac{|d-d_1|}{\|\mathbf{n}\|}$$

In Exercises 49–54, compute the angle between the two planes, defined as the angle θ (between 0 and π) between their normal vectors (Figure 11).

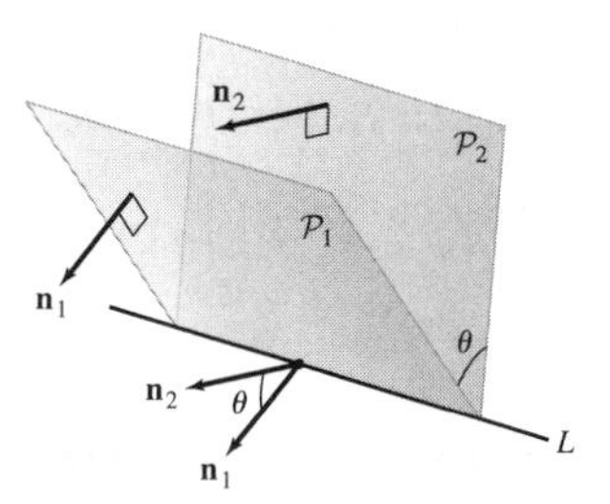

FIGURE 11 By definition, the angle between two planes is the angle between their normal vectors.

49. Planes with normals $\mathbf{n}_1 = \langle 1, 0, 1\rangle$, $\mathbf{n}_2 = \langle -1, 1, 1\rangle$

SOLUTION Using the formula for the angle between two vectors we get

$$\cos\theta = \frac{\mathbf{n}_1 \cdot \mathbf{n}_2}{\|\mathbf{n}_1\|\|\mathbf{n}_2\|} = \frac{\langle 1, 0, 1\rangle \cdot \langle -1, 1, 1\rangle}{\|\langle 1, 0, 1\rangle\|\|\langle -1, 1, 1\rangle\|} = \frac{-1+0+1}{\sqrt{1^2+0+1^2}\sqrt{(-1)^2+1^2+1^2}} = 0$$

The solution for $0 \le \theta < \pi$ is $\theta = \frac{\pi}{2}$.

51. Planes with normals $\mathbf{n}_1 = \langle 1, 0, 0 \rangle$, $\mathbf{n}_2 = \langle 0, 1, 0 \rangle$

SOLUTION Since $\mathbf{n}_1 \cdot \mathbf{n}_2 = \langle 1, 0, 0 \rangle \cdot \langle 0, 1, 0 \rangle = 1 \cdot 0 + 0 \cdot 1 + 0 \cdot 0 = 0$, the normals $\mathbf{n}_1$ and $\mathbf{n}_2$ are orthogonal, hence the angle between the planes is a right angle.

53. $x - 3y + z = 3$ and $2x - 3z = 4$

SOLUTION The planes $x - 3y + z = 3$ and $2x - 3z = 4$ have the normals $\mathbf{n}_1 = \langle 1, -3, 1 \rangle$ and $\mathbf{n}_2 = \langle 2, 0, -3 \rangle$ respectively. We use the formula for the angle between two vectors:

$$\cos\theta = \frac{\mathbf{n}_1 \cdot \mathbf{n}_2}{\|\mathbf{n}_1\|\|\mathbf{n}_2\|} = \frac{\langle 1, -3, 1 \rangle \cdot \langle 2, 0, -3 \rangle}{\|\langle 1, -3, 1 \rangle\|\|\langle 2, 0, -3 \rangle\|} = \frac{2 + 0 - 3}{\sqrt{1^2 + (-3)^2 + 1^2}\sqrt{2^2 + 0 + (-3)^2}} = \frac{-1}{\sqrt{11}\sqrt{13}} \approx -0.084$$

The solution for $0 \le \theta < \pi$ is $\theta = 1.655$ rad or $\theta = 94.80°$.

55. Find an equation of a plane making an angle of $\frac{\pi}{2}$ with the plane $3x + y - 4z = 2$.

SOLUTION The angle θ between two planes (chosen so that $0 \le \theta < \pi$) is defined as the angle between their normal vectors. The following vector is normal to the plane $3x + y - 4z = 2$:

$$\mathbf{n}_1 = \langle 3, 1, -4 \rangle$$

Let $\mathbf{n} \cdot \langle x, y, z \rangle = d$ denote the equation of a plane making an angle of $\frac{\pi}{2}$ with the given plane, where $\mathbf{n} = \langle a, b, c \rangle$. Since the two planes are perpendicular, the dot product of their normal vectors is zero. That is,

$$\mathbf{n} \cdot \mathbf{n}_1 = \langle a, b, c \rangle \cdot \langle 3, 1, -4 \rangle = 3a + b - 4c = 0 \quad \Rightarrow \quad b = -3a + 4c$$

Thus, the required planes (there is more than one plane) have the following normal vector:

$$\mathbf{n} = \langle a, -3a + 4c, c \rangle$$

We obtain the following equation:

$$\begin{aligned} \mathbf{n} \cdot \langle x, y, c \rangle &= d \\ \langle a, -3a + 4c, c \rangle \cdot \langle x, y, z \rangle &= d \\ ax + (4c - 3a)y + cz &= d \end{aligned}$$

Every choice of the values of a, c and d yields a plane with the desired property. For example, we set $a = c = d = 1$ to obtain

$$x + y + z = 1$$

57. Find a plane that is perpendicular to the two planes $x + y = 3$ and $x + 2y - z = 4$.

SOLUTION The vector forms of the equations of the planes are $\langle 1, 1, 0 \rangle \cdot \langle x, y, z \rangle = 3$ and $\langle 1, 2, -1 \rangle \cdot \langle x, y, z \rangle = 4$, hence the vectors $\mathbf{n}_1 = \langle 1, 1, 0 \rangle$ and $\mathbf{n}_2 = \langle 1, 2, -1 \rangle$ are normal to the planes. We denote the equation of the planes which are perpendicular to the two planes by

$$ax + by + cz = d \tag{1}$$

Then, the normal $\mathbf{n} = \langle a, b, c \rangle$ to the planes is orthogonal to the normals $\mathbf{n}_1$ and $\mathbf{n}_2$ of the given planes. Therefore, $\mathbf{n} \cdot \mathbf{n}_1 = 0$ and $\mathbf{n} \cdot \mathbf{n}_2 = 0$ which gives us

$$\langle a, b, c \rangle \cdot \langle 1, 1, 0 \rangle = 0, \quad \langle a, b, c \rangle \cdot \langle 1, 2, -1 \rangle = 0$$

We obtain the following equations:

$$\begin{cases} a + b = 0 \\ a + 2b - c = 0 \end{cases}$$

The first equation implies that $b = -a$. Substituting in the second equation we get $a - 2a - c = 0$, or $c = -a$. Substituting $b = -a$ and $c = -a$ in (1) gives (for $a \ne 0$):

$$ax - ay - az = d \quad \Rightarrow \quad x - y - z = \frac{d}{a}$$

$\frac{d}{a}$ is an arbitrary constant which we denote by f. The planes which are perpendicular to the given planes are, therefore,

$$x - y - z = f$$

59. Let $\mathcal{L}$ be the intersection of the planes $x - y - z = 1$ and $2x + 3y + z = 2$. Find parametric equations for the line $\mathcal{L}$. *Hint:* To find a point $\mathcal{L}$, substitute an arbitrary value for z (say, $z = 2$) and then solve the resulting pair of equations for x and y.

SOLUTION We use Exercise 56 to find a direction vector for the line of intersection $\mathcal{L}$ of the planes $x - y - z = 1$ and $2x + 3y + z = 2$. We identify the normals $\mathbf{n}_1 = \langle 1, -1, -1\rangle$ and $\mathbf{n}_2 = \langle 2, 3, 1\rangle$ to the two planes respectively. Hence, a direction vector for $\mathcal{L}$ is the cross product $\mathbf{v} = \mathbf{n}_1 \times \mathbf{n}_2$. We find it here:

$$\mathbf{v} = \mathbf{n}_1 \times \mathbf{n}_2 = \begin{vmatrix} \mathbf{i} & \mathbf{j} & \mathbf{k} \\ 1 & -1 & -1 \\ 2 & 3 & 1 \end{vmatrix} = 2\mathbf{i} - 3\mathbf{j} + 5\mathbf{k} = \langle 2, -3, 5\rangle$$

We now need to find a point on $\mathcal{L}$. We choose $z = 2$, substitute in the equations of the planes and solve the resulting equations for x and y. This gives

$$\begin{array}{c} x - y - 2 = 1 \\ 2x + 3y + 2 = 2 \end{array} \quad \text{or} \quad \begin{array}{c} x - y = 3 \\ 2x + 3y = 0 \end{array}$$

The 1st equation implies that $y = x - 3$. Substituting in the 2nd equation and solving for x gives

$$2x + 3(x - 3) = 0$$

$$5x = 9 \quad \Rightarrow \quad x = \frac{9}{5}, \quad y = \frac{9}{5} - 3 = -\frac{6}{5}$$

We conclude that the point $\left(\frac{9}{5}, -\frac{6}{5}, 2\right)$ is on $\mathcal{L}$. We now use the vector parametrization of a line to obtain the following parametrization for $\mathcal{L}$:

$$\mathbf{r}(t) = \left\langle \frac{9}{5}, -\frac{6}{5}, 2 \right\rangle + t\langle 2, -3, 5\rangle$$

This yields the parametric equations

$$x = \frac{9}{5} + 2t, \quad y = -\frac{6}{5} - 3t, \quad z = 2 + 5t$$

61. Find parametric equations for the line that is perpendicular to the plane $3x + 5y - 7z = 29$ and passes through the point $P_0 = (3, -1, 1)$.

SOLUTION Since the line $\mathcal{L}$ is perpendicular to the plane $3x + 5y - 7z = 29$, the direction vectors of $\mathcal{L}$ are parallel to the vector $\mathbf{n} = \langle 3, 5, -7\rangle$ normal to the plane. Hence we may choose $\mathbf{v} = \langle 3, 5, -7\rangle$ as a direction vector for $\mathcal{L}$. We now use the vector parametrization of the line to obtain the following parametrization of $\mathcal{L}$:

$$\mathbf{r}(t) = \langle 3, -1, 1\rangle + t\langle 3, 5, -7\rangle$$

yielding the parametric equations

$$x = 3 + 3t, \quad y = -1 + 5t, \quad z = 1 - 7t.$$

Further Insights and Challenges

In Exercises 62–72, if $\mathcal{P}$ is a plane and Q is a point not lying on $\mathcal{P}$, then the nearest point to Q on $\mathcal{P}$ is the unique point P on $\mathcal{P}$ such that $\overline{PQ}$ is orthogonal to $\mathcal{P}$ (Figure 12).

63. Use Eq. (6) to find the nearest point to $Q = (2, 1, 2)$ on the plane $x + y + z = 1$.

SOLUTION We identify $\mathbf{n} = \langle 1, 1, 1\rangle$ as a vector normal to the plane. By Eq. (6) the nearest point P to Q is determined by

$$\overrightarrow{OP} = \overrightarrow{OQ} + \left(\frac{d - \overrightarrow{OQ} \cdot \mathbf{n}}{\mathbf{n} \cdot \mathbf{n}}\right)\mathbf{n}$$

We substitute $\mathbf{n} = \langle 1, 1, 1\rangle$, $\overrightarrow{OQ} = \langle 2, 1, 2\rangle$ and $d = 1$ in this equation to obtain

$$\begin{aligned} \overrightarrow{OP} &= \langle 2, 1, 2\rangle + \frac{1 - \langle 2, 1, 2\rangle \cdot \langle 1, 1, 1\rangle}{\langle 1, 1, 1\rangle \cdot \langle 1, 1, 1\rangle}\langle 1, 1, 1\rangle = \langle 2, 1, 2\rangle + \frac{1 - (2 + 1 + 2)}{1 + 1 + 1}\langle 1, 1, 1\rangle \\ &= \langle 2, 1, 2\rangle - \frac{4}{3}\langle 1, 1, 1\rangle = \left\langle \frac{2}{3}, -\frac{1}{3}, \frac{2}{3}\right\rangle \end{aligned}$$

The terminal point $P = \left(\frac{2}{3}, -\frac{1}{3}, \frac{2}{3}\right)$ of $\overrightarrow{OP}$ is the nearest point to $Q = (2, 1, 2)$ on the plane.

65. In the notation of Exercise 62, the *distance* from Q to $\mathcal{P}$ is the length $\|\overrightarrow{PQ}\|$. Show that if $Q = (x_1, y_1, z_1)$, then

$$\text{Distance from } Q \text{ to } \mathcal{P} = \frac{|ax_1 + by_1 + cz_1 - d|}{\|\mathbf{n}\|} \qquad \boxed{7}$$

SOLUTION The distance l from Q to P is the length of the vector $\overrightarrow{PQ}$, that is,

$$\ell = \|\overrightarrow{PQ}\| = \|\overrightarrow{OQ} - \overrightarrow{OP}\| \tag{1}$$

By Eq. (6) in Exercise 62,

$$\overrightarrow{OP} - \overrightarrow{OQ} = \left(\frac{d - \overrightarrow{OQ} \cdot \mathbf{n}}{\|\mathbf{n}\|^2}\right)\mathbf{n}$$

Combining with (1) and noticing that $\frac{\mathbf{n}}{\|\mathbf{n}\|}$ is a unit vector, we have

$$\ell = \left\|\left(\frac{d - \overrightarrow{OQ} \cdot \mathbf{n}}{\|\mathbf{n}\|^2}\right)\mathbf{n}\right\| = \frac{|d - \overrightarrow{OQ} \cdot \mathbf{n}|}{\|\mathbf{n}\|} \cdot \left\|\frac{\mathbf{n}}{\|\mathbf{n}\|}\right\| = \frac{|d - \overrightarrow{OQ} \cdot \mathbf{n}|}{\|\mathbf{n}\|} \tag{2}$$

We compute the numerator in (2):

$$|d - \overrightarrow{OQ} \cdot \mathbf{n}| = |d - \langle x_1, y_1, z_1\rangle \cdot \langle a, b, c\rangle| = |d - (ax_1 + by_1 + cz_1)| = |ax_1 + by_1 + cz_1 - d|$$

Substituting into (2) we obtain the following distance:

$$\ell = \frac{|ax_1 + by_1 + cz_1 - d|}{\|\mathbf{n}\|}$$

67. Find the distance from $Q = (1, 2, 2)$ to the plane $\mathbf{n} \cdot \langle x, y, z\rangle = 3$, where $\mathbf{n} = \left\langle \frac{3}{5}, \frac{4}{5}, 0\right\rangle$.

SOLUTION We write the equation of the plane in scalar form:

$$\left\langle \frac{3}{5}, \frac{4}{5}, 0\right\rangle \cdot \langle x, y, z\rangle = 3$$

$$\frac{3}{5}x + \frac{4}{5}y + 0 = 3$$

$$\frac{3}{5}x + \frac{4}{5}y = 3$$

We use Eq. (7) in for the distance ℓ from $Q = (x_1, y_1, z_1)$ to the plane $ax + by + cz = d$:

$$\ell = \frac{|ax_1 + by_1 + cz_1 - d|}{\|\mathbf{n}\|} \tag{1}$$

In our example, $a = \frac{3}{5}$, $b = \frac{4}{5}$, $c = 0$, $d = 3$. Also $(x_1, y_1, z_1) = (1, 2, 2)$ and $\mathbf{n} = \left\langle \frac{3}{5}, \frac{4}{5}, 0\right\rangle$. Substituting these values in the formula (1) for the distance ℓ, we get

$$\ell = \frac{\left|\frac{3}{5} \cdot 1 + \frac{4}{5} \cdot 2 + 0 - 3\right|}{\sqrt{\left(\frac{3}{5}\right)^2 + \left(\frac{4}{5}\right)^2 + 0^2}} = \frac{\frac{4}{5}}{1} = \frac{4}{5}$$

We found that the distance from $Q = (1, 2, 2)$ to the given plane is $\ell = \frac{4}{5}$.

69. Show that the point Q on the plane $\mathbf{n} \cdot \langle x, y, z\rangle = d$ nearest the origin lies on the ray through $\mathbf{n}$. More precisely, show that Q is the terminal point of $(d/\|\mathbf{n}\|^2)\mathbf{n}$.

SOLUTION Using Eq. (6) in Exercise 62, the point Q is the terminal point of the following vector (note that we switched P and Q in the formula):

$$\overrightarrow{OQ} = \mathbf{0} + \left(\frac{d - \mathbf{0} \cdot \mathbf{n}}{\|\mathbf{n}\|^2}\right) \cdot \mathbf{n} = \frac{d}{\|\mathbf{n}\|^2}\mathbf{n}$$

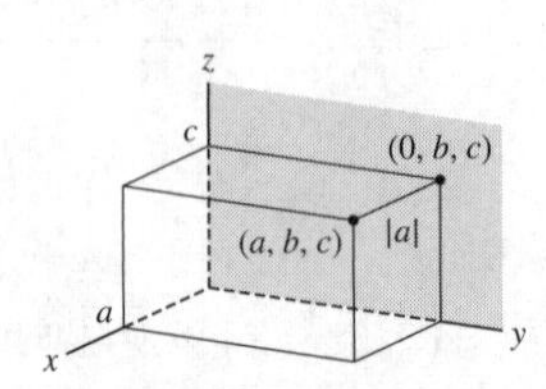

It follows that the point Q lies on the ray though $\mathbf{n}$. Alternatively, by definition of the nearest point, Q is the point on the line such that $\overrightarrow{OQ}$ is orthogonal to the plane. Since $\mathbf{n}$ is also orthogonal to the plane (and we assume that the initial point of $\mathbf{n}$ is at the origin), it follows that Q lies on the ray through $\mathbf{n}$.

71. Let $\mathbf{n} = \overrightarrow{OP}$, where $P = (x_0, y_0, z_0)$ is a point on the sphere $x^2 + y^2 + z^2 = r^2$, and let $\mathcal{P}$ be the plane with equation $\mathbf{n} \cdot \langle x, y, z \rangle = r^2$. Show that the point on $\mathcal{P}$ nearest the origin is P itself and conclude that $\mathcal{P}$ is tangent to the sphere at P (Figure 13).

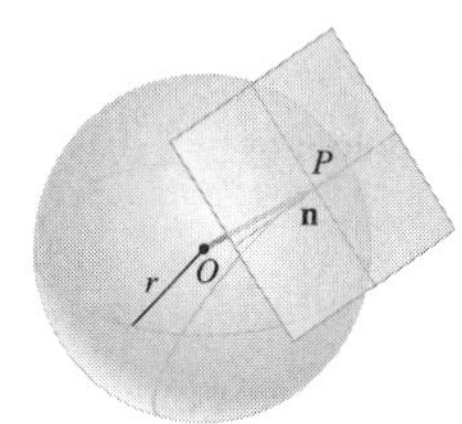

FIGURE 13 The terminal point of $\mathbf{n}$ lies on the sphere of radius r.

SOLUTION First notice that the terminal point P of $\mathbf{n}$ lies on the plane $\mathcal{P}$, since substituting $\langle x, y, z \rangle = \overrightarrow{OP} = \mathbf{n}$ in the equation of the plane gives

$$\mathbf{n} \cdot \langle x, y, z \rangle = \mathbf{n} \cdot \overrightarrow{OP} = \mathbf{n} \cdot \mathbf{n} = \|\mathbf{n}\|^2 = r^2$$

Since the vector $\overrightarrow{OP} = \mathbf{n}$ is orthogonal to the plane, that is, the radius $\overline{OP}$ is perpendicular to the plane, we conclude that P is the point on $\mathcal{P}$ closest to the origin and that $\mathcal{P}$ is tangent to the sphere of radius r centered at the origin. Of course, P is the tangency point.

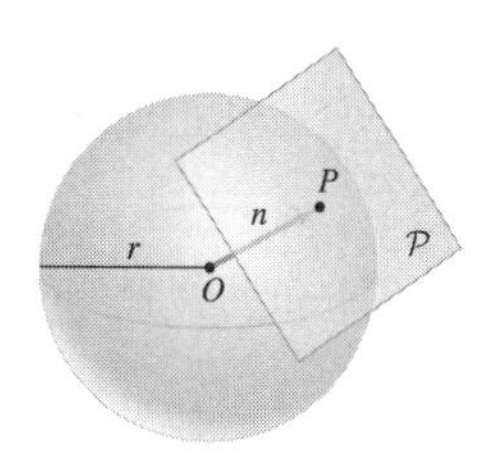

13.6 A Survey of Quadric Surfaces (ET Section 12.6)

Preliminary Questions

1. True or false: All traces of an ellipsoid are ellipses.

SOLUTION This statement is true, mostly. All traces of an ellipsoid $\left(\frac{x}{a}\right)^2 + \left(\frac{y}{b}\right)^2 + \left(\frac{z}{c}\right)^2 = 1$ are ellipses, except for the traces obtained by intersecting the ellipsoid with the planes $x = \pm a$, $y = \pm b$ and $z = \pm c$. These traces reduce to the single points $(\pm a, 0, 0)$, $(0, \pm b, 0)$ and $(0, 0, \pm c)$ respectively.

2. True or false: All traces of a hyperboloid are hyperbolas.

SOLUTION The statement is false. For a hyperbola in the standard orientation, the horizontal traces are ellipses (or perhaps empty for a hyperbola of two sheets), and the vertical traces are hyperbolas.

3. Which quadric surfaces have both hyperbolas and parabolas as traces?

SOLUTION The hyperbolic paraboloid $z = \left(\frac{x}{a}\right)^2 - \left(\frac{y}{b}\right)^2$ has vertical trace curves which are parabolas. If we set $x = x_0$ or $y = y_0$ we get

$$z = \left(\frac{x_0}{a}\right)^2 - \left(\frac{y}{b}\right)^2 \quad \Rightarrow \quad z = -\left(\frac{y}{b}\right)^2 + C$$

$$z = \left(\frac{x}{a}\right)^2 - \left(\frac{y_0}{b}\right)^2 \quad \Rightarrow \quad z = \left(\frac{x}{a}\right)^2 + C$$

The hyperbolic paraboloid has vertical traces which are hyperbolas, since for $z = z_0$, $(z_0 > 0)$, we get

$$z_0 = \left(\frac{x}{a}\right)^2 - \left(\frac{y}{b}\right)^2$$

4. Is there any quadric surface whose traces are all parabolas?

SOLUTION There is no quadric surface whose traces are all parabolas.

5. A surface is called **bounded** if there exists $M > 0$ such that every point on the surfaces lies at a distance of at most M from the origin. Which of the quadric surfaces are bounded?

SOLUTION The only quadric surface that is bounded is the ellipsoid

$$\left(\frac{x}{a}\right)^2 + \left(\frac{y}{b}\right)^2 + \left(\frac{z}{c}\right)^2 = 1.$$

All other quadric surfaces are not bounded, since at least one of the coordinates can increase or decrease without bound.

6. Give an equation for a quadric surface that consists of two separate components.

SOLUTION The hyperboloid of two sheets consists of two separate components. An example of a hyperboloid of two sheets is

$$-\left(\frac{x}{2}\right)^2 - \left(\frac{y}{3}\right)^2 + \left(\frac{z}{4}\right)^2 = 1$$

7. What is the definition of a parabolic cylinder?

SOLUTION A parabolic cylinder consists of all vertical lines passing through a parabola C in the xy-plane.

Exercises

In Exercises 1–6, state whether the given equation defines an ellipsoid or hyperboloid, and if a hyperboloid, whether it is of one or two sheets.

1. $\left(\frac{x}{2}\right)^2 + \left(\frac{y}{3}\right)^2 + \left(\frac{z}{5}\right)^2 = 1$

SOLUTION This equation is the equation of an ellipsoid.

3. $x^2 + 3y^2 + 9z^2 = 1$

SOLUTION We rewrite the equation as follows:

$$x^2 + \left(\frac{y}{\frac{1}{\sqrt{3}}}\right)^2 + \left(\frac{z}{\frac{1}{3}}\right)^2 = 1$$

This equation defines an ellipsoid.

5. $x^2 - 3y^2 + 9z^2 = 1$

SOLUTION We rewrite the equation in the form

$$x^2 - \left(\frac{y}{\frac{1}{\sqrt{3}}}\right)^2 + \left(\frac{z}{\frac{1}{3}}\right)^2 = 1$$

This is the equation of a hyperboloid of one sheet.

In Exercises 7–12, state whether the given equation defines an elliptic paraboloid, hyperbolic paraboloid, or elliptic cone.

7. $z = \left(\frac{x}{4}\right)^2 + \left(\frac{y}{3}\right)^2$

SOLUTION This equation defines an elliptic paraboloid.

9. $z = \left(\frac{x}{9}\right)^2 - \left(\frac{y}{12}\right)^2$

SOLUTION This equation defines a hyperbolic paraboloid.

11. $3x^2 - 7y^2 = z$

SOLUTION Rewriting the equation as

$$z = \left(\frac{x}{\frac{1}{\sqrt{3}}}\right)^2 - \left(\frac{y}{\frac{1}{\sqrt{7}}}\right)^2$$

we identify it as the equation of a hyperbolic paraboloid.

In Exercises 13–20, state the type of the quadric surface and describe the trace obtained by intersecting with the given plane.

13. $x^2 + \left(\frac{y}{4}\right)^2 + z^2 = 1, \quad y = 0$

SOLUTION The equation $x^2 + \left(\frac{y}{4}\right)^2 + z^2 = 1$ defines an ellipsoid. The xz-trace is obtained by substituting $y = 0$ in the equation of the ellipsoid. This gives the equation $x^2 + z^2 = 1$ which defines a circle in the xz-plane.

15. $x^2 + \left(\frac{y}{4}\right)^2 + z^2 = 1, \quad z = \frac{1}{4}$

SOLUTION The quadric surface is an ellipsoid, since its equation has the form $\left(\frac{x}{a}\right)^2 + \left(\frac{y}{b}\right)^2 + \left(\frac{z}{c}\right)^2 = 1$ for $a = 1$, $b = 4$, $c = 1$. To find the trace obtained by intersecting the ellipsoid with the plane $z = \frac{1}{4}$, we set $z = \frac{1}{4}$ in the equation of the ellipsoid. This gives

$$lx^2 + \left(\frac{y}{4}\right)^2 + \left(\frac{1}{4}\right)^2 = 1$$

$$x^2 + \frac{y^2}{16} = \frac{15}{16}$$

To get the standard form we divide by $\frac{15}{16}$ to obtain

$$\frac{x^2}{\frac{15}{16}} + \frac{y^2}{\frac{16 \cdot 15}{16}} = 1 \quad \Rightarrow \quad \left(\frac{x}{\frac{\sqrt{15}}{4}}\right)^2 + \left(\frac{y}{\sqrt{15}}\right)^2 = 1 \tag{1}$$

We conclude that the trace is an ellipse on the xy-plane, whose equation is given in (1).

17. $\left(\frac{x}{3}\right)^2 + \left(\frac{y}{5}\right)^2 - 5z^2 = 1, \quad y = 1$

SOLUTION Rewriting the equation in the form

$$\left(\frac{x}{3}\right)^2 + \left(\frac{y}{5}\right)^2 - \left(\frac{z}{\frac{1}{\sqrt{5}}}\right)^2 = 1$$

we identify it as the equation of a hyperboloid of one sheet. Substituting $y = 1$ we get

$$\frac{x^2}{9} + \frac{1}{25} - 5z^2 = 1$$

$$\frac{x^2}{9} - 5z^2 = \frac{24}{25}$$

$$\frac{25}{24 \cdot 9}x^2 - \frac{25 \cdot 5}{24}z^2 = 1$$

$$\left(\frac{x}{\frac{6\sqrt{6}}{5}}\right)^2 - \left(\frac{z}{\frac{2}{5}\sqrt{\frac{6}{5}}}\right)^2 = 1$$

Thus, the trace on the plane $y = 1$ is a hyperbola.

19. $y = 3x^2, \quad z = 27$

SOLUTION This equation defines a parabolic cylinder, consisting of all vertical lines passing through the parabola $y = 3x^2$ in the xy-plane. Hence, the trace of the cylinder on the plane $z = 27$ is the parabola $y = 3x^2$ on this plane, that is, the following set:

$$\left\{(x, y, z) : y = 3x^2, \ z = 27\right\}.$$

21. Match the ellipsoids in Figure 12 with the equation:

(a) $x^2 + 4y^2 + 4z^2 = 16$

(b) $4x^2 + y^2 + 4z^2 = 16$

(c) $4x^2 + 4y^2 + z^2 = 16$

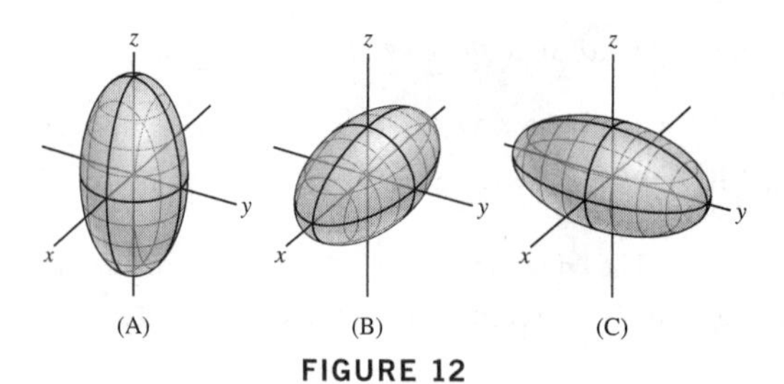

FIGURE 12

SOLUTION
(a) We rewrite the equation in the form

$$\left(\frac{x}{4}\right)^2 + \left(\frac{y}{2}\right)^2 + \left(\frac{z}{2}\right)^2 = 1$$

The ellipsoid intersects the x, y, and z axes at the points $(\pm 4, 0, 0)$, $(0, \pm 2, 0)$, and $(0, 0, \pm 2)$, hence (B) is the corresponding figure.
(b) We rewrite the equation in the form

$$\left(\frac{x}{2}\right)^2 + \left(\frac{y}{4}\right)^2 + \left(\frac{z}{2}\right)^2 = 1$$

The x, y, and z intercepts are $(\pm 2, 0, 0)$, $(0, \pm 4, 0)$, and $(0, 0, \pm 2)$ respectively, hence (C) is the correct figure.
(c) We write the equation in the form

$$\left(\frac{x}{2}\right)^2 + \left(\frac{y}{2}\right)^2 + \left(\frac{z}{4}\right)^2 = 1$$

The x, y, and z intercepts are $(\pm 2, 0, 0)$, $(0, \pm 2, 0)$, and $(0, 0, \pm 4)$ respectively, hence the corresponding figure is (A).

23. What is the equation of the surface obtained when the elliptic paraboloid $z = \left(\frac{x}{2}\right)^2 + \left(\frac{y}{4}\right)^2$ is rotated about the x-axis by $90°$? Refer to Figure 13.

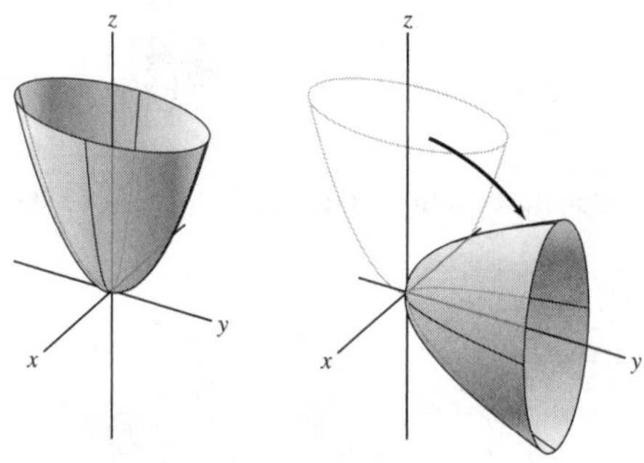

FIGURE 13

SOLUTION The axis of symmetry of the resulting surface is the y-axis rather than the z-axis. Interchanging y and z in the given equation gives the following equation of the rotated paraboloid:

$$y = \left(\frac{x}{2}\right)^2 + \left(\frac{z}{4}\right)^2$$

25. For which values of h is the intersection of the horizontal plane $x = h$ and the hyperboloid $\left(\frac{x}{2}\right)^2 - \left(\frac{y}{4}\right)^2 - \left(\frac{z}{9}\right)^2 = 1$ empty?

SOLUTION We set $x = h$ in the equation of the hyperboloid and transfer sides. We get

$$\left(\frac{h}{2}\right)^2 - \left(\frac{y}{4}\right)^2 - \left(\frac{z}{9}\right)^2 = 1$$

$$\left(\frac{y}{4}\right)^2 + \left(\frac{z}{9}\right)^2 = \left(\frac{h}{4}\right)^2 - 1$$

Since the sum of squares on the left-hand side is nonnegative, this equation has no solutions if the right-hand side is negative. That is, if

$$\frac{h^2}{4} - 1 < 0$$

$$h^2 < 4 \quad \Rightarrow \quad -2 < h < 2$$

We conclude that the given hyperboloid does not intersect the planes $x = h$ for $|h| < 2$.

In Exercises 27–32, sketch the given surface.

27. $x^2 + y^2 - z^2 = 1$

SOLUTION This equation defines a hyperboloid of one sheet. The trace on the plane $z = z_0$ is the circle $x^2 + y^2 = 1 + z_0^2$. The trace on the plane $y = y_0$ is the hyperbola $x^2 - z^2 = 1 - y_0^2$ and the trace on the plane $x = x_0$ is the hyperbola $y^2 - z^2 = 1 - x_0^2$. We obtain the following surface:

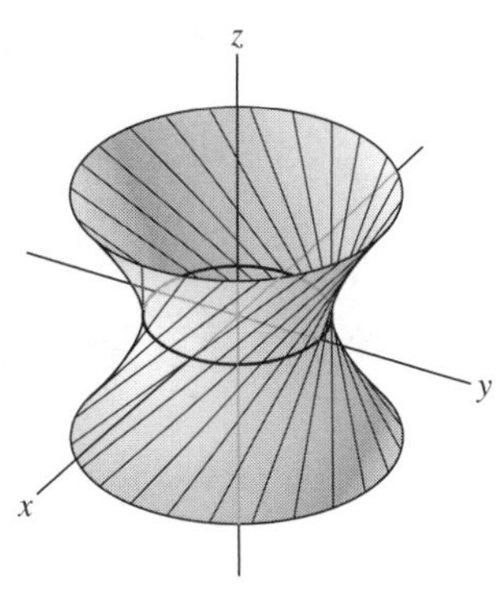

Graph of $x^2 + y^2 - z^2 = 1$

29. $z = \left(\frac{x}{4}\right)^2 + \left(\frac{y}{8}\right)^2$

SOLUTION This equation defines an elliptic paraboloid, as shown in the following figure:

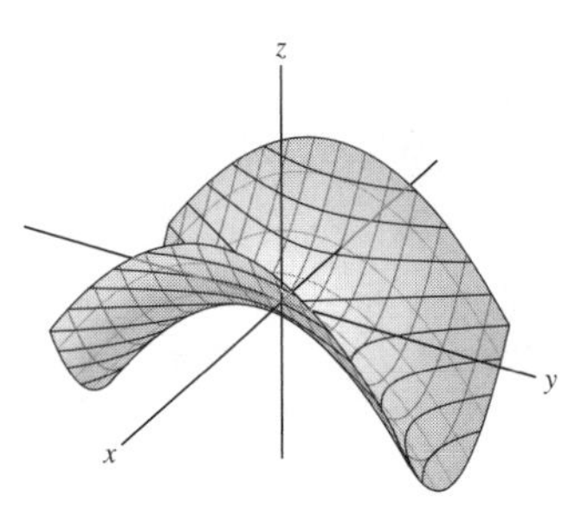

31. $z^2 = \left(\frac{x}{4}\right)^2 + \left(\frac{y}{8}\right)^2$

SOLUTION This equation defines the following elliptic cone:

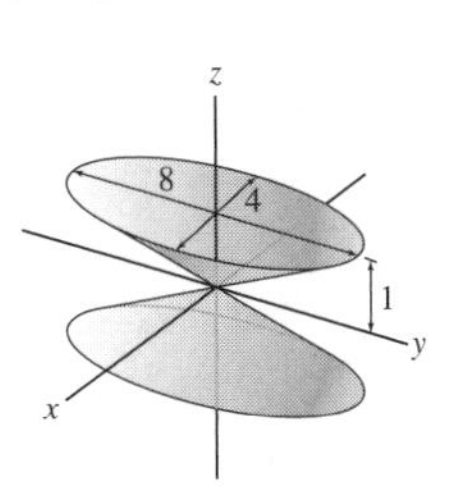

33. Find the equation of the ellipsoid passing through the points marked in Figure 14(A).

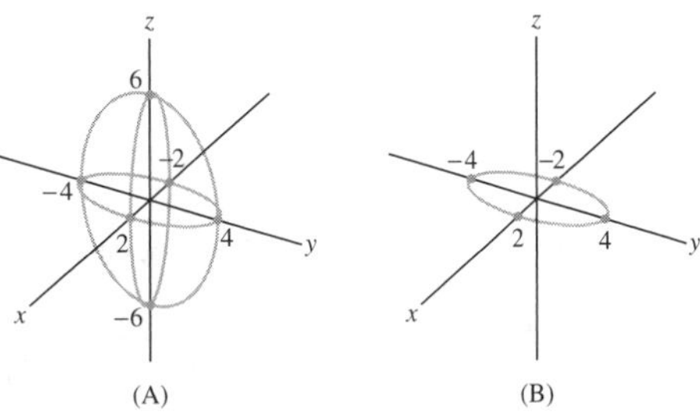

FIGURE 14

SOLUTION The equation of an ellipsoid is

$$\left(\frac{x}{a}\right)^2+\left(\frac{y}{b}\right)^2+\left(\frac{z}{c}\right)^2=1 \tag{1}$$

The x, y and z intercepts are $(\pm a, 0, 0)$, $(0, \pm b, 0)$ and $(0, 0, \pm c)$ respectively. The x, y and z intercepts of the desired ellipsoid are $(\pm 2, 0, 0)$, $(0, \pm 4, 0)$ and $(0, 0, \pm 6)$ respectively, hence $a = 2$, $b = 4$ and $c = 6$. Substituting into (1) we get

$$\left(\frac{x}{2}\right)^2+\left(\frac{y}{4}\right)^2+\left(\frac{z}{6}\right)^2=1.$$

35. Find the equation of the hyperboloid shown in Figure 15(A).

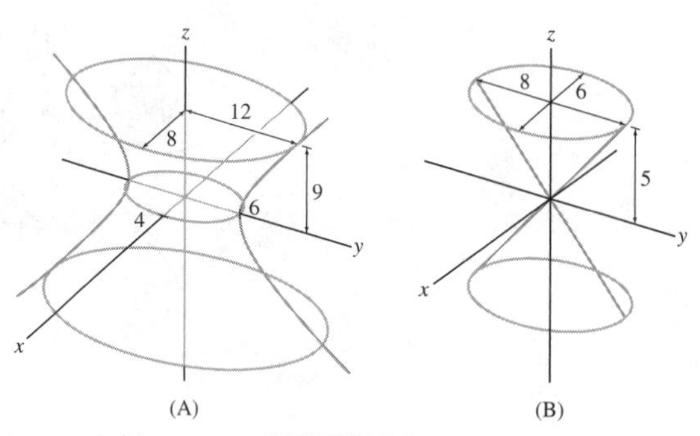

FIGURE 15

SOLUTION The hyperboloid in the figure is of one sheet and the intersections with the planes $z = z_0$ are ellipses. Hence, the equation of the hyperboloid has the form

$$\left(\frac{x}{a}\right)^2+\left(\frac{y}{b}\right)^2-\left(\frac{z}{c}\right)^2=1 \tag{1}$$

Substituting $z = 0$ we get

$$\left(\frac{x}{a}\right)^2+\left(\frac{y}{b}\right)^2=1$$

By the given information this ellipse has x and y intercepts at the points $(\pm 4, 0)$ and $(0, \pm 6)$ hence $a = 4$, $b = 6$. Substituting in (1) we get

$$\left(\frac{x}{4}\right)^2+\left(\frac{y}{6}\right)^2-\left(\frac{z}{c}\right)^2=1 \tag{2}$$

Substituting $z = 9$ we get

$$\frac{x^2}{16}+\frac{y^2}{36}-\frac{9^2}{c^2}=1$$

$$\frac{x^2}{16}+\frac{y^2}{36}=1+\frac{81}{c^2}=\frac{c^2+81}{c^2}$$

$$\frac{c^2x^2}{16(81+c^2)}+\frac{c^2y^2}{36(81+c^2)}=1$$

$$\left(\frac{x}{\frac{4}{c}\sqrt{81+c^2}}\right)^2+\left(\frac{y}{\frac{6}{c}\sqrt{81+c^2}}\right)^2=1$$

By the given information the following must hold:

$$\begin{matrix}\frac{4}{c}\sqrt{81+c^2}=8\\ \frac{6}{c}\sqrt{81+c^2}=12\end{matrix} \quad\Rightarrow\quad \frac{\sqrt{81+c^2}}{c}=2 \quad\Rightarrow\quad 81+c^2=4c^2 \quad\Rightarrow\quad 3c^2=81$$

Thus, $c = 3\sqrt{3}$, and by substituting in (2) we obtain the following equation:

$$\left(\frac{x}{4}\right)^2+\left(\frac{y}{6}\right)^2-\left(\frac{z}{3\sqrt{3}}\right)^2=1$$

37. Determine the vertical traces of elliptic and parabolic cylinders in standard form.

SOLUTION The vertical traces of elliptic or parabolic cylinders are one or two vertical lines, or an empty set.

39. Let $\mathcal{C}$ be an ellipse in a horizonal plane lying above the xy-plane. Which type of quadric surface is made up of all lines passing through the origin and a point on $\mathcal{C}$?

SOLUTION The quadric surface is the upper part of an elliptic cone.

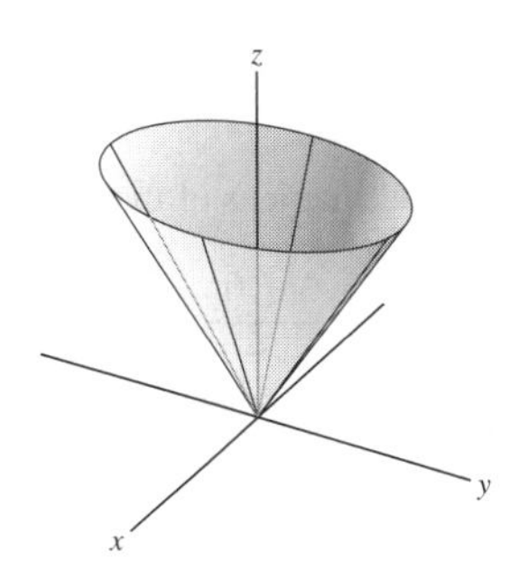

Further Insights and Challenges

41. Let $Q = (m, n, r)$ be a point on the ellipsoid with equation $\left(\frac{x}{a}\right)^2 + \left(\frac{y}{b}\right)^2 + \left(\frac{z}{c}\right)^2 = 1$ (Figure 16). Let $\mathcal{P}$ be the plane with equation

$$\frac{mx}{a^2} + \frac{ny}{b^2} + \frac{rz}{c^2} = 1$$

Show that Q lies on $\mathcal{P}$ and that the tangent lines to the trace curves through Q are each contained in $\mathcal{P}$. This shows that $\mathcal{P}$ is the tangent plane at Q, as defined in Section 15.4 (ET Section 14.4).

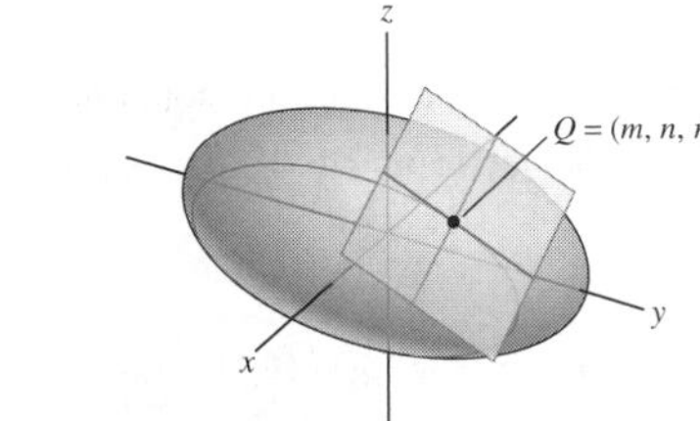

FIGURE 16 Ellipsoid $\left(\frac{x}{a}\right)^2 + \left(\frac{y}{b}\right)^2 + \left(\frac{z}{c}\right)^2 = 1$.

SOLUTION

(a) Since the point $Q = (m, n, r)$ is on the ellipsoid, the following holds:

$$\left(\frac{m}{a}\right)^2 + \left(\frac{n}{b}\right)^2 + \left(\frac{r}{c}\right)^2 = 1$$

$$\frac{m^2}{a^2} + \frac{n^2}{b^2} + \frac{r^2}{c^2} = 1 \quad \Rightarrow \quad \frac{m \cdot m}{a^2} + \frac{n \cdot n}{b^2} + \frac{r \cdot r}{c^2} = 1 \tag{1}$$

The meaning of (1) is that $x = m$, $y = n$, and $z = r$ satisfy the equation of the plane $\mathcal{P}$, that is, the point Q lies on $\mathcal{P}$. We now consider the three trace curves passing through $Q = (m, n, r)$. These are the intersections of the ellipsoid with the planes $x = m$, $y = n$, and $z = r$ respectively. That is,

$$x = m: \quad \left(\frac{m}{a}\right)^2 + \left(\frac{y}{b}\right)^2 + \left(\frac{z}{c}\right)^2 = 1 \quad \Rightarrow \quad \frac{y^2}{b^2} + \frac{z^2}{c^2} = 1 - \frac{m^2}{a^2}$$

$$y = n: \quad \left(\frac{x}{a}\right)^2 + \left(\frac{n}{b}\right)^2 + \left(\frac{z}{c}\right)^2 = 1 \quad \Rightarrow \quad \frac{x^2}{a^2} + \frac{z^2}{c^2} = 1 - \frac{n^2}{b^2} \tag{2}$$

$$z = r: \quad \left(\frac{x}{a}\right)^2 + \left(\frac{y}{b}\right)^2 + \left(\frac{r}{c}\right)^2 = 1 \quad \Rightarrow \quad \frac{x^2}{a^2} + \frac{y^2}{b^2} = 1 - \frac{r^2}{c^2}$$

Let the traces in (2) have the following parametric equations in respective order (where the parameters belong to certain intervals):

$$y = y(t),\ z = z(t) \text{ and } \left(y(t_0), z(t_0)\right) = (n, r)$$

$$x = x(t),\ z = z(t) \text{ and } \big(x(t_1), z(t_1)\big) = (m, r)$$
$$x = x(t),\ y = y(t) \text{ and } \big(x(t_2), y(t_2)\big) = (m, n)$$

The direction vectors of the tangent lines are, thus (in respective order)

$$\begin{aligned} \mathbf{u}_1 &= \langle 0, y'(t_0), z'(t_0)\rangle, \\ \mathbf{u}_2 &= \langle x'(t_1), 0, z'(t_1)\rangle, \\ \mathbf{u}_3 &= \langle x'(t_2), y'(t_2), 0\rangle. \end{aligned} \tag{3}$$

(b) By differentiating the equations of the trace curves in (2) with respect to the parameter t we get

$$\begin{aligned} \frac{2yy'(t)}{b^2} + \frac{2zz'(t)}{c^2} = 0 \quad &\Rightarrow \quad z'(t) = -\frac{c^2 y}{b^2 z} y'(t) \quad \Rightarrow \quad z'(t_0) = -\frac{c^2 n}{b^2 r} y'(t_0) \\ \frac{2xx'(t)}{a^2} + \frac{2zz'(t)}{c^2} = 0 \quad &\Rightarrow \quad z'(t) = -\frac{c^2 x}{a^2 z} x'(t) \quad \Rightarrow \quad z'(t_1) = -\frac{c^2 m}{a^2 r} x'(t_1) \\ \frac{2xx'(t)}{a^2} + \frac{2yy'(t)}{b^2} = 0 \quad &\Rightarrow \quad y'(t) = -\frac{b^2 x}{a^2 y} x'(t) \quad \Rightarrow \quad y'(t_2) = -\frac{b^2 m}{a^2 n} x'(t_2) \end{aligned} \tag{4}$$

Combining (3) and (4) gives

$$\mathbf{u}_1 = \left\langle 0, y'(t_0), -\frac{c^2 n}{b^2 r} y'(t_0) \right\rangle = y'(t_0) \left\langle 0, 1, -\frac{c^2 n}{b^2 r} \right\rangle$$
$$\mathbf{u}_2 = \left\langle x'(t_1), 0, -\frac{c^2 m}{a^2 r} x'(t_1) \right\rangle = x'(t_1) \left\langle 1, 0, -\frac{c^2 m}{a^2 r} \right\rangle$$
$$\mathbf{u}_3 = \left\langle x'(t_2), -\frac{b^2 m}{a^2 n} x'(t_2), 0 \right\rangle = x'(t_2) \left\langle 1, -\frac{b^2 m}{a^2 n}, 0 \right\rangle$$

(c) Now, to show that the tangent lines are contained in $\mathcal{P}$, we must show that their direction vectors are in $\mathcal{P}$, that is, they are orthogonal to the vector $\mathbf{n} = \left\langle \frac{m}{a^2}, \frac{n}{b^2}, \frac{r}{c^2} \right\rangle$ normal to the plane. We show this by proving that the following dot products are zero. That is,

$$\mathbf{u}_1 \cdot \mathbf{n} = y'(t_0) \left\langle 0, 1, \frac{-c^2 n}{b^2 r} \right\rangle \cdot \left\langle \frac{m}{a^2}, \frac{n}{b^2}, \frac{r}{c^2} \right\rangle = y'(t_0) \left(\frac{n}{b^2} - \frac{n}{b^2} \right) = 0$$
$$\mathbf{u}_2 \cdot \mathbf{n} = x'(t_1) \left\langle 1, 0, \frac{-c^2 m}{a^2 r} \right\rangle \cdot \left\langle \frac{m}{a^2}, \frac{n}{b^2}, \frac{r}{c^2} \right\rangle = x'(t_1) \left(\frac{m}{a^2} - \frac{m}{a^2} \right) = 0$$
$$\mathbf{u}_3 \cdot \mathbf{n} = x'(t_2) \left\langle 1, \frac{-b^2 m}{a^2 n}, 0 \right\rangle \cdot \left\langle \frac{m}{a^2}, \frac{n}{b^2}, \frac{r}{c^2} \right\rangle = x'(t_2) \left(\frac{m}{a^2} - \frac{m}{a^2} \right) = 0$$

Since the tangent lines intersect $\mathcal{P}$ (at the point P) and their direction vectors are in $\mathcal{P}$, it follows that the tangent lines are contained in the plane $\mathcal{P}$.

In Exercises 43–44, let $\mathcal{C}$ be a curve in $\mathbf{R}^3$ not passing through the origin. The cone on $\mathcal{C}$ is the surface consisting of all lines passing through the origin and a point on $\mathcal{C}$.

43. Show that the elliptic cone $\left(\frac{z}{c}\right)^2 = \left(\frac{x}{a}\right)^2 + \left(\frac{y}{b}\right)^2$ is, in fact, a cone on the ellipse $\mathcal{C}$ consisting of all points (x, y, c) such that $\left(\frac{x}{a}\right)^2 + \left(\frac{y}{b}\right)^2 = 1$.

SOLUTION

Step 1. We verify that the lines $\overline{OP}$ where P is a point (α, β, c) such that $\left(\frac{\alpha}{a}\right)^2 + \left(\frac{\beta}{b}\right)^2 = 1$ are contained in the elliptic cone $\left(\frac{x}{a}\right)^2 + \left(\frac{y}{b}\right)^2 = \left(\frac{z}{c}\right)^2$. The parametric equations of the line $\overline{OP}$ are

$$x = t\alpha, \quad y = t\beta, \quad z = tc$$

Substituting in the left hand side of the equation of the cone $\left(\frac{x}{a}\right)^2 + \left(\frac{y}{b}\right)^2 = \left(\frac{z}{c}\right)^2$ gives

$$\left(\frac{x}{a}\right)^2 + \left(\frac{y}{b}\right)^2 = \left(\frac{t\alpha}{a}\right)^2 + \left(\frac{t\beta}{b}\right)^2 = t^2 \left(\left(\frac{\alpha}{a}\right)^2 + \left(\frac{\beta}{b}\right)^2 \right) = t^2 \cdot 1 = \left(\frac{tc}{c}\right)^2 = \left(\frac{z}{c}\right)^2$$

Therefore, the line $\overline{OP}$ is contained in the elliptic cone.

Step 2. We show that every point (x_0, y_0, z_0) on the elliptic cone is contained in a certain line $\overline{OP}$ where P is a point on $\mathcal{C}$. Since (x_0, y_0, z_0) is on the cone, we have $\left(\frac{x_0}{a}\right)^2 + \left(\frac{y_0}{b}\right)^2 = \left(\frac{z_0}{c}\right)^2$, hence,

$$\left(\frac{x_0}{a\frac{z_0}{c}}\right)^2 + \left(\frac{y_0}{b\frac{z_0}{c}}\right)^2 = 1$$

or

$$\left(\frac{x_0\frac{c}{z_0}}{a}\right)^2 + \left(\frac{y_0\frac{c}{z_0}}{b}\right)^2 = 1 \tag{1}$$

We define P as the point $P = \left(\frac{x_0 c}{z_0}, \frac{y_0 c}{z_0}, c\right)$. By (1) P is on the ellipse $\mathcal{C}$. We show that (x_0, y_0, z_0) lies on the line through the origin and P. The parametric equations of this line are

$$x = t\frac{x_0 c}{z_0}, \quad y = t\frac{y_0 c}{z_0}, \quad z = tc$$

Now, (x_0, y_0, z_0) corresponds to the parameter $t = \frac{z_0}{c}$. This proves that any point $P = (x_0, y_0, z_0)$ on the cone is included in a certain line through the origin and a point on $\mathcal{C}$. By virtue of step 1 and step 2, we conclude that the cone on the ellipse $\mathcal{C}$ is equal to the elliptic cone.

13.7 Cylindrical and Spherical Coordinates (ET Section 12.7)

Preliminary Questions

1. Describe the surfaces $r = R$ in cylindrical coordinates and $\rho = R$ in spherical coordinates.

SOLUTION The surface $r = R$ consists of all points located at a distance R from the z-axis. This surface is the cylinder of radius R whose axis is the z-axis. The surface $\rho = R$ consists of all points located at a distance R from the origin. This is the sphere of radius R centered at the origin.

2. Which statement about the cylindrical coordinates is correct?

(a) If $\theta = 0$, then P lies on the z-axis.

(b) If $\theta = 0$, then P lies in the xz-plane.

SOLUTION The equation $\theta = 0$ defines the half-plane of all points that project onto the ray $\theta = 0$, that is, onto the nonnegative x-axis. This half plane is part of the (x, z)-plane, therefore if $\theta = 0$, then P lies in the (x, z)-plane.

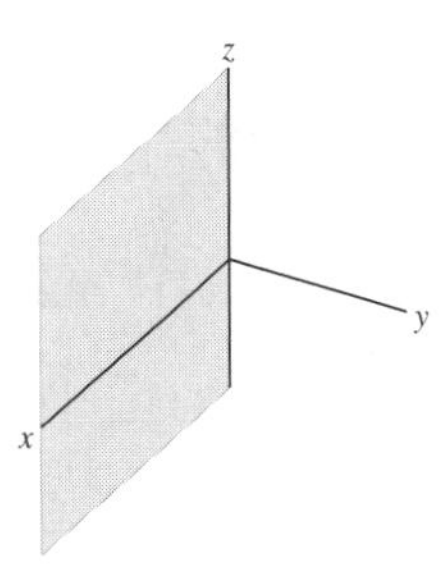

The half-plane $\theta = 0$

For instance, the point $P = (1, 0, 1)$ satisfies $\theta = 0$, but it does not lie on the z-axis. We conclude that statement (b) is correct and statement (a) is false.

3. Which statement about spherical coordinates is correct?

(a) If $\phi = 0$, then P lies on the z-axis.

(b) If $\phi = 0$, then P lies in the xy-plane.

SOLUTION The equation $\phi = 0$ describes the nonnegative z-axis. Therefore, if $\phi = 0$, P lies on the z-axis as stated in (a). Statement (b) is false, since the point $(0, 0, 1)$ satisfies $\phi = 0$, but it does not lie in the (x, y)-plane.

4. The level surface $\phi = \phi_0$ in spherical coordinates, usually a cone, reduces to a half-line for two values of ϕ_0. Which two values?

SOLUTION For $\phi_0 = 0$, the level surface $\phi = 0$ is the upper part of the z-axis. For $\phi_0 = \pi$, the level surface $\phi = \pi$ is the lower part of the z-axis. These are the two values of ϕ_0 where the level surface $\phi = \phi_0$ reduces to a half-line.

5. For which value of ϕ_0 is $\phi = \phi_0$ a plane? Which plane?

SOLUTION For $\phi_0 = \frac{\pi}{2}$, the level surface $\phi = \frac{\pi}{2}$ is the xy-plane.

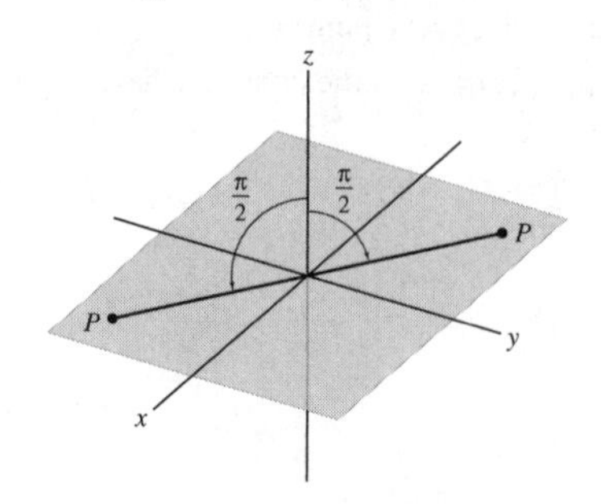

Exercises

In Exercises 1–4, convert from cylindrical to rectangular coordinates.

1. $(4, \pi, 4)$

SOLUTION By the given data $r = 4$, $\theta = \pi$ and $z = 4$. Hence,

$$\begin{aligned} x &= r\cos\theta = 4\cos\pi = 4\cdot(-1) = -4 \\ y &= r\sin\theta = 4\sin\pi = 4\cdot 0 \\ z &= 4 \end{aligned} \quad\Rightarrow\quad (x, y, z) = (-4, 0, 4)$$

3. $\left(0, \frac{\pi}{5}, \frac{1}{2}\right)$

SOLUTION We have $r = 0$, $\theta = \frac{\pi}{5}$, $z = \frac{1}{2}$. Thus,

$$\begin{aligned} x &= r\cos\theta = 0\cdot\cos\frac{\pi}{5} = 0 \\ y &= r\sin\theta = 0\cdot\sin\frac{\pi}{5} = 0 \\ z &= \frac{1}{2} \end{aligned} \quad\Rightarrow\quad (x, y, z) = \left(0, 0, \frac{1}{2}\right)$$

In Exercises 5–10, convert from rectangular to cylindrical coordinates.

5. $(1, -1, 1)$

SOLUTION We are given that $x = 1$, $y = -1$, $z = 1$. We find r:

$$r = \sqrt{x^2 + y^2} = \sqrt{1^2 + (-1)^2} = \sqrt{2}$$

Next we find θ. The point $(x, y) = (1, -1)$ lies in the fourth quadrant, hence,

$$\tan\theta = \frac{y}{x} = \frac{-1}{1} = -1, \quad \frac{3\pi}{2} \le \theta \le 2\pi \quad\Rightarrow\quad \theta = \frac{7\pi}{4}$$

We conclude that the cylindrical coordinates of the point are

$$(r, \theta, z) = \left(\sqrt{2}, \frac{7\pi}{4}, 1\right).$$

7. $(1, \sqrt{3}, 7)$

SOLUTION We have $x = 1$, $y = \sqrt{3}$, $z = 7$. We first find r:

$$r = \sqrt{x^2 + y^2} = \sqrt{1^2 + \left(\sqrt{3}\right)^2} = 2$$

Since the point $(x, y) = \left(1, \sqrt{3}\right)$ lies in the first quadrant, $0 \le \theta \le \frac{\pi}{2}$. Hence,

$$\tan\theta = \frac{y}{x} = \frac{\sqrt{3}}{1} = \sqrt{3}, \quad 0 \le \theta \le \frac{\pi}{2} \quad\Rightarrow\quad \theta = \frac{\pi}{3}$$

The cylindrical coordinates are thus

$$(r, \theta, z) = \left(2, \frac{\pi}{3}, 7\right).$$

9. $\left(\frac{5}{\sqrt{2}}, \frac{5}{\sqrt{2}}, 2\right)$

SOLUTION We have $x = \frac{5}{\sqrt{2}}$, $y = \frac{5}{\sqrt{2}}$, $z = 2$. We find r:

$$r = \sqrt{x^2 + y^2} = \sqrt{\left(\frac{5}{\sqrt{2}}\right)^2 + \left(\frac{5}{\sqrt{2}}\right)^2} = \sqrt{25} = 5$$

Since the point $(x, y) = \left(\frac{5}{\sqrt{2}}, \frac{5}{\sqrt{2}}\right)$ is in the first quadrant, $0 \le \theta \le \frac{\pi}{2}$, therefore,

$$\tan\theta = \frac{y}{x} = \frac{5/\sqrt{2}}{5/\sqrt{2}} = 1, \quad 0 \le \theta \le \frac{\pi}{2} \quad \Rightarrow \quad \theta = \frac{\pi}{4}$$

The corresponding cylindrical coordinates are

$$(r, \theta, z) = \left(5, \frac{\pi}{4}, 2\right).$$

In Exercises 11–16, describe the set in cylindrical coordinates.

11. $x^2 + y^2 \le 1$

SOLUTION The inequality describes a solid cylinder of radius 1 centered on the z-axis. Since $x^2 + y^2 = r^2$, this inequality can be written as $r^2 \le 1$.

13. $x^2 + y^2 + z^2 = 4, \quad x \ge 0, \quad y \ge 0, \quad z \ge 0$

SOLUTION We express z in terms of x and y. Since $z \ge 0$ we get

$$x^2 + y^2 + z^2 = 4 \quad \Rightarrow \quad z^2 = 4 - (x^2 + y^2) \quad \Rightarrow \quad z = \sqrt{4 - (x^2 + y^2)} \tag{1}$$

The cylindrical coordinates are (r, θ, z) where $x^2 + y^2 = r^2$. Substituting into (1) gives

$$z = \sqrt{4 - r^2} \tag{2}$$

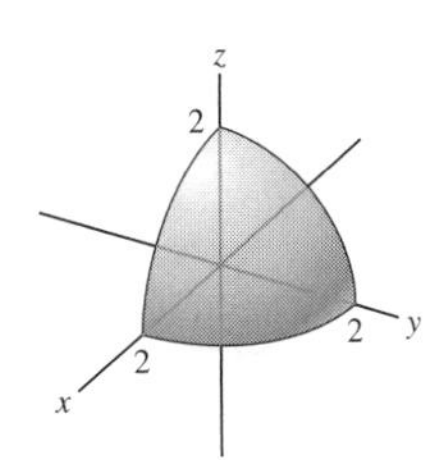

We find the interval for θ. The given set is the part of the sphere $x^2 + y^2 + z^2 = 4$ in the first octant.

$$\text{Hence, the angle } \theta \text{ is changing between 0 and } \frac{\pi}{2}. \tag{3}$$

We combine (2) and (3) to obtain the following representation:

$$z = \sqrt{4 - r^2}, \quad 0 \le \theta \le \frac{\pi}{2}.$$

15. $y^2 + z^2 \le 9, \quad x \ge y$

SOLUTION The region $x \ge y$ in the xy-plane is determined by the inequalities $\frac{5\pi}{4} \le \theta \le 2\pi$, $0 \le \theta \le \frac{\pi}{4}$ (and the origin). Since $y = r\sin\theta$, the region $y^2 + z^2 \le 9$ can be written as

$$r^2\sin^2\theta + z^2 \le 9$$

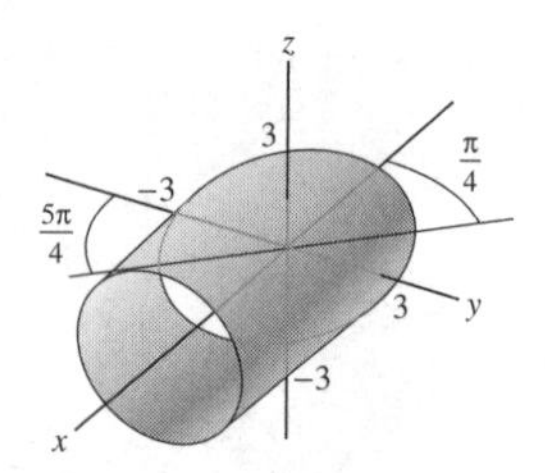

We obtain the following description in cylindrical coordinates:

$$r^2 \sin^2 \theta + z^2 \le 9, \quad 0 \le \theta \le \frac{\pi}{4} \quad \text{or} \quad \frac{5\pi}{4} \le \theta \le 2\pi.$$

In Exercises 17–19, sketch the level surface.

17. $r = 4$

SOLUTION The surface $r = 4$ consists of all points located at a distance 4 from the z-axis. It is a cylinder of radius 4 whose axis is the z-axis. The cylinder is shown in the following figure:

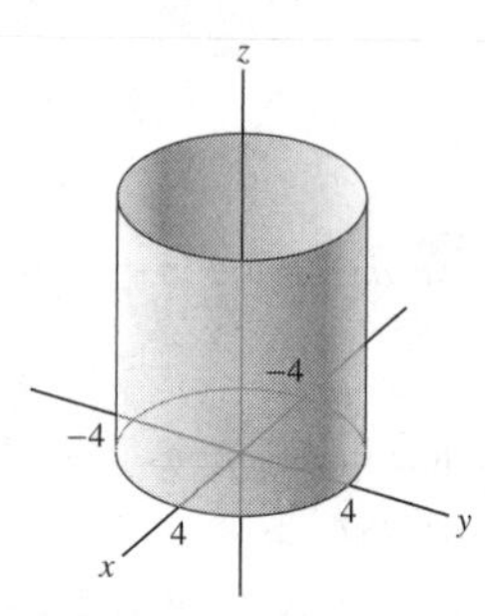

19. $z = -2$

SOLUTION $z = -2$ is the horizontal plane at height -2, shown in the following figure:

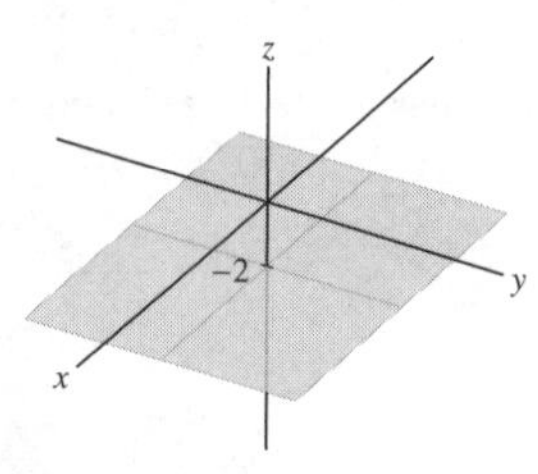

In Exercises 20–23, sketch the set (described in cylindrical coordinates).

21. $z^2 + r^2 \le 4$

SOLUTION The region $z^2 + r^2 \le 4$ is shown in the following figure:

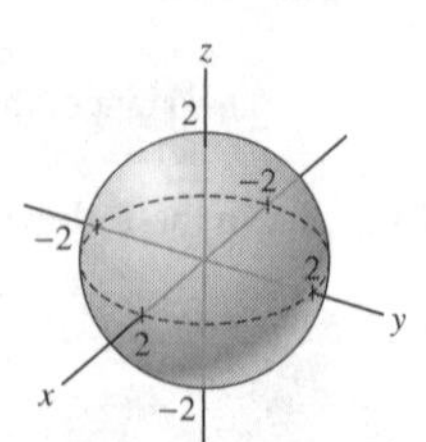

In rectangular coordinates the inequality is $z^2 + \left(x^2 + y^2\right) \le 4$, or $x^2 + y^2 + z^2 \le 4$, which is a ball of radius 2.

23. $r \le 3, \quad \pi \le \theta \le \frac{3\pi}{2}, \quad z = 4$

SOLUTION The region $r \le 3$, $\pi \le \theta \le \frac{3\pi}{2}$, $z = 4$ is shown in the following figure:

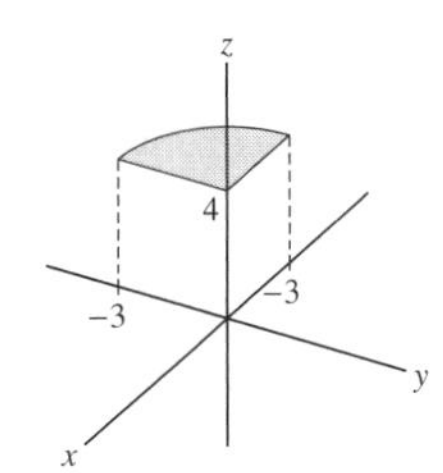

In Exercises 24–29, find an equation of the form $r = f(\theta, z)$ in cylindrical coordinates for the following surfaces.

25. $x^2 + y^2 + z^2 = 4$

SOLUTION Since $x^2 + y^2 = r^2$, we get

$$r^2 + z^2 = 4$$
$$r^2 = 4 - z^2 \quad \Rightarrow \quad r = \sqrt{4 - z^2}$$

.

27. $x^2 - y^2 = 4$

SOLUTION We substitute $x = r\cos\theta$, $y = r\sin\theta$ and use the trigonometric identity $\cos^2\theta - \sin^2\theta = \cos 2\theta$. This gives

$$r^2\cos^2\theta - r^2\sin^2\theta = 4$$
$$r^2\left(\cos^2\theta - \sin^2\theta\right) = 4$$
$$r^2\cos 2\theta = 4 \quad \Rightarrow \quad r = \frac{2}{\sqrt{\cos 2\theta}}$$

29. $z = 3xy$

SOLUTION We substitute $x = r\cos\theta$, $y = r\sin\theta$ and use the trigonometric identity $\sin\theta\cos\theta = \frac{1}{2}\sin 2\theta$. This gives

$$z = 3\,(r\cos\theta)\,(r\sin\theta) = 3r^2\cos\theta\sin\theta = \frac{3}{2}r^2\sin 2\theta$$

Thus,

$$r^2 = \frac{2z}{3\sin 2\theta} \quad \Rightarrow \quad r = \sqrt{\frac{2z}{3\sin 2\theta}}.$$

In Exercises 30–35, convert from spherical to rectangular coordinates.

31. $\left(2, \frac{\pi}{4}, \frac{\pi}{3}\right)$

SOLUTION We are given that $\rho = 2$, $\theta = \frac{\pi}{4}$, $\phi = \frac{\pi}{3}$. The relations between the spherical and the rectangular coordinates imply

$$x = \rho\sin\phi\cos\theta = 2\sin\frac{\pi}{3}\cos\frac{\pi}{4} = 2\cdot\frac{\sqrt{3}}{2}\cdot\frac{\sqrt{2}}{2} = \frac{\sqrt{6}}{2}$$
$$y = \rho\sin\phi\sin\theta = 2\sin\frac{\pi}{3}\sin\frac{\pi}{4} = 2\cdot\frac{\sqrt{3}}{2}\cdot\frac{\sqrt{2}}{2} = \frac{\sqrt{6}}{2} \quad \Rightarrow \quad (x, y, z) = \left(\frac{\sqrt{6}}{2}, \frac{\sqrt{6}}{2}, 1\right)$$
$$z = \rho\cos\phi = 2\cos\frac{\pi}{3} = 2\cdot\frac{1}{2} = 1$$

33. $\left(5, \frac{3\pi}{4}, \frac{\pi}{4}\right)$

SOLUTION We have $\rho = 5$, $\theta = \frac{3\pi}{4}$, $\phi = \frac{\pi}{4}$. Using the relations between spherical and rectangular coordinates we have

$$x = \rho\sin\phi\cos\theta = 5\sin\frac{\pi}{4}\cos\frac{3\pi}{4} = 5\cdot\frac{\sqrt{2}}{2}\cdot\left(-\frac{\sqrt{2}}{2}\right) = -2.5$$
$$y = \rho\sin\phi\sin\theta = 5\sin\frac{\pi}{4}\sin\frac{3\pi}{4} = 5\cdot\frac{\sqrt{2}}{2}\cdot\frac{\sqrt{2}}{2} = 2.5 \quad \Rightarrow \quad (x, y, z) = \left(-2.5, 2.5, 2.5\sqrt{2}\right)$$
$$z = \rho\cos\phi = 5\cos\frac{\pi}{4} = 5\cdot\frac{\sqrt{2}}{2} = 2.5\sqrt{2}$$

35. $(0.5, 3.7, 2)$

SOLUTION Using the relations between the spherical and the rectangular coordinates with $\rho = 0.5$, $\theta = 3.7$, and $\phi = 2$, we obtain

$$\begin{aligned} x &= \rho \sin\phi \cos\theta = 0.5 \sin 2 \cos 3.7 = -0.386 \\ y &= \rho \sin\phi \sin\theta = 0.5 \sin 2 \sin 3.7 = -0.241 \quad \Rightarrow \quad (x, y, z) = (-0.386, -0.241, -0.208) \\ z &= \rho \cos\phi = 0.5 \cos 2 = -0.208 \end{aligned}$$

In Exercises 36–41, convert from rectangular to spherical coordinates.

37. $\left(\frac{\sqrt{3}}{2}, \frac{3}{2}, 1\right)$

SOLUTION We have $x = \frac{\sqrt{3}}{2}$, $y = \frac{3}{2}$, and $z = 1$. The radial coordinate is

$$\rho = \sqrt{x^2 + y^2 + z^2} = \sqrt{\left(\frac{\sqrt{3}}{2}\right)^2 + \left(\frac{3}{2}\right)^2 + 1^2} = 2$$

The angular coordinate θ satisfies

$$\tan\theta = \frac{y}{x} = \frac{3/2}{\sqrt{3}/2} = \sqrt{3} \quad \Rightarrow \quad \theta = \frac{\pi}{3} \quad \text{or} \quad \theta = \frac{4\pi}{3}$$

Since the point $(x, y) = \left(\frac{\sqrt{3}}{2}, \frac{3}{2}\right)$ is in the first quadrant, the correct choice is $\theta = \frac{\pi}{3}$. The angle of declination ϕ satisfies

$$\cos\phi = \frac{z}{\rho} = \frac{1}{2}, \quad 0 \le \phi \le \pi \quad \Rightarrow \quad \phi = \frac{\pi}{3}$$

The spherical coordinates are thus

$$(\rho, \theta, \phi) = \left(2, \frac{\pi}{3}, \frac{\pi}{3}\right)$$

39. $(1, -1, 1)$

SOLUTION We have $x = 1$, $y = -1$, and $z = 1$. The radial coordinate is

$$\rho = \sqrt{1^2 + (-1)^2 + 1^2} = \sqrt{3}$$

The angular coordinate θ satisfies

$$\tan\theta = \frac{y}{x} = \frac{-1}{1} = -1 \quad \Rightarrow \quad \theta = \frac{3\pi}{4} \text{ or } \theta = \frac{7\pi}{4}$$

Since $(x, y) = (1, -1)$ is in the fourth quadrant, the angle is $\theta = \frac{7\pi}{4}$. The angle of declination satisfies

$$\cos\phi = \frac{z}{\rho} = \frac{1}{\sqrt{3}}, \quad 0 \le \phi \le \pi \quad \Rightarrow \quad \phi = 0.955$$

We conclude that

$$(\rho, \theta, \phi) = \left(\sqrt{3}, \frac{7\pi}{4}, 0.955\right).$$

41. $\left(\frac{\sqrt{2}}{2}, \frac{\sqrt{2}}{2}, \sqrt{3}\right)$

SOLUTION We are given that $x = y = \frac{\sqrt{2}}{2}$ and $z = \sqrt{3}$. Hence,

$$\rho = \sqrt{x^2 + y^2 + z^2} = \sqrt{\left(\frac{\sqrt{2}}{2}\right)^2 + \left(\frac{\sqrt{2}}{2}\right)^2 + \left(\sqrt{3}\right)^2} = 2$$

The angle θ satisfies $0 \le \theta \le \frac{\pi}{2}$ since $(x, y) = \left(\frac{\sqrt{2}}{2}, \frac{\sqrt{2}}{2}\right)$ is in the first quadrant. Also $\tan\theta = \frac{y}{x} = 1$, hence $\theta = \frac{\pi}{4}$. The angle of declination satisfies

$$\cos\phi = \frac{z}{\rho} = \frac{\sqrt{3}}{2}, \quad 0 \le \phi \le \pi \quad \Rightarrow \quad \phi = \frac{\pi}{6}$$

We conclude that

$$(\rho, \theta, \phi) = \left(2, \frac{\pi}{4}, \frac{\pi}{6}\right).$$

In Exercises 42–47, describe the given set in spherical coordinates.

43. $x^2 + y^2 + z^2 = 1, \quad z \ge 0$

SOLUTION Since $\rho^2 = x^2 + y^2 + z^2$ the equation becomes $\rho^2 = 1$ or $\rho = 1$. The inequality $z \ge 0$ implies that $\cos\phi = \frac{z}{\rho} \ge 0$. Also $0 \le \phi \le \pi$ by definition, hence $0 \le \phi \le \frac{\pi}{2}$. The spherical description of the set is thus

$$\rho = 1, \quad 0 \le \phi \le \frac{\pi}{2}.$$

45. $x^2 + y^2 + z^2 \le 1, \quad x = y, \quad x \ge 0, \quad y \ge 0$

SOLUTION Substituting $x^2 + y^2 + z^2 = \rho^2$ yields $\rho^2 \le 1$ or $0 \le \rho \le 1$. The inequalities $x \ge 0$, $y \ge 0$ determine the first quadrant which is also determined by

$$0 \le \theta \le \frac{\pi}{2} \qquad (1)$$

The line $y = x$ is determined by $\theta = \frac{\pi}{4}$ or $\theta = \frac{5\pi}{4}$ (and the origin). Combining with (1) we get $\theta = \frac{\pi}{4}$. We conclude that the description of the given set in spherical coordinates is

$$\left\{(\rho, \theta, \phi) : 0 \le \rho \le 1, \theta = \frac{\pi}{4}\right\}$$

47. $x^2 + y^2 = 3z^2$

SOLUTION We substitute the spherical coordinates $x = \rho\sin\phi\cos\theta$, $y = \rho\sin\phi\sin\theta$, $z = \rho\cos\phi$ in the given equation, and simplify. This gives

$$\rho^2\sin^2\phi\cos^2\theta + \rho^2\sin^2\phi\sin^2\theta = 3\rho^2\cos^2\phi$$
$$\rho^2\sin^2\phi\left(\cos^2\theta + \sin^2\theta\right) = 3\rho^2\cos^2\phi$$
$$\rho^2\sin^2\phi \cdot 1 = 3\rho^2\cos^2\phi$$

One solution is $\rho = 0$ (the origin). For $\rho \ne 0$ we divide both sides by ρ to obtain

$$\sin^2\phi = 3\cos^2\phi. \qquad (1)$$

When $\cos\phi = 0$, $\sin\phi \ne 0$. Hence the points where $\cos\phi = 0$ are not solutions. We, thus, can divide the two sides by $\cos^2\phi$ to obtain

$$\frac{\sin^2\phi}{\cos^2\phi} = 3 \quad \Rightarrow \quad \tan\phi = \sqrt{3} \quad \text{or} \quad \tan\phi = -\sqrt{3}.$$

The solutions for $0 \le \phi \le \pi$ are

$$\phi = \frac{\pi}{3} \quad \text{and} \quad \phi = \frac{2\pi}{3}. \qquad (2)$$

By (1) and (2) we obtain the following representation in spherical coordinates:

$$\phi = \frac{\pi}{3} \quad \text{or} \quad \phi = \frac{2\pi}{3}. \qquad (3)$$

Notice that by (3) we see that the set is the surface obtained while rotating a line that makes an angle of $\frac{\pi}{3}$ with the positive z-axis, about the z-axis. In other words, a double cone.

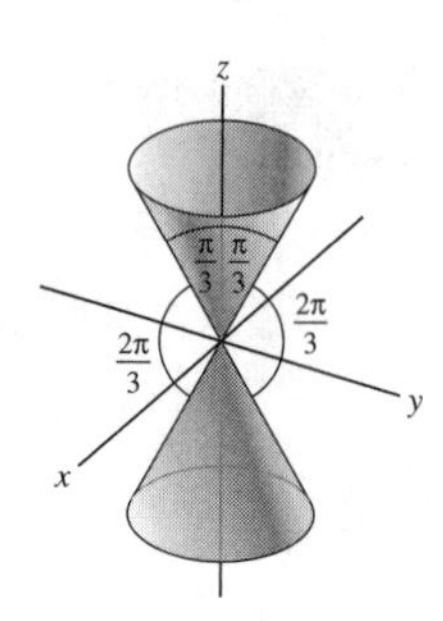

In Exercises 48–50, sketch the level surface.

49. $\theta = \dfrac{\pi}{3}$

SOLUTION The equation $\theta = \frac{\pi}{3}$ describes the following half plane:

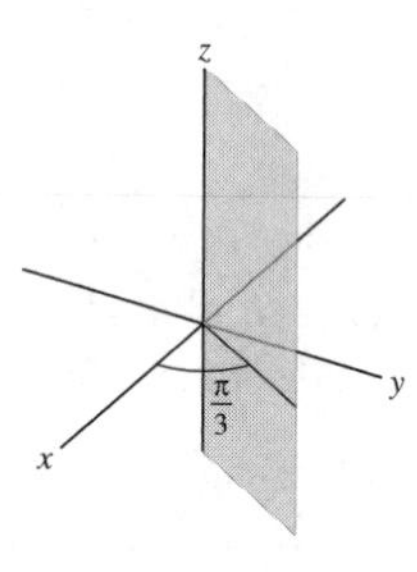

In Exercises 51–54, sketch the set of points.

51. $\rho = 2, \quad 0 \le \phi \le \dfrac{\pi}{2}$

SOLUTION The set

$$\rho = 2, \quad 0 \le \phi \le \frac{\pi}{2}$$

is shown in the following figure:

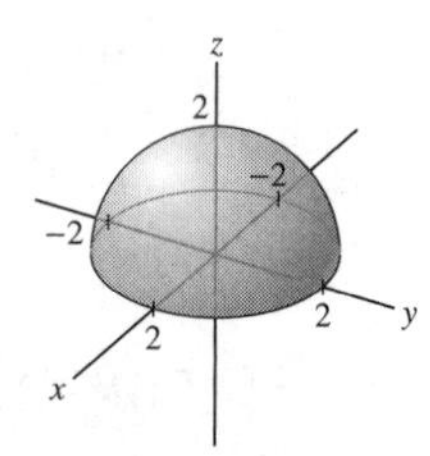

It is the upper half of the sphere with radius 2.

53. $\rho \le 2, \quad 0 \le \theta \le \dfrac{\pi}{2}, \quad \dfrac{\pi}{2} \le \phi \le \pi$

SOLUTION This set is the part of the ball of radius 2 which is below the first quadrant of the xy-plane, as shown in the following figure:

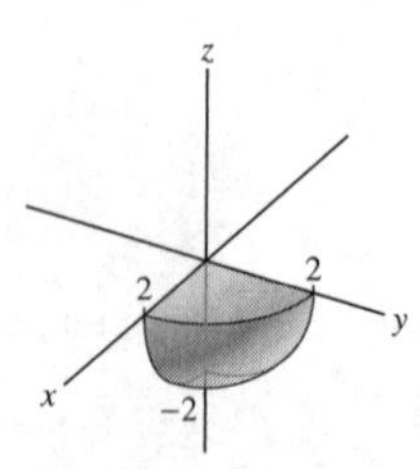

In Exercises 55–60, find an equation of the form $\rho = f(\theta, \phi)$ in spherical coordinates for the following surfaces.

55. $z = 2$

SOLUTION Since $z = \rho\cos\phi$, we have $\rho\cos\phi = 2$, or $\rho = \frac{2}{\cos\phi}$.

57. $x = z^2$

SOLUTION Substituting $x = \rho\cos\theta\sin\phi$ and $z = \rho\cos\phi$ we obtain

$$\rho\cos\theta\sin\phi = \rho^2\cos^2\phi$$
$$\cos\theta\sin\phi = \rho\cos^2\phi$$
$$\rho = \frac{\cos\theta\sin\phi}{\cos^2\phi} = \frac{\cos\theta\tan\phi}{\cos\phi}$$

59. $x^2 - y^2 = 4$

SOLUTION We substitute $x = \rho\cos\theta\sin\phi$ and $y = \rho\sin\theta\sin\phi$ to obtain

$$4 = \rho^2\cos^2\theta\sin^2\phi - \rho^2\sin^2\theta\sin^2\phi = \rho^2\sin^2\phi\left(\cos^2\theta - \sin^2\theta\right)$$

Using the identity $\cos^2\theta - \sin^2\theta = \cos 2\theta$ we get

$$4 = \rho^2\sin^2\phi\cos 2\theta$$
$$\rho^2 = \frac{4}{\sin^2\phi\cos 2\theta}$$

We take the square root of both sides. Since $0 < \phi < \pi$ we have $\sin\phi > 0$, hence,

$$\rho = \frac{2}{\sin\phi\sqrt{\cos 2\theta}}$$

61. Which of (a)–(c) is the equation of the cylinder of radius R in spherical coordinates? Refer to Figure 15.

(a) $R\rho = \sin\phi$ **(b)** $\rho\sin\phi = R$ **(c)** $\rho = R\sin\phi$

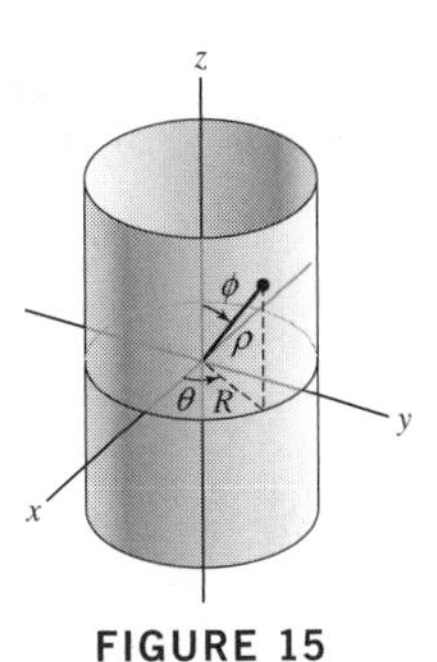

FIGURE 15

SOLUTION The equation of the cylinder of radius R in rectangular coordinates is $x^2 + y^2 = R^2$ (z is unlimited). Substituting the formulas for x and y in terms of ρ, θ and ϕ yields

$$R^2 = \rho^2\cos^2\theta\sin^2\phi + \rho^2\sin^2\theta\sin^2\phi = \rho^2\sin^2\phi\left(\cos^2\theta + \sin^2\theta\right) = \rho^2\sin^2\phi$$

Hence,

$$R^2 = \rho^2\sin^2\phi$$

We take the square root of both sides. Since $0 \le \phi \le \pi$, we have $\sin\phi \ge 0$, therefore,

$$R = \rho\sin\phi$$

Equation (b) is the correct answer.

63. Consider a rectangular coordinate system with origin at the center of the earth, z-axis through the North Pole, and x-axis through the prime meridian. Find the rectangular coordinates of Sydney, Australia (34° S, 151° E), and Bogota, Colombia (4° 32′ N, 74° 15′ W). A minute is $1/60°$. Assume that the earth is a sphere of radius $R = 6{,}367$ km.

SOLUTION We first find the angle (θ, ϕ) for the two towns. For Sydney $\theta = 151°$, since its longitude lies to the east of Greenwich, that is, in the positive θ direction. Sydney's latitude is south of the equator, hence $\phi = 90 + 34 = 124°$.

For Bogota, we have $\theta = 360° - 74°15' = 285°45'$, since $74°15'W$ refers to $74°15'$ in the negative θ direction. The latitude is north of the equator hence $\phi = 90° - 4°32' = 85°28'$.

We now use the formulas of x,y and z in terms of ρ, θ, ϕ to find the rectangular coordinates of the two towns. (Notice that $285°45' = 285.75°$ and $85°28' = 85.47°$).

Sydney:

$$x = \rho \cos\theta \sin\phi = 6367 \cos 151° \sin 124° = -4616.7$$
$$y = \rho \sin\theta \sin\phi = 6367 \sin 151° \sin 124° = 2559$$
$$z = \rho \cos\phi = 6367 \cos 124° = -3560.4$$

Bogota:

$$x = \rho \cos\theta \sin\phi = 6367 \cos 285.75° \sin 85.47° = 1722.9$$
$$y = \rho \sin\theta \sin\phi = 6367 \sin 285.75° \sin 85.47° = -6108.8$$
$$z = \rho \cos\phi = 6367 \cos 85.47° = 502.9$$

65. Find an equation of the form $z = f(r, \theta)$ in cylindrical coordinates for $z^2 = x^2 - y^2$.

SOLUTION In cylindrical coordinates, $x = r\cos\theta$ and $y = r\sin\theta$. Hence,

$$z^2 = x^2 - y^2 = r^2\cos^2\theta - r^2\sin^2\theta$$

We use the identity $\cos^2\theta - \sin^2\theta = \cos 2\theta$ to obtain

$$z^2 = r^2\cos 2\theta \quad \Rightarrow \quad z = \pm r\sqrt{\cos 2\theta}$$

67. Explain the following statement: If the equation of a surface in cylindrical or spherical coordinates does not involve the coordinate θ, then the surface is rotationally symmetric with respect to the z-axis.

SOLUTION Since the equation of the surface does not involve the coordinate θ, then for every point P on the surface ($P = (\rho_0, \theta_0, \phi_0)$ in spherical coordinates or $P = (r_0, \theta_0, z_0)$ in cylindrical coordinates) so also all the points (ρ_0, θ, ϕ_0) or (r_0, θ, z_0) are on the surface. That is, all the points obtained by rotating P around the z-axis are on the surface. Hence, the surface is rotationally symmetric with respect to the z-axis.

69. Find equations $r = g(\theta, z)$ (cylindrical) and $\rho = f(\theta, \phi)$ (spherical) for the hyperboloid $x^2 + y^2 = z^2 + 1$ (Figure 16). Do there exist points on the hyperboloid with $\phi = 0$ or π? Which values of ϕ occur for points on the hyperboloid?

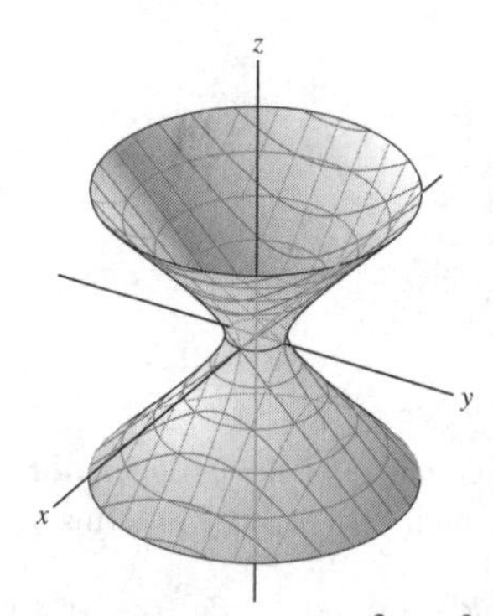

FIGURE 16 The hyperboloid $x^2 + y^2 = z^2 + 1$.

SOLUTION For the cylindrical coordinates (r, θ, z) we have $x^2 + y^2 = r^2$. Substituting into the equation $x^2 + y^2 = z^2 + 1$ gives

$$r^2 = z^2 + 1 \quad \Rightarrow \quad r = \sqrt{z^2 + 1}$$

For the spherical coordinates (ρ, θ, ϕ) we have $x = \rho\sin\phi\cos\theta$, $y = \rho\sin\phi\sin\theta$ and $z = \rho\cos\phi$. We substitute into the equation of the hyperboloid $x^2 + y^2 = z^2 + 1$ and simplify to obtain

$$\rho^2\sin^2\phi\cos^2\theta + \rho^2\sin^2\phi\sin^2\theta = \rho^2\cos^2\phi + 1$$
$$\rho^2\sin^2\phi\left(\cos^2\theta + \sin^2\theta\right) = \rho^2\cos^2\phi + 1$$
$$\rho^2\left(\sin^2\phi - \cos^2\phi\right) = 1$$

Using the trigonometric identity $\cos 2\phi = \cos^2\phi - \sin^2\phi$ we get

$$\rho^2 \cdot (-\cos 2\phi) = 1 \quad \Rightarrow \quad \rho = \sqrt{-\frac{1}{\cos 2\phi}}$$

For $\phi = 0$ and $\phi = \pi$ we have $\cos 2 \cdot 0 = 1$ and $\cos 2\pi = 1$. In both cases $-\frac{1}{\cos 2\phi} = -1 < 0$, hence there is no real value of ρ satisfying $\rho = \sqrt{-\frac{1}{\cos 2\phi}}$. We conclude that there are no points on the hyperboloid with $\phi = 0$ or π.

To obtain a real ρ such that $\rho = \sqrt{-\frac{1}{\cos 2\phi}}$, we must have $-\frac{1}{\cos 2\phi} > 0$. That is, $\cos 2\phi < 0$ (and of course $0 \le \phi \le \pi$). The corresponding values of ϕ are

$$\frac{\pi}{2} < 2\phi \le \frac{3\pi}{2} \quad \Rightarrow \quad \frac{\pi}{4} < \phi \le \frac{3\pi}{4}$$

Further Insights and Challenges

In Exercises 70–74, a ***great circle*** *on a sphere S with center O is a circle obtained by intersecting S with a plane that passes through O (Figure 17). If P and Q are not antipodal (on opposite sides), there is a unique great circle through P and Q on S (intersect S with the plane through O, P, and Q). The geodesic distance from P to Q is defined as the length of the smaller of the two circular arcs of this great circle.*

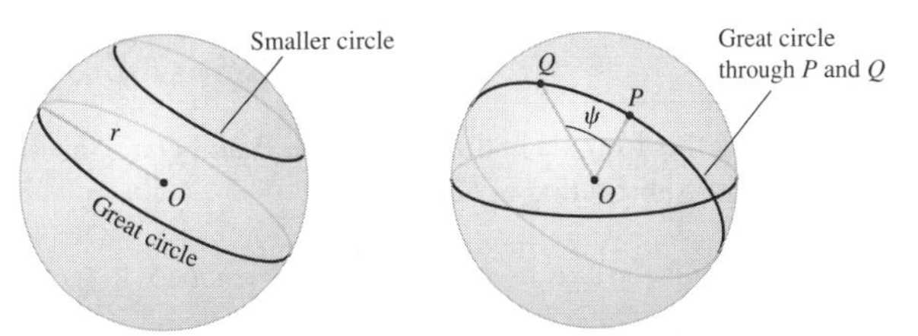

FIGURE 17

71. Show that the geodesic distance from $Q = (a, b, c)$ to the North Pole $P = (0, 0, R)$ is equal to $R \cos^{-1}\left(\frac{c}{R}\right)$.

SOLUTION Let ψ be the central angle between P and Q, that is, the angle between the vectors $\mathbf{v} = \overrightarrow{OP}$ and $\mathbf{u} = \overrightarrow{OQ}$. By Exercise 70 the geodesic distance from P to Q is $R\psi$. We find ψ. By the formula for the cosine of the angle between two vectors, we have

$$\cos \psi = \frac{\mathbf{u} \cdot \mathbf{v}}{\|\mathbf{u}\| \|\mathbf{v}\|} \tag{1}$$

We compute the values in this quotient:

$$\mathbf{u} \cdot \mathbf{v} = \langle 0, 0, R \rangle \cdot \langle a, b, c \rangle = 0 + 0 + Rc = Rc$$

$$\|\mathbf{v}\| = \|\overrightarrow{OP}\| = R$$

$$\|\mathbf{u}\| = \|\overrightarrow{OQ}\| = \sqrt{a^2 + b^2 + c^2} = R$$

Substituting in (1) we get

$$\cos \psi = \frac{Rc}{R^2} = \frac{c}{R} \quad \Rightarrow \quad \psi = \cos^{-1}\left(\frac{c}{R}\right)$$

The geodesic distance from Q to P is thus

$$R\psi = R \cos^{-1}\left(\frac{c}{R}\right)$$

73. Show that the central angle ψ between points P and Q on a sphere (of any radius) with angular coordinates (θ, ϕ) and (θ', ϕ') is equal to

$$\psi = \cos^{-1}\left(\sin \phi \sin \phi' \cos(\theta - \theta') + \cos \phi \cos \phi'\right)$$

Hint: Compute the dot product of $\overrightarrow{OP}$ and $\overrightarrow{OQ}$. Check this formula by computing the geodesic distance between the North and South Poles.

SOLUTION We denote the vectors $\mathbf{u} = \overrightarrow{OP}$ and $\mathbf{v} = \overrightarrow{OQ}$. By the formula for the angle between two vectors we have

$$\psi = \cos^{-1}\left(\frac{\mathbf{u} \cdot \mathbf{v}}{\|\mathbf{u}\| \|\mathbf{v}\|}\right)$$

Denoting by R the radius of the sphere, we have $\|\mathbf{u}\| = \|\mathbf{v}\| = R$, hence,

$$\psi = \cos^{-1}\left(\frac{\mathbf{u} \cdot \mathbf{v}}{R^2}\right) \tag{1}$$

The rectangular coordinates of $\mathbf{u}$ and $\mathbf{v}$ are

u	v
$x = R\sin\phi\cos\theta$	$x' = R\sin\phi'\cos\theta'$
$y = R\sin\phi\sin\theta$	$y' = R\sin\phi'\sin\theta'$
$z = R\cos\phi$	$z' = R\cos\phi'$

Hence,

$$\begin{aligned}\mathbf{u}\cdot\mathbf{v} &= R^2\sin\phi\cos\theta\sin\phi'\cos\theta' + R^2\sin\phi\sin\theta\sin\phi'\sin\theta' + R^2\cos\phi\cos\phi' \\ &= R^2\left[\sin\phi\sin\phi'\left(\cos\theta\cos\theta' + \sin\theta\sin\theta'\right) + \cos\phi\cos\phi'\right]\end{aligned}$$

We use the identity $\cos(\alpha-\beta) = \cos\alpha\cos\beta + \sin\alpha\sin\beta$ to obtain

$$\mathbf{u}\cdot\mathbf{v} = R^2\left(\sin\phi\sin\phi'\cos\left(\theta-\theta'\right) + \cos\phi\cos\phi'\right)$$

Substituting in (1) we obtain

$$\psi = \cos^{-1}\left(\sin\phi\sin\phi'\cos\left(\theta-\theta'\right) + \cos\phi\cos\phi'\right) \tag{2}$$

We now check this formula in the case where P and Q are the north and south poles respectively. In this case $\theta = \theta' = 0$, $\phi = 0$, $\phi' = \pi$. Substituting in (2) gives

$$\psi = \cos^{-1}\left(\sin 0\sin\pi\cos 0 + \cos 0\cos\pi\right) = \cos^{-1}(-1) = \pi$$

Using Exercise 70, the geodesic distance between the two poles is $R\psi = R\pi$, in accordance with the formula for the length of a semicircle.

CHAPTER REVIEW EXERCISES

In Exercises 1–6, let $\mathbf{v} = \langle -2, 5\rangle$, *and* $\mathbf{w} = \langle 3, -2\rangle$.

1. Calculate $5\mathbf{w} - 3\mathbf{v}$ and $5\mathbf{v} - 3\mathbf{w}$.

SOLUTION We use the definition of basic vector operations to compute the two linear combinations:

$$\begin{aligned}5\mathbf{w} - 3\mathbf{v} &= 5\langle 3, -2\rangle - 3\langle -2, 5\rangle = \langle 15, -10\rangle + \langle 6, -15\rangle = \langle 21, -25\rangle \\ 5\mathbf{v} - 3\mathbf{w} &= 5\langle -2, 5\rangle - 3\langle 3, -2\rangle = \langle -10, 25\rangle + \langle -9, 6\rangle = \langle -19, 31\rangle\end{aligned}$$

3. Find the unit vector in the direction of $\mathbf{v}$.

SOLUTION The unit vector in the direction of $\mathbf{v}$ is

$$\mathbf{e_v} = \frac{1}{\|\mathbf{v}\|}\mathbf{v}$$

We compute the length of $\mathbf{v}$:

$$\|\mathbf{v}\| = \sqrt{(-2)^2 + 5^2} = \sqrt{29}$$

Hence,

$$\mathbf{e_v} = \frac{\mathbf{v}}{\|\mathbf{v}\|} = \frac{\langle -2, 5\rangle}{\sqrt{29}} = \left\langle \frac{-2}{\sqrt{29}}, \frac{5}{\sqrt{29}}\right\rangle.$$

5. Express $\mathbf{i}$ as a linear combination $r\mathbf{v} + s\mathbf{w}$.

SOLUTION We use basic properties of vector algebra to write

$$\begin{aligned}\mathbf{i} &= r\mathbf{v} + s\mathbf{w} \\ \langle 1, 0\rangle &= r\langle -2, 5\rangle + s\langle 3, -2\rangle = \langle -2r + 3s, 5r - 2s\rangle\end{aligned} \tag{1}$$

The vector are equivalent, hence,

$$\begin{aligned}1 &= -2r + 3s \\ 0 &= 5r - 2s\end{aligned}$$

The second equation implies that $s = \frac{5}{2}r$. We substitute in the first equation and solve for r:

$$1 = -2r + 3 \cdot \frac{5}{2}r$$

$$1 = \frac{11}{2}r$$

$$r = \frac{2}{11} \quad \Rightarrow \quad s = \frac{5}{2} \cdot \frac{2}{11} = \frac{5}{11}$$

Substituting in (1) we obtain

$$\mathbf{i} = \frac{2}{11}\mathbf{v} + \frac{5}{11}\mathbf{w}.$$

7. If $P = (1, 4)$ and $Q = (-3, 5)$, what are the components of $\overrightarrow{PQ}$? What is the length of $\overrightarrow{PQ}$?

SOLUTION By the Definition of Components of a Vector we have

$$\overrightarrow{PQ} = \langle -3 - 1, 5 - 4 \rangle = \langle -4, 1 \rangle$$

The length of $\overrightarrow{PQ}$ is

$$\|\overrightarrow{PQ}\| = \sqrt{(-4)^2 + 1^2} = \sqrt{17}.$$

9. Find the vector with length 3 making an angle of $\frac{7\pi}{4}$ with the positive x-axis.

SOLUTION We denote the vector by $\mathbf{v} = \langle a, b \rangle$. $\mathbf{v}$ makes an angle $\theta = \frac{7\pi}{4}$ with the x-axis, and its length is 3, hence,

$$a = \|\mathbf{v}\| \cos\theta = 3\cos\frac{7\pi}{4} = \frac{3}{\sqrt{2}}$$

$$b = \|\mathbf{v}\| \sin\theta = 3\sin\frac{7\pi}{4} = -\frac{3}{\sqrt{2}}$$

That is,

$$\mathbf{v} = \langle a, b \rangle = \left\langle \frac{3}{\sqrt{2}}, -\frac{3}{\sqrt{2}} \right\rangle.$$

11. Find the value of β for which $\mathbf{w} = \langle -2, \beta \rangle$ is parallel to $\mathbf{v} = \langle 4, -3 \rangle$.

SOLUTION If $\mathbf{v} = \langle 4, -3 \rangle$ and $\mathbf{w} = \langle -2, \beta \rangle$ are parallel, there exists a scalar λ such that $\mathbf{w} = \lambda \mathbf{v}$. That is,

$$\langle -2, \beta \rangle = \lambda \langle 4, -3 \rangle = \langle 4\lambda, -3\lambda \rangle$$

yielding

$$-2 = 4\lambda \quad \text{and} \quad \beta = -3\lambda$$

These equations imply that $\lambda = -\frac{1}{2}$ and $\lambda = -\frac{\beta}{3}$. Equating the two expressions for λ gives

$$-\frac{1}{2} = -\frac{\beta}{3} \quad \text{or} \quad \beta = \frac{3}{2}.$$

13. Let $\mathbf{w} = \langle 2, -2, 1 \rangle$ and $\mathbf{v} = \langle 4, 5, -4 \rangle$. Solve for $\mathbf{u}$ if $\mathbf{v} + 5\mathbf{u} = 3\mathbf{w} - \mathbf{u}$.

SOLUTION Using vector algebra we have

$$\mathbf{v} + 5\mathbf{u} = 3\mathbf{w} - \mathbf{u}$$

$$6\mathbf{u} = 3\mathbf{w} - \mathbf{v}$$

$$\mathbf{u} = \frac{1}{2}\mathbf{w} - \frac{1}{6}\mathbf{v} = \left\langle 1, -1, \frac{1}{2} \right\rangle - \left\langle \frac{4}{6}, \frac{5}{6}, -\frac{4}{6} \right\rangle = \left\langle \frac{1}{3}, -\frac{11}{6}, \frac{7}{6} \right\rangle$$

15. Find a parametrization $\mathbf{r}_1(t)$ of the line passing through $(1, 4, 5)$ and $(-2, 3, -1)$. Then find a parametrization $\mathbf{r}_2(t)$ of the line parallel to $\mathbf{r}_1$ passing through $(1, 0, 0)$.

SOLUTION Since the points $P = (-2, 3, -1)$ and $Q = (1, 4, 5)$ are on the line l_1, the vector $\overrightarrow{PQ}$ is a direction vector for the line. We find this vector:

$$\overrightarrow{PQ} = \langle 1 - (-2), 4 - 3, 5 - (-1) \rangle = \langle 3, 1, 6 \rangle$$

Substituting $\mathbf{v} = \langle 3, 1, 6 \rangle$ and $P_0 = \langle 1, 4, 5 \rangle$ in the vector parametrization of the line we obtain the following equation for l_1:

$$\mathbf{r}_1(t) = \overrightarrow{OP_0} + t\mathbf{v}$$
$$\mathbf{r}_1(t) = \langle 1, 4, 5 \rangle + t\langle 3, 1, 6 \rangle = \langle 1 + 3t, 4 + t, 5 + 6t \rangle$$

The line l_2 is parallel to l_1, hence $\overrightarrow{PQ} = \langle 3, 1, 6 \rangle$ is also a direction vector for l_2. Substituting $\mathbf{v} = \langle 3, 1, 6 \rangle$ and $P_0 = (1, 0, 0)$ in the vector parametrization of the line we obtain the following equation for l_2:

$$\mathbf{r}_2(t) = \overrightarrow{OP_0} + t\mathbf{v}$$
$$\mathbf{r}_2(t) = \langle 1, 0, 0 \rangle + t\langle 3, 1, 6 \rangle = \langle 1 + 3t, t, 6t \rangle$$

17. Find a and b such that the lines $\mathbf{r}_1 = \langle 1, 2, 1 \rangle + t\langle 1, -1, 1 \rangle$ and $\mathbf{r}_2 = \langle 3, -1, 1 \rangle + t\langle a, b, -2 \rangle$ are parallel.

SOLUTION The lines are parallel if and only if the direction vectors $\mathbf{v}_1 = \langle 1, -1, 1 \rangle$ and $\mathbf{v}_2 = \langle a, b, -2 \rangle$ are parallel. That is, if and only if there exists a scalar λ such that:

$$\mathbf{v}_2 = \lambda \mathbf{v}_1$$
$$\langle a, b, -2 \rangle = \lambda \langle 1, -1, 1 \rangle = \langle \lambda, -\lambda, \lambda \rangle$$

We obtain the following equations:

$$\begin{aligned} a &= \lambda \\ b &= -\lambda \quad \Rightarrow \quad a = -2, \quad b = 2 \\ -2 &= \lambda \end{aligned}$$

19. Sketch the vector sum $\mathbf{v} = \mathbf{v}_1 - \mathbf{v}_2 + \mathbf{v}_3$ for the vectors in Figure 1(A).

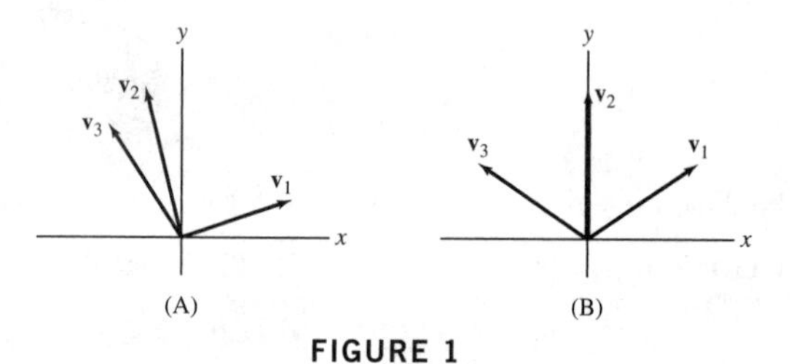

FIGURE 1

SOLUTION Using the Parallelogram Law we obtain the vector sum shown in the figure.

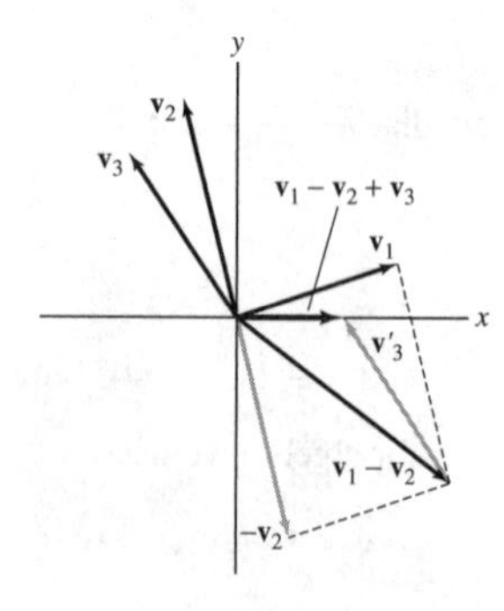

We first add $\mathbf{v}_1$ and $-\mathbf{v}_2$, then we add $\mathbf{v}_3$ to $\mathbf{v}_1 - \mathbf{v}_2$.

21. Use vectors to prove that the line connecting the midpoints of two sides of a triangle is parallel to the third side.

SOLUTION Let E and F be the midpoints of sides AC and BC in a triangle ABC (see figure).

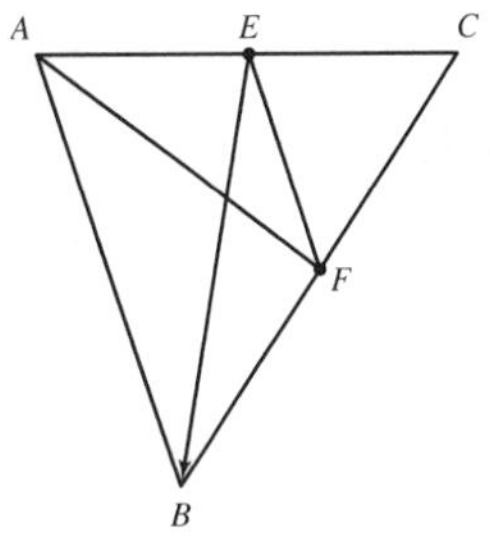

We must show that

$$\overrightarrow{EF} \parallel \overrightarrow{AB}$$

Using the Parallelogram Law we have

$$\overrightarrow{EF} = \overrightarrow{EA} + \overrightarrow{AB} + \overrightarrow{BF} \tag{1}$$

By the definition of the points E and F,

$$\overrightarrow{EA} = \frac{1}{2}\overrightarrow{CA}; \quad \overrightarrow{BF} = \frac{1}{2}\overrightarrow{BC}$$

We substitute (1) to obtain

$$\overrightarrow{EF} = \frac{1}{2}\overrightarrow{CA} + \overrightarrow{AB} + \frac{1}{2}\overrightarrow{BC} = \overrightarrow{AB} + \frac{1}{2}\left(\overrightarrow{CA} + \overrightarrow{BC}\right)$$
$$= \overrightarrow{AB} + \frac{1}{2}\left(\overrightarrow{BC} + \overrightarrow{CA}\right) = \overrightarrow{AB} + \frac{1}{2}\overrightarrow{BA} = \overrightarrow{AB} - \frac{1}{2}\overrightarrow{AB} = \frac{1}{2}\overrightarrow{AB}$$

Therefore, $\overrightarrow{EF}$ is a constant multiple of $\overrightarrow{AB}$, which implies that $\overrightarrow{EF}$ and $\overrightarrow{AB}$ are parallel vectors.

In Exercises 22–27, let $\mathbf{v} = \langle 1, 3, -2\rangle$ *and* $\mathbf{w} = \langle 2, -1, 4\rangle$.

23. Compute the angle between $\mathbf{v}$ and $\mathbf{w}$.

SOLUTION The cosine of the angle θ between $\mathbf{v}$ and $\mathbf{w}$ is

$$\cos\theta = \frac{\mathbf{v}\cdot\mathbf{w}}{\|\mathbf{v}\|\|\mathbf{w}\|} \tag{1}$$

We compute the lengths of the vectors:

$$\|\mathbf{v}\| = \|\langle 1, 3, -2\rangle\| = \sqrt{1^2 + 3^2 + (-2)^2} = \sqrt{14}$$
$$\|\mathbf{w}\| = \|\langle 2, -1, 4\rangle\| = \sqrt{2^2 + (-1)^2 + 4^2} = \sqrt{21}$$

In the previous exercise we found that $\mathbf{v}\cdot\mathbf{w} = -9$. Substituting these values in (1) gives

$$\cos\theta = \frac{-9}{\sqrt{14}\cdot\sqrt{21}} = \frac{-9}{7\sqrt{6}} \approx -0.5249$$

The solution for $0 \le \theta \le \pi$ is

$$\theta = 2.123 \text{ rad.}$$

25. Find the area of the parallelogram spanned by $\mathbf{v}$ and $\mathbf{w}$.

SOLUTION The parallelogram spanned by $\mathbf{v}$ and $\mathbf{w}$ has area $\|\mathbf{v}\times\mathbf{w}\|$. In the previous exercise, we found that $\mathbf{v}\times\mathbf{w} = \langle 10, -8, -7\rangle$. Therefore the area A of the parallelogram is

$$A = \|\mathbf{v}\times\mathbf{w}\| = \|\langle 10, -8, -7\rangle\| = \sqrt{10^2 + (-8)^2 + (-7)^2} = \sqrt{213} \approx 14.59$$

27. Find all the vectors orthogonal to both $\mathbf{v}$ and $\mathbf{w}$.

SOLUTION A vector $\mathbf{u} = \langle a, b, c\rangle$ is orthogonal to $\mathbf{v}$ and to $\mathbf{w}$ if the dot products $\mathbf{u}\cdot\mathbf{v}$ and $\mathbf{u}\cdot\mathbf{w}$ are zero. That is,

$$\mathbf{u}\cdot\mathbf{v} = 0 \quad \text{and} \quad \mathbf{u}\cdot\mathbf{w} = 0.$$

We compute the dot products:

$$\mathbf{u}\cdot\mathbf{v} = \langle a, b, c\rangle \cdot \langle 1, 3, -2\rangle = a + 3b - 2c$$

$$\mathbf{u}\cdot\mathbf{w} = \langle a, b, c\rangle \cdot \langle 2, -1, 4\rangle = 2a - b + 4c$$

We obtain the following equations:

$$a + 3b - 2c = 0$$
$$2a - b + 4c = 0$$

The first equation implies $a = 2c - 3b$. Substituting in the second equation and solving for b in terms of c gives

$$2(2c - 3b) - b + 4c = 0$$
$$4c - 6b - b + 4c = 0$$
$$8c - 7b = 0 \quad \Rightarrow \quad b = \frac{8}{7}c$$

We find a in terms of c, using the relation $a = 2c - 3b$:

$$a = 2c - 3\cdot\frac{8}{7}c = 2c - \frac{24}{7}c = -\frac{10}{7}c.$$

The solutions are, thus,

$$\mathbf{u} = \langle a, b, c\rangle = \left\langle -\frac{10}{7}c, \frac{8}{7}c, c\right\rangle = -\frac{c}{7}\langle 10, -8, -7\rangle$$

We conclude that the vectors orthogonal to $\mathbf{v}$ and $\mathbf{w}$ are all the vectors parallel to $\langle 10, -8, -7\rangle$.

29. A 50-kg wagon is pulled to the right by a force $\mathbf{F}_1$ making an angle of 30° with the ground. At the same time the wagon is pulled to the left by a horizontal force $\mathbf{F}_2$.

(a) Find the magnitude of $\mathbf{F}_1$ in terms of the magnitude of $\mathbf{F}_2$ if the wagon does not move.

(b) What is the maximal magnitude of $\mathbf{F}_1$ that can be applied to the wagon without lifting it?

SOLUTION

(a) By Newton's Law, at equilibrium, the total force acting on the wagon is zero.

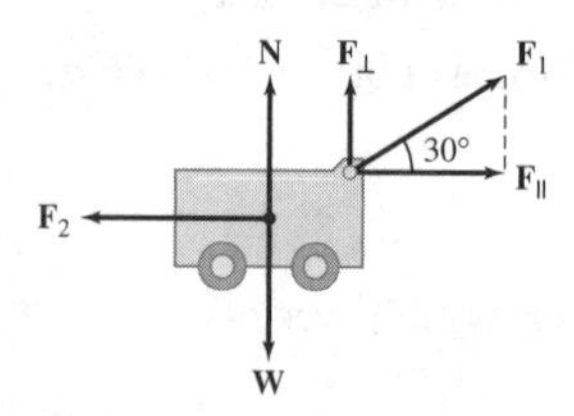

We resolve the force $\mathbf{F}_1$ into its components:

$$\mathbf{F}_1 = \mathbf{F}_{\|} + \mathbf{F}_{\perp}$$

where $\mathbf{F}_{\|}$ is the horizontal component and $\mathbf{F}_{\perp}$ is the vertical component. Since the wagon does not move, the magnitude of $\mathbf{F}_{\|}$ must be equal to the magnitude of $\mathbf{F}_2$. That is,

$$\|\mathbf{F}_{\|}\| = \|\mathbf{F}_1\|\cos 30^\circ = \|\mathbf{F}_2\|$$

The above equation gives:

$$\|\mathbf{F}_1\|\frac{\sqrt{3}}{2} = \|\mathbf{F}_2\| \quad \Rightarrow \quad \|\mathbf{F}_1\| = \frac{2\|\mathbf{F}_2\|}{\sqrt{3}}$$

(b) The maximum magnitude of force $\mathbf{F}_1$ that can be applied to the wagon without lifting the wagon is found by comparing the vertical forces:

$$\|\mathbf{F}_1\|\sin 30^\circ = 9.8\cdot 50$$
$$\|\mathbf{F}_1\|\cdot\frac{1}{2} = 9.8\cdot 50 \quad \Rightarrow \quad \|\mathbf{F}_1\| = 9.8\cdot 100 = 980\text{ N}$$

In Exercises 31–34, let $\mathbf{v} = \langle 1, 2, 4\rangle$, $\mathbf{u} = \langle 6, -1, 2\rangle$, *and* $\mathbf{w} = \langle 1, 0, -3\rangle$. *Calculate the given quantity.*

31. $\mathbf{v}\times\mathbf{w}$

SOLUTION We use the definition of the cross product as a determinant to compute $\mathbf{v} \times \mathbf{w}$:

$$\mathbf{v} \times \mathbf{w} = \begin{vmatrix} \mathbf{i} & \mathbf{j} & \mathbf{k} \\ 1 & 2 & 4 \\ 1 & 0 & -3 \end{vmatrix} = \begin{vmatrix} 2 & 4 \\ 0 & -3 \end{vmatrix}\mathbf{i} - \begin{vmatrix} 1 & 4 \\ 1 & -3 \end{vmatrix}\mathbf{j} + \begin{vmatrix} 1 & 2 \\ 1 & 0 \end{vmatrix}\mathbf{k}$$

$$= (-6-0)\mathbf{i} - (-3-4)\mathbf{j} + (0-2)\mathbf{k} = -6\mathbf{i} + 7\mathbf{j} - 2\mathbf{k} = \langle -6, 7, -2 \rangle$$

33. $\det \begin{pmatrix} \mathbf{u} \\ \mathbf{v} \\ \mathbf{w} \end{pmatrix}$

SOLUTION We compute the determinant:

$$\det \begin{pmatrix} \mathbf{u} \\ \mathbf{v} \\ \mathbf{w} \end{pmatrix} = \begin{vmatrix} 6 & -1 & 2 \\ 1 & 2 & 4 \\ 1 & 0 & -3 \end{vmatrix} = 6 \cdot \begin{vmatrix} 2 & 4 \\ 0 & -3 \end{vmatrix} + 1 \cdot \begin{vmatrix} 1 & 4 \\ 1 & -3 \end{vmatrix} + 2 \begin{vmatrix} 1 & 2 \\ 1 & 0 \end{vmatrix}$$

$$= 6 \cdot (-6-0) + 1 \cdot (-3-4) + 2 \cdot (0-2) = -47$$

35. Use the cross product to find the area of the triangle whose vertices are $(1, 3, -1)$, $(2, -1, 3)$, and $(4, 1, 1)$.

SOLUTION Let $A = (1, 3, -1)$, $B = (2, -1, 3)$ and $C = (4, 1, 1)$.

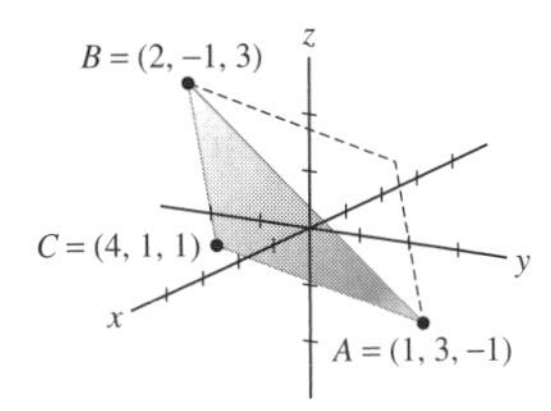

The area S of the triangle ABC is half the area of the parallelogram spanned by $\overrightarrow{AB}$ and $\overrightarrow{AC}$. Using the Formula for the Area of the Parallelogram, we conclude that the area of the triangle is:

$$S = \frac{1}{2} \left\| \overrightarrow{AB} \times \overrightarrow{AC} \right\| \tag{1}$$

We first compute the vectors $\overrightarrow{AB}$ and $\overrightarrow{AC}$:

$$\overrightarrow{AB} = \langle 2-1, -1-3, 3-(-1) \rangle = \langle 1, -4, 4 \rangle$$

$$\overrightarrow{AC} = \langle 4-1, 1-3, 1-(-1) \rangle = \langle 3, -2, 2 \rangle$$

We compute the cross product of the two vectors:

$$\overrightarrow{AB} \times \overrightarrow{AC} = \begin{vmatrix} \mathbf{i} & \mathbf{j} & \mathbf{k} \\ 1 & -4 & 4 \\ 3 & -2 & 2 \end{vmatrix} = \begin{vmatrix} -4 & 4 \\ -2 & 2 \end{vmatrix}\mathbf{i} - \begin{vmatrix} 1 & 4 \\ 3 & 2 \end{vmatrix}\mathbf{j} + \begin{vmatrix} 1 & -4 \\ 3 & -2 \end{vmatrix}\mathbf{k}$$

$$= (-8-(-8))\mathbf{i} - (2-12)\mathbf{j} + (-2-(-12))\mathbf{k}$$

$$= 10\mathbf{j} + 10\mathbf{k} = \langle 0, 10, 10 \rangle = 10\langle 0, 1, 1 \rangle$$

The length of $\overrightarrow{AB} \times \overrightarrow{AC}$ is, thus:

$$\left\| \overrightarrow{AB} \times \overrightarrow{AC} \right\| = \| 10\langle 0, 1, 1 \rangle \| = 10 \| \langle 0, 1, 1 \rangle \| = 10\sqrt{0^2 + 1^2 + 1^2} = 10\sqrt{2}$$

Substituting in (1) gives the following area:

$$S = \frac{1}{2} \cdot 10\sqrt{2} = 5\sqrt{2}.$$

37. Show that if the vectors $\mathbf{v}$, $\mathbf{w}$ are orthogonal, then $\|\mathbf{v} + \mathbf{w}\|^2 = \|\mathbf{v}\|^2 + \|\mathbf{w}\|^2$.

SOLUTION The vectors $\mathbf{v}$ and $\mathbf{w}$ are orthogonal, hence:

$$\mathbf{v} \cdot \mathbf{w} = 0 \tag{1}$$

Using the relation of the dot product with length and properties of the dot product we obtain:

$$\|\mathbf{v} + \mathbf{w}\|^2 = (\mathbf{v} + \mathbf{w}) \cdot (\mathbf{v} + \mathbf{w}) = \mathbf{v} \cdot (\mathbf{v} + \mathbf{w}) + \mathbf{w} \cdot (\mathbf{v} + \mathbf{w})$$

$$= \mathbf{v}\cdot\mathbf{v}+\mathbf{v}\cdot\mathbf{w}+\mathbf{w}\cdot\mathbf{v}+\mathbf{w}\cdot\mathbf{w} = \|\mathbf{v}\|^2 + 2\mathbf{v}\cdot\mathbf{w} + \|\mathbf{w}\|^2 \quad (2)$$

Combining (1) and (2) we get:

$$\|\mathbf{v}+\mathbf{w}\|^2 = \|\mathbf{v}\|^2 + \|\mathbf{w}\|^2.$$

39. Show that the equation $\langle 1, 2, 3\rangle \times \mathbf{v} = \langle -1, 2, a\rangle$ has no solution for $a \neq -1$.

SOLUTION By properties of the cross product, the vector $\langle -1, 2, a\rangle$ is orthogonal to $\langle 1, 2, 3\rangle$, hence the dot product of these vectors is zero. That is:

$$\langle -1, 2, a\rangle \cdot \langle 1, 2, 3\rangle = 0$$

We compute the dot product and solve for a:

$$-1 + 4 + 3a = 0$$
$$3a = -3 \quad \Rightarrow \quad a = -1$$

We conclude that if the given equation is solvable, then $a = -1$.

41. Use the identity

$$\mathbf{u} \times (\mathbf{v} \times \mathbf{w}) = (\mathbf{u}\cdot\mathbf{w})\,\mathbf{v} - (\mathbf{u}\cdot\mathbf{v})\,\mathbf{w}$$

to prove that

$$\mathbf{u} \times (\mathbf{v} \times \mathbf{w}) + \mathbf{v} \times (\mathbf{w} \times \mathbf{u}) + \mathbf{w} \times (\mathbf{u} \times \mathbf{v}) = \mathbf{0}$$

SOLUTION The given identity implies that:

$$\mathbf{u} \times (\mathbf{v} \times \mathbf{w}) = (\mathbf{u}\cdot\mathbf{w})\,\mathbf{v} - (\mathbf{u}\cdot\mathbf{v})\,\mathbf{w}$$
$$\mathbf{v} \times (\mathbf{w} \times \mathbf{u}) = (\mathbf{v}\cdot\mathbf{u})\,\mathbf{w} - (\mathbf{v}\cdot\mathbf{w})\,\mathbf{u}$$
$$\mathbf{w} \times (\mathbf{u} \times \mathbf{v}) = (\mathbf{w}\cdot\mathbf{v})\,\mathbf{u} - (\mathbf{w}\cdot\mathbf{u})\,\mathbf{v}$$

Adding the three equations and using the commutativity of the dot product we find that:

$$\mathbf{u} \times (\mathbf{v} \times \mathbf{w}) + \mathbf{v} \times (\mathbf{w} \times \mathbf{u}) + \mathbf{w} \times (\mathbf{u} \times \mathbf{v})$$
$$= (\mathbf{u}\cdot\mathbf{w} - \mathbf{w}\cdot\mathbf{u})\,\mathbf{v} + (\mathbf{v}\cdot\mathbf{u} - \mathbf{u}\cdot\mathbf{v})\,\mathbf{w} + (\mathbf{w}\cdot\mathbf{v} - \mathbf{v}\cdot\mathbf{w})\,\mathbf{u} = \mathbf{0}$$

43. Write the equation of the plane $\mathcal{P}$ with vector equation

$$\langle 1, 4, -3\rangle \cdot \langle x, y, z\rangle = 7$$

in the form

$$a\,(x - x_0) + b\,(y - y_0) + c\,(z - z_0) = 0$$

Hint: You must find a point $P = (x_0, y_0, z_0)$ on $\mathcal{P}$.

SOLUTION We identify the vector $\mathbf{n} = \langle a, b, c\rangle = \langle 1, 4, -3\rangle$ that is normal to the plane, hence we may choose,

$$a = 1, \quad b = 4, \quad c = -3.$$

We now must find a point in the plane. The point $(x_0, y_0, z_0) = (0, 1, -1)$, for instance, satisfies the equation of the plane, therefore the equation may be written in the form:

$$1(x - 0) + 4(y - 1) - 3(z - (-1)) = 0$$

or

$$(x - 0) + 4(y - 1) - 3(z + 1) = 0$$

45. Find the plane through $P = (4, -1, 9)$ containing the line $\mathbf{r}(t) = \langle 1, 4, -3\rangle + t\langle 2, 1, 1\rangle$.

SOLUTION Since the plane contains the line, the direction vector of the line, $\mathbf{v} = \langle 2, 1, 1\rangle$, is in the plane. To find another vector in the plane, we use the points $A = (1, 4, -3)$ and $B = (4, -1, 9)$ that lie in the plane, and compute the vector $\mathbf{u} = \overrightarrow{AB}$:

$$\mathbf{u} = \overrightarrow{AB} = \langle 4 - 1, -1 - 4, 9 - (-3)\rangle = \langle 3, -5, 12\rangle$$

We now compute the cross product $\mathbf{n} = \mathbf{v} \times \mathbf{u}$ that is normal to the plane:

$$\mathbf{n} = \mathbf{v} \times \mathbf{u} = \begin{vmatrix} \mathbf{i} & \mathbf{j} & \mathbf{k} \\ 2 & 1 & 1 \\ 3 & -5 & 12 \end{vmatrix} = \begin{vmatrix} 1 & 1 \\ -5 & 12 \end{vmatrix} \mathbf{i} - \begin{vmatrix} 2 & 1 \\ 3 & 12 \end{vmatrix} \mathbf{j} + \begin{vmatrix} 2 & 1 \\ 3 & -5 \end{vmatrix} \mathbf{k}$$

$$= (12+5)\mathbf{i} - (24-3)\mathbf{j} + (-10-3)\mathbf{k} = 17\mathbf{i} - 21\mathbf{j} - 13\mathbf{k} = \langle 17, -21, -13 \rangle$$

Finally, we use the vector form of the equation of the plane with $\mathbf{n} = \langle 17, -21, -13 \rangle$ and $P_0 = (4, -1, 9)$ to obtain the following equation:

$$\mathbf{n} \cdot \langle x, y, z \rangle = \mathbf{n} \cdot \langle x_0, y_0, z_0 \rangle$$

$$\langle 17, -21, -13 \rangle \cdot \langle x, y, z \rangle = \langle 17, -21, -13 \rangle \cdot \langle 4, -1, 9 \rangle$$

$$17x - 21y - 13z = 17 \cdot 4 + 21 - 13 \cdot 9 = -28$$

The equation of the plane is, thus,

$$17x - 21y - 13z = -28.$$

47. Find the trace of the plane $3x - 2y + 5z = 4$ in the xy-plane.

SOLUTION The xy-plane has equation $z = 0$, therefore the intersection of the plane $3x - 2y + 5z = 4$ with the xy-plane must satisfy both $z = 0$ and the equation of the plane. Therefore the trace has the following equation:

$$3x - 2y + 5 \cdot 0 = 4 \quad \Rightarrow \quad 3x - 2y = 4$$

We conclude that the trace of the plane in the xy-plane is the line $3x - 2y = 4$ in the xy-plane.

In Exercises 49–54, determine the type of the quadric surface.

49. $\left(\frac{x}{3}\right)^2 + \left(\frac{y}{4}\right)^2 + 2z^2 = 1$

SOLUTION Writing the equation in the form:

$$\left(\frac{x}{3}\right)^2 + \left(\frac{y}{4}\right)^2 + \left(\frac{z}{\frac{1}{\sqrt{2}}}\right)^2 = 1$$

we identify the quadric surface as an ellipsoid.

51. $\left(\frac{x}{3}\right)^2 + \left(\frac{y}{4}\right)^2 - 2z = 0$

SOLUTION We rewrite this equation as:

$$2z = \left(\frac{x}{3}\right)^2 + \left(\frac{y}{4}\right)^2$$

or

$$z = \left(\frac{x}{3\sqrt{2}}\right)^2 + \left(\frac{y}{4\sqrt{2}}\right)^2$$

This is the equation of an elliptic paraboloid.

53. $\left(\frac{x}{3}\right)^2 - \left(\frac{y}{4}\right)^2 - 2z^2 = 0$

SOLUTION This equation may be rewritten in the form

$$\left(\frac{x}{3}\right)^2 - \left(\frac{y}{4}\right)^2 = \left(\frac{z}{\frac{1}{\sqrt{2}}}\right)^2$$

we identify the quadric surface as an elliptic cone.

55. Determine the type of the quadric surface $ax^2 + by^2 - z^2 = 1$ if:

(a) $a < 0, \quad b < 0$

(b) $a > 0, \quad b > 0$

(c) $a > 0, \quad b < 0$

SOLUTION

(a) If $a < 0$, $b < 0$ then for all x, y and z we have $ax^2 + by^2 - z^2 < 0$, hence there are no points that satisfy $ax^2 + by^2 - z^2 = 1$. Therefore it is the empty set.

(b) For $a > 0$ and $b > 0$ we rewrite the equation as

$$\left(\frac{x}{\frac{1}{\sqrt{a}}}\right)^2 + \left(\frac{y}{\frac{1}{\sqrt{b}}}\right)^2 - z^2 = 1$$

which is the equation of a hyperboloid of one sheet.

(c) For $a > 0$, $b < 0$ we rewrite the equation in the form

$$\left(\frac{x}{\frac{1}{\sqrt{a}}}\right)^2 - \left(\frac{y}{\frac{1}{\sqrt{|b|}}}\right)^2 - z^2 = 1$$

which is the equation of a hyperboloid of two sheets.

57. Convert $(x, y, z) = (3, 4, -1)$ from rectangular to cylindrical and spherical coordinates.

SOLUTION In cylindrical coordinates (r, θ, z) we have

$$r = \sqrt{x^2 + y^2}, \quad \tan\theta = \frac{y}{x}$$

Therefore, $r = \sqrt{3^2 + 4^2} = 5$ and $\tan\theta = \frac{4}{3}$. The projection of the point $(3, 4, -1)$ onto the xy-plane is the point $(3, 4)$, in the first quadrant. Therefore, the corresponding value of θ is $\tan^{-1}\frac{4}{3} \approx 0.93$ rad. The cylindrical coordinates are, thus,

$$(r, \theta, z) = \left(5, \tan^{-1}\frac{4}{3}, -1\right)$$

The spherical coordinates (ρ, θ, ϕ) satisfy

$$\rho = \sqrt{x^2 + y^2 + z^2}, \quad \tan\theta = \frac{y}{x}, \quad \cos\phi = \frac{z}{\rho}$$

Therefore,

$$\rho = \sqrt{3^2 + 4^2 + (-1)^2} = \sqrt{26}$$

$$\tan\theta = \frac{4}{3}$$

$$\cos\phi = \frac{-1}{\sqrt{26}}$$

The angle θ is the same as in the cylindrical coordinates, that is, $\theta = \tan^{-1}\frac{4}{3}$. The angle ϕ is the solution of $\cos\phi = \frac{-1}{\sqrt{26}}$ that satisfies $0 \le \phi \le \pi$, that is, $\phi = \cos^{1}\left(\frac{-1}{\sqrt{26}}\right) \approx 1.77$ rad. The spherical coordinates are, thus,

$$(\rho, \theta, \phi) = \left(\sqrt{26}, \tan^{-1}\frac{4}{3}, \cos^{-1}\left(\frac{-1}{\sqrt{26}}\right)\right).$$

59. Convert the point $(\rho, \theta, \phi) = \left(3, \frac{\pi}{6}, \frac{\pi}{3}\right)$ from spherical to cylindrical coordinates.

SOLUTION By the given information, $\rho = 3$, $\theta = \frac{\pi}{6}$, and $\phi = \frac{\pi}{3}$. We must determine the cylindrical coordinates (r, θ, z). The angle θ is the same as in spherical coordinates. We find z using the relation $\cos\phi = \frac{z}{\rho}$, or $z = \rho\cos\phi$. We obtain

$$z = \rho\cos\phi = 3\cos\frac{\pi}{3} = 3 \cdot \frac{1}{2} = \frac{3}{2}$$

We find r using the relation $\rho^2 = x^2 + y^2 + z^2 = r^2 + z^2$, or $r = \sqrt{\rho^2 - z^2}$, we get

$$r = \sqrt{3^2 - \left(\frac{3}{2}\right)^2} = \sqrt{\frac{27}{4}} = \frac{3\sqrt{3}}{2}$$

Hence, in cylindrical coordinates we obtain the following description:

$$(r, \theta, z) = \left(\frac{3\sqrt{3}}{2}, \frac{\pi}{6}, \frac{3}{2}\right).$$

61. Sketch the graph of the cylindrical equation $z = 2r\cos\theta$ and write the equation in rectangular coordinates.

SOLUTION To obtain the equation in rectangular coordinates, we substitute $x = r\cos\theta$ in the equation $z = 2r\cos\theta$:

$$z = 2r\cos\theta = 2x \quad \Rightarrow \quad z = 2x$$

This is the equation of a plane normal to the xz-plane, whose intersection with the xz-plane is the line $z = 2x$. The graph of the plane is shown in the following figure (the same plane drawn twice, using the cylindrical coordinates' equation and using the rectangular coordinates' equation):

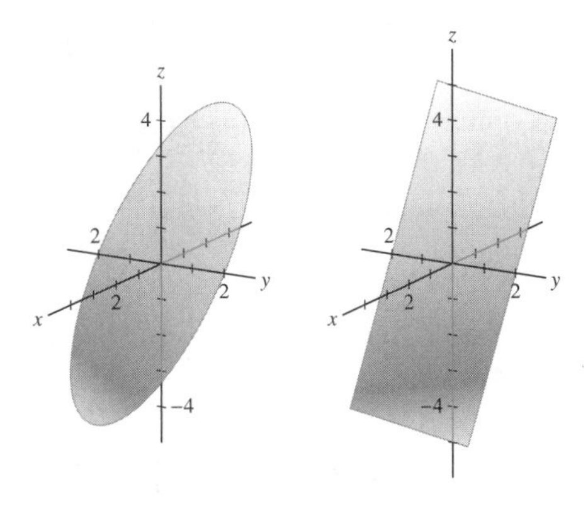

63. Show that the cylindrical equation

$$r^2(1 - 2\sin^2\theta) + z^2 = 1$$

is a hyperboloid of one sheet.

SOLUTION We rewrite the equation in the form

$$r^2 - 2(r\sin\theta)^2 + z^2 = 1$$

To write this equation in rectangular coordinates, we substitute $r^2 = x^2 + y^2$ and $r\sin\theta = y$. This gives

$$x^2 + y^2 - 2y^2 + z^2 = 1$$
$$x^2 - y^2 + z^2 = 1$$

We now can identify the surface as a hyperboloid of one sheet.

65. Describe how the surface with spherical equation

$$\rho^2(1 + A\cos^2\phi) = 1$$

depends on the constant A.

SOLUTION To identify the surface we convert the equation to rectangular coordinates. We write

$$\rho^2 + A\rho^2\cos^2\phi = 1$$

To obtain the following equation in terms of x, y, z only, we substitute $\rho^2 = x^2 + y^2 + z^2$ and $\rho\cos\phi = z$:

$$x^2 + y^2 + z^2 + Az^2 = 1$$
$$x^2 + y^2 + (1 + A)z^2 = 1 \tag{1}$$

Case 1: $A < -1$. Then $A + 1 < 0$ and the equation can be rewritten in the form

$$x^2 + y^2 - \left(\frac{z}{|1 + A|^{-1/2}}\right)^2 = 1$$

The corresponding surface is a hyperboloid of one sheet.

Case 2: $A = -1$. Equation (1) becomes:

$$x^2 + y^2 = 1$$

In R^3, this equation describes a cylinder with the z-axis as its central axis.

Case 3: $A > -1$. Then equation (1) can be rewritten as

$$x^2 + y^2 + \left(\frac{z}{(1+A)^{-1/2}}\right)^2 = 1$$

Then if $A = 0$ the equation $x^2 + y^2 + z^2 = 1$ describes the unit sphere in R^3. Otherwise, the surface is an ellipsoid.

67. Let c be a scalar, let $\mathbf{a}$ and $\mathbf{b}$ be vectors, and let $\mathbf{X} = \langle x, y, z \rangle$. Show that the equation $(\mathbf{X} - \mathbf{a})(\mathbf{X} - \mathbf{b}) = c^2$ defines a sphere with center $\mathbf{m} = \frac{1}{2}(\mathbf{a} + \mathbf{b})$ and radius R, where $R^2 = c^2 + \|\frac{1}{2}(\mathbf{a} - \mathbf{b})\|^2$.

SOLUTION We evaluate the following length:

$$\begin{aligned}
\|\mathbf{x} - \mathbf{m}\|^2 = \left\|\mathbf{x} - \frac{1}{2}(\mathbf{a} + \mathbf{b})\right\|^2 &= \left((\mathbf{x} - \mathbf{a}) + \frac{1}{2}(\mathbf{a} - \mathbf{b})\right) \cdot \left((\mathbf{x} - \mathbf{b}) - \frac{1}{2}(\mathbf{a} - \mathbf{b})\right) \\
&= (\mathbf{x} - \mathbf{a}) \cdot (\mathbf{x} - \mathbf{b}) - \frac{1}{2}(\mathbf{x} - \mathbf{a}) \cdot (\mathbf{a} - \mathbf{b}) + \frac{1}{2}(\mathbf{a} - \mathbf{b}) \cdot (\mathbf{x} - \mathbf{b}) - \frac{1}{4}(\mathbf{a} - \mathbf{b}) \cdot (\mathbf{a} - \mathbf{b}) \\
&= (\mathbf{x} - \mathbf{a}) \cdot (\mathbf{x} - \mathbf{b}) + \frac{1}{2}(\mathbf{a} - \mathbf{b}) \cdot (\mathbf{x} - \mathbf{b} - \mathbf{x} + \mathbf{a}) - \frac{1}{4}(\mathbf{a} - \mathbf{b}) \cdot (\mathbf{a} - \mathbf{b}) \\
&= (\mathbf{x} - \mathbf{a}) \cdot (\mathbf{x} - \mathbf{b}) + \frac{1}{2}(\mathbf{a} - \mathbf{b}) \cdot (\mathbf{a} - \mathbf{b}) - \frac{1}{4}(\mathbf{a} - \mathbf{b}) \cdot (\mathbf{a} - \mathbf{b}) \\
&= (\mathbf{x} - \mathbf{a}) \cdot (\mathbf{x} - \mathbf{b}) + \frac{1}{4}(\mathbf{a} - \mathbf{b}) \cdot (\mathbf{a} - \mathbf{b}) \\
&= (\mathbf{x} - \mathbf{a}) \cdot (\mathbf{x} - \mathbf{b}) + \left\|\frac{1}{2}(\mathbf{a} - \mathbf{b})\right\|^2
\end{aligned}$$

Since $R^2 = c^2 + \|\frac{1}{2}(\mathbf{a} - \mathbf{b})\|^2$ we get

$$\|\mathbf{x} - \mathbf{m}\|^2 = (\mathbf{x} - \mathbf{a}) \cdot (\mathbf{x} - \mathbf{b}) + R^2 - c^2$$

We conclude that if $(\mathbf{x} - \mathbf{a})(\mathbf{x} - \mathbf{b}) = c^2$ then $\|\mathbf{x} - \mathbf{m}\|^2 = R^2$. That is, the equation $(\mathbf{x} - \mathbf{a})(\mathbf{x} - \mathbf{b}) = c^2$ defines a sphere with center $\mathbf{m}$ and radius R.

14 CALCULUS OF VECTOR-VALUED FUNCTIONS

14.1 Vector-Valued Functions (ET Section 13.1)

Preliminary Questions

1. Which one of the following does *not* parametrize a line?

(a) $\mathbf{r}_1(t) = \langle 8 - t, 2t, 3t \rangle$

(b) $\mathbf{r}_2(t) = t^3\mathbf{i} - 7t^3\mathbf{j} + t^3\mathbf{k}$

(c) $\mathbf{r}_3(t) = \langle 8 - 4t^3, 2 + 5t^2, 9t^3 \rangle$

SOLUTION

(a) This is a parametrization of the line passing through the point $(8, 0, 0)$ in the direction parallel to the vector $\langle -1, 2, 3 \rangle$, since:

$$\langle 8 - t, 2t, 3t \rangle = \langle 8, 0, 0 \rangle + t \langle -1, 2, 3 \rangle$$

(b) Using the parameter $s = t^3$ we get:

$$\langle t^3\mathbf{i} - 7t^3\mathbf{j} + t^3\mathbf{k} \rangle = \langle s, -7s, s \rangle = s \langle 1, -7, 1 \rangle$$

This is a parametrization of the line through the origin, with the direction vector $\mathbf{v} = \langle -1, 7, 1 \rangle$.

(c) The parametrization $\langle 8 - 4t^3, 2 + 5t^2, 9t^3 \rangle$ does not parametrize a line. In particular, the points $(8, 2, 0)$ (at $t = 0$), $(4, 7, 9)$ (at $t = 1$), and $(-24, 22, 72)$ (at $t = 2$) are not colinear.

2. What is the projection of $\mathbf{r}(t) = t\mathbf{i} + t^4\mathbf{j} + e^t\mathbf{k}$ onto the xz-plane?

SOLUTION The projection of the path onto the xz-plane is the curve traced by $t\mathbf{i} + e^t\mathbf{k} = \langle t, 0, e^t \rangle$. This is the curve $z = e^x$ in the xz-plane.

3. Which projection of $\langle \cos t, \cos 2t, \sin t \rangle$ is a circle?

SOLUTION The parametric equations are

$$x = \cos t, \quad y = \cos 2t, \quad z = \sin t$$

The projection onto the xz-plane is $\langle \cos t, 0, \sin t \rangle$. Since $x^2 + z^2 = \cos^2 t + \sin^2 t = 1$, the projection is a circle in the xz-plane. The projection onto the xy-plane is traced by the curve $\langle \cos t, \cos 2t, 0 \rangle$. Therefore, $x = \cos t$ and $y = \cos 2t$. We express y in terms of x:

$$y = \cos 2t = 2\cos^2 t - 1 = 2x^2 - 1$$

The projection onto the xy-plane is a parabola. The projection onto the yz-plane is the curve $\langle 0, \cos 2t, \sin t \rangle$. Hence $y = \cos 2t$ and $z = \sin t$. We find y as a function of z:

$$y = \cos 2t = 1 - 2\sin^2 t = 1 - 2z^2$$

The projection onto the yz-plane is again a parabola.

4. What is the center of the circle with parametrization

$$\mathbf{r}(t) = (-2 + \cos t)\mathbf{i} + 2\mathbf{j} + (3 - \sin t)\mathbf{k}?$$

SOLUTION The parametric equations are

$$x = -2 + \cos t, \quad y = 2, \quad z = 3 - \sin t$$

Therefore, the curve is contained in the plane $y = 2$, and the following holds:

$$(x + 2)^2 + (z - 3)^2 = \cos^2 t + \sin^2 t = 1$$

We conclude that the curve $\mathbf{r}(t)$ is the circle of radius 1 in the plane $y = 2$ centered at the point $(-2, 2, 3)$.

5. How do the paths $\mathbf{r}_1(t) = \langle \cos t, \sin t \rangle$ and $\mathbf{r}_2(t) = \langle \sin t, \cos t \rangle$ around the unit circle differ?

SOLUTION The two paths describe the unit circle. However, as t increases from 0 to 2π, the point on the path $\sin t\mathbf{i} + \cos t\mathbf{j}$ moves in a clockwise direction, whereas the point on the path $\cos t\mathbf{i} + \sin t\mathbf{j}$ moves in a counterclockwise direction.

6. Which three of the following vector-valued functions parametrize the same space curve?

(a) $(-2 + \cos t)\mathbf{i} + 9\mathbf{j} + (3 - \sin t)\mathbf{k}$

(b) $(2 + \cos t)\mathbf{i} - 9\mathbf{j} + (-3 - \sin t)\mathbf{k}$

(c) $(-2 + \cos 3t)\mathbf{i} + 9\mathbf{j} + (3 - \sin 3t)\mathbf{k}$

(d) $(-2 - \cos t)\mathbf{i} + 9\mathbf{j} + (3 + \sin t)\mathbf{k}$

(e) $(2 + \cos t)\mathbf{i} + 9\mathbf{j} + (3 + \sin t)\mathbf{k}$

SOLUTION All the curves except for (b) lie in the vertical plane $y = 9$. We identify each one of the curves (a), (c), (d) and (e).

(a) The parametric equations are:

$$x = -2 + \cos t, \quad y = 9, \quad z = 3 - \sin t$$

Hence,

$$(x+2)^2 + (z-3)^2 = (\cos t)^2 + (-\sin t)^2 = 1$$

This is the circle of radius 1 in the plane $y = 9$, centered at $(-2, 9, 3)$.

(c) The parametric equations are:

$$x = -2 + \cos 3t, \quad y = 9, \quad z = 3 - \sin 3t$$

Hence,

$$(x+2)^2 + (z-3)^2 = (\cos 3t)^2 + (-\sin 3t)^2 = 1$$

This is the circle of radius 1 in the plane $y = 9$, centered at $(-2, 9, 3)$.

(d) In this curve we have:

$$x = -2 - \cos t, \quad y = 9, \quad z = 3 + \sin t$$

Hence,

$$(x+2)^2 + (z-3)^2 = (-\cos t)^2 + (\sin t)^2 = 1$$

Again, the circle of radius 1 in the plane $y = 9$, centered at $(-2, 9, 3)$.

(e) In this parametrization we have:

$$x = 2 + \cos t, \quad y = 9, \quad z = 3 + \sin t$$

Hence,

$$(x-2)^2 + (z-3)^2 = (\cos t)^2 + (\sin t)^2 = 1$$

This is the circle of radius 1 in the plane $y = 9$, centered at $(2, 9, 3)$.

We conclude that (a), (c) and (d) parametrize the same circle whereas (b) and (e) are different curves.

Exercises

1. What is the domain of $\mathbf{r}(t) = e^t\mathbf{i} + \frac{1}{t}\mathbf{j} + (t+1)^{-3}\mathbf{k}$?

SOLUTION $\mathbf{r}(t)$ is defined for $t \neq 0$ and $t \neq -1$, hence the domain of $\mathbf{r}(t)$ is:

$$D = \{t \in \mathbf{R} : t \neq 0, t \neq -1\}$$

3. Find a vector parametrization of the line through $P = (3, -5, 7)$ in the direction $\mathbf{v} = \langle 3, 0, 1 \rangle$.

SOLUTION We use the vector parametrization of the line to obtain:

$$\mathbf{r}(t) = \overrightarrow{OP} + t\mathbf{v} = \langle 3, -5, 7 \rangle + t\langle 3, 0, 1 \rangle = \langle 3 + 3t, -5, 7 + t \rangle$$

or in the form:

$$\mathbf{r}(t) = (3 + 3t)\mathbf{i} - 5\mathbf{j} + (7 + t)\mathbf{k}$$

5. Match the space curves in Figure 8 with their projections onto the xy-plane in Figure 9.

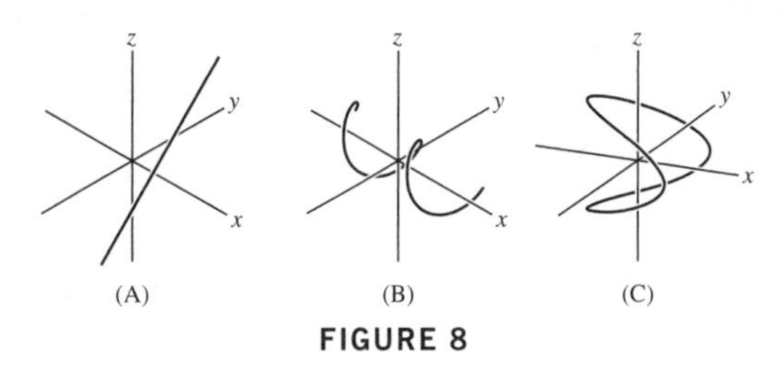

FIGURE 8

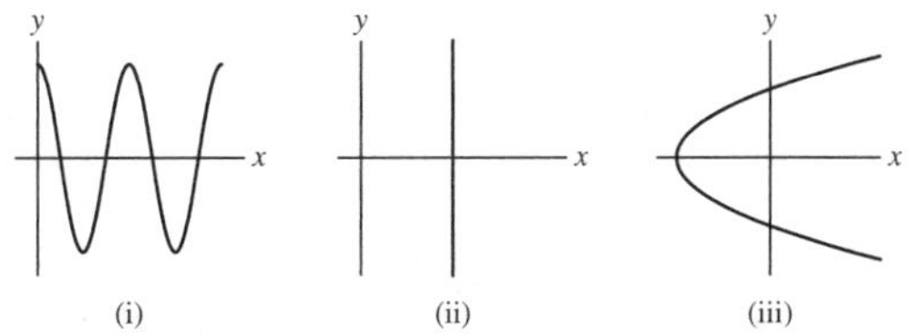

(i) (ii) (iii)

FIGURE 9

SOLUTION The projection of curve (C) onto the xy-plane is neither a segment nor a periodic wave. Hence, the correct projection is (iii), rather than the two other graphs. The projection of curve (A) onto the xy-plane is a vertical line, hence the corresponding projection is (ii). The projection of curve (B) onto the xy-plane is a periodic wave as illustrated in (i).

7. Match the vector-valued functions (a)–(f) with the space curves (i)–(vi) in Figure 10.

(a) $\mathbf{r}(t) = \langle t + 15, e^{0.08t} \cos t, e^{0.08t} \sin t \rangle$

(b) $\mathbf{r}(t) = \langle \cos t, \sin t, \sin 12t \rangle$

(c) $\mathbf{r}(t) = \left\langle t, t, \dfrac{25t}{1+t^2} \right\rangle$

(d) $\mathbf{r}(t) = \langle \cos^3 t, \sin^3 t, \sin 2t \rangle$

(e) $\mathbf{r}(t) = \langle t, t^2, 2t \rangle$

(f) $\mathbf{r}(t) = \langle \cos t, \sin t, \cos t \sin 12t \rangle$

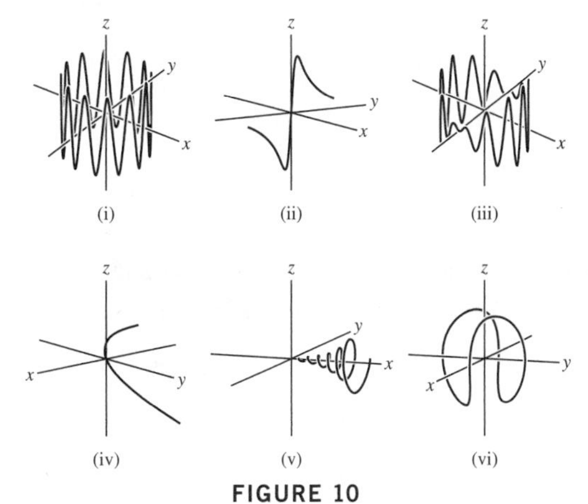

FIGURE 10

SOLUTION

(a) (v) **(b)** (i) **(c)** (ii)

(d) (vi) **(e)** (iv) **(f)** (iii)

9. Match the space curves (A)–(C) in Figure 11 with their projections (i)–(iii) onto the xy-plane.

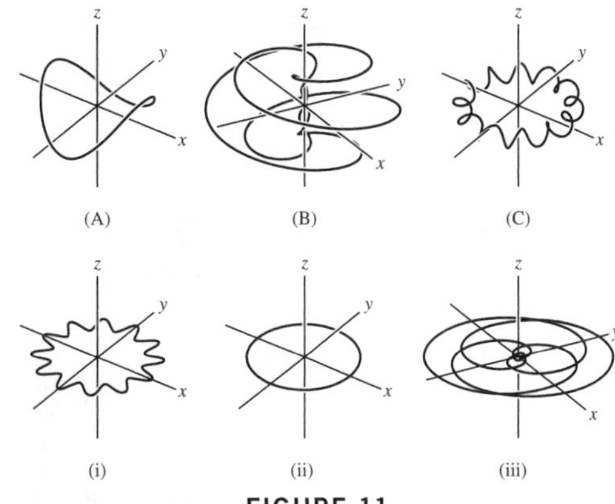

FIGURE 11

SOLUTION Observing the curves and the projections onto the xy-plane we conclude that: Projection (i) corresponds to curve (C); Projection (ii) corresponds to curve (A); Projection (iii) corresponds to curve (B).

In Exercises 10–13, the function $\mathbf{r}(t)$ *traces a circle. Determine the radius, center, and plane containing the circle.*

11. $\mathbf{r}(t) = 7\mathbf{i} + (12\cos t)\mathbf{j} + (12\sin t)\mathbf{k}$

SOLUTION We have:

$$x(t) = 7, \quad y(t) = 12\cos t, \quad z(t) = 12\sin t$$

Hence,

$$y(t)^2 + z(t)^2 = 144\cos^2 t + 144\sin^2 t = 144\left(\cos^2 t + \sin^2 t\right) = 144$$

This is the equation of a circle in the vertical plane $x = 7$. The circle is centered at the point $(7, 0, 0)$ and its radius is $\sqrt{144} = 12$.

13. $\mathbf{r}(t) = \langle 6 + 3\sin t, 9, 4 + 3\cos t\rangle$

SOLUTION Since $y(t) = 9$ the curve is contained in the vertical plane $y = 9$. By the given equations, $x(t) = 6 + 3\sin t$ and $z = 4 + 3\cos t$, hence:

$$\left(\frac{x-6}{3}\right)^2 + \left(\frac{z-4}{3}\right)^2 = \sin^2 t + \cos^2 t = 1$$

We conclude that the function traces a circle in the vertical plane $y = 9$, centered at the point $(6, 9, 4)$ and with radius 3.

15. Do either of $P = (4, 11, 20)$ or $Q = (-1, 6, 16)$ lie on the curve $\mathbf{r}(t) = \left\langle 1 + t, 2 + t^2, t^4\right\rangle$?

SOLUTION The point $P = (4, 11, 20)$ lies on the curve $\mathbf{r}(t) = \left\langle 1 + t, 2 + t^2, t^4\right\rangle$ if there exists a value of t such that $\overrightarrow{OP} = \mathbf{r}(t)$. That is,

$$\langle 4, 11, 20\rangle = \left\langle 1 + t, 2 + t^2, t^4\right\rangle$$

Equating like components we get:

$$\begin{aligned} 1 + t &= 4 \\ 2 + t^2 &= 11 \\ t^4 &= 20 \end{aligned}$$

The first equation implies that $t = 3$, but this value does not satisfy the third equation. We conclude that P does not lie on the curve. The point $Q = (-1, 6, 16)$ lies on the curve if there exists a value of t such that:

$$\langle -1, 6, 16\rangle = \left\langle 1 + t, 2 + t^2, t^4\right\rangle$$

or equivalently:

$$\begin{aligned} 1 + t &= -1 \\ 2 + t^2 &= 6 \\ t^4 &= 16 \end{aligned}$$

These equations have the solution $t = -2$, hence $Q = (-1, 6, 16)$ lies on the curve.

17. Find the points where the path $\mathbf{r}(t) = \langle \sin t, \cos t, \sin t\cos 2t\rangle$ intersects the xy-plane.

SOLUTION The curve intersects the xy-plane at the points where $z = 0$. That is, $\sin t\cos 2t = 0$ and so either $\sin t = 0$ or $\cos 2t = 0$. The solutions are, thus:

$$t = \pi k \text{ or } t = \frac{\pi}{4} + \frac{\pi k}{2}, \quad k = 0, \pm 1, \pm 2, \ldots$$

The values $t = \pi k$ yield the points: $(\sin \pi k, \cos \pi k, 0) = \left(0, (-1)^k, 0\right)$. The values $t = \frac{\pi}{4} + \frac{\pi k}{2}$ yield the points:

$$k = 0: \left(\sin\frac{\pi}{4}, \cos\frac{\pi}{4}, 0\right) = \left(\frac{1}{\sqrt{2}}, \frac{1}{\sqrt{2}}, 0\right)$$

$$k = 1: \left(\sin\frac{3\pi}{4}, \cos\frac{3\pi}{4}, 0\right) = \left(\frac{1}{\sqrt{2}}, -\frac{1}{\sqrt{2}}, 0\right)$$

$$k = 2: \left(\sin\frac{5\pi}{4}, \cos\frac{5\pi}{4}, 0\right) = \left(-\frac{1}{\sqrt{2}}, -\frac{1}{\sqrt{2}}, 0\right)$$

$$k = 3: \left(\sin\frac{7\pi}{4}, \cos\frac{7\pi}{4}, 0\right) = \left(-\frac{1}{\sqrt{2}}, \frac{1}{\sqrt{2}}, 0\right)$$

(Other values of k do not provide new points). We conclude that the curve intersects the xy-plane at the following points: $(0, 1, 0)$, $(0, -1, 0)$, $\left(\frac{1}{\sqrt{2}}, \frac{1}{\sqrt{2}}, 0\right)$, $\left(\frac{1}{\sqrt{2}}, -\frac{1}{\sqrt{2}}, 0\right)$, $\left(-\frac{1}{\sqrt{2}}, -\frac{1}{\sqrt{2}}, 0\right)$, $\left(-\frac{1}{\sqrt{2}}, \frac{1}{\sqrt{2}}, 0\right)$

19. Find a parametrization of the curve in Exercise 18 using trigonometric functions.

SOLUTION The curve in Exercise 18 is the intersection of the surfaces $y^2 - z^2 = x - 2$, $y^2 + z^2 = 9$. The circle $y^2 + z^2 = 9$ is parametrized by $y = 3\cos t$, $z = 3\sin t$. Substituting in the first equation and using the identity $\cos^2 t - \sin^2 t = \cos 2t$, gives:

$$x = 2 + y^2 - z^2 = 2 + (3\cos t)^2 - (3\sin t)^2 = 2 + 9\left(\cos^2 t - \sin^2 t\right) = 2 + 9\cos 2t$$

We obtain the following trigonometric parametrization:

$$\mathbf{r}(t) = \langle 2 + 9\cos 2t, 3\cos t, 3\sin t\rangle$$

21. Show that any point on $x^2 + y^2 = z^2$ can be written in the form $(z\cos\theta, z\sin\theta, z)$ for some θ. Use this to find a parametrization of Viviani's curve (Exercise 20) with θ as parameter.

SOLUTION We first verify that $x = z\cos\theta$, $y = z\sin\theta$, and $z = z$ satisfy the equation of the surface:

$$x^2 + y^2 = z^2\cos^2\theta + z^2\sin^2\theta = z^2\left(\cos^2\theta + \sin^2\theta\right) = z^2$$

We now show that if (x, y, z) satisfies $x^2 + y^2 = z^2$, then there exists a value of θ such that $x = z\cos\theta$, $y = z\sin\theta$. Since $x^2 + y^2 = z^2$, we have $|x| \le |z|$ and $|y| \le |z|$. If $z = 0$, then also $x = y = 0$ and any value of θ is adequate. If $z \ne 0$ then $\|\frac{x}{z}\| \le 1$ and $\|\frac{y}{z}\| \le 1$, hence there exists θ_0 such that $\frac{x}{z} = \cos\theta_0$. Hence,

$$\frac{y}{z} = \pm\sqrt{\frac{z^2 - x^2}{z^2}} = \pm\sqrt{1 - \left(\frac{x}{z}\right)^2} = \pm\sqrt{1 - \cos^2\theta_0} = \pm\sin\theta_0$$

If $\frac{x}{z}$ and $\frac{y}{z}$ are both positive, we choose θ_0 such that $0 < \theta_0 < \frac{\pi}{2}$. If $\frac{x}{z} > 0$ and $\frac{y}{z} < 0$ we choose θ_0 such that $\frac{3\pi}{2} < \theta_0 < 2\pi$. If $\frac{x}{z} < 0$ and $\frac{y}{z} < 0$ we choose θ_0 such that $\pi < \theta_0 < \frac{3\pi}{2}$, and if $\frac{x}{z} < 0$ and $\frac{y}{z} > 0$ we choose θ_0 such that $\frac{\pi}{2} < \theta_0 < \pi$. In either case we can represent the points on the surface as required. Viviani's curve is the intersection of the surfaces $x^2 + y^2 = z^2$ and $x = z^2$. The points on these surfaces are of the form:

$$\begin{aligned} x^2 + y^2 = z^2&: \quad (z\cos\theta, z\sin\theta, z) \\ x = z^2&: \quad (z^2, y, z) \end{aligned} \tag{1}$$

The points (x, y, z) on the intersection curve must satisfy the following equations:

$$\begin{cases} z^2 = z\cos\theta \\ y = z\sin\theta \end{cases}$$

The first equation implies that $z = 0$ or $z = \cos\theta$. The second equation implies that $y = 0$ or $y = \cos\theta\sin\theta = \frac{1}{2}\sin 2\theta$. The x coordinate is obtained by substituting $z = \cos\theta$ in $x = z\cos\theta$ (or in $x = z^2$). That is, $x = \cos^2\theta$. We obtain the following vector parametrization of the curve:

$$\mathbf{r}(t) = \left\langle \cos^2\theta, \frac{1}{2}\sin 2\theta, \cos\theta \right\rangle$$

23. CAS Use sine and cosine to parametrize the intersection of the surfaces $x^2 + y^2 = 1$ and $z = 4x^2$, and plot this curve using a CAS (Figure 13).

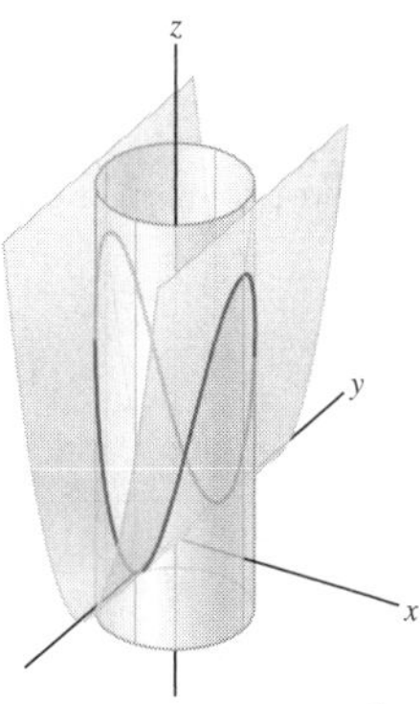

FIGURE 13 Intersection of the surfaces $x^2 + y^2 = 1$ and $z = 4x^2$.

SOLUTION The points on the cylinder $x^2 + y^2 = 1$ and on the parabolic cylinder $z = 4x^2$ can be written in the form:

$$x^2 + y^2 = 1: \quad (\cos t, \sin t, z)$$
$$z = 4x^2: \quad (x, y, 4x^2)$$

The points (x, y, z) on the intersection curve must satisfy the following equations:

$$\begin{array}{l} x = \cos t \\ y = \sin t \\ z = 4x^2 \end{array} \quad \Rightarrow \quad x = \cos t, \; y = \sin t, \; z = 4\cos^2 t$$

We obtain the vector parametrization:

$$\mathbf{r}(t) = \langle \cos t, \sin t, 4\cos^2 t \rangle$$

Using the CAS we obtain the following curve:

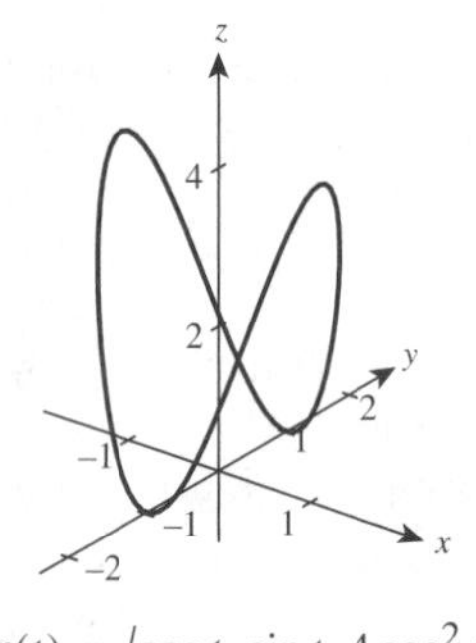

$\mathbf{r}(t) = \langle \cos t, \sin t, 4\cos^2 t \rangle$

In Exercises 25–34, find a parametrization of the curve.

25. The vertical line passing through the point $(3, 2, 0)$

SOLUTION The points of the vertical line passing through the point $(3, 2, 0)$ can be written as $(3, 2, z)$. Using $z = t$ as parameter we get the following parametrization:

$$\mathbf{r}(t) = \langle 3, 2, t \rangle, \quad -\infty < t < \infty$$

27. The line through the origin whose projection on the xy-plane is a line of slope 3 and on the yz-plane is a line of slope 5 (i.e., $\Delta y/\Delta z = 5$)

SOLUTION We denote by (x, y, z) the points on the line. The projection of the line on the xy-plane is the line through the origin having slope 3, that is the line $y = 3x$ in the xy-plane. The projection of the line on the yz-plane is the line through the origin with slope 5, that is the line $z = 5y$. Thus, the points on the desired line satisfy the following equalities:

$$\begin{array}{l} y = 3x \\ z = 5y \end{array} \quad \Rightarrow \quad y = 3x, \; z = 5 \cdot 3x = 15x$$

We conclude that the points on the line are all the points in the form $(x, 3x, 15x)$. Using $x = t$ as parameter we obtain the following parametrization:

$$\mathbf{r}(t) = \langle t, 3t, 15t \rangle.$$

29. The circle of radius 2 with center $(1, 2, 5)$ in a plane parallel to the yz-plane

SOLUTION The circle is parallel to the yz-plane and centered at $(1, 2, 5)$, hence the x-coordinates of the points on the circle are $x = 1$. The projection of the circle on the yz-plane is a circle of radius 2 centered at $(2, 5)$. This circle is parametrized by:

$$y = 2 + 2\cos t, \quad z = 5 + 2\sin t$$

We conclude that the points on the required circle can be written as $(1, 2 + 2\cos t, 5 + 2\sin t)$. This gives the following parametrization:

$$\mathbf{r}(t) = \langle 1, 2 + 2\cos t, 5 + 2\sin t \rangle.$$

31. The intersection of the plane $y = \frac{1}{2}$ with the sphere $x^2 + y^2 + z^2 = 1$

SOLUTION Substituting $y = \frac{1}{2}$ in the equation of the sphere gives:

$$x^2 + \left(\frac{1}{2}\right)^2 + z^2 = 1 \quad \Rightarrow \quad x^2 + z^2 = \frac{3}{4}$$

This circle in the horizontal plane $y = \frac{1}{2}$ has the parametrization $x = \frac{\sqrt{3}}{2}\cos t$, $z = \frac{\sqrt{3}}{2}\sin t$. Therefore, the points on the intersection of the plane $y = \frac{1}{2}$ and the sphere $x^2 + y^2 + z^2 = 1$, can be written in the form $\left(\frac{\sqrt{3}}{2}\cos t, \frac{1}{2}, \frac{\sqrt{3}}{2}\sin t\right)$, yielding the following parametrization:

$$\mathbf{r}(t) = \left\langle \frac{\sqrt{3}}{2}\cos t, \frac{1}{2}, \frac{\sqrt{3}}{2}\sin t \right\rangle.$$

33. The ellipse $\left(\frac{x}{2}\right)^2 + \left(\frac{z}{3}\right)^2 = 1$ in the xz-plane, translated to have center $(3, 1, 5)$ [Figure 14(A)]

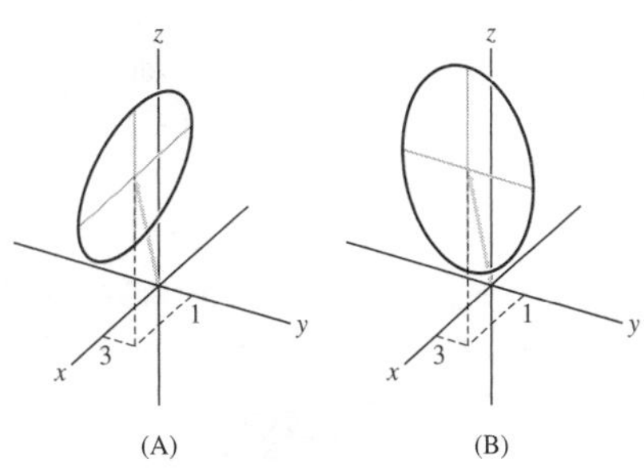

FIGURE 14 The ellipses described in Exercise 33 and 34.

SOLUTION The translated ellipse is in the vertical plane $y = 1$, hence the y-coordinate of the points on this ellipse is $y = 1$. The x and z coordinates satisfy the equation of the ellipse:

$$\left(\frac{x-3}{2}\right)^2 + \left(\frac{z-5}{3}\right)^2 = 1.$$

This ellipse is parametrized by the following equations:

$$x = 3 + 2\cos t, \quad z = 5 + 3\sin t.$$

Therefore, the points on the translated ellipse can be written as $(3 + 2\cos t, 1, 5 + 3\sin t)$. This gives the following parametrization:

$$\mathbf{r}(t) = \langle 3 + 2\cos t, 1, 5 + 3\sin t \rangle.$$

In Exercises 35–37, assume that two paths $\mathbf{r}_1(t)$ and $\mathbf{r}_2(t)$ intersect if there is a point P lying on both curves. We say that $\mathbf{r}_1(t)$ and $\mathbf{r}_2(t)$ collide *if $\mathbf{r}_1(t_0) = \mathbf{r}_2(t_0)$ at some time t_0.*

35. Which of the following are true?

(a) If $\mathbf{r}_1$ and $\mathbf{r}_2$ intersect, then they collide.

(b) If $\mathbf{r}_1$ and $\mathbf{r}_2$ collide, then they intersect.

(c) Intersection depends only on the underlying curves traced by $\mathbf{r}_1$ and $\mathbf{r}_2$ but collision depends on the actual parametrizations.

SOLUTION

(a) This statement is wrong. $\mathbf{r}_1(t)$ and $\mathbf{r}_2(t)$ may intersect but the point of intersection may correspond to different values of the parameters in the two curves, as illustrated in the following example:

$$\mathbf{r}_1(t) = \langle \cos t, \sin t \rangle \text{ (the unit circle)}$$

$$\mathbf{r}_2(s) = \langle s, 1 \rangle \text{ (the horizontal line } y = 1\text{)}$$

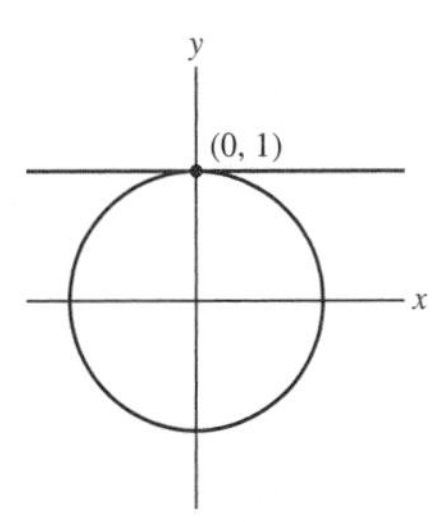

The point of intersection $(0, 1)$ corresponds to $t = \frac{\pi}{2}$ and $s = 0$.

(b) This statement is true. If $\mathbf{r}_1(t_0) = \mathbf{r}_2(t_0)$, then the head of the vector $\mathbf{r}_1(t_0)$ (or $\mathbf{r}_2(t_0)$) is a point of intersection of the two curves.

(c) The statement is true. Intersection is a geometric property of the curves and it is independent of the parametrization we choose for the curves. Collision depends on the actual parametrization. Notice that if we parametrize the line $y = 1$ in the example given in part (a) by $\mathbf{r}_3(s) = \langle s - \frac{\pi}{2}, 1\rangle$, then $\mathbf{r}_1\left(\frac{\pi}{2}\right) = \mathbf{r}_3\left(\frac{\pi}{2}\right)$ hence the two paths collide.

37. Determine whether $\mathbf{r}_1$ and $\mathbf{r}_2$ collide or intersect:

$$\mathbf{r}_1(t) = \langle t, t^2, t^3\rangle, \qquad \mathbf{r}_2(t) = \langle 4t + 6, 4t^2, 7 - t\rangle$$

SOLUTION The two paths collide if there exists a value of t such that:

$$\langle t, t^2, t^3\rangle = \langle 4t + 6, 4t^2, 7 - t\rangle$$

Equating corresponding components we obtain the following equations:

$$\begin{aligned} t &= 4t + 6 \\ t^2 &= 4t^2 \\ t^3 &= 7 - t \end{aligned}$$

The second equation implies that $t = 0$, but this value does not satisfy the other equations. Therefore, the equations have no solution, which means that the paths do not collide. The two paths intersect if there exist values of t and s such that:

$$\langle t, t^2, t^3\rangle = \langle 4s + 6, 4s^2, 7 - s\rangle$$

Or equivalently:

$$\begin{aligned} t &= 4s + 6 \\ t^2 &= 4s^2 \\ t^3 &= 7 - s \end{aligned} \qquad (1)$$

The second equation implies that $t_1 = 2s$ or $t_2 = -2s$. Substituting $t_1 = 2s$ and $t_2 = -2s$ in the first equation gives:

$$t_1 = 2s : 2s = 4s + 6 \quad\Rightarrow\quad 2s = -6 \quad\Rightarrow\quad s_1 = -3$$

$$t_2 = -2s : -2s = 4s + 6 \quad\Rightarrow\quad 6s = -6 \quad\Rightarrow\quad s_2 = -1$$

The solutions of the first two equations are thus

$$(t_1, s_1) = (-6, -3); \qquad (t_2, s_2) = (2, -1)$$

(t_1, s_1) does not satisfy the third equation whereas (t_2, s_2) does. We conclude that the equations in (1) have a solution $t = 2$, $s = -1$, hence the two paths intersect.

Further Insights and Challenges

39. Find the maximum height above the xy-plane of a point on $\mathbf{r}(t) = \langle e^t, \sin t, t(4 - t)\rangle$.

SOLUTION The height of a point is the value of the z-coordinate of the point. Therefore we need to maximize the function $z = t(4 - t)$. $z(t)$ is a quadratic function having the roots $t = 0$ and $t = 4$, hence the maximum value is obtained at the midpoint of the interval $0 \le t \le 4$, that is at $t = 2$. The corresponding value of z is:

$$z_{\max} = z(2) = 2(4 - 2) = 4$$

The point of maximum height is, thus,

$$(e^2, \sin 2, 4) \approx (7.39, 0.91, 4)$$

41. CAS Now reprove the result of Exercise 40 using vector geometry. Assume that the cylinder has equation $x^2 + y^2 = r^2$ and the plane has equation $z = ax + by$.

(a) Show that the upper and lower spheres in Figure 15 have centers

$$C_1 = \left(0, 0, r\sqrt{a^2 + b^2 + 1}\right)$$

$$C_2 = \left(0, 0, -r\sqrt{a^2 + b^2 + 1}\right)$$

(b) Show that the points where the plane is tangent to the sphere are

$$F_1 = \frac{r}{\sqrt{a^2+b^2+1}}\langle a, b, a^2+b^2\rangle$$

$$F_2 = \frac{-r}{\sqrt{a^2+b^2+1}}\langle a, b, a^2+b^2\rangle$$

Hint: Show that $\overline{C_1F_1}$ and $\overline{C_2F_2}$ have length r and are orthogonal to the plane.

(c) Verify, with the aid of a computer algebra system, that Eq. (2) holds with $K = 2r\sqrt{a^2+b^2+1}$. To simplify the algebra, observe that since a and b are arbitrary, it suffices to verify Eq. (2) for the point $P = (r, 0, ar)$.

SOLUTION

(a) and (b) Since F_1 is the tangency point of the sphere and the plane, the radius to F_1 is orthogonal to the plane. Therefore to show that the center of the sphere is at C_1 and the tangency point is the given point we must show that:

$$\|\overrightarrow{C_1F_1}\| = r \tag{1}$$

$$\overrightarrow{C_1F_1} \text{ is orthogonal to the plane.} \tag{2}$$

We compute the vector $\overrightarrow{C_1F_1}$:

$$\overrightarrow{C_1F_1} = \left\langle \frac{ra}{\sqrt{a^2+b^2+1}}, \frac{rb}{\sqrt{a^2+b^2+1}}, \frac{r(a^2+b^2)}{\sqrt{a^2+b^2+1}} - r\sqrt{a^2+b^2+1}\right\rangle = \frac{r}{\sqrt{a^2+b^2+1}}\langle a, b, -1\rangle$$

Hence,

$$\|\overrightarrow{C_1F_1}\| = \frac{r}{\sqrt{a^2+b^2+1}}\|\langle a, b, -1\rangle\| = \frac{r}{\sqrt{a^2+b^2+1}}\sqrt{a^2+b^2+(-1)^2} = r$$

We, thus, proved that (1) is satisfied. To show (2) we must show that $\overrightarrow{C_1F_1}$ is parallel to the normal vector $\langle a, b, -1\rangle$ to the plane $z = ax + by$ (i.e., $ax + by - z = 0$). The two vectors are parallel since by (1) $\overrightarrow{C_1F_1}$ is a constant multiple of $\langle a, b, -1\rangle$. In a similar manner one can show (1) and (2) for the vector $\overrightarrow{C_2F_2}$.

(c) This is an extremely challenging problem. As suggested in the book, we use $P = (r, 0, ar)$, and we also use the expressions for F_1 and F_2 as given above. This gives us:

$$PF_1 = \sqrt{\left(1 + 2a^2 + b^2 - 2a\sqrt{1+a^2+b^2}\right)r^2}$$

$$PF_2 = \sqrt{\left(1 + 2a^2 + b^2 + 2a\sqrt{1+a^2+b^2}\right)r^2}$$

Their sum is not very inspiring:

$$PF_1 + PF_2 = \sqrt{\left(1 + 2a^2 + b^2 - 2a\sqrt{1+a^2+b^2}\right)r^2} + \sqrt{\left(1 + 2a^2 + b^2 + 2a\sqrt{1+a^2+b^2}\right)r^2}$$

Let us look, instead, at $(PF_1 + PF_2)^2$, and show that this is equal to K^2. Since everything is positive, this will imply that $PF_1 + PF_2 = K$, as desired.

$$\begin{aligned}(PF_1 + PF_2)^2 &= 2r^2 + 4a^2r^2 + 2b^2r^2 + 2\sqrt{r^4 + 2b^2r^4 + b^4r^4}\\ &= 2r^2 + 4a^2r^2 + 2b^2r^2 + 2(1+b^2)r^2 = 4r^2(1+a^2+b^2) = K^2\end{aligned}$$

14.2 Calculus of Vector-Valued Functions (ET Section 13.2)

Preliminary Questions

1. State the three forms of the Product Rule for vector-valued functions.

SOLUTION The Product Rule for scalar multiple $f(t)$ of a vector-valued function $\mathbf{r}(t)$ states that:

$$\frac{d}{dt}f(t)\mathbf{r}(t) = f(t)\mathbf{r}'(t) + f'(t)\mathbf{r}(t)$$

The Product Rule for dot products states that:

$$\frac{d}{dt}\mathbf{r}_1(t)\cdot\mathbf{r}_2(t) = \mathbf{r}_1(t)\cdot\mathbf{r}_2'(t) + \mathbf{r}_1'(t)\cdot\mathbf{r}_2(t)$$

Finally, the Product Rule for cross product is

$$\frac{d}{dt}\mathbf{r}_1(t) \times \mathbf{r}_2(t) = \mathbf{r}_1(t) \times \mathbf{r}_2'(t) + \mathbf{r}_1'(t) \times \mathbf{r}_2(t).$$

In Questions 2–6, indicate whether true or false and if false, provide a correct statement.

2. The derivative of a vector-valued function is defined as the limit of the difference quotient, just as in the scalar-valued case.

SOLUTION The statement is true. The derivative of a vector-valued function $\mathbf{r}(t)$ is defined a limit of the difference quotient:

$$\mathbf{r}'(t) = \lim_{t \to 0} \frac{\mathbf{r}(t+h) - \mathbf{r}(t)}{h}$$

in the same way as in the scalar-valued case.

3. There are two Chain Rules for vector-valued functions, one for the composite of two vector-valued functions and one for the composite of a vector-valued and scalar-valued function.

SOLUTION This statement is false. A vector-valued function $\mathbf{r}(t)$ is a function whose domain is a set of real numbers and whose range consists of position vectors. Therefore, if $\mathbf{r}_1(t)$ and $\mathbf{r}_2(t)$ are vector-valued functions, the composition "$(\mathbf{r}_1 \cdot \mathbf{r}_2)(t) = \mathbf{r}_1(\mathbf{r}_2(t))$" has no meaning since $\mathbf{r}_2(t)$ is a vector and not a real number. However, for a scalar-valued function $f(t)$, the composition $\mathbf{r}(f(t))$ has a meaning, and there is a Chain Rule for differentiability of this vector-valued function.

4. The terms "velocity vector" and "tangent vector" for a path $\mathbf{r}(t)$ mean one and the same thing.

SOLUTION This statement is true.

5. The derivative of a vector-valued function is the slope of the tangent line, just as in the scalar case.

SOLUTION The statement is false. The derivative of a vector-valued function is again a vector-valued function, hence it cannot be the slope of the tangent line (which is a scalar). However, the derivative, $\mathbf{r}'(t_0)$ is the direction vector of the tangent line to the curve traced by $\mathbf{r}(t)$, at $\mathbf{r}(t_0)$.

6. The derivative of the cross product is the cross product of the derivatives.

SOLUTION The statement is false, since usually,

$$\frac{d}{dt}\mathbf{r}_1(t) \times \mathbf{r}_2(t) \neq \mathbf{r}_1'(t) \times \mathbf{r}_2'(t)$$

The correct statement is the Product Rule for Cross Products. That is,

$$\frac{d}{dt}\mathbf{r}_1(t) \times \mathbf{r}_2(t) = \mathbf{r}_1(t) \times \mathbf{r}_2'(t) + \mathbf{r}_1'(t) \times \mathbf{r}_2(t)$$

7. State whether the following derivatives of vector-valued functions $\mathbf{r}_1(t)$ and $\mathbf{r}_2(t)$ are scalars or vectors:

(a) $\frac{d}{dt}\mathbf{r}_1(t)$ **(b)** $\frac{d}{dt}(\mathbf{r}_1(t) \cdot \mathbf{r}_2(t))$

(c) $\frac{d}{dt}(\mathbf{r}_1(t) \times \mathbf{r}_2(t))$

SOLUTION (a) vector, (b) scalar, (c) vector.

Exercises

In Exercises 1–4, evaluate the limit.

1. $\lim_{t \to 3} \left\langle t^2, 4t, \frac{1}{t} \right\rangle$

SOLUTION By the theorem on vector-valued limits we have:

$$\lim_{t \to 3} \left\langle t^2, 4t, \frac{1}{t} \right\rangle = \left\langle \lim_{t \to 3} t^2, \lim_{t \to 3} 4t, \lim_{t \to 3} \frac{1}{t} \right\rangle = \left\langle 9, 12, \frac{1}{3} \right\rangle.$$

3. $\lim_{t \to 0} e^{2t}\mathbf{i} + \ln(t+1)\mathbf{j} + 4\mathbf{k}$

SOLUTION Computing the limit of each component, we obtain:

$$\lim_{t\to 0}\left(e^{2t}\mathbf{i}+\ln(t+1)\mathbf{j}+4\mathbf{k}\right)=\left(\lim_{t\to 0}e^{2t}\right)\mathbf{i}+\left(\lim_{t\to 0}\ln(t+1)\right)\mathbf{j}+\left(\lim_{t\to 0}4\right)\mathbf{k}=e^0\mathbf{i}+(\ln 1)\mathbf{j}+4\mathbf{k}=\mathbf{i}+4\mathbf{k}$$

5. Evaluate $\lim_{h\to 0}\dfrac{\mathbf{r}(t+h)-\mathbf{r}(t)}{h}$ for $\mathbf{r}(t)=\left\langle t^{-1},\sin t,4\right\rangle$.

SOLUTION This limit is the derivative $\frac{d\mathbf{r}}{dt}$. Using componentwise differentiation yields:

$$\lim_{h\to 0}\frac{\mathbf{r}(t+h)-\mathbf{r}(t)}{h}=\frac{d\mathbf{r}}{dt}=\left\langle\frac{d}{dt}\left(t^{-1}\right),\frac{d}{dt}(\sin t),\frac{d}{dt}(4)\right\rangle=\left\langle-\frac{1}{t^2},\cos t,0\right\rangle.$$

In Exercises 7–14, compute the derivative.

7. $\mathbf{r}(t)=\left\langle t,t^2,t^3\right\rangle$

SOLUTION Using componentwise differentiation we get:

$$\frac{d\mathbf{r}}{dt}=\left\langle\frac{d}{dt}(t),\frac{d}{dt}(t^2),\frac{d}{dt}(t^3)\right\rangle=\left\langle 1,2t,3t^2\right\rangle$$

9. $\mathbf{w}(s)=\left\langle e^s,e^{-2s}\right\rangle$

SOLUTION Componentwise differentiation gives:

$$\mathbf{w}'(s)=\left\langle (e^s)',\left(e^{-2s}\right)'\right\rangle=\left\langle e^s,-2e^{-2s}\right\rangle$$

11. $\mathbf{r}(t)=\left\langle t-t^{-1},4t^2,8\right\rangle$

SOLUTION We compute the derivative of each component to obtain:

$$\mathbf{r}'(t)=\left\langle\left(t-t^{-1}\right)',\left(4t^2\right)',(8)'\right\rangle=\left\langle 1+t^{-2},8t,0\right\rangle$$

13. $\mathbf{a}(\theta)=(\cos 2\theta)\mathbf{i}+(\sin 2\theta)\mathbf{j}+(\sin 4\theta)\mathbf{k}$

SOLUTION Using componentwise differentiation yields:

$$\mathbf{a}'(\theta)=(\cos 2\theta)'\mathbf{i}+(\sin 2\theta)'\mathbf{j}+(\sin 4\theta)'\mathbf{k}=(-2\sin 2\theta)\,\mathbf{i}+(2\cos 2\theta)\,\mathbf{j}+(4\cos 4\theta)\,\mathbf{k}$$

15. Calculate $\mathbf{r}'(t)$ and $\mathbf{r}''(t)$ for $\mathbf{r}(t)=\left\langle t,t^2,t^3\right\rangle$.

SOLUTION We perform the differentiation componentwise to obtain:

$$\mathbf{r}'(t)=\left\langle (t)',\left(t^2\right)',\left(t^3\right)'\right\rangle=\left\langle 1,2t,3t^2\right\rangle$$

We now differentiate the derivative vector to find the second derivative:

$$\mathbf{r}''(t)=\frac{d}{dt}\left\langle 1,2t,3t^2\right\rangle=\langle 0,2,6t\rangle\,.$$

17. Sketch the curve $\mathbf{r}_1(t)=\left\langle t,t^2\right\rangle$ together with its tangent vector at $t=1$. Then do the same for $\mathbf{r}_2(t)=\left\langle t^3,t^6\right\rangle$.

SOLUTION Note that $\mathbf{r}_1{}'(t)=\langle 1,2t\rangle$ and so $\mathbf{r}_1{}'(1)=\langle 1,2\rangle$. The graph of $\mathbf{r}_1(t)$ satisfies $y=x^2$. Likewise, $\mathbf{r}_2{}'(t)=\left\langle 3t^2,6t^5\right\rangle$ and so $\mathbf{r}_2{}'(1)=\langle 3,6\rangle$. The graph of $\mathbf{r}_2(t)$ also satisfies $y=x^2$. Both graphs and tangent vectors are given here.

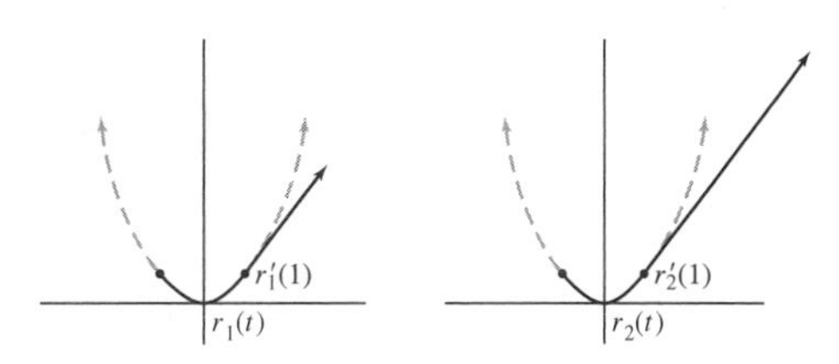

In Exercises 19–22, use the appropriate Product Rule to evaluate the derivative, where

$$\mathbf{r}_1(t)=\left\langle 8t,4,-t^3\right\rangle,\qquad \mathbf{r}_2(t)=\left\langle 0,e^t,-6\right\rangle$$

19. $\frac{d}{dt}(\mathbf{r}_1(t) \cdot \mathbf{r}_2(t))$

SOLUTION By the Product Rule for dot products we have:

$$\frac{d}{dt}\mathbf{r}_1 \cdot \mathbf{r}_2 = \mathbf{r}_1 \cdot \mathbf{r}_2{}' + \mathbf{r}_1{}' \cdot \mathbf{r}_2$$

We compute the derivatives of $\mathbf{r}_1$ and $\mathbf{r}_2$:

$$\mathbf{r}_1' = \frac{d}{dt}\langle 8t, 4, -t^3 \rangle = \langle 8, 0, -3t^2 \rangle$$

$$\mathbf{r}_2' = \frac{d}{dt}\langle 0, e^t, -6 \rangle = \langle 0, e^t, 0 \rangle$$

By (1) we have:

$$\frac{d}{dt}\mathbf{r}_1(t) \cdot \mathbf{r}_2(t) = \langle 8t, 4, -t^3 \rangle \cdot \langle 0, e^t, 0 \rangle + \langle 8, 0, -3t^2 \rangle \cdot \langle 0, e^t, -6 \rangle = 4e^t + 18t^2$$

21. $\frac{d}{dt}(\mathbf{r}_1(t) \times \mathbf{r}_2(t))$

SOLUTION We use the Product Rule for cross products:

$$\frac{d}{dt}\mathbf{r}_1 \times \mathbf{r}_2 = \mathbf{r}_1 \times \mathbf{r}_2' + \mathbf{r}_1' \times \mathbf{r}_2 = \langle 8t, 4, -t^3 \rangle \times \langle 0, e^t, 0 \rangle + \langle 8, 0, -3t^2 \rangle \times \langle 0, e^t, -6 \rangle$$

$$= \begin{vmatrix} \mathbf{i} & \mathbf{j} & \mathbf{k} \\ 8t & 4 & -t^3 \\ 0 & e^t & 0 \end{vmatrix} + \begin{vmatrix} \mathbf{i} & \mathbf{j} & \mathbf{k} \\ 8 & 0 & -3t^2 \\ 0 & e^t & -6 \end{vmatrix} = t^3 e^t \mathbf{i} + 8te^t \mathbf{k} + 3t^2 e^t \mathbf{i} + 48\mathbf{j} + 8e^t \mathbf{k}$$

$$= t^2 e^t (t+3)\mathbf{i} + 48\mathbf{j} + 8e^t(t+1)\mathbf{k} = \langle t^2 e^t (t+3), 48, 8e^t(t+1) \rangle$$

In Exercises 23–25, let

$$\mathbf{r}_1(t) = \langle t^2, t^3, 4t \rangle, \qquad \mathbf{r}_2(t) = \langle t^{-1}, 1+t, 2 \rangle$$

23. Let $F(t) = \mathbf{r}_1(t) \cdot \mathbf{r}_2(t)$.

(a) Calculate $F'(t)$ using the Product Rule.

(b) Expand the product $\mathbf{r}_1(t) \cdot \mathbf{r}_2(t)$ and differentiate. Compare with part (a).

SOLUTION

(a) By the Product Rule for dot products we have:

$$F'(t) = \mathbf{r}_1(t) \cdot \mathbf{r}_2'(t) + \mathbf{r}_1'(t) \cdot \mathbf{r}_2(t)$$

We compute the derivatives of $\mathbf{r}_1(t)$ and $\mathbf{r}_2(t)$:

$$\mathbf{r}_1'(t) = \frac{d}{dt}\langle t^2, t^3, 4t \rangle = \langle 2t, 3t^2, 4 \rangle$$

$$\mathbf{r}_2'(t) = \frac{d}{dt}\langle t^{-1}, 1+t, 2 \rangle = \langle -t^{-2}, 1, 0 \rangle$$

Thus,

$$F'(t) = \langle t^2, t^3, 4t \rangle \cdot \langle -t^{-2}, 1, 0 \rangle + \langle 2t, 3t^2, 4 \rangle \cdot \langle t^{-1}, 1+t, 2 \rangle$$

$$= \left(-1 + t^3 + 0\right) + \left(2 + 3t^2 + 3t^3 + 8\right) = 4t^3 + 3t^2 + 9$$

That is,

$$F'(t) = 4t^3 + 3t^2 + 9 \tag{1}$$

(b) We now first compute the product $\mathbf{r}_1(t) \cdot \mathbf{r}_2(t)$ and then differentiate the resulting function. This gives:

$$F(t) = \mathbf{r}_1(t) \cdot \mathbf{r}_2(t) = \langle t^2, t^3, 4t \rangle \cdot \langle t^{-1}, 1+t, 2 \rangle = t + t^3(1+t) + 8t = t^4 + t^3 + 9t$$

Differentiating $F(t)$ gives:

$$F'(t) = 4t^3 + 3t^2 + 9 \tag{2}$$

The derivatives in (1) and (2) are the same, as expected.

25. Find the rate of change of the angle θ between $\mathbf{r}_1(t)$ and $\mathbf{r}_2(t)$ at $t = 2$, assuming that t is measured in seconds.

SOLUTION Recall the formula for the dot product:

$$\mathbf{r}_1(t) \cdot \mathbf{r}_2(t) = \|\mathbf{r}_1(t)\| \|\mathbf{r}_2(t)\| \cos\theta$$

Thus,

$$\cos\theta = \frac{\mathbf{r}_1(t) \cdot \mathbf{r}_2(t)}{\|\mathbf{r}_1(t)\| \|\mathbf{r}_2(t)\|} = \frac{t + t^3 + t^4 + 8t}{\sqrt{t^4 + t^6 + 16t^2}\sqrt{t^{-2} + 5 + 2t + t^2}}$$
$$= \frac{t^4 + t^3 + 9t}{\sqrt{t^2 + t^4 + 16}\sqrt{1 + 5t^2 + 2t^3 + t^4}}$$
$$= (t^4 + t^3 + 9t)\left[(t^2 + t^4 + 16)(1 + 5t^2 + 2t^3 + t^4)\right]^{-1/2}$$

Taking the derivative, we find that (after a lot of work)

$$\frac{d}{dt}\cos\theta = \frac{144 + 48t^2 - 80t^3 - 36t^4 + 264t^5 - 6t^6 - 24t^8 + 5t^9}{\left(\left(16 + t^2 + t^4\right)\left(1 + 5t^2 + 2t^3 + t^4\right)\right)^{\frac{3}{2}}}$$

So, at $t = 2$, we get (after a lot of work)

$$\left.\frac{d}{dt}\cos\theta\right|_{t=2} = \frac{50}{159\sqrt{53}}$$

On the other hand, using the chain rule,

$$\frac{d}{dt}\cos\theta = -\sin\theta\frac{d}{dt}\theta$$

So, we have that

$$\frac{d}{dt}\theta = \frac{1}{-\sin\theta}\frac{d}{dt}\cos\theta$$

Thus, at $t = 2$,

$$\left.\frac{d}{dt}\theta\right|_{t=2} = \frac{1}{-\sin\theta}\left.\frac{d}{dt}\cos\theta\right|_{t=2}$$

We need only calculate $\sin\theta$ at $t = 2$. From above, we know that

$$\cos\theta = (t^4 + t^3 + 9t)\left[(t^2 + t^4 + 16)(1 + 5t^2 + 2t^3 + t^4)\right]^{-1/2}$$

so at $t = 2$, $\cos\theta = \frac{7}{\sqrt{53}}$. Since $\sin^2\theta + \cos^2\theta = 1$, we get that $\sin\theta = \frac{2}{\sqrt{53}}$. We can now conclude that

$$\left.\frac{d}{dt}\theta\right|_{t=2} = \frac{-\sqrt{53}}{2} \cdot \frac{50}{159\sqrt{53}} = \frac{-25}{159}$$

In Exercises 26–29, evaluate $\frac{d}{dt}\mathbf{r}(g(t))$ *using the Chain Rule.*

27. $\mathbf{r}(t) = \langle e^t, e^{2t}, 4\rangle, \quad g(t) = 4t + 9$

SOLUTION We first differentiate the two functions:

$$\mathbf{r}'(t) = \frac{d}{dt}\langle e^t, e^{2t}, 4\rangle = \langle e^t, 2e^{2t}, 0\rangle$$
$$g'(t) = \frac{d}{dt}(4t + 9) = 4$$

Using the Chain Rule we get:

$$\frac{d}{dt}\mathbf{r}(g(t)) = g'(t)\mathbf{r}'(g(t)) = 4\langle e^{4t+9}, 2e^{2(4t+9)}, 0\rangle = \langle 4e^{4t+9}, 8e^{8t+18}, 0\rangle$$

29. $\mathbf{r}(t) = \langle 3^t, \tan^{-1} t\rangle, \quad g(t) = \sin t$

SOLUTION We first compute the derivatives of the two functions:

$$\mathbf{r}'(t) = \frac{d}{dt}\langle 3^t, \tan^{-1} t\rangle = \left\langle 3^t \ln 3, \frac{1}{1+t^2}\right\rangle$$

$$g'(t) = \cos t$$

We now differentiate the composition function $\mathbf{r}(g(t))$ using the Chain Rule:

$$\frac{d}{dt}\mathbf{r}(g(t)) = g'(t)\mathbf{r}'(g(t)) = \cos t\left\langle 3^{\sin t}\ln 3, \frac{1}{1+\sin^2 t}\right\rangle = \left\langle 3^{\sin t}\cos t \ln 3, \frac{\cos t}{1+\sin^2 t}\right\rangle$$

31. Let $\mathbf{r}(t) = \langle t^2, 1-t, 4t\rangle$. Calculate the derivative of $\mathbf{r}(t)\cdot\mathbf{a}(t)$ at $t = 2$, assuming that $\mathbf{a}(2) = \langle 1, 3, 3\rangle$ and $\mathbf{a}'(2) = \langle -1, 4, 1\rangle$.

SOLUTION By the Product Rule for dot products we have

$$\frac{d}{dt}\mathbf{r}(t)\cdot\mathbf{a}(t) = \mathbf{r}(t)\cdot\mathbf{a}'(t) + \mathbf{r}'(t)\cdot\mathbf{a}(t)$$

At $t = 2$ we have

$$\frac{d}{dt}\mathbf{r}(t)\cdot\mathbf{a}(t)\bigg|_{t=2} = \mathbf{r}(2)\cdot\mathbf{a}'(2) + \mathbf{r}'(2)\cdot\mathbf{a}(2) \tag{1}$$

We compute the derivative $\mathbf{r}'(2)$:

$$\mathbf{r}'(t) = \frac{d}{dt}\langle t^2, 1-t, 4t\rangle = \langle 2t, -1, 4\rangle \quad\Rightarrow\quad \mathbf{r}'(2) = \langle 4, -1, 4\rangle \tag{2}$$

Also, $\mathbf{r}(2) = \langle 2^2, 1-2, 4\cdot 2\rangle = \langle 4, -1, 8\rangle$. Substituting the vectors in the equation above, we obtain:

$$\frac{d}{dt}\mathbf{r}(t)\cdot\mathbf{a}(t)\bigg|_{t=2} = \langle 4, -1, 8\rangle\cdot\langle -1, 4, 1\rangle + \langle 4, -1, 4\rangle\cdot\langle 1, 3, 3\rangle = (-4-4+8) + (4-3+12) = 13$$

The derivative of $\mathbf{r}(t)\cdot\mathbf{a}(t)$ at $t = 2$ is 13.

In Exercises 33–37, find a parametrization of the tangent line at the point indicated.

33. $\mathbf{r}(t) = \langle 1-t^2, 5t, 2t^3\rangle, \quad t = 2$

SOLUTION The tangent line is parametrized by:

$$\ell(t) = \mathbf{r}(2) + t\mathbf{r}'(2) \tag{1}$$

We compute the vectors in the above parametrization:

$$\mathbf{r}(2) = \langle 1-2^2, 5\cdot 2, 2\cdot 2^3\rangle = \langle -3, 10, 16\rangle$$

$$\mathbf{r}'(t) = \frac{d}{dt}\langle 1-t^2, 5t, 2t^3\rangle = \langle -2t, 5, 6t^2\rangle \quad\Rightarrow\quad \mathbf{r}'(2) = \langle -4, 5, 24\rangle$$

Substituting the vectors in (1) we obtain the following parametrization:

$$\ell(t) = \langle -3, 10, 16\rangle + t\langle -4, 5, 24\rangle = \langle -3-4t, 10+5t, 16+24t\rangle$$

35. $\mathbf{r}(s) = 4s^{-1}\mathbf{i} - 8s^{-3}\mathbf{k}, \quad s = 2$

SOLUTION The tangent line has the following parametrization:

$$\ell(s) = \mathbf{r}(2) + s\mathbf{r}'(2) \tag{1}$$

We compute the vectors $\mathbf{r}(2)$ and $\mathbf{r}'(2)$:

$$\mathbf{r}(2) = 4\cdot 2^{-1}\mathbf{i} - 8\cdot 2^{-3}\mathbf{k} = 2\mathbf{i} - \mathbf{k}$$

$$\mathbf{r}'(s) = \frac{d}{ds}\left(4s^{-1}\mathbf{i} - 8s^{-3}\mathbf{k}\right) = -4s^{-2}\mathbf{i} + 24s^{-4}\mathbf{k}$$

$$\Rightarrow \mathbf{r}'(2) = -4\cdot 2^{-2}\mathbf{i} + 24\cdot 2^{-4}\mathbf{k} = -\mathbf{i} + \frac{3}{2}\mathbf{k}$$

Substituting in (1) gives the following parametrization:

$$\ell(t) = 2\mathbf{i} - \mathbf{k} + s\left(-\mathbf{i} + \frac{3}{2}\mathbf{k}\right) = (2-s)\,\mathbf{i} + \left(\frac{3}{2}s - 1\right)\mathbf{k}$$

or in scalar form:

$$x = 2 - s, \quad y = 0, \quad z = \frac{3}{2}s - 1.$$

37. $\mathbf{r}(s) = \ln s\mathbf{i} + s^{-1}\mathbf{j} + 9s\mathbf{k}, \quad s = 1$

SOLUTION The tangent line has the following parametrization:

$$\ell(s) = \mathbf{r}(1) + s\mathbf{r}'(1) \qquad (1)$$

We compute the vectors $\mathbf{r}(1)$ and $\mathbf{r}'(1)$:

$$\mathbf{r}(1) = \ln 1\mathbf{i} + 1^{-1}\mathbf{j} + 9\cdot 1\mathbf{k} = \mathbf{j} + 9\mathbf{k}$$

$$\mathbf{r}'(s) = \frac{d}{ds}(\ln s\mathbf{i} + s^{-1}\mathbf{j} + 9s\mathbf{k}) = \frac{1}{s}\mathbf{i} - s^{-2}\mathbf{j} + 9\mathbf{k} \quad \Rightarrow \quad \mathbf{r}'(1) = \mathbf{i} - \mathbf{j} + 9\mathbf{k}$$

We substitute the vectors in (1) to obtain the following parametrization:

$$\ell(s) = \mathbf{j} + 9\mathbf{k} + s(\mathbf{i} - \mathbf{j} + 9\mathbf{k}) = s\mathbf{i} + (1-s)\mathbf{j} + (9+9s)\mathbf{k}$$

or in scalar form:

$$x = s, \quad y = 1 - s, \quad z = 9 + 9s.$$

39. Show, by finding a counterexample, that in general $\|\mathbf{r}'(t)\|$ need not equal $\|\mathbf{r}(t)\|'$.

SOLUTION Let $\mathbf{r}(t) = \langle 1, 1, t\rangle$. Then $\|\mathbf{r}(t)\| = \sqrt{1^2 + 1^2 + t^2} = \sqrt{2 + t^2}$, hence:

$$\|\mathbf{r}(t)\|' = \frac{d}{dt}\left(\sqrt{2+t^2}\right) = \frac{2t}{2\sqrt{2+t^2}} = \frac{t}{\sqrt{2+t^2}}$$

On the other hand, we have $\mathbf{r}'(t) = \langle 0, 0, 1\rangle$, hence:

$$\|\mathbf{r}'(t)\| = \sqrt{0^2 + 0^2 + 1^2} = 1.$$

We see that $\|\mathbf{r}'(t)\| \neq \|\mathbf{r}(t)\|'$.

In Exercises 40–45, evaluate the integrals.

41. $\int_1^4 \left(t^{-1}\mathbf{i} + 4\sqrt{t}\,\mathbf{j} - 8t^{3/2}\mathbf{k}\right) dt$

SOLUTION We perform the integration componentwise. Computing the integral of each component we get:

$$\int_1^4 t^{-1}\,dt = \ln t\Big|_1^4 = \ln 4 - \ln 1 = \ln 4$$

$$\int_1^4 4\sqrt{t}\,dt = 4\cdot\frac{2}{3}t^{3/2}\Big|_1^4 = \frac{8}{3}\left(4^{3/2} - 1\right) = \frac{56}{3}$$

$$\int_1^4 -8t^{3/2}\,dt = -\frac{16}{5}t^{5/2}\Big|_1^4 = -\frac{16}{5}\left(4^{5/2} - 1\right) = -\frac{496}{5}$$

Hence,

$$\int_1^4 \left(t^{-1}\mathbf{i} + 4\sqrt{t}\mathbf{j} - 8t^{3/2}\mathbf{k}\right) dt = (\ln 4)\,\mathbf{i} + \frac{56}{3}\mathbf{j} - \frac{496}{5}\mathbf{k}$$

43. $\int_{-2}^2 (u^3\mathbf{i} + u^5\mathbf{j} + u^7\mathbf{k})\,du$

SOLUTION We perform componentwise integration, but before doing so we notice that u^3, u^5 and u^7 are all odd functions, so their integrals over this symmetric region will all be zero! Thus, the answer is $0\mathbf{i} + 0\mathbf{j} + 0\mathbf{k}$.

45. $\int_0^t (3s\mathbf{i} + 6s^2\mathbf{j} + 9\mathbf{k})\, ds$

SOLUTION We first compute the integral of each component:

$$\int_0^t 3s\, ds = \frac{3}{2}s^2\Big|_0^t = \frac{3}{2}t^2$$

$$\int_0^t 6s^2\, ds = \frac{6}{3}s^3\Big|_0^t = 2t^3$$

$$\int_0^t 9\, ds = 9s\Big|_0^t = 9t$$

Hence,

$$\int_0^t \left(3s\mathbf{i} + 6s^2\mathbf{j} + 9\mathbf{k}\right) dt = \left(\int_0^t 3s\, ds\right)\mathbf{i} + \left(\int_0^t 6s^2\, ds\right)\mathbf{j} + \left(\int_0^t 9\, ds\right)\mathbf{k} = \left(\frac{3}{2}t^2\right)\mathbf{i} + (2t^3)\mathbf{j} + (9t)\mathbf{k}$$

In Exercises 46–53, find the general solution $\mathbf{r}(t)$ *of the differential equation and the solution with the given initial condition.*

47. $\mathbf{r}'(t) = \mathbf{i} - \mathbf{j}, \quad \mathbf{r}(0) = 2\mathbf{i} + 3\mathbf{k}$

SOLUTION The general solution is obtained by integrating $\mathbf{r}'(t)$:

$$\mathbf{r}(t) = \int (\mathbf{i} - \mathbf{j})\, dt = \left(\int 1\, dt\right)\mathbf{i} - \left(\int 1\, dt\right)\mathbf{j} = t\mathbf{i} - t\mathbf{j} + \mathbf{c} \tag{1}$$

Hence,

$$\mathbf{r}(0) = 0\mathbf{i} - 0\mathbf{j} + \mathbf{c} = \mathbf{c}$$

The solution with the initial condition $\mathbf{r}(0) = 2\mathbf{i} + 3\mathbf{k}$ must satisfy:

$$\mathbf{r}(0) = \mathbf{c} = 2\mathbf{i} + 3\mathbf{k}$$

Substituting in (1) yields the solution:

$$\mathbf{r}(t) = t\mathbf{i} - t\mathbf{j} + 2\mathbf{i} + 3\mathbf{k} = (t+2)\,\mathbf{i} - t\mathbf{j} + 3\mathbf{k}$$

49. $\mathbf{r}'(t) = \langle \sin 3t, \sin 3t, t\rangle, \quad \mathbf{r}(0) = \langle 0, 1, 8\rangle$

SOLUTION We first integrate the vector $\mathbf{r}'(t)$ to find the general solution:

$$\mathbf{r}(t) = \int \langle \sin 3t, \sin 3t, t\rangle\, dt = \left\langle \int \sin 3t\, dt, \int \sin 3t\, dt, \int t\, dt \right\rangle = \left\langle -\frac{1}{3}\cos 3t, -\frac{1}{3}\cos 3t, \frac{1}{2}t^2 \right\rangle + \mathbf{c} \tag{1}$$

Substituting the initial condition we obtain:

$$\mathbf{r}(0) = \left\langle -\frac{1}{3}\cos 0, -\frac{1}{3}\cos 0, \frac{1}{2}\cdot 0^2 \right\rangle + \mathbf{c} = \langle 0, 1, 8\rangle = \left\langle -\frac{1}{3}, -\frac{1}{3}, 0 \right\rangle + \mathbf{c} = \langle 0, 1, 8\rangle$$

Hence,

$$\mathbf{c} = \langle 0, 1, 8\rangle - \left\langle -\frac{1}{3}, -\frac{1}{3}, 0 \right\rangle = \left\langle \frac{1}{3}, \frac{4}{3}, 8 \right\rangle$$

Substituting in (1) we obtain the solution:

$$\mathbf{r}(t) = \left\langle -\frac{1}{3}\cos 3t, -\frac{1}{3}\cos 3t, \frac{1}{2}t^2 \right\rangle + \left\langle \frac{1}{3}, \frac{4}{3}, 8 \right\rangle = \left\langle \frac{1}{3}(1 - \cos 3t), \frac{1}{3}(4 - \cos 3t), 8 + \frac{1}{2}t^2 \right\rangle$$

51. $\mathbf{r}''(t) = 16\mathbf{k}, \quad \mathbf{r}(0) = \langle 1, 0, 0\rangle, \quad \mathbf{r}'(0) = \langle 0, 1, 0\rangle$

SOLUTION To find the general solution we first find $\mathbf{r}'(t)$ by integrating $\mathbf{r}''(t)$:

$$\mathbf{r}'(t) = \int \mathbf{r}''(t)\, dt = \int 16\mathbf{k}\, dt = (16t)\,\mathbf{k} + \mathbf{c}_1 \tag{1}$$

We now integrate $\mathbf{r}'(t)$ to find the general solution $\mathbf{r}(t)$:

$$\mathbf{r}(t) = \int \mathbf{r}'(t)\,dt = \int ((16t)\,\mathbf{k} + \mathbf{c}_1)\,dt = \left(\int 16(t)\,dt\right)\mathbf{k} + \mathbf{c}_1 t + \mathbf{c}_2 = (8t^2)\mathbf{k} + \mathbf{c}_1 t + \mathbf{c}_2 \qquad (2)$$

We substitute the initial conditions in (1) and (2). This gives:

$$\mathbf{r}'(0) = \mathbf{c}_1 = \langle 0, 1, 0\rangle = \mathbf{j}$$

$$\mathbf{r}(0) = 0\mathbf{k} + \mathbf{c}_1 \cdot 0 + \mathbf{c}_2 = \langle 1, 0, 0\rangle \quad \Rightarrow \quad \mathbf{c}_2 = \langle 1, 0, 0\rangle = \mathbf{i}$$

Combining with (2) we obtain the following solution:

$$\mathbf{r}(t) = (8t^2)\mathbf{k} + t\mathbf{j} + \mathbf{i} = \mathbf{i} + t\mathbf{j} + (8t^2)\mathbf{k}$$

53. $\mathbf{r}''(t) = \left\langle e^t, \sin t, \cos t\right\rangle$, $\mathbf{r}(0) = \langle 1, 0, 1\rangle$, $\mathbf{r}'(0) = \langle 0, 2, 2\rangle$

SOLUTION We perform integration componentwise on $\mathbf{r}''(t)$ to obtain:

$$\mathbf{r}'(t) = \int \left\langle e^t, \sin t, \cos t\right\rangle dt = \left\langle e^t, -\cos t, \sin t\right\rangle + \mathbf{c}_1 \qquad (1)$$

We now integrate $\mathbf{r}'(t)$ to obtain the general solution:

$$\mathbf{r}(t) = \int \left(\left\langle e^t, -\cos t, \sin t\right\rangle + \mathbf{c}_1\right) dt = \left\langle e^t, -\sin t, -\cos t\right\rangle + \mathbf{c}_1 t + \mathbf{c}_2 \qquad (2)$$

Now, we substitute the initial conditions $\mathbf{r}(0) = \langle 1, 0, 1\rangle$ and $\mathbf{r}'(0) = \langle 0, 2, 2\rangle$ into (1) and (2) and solve for the vectors $\mathbf{c}_1$ and $\mathbf{c}_2$. We obtain:

$$\mathbf{r}'(0) = \langle 1, -1, 0\rangle + \mathbf{c}_1 = \langle 0, 2, 2\rangle \quad \Rightarrow \quad \mathbf{c}_1 = \langle -1, 3, 2\rangle$$

$$\mathbf{r}(0) = \langle 1, 0, -1\rangle + \mathbf{c}_2 = \langle 1, 0, 1\rangle \quad \Rightarrow \quad \mathbf{c}_2 = \langle 0, 0, 2\rangle$$

Finally we combine the above to obtain the solution:

$$\mathbf{r}(t) = \left\langle e^t, -\sin t, -\cos t\right\rangle + \langle -1, 3, 2\rangle\, t + \langle 0, 0, 2\rangle = \left\langle e^t - t, -\sin t + 3t, -\cos t + 2t + 2\right\rangle$$

55. The path $\mathbf{r}(t)$ of a particle satisfies $\dfrac{d\mathbf{r}}{dt} = \left\langle 8, 5 - 3t, 4t^2\right\rangle$. Where is the particle located at $t = 4$ if $\mathbf{r}(0) = \langle 1, 6, 0\rangle$?

SOLUTION We first find the general solution by integrating the vector $\frac{d\mathbf{r}}{dt}$ componentwise. This gives:

$$\mathbf{r}(t) = \int \left\langle 8, 5 - 3t, 4t^2\right\rangle dt = \left\langle 8t, 5t - \frac{3}{2}t^2, \frac{4}{3}t^3\right\rangle + \mathbf{c} \qquad (1)$$

Substituting $t = 0$ we get:

$$\mathbf{r}(0) = \left\langle 8 \cdot 0, 5 \cdot 0 - \frac{3}{2} \cdot 0^2, \frac{4}{3} \cdot 0^3\right\rangle + \mathbf{c} = \mathbf{c}$$

The initial condition $\mathbf{r}(0) = \langle 1, 6, 0\rangle$ gives $\mathbf{c} = \langle 1, 6, 0\rangle$. Combining with (1) we obtain the following solution:

$$\mathbf{r}(t) = \left\langle 8t, 5t - \frac{3}{2}t^2, \frac{4}{3}t^3\right\rangle + \langle 1, 6, 0\rangle = \left\langle 1 + 8t, 6 + 5t - \frac{3}{2}t^2, \frac{4}{3}t^3\right\rangle$$

To find the particle's position at $t = 4$, we substitute $t = 4$ in $\mathbf{r}(t)$ obtaining:

$$\mathbf{r}(4) = \left\langle 1 + 8 \cdot 4, 6 + 5 \cdot 4 - \frac{3}{2} \cdot 4^2, \frac{4}{3} \cdot 4^3\right\rangle = \left\langle 33, 2, \frac{256}{3}\right\rangle$$

57. Find all solutions to $\mathbf{r}'(t) = \mathbf{v}$, where $\mathbf{v}$ is a constant vector in $\mathbf{R}^3$.

SOLUTION We denote the components of the constant vector $\mathbf{v}$ by $\mathbf{v} = \langle v_1, v_2, v_3\rangle$ and integrate to find the general solution. This gives:

$$\mathbf{r}(t) = \int \mathbf{v}\,dt = \int \langle v_1, v_2, v_3\rangle\,dt = \left\langle \int v_1\,dt, \int v_2\,dt, \int v_3\,dt\right\rangle$$

$$= \langle v_1 t + c_1, v_2 t + c_2, v_3 t + c_3\rangle = t\,\langle v_1, v_2, v_3\rangle + \langle c_1, c_2, c_3\rangle$$

We let $\mathbf{c} = \langle c_1, c_2, c_3\rangle$ and obtain:

$$\mathbf{r}(t) = t\mathbf{v} + \mathbf{c} = \mathbf{c} + t\mathbf{v}$$

Notice that the solutions are the vector parametrizations of all the lines with direction vector $\mathbf{v}$.

59. Find all solutions to $\mathbf{r}'(t) = 2\mathbf{r}(t)$ where $\mathbf{r}(t)$ is a vector-valued function in three-space.

SOLUTION We denote the components of $\mathbf{r}(t)$ by $\mathbf{r}(t) = \langle x(t), y(t), z(t)\rangle$. Then, $\mathbf{r}'(t) = \langle x'(t), y'(t), z'(t)\rangle$. Substituting in the differential equation we get:

$$\langle x'(t), y'(t), z'(t)\rangle = 2\langle x(t), y(t), z(t)\rangle$$

Equating corresponding components gives:

$$\begin{aligned} x'(t) &= 2x(t) & & x(t) = c_1 e^{2t} \\ y'(t) &= 2y(t) & \Rightarrow\ & y(t) = c_2 e^{2t} \\ z'(t) &= 2z(t) & & z(t) = c_3 e^{2t} \end{aligned}$$

We denote the constant vector by $\mathbf{c} = \langle c_1, c_2, c_3\rangle$ and obtain the following solutions:

$$\mathbf{r}(t) = \langle c_1 e^{2t}, c_2 e^{2t}, c_3 e^{2t}\rangle = e^{2t}\langle c_1, c_2, c_3\rangle = e^{2t}\mathbf{c}$$

61. Prove that $\mathbf{r}(t_0)$ and $\mathbf{r}'(t_0)$ are orthogonal at values $t = t_0$ where $\|\mathbf{r}(t)\|$ takes on a local minimum or maximum value. Explain how this result is related to Figure 7. *Hint:* In the figure, $\|\mathbf{r}(t_0)\|$ is a minimum and the path $\mathbf{r}(t)$ intersects the sphere of radius $\|\mathbf{r}(t_0)\|$ in a single point (and hence is tangent at that point).

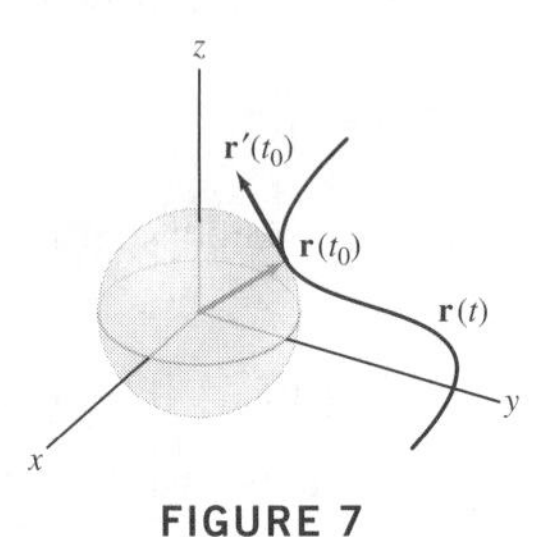

FIGURE 7

SOLUTION Suppose that $\|\mathbf{r}(t)\|$ takes on a minimum or maximum value at $t = t_0$. Hence, $\|\mathbf{r}(t)\|^2$ also takes on a minimum or maximum value at $t = t_0$, therefore $\frac{d}{dt}\|\mathbf{r}(t)\|^2\big|_{t=t_0} = 0$. Using the Product Rule for dot products we get

$$\frac{d}{dt}\|\mathbf{r}(t)\|^2\bigg|_{t=t_0} = \frac{d}{dt}\mathbf{r}(t)\cdot\mathbf{r}(t)\bigg|_{t=t_0} = \mathbf{r}(t_0)\cdot\mathbf{r}'(t_0) + \mathbf{r}'(t_0)\cdot\mathbf{r}(t_0) = 2\mathbf{r}(t_0)\cdot\mathbf{r}'(t_0) = 0$$

Thus $\mathbf{r}(t_0)\cdot\mathbf{r}'(t_0) = 0$, which implies the orthogonality of $\mathbf{r}(t_0)$ and $\mathbf{r}'(t_0)$. In Figure 7, $\|\mathbf{r}(t_0)\|$ is a minimum and the path intersects the sphere of radius $\|\mathbf{r}(t_0)\|$ at a single point. Therefore, the point of intersection is a tangency point which implies that $\mathbf{r}'(t_0)$ is tangent to the sphere at t_0. We conclude that $\mathbf{r}(t_0)$ and $\mathbf{r}'(t_0)$ are orthogonal.

Further Insights and Challenges

63. In this exercise, we verify that the definition of the tangent line using vector-valued functions agrees with the usual definition in terms of the scalar derivative in the case of a plane curve. Suppose that $\mathbf{r}(t) = \langle x(t), y(t)\rangle$ traces a plane curve $\mathcal{C}$.

(a) Show that $\dfrac{dy}{dx} = \dfrac{y'(t)}{x'(t)}$ at any point such that $x'(t) \neq 0$. *Hint:* By the Chain Rule, $\dfrac{dy}{dt} = \dfrac{dy}{dx}\dfrac{dx}{dt}$.

(b) Show that if $x'(t_0) \neq 0$, then the line $\mathbf{L}(t) = \mathbf{r}(t_0) + t\mathbf{r}'(t_0)$ passes through $\mathbf{r}(t_0)$ and has slope $\dfrac{dy}{dx}\bigg|_{t=t_0}$.

SOLUTION

(a) By the Chain Rule we have

$$\frac{dy}{dt} = \frac{dy}{dx}\cdot\frac{dx}{dt}$$

Hence, at the points where $\frac{dx}{dt} \neq 0$ we have:

$$\frac{dy}{dx} = \frac{\frac{dy}{dt}}{\frac{dx}{dt}} = \frac{y'(t)}{x'(t)}$$

(b) The line $\ell(t) = \langle a, b\rangle + t\mathbf{r}'(t_0)$ passes through (a, b) at $t = 0$. It holds that:

$$\ell(0) = \langle a, b\rangle + 0\mathbf{r}'(t_0) = \langle a, b\rangle$$

That is, (a, b) is the terminal point of the vector $\ell(0)$, hence the line passes through (a, b). The line has the direction vector $\mathbf{r}'(t_0) = \langle x'(t_0), y'(t_0)\rangle$, therefore the slope of the line is $\frac{y'(t_0)}{x'(t_0)}$ which is equal to $\frac{dy}{dx}\big|_{t=t_0}$ by part (a).

65. Verify the Chain Rule for vector-valued functions.

SOLUTION Let $g(t)$ and $\mathbf{r}(t) = \langle x(t), y(t), z(t)\rangle$ be differentiable scalar and vector valued functions respectively. We must show that:

$$\frac{d}{dt}\mathbf{r}(g(t)) = g'(t)\mathbf{r}'(g(t)).$$

We have

$$\mathbf{r}(g(t)) = \langle x(g(t)), y(g(t)), z(g(t))\rangle$$

We differentiate the vector componentwise, using the Chain Rule for scalar functions. This gives:

$$\frac{d}{dt}\mathbf{r}(g(t)) = \left\langle \frac{d}{dt}(x(g(t))), \frac{d}{dt}(y(g(t))), \frac{d}{dt}(z(g(t)))\right\rangle = \left\langle g'(t)x'(g(t)), g'(t)y'(g(t)), g'(t)z'(g(t))\right\rangle$$
$$= g'(t)\left\langle x'(g(t)), y'(g(t)), z'(g(t))\right\rangle = g'(t)\mathbf{r}'(g(t))$$

67. Prove that $\frac{d}{dt}(\mathbf{r}\cdot(\mathbf{r}'\times\mathbf{r}'')) = \mathbf{r}\cdot(\mathbf{r}'\times\mathbf{r}''')$

SOLUTION We use the Product Rule for dot products to obtain:

$$\frac{d}{dt}\left(\mathbf{r}\cdot\left(\mathbf{r}'\times\mathbf{r}''\right)\right) = \mathbf{r}\cdot\frac{d}{dt}\left(\mathbf{r}'\times\mathbf{r}''\right) + \mathbf{r}'\cdot\left(\mathbf{r}'\times\mathbf{r}''\right) \tag{1}$$

By the Product Rule for cross products and properties of cross products, we have:

$$\frac{d}{dt}\left(\mathbf{r}'\times\mathbf{r}''\right) = \mathbf{r}'\times\mathbf{r}''' + \mathbf{r}''\times\mathbf{r}'' = \mathbf{r}'\times\mathbf{r}''' + \mathbf{0} = \mathbf{r}'\times\mathbf{r}''' \tag{2}$$

Substituting (2) into (1) yields:

$$\frac{d}{dt}\left(\mathbf{r}\cdot\left(\mathbf{r}'\times\mathbf{r}''\right)\right) = \mathbf{r}\cdot\left(\mathbf{r}'\times\mathbf{r}'''\right) + \mathbf{r}'\cdot\left(\mathbf{r}'\times\mathbf{r}''\right) \tag{3}$$

Since $\mathbf{r}'\times\mathbf{r}''$ is orthogonal to $\mathbf{r}'$, the dot product $\mathbf{r}'\cdot\left(\mathbf{r}'\times\mathbf{r}''\right) = 0$. So (3) gives:

$$\frac{d}{dt}\left(\mathbf{r}\cdot\left(\mathbf{r}'\times\mathbf{r}''\right)\right) = \mathbf{r}\cdot\left(\mathbf{r}'\times\mathbf{r}'''\right) + 0 = \mathbf{r}\cdot\left(\mathbf{r}'\times\mathbf{r}'''\right)$$

Exercises 68–71 establish additional properties of vector-valued integrals. Assume that all functions are integrable.

69. Prove the Substitution Rule [where $g(t)$ is a differentiable scalar function]:

$$\int_a^b \mathbf{r}(g(t))g'(t)\,dt = \int_{g^{-1}(a)}^{g^{-1}(b)} \mathbf{r}(u)\,du$$

SOLUTION (Note that an early edition of the textbook had the integral limits as $g(a)$ and $g(b)$; they should actually be $g^{-1}(a)$ and $g^{-1}(b)$.) We denote the components of the vector-valued function by $\mathbf{r}(t)\,dt = \langle x(t), y(t), z(t)\rangle$. Using componentwise integration we have:

$$\int_a^b \mathbf{r}(t)\,dt = \left\langle \int_a^b x(t)\,dt, \int_a^b y(t)\,dt, \int_a^b z(t)\,dt\right\rangle$$

Write $\int_a^b x(t)\,dt$ as $\int_a^b x(s)\,ds$. Let $s = g(t)$, so $ds = g'(t)\,dt$. The substitution gives us $\int_{g^{-1}(a)}^{g^{-1}(b)} x(g(t))g'(t)\,dt$. A similar procedure for the other two integrals gives us:

$$\int_a^b \mathbf{r}(t)\,dt = \left\langle \int_{g^{-1}(a)}^{g^{-1}(b)} x(g(t))\,g'(t)\,dt, \int_{g^{-1}(a)}^{g^{-1}(b)} y(g(t))\,g'(t)\,dt, \int_{g^{-1}(a)}^{g^{-1}(b)} z(g(t))\,g'(t)\,dt\right\rangle$$
$$= \int_{g^{-1}(a)}^{g^{-1}(b)} \left\langle x(g(t))\,g'(t), y(g(t))\,g'(t), z(g(t))\,g'(t)\right\rangle dt$$
$$= \int_{g^{-1}(a)}^{g^{-1}(b)} \langle x(g(t)), y(g(t)), z(g(t))\rangle\, g'(t)\,dt = \int_{g^{-1}(a)}^{g^{-1}(b)} \mathbf{r}(g(t))\,g'(t)\,dt$$

71. Show that if $\|\mathbf{r}(t)\| \le K$ for $t \in [a, b]$, then

$$\left\| \int_a^b \mathbf{r}(t)\, dt \right\| \le K(b-a)$$

SOLUTION Think of $\mathbf{r}(t)$ as a velocity vector. Then, $\int_a^b \mathbf{r}(t)\, dt$ gives the displacement vector from the location at time $t = a$ to the time $t = b$, and so $\left\| \int_a^b \mathbf{r}(t)\, dt \right\|$ gives the length of this displacement vector. But, since speed is $\|\mathbf{r}(t)\|$ which is less than or equal to K, then in the interval $a \le t \le b$, the object can move a total distance not more than $K(b-a)$. Thus, the length of the displacement vector is $\le K(b-a)$, which gives us $\left\| \int_a^b \mathbf{r}(t)\, dt \right\| \le K(b-a)$, as desired.

14.3 Arc Length and Speed (ET Section 13.3)

Preliminary Questions

1. At a given instant, a car on a roller coaster has velocity vector $\mathbf{r}' = \langle 25, -35, 10 \rangle$ (in miles per hour). What would the velocity vector be if the speed were doubled? What would it be if the car's direction were reversed but its speed remained unchanged?

SOLUTION The speed is doubled but the direction is unchanged, hence the new velocity vector has the form:

$$\lambda \mathbf{r}' = \lambda \langle 25, -35, 10 \rangle \quad \text{for } \lambda > 0$$

We use $\lambda = 2$, and so the new velocity vector is $\langle 50, -70, 20 \rangle$. If the direction is reversed but the speed is unchanged, the new velocity vector is:

$$-\mathbf{r}' = \langle -25, 35, -10 \rangle .$$

2. Two cars travel in the same direction along the same roller coaster (at different times). Which of the following statements about their velocity vectors at a given point P on the roller coaster are true?

(a) The velocity vectors are identical.

(b) The velocity vectors point in the same direction but may have different lengths.

(c) The velocity vectors may point in opposite directions.

SOLUTION

(a) The length of the velocity vector is the speed of the particle. Therefore, if the speeds of the cars are different the velocities are not identical. The statement is false.

(b) The velocity vector is tangent to the curve. Since the cars travel in the same direction, their velocity vectors point in the same direction. The statement is true.

(c) Since the cars travel in the same direction, the velocity vectors point in the same direction. The statement is false.

3. A mosquito flies along a parabola with speed $v(t) = t^2$. Let $L(t)$ be the total distance traveled at time t.

(a) How fast is $L(t)$ changing at $t = 2$?

(b) Is $L(t)$ equal to the mosquito's distance from the origin?

SOLUTION

(a) By the Arc Length Formula, we have:

$$L(t) = \int_{t_0}^{t} \|\mathbf{r}'(t)\|\, dt = \int_{t_0}^{t} v(t)\, dt$$

Therefore,

$$L'(t) = v(t)$$

To find the rate of change of $L(t)$ at $t = 2$ we compute the derivative of $L(t)$ at $t = 2$, that is,

$$L'(2) = v(2) = 2^2 = 4$$

(b) $L(t)$ is the distance along the path traveled by the mosquito. This distance is usually different from the mosquito's distance from the origin, which is the length of $\mathbf{r}(t)$.

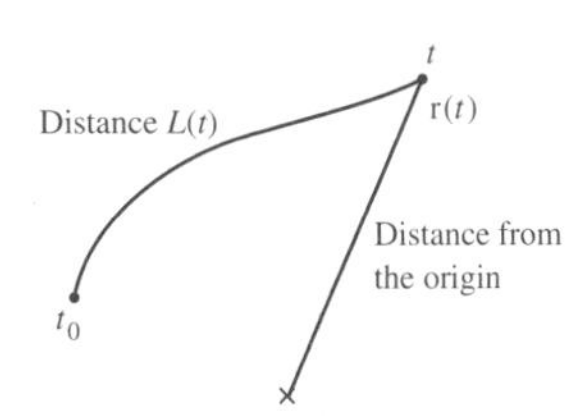

4. What is the length of the path traced by $\mathbf{r}(t)$ for $4 \le t \le 10$ if $\mathbf{r}(t)$ is an arc length parametrization?

SOLUTION Since $\mathbf{r}(t)$ is an arc length parametrization, the length of the path for $4 \le t \le 10$ is equal to the length of the time interval $4 \le t \le 10$, which is 6.

Exercises

In Exercises 1–6, compute the length of the curve over the given interval.

1. $\mathbf{r}(t) = \langle 3t, 4t - 3, 6t + 1 \rangle, \quad 0 \le t \le 3$

SOLUTION We have $x(t) = 3t$, $y(t) = 4t - 3$, $z(t) = 6t + 1$ hence

$$x'(t) = 3, \quad y'(t) = 4, \quad z'(t) = 6.$$

We use the Arc Length Formula to obtain:

$$L = \int_0^3 \|\mathbf{r}'(t)\|\, dt = \int_0^3 \sqrt{x'(t)^2 + y'(t)^2 + z'(t)^2}\, dt = \int_0^3 \sqrt{3^2 + 4^2 + 6^2}\, dt = 3\sqrt{61}$$

3. $\mathbf{r}(t) = \langle 2t, \ln t, t^2 \rangle, \quad 1 \le t \le 4$

SOLUTION The derivative of $\mathbf{r}(t)$ is $\mathbf{r}'(t) = \left\langle 2, \frac{1}{t}, 2t \right\rangle$. We use the Arc Length Formula to obtain:

$$L = \int_1^4 \|\mathbf{r}'(t)\|\, dt = \int_1^4 \sqrt{2^2 + \left(\frac{1}{t}\right)^2 + (2t)^2}\, dt = \int_1^4 \sqrt{4t^2 + 4 + \frac{1}{t^2}}\, dt = \int_1^4 \sqrt{\left(2t + \frac{1}{t}\right)^2}\, dt$$
$$= \int_1^4 \left(2t + \frac{1}{t}\right) dt = t^2 + \ln t \Big|_1^4 = (16 + \ln 4) - (1 + \ln 1) = 15 + \ln 4$$

5. $\mathbf{r}(t) = t\mathbf{i} + 2t\mathbf{j} + (t^2 - 3)\mathbf{k}, \quad 0 \le t \le 2$. *Hint:*

$$\int \sqrt{t^2 + a^2}\, dt = \frac{1}{2} t \sqrt{t^2 + a^2} + \frac{1}{2} a^2 \ln\left(t + \sqrt{t^2 + a^2}\right)$$

SOLUTION The derivative of $\mathbf{r}(t)$ is $\mathbf{r}'(t) = \mathbf{i} + 2\mathbf{j} + 2t\mathbf{k}$. Using the Arc Length Formula we get:

$$L = \int_0^2 \|\mathbf{r}'(t)\|\, dt = \int_0^2 \sqrt{1^2 + (2)^2 + (2t)^2}\, dt = \int_0^2 \sqrt{4t^2 + 5}\, dt$$

We substitute $u = 2t$, $du = 2\, dt$ and use the given integration formula. This gives:

$$L = \frac{1}{2} \int_0^4 \sqrt{u^2 + 5}\, du = \frac{1}{4} u \sqrt{u^2 + 5} + \frac{1}{4} \cdot 5 \ln\left(u + \sqrt{u^2 + 5}\right)\Big|_0^4$$
$$= \frac{1}{4} \cdot 4\sqrt{4^2 + 5} + \frac{5}{4} \ln\left(4 + \sqrt{4^2 + 5}\right) - \frac{5}{4} \ln \sqrt{5} = \sqrt{21} + \frac{5}{4} \ln\left(4 + \sqrt{21}\right) - \frac{5}{4} \ln \sqrt{5}$$
$$= \sqrt{21} + \frac{5}{4} \ln \frac{4 + \sqrt{21}}{\sqrt{5}} \approx 6.26$$

7. Compute $s(t) = \int_0^t \|\mathbf{r}'(u)\|\, du$ for $\mathbf{r}(t) = \langle t^2, 2t^2, t^3 \rangle$.

SOLUTION The derivative of $\mathbf{r}(t)$ is $\mathbf{r}'(t) = \langle 2t, 4t, 3t^2 \rangle$. Hence,

$$\|\mathbf{r}'(t)\| = \sqrt{(2t)^2 + (4t)^2 + (3t^2)^2} = \sqrt{4t^2 + 16t^2 + 9t^4} = \sqrt{20 + 9t^2}\, t$$

Hence,

$$s(t) = \int_0^t \|\mathbf{r}'(u)\|\, du = \int_0^t \sqrt{20 + 9u^2} u\, du$$

We compute the integral using the substitution $v = 20 + 9u^2$, $dv = 18u\, du$. This gives:

$$s(t) = \frac{1}{18} \int_{20}^{20+9t^2} v^{1/2}\, dv = \frac{1}{18} \cdot \frac{2}{3} v^{3/2} \Big|_{20}^{20+9t^2} = \frac{1}{27} \left((20 + 9t^2)^{3/2} - 20^{3/2} \right).$$

In Exercises 8–11, find the speed at the given value of t.

9. $\mathbf{r}(t) = \langle e^{t-3}, 12, 3t^{-1} \rangle$, $t = 3$

SOLUTION The velocity vector is $\mathbf{r}'(t) = \langle e^{t-3}, 0, -3t^{-2} \rangle$ and at $t = 3$, $\mathbf{r}'(3) = \langle e^{3-3}, 0, -3 \cdot 3^{-2} \rangle = \langle 1, 0, -\frac{1}{3} \rangle$. The speed is the magnitude of the velocity vector, that is,

$$v(3) = \|\mathbf{r}'(3)\| = \sqrt{1^2 + 0^2 + \left(-\frac{1}{3}\right)^2} = \sqrt{\frac{10}{9}} \approx 1.05$$

11. $\mathbf{r}(t) = \langle \cosh t, \sinh t, t \rangle$, $t = 0$

SOLUTION The velocity vector is $\mathbf{r}'(t) = \langle \sinh t, \cosh t, 1 \rangle$. At $t = 0$ the velocity is $\mathbf{r}'(0) = \langle \sinh(0), \cosh(0), 1 \rangle = \langle 0, 1, 1 \rangle$, hence the speed is

$$v(0) = \|\mathbf{r}'(0)\| = \sqrt{0^2 + 1^2 + 1^2} = \sqrt{2}.$$

13. A bee with velocity vector $\mathbf{r}'(t)$ starts out at the origin at $t = 0$ and flies around for T seconds. Where is the bee located at time T if $\int_0^T \mathbf{r}'(u)\, du = \mathbf{0}$? What does the quantity $\int_0^T \|\mathbf{r}'(u)\|\, du$ represent?

SOLUTION By the Fundamental Theorem for vector-valued functions, $\int_0^T \mathbf{r}'(u)\, du = \mathbf{r}(T) - \mathbf{r}(0)$, hence by the given information $\mathbf{r}(T) = \mathbf{r}(0)$. It follows that at time T the bee is located at the starting point which is at the origin. The integral $\int_0^T \|\mathbf{r}'(u)\|\, du$ is the length of the path traveled by the bee in the time interval $0 \le t \le T$. Notice that there is a difference between the displacement and the actual length traveled.

15. Let $\mathbf{r}(t) = \langle 3t + 1, 4t - 5, 2t \rangle$.

(a) Calculate $s(t) = \int_0^t \|\mathbf{r}'(u)\|\, du$ as a function of t.

(b) Find the inverse $\varphi(s) = t(s)$ and show that $\mathbf{r}_1(s) = \mathbf{r}(\varphi(s))$ is an arc length parametrization.

SOLUTION

(a) We differentiate $\mathbf{r}(t)$ componentwise and then compute the norm of the derivative vector. This gives:

$$\mathbf{r}'(t) = \langle 3, 4, 2 \rangle$$

$$\|\mathbf{r}'(t)\| = \sqrt{3^2 + 4^2 + 2^2} = \sqrt{29}$$

We compute $s(t)$:

$$s(t) = \int_0^t \|\mathbf{r}'(u)\|\, du = \int_0^t \sqrt{29}\, du = \sqrt{29}\, u \Big|_0^t = \sqrt{29} t$$

(b) We find the inverse $\varphi(s) = t(s)$ by solving $s = \sqrt{29} t$ for t. We obtain:

$$s = \sqrt{29} t \quad \Rightarrow \quad t = \varphi(s) = \frac{s}{\sqrt{29}}$$

We obtain the following arc length parametrization:

$$\mathbf{r}_1(s) = \mathbf{r}\left(\frac{s}{\sqrt{29}}\right) = \left\langle \frac{3s}{\sqrt{29}} + 1, \frac{4s}{\sqrt{29}} - 5, \frac{2s}{\sqrt{29}} \right\rangle$$

To verify that $\mathbf{r}_1(s)$ is an arc length parametrization we must show that $\|\mathbf{r}_1'(s)\| = 1$. We compute $\mathbf{r}_1'(s)$:

$$\mathbf{r}_1'(s) = \frac{d}{ds} \left\langle \frac{3s}{\sqrt{29}} + 1, \frac{4s}{\sqrt{29}} - 5, \frac{2s}{\sqrt{29}} \right\rangle = \left\langle \frac{3}{\sqrt{29}}, \frac{4}{\sqrt{29}}, \frac{2}{\sqrt{29}} \right\rangle = \frac{1}{\sqrt{29}} \langle 3, 4, 2 \rangle$$

Thus,

$$\|\mathbf{r}_1'(s)\| = \frac{1}{\sqrt{29}} \| \langle 3, 4, 2 \rangle \| = \frac{1}{\sqrt{29}} \sqrt{3^2 + 4^2 + 2^2} = \frac{1}{\sqrt{29}} \cdot \sqrt{29} = 1$$

17. Find a path that traces the circle in the plane $y = 10$ with radius 4 and center $(2, 10, -3)$ with constant speed 8.

SOLUTION We start with the following parametrization of the circle:

$$\mathbf{r}(t) = \langle 2, 10, -3 \rangle + 4 \langle \cos t, 0, \sin t \rangle = \langle 2 + 4\cos t, 10, -3 + 4\sin t \rangle$$

We need to reparametrize the curve by making a substitution $t = \varphi(s)$, so that the new parametrization $\mathbf{r}_1(s) = \mathbf{r}(\varphi(s))$ satisfies $\|\mathbf{r}_1'(s)\| = 8$ for all s. We find $\mathbf{r}_1'(s)$ using the Chain Rule:

$$\mathbf{r}_1'(s) = \frac{d}{ds}\mathbf{r}(\varphi(s)) = \varphi'(s)\mathbf{r}'(\varphi(s)) \tag{1}$$

Next, we differentiate $\mathbf{r}(t)$ and then replace t by $\varphi(s)$:

$$\mathbf{r}'(t) = \langle -4\sin t, 0, 4\cos t \rangle$$

$$\mathbf{r}'(\varphi(s)) = \langle -4\sin\varphi(s), 0, 4\cos\varphi(s) \rangle$$

Substituting in (1) we get:

$$\mathbf{r}_1'(s) = \varphi'(s)\langle -4\sin\varphi(s), 0, 4\cos\varphi(s) \rangle = -4\varphi'(s)\langle \sin\varphi(s), 0, -\cos\varphi(s) \rangle$$

Hence,

$$\|\mathbf{r}_1'(s)\| = 4|\varphi'(s)|\sqrt{(\sin\varphi(s))^2 + (-\cos\varphi(s))^2} = 4|\varphi'(s)|$$

To satisfy $\|\mathbf{r}_1'(s)\| = 8$ for all s, we choose $\varphi'(s) = 2$. We may take the antiderivative $\varphi(s) = 2 \cdot s$, and obtain the following parametrization:

$$\mathbf{r}_1(s) = \mathbf{r}(\varphi(s)) = \mathbf{r}(2s) = \langle 2 + 4\cos(2s), 10, -3 + 4\sin(2s) \rangle.$$

This is a parametrization of the given circle, with constant speed 8.

19. Find an arc length parametrization of $\mathbf{r}(t) = \langle e^t \sin t, e^t \cos t, e^t \rangle$.

SOLUTION An arc length parametrization is $\mathbf{r}_1(s) = \mathbf{r}(\varphi(s))$ where $t = \varphi(s)$ is the inverse of the arc length function. We compute the arc length function:

$$s(t) = \int_0^t \|\mathbf{r}'(u)\|\, du \tag{1}$$

Differentiating $\mathbf{r}(t)$ and computing the norm of $\mathbf{r}'(t)$ gives:

$$\mathbf{r}'(t) = \langle e^t \sin t + e^t \cos t, e^t \cos t - e^t \sin t, e^t \rangle = e^t \langle \sin t + \cos t, \cos t - \sin t, 1 \rangle$$

$$\begin{aligned}
\|\mathbf{r}'(t)\| &= e^t \sqrt{(\sin t + \cos t)^2 + (\cos t - \sin t)^2 + 1^2} \\
&= e^t (\sin^2 t + 2\sin t\cos t + \cos^2 t + \cos^2 t - 2\sin t\cos t + \sin^2 t + 1)^{1/2} \\
&= e^t\sqrt{2(\sin^2 t + \cos^2 t) + 1} = e^t\sqrt{2 \cdot 1 + 1} = \sqrt{3}\,e^t
\end{aligned} \tag{2}$$

Substituting (2) into (1) gives:

$$s(t) = \int_0^t \sqrt{3}\,e^u\,du = \sqrt{3}\,e^u\Big|_0^t = \sqrt{3}(e^t - e^0) = \sqrt{3}(e^t - 1)$$

We find the inverse function of $s(t)$ by solving $s = \sqrt{3}\left(e^t - 1\right)$ for t. We obtain:

$$s = \sqrt{3}(e^t - 1)$$

$$\frac{s}{\sqrt{3}} = e^t - 1$$

$$e^t = 1 + \frac{s}{\sqrt{3}} \quad \Rightarrow \quad t = \varphi(s) = \ln\left(1 + \frac{s}{\sqrt{3}}\right)$$

An arc length parametrization for $\mathbf{r}_1(s) = \mathbf{r}(\varphi(s))$ is:

$$\left\langle e^{\ln(1+s/(\sqrt{3}))}\sin\left(\ln\left(1 + \frac{s}{\sqrt{3}}\right)\right), e^{\ln(1+s/(\sqrt{3}))}\cos\left(\ln\left(1 + \frac{s}{\sqrt{3}}\right)\right), e^{\ln(1+s/(\sqrt{3}))}\right\rangle$$

$$= \left(1 + \frac{s}{\sqrt{3}}\right)\left\langle \sin\left(\ln\left(1 + \frac{s}{\sqrt{3}}\right)\right), \cos\left(\ln\left(1 + \frac{s}{\sqrt{3}}\right)\right), 1\right\rangle$$

21. Express the arc length L of $y = x^3$ for $0 \le x \le 8$ as an integral in two ways, using the parametrizations $\mathbf{r}_1(t) = \langle t, t^3 \rangle$ and $\mathbf{r}_2(t) = \langle t^3, t^9 \rangle$. Do not evaluate the integrals, but use substitution to show that they yield the same result.

SOLUTION For $\mathbf{r}_1(t) = \langle t, t^3 \rangle$ we have $\mathbf{r}_1'(t) = \langle 1, 3t^2 \rangle$ hence $\|\mathbf{r}_1'(t)\| = \sqrt{1+9t^4}$. For $\mathbf{r}_2(t) = \langle t^3, t^9 \rangle$ we have $\mathbf{r}_2'(t) = \langle 3t^2, 9t^8 \rangle$ hence $\|\mathbf{r}_2'(t)\| = \sqrt{9t^4 + 81t^{16}}$. The length L may be computed using the two parametrizations by the following integrals (notice that in the second parametrization $0 \le t^3 \le 8$ hence $0 \le t \le 2$).

$$L = \int_0^8 \|\mathbf{r}_1'(t)\|\, dt = \int_0^8 \sqrt{1+9t^4}\, dt \tag{1}$$

$$L = \int_0^2 \|\mathbf{r}_2'(t)\|\, dt = \int_0^2 \sqrt{1+9t^{12}}\, 3t^2\, dt \tag{2}$$

We use the substitution $u = t^3$, $du = 3t^2\, dt$ in the second integral to obtain:

$$\int_0^2 \sqrt{1+9t^{12}}\, 3t^2\, dt = \int_0^8 \sqrt{1+9u^4}\, du$$

This integral is the same as the integral in (1), in accordance with the well known property: the arc length is independent of the parametrization we choose for the curve.

23. Consider the two springs in Figure 5. One has radius 5 cm, height 4 cm, and makes three complete turns. The other has height 3 cm, radius 4 cm, and makes five complete turns.

(a) Take a guess as to which spring uses more wire.

(b) Compute the lengths of the two springs (use Exercise 22) and compare.

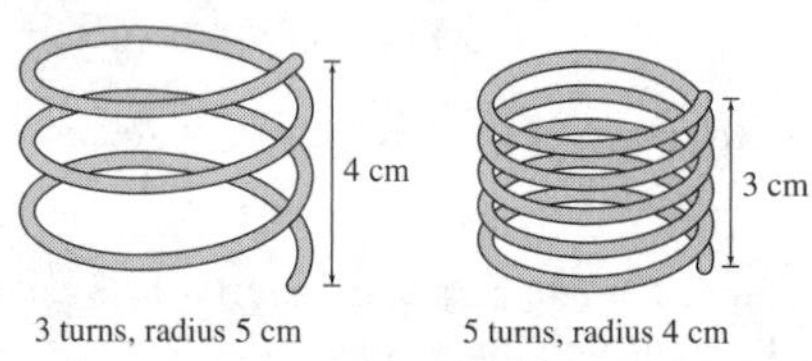

FIGURE 5 Which spring uses more wire?

SOLUTION

(a) The second wire seems to use more wire than the first one.

(b) Setting $R = 5$, $h = 4$ and $N = 3$ in the parametrization in Exercise 22 gives:

$$\mathbf{r}_1(t) = \left\langle 5\cos\frac{2\pi \cdot 3t}{4}, 5\sin\frac{2\pi \cdot 3t}{4}, t \right\rangle = \left\langle 5\cos\frac{3\pi t}{2}, 5\sin\frac{3\pi t}{2}, t \right\rangle, \quad 0 \le t \le 4$$

Setting $R = 4$, $h = 3$ and $N = 5$ in this parametrization we get:

$$\mathbf{r}_2(t) = \left\langle 4\cos\frac{2\pi \cdot 5t}{3}, 4\sin\frac{2\pi \cdot 5t}{3}, t \right\rangle = \left\langle 4\cos\frac{10\pi t}{3}, 4\sin\frac{10\pi t}{3}, t \right\rangle, \quad 0 \le t \le 3$$

We find the derivatives of the two vectors and their lengths:

$$\mathbf{r}_1'(t) = \left\langle -\frac{15\pi}{2}\sin\frac{3\pi t}{2}, \frac{15\pi}{2}\cos\frac{3\pi t}{2}, 1 \right\rangle \quad \Rightarrow \quad \|\mathbf{r}_1'(t)\| = \sqrt{\frac{225\pi^2}{4} + 1} = \frac{1}{2}\sqrt{225\pi^2 + 4}$$

$$\mathbf{r}_2'(t) = \left\langle -\frac{40\pi}{3}\sin\frac{10\pi t}{3}, \frac{40\pi}{3}\cos\frac{10\pi t}{3}, 1 \right\rangle \quad \Rightarrow \quad \|\mathbf{r}_2'(t)\| = \sqrt{\frac{1{,}600\pi^2}{9} + 1} = \frac{1}{3}\sqrt{1{,}600\pi^2 + 9}$$

Using the Arc Length Formula we obtain the following lengths:

$$L_1 = \int_0^4 \frac{1}{2}\sqrt{225\pi^2 + 4}\, dt = 2\sqrt{225\pi^2 + 4} \approx 94.3$$

$$L_2 = \int_0^3 \frac{1}{3}\sqrt{1{,}600\pi^2 + 9}\, dt = \sqrt{1{,}600\pi^2 + 9} \approx 125.7$$

We see that the second spring uses more wire than the first one.

25. Evaluate $s(t) = \int_{-\infty}^{t} \|\mathbf{r}'(u)\|\, du$ for the **Bernoulli spiral** $\mathbf{r}(t) = \langle e^t \cos 4t, e^t \sin 4t \rangle$ (Figure 6). It is convenient to take $-\infty$ as the lower limit since $s(-\infty) = 0$. Then:

(a) Use s to obtain an arc length parametrization of $\mathbf{r}(t)$.

(b) Prove that the angle between the position vector and the tangent vector is constant.

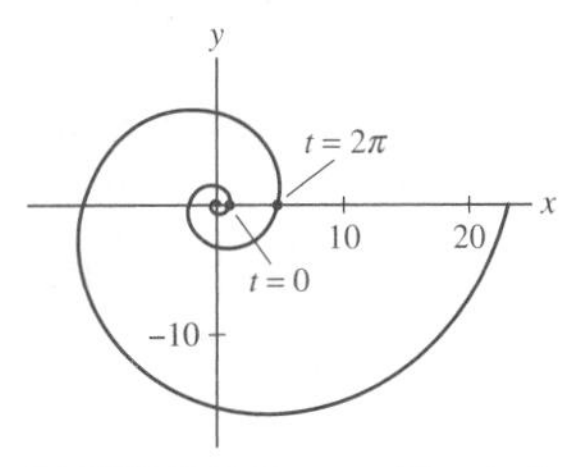

FIGURE 6 Bernoulli spiral.

SOLUTION

(a) We differentiate $\mathbf{r}(t)$ and compute the norm of the derivative vector. This gives:

$$\mathbf{r}'(t) = \left\langle e^t \cos 4t - 4e^t \sin 4t, e^t \sin 4t + 4e^t \cos 4t \right\rangle = e^t \left\langle \cos 4t - 4\sin 4t, \sin 4t + 4\cos 4t \right\rangle$$

$$\begin{aligned}\|\mathbf{r}'(t)\| &= e^t \sqrt{(\cos 4t - 4\sin 4t)^2 + (\sin 4t + 4\cos 4t)^2} \\ &= e^t \left(\cos^2 4t - 8\cos 4t \sin 4t + 16\sin^2 4t + \sin^2 4t + 8\sin 4t \cos 4t + 16\cos^2 4t\right)^{1/2} \\ &= e^t \sqrt{\cos^2 4t + \sin^2 4t + 16\left(\sin^2 4t + \cos^2 4t\right)} = e^t \sqrt{1 + 16 \cdot 1} = \sqrt{17}e^t\end{aligned}$$

We now evaluate the improper integral:

$$\begin{aligned}s(t) &= \int_{-\infty}^{t} \|\mathbf{r}'(u)\|\, du = \lim_{R\to-\infty} \int_R^t \sqrt{17}e^u\, du = \lim_{R\to-\infty} \sqrt{17}e^u \Big|_R^t = \lim_{R\to-\infty} \sqrt{17}(e^t - e^R) \\ &= \sqrt{17}(e^t - 0) = \sqrt{17}e^t\end{aligned}$$

An arc length parametrization of $\mathbf{r}(t)$ is $\mathbf{r}_1(s) = \mathbf{r}\left(\varphi(s)\right)$ where $t = \varphi(s)$ is the inverse function of $s(t)$. We find $t = \varphi(s)$ by solving $s = \sqrt{17}e^t$ for t:

$$s = \sqrt{17}e^t \quad \Rightarrow \quad e^t = \frac{s}{\sqrt{17}} \Rightarrow t = \varphi(s) = \ln \frac{s}{\sqrt{17}}$$

An arc length parametrization of $\mathbf{r}(t)$ is:

$$\begin{aligned}\mathbf{r}_1(s) = \mathbf{r}\left(\varphi(s)\right) &= \left\langle e^{\ln(s/(\sqrt{17}))} \cos\left(4\ln\frac{s}{\sqrt{17}}\right), e^{\ln(s/(\sqrt{17}))} \sin\left(4\ln\frac{s}{\sqrt{17}}\right)\right\rangle \\ &= \frac{s}{\sqrt{17}} \left\langle \cos\left(4\ln\frac{s}{\sqrt{17}}\right), \sin\left(4\ln\frac{s}{\sqrt{17}}\right)\right\rangle \qquad (1)\end{aligned}$$

(b) The cosine of the angle θ between the position vector $\mathbf{r}_1(s)$ and the tangent vector $\mathbf{r}_1'(s)$ is:

$$\cos\theta = \frac{\mathbf{r}_1(s) \cdot \mathbf{r}_1'(s)}{\|\mathbf{r}_1(s)\|\,\|\mathbf{r}_1'(s)\|}$$

Since for the arc length parametrization $\|\mathbf{r}_1'(s)\| = 1$, we obtain:

$$\cos\theta = \frac{\mathbf{r}_1(s) \cdot \mathbf{r}_1'(s)}{\|\mathbf{r}_1(s)\|} \qquad (2)$$

We compute the dot product in (2). We first compute $\mathbf{r}_1'(s)$ from (1):

$$\begin{aligned}\mathbf{r}_1'(s) &= \frac{1}{\sqrt{17}} \left\langle \cos\left(4\ln\frac{s}{\sqrt{17}}\right), \sin\left(4\ln\frac{s}{\sqrt{17}}\right)\right\rangle \\ &\quad + \frac{s}{\sqrt{17}} \left\langle -\sin\left(4\ln\frac{s}{\sqrt{17}}\right) \cdot \frac{4\sqrt{17}}{s} \cdot \frac{1}{\sqrt{17}}, \cos\left(4\ln\frac{s}{\sqrt{17}}\right) \cdot \frac{4\sqrt{17}}{s} \cdot \frac{1}{\sqrt{17}}\right\rangle \\ &= \frac{1}{\sqrt{17}} \left\langle \cos\left(4\ln\frac{s}{\sqrt{17}}\right), \sin\left(4\ln\frac{s}{\sqrt{17}}\right)\right\rangle + \frac{4}{\sqrt{17}} \left\langle -\sin\left(4\ln\frac{s}{\sqrt{17}}\right), \cos\left(4\ln\frac{s}{\sqrt{17}}\right)\right\rangle \\ &= \frac{1}{\sqrt{17}} \left\langle \cos\left(4\ln\frac{s}{\sqrt{17}}\right) - 4\sin\left(4\ln\frac{s}{\sqrt{17}}\right), \sin\left(4\ln\frac{s}{\sqrt{17}}\right) + 4\cos\left(4\ln\frac{s}{\sqrt{17}}\right)\right\rangle\end{aligned}$$

Thus,

$$\mathbf{r}_1(s)\cdot\mathbf{r}_1'(s)=\frac{s}{\sqrt{17}}\cdot\frac{1}{\sqrt{17}}\left\{\cos\left(4\ln\frac{s}{\sqrt{17}}\right)\left(\cos\left(4\ln\frac{s}{\sqrt{17}}\right)-4\sin\left(4\ln\frac{s}{\sqrt{17}}\right)\right)\right.$$
$$\left.+\sin\left(4\ln\frac{s}{\sqrt{17}}\right)\left(\sin\left(4\ln\frac{s}{\sqrt{17}}\right)+4\cos\left(4\ln\frac{s}{\sqrt{17}}\right)\right)\right\}$$
$$=\frac{s}{17}\left(\cos^2\left(4\ln\frac{s}{\sqrt{17}}\right)-4\cos\left(4\ln\frac{s}{\sqrt{17}}\right)\sin\left(4\ln\frac{s}{\sqrt{17}}\right)\right.$$
$$\left.+\sin^2\left(4\ln\frac{s}{\sqrt{17}}\right)+4\sin\left(4\ln\frac{s}{\sqrt{17}}\right)\cos\left(4\ln\frac{s}{\sqrt{17}}\right)\right)$$
$$=\frac{s}{17}\left(\cos^2\left(4\ln\frac{s}{\sqrt{17}}\right)+\sin^2\left(4\ln\frac{s}{\sqrt{17}}\right)\right)=\frac{s}{17}\cdot 1=\frac{s}{17} \tag{3}$$

We now compute $\|\mathbf{r}_1(s)\|$ from (1) (Notice that $s(t)>0$ for all t):

$$\|\mathbf{r}_1(s)\|=\frac{|s|}{\sqrt{17}}\sqrt{\cos^2\left(4\ln\frac{s}{\sqrt{17}}\right)+\sin^2\left(4\ln\frac{s}{\sqrt{17}}\right)}=\frac{s}{\sqrt{17}}\cdot 1=\frac{s}{\sqrt{17}} \tag{4}$$

Combining (2), (3) and (4) yields:

$$\cos\theta=\frac{\frac{s}{17}}{\frac{s}{\sqrt{17}}}=\frac{1}{\sqrt{17}}$$

The solution for $0\le\theta\le\pi$ is $\theta=1.326$ rad. Thus, the angle between $\mathbf{r}_1(s)$ and $\mathbf{r}_1'(s)$ is constant.

Further Insights and Challenges

27. Show that path $\mathbf{r}(t)=\left\langle\frac{1-t^2}{1+t^2},\frac{2t}{1+t^2}\right\rangle$ parametrizes the unit circle with the point $(-1,0)$ excluded for $-\infty<t<\infty$. Use this parametrization to compute the length of the unit circle as an improper integral. *Hint:* The expression for $\|\mathbf{r}'(t)\|$ simplifies.

SOLUTION We have $x(t)=\frac{1-t^2}{1+t^2}$, $y(t)=\frac{2t}{1+t^2}$. Hence,

$$x^2(t)+y^2(t)=\left(\frac{1-t^2}{1+t^2}\right)^2+\left(\frac{2t}{1+t^2}\right)^2=\frac{1-2t^2+t^4+4t^2}{\left(1+t^2\right)^2}=\frac{1+2t^2+t^4}{\left(1+t^2\right)^2}=\frac{\left(1+t^2\right)^2}{\left(1+t^2\right)^2}=1$$

It follows that the path $\mathbf{r}(t)$ lies on the unit circle. We now show that the entire circle is indeed parametrized by $\mathbf{r}(t)$ as t moves from $-\infty$ to ∞. First, note that $x'(t)$ can be written as $\left[-2t(1+t^2)-2t(1-t^2)\right]/(1+t^2)^2$ which is $-4t/(1+t^2)^2$. So, for t negative, $x(t)$ is an increasing function, $y(t)$ is negative, and since $\lim_{t\to-\infty}x(t)=-1$ and $\lim_{t\to 0}x(t)=1$, we conclude that $\mathbf{r}(t)$ does indeed parametrize the lower half of the circle for negative t. A similar argument proves that we get the upper half of the circle for positive t. We now compute $\mathbf{r}'(t)$ and its length:

$$\mathbf{r}'(t)=\left\langle\frac{-2t(1+t^2)-2t(1-t^2)}{(1+t^2)^2},\frac{2(1+t^2)-2t\cdot 2t}{(1+t^2)^2}\right\rangle$$
$$=\left\langle-\frac{4t}{(1+t^2)^2},\frac{2-2t^2}{(1+t^2)^2}\right\rangle=\frac{1}{\left(1+t^2\right)^2}\left\langle-4t,2(1-t^2)\right\rangle$$
$$\|\mathbf{r}'(t)\|=\frac{1}{(1+t^2)^2}\sqrt{16t^2+4(1-t^2)^2}=\frac{2}{(1+t^2)^2}\sqrt{t^4+2t^2+1}$$
$$=\frac{2}{(1+t^2)^2}\sqrt{(t^2+1)^2}=\frac{2(t^2+1)}{\left(1+t^2\right)^2}=\frac{2}{1+t^2}$$

That is,

$$\|\mathbf{r}'(t)\|=\frac{2}{1+t^2}$$

We now use the Arc Length Formula to compute the length of the circle:

$$L=\int_{-\infty}^{\infty}\|\mathbf{r}'(t)\|\,dt=2\int_{-\infty}^{\infty}\frac{dt}{1+t^2}=2\left(\lim_{R\to\infty}\tan^{-1}R-\lim_{R\to-\infty}\tan^{-1}R\right)=2\left(\frac{\pi}{2}-\left(-\frac{\pi}{2}\right)\right)=2\pi$$

29. The curve $\mathbf{r}(t) = \langle t - \tanh t, \operatorname{sech} t\rangle$ is called a **tractrix**.

(a) Show that the arc length function $s(t) = \int_0^t \|\mathbf{r}'(u)\|\,du$ is equal to $s(t) = \ln(\cosh t)$.

(b) Show that $t = \varphi(s) = \ln(e^s + \sqrt{e^{2s} - 1})$ is an inverse of $s(t)$ and verify that

$$\mathbf{r}_1(s) = \left\langle \tanh^{-1}\left(\sqrt{1 - e^{-2s}}\right) - \sqrt{1 - e^{-2s}}, e^{-s}\right\rangle$$

is an arc length parametrization of the tractrix.

(c) Plot the tractrix if you have a computer algebra system.

SOLUTION

(a) We compute the derivative vector and its length:

$$\mathbf{r}'(t) = \langle 1 - \operatorname{sech}^2 t, -\operatorname{sech} t \tanh t\rangle$$

$$\|\mathbf{r}'(t)\| = \sqrt{(1 - \operatorname{sech}^2 t) + \operatorname{sech}^2 t \tanh^2 t} = \sqrt{1 - 2\operatorname{sech}^2 t + \operatorname{sech}^4 t + \operatorname{sech}^2 t \tanh^2 t}$$

$$= \sqrt{-\operatorname{sech}^2 t(2 - \tanh^2 t) + 1 + \operatorname{sech}^4 t}$$

We use the identity $1 - \tanh^2 t = \operatorname{sech}^2 t$ to write:

$$\|\mathbf{r}'(t)\| = \sqrt{-\operatorname{sech}^2 t(1 + \operatorname{sech}^2 t) + 1 + \operatorname{sech}^4 t} = \sqrt{-\operatorname{sech}^2 t - \operatorname{sech}^4 t + 1 + \operatorname{sech}^4 t}$$

$$= \sqrt{1 - \operatorname{sech}^2 t} = \sqrt{\tanh^2 t} = |\tanh t|$$

For $t \ge 0$, $\tanh t \ge 0$ hence, $\|\mathbf{r}'(t)\| = \tanh t$. We now apply the Arc Length Formula to obtain:

$$s(t) = \int_0^t \|\mathbf{r}'(u)\|\,du = \int_0^t (\tanh u)\,du = \ln(\cosh u)\Big|_0^t = \ln(\cosh t) - \ln(\cosh 0) = \ln(\cosh t) - \ln 1 = \ln(\cosh t)$$

That is:

$$s(t) = \ln(\cosh t)$$

(b) We show that the function $t = \varphi(s) = \ln\left(e^s + \sqrt{e^{2s} - 1}\right)$ is an inverse of $s(t)$. First we note that $s'(t) = \tanh t$, hence $s'(t) > 0$ for $t > 0$, which implies that $s(t)$ has an inverse function for $t \ge 0$. Therefore, it suffices to verify that $\varphi(s(t)) = t$. We have:

$$\varphi(s(t)) = \ln\left(e^{\ln(\cosh t)} + \sqrt{e^{2\ln(\cosh t)} - 1}\right) = \ln\left(\cosh t + \sqrt{\cosh^2 t - 1}\right)$$

Since $\cosh^2 t - 1 = \sinh^2 t$ we obtain (for $t \ge 0$):

$$\varphi(s(t)) = \ln\left(\cosh t + \sqrt{\sinh^2 t}\right) = \ln(\cosh t + \sinh t) = \ln\left(\frac{e^t + e^{-t}}{2} + \frac{e^t - e^{-t}}{2}\right) = \ln\left(e^t\right) = t$$

We thus proved that $t = \varphi(s)$ is an inverse of $s(t)$. Therefore, the arc length parametrization is obtained by substituting $t = \varphi(s)$ in $\mathbf{r}(t) = \langle t - \tanh t, \operatorname{sech} t\rangle$. We compute t, $\tanh t$ and $\operatorname{sech} t$ in terms of s. We have:

$$s = \ln(\cosh t) \quad\Rightarrow\quad e^s = \cosh t \quad\Rightarrow\quad \operatorname{sech} t = e^{-s}$$

Also:

$$\tanh^2 t = 1 - \operatorname{sech}^2 t = 1 - e^{-2s} \quad\Rightarrow\quad \tanh t = \sqrt{1 - e^{-2s}} \quad\Rightarrow\quad t = \tanh^{-1}\sqrt{1 - e^{-2s}}$$

Substituting in $\mathbf{r}(t)$ gives:

$$\mathbf{r}_1(s) = \langle t - \tanh t, \operatorname{sech} t\rangle = \left\langle \tanh^{-1}\sqrt{1 - e^{-2s}} - \sqrt{1 - e^{-2s}}, e^{-s}\right\rangle$$

(c) The tractrix is shown in the following figure:

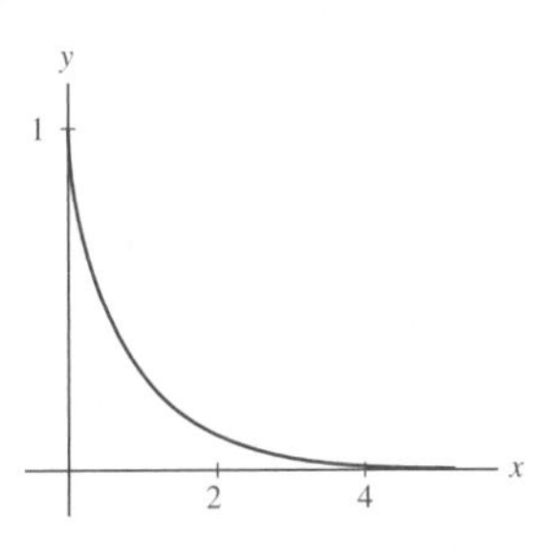

14.4 Curvature (ET Section 13.4)

Preliminary Questions

1. What is the unit tangent vector of a line with direction vector $\mathbf{v} = \langle 2, 1, -2 \rangle$?

SOLUTION A line with direction vector $\mathbf{v}$ has the parametrization:

$$\mathbf{r}(t) = \overrightarrow{OP_0} + t\mathbf{v}$$

hence, since $\overrightarrow{OP_0}$ and $\mathbf{v}$ are constant vectors, we have:

$$\mathbf{r}'(t) = \mathbf{v}$$

Therefore, since $\|\mathbf{v}\| = 3$, the unit tangent vector is:

$$\mathbf{T}(t) = \frac{\mathbf{r}'(t)}{\|\mathbf{r}'(t)\|} = \frac{\mathbf{v}}{\|\mathbf{v}\|} = \langle 2/3, 1/3, -2/3 \rangle$$

2. What is the curvature of a circle of radius 4?

SOLUTION The curvature of a circle of radius R is $\frac{1}{R}$, hence the curvature of a circle of radius 4 is $\frac{1}{4}$.

3. Which has larger curvature, a circle of radius 2 or a circle of radius 4?

SOLUTION The curvature of a circle of radius 2 is $\frac{1}{2}$, and it is larger than the curvature of a circle of radius 4, which is $\frac{1}{4}$.

4. What is the curvature of $\mathbf{r}(t) = \langle 2 + 3t, 7t, 5 - t \rangle$?

SOLUTION $\mathbf{r}(t)$ parametrizes the line $\langle 2, 0, 5 \rangle + t\,\langle 3, 7, -1 \rangle$, and a line has zero curvature.

5. What is the curvature at a point where $\mathbf{T}'(s) = \langle 1, 2, 3 \rangle$ in an arc length parametrization $\mathbf{r}(s)$?

SOLUTION The curvature is given by the formula:

$$\kappa(t) = \frac{\|\mathbf{T}'(t)\|}{\|\mathbf{r}'(t)\|}$$

In an arc length parametrization, $\|\mathbf{r}'(t)\| = 1$ for all t, hence the curvature is $\kappa(t) = \|\mathbf{T}'(t)\|$. Using the given information we obtain the following curvature:

$$\kappa = \|\,\langle 1, 2, 3 \rangle\,\| = \sqrt{1^2 + 2^2 + 3^2} = \sqrt{14}$$

6. What is the radius of curvature of a circle of radius 4?

SOLUTION The definition of the osculating circle implies that the osculating circles at the points of a circle, is the circle itself. Therefore, the radius of curvature is the radius of the circle, that is, 4.

7. What is the radius of curvature at P if $\kappa_P = 9$?

SOLUTION The radius of curvature is the reciprocal of the curvature, hence the radius of curvature at P is:

$$R = \frac{1}{\kappa_P} = \frac{1}{9}$$

Exercises

In Exercises 1–6, calculate $\mathbf{r}'(t)$ *and* $\mathbf{T}(t)$, *and evaluate* $\mathbf{T}(1)$.

1. $\mathbf{r}(t) = \left\langle 12t^3, 18t^2, 9t^4 \right\rangle$

SOLUTION The derivative vector is:

$$\mathbf{r}'(t) = \left\langle 36t^2, 36t, 36t^3 \right\rangle = 36\left\langle t^2, t, t^3 \right\rangle \quad \Rightarrow \quad \|\mathbf{r}'(t)\| = 36\sqrt{t^4 + t^2 + t^6}$$

The unit vector is thus:

$$\mathbf{T}(t) = \frac{\mathbf{r}'(t)}{\|\mathbf{r}'(t)\|} = \frac{36\langle t^2, t, t^3 \rangle}{36\sqrt{t^4 + t^2 + t^6}} = \frac{t\langle t, 1, t^2 \rangle}{|t|\sqrt{1 + t^2 + t^4}} = \frac{t}{|t|\sqrt{1 + t^2 + t^4}}\langle t, 1, t^2 \rangle$$

At $t = 1$ we have:

$$\mathbf{T}(1) = \frac{1}{\sqrt{1 + 1^2 + 1^4}}\langle 1, 1, 1^2 \rangle = \frac{1}{\sqrt{3}}\langle 1, 1, 1 \rangle = \left\langle \frac{1}{\sqrt{3}}, \frac{1}{\sqrt{3}}, \frac{1}{\sqrt{3}} \right\rangle.$$

3. $\mathbf{r}(t) = \langle 3 + 4t, 3 - 5t, 9t \rangle$

SOLUTION We first find the vector $\mathbf{r}'(t)$ and its length:

$$\mathbf{r}'(t) = \langle 4, -5, 9 \rangle \quad \Rightarrow \quad \|\mathbf{r}'(t)\| = \sqrt{4^2 + (-5)^2 + 9^2} = \sqrt{122}$$

The unit tangent vector is therefore:

$$\mathbf{T}(t) = \frac{\mathbf{r}'(t)}{\|\mathbf{r}'(t)\|} = \frac{1}{\sqrt{122}} \langle 4, -5, 9 \rangle = \left\langle \frac{4}{\sqrt{122}}, -\frac{5}{\sqrt{122}}, \frac{9}{\sqrt{122}} \right\rangle$$

We see that the unit tangent vector is constant, since the curve is a straight line.

5. $\mathbf{r}(t) = \langle 4t^2, 9t \rangle$

SOLUTION We differentiate $\mathbf{r}(t)$ to obtain:

$$\mathbf{r}'(t) = \langle 8t, 9 \rangle \quad \Rightarrow \quad \|\mathbf{r}'(t)\| = \sqrt{(8t)^2 + 9^2} = \sqrt{64t^2 + 81}$$

We now find the unit tangent vector:

$$\mathbf{T}(t) = \frac{\mathbf{r}'(t)}{\|\mathbf{r}'(t)\|} = \frac{1}{\sqrt{64t^2 + 81}} \langle 8t, 9 \rangle$$

For $t = 1$ we obtain the vector:

$$\mathbf{T}(t) = \frac{1}{\sqrt{64 + 81}} \langle 8, 9 \rangle = \left\langle \frac{8}{\sqrt{145}}, \frac{9}{\sqrt{145}} \right\rangle.$$

In Exercises 7–12, use Eq. (3) to calculate $\kappa(t)$.

7. $\mathbf{r}(t) = \langle 1, e^t, t \rangle$

SOLUTION We compute the first and the second derivatives of $\mathbf{r}(t)$:

$$\mathbf{r}'(t) = \langle 0, e^t, 1 \rangle, \quad \mathbf{r}''(t) = \langle 0, e^t, 0 \rangle.$$

Next, we find the cross product $\mathbf{r}'(t) \times \mathbf{r}''(t)$:

$$\mathbf{r}'(t) \times \mathbf{r}''(t) = \begin{vmatrix} \mathbf{i} & \mathbf{j} & \mathbf{k} \\ 0 & e^t & 1 \\ 0 & e^t & 0 \end{vmatrix} = \begin{vmatrix} e^t & 1 \\ e^t & 0 \end{vmatrix} \mathbf{i} - \begin{vmatrix} 0 & 1 \\ 0 & 0 \end{vmatrix} \mathbf{j} + \begin{vmatrix} 0 & e^t \\ 0 & e^t \end{vmatrix} \mathbf{k} = -e^t \mathbf{i} = \langle -e^t, 0, 0 \rangle$$

We need to find the lengths of the following vectors:

$$\|\mathbf{r}'(t) \times \mathbf{r}''(t)\| = \left| \langle -e^t, 0, 0 \rangle \right| = e^t$$

$$\|\mathbf{r}'(t)\| = \sqrt{0^2 + (e^t)^2 + 1^2} = \sqrt{1 + e^{2t}}$$

We now use the formula for curvature to calculate $\kappa(t)$:

$$\kappa(t) = \frac{\|\mathbf{r}'(t) \times \mathbf{r}''(t)\|}{\|\mathbf{r}'(t)\|^3} = \frac{e^t}{\left(\sqrt{1 + e^{2t}}\right)^3} = \frac{e^t}{\left(1 + e^{2t}\right)^{3/2}}$$

9. $\mathbf{r}(t) = \langle 4\cos t, t, 4\sin t \rangle$

SOLUTION By the formula for curvature we have:

$$\kappa(t) = \frac{\|\mathbf{r}'(t) \times \mathbf{r}''(t)\|}{\|\mathbf{r}'(t)\|^3} \tag{1}$$

First we find $\mathbf{r}'(t)$ and $\mathbf{r}''(t)$:

$$\mathbf{r}'(t) = \langle -4\sin t, 1, 4\cos t \rangle$$

$$\mathbf{r}''(t) = \langle -4\cos t, 0, -4\sin t \rangle$$

We compute the cross product:

$$\mathbf{r}'(t) \times \mathbf{r}''(t) = \begin{vmatrix} \mathbf{i} & \mathbf{j} & \mathbf{k} \\ -4\sin t & 1 & 4\cos t \\ -4\cos t & 0 & -4\sin t \end{vmatrix}$$

$$= \begin{vmatrix} 1 & 4\cos t \\ 0 & -4\sin t \end{vmatrix} \mathbf{i} - \begin{vmatrix} -4\sin t & 4\cos t \\ -4\cos t & -4\sin t \end{vmatrix} \mathbf{j} + \begin{vmatrix} -4\sin t & 1 \\ -4\cos t & 0 \end{vmatrix} \mathbf{k}$$

$$= -4\sin t\mathbf{i} - \left(16\sin^2 t + 16\cos^2 t\right)\mathbf{j} + 4\cos t\mathbf{k}$$

$$= -4\sin t\mathbf{i} - 16\mathbf{j} + 4\cos t\mathbf{k} = 4\langle -\sin t, -4, \cos t\rangle$$

We compute the lengths of the following vectors:

$$\|\mathbf{r}'(t) \times \mathbf{r}''(t)\| = 4\sqrt{(-\sin t)^2 + (-4)^2 + \cos^2 t} = 4\sqrt{\sin^2 t + 16 + \cos^2 t} = 4\sqrt{17}\|\mathbf{r}'(t)\|$$

$$= \sqrt{(-4\sin t)^2 + 1^2 + (4\cos t)^2} = \sqrt{16\sin^2 t + 1 + 16\cos^2 t} = \sqrt{17}$$

Substituting in (1) gives the following curvature:

$$\kappa(t) = \frac{4\sqrt{17}}{\left(\sqrt{17}\right)^3} = \frac{4\sqrt{17}}{17\sqrt{17}} = \frac{4}{17}$$

We see that this curve has constant curvature.

11. $\mathbf{r}(t) = \langle t^{-1}, 1, t\rangle$

SOLUTION By the formula for curvature we have:

$$\kappa(t) = \frac{\|\mathbf{r}'(t) \times \mathbf{r}''(t)\|}{\|\mathbf{r}'(t)\|^3} \tag{1}$$

We now find $\mathbf{r}'(t)$, $\mathbf{r}''(t)$ and their cross product. This gives:

$$\mathbf{r}'(t) = \langle -t^{-2}, 0, 1\rangle, \quad \mathbf{r}''(t) = \langle 2t^{-3}, 0, 0\rangle$$

$$\mathbf{r}'(t) \times \mathbf{r}''(t) = \left(-t^{-2}\mathbf{i} + \mathbf{k}\right) \times 2t^{-3}\mathbf{i} = 2t^{-3}\mathbf{k} \times \mathbf{i} = 2t^{-3}\mathbf{j}$$

We compute the lengths of the vector in (1):

$$\|\mathbf{r}'(t) \times \mathbf{r}''(t)\| = \|2t^{-3}\mathbf{j}\| = 2|t^{-3}|$$

$$\|\mathbf{r}'(t)\| = \sqrt{\left((-t)^{-2}\right)^2 + 0^2 + 1^2} = \sqrt{t^{-4} + 1}$$

Substituting in (1) we obtain the following curvature:

$$\kappa(t) = \frac{2|t|^{-3}}{\left(\sqrt{t^{-4} + 1}\right)^3} = \frac{2|t|^{-3}}{\left(t^{-4} + 1\right)^{3/2}}$$

We multiply through by $|t|^{4 \cdot 3/2} = |t|^6$ to obtain:

$$\kappa(t) = \frac{2|t|^3}{\left(1 + |t|^4\right)^{3/2}}$$

In Exercises 13–16, find the curvature of the plane curve at the point indicated.

13. $y = e^t, \quad t = 3$

SOLUTION We use the curvature of a graph in the plane:

$$\kappa(t) = \frac{|f''(t)|}{\left(1 + f'(t)^2\right)^{3/2}}$$

In our case $f(t) = e^t$, hence $f'(t) = f''(t) = e^t$ and we obtain:

$$\kappa(t) = \frac{e^t}{\left(1 + e^{2t}\right)^{3/2}} \quad \Rightarrow \quad \kappa(3) = \frac{e^3}{\left(1 + e^6\right)^{3/2}} \approx 0.0025$$

15. $y = t^4, \quad t = 2$

SOLUTION By the curvature of a graph in the plane, we have:

$$\kappa(t) = \frac{|f''(t)|}{\left(1 + f'(t)^2\right)^{3/2}}$$

In this case $f(t) = t^4$, $f'(t) = 4t^3$, $f''(t) = 12t^2$. Hence,

$$\kappa(t) = \frac{12t^2}{\left(1 + \left(4t^3\right)^2\right)^{3/2}} = \frac{12t^2}{\left(1 + 16t^6\right)^{3/2}}$$

At $t = 2$ we obtain the following curvature:

$$\kappa(2) = \frac{12 \cdot 2^2}{(1 + 16 \cdot 2^6)^{3/2}} = \frac{48}{(1{,}025)^{3/2}} \approx 0.0015.$$

17. Find the curvature of $\mathbf{r}(t) = \langle 2\sin t, \cos 3t, t\rangle$ at $t = \frac{\pi}{3}$ and $t = \frac{\pi}{2}$ (Figure 15).

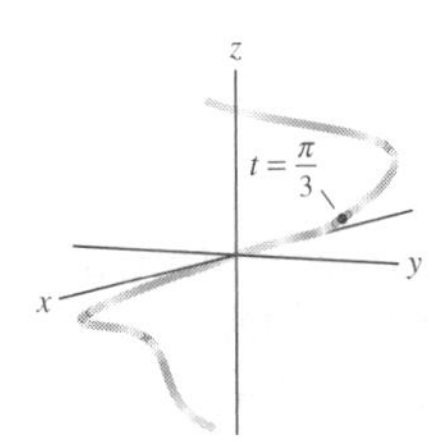

FIGURE 15 The curve $\mathbf{r}(t) = \langle 2\sin t, \cos 3t, t\rangle$.

SOLUTION By the formula for curvature we have:

$$\kappa(t) = \frac{\|\mathbf{r}'(t) \times \mathbf{r}''(t)\|}{\|\mathbf{r}'(t)\|^3} \tag{1}$$

We compute the first and second derivatives:

$$\mathbf{r}'(t) = \langle 2\cos t, -3\sin 3t, 1\rangle, \quad \mathbf{r}''(t) = \langle -2\sin t, -9\cos 3t, 0\rangle$$

At the points $t = \frac{\pi}{3}$ and $t = \frac{\pi}{2}$ we have:

$$\mathbf{r}'\left(\frac{\pi}{3}\right) = \left\langle 2\cos\frac{\pi}{3}, -3\sin\frac{3\pi}{3}, 1\right\rangle = \left\langle 2\cos\frac{\pi}{3}, -3\sin\pi, 1\right\rangle = \langle 1, 0, 1\rangle$$

$$\mathbf{r}''\left(\frac{\pi}{3}\right) = \left\langle -2\sin\frac{\pi}{3}, -9\cos\frac{3\pi}{3}, 0\right\rangle = \left\langle -\sqrt{3}, 9, 0\right\rangle$$

$$\mathbf{r}'\left(\frac{\pi}{2}\right) = \left\langle 2\cos\frac{\pi}{2}, -3\sin\frac{3\pi}{2}, 1\right\rangle = \langle 0, 3, 1\rangle$$

$$\mathbf{r}''\left(\frac{\pi}{2}\right) = \left\langle -2\sin\frac{\pi}{2}, -9\cos\frac{3\pi}{2}, 0\right\rangle = \langle -2, 0, 0\rangle$$

We compute the cross products required to use (1):

$$\mathbf{r}'\left(\frac{\pi}{3}\right) \times \mathbf{r}''\left(\frac{\pi}{3}\right) = \begin{vmatrix} \mathbf{i} & \mathbf{j} & \mathbf{k} \\ 1 & 0 & 1 \\ -\sqrt{3} & 9 & 0 \end{vmatrix} = \begin{vmatrix} 0 & 1 \\ 9 & 0 \end{vmatrix}\mathbf{i} - \begin{vmatrix} 1 & 1 \\ -\sqrt{3} & 0 \end{vmatrix}\mathbf{j} + \begin{vmatrix} 1 & 0 \\ -\sqrt{3} & 9 \end{vmatrix}\mathbf{k} = -9\mathbf{i} - \sqrt{3}\mathbf{j} + 9\mathbf{k}$$

$$\mathbf{r}'\left(\frac{\pi}{2}\right) \times \mathbf{r}''\left(\frac{\pi}{2}\right) = \begin{vmatrix} \mathbf{i} & \mathbf{j} & \mathbf{k} \\ 0 & 3 & 1 \\ -2 & 0 & 0 \end{vmatrix} = \begin{vmatrix} 3 & 1 \\ 0 & 0 \end{vmatrix}\mathbf{i} - \begin{vmatrix} 0 & 1 \\ -2 & 0 \end{vmatrix}\mathbf{j} + \begin{vmatrix} 0 & 3 \\ -2 & 0 \end{vmatrix}\mathbf{k} = -2\mathbf{j} + 6\mathbf{k}$$

Hence,

$$\left\|\mathbf{r}'\left(\frac{\pi}{3}\right) \times \mathbf{r}''\left(\frac{\pi}{3}\right)\right\| = \sqrt{(-9)^2 + \left(-\sqrt{3}\right)^2 + 9^2} = \sqrt{165}$$

$$\left\|\mathbf{r}'\left(\frac{\pi}{3}\right)\right\| = \sqrt{1^2 + 0^2 + 1^2} = \sqrt{2}$$

At $t = \frac{\pi}{2}$ we have:

$$\left\| \mathbf{r}'\left(\frac{\pi}{2}\right) \times \mathbf{r}''\left(\frac{\pi}{2}\right) \right\| = \sqrt{(-2)^2 + 6^2} = \sqrt{40} = 2\sqrt{10}$$

$$\left\| \mathbf{r}'\left(\frac{\pi}{2}\right) \right\| = \sqrt{0^2 + 3^2 + 1^2} = \sqrt{10}$$

Substituting the values for $t = \frac{\pi}{3}$ and $t = \frac{\pi}{2}$ in (1) we obtain the following curvatures:

$$\kappa\left(\frac{\pi}{3}\right) = \frac{\sqrt{165}}{\left(\sqrt{2}\right)^3} = \frac{\sqrt{165}}{2\sqrt{2}} \approx 4.54$$

$$\kappa\left(\frac{\pi}{2}\right) = \frac{2\sqrt{10}}{\left(\sqrt{10}\right)^3} = \frac{2\sqrt{10}}{10\sqrt{10}} = 0.2$$

19. Show that curvature at an inflection point of a plane curve $y = f(x)$ is zero.

SOLUTION The curvature of the graph $y = f(x)$ in the plane is the following function:

$$\kappa(x) = \frac{|f''(x)|}{\left(1 + f'(x)^2\right)^{3/2}} \tag{1}$$

At an inflection point the second derivative changes its sign. Therefore, if f'' is continuous at the inflection point, it is zero at this point, hence by (1) the curvature at this point is zero.

21. Find the value of α such that the curvature of $y = e^{\alpha x}$ at $x = 0$ is as large as possible.

SOLUTION Using the curvature of a graph in the plane we have:

$$\kappa(x) = \frac{|y''(x)|}{\left(1 + y'(x)^2\right)^{3/2}} \tag{1}$$

In our case $y'(x) = \alpha e^{\alpha x}$, $y''(x) = \alpha^2 e^{\alpha x}$. Substituting in (1) we obtain

$$\kappa(x) = \frac{\alpha^2 e^{\alpha x}}{\left(1 + \alpha^2 e^{2\alpha x}\right)^{3/2}}$$

The curvature at the origin is thus

$$\kappa(0) = \frac{\alpha^2 e^{\alpha \cdot 0}}{\left(1 + \alpha^2 e^{2\alpha \cdot 0}\right)^{3/2}} = \frac{\alpha^2}{\left(1 + \alpha^2\right)^{3/2}}$$

Since $\kappa(0)$ and $\kappa^2(0)$ have their maximum values at the same values of α, we may maximize the function:

$$g(\alpha) = \kappa^2(0) = \frac{\alpha^4}{(1 + \alpha^2)^3}$$

We find the stationary points:

$$g'(\alpha) = \frac{4\alpha^3(1 + \alpha^2)^3 - \alpha^4(3)(1 + \alpha^2)^2 2\alpha}{(1 + \alpha^2)^6} = \frac{2\alpha^3(1 + \alpha^2)^2(2 - \alpha^2)}{(1 + \alpha^2)^6} = 0$$

The stationary points are the solutions of the following equation:

$$2\alpha^3(1 + \alpha^2)^2(2 - \alpha^2) = 0$$

$$\swarrow \qquad\qquad \searrow$$

$$\alpha^3 = 0 \quad \text{or} \quad 2 - \alpha^2 = 0$$

$$\alpha = 0 \qquad\qquad \alpha = \pm\sqrt{2}$$

Since $g(\alpha) \ge 0$ and $g(0) = 0$, $\alpha = 0$ is a minimum point. Also, $g'(\alpha)$ is positive immediately to the left of $\sqrt{2}$ and negative to the right. Hence, $\alpha = \sqrt{2}$ is a maximum point. Since $g(\alpha)$ is an even function, $\alpha = -\sqrt{2}$ is a maximum point as well. Conclusion: $\kappa(x)$ takes its maximum value at the origin when $\alpha = \pm\sqrt{2}$.

23. Show that the curvature function of the parametrization $\mathbf{r}(t) = \langle a\cos t, b\sin t \rangle$ of the ellipse $\left(\frac{x}{a}\right)^2 + \left(\frac{y}{b}\right)^2 = 1$ is

$$\kappa(t) = \frac{ab}{(b^2\cos^2 t + a^2\sin^2 t)^{3/2}}$$

8

SOLUTION The curvature is the following function:

$$\kappa(t) = \frac{\|\mathbf{r}'(t) \times \mathbf{r}''(t)\|}{\|\mathbf{r}'(t)\|^3} \tag{1}$$

We compute the derivatives and their cross product:

$$\mathbf{r}'(t) = \langle -a\sin t, b\cos t\rangle, \mathbf{r}''(t) = \langle -a\cos t, -b\sin t\rangle$$

$$\begin{aligned}\mathbf{r}'(t) \times \mathbf{r}''(t) &= (-a\sin t\mathbf{i} + b\cos t\mathbf{j}) \times (-a\cos t\mathbf{i} - b\sin t\mathbf{j})\\ &= ab\sin^2 t\mathbf{k} + ab\cos^2 t\mathbf{k} = ab\left(\sin^2 t + \cos^2 t\right)\mathbf{k} = ab\mathbf{k}\end{aligned}$$

Thus,

$$\|\mathbf{r}'(t) \times \mathbf{r}''(t)\| = \|ab\mathbf{k}\| = ab$$

$$\|\mathbf{r}'(t)\| = \sqrt{(-a\sin t)^2 + (b\cos t)^2} = \sqrt{a^2\sin^2 t + b^2\cos^2 t}$$

Substituting in (1) we obtain the following curvature:

$$\kappa(t) = \frac{ab}{\left(\sqrt{a^2\sin^2 t + b^2\cos^2 t}\right)^3} = \frac{ab}{\left(a^2\sin^2 t + b^2\cos^2 t\right)^{3/2}}$$

25. In the notation of Exercise 23, assume that $a \ge b$. Show that $b/a^2 \le \kappa(t) \le a/b^2$ for all t.

SOLUTION In Exercise 23 we showed that the curvature of the ellipse $\mathbf{r}(t) = \langle a\cos t, b\sin t\rangle$ is the following function:

$$\kappa(t) = \frac{ab}{\left(b^2\cos^2 t + a^2\sin^2 t\right)^{3/2}}$$

Since $a \ge b > 0$ the quotient becomes greater if we replace a by b in the denominator, and it becomes smaller if we replace b by a in the denominator. We use the identity $\cos^2 t + \sin^2 t = 1$ to obtain:

$$\frac{ab}{\left(a^2\cos^2 t + a^2\sin^2 t\right)^{3/2}} \le \kappa(t) \le \frac{ab}{\left(b^2\cos^2 t + b^2\sin^2 t\right)^{3/2}}$$

$$\frac{ab}{\left(a^2\left(\cos^2 t + \sin^2 t\right)\right)^{3/2}} \le \kappa(t) \le \frac{ab}{\left(b^2\left(\cos^2 t + \sin^2 t\right)\right)^{3/2}}$$

$$\frac{ab}{a^3} = \frac{ab}{\left(a^2\right)^{3/2}} \le \kappa(t) \le \frac{ab}{\left(b^2\right)^{3/2}} = \frac{ab}{b^3}$$

$$\frac{b}{a^2} \le \kappa(t) \le \frac{a}{b^2}$$

In Exercises 27–30, use Eq. (9) to compute the curvature at the given point.

27. $\left\langle t^2, t^3\right\rangle$, $t = 2$

SOLUTION For the given parametrization, $x(t) = t^2$, $y(t) = t^3$, hence

$$\begin{aligned}x'(t) &= 2t\\ x''(t) &= 2\\ y'(t) &= 3t^2\\ y''(t) &= 6t\end{aligned}$$

At the point $t = 2$ we have

$$x'(2) = 4, \quad x''(2) = 2, \quad y'(2) = 3 \cdot 2^2 = 12, \quad y''(2) = 12$$

Substituting in Eq. (9) we get

$$\kappa(2) = \frac{|x'(2)y''(2) - x''(2)y'(2)|}{\left(x'(2)^2 + y'(2)^2\right)^{3/2}} = \frac{|4 \cdot 12 - 2 \cdot 12|}{\left(4^2 + 12^2\right)^{3/2}} = \frac{24}{160^{3/2}} \approx 0.012$$

29. $\langle t\cos t, \sin t\rangle$, $t=\pi$

SOLUTION We have $x(t)=t\cos t$ and $y(t)=\sin t$, hence:

$$x'(t)=\cos t - t\sin t \quad\Rightarrow\quad x'(\pi)=\cos\pi-\pi\sin\pi=-1$$

$$x''(t)=-\sin t-(\sin t+t\cos t)=-2\sin t-t\cos t \quad\Rightarrow\quad x''(\pi)=-2\sin\pi-\pi\cos\pi=\pi$$

$$y'(t)=\cos t \quad\Rightarrow\quad y'(\pi)=\cos\pi=-1$$

$$y''(t)=-\sin t \quad\Rightarrow\quad y''(\pi)=-\sin\pi=0$$

Substituting in Eq. (9) gives the following curvature:

$$\kappa(\pi)=\frac{|x'(\pi)y''(\pi)-x''(\pi)y'(\pi)|}{\left(x'(\pi)^2+y'(\pi)^2\right)^{3/2}}=\frac{|-1\cdot 0-\pi\cdot(-1)|}{\left((-1)^2+(-1)^2\right)^{3/2}}=\frac{\pi}{2\sqrt{2}}\approx 1.11$$

31. Let $s(t)=\int_{-\infty}^{t}\|\mathbf{r}'(u)\|\,du$ for the Bernoulli spiral $\mathbf{r}(t)=\langle e^t\cos 4t, e^t\sin 4t\rangle$ (see Exercise 25 in Section 14.3). Show that the radius of curvature is proportional to $s(t)$.

SOLUTION The radius of curvature is the reciprocal of the curvature:

$$R(t)=\frac{1}{\kappa(t)}$$

We compute the curvature using the equality given in Exercise 25 in Section 3:

$$\kappa(t)=\frac{|x'(t)y''(t)-x''(t)y'(t)|}{\left(x'(t)^2+y'(t)^2\right)^{3/2}} \tag{1}$$

In our case, $x(t)=e^t\cos 4t$ and $y(t)=e^t\sin 4t$. Hence:

$$x'(t)=e^t\cos 4t-4e^t\sin 4t=e^t(\cos 4t-4\sin 4t)$$

$$x''(t)=e^t(\cos 4t-4\sin 4t)+e^t(-4\sin 4t-16\cos 4t)=-e^t(15\cos 4t+8\sin 4t)$$

$$y'(t)=e^t\sin 4t+4e^t\cos 4t=e^t(\sin 4t+4\cos 4t)$$

$$y''(t)=e^t(\sin 4t+4\cos 4t)+e^t(4\cos 4t-16\sin 4t)=e^t(8\cos 4t-15\sin 4t)$$

We compute the numerator in (1):

$$\begin{aligned}x'(t)y''(t)-x''(t)y'(t)&=e^{2t}(\cos 4t-4\sin 4t)\cdot(8\cos 4t-15\sin 4t)+e^{2t}(15\cos 4t+8\sin 4t)\cdot(\sin 4t+4\cos 4t)\\&=e^{2t}\left(68\cos^2 4t+68\sin^2 4t\right)=68e^{2t}\end{aligned}$$

We compute the denominator in (1):

$$\begin{aligned}x'(t)^2+y'(t)^2&=e^{2t}(\cos 4t-4\sin 4t)^2+e^{2t}(\sin 4t+4\cos 4t)^2\\&=e^{2t}\left(\cos^2 4t-8\cos 4t\sin 4t+16\sin^2 4t+\sin^2 4t+8\sin 4t\cos 4t+16\cos^2 4t\right)\\&=e^{2t}\left(\cos^2 4t+\sin^2 4t+16\left(\sin^2 4t+\cos^2 4t\right)\right)\\&=e^{2t}(1+16\cdot 1)=17e^{2t}\end{aligned} \tag{2}$$

Hence

$$\left(x'(t)^2+y'(t)^2\right)^{3/2}=17^{3/2}e^{3t}$$

Substituting in (2) we have

$$\kappa(t)=\frac{68e^{2t}}{17^{3/2}e^{3t}}=\frac{4}{\sqrt{17}}e^{-t} \quad\Rightarrow\quad R=\frac{\sqrt{17}}{4}e^t \tag{3}$$

On the other hand, by the Fundamental Theorem and (2) we have

$$s'(t)=\|\mathbf{r}'(t)\|=\sqrt{x'(t)^2+y'(t)^2}=\sqrt{17e^{2t}}=\sqrt{17}e^t$$

We integrate to obtain

$$s(t)=\int\sqrt{17}\,e^t\,dt=\sqrt{17}\,e^t+C \tag{4}$$

Since $s(t) = \int_{-\infty}^{t} \|\mathbf{r}'(u)\|\, du$, we have $\lim_{t\to-\infty} s(t) = 0$, hence by (4):

$$0 = \lim_{t\to-\infty} \left(\sqrt{17}e^t + C\right) = 0 + C = C.$$

Substituting $C = 0$ in (4) we get:

$$s(t) = \sqrt{17}e^t \tag{5}$$

Combining (3) and (5) gives:

$$R(t) = \frac{1}{4}s(t)$$

which means that the radius of curvature is proportional to $s(t)$.

33. CAS Plot and compute the curvature $\kappa(t)$ of the **clothoid** $\mathbf{r}(t) = \langle x(t), y(t)\rangle$, where

$$x(t) = \int_0^t \sin\frac{u^3}{3}\, du, \qquad y(t) = \int_0^t \cos\frac{u^3}{3}\, du$$

SOLUTION We use the following formula for the curvature (given earlier):

$$\kappa(t) = \frac{|x'(t)y''(t) - x''(t)y'(t)|}{\left(x'(t)^2 + y'(t)^2\right)^{3/2}} \tag{1}$$

We compute the first and second derivatives of $x(t)$ and $y(t)$. Using the Fundamental Theorem and the Chain Rule we get:

$$x'(t) = \sin\frac{t^3}{3}$$

$$x''(t) = \frac{3t^2}{3}\cos\frac{t^3}{3} = t^2\cos\frac{t^3}{3}$$

$$y'(t) = \cos\frac{t^3}{3}$$

$$y''(t) = \frac{3t^2}{3}\left(-\sin\frac{t^3}{3}\right) = -t^2\sin\frac{t^3}{3}$$

Substituting in (1) gives the following curvature function:

$$\kappa(t) = \frac{\left|\sin\frac{t^3}{3}\left(-t^2\sin\frac{t^3}{3}\right) - t^2\cos\frac{t^3}{3}\cos\frac{t^3}{3}\right|}{\left(\left(\sin\frac{t^3}{3}\right)^2 + \left(\cos\frac{t^3}{3}\right)^2\right)^{3/2}} = \frac{t^2\left(\sin^2\frac{t^3}{3} + \cos^2\frac{t^3}{3}\right)}{1^{3/2}} = t^2$$

That is, $\kappa(t) = t^2$. Here is a plot of the curvature as a function of t:

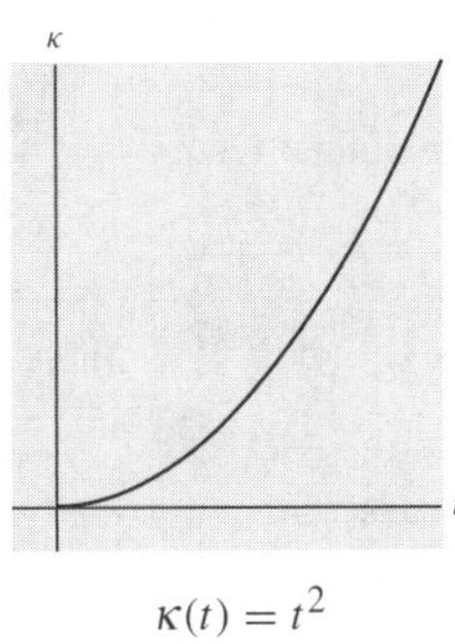

$\kappa(t) = t^2$

35. Find the unit normal vector $\mathbf{N}(t)$ to $\mathbf{r}(t) = \langle 4, \sin 2t, \cos 2t\rangle$.

SOLUTION We first find the unit tangent vector:

$$\mathbf{T}(t) = \frac{\mathbf{r}'(t)}{\|\mathbf{r}'(t)\|} \tag{1}$$

We have

$$\mathbf{r}'(t) = \frac{d}{dt}\langle 4, \sin 2t, \cos 2t\rangle = \langle 0, 2\cos 2t, -2\sin 2t\rangle = 2\langle 0, \cos 2t, -\sin 2t\rangle$$

$$\|\mathbf{r}'(t)\| = 2\sqrt{0^2 + \cos^2 2t + (-\sin 2t)^2} = 2\sqrt{0+1} = 2$$

Substituting in (1) gives:

$$\mathbf{T}(t) = \frac{2\langle 0, \cos 2t, -\sin 2t\rangle}{2} = \langle 0, \cos 2t, -\sin 2t\rangle$$

The normal vector is the following vector:

$$\mathbf{N}(t) = \frac{\mathbf{T}'(t)}{\|\mathbf{T}'(t)\|} \tag{2}$$

We compute the derivative of the unit tangent vector and its length:

$$\mathbf{T}'(t) = \frac{d}{dt}\langle 0, \cos 2t, -\sin 2t\rangle = \langle 0, -2\sin 2t, -2\cos 2t\rangle = -2\langle 0, \sin 2t, \cos 2t\rangle$$

$$\|\mathbf{T}'(t)\| = 2\sqrt{0^2 + \sin^2 2t + \cos^2 2t} = 2\sqrt{0+1} = 2$$

Substituting in (2) we obtain:

$$\mathbf{N}(t) = \frac{-2\langle 0, \sin 2t, \cos 2t\rangle}{2} = \langle 0, -\sin 2t, -\cos 2t\rangle$$

37. Find the normal vectors to $\mathbf{r}(t) = \langle t, \cos t\rangle$ at $t = \frac{\pi}{4}$ and $t = \frac{3\pi}{4}$.

SOLUTION The normal vector to $\mathbf{r}(t) = \langle t, \cos t\rangle$ is $\mathbf{T}'(t)$, where $\mathbf{T}(t) = \frac{\mathbf{r}'(t)}{\|\mathbf{r}'(t)\|}$ is the unit tangent vector. We have

$$\mathbf{r}'(t) = \langle 1, -\sin t\rangle \quad \Rightarrow \quad \|\mathbf{r}'(t)\| = \sqrt{1^2 + (\sin t)^2} = \sqrt{1+\sin^2 t}$$

Hence,

$$\mathbf{T}(t) = \frac{1}{\sqrt{1+\sin^2 t}}\langle 1, -\sin t\rangle$$

We compute the derivative of $\mathbf{T}(t)$ to find the normal vector.We use the Product Rule and the Chain Rule to obtain:

$$\begin{aligned}
\mathbf{T}'(t) &= \frac{1}{\sqrt{1+\sin^2 t}}\frac{d}{dt}\langle 1, -\sin t\rangle + \left(\frac{1}{\sqrt{1+\sin^2 t}}\right)'\langle 1, -\sin t\rangle \\
&= \frac{1}{\sqrt{1+\sin^2 t}}\langle 0, -\cos t\rangle - \frac{1}{1+\sin^2 t}\cdot\frac{2\sin t\cos t}{2\sqrt{1+\sin^2 t}}\langle 1, -\sin t\rangle \\
&= \frac{1}{\sqrt{1+\sin^2 t}}\langle 0, -\cos t\rangle - \frac{\sin 2t}{2\left(1+\sin^2 t\right)^{3/2}}\langle 1, -\sin t\rangle
\end{aligned}$$

At $t = \frac{\pi}{4}$ we obtain the normal vector:

$$\mathbf{T}'\left(\frac{\pi}{4}\right) = \frac{1}{\sqrt{1+\frac{1}{2}}}\left\langle 0, -\frac{1}{\sqrt{2}}\right\rangle - \frac{1}{2\left(1+\frac{1}{2}\right)^{3/2}}\left\langle 1, -\frac{1}{\sqrt{2}}\right\rangle = \left\langle 0, -\frac{1}{\sqrt{3}}\right\rangle - \left\langle \frac{\sqrt{2}}{3\sqrt{3}}, \frac{-1}{3\sqrt{3}}\right\rangle = \left\langle \frac{-\sqrt{2}}{3\sqrt{3}}, \frac{-2}{3\sqrt{3}}\right\rangle$$

At $t = \frac{3\pi}{4}$ we obtain:

$$\mathbf{T}'\left(\frac{3\pi}{4}\right) = \frac{1}{\sqrt{1+\frac{1}{2}}}\left\langle 0, \frac{1}{\sqrt{2}}\right\rangle - \frac{-1}{2\left(1+\frac{1}{2}\right)^{3/2}}\left\langle 1, -\frac{1}{\sqrt{2}}\right\rangle = \left\langle 0, \frac{1}{\sqrt{3}}\right\rangle + \left\langle \frac{\sqrt{2}}{3\sqrt{3}}, \frac{-1}{3\sqrt{3}}\right\rangle = \left\langle \frac{\sqrt{2}}{3\sqrt{3}}, \frac{2}{3\sqrt{3}}\right\rangle$$

39. Find the unit normal to the clothoid (Exercise 33) at $t = \pi^{1/3}$.

SOLUTION The Clothoid is the plane curve $\mathbf{r}(t) = \langle x(t), y(t)\rangle$ with

$$x(t) = \int_0^t \sin\frac{u^3}{3}\,du, \quad y(t) = \int_0^t \cos\frac{u^3}{3}\,du$$

The unit normal is the following vector:

$$\mathbf{N}(t) = \frac{\mathbf{T}'(t)}{\|\mathbf{T}'(t)\|} \tag{1}$$

We first find the unit tangent vector $\mathbf{T}(t) = \frac{\mathbf{r}'(t)}{\|\mathbf{r}'(t)\|}$. By the Fundamental Theorem we have

$$\mathbf{r}'(t) = \left\langle \sin\frac{t^3}{3}, \cos\frac{t^3}{3}\right\rangle \quad\Rightarrow\quad \|\mathbf{r}'(t)\| = \sqrt{\sin^2\frac{t^3}{3} + \cos^2\frac{t^3}{3}} = \sqrt{1} = 1$$

Hence,

$$\mathbf{T}(t) = \left\langle \sin\frac{t^3}{3}, \cos\frac{t^3}{3}\right\rangle$$

We now differentiate $\mathbf{T}(t)$ using the Chain Rule to obtain:

$$\mathbf{T}'(t) = \left\langle \frac{3t^2}{3}\cos\frac{t^3}{3}, \frac{-3t^2}{3}\sin\frac{t^3}{3}\right\rangle = t^2\left\langle \cos\frac{t^3}{3}, -\sin\frac{t^3}{3}\right\rangle$$

Hence,

$$\|\mathbf{T}'(t)\| = t^2\sqrt{\cos^2\frac{t^3}{3} + \left(-\sin\frac{t^3}{3}\right)^2} = t^2$$

Substituting in (1) we obtain the following unit normal:

$$\mathbf{N}(t) = \left\langle \cos\frac{t^3}{3}, -\sin\frac{t^3}{3}\right\rangle$$

At the point $T = \pi^{1/3}$ the unit normal is

$$\mathbf{N}(\pi^{1/3}) = \left\langle \cos\frac{(\pi^{1/3})^3}{3}, -\sin\frac{(\pi^{1/3})^3}{3}\right\rangle = \left\langle \cos\frac{\pi}{3}, -\sin\frac{\pi}{3}\right\rangle = \left\langle \frac{1}{2}, -\frac{\sqrt{3}}{2}\right\rangle$$

In Exercises 41–46, use Eq. (10) to find **N** *at the point indicated.*

41. $\langle 1+t^2, 2t, t^3\rangle, \quad t = 1$

SOLUTION We compute the values in formula (10). In our case

$$\mathbf{r}(t) = \langle 1+t^2, 2t, t^3\rangle$$

Hence,

$$\mathbf{r}'(t) = \langle 2t, 2, 3t^2\rangle$$
$$\mathbf{r}''(t) = \langle 2, 0, 6t\rangle$$
$$v(t) = \|\mathbf{r}'(t)\| = \sqrt{(2t)^2 + 2^2 + (3t^2)^2} = \sqrt{4t^2+4+9t^4} = \sqrt{4+4t^2+9t^4}$$
$$v'(t) = \frac{8t+36t^3}{2\sqrt{4+4t^2+9t^4}} = \frac{4t+18t^3}{\sqrt{4+4t^2+9t^4}}$$

At the point $t = 1$, we have

$$\mathbf{r}''(1) = \langle 2, 0, 6\rangle, \quad v'(1) = \frac{22}{\sqrt{17}}, \quad T(1) = \frac{\mathbf{r}'(1)}{\|\mathbf{r}'(1)\|} = \frac{\langle 2, 2, 3\rangle}{\sqrt{17}}$$

Hence,

$$\mathbf{r}''(1) - v'(1)\mathbf{T}(1) = \langle 2, 0, 6\rangle - \frac{22}{\sqrt{17}}\cdot\frac{1}{\sqrt{17}}\langle 2, 2, 3\rangle = \langle 2, 0, 6\rangle - \left\langle \frac{44}{17}, \frac{44}{17}, \frac{66}{17}\right\rangle$$

$$= \left\langle -\frac{10}{17}, -\frac{44}{17}, \frac{36}{17} \right\rangle = \frac{1}{17} \langle -10, -44, 36 \rangle$$

$$\|\mathbf{r}''(1) - v'(1)\mathbf{T}(1)\| = \frac{1}{17}\sqrt{(-10)^2 + (-44)^2 + 36^2} = \frac{1}{17}\sqrt{3{,}332} = \frac{14\sqrt{17}}{17} = \frac{14}{\sqrt{17}}$$

Substituting in equation (10) we get

$$\mathbf{N}(1) = \frac{\mathbf{r}''(1) - v'(1)\mathbf{T}(1)}{\|\mathbf{r}''(1) - v'(1)\mathbf{T}(1)\|} = \frac{\frac{1}{17}\langle -10, -44, 36 \rangle}{\frac{14}{\sqrt{17}}} = \frac{1}{7\sqrt{17}} \langle -5, -22, 18 \rangle$$

43. $\langle t - \sin t, 1 - \cos t \rangle, \quad t = \pi$

SOLUTION We use the following equality:

$$\mathbf{N}(t) = \frac{\mathbf{r}''(t) - v'(t)\mathbf{T}(t)}{\|\mathbf{r}''(t) - v'(t)\mathbf{T}(t)\|} \tag{1}$$

We compute the vectors in the above equality. For $r(t) = \langle t - \sin t, 1 - \cos t \rangle$ we have

$$\mathbf{r}'(t) = \langle 1 - \cos t, \sin t \rangle$$

$$\mathbf{r}''(t) = \langle \sin t, \cos t \rangle$$

$$v(t) = \|\mathbf{r}'(t)\| = \sqrt{(1 - \cos t)^2 + \sin^2 t} = \sqrt{1 - 2\cos t + \cos^2 t + \sin^2 t}$$

$$= \sqrt{1 - 2\cos t + 1} = \sqrt{2(1 - \cos t)} = \sqrt{2 \cdot 2\sin^2 \frac{t}{2}} = 2\left|\sin \frac{t}{2}\right|$$

For $0 \le t \le 2\pi$, $\sin \frac{t}{2} \ge 0$, hence $v(t) = 2\sin \frac{t}{2}$. Therefore,

$$v'(t) = 2 \cdot \frac{1}{2}\cos \frac{t}{2} = \cos \frac{t}{2}, \quad 0 \le t \le 2\pi$$

At the point $t = \pi$ we have

$$\mathbf{r}''(\pi) = \langle \sin \pi, \cos \pi \rangle = \langle 0, -1 \rangle$$

$$v'(\pi) = \cos \frac{\pi}{2} = 0$$

$$\mathbf{r}'(\pi) = \langle 1 - \cos \pi, \sin \pi \rangle = \langle 2, 0 \rangle$$

$$\mathbf{T}(\pi) = \frac{\mathbf{r}'(\pi)}{\|\mathbf{r}'(\pi)\|} = \frac{\langle 2, 0 \rangle}{2} = \langle 1, 0 \rangle$$

We now substitute these values in (1) to obtain the following unit normal:

$$\mathbf{N}(\pi) = \frac{\langle 0, -1 \rangle - 0\langle 1, 0 \rangle}{\|\langle 0, -1 \rangle - 0\langle 1, 0 \rangle\|} = \frac{\langle 0, -1 \rangle}{1} = \langle 0, -1 \rangle$$

45. $\langle t^{-1}, t, t^2 \rangle, \quad t = -1$

SOLUTION We use the equality

$$\mathbf{N}(t) = \frac{\mathbf{r}''(t) - v'(t)\mathbf{T}(t)}{\|\mathbf{r}''(t) - v'(t)\mathbf{T}(t)\|} \tag{1}$$

We compute the vectors in the above equality. For $\mathbf{r}(t) = \langle t^{-1}, t, t^2 \rangle$ we have

$$\mathbf{r}'(t) = \langle -t^{-2}, 1, 2t \rangle$$

$$\mathbf{r}''(t) = \langle 2t^{-3}, 0, 2 \rangle$$

$$v(t) = \|\mathbf{r}'(t)\| = \sqrt{t^{-4} + 1 + 4t^2}$$

$$v'(t) = \frac{-4t^{-5} + 8t}{2\sqrt{t^{-4} + 1 + 4t^2}} = \frac{-2t^{-5} + 4t}{\sqrt{t^{-4} + 1 + 4t^2}}$$

At the point $t = -1$ we get

$$\mathbf{r}'(-1) = \langle -1, 1, -2 \rangle, \quad \mathbf{r}''(-1) = \langle -2, 0, 2 \rangle, \quad v'(-1) = \frac{2 - 4}{\sqrt{1 + 1 + 4}} = \frac{-2}{\sqrt{6}},$$

$$\mathbf{T}(-1) = \frac{\mathbf{r}'(-1)}{\|\mathbf{r}'(-1)\|} = \frac{\langle -1, 1, -2\rangle}{\sqrt{6}}$$

Hence,

$$\mathbf{r}''(-1) - v'(-1)\mathbf{T}(-1) = \langle -2, 0, 2\rangle + \frac{2}{\sqrt{6}} \cdot \frac{1}{\sqrt{6}} \langle -1, 1, -2\rangle = \left\langle -\frac{7}{3}, \frac{1}{3}, \frac{4}{3}\right\rangle = \frac{1}{3}\langle -7, 1, 4\rangle$$

$$\|\mathbf{r}''(-1) - v'(-1)\mathbf{T}(-1)\| = \frac{1}{3}\sqrt{(-7)^2 + 1^2 + 4^2} = \frac{1}{3}\sqrt{66}$$

Substituting in (1) gives the following unit normal:

$$\mathbf{N}(-1) = \frac{\frac{1}{3}\langle -7, 1, 4\rangle}{\frac{1}{3}\sqrt{66}} = \frac{1}{\sqrt{66}}\langle -7, 1, 4\rangle$$

47. Let $f(x) = x^2$. Show that the center of the osculating circle at (x_0, x_0^2) is given by $\left(-4x_0^3, \frac{1}{2} + 3x_0^2\right)$.

SOLUTION We parametrize the curve by $\mathbf{r}(x) = \langle x, x^2\rangle$. The center Q of the osculating circle at $x = x_0$ has the position vector

$$\overrightarrow{OQ} = \mathbf{r}(x_0) + \kappa(x_0)^{-1}\mathbf{N}(x_0) \tag{1}$$

We first find the curvature, using the formula for the curvature of a graph in the plane. We have $f'(x) = 2x$ and $f''(x) = 2$, hence,

$$\kappa(x) = \frac{|f''(x)|}{(1 + f'(x)^2)^{3/2}} = \frac{2}{(1 + 4x^2)^{3/2}} \quad \Rightarrow \quad \kappa(x_0)^{-1} = \frac{1}{2}(1 + 4x_0^2)^{3/2}$$

To find the unit normal vector $\mathbf{N}(x_0)$ we use the following considerations:

- The tangent vector is $\mathbf{r}'(x_0) = \langle 1, 2x_0\rangle$, hence the vector $\langle -2x_0, 1\rangle$ is orthogonal to $\mathbf{r}'(x_0)$ (since their dot product is zero). Hence $\mathbf{N}(x_0)$ is one of the two unit vectors $\pm\frac{1}{\sqrt{1+4x_0^2}}\langle -2x_0, 1\rangle$.
- The graph of $f(x) = x^2$ shows that the unit normal vector points in the positive y-direction, hence, the appropriate choice is:

$$\mathbf{N}(x_0) = \frac{1}{\sqrt{1 + 4x_0^2}}\langle -2x_0, 1\rangle \tag{2}$$

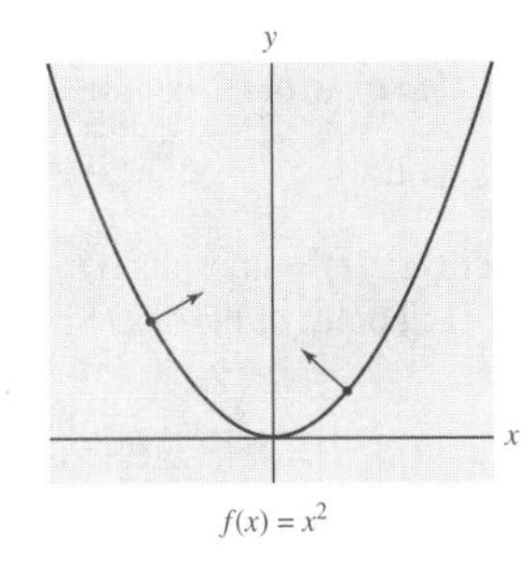

$f(x) = x^2$

We now substitute (2), (3), and $\mathbf{r}(x_0) = \langle x_0, x_0^2\rangle$ in (1) to obtain

$$\overrightarrow{OQ} = \langle x_0, x_0^2\rangle + \frac{1}{2}(1 + 4x_0^2)^{3/2} \cdot \frac{1}{\sqrt{1 + 4x_0^2}}\langle -2x_0, 1\rangle = \langle x_0, x_0^2\rangle + \frac{1}{2}(1 + 4x_0^2)\langle -2x_0, 1\rangle$$

$$= \langle x_0, x_0^2\rangle + \left\langle -x_0 - 4x_0^3, \frac{1}{2}(1 + 4x_0^2)\right\rangle = \left\langle -4x_0^3, \frac{1}{2} + 3x_0^2\right\rangle$$

The center of the osculating circle is the terminal point of $\overrightarrow{OQ}$, that is,

$$Q = \left(-4x_0^3, \frac{1}{2} + 3x_0^2\right)$$

49. Find a parametrization of the osculating circle to $y = \sin x$ at $x = \frac{\pi}{2}$.

SOLUTION We use the parametrization $\mathbf{r}(x) = \langle x, \sin x \rangle$. The radius of the osculating circle is the radius of curvature $R = \frac{1}{\kappa(\frac{\pi}{2})}$ and the center is the terminal point of the following vector:

$$\overrightarrow{OQ} = \mathbf{r}\left(\frac{\pi}{2}\right) + R\mathbf{N}\left(\frac{\pi}{2}\right)$$

We first compute the curvature. Since $y'(x) = \cos x$ and $y''(x) = -\sin x$, we have:

$$\kappa(x) = \frac{|y''(x)|}{\left(1 + y'(x)^2\right)^{3/2}} = \frac{|-\sin x|}{\left(1 + \cos^2 x\right)^{3/2}} \quad \Rightarrow \quad \kappa\left(\frac{\pi}{2}\right) = \frac{\sin \frac{\pi}{2}}{\left(1 + \cos^2 \frac{\pi}{2}\right)^{3/2}} = \frac{1}{1} = 1$$

We compute the unit normal vector $\mathbf{N}(x)$. $\mathbf{N}(x)$ is a unit vector orthogonal to the tangent vector $\mathbf{r}'(x) = \langle 1, \cos x \rangle$. We observe that $\langle -\cos x, 1 \rangle$ is orthogonal to $\mathbf{r}'(x)$, since their dot product is zero. Therefore, $\mathbf{N}(x)$ is the unit vector in the direction of either $\langle -\cos x, 1 \rangle$ or $-\langle -\cos x, 1 \rangle$, depending on the graph. Considering the accompanying figure, we see that the unit normal vector at $x = \pi/2$ points to the negative y-direction. Thus,

$$\mathbf{N}(x) = \frac{\langle \cos x, -1 \rangle}{\| \langle \cos x, -1 \rangle \|} = \frac{\langle \cos x, -1 \rangle}{\sqrt{\cos^2 x + (-1)^2}} \quad \Rightarrow \quad \mathbf{N}\left(\frac{\pi}{2}\right) = \langle 0, -1 \rangle$$

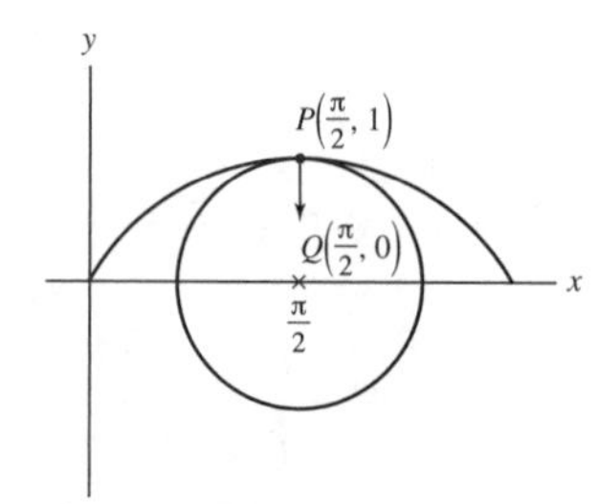

We now find the center of the osculating circle. We substitute $R = \frac{1}{\kappa(\frac{\pi}{2})} = 1$, $\mathbf{N}\left(\frac{\pi}{2}\right) = \langle 0, -1 \rangle$, and $\mathbf{r}\left(\frac{\pi}{2}\right) = \left\langle \frac{\pi}{2}, \sin \frac{\pi}{2} \right\rangle = \left\langle \frac{\pi}{2}, 1 \right\rangle$ into (1) to obtain

$$\overrightarrow{OQ} = \left\langle \frac{\pi}{2}, 1 \right\rangle + 1 \cdot \langle 0, -1 \rangle = \left\langle \frac{\pi}{2}, 0 \right\rangle$$

The osculating circle is the circle with center at the point $\left(\frac{\pi}{2}, 0\right)$ and radius 1, so it has the following parametrization:

$$\mathbf{c}(t) = \left\langle \frac{\pi}{2}, 0 \right\rangle + 1 \cdot \langle \cos t, \sin t \rangle = \left\langle \frac{\pi}{2}, 0 \right\rangle + \langle \cos t, \sin t \rangle$$

In Exercises 51–55, find a parametrization of the osculating circle at the point indicated.

51. $\langle \cos t, \sin t \rangle, \quad t = \frac{\pi}{4}$

SOLUTION The curve $\mathbf{r}(t) = \langle \cos t, \sin t \rangle$ is the unit circle. By the definition of the osculating circle, it follows that the osculating circle at each point of the circle is the circle itself. Therefore the osculating circle to the unit circle at $t = \frac{\pi}{4}$ is the unit circle itself.

53. $\langle t - \sin t, 1 - \cos t \rangle, \quad t = \pi$ (use Exercise 43)

SOLUTION

Step 1. Find κ and $\mathbf{N}$. In Exercise 43 we found that:

$$\mathbf{N}(\pi) = \langle 0, -1 \rangle \tag{1}$$

To find κ we use the formula for curvature:

$$\kappa(\pi) = \frac{\|\mathbf{r}'(\pi) \times \mathbf{r}''(\pi)\|}{\|\mathbf{r}'(\pi)\|^3} \tag{2}$$

For $\mathbf{r}(t) = \langle t - \sin t, 1 - \cos t \rangle$ we have:

$$\mathbf{r}'(t) = \langle 1 - \cos t, \sin t \rangle \quad \Rightarrow \quad \mathbf{r}'(\pi) = \langle 1 - \cos \pi, \sin \pi \rangle = \langle 2, 0 \rangle$$
$$\mathbf{r}''(t) = \langle \sin t, \cos t \rangle \quad \Rightarrow \quad \mathbf{r}''(\pi) = \langle \sin \pi, \cos \pi \rangle = \langle 0, -1 \rangle$$

Hence,

$$\mathbf{r}'(\pi) \times \mathbf{r}''(\pi) = 2\mathbf{i} \times (-\mathbf{j}) = -2\mathbf{k}$$

$$\|\mathbf{r}'(\pi) \times \mathbf{r}''(\pi)\| = \|-2\mathbf{k}\| = 2 \quad \text{and} \quad \|\mathbf{r}'(\pi)\| = \|\langle 2, 0\rangle\| = 2$$

Substituting in (2) we get:

$$\kappa(\pi) = \frac{2}{2^3} = \frac{1}{4} \tag{3}$$

Step 2. Find the center of the osculating circle. The center Q of the osculating circle at $\mathbf{r}(\pi) = \langle \pi, 2\rangle$ has position vector

$$\overrightarrow{OQ} = \mathbf{r}(\pi) + \kappa(\pi)^{-1} N(\pi)$$

Substituting (1), (3) and $\mathbf{r}(\pi) = \langle \pi, 2\rangle$ we get:

$$\overrightarrow{OQ} = \langle \pi, 2\rangle + \left(\frac{1}{4}\right)^{-1} \langle 0, -1\rangle = \langle \pi, 2\rangle + \langle 0, -4\rangle = \langle \pi, -2\rangle$$

Step 3. Parametrize the osculating circle. The osculating circle has radius $R = \frac{1}{\kappa(\pi)}$ and it is centered at $(\pi, -2)$, hence it has the following parametrization:

$$\mathbf{c}(t) = \langle \pi, -2\rangle + 4\langle \cos t, \sin t\rangle$$

55. $\mathbf{r}(t) = \langle \cosh t, \sinh t, t\rangle, \quad t = 0$ (use Exercise 12)

SOLUTION

Step 1. Find κ and $\mathbf{N}$. In Exercise 12 we found that:

$$\kappa(t) = \frac{1}{2\cosh^2 t} \quad \Rightarrow \quad \kappa(0) = \frac{1}{2\cosh^2 0} = \frac{1}{2} \tag{1}$$

We now must find the unit normal $\mathbf{N}$. We have:

$$\mathbf{r}'(t) = \langle \sinh t, \cosh t, 1\rangle$$

$$\|\mathbf{r}'(t)\| = \sqrt{\sinh^2 t + \cosh^2 t + 1} = \sqrt{\cosh^2 t - 1 + \cosh^2 t + 1} = \sqrt{2\cosh^2 t} = \sqrt{2}\cosh t$$

$$\mathbf{T}(t) = \frac{\mathbf{r}'(t)}{\|\mathbf{r}'(t)\|} = \frac{1}{\sqrt{2}\cosh t}\langle \sinh t, \cosh t, 1\rangle = \frac{1}{\sqrt{2}}\langle \tanh t, 1, \operatorname{sech} t\rangle$$

$$\mathbf{T}'(t) = \frac{1}{\sqrt{2}}\left\langle \operatorname{sech}^2 t, 0, -\operatorname{sech} t \tanh t\right\rangle$$

We compute the length of $\mathbf{T}'(t)$. Using the identity $\tanh^2 t + \operatorname{sech}^2 t = 1$ we get:

$$\|\mathbf{T}'(t)\| = \frac{1}{\sqrt{2}}\sqrt{\operatorname{sech}^4 t + 0 + \operatorname{sech}^2 t \tanh^2 t} = \frac{1}{\sqrt{2}}\sqrt{\operatorname{sech}^2 t\left(\tanh^2 t + \operatorname{sech}^2 t\right)} = \frac{1}{\sqrt{2}}\sqrt{\operatorname{sech}^2 t \cdot 1} = \frac{\operatorname{sech} t}{\sqrt{2}}$$

Hence,

$$\mathbf{N}(t) = \frac{\mathbf{T}'(t)}{\|\mathbf{T}'(t)\|} = \frac{\sqrt{2}}{\operatorname{sech} t}\frac{1}{\sqrt{2}} \cdot \left\langle \operatorname{sech}^2 t, 0, -\operatorname{sech} t \tanh t\right\rangle = \frac{1}{2}\langle \operatorname{sech} t, 0, -\tanh t\rangle$$

At the point $t = 0$ we have $\operatorname{sech} 0 = 1$, $\tanh 0 = 0$, hence

$$\mathbf{N}(0) = \langle 1, 0, 0\rangle \tag{2}$$

Step 2. Find the center of the osculating circle. The center Q of the osculating circle at $\mathbf{r}(0) = \langle 1, 0, 0\rangle$ has position vector:

$$\overrightarrow{OQ} = \mathbf{r}(0) + \kappa(0)^{-1}\mathbf{N}(0)$$

Substituting (1), (2) and $\mathbf{r}(0) = \langle 1, 0, 0\rangle$ we get:

$$\overrightarrow{OQ} = \langle 1, 0, 0\rangle + 2 \cdot \langle 1, 0, 0\rangle = \langle 3, 0, 0\rangle$$

Step 3. Parametrize the osculating circle. The osculating circle is centered at $Q = (3, 0, 0)$ and has radius $R = \frac{1}{\kappa(0)} = 2$, hence it has the following parametrization:

$$\mathbf{c}(t) = \langle 3, 0, 0\rangle + 2\mathbf{N}\cos t + 2\mathbf{T}\sin t = \langle 3, 0, 0\rangle + 2\langle 1, 0, 0\rangle\cos t + \frac{2}{\sqrt{2}}\langle 0, 1, 1\rangle\sin t$$

57. Figure 17 shows the graph of the half-ellipse $y = \pm\sqrt{2rx - px^2}$, where r and p are positive constants. Show that the radius of curvature at the origin is equal to r. *Hint:* One way of proceeding is to write the ellipse in the form of Exercise 23 and apply Eq. (8).

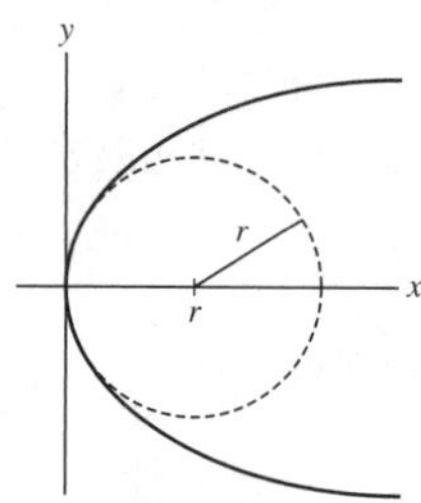

FIGURE 17 The curve $y = \sqrt{2rx - px^2}$ and the osculating circle at the origin.

SOLUTION The radius of curvature is the reciprocal of the curvature. We thus must find the curvature at the origin. We use the following simple variant of the formula for the curvature of a graph in the plane:

$$\kappa(y) = \frac{|x''(y)|}{\left(1 + x'(y)^2\right)^{3/2}} \tag{1}$$

(The traditional formula of $\kappa(x) = \frac{|y''(x)|}{\left(1+y'(x)^2\right)^{3/2}}$ is inappropriate for this problem, as $y'(x)$ is undefined at $x = 0$.) We find x in terms of y:

$$y = \sqrt{2rx - px^2}$$
$$y^2 = 2rx - px^2$$
$$px^2 - 2rx + y^2 = 0$$

We solve for x and obtain:

$$x = \pm\frac{1}{p}\sqrt{r^2 - py^2} + \frac{r}{p}, \quad y \geq 0.$$

We find x' and x'':

$$x' = \pm\frac{-2py}{2p\sqrt{r^2 - py^2}} = \pm\frac{y}{\sqrt{r^2 - py^2}}$$

$$x'' = \pm\frac{1 \cdot \sqrt{r^2 - py^2} - y \cdot \frac{-py}{\sqrt{r^2-py^2}}}{r^2 - py^2} = \pm\frac{r^2 - py^2 + py^2}{\left(r^2 - py^2\right)^{3/2}} = \pm\frac{r^2}{\left(r^2 - py^2\right)^{3/2}}$$

At the origin we get:

$$x'(0) = 0, \quad x''(0) = \frac{\pm r^2}{(r^2)^{3/2}} = \frac{\pm 1}{r}$$

Substituting in (1) gives the following curvature at the origin:

$$\kappa(0) = \frac{|x''(0)|}{(1 + x'(0)^2)^{3/2}} = \frac{|\frac{\pm 1}{r}|}{(1+0)^{3/2}} = \frac{1}{|r|} = \frac{1}{r}$$

We conclude that the radius of curvature at the origin is

$$R = \frac{1}{\kappa(0)} = r$$

59. The **angle of inclination** of a plane curve with parametrization $\mathbf{r}(t)$ is defined as the angle $\theta(t)$ between the unit tangent vector $\mathbf{T}(t)$ and the x-axis (Figure 19). Show that $\|\mathbf{T}'(t)\| = |\theta'(t)|$ and conclude that if $\mathbf{r}(s)$ is a parametrization by arc length, then

$$\kappa(s) = \left|\frac{d\theta}{ds}\right|$$

Hint: Observe that $\mathbf{T}(t) = \langle\cos\theta(t), \sin\theta(t)\rangle$.

11

FIGURE 19 The curvature is the rate of change of $\theta(t)$.

SOLUTION Since $\mathbf{T}(t)$ is a unit vector that makes an angle $\theta(t)$ with the positive x-axis, we have

$$\mathbf{T}(t) = \langle \cos\theta(t), \sin\theta(t) \rangle .$$

Differentiating this vector using the Chain Rule gives:

$$\mathbf{T}'(t) = \left\langle -\theta'(t)\sin\theta(t), \theta'(t)\cos\theta(t) \right\rangle = \theta'(t)\left\langle -\sin\theta(t), \cos\theta(t) \right\rangle$$

We compute the norm of the vector $\mathbf{T}'(t)$:

$$\|\mathbf{T}'(t)\| = \|\theta'(t)\langle -\sin\theta(t), \cos\theta(t)\rangle\| = |\theta'(t)|\sqrt{(-\sin\theta(t))^2 + (\cos\theta(t))^2} = |\theta'(t)| \cdot 1 = |\theta'(t)|$$

When $\mathbf{r}(s)$ is a parametrization by arc length we have:

$$\kappa(s) = \left\|\frac{d\mathbf{T}}{ds}\right\| = \left\|\frac{d\mathbf{T}}{dt}\right\| \left|\frac{dt}{d\theta}\frac{d\theta}{ds}\right| = |\theta'(t)| \frac{1}{|\theta'(t)|}\left|\frac{d\theta}{ds}\right| = \left|\frac{d\theta}{ds}\right|$$

as desired.

61. Verify Eq. (11) for a circle of radius R. Suppose that a particle traverses the circle at unit speed. Show that the change in the angle during an interval of length Δt is $\Delta\theta = \Delta t/R$ and conclude that $\theta'(s) = 1/R$.

SOLUTION The particle traverses the circle at unit speed hence the parametrization is the arc length parametrization of the circle. That is,

$$\mathbf{r}(s) = R\left\langle \cos\frac{s}{R}, \sin\frac{s}{R} \right\rangle$$

The angle is $\theta(s) = \frac{s}{R}$, hence the change in the angle during an interval of length Δs is:

$$\Delta\theta = \theta(s + \Delta s) - \theta(s) = \frac{s + \Delta s}{R} - \frac{s}{R} = \frac{\Delta s}{R}$$

Therefore, $\frac{\Delta\theta}{\Delta s} = \frac{1}{R}$ and we obtain the following derivative:

$$\theta'(s) = \lim_{\Delta s \to 0} \frac{\Delta\theta}{\Delta s} = \lim_{\Delta s \to 0} \frac{1}{R} = \frac{1}{R}$$

The curvature of a circle of radius R is $\frac{1}{R}$, therefore, we have:

$$\frac{d\theta}{ds} = \frac{1}{R}.$$

This equality verifies Eq. (11), for this case.

63. Use the parametrization $\mathbf{r}(\theta) = \langle f(\theta)\cos\theta, f(\theta)\sin\theta \rangle$ to show that a curve $r = f(\theta)$ in polar coordinates has curvature

$$\kappa(\theta) = \frac{|f(\theta)^2 + 2f'(\theta)^2 - 2f(\theta)f''(\theta)|}{(f(\theta)^2 + f'(\theta)^2)^{3/2}} \qquad \boxed{12}$$

SOLUTION By the formula for curvature we have

$$\kappa(\theta) = \frac{\|\mathbf{r}'(\theta) \times \mathbf{r}''(\theta)\|}{\|\mathbf{r}'(\theta)\|^3} \qquad (1)$$

We differentiate $\mathbf{r}(\theta)$ and $\mathbf{r}'(\theta)$:

$$\begin{aligned}
\mathbf{r}'(\theta) &= \left\langle f'(\theta)\cos\theta - f(\theta)\sin\theta, f'(\theta)\sin\theta + f(\theta)\cos\theta \right\rangle \\
\mathbf{r}''(\theta) &= \left\langle f''(\theta)\cos\theta - f'(\theta)\sin\theta - f'(\theta)\sin\theta - f(\theta)\cos\theta,\right. \\
&\qquad \left. f''(\theta)\sin\theta + f'(\theta)\cos\theta + f'(\theta)\cos\theta - f(\theta)\sin\theta \right\rangle \\
&= \left\langle \left(f''(\theta) - f(\theta)\right)\cos\theta - 2f'(\theta)\sin\theta, \left(f''(\theta) - f(\theta)\right)\sin\theta + 2f'(\theta)\cos\theta \right\rangle
\end{aligned}$$

Hence,

$$\begin{aligned}
\mathbf{r}'(\theta)\times\mathbf{r}''(\theta) &= \left(f'(\theta)\cos\theta - f(\theta)\sin\theta\right)\cdot\left(\left(f''(\theta)-f(\theta)\right)\sin\theta + 2f'(\theta)\cos\theta\right)\mathbf{k}\\
&\quad - \left(f'(\theta)\sin\theta + f(\theta)\cos\theta\right)\cdot\left(\left(f''(\theta)-f(\theta)\right)\cos\theta - 2f'(\theta)\sin\theta\right)\mathbf{k}\\
&= \Big\{f'(\theta)\left(f''(\theta)-f(\theta)\right)\cos\theta\sin\theta - f(\theta)\left(f''(\theta)-f(\theta)\right)\sin^2\theta + 2f'^2(\theta)\cos^2\theta\\
&\quad - 2f(\theta)f'(\theta)\sin\theta\cos\theta\Big(-f'(\theta)\left(f''(\theta)-f(\theta)\right)\sin\theta\cos\theta - f(\theta)\left(f''(\theta)-f(\theta)\right)\cos^2\theta\\
&\quad + 2f'(\theta)^2\sin^2\theta + 2f(\theta)f'(\theta)\cos\theta\sin\theta\Big)\Big\}\mathbf{k}\\
&= \Big(-f(\theta)\left(f''(\theta)-f(\theta)\right)\left(\sin^2\theta+\cos^2\theta\right) + 2f'^2(\theta)\left(\cos^2\theta+\sin^2\theta\right)\Big)\mathbf{k}\\
&= \Big(-f(\theta)\left(f''(\theta)-f(\theta)\right) + 2f'^2(\theta)\Big)\mathbf{k}\\
&= \Big(-f(\theta)f''(\theta) + f^2(\theta) + 2f'^2(\theta)\Big)\mathbf{k}
\end{aligned}$$

The length of the cross product is:

$$\|\mathbf{r}'(\theta)\times\mathbf{r}''(\theta)\| = |f^2(\theta) + 2f'^2(\theta) - f(\theta)f''(\theta)| \tag{2}$$

We compute the length of $\mathbf{r}'(\theta)$:

$$\begin{aligned}
\|\mathbf{r}'(\theta)\|^2 &= \left(f'(\theta)\cos\theta - f(\theta)\sin\theta\right)^2 + \left(f'(\theta)\sin\theta + f(\theta)\cos\theta\right)^2\\
&= f'^2(\theta)\cos^2\theta - 2f'(\theta)f(\theta)\cos\theta\sin\theta + f^2(\theta)\sin^2\theta + f'^2(\theta)\sin^2\theta\\
&\quad + 2f'(\theta)f(\theta)\sin\theta\cos\theta + f^2(\theta)\cos^2\theta\\
&= f'^2(\theta)\left(\cos^2\theta+\sin^2\theta\right) + f^2(\theta)\left(\sin^2\theta+\cos^2\theta\right) = f'^2(\theta) + f^2(\theta)
\end{aligned}$$

Hence,

$$\|\mathbf{r}'(\theta)\| = \sqrt{f'^2(\theta) + f^2(\theta)} \tag{3}$$

Substituting (2) and (3) in (1) gives:

$$\kappa(\theta) = \frac{|f^2(\theta) + 2f'^2(\theta) - f(\theta)f''(\theta)|}{\left(f'^2(\theta) + f^2(\theta)\right)^{3/2}}$$

In Exercises 64–66, use Eq. (12) to find the curvature of the curve given in polar form.

65. $f(\theta) = \theta$

SOLUTION We have $f'(\theta) = 1$, $f''(\theta) = 0$. The numerator and denominator in Eq. (12) are thus:

$$\begin{aligned}
f(\theta)^2 + 2f'(\theta) - f(\theta)f''(\theta) &= \theta^2 + 2\cdot 1 - 0 = \theta^2 + 2\\
\left(f(\theta)^2 + f'(\theta)^2\right)^{3/2} &= \left(\theta^2+1\right)^{3/2}
\end{aligned}$$

Hence,

$$\kappa(\theta) = \frac{\theta^2+2}{\left(\theta^2+1\right)^{3/2}}$$

67. Use Eq. (12) to find the curvature of the general Bernoulli spiral $r = ae^{b\theta}$ in polar form (a and b are constants).

SOLUTION By Eq. (12):

$$\kappa(\theta) = \frac{|f(\theta)^2 + 2f'(\theta)^2 - f(\theta)f''(\theta)|}{\left(f(\theta)^2 + f'^2(\theta)\right)^{3/2}}$$

In our case $f(\theta) = ae^{b\theta}$ hence $f'(\theta) = abe^{b\theta}$ and $f''(\theta) = ab^2e^{b\theta}$. We compute the numerator of $\kappa(\theta)$:

$$f(\theta)^2 + 2f'(\theta)^2 - f(\theta)f''(\theta) = a^2e^{2b\theta} + 2a^2b^2e^{2b\theta} - ae^{b\theta}\cdot ab^2e^{b\theta} = a^2e^{2b\theta} + 2a^2b^2e^{2b\theta} - a^2b^2e^{2b\theta}$$

$$= a^2e^{2b\theta} + a^2b^2e^{2b\theta} = a^2(1+b^2)e^{2b\theta}$$

We compute the denominator of $\kappa(\theta)$:

$$\left(f(\theta)^2 + f'(\theta)^2\right)^{3/2} = \left(a^2e^{2b\theta} + a^2b^2e^{2b\theta}\right)^{3/2} = \left(a^2e^{2b\theta}\left(1+b^2\right)\right)^{3/2} = a^3e^{3b\theta}\left(1+b^2\right)^{3/2}$$

Therefore:

$$\kappa(\theta) = \frac{a^2(1+b^2)e^{2b\theta}}{a^3(1+b^2)^{3/2}e^{3b\theta}} = \frac{1}{a\sqrt{1+b^2}}e^{-b\theta}$$

69. Show that

$$\gamma(s) = \frac{1}{\kappa}\mathbf{N} + \frac{1}{\kappa}\left((\sin\kappa s)\mathbf{T} - (\cos\kappa s)\mathbf{N}\right)$$

is an arc length parametrization of the osculating circle.

SOLUTION Let P be a fixed point on the curve C, $\mathbf{T}$ and $\mathbf{N}$ are the unit tangent and the unit normal to the curve at P. We place the xy-coordinate system so that the origin is at P and the x and y axes are in the directions of $\mathbf{T}$ and $\mathbf{N}$, respectively. We next show that $\gamma(s)$ is an arc length parametrization of the osculating circle at P.

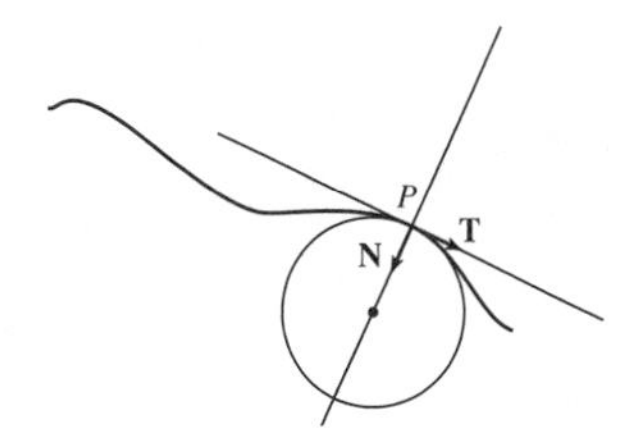

We compute the following expression:

$$\left\|\gamma(s) - \frac{1}{\kappa}\mathbf{N}\right\|^2 = \frac{1}{\kappa^2}\|\,(\sin\kappa s)\,\mathbf{T} - (\cos\kappa s)\,\mathbf{N}\|^2 = \frac{1}{\kappa^2}\left((\sin\kappa s)\,\mathbf{T} - (\cos\kappa s)\,\mathbf{N}\right)\cdot\left((\sin\kappa s)\,\mathbf{T} - (\cos\kappa s)\,\mathbf{N}\right)$$

$$= \frac{1}{\kappa^2}\left(\sin^2\kappa s\mathbf{T}\cdot\mathbf{T} - (\sin\kappa s\cos\kappa s)\,\mathbf{T}\cdot\mathbf{N} - (\cos\kappa s\sin\kappa s)\,\mathbf{N}\cdot\mathbf{T} + \left(\cos^2\kappa s\right)\mathbf{N}\cdot\mathbf{N}\right)$$

The vectors $\mathbf{T}$ and $\mathbf{N}$ are orthogonal unit vectors, hence $\mathbf{T}\cdot\mathbf{N} = \mathbf{N}\cdot\mathbf{T} = 0$ and $\mathbf{T}\cdot\mathbf{T} = \|\mathbf{T}\|^2 = 1$, $\mathbf{N}\cdot\mathbf{N} = \|\mathbf{N}\|^2 = 1$. We use the identity $\sin^2(\kappa s) + \cos^2(\kappa s) = 1$ to obtain

$$\left\|\gamma(s) - \frac{1}{\kappa}\mathbf{N}\right\|^2 = \frac{1}{\kappa^2}\left(\sin^2\kappa s + \cos^2\kappa s\right) = \frac{1}{\kappa^2}$$

That is,

$$\left\|\gamma(s) - \frac{1}{\kappa}\mathbf{N}\right\| = \frac{1}{\kappa} \tag{1}$$

Notice that κ, $\mathbf{N}$, and $\mathbf{T}$ are fixed and only s is changing in $\gamma(s)$. It follows by (1) that $\gamma(s)$ is a circle of radius $\frac{1}{\kappa}$ centered at $\frac{1}{\kappa}\mathbf{N}$. The curvature of the circle is the reciprocal of the radius, which is κ (the curvature of C at the point P). We thus showed that the circle $\gamma(s)$ satisfies the second condition in the definition of the osculating circle. We now show that the first condition is satisfied as well.

The center of the circle is the terminal point of the vector $\frac{1}{\kappa}\mathbf{N}$, which is in the direction of $\mathbf{N}$ and orthogonal to $\mathbf{T}$. This shows that $\mathbf{T}$ and $\mathbf{N}$ are the unit tangent and unit normal to the circle at P. Finally, we verify that the given parametrization is the arc length parametrization, by showing that $\|\gamma'(s)\| = 1$. Differentiating $\gamma(s)$ with respect to s gives (notice that κ, $\mathbf{T}$, and $\mathbf{N}$ are fixed):

$$\gamma'(s) = \frac{1}{\kappa}\left((\kappa\cos\kappa s)\,\mathbf{T} + (\kappa\sin\kappa s)\,\mathbf{N}\right) = (\cos\kappa s)\mathbf{T} + (\sin\kappa s)\mathbf{N}$$

Hence, since $\mathbf{T}\cdot\mathbf{T} = \mathbf{N}\cdot\mathbf{N} = 1$ and $\mathbf{T}\cdot\mathbf{N} = \mathbf{N}\cdot\mathbf{T} = 0$ we get:

$$\|\gamma'(s)\|^2 = \left((\cos\kappa s)\mathbf{T} + (\sin\kappa s)\mathbf{N}\right)\cdot\left((\cos\kappa s)\mathbf{T} + (\sin\kappa s)\mathbf{N}\right)$$

$$= \left(\cos^2\kappa s\right)\mathbf{T}\cdot\mathbf{T} + (\cos\kappa s)(\sin\kappa s)\mathbf{T}\cdot\mathbf{N} + (\sin\kappa s\cos\kappa s)\mathbf{N}\cdot\mathbf{T} + \left(\sin^2\kappa s\right)\mathbf{N}\cdot\mathbf{N}$$

$$= \cos^2\kappa s + \sin^2\kappa s = 1$$

Hence

$$\|\gamma'(s)\| = 1$$

71. Let $\mathbf{r}(t) = \langle x(t), y(t), z(t)\rangle$ be a path with curvature $\kappa(t)$ and define the scaled path $\mathbf{r}_1(t) = \langle \lambda x(t), \lambda y(t), \lambda z(t)\rangle$, where $\lambda \neq 0$ is a constant. Prove that curvature varies inversely with the scale factor, that is, the curvature $\kappa_1(t)$ of $\mathbf{r}_1(t)$ is $\kappa_1(t) = \lambda^{-1}\kappa(t)$. This explains why the curvature of a circle of radius R is proportional to $1/R$ (in fact, it is equal to $1/R$). *Hint:* Use Eq. (3).

SOLUTION The resulting curvature k_1 and the original curvature κ are:

$$\kappa_1(t) = \frac{\|\mathbf{r}_1'(t) \times \mathbf{r}_1''(t)\|}{\|\mathbf{r}_1'(t)\|^3}, \qquad \kappa(t) = \frac{\|\mathbf{r}'(t) \times \mathbf{r}''(t)\|}{\|\mathbf{r}'(t)\|^3}$$

We have

$$\mathbf{r}_1'(t) = \frac{d}{dt}(\lambda \mathbf{r}(t)) = \lambda \mathbf{r}'(t)$$

$$\mathbf{r}_1''(t) = \frac{d}{dt}\left(r_1'(t)\right) = \frac{d}{dt}\left(\lambda \mathbf{r}'(t)\right) = \lambda \mathbf{r}''(t)$$

Hence,

$$\|\mathbf{r}_1'(t) \times \mathbf{r}_1''(t)\| = \|\lambda \mathbf{r}'(t) \times \lambda \mathbf{r}''(t)\| = \lambda^2 \|\mathbf{r}'(t) \times \mathbf{r}''(t)\|$$

$$\|\mathbf{r}_1'(t)\| = \|\lambda \mathbf{r}'(t)\| = |\lambda| \|\mathbf{r}'(t)\|$$

Substituting in (1) we get:

$$\kappa_1(t) = \frac{\lambda^2 \|\mathbf{r}'(t) \times \mathbf{r}''(t)\|}{|\lambda|^3 \|\mathbf{r}'(t)\|^3} = \frac{1}{|\lambda|} \frac{\|\mathbf{r}'(t) \times \mathbf{r}''(t)\|}{\|\mathbf{r}'(t)\|^3} = \frac{1}{|\lambda|}\kappa(t)$$

We conclude that the resulting curvature is:

$$\kappa_1(t) = \frac{1}{|\lambda|}\kappa(t)$$

Multiplying the vector by λ causes the curvature to be divided by $|\lambda|$.

Further Insights and Challenges

73. Let $\mathbf{r}(s)$ be an arc length parametrization of a closed curve $\mathcal{C}$ of length L. We call $\mathcal{C}$ an **oval** if $d\theta/ds > 0$ (see Exercise 59). Observe that $-\mathbf{N}$ points to the *outside* of $\mathcal{C}$. For $k > 0$, the curve $\mathcal{C}_1$ defined by $\mathbf{r}_1(s) = \mathbf{r}(s) - k\mathbf{N}$ is called the expansion of $c(s)$ in the normal direction.

(a) Show that $\|\mathbf{r}_1'(s)\| = \|\mathbf{r}'(s)\| + k\kappa(s)$.

(b) As P moves around the oval counterclockwise, θ increases by 2π [Figure 20(A)]. Use this and a change of variables to prove that $\int_0^L \kappa(s)\,ds = 2\pi$.

(c) Show that $\mathcal{C}_1$ has length $L + 2\pi k$.

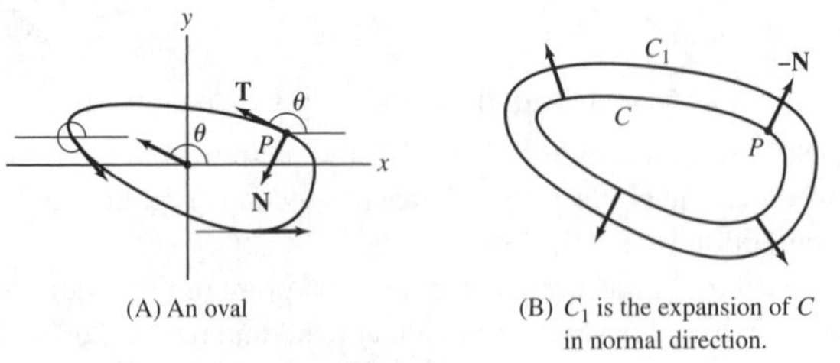

FIGURE 20 As P moves around the oval, θ increases by 2π.

SOLUTION

(a) Since $\mathbf{r}_1(s) = \mathbf{r}(s) - k\mathbf{N}$ we have

$$\mathbf{r}_1'(s) = \mathbf{r}'(s) - k\frac{d\mathbf{N}}{ds} \tag{1}$$

We compute $\frac{d\mathbf{N}}{ds}$ using the Chain Rule:

$$\frac{d\mathbf{N}}{ds} = \frac{d\mathbf{N}}{d\theta} \cdot \frac{d\theta}{ds} \tag{2}$$

By Exercise 59 and since $\mathcal{C}$ is oval we have:

$$\kappa(s) = \left|\frac{d\theta}{ds}\right| = \frac{d\theta}{ds} \tag{3}$$

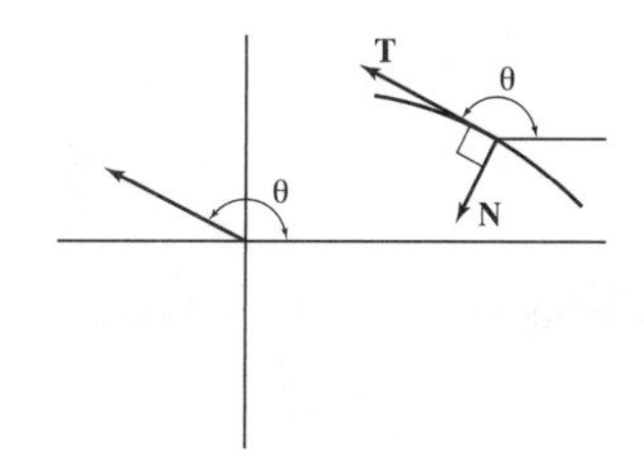

Also, as illustrated in the figure, the following holds:

$$\mathbf{N} = \left\langle \cos\left(\frac{\pi}{2} + \theta\right), \sin\left(\frac{\pi}{2} + \theta\right) \right\rangle = \langle -\sin\theta, \cos\theta \rangle$$

Hence:

$$\frac{d\mathbf{N}}{d\theta} = \langle -\cos\theta, -\sin\theta \rangle = -\langle \cos\theta, \sin\theta \rangle = -\mathbf{T} \tag{4}$$

Substituting (3) and (4) in (2) yields:

$$\frac{d\mathbf{N}}{ds} = -\kappa(s)\mathbf{T}(s)$$

Substituting in (1) we obtain:

$$\mathbf{r}_1'(s) = \mathbf{r}'(s) + k\kappa(s)\mathbf{T}(s)$$

In the arc length parametrization, $\mathbf{T}(s) = \mathbf{r}'(s)$, therefore:

$$\mathbf{r}_1'(s) = \mathbf{r}'(s) + k\kappa(s)\mathbf{r}'(s) = \mathbf{r}'(s)\,(1 + k\kappa(s))$$

Computing the length and using $\|\mathbf{r}'(s)\| = 1$ we obtain:

$$\|\mathbf{r}_1'(s)\| = \|\mathbf{r}'(s)\|\,(1 + k\kappa(s)) = \|\mathbf{r}'(s)\| + \|\mathbf{r}'(s)\| \cdot k\kappa(s) = \|\mathbf{r}'(s)\| + k\kappa(s)$$

(b) In Exercise 59 we showed that:

$$\kappa(s) = \left|\frac{d\theta}{ds}\right|$$

Since $\frac{d\theta}{ds} > 0$ we have $\kappa(s) = \frac{d\theta}{ds}$. As P moves around the oval, θ increases by 2π, hence $\theta\,(s = L) - \theta\,(s = 0) = 2\pi$. Using these considerations we get:

$$\int_0^L \kappa(s)\,ds = \int_{\theta(0)}^{\theta(L)} \frac{d\theta}{ds}\,ds = \int_{\theta(0)}^{\theta(L)} d\theta = \theta(L) - \theta(0) = 2\pi.$$

(c) We use the Arc Length Formula and the equality in part (a) to write the length L_1 of C_1 as the following integral:

$$L_1 = \int_0^L \|\mathbf{r}_1'(s)\|\,ds = \int_0^L \|\mathbf{r}'(s)\|\,ds + k\int_0^L \kappa(s)\,ds$$

By the Arc Length Formula, the first integral is the length L of C. The second integral was computed in part (b). Therefore we get:

$$L_1 = L + k \cdot 2\pi = L + 2\pi k.$$

In Exercises 74–81, let **B** *denote the* binormal vector *at a point on a space curve* C*, defined by* $\mathbf{B} = \mathbf{T} \times \mathbf{N}$.

75. Follow steps (a)–(c) to prove that there is a number τ (lowercase Greek tau) called the **torsion** such that

$$\frac{d\mathbf{B}}{ds} = -\tau\mathbf{N}$$

13

(a) Show that $\dfrac{d\mathbf{B}}{ds} = \mathbf{T} \times \dfrac{d\mathbf{N}}{ds}$ and conclude that $d\mathbf{B}/ds$ is orthogonal to $\mathbf{T}$.

(b) Differentiate $\mathbf{B} \cdot \mathbf{B} = 1$ with respect to s to show that $d\mathbf{B}/ds$ is orthogonal to $\mathbf{B}$.

(c) Conclude that $d\mathbf{B}/ds$ is a multiple of $\mathbf{N}$.

SOLUTION

(a) Using the Product Rule for cross product we have:

$$\frac{d\mathbf{B}}{ds} = \frac{d}{ds}(\mathbf{T} \times \mathbf{N}) = \frac{d\mathbf{T}}{ds} \times \mathbf{N} + \mathbf{T} \times \frac{d\mathbf{N}}{ds}$$

$\mathbf{N}$ is a unit vector in the direction of $\frac{d\mathbf{T}}{ds}$, hence $\frac{d\mathbf{T}}{ds} \times \mathbf{N} = \mathbf{0}$, so we obtain:

$$\frac{d\mathbf{B}}{ds} = \mathbf{T} \times \frac{d\mathbf{N}}{ds}$$

By properties of cross products we conclude that $\frac{d\mathbf{B}}{ds}$ is orthogonal to $\mathbf{T}$.

(b) We differentiate $\mathbf{B} \cdot \mathbf{B} = 1$ using the Product Rule for dot products:

$$\mathbf{B} \cdot \frac{d\mathbf{B}}{ds} + \frac{d\mathbf{B}}{ds} \cdot \mathbf{B} = 0$$

$$2\mathbf{B} \cdot \frac{d\mathbf{B}}{ds} = 0 \quad \Rightarrow \quad \mathbf{B} \cdot \frac{d\mathbf{B}}{ds} = 0$$

Since the dot product of $\mathbf{B}$ and $\frac{d\mathbf{B}}{ds}$ is zero, the two vectors are orthogonal.

(c) In parts (a) and (b) we showed that $\frac{d\mathbf{B}}{ds}$ is orthogonal to $\mathbf{B}$ and $\mathbf{T}$. It follows that $\frac{d\mathbf{B}}{ds}$ is parallel to any other vector that is orthogonal to $\mathbf{B}$ and $\mathbf{T}$. We show that $\mathbf{N}$ is such a vector.

Since $\mathbf{B} = \mathbf{T} \times \mathbf{N}$, the vectors $\mathbf{N}$ and $\mathbf{B}$ are orthogonal. The unit normal $\mathbf{N}$ is also orthogonal to the unit tangent $\mathbf{T}$. We conclude that $\frac{d\mathbf{B}}{ds}$ and $\mathbf{N}$ are parallel, hence there exists a number $(-\tau)$ such that:

$$\frac{d\mathbf{B}}{ds} = -\tau\mathbf{N}.$$

77. Torsion means "twisting." Is this an appropriate name for τ? Explain by interpreting τ geometrically.

SOLUTION $\mathbf{B}$ is the unit normal to the osculating plane at a point P on the curve. As P moves along the curve, the unit normal $\mathbf{B}$ is changing by $\frac{d\mathbf{B}}{ds} = -\tau\mathbf{N}$. Geometrically the osculating plane is "twisted" and τ is a measure for this twisting.

79. Follow steps (a)–(b) to prove

$$\frac{d\mathbf{N}}{ds} = -\kappa\mathbf{T} + \tau\mathbf{B} \qquad \boxed{15}$$

(a) Show that $d\mathbf{N}/ds$ is orthogonal to $\mathbf{N}$. Conclude that $d\mathbf{N}/ds$ lies in the plane spanned by $\mathbf{T}$ and $\mathbf{B}$, and hence, $d\mathbf{N}/ds = a\mathbf{T} + b\mathbf{B}$ for some scalars a, b.

(b) Use $\mathbf{N} \cdot \mathbf{T} = 0$ to show that $\mathbf{T} \cdot \frac{d\mathbf{N}}{ds} = -\mathbf{N} \cdot \frac{d\mathbf{T}}{ds}$ and compute a. Compute b similarly.

Equations (13) and (15) together with $d\mathbf{T}/dt = \kappa\mathbf{N}$ are called the **Frenet formulas** and were discovered by the French geometer Jean Frenet (1816–1900).

SOLUTION

(a) We first show that $\frac{d\mathbf{N}}{ds}$ is orthogonal to $\mathbf{N}$. Earlier we showed that $\frac{d\mathbf{B}}{ds} = \mathbf{T} \times \frac{d\mathbf{N}}{ds}$ and $\frac{d\mathbf{B}}{ds} = -\tau\mathbf{N}$, hence:

$$-\tau\mathbf{N} = \mathbf{T} \times \frac{d\mathbf{N}}{ds}$$

By properties of the cross product, this equality implies that $\frac{d\mathbf{N}}{ds}$ is orthogonal to $-\tau\mathbf{N}$, hence it is orthogonal to $\mathbf{N}$. Now, $\mathbf{N}$ is orthogonal to $\mathbf{T}$ and $\mathbf{B}$, hence $\mathbf{N}$ is normal to the plane spanned by $\mathbf{T}$ and $\mathbf{B}$. Therefore, since $\mathbf{N}$ is orthogonal to $\frac{d\mathbf{N}}{ds}$, this last vector lies in the plane spanned by $\mathbf{T}$ and $\mathbf{B}$, that is, there exist scalars a and b such that:

$$\frac{d\mathbf{N}}{ds} = a\mathbf{T} + b\mathbf{B}$$

(b) By the orthogonality of $\mathbf{N}$ and $\mathbf{T}$ we have:

$$\mathbf{N} \cdot \mathbf{T} = 0$$

Differentiating this equality, using the product rule for dot product we get:

$$\mathbf{N} \cdot \frac{d\mathbf{T}}{ds} + \frac{d\mathbf{N}}{ds} \cdot \mathbf{T} = 0 \quad \Rightarrow \quad \mathbf{T} \cdot \frac{d\mathbf{N}}{ds} = -\mathbf{N} \cdot \frac{d\mathbf{T}}{ds}$$

To compute a, we substitute $\frac{d\mathbf{N}}{ds} = a\mathbf{T} + b\mathbf{B}$ and use $\mathbf{T} \cdot \mathbf{T} = \|\mathbf{T}\|^2 = 1$ and $\mathbf{T} \cdot \mathbf{B} = 0$. This gives:

$$\mathbf{T} \cdot (a\mathbf{T} + b\mathbf{B}) = -\mathbf{N} \cdot \frac{d\mathbf{T}}{ds}$$

$$a\mathbf{T}\cdot\mathbf{T}+b\mathbf{T}\cdot\mathbf{B}=-\mathbf{N}\cdot\frac{d\mathbf{T}}{ds}$$

$$a\cdot 1+b\cdot 0=-\mathbf{N}\cdot\frac{d\mathbf{T}}{ds}\quad\Rightarrow\quad a=-\mathbf{N}\cdot\frac{d\mathbf{T}}{ds}\tag{1}$$

To find b we differentiate the equality $\mathbf{N}\cdot\mathbf{B}=0$ (notice that by $\mathbf{B}=\mathbf{T}\times\mathbf{N}$ follows the orthogonality of $\mathbf{N}$ and $\mathbf{B}$). We get:

$$\mathbf{N}\cdot\frac{d\mathbf{B}}{ds}+\frac{d\mathbf{N}}{ds}\cdot\mathbf{B}=0\quad\Rightarrow\quad\frac{d\mathbf{N}}{ds}\cdot\mathbf{B}=-\mathbf{N}\cdot\frac{d\mathbf{B}}{ds}$$

We now substitute $\frac{d\mathbf{N}}{ds}=a\mathbf{T}+b\mathbf{B}$ and we use $\mathbf{B}\cdot\mathbf{B}=\|\mathbf{B}\|^2=1$ and $\mathbf{T}\cdot\mathbf{B}=0$ to obtain:

$$(a\mathbf{T}+b\mathbf{B})\cdot\mathbf{B}=-\mathbf{N}\cdot\frac{d\mathbf{B}}{ds}$$

$$a\mathbf{T}\cdot\mathbf{B}+b\mathbf{B}\cdot\mathbf{B}=-\mathbf{N}\cdot\frac{d\mathbf{B}}{ds}$$

$$a\cdot 0+b\cdot 1=-\mathbf{N}\cdot\frac{d\mathbf{B}}{ds}\quad\Rightarrow\quad b=-\mathbf{N}\cdot\frac{d\mathbf{B}}{ds}$$

Since $\frac{d\mathbf{B}}{ds}=-\tau\mathbf{N}$ we may write:

$$b=-\mathbf{N}\cdot(-\tau\mathbf{N})=\tau\mathbf{N}\cdot\mathbf{N}=\tau\|\mathbf{N}\|^2=\tau\tag{2}$$

Also for the arc length parametrization $\frac{d\mathbf{T}}{ds}=\kappa(s)\mathbf{N}$, hence by (1):

$$a=-\mathbf{N}\cdot\kappa(s)\mathbf{N}=-\kappa(s)\mathbf{N}\cdot\mathbf{N}=-\kappa(s)\|\mathbf{N}\|^2=-\kappa(s)\tag{3}$$

We combine (2), (3), and part (a) to conclude:

$$\frac{d\mathbf{N}}{ds}=-\kappa\mathbf{T}+\tau\mathbf{B}.$$

81. The vector $\mathbf{N}$ can be computed using $\mathbf{N}=\mathbf{B}\times\mathbf{T}$ [Eq. (14)] with $\mathbf{B}$, as in Eq. (16). Use this method to find $\mathbf{N}$ in the following cases:

(a) $\mathbf{r}(t)=\langle\cos t,t,t^2\rangle$ at $t=0$

(b) $\mathbf{r}(t)=\langle t^2,t^{-1},t\rangle$ at $t=1$

SOLUTION

(a) We first compute the vector $\mathbf{B}$ using Eq. (16):

$$\mathbf{B}=\frac{\mathbf{r}'\times\mathbf{r}''}{\|\mathbf{r}'\times\mathbf{r}''\|}\tag{1}$$

Differentiating $\mathbf{r}(t)=\langle\cos t,t,t^2\rangle$ gives

$$\begin{array}{l}\mathbf{r}'(t)=\langle-\sin t,1,2t\rangle\\ \mathbf{r}''(t)=\langle-\cos t,0,2\rangle\end{array}\quad\Rightarrow\quad\begin{array}{l}\mathbf{r}'(0)=\langle 0,1,0\rangle\\ \mathbf{r}''(0)=\langle -1,0,2\rangle\end{array}$$

We compute the cross product:

$$\mathbf{r}'(0)\times\mathbf{r}''(0)=\mathbf{j}\times(-\mathbf{i}+2\mathbf{k})=-\mathbf{j}\times\mathbf{i}+2\mathbf{j}\times\mathbf{k}=\mathbf{k}+2\mathbf{i}=\langle 2,0,1\rangle$$

$$\|\mathbf{r}'(0)\times\mathbf{r}''(0)\|=\sqrt{2^2+0^2+1^2}=\sqrt{5}$$

Substituting in (1) we obtain:

$$\mathbf{B}(0)=\frac{\langle 2,0,1\rangle}{\sqrt{5}}=\frac{1}{\sqrt{5}}\langle 2,0,1\rangle$$

We now compute $\mathbf{T}(0)$:

$$\mathbf{T}(0)=\frac{\mathbf{r}'(0)}{\|\mathbf{r}'(0)\|}=\frac{\langle 0,1,0\rangle}{\|\langle 0,1,0\rangle\|}=\langle 0,1,0\rangle$$

Finally we find $\mathbf{N}=\mathbf{B}\times\mathbf{T}$:

$$\mathbf{N}(0)=\frac{1}{\sqrt{5}}\langle 2,0,1\rangle\times\langle 0,1,0\rangle=\frac{1}{\sqrt{5}}(2\mathbf{i}+\mathbf{k})\times\mathbf{j}=\frac{1}{\sqrt{5}}(2\mathbf{i}\times\mathbf{j}+\mathbf{k}\times\mathbf{j})=\frac{1}{\sqrt{5}}(2\mathbf{k}-\mathbf{i})=\frac{1}{\sqrt{5}}\langle -1,0,2\rangle$$

(b) Differentiating $\mathbf{r}(t) = \langle t^2, t^{-1}, t\rangle$ gives

$$\begin{aligned}\mathbf{r}'(t) &= \langle 2t, -t^{-2}, 1\rangle \\ \mathbf{r}''(t) &= \langle 2, 2t^{-3}, 0\rangle\end{aligned} \quad\Rightarrow\quad \begin{aligned}\mathbf{r}'(1) &= \langle 2, -1, 1\rangle \\ \mathbf{r}''(1) &= \langle 2, 2, 0\rangle\end{aligned}$$

We compute the cross product:

$$\mathbf{r}'(1)\times\mathbf{r}''(1) = \begin{vmatrix} \mathbf{i} & \mathbf{j} & \mathbf{k} \\ 2 & -1 & 1 \\ 2 & 2 & 0 \end{vmatrix} = \begin{vmatrix} -1 & 1 \\ 2 & 0 \end{vmatrix}\mathbf{i} - \begin{vmatrix} 2 & 1 \\ 2 & 0 \end{vmatrix}\mathbf{j} + \begin{vmatrix} 2 & -1 \\ 2 & 2 \end{vmatrix}\mathbf{k} = -2\mathbf{i} + 2\mathbf{j} + 6\mathbf{k} = \langle -2, 2, 6\rangle$$

$$\|\mathbf{r}'(1)\times\mathbf{r}''(1)\| = \sqrt{(-2)^2 + 2^2 + 6^2} = \sqrt{44} = 2\sqrt{11}$$

Substituting in (1) gives:

$$\mathbf{B}(1) = \frac{\langle -2, 2, 6\rangle}{2\sqrt{11}} = \frac{1}{\sqrt{11}}\langle -1, 1, 3\rangle$$

We now find $\mathbf{T}(1)$:

$$\mathbf{T}(1) = \frac{\mathbf{r}'(1)}{\|\mathbf{r}'(1)\|} = \frac{\langle 2, -1, 1\rangle}{\sqrt{4+1+1}} = \frac{1}{\sqrt{6}}\langle 2, -1, 1\rangle$$

Finally we find $\mathbf{N}(1)$ by computing the following cross product:

$$\mathbf{N}(1) = \mathbf{B}(1)\times\mathbf{T}(1) = \frac{1}{\sqrt{11}}\langle -1, 1, 3\rangle \times \frac{1}{\sqrt{6}}\langle 2, -1, 1\rangle = \frac{1}{\sqrt{66}}\begin{vmatrix} \mathbf{i} & \mathbf{j} & \mathbf{k} \\ -1 & 1 & 3 \\ 2 & -1 & 1 \end{vmatrix}$$

$$= \frac{1}{\sqrt{66}}\left\{\begin{vmatrix} 1 & 3 \\ -1 & 1 \end{vmatrix}\mathbf{i} - \begin{vmatrix} -1 & 3 \\ 2 & 1 \end{vmatrix}\mathbf{j} + \begin{vmatrix} -1 & 1 \\ 2 & -1 \end{vmatrix}\mathbf{k}\right\} = \frac{1}{\sqrt{66}}(4\mathbf{i} + 7\mathbf{j} - \mathbf{k}) = \frac{1}{\sqrt{66}}\langle 4, 7, -1\rangle$$

14.5 Motion in Three-Space (ET Section 13.5)

Preliminary Questions

1. If a particle travels with constant speed, must its acceleration vector be zero? Explain.

SOLUTION If the speed of the particle is constant, the tangential component, $a_T(t) = v'(t)$, of the acceleration is zero. However, the normal component, $a_N(t) = \kappa(t)v(t)^2$ is not necessarily zero, since the particle may change its direction.

2. For a particle in uniform circular motion around a circle, which of the vectors $\mathbf{v}(t)$ or $\mathbf{a}(t)$ always points toward the center of the circle?

SOLUTION For a particle in uniform circular motion around a circle, the acceleration vector $\mathbf{a}(t)$ points towards the center of the circle, whereas $\mathbf{v}(t)$ is tangent to the circle.

3. Two objects travel to the right along the parabola $y = x^2$ with nonzero speed. Which of the following must be true?

(a) Their velocity vectors point in the same direction.

(b) Their velocity vectors have the same length.

(c) Their acceleration vectors point in the same direction.

SOLUTION

(a) The velocity vector points in the direction of motion, hence the velocities of the two objects point in the same direction.

(b) The length of the velocity vector is the speed. Since the speeds are not necessarily equal, the velocity vectors may have different lengths.

(c) The acceleration is determined by the tangential component $v'(t)$ and the normal component $\kappa(t)v(t)^2$. Since v and v' may be different for the two objects, the acceleration vectors may have different directions.

4. Use the decomposition of acceleration into tangential and normal components to explain the following statement: If the speed is constant, then the acceleration and velocity vectors are orthogonal.

SOLUTION If the speed is constant, $v'(t) = 0$. Therefore, the acceleration vector has only the normal component:

$$\mathbf{a}(t) = a_N(t)\mathbf{N}(t)$$

The velocity vector always points in the direction of motion. Since the vector $\mathbf{N}(t)$ is orthogonal to the direction of motion, the vectors $\mathbf{a}(t)$ and $\mathbf{v}(t)$ are orthogonal.

5. If a particle travels along a straight line, then the acceleration and velocity vectors are (choose the correct statement):

(a) Orthogonal **(b)** Parallel

SOLUTION Since a line has zero curvature, the normal component of the acceleration is zero, hence $\mathbf{a}(t)$ has only the tangential component. The velocity vector is always in the direction of motion, hence the acceleration and the velocity vectors are parallel to the line. We conclude that (b) is the correct statement.

6. What is the length of the acceleration vector of a particle traveling around a circle of radius 2 cm with constant velocity 4 cm/s?

SOLUTION The acceleration vector is given by the following decomposition:

$$\mathbf{a}(t) = v'(t)\mathbf{T}(t) + \kappa(t)v(t)^2\mathbf{N}(t) \tag{1}$$

In our case $v(t) = 4$ is constant hence $v'(t) = 0$. In addition, the curvature of a circle of radius 2 is $\kappa(t) = \frac{1}{2}$. Substituting $v(t) = 4$, $v'(t) = 0$ and $\kappa(t) = \frac{1}{2}$ in (1) gives:

$$\mathbf{a}(t) = \frac{1}{2} \cdot 4^2 N(t) = 8N(t)$$

The length of the acceleration vector is, thus,

$$\|\mathbf{a}(t)\| = 8 \text{ cm/s}^2$$

7. Two cars are racing around a circular track. If, at a certain moment, both of their speedometers read 110 mph. then the two cars have the same (choose one):

(a) $a_{\mathbf{T}}$ **(b)** $a_{\mathbf{N}}$

SOLUTION The tangential acceleration a_T and the normal acceleration a_N are the following values:

$$a_T(t) = v'(t); \quad a_N(t) = \kappa(t)v(t)^2$$

At the moment where both speedometers read 110 mph, the speeds of the two cars are $v = 110$ mph. Since the track is circular, the curvature $\kappa(t)$ is constant, hence the normal accelerations of the two cars are equal at this moment. Statement (b) is correct.

Exercises

1. Use the table here to calculate the difference quotients $\dfrac{\mathbf{r}(1+h) - \mathbf{r}(1)}{h}$ for $h = -0.2, -0.1, 0.1, 0.2$. Then estimate the velocity and speed at $t = 1$.

$\mathbf{r}(0.8)$	$\langle 1.557, 2.459, -1.970\rangle$
$\mathbf{r}(0.9)$	$\langle 1.559, 2.634, -1.740\rangle$
$\mathbf{r}(1)$	$\langle 1.540, 2.841, -1.443\rangle$
$\mathbf{r}(1.1)$	$\langle 1.499, 3.078, -1.035\rangle$
$\mathbf{r}(1.2)$	$\langle 1.435, 3.342, -0.428\rangle$

SOLUTION

$(h = -0.2)$

$$\frac{\mathbf{r}(1-0.2) - \mathbf{r}(1)}{-0.2} = \frac{\mathbf{r}(0.8) - \mathbf{r}(1)}{-0.2} = \frac{\langle 1.557, 2.459, -1.970\rangle - \langle 1.540, 2.841, -1.443\rangle}{-0.2}$$
$$= \frac{\langle 0.017, -0.382, -0.527\rangle}{-0.2} = \langle -0.085, 1.91, 2.635\rangle$$

$(h = -0.1)$

$$\frac{\mathbf{r}(1-0.1) - \mathbf{r}(1)}{-0.1} = \frac{\mathbf{r}(0.9) - \mathbf{r}(1)}{-0.1} = \frac{\langle 1.559, 2.634, -1.740\rangle - \langle 1.540, 2.841, -1.443\rangle}{-0.1}$$
$$= \frac{\langle 0.019, -0.207, -0.297\rangle}{-0.1} = \langle -0.19, 2.07, 2.97\rangle$$

$(h = 0.1)$

$$\frac{\mathbf{r}(1+0.1) - \mathbf{r}(1)}{0.1} = \frac{\mathbf{r}(1.1) - \mathbf{r}(1)}{0.1} = \frac{\langle 1.499, 3.078, -1.035\rangle - \langle 1.540, 2.841, -1.443\rangle}{0.1}$$
$$= \frac{\langle -0.041, 0.237, 0.408\rangle}{0.1} = \langle -0.41, 2.37, 4.08\rangle$$

$(h = 0.2)$

$$\frac{\mathbf{r}(1+0.2)-\mathbf{r}(1)}{0.2} = \frac{\mathbf{r}(1.2)-\mathbf{r}(1)}{0.2} = \frac{\langle 1.435, 3.342, -0.428\rangle - \langle 1.540, 2.841, -1.443\rangle}{0.2}$$

$$= \frac{\langle -0.105, 0.501, 1.015\rangle}{0.2} = \langle -0.525, 2.505, 5.075\rangle$$

The velocity vector is defined by:

$$\mathbf{v}(t) = \mathbf{r}'(t) = \lim_{h\to 0} \frac{\mathbf{r}(t+h)-\mathbf{r}(t)}{h}$$

We may estimate the velocity at $t = 1$ by:

$$\mathbf{v}(1) \approx \langle -0.3, 2.2, 3.5\rangle$$

and the speed by:

$$v(1) = \|\mathbf{v}(1)\| \approx \sqrt{0.3^2 + 2.2^2 + 3.5^2} \cong 4.1$$

In Exercises 3–6, calculate the velocity and acceleration vectors and the speed at the time indicated.

3. $\mathbf{r}(t) = \langle t^3, 1-t, 4t^2\rangle, \quad t = 1$

SOLUTION In this case $\mathbf{r}(t) = \langle t^3, 1-t, 4t^2\rangle$ hence:

$$\mathbf{v}(t) = \mathbf{r}'(t) = \langle 3t^2, -1, 8t\rangle \quad \Rightarrow \quad \mathbf{v}(1) = \langle 3, -1, 8\rangle$$

$$\mathbf{a}(t) = \mathbf{r}''(t) = \langle 6t, 0, 8\rangle \quad \Rightarrow \quad \mathbf{a}(1) = \langle 6, 0, 8\rangle$$

The speed is the magnitude of the velocity vector, that is,

$$v(1) = \|\mathbf{v}(1)\| = \sqrt{3^2 + (-1)^2 + 8^2} = \sqrt{74}$$

5. $\mathbf{r}(\theta) = \langle \sin\theta, \cos\theta, \cos 3\theta\rangle, \quad \theta = \frac{\pi}{3}$

SOLUTION Differentiating $\mathbf{r}(\theta) = \langle \sin\theta, \cos\theta, \cos 3\theta\rangle$ gives:

$$\mathbf{v}(\theta) = \mathbf{r}'(\theta) = \langle \cos\theta, -\sin\theta, -3\sin 3\theta\rangle$$

$$\Rightarrow \mathbf{v}\left(\frac{\pi}{3}\right) = \left\langle \cos\frac{\pi}{3}, -\sin\frac{\pi}{3}, -3\sin\pi\right\rangle = \left\langle \frac{1}{2}, -\frac{\sqrt{3}}{2}, 0\right\rangle$$

$$\mathbf{a}(\theta) = \mathbf{r}''(\theta) = \langle -\sin\theta, -\cos\theta, -9\cos 3\theta\rangle$$

$$\Rightarrow \mathbf{a}\left(\frac{\pi}{3}\right) = \left\langle -\sin\frac{\pi}{3}, -\cos\frac{\pi}{3}, -9\cos\pi\right\rangle = \left\langle -\frac{\sqrt{3}}{2}, -\frac{1}{2}, 9\right\rangle$$

The speed is the magnitude of the velocity vector, that is:

$$v\left(\frac{\pi}{3}\right) = \left\|\mathbf{v}\left(\frac{\pi}{3}\right)\right\| = \sqrt{\left(\frac{1}{2}\right)^2 + \left(-\frac{\sqrt{3}}{2}\right)^2 + 0^2} = 1$$

7. Find $\mathbf{a}(t)$ for a particle moving around a circle of radius 8 cm at a constant speed of $v = 4$ cm/s (see Example 2). Draw the path and acceleration vector at $t = \frac{\pi}{4}$.

SOLUTION The position vector is:

$$\mathbf{r}(t) = 8\langle \cos\omega t, \sin\omega t\rangle$$

Hence,

$$\mathbf{v}(t) = \mathbf{r}'(t) = 8\langle -\omega\sin\omega t, \omega\cos\omega t\rangle = 8\omega\langle -\sin\omega t, \cos\omega t\rangle \tag{1}$$

We are given that the speed of the particle is $v = 4$cm/s. The speed is the magnitude of the velocity vector, hence:

$$v = 8\omega\sqrt{(-\sin\omega t)^2 + \cos^2\omega t} = 8\omega = 4 \quad \Rightarrow \quad \omega = \frac{1}{2} \text{ rad/s}$$

Substituting in (2) we get:

$$\mathbf{v}(t) = 4\left\langle -\sin\frac{t}{2}, \cos\frac{t}{2}\right\rangle$$

We now find $\mathbf{a}(t)$ by differentiating the velocity vector. This gives

$$\mathbf{a}(t) = \mathbf{v}'(t) = 4\left\langle -\frac{1}{2}\cos\frac{t}{2}, -\frac{1}{2}\sin\frac{t}{2}\right\rangle = -2\left\langle \cos\frac{t}{2}, \sin\frac{t}{2}\right\rangle$$

The path of the particle is $\mathbf{r}(t) = 8\left\langle \cos\frac{t}{2}, \sin\frac{t}{2}\right\rangle$ and the acceleration vector at $t = \frac{\pi}{4}$ is:

$$\mathbf{a}\left(\frac{\pi}{4}\right) = -2\left\langle \cos\frac{\pi}{8}, \sin\frac{\pi}{8}\right\rangle \approx \langle -1.85, -0.77\rangle$$

The path $\mathbf{r}(t)$ and the acceleration vector at $t = \frac{\pi}{4}$ are shown in the following figure:

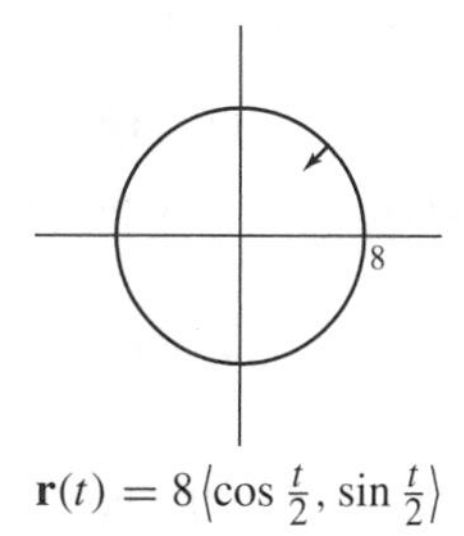

$\mathbf{r}(t) = 8\left\langle \cos\frac{t}{2}, \sin\frac{t}{2}\right\rangle$

9. Sketch the path $\mathbf{r}(t) = \left\langle t^2, t^3\right\rangle$ together with the velocity and acceleration vectors at $t = 1$.

SOLUTION We compute the velocity and acceleration vectors at $t = 1$:

$$\mathbf{v}(t) = \mathbf{r}'(t) = \left\langle 2t, 3t^2\right\rangle \quad \Rightarrow \quad \mathbf{v}(1) = \langle 2, 3\rangle$$
$$\mathbf{a}(t) = \mathbf{v}'(t) = \langle 2, 6t\rangle \quad \Rightarrow \quad \mathbf{a}(1) = \langle 2, 6\rangle$$

The following figure shows the path $\mathbf{r}(t) = \left\langle t^2, t^3\right\rangle$ and the vectors $\mathbf{v}(1)$ and $\mathbf{a}(1)$:

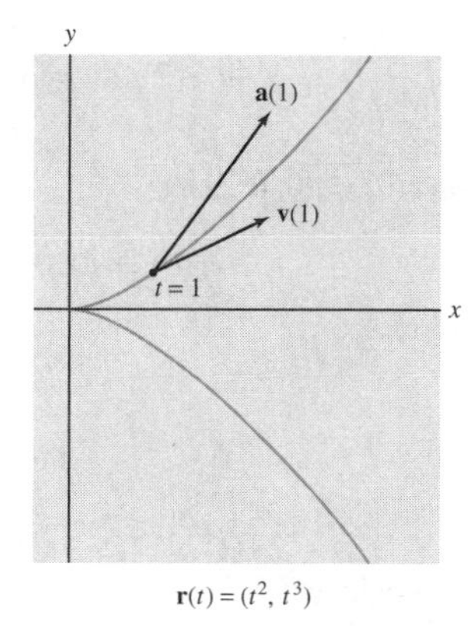

$\mathbf{r}(t) = (t^2, t^3)$

In Exercises 11–14, find $\mathbf{v}(t)$ given $\mathbf{a}(t)$ and the initial velocity.

11. $\mathbf{a}(t) = \left\langle t, 4\right\rangle$, $\quad \mathbf{v}(0) = \left\langle \frac{1}{3}, -2\right\rangle$

SOLUTION We find $\mathbf{v}(t)$ by integrating $\mathbf{a}(t)$:

$$\mathbf{v}(t) = \int_0^t \mathbf{a}(u)du = \int_0^t \langle u, 4\rangle\, du = \left\langle \frac{1}{2}u^2, 4u\right\rangle\Big|_0^t + \mathbf{v}_0 = \left\langle \frac{t^2}{2}, 4t\right\rangle + \mathbf{v}_0$$

The initial condition gives:

$$\mathbf{v}(0) = \langle 0, 0\rangle + \mathbf{v}_0 = \left\langle \frac{1}{3}, -2\right\rangle \quad \Rightarrow \quad \mathbf{v}_0 = \left\langle \frac{1}{3}, -2\right\rangle$$

Hence,

$$\mathbf{v}(t) = \left\langle \frac{t^2}{2}, 4t\right\rangle + \left\langle \frac{1}{3}, -2\right\rangle = \left\langle \frac{3t^2+2}{6}, 4t-2\right\rangle$$

13. $\mathbf{a}(t) = \mathbf{k}, \quad \mathbf{v}(0) = \mathbf{i}$

SOLUTION We compute $\mathbf{v}(t)$ by integrating the acceleration vector:

$$\mathbf{v}(t) = \int_0^t \mathbf{a}(u)\,du = \int_0^t \mathbf{k}\,du = \mathbf{k}u\Big|_0^t + \mathbf{v}_0 = t\mathbf{k} + \mathbf{v}_0 \tag{1}$$

Substituting the initial condition gives:

$$\mathbf{v}(0) = 0\mathbf{k} + \mathbf{v}_0 = \mathbf{i} \quad \Rightarrow \quad \mathbf{v}_0 = \mathbf{i}$$

Combining with (1) we obtain:

$$\mathbf{v}(t) = \mathbf{i} + t\mathbf{k}$$

In Exercises 15–18, find $\mathbf{r}(t)$ *and* $\mathbf{v}(t)$ *given* $\mathbf{a}(t)$ *and the initial velocity and position.*

15. $\mathbf{a}(t) = \langle t, 4\rangle, \quad \mathbf{v}(0) = \langle 3, -2\rangle, \quad \mathbf{r}(0) = \langle 0, 0\rangle$

SOLUTION We first integrate $\mathbf{a}(t)$ to find the velocity vector:

$$\mathbf{v}(t) = \int_0^t \langle u, 4\rangle\,du = \left\langle \frac{u^2}{2}, 4u\right\rangle\Big|_0^t + \mathbf{v}_0 = \left\langle \frac{t^2}{2}, 4t\right\rangle + \mathbf{v}_0 \tag{1}$$

The initial condition $\mathbf{v}(0) = \langle 3, -2\rangle$ gives:

$$\mathbf{v}(0) = \langle 0, 0\rangle + \mathbf{v}_0 = \langle 3, -2\rangle \quad \Rightarrow \quad \mathbf{v}_0 = \langle 3, -2\rangle$$

Substituting in (1) we get:

$$\mathbf{v}(t) = \left\langle \frac{t^2}{2}, 4t\right\rangle + \langle 3, -2\rangle = \left\langle \frac{t^2}{2} + 3, 4t - 2\right\rangle$$

We now integrate the velocity vector to find $\mathbf{r}(t)$:

$$\mathbf{r}(t) = \int_0^t \left\langle \frac{u^2}{2} + 3, 4u - 2\right\rangle du = \left\langle \frac{u^3}{6} + 3u, 2u^2 - 2u\right\rangle\Big|_0^t + \mathbf{r}_0 = \left\langle \frac{t^3}{6} + 3t, 2t^2 - 2t\right\rangle + \mathbf{r}_0$$

The initial condition $\mathbf{r}(0) = \langle 0, 0\rangle$ gives:

$$\mathbf{r}(0) = \langle 0, 0\rangle + \mathbf{r}_0 = \langle 0, 0\rangle \quad \Rightarrow \quad \mathbf{r}_0 = \langle 0, 0\rangle$$

Hence,

$$\mathbf{r}(t) = \left\langle \frac{t^3}{6} + 3t, 2t^2 - 2t\right\rangle$$

17. $\mathbf{a}(t) = t\mathbf{k}, \quad \mathbf{v}(0) = \mathbf{i}, \mathbf{r}(0) = \mathbf{j}$

SOLUTION Integrating the acceleration vector gives:

$$\mathbf{v}(t) = \int_0^t u\mathbf{k}\,du = \frac{u^2}{2}\mathbf{k}\Big|_0^t + \mathbf{v}_0 = \frac{t^2}{2}\mathbf{k} + \mathbf{v}_0 \tag{1}$$

The initial condition for $\mathbf{v}(t)$ gives:

$$\mathbf{v}(0) = \frac{0^2}{2}\mathbf{k} + \mathbf{v}_0 = \mathbf{i} \quad \Rightarrow \quad \mathbf{v}_0 = \mathbf{i}$$

We substitute in (1):

$$\mathbf{v}(t) = \frac{t^2}{2}\mathbf{k} + \mathbf{i} = \mathbf{i} + \frac{t^2}{2}\mathbf{k}$$

We now integrate $\mathbf{v}(t)$ to find $\mathbf{r}(t)$:

$$\mathbf{r}(t) = \int_0^t \left(\mathbf{i} + \frac{u^2}{2}\mathbf{k}\right) du = u\mathbf{i} + \frac{u^3}{6}\mathbf{k}\Big|_0^t + \mathbf{r}_0 = t\mathbf{i} + \frac{t^3}{6}\mathbf{k} + \mathbf{r}_0 \tag{2}$$

The initial condition for $\mathbf{r}(t)$ gives:

$$\mathbf{r}(0) = 0\mathbf{i} + 0\mathbf{k} + \mathbf{r}_0 = \mathbf{j} \quad \Rightarrow \quad \mathbf{r}_0 = \mathbf{j}$$

Combining with (2) gives the position vector:

$$\mathbf{r}(t) = t\mathbf{i} + \mathbf{j} + \frac{t^3}{6}\mathbf{k}$$

19. A bullet is fired from the ground at an angle of 45°. What initial speed must the bullet have in order to hit the top of a 400-ft tower located 600 ft away?

SOLUTION We place the gun at the origin and let $\mathbf{r}(t)$ be the bullet's position vector.

Step 1. Use Newton's Law. The net force vector acting on the bullet is the force of gravity $\mathbf{F} = \langle 0, -gm\rangle = m\langle 0, -g\rangle$. By Newton's Second Law, $\mathbf{F} = m\mathbf{r}''(t)$, hence:

$$m\langle 0, -g\rangle = m\mathbf{r}''(t) \quad \Rightarrow \quad \mathbf{r}''(t) = \langle 0, -g\rangle$$

We compute the position vector by integrating twice:

$$\mathbf{r}'(t) = \int_0^t \mathbf{r}''(u)\,du = \int_0^t \langle 0, -g\rangle\,du = \langle 0, -gt\rangle + \mathbf{v}_0$$

$$\mathbf{r}(t) = \int_0^t \mathbf{r}'(u)\,du = \int_0^t (\langle 0, -gu\rangle + \mathbf{v}_0)\,du = \left\langle 0, -g\frac{t^2}{2}\right\rangle + \mathbf{v}_0 t + \mathbf{r}_0$$

That is,

$$\mathbf{r}(t) = \left\langle 0, \frac{-g}{2}t^2\right\rangle + \mathbf{v}_0 t + \mathbf{r}_0 \tag{1}$$

Since the gun is at the origin, $\mathbf{r}_0 = \mathbf{0}$. The bullet is fired at an angle of 45°, hence the initial velocity $\mathbf{v}_0$ points in the direction of the unit vector $\langle\cos 45°, \sin 45°\rangle = \left\langle\frac{\sqrt{2}}{2}, \frac{\sqrt{2}}{2}\right\rangle$ therefore, $\mathbf{v}_0 = v_0\left\langle\frac{\sqrt{2}}{2}, \frac{\sqrt{2}}{2}\right\rangle$. Substituting these initial values in (1) gives:

$$\mathbf{r}(t) = \left\langle 0, \frac{-g}{2}t^2\right\rangle + tv_0\left\langle\frac{\sqrt{2}}{2}, \frac{\sqrt{2}}{2}\right\rangle$$

Step 2. Solve for v_0. The position vector of the top of the tower is $\langle 600, 400\rangle$, hence at the moment of hitting the tower we have,

$$\mathbf{r}(t) = \left\langle 0, \frac{-g}{2}t^2\right\rangle + tv_0\left\langle\frac{\sqrt{2}}{2}, \frac{\sqrt{2}}{2}\right\rangle = \langle 600, 400\rangle$$

$$\left\langle tv_0\frac{\sqrt{2}}{2}, \frac{-g}{2}t^2 + \frac{\sqrt{2}}{2}tv_0\right\rangle = \langle 600, 400\rangle$$

Equating components, we get the equations:

$$\begin{cases} tv_0\dfrac{\sqrt{2}}{2} = 600 \\ -\dfrac{g}{2}t^2 + \dfrac{\sqrt{2}}{2}tv_0 = 400 \end{cases}$$

The first equation implies that $t = \frac{1{,}200}{\sqrt{2}v_0}$. We substitute in the second equation and solve for v_0 (we use $g = 32\text{ ft/s}^2$):

$$-16\left(\frac{1{,}200}{\sqrt{2}v_0}\right)^2 + \frac{\sqrt{2}}{2}\left(\frac{1{,}200}{\sqrt{2}v_0}\right)v_0 = 400$$

$$-8\left(\frac{1{,}200}{v_0}\right)^2 + 600 = 400$$

$$\left(\frac{1{,}200}{v_0}\right)^2 = 25 \quad \Rightarrow \quad \frac{1{,}200}{v_0} = 5 \quad \Rightarrow \quad v_0 = 240\text{ ft/s}$$

The initial speed of the bullet must be $v_0 = 240$ ft/s.

21. A projectile fired at an angle of 60° lands 1,200 ft away. What was its initial speed?

SOLUTION We place the gun at the origin and let $\mathbf{r}(t)$ be the projectile's position vector. The net force acting on the projectile is:

$$\mathbf{F} = \langle 0, -mg \rangle = -m \langle 0, g \rangle = -m \langle 0, 32 \rangle$$

By Newton's Second Law, $\mathbf{F} = m\mathbf{r}''(t)$, hence:

$$-m \langle 0, 32 \rangle = m\mathbf{r}''(t) \quad \Rightarrow \quad \mathbf{r}''(t) = -\langle 0, 32 \rangle$$

We integrate twice to obtain:

$$\mathbf{v}(t) = \int_0^t \mathbf{r}''(u)\, du = \int_0^t \langle 0, -32 \rangle\, du = \langle 0, -32t \rangle + \mathbf{v}_0$$

$$\mathbf{r}(t) = \int_0^t \mathbf{v}(u)\, du = \int_0^t (\langle 0, -32u \rangle + \mathbf{v}_0)\, du = \left\langle 0, -16t^2 \right\rangle + \mathbf{v}_0 t + \mathbf{r}_0 \qquad (1)$$

Since the gun is at the origin, $\mathbf{r}_0 = 0$. The firing is at an angle of 60°, hence the initial velocity points in the direction of the unit vector $\left\langle \cos 60°, \sin 60° \right\rangle = \left\langle \frac{1}{2}, \frac{\sqrt{3}}{2} \right\rangle$. Hence,

$$\mathbf{v}_0 = v_0 \left\langle \frac{1}{2}, \frac{\sqrt{3}}{2} \right\rangle = \left\langle \frac{v_0}{2}, \frac{\sqrt{3}v_0}{2} \right\rangle$$

Substituting the initial vectors in (1) we get:

$$\mathbf{r}(t) = \left\langle 0, -16t^2 \right\rangle + \left\langle \frac{v_0}{2}, \frac{\sqrt{3}v_0}{2} \right\rangle t = \left\langle \frac{v_0 t}{2}, -16t^2 + \frac{\sqrt{3}v_0}{2} t \right\rangle$$

At the moment of landing, the x-component of $\mathbf{r}(t)$ is 1200, and the y-component is zero. The corresponding equations are:

$$\begin{cases} \dfrac{v_0}{2} t = 1{,}200 \\ -16t^2 + \dfrac{\sqrt{3}v_0}{2} t = 0 \end{cases}$$

The first equation implies that $v_0 t = 2{,}400$. We substitute in the second equation and solve for t:

$$-16t^2 + \frac{\sqrt{3}}{2} \cdot 2{,}400 = 0$$

$$16t^2 = 1{,}200\sqrt{3}$$

$$t^2 \approx 129.9 \quad \Rightarrow \quad t \approx 11.4 \text{ s}$$

By $v_0 t = 2{,}400$ we get $v_0 = \frac{2{,}400}{11.4} \approx 210.53$ ft/s.

23. One player throws a baseball to another player standing 80 ft away with initial speed 60 ft/s. Use the result of Exercise 22 to find two angles θ at which the ball can be released. Which angle gets the ball there faster?

SOLUTION We suppose that the baseball is thrown from the origin, and that $\mathbf{r}(t)$ is the baseball's position vector. By Exercise 22 the total distance travelled by the ball is $\frac{v_0^2}{g} \sin 2\theta$. Using the given information we obtain the following equation:

$$\frac{60^2}{32} \sin 2\theta = 80$$

$$\sin 2\theta = \frac{32 \cdot 80}{60^2} \approx 0.711$$

The solutions for $0 \le \theta \le 90°$ are:

$$\begin{matrix} 2\theta \approx 45.32° \\ \theta \approx 22.66° \end{matrix} \quad \text{or} \quad \begin{matrix} 2\theta \approx 134.68° \\ \theta \approx 67.34° \end{matrix}$$

By Newton's Second Law we have:

$$\mathbf{F} = m \langle 0, -g \rangle = m\mathbf{r}''(t) \quad \Rightarrow \quad \mathbf{r}''(t) = \langle 0, -g \rangle = \langle 0, -32 \rangle$$

Integrating gives:

$$\mathbf{v}(t) = \int_0^t \mathbf{r}''(u)\,du = \int_0^t \langle 0, -32\rangle\,du = \langle 0, -32t\rangle + \mathbf{v}_0 \tag{1}$$

The initial velocity points in the direction of the unit vector $\langle \cos\theta, \sin\theta\rangle$ and its magnitude is the initial speed $v_0 = 60$. Hence, $\mathbf{v}_0 = 60\langle \cos\theta, \sin\theta\rangle$. Substituting in (1) we get:

$$\mathbf{v}(t) = \langle 0, -32t\rangle + 60\langle \cos\theta, \sin\theta\rangle \tag{2}$$

Integrating this vector with respect to t and using $\mathbf{r}_0 = \mathbf{0}$ we obtain:

$$\mathbf{r}(t) = \int_0^t \mathbf{v}(u)\,du = \int_0^t \left(\langle 0, -32u\rangle + 60\langle \cos\theta, \sin\theta\rangle\right)du = \left\langle 0, -16t^2\right\rangle + 60t\langle \cos\theta, \sin\theta\rangle$$

At the final time $x(t) = 80$. This gives:

$$x(t) = 60t\cos\theta = 80 \quad\Rightarrow\quad t = \frac{4}{3\cos\theta}$$

Since we want to minimize t we need to maximize $\cos\theta$, hence, to minimize θ. Therefore, $\theta = 22.66^\circ$ will get the ball faster to the other player.

25. At a certain moment, a moving particle has velocity $\mathbf{v} = \langle 2, 2, -1\rangle$ and $\mathbf{a} = \langle 0, 4, 3\rangle$. Find $\mathbf{T}$, $\mathbf{N}$, and the decomposition of $\mathbf{a}$ into tangential and normal components.

SOLUTION We go through the following steps:

Step 1. Compute $\mathbf{T}$ and $a_{\mathbf{T}}$. The unit tangent is the following vector:

$$\mathbf{T} = \frac{\mathbf{v}}{\|\mathbf{v}\|} = \frac{\langle 2, 2, -1\rangle}{\sqrt{2^2 + 2^2 + (-1)^2}} = \frac{1}{3}\langle 2, 2, -1\rangle \tag{1}$$

The tangential component of $\mathbf{a} = \langle 0, 4, 3\rangle$ is:

$$a_{\mathbf{T}} = \mathbf{a}\cdot\mathbf{T} = \langle 0, 4, 3\rangle \cdot \frac{1}{3}\langle 2, 2, -1\rangle = \frac{1}{3}(0 + 8 - 3) = \frac{5}{3}$$

Step 2. Compute $a_{\mathbf{N}}$ and $\mathbf{N}$. Since $a_{\mathbf{N}}\mathbf{N} = \mathbf{a} - a_{\mathbf{T}}\mathbf{T}$, we have:

$$a_{\mathbf{N}}\mathbf{N} = \langle 0, 4, 3\rangle - \frac{5}{3}\cdot\frac{1}{3}\langle 2, 2, -1\rangle = \langle 0, 4, 3\rangle - \left\langle \frac{10}{9}, \frac{10}{9}, -\frac{5}{9}\right\rangle = \frac{1}{9}\langle -10, 26, 32\rangle \tag{2}$$

The unit normal $\mathbf{N}$ is a unit vector, therefore:

$$a_{\mathbf{N}} = \|a_{\mathbf{N}}\mathbf{N}\| = \frac{1}{9}\sqrt{(-10)^2 + 26^2 + 32^2} = \frac{1}{9}\cdot 30\sqrt{2} = \frac{10\sqrt{2}}{3} \tag{3}$$

We compute $\mathbf{N}$, using (3) and (4):

$$\mathbf{N} = \frac{a_{\mathbf{N}}\mathbf{N}}{a_{\mathbf{N}}} = \frac{\frac{1}{9}\langle -10, 26, 32\rangle}{\frac{10\sqrt{2}}{3}} = \frac{1}{15\sqrt{2}}\langle -5, 13, 16\rangle$$

Step 3. Write the decomposition. Using (1)–(4) we obtain the following decomposition:

$$\mathbf{a} = a_{\mathbf{T}}\mathbf{T} + a_{\mathbf{N}}\mathbf{N}$$

$$\langle 0, 4, 3\rangle = \frac{5}{3}\mathbf{T} + \frac{10\sqrt{2}}{3}\mathbf{N},$$

where $\mathbf{T} = \left\langle \frac{2}{3}, \frac{2}{3}, -\frac{1}{3}\right\rangle$ and $\mathbf{N} = \frac{1}{15\sqrt{2}}\langle -5, 13, 16\rangle$.

27. A particle follows a path $\mathbf{r}(t)$ for $0 \le t \le T$, beginning at the origin O. The vector $\overline{\mathbf{v}} = \frac{1}{T}\int_0^T \mathbf{r}'(t)\,dt$ is called the **average velocity** vector. Suppose that $\overline{\mathbf{v}} = \mathbf{0}$. Answer and explain the following:

(a) Where is the particle located at time T if $\overline{\mathbf{v}} = \mathbf{0}$?

(b) Is the particle's average speed necessarily equal to zero?

SOLUTION

(a) If the average velocity is 0, then the particle must be back at its original position at time $t = T$. This is perhaps best seen by noting that $\overline{\mathbf{v}} = \frac{1}{T}\int_0^T \mathbf{r}'(t)\,dt = \mathbf{r}(t)\Big|_0^T$.

(b) The average speed need not be zero! Consider a particle moving at constant speed around a circle, with position vector $\mathbf{r}(t) = \langle \cos t, \sin t \rangle$. From 0 to 2π, this has average velocity of 0, but constant average speed of 1.

29. A space shuttle orbits the earth at an altitude 200 miles above the earth's surface, with constant speed $v = 17{,}000$ mph. Find the magnitude of the shuttle's acceleration (in ft/s^2), assuming that the radius of the earth is 4,000 miles (Figure 11).

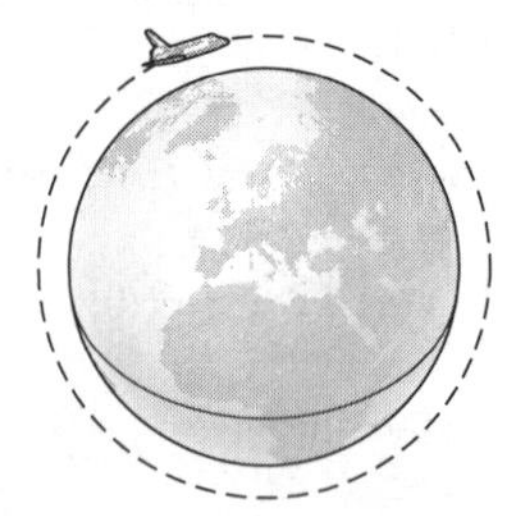

FIGURE 11 Space shuttle orbit.

SOLUTION The shuttle is in a uniform circular motion, therefore the tangential component of its acceleration is zero, and the acceleration can be written as:

$$\mathbf{a} = \kappa v^2 \mathbf{N} \qquad (1)$$

The radius of motion is $4000 + 200 = 4200$ miles hence the curvature is $\kappa = \frac{1}{4200}$. Also by the given information the constant speed is $v = 17{,}000$ mph. Substituting these values in (1) we get:

$$\mathbf{a} = \frac{1}{4200} \cdot 17{,}000^2 \mathbf{N} = 6.88 \cdot 10^4 \text{ miles/h}^2$$

The magnitude of the shuttle's acceleration is thus:

$$\|\mathbf{a}\| = 6.88 \cdot 10^4 \text{ miles/h}^2$$

In units of ft/s^2 we obtain

$$\|\mathbf{a}\| = \frac{6.88 \cdot 10^4 \cdot 5.28 \cdot 10^3}{3600^2} = 28.03 \text{ ft/s}^2$$

In Exercises 30–33, use (4) and (5) to find $a_{\mathbf{T}}$ and $a_{\mathbf{N}}$ as a function of t or at the point indicated.

31. $\mathbf{r}(t) = \langle t, \cos t, \sin t \rangle$

SOLUTION We find $a_{\mathbf{T}}$ and $a_{\mathbf{N}}$ using the following equalities:

$$a_{\mathbf{T}} = \mathbf{a} \cdot \mathbf{T}, \; a_{\mathbf{N}} = \frac{\|\mathbf{a} \times \mathbf{v}\|}{\|\mathbf{v}\|}.$$

We compute $\mathbf{v}$ and $\mathbf{a}$ by differentiating $\mathbf{r}$ twice:

$$\mathbf{v}(t) = \mathbf{r}'(t) = \langle 1, -\sin t, \cos t \rangle \quad \Rightarrow \quad \|\mathbf{v}(t)\| = \sqrt{1 + (-\sin t)^2 + \cos^2 t} = \sqrt{2}$$

$$\mathbf{a}(t) = \mathbf{r}''(t) = \langle 0, -\cos t, -\sin t \rangle$$

The unit tangent vector $\mathbf{T}$ is, thus:

$$\mathbf{T}(t) = \frac{\mathbf{v}(t)}{\|\mathbf{v}(t)\|} = \frac{1}{\sqrt{2}} \langle 1, -\sin t, \cos t \rangle$$

Since the speed is constant ($v = \|\mathbf{v}(t)\| = \sqrt{2}$), the tangential component of the acceleration is zero, that is:

$$a_{\mathbf{T}} = 0$$

To find $a_{\mathbf{N}}$ we first compute the following cross product:

$$\mathbf{a} \times \mathbf{v} = \begin{vmatrix} \mathbf{i} & \mathbf{j} & \mathbf{k} \\ 0 & -\cos t & -\sin t \\ 1 & -\sin t & \cos t \end{vmatrix} = \begin{vmatrix} -\cos t & -\sin t \\ -\sin t & \cos t \end{vmatrix} \mathbf{i} - \begin{vmatrix} 0 & -\sin t \\ 1 & \cos t \end{vmatrix} \mathbf{j} + \begin{vmatrix} 0 & -\cos t \\ 1 & -\sin t \end{vmatrix} \mathbf{k}$$

$$= -\left(\cos^2 t + \sin^2 t\right) \mathbf{i} - \sin t \mathbf{j} + \cos t \mathbf{k} = -\mathbf{i} - \sin t \mathbf{j} + \cos t \mathbf{k} = \langle -1, -\sin t, \cos t \rangle$$

Hence,

$$a_{\mathbf{N}} = \frac{\|\mathbf{a} \times \mathbf{v}\|}{\|\mathbf{v}\|} = \frac{\sqrt{(-1)^2 + (-\sin t)^2 + \cos^2 t}}{\sqrt{2}} = \frac{\sqrt{2}}{\sqrt{2}} = 1.$$

33. $\mathbf{r}(t) = \langle e^t, t, e^{-t} \rangle, \quad t = 0$

SOLUTION We use the following equalities:

$$a_{\mathbf{T}} = \mathbf{a} \cdot \mathbf{T}, \quad a_{\mathbf{N}} = \frac{\|\mathbf{a} \times \mathbf{v}\|}{\|\mathbf{v}\|}$$

We first find $\mathbf{v}$ and $\mathbf{a}$ by twice differentiating $\mathbf{r}$. This gives:

$$\mathbf{v}(t) = \mathbf{r}'(t) = \langle e^t, 1, -e^{-t} \rangle \quad \Rightarrow \quad \mathbf{v}(0) = \langle 1, 1, -1 \rangle, \quad \|\mathbf{v}(0)\| = \sqrt{3}$$

$$\mathbf{a}(t) = \mathbf{r}''(t) = \langle e^t, 0, e^{-t} \rangle \quad \Rightarrow \quad \mathbf{a}(0) = \langle 1, 0, 1 \rangle$$

The unit tangent at $t = 0$ is, thus:

$$\mathbf{T}(0) = \frac{\mathbf{v}(0)}{\|\mathbf{v}(0)\|} = \frac{1}{\sqrt{3}} \langle 1, 1, -1 \rangle$$

We now find $a_{\mathbf{T}}$ at the point $t = 0$:

$$a_{\mathbf{T}} = \mathbf{a} \cdot \mathbf{T} = \langle 1, 0, 1 \rangle \cdot \frac{1}{\sqrt{3}} \langle 1, 1, -1 \rangle = \frac{1}{\sqrt{3}} (1 + 0 - 1) = 0$$

To find $a_{\mathbf{N}}$ we first compute the cross product:

$$\mathbf{a} \times \mathbf{v} = \langle 1, 0, 1 \rangle \times \langle 1, 1, -1 \rangle = \begin{vmatrix} \mathbf{i} & \mathbf{j} & \mathbf{k} \\ 1 & 0 & 1 \\ 1 & 1 & -1 \end{vmatrix} = \begin{vmatrix} 0 & 1 \\ 1 & -1 \end{vmatrix} \mathbf{i} - \begin{vmatrix} 1 & 1 \\ 1 & -1 \end{vmatrix} \mathbf{j} + \begin{vmatrix} 1 & 0 \\ 1 & 1 \end{vmatrix} \mathbf{k}$$

$$= -1\mathbf{i} + 2\mathbf{j} + 1\mathbf{k} = \langle -1, 2, 1 \rangle$$

Hence,

$$a_{\mathbf{N}} = \frac{\|\mathbf{a} \times \mathbf{v}\|}{\|\mathbf{v}\|} = \frac{\|\langle -1, 2, 1 \rangle\|}{\sqrt{3}} = \frac{\sqrt{6}}{\sqrt{3}} = \sqrt{2}$$

In Exercise 34–41, use (4) and (6) to find the decomposition of $\mathbf{a}(t)$ *into tangential and normal components at the point indicated, as in Example 6.*

35. $\mathbf{r}(t) = \langle t, e^t, te^t \rangle, \quad t = 0$

SOLUTION We have $\mathbf{r}(t) = \langle t, e^t, te^t \rangle$. We find the decomposition of $\mathbf{a}(t)$ into tangential and normal components, using the following steps.

Step 1. Compute $\mathbf{T}$ and $a_{\mathbf{T}}$. Differentiating $\mathbf{r}(t)$ twice we obtain:

$$\mathbf{v}(t) = \mathbf{r}'(t) = \langle 1, e^t, e^t + te^t \rangle = \langle 1, e^t, (1+t)e^t \rangle$$

$$\Rightarrow \quad \mathbf{v}(0) = \langle 1, 1, 1 \rangle, \quad \|\mathbf{v}(0)\| = \sqrt{1^2 + 1^2 + 1^2} = \sqrt{3}$$

$$\mathbf{a}(t) = \mathbf{r}''(t) = \langle 0, e^t, e^t + (1+t)e^t \rangle = \langle 0, e^t, (2+t)e^t \rangle \quad \Rightarrow \quad \mathbf{a}(0) = \langle 0, 1, 2 \rangle$$

We compute the unit normal vector at $t = 0$:

$$\mathbf{T} = \frac{\mathbf{v}}{\|\mathbf{v}\|} = \frac{\langle 1, 1, 1 \rangle}{\sqrt{3}} = \frac{1}{\sqrt{3}} \langle 1, 1, 1 \rangle$$

Using Eq. (4) we obtain:

$$a_{\mathbf{T}} = a \cdot \mathbf{T} = \langle 0, 1, 2 \rangle \cdot \frac{1}{\sqrt{3}} \langle 1, 1, 1 \rangle = \frac{1}{\sqrt{3}} (0 + 1 + 2) = \frac{3}{\sqrt{3}} = \sqrt{3}$$

Step 2. Compute $a_{\mathbf{N}}$ and $\mathbf{N}$. By Eq. (6) we have:

$$a_{\mathbf{N}}\mathbf{N} = \mathbf{a} - a_{\mathbf{T}}\mathbf{T} = \langle 0, 1, 2 \rangle - \sqrt{3} \cdot \frac{1}{\sqrt{3}} \langle 1, 1, 1 \rangle = \langle -1, 0, 1 \rangle$$

Since $\mathbf{N}$ is a unit vector, the following holds:

$$a_{\mathbf{N}} = \|a_{\mathbf{N}}\mathbf{N}\| = \sqrt{(-1)^2 + 0^2 + 1^2} = \sqrt{2}$$

$$\mathbf{N} = \frac{a_{\mathbf{N}}\mathbf{N}}{a_{\mathbf{N}}} = \frac{\langle -1, 0, 1\rangle}{\sqrt{2}} = \left\langle -\frac{1}{\sqrt{2}}, 0, \frac{1}{\sqrt{2}}\right\rangle$$

Step 3. Write the decomposition. Using the results obtained in the previous steps, we obtain the following decomposition of $\mathbf{a}(0)$:

$$\mathbf{a}(0) = a_{\mathbf{T}}(0)\mathbf{T}(0) + a_{\mathbf{N}}(0)\mathbf{N}(0) = \sqrt{3}\mathbf{T} + \sqrt{2}\mathbf{N}$$

where $\mathbf{T} = \mathbf{T}(0) = \frac{1}{\sqrt{3}}\langle 1, 1, 1\rangle$ and $\mathbf{N} = \mathbf{N}(0) = \left\langle -\frac{1}{\sqrt{2}}, 0, \frac{1}{\sqrt{2}}\right\rangle$.

37. $\mathbf{r}(t) = \left\langle t, \frac{1}{2}t^2, \frac{1}{6}t^3\right\rangle, \quad t = 4$

SOLUTION

(a) By Eq. (4) we have:

$$a_{\mathbf{T}} = \mathbf{a}\cdot\mathbf{T} = \frac{\mathbf{a}\cdot\mathbf{v}}{\|\mathbf{v}\|}$$

We compute $\mathbf{v}$ and $\mathbf{a}$ by differentiating $\mathbf{r}(t) = \left\langle t, \frac{1}{2}t^2, \frac{1}{6}t^3\right\rangle$ twice. We obtain:

$$\mathbf{v}(t) = \mathbf{r}'(t) = \left\langle 1, t, \frac{1}{2}t^2\right\rangle \quad\Rightarrow\quad \mathbf{v}(4) = \langle 1, 4, 8\rangle,$$

$$\|\mathbf{v}(4)\| = \sqrt{1^2 + 4^2 + 8^2} = 9$$

$$\mathbf{a}(t) = \mathbf{r}''(t) = \langle 0, 1, t\rangle \quad\Rightarrow\quad \mathbf{a}(4) = \langle 0, 1, 4\rangle$$

Hence,

$$a_{\mathbf{T}}(4) = \frac{\mathbf{a}(4)\cdot\mathbf{v}(4)}{\|\mathbf{v}(4)\|} = \frac{\langle 0, 1, 4\rangle\cdot\langle 1, 4, 8\rangle}{9} = \frac{36}{9} = 4$$

(b) By Eq. (6) we have:

$$a_{\mathbf{N}}\mathbf{N} = \mathbf{a} - a_{\mathbf{T}}\mathbf{T}$$

Since $\mathbf{T}(4) = \frac{\mathbf{v}(4)}{\|\mathbf{v}(4)\|} = \frac{\langle 1,4,8\rangle}{9} = \left\langle \frac{1}{9}, \frac{4}{9}, \frac{8}{9}\right\rangle$ and by part (a) $a_{\mathbf{T}}(4) = 4$, and $\mathbf{a}(4) = \langle 0, 1, 4\rangle$, we have:

$$a_{\mathbf{N}}\mathbf{N} = \langle 0, 1, 4\rangle - 4\cdot\left\langle \frac{1}{9}, \frac{4}{9}, \frac{8}{9}\right\rangle = \left\langle -\frac{4}{9}, -\frac{7}{9}, \frac{4}{9}\right\rangle$$

We see that $a_{\mathbf{N}}\mathbf{N}$ is a unit vector and conclude that:

$$\mathbf{N}(4) = a_{\mathbf{N}}\mathbf{N} = \left\langle -\frac{4}{9}, -\frac{7}{9}, \frac{4}{9}\right\rangle \quad\text{and}\quad a_{\mathbf{N}}(4) = 1.$$

(c) The acceleration vector $\mathbf{a}(4)$ can be written as the following decomposition:

$$\mathbf{a}(4) = a_{\mathbf{T}}(4)\mathbf{T}(4) + a_{\mathbf{N}}(4)\mathbf{N}(4)$$

In parts (a) and (b) we found that $a_{\mathbf{T}}(4) = 4$, $\mathbf{T}(4) = \left\langle \frac{1}{9}, \frac{4}{9}, \frac{8}{9}\right\rangle$, $a_{\mathbf{N}}(4) = 1$, and $\mathbf{N}(4) = \left\langle -\frac{4}{9}, -\frac{7}{9}, \frac{4}{9}\right\rangle$. This gives the following decomposition:

$$\mathbf{a}(4) = 4\mathbf{T}(4) + 1\mathbf{N}(4) = 4\mathbf{T}(4) + \mathbf{N}(4) = \left\langle \frac{4}{9}, \frac{16}{9}, \frac{32}{9}\right\rangle + \left\langle -\frac{4}{9}, -\frac{7}{9}, \frac{4}{9}\right\rangle = \langle 0, 1, 4\rangle$$

39. $\mathbf{r}(\theta) = \langle \cos\theta, \sin\theta, \theta\rangle, \quad \theta = 0$

SOLUTION We have $\mathbf{r}(t) = \langle \cos t, \sin t, t\rangle$. We find the decomposition of $\mathbf{a}(0)$, using the following steps:

Step 1. Compute $\mathbf{T}$ and $a_{\mathbf{T}}$. We differentiate $\mathbf{r}(t)$ twice to obtain:

$$\mathbf{v}(t) = \mathbf{r}'(t) = \langle -\sin t, \cos t, 1\rangle$$

$$\mathbf{a}(t) = \mathbf{r}''(t) = \langle -\cos t, -\sin t, 0\rangle \quad\Rightarrow\quad \mathbf{a}(0) = \langle -1, 0, 0\rangle$$

Computing the magnitude of $\mathbf{v}(t)$ (the speed) we obtain:

$$v(t) = \|\mathbf{v}(t)\| = \sqrt{(-\sin t)^2 + (\cos t)^2 + 1^2} = \sqrt{1+1} = \sqrt{2}$$

Since the speed is constant, the tangential component of the acceleration $\mathbf{a}(t)$ is zero. That is:

$$a_{\mathbf{T}}(t) = 0$$

Step 2. Compute $a_{\mathbf{N}}$ and $\mathbf{N}$. Since $a_{\mathbf{T}}(t) = 0$, we obtain the decomposition:

$$\mathbf{a}(t) = a_{\mathbf{N}}(t)\mathbf{N}(t) \quad \Rightarrow \quad \mathbf{a}(0) = a_{\mathbf{N}}(0)\mathbf{N}(0) \tag{1}$$

Notice that the vector $\mathbf{a}(0) = \langle -1, 0, 0\rangle$ is already a unit vector, hence (1) implies that $\mathbf{N}(0) = \mathbf{a}(0)$ and $a_{\mathbf{N}}(0) = 1$. Hence the required decomposition reduces to:

$$\mathbf{a}(0) = 1 \cdot \mathbf{N}(0) = 1 \cdot \mathbf{N} \quad \text{where} \quad \mathbf{N} = \mathbf{a}(0) = \langle -1, 0, 0\rangle$$

41. $\mathbf{r}(t) = \langle t, \cos t, t\sin t\rangle, \quad t = \frac{\pi}{2}$

SOLUTION In this case, $\mathbf{r}(t) = \langle t, \cos t, t\sin t\rangle$. We find the decomposition of $\mathbf{a}\left(\frac{\pi}{2}\right)$, using the following steps:

Step 1. Compute $\mathbf{T}$ and $a_{\mathbf{T}}$. We differentiate $\mathbf{r}(t)$ twice to obtain:

$$\begin{aligned}
\mathbf{v}(t) &= \mathbf{r}'(t) = \langle 1, -\sin t, \sin t + t\cos t\rangle \\
&\Rightarrow \mathbf{v}\left(\frac{\pi}{2}\right) = \langle 1, -1, 1\rangle, \\
\left\|\mathbf{v}\left(\frac{\pi}{2}\right)\right\| &= \sqrt{3} \\
\mathbf{a}(t) &= \mathbf{r}''(t) = \langle 0, -\cos t, \cos t + \cos t - t\sin t\rangle = \langle 0, -\cos t, 2\cos t - t\sin t\rangle \\
&\Rightarrow \mathbf{a}\left(\frac{\pi}{2}\right) = \left\langle 0, 0, -\frac{\pi}{2}\right\rangle
\end{aligned}$$

The unit tangent at $t = \frac{\pi}{2}$ is, thus:

$$\mathbf{T}\left(\frac{\pi}{2}\right) = \frac{\mathbf{v}\left(\frac{\pi}{2}\right)}{\left\|\mathbf{v}\left(\frac{\pi}{2}\right)\right\|} = \frac{\langle 1, -1, 1\rangle}{\sqrt{3}} = \frac{1}{\sqrt{3}}\langle 1, -1, 1\rangle$$

Using Eq. (4) we get at $t = \frac{\pi}{2}$:

$$a_{\mathbf{T}} = \mathbf{a}\cdot\mathbf{T} = \left\langle 0, 0, -\frac{\pi}{2}\right\rangle \cdot \frac{1}{\sqrt{3}}\langle 1, -1, 1\rangle = \frac{1}{\sqrt{3}}\cdot\left(-\frac{\pi}{2}\right) = -\frac{\pi}{2\sqrt{3}}$$

Step 2. Compute $a_{\mathbf{N}}$ and $\mathbf{N}$. By Eq. (6) we have at $t = \frac{\pi}{2}$:

$$\begin{aligned}
a_{\mathbf{N}}\mathbf{N} &= \mathbf{a} - a_{\mathbf{T}}\mathbf{T} = \left\langle 0, 0, -\frac{\pi}{2}\right\rangle - \left(-\frac{\pi}{2\sqrt{3}}\right)\cdot\frac{1}{\sqrt{3}}\langle 1, -1, 1\rangle = \left\langle 0, 0, -\frac{\pi}{2}\right\rangle + \frac{\pi}{6}\langle 1, -1, 1\rangle \\
&= \left\langle \frac{\pi}{6}, -\frac{\pi}{6}, -\frac{\pi}{3}\right\rangle = \frac{\pi}{6}\langle 1, -1, -2\rangle
\end{aligned}$$

Since $\mathbf{N}$ is a unit vector we get:

$$\mathbf{a_N} = \|a_{\mathbf{N}}\mathbf{N}\| = \frac{\pi}{6}\sqrt{1^2 + (-1)^2 + (-2)^2} = \frac{\pi}{\sqrt{6}}$$

$$\mathbf{N} = \frac{a_{\mathbf{N}}\mathbf{N}}{a_{\mathbf{N}}} = \frac{\frac{\pi}{6}\langle 1, -1, -2\rangle}{\frac{\pi}{\sqrt{6}}} = \frac{1}{\sqrt{6}}\langle 1, -1, -2\rangle$$

Step 3. Write the decomposition. Using the results obtained in the previous steps we obtain the following decomposition of $\mathbf{a}\left(\frac{\pi}{2}\right)$:

$$a = a_{\mathbf{T}}\mathbf{T} + a_{\mathbf{N}}\mathbf{N} = -\frac{\pi}{2\sqrt{3}}\mathbf{T} + \frac{\pi}{\sqrt{6}}\mathbf{N}$$

where $\mathbf{T} = \frac{1}{\sqrt{3}}\langle 1, -1, 1\rangle$ and $\mathbf{N} = \frac{1}{\sqrt{6}}\langle 1, -1, -2\rangle$.

43. Find the components $a_{\mathbf{T}}$ and $a_{\mathbf{N}}$ of the acceleration vector of a particle moving along a circular path of radius $R = 100$ cm with constant velocity $v_0 = 5$ cm/s.

SOLUTION Since the particle moves with constant speed, we have $v'(t) = 0$, hence:

$$a_{\mathbf{T}} = v'(t) = 0$$

The normal component of the acceleration is $a_{\mathbf{N}} = \kappa(t)v(t)^2$. The curvature of a circular path of radius $R = 100$ is $\kappa(t) = \frac{1}{R} = \frac{1}{100}$, and the velocity is the constant value $v(t) = v_0 = 5$. Hence,

$$a_{\mathbf{N}} = \frac{1}{R}v_0^2 = \frac{25}{100} = 0.25 \text{ cm/s}^2$$

45. A car proceeds along a circular path of radius $R = 1{,}000$ ft centered at the origin. Starting at rest, its speed increases at a rate of t ft/s^2. Find the acceleration vector $\mathbf{a}$ at time $t = 3$ s and determine its decomposition into normal and tangential components.

SOLUTION The acceleration vector can be decomposed into tangential and normal directions as follows:

$$\mathbf{a}(t) = a_{\mathbf{T}}(t)\mathbf{T}(t) + a_{\mathbf{N}}(t)\mathbf{N}(t) \tag{1}$$

where

$$a_{\mathbf{T}}(t) = v'(t) \quad \text{and} \quad a_{\mathbf{N}}(t) = \kappa(t)v(t)^2 \tag{2}$$

Since the speed $v(t)$ is increasing at a rate of t ft/s^2, we have $v'(t) = t$. The car starts at rest hence the initial speed is $v_0 = 0$. We now integrate to find $v(t)$:

$$v(t) = \int_0^t v'(u)\,du = \int_0^t u\,du = \frac{1}{2}t^2 + v_0 = \frac{1}{2}t^2 + 0 = \frac{1}{2}t^2$$

The curvature of the circular path is $\kappa(t) = \frac{1}{R} = \frac{1}{1000}$. Substituting $v'(t) = t$, $\kappa = \frac{1}{1000}$, and $v(t) = \frac{1}{2}t^2$ in (2) gives:

$$a_{\mathbf{T}}(t) = t, \quad a_{\mathbf{N}}(t) = \frac{1}{1000}\left(\frac{1}{2}t^2\right)^2 = \frac{1}{4000}t^4$$

Combining with (1) gives the following decomposition:

$$a(t) = t\mathbf{T}(t) + \frac{1}{4000}t^4\mathbf{N}(t) \tag{3}$$

We now find the unit tangent $\mathbf{T}(t)$ and the unit normal $\mathbf{N}(t)$.

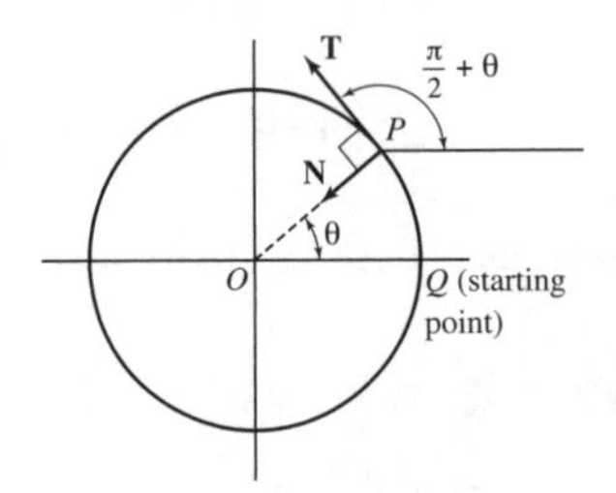

We have (see figure):

$$\mathbf{T} = \left\langle \cos\left(\frac{\pi}{2} + \theta\right), \sin\left(\frac{\pi}{2} + \theta\right)\right\rangle = \langle -\sin\theta, \cos\theta\rangle \tag{4}$$

$$\mathbf{N} = \langle \cos(\pi + \theta), \sin(\pi + \theta)\rangle = \langle -\cos\theta, -\sin\theta\rangle \tag{5}$$

We use the arc length formula to find θ:

$$\widehat{PQ} = \int_0^t \|r'(u)\|\,du = \int_0^t v(u)\,du = \int_0^t \frac{1}{2}u^2\,du = \frac{t^3}{6}$$

In addition, $\widehat{PQ} = R\theta = 1000\theta$. Hence,

$$1{,}000\theta = \frac{t^3}{6} \quad \Rightarrow \quad \theta = \frac{t^3}{6{,}000}$$

Substituting in (4) and (5) yields:

$$\mathbf{T} = \left\langle -\sin\frac{t^3}{6{,}000}, \cos\frac{t^3}{6{,}000}\right\rangle; \quad \mathbf{N} = \left\langle -\cos\frac{t^3}{6{,}000}, -\sin\frac{t^3}{6{,}000}\right\rangle \tag{6}$$

We now combine (3) and (6) to obtain the following decomposition:

$$\mathbf{a}(t) = t\left\langle -\sin\frac{t^3}{6{,}000}, \cos\frac{t^3}{6{,}000}\right\rangle + \frac{1}{4{,}000}t^4\left\langle -\cos\frac{t^3}{6{,}000}, -\sin\frac{t^3}{6{,}000}\right\rangle$$

At $t = 3$ we get:

$$a_{\mathbf{T}} = 3a_{\mathbf{N}} = \frac{3^4}{4{,}000} \approx 0.02025$$

$$\mathbf{T} = \left\langle -\sin\frac{3^3}{6{,}000}, \cos\frac{3^3}{6{,}000}\right\rangle \approx \langle -0.0045, 0.9999\rangle$$

$$\mathbf{N} = \left\langle -\cos\frac{3^3}{6{,}000}, -\sin\frac{3^3}{6{,}000}\right\rangle \approx \langle -0.9999, -0.0045\rangle$$

47. Suppose that $\mathbf{r} = \mathbf{r}(t)$ lies on a sphere of radius R for all t. Let $\mathbf{J} = \mathbf{r} \times \mathbf{r}'$. Show that $\mathbf{r}' = (\mathbf{J} \times \mathbf{r})/\|\mathbf{r}\|^2$. *Hint:* Observe that $\mathbf{r}$ and $\mathbf{r}'$ are perpendicular.

SOLUTION

(a) Solution 1. Since $\mathbf{r} = \mathbf{r}(t)$ lies on the sphere, the vectors $\mathbf{r} = \mathbf{r}(t)$ and $\mathbf{r}' = \mathbf{r}'(t)$ are orthogonal, therefore:

$$\mathbf{r} \cdot \mathbf{r}' = 0 \tag{1}$$

We use the following well-known equality:

$$\mathbf{a} \times (\mathbf{b} \times \mathbf{c}) = (\mathbf{a} \cdot \mathbf{c})\,\mathbf{b} - (\mathbf{a} \cdot \mathbf{b}) \cdot \mathbf{c}$$

Using this equality and (1) we obtain:

$$\begin{aligned}\mathbf{J} \times \mathbf{r} &= (\mathbf{r} \times \mathbf{r}') \times \mathbf{r} = -\mathbf{r} \times (\mathbf{r} \times \mathbf{r}') = -\left((\mathbf{r} \cdot \mathbf{r}')\,\mathbf{r} - (\mathbf{r} \cdot \mathbf{r})\,\mathbf{r}'\right) \\ &= -(\mathbf{r} \cdot \mathbf{r}')\,\mathbf{r} + \|\mathbf{r}\|^2\mathbf{r}' = 0\mathbf{r} + \|\mathbf{r}\|^2\mathbf{r}' = \|\mathbf{r}\|^2\mathbf{r}'\end{aligned}$$

Divided by the scalar $\|\mathbf{r}\|^2$ we obtain:

$$\mathbf{r}' = \frac{\mathbf{J} \times \mathbf{r}}{\|\mathbf{r}\|^2}$$

(b) Solution 2. The cross product $\mathbf{J} = \mathbf{r} \times \mathbf{r}'$ is orthogonal to $\mathbf{r}$ and $\mathbf{r}'$. Also, $\mathbf{r}$ and $\mathbf{r}'$ are orthogonal, hence the vectors $\mathbf{r}$, $\mathbf{r}'$ and $\mathbf{J}$ are mutually orthogonal. Now, since $\mathbf{r}'$ is orthogonal to $\mathbf{r}$ and $\mathbf{J}$, the right-hand rule implies that $\mathbf{r}'$ points in the direction of $\mathbf{J} \times \mathbf{r}$. Therefore, for some $\alpha > 0$ we have:

$$\mathbf{r}' = \alpha\mathbf{J} \times \mathbf{r} = \|\mathbf{r}'\| \cdot \frac{\mathbf{J} \times \mathbf{r}}{\|\mathbf{J} \times \mathbf{r}\|} \tag{2}$$

By properties of the cross product and since $\mathbf{J}$, $\mathbf{r}$, and $\mathbf{r}'$ are mutually orthogonal we have:

$$\|\mathbf{J} \times \mathbf{r}\| = \|\mathbf{J}\|\,\|\mathbf{r}\| = \|\mathbf{r} \times \mathbf{r}'\|\,\|\mathbf{r}\| = \|\mathbf{r}\|\,\|\mathbf{r}'\|\,\|\mathbf{r}\| = \|\mathbf{r}\|^2\|\mathbf{r}'\|$$

Substituting in (2) we get:

$$\mathbf{r}' = \|\mathbf{r}'\|\frac{\mathbf{J} \times \mathbf{r}}{\|\mathbf{r}\|^2\|\mathbf{r}'\|} = \frac{\mathbf{J} \times \mathbf{r}}{\|\mathbf{r}\|^2}$$

Further Insights and Challenges

49. The orbit of a planet is an ellipse with the sun at one focus. The sun's gravitational force acts along the radial line from the planet to the sun (the dashed lines in Figure 13), and by Newton's Second Law, the acceleration vector points in the same direction. Explain in words why the planet must slow down in the upper half of the orbit (as it moves away from the sun) and speed up in the lower half. Kepler's Second Law, discussed in the next section, gives a more precise version of this qualitative conclusion. *Hint:* Consider the decomposition of $\mathbf{a}$ into normal and tangential components.

FIGURE 13 Elliptical orbit of a planet around the sun.

SOLUTION In the upper half of the orbit, as the planet moves away from the sun the acceleration vector has a negative component in the tangential direction **T**, so the particle's velocity is decreasing (since $a_{\mathbf{T}}(t) = v'(t) < 0$).

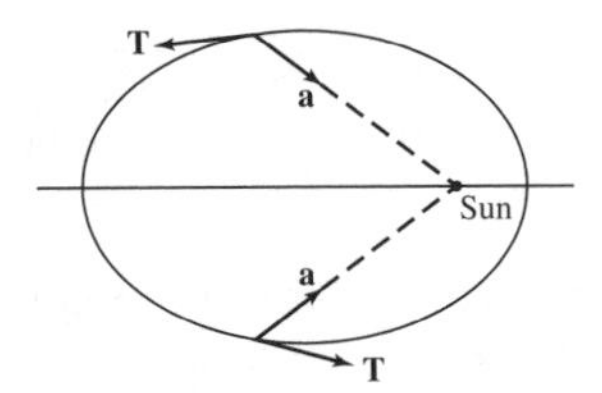

However, in the lower half of the orbit, as the planet gets closer to the sun, the acceleration has a positive component in the tangential direction, that is, $a_{\mathbf{T}}(t) = v'(t) > 0$. Therefore the velocity $v(t)$ is increasing.

In Exercises 50–54, consider a car of mass m traveling along a curved but level road. To avoid skidding, the road must supply a frictional force $\mathbf{F} = m\mathbf{a}$, *where* **a** *is the car's acceleration vector. The maximum magnitude of the frictional force is* μmg, *where* μ *is the coefficient of friction and* $g = 32$ ft/s^2. *Let* v *be the car's speed in feet per second.*

51. Suppose that the maximum radius of curvature along a curved highway is $R = 600$ ft. How fast can a car travel (at constant speed) along the highway without skidding if the coefficient of friction is $\mu = 0.5$?

SOLUTION In Exercise 50 we showed that the car will not skid if the following inequality is satisfied:

$$(v')^2 + \frac{v^4}{R^2} < \mu^2 g^2$$

We compute the constant speed v for which the car can travel without skidding. In case of constant speed, $v' = 0$. We substitute $R = 600$, $\mu = 0.5$ and $g = 32$ and solve for v. This gives:

$$\frac{v^4}{600^2} < 0.5^2 \cdot 32^2$$

$$v^4 < 9{,}600^2 \quad \Rightarrow \quad v < 98 \text{ ft/s}$$

The maximum speed (in case of constant speed) is about 98 ft/s.

53. You want to reverse your direction in the shortest possible time by driving around a semicircular bend (Figure 14). If you travel at the maximum possible *constant speed* v that will not cause skidding, is it faster to hug the inside curve (radius r) or the outside curb (radius R)? *Hint:* Use Eq. (7) to show that at maximum speed, the time required to drive around the semicircle is proportional to the square root of the radius.

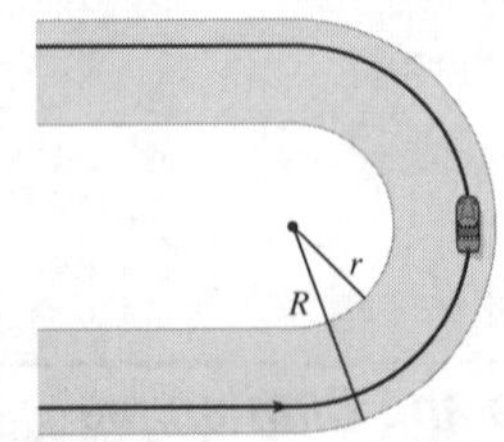

FIGURE 14 Car going around the bend.

SOLUTION In Exercise 50 we showed that the car will not skid if the following inequality is satisfied:

$$(v')^2 + \frac{v^4}{R^2} < \mu^2 g^2$$

In case of constant speed, $v' = 0$, so the inequality becomes:

$$\frac{v^4}{R^2} < \mu^2 g^2$$

We solve for v:

$$v^4 < (\mu g R)^2 \quad \Rightarrow \quad v < \sqrt{\mu g R}$$

The maximum speed in which skidding does not occur is, thus,

$$v \approx \sqrt{\mu g R} \tag{1}$$

If T is the time required to drive around the semicircle of radius R at the constant speed v, then the length of the semicircle can be written as:

$$\pi R = \int_0^T \|\mathbf{r}'(t)\|\, dt = \int_0^T v\, dt = vT$$

Hence,

$$T = \frac{\pi R}{v} \tag{2}$$

Combining (1) and (2) gives:

$$T \approx \frac{\pi R}{\sqrt{\mu g R}} \approx \frac{\pi}{\sqrt{\mu g}}\sqrt{R}$$

We conclude that it is faster to hug the inside curve of radius r ($r < R$), rather than the outside curve of radius R.

14.6 Planetary Motion According to Kepler and Newton (ET Section 13.6)

Preliminary Questions

1. Describe the relation between the vector $\mathbf{J} = \mathbf{r} \times \mathbf{r}'$ and the rate at which the radial vector sweeps out area.

SOLUTION The rate at which the radial vector sweeps out area equals half the magnitude of the vector $\mathbf{J}$. This relation is expressed in the formula:

$$\frac{dA}{dt} = \frac{1}{2}\|\mathbf{J}\|.$$

2. Equation (1) shows that $\mathbf{r}''$ is proportional to $\mathbf{r}$. Explain how this fact is used to prove Kepler's Second Law.

SOLUTION In the proof of Kepler's Second Law it is shown that the rate at which area is swept out is

$$\frac{dA}{dt} = \frac{1}{2}\|\mathbf{J}\|, \quad \text{where} \quad \mathbf{J} = \mathbf{r}(t) \times \mathbf{r}'(t)$$

To show that $\|\mathbf{J}\|$ is constant, show that $\mathbf{J}$ is constant. This is done using the proportionality of $\mathbf{r}''$ and $\mathbf{r}$ which implies that $\mathbf{r}(t) \times \mathbf{r}''(t) = 0$. Using this we get:

$$\frac{d\mathbf{J}}{dt} = \frac{d}{dt}(\mathbf{r} \times \mathbf{r}') = \mathbf{r} \times \mathbf{r}'' + \mathbf{r}' \times \mathbf{r}' = \mathbf{0} + \mathbf{0} = \mathbf{0} \Rightarrow \mathbf{J} = \text{const}$$

3. How is the period T affected if the semimajor axis a is increased four-fold?

SOLUTION Kepler's Third Law states that the period T of the orbit is given by:

$$T^2 = \left(\frac{4\pi^2}{GM}\right)a^3$$

or

$$T = \frac{2\pi}{\sqrt{GM}}a^{3/2}$$

If a is increased four-fold the period becomes:

$$\frac{2\pi}{\sqrt{GM}}(4a)^{3/2} = 8 \cdot \frac{2\pi}{\sqrt{GM}}a^{3/2}$$

That is, the period is increased eight-fold.

Exercises

1. Kepler's Third Law states that T^2/a^3 has the same value for each planetary orbit. Do the data in the following table support this conclusion? Estimate the length of Jupiter's period, assuming that $a = 77.8 \times 10^{10}$ m.

Planet	Mercury	Venus	Earth	Mars
a (10^{10} m)	5.79	10.8	15.0	22.8
T (years)	0.241	0.615	1.00	1.88

SOLUTION Using the given data we obtain the following values of T^2/a^3, where a, as always, is measured not in meters but in 10^{10} m:

Planet	Mercury	Venus	Earth	Mars
T^2/a^3	$2.99 \cdot 10^{-4}$	$3 \cdot 10^{-4}$	$2.96 \cdot 10^{-4}$	$2.98 \cdot 10^{-4}$

The data on the planets supports Kepler's prediction. We estimate Jupiter's period (using the given a) as $T \approx \sqrt{a^3 \cdot 3 \cdot 10^{-4}} \approx 11.9$ years.

3. The earth's orbit is nearly circular with radius $R = 93 \times 10^6$ miles (the eccentricity is $e = 0.017$). Find the rate at which the earth's radial vector sweeps out area in units of ft^2/s. What is the magnitude of the vector $\mathbf{J} = \mathbf{r} \times \mathbf{r}'$ for the earth (in units of squared feet per second)?

SOLUTION The rate at which the earth's radial vector sweeps out area is

$$\frac{dA}{dt} = \frac{1}{2}\|\mathbf{J}\|; \quad \mathbf{J} = \mathbf{r}(t) \times \mathbf{r}'(t) \tag{1}$$

Since $\mathbf{J}$ is a constant vector, its length is constant. Moreover, if we assume that the orbit is circular then $\mathbf{r}(t)$ lies on a circle, and therefore $\mathbf{r}(t)$ and $\mathbf{r}'(t)$ are orthogonal. Using properties of the cross product we get:

$$\|\mathbf{J}\| = \|\mathbf{r}(t) \times \mathbf{r}'(t)\| = \|\mathbf{r}(t)\|\|\mathbf{r}'(t)\| = R\|\mathbf{r}'(t)\| = \text{const}$$

We conclude that the speed $v = \|\mathbf{r}'(t)\|$ is constant. We find the speed using the following equality:

$$2\pi R = vT \Rightarrow v = \frac{2\pi R}{T}.$$

Therefore,

$$\|\mathbf{J}\| = R \cdot \frac{2\pi R}{T} = \frac{2\pi R^2}{T}.$$

Substituting in (1) we get:

$$\frac{dA}{dt} = \frac{1}{2} \cdot \frac{2\pi R^2}{T} = \frac{\pi R^2}{T}.$$

For $R = 93 \times 10^6$ miles $= 4.9 \times 10^{11}$ ft and $T = 365 \times 24 \times 3{,}600 = 31{,}536{,}000$ s we obtain:

$$\|\mathbf{J}\| = \frac{2\pi \cdot (4.9 \cdot 10^{11})^2}{31{,}536{,}000} = 4.78 \times 10^{16}\ \text{ft}^2/\text{s}$$

$$\frac{dA}{dt} = 2.39 \times 10^{16}\ \text{ft}^2/\text{s}$$

5. Ganymede, one of Jupiter's moons discovered by Galileo, has an orbital period of 7.154 days and a semimajor axis of 1.07×10^9 m. Use Exercise 4 to estimate the mass of Jupiter.

SOLUTION By Exercise 4, the mass of Jupiter can be computed using the following equality:

$$M = \frac{4\pi^2}{G}\frac{a^3}{T^2}$$

We substitute the given data $T = 7.154 \cdot 24 \cdot 60^2 = 618{,}105.6$ $a = 1.07 \times 10^9$ m and $G = 6.67300 \times 10^{-11}\text{m}^3\text{kg}^{-1}\text{s}^{-1}$, to obtain:

$$M = \frac{4\pi^2 \cdot \left(1.07 \times 10^9\right)^3}{6.67300 \times 10^{-11} \cdot (618{,}105.6)^2} \approx 1.897 \times 10^{27}\ \text{kg}.$$

7. Use the fact that **J** is constant to show that a planet in a circular orbit travels at constant speed.

SOLUTION It is shown in the proof of Kepler's Second Law that the vector $\mathbf{J} = \mathbf{r}(t) \times \mathbf{r}'(t)$ is constant, hence its length is constant:

$$\|\mathbf{J}\| = \|\mathbf{r}(t) \times \mathbf{r}'(t)\| = \text{const} \tag{1}$$

We consider the orbit as a circle of radius R, therefore, $\mathbf{r}(t)$ and $\mathbf{r}'(t)$ are orthogonal and $\|\mathbf{r}(t)\| = R$. By (1) and using properties of the cross product we obtain:

$$\|\mathbf{r}(t) \times \mathbf{r}'(t)\| = \|\mathbf{r}(t)\| \|\mathbf{r}'(t)\| \sin\frac{\pi}{2} = R \cdot \|\mathbf{r}'(t)\| = \text{const}$$

We conclude that $\|\mathbf{r}'(t)\|$ is constant, that is the speed $v = \|\mathbf{r}'(t)\|$ of the planet is constant.

9. Show directly that the circular orbit

$$\mathbf{r}(t) = \langle R\cos\omega t, R\sin\omega t\rangle$$

satisfies the differential equation, Eq. (2), provided that $\omega^2 = kR^{-3}$. Then deduce Kepler's Third Law $T^2 = \left(\frac{4\pi^2}{k}\right) R^3$ for this orbit.

SOLUTION Note that $\|\mathbf{r}\| = R$, and note that

$$\mathbf{r}' = \langle -R\omega\sin\omega t, R\omega\cos\omega t\rangle \quad \text{and} \quad \mathbf{r}'' = \left\langle -R\omega^2\cos\omega t, -R\omega^2\sin\omega t\right\rangle$$

We rewrite this as:

$$\mathbf{r}'' = -\omega^2 \langle R\cos\omega t, R\sin\omega t\rangle = -\omega^2\mathbf{r}$$

Since $\omega^2 = k/R^3$ and $R = \|\mathbf{r}\|$, we get $\mathbf{r}'' = \frac{-k}{\|\mathbf{r}\|^3}\mathbf{r}$, as desired. Since $T = \frac{2\pi}{\omega}$ then $T^2 = \frac{4\pi^2}{\omega^2} = \frac{4\pi^2 R^3}{k}$, as desired.

11. Use the results of Exercises 8 and 10 to find the velocity of a satellite in geosynchronous orbit.

SOLUTION In Exercise 8 we showed that the velocity of a planet in a circular orbit of radius a is:

$$v = \frac{2\pi a}{T} \tag{1}$$

A geosynchronous orbit has period $T = 24$ hours and in Exercise 10 we found that $a = 42{,}246$ km. Substituting in (1) we get:

$$v = \frac{2\pi \cdot 42{,}246}{24} = 11{,}060 \text{ km/h}$$

13. Mass of the Milky Way The sun revolves around the center of mass of the Milky Way galaxy in an orbit that is approximately circular, of radius $a \approx 2.8 \times 10^{17}$ km and velocity $v \approx 250$ km/s. Use the result of Exercise 12 to estimate the mass of the portion of the Milky Way inside the sun's orbit (place all of this mass at the center of the orbit).

SOLUTION Let M be the mass of the portion of the Milky Way inside the sun's orbit, assuming that all this mass is at the center of the sun's orbit. By Exercise 12, the following equality holds:

$$M = \frac{av^2}{G}.$$

We substitute the values $a = 2.8 \times 10^{20}$ m, $v = 250 \times 10^3$ m/s and $G = 6.673 \times 10^{-11}$ m^{3}kg^{-1}s^{-1} and compute the mass M. This gives:

$$M = \frac{2.8 \cdot 10^{20} \cdot (250 \cdot 10^3)^2}{6.673 \cdot 10^{-11}} = 2.6225 \times 10^{41} \text{ kg}.$$

The mass of the sun is 1.989×10^{30} kg, hence M is 1.32×10^{11} times the mass of the sun (132 billions times the mass of the sun).

15. Show that the total energy (11) of a planet in a circular orbit of radius R is $E = -GMm/(2R)$. *Hint:* Use Exercise 8.

SOLUTION The total energy of a planet in a circular orbit of radius R is

$$E = \frac{1}{2}mv^2 - \frac{GMm}{\|\mathbf{r}\|} = \frac{1}{2}mv^2 - \frac{GMm}{R} \tag{1}$$

In Exercise 8 we showed that

$$v^2 = \frac{GM}{R} \tag{2}$$

Substituting (2) in (1) we obtain:

$$E = \frac{1}{2}m\frac{GM}{R} - \frac{GMm}{R} = -\frac{1}{2}\frac{GMm}{R} = -\frac{GMm}{2R}.$$

In Exercises 16–20, we consider a planetary orbit with orbital parameters p and e. The perihelion and aphelion of the orbits are the points on the orbit closest to and farthest from the sun (Figure 7). Denote the distances from the sun at the perihelion and aphelion by r_{per} and r_{ap} and the speeds of the planet at the perihelion and aphelion by v_{per} and v_{ap}.

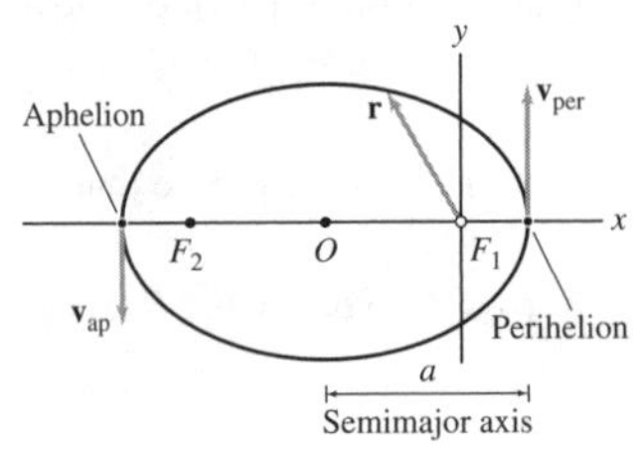

FIGURE 7 $\mathbf{r}$ and $\mathbf{r}'$ are perpendicular at the perihelion and aphelion.

17. Compute r_{per} and r_{ap} for the orbit of Mercury, which has eccentricity $e = 0.244$ (see the table in Exercise 1 for the semimajor axis).

SOLUTION The length of the semi-major axis of the orbit of mercury is $a = 5.79 \cdot 10^7$ km. We substitute a and $e = 0.244$ in the formulas for r_{per} and r_{ap} obtained in Exercise 16, to obtain the shortest and longest distances respectively. This gives:

$$r_{\text{per}} = a(1-e) = 5.79 \cdot 10^7(1-0.244) = 4.377 \cdot 10^7 \text{ km}$$

$$r_{\text{ap}} = a(1+e) = 5.79 \cdot 10^7(1+0.244) = 7.203 \cdot 10^7 \text{ km}.$$

19. Prove that $v_{\text{per}}(1-e) = v_{\text{ap}}(1+e)$. *Hint:* $\mathbf{r} \times \mathbf{r}'$ is constant by Eq. (5). Compute this cross product at the perihelion and aphelion, noting that $\mathbf{r}$ is perpendicular to $\mathbf{r}'$ at these two points.

SOLUTION Since the vector $\mathbf{J}(t) = \mathbf{r}(t) \times \mathbf{r}'(t)$ is constant, it is the same vector at the perigee and at the apogee, hence we may equate the length of $\mathbf{J}(t)$ at these two points. Since at the perigee and at the apogee $\mathbf{r}(t)$ and $\mathbf{r}'(t)$ are orthogonal we have by properties of the cross product:

$$\|\mathbf{r}_{\text{ap}} \times \mathbf{r}'_{\text{ap}}\| = \|\mathbf{r}_{\text{ap}}\|\|\mathbf{r}'_{\text{ap}}\| = r_{\text{ap}}v_{\text{ap}}$$

$$\|\mathbf{r}_{\text{per}} \times \mathbf{r}'_{\text{per}}\| = \|\mathbf{r}_{\text{per}}\|\|\mathbf{r}'_{\text{per}}\| = r_{\text{per}}v_{\text{per}}$$

Equating the two values gives:

$$r_{\text{ap}}v_{\text{ap}} = r_{\text{per}}v_{\text{per}} \tag{1}$$

In Exercise 16 we showed that $r_{\text{per}} = a(1-e)$ and $r_{\text{ap}} = a(1+e)$. Substituting in (1) we obtain:

$$a(1+e)v_{\text{ap}} = a(1-e)v_{\text{per}}$$

$$(1+e)v_{\text{ap}} = (1-e)v_{\text{per}}$$

21. Show that the total mechanical energy E of a planet in an elliptical orbit with semimajor axis a is $E = -\dfrac{GMm}{2a}$. *Hint:* Use Exercise 20 to compute the total energy at the perihelion.

SOLUTION The total energy of a planet of mass m orbiting a sun of mass M with position $\mathbf{r}$ and speed $v = \|\mathbf{r}'\|$ is (given in Exercise 14):

$$E = \frac{1}{2}mv^2 - \frac{GMm}{\|\mathbf{r}\|} \tag{1}$$

The energy E is conserved, so we can compute it using any point on the elliptical orbit, for instance the perihelion. By Exercise 16 and Exercise 20 we have:

$$r_{\text{per}} = a(1-e)$$
$$v_{\text{per}} = \sqrt{\frac{GM}{a}\frac{1+e}{1-e}} \tag{2}$$

Substituting (2) into (1) gives:

$$E = \frac{1}{2}m \cdot \frac{GM}{a}\frac{1+e}{1-e} - \frac{GMm}{a(1-e)} = \frac{GMm}{a(1-e)}\left(\frac{1+e}{2} - 1\right) = \frac{GMm}{a(1-e)}\frac{1+e-2}{2} = \frac{GMm}{a(1-e)}\frac{e-1}{2} = -\frac{GMm}{2a}$$

23. Two space shuttles A and B orbit the earth along the solid trajectory in Figure 8. Hoping to catch up to B, the pilot of A applies a forward thrust to increase her shuttle's kinetic energy. Use Exercise 21 to show that shuttle A will move off into a larger orbit as shown in the figure. Then use Kepler's Third Law to show that A's orbital period T will increase (and she will fall farther and farther behind B)!

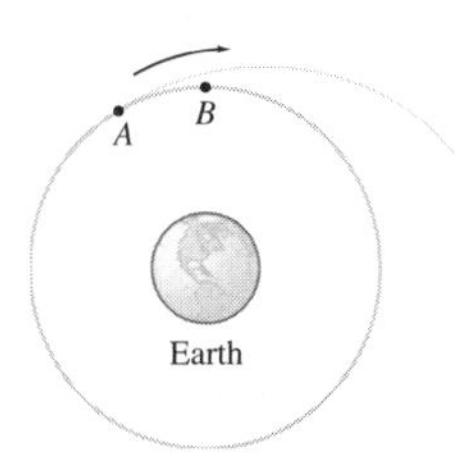

FIGURE 8

SOLUTION In Exercise 21 we showed that the total mechanical energy E of a planet in an elliptical orbit with semi-major axis a is

$$E = \frac{-GMm}{2a} \tag{1}$$

Since E is increased, a is increased, resulting in moving to an elliptic orbit as the dashed orbit in the figure. Now, by Kepler's Third Law,

$$T^2 = \left(\frac{4\pi^2}{GM}\right)a^3$$

We conclude that the orbital period T of shuttle A is also increasing, which means that A will get further and further behind B.

Further Insights and Challenges

In Exercises 25–26, we prove Kepler's Third Law. Figure 9 shows an elliptical orbit with polar equation

$$r = \frac{p}{1+e\cos\theta}$$

where $p = \|\mathbf{J}\|^2/k$. Let a and b be the semimajor and semiminor axes, respectively. The origin is located at F_1.

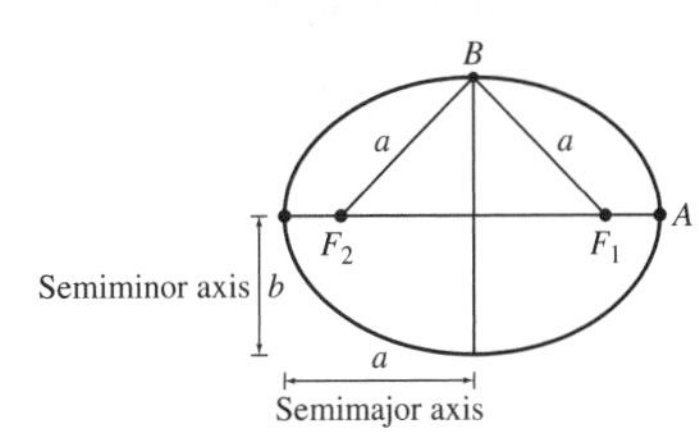

FIGURE 9

25. The goal of this exercise is to show that $b = \sqrt{pa}$.

(a) Show that $F_1A + F_2A = 2a$. Conclude that $F_1B + F_2B = 2a$ and hence $F_1B = F_2B = a$.

(b) Show that $F_1A = \dfrac{p}{1+e}$ and $F_2A = \dfrac{p}{1-e}$, and conclude that $a = \dfrac{p}{1-e^2}$.

(c) Use the Pythagorean Theorem to prove that

$$b = \frac{p}{\sqrt{1-e^2}} = \sqrt{pa}$$

SOLUTION

(a) Since $CF_2 = AF_1$, we have:

$$F_2A = CA - CF_2 = 2a - F_1A$$

Therefore,

$$F_1A + F_2A = 2a \tag{1}$$

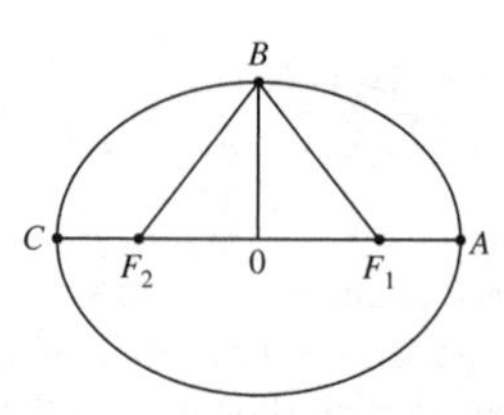

The ellipse is the set of all points such that the sum of the distances to the two foci F_1 and F_2 is constant. Therefore,

$$F_1A + F_2A = F_1B + F_2B \tag{2}$$

Combining (1) and (2), we obtain:

$$F_1B + F_2B = 2a \tag{3}$$

The triangle F_2BF_1 is isosceles, hence $F_2B = F_1B$ and so we conclude that

$$F_1B = F_2B = a$$

(b) The polar equation of the ellipse, where the focus F_1 is at the origin is

$$r = \frac{p}{1 + e\cos\theta}$$

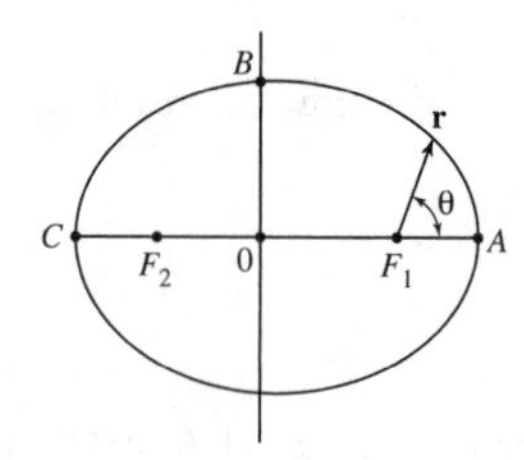

The point A corresponds to $\theta = 0$, hence,

$$F_1A = \frac{p}{1 + e\cos 0} = \frac{p}{1 + e} \tag{4}$$

The point C corresponds to $\theta = \pi$ hence,

$$F_1C = \frac{p}{1 + e\cos\pi} = \frac{p}{1 - e}$$

We now find F_2A. Using the equality $CF_2 = AF_1$ we get:

$$F_2A = F_2F_1 + F_1A = F_2F_1 + F_2C = F_1C = \frac{p}{1 - e}$$

That is,

$$F_2A = \frac{p}{1 - e} \tag{5}$$

Combining (1), (4), and (5) we obtain:

$$\frac{p}{1 + e} + \frac{p}{1 - e} = 2a$$

Hence,

$$a = \frac{1}{2}\left(\frac{p}{1 + e} + \frac{p}{1 - e}\right) = \frac{p(1 - e) + p(1 + e)}{2(1 + e)(1 - e)} = \frac{2p}{2\left(1 - e^2\right)} = \frac{p}{1 - e^2}$$

(c) We use Pythagoras' Theorem for the triangle OBF_1:

$$OB^2 + OF_1^2 = BF_1^2 \tag{6}$$

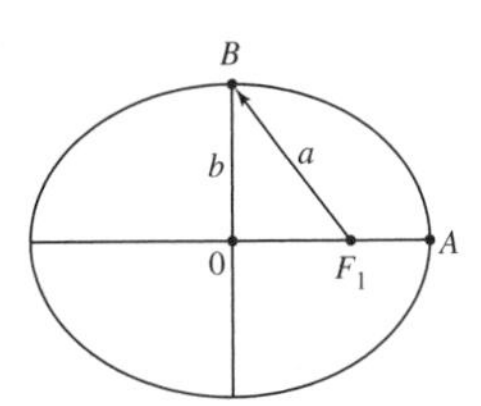

Using (4) we have

$$OF_1 = a - F_1A = a - \frac{p}{1+e}$$

Also $OB = b$ and $BF_1 = a$, hence (6) gives:

$$b^2 + \left(a - \frac{p}{1+e}\right)^2 = a^2$$

We solve for b:

$$b^2 + a^2 - \frac{2ap}{1+e} + \frac{p^2}{(1+e)^2} = a^2$$

$$b^2 - \frac{2ap}{1+e} + \frac{p^2}{(1+e)^2} = 0$$

In part (b) we showed that $a = \frac{p}{1-e^2}$. We substitute to obtain:

$$b^2 - \frac{2p}{1+e} \cdot \frac{p}{1-e^2} + \frac{p^2}{(1+e)^2} = 0$$

$$b^2 = \frac{2p^2}{(1+e)^2(1-e)} - \frac{p^2}{(1+e)^2} = \frac{2p^2 - p^2(1-e)}{(1+e)^2(1-e)}$$

$$= \frac{p^2(1+e)}{(1+e)^2(1-e)} = \frac{p^2}{1-e^2}$$

Hence,

$$b = \frac{p}{\sqrt{1-e^2}}$$

Since $1 - e^2 = \frac{p}{a}$ we also have

$$b = \frac{p}{\sqrt{\frac{p}{a}}} = \sqrt{ap}$$

27. Let $\mathbf{e_r} = \langle \cos\theta, \sin\theta \rangle$ and $\mathbf{e}_\theta = \langle -\sin\theta, \cos\theta \rangle$. Write the position vector of a planet as $\mathbf{r} = r\mathbf{e_r}$, where $r = \|\mathbf{r}\|$.

(a) Show that $\dfrac{d\mathbf{e}_\theta}{d\theta} = -\mathbf{e_r}$.

(b) Write Eq. (2) in the form $\dfrac{d\mathbf{v}}{dt} = -\dfrac{k}{r^2}\mathbf{e_r}$ and use the Chain Rule to show that

$$\frac{d\mathbf{v}}{d\theta}\frac{d\theta}{dt} = \frac{k}{r^2}\frac{d\mathbf{e}_\theta}{d\theta}$$

(c) Show that $\dfrac{d\mathbf{v}}{d\theta} = \dfrac{k}{\|\mathbf{J}\|}\dfrac{d\mathbf{e}_\theta}{d\theta}$. *Hint:* Use Exercise 24 to show that $\dfrac{d\theta}{dt} = \|\mathbf{J}\|/r^2$.

(d) Conclude that there is a constant vector $\mathbf{w}$ such that

$$\mathbf{v}(\theta) = \frac{k}{\|\mathbf{J}\|}\mathbf{e}_\theta + \mathbf{w}$$

This shows that as θ varies from 0 to 2π, the velocity vector $\mathbf{v}$ traces out a circle of radius $k/\|\mathbf{J}\|$ with center at the terminal point of $\mathbf{w}$ (Figure 10).

Planetary orbit Velocity circle

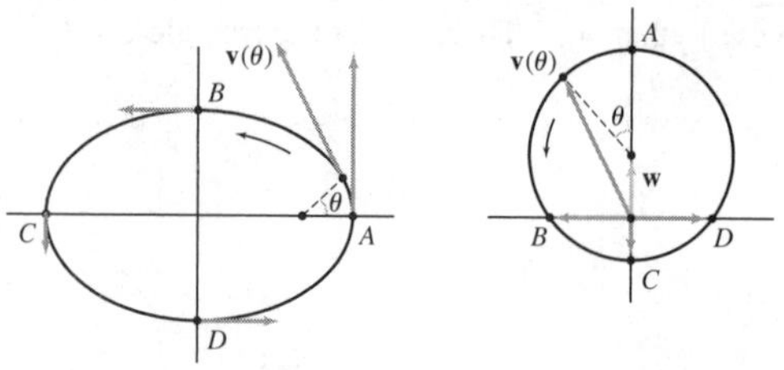

FIGURE 10 The terminal point of the velocity vector traces out the circle as the planet travels along its orbit.

SOLUTION

(a) Differentiating the vector $\mathbf{e}_\theta = \langle -\sin\theta, \cos\theta\rangle$ with respect to θ gives:

$$\frac{d\mathbf{e}_\theta}{d\theta} = \frac{d}{d\theta}\langle -\sin\theta, \cos\theta\rangle = \langle -\cos\theta, -\sin\theta\rangle = -\langle \cos\theta, \sin\theta\rangle = -\mathbf{e_r}$$

(b) Eq. (2) is the following equality:

$$\mathbf{r}''(t) = \frac{-k}{\|\mathbf{r}\|^3}\mathbf{r}(t) = \frac{-GM}{\|\mathbf{r}\|^3}\mathbf{r}(t)$$

Writing $\mathbf{r} = \|\mathbf{r}\|\mathbf{e_r}$ and $\mathbf{r}''(t) = \frac{d}{dt}\mathbf{r}'(t) = \frac{d\mathbf{v}}{dt}$ we get:

$$\frac{d\mathbf{v}}{dt} = \frac{-GM}{\|\mathbf{r}\|^3}\|\mathbf{r}\|\mathbf{e_r} = \frac{-GM}{\|\mathbf{r}\|^2}\mathbf{e_r} = \frac{-GM}{r^2}\mathbf{e_r}$$

That is,

$$\frac{d\mathbf{v}}{dt} = -\frac{GM}{r^2}\mathbf{e_r} \tag{1}$$

By the Chain Rule, $\frac{d\mathbf{v}}{dt} = \frac{d\mathbf{v}}{d\theta}\cdot\frac{d\theta}{dt}$ and by part (a) $\frac{d\mathbf{e}_\theta}{d\theta} = -\mathbf{e_r}$. Substituting in (1), we get:

$$\frac{d\mathbf{v}}{d\theta}\frac{d\theta}{dt} = \frac{GM}{r^2}\frac{d\mathbf{e}_\theta}{d\theta}$$

(c) In Exercise 24 we showed that

$$2\frac{dA}{dt} = \|\mathbf{r}\times\mathbf{r}'\| = \|\mathbf{J}\| \Rightarrow \frac{dA}{dt} = \frac{1}{2}\|\mathbf{J}\|$$

and

$$\frac{dA}{dt} = \frac{1}{2}r^2\frac{d\theta}{dt}$$

Combining the two equalities we get:

$$\frac{1}{2}r^2\frac{d\theta}{dt} = \frac{1}{2}\|\mathbf{J}\| \Rightarrow \frac{d\theta}{dt} = \frac{\|\mathbf{J}\|}{r^2}$$

Substituting in the equality obtained in part (b) we obtain:

$$\frac{d\mathbf{v}}{d\theta}\frac{\|\mathbf{J}\|}{r^2} = \frac{GM}{r^2}\frac{d\mathbf{e}_\theta}{d\theta}$$

Denoting $C = \frac{GM}{\|\mathbf{J}\|}$ we obtain:

$$\frac{d\mathbf{v}}{d\theta} = \frac{GM}{\|\mathbf{J}\|}\frac{d\mathbf{e}_\theta}{d\theta} = C\frac{d\mathbf{e}_\theta}{d\theta}$$

(d) Integrating the two sides of $\frac{d\mathbf{v}}{d\theta} = C\frac{d\mathbf{e}_\theta}{d\theta}$ we have

$$\mathbf{v}(\theta) = \int \frac{d\mathbf{v}}{d\theta}d\theta = C\int\frac{d\mathbf{e}_\theta}{d\theta}d\theta = C\mathbf{e}_\theta + \mathbf{u}$$

where $\mathbf{u}$ is a constant vector. Notice that $\|\mathbf{v}(\theta) - \mathbf{u}\| = \|C\mathbf{e}_\theta\| = |C|$, which is the equation of a circle of radius C (recall, $C = GM/\|\mathbf{J}\| = k/\|\mathbf{J}\|$) centered at the terminal point of $\mathbf{u}$.

CHAPTER REVIEW EXERCISES

1. Determine the domains of the vector-valued functions.

(a) $\mathbf{r}_1(t) = \langle t^{-1}, (t+1)^{-1}, \sin^{-1} t \rangle$

(b) $\mathbf{r}_2(t) = \langle \sqrt{8-t^3}, \ln t, e^{\sqrt{t}} \rangle$

SOLUTION

(a) We find the domain of $\mathbf{r}_1(t) = \langle t^{-1}, (t+1)^{-1}, \sin^{-1} t \rangle$. The function t^{-1} is defined for $t \neq 0$. $(t+1)^{-1}$ is defined for $t \neq -1$ and $\sin^{-1} t$ is defined for $-1 \le t \le 1$. Hence, the domain of $\mathbf{r}_1(t)$ is defined by the following inequalities:

$$\begin{aligned} t &\neq 0 \\ t &\neq -1 \quad \Rightarrow \quad -1 < t < 0 \quad \text{or} \quad 0 < t \le 1 \\ -1 \le t &\le 1 \end{aligned}$$

(b) We find the domain of $\mathbf{r}_2(t) = \langle \sqrt{8-t^3}, \ln t, e^{\sqrt{t}} \rangle$. The domain of $\sqrt{8-t^3}$ is $8 - t^3 \ge 0$. The domain of $\ln t$ is $t > 0$ and $e^{\sqrt{t}}$ is defined for $t \ge 0$. Hence, the domain of $\mathbf{r}_2(t)$ is defined by the following inequalities:

$$\begin{aligned} 8 - t^3 &\ge 0 & & t^3 \le 8 & & \\ t &> 0 \quad \Rightarrow & & & \Rightarrow \quad & 0 < t \le 2 \\ t &\ge 0 & & t > 0 & & \end{aligned}$$

3. Find a vector parametrization of the intersection of the surfaces $x^2 + y^4 + 2z^3 = 6$ and $x = y^2$ in $\mathbf{R}^3$.

SOLUTION We need to find a vector parametrization $\mathbf{r}(t) = \langle x(t), y(t), z(t) \rangle$ for the intersection curve. Using $t = y$ as a parameter, we have $x = t^2$ and $y = t$. We substitute in the equation of the surface $x^2 + y^4 + 2z^3 = 6$ and solve for z in terms of t. This gives:

$$\begin{aligned} t^4 + t^4 + 2z^3 &= 6 \\ 2t^4 + 2z^3 &= 6 \\ z^3 &= 3 - t^4 \quad \Rightarrow \quad z = \sqrt[3]{3 - t^4} \end{aligned}$$

We obtain the following parametrization of the intersection curve:

$$\mathbf{r}(t) = \left\langle t^2, t, \sqrt[3]{3 - t^4} \right\rangle.$$

In Exercises 5–10, calculate the derivative indicated.

5. $\mathbf{r}'(t)$, where $\mathbf{r}(t) = \langle 1 - t, t^{-2}, \ln t \rangle$

SOLUTION We use the Theorem on Componentwise Differentiation to compute the derivative $\mathbf{r}'(t)$. We get

$$\mathbf{r}'(t) = \langle (1-t)', (t^{-2})', (\ln t)' \rangle = \left\langle -1, -2t^{-3}, \frac{1}{t} \right\rangle$$

7. $\mathbf{r}'(0)$, where $\mathbf{r}(t) = \langle e^{2t}, e^{-4t^2}, e^{6t} \rangle$

SOLUTION We differentiate $\mathbf{r}(t)$ componentwise to find $\mathbf{r}'(t)$:

$$\mathbf{r}'(t) = \left\langle (e^{2t})', \left(e^{-4t^2}\right)', (e^{6t})' \right\rangle = \left\langle 2e^{2t}, -8te^{-4t^2}, 6e^{6t} \right\rangle$$

The derivative $\mathbf{r}'(0)$ is obtained by setting $t = 0$ in $\mathbf{r}'(t)$. This gives

$$\mathbf{r}'(0) = \left\langle 2e^{2\cdot 0}, -8 \cdot 0e^{-4\cdot 0^2}, 6e^{6\cdot 0} \right\rangle = \langle 2, 0, 6 \rangle$$

9. $\frac{d}{dt} e^t \langle 1, t, t^2 \rangle$

SOLUTION Using the Product Rule for differentiation gives

$$\begin{aligned} \frac{d}{dt} e^t \langle 1, t, t^2 \rangle &= e^t \frac{d}{dt} \langle 1, t, t^2 \rangle + (e^t)' \langle 1, t, t^2 \rangle = e^t \langle 0, 1, 2t \rangle + e^t \langle 1, t, t^2 \rangle \\ &= e^t \left(\langle 0, 1, 2t \rangle + \langle 1, t, t^2 \rangle \right) = e^t \langle 1, 1 + t, 2t + t^2 \rangle \end{aligned}$$

In Exercises 11–14, calculate the derivative at $t = 3$ assuming that

$$\mathbf{r}_1(3) = \langle 1, 1, 0 \rangle, \qquad \mathbf{r}_2(3) = \langle 1, 1, 0 \rangle$$
$$\mathbf{r}_1'(3) = \langle 0, 0, 1 \rangle, \qquad \mathbf{r}_2'(3) = \langle 0, 2, 4 \rangle$$

11. $\frac{d}{dt}(6\mathbf{r}_1(t) - 4 \cdot \mathbf{r}_2(t))$

SOLUTION Using Differentiation Rules we obtain:

$$\frac{d}{dt}(6\mathbf{r}_1(t) - 4\mathbf{r}_2(t))\bigg|_{t=3} = 6\mathbf{r}_1'(3) - 4\mathbf{r}_2'(3) = 6 \cdot \langle 0, 0, 1 \rangle - 4 \cdot \langle 0, 2, 4 \rangle$$
$$= \langle 0, 0, 6 \rangle - \langle 0, 8, 16 \rangle = \langle 0, -8, -10 \rangle$$

13. $\frac{d}{dt}\left(\mathbf{r}_1(t) \cdot \mathbf{r}_2(t)\right)$

SOLUTION Using Product Rule for Dot Products we obtain:

$$\frac{d}{dt}\mathbf{r}_1(t) \cdot \mathbf{r}_2(t) = \mathbf{r}_1(t) \cdot \mathbf{r}_2'(t) + \mathbf{r}_1'(t) \cdot \mathbf{r}_2(t)$$

Setting $t = 3$ gives:

$$\frac{d}{dt}\mathbf{r}_1(t) \cdot \mathbf{r}_2(t)\bigg|_{t=3} = \mathbf{r}_1(3) \cdot \mathbf{r}_2'(3) + \mathbf{r}_1'(3) \cdot \mathbf{r}_2(3) = \langle 1, 1, 0 \rangle \cdot \langle 0, 2, 4 \rangle + \langle 0, 0, 1 \rangle \cdot \langle 1, 1, 0 \rangle = 2 + 0 = 2$$

15. Calculate $\int_0^3 \left\langle 4t + 3, t^2, -4t^3 \right\rangle dt$.

SOLUTION By the definition of vector-valued integration, we have

$$\int_0^3 \left\langle 4t + 3, t^2, -4t^3 \right\rangle dt = \left\langle \int_0^3 (4t + 3)\, dt, \int_0^3 t^2\, dt, \int_0^3 -4t^3\, dt \right\rangle \tag{1}$$

We compute the integrals on the right-hand side:

$$\int_0^3 (4t + 3)\, dt = 2t^2 + 3t\bigg|_0^3 = 2 \cdot 9 + 3 \cdot 3 - 0 = 27$$
$$\int_0^3 t^2\, dt = \frac{t^3}{3}\bigg|_0^3 = \frac{3^3}{3} = 9$$
$$\int_0^3 -4t^3\, dt = -t^4\bigg|_0^3 = -3^4 = -81$$

Substituting in (1) gives the following integral:

$$\int_0^3 \left\langle 4t + 3, t^2, -4t^3 \right\rangle dt = \langle 27, 9, -81 \rangle$$

17. Find the unit tangent vector to $\mathbf{r}(t) = \left\langle \sin t, t, \cos t \right\rangle$ at $t = \pi$.

SOLUTION The unit tangent vector at $t = \pi$ is

$$\mathbf{T}(\pi) = \frac{\mathbf{r}'(\pi)}{\|\mathbf{r}'(\pi)\|} \tag{1}$$

We differentiate $\mathbf{r}(t)$ componentwise to obtain:

$$\mathbf{r}'(t) = \langle \cos t, 1, -\sin t \rangle$$

Therefore,

$$\mathbf{r}'(\pi) = \langle \cos \pi, 1, -\sin \pi \rangle = \langle -1, 1, 0 \rangle$$

We compute the length of $\mathbf{r}'(\pi)$:

$$\|\mathbf{r}'(\pi)\| = \sqrt{(-1)^2 + 1^2 + 0^2} = \sqrt{2}$$

Substituting in (1) gives:

$$\mathbf{T}(\pi) = \left\langle \frac{-1}{\sqrt{2}}, \frac{1}{\sqrt{2}}, 0 \right\rangle$$

19. A particle located at $(1, 1, 0)$ at time $t = 0$ follows a path whose velocity vector is $\mathbf{v}(t) = \left\langle 1, t, 2t^2 \right\rangle$. Find the particle's location at $t = 2$.

SOLUTION We first find the path $\mathbf{r}(t)$ by integrating the velocity vector $\mathbf{v}(t)$:

$$\mathbf{r}(t) = \int \left\langle 1, t, 2t^2 \right\rangle dt = \left\langle \int 1\, dt, \int t\, dt, \int 2t^2\, dt \right\rangle = \left\langle t + c_1, \frac{1}{2}t^2 + c_2, \frac{2}{3}t^3 + c_3 \right\rangle$$

Denoting by $\mathbf{c} = \langle c_1, c_2, c_3 \rangle$ the constant vector, we obtain:

$$\mathbf{r}(t) = \left\langle t, \frac{1}{2}t^2, \frac{2}{3}t^3 \right\rangle + \mathbf{c} \tag{1}$$

To find the constant vector $\mathbf{c}$, we use the given information on the initial position of the particle. At time $t = 0$ it is at the point $(1, 1, 0)$. That is, by (1):

$$\mathbf{r}(0) = \langle 0, 0, 0 \rangle + \mathbf{c} = \langle 1, 1, 0 \rangle$$

or,

$$\mathbf{c} = \langle 1, 1, 0 \rangle$$

We substitute in (1) to obtain:

$$\mathbf{r}(t) = \left\langle t, \frac{1}{2}t^2, \frac{2}{3}t^3 \right\rangle + \langle 1, 1, 0 \rangle = \left\langle t + 1, \frac{1}{2}t^2 + 1, \frac{2}{3}t^3 \right\rangle$$

Finally, we substitute $t = 2$ to obtain the particle's location at $t = 2$:

$$\mathbf{r}(2) = \left\langle 2 + 1, \frac{1}{2} \cdot 2^2 + 1, \frac{2}{3} \cdot 2^3 \right\rangle = \left\langle 3, 3, \frac{16}{3} \right\rangle$$

At time $t = 2$ the particle is located at the point

$$\left(3, 3, \frac{16}{3} \right)$$

21. Compute the length of the path $\mathbf{r}(t) = \left\langle \sin 2t, \cos 2t, 3t - 1 \right\rangle$ for $1 \le t \le 3$.

SOLUTION We use the formula for the arc length:

$$L = \int_1^3 \|\mathbf{r}'(t)\|\, dt \tag{1}$$

We compute the derivative vector $\mathbf{r}'(t)$ and its length:

$$\mathbf{r}'(t) = \langle 2\cos 2t, -2\sin 2t, 3 \rangle$$

$$\|\mathbf{r}'(t)\| = \sqrt{(2\cos 2t)^2 + (-2\sin 2t)^2 + 3^2} = \sqrt{4\cos^2 2t + 4\sin^2 2t + 9}$$

$$= \sqrt{4\left(\cos^2 2t + \sin^2 2t\right) + 9} = \sqrt{4 \cdot 1 + 9} = \sqrt{13}$$

We substitute in (1) and compute the integral to obtain the following length:

$$L = \int_1^3 \sqrt{13}\, dt = \sqrt{13}t \Big|_1^3 = 2\sqrt{13}.$$

23. A string in the shape of a helix has a height of 20 cm and makes four full rotations over a circle of radius 5 cm. Find a parametrization $\mathbf{r}(t)$ of the string and compute its length.

SOLUTION Since the radius is 5 cm and the height is 20 cm, the helix is traced by a parametrization of the form:

$$\mathbf{r}(t) = \langle 5\cos at, 5\sin at, t\rangle, \quad 0 \le t \le 20$$

Since the helix makes exactly 4 full rotations, we have:

$$a \cdot 20 = 4 \cdot 2\pi \quad \Rightarrow \quad a = \frac{2\pi}{5}$$

The parametrization of the helix is, thus:

$$\mathbf{r}(t) = \left\langle 5\cos\frac{2\pi t}{5}, 5\sin\frac{2\pi t}{5}, t\right\rangle, \quad 0 \le t \le 20$$

The helix is shown in the following figure:

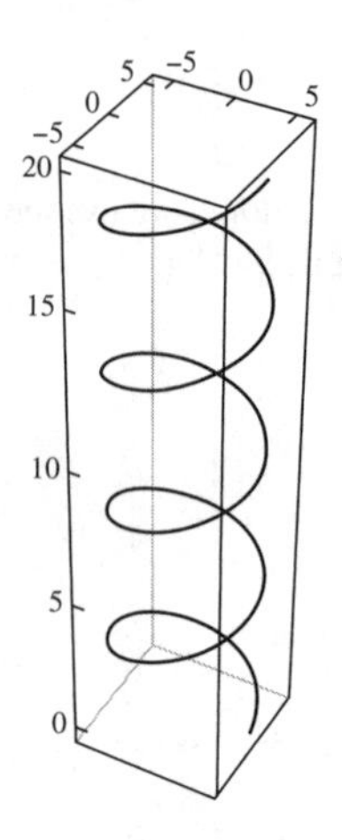

To find the length of the helix, we use the arc length formula:

$$L = \int_0^{20} \|\mathbf{r}'(t)\|\, dt \tag{1}$$

We find $\mathbf{r}'(t)$ and its length:

$$\mathbf{r}'(t) = \left\langle -5 \cdot \frac{2\pi}{5}\sin\frac{2\pi t}{5}, 5 \cdot \frac{2\pi}{5}\cos\frac{2\pi t}{5}, 1\right\rangle = \left\langle -2\pi\sin\frac{2\pi t}{5}, 2\pi\cos\frac{2\pi t}{5}, 1\right\rangle$$

$$\|\mathbf{r}'(t)\| = \sqrt{4\pi^2\sin^2\frac{2\pi t}{5} + 4\pi^2\cos^2\frac{2\pi t}{5} + 1} = \sqrt{4\pi^2\left(\sin^2\frac{2\pi t}{5} + \cos^2\frac{2\pi t}{5}\right) + 1} = \sqrt{1 + 4\pi^2}$$

Substituting in (1) we get:

$$L = \int_0^{20} \sqrt{1 + 4\pi^2}\, dt = 20\sqrt{1 + 4\pi^2} \approx 127.2$$

25. Calculate the curvature $\kappa(t)$ for $\mathbf{r}(t) = \langle t^{-1}, \ln t, t\rangle$ and find the unit tangent and normal vectors at $t = 1$.

SOLUTION The unit normal vector is defined by:

$$\mathbf{N}(t) = \frac{\mathbf{T}'(t)}{\|\mathbf{T}'(t)\|} \tag{1}$$

Since $\mathbf{T}(t) = \frac{\mathbf{r}'(t)}{\|\mathbf{r}'(t)\|}$, we must find $\mathbf{r}'(t)$ and its length:

$$\mathbf{r}'(t) = \frac{d}{dt}\langle t^{-1}, \ln t, t\rangle = \left\langle -t^{-2}, \frac{1}{t}, 1\right\rangle = \frac{1}{t^2}\langle -1, t, t^2\rangle$$

$$\|\mathbf{r}'(t)\| = \frac{1}{t^2}\sqrt{1 + t^2 + t^4} \tag{2}$$

Hence:

$$\mathbf{T}(t) = \frac{\mathbf{r}'(t)}{\|\mathbf{r}'(t)\|} = \left\langle \frac{-1}{\sqrt{1+t^2+t^4}}, \frac{t}{\sqrt{1+t^2+t^4}}, \frac{t^2}{\sqrt{1+t^2+t^4}}\right\rangle$$

Setting $t = 1$ gives:

$$\mathbf{T}(1) = \left\langle -\frac{1}{\sqrt{3}}, \frac{1}{\sqrt{3}}, \frac{1}{\sqrt{3}} \right\rangle$$

We now compute $\mathbf{T}'(t)$ and its length:

$$\mathbf{T}'(t) = \frac{d}{dt}\left\langle \frac{-1}{\sqrt{1+t^2+t^4}}, \frac{t}{\sqrt{1+t^2+t^4}}, \frac{t^2}{\sqrt{1+t^2+t^4}} \right\rangle$$

$$= \left\langle \frac{\frac{2t+4t^3}{2\sqrt{1+t^2+t^4}}}{1+t^2+t^4}, \frac{1\cdot\sqrt{1+t^2+t^4} - t\cdot\frac{2t+4t^3}{2\sqrt{1+t^2+t^4}}}{1+t^2+t^4}, \frac{2t\sqrt{1+t^2+t^4} - t^2\cdot\frac{2t+4t^3}{2\sqrt{1+t^2+t^4}}}{1+t^2+t^4} \right\rangle$$

$$= \left\langle \frac{t+2t^3}{\left(1+t^2+t^4\right)^{3/2}}, \frac{1-t^4}{\left(1+t^2+t^4\right)^{3/2}}, \frac{2t+t^3}{\left(1+t^2+t^4\right)^{3/2}} \right\rangle$$

$$\|\mathbf{T}'(t)\| = \frac{1}{\left(1+t^2+t^4\right)^{3/2}}\sqrt{\left(t+2t^3\right)^2 + \left(1-t^4\right)^2 + \left(2t+t^3\right)^2} = \frac{\sqrt{t^8+5t^6+6t^4+5t^2+1}}{\left(1+t^2+t^4\right)^{3/2}}$$

Substituting in (1) we get:

$$\mathbf{N}(t) = \frac{1}{\sqrt{t^8+5t^6+6t^4+5t^2+1}}\left\langle t+2t^3, 1-t^4, 2t+t^3 \right\rangle \quad \Rightarrow \quad \mathbf{N}(1) = \frac{3}{\sqrt{18}}\langle 1, 0, 1\rangle$$

To find the curvature we use the following formula:

$$\kappa(t) = \frac{\|\mathbf{r}'(t)\times\mathbf{r}''(t)\|}{\|\mathbf{r}'(t)\|^3} \tag{3}$$

We first find $\mathbf{r}''(t)$:

$$\mathbf{r}''(t) = \frac{d}{dt}\left\langle -t^{-2}, t^{-1}, 1\right\rangle = \left\langle 2t^{-3}, -t^{-2}, 0\right\rangle = \left\langle \frac{2}{t^3}, -\frac{1}{t^2}, 0\right\rangle$$

We compute the cross product:

$$\mathbf{r}'(t)\times\mathbf{r}''(t) = \begin{vmatrix} \mathbf{i} & \mathbf{j} & \mathbf{k} \\ -\frac{1}{t^2} & \frac{1}{t} & 1 \\ \frac{2}{t^3} & -\frac{1}{t^2} & 0 \end{vmatrix} = \frac{1}{t^2}\mathbf{i} + \frac{2}{t^3}\mathbf{j} + \left(\frac{1}{t^4} - \frac{2}{t^4}\right)\mathbf{k} = \frac{1}{t^2}\mathbf{i} + \frac{2}{t^3}\mathbf{j} - \frac{1}{t^4}\mathbf{k}$$

Hence:

$$\|\mathbf{r}'(t)\times\mathbf{r}''(t)\| = \sqrt{\left(\frac{1}{t^2}\right)^2 + \left(\frac{2}{t^3}\right)^2 + \left(-\frac{1}{t^4}\right)^2} = \sqrt{\frac{1}{t^4} + \frac{4}{t^6} + \frac{1}{t^8}} = \frac{1}{t^4}\sqrt{t^4+4t^2+1} \tag{4}$$

We now substitute (2) and (4) in (3) to obtain the following curvature:

$$\kappa(t) = \frac{t^2\sqrt{t^4+4t^2+1}}{\left(\sqrt{t^4+t^2+1}\right)^3}$$

Setting $t = 1$ we get:

$$\kappa(1) = \frac{\sqrt{2}}{3}$$

In Exercises 27–30, let $\mathbf{r}(t) = \left\langle t, e^{-t^2}\right\rangle$.

27. Compute the curvature function $\kappa(t)$.

SOLUTION The curvature is the following function:

$$\kappa(t) = \frac{\|\mathbf{r}'(t)\times\mathbf{r}''(t)\|}{\|\mathbf{r}'(t)\|^3} \tag{1}$$

We find the derivatives of $\mathbf{r}(t)$:

$$\mathbf{r}'(t) = \left\langle 1, -2te^{-t^2} \right\rangle$$

$$\mathbf{r}''(t) = \left\langle 0, -2e^{-t^2} - 2t \cdot (-2t)e^{-t^2} \right\rangle = \left\langle 0, (4t^2 - 2)e^{-t^2} \right\rangle$$

Therefore:

$$\mathbf{r}'(t) \times \mathbf{r}''(t) = \left(\mathbf{i} - 2te^{-t^2}\mathbf{j}\right) \times (4t^2 - 2)e^{-t^2}\mathbf{j} = (4t^2 - 2)e^{-t^2}\mathbf{k} = \left\langle 0, 0, (4t^2 - 2)e^{-t^2} \right\rangle$$

We compute the lengths of the vectors in (1):

$$\|\mathbf{r}'(t)\| = \sqrt{1^2 + \left(-2te^{-t^2}\right)^2} = \sqrt{1 + 4t^2e^{-2t^2}}$$

$$\|\mathbf{r}'(t) \times \mathbf{r}''(t)\| = |4t^2 - 2|e^{-t^2}$$

Substituting in (1) gives the following curvature:

$$\kappa(t) = \frac{|4t^2 - 2|e^{-t^2}}{\left(1 + 4t^2e^{-2t^2}\right)^{3/2}}$$

29. Find the unit tangent and normal vectors at $t = 0$ and $t = 1$.

SOLUTION The unit tangent vector is defined by:

$$\mathbf{T}(t) = \frac{\mathbf{r}'(t)}{\|\mathbf{r}'(t)\|} \qquad (1)$$

The derivative of $\mathbf{r}(t) = \left\langle t, e^{-t^2} \right\rangle$ is:

$$\mathbf{r}'(t) = \left\langle 1, -2te^{-t^2} \right\rangle$$

At $t = 0$ and $t = 1$ we have:

$$\mathbf{r}'(0) = \left\langle 1, -2 \cdot 0 \cdot e^{-0^2} \right\rangle = \langle 1, 0 \rangle$$

$$\|\mathbf{r}'(0)\| = \sqrt{1^2 + 0^2} = 1$$

$$\mathbf{r}'(1) = \left\langle 1, -2 \cdot 1 \cdot e^{-1^2} \right\rangle = \left\langle 1, -\frac{2}{e} \right\rangle$$

$$\|\mathbf{r}'(1)\| = \sqrt{1^2 + \left(-\frac{2}{e}\right)^2} = \sqrt{1 + \frac{4}{e^2}} = \frac{1}{e}\sqrt{e^2 + 4}$$

Substituting in (1) we obtain:

$$\mathbf{T}(0) = \frac{\mathbf{r}'(0)}{\|\mathbf{r}'(0)\|} = \langle 1, 0 \rangle$$

$$\mathbf{T}(1) = \frac{\mathbf{r}'(1)}{\|\mathbf{r}'(1)\|} = \frac{\left\langle 1, -\frac{2}{e} \right\rangle}{\frac{1}{e}\sqrt{e^2 + 4}} = \left\langle \frac{e}{\sqrt{e^2 + 4}}, \frac{-2}{\sqrt{e^2 + 4}} \right\rangle$$

The unit normal vector is a unit vector orthogonal to $\mathbf{r}'(t) = \left\langle 1, -2te^{-t^2} \right\rangle$. We observe that $\left\langle 2te^{-t^2}, 1 \right\rangle$ is orthogonal to $\mathbf{r}'(t)$, since their dot product is zero, hence $\mathbf{N}(t)$ is a unit vector in the direction of $\left\langle 2te^{-t^2}, 1 \right\rangle$ or $-\left\langle 2te^{-t^2}, 1 \right\rangle$.

Recall that $\mathbf{N}(t)$ points in the direction of bending. A graph of $\mathbf{r}(t)$ (which is also the graph of $y = e^{-x^2}$) helps us find the appropriate direction. It is easy to show (by taking y' and y'') that the inflection points occur at $x = \pm 1/\sqrt{2}$. Thus, we get the following picture:

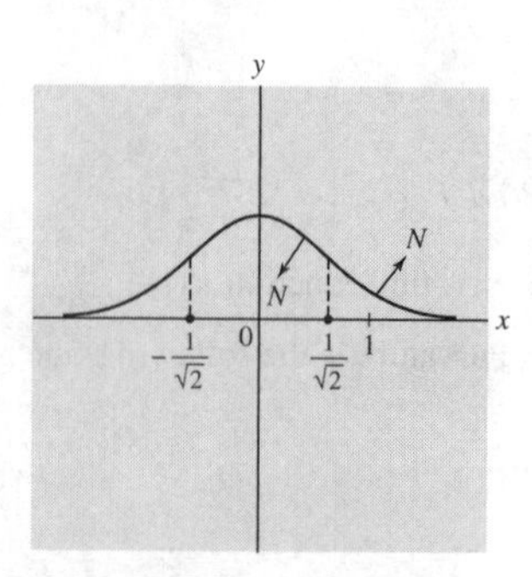

We conclude that:

$$\mathbf{N}(t) = \begin{cases} \dfrac{\left\langle -2te^{-t^2}, -1 \right\rangle}{\sqrt{1+4t^2e^{-2t^2}}}, & |t| < \dfrac{1}{\sqrt{2}} \\ \dfrac{\left\langle 2te^{-t^2}, 1 \right\rangle}{\sqrt{1+4t^2e^{-2t^2}}}, & |t| > \dfrac{1}{\sqrt{2}} \end{cases}$$

For $t = 0$ and $t = 1$ we get:

$$\mathbf{N}(0) = \langle 0, -1 \rangle, \quad \mathbf{N}(1) = \frac{\left\langle 2e^{-1}, 1 \right\rangle}{\sqrt{1+4e^{-2}}} = \left\langle \frac{2}{\sqrt{e^2+4}}, \frac{e}{\sqrt{e^2+4}} \right\rangle$$

31. Find the curvature $\kappa(t)$ and unit normal vector $\mathbf{N}(t)$ for $\mathbf{r}(t) = \left\langle \sin t, \sin t, \cos t \right\rangle$.

SOLUTION The curvature is the following function:

$$\kappa(t) = \frac{\|\mathbf{r}'(t) \times \mathbf{r}''(t)\|}{\|\mathbf{r}'(t)\|^3} \tag{1}$$

We compute the derivatives of $\mathbf{r}(t) = \langle \sin t, \sin t, \cos t \rangle$:

$$\mathbf{r}'(t) = \langle \cos t, \cos t, -\sin t \rangle,$$
$$\mathbf{r}''(t) = \langle -\sin t, -\sin t, -\cos t \rangle$$

We calculate the cross product:

$$\begin{aligned} \mathbf{r}'(t) \times \mathbf{r}''(t) &= \begin{vmatrix} \mathbf{i} & \mathbf{j} & \mathbf{k} \\ \cos t & \cos t & -\sin t \\ -\sin t & -\sin t & -\cos t \end{vmatrix} \\ &= \left(-\cos^2 t - \sin^2 t\right)\mathbf{i} - \left(-\cos^2 t - \sin^2 t\right)\mathbf{j} + (-\cos t \sin t + \cos t \sin t)\mathbf{k} \\ &= -\mathbf{i} + \mathbf{j} = \langle -1, 1, 0 \rangle \end{aligned}$$

We find the lengths of the vectors in (1):

$$\|\mathbf{r}'(t) \times \mathbf{r}''(t)\| = \| \langle -1, 1, 0 \rangle \| = \sqrt{2}$$
$$\left\|\mathbf{r}'(t)\right\| = \sqrt{\cos^2 t + \cos^2 t + \sin^2 t} = \sqrt{1+\cos^2 t}$$

Substituting in (1) gives the following curvature:

$$\kappa(t) = \frac{\sqrt{2}}{\left(1+\cos^2 t\right)^{3/2}}$$

The unit normal is the following vector:

$$\mathbf{N}(t) = \frac{\mathbf{T}'(t)}{\|\mathbf{T}'(t)\|} \tag{2}$$

We find the unit tangent vector:

$$\mathbf{T}(t) = \frac{\mathbf{r}'(t)}{\|\mathbf{r}'(t)\|} = \frac{1}{\sqrt{1+\cos^2 t}} \langle \cos t, \cos t, -\sin t \rangle$$

We differentiate $\mathbf{T}(t)$ using the Product Rule:

$$\begin{aligned} \mathbf{T}'(t) &= \frac{1}{\sqrt{1+\cos^2 t}} \langle -\sin t, -\sin t, -\cos t \rangle + \frac{\frac{2\cos t \sin t}{2\sqrt{1+\cos^2 t}}}{1+\cos^2 t} \langle \cos t, \cos t, -\sin t \rangle \\ &= \frac{1}{\sqrt{1+\cos^2 t}} \langle -\sin t, -\sin t, -\cos t \rangle + \frac{\cos t \sin t}{\left(1+\cos^2 t\right)^{3/2}} \langle \cos t, \cos t, -\sin t \rangle \\ &= \frac{1}{\left(1+\cos^2 t\right)^{3/2}} \left[(1+\cos^2 t) \langle -\sin t, -\sin t, -\cos t \rangle + \sin t \cos t \langle \cos t, \cos t, -\sin t \rangle \right] \\ &= \frac{1}{\left(1+\cos^2 t\right)^{3/2}} \cdot \left\langle -\sin t, -\sin t, -\cos t - \cos^3 t - \sin^2 t \cos t \right\rangle \end{aligned}$$

$$= \frac{1}{\left(1+\cos^2 t\right)^{3/2}} \cdot \langle -\sin t, -\sin t, -2\cos t \rangle$$

Hence:

$$\|\mathbf{T}'(t)\| = \frac{1}{\left(1+\cos^2 t\right)^{3/2}} \sqrt{2\sin^2 + 4\cos^2 t} = \frac{\sqrt{2}\sqrt{1+\cos^2 t}}{\left(1+\cos^2 t\right)^{3/2}} = \frac{\sqrt{2}}{1+\cos^2 t}$$

Thus,

$$\mathbf{N}(t) = \frac{-1}{\sqrt{2}} \cdot \frac{1}{\sqrt{1+\cos^2 t}} \cdot \langle \sin t, \sin t, 2\cos t \rangle$$

In Exercises 33–34, write the acceleration vector **a** *at the point indicated as a sum of tangential and normal components.*

33. $\mathbf{r}(\theta) = \langle \cos\theta, \sin 2\theta \rangle, \quad \theta = \frac{\pi}{4}$

SOLUTION

Step 1. Compute **T** and $a_{\mathbf{T}}$. Since $\mathbf{r}(\theta) = \langle \cos\theta, \sin 2\theta \rangle$, we have $\mathbf{r}'(\theta) = \langle -\sin\theta, 2\cos 2\theta \rangle$ and $\mathbf{r}''(\theta) = \langle -\cos\theta, -4\sin 2\theta \rangle$. Thus, at $\theta = \frac{\pi}{4}$,

$$\mathbf{v} = \mathbf{r}'\left(\frac{\pi}{4}\right) = \left\langle -\sin\frac{\pi}{4}, 2\cos\frac{\pi}{2} \right\rangle = \left\langle -\frac{1}{\sqrt{2}}, 0 \right\rangle$$

$$\mathbf{T} = \frac{\mathbf{v}}{\|\mathbf{v}\|} = \frac{\left\langle -\frac{1}{\sqrt{2}}, 0 \right\rangle}{\sqrt{\left(\frac{1}{\sqrt{2}}\right)^2 + 0}} = \sqrt{2}\left\langle -\frac{1}{\sqrt{2}}, 0 \right\rangle = \langle -1, 0 \rangle$$

$$\mathbf{a} = \mathbf{r}''\left(\frac{\pi}{4}\right) = \left\langle -\cos\frac{\pi}{4}, -4\sin\frac{\pi}{2} \right\rangle = \left\langle -\frac{1}{\sqrt{2}}, -4 \right\rangle$$

$$a_{\mathbf{T}} = \mathbf{a} \cdot \mathbf{T} = \left\langle -\frac{1}{\sqrt{2}}, -4 \right\rangle \cdot \langle -1, 0 \rangle = \frac{1}{\sqrt{2}}$$

Step 2. Compute $a_{\mathbf{N}}$ and **N**. We have

$$a_{\mathbf{N}}\mathbf{N} = \mathbf{a} - a_{\mathbf{T}}\mathbf{T} = \left\langle -\frac{1}{\sqrt{2}}, -4 \right\rangle - \frac{1}{\sqrt{2}}\langle -1, 0 \rangle = \langle 0, -4 \rangle$$

Since **N** is a unit vector, we can find $a_{\mathbf{N}}$ by:

$$a_{\mathbf{N}} = \|a_{\mathbf{N}}\mathbf{N}\| = \|\langle 0, -4 \rangle\| = 4$$

Therefore,

$$\mathbf{N} = \frac{a_{\mathbf{N}}\mathbf{N}}{a_{\mathbf{N}}} = \frac{\langle 0, -4 \rangle}{4} = \langle 0, -1 \rangle$$

Step 3. Write the decomposition. We found that $a_{\mathbf{T}} = \frac{1}{\sqrt{2}}$ and $a_{\mathbf{N}} = 4$, hence the decomposition of **a** is

$$\mathbf{a} = a_{\mathbf{T}}\mathbf{T} + a_{\mathbf{N}}\mathbf{N} = \frac{1}{\sqrt{2}}\mathbf{T} + 4\mathbf{N}$$

where $\mathbf{T} = \langle -1, 0 \rangle$ and $\mathbf{N} = \langle 0, -1 \rangle$.

35. Find the osculating circle to the curve $y = e^{-x^2}$ at $x = 0$.

SOLUTION We first find the curvature, using the formula for the curvature of a graph in the plane:

$$\kappa(x) = \frac{|y''(x)|}{\left(1 + y'(x)^2\right)^{3/2}} \tag{1}$$

Since $y = e^{-x^2}$, we have:

$$y'(x) = -2xe^{-x^2},$$

$$y''(x) = -2e^{-x^2} + 4x^2e^{-x^2} = 2\left(2x^2 - 1\right)e^{-x^2}$$

Substituting in (1) we obtain the following curvature at $x = 0$:

$$\kappa(x) = \frac{|y''(0)|}{\left(1 + y'(0)^2\right)^{3/2}} = \frac{|-2|}{(1+0)^{3/2}} = 2 \tag{2}$$

We now find the osculating circle at $x = 0$. The radius of the osculating circle is $R = \frac{1}{\kappa(0)} = \frac{1}{2}$, and since the graph of $y = e^{-x^2}$ is symmetric about the y-axis, so also must be the osculating circle (at $x = 0$). The circle touches the graph at $(0, 1)$, is symmetric about the y-axis, and has radius $\frac{1}{2}$, so it must be centered at $\left(0, \frac{1}{2}\right)$. Therefore it is parametrized by,

$$\mathbf{c}(t) = \left\langle 0, \frac{1}{2} \right\rangle + \frac{1}{2}\langle \cos t, \sin t \rangle, \quad 0 \le t \le 2\pi$$

37. If a planet is in "orbit" around a sun whose mass is zero, Newton's Laws imply that the position vector of the plane satisfies $\mathbf{r}''(t) = \mathbf{0}$. Show that in this case, the orbit is the straight line with parametrization $\mathbf{r}(t) = \mathbf{r} + t\mathbf{v}$, where $\mathbf{r} = \mathbf{r}(0)$ and $\mathbf{v} = \mathbf{r}'(0)$ (Figure 1).

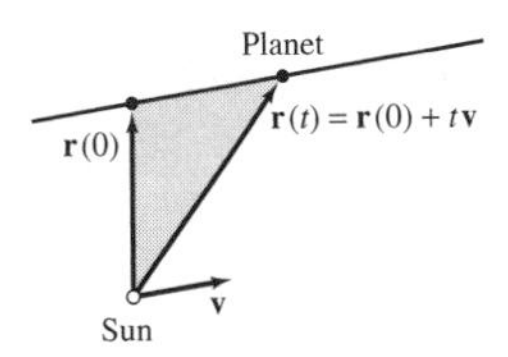

FIGURE 1

SOLUTION Integrating $\mathbf{r}''(t) = \mathbf{0}$ gives:

$$\mathbf{r}'(t) = \mathbf{c}$$

The constant $\mathbf{c}$ is $\mathbf{r}'(0) = \mathbf{v}(0)$. That is,

$$\mathbf{r}'(t) = \mathbf{v}$$

We integrate again:

$$\mathbf{r}(t) = \mathbf{v}t + \mathbf{d}$$

The constant $\mathbf{d}$ is $\mathbf{r} = \mathbf{r}(0)$. Hence, $\mathbf{r}(t) = \mathbf{r} + t\mathbf{v}$, where $\mathbf{r} = \mathbf{r}(0)$ and $\mathbf{v} = \mathbf{r}'(0)$.

39. Suppose that the planetary orbit in Figure 2 is an ellipse with eccentricity e (by definition, $e = c/a$). Use Kepler's Second Law to show that if the period of the orbit is T, the time it takes for a planet to travel from A' to B' is equal to

$$\left(\frac{1}{4} + \frac{e}{2\pi}\right) T$$

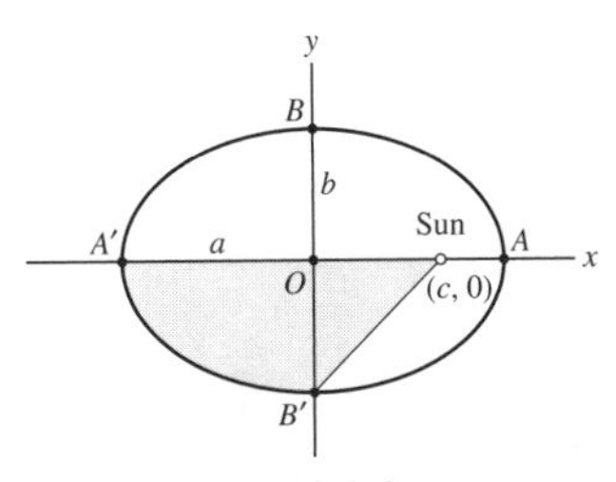

FIGURE 2

SOLUTION By the Law of Equal Areas, the position vector pointing from the sun to the planet sweeps out equal areas in equal times. We denote by S_1 the area swept by the position vector when the planet moves from A' to B', and t is the desired time. Since the position vector sweeps out the whole area of the ellipse (πab) in time T, the Law of Equal Areas implies that:

$$\frac{S_1}{\pi ab} = \frac{t}{T} \quad \Rightarrow \quad t = \frac{TS_1}{\pi ab} \tag{1}$$

We now find the area S_1 as the sum of the area of a quarter of the ellipse and the area of the triangle ODB. That is,

$$S_1 = \frac{\pi ab}{4} + \frac{\overline{OD} \cdot \overline{OB}'}{2} = \frac{\pi ab}{4} + \frac{cb}{2} = \frac{b}{4}(\pi a + 2c)$$

Substituting in (1) we get:

$$t = \frac{Tb(\pi a + 2c)}{4\pi ab} = \frac{T(\pi a + 2c)}{4\pi a} = T\left(\frac{1}{4} + \frac{1}{2\pi}\frac{c}{a}\right) = T\left(\frac{1}{4} + \frac{e}{2\pi}\right)$$

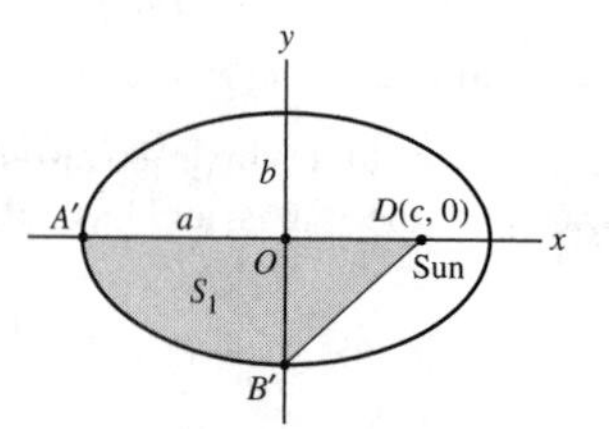

15 DIFFERENTIATION IN SEVERAL VARIABLES

15.1 Functions of Two or More Variables (ET Section 14.1)

Preliminary Questions

1. What is the difference between a horizontal trace and a level curve? How are they related?

SOLUTION A horizontal trace at height c consists of all points (x, y, c) such that $f(x, y) = c$. A level curve is the curve $f(x, y) = c$ in the xy-plane. The horizontal trace is in the $z = c$ plane. The two curves are related in the sense that the level curve is the projection of the horizontal trace on the xy-plane. The two curves have the same shape but they are located in parallel planes.

2. Describe the trace of $f(x, y) = x^2 - \sin(x^3 y)$ in the xz-plane.

SOLUTION The intersection of the graph of $f(x, y) = x^2 - \sin(x^3 y)$ with the xz-plane is obtained by setting $y = 0$ in the equation $z = x^2 - \sin(x^3 y)$. We get the equation $z = x^2 - \sin 0 = x^2$. This is the parabola $z = x^2$ in the xz-plane.

3. Is it possible for two different level curves of a function to intersect? Explain.

SOLUTION Two different level curves of $f(x, y)$ are the curves in the xy-plane defined by equations $f(x, y) = c_1$ and $f(x, y) = c_2$ for $c_1 \neq c_2$. If the curves intersect at a point (x_0, y_0), then $f(x_0, y_0) = c_1$ and $f(x_0, y_0) = c_2$, which implies that $c_1 = c_2$. Therefore, two different level curves of a function do not intersect.

4. Describe the contour map of $f(x, y) = x$ with contour interval 1.

SOLUTION The level curves of the function $f(x, y) = x$ are the vertical lines $x = c$. Therefore, the contour map of f with contour interval 1 consists of vertical lines so that every two adjacent lines are distanced one unit from another.

5. How will the contour maps of $f(x, y) = x$ and $g(x, y) = 2x$ with contour interval 1 look different?

SOLUTION The level curves of $f(x, y) = x$ are the vertical lines $x = c$, and the level curves of $g(x, y) = 2x$ are the vertical lines $2x = c$ or $x = \frac{c}{2}$. Therefore, the contour map of $f(x, y) = x$ with contour interval 1 consists of vertical lines with distance one unit between adjacent lines, whereas in the contour map of $g(x, y) = 2x$ (with contour interval 1) the distance between two adjacent vertical lines is $\frac{1}{2}$.

Exercises

In Exercises 1–4, evaluate the function at the specified points.

1. $f(x, y) = x + yx^3, \quad (2, 2), (-1, 4), (6, \frac{1}{2})$

SOLUTION We substitute the values for x and y in $f(x, y)$ and compute the values of f at the given points. This gives

$$f(2, 2) = 2 + 2 \cdot 2^3 = 18$$
$$f(-1, 4) = -1 + 4 \cdot (-1)^3 = -5$$
$$f\left(6, \tfrac{1}{2}\right) = 6 + \tfrac{1}{2} \cdot 6^3 = 114$$

3. $h(x, y, z) = xyz^{-2}, \quad (3, 8, 2), (3, -2, -6)$

SOLUTION Substituting $(x, y, z) = (3, 8, 2)$ and $(x, y, z) = (3, -2, -6)$ in the function, we obtain

$$h(3, 8, 2) = 3 \cdot 8 \cdot 2^{-2} = 3 \cdot 8 \cdot \frac{1}{4} = 6$$
$$h(3, -2, -6) = 3 \cdot (-2) \cdot (-6)^{-2} = -6 \cdot \frac{1}{36} = -\frac{1}{6}$$

In Exercises 5–16, sketch the domain of the function.

5. $f(x, y) = 4x - 7y$

SOLUTION The function is defined for all x and y, hence the domain is the entire xy-plane.

7. $f(x, y) = \ln(y - 2x)$

SOLUTION The function is defined if $y - 2x > 0$ or $y > 2x$. This is the region in the xy-plane that is above the line $y = 2x$.

$$D = \{(x, y) : y > 2x\}$$

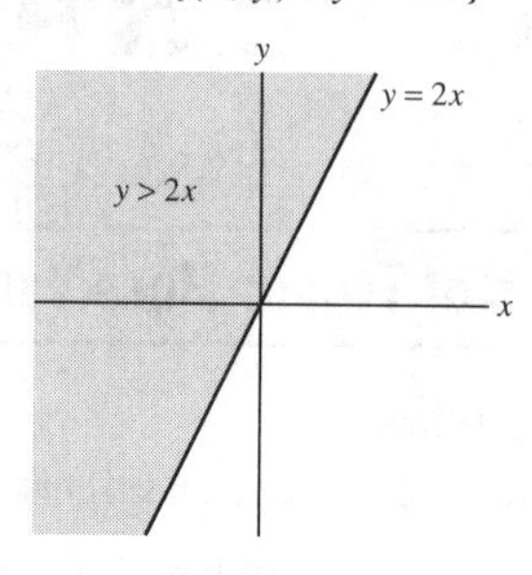

9. $G(x, t) = e^{1/(x+t)}$

SOLUTION The function is defined if $x + t \neq 0$, that is, $x \neq -t$. The domain is the same as the domain of $h(x, t)$ in the previous exercise.

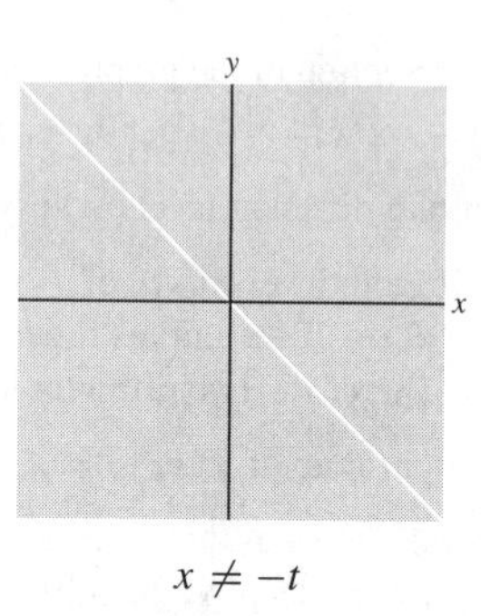

$x \neq -t$

11. $f(x, y) = \sin \dfrac{y}{x}$

SOLUTION The function is defined for all $x \neq 0$. The domain is the xy-plane with the y-axis excluded.

$$D = \{(x, y) : x \neq 0\}$$

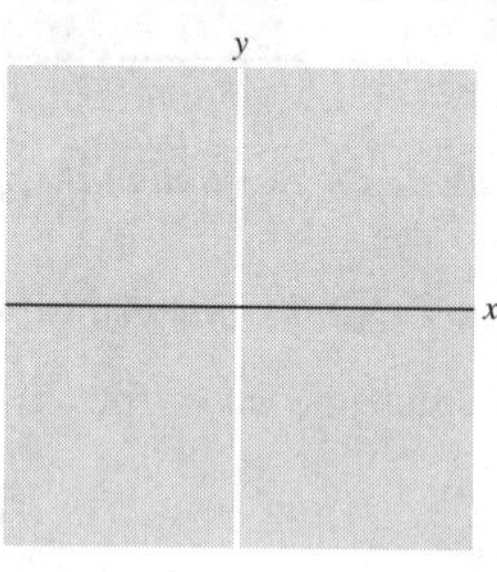

$x \neq 0$

13. $F(I, R) = \sqrt{IR}$

SOLUTION The function is defined if $IR \geq 0$. Therefore the domain is the first and the third quadrants of the IR-plane including both axes.

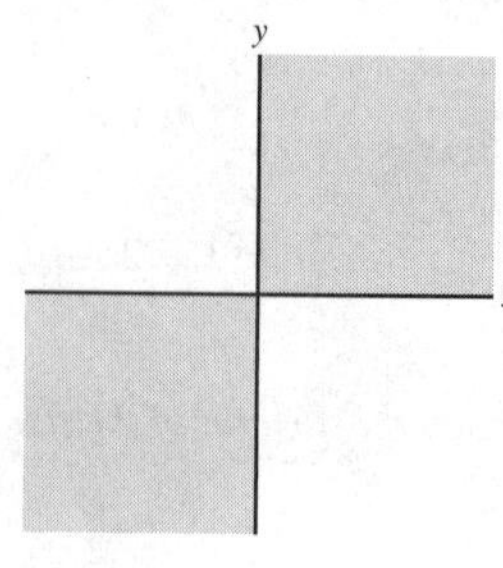

$IR \geq 0$

15. $g(r,t) = \dfrac{1}{r^2 - t}$

SOLUTION The function is defined if $r^2 - t \neq 0$, that is, $t \neq r^2$. The domain is the rt-plane with the parabola $t = r^2$ excluded.

$$D = \left\{(r,t) : t \neq r^2\right\}$$

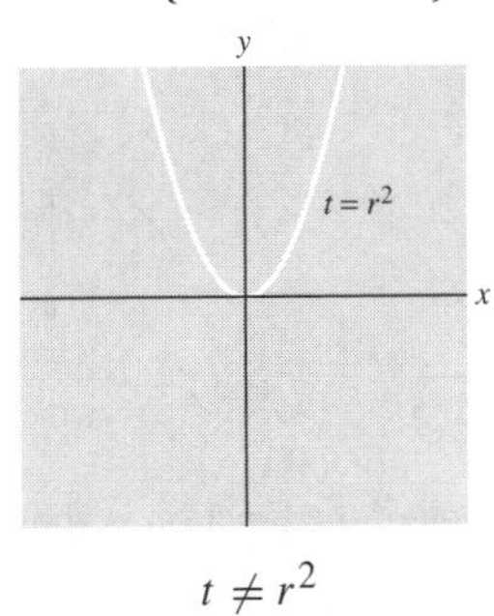

$t \neq r^2$

In Exercises 17–19, describe the domain and range of the function.

17. $f(x,y,z) = xz + e^y$

SOLUTION The domain of f is the entire (x,y,z)-space. Since f takes all the real values, the range is the entire real line.

19. $f(x,y,z) = \sqrt{9 - x^2 - y^2 - z^2}$

SOLUTION The function is defined if $9 - x^2 - y^2 - z^2 \geq 0$, i.e, $x^2 + y^2 + z^2 \leq 9$. Therefore, the domain is the set $\{(x,y,z) : x^2 + y^2 + z^2 \leq 9\}$, which is the ball of radius 3 centered at the origin, including the sphere $x^2 + y^2 + z^2 = 9$. To find the range, we examine the values ω taken by f. For ω in the range, the following equation has solutions x, y, z:

$$\sqrt{9 - x^2 - y^2 - z^2} = \omega \qquad (1)$$

Raising to the power of two and transfering sides gives

$$9 - x^2 - y^2 - z^2 = \omega^2$$
$$x^2 + y^2 + z^2 = 9 - \omega^2$$

The left-hand side is nonnegative, hence also $9 - \omega^2 \geq 0$ or $\omega^2 \leq 9$. Therefore, $-3 \leq \omega \leq 3$. By (1), $\omega \geq 0$, hence we must satisfy $0 \leq \omega \leq 3$. We obtain the following range:

$$\{\omega \in R : 0 \leq \omega \leq 3\}.$$

In Exercises 21–30, sketch the graph and describe the vertical and horizontal traces.

21. $f(x,y) = 12 - 3x - 4y$

SOLUTION The graph of $f(x,y) = 12 - 3x - 4y$ is shown in the figure:

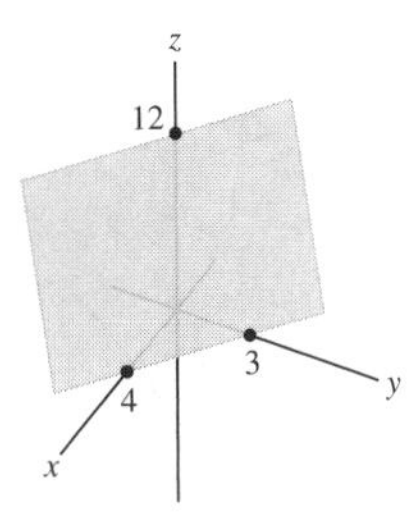

The horizontal trace at height c is the line $12 - 3x - 4y = c$ or $3x + 4y = 12 - c$ in the plane $z = c$.

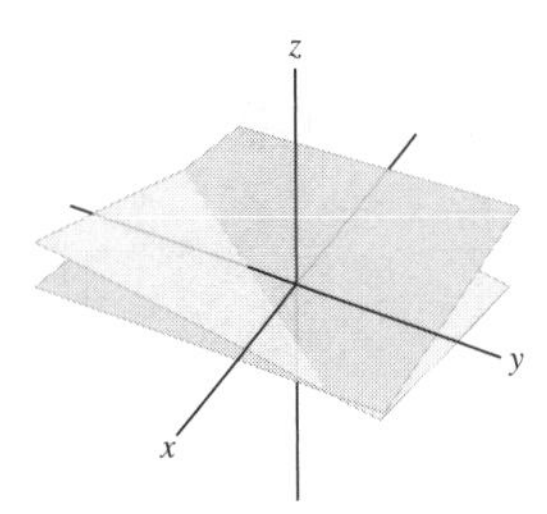

The vertical traces obtained by setting $x = a$ or $y = a$ are the lines $z = (12 - 3a) - 4y$ and $z = -3x + (12 - 4a)$ in the planes $x = a$ and $y = a$, respectively.

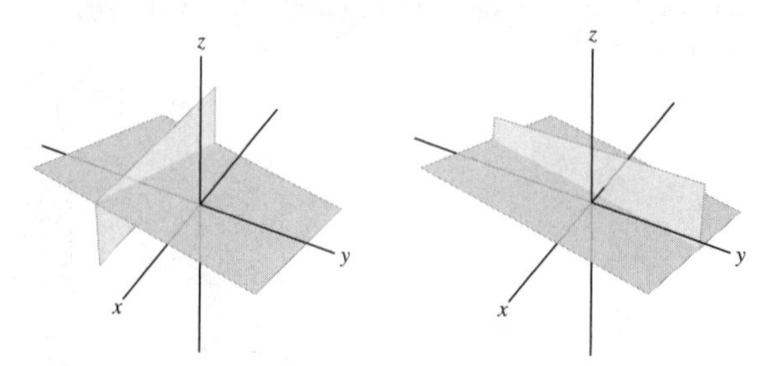

23. $f(x, y) = x^2 + 4y^2$

SOLUTION The graph of the function is shown in the figure:

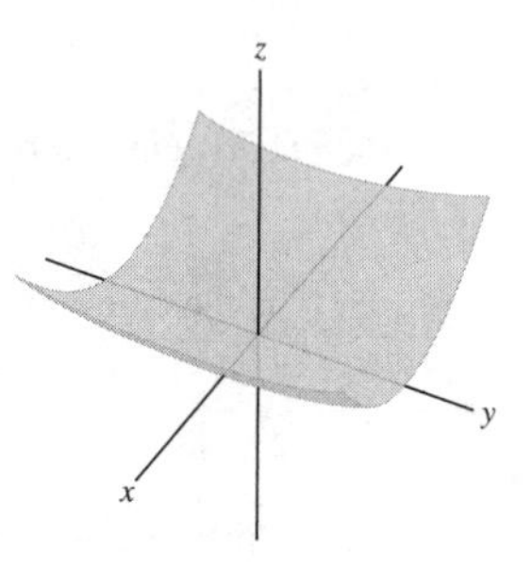

The horizontal trace at height c is the curve $x^2 + 4y^2 = c$, where $c \geq 0$ (if $c = 0$, it is the origin). The horizontal traces are ellipses for $c > 0$.

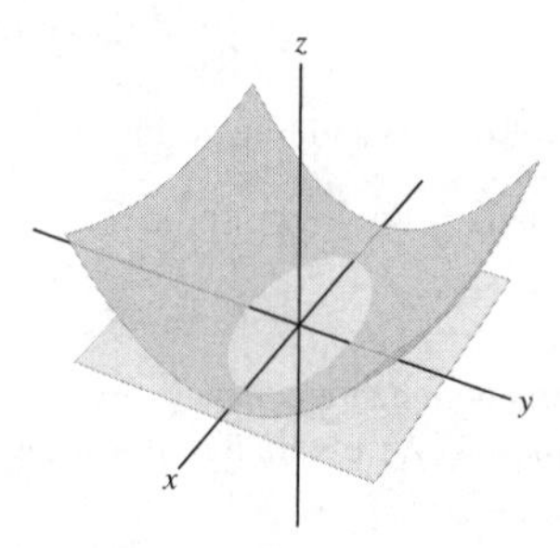

The vertical trace in the plane $x = a$ is the parabola $z = a^2 + 4y^2$, and the vertical trace in the plane $y = a$ is the parabola $z = x^2 + 4a^2$.

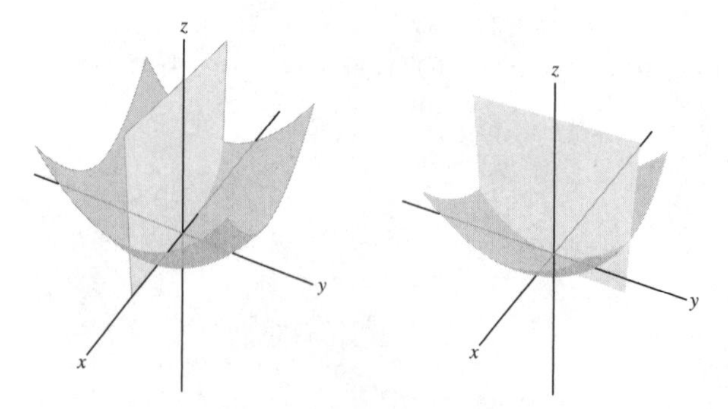

25. $f(x, y) = \dfrac{1}{x^2 + y^2 + 1}$

SOLUTION The graph of the function is shown in the figure:

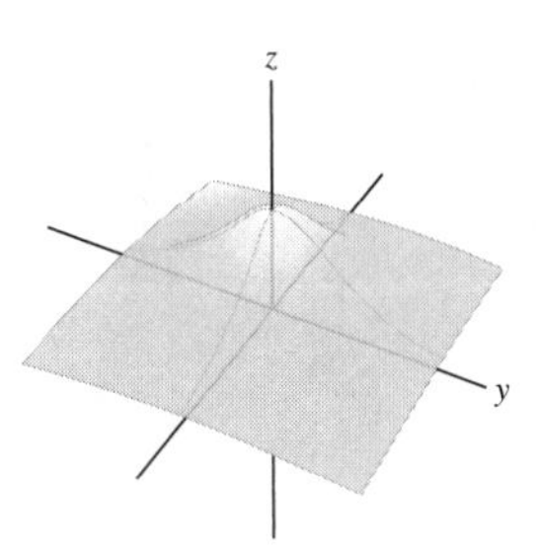

The horizontal trace at height c is the following curve in the plane $z = c$:

$$\frac{1}{x^2 + y^2 + 1} = c \quad \Rightarrow \quad x^2 + y^2 + 1 = \frac{1}{c} \quad \Rightarrow \quad x^2 + y^2 = \frac{1}{c} - 1$$

For $0 < c < 1$ it is a circle of radius $\sqrt{\frac{1}{c} - 1}$ centered at $(0, 0)$, and for $c = 1$ it is the origin.

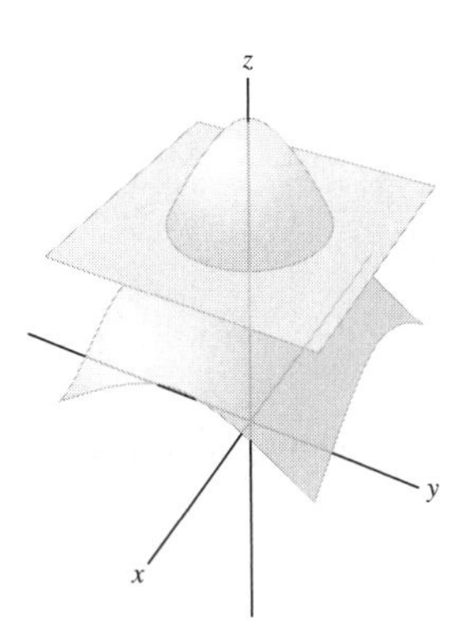

The vertical trace in the plane $x = a$ is the following curve in the plane $x = a$:

$$z = \frac{1}{a^2 + y^2 + 1} \quad \Rightarrow \quad z = \frac{1}{(1 + a^2) + y^2}$$

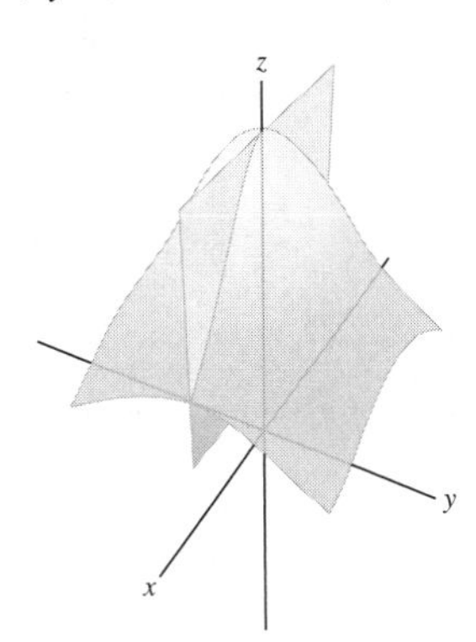

The vertical trace in the plane $y = a$ is the curve $z = \dfrac{1}{x^2 + a^2 + 1}$ in this plane.

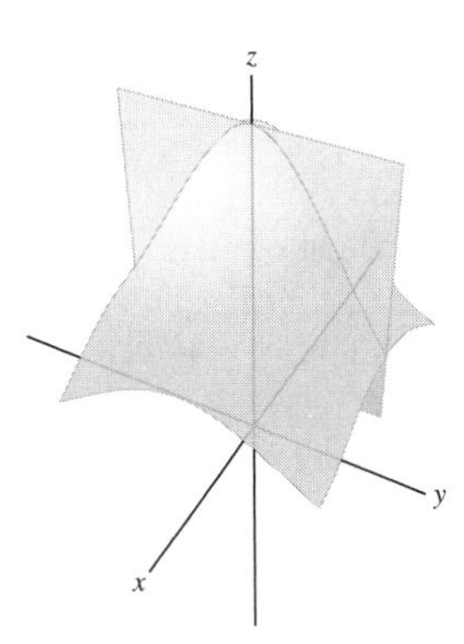

27. $f(x, y) = x + |y|$

SOLUTION The graph of f is shown in the figure:

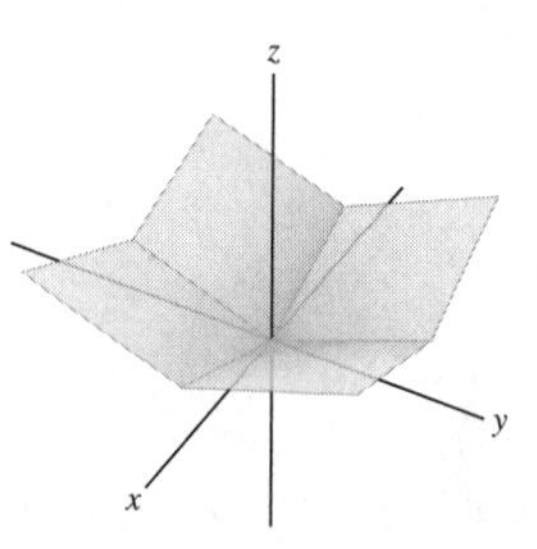

The horizontal trace at height c is $x + |y| = c$, or in other words, $x = c - |y|$.

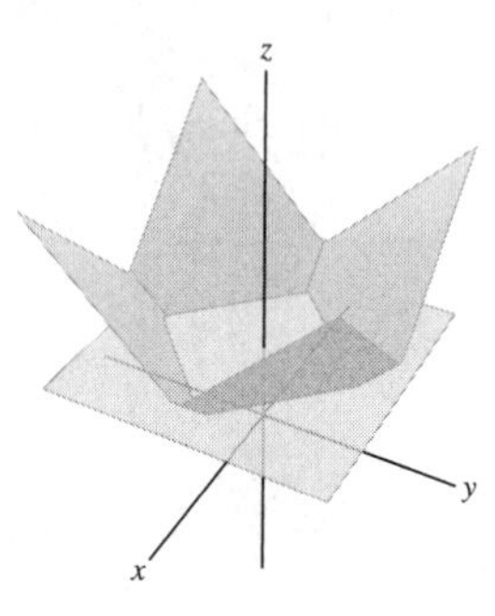

The vertical traces obtained by setting $x = a$ or $y = a$ are $z = a + |y|$ or $z = x + |a|$ in the planes $x = a$ and $y = a$, respectively.

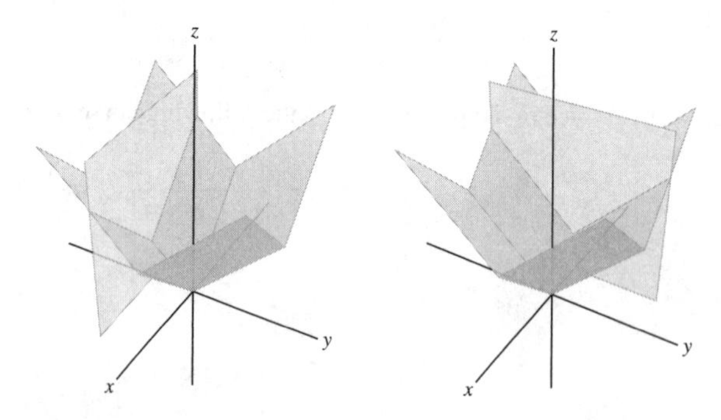

29. $f(x, y) = \sqrt{1 - x^2 - y^2}$

SOLUTION The graph of f is shown in the figure below:

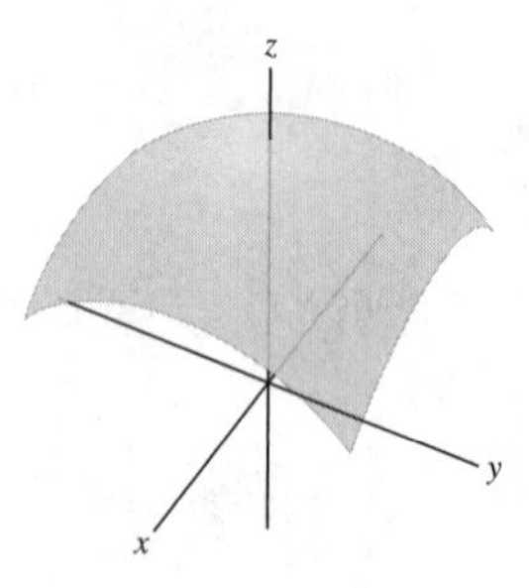

The horizontal trace at height $c \geq 0$ is

$$\sqrt{1 - x^2 - y^2} = c \quad \Rightarrow \quad 1 - x^2 - y^2 = c^2 \quad \Rightarrow \quad x^2 + y^2 = 1 - c^2$$

For $0 \leq c < 1$ it is the circle of radius $\sqrt{1 - c^2}$ centered at the origin, and for $c = 1$ it is the point at the origin.

The vertical traces in the planes $x = a$ and $y = a$ (for $|a| \le 1$) are $z = \sqrt{1 - a^2 - y^2}$ and $z = \sqrt{1 - a^2 - x^2}$ in these planes.

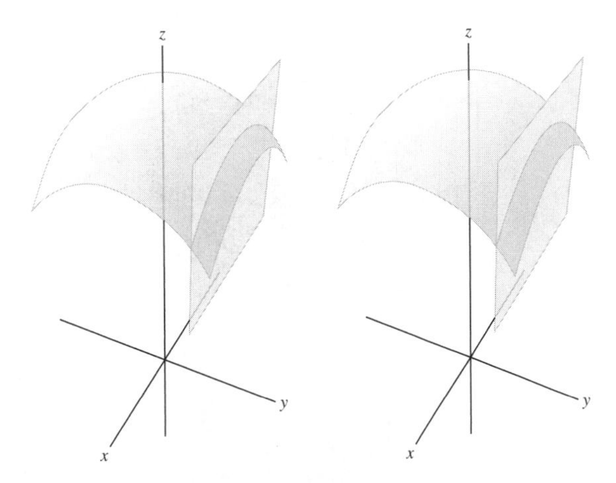

31. Sketch the contour maps of $f(x, y) = x + y$ with contour intervals $m = 1$ and 2.

SOLUTION The level curves are $x + y = c$ or $y = c - x$. Using contour interval $m = 1$, we plot $y = c - x$ for various values of c.

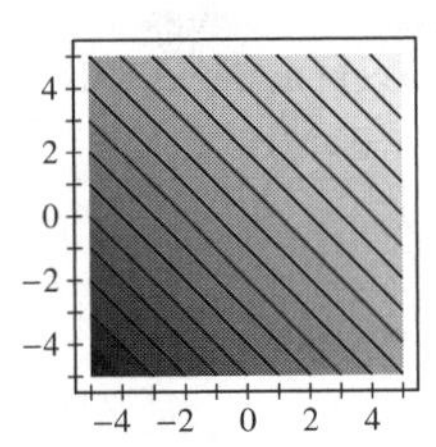

Using contour interval $m = 2$, we plot $y = c - x$ for various values of c.

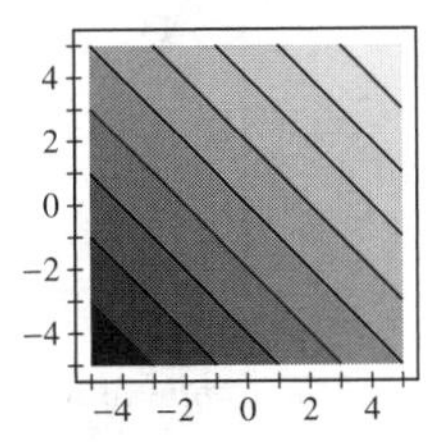

33. The function $f(x, t) = t^{-1/2}e^{-x^2/t}$, whose graph is shown in Figure 22, models the temperature along a metal bar after an intense burst of heat is applied at its center point.

(a) Sketch the graphs of the vertical traces at times $t = 0.5, 1, 1.5, 2$. What do these traces tell us about the way heat diffuses through the bar?

(b) Sketch the vertical trace at $x = 0$. Describe how temperature varies in time at the center point.

(c) Sketch the vertical trace $x = c$ for $c = \pm 0.2, \pm 0.4$. Describe how temperature varies in time at points near the center.

FIGURE 22 Graph of $f(x, t) = t^{-1/2}e^{-x^2/t}$ beginning shortly after $t = 0$.

SOLUTION

(a) The vertical traces at times $t = 0.5, 1, 1.5, 2$ are

$$z = \sqrt{2}e^{-2x^2} \text{ in the plane } t = 0.5$$
$$z = e^{-x^2} \text{ in the plane } t = 1$$
$$z = \frac{1}{\sqrt{3/2}}e^{-2x^2/3} \text{ in the plane } t = 1.5$$
$$z = \frac{1}{\sqrt{2}}e^{-x^2/2} \text{ in the plane } t = 2$$

These vertical traces are shown in the following figure:

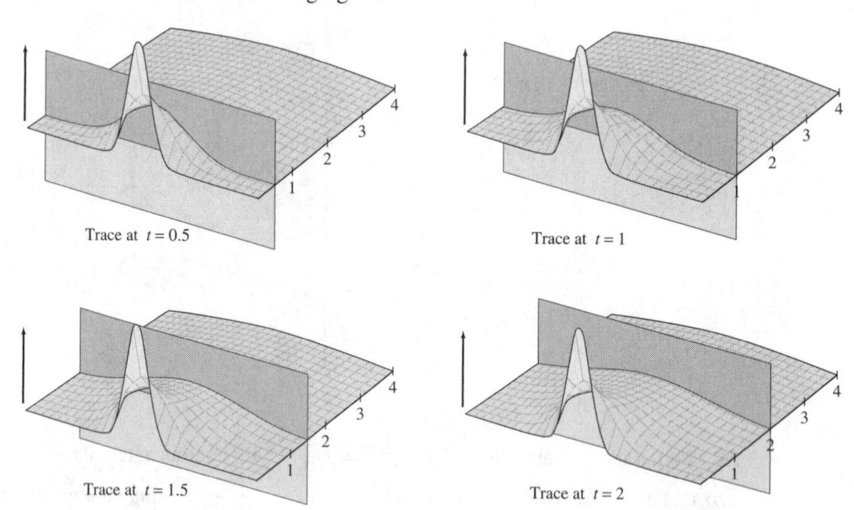

At each time the temperature decreases as we move away from the center point. Also, as t increases, the temperature at each point in the bar (except at the middle) increases and then decreases (as can be seen in Figure 22).

(b) The vertical trace at $x = 0$ is $z = t^{-1/2} = \frac{1}{\sqrt{t}}$. This trace is shown in the following figure:

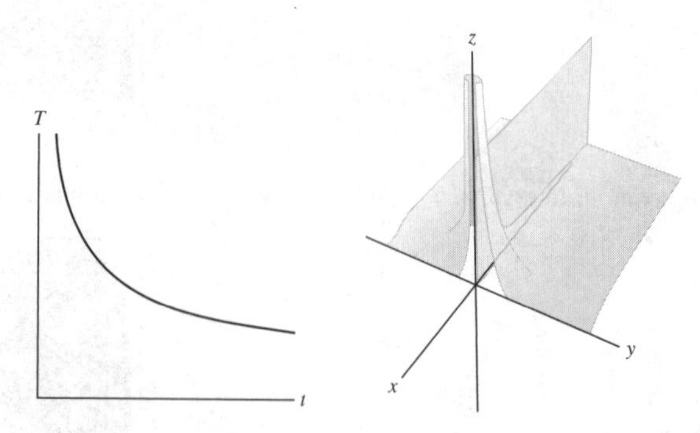

As suggested by the graph, the temperature is decreasing with time at the center point, according to $\frac{1}{\sqrt{t}}$.

(c) The vertical traces $x = c$ for the given values of c are:

$$z = \frac{1}{\sqrt{t}}e^{-\frac{0.04}{t}} \text{ in the planes } x = 0.2 \text{ and } x = -0.2$$
$$z = \frac{1}{\sqrt{t}}e^{-\frac{0.16}{t}} \text{ in the planes } x = 0.4 \text{ and } x = -0.4.$$

We see that for small values of t the temperature increases quickly and then slowly decreases as t increases.

In Exercises 34–43, draw a contour map of $f(x, y)$ with an appropriate contour interval, showing at least six level curves.

35. $f(x, y) = e^{y/x}$

SOLUTION The level curves are

$$e^{\frac{y}{x}} = c \quad \Rightarrow \quad \frac{y}{x} = \ln c \quad \Rightarrow \quad y = x \ln c$$

For $c > 1$, the level curves are the lines $y = x \ln c$, and for $c = 1$ the level curve is the x-axis. We use the contour interval $m = 3$ and show a contour map with the level curves $c = 1, 4, 7, 10, 13, 16$:

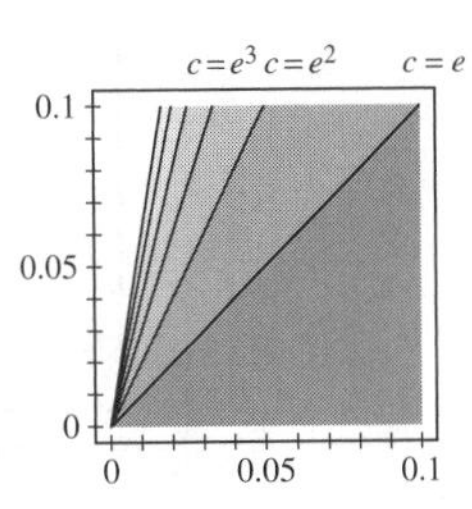

37. $f(x, y) = xy$

SOLUTION The level curves are $xy = c$ or $y = \frac{c}{x}$. These are hyperbolas in the xy-plane. We draw a contour map of the function using contour interval $m = 1$ and $c = 0, \pm1, \pm2, \pm3$:

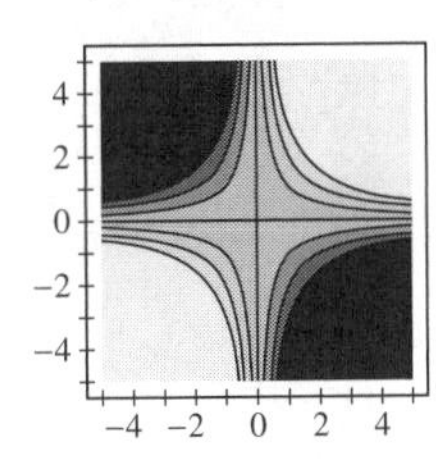

39. $f(x, y) = x + 2y - 1$

SOLUTION The level curves are the lines $x + 2y - 1 = c$ or $y = -\frac{x}{2} + \frac{c+1}{2}$. We draw a contour map using the contour interval $m = 4$ and $c = -9, -5, -1, 3, 7, 11$. The corresponding level curves are:

$$\underset{c=-9}{y = -\frac{x}{2} - 4}, \quad \underset{c=-5}{y = -\frac{x}{2} - 2}, \quad \underset{c=-1}{y = -\frac{x}{2}}, \quad \underset{c=3}{y = -\frac{x}{2} + 2},$$

$$\underset{c=7}{y = -\frac{x}{2} + 4}, \quad \underset{c=11}{y = -\frac{x}{2} + 6}$$

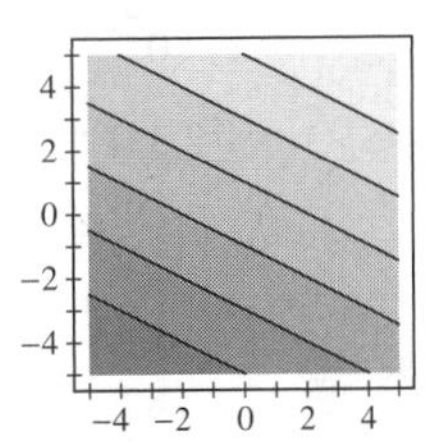

41. $f(x, y) = x^2$

SOLUTION The level curves are $x^2 = c$. For $c > 0$ these are the two vertical lines $x = \sqrt{c}$ and $x = -\sqrt{c}$ and for $c = 0$ it is the y-axis. We draw a contour map using contour interval $m = 4$ and $c = 0, 4, 8, 12, 16, 20$:

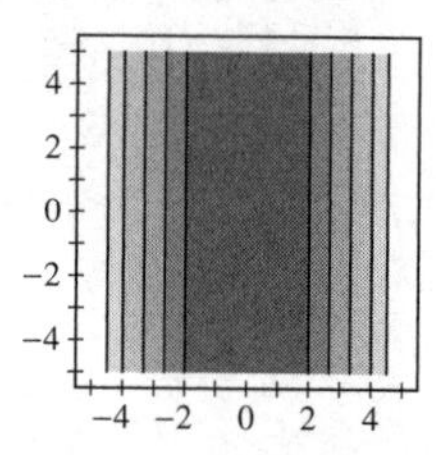

43. $f(x, y) = \dfrac{10}{1 + x^2 + y^2}$

SOLUTION The level curves are:

$$\frac{10}{1 + x^2 + y^2} = c \quad \Rightarrow \quad 1 + x^2 + y^2 = \frac{10}{c} \quad \Rightarrow \quad x^2 + y^2 = \frac{10}{c} - 1$$

For $0 < c < 10$ these are circles of radius $\sqrt{\frac{10}{c} - 1}$ centered at the origin. For $c = 10$ it is the point at the origin. We plot a contour map with contour interval $m = 1.5$ for $c = 10, 8.5, 7, 5.5, 4, 2.5$.

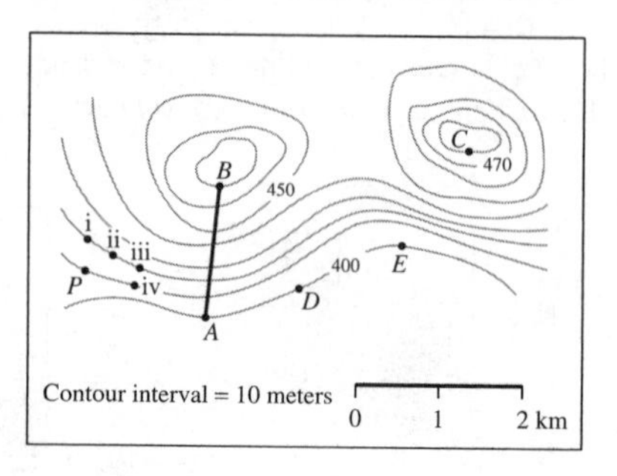

45. Which linear function has the contour map shown in Figure 24 (with level curve $c = 0$ as indicated), assuming that the contour interval is $m = 6$? What if $m = 3$?

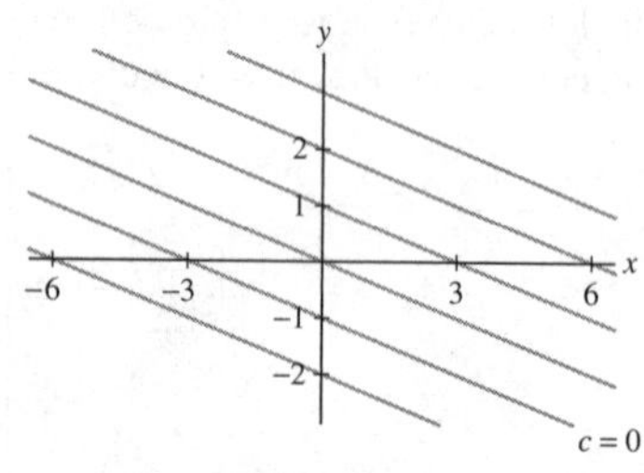

FIGURE 24

SOLUTION We denote the linear function by

$$f(x, y) = \alpha x + \beta y + \gamma \tag{1}$$

The level curves of f are

$$\alpha x + \beta y + \gamma = c \tag{2}$$

By the given information, the level curve for $c = 0$ is the line passing through the points $(0, -1)$ and $(-3, 0)$. We find the equation of this line:

$$y = \frac{0 - (-1)}{-3 - 0}(x + 3) \quad \Rightarrow \quad y = -\frac{1}{3}(x + 3) \quad \Rightarrow \quad x + 3y + 3 = 0$$

Setting $c = 0$ in (2) gives $\alpha x + \beta y + \gamma = 0$. Hence,

$$\frac{\alpha}{1} = \frac{\beta}{3} = \frac{\gamma}{3} \quad \Rightarrow \quad \beta = 3\alpha, \quad \gamma = 3\alpha$$

Substituting into (2) gives

$$\alpha x + 3\alpha y + 3\alpha = c$$

or

$$x + 3y + 3 = \frac{c}{\alpha} \tag{3}$$

Case 1: $m = 6$. In this case, the closest line above $c = 0$ corresponds to $c = 6$. This line is parallel to $x + 3y + 3 = 0$ and passes through the origin, hence its equation is $x + 3y = 0$. Setting $c = 6$ in (3) gives

$$x + 3y + 3 = \frac{6}{\alpha} \quad \text{or} \quad x + 3y = \frac{6}{\alpha} - 3$$

This line coincides with the line $x + 3y = 0$ only if $\frac{6}{\alpha} - 3 = 0$, that is, $\alpha = 2$. Substituting $\alpha = 2$, $\beta = 3\alpha = 6$, and $\gamma = 3\alpha = 6$ in (1) gives

$$f(x, y) = 2x + 6y + 6$$

Case 2: $m = 3$. In this case the closest line above $c = 0$ corresponds to $c = 3$. Substituting $c = 3$ in (3) gives the level curve $x + 3y + 3 = \frac{3}{\alpha}$ or $x + 3y = \frac{3}{\alpha} - 3$. This line passes through the origin if $\frac{3}{\alpha} - 3 = 0$, that is, $\alpha = 1$. Hence, $\alpha = 1$, $\beta = 3\alpha = 3$, $\gamma = 3\alpha = 3$. Substituting in (1) gives

$$f(x, y) = x + 3y + 3$$

In Exercises 46–49, $f(S, T)$ denotes the density of seawater at salinity level S (parts per thousand) and temperature T (degrees Celsius). Refer to the contour map of $f(S, T)$ in Figure 25.

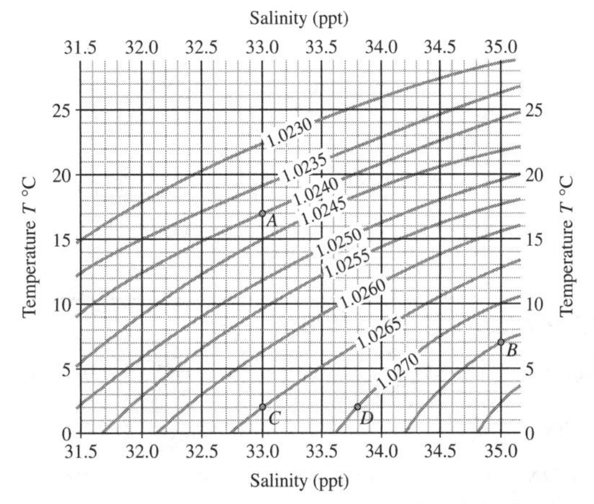

FIGURE 25 Contour map of seawater density $f(S, T)$ (kg/m^3).

47. Calculate the average ROC of density with respect to salinity from C to D.

SOLUTION For fixed temperature, the segment $\overline{CD}$ spans one level curve and the level curve of D is to the right of the level curve of C. Therefore, the change in density from C to D is $\Delta d = 0.0005$ kg/m^3. The salinity at D is greater than the salinity at C and $\Delta s = 0.8$ ppt. Therefore,

$$\text{Average ROC from } C \text{ to } D = \frac{\Delta d}{\Delta s} = \frac{0.0005}{0.8} = 0.000625 \text{ kg/m}^3 \cdot \text{ppt}.$$

49. Does water density appear to be more sensitive to a change in temperature at point A or point B?

SOLUTION The two adjacent level curves are closer to the level curve of A than the corresponding two adjacent level curves are to the level curve of B. This suggests that water density is more sensitive to a change in temperature at A than at B.

In Exercises 50–53, refer to Figure 26.

51. Estimate the average ROC from A and B and from A to C.

SOLUTION The change in elevation from A to B is 70 m. The scale shows that $\overline{AB}$ is approximately 1500 m. Therefore,

$$\text{Average ROC from } A \text{ to } B = \frac{70}{1500} \approx 0.047.$$

The change in elevation from A to C is obtained by multiplying the number of level curves between A and C, which is 8, by the contour interval 10 meters, giving $8 \cdot 10 = 80$ m. Using the scale, we approximate the distance $\overline{AC}$ by 3500 km. Therefore,

$$\text{Average ROC from } A \text{ to } C = \frac{80}{3500} \approx 0.023.$$

53. Sketch the paths of steepest ascent beginning at D and E.

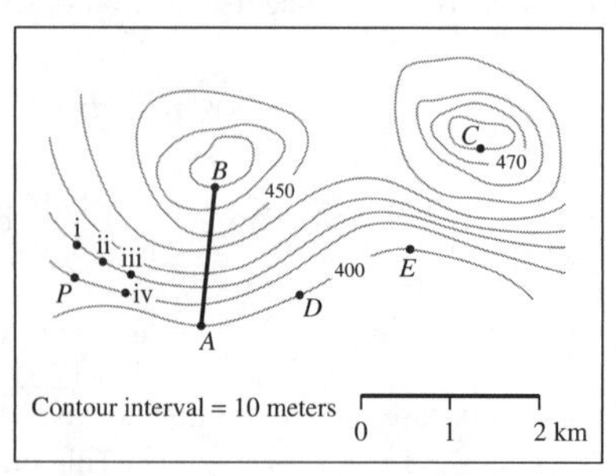

FIGURE 26 Contour map of mountain.

SOLUTION Starting at D or E, we draw a path that everywhere along the way points on the steepest direction, that is, moves as straight as possible from one level curve to the next. We obtain the following paths:

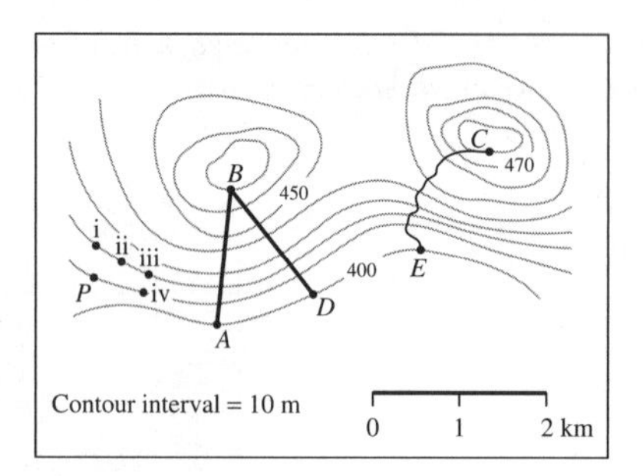

Further Insights and Challenges

55. Let $f(x, y) = \dfrac{x}{\sqrt{x^2 + y^2}}$ for $(x, y) \neq 0$. Write f as a function $f(r, \theta)$ in polar coordinates and use this to find the level curves of f.

SOLUTION In polar coordinates $x = r\cos\theta$ and $r = \sqrt{x^2 + y^2}$. Hence,

$$f(r, \theta) = \frac{r\cos\theta}{r} = \cos\theta.$$

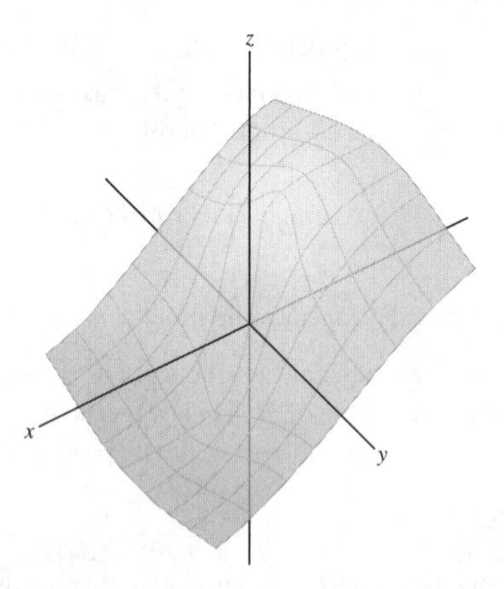

The level curves are the curves $\cos\theta = c$ in the $r\theta$-plane, for $|c| \leq 1$. For $-1 < c < 1$, $c \neq 0$, the level curves $\cos\theta = c$ are the two rays $\theta = \cos^{-1} c$ and $\theta = -\cos^{-1} c$.

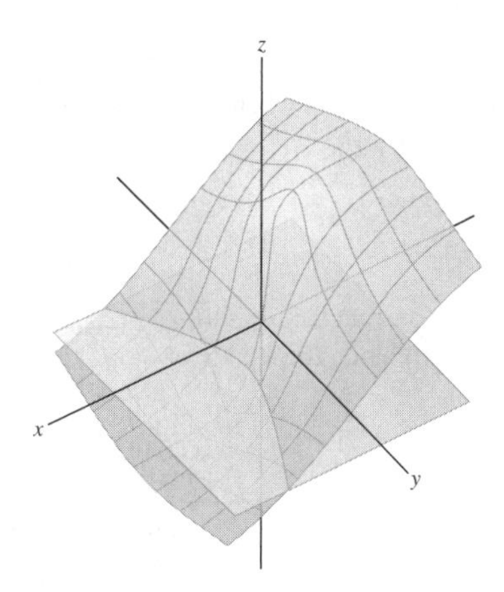

For $c = 0$, the level curve $\cos\theta = 0$ is the y-axis; for $c = 1$ the level curve $\cos\theta = 1$ is the nonnegative x-axis.

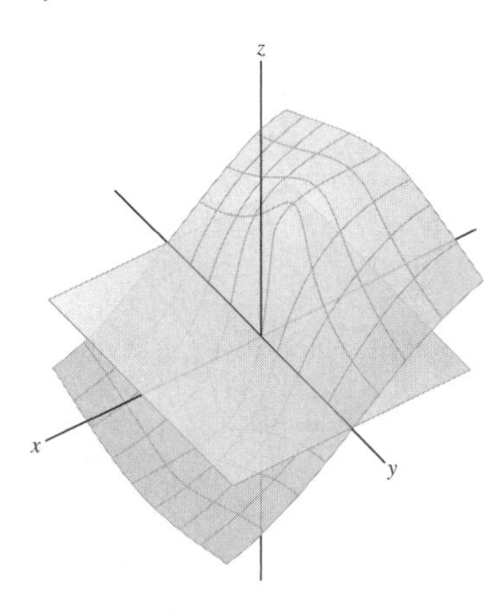

For $c = -1$, the level curve $\cos\theta = -1$ is the negative x-axis.

15.2 Limits and Continuity in Several Variables (ET Section 14.2)

Preliminary Questions

1. What is the difference between $D(P, r)$ and $D^*(P, r)$?

SOLUTION $D(P, r)$ is the open disk of radius r and center (a, b). It consists of all points distanced less than r from P, hence $D(P, r)$ includes the point P. $D^*(P, r)$ consists of all points in $D(P, r)$ other than P itself.

2. Suppose that $f(x, y)$ is continuous at $(2, 3)$ and that $f(2, y) = y^3$ for $y \neq 3$. What is the value $f(2, 3)$?

SOLUTION $f(x, y)$ is continuous at $(2, 3)$, hence the following holds:

$$f(2,3) = \lim_{(x,y)\to(2,3)} f(x,y)$$

Since the limit exists, we may compute it by approaching $(2, 3)$ along the vertical line $x = 2$. This gives

$$f(2,3) = \lim_{(x,y)\to(2,3)} f(x,y) = \lim_{y\to 3} f(2,y) = \lim_{y\to 3} y^3 = 3^3 = 27$$

We conclude that $f(2, 3) = 27$.

3. Suppose that $Q(x, y)$ is a function such that $\dfrac{1}{Q(x,y)}$ is continuous for all (x, y). Which of the following statements are true?

(a) $Q(x, y)$ is continuous at all (x, y).

(b) $Q(x, y)$ is continuous for $(x, y) \neq (0, 0)$.

(c) $Q(x, y) \neq 0$ for all (x, y).

SOLUTION All three statements are true. Let $f(x, y) = \frac{1}{Q(x,y)}$. Hence $Q(x, y) = \frac{1}{f(x,y)}$.

(a) Since f is continuous, Q is continuous whenever $f(x, y) \neq 0$. But by the definition of f it is never zero, therefore Q is continuous at all (x, y).

(b) Q is continuous everywhere including at $(0, 0)$.

(c) Since $f(x, y) = \frac{1}{Q(x,y)}$ is continuous, the denominator is never zero, that is, $Q(x, y) \neq 0$ for all (x, y). Moreover, there are no points where $Q(x, y) = 0$. (The equality $Q(x, y) = (0, 0)$ is meaningless since the range of Q consists of real numbers.)

4. Suppose that $f(x, 0) = 3$ for all $x \neq 0$ and $f(0, y) = 5$ for all $y \neq 0$. What can you conclude about $\lim_{(x,y)\to(0,0)} f(x, y)$?

SOLUTION We show that the limit $\lim_{(x,y)\to(0,0)} f(x, y)$ does not exist. Indeed, if the limit exists, it may be computed by approaching $(0, 0)$ along the x-axis or along the y-axis. We compute these two limits:

$$\lim_{\substack{(x,y)\to(0,0)\\ \text{along } y=0}} f(x, y) = \lim_{x\to 0} f(x, 0) = \lim_{x\to 0} 3 = 3$$

$$\lim_{\substack{(x,y)\to(0,0)\\ \text{along } x=0}} f(x, y) = \lim_{y\to 0} f(0, y) = \lim_{y\to 0} 5 = 5$$

Since the limits are different, $f(x, y)$ does not approach one limit as $(x, y) \to (0, 0)$, hence the limit $\lim_{(x,y)\to(0,0)} f(x, y)$ does not exist.

Exercises

In Exercises 1–10, use continuity to evaluate the limit.

1. $\lim_{(x,y)\to(1,2)} (x^2 + y)$

SOLUTION Since the function $x^2 + y$ is continuous, we evaluate the limit by substitution:

$$\lim_{(x,y)\to(1,2)} (x^2 + y) = 1^2 + 2 = 3$$

3. $\lim_{(x,y)\to(3,-1)} (3x^2y - 2xy^3)$

SOLUTION The function $3x^2y - 2xy^3$ is continuous everywhere because it is a polynomial. We evaluate the limit by substitution:

$$\lim_{(x,y)\to(3,-1)} (3x^2y - 2xy^3) = 3 \cdot 3^2 \cdot (-1) - 2 \cdot 3 \cdot (-1)^3 = -27 + 6 = -21$$

5. $\lim_{(x,y)\to(1,2)} \frac{x^2 + y}{x + y^2}$

SOLUTION The rational function $\frac{x^2+y}{x+y^2}$ is continuous at $(1, 2)$ since the denominator is not zero at this point. We compute the limit using substitution:

$$\lim_{(x,y)\to(1,2)} \frac{x^2 + y}{x + y^2} = \frac{1^2 + 2}{1 + 2^2} = \frac{3}{5}$$

7. $\lim_{(x,y)\to(4,4)} \arctan \frac{y}{x}$

SOLUTION Since the function $\arctan \frac{y}{x}$ is continuous at the point $(4, 4)$, we compute the limit by substitution:

$$\lim_{(x,y)\to(4,4)} \arctan \frac{y}{x} = \arctan \frac{4}{4} = \arctan 1 = \frac{\pi}{4}$$

9. $\lim_{(x,y)\to(1,1)} \frac{e^{x^2} - e^{-y^2}}{x + y}$

SOLUTION The function is the quotient of two continuous functions, and the denominator is not zero at the point $(1, 1)$. Therefore, the function is continuous at this point, and we may compute the limit by substitution:

$$\lim_{(x,y)\to(1,1)} \frac{e^{x^2} - e^{-y^2}}{x + y} = \frac{e^{1^2} - e^{-1^2}}{1 + 1} = \frac{e - \frac{1}{e}}{2} = \frac{1}{2}(e - e^{-1})$$

In Exercises 11–18, evaluate the limit or determine that it does not exist.

11. $\displaystyle\lim_{(x,y)\to(0,1)} \frac{x}{y}$

SOLUTION The function $\frac{x}{y}$ is continuous at the point (0, 1), hence we compute the limit by substitution:

$$\lim_{(x,y)\to(0,1)} \frac{x}{y} = \frac{0}{1} = 0$$

13. $\displaystyle\lim_{(x,y)\to(1,1)} e^{xy}\ln(1+xy)$

SOLUTION The function is continuous at (1, 1), hence we compute the limit using substitution:

$$\lim_{(x,y)\to(1,1)} e^{xy}\ln(1+xy) = e^{1\cdot 1}\ln(1+1\cdot 1) = e\ln 2$$

15. $\displaystyle\lim_{(x,y)\to(-1,-2)} x^2|y|^3$

SOLUTION The function $x^2|y|^3$ is continuous everywhere. We compute the limit using substitution:

$$\lim_{(x,y)\to(-1,-2)} x^2|y|^3 = (-1)^2|-2|^3 = 8$$

17. $\displaystyle\lim_{(x,y)\to(4,2)} \frac{y-2}{\sqrt{x^2-4}}$

SOLUTION The function is continuous at the point (4, 2), since it is the quotient of two continuous functions and the denominator is not zero at (4, 2). We compute the limit by substitution:

$$\lim_{(x,y)\to(4,2)} \frac{y-2}{\sqrt{x^2-4}} = \frac{2-2}{\sqrt{4^2-4}} = \frac{0}{\sqrt{12}} = 0$$

In Exercises 19–22, assume that

$$\lim_{(x,y)\to(2,5)} f(x,y) = 3, \qquad \lim_{(x,y)\to(2,5)} g(x,y) = 7$$

19. $\displaystyle\lim_{(x,y)\to(2,5)} \big(f(x,y)+4g(x,y)\big)$

SOLUTION Using the Sum Law and the Constant Multiples Law we get

$$\lim_{(x,y)\to(2,5)} (f(x,y)+4g(x,y)) = \lim_{(x,y)\to(2,5)} f(x,y) + 4\lim_{(x,y)\to(2,5)} g(x,y) = 3+4\cdot 7 = 31$$

21. $\displaystyle\lim_{(x,y)\to(2,5)} e^{f(x,y)}$

SOLUTION $e^{f(x,y)}$ is the composition of $G(u)=e^u$ and $u=f(x,y)$. Since e^u is continuous, we may evaluate the limit as follows:

$$\lim_{(x,y)\to(2,5)} e^{f(x,y)} = e^{\lim_{(x,y)\to(2,5)} f(x,y)} = e^3$$

In Exercises 23–32, evaluate the limit or determine that the limit does not exist. You may evaluate the limit of a product function as a product of limits as in Example 3.

23. $\displaystyle\lim_{(x,y)\to(0,0)} \frac{(\sin x)(\sin y)}{xy}$

SOLUTION We evaluate the limit as a product of limits:

$$\lim_{(x,y)\to(0,0)} \frac{(\sin x)(\sin y)}{xy} = \left(\lim_{x\to 0}\frac{\sin x}{x}\right)\left(\lim_{y\to 0}\frac{\sin y}{y}\right) = 1\cdot 1 = 1.$$

25. $\displaystyle\lim_{(z,w)\to(-1,2)} (z^2w-9z)$

SOLUTION The function is continuous everywhere since it is a polynomial. Therefore we use substitution to evaluate the limit:

$$\lim_{(z,\omega)\to(-1,2)} (z^2\omega - 9z) = (-1)^2 \cdot 2 - 9 \cdot (-1) = 11.$$

27. $\displaystyle\lim_{(h,k)\to(2,0)} h^4 \frac{(2+k)^2-4}{k}$

SOLUTION We write the limit as a product of limits:

$$\lim_{(h,k)\to(2,0)} h^4 \frac{(2+k)^2-4}{k} = \lim_{h\to 2} h^4 \lim_{k\to 0} \frac{(2+k)^2-4}{k} = 4 \cdot \lim_{k\to 0} \frac{(2+k)^2-4}{k} \tag{1}$$

We show that this limit exists. Notice that the limit is the derivative of $f(x) = x^2$ at $x = 2$. That is,

$$\lim_{k\to 0} \frac{(2+k)^2-4}{k} = \lim_{k\to 0} \frac{f(2+k)-f(2)}{k} = f'(2) = 2x\Big|_{x=2} = 4$$

Combining with (1), we conclude that the given limit is $4 \cdot 4 = 16$.

29. $\displaystyle\lim_{(x,y)\to(0,0)} e^{1/x} \tan^{-1}\frac{1}{y}$

SOLUTION We show that the limit along the line $y = x$ does not exist. We have,

$$\lim_{\substack{(x,y)\to(0,0)\\ \text{along } y=x}} e^{1/x} \tan^{-1}\frac{1}{y} = \lim_{x\to 0} e^{1/x} \tan^{-1}\frac{1}{x} \tag{1}$$

We consider the two one-sided limits. As $x \to 0+$, $e^{1/x}$ is increasing without bound and $\lim_{x\to 0+} \tan^{-1}\frac{1}{x} = \frac{\pi}{2}$. Therefore,

$$\lim_{x\to 0+} e^{1/x} \tan^{-1}\frac{1}{x} = +\infty$$

As $x \to 0-$, we have

$$\lim_{x\to 0-} e^{1/x} \tan^{-1}\frac{1}{x} = \left(\lim_{x\to 0-} e^{1/x}\right)\left(\lim_{x\to 0-} \tan^{-1}\frac{1}{x}\right) = 0 \cdot \left(\frac{-\pi}{2}\right) = 0$$

Since the one-sided limits are not equal, the limit in (1) does not exist. Therefore, the given limit does exist.

31. $\displaystyle\lim_{(x,y)\to(0,0)} x \ln y$

SOLUTION We first compute the limit as (x, y) approaches the origin along the positive y-axis ($x = 0$, $y > 0$):

$$\lim_{\substack{(x,y)\to(0,0)\\ \text{along the positive}\\ y\text{-axis}}} x \ln y = \lim_{y\to 0+} 0 \cdot \ln y = \lim_{y\to 0} 0 = 0$$

We now compute the limit as (x, y) approaches the origin along the curve $y = e^{-1/x}$, $x > 0$ (notice that $y \to 0+$ as $x \to 0+$):

$$\lim_{\substack{(x,y)\to(0,0)\\ \text{along } y=e^{-1/x},\ x>0}} x \ln y = \lim_{x\to 0+} x \ln e^{-1/x} = \lim_{x\to 0+} x \cdot \left(-\frac{1}{x}\right) = \lim_{x\to 0+} (-1) = -1$$

Since the limits along the two paths are not equal, the limit itself does not exist.

33. Let $f(x, y) = \dfrac{x^3 + y^3}{x^2 + y^2}$.

(a) Show that

$$|x^3| \le |x|(x^2 + y^2), \quad |y^3| \le |y|(x^2 + y^2)$$

(b) Show that $|f(x, y)| \le |x| + |y|$.

(c) Use (b) and the formal definition of the limit to prove that $\displaystyle\lim_{(x,y)\to(0,0)} f(x, y) = 0$.

(d) Verify the conclusion of (c) again using polar coordinates as in Example 6.

SOLUTION

(a) Since $|x|y^2 \ge 0$, we have

$$|x^3| \le |x^3| + |x|y^2 = |x|^3 + |x|y^2 = |x|(x^2 + y^2)$$

Similarly, since $|y|x^2 \ge 0$, we have

$$|y^3| \le |y^3| + |y|x^2 = |y|^3 + |y|x^2 = |y|(x^2 + y^2)$$

(b) We use the triangle inequality to write

$$|f(x, y)| = \frac{|x^3 + y^3|}{x^2 + y^2} \le \frac{|x^3| + |y^3|}{x^2 + y^2}$$

We continue using the inequality in part (a):

$$|f(x, y)| \le \frac{|x|(x^2 + y^2) + |y|(x^2 + y^2)}{x^2 + y^2} = \frac{(|x| + |y|)(x^2 + y^2)}{x^2 + y^2} = |x| + |y|$$

That is,

$$|f(x, y)| \le |x| + |y|$$

(c) In part (b) we showed that

$$|f(x, y)| \le |x| + |y| \tag{1}$$

Let $\epsilon > 0$. Then if $|x| < \frac{\epsilon}{2}$ and $|y| < \frac{\epsilon}{2}$, we have by (1)

$$|f(x, y) - 0| \le |x| + |y| < \frac{\epsilon}{2} + \frac{\epsilon}{2} = \epsilon \tag{2}$$

Notice that if $x^2 + y^2 < \frac{\epsilon^2}{4}$, then $x^2 < \frac{\epsilon^2}{4}$ and $y^2 < \frac{\epsilon^2}{4}$. Hence $|x| < \frac{\epsilon}{2}$ and $|y| < \frac{\epsilon}{2}$, so (1) holds. In other words, using $D^\star\left(\frac{\epsilon}{2}\right)$ to represent the punctured disc of radius $\epsilon/2$ centered at the origin, we have

$$(x, y) \in D^\star\left(\frac{\epsilon}{2}\right) \Rightarrow |x| < \frac{\epsilon}{2}$$

and

$$|y| < \frac{\epsilon}{2} \Rightarrow |f(x, y) - 0| < \epsilon$$

We conclude by the limit definition that

$$\lim_{(x,y)\to(0,0)} f(x, y) = 0$$

(d) We verify the limit using polar coordinates:

$$x = r\cos\theta, \quad y = r\sin\theta$$

Then, $(x, y) \to (0, 0)$ if and only if $x^2 + y^2 \to 0$, that is, if and only if $r \to 0+$. We have

$$x^2 + y^2 = r^2$$
$$x^3 + y^3 = r^3\cos^3\theta + r^3\sin^3\theta = r^3(\cos^3\theta + \sin^3\theta)$$

Hence,

$$f(x, y) = \frac{x^3 + y^3}{x^2 + y^2} = \frac{r^3(\cos^3\theta + \sin^3\theta)}{r^2} = r(\cos^3\theta + \sin^3\theta) \tag{3}$$

Since $|\cos^3\theta + \sin^3\theta| \le |\cos^3\theta| + |\sin^3\theta| \le 2$, we have

$$0 \le |r(\cos^3\theta + \sin^3\theta)| \le 2r \tag{4}$$

As r approaches zero, $2r$ approaches zero, hence (4) and the Squeeze Theorem imply that

$$\lim_{r\to 0+} r(\cos^3\theta + \sin^3\theta) = 0 \tag{5}$$

By (3) and (5) we conclude that

$$\lim_{(x,y)\to(0,0)} f(x, y) = \lim_{r\to 0+} r(\cos^3\theta + \sin^3\theta) = 0$$

35. Show that $\lim_{(x,y)\to(0,0)} \frac{|x|}{|x|+|y|}$ does not exist. *Hint:* Consider the limits along the lines $y = mx$.

SOLUTION We compute the limit as (x, y) approaches the origin along the line $y = mx$, for a fixed value of m. Substituting $y = mx$ in the function $f(x, y) = \frac{|x|}{|x|+|y|}$, we get for $x \neq 0$:

$$f(x, mx) = \frac{|x|}{|x| + m|x|} = \frac{|x|}{|x|(1+m)} = \frac{1}{1+m}$$

As (x, y) approaches $(0, 0)$, $(x, y) \neq (0, 0)$. Therefore $x \neq 0$ on the line $y = mx$. Thus,

$$\lim_{\substack{(x,y)\to(0,0)\\ \text{along } y=mx}} f(x, y) = \lim_{x\to 0} \frac{1}{1+m} = \frac{1}{1+m}$$

We see that the limits along the lines $y = mx$ are different, hence $f(x, y)$ does not approach one limit as $(x, y) \to (0, 0)$. We conclude that the given limit does not exist.

37. Let $a, b \geq 0$. Show that $\lim_{(x,y)\to(0,0)} \frac{x^a y^b}{x^2 + y^2} = 0$ if $a + b > 2$ and the limit does not exist if $a + b \leq 2$.

SOLUTION We first show that the limit is zero if $a + b > 2$. We compute the limit using the polar coordinates $x = r\cos\theta$, $y = r\sin\theta$. Then $(x, y) \to (0, 0)$ if and only if $x^2 + y^2 \to 0$, that is, if and only if $r \to 0+$. Therefore,

$$\lim_{(x,y)\to(0,0)} \frac{x^a y^b}{x^2 + y^2} = \lim_{r\to 0+} \frac{(r\cos\theta)^a (r\sin\theta)^b}{r^2} = \lim_{r\to 0+} \frac{r^{a+b}\cos^a\theta \sin^b\theta}{r^2}$$

$$= \lim_{r\to 0+} (r^{a+b-2}\cos^a\theta\sin^b\theta) \qquad (1)$$

The following inequality holds:

$$0 \leq |r^{a+b-2}\cos^a\theta\sin^b\theta| \leq r^{a+b-2} \qquad (2)$$

Since $a + b > 2$, $\lim_{r\to 0+} r^{a+b-2} = 0$, therefore (2) and the Squeeze Theorem imply that

$$\lim_{r\to 0} (r^{a+b-2}\cos^a\theta\sin^b\theta) = 0 \qquad (3)$$

We combine (1) and (3) to conclude that if $a + b > 2$, then

$$\lim_{(x,y)\to(0,0)} \frac{x^a y^b}{x^2 + y^2} = 0$$

We now consider the case $a + b < 2$. We examine the limit as (x, y) approaches the origin along the line $y = x$. Along this line, $\theta = \frac{\pi}{4}$, therefore (1) gives

$$\lim_{(x,y)\to(0,0)} \frac{x^a y^b}{x^2 + y^2} = \lim_{r\to 0+} \left(r^{a+b-2}\cos^a\frac{\pi}{4}\sin^b\frac{\pi}{4}\right) = \lim_{r\to 0+} \left(r^{a+b-2}\cdot\left(\frac{1}{\sqrt{2}}\right)^a\cdot\left(\frac{1}{\sqrt{2}}\right)^b\right) = \lim_{r\to 0+} \frac{r^{a+b-2}}{(\sqrt{2})^{a+b}}$$

Since $a + b < 2$, we have $a + b - 2 < 0$ therefore $\lim_{r\to 0+} r^{a+b-2}$ does not exist. It follows that if $a + b < 2$, the given limit does not exist. Finally we examine the case $a + b = 2$. By (1) we get

$$\lim_{(x,y)\to(0,0)} \frac{x^a y^b}{x^2 + y^2} = \lim_{r\to 0+} (r^0\cos^a\theta\sin^b\theta) = \lim_{r\to 0+} \cos^a\theta\sin^b\theta = \cos^a\theta\sin^b\theta$$

We see that the function does not approach one limit. For example, approaching the origin along the lines $y = x$ (i.e., $\theta = \frac{\pi}{4}$) and $y = 0$ (i.e., $\theta = 0$) gives two different limits $\cos^a\frac{\pi}{4}\sin^b\frac{\pi}{4} = \left(\frac{\sqrt{2}}{2}\right)^{a+b}$ and $\cos^a 0\sin^b 0 = 0$. We conclude that if $a + b = 2$, the limit does not exist.

39. Figure 8 shows the contour maps of two functions with contour interval $m = 2$. Explain why the limit $\lim_{(x,y)\to P} f(x, y)$ does not exist. Does $\lim_{(x,y)\to P} g(x, y)$ appear to exist in (B)? If so, what is its limit?

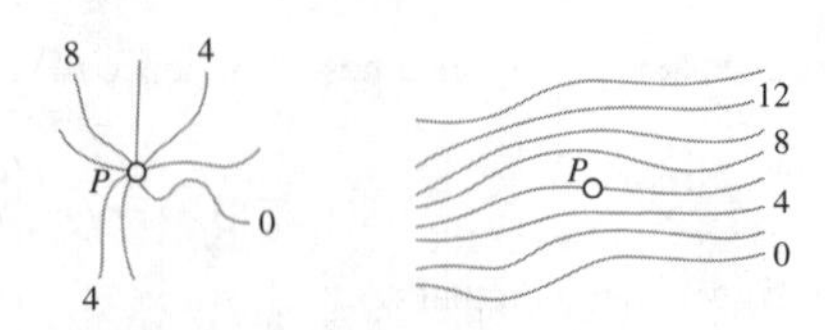

(A) Contour map of $f(x, y)$ (B) Contour map of $g(x, y)$

FIGURE 8

SOLUTION As (x, y) approaches arbitrarily close to P, the function $f(x, y)$ takes the values 0, 4, and 8. Therefore $f(x, y)$ does not approach one limit as $(x, y) \to P$. Rather, the limit depends on the contour along which (x, y) is approaching P. This implies that the limit $\lim_{(x,y)\to P} f(x, y)$ does not exist. In (B) the limit $\lim_{(x,y)\to P} g(x, y)$ appears to exist. If it exists, it must be 6, which is the level curve of P.

Further Insights and Challenges

41. CAS The function $f(x, y) = \dfrac{\sin(xy)}{xy}$ is defined for $xy \neq 0$.

(a) Is it possible to extend the domain of $f(x, y)$ to all of $\mathbf{R}^2$ so that the result is a continuous function?

(b) Use a computer algebra system to plot $f(x, y)$. Does the result support your conclusion in (a)?

SOLUTION

(a) We define $f(x, y)$ on the x- and y-axes by $f(x, y) = 1$ if $xy = 0$. We now show that f is continuous. f is continuous at the points where $xy \neq 0$. We next show continuity at $(x_0, 0)$ (including $x_0 = 0$). For the points $(0, y_0)$, the proof is similar and hence will be omitted. To prove continuity at $P = (x_0, 0)$ we have to show that

$$\lim_{(x,y)\to P} f(x, y) = \lim_{(x,y)\to P} \frac{\sin xy}{xy} = 1 \tag{1}$$

Let us denote $u = xy$. As $(x, y) \to (x_0, 0)$, $u = x \cdot y \to x_0 \cdot 0 = 0$. Thus,

$$\lim_{(x,y)\to P} f(x, y) = \lim_{(x,y)\to(x_0,0)} \frac{\sin xy}{xy} = \lim_{u\to 0} \frac{\sin u}{u} = 1 = f(x_0, 0).$$

(b) The following figure shows the graph of $f(x, y) = \frac{\sin xy}{xy}$:

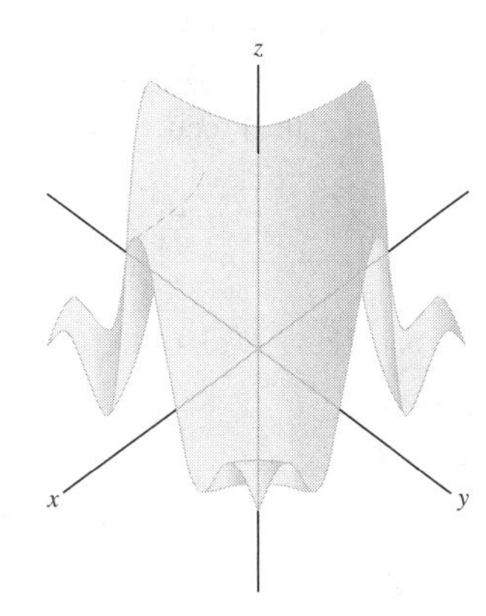

The graph shows that, near the axes, the values of $f(x, y)$ are approaching 1, as shown in part (a).

43. The function $f(x, y) = \dfrac{x^2 y}{x^4 + y^2}$ provides an interesting example where the limit as $(x, y) \to (0, 0)$ does not exist, even though the limit along every line $y = mx$ exists and is zero (Figure 9).

(a) Show that the limit along any line $y = mx$ exists and is equal to 0.

(b) Calculate $f(x, y)$ at the points $(10^{-1}, 10^{-2})$, $(10^{-5}, 10^{-10})$, $(10^{-20}, 10^{-40})$. Do not use a calculator.

(c) Show that $\lim_{(x,y)\to(0,0)} f(x, y)$ does not exist. *Hint:* Compute the limit along the parabola $y = x^2$.

SOLUTION

(a) Substituting $y = mx$ in $f(x, y) = \frac{x^2 y}{x^4+y^2}$, we get

$$f(x, mx) = \frac{x^2 \cdot mx}{x^4 + (mx)^2} = \frac{mx^3}{x^2(x^2 + m^2)} = \frac{mx}{x^2 + m^2}$$

We compute the limit as $x \to 0$ by substitution:

$$\lim_{x\to 0} f(x, mx) = \lim_{x\to 0} \frac{mx}{x^2 + m^2} = \frac{m \cdot 0}{0^2 + m^2} = 0$$

(b) We compute $f(x, y)$ at the given points:

$$f(10^{-1}, 10^{-2}) = \frac{10^{-2} \cdot 10^{-2}}{10^{-4} + 10^{-4}} = \frac{10^{-4}}{2 \cdot 10^{-4}} = \frac{1}{2}$$

$$f(10^{-5}, 10^{-10}) = \frac{10^{-10} \cdot 10^{-10}}{10^{-20} + 10^{-20}} = \frac{10^{-20}}{2 \cdot 10^{-20}} = \frac{1}{2}$$

$$f(10^{-20}, 10^{-40}) = \frac{10^{-40} \cdot 10^{-40}}{10^{-80} + 10^{-80}} = \frac{10^{-80}}{2 \cdot 10^{-80}} = \frac{1}{2}$$

(c) We compute the limit as (x, y) approaches the origin along the parabola $y = x^2$ (by part (b), the limit appears to be $\frac{1}{2}$). We substitute $y = x^2$ in the function and compute the limit as $x \to 0$. This gives

$$\lim_{\substack{(x,y)\to 0 \\ \text{along } y=x^2}} f(x, y) = \lim_{x\to 0} f(x, x^2) = \lim_{x\to 0} \frac{x^2 \cdot x^2}{x^4 + (x^2)^2} = \lim_{x\to 0} \frac{x^4}{2x^4} = \lim_{x\to 0} \frac{1}{2} = \frac{1}{2}$$

However, in part (a), we showed that the limit along the lines $y = mx$ is zero. Therefore $f(x, y)$ does not approach one limit as $(x, y) \to (0, 0)$, so the limit $\lim_{(x,y)\to(0,0)} f(x, y)$ does not exist.

15.3 Partial Derivatives (ET Section 14.3)

Preliminary Questions

1. Patricia derived the following *incorrect* formula by misapplying the Product Rule:

$$\frac{\partial}{\partial x}(x^2 y^2) = x^2(2y) + y^2(2x)$$

What was her mistake and what is the correct calculation?

SOLUTION To compute the partial derivative with respect to x, we treat y as a constant. Therefore the Constant Multiple Rule must be used rather than the Product Rule. The correct calculation is:

$$\frac{\partial}{\partial x}(x^2 y^2) = y^2 \frac{\partial}{\partial x}(x^2) = y^2 \cdot 2x = 2xy^2.$$

2. Explain why it is not necessary to use the Quotient Rule to compute $\frac{\partial}{\partial x}\left(\frac{x+y}{y+1}\right)$. Should the Quotient Rule be used to compute $\frac{\partial}{\partial y}\left(\frac{x+y}{y+1}\right)$?

SOLUTION In differentiating with respect to x, y is considered a constant. Therefore in this case the Constant Multiple Rule can be used to obtain

$$\frac{\partial}{\partial x}\left(\frac{x+y}{y+1}\right) = \frac{1}{y+1}\frac{\partial}{\partial x}(x+y) = \frac{1}{y+1} \cdot 1 = \frac{1}{y+1}.$$

As for the second part, since y appears in both the numerator and the denominator, the Quotient Rule is indeed needed.

3. Which of the following partial derivatives should be evaluated without using the Quotient Rule?

(a) $\frac{\partial}{\partial x}\frac{xy}{y^2+1}$ **(b)** $\frac{\partial}{\partial y}\frac{xy}{y^2+1}$ **(c)** $\frac{\partial}{\partial x}\frac{y^2}{y^2+1}$

SOLUTION

(a) This partial derivative does not require use of the Quotient Rule, since the Constant Multiple Rule gives

$$\frac{\partial}{\partial x}\left(\frac{xy}{y^2+1}\right) = \frac{y}{y^2+1}\frac{\partial}{\partial x}(x) = \frac{y}{y^2+1} \cdot 1 = \frac{y}{y^2+1}.$$

(b) This partial derivative requires use of the Quotient Rule.

(c) Since y is considered a constant in differentiating with respect to x, we do not need the Quotient Rule to state that $\frac{\partial}{\partial x}\left(\frac{y^2}{y^2+1}\right) = 0.$

4. What is f_x, where $f(x, y, z) = (\sin yz)e^{z^3 - z^{-1}\sqrt{y}}$?

SOLUTION In differentiating with respect to x, we treat y and z as constants. Therefore, the whole expression for $f(x, y, z)$ is treated as constant, so the derivative is zero:

$$\frac{\partial}{\partial x}\left(\sin yz e^{z^3 - z^{-1}\sqrt{y}}\right) = 0.$$

5. Which of the following partial derivatives are equal to f_{xxy}?

(a) f_{xyx} **(b)** f_{yyx} **(c)** f_{xyy} **(d)** f_{yxx}

SOLUTION f_{xxy} involves two differentiations with respect to x and one differentiation with respect to y. Therefore, if f satisfies the assumptions of Clairaut's Theorem, then

$$f_{xxy} = f_{xyx} = f_{yxx}$$

Exercises

1. Use the limit definition of the partial derivative to verify the formulas

$$\frac{\partial}{\partial x}xy^2 = y^2, \qquad \frac{\partial}{\partial y}xy^2 = 2xy$$

SOLUTION Using the limit definition of the partial derivative, we have

$$\frac{\partial}{\partial x}xy^2 = \lim_{h\to 0}\frac{(x+h)y^2 - xy^2}{h} = \lim_{h\to 0}\frac{xy^2 + hy^2 - xy^2}{h} = \lim_{h\to 0}\frac{hy^2}{h} = \lim_{h\to 0} y^2 = y^2$$

$$\frac{\partial}{\partial y}xy^2 = \lim_{k\to 0}\frac{x(y+k)^2 - xy^2}{k} = \lim_{k\to 0}\frac{x(y^2+2yk+k^2) - xy^2}{k} = \lim_{k\to 0}\frac{xy^2 + 2xyk + xk^2 - xy^2}{k}$$

$$= \lim_{k\to 0}\frac{k(2xy+xk)}{k} = \lim_{k\to 0}(2xy+k) = 2xy + 0 = 2xy$$

3. Use the Quotient Rule to compute $\frac{\partial}{\partial y}\frac{y}{x+y}$.

SOLUTION Using the Quotient Rule we obtain

$$\frac{\partial}{\partial y}\frac{y}{x+y} = \frac{(x+y)\frac{\partial}{\partial y}(y) - y\frac{\partial}{\partial y}(x+y)}{(x+y)^2} = \frac{(x+y)\cdot 1 - y\cdot 1}{(x+y)^2} = \frac{x}{(x+y)^2}$$

5. Calculate $f_z(2, 3, 1)$, where $f(x, y, z) = xyz$.

SOLUTION We first find the partial derivative $f_z(x, y, z)$:

$$f_z(x, y, z) = \frac{\partial}{\partial z}(xyz) = xy$$

Substituting the given point we get

$$f_z(2, 3, 1) = 2\cdot 3 = 6$$

7. The plane $y = 1$ intersects the surface $z = x^4 + 6xy - y^4$ in a certain curve (Figure 7). Find the slope of the tangent line to this curve at the point $P = (1, 1, 6)$.

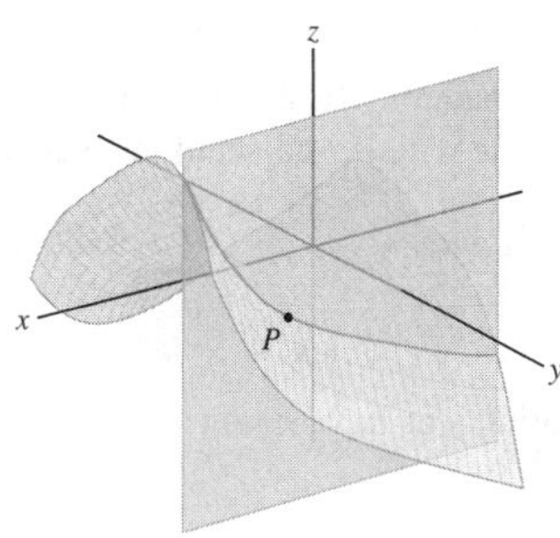

FIGURE 7 Graph of $f(x, y) = x^4 + 6xy - y^4$.

SOLUTION The slope of the tangent line to the curve $z = z(x, 1) = x^4 + 6x - 1$, obtained by intersecting the surface $z = x^4 + 6xy - y^4$ with the plane $y = 1$, is the partial derivative $\frac{\partial z}{\partial x}(1, 1)$.

$$\frac{\partial z}{\partial x} = \frac{\partial}{\partial x}(x^4 + 6xy - y^4) = 4x^3 + 6y$$

$$m = \frac{\partial z}{\partial x}(1, 1) = 4\cdot 1^3 + 6\cdot 1 = 10$$

In Exercises 9–11, refer to Figure 8.

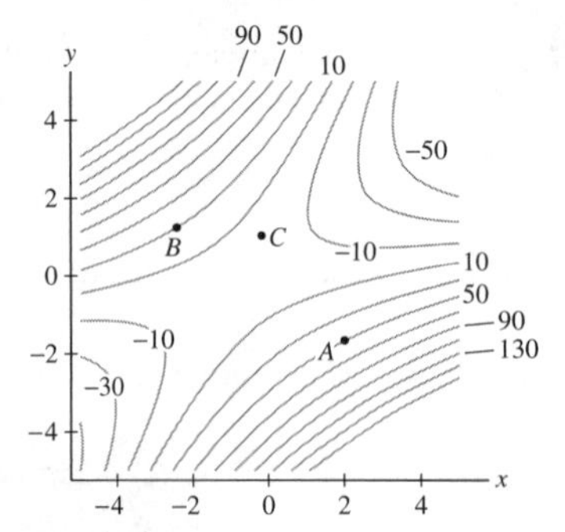

FIGURE 8 Contour map of $f(x, y)$.

9. Estimate f_x and f_y at point A.

SOLUTION To estimate f_x we move horizontally to the next level curve in the direction of growing x, to a point A'. The change in f from A to A' is the contour interval, $\Delta f = 70 - 50 = 20$.

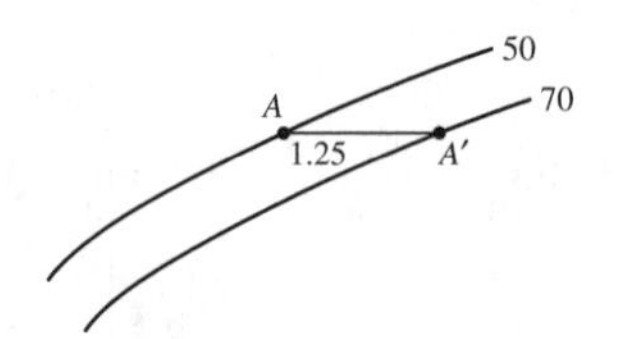

The distance between A and A' is approximately $\Delta x \approx 1.25$. Hence,

$$f_x(A) \approx \frac{\Delta f}{\Delta x} = \frac{20}{1.25} = 16$$

To estimate f_y we move vertically from A to a point A'' on the next level curve in the direction of growing y. The change in f from A to A'' is $\Delta f = 30 - 50 = -20$.

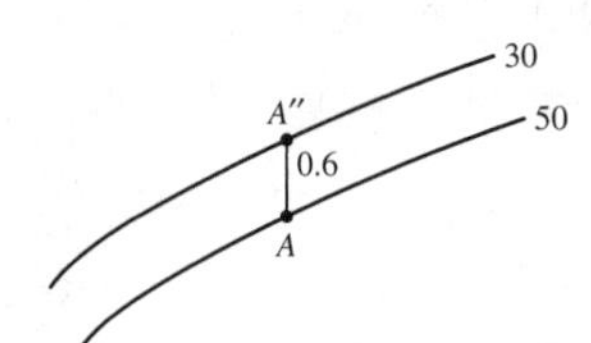

The distance between A and A'' is $\Delta y \approx 0.6$. Hence,

$$f_y(A) \approx \frac{\Delta f}{\Delta y} = \frac{-20}{0.6} \approx -33.3.$$

11. At which of A, B, or C is f_y smallest?

SOLUTION We consider vertical lines through A, B, and C. The distance between each point A, B, C and the intersection of the vertical line with the adjacent level curves is the largest at C. It means that f_y is smallest at C.

In Exercises 12–39, compute the partial derivatives.

13. $z = x^4y^3$

SOLUTION Treating y as a constant (to find z_x) and x as a constant (to find z_y) and using Rules for Differentiation, we get,

$$\frac{\partial}{\partial x}(x^4y^3) = y^3\frac{\partial}{\partial x}(x^4) = y^3 \cdot 4x^3 = 4x^3y^3$$

$$\frac{\partial}{\partial y}(x^4y^3) = x^4\frac{\partial}{\partial y}(y^3) = x^4 \cdot 3y^2 = 3x^4y^2$$

15. $V = \pi r^2 h$

SOLUTION We find $\frac{\partial V}{\partial r}$ and $\frac{\partial V}{\partial h}$:

$$\frac{\partial V}{\partial r} = \frac{\partial}{\partial r}(\pi r^2 h) = \pi h \frac{\partial}{\partial r}(r^2) = \pi h \cdot 2r = 2\pi hr$$

$$\frac{\partial V}{\partial h} = \frac{\partial}{\partial h}(\pi r^2 h) = \pi r^2$$

17. $z = \frac{x}{x-y}$

SOLUTION We differentiate with respect to x, using the Quotient Rule. We get

$$\frac{\partial}{\partial x}\left(\frac{x}{x-y}\right) = \frac{(x-y)\frac{\partial}{\partial x}(x) - x\frac{\partial}{\partial x}(x-y)}{(x-y)^2} = \frac{(x-y)\cdot 1 - x \cdot 1}{(x-y)^2} = \frac{-y}{(x-y)^2}$$

We now differentiate with respect to y, using the Chain Rule:

$$\frac{\partial}{\partial y}\left(\frac{x}{x-y}\right) = x\frac{\partial}{\partial y}\left(\frac{1}{x-y}\right) = x \cdot \frac{-1}{(x-y)^2}\frac{\partial}{\partial y}(x-y) = x \cdot \frac{-1}{(x-y)^2} \cdot (-1) = \frac{x}{(x-y)^2}$$

19. $z = \frac{x}{\sqrt{x^2+y^2}}$

SOLUTION We compute $\frac{\partial z}{\partial x}$ using the Quotient Rule and the Chain Rule:

$$\frac{\partial z}{\partial x} = \frac{1 \cdot \sqrt{x^2+y^2} - x\frac{\partial}{\partial x}\sqrt{x^2+y^2}}{\left(\sqrt{x^2+y^2}\right)^2} = \frac{\sqrt{x^2+y^2} - x \cdot \frac{2x}{2\sqrt{x^2+y^2}}}{x^2+y^2} = \frac{x^2+y^2-x^2}{(x^2+y^2)^{3/2}} = \frac{y^2}{(x^2+y^2)^{3/2}}$$

We compute $\frac{\partial z}{\partial y}$ using the Chain Rule:

$$\frac{\partial z}{\partial y} = x\frac{\partial}{\partial y}(x^2+y^2)^{-1/2} = x \cdot \left(-\frac{1}{2}\right)(x^2+y^2)^{-3/2} \cdot 2y = \frac{-xy}{(x^2+y^2)^{3/2}}$$

21. $z = \sin(u^2 v)$

SOLUTION By the Chain Rule,

$$\frac{d}{du}\sin\omega = \cos\omega\frac{d\omega}{du} \quad \text{and} \quad \frac{d}{dv}\sin\omega = \cos\omega\frac{d\omega}{dv}.$$

Applying this with $\omega = u^2 v$ gives

$$\frac{\partial}{\partial u}\sin(u^2v) = \cos(u^2v)\frac{\partial}{\partial u}(u^2v) = \cos(u^2v)\cdot 2uv = 2uv\cos(u^2v)$$

$$\frac{\partial}{\partial v}\sin(u^2v) = \cos(u^2v)\frac{\partial}{\partial v}(u^2v) = \cos(u^2v)\cdot u^2 = u^2\cos(u^2v)$$

23. $S = \tan^{-1}(wz)$

SOLUTION By the Chain Rule,

$$\frac{d}{dw}\tan^{-1}u = \frac{1}{1+u^2}\frac{du}{dw} \quad \text{and} \quad \frac{d}{dz}\tan^{-1}u = \frac{1}{1+u^2}\frac{du}{dz}$$

Using this rule with $u = wz$ gives

$$\frac{dS}{dw} = \frac{\partial}{\partial w}\tan^{-1}(wz) = \frac{1}{1+(wz)^2}\frac{\partial}{\partial w}(wz) = \frac{z}{1+w^2z^2}$$

$$\frac{dS}{dz} = \frac{\partial}{\partial z}\tan^{-1}(wz) = \frac{1}{1+(wz)^2}\frac{\partial}{\partial z}(wz) = \frac{w}{1+w^2z^2}$$

25. $z = \ln(x^2+y^2)$

SOLUTION Using the Chain Rule we have

$$\frac{\partial z}{\partial x} = \frac{1}{x^2+y^2}\frac{\partial}{\partial x}(x^2+y^2) = \frac{1}{x^2+y^2}\cdot 2x = \frac{2x}{x^2+y^2}$$

$$\frac{\partial z}{\partial y} = \frac{1}{x^2+y^2}\frac{\partial}{\partial y}(x^2+y^2) = \frac{1}{x^2+y^2}\cdot 2y = \frac{2y}{x^2+y^2}$$

27. $z = e^{xy}$

SOLUTION We use the Chain Rule, $\frac{d}{dx}e^u = e^u\frac{du}{dx}$; $\frac{d}{dy}e^u = e^u\frac{du}{dy}$ with $u = xy$ to obtain

$$\frac{\partial}{\partial x}e^{xy} = e^{xy}\frac{\partial}{\partial x}(xy) = e^{xy}y = ye^{xy}$$

$$\frac{\partial}{\partial y}e^{xy} = e^{xy}\frac{\partial}{\partial y}(xy) = e^{xy}x = xe^{xy}$$

29. $z = e^{-x^2-y^2}$

SOLUTION We use the Chain Rule to find $\frac{\partial z}{\partial x}$ and $\frac{\partial z}{\partial y}$:

$$\frac{\partial z}{\partial x} = e^{-x^2-y^2}\frac{\partial}{\partial x}(-x^2-y^2) = e^{-x^2-y^2}\cdot(-2x) = -2xe^{-x^2-y^2}$$

$$\frac{\partial z}{\partial y} = e^{-x^2-y^2}\frac{\partial}{\partial y}(-x^2-y^2) = e^{-x^2-y^2}\cdot(-2y) = -2ye^{-x^2-y^2}$$

31. $z = y^x$

SOLUTION To find $\frac{\partial z}{\partial y}$, we use the Power Rule for differentiation:

$$\frac{\partial z}{\partial y} = xy^{x-1}$$

To find $\frac{\partial z}{\partial x}$, we use the derivative of the exponent function:

$$\frac{\partial z}{\partial x} = y^x \ln y$$

33. $z = \sinh(x^2y)$

SOLUTION By the Chain Rule, $\frac{d}{dx}\sinh u = \cosh u\frac{du}{dx}$ and $\frac{d}{dy}\sinh u = \cosh u\frac{du}{dy}$. We use the Chain Rule with $u = x^2y$ to obtain

$$\frac{\partial}{\partial x}\sinh(x^2y) = \cosh(x^2y)\frac{\partial}{\partial x}(x^2y) = 2xy\cosh(x^2y)$$

$$\frac{\partial}{\partial y}\sinh(x^2y) = \cosh(x^2y)\frac{\partial}{\partial y}(x^2y) = x^2\cosh(x^2y)$$

35. $w = \dfrac{x}{y+z}$

SOLUTION We have

$$\frac{\partial w}{\partial x} = \frac{\partial}{\partial x}\left(\frac{x}{y+z}\right) = \frac{1}{y+z}\frac{\partial}{\partial x}(x) = \frac{1}{y+z}$$

To find $\frac{\partial w}{\partial y}$ and $\frac{\partial w}{\partial z}$, we use the Chain Rule:

$$\frac{\partial w}{\partial y} = x\frac{\partial}{\partial y}\left(\frac{1}{y+z}\right) = x\cdot\frac{-1}{(y+z)^2}\frac{\partial}{\partial y}(y+z) = x\cdot\frac{-1}{(y+z)^2}\cdot 1 = \frac{-x}{(y+z)^2}$$

$$\frac{\partial w}{\partial z} = x\frac{\partial}{\partial z}\left(\frac{1}{y+z}\right) = x\cdot\frac{-1}{(y+z)^2}\frac{\partial}{\partial z}(y+z) = x\cdot\frac{-1}{(y+z)^2}\cdot 1 = \frac{-x}{(y+z)^2}$$

37. $w = \dfrac{x}{(x^2+y^2+z^2)^{3/2}}$

SOLUTION To find $\frac{\partial w}{\partial x}$, we use the Quotient Rule and the Chain Rule:

$$\frac{\partial w}{\partial x} = \frac{1\cdot(x^2+y^2+z^2)^{3/2} - x\cdot\frac{3}{2}(x^2+y^2+z^2)^{1/2}\cdot 2x}{(x^2+y^2+z^2)^3} = (x^2+y^2+z^2)^{1/2}\frac{(x^2+y^2+z^2)-x\cdot 3x}{(x^2+y^2+z^2)^3}$$

$$= \frac{x^2+y^2+z^2-3x^2}{(x^2+y^2+z^2)^{5/2}} = \frac{y^2+z^2-2x^2}{(x^2+y^2+z^2)^{5/2}}$$

We now use the Chain Rule to compute $\frac{\partial w}{\partial y}$ and $\frac{\partial w}{\partial z}$:

$$\frac{\partial w}{\partial y} = x\frac{\partial}{\partial y}\frac{1}{(x^2+y^2+z^2)^{3/2}} = x\frac{\partial}{\partial y}(x^2+y^2+z^2)^{-3/2}$$

$$= x\cdot\left(-\frac{3}{2}\right)(x^2+y^2+z^2)^{-5/2}\cdot 2y = -\frac{3xy}{(x^2+y^2+z^2)^{5/2}}$$

$$\frac{\partial w}{\partial z} = x\frac{\partial}{\partial z}\frac{1}{(x^2+y^2+z^2)^{3/2}} = x\frac{\partial}{\partial z}(x^2+y^2+z^2)^{-3/2}$$

$$= x\cdot\left(-\frac{3}{2}\right)(x^2+y^2+z^2)^{-5/2}\cdot 2z = -\frac{3xz}{(x^2+y^2+z^2)^{5/2}}$$

39. $U = \frac{e^{-rt}}{r}$

SOLUTION We have

$$\frac{\partial U}{\partial r} = \frac{-te^{-rt}\cdot r - e^{-rt}\cdot 1}{r^2} = \frac{-(1+rt)e^{-rt}}{r^2}$$

and also

$$\frac{\partial U}{\partial t} = \frac{-re^{-rt}}{r} = -e^{-rt}$$

In Exercises 40–44, compute the given partial derivatives.

41. $f(x, y) = \sin(x^2 - y)$, $f_y(0, \pi)$

SOLUTION We differentiate with respect to y, using the Chain Rule. This gives

$$f_y(x, y) = \cos(x^2-y)\frac{\partial}{\partial y}(x^2-y) = \cos(x^2-y)\cdot(-1) = -\cos(x^2-y)$$

Evaluating at $(0, \pi)$ we obtain

$$f_y(0, \pi) = -\cos(0^2-\pi) = -\cos(-\pi) = -\cos\pi = 1.$$

43. $h(x, z) = e^{xz-x^2z^3}$, $h_z(2, 1)$

SOLUTION We first find the partial derivative $h_z(x, z)$ using the Chain Rule:

$$h_z(x, z) = e^{xz-x^2z^3}\frac{\partial}{\partial z}(xz-x^2z^3) = e^{xz-x^2z^3}(x - x^2\cdot 3z^2) = (x-3x^2z^2)e^{xz-x^2z^3}$$

At the point $(2, 1)$ we get

$$h_z(2, 1) = (2 - 3\cdot 2^2\cdot 1^2)e^{2\cdot 1 - 2^2\cdot 1^3} = -10e^{-2}.$$

45. Calculate $\frac{\partial W}{\partial E}$ and $\frac{\partial W}{\partial T}$, where $W = e^{-E/kT}$.

SOLUTION We use the Chain Rule

$$\frac{d}{dE}e^u = e^u\frac{du}{dE} \quad \text{and} \quad \frac{d}{dT}e^u = e^u\frac{du}{dT}$$

with $u = -\frac{E}{kT}$, to obtain

$$\frac{\partial W}{\partial E} = e^{-E/kT}\frac{\partial}{\partial E}\left(-\frac{E}{kT}\right) = e^{-E/kT}\left(-\frac{1}{kT}\right) = -\frac{1}{kT}e^{-E/kT}$$

$$\frac{\partial W}{\partial T} = e^{-E/kT}\frac{\partial}{\partial T}\left(-\frac{E}{kT}\right) = e^{-E/kT}\cdot\left(-\frac{E}{k}\right)\frac{\partial}{\partial T}\left(\frac{1}{T}\right) = e^{-E/kT}\left(-\frac{E}{k}\right)\left(-\frac{1}{T^2}\right) = \frac{E}{kT^2}e^{-E/kT}$$

47. The volume of a right-circular cone of radius r and height h is $V = \frac{\pi}{3}r^2h$. Calculate $\frac{\partial V}{\partial r}$ and $\frac{\partial V}{\partial h}$.

SOLUTION We obtain the following derivatives:

$$\frac{\partial V}{\partial r} = \frac{\partial}{\partial r}\left(\frac{\pi}{3}r^2h\right) = \frac{\pi h}{3}\frac{\partial}{\partial r}r^2 = \frac{\pi h}{3}\cdot 2r = \frac{2\pi hr}{3}$$

$$\frac{\partial V}{\partial h} = \frac{\partial}{\partial h}\left(\frac{\pi}{3}r^2h\right) = \frac{\pi}{3}r^2$$

49. Use the contour map of $f(x, y)$ in Figure 9 to explain why the following are true:

(a) f_x and f_y are both larger at P than at Q.

(b) $f_x(x, y)$ is decreasing as a function of y, that is, for any x, $f_x(x, b_1) > f_x(x, b_2)$ if $b_1 < b_2$.

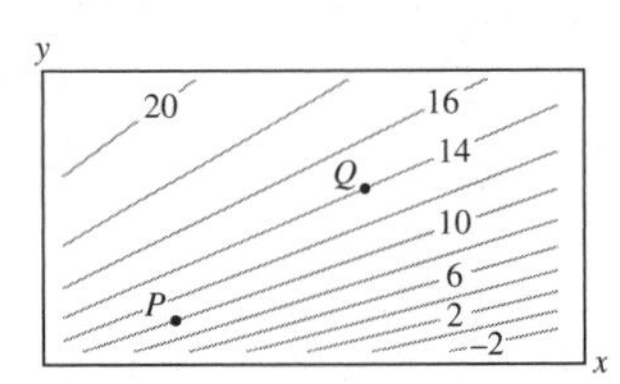

FIGURE 9 Contour interval 2.

SOLUTION

(a) Since the vertical and horizontal lines through P meet more level curves than vertical and horizontal lines through Q, f is increasing more rapidly in the y and x direction at P than at Q. Therefore, f_x and f_y are both larger at P than at Q.

(b) For any fixed value of x, a horizontal line through (x, b_2) meets fewer level curves than a horizontal line through (x, b_1), for $b_1 < b_2$. Thus, f_x is decreasing as a function of y.

51. Seawater Density Refer to Example 6 and Figure 4.

(a) Estimate $\frac{\partial \rho}{\partial T}$ and $\frac{\partial \rho}{\partial S}$ at points B and C.

(b) Which has a greater effect on density ρ at B, a 1° increase in temperature or a 1-ppt increase in salinity?

(c) True or false: The density of warm seawater is more sensitive to a change in temperature than the density of cold seawater? Explain.

SOLUTION

(a) We estimate $\frac{\partial \rho}{\partial T}$ at B. Since T varies in the vertical direction, we move vertically from B to a point B' on the next level curve, in the direction of increasing T (upward). The change in ρ from B to B' is $\Delta\rho = 1.0235 - 1.0240 = -0.0005$. The change in T from B to B' is $\Delta T = 2.5°$. Hence,

$$\left.\frac{\partial \rho}{\partial T}\right|_B \approx \frac{\Delta\rho}{\Delta T} \approx \frac{-0.0005}{2.5} = -2\cdot 10^{-4}$$

B′ 1.0235
1.0240
B

We now compute $\left.\frac{\partial \rho}{\partial T}\right|_C$. Following the same procedure as before, we find that

$$\left.\frac{\partial \rho}{\partial T}\right|_C \approx \frac{\Delta\rho}{\Delta T} \approx \frac{-0.0005}{2.5} = -0.0002$$

We now estimate $\frac{\partial \rho}{\partial S}$ at C. Since S varies in the horizontal direction, we move horizontally from C to a point C' on the next level curve, in the direction of increasing S (to the right). The change in ρ from C to C' is $\Delta\rho = 1.0255 - 1.0250 = 0.0005$. The change in S from C to C' is $\Delta S = 0.5$. Hence,

$$\left.\frac{\partial \rho}{\partial S}\right|_C \approx \frac{\Delta\rho}{\Delta S} = \frac{0.0005}{0.5} = 0.001.$$

Finally, we now estimate $\frac{\partial \rho}{\partial S}$ at B, using the same technique. We get

$$\left.\frac{\partial \rho}{\partial S}\right|_B \approx \frac{\Delta \rho}{\Delta S} \approx \frac{0.0005}{0.6} = 0.0008$$

(b) The change $\Delta \rho$ in the density as a result of a change ΔT in temperature (where S remains unchanged), and as a result of a change ΔS in salinity (where T remains unchanged), can be estimated by

$$\Delta \rho \approx \left.\frac{\partial \rho}{\partial T}\right|_C \Delta T; \quad \Delta \rho \approx \left.\frac{\partial \rho}{\partial S}\right|_C \Delta S \tag{1}$$

We substitute $\Delta T = 1°$, $\Delta S = 1$ ppt, $\left.\frac{\partial \rho}{\partial T}\right|_C \approx -0.0002$, and $\left.\frac{\partial \rho}{\partial S}\right|_C \approx 0.001$ (obtained in part (a)) in (1), to obtain

$\Delta \rho \approx -0.0002$ for the change in temperature
$\Delta \rho \approx 0.001$ for the change in salinity.

We conclude that the effect caused by the change in salinity is greater than the effect caused by the change in temperature, at the point C.

(c) At low temperatures the vertical distances between adjacent level curves are greater than at higher temperatures. This means that the density of cold water is less sensitive to a change in the temperature than the density of warm water. Thus, the statement is true.

In Exercises 53–58, compute the derivative indicated.

53. $f(x, y) = 3x^2y - 6xy^4$, $\frac{\partial^2 f}{\partial x^2}$ and $\frac{\partial^2 f}{\partial y^2}$

SOLUTION We first compute the partial derivatives $\frac{\partial f}{\partial x}$ and $\frac{\partial f}{\partial y}$:

$$\frac{\partial f}{\partial x} = 6xy - 6y^4; \quad \frac{\partial f}{\partial y} = 3x^2 - 6x \cdot 4y^3 = 3x^2 - 24xy^3$$

We now differentiate $\frac{\partial f}{\partial x}$ with respect to x and $\frac{\partial f}{\partial y}$ with respect to y. We get

$$\frac{\partial^2 f}{\partial x^2} = \frac{\partial}{\partial x} f_x = 6y; \quad \frac{\partial^2 f}{\partial y^2} = \frac{\partial}{\partial y} f_y = -24x \cdot 3y^2 = -72xy^2.$$

55. $g(x, y) = \frac{xy}{x - y}$, $\frac{\partial^2 g}{\partial x\, \partial y}$

SOLUTION By definition we have

$$\frac{\partial^2 g}{\partial x \partial y} = g_{yx} = \frac{\partial}{\partial x}\left(\frac{\partial g}{\partial y}\right)$$

Thus, we must find $\frac{\partial g}{\partial y}$:

$$\frac{\partial g}{\partial y} = x\frac{\partial}{\partial y}\left(\frac{y}{x - y}\right) = x\frac{1 \cdot (x - y) - y \cdot (-1)}{(x - y)^2} = \frac{x^2}{(x - y)^2}$$

Differentiating $\frac{\partial g}{\partial y}$ with respect to x, using the Quotient Rule, we obtain

$$\frac{\partial^2 g}{\partial x \partial y} = \frac{\partial}{\partial x}\left(\frac{\partial g}{\partial y}\right) = \frac{\partial}{\partial x}\frac{x^2}{(x - y)^2} = \frac{2x(x - y)^2 - x^2 \cdot 2(x - y)}{(x - y)^4} = -\frac{2xy}{(x - y)^3}$$

57. $h(u, v) = \frac{u}{u + v}$, $h_{uv}(u, v)$

SOLUTION By definition $h_{uv} = \frac{\partial}{\partial v}\left(\frac{\partial h}{\partial u}\right)$. We compute $\frac{\partial h}{\partial u}$ using the Quotient Rule:

$$\frac{\partial h}{\partial u} = \frac{\partial}{\partial u}\frac{u}{u + v} = \frac{1 \cdot (u + v) - u \cdot 1}{(u + v)^2} = \frac{v}{(u + v)^2}$$

We now use again the Quotient Rule to differentiate $\frac{\partial h}{\partial u}$ with respect to v. We obtain

$$h_{uv}(u, v) = \frac{\partial}{\partial v}\frac{v}{(u + v)^2} = \frac{1 \cdot (u + v)^2 - v \cdot 2(u + v)}{(u + v)^4} = \frac{u - v}{(u + v)^3}$$

In Exercises 59–60, use Table 1 to answer the following questions.

TABLE 1 Seawater Density ρ as a Function of Temperature T and Salinity S

T \ S	30	31	32	33	34	35	36
12	22.75	23.51	24.27	25.07	25.82	26.6	27.36
10	23.07	23.85	24.62	25.42	26.17	26.99	27.73
8	23.36	24.15	24.93	25.73	26.5	27.28	29.09
6	23.62	24.44	25.22	26	26.77	27.55	28.35
4	23.85	24.62	25.42	26.23	27	27.8	28.61
2	24	24.78	25.61	26.38	27.18	28.01	28.78
0	24.11	24.92	25.72	26.5	27.34	28.12	28.91

59. Estimate $\frac{\partial \rho}{\partial T}$ and $\frac{\partial \rho}{\partial S}$ at $(S, T) = (34, 2)$ and $(35, 10)$.

SOLUTION We estimate $\frac{\partial \rho}{\partial T}$ at the given points using the values in Table 1 and the following approximation:

$$\frac{\partial \rho}{\partial T}(34, 2) \approx \frac{\rho(34, 2+2) - \rho(34, 2)}{2} = \frac{\rho(34, 4) - \rho(34, 2)}{2} = \frac{27 - 27.18}{2} = -0.09$$

$$\frac{\partial \rho}{\partial T}(35, 10) \approx \frac{\rho(35, 10+2) - \rho(35, 10)}{2} = \frac{\rho(35, 12) - \rho(35, 10)}{2} = \frac{26.6 - 26.99}{2} = -0.195$$

We estimate the partial derivative $\frac{\partial \rho}{\partial S}$ at the given points:

$$\frac{\partial \rho}{\partial S}(34, 2) \approx \frac{\rho(34+1, 2) - \rho(34, 2)}{1} = \frac{\rho(35, 2) - \rho(34, 2)}{1} = 28.01 - 27.18 = 0.83$$

$$\frac{\partial \rho}{\partial S}(35, 10) \approx \frac{\rho(35+1, 10) - \rho(35, 10)}{1} = \frac{\rho(36, 10) - \rho(35, 10)}{1} = 27.73 - 26.99 = 0.74$$

61. Compute f_{xyz} for

$$f(x, y, z) = \sin(yx) + \tan\left(\frac{z + z^{-1}}{x - x^{-1}}\right)$$

Hint: Use a well-chosen order of differentiation on each term.

SOLUTION At the points where the derivatives are continuous, the partial derivative f_{xyz} may be performed in any order. To simplify the computation we first differentiate with respect to y. This gives

$$f_y(x, y, z) = \frac{\partial}{\partial y}\sin(yx) + 0 = \cos(yx)\frac{\partial}{\partial y}(yx) = x\cos(yx)$$

We now differentiate f_y with respect to z:

$$f_{yz}(x, y, z) = \frac{\partial}{\partial z}\big(x\cos(yx)\big) = 0$$

Hence,

$$f_{yzx}(x, y, z) = 0$$

We conclude that at the points where the partial derivatives are continuous,

$$f_{xyz}(x, y, z) = f_{yzx}(x, y, z) = 0.$$

In Exercises 63–70, compute the derivative indicated.

63. $f(u, v) = \cos(u + v^2)$, $\quad f_{uuv}$

SOLUTION Using the Chain Rule, we have

$$f_u = \frac{\partial}{\partial u}\cos(u+v^2) = -\sin(u+v^2)\cdot\frac{\partial}{\partial u}(u+v^2) = -\sin(u+v^2)$$

$$f_{uu} = \frac{\partial}{\partial u}\left(-\sin(u+v^2)\right) = -\cos(u+v^2)$$

$$f_{uuv} = \frac{\partial}{\partial v}\left(-\cos(u+v^2)\right) = \sin(u+v^2)\cdot\frac{\partial}{\partial v}(u+v^2) = 2v\sin(u+v^2)$$

65. $F(r,s,t) = r(s^2+t^2)$, F_{rst}

SOLUTION For $F(r,s,t) = r(s^2+t^2)$, we have

$$F_r = s^2+t^2$$

$$F_{rs} = 2s$$

$$F_{rst} = 0$$

67. $u(x,t) = t^{-1/2}e^{-(x^2/4t)}$, u_{xx}

SOLUTION Using the Chain Rule we obtain

$$u_x = t^{-1/2}\frac{\partial}{\partial x}(e^{-x^2/4t}) = t^{-1/2}\cdot e^{-x^2/4t}\frac{\partial}{\partial x}\left(-\frac{x^2}{4t}\right) = t^{-1/2}\cdot e^{-x^2/4t}\cdot\frac{-2x}{4t} = -\frac{1}{2}xt^{-3/2}e^{-x^2/4t}$$

We now differentiate u_x with respect to x, using the Product Rule and the Chain Rule:

$$u_{xx} = -\frac{1}{2}t^{-3/2}\frac{\partial}{\partial x}(xe^{-x^2/4t}) = -\frac{1}{2}t^{-3/2}\left(1\cdot e^{-x^2/4t} + x\cdot e^{-x^2/4t}\cdot\frac{-2x}{4t}\right)$$

$$= -\frac{1}{2}t^{-3/2}\left(e^{-x^2/4t} - \frac{x^2}{2t}e^{-x^2/4t}\right) = -\frac{1}{2}t^{-3/2}e^{-x^2/4t}\left(1-\frac{x^2}{2t}\right)$$

69. $g(x,y,z) = \sqrt{x^2+y^2+z^2}$, g_{xyz}

SOLUTION Differentiating with respect to x, using the Chain Rule, we get

$$g_x = \frac{\partial}{\partial x}\sqrt{x^2+y^2+z^2} = \frac{1}{2\sqrt{x^2+y^2+z^2}}\frac{\partial}{\partial x}(x^2+y^2+z^2) = \frac{1}{2\sqrt{x^2+y^2+z^2}}\cdot 2x = \frac{x}{\sqrt{x^2+y^2+z^2}}$$

We now differentiate g_x with respect to y, using the Chain Rule. This gives

$$g_{xy} = x\frac{\partial}{\partial y}(x^2+y^2+z^2)^{-1/2} = x\cdot\left(-\frac{1}{2}\right)(x^2+y^2+z^2)^{-3/2}\cdot 2y = \frac{-xy}{(x^2+y^2+z^2)^{3/2}}$$

Finally, we differentiate g_{xy} with respect to z, obtaining

$$g_{xyz} = -xy\frac{\partial}{\partial z}(x^2+y^2+z^2)^{-3/2} = -xy\cdot\left(-\frac{3}{2}\right)(x^2+y^2+z^2)^{-5/2}\cdot 2z = \frac{3xyz}{(x^2+y^2+z^2)^{5/2}}$$

71. Find (by guessing) a function such that $\frac{\partial f}{\partial x} = 2xy$ and $\frac{\partial f}{\partial y} = x^2$.

SOLUTION The function $f(x,y) = x^2y$ satisfies $\frac{\partial f}{\partial y} = x^2$ and $\frac{\partial f}{\partial x} = 2xy$.

73. Assume that f_{xy} and f_{yx} are continuous and that f_{yxx} exists. Show that f_{xyx} also exists and that $f_{yxx} = f_{xyx}$.

SOLUTION Since f_{xy} and f_{yx} are continuous, Clairaut's Theorem implies that

$$f_{xy} = f_{yx} \tag{1}$$

We are given that f_{yxx} exists. Using (1) we get

$$f_{yxx} = \frac{\partial}{\partial x}\frac{\partial}{\partial x}f_y = \frac{\partial}{\partial x}f_{yx} = \frac{\partial}{\partial x}f_{xy} = f_{xyx}$$

Therefore, f_{xyx} also exists and $f_{yxx} = f_{xyx}$.

75. Find all values of A and B such that $f(x,t) = e^{Ax+Bt}$ satisfies Eq. (3).

SOLUTION We compute the following partials, using the Chain Rule:

$$\frac{\partial f}{\partial t} = \frac{\partial}{\partial t}\left(e^{Ax+Bt}\right) = e^{Ax+Bt}\frac{\partial}{\partial t}(Ax+Bt) = Be^{Ax+Bt}$$

$$\frac{\partial f}{\partial x} = \frac{\partial}{\partial x}\left(e^{Ax+Bt}\right) = e^{Ax+Bt}\frac{\partial}{\partial x}(Ax+Bt) = Ae^{Ax+Bt}$$

$$\frac{\partial^2 f}{\partial x^2} = \frac{\partial}{\partial x}\left(Ae^{Ax+Bt}\right) = A\frac{\partial}{\partial x}\left(e^{Ax+Bt}\right) = Ae^{Ax+Bt}\frac{\partial}{\partial x}(Ax+Bt) = A^2e^{Ax+Bt}$$

Substituting these partials in the differential equation (3), we get

$$Be^{Ax+Bt} = A^2e^{Ax+Bt}$$

We divide by the nonzero e^{Ax+Bt} to obtain

$$B = A^2$$

We conclude that $f(x,t) = e^{Ax+Bt}$ satisfies equation (5) if and only if $B = A^2$, where A is arbitrary.

In Exercises 76–79, the ***Laplace operator*** Δ *is defined by* $\Delta f = f_{xx} + f_{yy}$. *A function* $u(x,y)$ *satisfying the Laplace equation* $\Delta u = 0$ *is called* ***harmonic***.

77. Find all harmonic polynomials $u(x,y)$ of degree three, that is, $u(x,y) = ax^3 + bx^2y + cxy^2 + dy^3$.

SOLUTION We compute the first-order partials u_x and u_y and the second-order partials u_{xx} and u_{yy} of the given polynomial $u(x,y)$. This gives

$$u_x = 3ax^2 + 2bxy + cy^2$$

$$u_y = bx^2 + 2cxy + 3dy^2$$

$$u_{xx} = 6ax + 2by$$

$$u_{yy} = 2cx + 6dy$$

The polynomial is harmonic if $u_{xx} + u_{yy} = 0$, that is, if for all x and y

$$6ax + 2by + 2cx + 6dy = 0$$

This equality holds for all x and y if and only if the coefficients of x and y are both zero. That is, $6a + 2c = 0$ (so $c = -3a$) and $2b + 6d = 0$ (so $b = -3d$). We conclude that the harmonic polynomials in the given form are

$$u(x,y) = ax^3 - 3dx^2y - 3axy^2 + dy^3$$

79. Find all constants a, b such that $u(x,y) = \cos(ax)e^{by}$ is harmonic.

SOLUTION To determine if the functions $\cos(ax)e^{by}$ are harmonic, we compute the following derivatives:

$$(\cos ax)' = -a\sin ax \Rightarrow (\cos ax)'' = -a^2\cos ax$$

$$(e^{by})' = be^{by} \Rightarrow (e^{by})'' = b^2e^{by} = a^2e^{by}$$

Thus, we can conclude

$$u_{xx} = \frac{\partial^2}{\partial x^2}\cos(ax)e^{by} = -a^2\cos(ax)e^{by} = -a^2u$$

$$u_{yy} = \frac{\partial^2}{\partial y^2}\cos(ax)e^{by} = b^2\cos(ax)e^{by} = b^2u$$

Thus, $u_{xx} + u_{yy} = (b^2 - a^2)u$, which equals 0 if and only if $a^2 = b^2$.

Further Insights and Challenges

81. CAS **Assumptions Matter** This exercise shows that the hypotheses of Clairaut's Theorem are needed. Let $f(x,y) = xy\dfrac{x^2 - y^2}{x^2 + y^2}$ for $(x,y) \neq (0,0)$ and $f(0,0) = 0$.

(a) Use a computer algebra system to verify the following formulas for $(x, y) \neq (0, 0)$:

$$f_x(x, y) = \frac{y(x^4 + 4x^2y^2 - y^4)}{(x^2 + y^2)^2}$$

$$f_y(x, y) = \frac{x(x^4 - 4x^2y^2 - y^4)}{(x^2 + y^2)^2}$$

(b) Use the limit definition of the partial derivative to show that $f_x(0, 0) = f_y(0, 0) = 0$ and that $f_{yx}(0, 0)$ and $f_{xy}(0, 0)$ both exist but are not equal.

(c) Use a computer algebra system to show that for $(x, y) \neq (0, 0)$:

$$f_{xy}(x, y) = f_{yx}(x, y) = \frac{x^6 + 9x^4y^2 - 9x^2y^4 - y^6}{(x^2 + y^2)^3}$$

Show that f_{xy} is not continuous at $(0, 0)$. *Hint:* Show that $\lim_{h\to 0} f_{xy}(h, 0) \neq \lim_{h\to 0} f_{xy}(0, h)$.

(d) Explain why the result of (b) does not contradict Clairaut's Theorem.

SOLUTION

(a) We use a CAS to verify the following partials for $(x, y) \neq (0, 0)$:

$$f_x(x, y) = \frac{y(x^4 + 4x^2y^2 - y^4)}{(x^2 + y^2)^2}$$

$$f_y(x, y) = \frac{x(x^4 - 4x^2y^2 - y^4)}{(x^2 + y^2)^2}$$

(b) Using the limit definition of the partial derivatives at the point $(0, 0)$ we have

$$f_x(0, 0) = \lim_{h\to 0} \frac{f(h, 0) - f(0, 0)}{h} = \lim_{h\to 0} \frac{h \cdot 0 \frac{h^2 - 0^2}{h^2 + 0^2} - 0}{h} = \lim_{h\to 0} \frac{0}{h} = 0$$

$$f_y(0, 0) = \lim_{k\to 0} \frac{f(0, k) - f(0, 0)}{k} = \lim_{k\to 0} \frac{0 \cdot k \frac{0^2 - k^2}{0^2 + k^2} - 0}{k} = \lim_{k\to 0} \frac{0}{k} = 0$$

We now use the derivatives in part (a) and the limit definition of the partial derivatives to compute $f_{yx}(0, 0)$ and $f_{xy}(0, 0)$. By the formulas in part (a), we have

$$f_y(0, 0) = 0, \quad f_y(h, 0) = \frac{h(h^4 - 0 - 0)}{(h^2 + 0)^2} = h$$

$$f_x(0, 0) = 0, \quad f_x(0, k) = \frac{k(0 + 0 - k^4)}{(0^2 + k^2)^2} = -k$$

Thus,

$$f_{yx}(0, 0) = \left.\frac{\partial}{\partial x} f_y\right|_{(0,0)} = \lim_{h\to 0} \frac{f_y(h, 0) - f_y(0, 0)}{h} = \lim_{h\to 0} \frac{h - 0}{h} = \lim_{h\to 0} 1 = 1$$

$$f_{xy}(0, 0) = \left.\frac{\partial}{\partial y} f_x\right|_{(0,0)} = \lim_{k\to 0} \frac{f_x(0, k) - f_x(0, 0)}{k} = \lim_{k\to 0} \frac{-k - 0}{k} = \lim_{k\to 0} (-1) = -1$$

We see that the mixed partials at the point $(0, 0)$ exist but are not equal.

(c) We verify, using a CAS, that for $(x, y) \neq (0, 0)$ the following derivatives hold:

$$f_{xy}(x, y) = f_{yx}(x, y) = \frac{x^6 + 9x^4y^2 - 9x^2y^4 - y^6}{(x^2 + y^2)^3}$$

To show that f_{xy} is not continuous at $(0, 0)$, we show that the limit $\lim_{(x,y)\to(0,0)} f_{xy}(x, y)$ does not exist. We compute the limit as (x, y) approaches the origin along the x-axis. Along this axis, $y = 0$; hence,

$$\lim_{\substack{(x,y)\to(0,0)\\ \text{along the } x\text{-axis}}} f_{xy}(x, y) = \lim_{h\to 0} f_{xy}(h, 0) = \lim_{h\to 0} \frac{h^6 + 9h^4 \cdot 0 - 9h^2 \cdot 0 - 0}{(0 + h^2)^3} = \lim_{h\to 0} 1 = 1$$

We compute the limit as (x, y) approaches the origin along the y-axis. Along this axis, $x = 0$, hence,

$$\lim_{\substack{(x,y)\to(0,0)\\ \text{along the } y\text{-axis}}} f_{xy}(x, y) = \lim_{h\to 0} f_{xy}(0, h) = \lim_{h\to 0} \frac{0 + 0 + 0 - h^6}{(0 + h^2)^3} = \lim_{h\to 0} (-1) = -1$$

Since the limits are not equal $f(x, y)$ does not approach one value as $(x, y) \to (0, 0)$, hence the limit $\lim_{(x,y)\to(0,0)} f_{xy}(x, y)$ does not exist, and $f_{xy}(x, y)$ is not continuous at the origin.

(d) The result of part (b) does not contradict Clairaut's Theorem since f_{xy} is not continuous at the origin. The continuity of the mixed derivative is essential in Clairaut's Theorem.

15.4 Differentiability, Linear Approximation, and Tangent Planes (ET Section 14.4)

Preliminary Questions

1. How is the linearization of $f(x, y)$ at (a, b) defined?

SOLUTION The linearization of $f(x, y)$ at (a, b) is the linear function

$$L(x, y) = f(a, b) + f_x(a, b)(x - a) + f_y(a, b)(y - b)$$

This function is the equation of the tangent plane to the surface $z = f(x, y)$ at $\left(a, b, f(a, b)\right)$.

2. Define local linearity for functions of two variables.

SOLUTION $f(x, y)$ is locally linear at (a, b) if the linear approximation $L(x, y)$ at (a, b) approximates $f(x, y)$ at (a, b) to first order. That is, if there exists a function $\epsilon(x, y)$ satisfying $\lim_{(x,y)\to(a,b)} \epsilon(x, y) = 0$ such that

$$f(x, y) - L(x, y) = \epsilon(x, y)\sqrt{(x - a)^2 + (y - b)^2}$$

for all (x, y) in an open disk D containing (a, b).

In Questions 3–5, assume that

$$f(2, 3) = 8, \qquad f_x(2, 3) = 5, \qquad f_y(2, 3) = 7$$

3. Which of (a)–(b) is the linearization of f at $(2, 3)$?

(a) $L(x, y) = 8 + 5x + 7y$

(b) $L(x, y) = 8 + 5(x - 2) + 7(y - 3)$

SOLUTION The linearization of f at $(2, 3)$ is the following linear function:

$$L(x, y) = f(2, 3) + f_x(2, 3)(x - 2) + f_y(2, 3)(y - 3)$$

That is,

$$L(x, y) = 8 + 5(x - 2) + 7(y - 3) = -23 + 5x + 7y$$

The function in (b) is the correct answer.

4. Estimate $f(2, 3.1)$.

SOLUTION We use the linear approximation

$$f(a + h, b + k) \approx f(a, b) + f_x(a, b)h + f_y(a, b)k$$

We let $(a, b) = (2, 3)$, $h = 0$, $k = 3.1 - 3 = 0.1$. Then,

$$f(2, 3.1) \approx f(2, 3) + f_x(2, 3) \cdot 0 + f_y(2, 3) \cdot 0.1 = 8 + 0 + 7 \cdot 0.1 = 8.7$$

We get the estimation $f(2, 3.1) \approx 8.7$.

5. Estimate Δf at $(2, 3)$ if $\Delta x = -0.3$ and $\Delta y = 0.2$.

SOLUTION The change in f can be estimated by the linear approximation as follows:

$$\Delta f \approx f_x(a, b)\Delta x + f_y(a, b)\Delta y$$

$$\Delta f \approx f_x(2, 3) \cdot (-0.3) + f_y(2, 3) \cdot 0.2$$

or

$$\Delta f \approx 5 \cdot (-0.3) + 7 \cdot 0.2 = -0.1$$

The estimated change is $\Delta f \approx -0.1$.

6. Which theorem allows us to conclude that $f(x, y) = x^3y^8$ is differentiable?

SOLUTION The function $f(x, y) = x^3y^8$ is a polynomial, hence $f_x(x, y)$ and $f_y(x, y)$ exist and are continuous. Therefore the Criterion for Differentiability implies that f is differentiable everywhere.

Exercises

1. Let $f(x, y) = x^2y^3$.

(a) Find the linearization of f at $(a, b) = (2, 1)$.

(b) Use the linear approximation to estimate $f(2.01, 1.02)$ and $f(1.97, 1.01)$, and compare your estimates with the values obtained using a calculator.

SOLUTION

(a) We compute the value of the function and its partial derivatives at $(a, b) = (2, 1)$:

$$\begin{aligned} f(x, y) &= x^2y^3 & & f(2, 1) = 4 \\ f_x(x, y) &= 2xy^3 & \Rightarrow\quad & f_x(2, 1) = 4 \\ f_y(x, y) &= 3x^2y^2 & & f_y(2, 1) = 12 \end{aligned}$$

The linear approximation is therefore

$$f(x, y) \approx f(2, 1) + f_x(2, 1)(x - 2) + f_y(2, 1)(y - 1)$$
$$f(x, y) \approx 4 + 4(x - 2) + 12(y - 1) = -16 + 4x + 12y$$

(b) For $h = x - 2$ and $k = y - 1$ we have the following form of the linear approximation at $(a, b) = (2, 1)$:

$$f(x, y) \approx f(2, 1) + f_x(2, 1)h + f_y(2, 1)k = 4 + 4h + 12k$$

To compute $f(2.01, 1.02)$ we set $h = 2.01 - 2 = 0.01$, $k = 1.02 - 1 = 0.02$ to obtain

$$f(2.01, 1.02) \approx 4 + 4 \cdot 0.01 + 12 \cdot 0.02 = 4.28$$

The actual value is

$$f(2.01, 1.02) = 2.01^2 \cdot 1.02^3 = 4.2874$$

To compute $f(1.97, 1.01)$ we set $h = 1.97 - 2 = -0.03$, $k = 1.01 - 1 = 0.01$ to obtain

$$f(1.97, 1.01) \approx 4 + 4 \cdot (-0.03) + 12 \cdot 0.01 = 4.$$

The actual value is

$$f(1.97, 1.01) = 1.97^2 \cdot 1.01^3 = 3.998.$$

3. Let $f(x, y) = x^3y^{-4}$. Use Eq. (5) to estimate the change $\Delta f = f(2.03, 0.9) - f(2, 1)$.

SOLUTION We compute the function and its partial derivatives at $(a, b) = (2, 1)$:

$$\begin{aligned} f(x, y) &= x^3y^{-4} & & f(2, 1) = 8 \\ f_x(x, y) &= 3x^2y^{-4} & \Rightarrow\quad & f_x(2, 1) = 12 \\ f_y(x, y) &= -4x^3y^{-5} & & f_y(2, 1) = -32 \end{aligned}$$

Also, $\Delta x = 2.03 - 2 = 0.03$ and $\Delta y = 0.9 - 1 = -0.1$. Therefore,

$$\Delta f = f(2.03, 0.9) - f(2, 1) \approx f_x(2, 1)\Delta x + f_y\Delta y = 12 \cdot 0.03 + (-32) \cdot (-0.1) = 3.56$$
$$\Delta f \approx 3.56$$

5. Use the linear approximation of $f(x, y) = e^{x^2+y}$ at $(0, 0)$ to estimate $f(0.01, -0.02)$. Compare with the value obtained using a calculator.

SOLUTION The linear approximation of f at the point $(0, 0)$ is

$$f(h, k) \approx f(0, 0) + f_x(0, 0)h + f_y(0, 0)k \tag{1}$$

We first must compute f and its partial derivative at the point $(0, 0)$. Using the Chain Rule we obtain

$$\begin{aligned} f(x, y) &= e^{x^2+y} & & f(0, 0) = e^0 = 1 \\ f_x(x, y) &= 2xe^{x^2+y} & \Rightarrow\quad & f_x(0, 0) = 2 \cdot 0 \cdot e^0 = 0 \\ f_y(x, y) &= e^{x^2+y} & & f_y(0, 0) = e^0 = 1 \end{aligned}$$

We substitute these values and $h = 0.01$, $k = -0.02$ in (1) to obtain

$$f(0.01, -0.02) \approx 1 + 0 \cdot 0.01 + 1 \cdot (-0.02) = 0.98$$

The actual value is $f(0.01, -0.02) = e^{0.01^2 - 0.02} \approx 0.9803$.

7. Use Eq. (3) to find an equation of the tangent plane to the graph of $f(x, y) = 2x^2 - 4xy^2$ at the point $(-1, 2)$.

SOLUTION The equation of the tangent plane at the point $(-1, 2, 18)$ is

$$z = f(-1, 2) + f_x(-1, 2)(x + 1) + f_y(-1, 2)(y - 2) \tag{1}$$

We compute the function and its partial derivatives at the point $(-1, 2)$:

$$\begin{aligned} f(x, y) &= 2x^2 - 4xy^2 & & f(-1, 2) = 18 \\ f_x(x, y) &= 4x - 4y^2 & \Rightarrow \quad & f_x(-1, 2) = -20 \\ f_y(x, y) &= -8xy & & f_y(-1, 2) = 16 \end{aligned}$$

Substituting in (1) we obtain the following equation of the tangent plane:

$$z = 18 - 20(x + 1) + 16(y - 2) = -34 - 20x + 16y$$

That is,

$$z = -34 - 20x + 16y$$

9. Find the linear approximation to $f(x, y, z) = \dfrac{xy}{z}$ at the point $(2, 1, 2)$.

SOLUTION The linear approximation to f at the point $(2, 1, 2)$ is:

$$f(x, y, z) \approx f(2, 1, 2) + f_x(2, 1, 2)(x - 2) + f_y(2, 1, 2)(y - 1) + f_z(2, 1, 2)(z - 2) \tag{1}$$

We compute the values of f and its partial derivatives at $(2, 1, 2)$:

$$\begin{aligned} f(x, y, z) &= \frac{xy}{z} & & f(2, 1, 2) = 1 \\ f_x(x, y, z) &= \frac{y}{z} & \Rightarrow \quad & f_x(2, 1, 2) = \frac{1}{2} \\ f_y(x, y, z) &= \frac{x}{z} & & f_y(2, 1, 2) = 1 \\ f_z(x, y, z) &= -\frac{xy}{z^2} & & f_z(2, 1, 2) = -\frac{1}{2} \end{aligned}$$

We substitute these values in (1) to obtain the following linear approximation:

$$\begin{aligned} \frac{xy}{z} &\approx 1 + \frac{1}{2}(x - 2) + 1 \cdot (y - 1) - \frac{1}{2}(z - 2) \\ \frac{xy}{z} &\approx \frac{1}{2}x + y - \frac{1}{2}z \end{aligned}$$

In Exercises 11–18, find an equation of the tangent plane at the given point.

11. $f(x, y) = x^2y + xy^3$, $(2, 1)$

SOLUTION The equation of the tangent plane at $(2, 1)$ is

$$z = f(2, 1) + f_x(2, 1)(x - 2) + f_y(2, 1)(y - 1) \tag{1}$$

We compute the values of f and its partial derivatives at $(2, 1)$:

$$\begin{aligned} f(x, y) &= x^2y + xy^3 & & f(2, 1) = 6 \\ f_x(x, y) &= 2xy + y^3 & \Rightarrow \quad & f_x(2, 1) = 5 \\ f_y(x, y) &= x^2 + 3xy^2 & & f_y(2, 1) = 10 \end{aligned}$$

We now substitute these values in (1) to obtain the following equation of the tangent plane:

$$z = 6 + 5(x - 2) + 10(y - 1) = 5x + 10y - 14$$

That is,

$$z = 5x + 10y - 14.$$

13. $f(x, y) = x^2 + y^{-2}$, $(4, 1)$

SOLUTION The equation of the tangent plane at $(4, 1)$ is

$$z = f(4,1) + f_x(4,1)(x-4) + f_y(4,1)(y-1) \tag{1}$$

We compute the values of f and its partial derivatives at $(4, 1)$:

$$\begin{aligned} f(x,y) &= x^2 + y^{-2} & & f(4,1) = 17 \\ f_x(x,y) &= 2x & \Rightarrow \quad & f_x(4,1) = 8 \\ f_y(x,y) &= -2y^{-3} & & f_y(4,1) = -2 \end{aligned}$$

Substituting in (1) we obtain the following equation of the tangent plane:

$$z = 17 + 8(x-4) - 2(y-1) = 8x - 2y - 13.$$

15. $F(r, s) = r^2 s^{-1/2} + s^{-3}$, $(2, 1)$

SOLUTION The equation of the tangent plane at $(2, 1)$ is

$$z = f(2,1) + f_r(2,1)(r-2) + f_s(2,1)(s-1) \tag{1}$$

We compute f and its partial derivatives at $(2, 1)$:

$$\begin{aligned} f(r,s) &= r^2 s^{-1/2} + s^{-3} & & f(2,1) = 5 \\ f_r(r,s) &= 2rs^{-1/2} & \Rightarrow \quad & f_r(2,1) = 4 \\ f_s(r,s) &= -\frac{1}{2}r^2 s^{-3/2} - 3s^{-4} & & f_s(2,1) = -5 \end{aligned}$$

We substitute these values in (1) to obtain the following equation of the tangent plane:

$$z = 5 + 4(r-2) - 5(s-1) = 4r - 5s + 2.$$

17. $f(x, y) = e^x \ln y$, $(0, 1)$

SOLUTION The equation of the tangent plane at $(0, 1)$ is

$$z = f(0,1) + f_x(0,1)x + f_y(0,1)(y-1) \tag{1}$$

We compute the values of f and its partial derivatives at $(0, 1)$:

$$\begin{aligned} f(x,y) &= e^x \ln y & & f(0,1) = 0 \\ f_x(x,y) &= e^x \ln y & \Rightarrow \quad & f_x(0,1) = 0 \\ f_y(x,y) &= \frac{e^x}{y} & & f_y(0,1) = 1 \end{aligned}$$

Substituting in (1) gives the following equation of the tangent plane:

$$z = 0 + 0x + 1(y-1) = y - 1$$

That is, $z = y - 1$.

19. Find the points on the graph of $z = 3x^2 - 4y^2$ at which the vector $\mathbf{n} = \langle 3, 2, 2 \rangle$ is normal to the tangent plane.

SOLUTION The equation of the tangent plane at the point $(a, b, f(a, b))$ on the graph of $z = f(x, y)$ is

$$z = f(a,b) + f_x(a,b)(x-a) + f_y(a,b)(y-b)$$

or

$$f_x(a,b)(x-a) + f_y(a,b)(y-b) - z + f(a,b) = 0$$

Therefore, the following vector is normal to the plane:

$$\mathbf{v} = \langle f_x(a,b), f_y(a,b), -1 \rangle$$

We compute the partial derivatives of the function $f(x, y) = 3x^2 - 4y^2$:

$$\begin{aligned} f_x(x,y) &= 6x & \Rightarrow \quad & f_x(a,b) = 6a \\ f_y(x,y) &= -8y & \Rightarrow \quad & f_y(a,b) = -8b \end{aligned}$$

Therefore, the vector $\mathbf{v} = \langle 6a, -8b, -1 \rangle$ is normal to the tangent plane at (a, b). Since we want $\mathbf{n} = \langle 3, 2, 2 \rangle$ to be normal to the plane, the vectors $\mathbf{v}$ and $\mathbf{n}$ must be parallel. That is, the following must hold:

$$\frac{6a}{3} = \frac{-8b}{2} = -\frac{1}{2}$$

which implies that $a = -\frac{1}{4}$ and $b = \frac{1}{8}$. We compute the z-coordinate of the point:

$$z = 3 \cdot \left(-\frac{1}{4}\right)^2 - 4\left(\frac{1}{8}\right)^2 = \frac{1}{8}$$

The point on the graph at which the vector $\mathbf{n} = \langle 3, 2, 2 \rangle$ is normal to the tangent plane is $\left(-\frac{1}{4}, \frac{1}{8}, \frac{1}{8}\right)$.

21. CAS Use a computer algebra system to plot the graph of $f(x, y) = x^2 + 3xy + y$ together with the tangent plane at $(x, y) = (1, 1)$ on the same screen.

SOLUTION The equation of the tangent plane at the point $(1, 1)$ is

$$z = f(1, 1) + f_x(1, 1)(x - 1) + f_y(1, 1)(y - 1)$$
$$z = 5 + 5(x - 1) + 4(1, 1)(y - 1)$$
$$z = 5x + 4y - 4$$

A sketch is shown here:

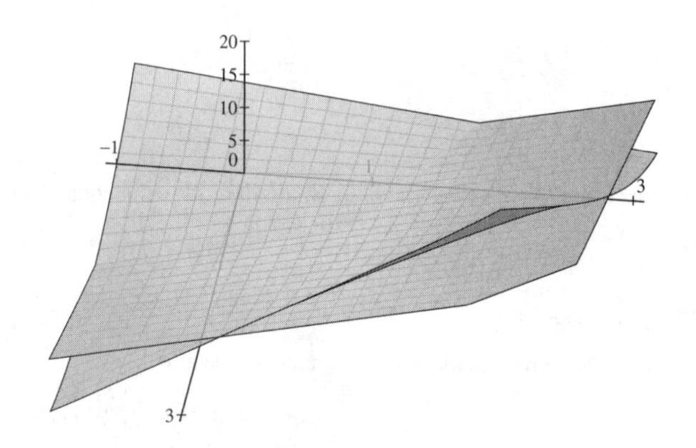

23. The following values are given:

$$f(1, 2) = 10, \qquad f(1.1, 2.01) = 10.3, \qquad f(1.04, 2.1) = 9.7$$

Find an approximation to the equation of the tangent plane to the graph of f at $(1, 2, 10)$.

SOLUTION The equation of the tangent plane at the point $(1, 2)$ is

$$z = f(1, 2) + f_x(1, 2)(x - 1) + f_y(1, 2)(y - 2)$$
$$z = 10 + f_x(1, 2)(x - 1) + f_y(1, 2)(y - 2) \tag{1}$$

Since the values of the partial derivatives at $(1, 2)$ are not given, we approximate them as follows:

$$f_x(1, 2) \approx \frac{f(1.1, 2) - f(1, 2)}{0.1} \approx \frac{f(1.1, 2.01) - f(1, 2)}{0.1} = 3$$
$$f_y(1, 2) \approx \frac{f(1, 2.1) - f(1, 2)}{0.1} \approx \frac{f(1.04, 2.1) - f(1, 2)}{0.1} = -3$$

Substituting in (1) gives the following approximation to the equation of the tangent plane:

$$z = 10 + 3(x - 1) - 3(y - 2)$$

That is, $z = 3x - 3y + 13$.

In Exercises 25–30, use the linear approximation to estimate the value. Compare with the value given by a calculator.

25. $(2.01)^3(1.02)^2$

SOLUTION The number $(2.01)^3(1.02)^2$ is a value of the function $f(x, y) = x^3y^2$. We use the linear approximation at $(2, 1)$, which is

$$f(2 + h, 1 + k) \approx f(2, 1) + f_x(2, 1)h + f_y(2, 1)k \tag{1}$$

We compute the value of the function and its partial derivatives at $(2, 1)$:

$$\begin{aligned} f(x, y) &= x^3y^2 & & f(2, 1) = 8 \\ f_x(x, y) &= 3x^2y^2 & \Rightarrow \quad & f_x(2, 1) = 12 \\ f_y(x, y) &= 2x^3y & & f_y(2, 1) = 16 \end{aligned}$$

Substituting these values and $h = 0.01$, $k = 0.02$ in (1) gives the approximation

$$(2.01)^3(1.02)^2 \approx 8 + 12 \cdot 0.01 + 16 \cdot 0.02 = 8.44$$

The value given by a calculator is 8.4487. The error is 0.0087 and the percentage error is

$$\text{Percentage error} \approx \frac{0.0087 \cdot 100}{8.4487} = 0.103\%$$

27. $\sqrt{3.01^2 + 3.99^2}$

SOLUTION This is a value of the function $f(x, y) = \sqrt{x^2 + y^2}$. We use the linear approximation at the point $(3, 4)$, which is

$$f(3 + h, 4 + k) \approx f(3, 4) + f_x(3, 4)h + f_y(3, 4)k \tag{1}$$

Using the chain Rule gives the following partial derivatives:

$$\begin{aligned} f(x, y) &= \sqrt{x^2 + y^2} & & f(3, 4) = 5 \\ f_x(x, y) &= \frac{2x}{2\sqrt{x^2 + y^2}} = \frac{x}{\sqrt{x^2 + y^2}} & \Rightarrow \quad & f_x(3, 4) = \frac{3}{5} \\ f_y(x, y) &= \frac{2y}{2\sqrt{x^2 + y^2}} = \frac{y}{\sqrt{x^2 + y^2}} & & f_y(3, 4) = \frac{4}{5} \end{aligned}$$

Substituting these values and $h = 0.01$, $k = -0.01$ in (1) gives the following approximation:

$$\sqrt{3.01^2 + 3.99^2} \approx 5 + \frac{3}{5} \cdot 0.01 + \frac{4}{5} \cdot (-0.01) = 4.998$$

The value given by a calculator is $\sqrt{3.01^2 + 3.99^2} \approx 4.99802$. The error is 0.00002 and the percentage error is at most

$$\text{Percentage error} \approx \frac{0.00002 \cdot 100}{4.99802} = 0.0004002\%$$

29. $\sqrt{(1.9)(2.02)(4.05)}$

SOLUTION We use the linear approximation of the function $f(x, y, z) = \sqrt{xyz}$ at the point $(2, 2, 4)$, which is

$$f(2 + h, 2 + k, 4 + l) \approx f(2, 2, 4) + f_x(2, 2, 4)h + f_y(2, 2, 4)k + f_z(2, 2, 4)l \tag{1}$$

We compute the values of the function and its partial derivatives at $(2, 2, 4)$:

$$\begin{aligned} f(x, y, z) &= \sqrt{xyz} & & f(2, 2, 4) = 4 \\ f_x(x, y, z) &= \frac{yz}{2\sqrt{xyz}} = \frac{1}{2}\sqrt{\frac{yz}{x}} & \Rightarrow \quad & f_x(2, 2, 4) = 1 \\ f_y(x, y, z) &= \frac{xz}{2\sqrt{xyz}} = \frac{1}{2}\sqrt{\frac{xz}{y}} & & f_y(2, 2, 4) = 1 \\ f_z(x, y, z) &= \frac{xy}{2\sqrt{xyz}} = \frac{1}{2}\sqrt{\frac{xy}{z}} & & f_z(2, 2, 4) = \frac{1}{2} \end{aligned}$$

Substituting these values and $h = -0.1$, $k = 0.02$, $l = 0.05$ in (1) gives the following approximation:

$$\sqrt{(1.9)(2.02)(4.05)} = 4 + 1 \cdot (-0.1) + 1 \cdot 0.02 + \frac{1}{2}(0.05) = 3.945$$

The value given by a calculator is:

$$\sqrt{(1.9)(2.02)(4.05)} \approx 3.9426$$

31. Estimate $f(2.1, 3.8)$ given that $f(2,4) = 5$, $f_x(2,4) = 0.3$, and $f_y(2,4) = -0.2$.

SOLUTION We use the linear approximation of f at the point $(2, 4)$, which is

$$f(2+h, 4+k) \approx f(2,4) + f_x(2,4)h + f_y(2,4)k$$

Substituting the given values and $h = 0.1$, $k = -0.2$ we obtain the following approximation:

$$f(2.1, 3.8) \approx 5 + 0.3 \cdot 0.1 + 0.2 \cdot 0.2 = 5.07.$$

In Exercises 32–34, let $I = W/H^2$ denote the BMI described in Example 6.

33. Suppose that $(W, H) = (34, 1.3)$ and W increases to 35. Use the linear approximation to estimate the increase in H required to keep I constant.

SOLUTION The linear approximation of $I = \frac{W}{H^2}$ at the point $(34, 1.3)$ is:

$$\Delta I = I(34+h, 1.3+k) - I(34, 1.3) \approx \frac{\partial I}{\partial W}(34, 1.3)h + \frac{\partial I}{\partial H}(34, 1.3)k \tag{1}$$

In the earlier exercise, we found that

$$\frac{\partial I}{\partial W}(34, 1.3) = 0.5917, \qquad \frac{\partial I}{\partial H}(34, 1.3) = -30.9513$$

We substitute these derivatives and $h = 1$ in (1), equate the resulting expression to zero and solve for k. This gives:

$$\Delta I \approx 0.5917 \cdot 1 - 30.9513 \cdot k = 0$$

$$0.5917 = 30.9513k \quad \Rightarrow \quad k = 0.0191$$

That is, for an increase in weight of 1 kg, the increase in height must be approximately 0.0191 meters (or 1.91 centimeters) in order to keep I constant.

35. The volume of a cylinder of radius r and height h is $V = \pi r^2 h$.

(a) Use the linear approximation to show that

$$\frac{\Delta V}{V} \approx \frac{2\Delta r}{r} + \frac{\Delta h}{h}$$

(b) Calculate the percentage increase in V if r and h are each increased by 2%.

(c) The volume of a certain cylinder V is determined by measuring r and h. Which will lead to a greater error in V: a 1% error in r or a 1% error in h?

SOLUTION

(a) The linear approximation is

$$\Delta V \approx V_r \Delta r + V_h \Delta h \tag{1}$$

We compute the partial derivatives of $V = \pi r^2 h$:

$$V_r = \pi h \frac{\partial}{\partial r} r^2 = 2\pi h r$$

$$V_h = \pi r^2 \frac{\partial}{\partial h} h = \pi r^2$$

Substituting in (1) gives

$$\Delta V \approx 2\pi h r \Delta r + \pi r^2 \Delta h$$

We divide by $V = \pi r^2 h$ to obtain

$$\frac{\Delta V}{V} \approx \frac{2\pi h r \Delta r}{V} + \frac{\pi r^2 \Delta h}{V} = \frac{2\pi h r \Delta r}{\pi r^2 h} + \frac{\pi r^2 \Delta h}{\pi r^2 h} = \frac{2\Delta r}{r} + \frac{\Delta h}{h}$$

That is,

$$\frac{\Delta V}{V} \approx \frac{2\Delta r}{r} + \frac{\Delta h}{h}$$

(b) The percentage increase in V is, by part (a),

$$\frac{\Delta V}{V} \cdot 100 \approx 2\frac{\Delta r}{r} \cdot 100 + \frac{\Delta h}{h} \cdot 100$$

We are given that $\frac{\Delta r}{r} \cdot 100 = 2$ and $\frac{\Delta h}{h} \cdot 100 = 2$, hence the percentage increase in V is

$$\frac{\Delta V}{V} \cdot 100 = 2 \cdot 2 + 2 = 6\%$$

(c) The percentage error in V is

$$\frac{\Delta V}{V} \cdot 100 = 2\frac{\Delta r}{r} \cdot 100 + \frac{\Delta h}{h} \cdot 100$$

A 1% error in r implies that $\frac{\Delta r}{r} \cdot 100 = 1$. Assuming that there is no error in h, we get

$$\frac{\Delta V}{V} \cdot 100 = 2 \cdot 1 + 0 = 2\%$$

A 1% in h implies that $\frac{\Delta h}{h} \cdot 100 = 1$. Assuming that there is no error in r, we get

$$\frac{\Delta V}{V} \cdot 100 = 0 + 1 = 1\%$$

We conclude that a 1% error in r leads to a greater error in V than a 1% error in h.

37. The monthly payment for a home loan is given by a function $f(P, r, N)$, where P is the principal (the initial size of the loan), r the interest rate, and N the length of the loan in months. Interest rates are expressed as a decimal: A 6% interest rate is denoted by $r = 0.06$. If $P = \$100{,}000$, $r = 0.06$, and $N = 240$ (a 20-year loan), then the monthly payment is $f(100{,}000, 0.06, 240) = 716.43$. Furthermore, with these values, we have

$$\frac{\partial f}{\partial P} = 0.0071, \qquad \frac{\partial f}{\partial r} = 5{,}769, \qquad \frac{\partial f}{\partial N} = -1.5467$$

Estimate:

(a) The change in monthly payment per \$1,000 increase in loan principal.

(b) The change in monthly payment if the interest rate increases to $r = 6.5\%$ and $r = 7\%$.

(c) The change in monthly payment if the length of the loan increases to 24 years.

SOLUTION

(a) The linear approximation to $f(P, r, N)$ is

$$\Delta f \approx \frac{\partial f}{\partial P}\Delta P + \frac{\partial f}{\partial r}\Delta r + \frac{\partial f}{\partial N}\Delta N$$

We are given that $\frac{\partial f}{\partial P} = 0.0071$, $\frac{\partial f}{\partial r} = 5769$, $\frac{\partial f}{\partial N} = -1.5467$, and $\Delta P = 1000$. Assuming that $\Delta r = 0$ and $\Delta N = 0$, we get

$$\Delta f \approx 0.0071 \cdot 1000 = 7.1$$

The change in monthly payment per thousand dollar increase in loan principal is \$7.1.

(b) By the given data, we have

$$\Delta f \approx 0.0071\Delta P + 5769\Delta r - 1.5467\Delta N \tag{1}$$

The interest rate 6.5% corresponds to $r = 0.065$, and the interest rate 7% corresponds to $r = 0.07$. In the first case $\Delta r = 0.065 - 0.06 = 0.005$ and in the second case $\Delta r = 0.07 - 0.06 = 0.01$. Substituting in (1), assuming that $\Delta P = 0$ and $\Delta N = 0$, gives

$$\Delta f = 5769 \cdot 0.005 = \$28.845$$

$$\Delta f = 5769 \cdot 0.01 = \$57.69$$

(c) We substitute $\Delta N = (24 - 20) \cdot 12 = 48$ months and $\Delta r = \Delta N = 0$ in (1) to obtain

$$\Delta f \approx -1.5467 \cdot 48 = -74.2416$$

The monthly payment will be reduced by \$74.2416.

39. The volume V of a cylinder is computed using the values 3.5 m for diameter and 6.2 m for height. Use the linear approximation to estimate the maximum error in V if each of these values has a possible error of at most 5%.

SOLUTION We denote by d and h the diameter and height of the cylinder, respectively. By the Formula for the Volume of a Cylinder we have

$$V = \pi\left(\frac{d}{2}\right)^2 h = \frac{\pi}{4}d^2 h$$

The linear approximation is

$$\Delta V \approx \frac{\partial V}{\partial d}\Delta d + \frac{\partial V}{\partial h}\Delta h \tag{1}$$

We compute the partial derivatives at $(d, h) = (3.5, 6.2)$:

$$\begin{aligned} \frac{\partial V}{\partial d}(d,h) &= \frac{\pi}{4}h \cdot 2d = \frac{\pi}{2}hd \\ \frac{\partial V}{\partial h}(d,h) &= \frac{\pi}{4}d^2 \end{aligned} \quad \Rightarrow \quad \begin{aligned} \frac{\partial V}{\partial d}(3.5, 6.2) &\approx 34.086 \\ \frac{\partial V}{\partial h}(3.5, 6.2) &= 9.621 \end{aligned}$$

Substituting these derivatives in (1) gives

$$\Delta V \approx 34.086\Delta d + 9.621\Delta h \tag{2}$$

We are given that the errors in the measurements of d and h are at most 5%. Hence,

$$\frac{\Delta d}{3.5} = 0.05 \quad \Rightarrow \quad \Delta d = 0.175$$

$$\frac{\Delta h}{6.2} = 0.05 \quad \Rightarrow \quad \Delta h = 0.31$$

Substituting in (2) we obtain

$$\Delta V \approx 34.086 \cdot 0.175 + 9.621 \cdot 0.31 \approx 8.948$$

The error in V is approximately 8.948 meters. The percentage error is at most

$$\frac{\Delta V \cdot 100}{V} = \frac{8.948 \cdot 100}{\frac{\pi}{4} \cdot 3.5^2 \cdot 6.2} = 15\%$$

Further Insights and Challenges

41. Assume that $f(0) = 0$. By the discussion in this section, $f(x)$ is differentiable at $x = 0$ if there is a constant M such that

$$f(h) = Mh + h\epsilon(h) \tag{7}$$

where $\lim_{h\to 0} \epsilon(h) = 0$ [in this case, $M = f'(0)$].

(a) Use this definition to verify that $f(x) = 1/(x+1)$ is differentiable with $M = -1$ and $E(h) = h/(h+1)$.

(b) Use this definition to verify that $f(x) = x^{3/2}$ is differentiable with $M = 0$.

(c) Show that $f(x) = x^{1/2}$ is not differentiable at $x = 0$ by showing that if M is any constant, and if we write $\sqrt{h} = Mh + h\epsilon(h)$, then the function $\epsilon(h)$ does not approach zero as $h \to 0$.

SOLUTION

(a) We define the function

$$F(x) = f(x) - 1 = \frac{1}{x+1} - 1 = \frac{-x}{x+1}$$

Then $F(0) = 0$, and F is differentiable at $x = 0$ if and only if f is differentiable at $x = 0$. We show that the definition (7) is satisfied by F. Differentiating F gives

$$F'(x) = \frac{d}{dx}\left(\frac{1}{x+1} - 1\right) = \frac{-1}{(x+1)^2} \quad \Rightarrow \quad M = F'(0) = -1$$

The function $E(h) = \frac{h}{h+1}$ satisfies $\lim_{h\to 0} E(h) = 0$, and the following equality holds:

$$Mh + hE(h) = -1 \cdot h + h \cdot \frac{h}{h+1} = \frac{-h(h+1) + h^2}{h+1} = \frac{-h}{h+1} = F(h)$$

That is, $F(x)$ is differentiable at $x = 0$, hence also $f(x)$ is differentiable at $x = 0$.

(b) We have $f(0)=0$, hence f is differentiable at $x=0$ if (7) holds. We compute M:

$$f'(x)=\frac{3}{2}x^{1/2} \quad \Rightarrow \quad M=f'(0)=0$$

We define the function $E(h)$ by

$$E(h)=\frac{f(h)}{h}=\frac{h^{3/2}}{h}=h^{1/2}$$

Then $\lim_{h\to 0} E(h)=0$ and equality (7) holds, since

$$f(h)=hE(h)=0h+hE(h)=Mh+hE(h)$$

(c) We show that $f(x)=x^{1/2}$ is not differentiable at $x=0$, by showing that for any constant M, the function $E(h)$ defined by Eq. (7) does not approach zero as $h\to 0$. For $f(h)=h^{1/2}$, Eq. (7) becomes

$$\sqrt{h}=Mh+hE(h)$$

We solve for $E(h)$ to obtain

$$\sqrt{h}-Mh=hE(h)$$

$$E(h)=\frac{\sqrt{h}-Mh}{h}=\frac{1}{\sqrt{h}}-M$$

As $h\to 0^+$, $E(h)$ is increasing without bound, therefore $E(h)$ does not approach zero as $h\to 0$. We conclude that $f(x)=\sqrt{x}$ is not differentiable at $x=0$.

15.5 The Gradient and Directional Derivatives (ET Section 14.5)

Preliminary Questions

1. Which of the following is a possible value of the gradient ∇f of a function $f(x,y)$ of two variables?

(a) 5 **(b)** $\langle 3,4\rangle$ **(c)** $\langle 3,4,5\rangle$

SOLUTION The gradient of $f(x,y)$ is a vector with two components, hence the possible value of the gradient $\nabla f=\left\langle \frac{\partial f}{\partial x},\frac{\partial f}{\partial y}\right\rangle$ is (b).

2. True or false: A differentiable function increases at the rate $\|\nabla f_P\|$ in the direction of ∇f_P?

SOLUTION The statement is true. The value $\|\nabla f_P\|$ is the rate of increase of f in the direction ∇f_P.

3. Describe the two main geometric properties of the gradient ∇f.

SOLUTION The gradient of f points in the direction of maximum rate of increase of f and is normal to the level curve (or surface) of f.

4. Express the partial derivative $\frac{\partial f}{\partial x}$ as a directional derivative $D_{\mathbf{u}}f$ for some unit vector $\mathbf{u}$.

SOLUTION The partial derivative $\frac{\partial f}{\partial x}$ is the following limit:

$$\frac{\partial f}{\partial x}(a,b)=\lim_{t\to 0}\frac{f(a+t,b)-f(a,b)}{t}$$

Considering the unit vector $\mathbf{i}=\langle 1,0\rangle$, we find that

$$\frac{\partial f}{\partial x}(a,b)=\lim_{t\to 0}\frac{f(a+t\cdot 1,b+t\cdot 0)-f(a,b)}{t}=D_{\mathbf{i}}f(a,b)$$

We see that the partial derivative $\frac{\partial f}{\partial x}$ can be viewed as a directional derivative in the direction of $\mathbf{i}$.

5. You are standing at point where the temperature gradient vector is pointing in the northeast (NE) direction. In which direction(s) should you walk to avoid a change in temperature?

(a) NE **(b)** NW **(c)** SE **(d)** SW

SOLUTION The rate of change of the temperature T at a point P in the direction of a unit vector $\mathbf{u}$, is the directional derivative $D_{\mathbf{u}}T(P)$, which is given by the formula

$$D_{\mathbf{u}}T(P)=\|\nabla f_P\|\cos\theta$$

To avoid a change in temperature, we must choose the direction $\mathbf{u}$ so that $D_{\mathbf{u}}T(P)=0$, that is, $\cos\theta=0$, so $\theta=\frac{\pi}{2}$ or $\theta=\frac{3\pi}{2}$. Since the gradient at P is pointing NE, we should walk NW or SE to avoid a change in temperature. Thus, the answer is (b) and (c).

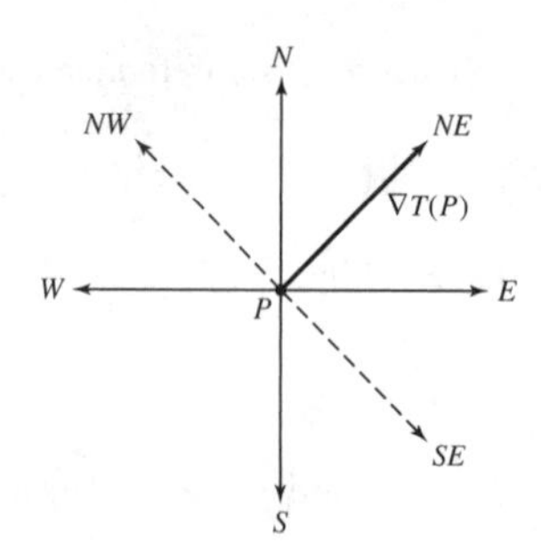

6. What is the rate of change of $f(x, y)$ at $(0, 0)$ in the direction making an angle of 45° with the x-axis if $\nabla f(0, 0) = \langle 2, 4\rangle$?

SOLUTION By the formula for directional derivatives, and using the unit vector $\langle 1/\sqrt{2}, 1/\sqrt{2}\rangle$, we get $\langle 2, 4\rangle \cdot \langle 1/\sqrt{2}, 1/\sqrt{2}\rangle = 6/\sqrt{2} = 3\sqrt{2}$.

Exercises

1. Let $f(x, y) = xy^2$ and $\mathbf{c}(t) = (\frac{1}{2}t^2, t^3)$.

(a) Calculate $\nabla f \cdot \mathbf{c}'(t)$.

(b) Use the Chain Rule for Paths to evaluate $\frac{d}{dt} f(\mathbf{c}(t))$ at $t = 1$ and $t = -1$.

SOLUTION

(a) We compute the partial derivatives of $f(x, y) = xy^2$:

$$\frac{\partial f}{\partial x} = y^2, \quad \frac{\partial f}{\partial y} = 2xy$$

The gradient vector is thus

$$\nabla f = \left\langle y^2, 2xy\right\rangle.$$

Also,

$$\mathbf{c}'(t) = \left\langle \left(\frac{1}{2}t^2\right)', \left(t^3\right)'\right\rangle = \left\langle t, 3t^2\right\rangle$$

$$\nabla f \cdot \mathbf{c}'(t) = \left\langle y^2, 2xy\right\rangle \cdot \left\langle t, 3t^2\right\rangle = y^2 t + 6xyt^2.$$

(b) Using the Chain Rule gives

$$\frac{d}{dt} f(\mathbf{c}(t)) = \frac{d}{dt}\left(\frac{1}{2}t^2 \cdot t^6\right) = \frac{d}{dt}\left(\frac{1}{2}t^8\right) = 4t^7$$

Substituting $x = \frac{1}{2}t^2$ and $y = t^3$, we obtain

$$\frac{d}{dt} f(\mathbf{c}(t)) = t^6 \cdot t + 2 \cdot \frac{1}{2}t^2 \cdot 3 \cdot t^3 \cdot t^2 = 4t^7$$

At the point $t = 1$ and $t = -1$, we get

$$\frac{d}{dt}(f(\mathbf{c}(t)))\bigg|_{t=1} = 4 \cdot 1^7 = 4, \quad \frac{d}{dt}(f(\mathbf{c}(t)))\bigg|_{t=-1} = 4 \cdot (-1)^7 = -4.$$

3. Let $f(x, y) = x^2 + y^2$ and $\mathbf{c}(t) = (\cos t, \sin t)$.

(a) Find $\frac{d}{dt} f(\mathbf{c}(t))$ without making any calculations. Explain.

(b) Verify your answer to (a) using the Chain Rule.

SOLUTION

(a) The level curves of $f(x, y)$ are the circles $x^2 + y^2 = \mathbf{c}^2$. Since $\mathbf{c}(t)$ is a parametrization of the unit circle, f has constant value 1 on $\mathbf{c}$. That is, $f(\mathbf{c}(t)) = 1$, which implies that $\frac{d}{dt} f(\mathbf{c}(t)) = 0$.

(b) We now find $\frac{d}{dt} f(\mathbf{c}(t))$ using the Chain Rule:

$$\frac{d}{dt} f(\mathbf{c}(t)) = \frac{\partial f}{\partial x}\frac{dx}{dt} + \frac{\partial f}{\partial y}\frac{dy}{dt} \tag{1}$$

We compute the derivatives involved in (1):

$$\frac{\partial f}{\partial x} = \frac{\partial}{\partial x}\left(x^2+y^2\right) = 2x, \quad \frac{\partial f}{\partial y} = \frac{\partial}{\partial y}\left(x^2+y^2\right) = 2y$$

$$\frac{dx}{dt} = \frac{d}{dt}(\cos t) = -\sin t, \quad \frac{dy}{dt} = \frac{d}{dt}(\sin t) = \cos t$$

Substituting the derivatives in (1) gives

$$\frac{d}{dt} f(\mathbf{c}(t)) = 2x(-\sin t) + 2y\cos t$$

Finally, we substitute $x = \cos t$ and $y = \sin t$ to obtain

$$\frac{d}{dt} f(\mathbf{c}(t)) = -2\cos t \sin t + 2\sin t\cos t = 0.$$

In Exercises 5–8, calculate the gradient.

5. $f(x,y) = \cos(x^2+y)$

SOLUTION We find the partial derivatives using the Chain Rule:

$$\frac{\partial f}{\partial x} = -\sin\left(x^2+y\right)\frac{\partial}{\partial x}\left(x^2+y\right) = -2x\sin\left(x^2+y\right)$$

$$\frac{\partial f}{\partial y} = -\sin\left(x^2+y\right)\frac{\partial}{\partial y}\left(x^2+y\right) = -\sin\left(x^2+y\right)$$

The gradient vector is thus

$$\nabla f = \left\langle \frac{\partial f}{\partial x}, \frac{\partial f}{\partial y}\right\rangle = \left\langle -2x\sin\left(x^2+y\right), -\sin\left(x^2+y\right)\right\rangle = -\sin\left(x^2+y\right)\langle 2x, 1\rangle$$

7. $h(x,y,z) = xyz^{-3}$

SOLUTION We compute the partial derivatives of $h(x,y,z) = xyz^{-3}$, obtaining

$$\frac{\partial h}{\partial x} = yz^{-3}, \quad \frac{\partial h}{\partial y} = xz^{-3}, \quad \frac{\partial h}{\partial z} = xy\cdot\left(-3z^{-4}\right) = -3xyz^{-4}$$

The gradient vector is thus

$$\nabla h = \left\langle \frac{\partial h}{\partial x}, \frac{\partial h}{\partial y}, \frac{\partial h}{\partial z}\right\rangle = \left\langle yz^{-3}, xz^{-3}, -3xyz^{-4}\right\rangle.$$

In Exercises 9–20, use the Chain Rule to calculate $\frac{d}{dt} f(\mathbf{c}(t))$.

9. $f(x,y) = 3x - 7y, \quad \mathbf{c}(t) = (\cos t, \sin t), \quad t = 0$

SOLUTION By the Chain Rule for paths, we have

$$\frac{d}{dt} f(\mathbf{c}(t)) = \nabla f_{\mathbf{c}(t)} \cdot \mathbf{c}'(t) \tag{1}$$

We compute the gradient and the derivative $\mathbf{c}'(t)$:

$$\nabla f = \left\langle \frac{\partial f}{\partial x}, \frac{\partial f}{\partial y}\right\rangle = \langle 3, -7\rangle, \quad \mathbf{c}'(t) = \langle -\sin t, \cos t\rangle$$

We determine these vectors at $t = 0$:

$$\mathbf{c}'(0) = \langle -\sin 0, \cos 0\rangle = \langle 0, 1\rangle$$

and since the gradient is a constant vector, we have

$$\nabla f_{\mathbf{c}(0)} = \nabla f_{(1,0)} = \langle 3, -7\rangle$$

Substituting these vectors in (1) gives

$$\frac{d}{dt} f(\mathbf{c}(t))\Big|_{t=0} = \langle 3, -7\rangle \cdot \langle 0, 1\rangle = 0 - 7 = -7$$

11. $f(x, y) = x^2 - 3xy, \quad \mathbf{c}(t) = (\cos t, \sin t), \quad t = 0$

SOLUTION By the Chain Rule For Paths we have

$$\frac{d}{dt} f(\mathbf{c}(t)) = \nabla f_{\mathbf{c}(t)} \cdot \mathbf{c}'(t) \tag{1}$$

We compute the gradient and $\mathbf{c}'(t)$:

$$\nabla f = \left\langle \frac{\partial f}{\partial x}, \frac{\partial f}{\partial y} \right\rangle = \langle 2x - 3y, -3x\rangle$$
$$\mathbf{c}'(t) = \langle -\sin t, \cos t\rangle$$

At the point $t = 0$ we have

$$\mathbf{c}(0) = (\cos 0, \sin 0) = (1, 0)$$
$$\mathbf{c}'(0) = \langle -\sin 0, \cos 0\rangle = \langle 0, 1\rangle$$
$$\nabla f\Big|_{\mathbf{c}(0)} = \nabla f_{(1,0)} = \langle 2 \cdot 1 - 3 \cdot 0, -3 \cdot 1\rangle = \langle 2, -3\rangle$$

Substituting in (1) we obtain

$$\frac{d}{dt} f(\mathbf{c}(t))\Big|_{t=0} = \langle 2, -3\rangle \cdot \langle 0, 1\rangle = -3$$

13. $f(x, y) = \sin(xy), \quad \mathbf{c}(t) = (e^{2t}, e^{3t}), \quad t = 0$

SOLUTION By the Chain Rule for Paths we have

$$\frac{d}{dt} f(\mathbf{c}(t)) = \nabla f_{\mathbf{c}(t)} \cdot \mathbf{c}'(t) \tag{1}$$

We compute the gradient and $\mathbf{c}'(t)$:

$$\nabla f = \left\langle \frac{\partial f}{\partial x}, \frac{\partial f}{\partial y} \right\rangle = \langle y\cos(xy), x\cos(xy)\rangle$$
$$\mathbf{c}'(t) = \left\langle 2e^{2t}, 3e^{3t} \right\rangle$$

At the point $t = 0$ we have

$$\mathbf{c}(0) = \left(e^0, e^0\right) = (1, 1)$$
$$\mathbf{c}'(0) = \left\langle 2e^0, 3e^0 \right\rangle = \langle 2, 3\rangle$$
$$\nabla f_{\mathbf{c}(0)} = \nabla f_{(1,1)} = \langle \cos 1, \cos 1\rangle$$

Substituting the vectors in (1) we get

$$\frac{d}{dt} f(\mathbf{c}(t))\Big|_{t=0} = \langle \cos 1, \cos 1\rangle \cdot \langle 2, 3\rangle = 5\cos 1$$

15. $f(x, y) = x - xy, \quad \mathbf{c}(t) = (t^2, t^2 - 4t), \quad t = 4$

SOLUTION We compute the gradient and $\mathbf{c}'(t)$:

$$\nabla f = \left\langle \frac{\partial f}{\partial x}, \frac{\partial f}{\partial y} \right\rangle = \langle 1 - y, -x\rangle$$
$$\mathbf{c}'(t) = (2t, 2t - 4)$$

At the point $t = 4$ we have

$$\mathbf{c}(4) = \left(4^2, 4^2 - 4 \cdot 4\right) = (16, 0)$$

$$\mathbf{c}'(4) = \langle 2\cdot 4, 2\cdot 4 - 4\rangle = \langle 8, 4\rangle$$

$$\nabla f_{\mathbf{c}(4)} = \nabla f_{(16,0)} = \langle 1-0, -16\rangle = \langle 1, -16\rangle$$

We now use the Chain Rule for Paths to compute the following derivative:

$$\left.\frac{d}{dt} f(\mathbf{c}(t))\right|_{t=4} = \nabla f_{\mathbf{c}(4)} \cdot \mathbf{c}'(4) = \langle 1, -16\rangle \cdot \langle 8, 4\rangle = 8 - 64 = -56$$

17. $f(x, y) = \ln x + \ln y, \quad \mathbf{c}(t) = (\cos t, t^2), \quad t = \frac{\pi}{4}$

SOLUTION We compute the gradient and $\mathbf{c}'(t)$:

$$\nabla f = \left\langle \frac{\partial f}{\partial x}, \frac{\partial f}{\partial y}\right\rangle = \left\langle \frac{1}{x}, \frac{1}{y}\right\rangle$$

$$\mathbf{c}'(t) = \langle -\sin t, 2t\rangle$$

At the point $t = \frac{\pi}{4}$ we have

$$\mathbf{c}\left(\frac{\pi}{4}\right) = \left(\cos\frac{\pi}{4}, \left(\frac{\pi}{4}\right)^2\right) = \left(\frac{\sqrt{2}}{2}, \frac{\pi^2}{16}\right)$$

$$\mathbf{c}'\left(\frac{\pi}{4}\right) = \left\langle -\sin\frac{\pi}{4}, \frac{2\pi}{4}\right\rangle = \left\langle -\frac{\sqrt{2}}{2}, \frac{\pi}{2}\right\rangle$$

$$\nabla f_{\mathbf{c}\left(\frac{\pi}{4}\right)} = \nabla f_{\left(\frac{\sqrt{2}}{2}, \frac{\pi^2}{16}\right)} = \left\langle \sqrt{2}, \frac{16}{\pi^2}\right\rangle$$

Using the Chain Rule for Paths we obtain the following derivative:

$$\left.\frac{d}{dt} f(\mathbf{c}(t))\right|_{t=\frac{\pi}{4}} = \nabla f_{\mathbf{c}\left(\frac{\pi}{4}\right)} \cdot \mathbf{c}'\left(\frac{\pi}{4}\right) = \left\langle \sqrt{2}, \frac{16}{\pi^2}\right\rangle \cdot \left\langle -\frac{\sqrt{2}}{2}, \frac{\pi}{2}\right\rangle = -1 + \frac{8}{\pi} \approx 1.546$$

19. $g(x, y, z) = xyz^{-1}, \quad \mathbf{c}(t) = (e^t, t, t^2), \quad t = 1$

SOLUTION By the Chain Rule for Paths we have

$$\frac{d}{dt} g(\mathbf{c}(t)) = \nabla g_{\mathbf{c}(t)} \cdot \mathbf{c}'(t) \tag{1}$$

We compute the gradient and $\mathbf{c}'(t)$:

$$\nabla g = \left\langle \frac{\partial g}{\partial x}, \frac{\partial g}{\partial y}, \frac{\partial g}{\partial z}\right\rangle = \left\langle yz^{-1}, xz^{-1}, -xyz^{-2}\right\rangle$$

$$\mathbf{c}'(t) = \left\langle e^t, 1, 2t\right\rangle$$

At the point $t = 1$ we have

$$\mathbf{c}(1) = (e, 1, 1)$$

$$\mathbf{c}'(1) = \langle e, 1, 2\rangle$$

$$\nabla g_{\mathbf{c}(1)} = \nabla g_{(e,1,1)} = \langle 1, e, -e\rangle$$

Substituting the vectors in (1) gives the following derivative:

$$\left.\frac{d}{dt} g(\mathbf{c}(t))\right|_{t=1} = \langle 1, e, -e\rangle \cdot \langle e, 1, 2\rangle = e + e - 2e = 0$$

In Exercises 21–30, calculate the directional derivative in the direction of $\mathbf{v}$ *at the given point. Remember to normalize the direction vector or use Eq. (2).*

21. $f(x, y) = x^2 + y^3, \quad \mathbf{v} = \langle 4, 3\rangle, \quad P = (1, 2)$

SOLUTION We first normalize the direction vector **v**:

$$\mathbf{u} = \frac{\mathbf{v}}{\|\mathbf{v}\|} = \frac{\langle 4, 3\rangle}{\sqrt{4^2+3^2}} = \left\langle \frac{4}{5}, \frac{3}{5}\right\rangle$$

We compute the gradient of $f(x, y) = x^2 + y^3$ at the given point:

$$\nabla f = \left\langle \frac{\partial f}{\partial x}, \frac{\partial f}{\partial y}\right\rangle = \left\langle 2x, 3y^2\right\rangle \quad \Rightarrow \quad \nabla f_{(1,2)} = \langle 2, 12\rangle$$

Using the Theorem on Evaluating Directional Derivatives, we get

$$D_{\mathbf{u}} f(1, 2) = \nabla f_{(1,2)} \cdot \mathbf{u} = \langle 2, 12\rangle \cdot \left\langle \frac{4}{5}, \frac{3}{5}\right\rangle = \frac{8}{5} + \frac{36}{5} = \frac{44}{5} = 8.8$$

23. $f(x, y) = x^2 y^3, \quad \mathbf{v} = \mathbf{i} + \mathbf{j}, \quad P = (\frac{1}{6}, 3)$

SOLUTION We normalize **v** to obtain a unit vector **u** in the direction of **v**:

$$\mathbf{u} = \frac{\mathbf{v}}{\|\mathbf{v}\|} = \frac{1}{\sqrt{2}}(\mathbf{i} + \mathbf{j}) = \frac{1}{\sqrt{2}}\mathbf{i} + \frac{1}{\sqrt{2}}\mathbf{j}$$

We compute the gradient of $f(x, y) = x^2 y^3$ at the point P:

$$\nabla f = \left\langle \frac{\partial f}{\partial x}, \frac{\partial f}{\partial y}\right\rangle = \left\langle 2xy^3, 3x^2y^2\right\rangle \quad \Rightarrow \quad \nabla f_{\left(\frac{1}{6},3\right)} = \left\langle 2 \cdot \frac{1}{6} \cdot 3^3, 3 \cdot \frac{1}{6^2} \cdot 3^2\right\rangle = \left\langle 9, \frac{3}{4}\right\rangle = 9\mathbf{i} + \frac{3}{4}\mathbf{j}$$

The directional derivative in the direction **v** is thus

$$D_{\mathbf{u}} f\left(\frac{1}{6}, 3\right) = \nabla f_{\left(\frac{1}{6},3\right)} \cdot \mathbf{u} = \left(9\mathbf{i} + \frac{3}{4}\mathbf{j}\right) \cdot \left(\frac{1}{\sqrt{2}}\mathbf{i} + \frac{1}{\sqrt{2}}\mathbf{j}\right) = \frac{9}{\sqrt{2}} + \frac{3}{4\sqrt{2}} = \frac{39}{4\sqrt{2}}$$

25. $f(x, y) = \tan^{-1}(xy), \quad \mathbf{v} = \langle 1, 1\rangle, \quad P = (3, 4)$

SOLUTION We first normalize **v** to obtain a unit vector **u** in the direction **v**:

$$\mathbf{u} = \frac{\mathbf{v}}{\|\mathbf{v}\|} = \frac{1}{\sqrt{2}}\langle 1, 1\rangle$$

We compute the gradient of $f(x, y) = \tan^{-1}(xy)$ at the point $P = (3, 4)$:

$$\nabla f = \left\langle \frac{\partial f}{\partial x}, \frac{\partial f}{\partial y}\right\rangle = \left\langle \frac{y}{1+(xy)^2}, \frac{x}{1+(xy)^2}\right\rangle = \frac{1}{1+x^2y^2}\langle y, x\rangle$$

$$\nabla f_{(3,4)} = \frac{1}{1+3^2\cdot 4^2}\langle 4, 3\rangle = \frac{1}{145}\langle 4, 3\rangle$$

Therefore, the directional derivative in the direction **v** is

$$D_{\mathbf{u}} f(3, 4) = \nabla f_{(3,4)} \cdot \mathbf{u} = \frac{1}{145}\langle 4, 3\rangle \cdot \frac{1}{\sqrt{2}}\langle 1, 1\rangle = \frac{1}{145\sqrt{2}}(4+3) = \frac{7}{145\sqrt{2}} = \frac{7\sqrt{2}}{290}$$

27. $f(x, y) = \ln(x^2 + y^2), \quad \mathbf{v} = 3\mathbf{i} - 2\mathbf{j}, \quad P = (1, 0)$

SOLUTION We normalize **v** to obtain a unit vector **u** in the direction **v**:

$$\mathbf{u} = \frac{\mathbf{v}}{\|\mathbf{v}\|} = \frac{1}{\sqrt{3^2+(-2)^2}}(3\mathbf{i} - 2\mathbf{j}) = \frac{1}{\sqrt{13}}(3\mathbf{i} - 2\mathbf{j})$$

We compute the gradient of $f(x, y) = \ln\left(x^2 + y^2\right)$ at the point $P = (1, 0)$:

$$\nabla f = \left\langle \frac{\partial f}{\partial x}, \frac{\partial f}{\partial y}\right\rangle = \left\langle \frac{2x}{x^2+y^2}, \frac{2y}{x^2+y^2}\right\rangle = \frac{2}{x^2+y^2}\langle x, y\rangle$$

$$\nabla f_{(1,0)} = \frac{2}{1^2+0^2}\langle 1, 0\rangle = \langle 2, 0\rangle = 2\mathbf{i}$$

The directional derivative in the direction **v** is thus

$$D_{\mathbf{u}} f(1, 0) = \nabla f_{(1,0)} \cdot \mathbf{u} = 2\mathbf{i} \cdot \frac{1}{\sqrt{13}}(3\mathbf{i} - 2\mathbf{j}) = \frac{6}{\sqrt{13}}$$

29. $g(x, y, z) = xe^{-yz}$, $\mathbf{v} = \langle 1, 1, 1 \rangle$, $P = (1, 2, 0)$

SOLUTION We first compute a unit vector **u** in the direction **v**:

$$\mathbf{u} = \frac{\mathbf{v}}{\|\mathbf{v}\|} = \frac{\langle 1, 1, 1 \rangle}{\sqrt{1^2 + 1^2 + 1^2}} = \frac{1}{\sqrt{3}} \langle 1, 1, 1 \rangle$$

We find the gradient of $f(x, y, z) = xe^{-yz}$ at the point $P = (1, 2, 0)$:

$$\nabla f = \left\langle \frac{\partial f}{\partial x}, \frac{\partial f}{\partial y}, \frac{\partial f}{\partial z} \right\rangle = \left\langle e^{-yz}, -xze^{-yz}, -xye^{-yz} \right\rangle = e^{-yz} \langle 1, -xz, -xy \rangle$$

$$\nabla f_{(1,2,0)} = e^0 \langle 1, 0, -2 \rangle = \langle 1, 0, -2 \rangle$$

The directional derivative in the direction **v** is thus

$$D_{\mathbf{u}} f(1, 2, 0) = \nabla f_{(1,2,0)} \cdot \mathbf{u} = \langle 1, 0, -2 \rangle \cdot \frac{1}{\sqrt{3}} \langle 1, 1, 1 \rangle = \frac{1}{\sqrt{3}} (1 + 0 - 2) = -\frac{1}{\sqrt{3}}$$

31. Find the directional derivative of $f(x, y) = x^2 + 4y^2$ at $P = (3, 2)$ in the direction pointing to the origin.

SOLUTION The direction vector is $\mathbf{v} = \overrightarrow{PO} = \langle -3, -2 \rangle$. A unit vector **u** in the direction **v** is obtained by normalizing **v**. That is,

$$\mathbf{u} = \frac{\mathbf{v}}{\|\mathbf{v}\|} = \frac{\langle -3, -2 \rangle}{\sqrt{3^2 + 2^2}} = \frac{-1}{\sqrt{13}} \langle 3, 2 \rangle$$

We compute the gradient of $f(x, y) = x^2 + 4y^2$ at the point $P = (3, 2)$:

$$\nabla f = \left\langle \frac{\partial f}{\partial x}, \frac{\partial f}{\partial y} \right\rangle = \langle 2x, 8y \rangle \quad \Rightarrow \quad \nabla f_{(3,2)} = \langle 6, 16 \rangle$$

The directional derivative is thus

$$D_{\mathbf{u}} f(3, 2) = \nabla f_{(3,2)} \cdot \mathbf{u} = \langle 6, 16 \rangle \cdot \frac{-1}{\sqrt{13}} \langle 3, 2 \rangle = \frac{-50}{\sqrt{13}}$$

33. A bug located at $(3, 9, 4)$ begins walking in a straight line toward $(5, 7, 3)$. At what rate is the bug's temperature changing if the temperature is $T(x, y, z) = xe^{y-z}$? Units are in meters and degrees Celsius.

SOLUTION The bug is walking in a straight line from the point $P = (3, 9, 4)$ towards $Q = (5, 7, 3)$, hence the rate of change in the temperature is the directional derivative in the direction of $\mathbf{v} = \overrightarrow{PQ}$. We first normalize **v** to obtain

$$\mathbf{v} = \overrightarrow{PQ} = \langle 5 - 3, 7 - 9, 3 - 4 \rangle = \langle 2, -2, -1 \rangle$$

$$\mathbf{u} = \frac{\mathbf{v}}{\|\mathbf{v}\|} = \frac{\langle 2, -2, -1 \rangle}{\sqrt{4 + 4 + 1}} = \frac{1}{3} \langle 2, -2, -1 \rangle$$

We compute the gradient of $T(x, y, z) = xe^{y-z}$ at $P = (3, 9, 4)$:

$$\nabla T = \left\langle \frac{\partial T}{\partial x}, \frac{\partial T}{\partial y}, \frac{\partial T}{\partial z} \right\rangle = \left\langle e^{y-z}, xe^{y-z}, -xe^{y-z} \right\rangle = e^{y-z} \langle 1, x, -x \rangle$$

$$\nabla T_{(3,9,4)} = e^{9-4} \langle 1, 3, -3 \rangle = e^5 \langle 1, 3, -3 \rangle$$

The rate of change of the bug's temperature at the starting point P is the directional derivative

$$D_{\mathbf{u}} f(P) = \nabla T_{(3,9,4)} \cdot \mathbf{u} = e^5 \langle 1, 3, -3 \rangle \cdot \frac{1}{3} \langle 2, -2, -1 \rangle = -\frac{e^5}{3} \approx -49.47$$

The answer is -49.47 degrees Celsius per meter.

35. Let $f(x, y) = xe^{x^2 - y}$ and $P = (1, 1)$.

(a) Calculate $\|\nabla f_P\|$.

(b) Find the rate of change of f in the direction ∇f_P.

(c) Find the rate of change of f in the direction of a vector making an angle of 45° with ∇f_P.

SOLUTION

(a) We compute the gradient of $f(x, y) = xe^{x^2-y}$. The partial derivatives are

$$\frac{\partial f}{\partial x} = 1 \cdot e^{x^2-y} + xe^{x^2-y} \cdot 2x = e^{x^2-y}\left(1 + 2x^2\right)$$
$$\frac{\partial f}{\partial y} = -xe^{x^2-y}$$

The gradient vector is thus

$$\nabla f = \left\langle \frac{\partial f}{\partial x}, \frac{\partial f}{\partial y} \right\rangle = \left\langle e^{x^2-y}\left(1 + 2x^2\right), -xe^{x^2-y} \right\rangle = e^{x^2-y}\left\langle 1 + 2x^2, -x \right\rangle$$

At the point $P = (1, 1)$ we have

$$\nabla f_P = e^0 \langle 1 + 2, -1 \rangle = \langle 3, -1 \rangle \quad \Rightarrow \quad \|\nabla f_P\| = \sqrt{3^2 + (-1)^2} = \sqrt{10}$$

(b) The rate of change of f in the direction of the gradient vector is the length of the gradient, that is, $\|\nabla f_P\| = \sqrt{10}$.

(c) Let $\mathbf{e_v}$ be the unit vector making an angle of 45° with ∇f_P. The rate of change of f in the direction of $\mathbf{e_v}$ is the directional derivative of f in the direction $\mathbf{e_v}$, which is the following dot product:

$$D_{\mathbf{e_v}} f(P) = \nabla f_P \cdot \mathbf{e_v} = \|\nabla f_P\| \|\mathbf{e_v}\| \cos 45^\circ = \sqrt{10} \cdot 1 \cdot \frac{1}{\sqrt{2}} = \sqrt{5} \approx 2.236$$

37. Let $T(x, y)$ be the temperature at location (x, y). Assume that $\nabla T = \langle y - 4, x + 2y \rangle$. Let $\mathbf{c}(t) = (t^2, t)$ be a path in the plane. Find the values of t such that

$$\frac{d}{dt} T(\mathbf{c}(t)) = 0$$

SOLUTION By the Chain Rule for Paths we have

$$\frac{d}{dt} T(\mathbf{c}(t)) = \nabla T_{\mathbf{c}(t)} \cdot \mathbf{c}'(t) \tag{1}$$

We compute the gradient vector ∇T for $x = t^2$ and $y = t$:

$$\nabla T = \left\langle t - 4, t^2 + 2t \right\rangle$$

Also $\mathbf{c}'(t) = \langle 2t, 1 \rangle$. Substituting in (1) gives

$$\frac{d}{dt} T(\mathbf{c}(t)) = \left\langle t - 4, t^2 + 2t \right\rangle \cdot \langle 2t, 1 \rangle = (t - 4) \cdot 2t + \left(t^2 + 2t\right) \cdot 1 = 3t^2 - 6t$$

We are asked to find the values of t such that

$$\frac{d}{dt} T(\mathbf{c}(t)) = 3t^2 - 6t = 0$$

We solve to obtain

$$3t^2 - 6t = 3t(t - 2) = 0 \quad \Rightarrow \quad t_1 = 0, \quad t_2 = 2$$

39. Find a vector normal to the surface $3z^3 + x^2y - y^2x = 1$ at $P = (1, -1, 1)$.

SOLUTION The gradient is normal to the level surfaces, that is ∇f_P is normal to the level surface $f(x, y, z) = 3z^3 + x^2y - y^2x = 1$. We compute the gradient vector at $P = (1, -1, 1)$:

$$\nabla f = \left\langle \frac{\partial f}{\partial x}, \frac{\partial f}{\partial y}, \frac{\partial f}{\partial z} \right\rangle = \left\langle 2xy - y^2, x^2 - 2yx, 9z^2 \right\rangle$$
$$\nabla f_P = \langle -3, 3, 9 \rangle$$

In Exercises 41–44, find an equation of the tangent plane to the surface at the given point.

41. $x^2 + 3y^2 + 4z^2 = 20, \quad P = (2, 2, 1)$

SOLUTION The equation of the tangent plane is

$$\nabla f_P \cdot \langle x-2, y-2, z-1\rangle = 0 \tag{1}$$

We compute the gradient of $f(x, y, z) = x^2 + 3y^2 + 4z^2$ at $P = (2, 2, 1)$:

$$\nabla f = \left\langle \frac{\partial f}{\partial x}, \frac{\partial f}{\partial y}, \frac{\partial f}{\partial z} \right\rangle = \langle 2x, 6y, 8z\rangle$$

At the point P we have

$$\nabla f_P = \langle 2\cdot 2, 6\cdot 2, 8\cdot 1\rangle = \langle 4, 12, 8\rangle$$

Substituting in (1) we obtain the following equation of the tangent plane:

$$\langle 4, 12, 8\rangle \cdot \langle x-2, y-2, z-1\rangle = 0$$
$$4(x-2) + 12(y-2) + 8(z-1) = 0$$
$$x - 2 + 3(y-2) + 2(z-1) = 0$$

or

$$x + 3y + 2z = 10$$

43. $x^2 + z^2 e^{y-x} = 13, \quad P = \left(2, 3, \frac{3}{\sqrt{e}}\right)$

SOLUTION We compute the gradient of $f(x, y, z) = x^2 + z^2 e^{y-x}$ at the point $P = \left(2, 3, \frac{3}{\sqrt{e}}\right)$:

$$\nabla f = \left\langle \frac{\partial f}{\partial x}, \frac{\partial f}{\partial y}, \frac{\partial f}{\partial z} \right\rangle = \left\langle 2x - z^2 e^{y-x}, z^2 e^{y-x}, 2ze^{y-x}\right\rangle$$

At the point $P = \left(2, 3, \frac{3}{\sqrt{e}}\right)$ we have

$$\nabla f_P = \left\langle 4 - \frac{9}{e}\cdot e, \frac{9}{e}\cdot e, 2\cdot\frac{3}{\sqrt{e}}\cdot e\right\rangle = \left\langle -5, 9, 6\sqrt{e}\right\rangle$$

The equation of the tangent plane at P is

$$\nabla f_P \cdot \left\langle x-2, y-3, z - \frac{3}{\sqrt{e}}\right\rangle = 0$$

That is,

$$-5(x-2) + 9(y-3) + 6\sqrt{e}\left(z - \frac{3}{\sqrt{e}}\right) = 0$$

or

$$-5x + 9y + 6\sqrt{e}z = 35$$

45. Verify what is clear from Figure 17: Every tangent plane to the cone $x^2 + y^2 - z^2 = 0$ passes through the origin.

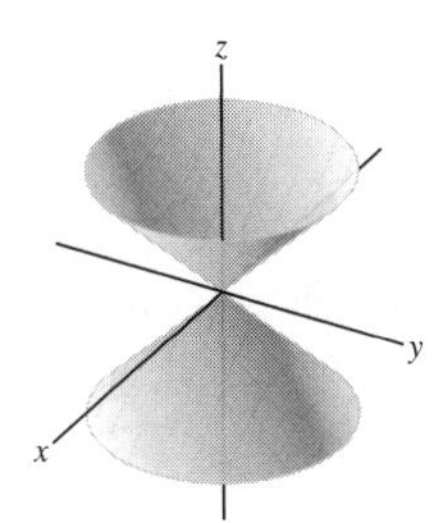

FIGURE 17 Graph of $x^2 + y^2 - z^2 = 0$.

SOLUTION The equation of the tangent plane to the surface $f(x, y, z) = x^2 + y^2 - z^2 = 0$ at the point $P = (x_0, y_0, z_0)$ on the surface is

$$\nabla f_P \cdot \langle x - x_0, y - y_0, z - z_0\rangle \tag{1}$$

We compute the gradient of $f(x, y, z) = x^2 + y^2 - z^2$ at P:

$$\nabla f = \left\langle \frac{\partial f}{\partial x}, \frac{\partial f}{\partial y}, \frac{\partial f}{\partial z} \right\rangle = \langle 2x, 2y, -2z \rangle$$

Hence,

$$\nabla f_P = \langle 2x_0, 2y_0, -2z_0 \rangle$$

Substituting in (1) we obtain the following equation of the tangent plane:

$$\begin{aligned} \langle 2x_0, 2y_0, -2z_0 \rangle \cdot \langle x - x_0, y - y_0, z - z_0 \rangle &= 0 \\ x_0(x - x_0) + y_0(y - y_0) - z_0(z - z_0) &= 0 \\ x_0x + y_0y - z_0z &= x_0^2 + y_0^2 - z_0^2 \end{aligned}$$

Since $P = (x_0, y_0, z_0)$ is on the surface, we have $x_0^2 + y_0^2 - z_0^2 = 0$. The equation of the tangent plane is thus

$$x_0x + y_0y - z_0z = 0$$

This plane passes through the origin.

47. Find a function $f(x, y, z)$ such that ∇f is the constant vector $\langle 1, 3, 1 \rangle$.

SOLUTION The gradient of $f(x, y, z)$ must satisfy the equality

$$\nabla f = \left\langle \frac{\partial f}{\partial x}, \frac{\partial f}{\partial y}, \frac{\partial f}{\partial z} \right\rangle = \langle 1, 3, 1 \rangle$$

Equating corresponding components gives

$$\frac{\partial f}{\partial x} = 1$$

$$\frac{\partial f}{\partial y} = 3$$

$$\frac{\partial f}{\partial z} = 1$$

One of the functions that satisfies these equalities is

$$f(x, y, z) = x + 3y + z$$

49. Find a function $f(x, y, z)$ such that $\nabla f = \langle x, y^2, z^3 \rangle$.

SOLUTION The following equality must hold:

$$\nabla f = \left\langle \frac{\partial f}{\partial x}, \frac{\partial f}{\partial y}, \frac{\partial f}{\partial z} \right\rangle = \left\langle x, y^2, z^3 \right\rangle$$

That is,

$$\frac{\partial f}{\partial x} = x$$

$$\frac{\partial f}{\partial y} = y^2$$

$$\frac{\partial f}{\partial z} = z^3$$

One of the functions that satisfies these equalities is

$$f(x, y, z) = \frac{1}{2}x^2 + \frac{1}{3}y^3 + \frac{1}{4}z^4$$

51. Find a function $f(x, y)$ such that $\nabla f = \langle y, x \rangle$.

SOLUTION We must find a function $f(x, y)$ such that

$$\nabla f = \left\langle \frac{\partial f}{\partial x}, \frac{\partial f}{\partial y} \right\rangle = \langle y, x \rangle$$

That is,

$$\frac{\partial f}{\partial x} = y, \quad \frac{\partial f}{\partial y} = x$$

We integrate the first equation with respect to x. Since y is treated as a constant, the constant of integration is a function of y. We get

$$f(x, y) = \int y\, dx = yx + g(y) \tag{1}$$

We differentiate f with respect to y and substitute in the second equation. This gives

$$\frac{\partial f}{\partial y} = \frac{\partial}{\partial y}(yx + g(y)) = x + g'(y)$$

Hence,

$$x + g'(y) = x \quad \Rightarrow \quad g'(y) = 0 \quad \Rightarrow \quad g(y) = C$$

Substituting in (1) gives

$$f(x, y) = yx + C$$

One of the solutions is $f(x, y) = yx$ (obtained for $C = 0$).

53. Let $\Delta f = f(a + h, b + k) - f(a, b)$ be the change in f at $P = (a, b)$. Set $\Delta \mathbf{v} = \langle h, k\rangle$. Show that the linear approximation can be written

$$\Delta f \approx \nabla f_P \cdot \Delta \mathbf{v} \tag{6}$$

SOLUTION The linear approximation is

$$\Delta f \approx f_x(a, b)h + f_y(a, b)k = \langle f_x(a, b), f_y(a, b)\rangle \cdot \langle h, k\rangle = \nabla f_P \cdot \Delta \mathbf{v}$$

55. Find the unit vector $\mathbf{n}$ normal to the surface $z^2 - 2x^4 - y^4 = 16$ at $P = (2, 2, 8)$ that points in the direction of the xy-plane.

SOLUTION The gradient vector ∇f_P is normal to the surface $f(x, y, z) = z^2 - 2x^4 - y^4 = 16$ at P. We find this vector:

$$\nabla f = \left\langle \frac{\partial f}{\partial x}, \frac{\partial f}{\partial y}, \frac{\partial f}{\partial z} \right\rangle = \left\langle -8x^3, -4y^3, 2z \right\rangle \quad \Rightarrow \quad \nabla f_{(2,2,8)} = \left\langle -8 \cdot 2^3, -4 \cdot 2^3, 2 \cdot 8 \right\rangle = \langle -64, -32, 16\rangle$$

We normalize to obtain a unit vector normal to the surface:

$$\frac{\nabla f_P}{\|\nabla f_P\|} = \frac{\langle -64, -32, 16\rangle}{\sqrt{(-64)^2 + 32^2 + 16^2}} = \frac{\langle -64, -32, 16\rangle}{16\sqrt{21}} = \frac{1}{\sqrt{21}}\langle -4, -2, 1\rangle$$

There are two unit normals to the surface at P, namely,

$$\mathbf{n} = \pm \frac{1}{\sqrt{21}}\langle -4, -2, 1\rangle$$

We need to find the normal that points in the direction of the xy-plane. Since the point $P = (2, 2, 8)$ is above the xy-plane, the normal we need has negative z-component. Hence,

$$\mathbf{n} = \frac{1}{\sqrt{21}}\langle 4, 2, -1\rangle$$

57. Let $f(x, y) = \tan^{-1}\frac{x}{y}$ and $\mathbf{u} = \left\langle \frac{\sqrt{2}}{2}, \frac{\sqrt{2}}{2} \right\rangle$.

(a) Calculate the gradient of f.

(b) Calculate $D_{\mathbf{u}} f(1, 1)$ and $D_{\mathbf{u}} f(\sqrt{3}, 1)$.

(c) Show that the lines $y = mx$ for $m \neq 0$ are level curves for f.

(d) Verify that ∇f_P is orthogonal to the level curve through P for $P = (x, y) \neq (0, 0)$.

SOLUTION

(a) We compute the partial derivatives of $f(x, y) = \tan^{-1}\frac{x}{y}$. Using the Chain Rule we get

$$\frac{\partial f}{\partial x} = \frac{1}{1+\left(\frac{x}{y}\right)^2} \cdot \frac{1}{y} = \frac{y}{x^2+y^2}$$

$$\frac{\partial f}{\partial y} = \frac{1}{1+\left(\frac{x}{y}\right)^2} \cdot \left(-\frac{x}{y^2}\right) = -\frac{x}{x^2+y^2}$$

The gradient of f is thus

$$\nabla f = \left\langle \frac{y}{x^2+y^2}, -\frac{x}{x^2+y^2} \right\rangle = \frac{1}{x^2+y^2}\langle y, -x\rangle$$

(b) By the Theorem on Evaluating Directional Derivatives,

$$D_{\mathbf{u}} f(a, b) = \nabla f_{(a,b)} \cdot \mathbf{u} \tag{1}$$

We find the values of the gradient at the two points:

$$\nabla f_{(1,1)} = \frac{1}{1^2+1^2}\langle 1, -1\rangle = \frac{1}{2}\langle 1, -1\rangle$$

$$\nabla f_{\left(\sqrt{3},1\right)} = \frac{1}{\left(\sqrt{3}\right)^2+1^2}\left\langle 1, -\sqrt{3}\right\rangle = \frac{1}{4}\left\langle 1, -\sqrt{3}\right\rangle$$

Substituting in (1) we obtain the following directional derivatives

$$D_{\mathbf{u}} f(1, 1) = \nabla f_{(1,1)} \cdot \mathbf{u} = \frac{1}{2}\langle 1, -1\rangle \cdot \left\langle \frac{\sqrt{2}}{2}, \frac{\sqrt{2}}{2} \right\rangle = 0$$

$$D_{\mathbf{u}} f\left(\sqrt{3}, 1\right) = \nabla f_{\left(\sqrt{3},1\right)} \cdot \mathbf{u} = \frac{1}{4}\left\langle 1, -\sqrt{3}\right\rangle \cdot \left\langle \frac{\sqrt{2}}{2}, \frac{\sqrt{2}}{2} \right\rangle = \frac{\sqrt{2}}{8}\left\langle 1, -\sqrt{3}\right\rangle \cdot \langle 1, 1\rangle$$

$$= \frac{\sqrt{2}}{8}\left(1-\sqrt{3}\right) = \frac{\sqrt{2}-\sqrt{6}}{8}$$

(c) Note that f is not defined for $y = 0$. For $x = 0$, the level curve of f is the y-axis, and the gradient vector is $\langle \frac{1}{y}, 0\rangle$, which is perpendicular to the y-axis. For $y \neq 0$ and $x \neq 0$, the level curves of f are the curves where $f(x, y)$ is constant. That is,

$$\tan^{-1}\frac{x}{y} = k$$

$$\frac{x}{y} = \tan k \qquad (\text{for } k \neq 0)$$

$$y = \frac{1}{\tan k}x$$

We conclude that the lines $y = mx$, $m \neq 0$, are level curves for f.

(d) By part (c), the level curve through $P = (x_0, y_0)$ is the line $y = \frac{y_0}{x_0}x$. This line has a direction vector $\left\langle 1, \frac{y_0}{x_0}\right\rangle$. The gradient at P is, by part (a), $\nabla f_P = \frac{1}{x_0^2+y_0^2}\langle y_0, -x_0\rangle$. We verify that the two vectors are orthogonal:

$$\left(1, \frac{y_0}{x_0}\right) \cdot \nabla f_P = \left\langle 1, \frac{y_0}{x_0}\right\rangle \cdot \frac{1}{x_0^2+y_0^2}\langle y_0, -x_0\rangle = \frac{1}{x_0^2+y_0^2}\left(y_0 - \frac{x_0 y_0}{x_0}\right) = 0$$

Since the dot products is zero, the two vectors are orthogonal as expected (Theorem 6).

59. Let C be the curve of intersection of the two surfaces $x^2+y^2+z^2 = 3$ and $(x-2)^2+(y-2)^2+z^2 = 3$. Use the result of Exercise 58 to find parametric equations of the tangent line to C at $P = (1, 1, 1)$.

SOLUTION The parametric equations of the tangent line to C at $P = (1, 1, 1)$ are

$$x = 1 + at, \quad y = 1 + bt, \quad z = 1 + ct \tag{1}$$

where $\mathbf{v} = \langle a, b, c\rangle$ is a direction vector for the line. By Exercise 58 $\mathbf{v}$ may be chosen as the following cross product:

$$\mathbf{v} = \nabla F_P \times \nabla G_P \tag{2}$$

where $F(x, y, z) = x^2 + y^2 + z^2$ and $G(x, y, z) = (x-2)^2 + (y-2)^2 + z^2$. We compute ∇F_P and ∇G_P:

$$\begin{aligned} F_x(x, y, z) &= 2x \\ F_y(x, y, z) &= 2y \quad \Rightarrow \quad \nabla F_P = \langle 2\cdot 1, 2\cdot 1, 2\cdot 1\rangle = \langle 2, 2, 2\rangle \\ F_z(x, y, z) &= 2z \end{aligned}$$

$$\begin{aligned} G_x(x, y, z) &= 2(x-2) \\ G_y(x, y, z) &= 2(y-2) \quad \Rightarrow \quad \nabla G_P = \langle 2(1-2), 2(1-2), 2\cdot 1\rangle = \langle -2, -2, 2\rangle \\ G_z(x, y, z) &= 2z \end{aligned}$$

Hence,

$$\mathbf{v} = \langle 2, 2, 2\rangle \times \langle -2, -2, 2\rangle = \begin{vmatrix} \mathbf{i} & \mathbf{j} & \mathbf{k} \\ 2 & 2 & 2 \\ -2 & -2 & 2 \end{vmatrix} = (4+4)\mathbf{i} - (4+4)\mathbf{j} + (-4+4)\mathbf{k} = 8\mathbf{i} - 8\mathbf{j} = \langle 8, -8, 0\rangle$$

Therefore, $\mathbf{v} = \langle a, b, c\rangle = \langle 8, -8, 0\rangle$, yielding $a = 8$, $b = -8$, $c = 0$. Substituting in (1) gives the following equations of the tangent line: $x = 1 + 8t$, $y = 1 - 8t$, $z = 1$.

61. Verify the linearity relations for gradients:

(a) $\nabla(f + g) = \nabla f + \nabla g$

(b) $\nabla(cf) = c\nabla f$

SOLUTION

(a) We use the linearity relations for partial derivative to write

$$\begin{aligned} \nabla(f+g) &= \langle (f+g)_x, (f+g)_y, (f+g)_z\rangle = \langle f_x + g_x, f_y + g_y, f_z + g_z\rangle \\ &= \langle f_x, f_y, f_z\rangle + \langle g_x, g_y, g_z\rangle = \nabla f + \nabla g \end{aligned}$$

(b) We use the linearity properties of partial derivatives to write

$$\nabla(cf) = \langle (cf)_x, (cf)_y, (cf)_z\rangle = \langle cf_x, cf_y, cf_z\rangle = c\langle f_x, f_y, f_z\rangle = c\nabla f$$

63. Prove the Product Rule for gradients (Theorem 1).

SOLUTION We must show that if $f(x, y, z)$ and $g(x, y, z)$ are differentiable, then

$$\nabla(fg) = f\nabla g + g\nabla f$$

Using the Product Rule for partial derivatives we get

$$\begin{aligned} \nabla(fg) &= \langle (fg)_x, (fg)_y, (fg)_z\rangle = \langle f_x g + f g_x, f_y g + f g_y, f_z g + f g_z\rangle \\ &= \langle f_x g, f_y g, f_z g\rangle + \langle f g_x, f g_y, f g_z\rangle = \langle f_x, f_y, f_z\rangle g + f\langle g_x, g_y, g_z\rangle = g\nabla f + f\nabla g \end{aligned}$$

Further Insights and Challenges

65. Let $f(x, y) = (xy)^{1/3}$.

(a) Use the limit definition to show that $f_x(0, 0) = f_y(0, 0) = 0$.

(b) Use the limit definition to show that the directional derivative $D_{\mathbf{u}} f(0, 0)$ does not exist for any unit vector $\mathbf{u}$ other than $\mathbf{i}$ and $\mathbf{j}$.

(c) Is f differentiable at $(0, 0)$?

SOLUTION

(a) By the limit definition and since $f(0, 0) = 0$, we have

$$f_x(0, 0) = \lim_{h\to 0} \frac{f(h, 0) - f(0, 0)}{h} = \lim_{h\to 0} \frac{(h\cdot 0)^{1/3} - 0}{h} = \lim_{h\to 0} \frac{0}{h} = 0$$

$$f_y(0, 0) = \lim_{h\to 0} \frac{f(0, h) - f(0, 0)}{h} = \lim_{h\to 0} \frac{(0\cdot h)^{1/3} - 0}{h} = \lim_{h\to 0} \frac{0}{h} = 0$$

(b) By the limit definition of the directional derivative, and for $\mathbf{u} = \langle u_1, u_2\rangle$ a unit vector, we have

$$D_{\mathbf{u}} f(0, 0) = \lim_{t\to 0} \frac{f(tu_1, tu_2) - f(0, 0)}{t} = \lim_{t\to 0} \frac{\left(t^2 u_1 u_2\right)^{1/3} - 0}{t} = \lim_{t\to 0} \frac{u_1 u_2}{t^{1/3}}$$

This limit does not exist unless $u_1 = 0$ or $u_2 = 0$. $u_1 = 0$ corresponds to the unit vector $\mathbf{j}$, and $u_2 = 0$ corresponds to the unit vector $\mathbf{i}$.

(c) If f was differentiable at $(0, 0)$, then $D_{\mathbf{u}} f(0, 0)$ would exist for any vector $\mathbf{u}$. Therefore, using the result obtained in part (b), f is not differentiable at $(0, 0)$.

67. This exercise shows that there exists a function which is not differentiable at $(0, 0)$ even though all directional derivatives at $(0, 0)$ exist. Define $f(x, y) = \dfrac{x^2 y}{x^2 + y^2}$ for $(x, y) \neq 0$ and $f(0, 0) = 0$.

(a) Use the limit definition to show that $D_{\mathbf{v}} f(0, 0)$ exists for all vectors $\mathbf{v}$. Show that $f_x(0, 0) = f_y(0, 0) = 0$.

(b) Prove that f is *not* differentiable at $(0, 0)$ by showing that Eq. (7) does not hold.

SOLUTION

(a) Let $\mathbf{v} \neq \mathbf{0}$ be the vector $\mathbf{v} = \langle v_1, v_2 \rangle$. By the definition of the derivative $D_{\mathbf{v}} f(0, 0)$, we have

$$D_{\mathbf{v}} f(0, 0) = \lim_{t \to 0} \frac{f(tv_1, tv_2) - f(0, 0)}{t} = \lim_{t \to 0} \frac{\frac{(tv_1)^2 tv_2}{(tv_1)^2 + (tv_2)^2} - 0}{t}$$

$$= \lim_{t \to 0} \frac{t^3 v_1^2 v_2}{t^3 \left(v_1^2 + v_2^2\right)} = \lim_{t \to 0} \frac{v_1^2 v_2}{v_1^2 + v_2^2} = \frac{v_1^2 v_2}{v_1^2 + v_2^2} \qquad (1)$$

Therefore $D_{\mathbf{v}} f(0, 0)$ exists for all vectors $\mathbf{v}$.

(b) In Exercise 66 we showed that if $f(x, y)$ is differentiable at $(0, 0)$ and $f(0, 0) = 0$, then

$$\lim_{(x,y) \to (0,0)} \frac{f(x, y) - f_x(0, 0)x - f_y(0, 0)y}{\sqrt{x^2 + y^2}} = 0$$

We now show that f does not satisfy the above equation. We first compute the partial derivatives $f_x(0, 0)$ and $f_y(0, 0)$. The partial derivatives f_x and f_y are the directional derivatives in the directions of $\mathbf{v} = \langle 1, 0 \rangle$ and $\mathbf{v} = \langle 0, 1 \rangle$, respectively. Substituting $v_1 = 1$, $v_2 = 0$ in (1) gives

$$f_x(0, 0) = \frac{1^2 \cdot 0}{1^2 + 0^2} = 0$$

Substituting $v_1 = 0$, $v_2 = 1$ in (1) gives

$$f_y(0, 0) = \frac{0^2 \cdot 1}{0^2 + 1^2} = 0$$

Also $f(0, 0) = 0$, therefore for $(x, y) \neq (0, 0)$ we have

$$\lim_{(x,y) \to (0,0)} \frac{f(x, y) - f_x(0, 0)x - f_y(0, 0)y}{\sqrt{x^2 + y^2}} = \lim_{(x,y) \to (0,0)} \frac{\frac{x^2 y}{x^2 + y^2} - 0x - 0y}{\sqrt{x^2 + y^2}} = \lim_{(x,y) \to (0,0)} \frac{x^2 y}{(x^2 + y^2)^{\frac{3}{2}}}$$

We compute the limit along the line $y = \sqrt{3}x$:

$$\lim_{\substack{(x,y) \to (0,0) \\ \text{along } y = \sqrt{3}x}} \frac{x^2 y}{(x^2 + y^2)^{3/2}} = \lim_{x \to 0} \frac{x^2 \sqrt{3}x}{\left(x^2 + \left(\sqrt{3}x\right)^2\right)^{3/2}} = \lim_{x \to 0} \frac{\sqrt{3}x^3}{(4x^2)^{3/2}} = \lim_{x \to 0} \frac{\sqrt{3}x^3}{8x^3} = \frac{\sqrt{3}}{8} \neq 0$$

Since this limit is not zero, f does not satisfy Eq. (7), hence f is not differentiable at $(0, 0)$.

69. Prove the following Quotient Rule where f, g are differentiable:

$$\nabla \left(\frac{f}{g} \right) = \frac{g \nabla f - f \nabla g}{g^2}$$

SOLUTION The Quotient Rule is valid for partial derivatives, therefore

$$\nabla \left(\frac{f}{g} \right) = \left\langle \frac{\partial}{\partial x} \left(\frac{f}{g} \right), \frac{\partial}{\partial y} \left(\frac{f}{g} \right), \frac{\partial}{\partial z} \left(\frac{f}{g} \right) \right\rangle = \left\langle \frac{g \frac{\partial f}{\partial x} - f \frac{\partial g}{\partial x}}{g^2}, \frac{g \frac{\partial f}{\partial y} - f \frac{\partial g}{\partial y}}{g^2}, \frac{g \frac{\partial f}{\partial z} - f \frac{\partial g}{\partial z}}{g^2} \right\rangle$$

$$= \left\langle \frac{g \frac{\partial f}{\partial x}}{g^2}, \frac{g \frac{\partial f}{\partial y}}{g^2}, \frac{g \frac{\partial f}{\partial z}}{g^2} \right\rangle - \left\langle \frac{f \frac{\partial g}{\partial x}}{g^2}, \frac{f \frac{\partial g}{\partial y}}{g^2}, \frac{f \frac{\partial g}{\partial z}}{g^2} \right\rangle = \frac{g}{g^2} \left\langle \frac{\partial f}{\partial x}, \frac{\partial f}{\partial y}, \frac{\partial f}{\partial z} \right\rangle - \frac{f}{g^2} \left\langle \frac{\partial g}{\partial x}, \frac{\partial g}{\partial y}, \frac{\partial g}{\partial z} \right\rangle$$

$$= \frac{g}{g^2} \nabla f - \frac{f}{g^2} \nabla g = \frac{g \nabla f - f \nabla g}{g^2}$$

In Exercises 70–72, a path $\mathbf{c}(t) = (x(t), y(t))$ follows the gradient of a function $f(x, y)$ if the tangent vector $\mathbf{c}'(t)$ points in the direction of ∇f for all t. In other words, $\mathbf{c}'(t) = k(t)\nabla f_{\mathbf{c}(t)}$ for some positive function $k(t)$. Note that in this case, $\mathbf{c}(t)$ crosses each level curve of $f(x, y)$ at a right angle.

71. Find a path of the form $\mathbf{c}(t) = (t, g(t))$ passing through $(1, 2)$ that follows the gradient of $f(x, y) = 2x^2 + 8y^2$ (Figure 18). *Hint:* Use Separation of Variables.

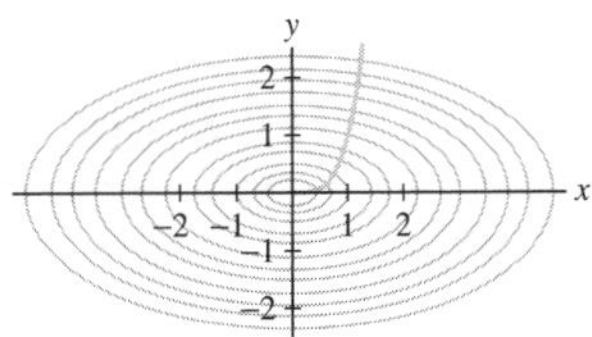

FIGURE 18 The path $\mathbf{c}(t)$ is orthogonal to the level curves of $f(x, y) = 2x^2 + 8y^2$.

SOLUTION By the previous exercise, if $\mathbf{c}(t) = (x(t), y(t))$ follows the gradient of f, then

$$\frac{dy}{dx} = \frac{y'(t)}{x'(t)} = \frac{f_y}{f_x} \tag{1}$$

We find the partial derivatives of f:

$$f_y = \frac{\partial}{\partial y}\left(2x^2 + 8y^2\right) = 16y, \quad f_x = \frac{\partial}{\partial x}\left(2x^2 + 8y^2\right) = 4x$$

Substituting in (1) we get

$$\frac{dy}{dx} = \frac{16y}{4x} = \frac{4y}{x}$$

We solve the differential equation using separation of variables. We obtain

$$\frac{dy}{y} = 4\frac{dx}{x}$$

$$\int \frac{dy}{y} = 4\int \frac{dx}{x}$$

$$\ln y = 4\ln x + c = \ln x^4 + c$$

or

$$y = e^{\ln x^4 + c} = e^c x^4$$

Denoting $k = e^c$, we obtain the following solution:

$$y = kx^4$$

The corresponding path may be parametrized using the parameter $x = t$ as

$$\mathbf{c}(t) = \left(t, kt^4\right) \tag{2}$$

Since we want the path to pass through $(1, 2)$, there must be a solution t for the equation

$$\left(t, kt^4\right) = (1, 2)$$

or

$$\begin{aligned} t &= 1 \\ kt^4 &= 2 \end{aligned} \quad \Rightarrow \quad k \cdot 1^4 = 2 \quad \Rightarrow \quad k = 2$$

Substituting in (2) we obtain the following path:

$$\mathbf{c}(t) = \left(t, 2t^4\right)$$

We now show that $\mathbf{c}$ follows the gradient of $f(x, y) = 2x^2 + 8y^2$. We have

$$\mathbf{c}'(t) = \left(1, 8t^3\right) \quad \text{and} \quad \nabla f = \left\langle f_x, f_y \right\rangle = \langle 4x, 16y \rangle$$

Therefore, $\nabla f_{\mathbf{c}(t)} = \left\langle 4t, 16 \cdot 2t^4 \right\rangle = \left\langle 4t, 32t^4 \right\rangle$, so we obtain

$$\mathbf{c}'(t) = \left(1, 8t^3\right) = \frac{1}{4t}\left\langle 4t, 32t^4 \right\rangle = \frac{1}{4t}\nabla f_{\mathbf{c}(t)}, \quad t \neq 0$$

For $t = 0$, $\nabla f_{\mathbf{c}(0)} = \nabla f_{(0,0)} = \langle 0, 0 \rangle$ and $\mathbf{c}'(0) = \langle 1, 0 \rangle$. We conclude that $\mathbf{c}$ follows the gradient of f for $t \neq 0$.

15.6 The Chain Rule (ET Section 14.6)

Preliminary Questions

1. Consider a function $f(x, y)$ where $x = uv$ and $y = u/v$.

(a) What are the primary derivatives of f?

(b) What are the independent variables?

SOLUTION

(a) The primary derivatives of f are $\frac{\partial f}{\partial x}$ and $\frac{\partial f}{\partial y}$.

(b) The independent variables are u and v, on which x and y depend.

In Questions 2–4, suppose that $f(u, v) = ue^v$, where $u = rs$ and $v = r + s$.

2. The composite function $f(u, v)$ is equal to:

(a) rse^{r+s} **(b)** re^s **(c)** rse^{rs}

SOLUTION The composite function $f(u, v)$ is obtained by replacing u and v in the formula for $f(u, v)$ by the corresponding functions $u = rs$ and $v = r + s$. This gives

$$f\big(u(r, s), v(r, s)\big) = u(r, s)e^{v(r,s)} = rse^{r+s}$$

Answer (a) is the correct answer.

3. What is the value of $f(u, v)$ at $(r, s) = (1, 1)$?

SOLUTION We compute $u = rs$ and $v = r + s$ at the point $(r, s) = (1, 1)$:

$$u(1, 1) = 1 \cdot 1 = 1; \quad v(1, 1) = 1 + 1 = 2$$

Substituting in $f(u, v) = ue^v$, we get

$$f(u, v)\Big|_{(r,s)=(1,1)} = 1 \cdot e^2 = e^2.$$

4. According to the Chain Rule, $\frac{\partial f}{\partial r}$ is equal to (choose correct answer):

(a) $\frac{\partial f}{\partial x}\frac{\partial x}{\partial r} + \frac{\partial f}{\partial x}\frac{\partial x}{\partial s}$

(b) $\frac{\partial f}{\partial x}\frac{\partial x}{\partial r} + \frac{\partial f}{\partial y}\frac{\partial y}{\partial r}$

(c) $\frac{\partial f}{\partial r}\frac{\partial r}{\partial x} + \frac{\partial f}{\partial s}\frac{\partial s}{\partial x}$

SOLUTION For a function $f(x, y)$ where $x = x(r, s)$ and $y = y(r, s)$, the Chain Rule states that the partial derivative $\frac{\partial f}{\partial r}$ is as given in (b). That is,

$$\frac{\partial f}{\partial x}\frac{\partial x}{\partial r} + \frac{\partial f}{\partial y}\frac{\partial y}{\partial r}$$

5. Suppose that x, y, z are functions of the independent variables u, v, w. Given a function $f(x, y, z)$, which of the following terms appear in the Chain Rule expression for $\frac{\partial f}{\partial w}$?

(a) $\frac{\partial f}{\partial v}\frac{\partial x}{\partial v}$ **(b)** $\frac{\partial f}{\partial w}\frac{\partial w}{\partial x}$

(c) $\frac{\partial f}{\partial z}\frac{\partial z}{\partial w}$ **(d)** $\frac{\partial f}{\partial v}\frac{\partial v}{\partial w}$

SOLUTION By the Chain Rule, the derivative $\frac{\partial f}{\partial w}$ is

$$\frac{\partial f}{\partial w} = \frac{\partial f}{\partial x}\frac{\partial x}{\partial w} + \frac{\partial f}{\partial y}\frac{\partial y}{\partial w} + \frac{\partial f}{\partial z}\frac{\partial z}{\partial w}$$

Therefore (c) is the only correct answer.

6. With notation as in the previous question, does $\frac{\partial x}{\partial v}$ appear in the Chain Rule expression for $\frac{\partial f}{\partial u}$?

SOLUTION The Chain Rule expression for $\frac{\partial f}{\partial u}$ is

$$\frac{\partial f}{\partial u} = \frac{\partial f}{\partial x}\frac{\partial x}{\partial u} + \frac{\partial f}{\partial y}\frac{\partial y}{\partial u} + \frac{\partial f}{\partial z}\frac{\partial z}{\partial u}$$

The derivative $\frac{\partial x}{\partial v}$ does not appear in differentiating f with respect to the independent variable u.

Exercises

1. Let $f(x, y, z) = x^2y^3 + z^4$ and $x = s^2$, $y = st^2$, and $z = s^2t$.

(a) Calculate the primary derivatives $\frac{\partial f}{\partial x}, \frac{\partial f}{\partial y}, \frac{\partial f}{\partial z}$.

(b) Calculate $\frac{\partial x}{\partial s}, \frac{\partial y}{\partial s}, \frac{\partial z}{\partial s}$.

(c) Compute $\frac{\partial f}{\partial s}$ using the Chain Rule:

$$\frac{\partial f}{\partial s} = \frac{\partial f}{\partial x}\frac{\partial x}{\partial s} + \frac{\partial f}{\partial y}\frac{\partial y}{\partial s} + \frac{\partial f}{\partial z}\frac{\partial z}{\partial s}$$

Express the answer in terms of the independent variables s, t.

SOLUTION

(a) The primary derivatives of $f(x, y, z) = x^2y^3 + z^4$ are

$$\frac{\partial f}{\partial x} = 2xy^3, \quad \frac{\partial f}{\partial y} = 3x^2y^2, \quad \frac{\partial f}{\partial z} = 4z^3$$

(b) The partial derivatives of x, y, and z with respect to s are

$$\frac{\partial x}{\partial s} = 2s, \quad \frac{\partial y}{\partial s} = t^2, \quad \frac{\partial z}{\partial s} = 2st$$

(c) We use the Chain Rule and the partial derivatives computed in parts (a) and (b) to find the following derivative:

$$\frac{\partial f}{\partial s} = \frac{\partial f}{\partial x}\frac{\partial x}{\partial s} + \frac{\partial f}{\partial y}\frac{\partial y}{\partial s} + \frac{\partial f}{\partial z}\frac{\partial z}{\partial s} = 2xy^3 \cdot 2s + 3x^2y^2t^2 + 4z^3 \cdot 2st = 4xy^3s + 3x^2y^2t^2 + 8z^3st$$

To express the answer in terms of the independent variables s, t we substitute $x = s^2$, $y = st^2$, $z = s^2t$. This gives

$$\frac{\partial f}{\partial s} = 4s^2(st^2)^3s + 3(s^2)^2(st^2)^2t^2 + 8(s^2t)^3st = 4s^6t^6 + 3s^6t^6 + 8s^7t^4 = 7s^6t^6 + 8s^7t^4.$$

In Exercises 3–10, use the Chain Rule to calculate the partial derivatives. Express the answer in terms of the independent variables.

3. $\frac{\partial f}{\partial s}, \frac{\partial f}{\partial r}$: $f(x, y, z) = xy + z^2$, $x = s^2$, $y = 2rs$, $z = r^2$

SOLUTION We perform the following steps:

Step 1. Compute the primary derivatives. The primary derivatives of $f(x, y, z) = xy + z^2$ are

$$\frac{\partial f}{\partial x} = y, \quad \frac{\partial f}{\partial y} = x, \quad \frac{\partial f}{\partial z} = 2z$$

Step 2. Apply the Chain Rule. By the Chain Rule,

$$\frac{\partial f}{\partial s} = \frac{\partial f}{\partial x} \cdot \frac{\partial x}{\partial s} + \frac{\partial f}{\partial y} \cdot \frac{\partial y}{\partial s} + \frac{\partial f}{\partial z} \cdot \frac{\partial z}{\partial s} \tag{1}$$

$$\frac{\partial f}{\partial r} = \frac{\partial f}{\partial x} \cdot \frac{\partial x}{\partial r} + \frac{\partial f}{\partial y} \cdot \frac{\partial y}{\partial r} + \frac{\partial f}{\partial z} \cdot \frac{\partial z}{\partial r} \tag{2}$$

We compute the partial derivatives of x, y, z with respect to s and r:

$$\frac{\partial x}{\partial s} = 2s, \quad \frac{\partial y}{\partial s} = 2r, \quad \frac{\partial z}{\partial s} = 0.$$

$$\frac{\partial x}{\partial r} = 0, \quad \frac{\partial y}{\partial r} = 2s, \quad \frac{\partial z}{\partial r} = 2r.$$

Substituting these derivatives and the primary derivatives computed in step 1 in (1) and (2), we get

$$\frac{\partial f}{\partial s} = y \cdot 2s + x \cdot 2r + 2z \cdot 0 = 2ys + 2xr$$

$$\frac{\partial f}{\partial r} = y \cdot 0 + x \cdot 2s + 2z \cdot 2r = 2xs + 4zr$$

Step 3. Express the answer in terms of r and s. We substitute $x = s^2$, $y = 2rs$, and $z = r^2$ in $\frac{\partial f}{\partial s}$ and $\frac{\partial f}{\partial r}$ in step 2, to obtain

$$\frac{\partial f}{\partial s} = 2rs \cdot 2s + s^2 \cdot 2r = 4rs^2 + 2rs^2 = 6rs^2.$$

$$\frac{\partial f}{\partial r} = 2s^2 \cdot s + 4r^2 \cdot r = 2s^3 + 4r^3.$$

5. $\frac{\partial g}{\partial u}, \frac{\partial g}{\partial v}$: $g(x, y) = \cos(x^2 - y^2)$, $x = 2u - 3v$, $y = -5u + 8v$

SOLUTION We use the following steps:

Step 1. Compute the primary derivatives. The primary derivatives of $g(x, y) = \cos(x^2 - y^2)$ are

$$\frac{\partial g}{\partial x} = -2x \sin(x^2 - y^2), \quad \frac{\partial g}{\partial y} = 2y \sin(x^2 - y^2)$$

Step 2. Apply the Chain Rule. By the Chain Rule,

$$\frac{\partial g}{\partial u} = \frac{\partial g}{\partial x}\frac{\partial x}{\partial u} + \frac{\partial g}{\partial y}\frac{\partial y}{\partial u} = -2x \sin(x^2 - y^2)\frac{\partial x}{\partial u} + 2y \sin(x^2 - y^2)\frac{\partial y}{\partial u}$$

$$\frac{\partial g}{\partial v} = \frac{\partial g}{\partial x}\frac{\partial x}{\partial v} + \frac{\partial g}{\partial y}\frac{\partial y}{\partial v} = -2x \sin(x^2 - y^2)\frac{\partial x}{\partial v} + 2y \sin(x^2 - y^2)\frac{\partial y}{\partial v}$$

We find the partial derivatives of x and y:

$$\frac{\partial x}{\partial u} = 2, \quad \frac{\partial y}{\partial u} = -5$$

$$\frac{\partial x}{\partial v} = -3, \quad \frac{\partial y}{\partial v} = 8$$

We substitute these derivatives to obtain

$$\frac{\partial g}{\partial u} = -4x \sin(x^2 - y^2) - 10y \sin(x^2 - y^2) = -(4x + 10y) \sin(x^2 - y^2) \tag{1}$$

$$\frac{\partial g}{\partial v} = 6x \sin(x^2 - y^2) + 16y \sin(x^2 - y^2) = (6x + 16y) \sin(x^2 - y^2) \tag{2}$$

Step 3. Express the answer in terms of u and v. We substitute $x = 2u - 3v$, $y = -5u + 8v$ in (1) and (2), to obtain

$$\frac{\partial g}{\partial u} = -(8u - 12v - 50u + 80v) \cdot \sin\left((2u - 3v)^2 - (-5u + 8v)^2\right)$$
$$= (42u - 68v) \sin\left(68uv - 21u^2 - 55v^2\right)$$

$$\frac{\partial g}{\partial v} = (12u - 18v - 80u + 128v) \cdot \sin\left((2u - 3v)^2 - (-5u + 8v)^2\right)$$
$$= (110v - 68u) \sin\left(68uv - 21u^2 - 55v^2\right)$$

7. $\frac{\partial F}{\partial x}, \frac{\partial F}{\partial y}$: $F(u, v) = e^{u+v}$, $u = x^2$, $v = xy$

SOLUTION We use the following steps:

Step 1. Compute the primary derivatives. The primary derivatives of $F(u, v) = e^{u+v}$ are

$$\frac{\partial f}{\partial u} = e^{u+v}, \quad \frac{\partial f}{\partial v} = e^{u+v}$$

Step 2. Apply the Chain Rule. By the Chain Rule,

$$\frac{\partial F}{\partial x} = \frac{\partial F}{\partial u}\frac{\partial u}{\partial x} + \frac{\partial F}{\partial v}\frac{\partial v}{\partial x} = e^{u+v}\frac{\partial u}{\partial x} + e^{u+v}\frac{\partial v}{\partial x} = e^{u+v}\left(\frac{\partial u}{\partial x} + \frac{\partial v}{\partial x}\right)$$

$$\frac{\partial F}{\partial y} = \frac{\partial F}{\partial u}\frac{\partial u}{\partial y} + \frac{\partial F}{\partial v}\frac{\partial v}{\partial y} = e^{u+v}\frac{\partial u}{\partial y} + e^{u+v}\frac{\partial v}{\partial y} = e^{u+v}\left(\frac{\partial u}{\partial y} + \frac{\partial v}{\partial y}\right)$$

We compute the partial derivatives of u and v with respect to x and y:

$$\frac{\partial u}{\partial x} = 2x, \quad \frac{\partial v}{\partial x} = y$$

$$\frac{\partial u}{\partial y} = 0, \quad \frac{\partial v}{\partial y} = x$$

We substitute to obtain

$$\frac{\partial F}{\partial x} = (2x + y)e^{u+v} \tag{1}$$

$$\frac{\partial F}{\partial y} = xe^{u+v} \tag{2}$$

Step 3. Express the answer in terms of x and y. We substitute $u = x^2$, $v = xy$ in (1) and (2), obtaining

$$\frac{\partial F}{\partial x} = (2x + y)e^{x^2+xy}, \quad \frac{\partial F}{\partial y} = xe^{x^2+xy}.$$

9. $\frac{\partial f}{\partial x}, \frac{\partial f}{\partial y}$: $f(r, \theta) = r\sin^2\theta$, $x = r\cos\theta$, $y = r\sin\theta$

SOLUTION We use the following steps:

Step 1. Compute the primary derivatives. The primary derivatives of $f(r, \theta) = r\sin^2\theta$ are

$$\frac{\partial f}{\partial r} = \sin^2\theta, \quad \frac{\partial f}{\partial \theta} = r \cdot 2\sin\theta\cos\theta = r\sin 2\theta$$

Step 2. Apply the Chain Rule. By the Chain Rule,

$$\frac{\partial f}{\partial x} = \frac{\partial f}{\partial r}\frac{\partial r}{\partial x} + \frac{\partial f}{\partial \theta}\frac{\partial \theta}{\partial x} = \sin^2\theta\frac{\partial r}{\partial x} + r\sin 2\theta\frac{\partial \theta}{\partial x} \tag{1}$$

$$\frac{\partial f}{\partial y} = \frac{\partial f}{\partial r}\frac{\partial r}{\partial y} + \frac{\partial f}{\partial \theta}\frac{\partial \theta}{\partial y} = \sin^2\theta\frac{\partial r}{\partial y} + r\sin 2\theta\frac{\partial \theta}{\partial y} \tag{2}$$

We compute the partial derivatives of r and θ with respect to x and y. Since $r^2 = x^2 + y^2$, we have

$$2r\frac{\partial r}{\partial x} = 2x \quad \Rightarrow \quad \frac{\partial r}{\partial x} = \frac{x}{r}$$

$$2r\frac{\partial r}{\partial y} = 2y \quad \Rightarrow \quad \frac{\partial r}{\partial y} = \frac{y}{r}$$

By the relation $\tan\theta = \frac{y}{x}$, we get

$$\frac{1}{\cos^2\theta}\frac{\partial \theta}{\partial x} = -\frac{y}{x^2} \quad \Rightarrow \quad \frac{\partial \theta}{\partial x} = -\frac{y\cos^2\theta}{x^2}$$

$$\frac{1}{\cos^2\theta}\frac{\partial \theta}{\partial y} = \frac{1}{x} \quad \Rightarrow \quad \frac{\partial \theta}{\partial y} = \frac{\cos^2\theta}{x}$$

Substituting these derivatives in (1) and (2) gives

$$\frac{\partial f}{\partial x} = \sin^2\theta \cdot \frac{x}{r} + r\sin 2\theta\left(-\frac{y\cos^2\theta}{x^2}\right) = \frac{x\sin^2\theta}{r} - \frac{yr\sin 2\theta\cos^2\theta}{x^2} \tag{3}$$

$$\frac{\partial f}{\partial y} = \sin^2\theta \cdot \frac{y}{r} + r\sin 2\theta\frac{\cos^2\theta}{x} = \frac{y\sin^2\theta}{r} + \frac{r\sin 2\theta\cos^2\theta}{x} \tag{4}$$

Step 3. Express answer in terms of x and y. We express r, $\cos\theta$, and $\sin 2\theta$ in terms of x and y. Since $r = \sqrt{x^2+y^2}$, we have

$$\cos\theta = \frac{x}{r} = \frac{x}{\sqrt{x^2+y^2}}$$

$$\sin\theta = \frac{y}{r} = \frac{y}{\sqrt{x^2+y^2}}$$

$$\sin 2\theta = 2\sin\theta\cos\theta = \frac{2xy}{x^2+y^2}$$

We substitute in (3) and (4) to obtain

$$\frac{\partial f}{\partial x} = \frac{x}{\sqrt{x^2+y^2}}\cdot\frac{y^2}{x^2+y^2} - \frac{y\sqrt{x^2+y^2}}{x^2}\cdot\frac{2xy}{x^2+y^2}\cdot\frac{x^2}{x^2+y^2}$$

$$= \frac{xy^2}{(x^2+y^2)^{3/2}} - \frac{2y^2x}{(x^2+y^2)^{3/2}} = \frac{-xy^2}{(x^2+y^2)^{3/2}}$$

$$\frac{\partial f}{\partial y} = \frac{y}{\sqrt{x^2+y^2}}\cdot\frac{y^2}{x^2+y^2} + \frac{\sqrt{x^2+y^2}}{x}\cdot\frac{2xy}{x^2+y^2}\cdot\frac{x^2}{x^2+y^2}$$

$$= \frac{y^3}{(x^2+y^2)^{3/2}} + \frac{2x^2y}{(x^2+y^2)^{3/2}} = \frac{y(y^2+2x^2)}{(x^2+y^2)^{3/2}}$$

In Exercises 11–16, use the Chain Rule to evaluate the partial derivative at the point specified.

11. $\frac{\partial f}{\partial u}$ and $\frac{\partial f}{\partial v}$ at $(u, v) = (-1, -1)$, where $f(x, y, z) = x^3 + yz^2$, $x = u^2 + v$, $y = u + v^2$, $z = uv$.

SOLUTION The primary derivatives of $f(x, y, z) = x^3 + yz^2$ are

$$\frac{\partial f}{\partial x} = 3x^2, \quad \frac{\partial f}{\partial y} = z^2, \quad \frac{\partial f}{\partial z} = 2yz$$

By the Chain Rule we have

$$\frac{\partial f}{\partial u} = \frac{\partial f}{\partial x}\frac{\partial x}{\partial u} + \frac{\partial f}{\partial y}\frac{\partial y}{\partial u} + \frac{\partial f}{\partial z}\frac{\partial z}{\partial u} = 3x^2\frac{\partial x}{\partial u} + z^2\frac{\partial y}{\partial u} + 2yz\frac{\partial z}{\partial u} \tag{1}$$

$$\frac{\partial f}{\partial v} = \frac{\partial f}{\partial x}\frac{\partial x}{\partial v} + \frac{\partial f}{\partial y}\frac{\partial y}{\partial v} + \frac{\partial f}{\partial z}\frac{\partial z}{\partial v} = 3x^2\frac{\partial x}{\partial v} + z^2\frac{\partial y}{\partial v} + 2yz\frac{\partial z}{\partial v} \tag{2}$$

We compute the partial derivatives of x, y, and z with respect to u and v:

$$\frac{\partial x}{\partial u} = 2u, \quad \frac{\partial y}{\partial u} = 1, \quad \frac{\partial z}{\partial u} = v$$

$$\frac{\partial x}{\partial v} = 1, \quad \frac{\partial y}{\partial v} = 2v, \quad \frac{\partial z}{\partial v} = u$$

Substituting in (1) and (2) we get

$$\frac{\partial f}{\partial u} = 6x^2u + z^2 + 2yzv \tag{3}$$

$$\frac{\partial f}{\partial v} = 3x^2 + 2vz^2 + 2yzu \tag{4}$$

We determine (x, y, z) for $(u, v) = (-1, -1)$:

$$x = (-1)^2 - 1 = 0, \quad y = -1 + (-1)^2 = 0, \quad z = (-1)\cdot(-1) = 1.$$

Finally, we substitute $(x, y, z) = (0, 0, 1)$ and $(u, v) = (-1, -1)$ in (3), (4) to obtain the following derivatives:

$$\left.\frac{\partial f}{\partial u}\right|_{(u,v)=(-1,-1)} = 6\cdot 0^2\cdot(-1) + 1^2 + 2\cdot 0\cdot 1\cdot(-1) = 1$$

$$\left.\frac{\partial f}{\partial v}\right|_{(u,v)=(-1,-1)} = 3\cdot 0^2 + 2\cdot(-1)\cdot 1^2 + 2\cdot 0\cdot 1\cdot(-1) = -2$$

13. $\frac{\partial g}{\partial \theta}$ at $(r, \theta) = (2\sqrt{2}, \frac{\pi}{4})$, where $g(x, y) = \frac{1}{x + y^2}$, $x = r\sin\theta$, $y = r\cos\theta$.

SOLUTION We compute the primary derivatives of $g(x, y) = \frac{1}{x+y^2}$:

$$\frac{\partial g}{\partial x} = -\frac{1}{(x+y^2)^2}, \quad \frac{\partial g}{\partial y} = -\frac{2y}{(x+y^2)^2}$$

By the Chain Rule we have

$$\frac{\partial g}{\partial \theta} = \frac{\partial g}{\partial x}\frac{\partial x}{\partial \theta} + \frac{\partial g}{\partial y}\frac{\partial y}{\partial \theta} = -\frac{1}{(x+y^2)^2}\frac{\partial x}{\partial \theta} - \frac{2y}{(x+y^2)^2}\frac{\partial y}{\partial \theta} = -\frac{1}{(x+y^2)^2}\left(\frac{\partial x}{\partial \theta} + 2y\frac{\partial y}{\partial \theta}\right)$$

We find the partial derivatives $\frac{\partial x}{\partial \theta}, \frac{\partial y}{\partial \theta}$:

$$\frac{\partial x}{\partial \theta} = r\cos\theta, \quad \frac{\partial y}{\partial \theta} = -r\sin\theta$$

Hence,

$$\frac{\partial g}{\partial \theta} = -\frac{r}{(x+y^2)^2}(\cos\theta - 2y\sin\theta) \tag{1}$$

At the point $(r, \theta) = \left(2\sqrt{2}, \frac{\pi}{4}\right)$, we have $x = 2\sqrt{2}\sin\frac{\pi}{4} = 2$ and $y = 2\sqrt{2}\cos\frac{\pi}{4} = 2$. Substituting $(r, \theta) = \left(2\sqrt{2}, \frac{\pi}{4}\right)$ and $(x, y) = (2, 2)$ in (1) gives the following derivative:

$$\left.\frac{\partial g}{\partial \theta}\right|_{(r,\theta)=\left(2\sqrt{2}, \frac{\pi}{4}\right)} = \frac{-2\sqrt{2}}{(2+2^2)^2}\left(\cos\frac{\pi}{4} - 4\sin\frac{\pi}{4}\right) = \frac{-\sqrt{2}}{18}\left(\frac{1}{\sqrt{2}} - \frac{4}{\sqrt{2}}\right) = \frac{1}{6}.$$

15. $\frac{\partial g}{\partial u}$ at $(u, v) = (0, 1)$, where $g(x, y) = x^2 - y^2$, $x = e^u\cos v$, $y = e^u\sin v$.

SOLUTION The primary derivatives of $g(x, y) = x^2 - y^2$ are

$$\frac{\partial g}{\partial x} = 2x, \quad \frac{\partial g}{\partial y} = -2y$$

By the Chain Rule we have

$$\frac{\partial g}{\partial u} = \frac{\partial g}{\partial x}\cdot\frac{\partial x}{\partial u} + \frac{\partial g}{\partial y}\cdot\frac{\partial y}{\partial u} = 2x\frac{\partial x}{\partial u} - 2y\frac{\partial y}{\partial u} \tag{1}$$

We find $\frac{\partial x}{\partial u}$ and $\frac{\partial y}{\partial u}$:

$$\frac{\partial x}{\partial u} = e^u\cos v, \quad \frac{\partial y}{\partial u} = e^u\sin v$$

Substituting in (1) gives

$$\frac{\partial g}{\partial u} = 2xe^u\cos v - 2ye^u\sin v = 2e^u(x\cos v - y\sin v) \tag{2}$$

We determine (x, y) for $(u, v) = (0, 1)$:

$$x = e^0\cos 1 = \cos 1, \quad y = e^0\sin 1 = \sin 1$$

Finally, we substitute $(u, v) = (0, 1)$ and $(x, y) = (\cos 1, \sin 1)$ in (2) and use the identity $\cos^2\alpha - \sin^2\alpha = \cos 2\alpha$, to obtain the following derivative:

$$\left.\frac{\partial g}{\partial u}\right|_{(u,v)=(0,1)} = 2e^0\left(\cos^2 1 - \sin^2 1\right) = 2\cdot\cos 2\cdot 1 = 2\cos 2$$

17. The temperature at a point (x, y) is $T(x, y) = 20 + 0.1(x^2 - xy)$ (degrees Celsius). A particle moves clockwise along the unit circle at unit speed (1 cm/s), starting at $(1, 0)$ at time $t = 0$. How fast is the particle's temperature changing at time $t = \pi$?

SOLUTION The particle moves along the unit circle at unit speed, hence its trajectory is parametrized by the arc length parametrization of the unit circle. That is,

$$c(t) = (\cos t, \sin t) \quad \Rightarrow \quad x(t) = \cos t, \quad y(t) = \sin t$$

We need to find $\frac{dT}{dt}\big|_{t=\pi}$. We first compute the primary derivatives of $T(x, y) = 20 + 0.1(x^2 - xy)$:

$$\frac{\partial T}{\partial x} = 0.2x - 0.1y, \qquad \frac{\partial T}{\partial y} = -0.1x$$

By the Chain Rule we have

$$\frac{dT}{dt} = \frac{\partial T}{\partial x}\frac{dx}{dt} + \frac{\partial T}{\partial y}\frac{dy}{dt} = (0.2x - 0.1y)\frac{dx}{dt} - 0.1x\frac{dy}{dt} \tag{1}$$

We compute $\frac{dx}{dt}$ and $\frac{dy}{dt}$:

$$\frac{dx}{dt} = -\sin t, \qquad \frac{dy}{dt} = \cos t$$

Substituting in (1) gives

$$\frac{dT}{dt} = -(0.2x - 0.1y)\sin t - 0.1x\cos t \tag{2}$$

We determine (x, y) for $t = \pi$:

$$x = \cos\pi = -1, \qquad y = \sin\pi = 0$$

Substituting $(x, y) = (-1, 0)$ and $t = \pi$ in (2), we obtain the following derivative:

$$\frac{dT}{dt} = -(-0.2 - 0)\sin\pi - 0.1\cdot(-1)\cos\pi = 0 + 0.1\cdot(-1) = -0.1$$

We conclude that at time $t = \pi$, the particle's temperature is decreasing at a rate of 0.1 degrees per second.

19. Let $u(r, \theta) = r^2\cos^2\theta$. Use Eq. (7) to compute $\|\nabla u\|^2$. Then compute $\|\nabla u\|^2$ directly by observing that $u(x, y) = x^2$ and compare.

SOLUTION By Eq. (7) we have

$$\|\nabla u\|^2 = u_r^2 + \frac{1}{r^2}u_\theta^2$$

We compute the partial derivatives of $u(r, \theta) = r^2\cos^2\theta$:

$$u_r = 2r\cos^2\theta, \qquad u_\theta = r^2\cdot 2\cos\theta(-\sin\theta) = -2r^2\cos\theta\sin\theta$$

Substituting in Eq. (7) we get

$$\|\nabla u\|^2 = \left(2r\cos^2\theta\right)^2 + \frac{1}{r^2}\left(-2r^2\cos\theta\sin\theta\right)^2 = 4r^2\cos^4\theta + 4r^2\cos^2\theta\sin^2\theta$$
$$= 4r^2\cos^2\theta(\cos^2\theta + \sin^2\theta) = 4r^2\cos^2\theta$$

That is,

$$\|\nabla u\|^2 = 4r^2\cos^2\theta \tag{1}$$

We now compute $\|\nabla u\|^2$ directly. We first express $u(r, \theta)$ as a function of x and y. Since $x = r\cos\theta$, we have

$$u(x, y) = x^2$$

Hence $u_x = 2x$, $u_y = 0$, so we obtain

$$\|\nabla u\|^2 = u_x^2 + u_y^2 = (2x)^2 + 0^2 = 4x^2 = 4(r\cos\theta)^2 = 4r^2\cos^2\theta$$

The answer agrees with the result in (1), as expected.

21. Express the derivatives $\frac{\partial f}{\partial \rho}, \frac{\partial f}{\partial \theta}, \frac{\partial f}{\partial \phi}$ in terms of $\frac{\partial f}{\partial x}, \frac{\partial f}{\partial y}, \frac{\partial f}{\partial z}$, where (ρ, θ, ϕ) are spherical coordinates.

SOLUTION The spherical coordinates are

$$x = \rho\sin\phi\cos\theta, \qquad y = \rho\sin\phi\sin\theta, \qquad z = \rho\cos\phi \tag{1}$$

We apply the Chain Rule to write

$$\frac{\partial f}{\partial \rho} = \frac{\partial f}{\partial x}\frac{\partial x}{\partial \rho} + \frac{\partial f}{\partial y}\frac{\partial y}{\partial \rho} + \frac{\partial f}{\partial z}\frac{\partial z}{\partial \rho}$$

$$\frac{\partial f}{\partial \theta} = \frac{\partial f}{\partial x}\frac{\partial x}{\partial \theta} + \frac{\partial f}{\partial y}\frac{\partial y}{\partial \theta} + \frac{\partial f}{\partial z}\frac{\partial z}{\partial \theta}$$

$$\frac{\partial f}{\partial \phi} = \frac{\partial f}{\partial x}\frac{\partial x}{\partial \phi} + \frac{\partial f}{\partial y}\frac{\partial y}{\partial \phi} + \frac{\partial f}{\partial z}\frac{\partial z}{\partial \phi} \tag{2}$$

We use (1) to compute the partial derivatives of x, y, and z with respect to ρ, θ, and ϕ. This gives

$$\frac{\partial x}{\partial \theta} = -\rho \sin\phi \sin\theta, \quad \frac{\partial y}{\partial \theta} = \rho \sin\phi \cos\theta, \quad \frac{\partial z}{\partial \theta} = 0$$

$$\frac{\partial x}{\partial \phi} = \rho \cos\phi \cos\theta, \quad \frac{\partial y}{\partial \phi} = \rho \cos\phi \sin\theta, \quad \frac{\partial z}{\partial \phi} = -\rho \sin\phi$$

$$\frac{\partial x}{\partial \rho} = \sin\phi \cos\theta, \quad \frac{\partial y}{\partial \rho} = \sin\phi \sin\theta, \quad \frac{\partial z}{\partial \rho} = \cos\phi$$

Substituting these derivatives in (2), we get

$$\frac{\partial f}{\partial \rho} = (\sin\phi\cos\theta)\frac{\partial f}{\partial x} + (\sin\phi\sin\theta)\frac{\partial f}{\partial y} + (\cos\phi)\frac{\partial f}{\partial z}$$

$$\frac{\partial f}{\partial \phi} = (\rho\cos\phi\cos\theta)\frac{\partial f}{\partial x} + (\rho\cos\phi\sin\theta)\frac{\partial f}{\partial y} - (\rho\sin\phi)\frac{\partial f}{\partial z}$$

$$\frac{\partial f}{\partial \theta} = (-\rho\sin\phi\sin\theta)\frac{\partial f}{\partial x} + (\rho\sin\phi\cos\theta)\frac{\partial f}{\partial y}$$

23. Calculate $\frac{\partial z}{\partial x}$ and $\frac{\partial z}{\partial y}$ at the points (3, 2, 1) and (3, 2, −1), where z is defined implicitly by the equation $z^4 + z^2x^2 - y - 8 = 0$.

SOLUTION For $F(x, y, z) = z^4 + z^2x^2 - y - 8 = 0$, we use the following equalities, (Eq. (6)):

$$\frac{\partial z}{\partial x} = -\frac{F_x}{F_z}, \quad \frac{\partial z}{\partial y} = -\frac{F_y}{F_z} \tag{1}$$

The partial derivatives of F are

$$F_x = 2z^2x, \quad F_y = -1, \quad F_z = 4z^3 + 2zx^2$$

Substituting in (1) gives

$$\frac{\partial z}{\partial x} = -\frac{2z^2x}{4z^3 + 2zx^2} = -\frac{zx}{2z^2 + x^2}$$

$$\frac{\partial z}{\partial y} = \frac{1}{4z^3 + 2zx^2}$$

At the point (3, 2, 1), we have

$$\left.\frac{\partial z}{\partial x}\right|_{(3,2,1)} = -\frac{1\cdot 3}{2\cdot 1^2 + 3^2} = -\frac{3}{11}, \quad \left.\frac{\partial z}{\partial y}\right|_{(3,2,1)} = \frac{1}{4\cdot 1^3 + 2\cdot 1\cdot 3^2} = \frac{1}{22}$$

At the point (3, 2, −1), we have

$$\left.\frac{\partial z}{\partial x}\right|_{(3,2,-1)} = -\frac{-3}{2\cdot(-1)^2 + 3^2} = \frac{3}{11}$$

$$\left.\frac{\partial z}{\partial y}\right|_{(3,2,-1)} = \frac{1}{4\cdot(-1)^3 + 2\cdot(-1)\cdot 3^2} = -\frac{1}{22}$$

In Exercises 24–29, calculate the derivative using implicit differentiation.

25. $\frac{\partial w}{\partial z}$, $\quad x^2w + w^3 + wz^2 + 3yz = 0$

SOLUTION We find the partial derivatives F_w and F_z of

$$F(x, w, z) = x^2w + w^3 + wz^2 + 3yz$$

$$F_w = x^2 + 3w^2 + z^2, \quad F_z = 2wz + 3y$$

Using Eq. (6) we get

$$\frac{\partial w}{\partial z} = -\frac{F_z}{F_w} = -\frac{2wz + 3y}{x^2 + 3w^2 + z^2}.$$

27. $\dfrac{\partial r}{\partial t}$ and $\dfrac{\partial t}{\partial r}$, $r^2 = te^{s/r}$

SOLUTION We use the formulas obtained by implicit differentiation of $F(r, s, t) = r^2 - te^{s/r}$ (Eq. (6)):

$$\frac{\partial r}{\partial t} = -\frac{F_t}{F_r}, \quad \frac{\partial t}{\partial r} = -\frac{F_r}{F_t} \tag{1}$$

The partial derivatives of F are

$$F_r = 2r - te^{s/r}\left(-\frac{s}{r^2}\right) = 2r + \frac{st}{r^2}e^{s/r}$$

$$F_t = -e^{s/r}$$

Substituting in (1) gives

$$\frac{\partial r}{\partial t} = \frac{e^{s/r}}{2r + \frac{st}{r^2}e^{s/r}} = \frac{r^2 e^{s/r}}{2r^3 + ste^{s/r}}$$

$$\frac{\partial t}{\partial r} = \frac{2r + \frac{st}{r^2}e^{s/r}}{e^{s/r}} = \frac{2r^3 + ste^{s/r}}{r^2 e^{s/r}} = 2re^{-s/r} + \frac{st}{r^2}$$

29. $\dfrac{\partial U}{\partial T}$ and $\dfrac{\partial T}{\partial U}$, $(TU - V)^2 \ln(W - UV) = 1$ at $(T, U, V, W) = (1, 1, 2, 4)$

SOLUTION Using the formulas obtained by implicit differentiation (Eq. (6)) we have,

$$\frac{\partial U}{\partial T} = -\frac{F_T}{F_U}, \quad \frac{\partial T}{\partial U} = -\frac{F_U}{F_T} \tag{1}$$

We compute the partial derivatives of $F(T, U, V, W) = (TU - V)^2 \ln(W - UV) - 1$:

$$F_T = 2U(TU - V)\ln(W - UV)$$

$$F_U = 2T(TU - V)\ln(W - UV) + (TU - V)^2 \cdot \frac{-V}{W - UV} = (TU - V)\left(2T\ln(W - UV) - \frac{V(TU - V)}{W - UV}\right)$$

At the point $(T, U, V, W) = (1, 1, 2, 4)$ we have

$$F_T = 2(1 - 2)\ln(4 - 2) = -2\ln 2$$

$$F_U = (1 - 2)\left(2\ln(4 - 2) - \frac{2(1 - 2)}{4 - 2}\right) = (-2\ln 2 - 1) = -1 - 2\ln 2$$

Substituting in (1) we obtain

$$\left.\frac{\partial U}{\partial T}\right|_{(1,1,2,4)} = -\frac{2\ln 2}{1 + 2\ln 2}, \quad \left.\frac{\partial T}{\partial U}\right|_{(1,1,2,4)} = -\frac{1 + 2\ln 2}{2\ln 2}.$$

31. Let $f(x, y, z) = F(r)$, where $r = \sqrt{x^2 + y^2 + z^2}$. Show that

$$\nabla f = F'(r)e_{\mathbf{r}}$$

8

where $e_{\mathbf{r}} = \dfrac{\mathbf{r}}{\|\mathbf{r}\|}$ and $\mathbf{r} = \langle x, y, z\rangle$.

SOLUTION The gradient of f is the following vector:

$$\nabla f = \left\langle \frac{\partial f}{\partial x}, \frac{\partial f}{\partial y}, \frac{\partial f}{\partial z}\right\rangle$$

We must express this vector in terms of $\mathbf{r}$ and r. Using the Chain Rule, we have

$$\frac{\partial f}{\partial x} = F'(r)\frac{\partial r}{\partial x} = F'(r) \cdot \frac{2x}{2\sqrt{x^2 + y^2 + z^2}} = F'(r) \cdot \frac{x}{r}$$

$$\frac{\partial f}{\partial y} = F'(r)\frac{\partial r}{\partial y} = F'(r) \cdot \frac{2y}{2\sqrt{x^2+y^2+z^2}} = F'(r) \cdot \frac{y}{r}$$

$$\frac{\partial f}{\partial z} = F'(r)\frac{\partial r}{\partial z} = F'(r) \cdot \frac{2z}{2\sqrt{x^2+y^2+z^2}} = F'(r) \cdot \frac{z}{r}$$

Hence,

$$\nabla f = \left\langle F'(r)\frac{x}{r}, F'(r)\frac{y}{r}, F'(r)\frac{z}{r} \right\rangle = \frac{F'(r)}{r}\langle x, y, z\rangle = F'(r)\frac{\mathbf{r}}{\|\mathbf{r}\|} = F'(r)e_{\mathbf{r}}$$

33. Use Eq. (8) to compute $\nabla\left(\frac{1}{r}\right)$ and $\nabla(\ln r)$.

SOLUTION To compute $\nabla(\frac{1}{r})$ using Eq. (8), we let $F(r) = \frac{1}{r}$:

$$F'(r) = -\frac{1}{r^2}$$

We obtain

$$\nabla\left(\frac{1}{r}\right) = F'(r)\mathbf{e_r} = -\frac{1}{r^2} \cdot \frac{\mathbf{r}}{\|\mathbf{r}\|} = -\frac{1}{r^3}\mathbf{r}$$

To compute $\nabla(\ln r)$ we let $F(r) = \ln r$, hence $F'(r) = \frac{1}{r}$. Thus,

$$\nabla(\ln r) = F'(r)\mathbf{e_r} = \frac{1}{r} \cdot \frac{\mathbf{r}}{\|\mathbf{r}\|} = \frac{\mathbf{r}}{r^2}$$

35. Jessica and Matthew are running toward the point P along the straight paths that make a fixed angle of θ (Figure 3). Suppose that Matthew runs with velocity v_a m/s and Jessica with velocity v_b m/s. Let $f(x, y)$ be the distance from Matthew to Jessica when Matthew is x meters from P and Jessica is y meters from P.

(a) Show that $f(x, y) = \sqrt{x^2 + y^2 - 2xy\cos\theta}$.

(b) Assume that $\theta = \pi/3$. Use the Chain Rule to determine the rate at which the distance between Matthew and Jessica is changing when $x = 30$, $y = 20$, $v_a = 4$ m/s, and $v_b = 3$ m/s.

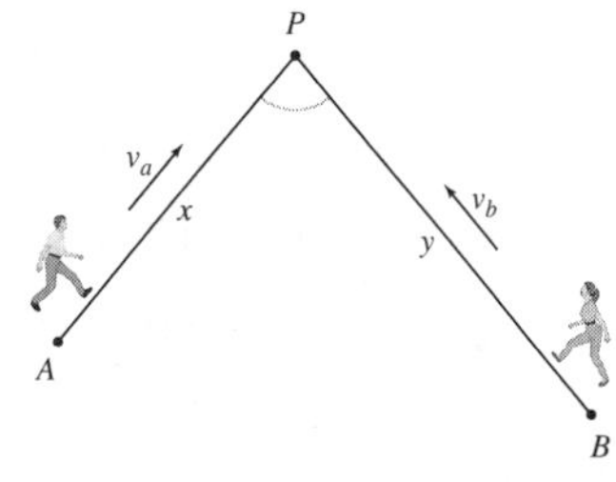

FIGURE 3

SOLUTION

(a) This is a simple application of the Law of Cosines. Connect points A and B in the diagram to form a line segment that we will call f. Then, the Law of Cosines says that $f^2 = x^2 + y^2 - 2xy\cos\theta$. By taking square roots, we find that $f = \sqrt{x^2 + y^2 - 2xy\cos\theta}$.

(b) Using the chain rule,

$$\frac{df}{dt} = \frac{\partial f}{\partial x}\frac{dx}{dt} + \frac{\partial f}{\partial y}\frac{dy}{dt}$$

so we get

$$\frac{df}{dt} = \frac{(x - y\cos\theta)dx/dt}{\sqrt{x^2+y^2-2xy\cos\theta}} + \frac{(y - x\cos\theta)dy/dt}{\sqrt{x^2+y^2-2xy\cos\theta}}$$

and using $x = 30$, $y = 20$, and $dx/dt = 4$, $dy/dt = 3$, we get

$$\frac{df}{dt} = \frac{180 - 170\cos\theta}{\sqrt{1300 - 1200\cos\theta}}$$

Further Insights and Challenges

In Exercises 37–40, a function $f(x, y, z)$ is called **homogeneous of degree n** *if $f(\lambda x, \lambda y, \lambda z) = \lambda^n f(x, y, z)$ for all $\lambda \in \mathbf{R}$.*

37. Show that the following functions are homogeneous and determine their degree.

(a) $f(x, y, z) = x^2y + xyz$

(b) $f(x, y, z) = 3x + 2y - 8z$

(c) $f(x, y, z) = \ln\left(\frac{xy}{z^2}\right)$

(d) $f(x, y, z) = z^4$

SOLUTION

(a) For $f(x, y, z) = x^2y + xyz$ we have

$$f(\lambda x, \lambda y, \lambda z) = (\lambda x)^2(\lambda y) + (\lambda x)(\lambda y)(\lambda z) = \lambda^3x^2y + \lambda^3xyz = \lambda^3(x^2y + xyz) = \lambda^3 f(x, y, z)$$

Hence, f is homogeneous of degree 3.

(b) For $f(x, y, z) = 3x + 2y - 8z$ we have

$$f(\lambda x, \lambda y, \lambda z) = 3(\lambda x) + 2(\lambda y) - 8(\lambda z) = \lambda(3x + 2y - 8z) = \lambda f(x, y, z)$$

Hence, f is homogeneous of degree 1.

(c) For $f(x, y, z) = \ln\left(\frac{xy}{z^2}\right)$ we have, for $\lambda \neq 0$,

$$f(\lambda x, \lambda y, \lambda z) = \ln\left(\frac{(\lambda x)(\lambda y)}{(\lambda z)^2}\right) = \ln\left(\frac{\lambda^2xy}{\lambda^2z^2}\right) = \ln\left(\frac{xy}{z^2}\right) = f(x, y, z) = \lambda^0 f(x, y, z)$$

Thus, f is homogeneous of degree 0.

(d) For $f(z) = z^4$ we have

$$f(\lambda z) = (\lambda z)^4 = \lambda^4z^4 = \lambda^4 f(z)$$

Hence, f is homogeneous of degree 4.

39. Prove that if $f(x, y, z)$ is homogeneous of degree n, then

$$x\frac{\partial f}{\partial x} + y\frac{\partial f}{\partial y} + z\frac{\partial f}{\partial z} = nf$$

9

Hint: Let $F(t) = f(tx, ty, tz)$ and calculate $F'(1)$ using the Chain Rule.

SOLUTION We use the Chain Rule to differentiate the function $F(t) = f(tx, ty, tz)$ with respect to t. This gives

$$F'(t) = \frac{\partial f}{\partial x} \cdot \frac{\partial(tx)}{\partial t} + \frac{\partial f}{\partial y} \cdot \frac{\partial(ty)}{\partial t} + \frac{\partial f}{\partial z} \cdot \frac{\partial(tz)}{\partial t} = x\frac{\partial f}{\partial x} + y\frac{\partial f}{\partial y} + z\frac{\partial f}{\partial z} \tag{1}$$

On the other hand, since f is homogeneous of degree n, we have

$$F(t) = f(tx, ty, tz) = t^n f(x, y, z)$$

Differentiating with respect to t we get

$$F'(t) = nt^{n-1} f(x, y, z) \tag{2}$$

By (1) and (2) we obtain

$$x\frac{\partial f}{\partial x} + y\frac{\partial f}{\partial y} + z\frac{\partial f}{\partial z} = nt^{n-1} f(x, y, z)$$

Substituting $t = 1$ gives

$$x\frac{\partial f}{\partial x} + y\frac{\partial f}{\partial y} + z\frac{\partial f}{\partial z} = nf$$

41. Suppose that $x = g(t, s)$, $y = h(t, s)$. Show that f_{tt} is equal to

$$f_{xx}\left(\frac{\partial x}{\partial t}\right)^2 + 2f_{xy}\left(\frac{\partial x}{\partial t}\right)\left(\frac{\partial y}{\partial t}\right) + f_{yy}\left(\frac{\partial y}{\partial t}\right)^2 + f_x\frac{\partial^2 x}{\partial t^2} + f_y\frac{\partial^2 y}{\partial t^2}$$

10

SOLUTION We are given that $x = g(t,s)$, $y = h(t,s)$. We must compute f_{tt} for a function $f(x,y)$. We first compute f_t using the Chain Rule:

$$f_t = f_x \frac{\partial x}{\partial t} + f_y \frac{\partial y}{\partial t}$$

To find f_{tt} we differentiate the two sides with respect to t using the Product Rule. This gives

$$f_{tt} = \frac{\partial}{\partial t}(f_x)\frac{\partial x}{\partial t} + f_x \frac{\partial^2 x}{\partial t^2} + \frac{\partial}{\partial t}(f_y)\frac{\partial y}{\partial t} + f_y \frac{\partial^2 y}{\partial t^2} \tag{1}$$

By the Chain Rule,

$$\frac{\partial}{\partial t}(f_x) = f_{xx}\frac{\partial x}{\partial t} + f_{xy}\frac{\partial y}{\partial t}$$
$$\frac{\partial}{\partial t}(f_y) = f_{yx}\frac{\partial x}{\partial t} + f_{yy}\frac{\partial y}{\partial t}$$

Substituting in (1) we obtain

$$f_{tt} = \left(f_{xx}\frac{\partial x}{\partial t} + f_{xy}\frac{\partial y}{\partial t}\right)\frac{\partial x}{\partial t} + f_x\frac{\partial^2 x}{\partial t^2} + \left(f_{yx}\frac{\partial x}{\partial t} + f_{yy}\frac{\partial y}{\partial t}\right)\frac{\partial y}{\partial t} + f_y\frac{\partial^2 y}{\partial t^2}$$
$$= f_{xx}\left(\frac{\partial x}{\partial t}\right)^2 + f_{xy}\left(\frac{\partial y}{\partial t}\right)\left(\frac{\partial x}{\partial t}\right) + f_x\frac{\partial^2 x}{\partial t^2} + f_{yx}\left(\frac{\partial x}{\partial t}\right)\left(\frac{\partial y}{\partial t}\right) + f_{yy}\left(\frac{\partial y}{\partial t}\right)^2 + f_y\frac{\partial^2 y}{\partial t^2}$$

If f_{xy} and f_{yx} are continuous, Clairaut's Theorem implies that $f_{xy} = f_{yx}$. Hence,

$$f_{tt} = f_{xx}\left(\frac{\partial x}{\partial t}\right)^2 + 2f_{xy}\left(\frac{\partial x}{\partial t}\right)\left(\frac{\partial y}{\partial t}\right) + f_{yy}\left(\frac{\partial y}{\partial t}\right)^2 + f_x\frac{\partial^2 x}{\partial t^2} + f_y\frac{\partial^2 y}{\partial t^2}$$

43. Prove that if $g(r)$ is a function of r as in Exercise 42, then

$$\frac{\partial^2 g}{\partial x_1^2} + \cdots + \frac{\partial^2 g}{\partial x_n^2} = g_{rr} + \frac{n-1}{r} g_r$$

SOLUTION In Exercise 42 we showed that

$$\frac{\partial^2 g}{\partial {x_i}^2} = \frac{x_i^2}{r^2} g_{rr} + \frac{r^2 - x_i^2}{r^3} g_r$$

Hence,

$$\frac{\partial^2 g}{\partial {x_i}^2} + \cdots + \frac{\partial^2 g}{\partial {x_n}^2} = \left(\frac{x_1^2}{r^2} g_{rr} + \frac{r^2 - x_1^2}{r^3} g_r\right) + \cdots + \left(\frac{x_n^2}{r^2} g_{rr} + \frac{r^2 - x_n^2}{r^3} g_r\right)$$
$$= \frac{x_1^2 + \cdots + x_n^2}{r^2} g_{rr} + \frac{1}{r^3} g_r \left((r^2 - x_1^2) + \cdots + (r^2 - x_n^2)\right)$$
$$= \frac{r^2}{r^2} g_{rr} + \frac{1}{r^3} g_r \left(nr^2 - (x_1^2 + \cdots + x_n^2)\right)$$
$$= g_{rr} + \frac{1}{r^3} g_r (nr^2 - r^2) = g_{rr} + \frac{r^2}{r^3} g_r (n-1) = g_{rr} + \frac{n-1}{r} g_r$$

In Exercises 44–48, the ***Laplace operator*** *is defined by* $\Delta f = f_{xx} + f_{yy}$. *A function* $f(x,y)$ *satisfying the Laplace equation* $\Delta f = 0$ *is called* ***harmonic***. *A function* $f(x,y)$ *is called* ***radial*** *if* $f(x,y) = g(r)$, *where* $r = \sqrt{x^2 + y^2}$.

45. Use Eq. (11) to show that $f(x,y) = \ln r$ is harmonic.

SOLUTION We must show that $f(r,\theta) = \ln r$ satisfies

$$\Delta f = f_{rr} + \frac{1}{r^2} f_{\theta\theta} + \frac{1}{r} f_r = 0$$

We compute the derivatives of $f(r,\theta) = \ln r$:

$$f_r = \frac{1}{r}, \quad f_{rr} = -\frac{1}{r^2}, \quad f_\theta = 0, \quad f_{\theta\theta} = 0$$

Hence,

$$\Delta f = f_{rr} + \frac{1}{r^2} f_{\theta\theta} + \frac{1}{r} f_r = -\frac{1}{r^2} + \frac{1}{r^2} \cdot 0 + \frac{1}{r} \cdot \frac{1}{r} = -\frac{1}{r^2} + \frac{1}{r^2} = 0$$

Since $\Delta f = 0$, f is harmonic.

47. Verify that $f(x, y) = \tan^{-1} \frac{y}{x}$ is harmonic using both the rectangular and polar expressions for Δf.

SOLUTION

(a) Using the rectangular expression for Δf:

$$\Delta f = f_{xx} + f_{yy}$$

We compute the partial derivatives of $f(x, y) = \tan^{-1} \left(\frac{y}{x}\right)$. Using the Chain Rule we get

$$f_x = \frac{1}{1 + \left(\frac{y}{x}\right)^2} \cdot \left(-\frac{y}{x^2}\right) = -\frac{y}{x^2 + y^2}$$

$$f_y = \frac{1}{1 + \left(\frac{y}{x}\right)^2} \cdot \frac{1}{x} = \frac{x}{x^2 + y^2}$$

$$f_{xx} = -\frac{-y}{\left(x^2 + y^2\right)^2} \cdot 2x = \frac{2xy}{\left(x^2 + y^2\right)^2}$$

$$f_{yy} = \frac{-x}{\left(x^2 + y^2\right)^2} \cdot 2y = \frac{-2xy}{\left(x^2 + y^2\right)^2}$$

Hence,

$$f_{xx} + f_{yy} = \frac{2xy}{\left(x^2 + y^2\right)^2} - \frac{2xy}{\left(x^2 + y^2\right)^2} = 0$$

(b) Using the polar expression for Δf,

$$\Delta f = f_{rr} + \frac{1}{r^2} f_{\theta\theta} + \frac{1}{r} f_r \qquad (1)$$

Since $\frac{y}{x} = \frac{r \sin\theta}{r \cos\theta} = \tan\theta$, we have $f(x, y) = \tan^{-1}\left(\frac{y}{x}\right) = \tan^{-1}(\theta) = \theta$. We compute the partial derivatives:

$$f_r = 0, \quad f_\theta = 1, \quad f_{rr} = 0, \quad f_{\theta\theta} = 0.$$

Substituting in (1), we get

$$\Delta f = 0 + \frac{1}{r^2} \cdot 0 + \frac{1}{r} \cdot 0 = 0$$

49. Figure 4 shows the graph of the equation

$$F(x, y, z) = x^2 + y^2 - z^2 - 12x - 8z - 4 = 0$$

(a) Use the quadratic formula to solve for z as a function of x and y. This gives two formulas, depending on the choice of a sign.

(b) Which formula defines the portion of the surface satisfying $z \geq -4$? Which formula defines the portion satisfying $z \leq -4$?

(c) Calculate $\frac{\partial z}{\partial x}$ using the formula $z = f(x, y)$ (for both choices of sign) and again via implicit differentiation. Verify that the two answers agree.

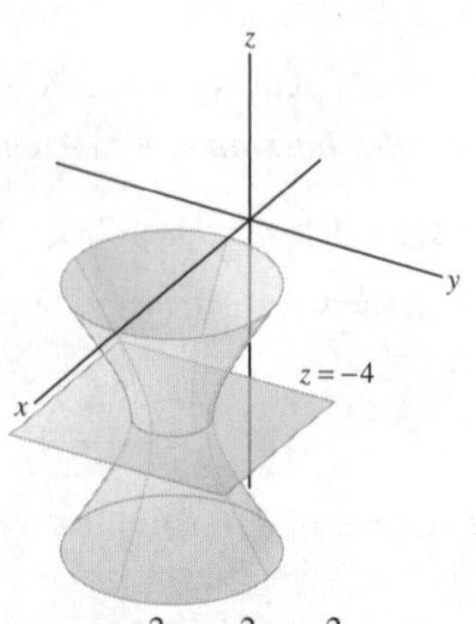

FIGURE 4 Graph of $x^2 + y^2 - z^2 - 12x - 8z - 4 = 0$.

SOLUTION

(a) We rewrite $F(x, y, z) = 0$ as a quadratic equation in the variable z:

$$z^2 + 8z + \left(4 + 12x - x^2 - y^2\right) = 0$$

We solve for z. The discriminant is

$$8^2 - 4\left(4 + 12x - x^2 - y^2\right) = 4x^2 + 4y^2 - 48x + 48 = 4\left(x^2 + y^2 - 12x + 12\right)$$

Hence,

$$z_{1,2} = \frac{-8 \pm \sqrt{4\left(x^2 + y^2 - 12x + 12\right)}}{2} = -4 \pm \sqrt{x^2 + y^2 - 12x + 12}$$

We obtain two functions:

$$z = -4 + \sqrt{x^2 + y^2 - 12x + 12}, \quad z = -4 - \sqrt{x^2 + y^2 - 12x + 12}$$

(b) The formula with the positive root defines the portion of the surface satisfying $z \geq -4$, and the formula with the negative root defines the portion satisfying $z \leq -4$.

(c) Differentiating $z = -4 + \sqrt{x^2 + y^2 - 12x + 12}$ with respect to x, using the Chain Rule, gives

$$\frac{\partial z}{\partial x} = \frac{2x - 12}{2\sqrt{x^2 + y^2 - 12x + 12}} = \frac{x - 6}{\sqrt{x^2 + y^2 - 12x + 12}} \tag{1}$$

Alternatively, using the formula for $\frac{\partial z}{\partial x}$ obtained by implicit differentiation gives

$$\frac{\partial z}{\partial x} = -\frac{F_x}{F_z} \tag{2}$$

We find the partial derivatives of $F(x, y, z) = x^2 + y^2 - z^2 - 12x - 8z - 4$:

$$F_x = 2x - 12, \quad F_z = -2z - 8$$

Substituting in (2) gives

$$\frac{\partial z}{\partial x} = -\frac{2x - 12}{-2z - 8} = \frac{x - 6}{z + 4}$$

This result is the same as the result in (1), since $z = -4 + \sqrt{x^2 + y^2 - 12x + 12}$ implies that

$$\sqrt{x^2 + y^2 - 12x + 12} = z + 4$$

For $z = -4 - \sqrt{x^2 + y^2 - 12x + 12}$, differentiating with respect to x gives

$$\frac{\partial z}{\partial x} = -\frac{2x - 12}{2\sqrt{x^2 + y^2 - 12x + 12}} = \frac{x - 6}{-\sqrt{x^2 + y^2 - 12x + 12}} = \frac{x - 6}{z + 4}$$

which is equal to $-\frac{F_x}{F_z}$ computed above.

15.7 Optimization in Several Variables (ET Section 14.7)

Preliminary Questions

1. The functions $f(x, y) = x^2 + y^2$ and $g(x, y) = x^2 - y^2$ both have a critical point at $(0, 0)$. How is the behavior of the two functions at the critical point different?

SOLUTION Let $f(x, y) = x^2 + y^2$ and $g(x, y) = x^2 - y^2$. In the domain $\mathbf{R}^2$, the partial derivatives of f and g are

$$f_x = 2x, \quad f_{xx} = 2, \quad f_y = 2y, \quad f_{yy} = 2, \quad f_{xy} = 0$$

$$g_x = 2x, \quad g_{xx} = 2, \quad g_y = -2y, \quad g_{yy} = -2, \quad g_{xy} = 0$$

Therefore, $f_x = f_y = 0$ at $(0, 0)$ and $g_x = g_y = 0$ at $(0, 0)$. That is, the two functions have one critical point, which is the origin. Since the discriminant of f is $D = 4 > 0$, $f_{xx} > 0$, and the discriminant of g is $D = -4 < 0$, f has a local minimum (which is also a global minimum) at the origin, whereas g has a saddle point there. Moreover, since $\lim_{y \to \infty} g(0, y) = -\infty$ and $\lim_{x \to \infty} g(x, 0) = \infty$, g does not have global extrema on the plane. Similarly, f does not have a global maximum but does have a global minimum, which is $f(0, 0) = 0$.

2. Identify the points indicated in the contour maps as local minima, maxima, saddle points, or neither (Figure 14).

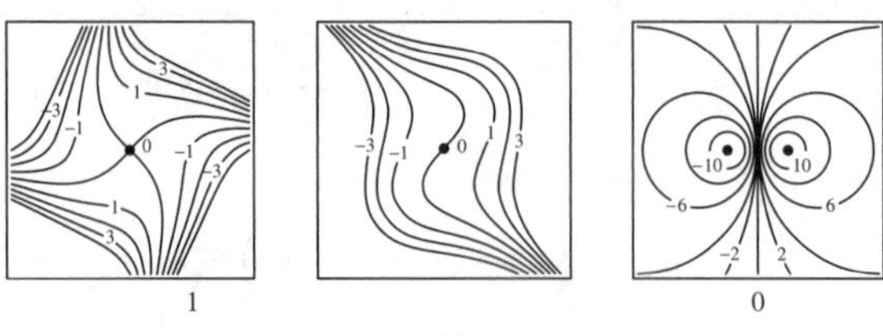

FIGURE 14

SOLUTION If $f(P)$ is a local minimum or maximum, then the nearby level curves are closed curves encircling P. In Figure (C), f increases in all directions emanating from P and decreases in all directions emanating from Q. Hence, f has a local minimum at P and local maximum at Q.

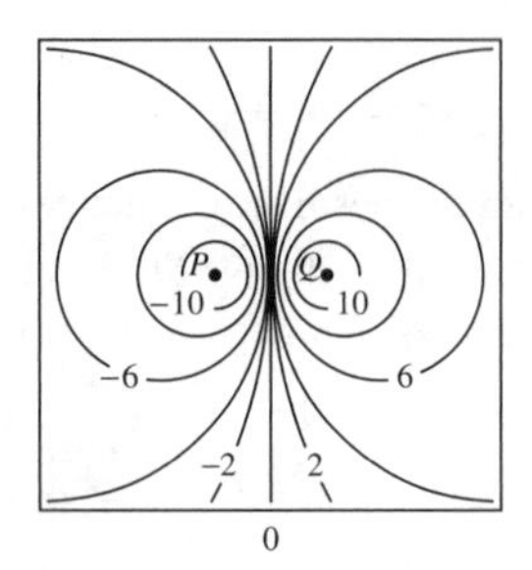

In Figure (A), the level curves through the point R consist of two intersecting lines that divide the neighborhood near R into four regions. f is decreasing in some directions and increasing in other directions. Therefore, R is a saddle point.

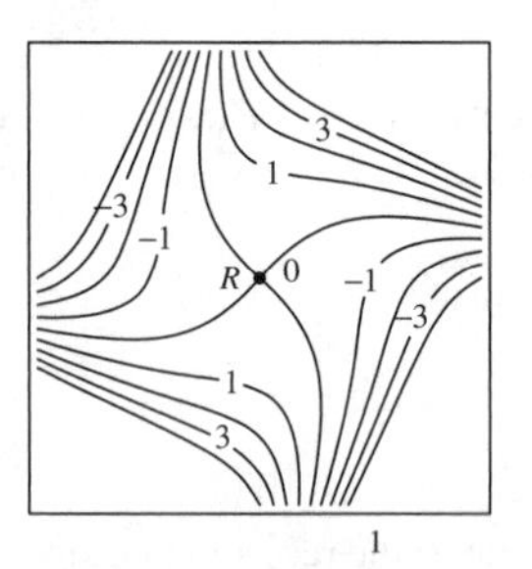

Figure (A)

Point S in Figure (B) is neither a local extremum nor a saddle point of f.

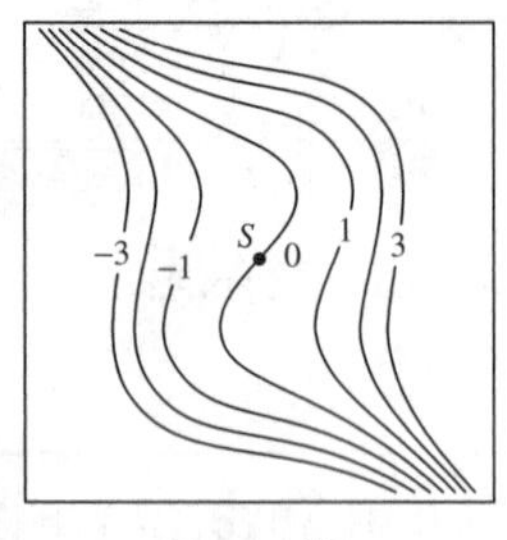

Figure (B)

3. Let $f(x, y)$ be a continuous function on a domain $\mathcal{D}$ in $\mathbf{R}^2$. Determine which of the following statements are true:

(a) If $\mathcal{D}$ is closed and bounded, then f takes on a maximum value on $\mathcal{D}$.

(b) If $\mathcal{D}$ is neither closed nor bounded, then f does not take on a maximum value of $\mathcal{D}$.

(c) $f(x, y)$ need not have a maximum value on $\mathcal{D} = \{(x, y) : 0 \le x, y \le 1\}$.

(d) A continuous function takes on neither a minimum nor a maximum value on the open quadrant $\{(x, y) : x > 0, y > 0\}$.

SOLUTION

(a) This statement is true. It follows by the Theorem on Existence of Global Extrema.

(b) The statement is false. Consider the constant function $f(x, y) = 2$ in the following domain:

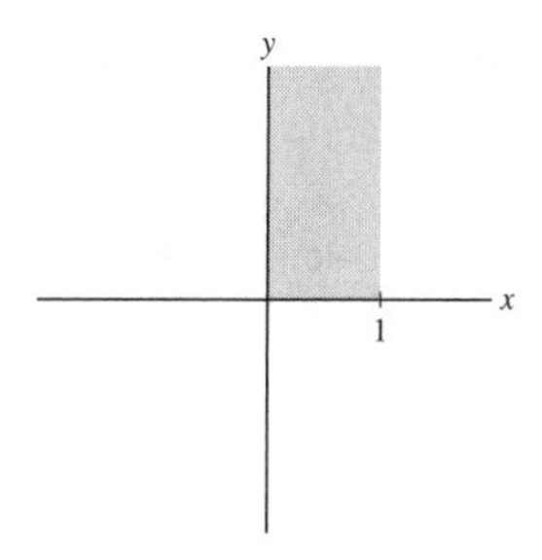

$$D = \{(x, y) : 0 < x \le 1,\ 0 \le y < \infty\}$$

Obviously f is continuous and D is neither closed nor bounded. However, f takes on a maximum value (which is 2) on D.

(c) The domain $D = \{(x, y) : 0 \le x, y \le 1\}$ is the following rectangle:

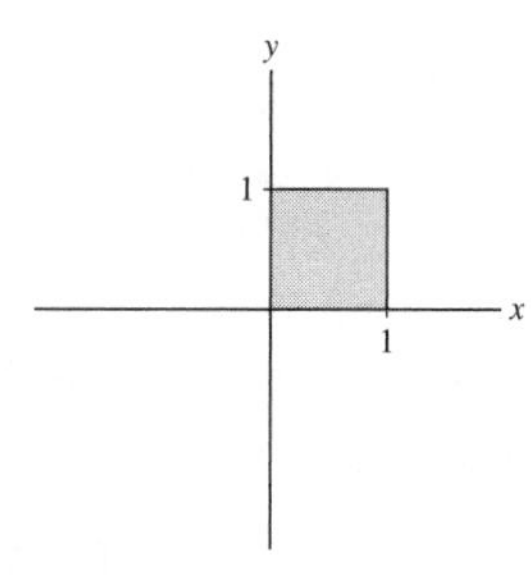

$$D = \{(x, y) : 0 \le x, y \le 1\}$$

D is closed and bounded, hence f takes on a maximum value on D. Thus the statement is false.

(d) The statement is false. The constant function $f(x, y) = c$ takes on minimum and maximum values on the open quadrant.

Exercises

1. Let $P = (a, b)$ be a critical point of $f(x, y) = x^2 + y^4 - 4xy$.

(a) First use $f_x(, x, y) = 0$ to show that $a = 2b$. Then use $f_y(x, y) = 0$ to show that $P = (0, 0)$, $(2\sqrt{2}, \sqrt{2})$, or $(-2\sqrt{2}, -\sqrt{2})$.

(b) Referring to Figure 15, determine the local minima and saddle points of $f(x, y)$ and find the absolute minimum value of $f(x, y)$.

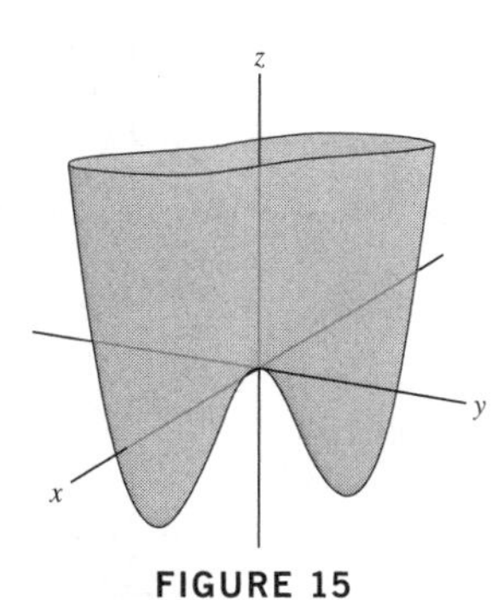

FIGURE 15

SOLUTION

(a) We find the partial derivatives:

$$f_x(x, y) = \frac{\partial}{\partial x}\left(x^2 + y^4 - 4xy\right) = 2x - 4y$$

$$f_y(x, y) = \frac{\partial}{\partial y}\left(x^2 + y^4 - 4xy\right) = 4y^3 - 4x$$

Since $P = (a, b)$ is a critical point, $f_x(a, b) = 0$. That is,

$$2a - 4b = 0 \quad \Rightarrow \quad a = 2b$$

Also $f_y(a,b) = 0$, hence,

$$4b^3 - 4a = 0 \quad \Rightarrow \quad a = b^3$$

We obtain the following equations for the critical points (a, b):

$$\begin{cases} a = 2b \\ a = b^3 \end{cases}$$

Equating the two equations, we get

$$2b = b^3$$

$$b^3 - 2b = b(b^2 - 2) = 0 \quad \Rightarrow \quad \begin{cases} b_1 = 0 \\ b_2 = \sqrt{2} \\ b_3 = -\sqrt{2} \end{cases}$$

Since $a = 2b$, we have $a_1 = 0$, $a_2 = 2\sqrt{2}$, $a_3 = -2\sqrt{2}$. The critical points are thus

$$P_1 = (0,0), \quad P_2 = \left(2\sqrt{2}, \sqrt{2}\right), \quad P_3 = \left(-2\sqrt{2}, -\sqrt{2}\right)$$

(b) Referring to Figure 14, we see that $P_1 = (0,0)$ is a saddle point and $P_2 = \left(2\sqrt{2}, \sqrt{2}\right)$, $P_3 = \left(-2\sqrt{2}, -\sqrt{2}\right)$ are local minima. The absolute minimum value of f is -4.

3. Find the critical points of

$$f(x, y) = 8y^4 + x^2 + xy - 3y^2 - y^3$$

Use the contour map in Figure 17 to determine their nature (minimum, maximum, saddle point).

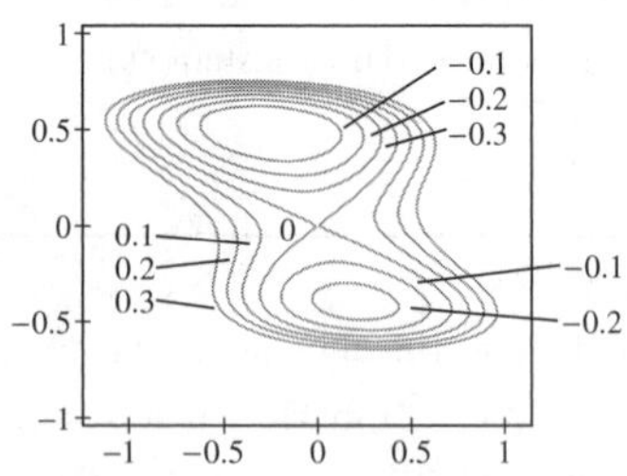

FIGURE 17 Contour map of $f(x, y) = 8y^4 + x^2 + xy - 3y^2 - y^3$.

SOLUTION The critical points are the solutions of $f_x = 0$ and $f_y = 0$. That is,

$$f_x(x, y) = 2x + y = 0$$
$$f_y(x, y) = 32y^3 + x - 6y - 3y^2 = 0$$

The first equation gives $y = -2x$. We substitute in the second equation and solve for x. This gives

$$32(-2x)^3 + x - 6(-2x) - 3(-2x)^2 = 0$$
$$-256x^3 + 13x - 12x^2 = 0$$
$$-x(256x^2 + 12x - 13) = 0$$

Hence $x = 0$ or $256x^2 + 12x - 13 = 0$. Solving the quadratic,

$$x_{1,2} = \frac{-12 \pm \sqrt{12^2 - 4 \cdot 256 \cdot (-13)}}{512} = \frac{-12 \pm 116}{512} \quad \Rightarrow \quad x = \frac{13}{64} \quad \text{or} \quad -\frac{1}{4}$$

Substituting in $y = -2x$ gives the y-coordinates of the critical points. The critical points are thus

$$(0,0), \quad \left(\frac{13}{64}, -\frac{13}{32}\right), \quad \left(-\frac{1}{4}, \frac{1}{2}\right)$$

We now use the contour map to determine the type of each critical point. The level curves through $(0,0)$ consist of two intersecting lines that divide the neighborhood near $(0,0)$ into four regions. The function is decreasing in the y direction and increasing in the x-direction. Therefore, $(0,0)$ is a saddle point. The level curves near the critical points $\left(\frac{13}{64}, -\frac{13}{32}\right)$ and $\left(-\frac{1}{4}, \frac{1}{2}\right)$ are closed curves encircling the points, hence these are local minima or maxima. The graph shows that $\left(\frac{13}{64}, -\frac{13}{32}\right)$ is a local maximum and $\left(-\frac{1}{4}, \frac{1}{2}\right)$ is a local minimum.

In Exercises 5–20, find the critical points of the function. Then use the Second Derivative Test to determine whether they are local minima or maxima (or state that the test fails).

5. $f(x, y) = x^2 + y^2 - xy + x$

SOLUTION

Step 1. Find the critical points. We set the first-order partial derivatives of $f(x, y) = x^2 + y^2 - xy + x$ equal to zero and solve:

$$f_x(x, y) = 2x - y + 1 = 0 \tag{1}$$

$$f_y(x, y) = 2y - x = 0 \tag{2}$$

Equation (2) implies that $x = 2y$. Substituting in (1) and solving for y gives

$$2 \cdot 2y - y + 1 = 0 \quad \Rightarrow \quad 3y = -1 \quad \Rightarrow \quad y = -\frac{1}{3}$$

The corresponding value of x is $x = 2 \cdot \left(-\frac{1}{3}\right) = -\frac{2}{3}$. The critical point is $\left(-\frac{2}{3}, -\frac{1}{3}\right)$.

Step 2. Compute the Discriminant. We find the second-order partials:

$$f_{xx}(x, y) = 2, \quad f_{yy}(x, y) = 2, \quad f_{xy}(x, y) = -1$$

The discriminant is

$$D(x, y) = f_{xx} f_{yy} - f_{xy}^2 = 2 \cdot 2 - (-1)^2 = 3$$

Step 3. Applying the Second Derivative Test. We have

$$D\left(-\frac{2}{3}, -\frac{1}{3}\right) = 3 > 0 \quad \text{and} \quad f_{xx}\left(-\frac{2}{3}, -\frac{1}{3}\right) = 2 > 0$$

The Second Derivative Test implies that $f\left(-\frac{2}{3}, -\frac{1}{3}\right)$ is a local minimum.

7. $f(x, y) = x^3 y + 12x^2 - 8y$

SOLUTION

Step 1. Find the critical points. We set the first-order partial derivatives of $f(x, y) = x^3 y + 12x^2 - 8y$ equal to zero and solve:

$$f_x(x, y) = 3x^2 y + 24x = 3x(xy + 8) = 0 \tag{1}$$

$$f_y(x, y) = x^3 - 8 = 0 \tag{2}$$

Equation (2) implies that $x = 2$. We substitute in equation (1) and solve for y to obtain

$$6(2y + 8) = 0 \quad \text{or} \quad y = -4$$

The critical point is $(2, -4)$.

Step 2. Compute the Discriminant. We find the second-order partials:

$$f_{xx}(x, y) = 6xy + 24, \quad f_{yy} = 0, \quad f_{xy} = 3x^2$$

The discriminant is thus

$$D(x, y) = f_{xx} f_{yy} - f_{xy}^2 = -9x^4$$

Step 3. Apply the Second Derivative Test. We have

$$D(2, -4) = -9 \cdot 2^4 < 0$$

Hence $(2, -4)$ is a saddle point.

9. $f(x, y) = 4x - 3x^3 - 2xy^2$

SOLUTION

Step 1. Find the critical points. We set the first-order derivatives of $f(x, y) = 4x - 3x^3 - 2xy^2$ equal to zero and solve:

$$f_x(x, y) = 4 - 9x^2 - 2y^2 = 0 \quad (1)$$

$$f_y(x, y) = -4xy = 0 \quad (2)$$

Equation (2) implies that $x = 0$ or $y = 0$. If $x = 0$, then equation (1) gives

$$4 - 2y^2 = 0 \quad \Rightarrow \quad y^2 = 2 \quad \Rightarrow \quad y = \sqrt{2}, \quad y = -\sqrt{2}$$

If $y = 0$, then equation (1) gives

$$4 - 9x^2 = 0 \quad \Rightarrow \quad 9x^2 = 4 \quad \Rightarrow \quad x = \frac{2}{3}, \quad x = -\frac{2}{3}$$

The critical points are therefore

$$\left(0, \sqrt{2}\right), \quad \left(0, -\sqrt{2}\right), \quad \left(\frac{2}{3}, 0\right), \quad \left(-\frac{2}{3}, 0\right)$$

Step 2. Compute the discriminant. The second-order partials are

$$f_{xx}(x, y) = -18x, \quad f_{yy}(x, y) = -4x, \quad f_{xy} = -4y$$

The discriminant is thus

$$D(x, y) = f_{xx}f_{yy} - f_{xy}^2 = -18x \cdot (-4x) - (-4y)^2 = 72x^2 - 16y^2$$

Step 3. Apply the Second Derivative Test. We have

$$D\left(0, \sqrt{2}\right) = -32 < 0$$

$$D\left(0, -\sqrt{2}\right) = -32 < 0$$

$$D\left(\frac{2}{3}, 0\right) = 72 \cdot \frac{4}{9} = 32 > 0,$$

$$f_{xx}\left(\frac{2}{3}, 0\right) = -18 \cdot \frac{2}{3} = -12 < 0$$

$$D\left(-\frac{2}{3}, 0\right) = 72 \cdot \frac{4}{9} = 32 > 0,$$

$$f_{xx}\left(-\frac{2}{3}, 0\right) = -18 \cdot \left(-\frac{2}{3}\right) = 12 > 0$$

The Second Derivative Test implies that the points $\left(0, \pm\sqrt{2}\right)$ are the saddle points, $f\left(\frac{2}{3}, 0\right)$ is a local maximum, and $f\left(-\frac{2}{3}, 0\right)$ is a local minimum.

11. $f(x, y) = x^4 + y^4 - 4xy$

SOLUTION

Step 1. Find the critical points. We set the first-order derivatives of $f(x, y) = x^4 + y^4 - 4xy$ equal to zero and solve:

$$f_x(x, y) = 4x^3 - 4y = 0, \quad f_y(x, y) = 4y^3 - 4x = 0 \quad (1)$$

Equation (1) implies that $y = x^3$. Substituting in (2) and solving for x, we obtain

$$\left(x^3\right)^3 - x = x^9 - x = x(x^8 - 1) = 0 \quad \Rightarrow \quad x = 0, \quad x = 1, \quad x = -1$$

The corresponding y coordinates are

$$y = 0^3 = 0, \quad y = 1^3 = 1, \quad y = (-1)^3 = -1$$

The critical points are therefore

$$(0, 0), \quad (1, 1), \quad (-1, -1)$$

Step 2. Compute the discriminant. We find the second-order partials:

$$f_{xx}(x, y) = 12x^2, \quad f_{yy}(x, y) = 12y^2, \quad f_{xy}(x, y) = -4$$

The discriminant is thus

$$D(x, y) = f_{xx} f_{yy} - f_{xy}^2 = 12x^2 \cdot 12y^2 - (-4)^2 = 144x^2y^2 - 16$$

Step 3. Apply the Second Derivative Test. We have

$$D(0, 0) = -16 < 0$$

$$D(1, 1) = 144 - 16 = 128 > 0, \quad f_{xx}(1, 1) = 12 > 0$$

$$D(-1, -1) = 144 - 16 = 128 > 0, \quad f_{xx}(-1, -1) = 12 > 0$$

We conclude that $(0, 0)$ is a saddle point, whereas $f(1, 1)$ and $f(-1, -1)$ are local minima.

13. $f(x, y) = xye^{-x^2-y^2}$

SOLUTION

Step 1. Find the critical points. We compute the partial derivatives of $f(x, y) = xye^{-x^2-y^2}$, using the Product Rule and the Chain Rule:

$$f_x(x, y, z) = y\left(1 \cdot e^{-x^2-y^2} + xe^{-x^2-y^2} \cdot (-2x)\right) = ye^{-x^2-y^2}\left(1 - 2x^2\right)$$

$$f_y(x, y, z) = x\left(1 \cdot e^{-x^2-y^2} + ye^{-x^2-y^2} \cdot (-2y)\right) = xe^{-x^2-y^2}\left(1 - 2y^2\right)$$

We set the partial derivatives equal to zero and solve to find the critical points. This gives

$$ye^{-x^2-y^2}\left(1 - 2x^2\right) = 0$$

$$xe^{-x^2-y^2}\left(1 - 2y^2\right) = 0$$

Since $e^{-x^2-y^2} \neq 0$, the first equation gives $y = 0$ or $1 - 2x^2 = 0$, that is, $y = 0$, $x = \frac{1}{\sqrt{2}}$, $x = -\frac{1}{\sqrt{2}}$. We substitute each of these values in the second equation and solve to obtain

$$y = 0: \quad xe^{-x^2} = 0 \quad \Rightarrow \quad x = 0$$

$$x = \frac{1}{\sqrt{2}}: \quad \frac{1}{\sqrt{2}}e^{-\frac{1}{2}-y^2}\left(1 - 2y^2\right) = 0 \quad \Rightarrow \quad 1 - 2y^2 = 0 \quad \Rightarrow \quad y = \pm\frac{1}{\sqrt{2}}$$

$$x = -\frac{1}{\sqrt{2}}: \quad -\frac{1}{\sqrt{2}}e^{-\frac{1}{2}-y^2}\left(1 - 2y^2\right) = 0 \quad \Rightarrow \quad 1 - 2y^2 = 0 \quad \Rightarrow \quad y = \pm\frac{1}{\sqrt{2}}$$

We obtain the following critical points: $(0, 0)$,

$$\left(\frac{1}{\sqrt{2}}, \frac{1}{\sqrt{2}}\right), \quad \left(\frac{1}{\sqrt{2}}, -\frac{1}{\sqrt{2}}\right), \quad \left(-\frac{1}{\sqrt{2}}, \frac{1}{\sqrt{2}}\right), \quad \left(-\frac{1}{\sqrt{2}}, -\frac{1}{\sqrt{2}}\right)$$

Step 2. Compute the second-order partials.

$$f_{xx}(x, y) = y\frac{\partial}{\partial x}\left(e^{-x^2-y^2}\left(1 - 2x^2\right)\right) = y\left(e^{-x^2-y^2}(-2x)\left(1 - 2x^2\right) + e^{-x^2-y^2}(-4x)\right)$$

$$= -2xye^{-x^2-y^2}\left(3 - 2x^2\right)$$

$$f_{yy}(x, y) = x\frac{\partial}{\partial y}\left(e^{-x^2-y^2}\left(1 - 2y^2\right)\right) = x\left(e^{-x^2-y^2}(-2y)\left(1 - 2y^2\right) + e^{-x^2-y^2}(-4y)\right)$$

$$= -2yxe^{-x^2-y^2}\left(3 - 2y^2\right)$$

$$f_{xy}(x, y) = \frac{\partial}{\partial y}f_x = \left(1 - 2x^2\right)\frac{\partial}{\partial y}\left(ye^{-x^2-y^2}\right) = \left(1 - 2x^2\right)\left(1 \cdot e^{-x^2-y^2} + ye^{-x^2-y^2}(-2y)\right)$$

$$= e^{-x^2-y^2}\left(1 - 2x^2\right)\left(1 - 2y^2\right)$$

The discriminant is

$$D(x, y) = f_{xx} f_{yy} - f_{xy}^2$$

Step 3. Apply the Second Derivative Test. We construct the following table:

Critical Point	f_{xx}	f_{yy}	f_{xy}	D	Type
$(0, 0)$	0	0	1	-1	$D < 0$, saddle point
$\left(\frac{1}{\sqrt{2}}, \frac{1}{\sqrt{2}}\right)$	$-\frac{2}{e}$	$-\frac{2}{e}$	0	$\frac{4}{e^2}$	$D > 0$, $f_{xx} < 0$ local maximum
$\left(\frac{1}{\sqrt{2}}, -\frac{1}{\sqrt{2}}\right)$	$\frac{2}{e}$	$\frac{2}{e}$	0	$\frac{4}{e^2}$	$D > 0$, $f_{xx} > 0$ local minimum
$\left(-\frac{1}{\sqrt{2}}, \frac{1}{\sqrt{2}}\right)$	$\frac{2}{e}$	$\frac{2}{e}$	0	$\frac{4}{e^2}$	$D > 0$, $f_{xx} > 0$ local minimum
$\left(-\frac{1}{\sqrt{2}}, -\frac{1}{\sqrt{2}}\right)$	$-\frac{2}{e}$	$-\frac{2}{e}$	0	$\frac{4}{e^2}$	$D > 0$, $f_{xx} < 0$ local maximum

15. $f(x, y) = e^x - xe^y$

SOLUTION

Step 1. Find the critical points. We set the first-order derivatives of $f(x, y) = e^x - xe^y$ equal to zero and solve:

$$f_x(x, y) = e^x - e^y = 0$$

$$f_y(x, y) = -xe^y = 0$$

Since $e^y \neq 0$, the second equation gives $x = 0$. Substituting in the first equation, we get

$$e^0 - e^y = 1 - e^y = 0 \quad \Rightarrow \quad e^y = 1 \quad \Rightarrow \quad y = 0$$

The critical point is $(0, 0)$.

Step 2. Compute the discriminant. We find the second-order partial derivatives:

$$f_{xx}(x, y) = \frac{\partial}{\partial x}\left(e^x - e^y\right) = e^x$$

$$f_{yy}(x, y) = \frac{\partial}{\partial y}\left(-xe^y\right) = -xe^y$$

$$f_{xy}(x, y) = \frac{\partial}{\partial y}\left(e^x - e^y\right) = -e^y$$

The discriminant is

$$D(x, y) = f_{xx}f_{yy} - f_{xy}^2 = -xe^{x+y} - e^{2y}$$

Step 3. Apply the Second Derivative Test. We have

$$D(0, 0) = 0 - e^0 = -1 < 0$$

The point $(0, 0)$ is a saddle point.

17. $f(x, y) = \ln x + 2\ln y - x - 4y$

SOLUTION

Step 1. Find the critical points. We set the first-order partials of $f(x, y) = \ln x + 2\ln y - x - 4y$ equal to zero and solve:

$$f_x(x, y) = \frac{1}{x} - 1 = 0, \quad f_y(x, y) = \frac{2}{y} - 4 = 0$$

The first equation gives $x = 1$, and the second equation gives $y = \frac{1}{2}$. We obtain the critical point $\left(1, \frac{1}{2}\right)$. Notice that f_x and f_y do not exist if $x = 0$ or $y = 0$, respectively, but these are not critical points since they are not in the domain of f. The critical point is thus $\left(1, \frac{1}{2}\right)$.

Step 2. Compute the discriminant. We find the second-order partials:

$$f_{xx}(x, y) = -\frac{1}{x^2}, \quad f_{yy}(x, y) = -\frac{2}{y^2}, \quad f_{xy}(x, y) = 0$$

The discriminant is

$$D(x, y) = f_{xx}f_{yy} - f_{xy}^2 = \frac{2}{x^2y^2}$$

Step 3. Apply the Second Derivative Test. We have

$$D\left(1, \frac{1}{2}\right) = \frac{2}{1^2 \cdot \left(\frac{1}{2}\right)^2} = 8 > 0, \quad f_{xx}\left(1, \frac{1}{2}\right) = -\frac{1}{1^2} = -1 < 0$$

We conclude that $f\left(1, \frac{1}{2}\right)$ is a local maximum.

19. $f(x, y) = x - y^2 - \ln(x + y)$

SOLUTION

Step 1. Find the critical points. We set the partial derivatives of $f(x, y) = x - y^2 - \ln(x + y)$ equal to zero and solve.

$$f_x(x, y) = 1 - \frac{1}{x+y} = 0, \quad f_y(x, y) = -2y - \frac{1}{x+y} = 0$$

The first equation implies that $\frac{1}{x+y} = 1$. Substituting in the second equation gives

$$-2y - 1 = 0 \quad \Rightarrow \quad 2y = -1 \quad \Rightarrow \quad y = -\frac{1}{2}$$

We substitute $y = -\frac{1}{2}$ in the first equation and solve for x:

$$1 - \frac{1}{x - \frac{1}{2}} = 0 \quad \Rightarrow \quad x - \frac{1}{2} = 1 \quad \Rightarrow \quad x = \frac{3}{2}$$

We obtain the critical point $\left(\frac{3}{2}, -\frac{1}{2}\right)$. Notice that although f_x and f_y do not exist where $x + y = 0$, these are not critical points since f is not defined at these points.

Step 2. Compute the discriminant. We compute the second-order partial derivatives:

$$f_{xx}(x, y) = \frac{\partial}{\partial x}\left(1 - \frac{1}{x+y}\right) = \frac{1}{(x+y)^2}$$

$$f_{yy}(x, y) = \frac{\partial}{\partial y}\left(-2y - \frac{1}{x+y}\right) = -2 + \frac{1}{(x+y)^2}$$

$$f_{xy}(x, y) = \frac{\partial}{\partial y}\left(1 - \frac{1}{x+y}\right) = \frac{1}{(x+y)^2}$$

The discriminant is

$$D(x, y) = f_{xx}f_{yy} - f_{xy}^2 = \frac{1}{(x+y)^2}\left(-2 + \frac{1}{(x+y)^2}\right) - \frac{1}{(x+y)^4} = \frac{-2}{(x+y)^2}$$

Step 3. Apply the Second Derivative Test. We have

$$D\left(\frac{3}{2}, -\frac{1}{2}\right) = \frac{-2}{\left(\frac{3}{2} - \frac{1}{2}\right)^2} = -2 < 0$$

We conclude that $\left(\frac{3}{2}, -\frac{1}{2}\right)$ is a saddle point.

21. Show that $f(x, y) = \sqrt{x^2 + y^2}$ has one critical point P and that f is nondifferentiable at P. Show that $f(P)$ is an absolute minimum value.

SOLUTION Since $f(x, y) = \sqrt{x^2 + y^2} \geq 0$ and $f(0, 0) = 0$, $f(0, 0)$ is an absolute minimum value. To find the critical point of f we first find the first derivatives:

$$f_x(x, y) = \frac{\partial}{\partial x}\left(\sqrt{x^2 + y^2}\right) = \frac{2x}{2\sqrt{x^2 + y^2}} = \frac{x}{\sqrt{x^2 + y^2}}$$

$$f_y(x, y) = \frac{\partial}{\partial y}\left(\sqrt{x^2 + y^2}\right) = \frac{2y}{2\sqrt{x^2 + y^2}} = \frac{y}{\sqrt{x^2 + y^2}}$$

Since f_x and f_y do not exist at $(0, 0)$ and the equations $f_x(x, y) = 0$ and $f_y(x, y) = 0$ have no solutions, the only critical point is $P = (0, 0)$, a point where f is non-differentiable.

23. CAS Use a computer algebra system to find numerical approximations to the critical points of

$$f(x, y) = (1 - x + x^2)e^{y^2} + (1 - y + y^2)e^{x^2}$$

Use Figure 18 to determine whether they correspond to local minima or maxima.

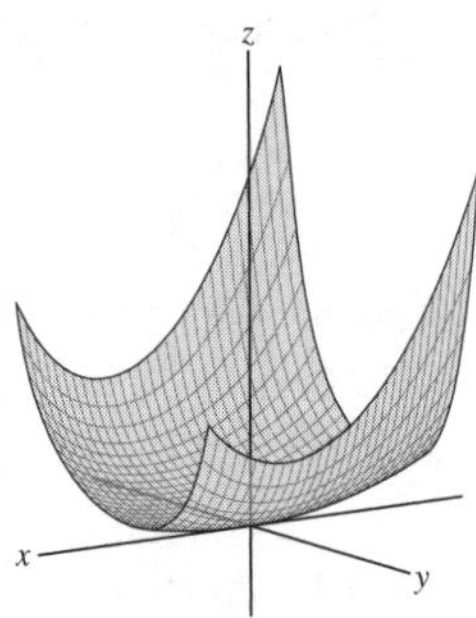

FIGURE 18 Plot of the function $f(x, y) = (1 - x + x^2)e^{y^2} + (1 - y + y^2)e^{x^2}$.

SOLUTION The critical points are the solutions of $f_x(x, y) = 0$ and $f_y(x, y) = 0$. We compute the partial derivatives:

$$f_x(x, y) = (-1 + 2x)e^{y^2} + \left(1 - y + y^2\right)e^{x^2} \cdot 2x$$

$$f_y(x, y) = \left(1 - x + x^2\right)e^{y^2} \cdot 2y + (-1 + 2y)e^{x^2}$$

Hence, the critical points are the solutions of the following equations:

$$(2x - 1)e^{y^2} + 2x\left(1 - y + y^2\right)e^{x^2} = 0$$

$$(2y - 1)e^{x^2} + 2y\left(1 - x + x^2\right)e^{y^2} = 0$$

Using a CAS we obtain the following solution: $x = y = 0.27788$, which from the figure is a local minimum.

25. Which of the following domains are closed and which are bounded?

(a) $\{(x, y) \in \mathbf{R}^2 : x^2 + y^2 \le 1\}$

(b) $\{(x, y) \in \mathbf{R}^2 : x^2 + y^2 < 1\}$

(c) $\{(x, y) \in \mathbf{R}^2 : x \ge 0\}$

(d) $\{(x, y) \in \mathbf{R}^2 : x > 0, y > 0\}$

(e) $\{(x, y) \in \mathbf{R}^2 : 1 \le x \le 4, 5 \le y \le 10\}$

(f) $\{(x, y) \in \mathbf{R}^2 : x > 0, x^2 + y^2 \le 10\}$

SOLUTION

(a) $\{(x, y) \in \mathbf{R}^2 : x^2 + y^2 \le 1\}$: This domain is bounded since it is contained, for instance, in the disk $x^2 + y^2 < 2$. The domain is also closed since it contains all of its boundary points, which are the points on the unit circle $x^2 + y^2 = 1$.

(b) $\{(x, y) \in \mathbf{R}^2 : x^2 + y^2 < 1\}$: The domain is contained in the disk $x^2 + y^2 < 1$, hence it is bounded. It is not closed since its boundary $x^2 + y^2 = 1$ is not contained in the domain.

(c) $\{(x, y) \in \mathbf{R}^2 : x \ge 0\}$:

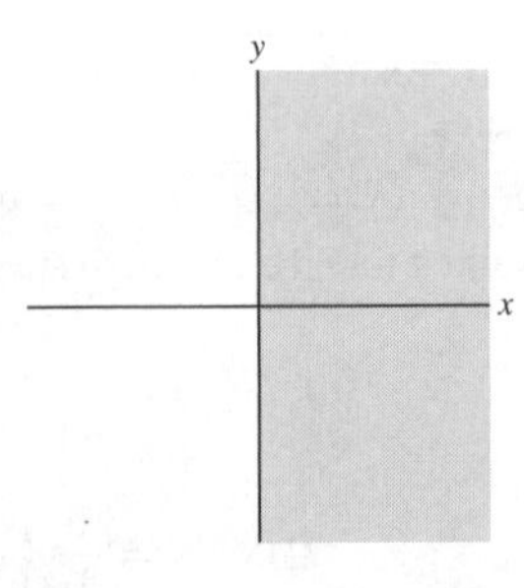

This domain is not contained in any disk, hence it is not bounded. However, the domain contains its boundary $x = 0$ (the y-axis), hence it is closed.

(d) $\{(x, y) \in \mathbf{R}^2 : x > 0, y > 0\}$:

The domain is not contained in any disk, hence it is not bounded. The boundary is the positive x and y axes, and it is not contained in the domain, therefore the domain is not closed.

(e) $\{(x, y) \in \mathbf{R}^2 : 1 \le x \le 4, 5 \le y \le 10\}$:

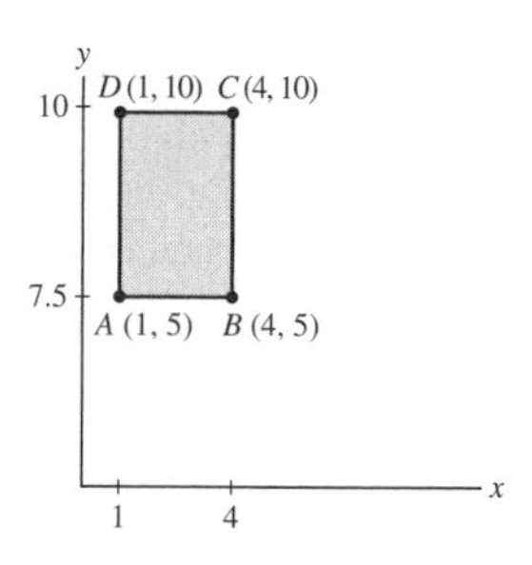

This domain is contained in the disk $x^2 + y^2 \le 11^2$, hence it is bounded. Moreover, the domain contains its boundary, which consists of the segments AB, BC, CD, AD shown in the figure, therefore the domain is closed.

(f) $\{(x, y) \in \mathbf{R}^2 : x > 0, x^2 + y^2 \le 10\}$:

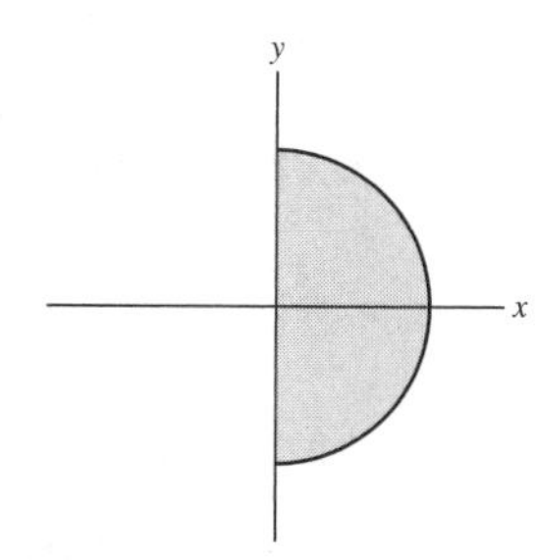

This domain is bounded since it is contained in the disk $x^2 + y^2 \le 10$. It is not closed since the part $\{(0, y) \in \mathbf{R}^2 : |y| \le \sqrt{10}\}$ of its boundary is not contained in the domain.

In Exercises 26–29, determine the global extreme values of the function on the given set without using calculus.

27. $f(x, y) = 2x - y, \quad 0 \le x \le 1, 0 \le y \le 3$

SOLUTION f is maximum when x is maximum and y is minimum, that is $x = 1$ and $y = 0$. f is minimum when x is minimum and y is maximum, that is, $x = 0$, $y = 3$. Therefore, the global maximum of f in the set is $f(1, 0) = 2 \cdot 1 - 0 = 2$ and the global minimum is $f(0, 3) = 2 \cdot 0 - 3 = -3$.

29. $f(x, y) = e^{-x^2-y^2}, \quad x^2 + y^2 \le 1$

SOLUTION The function $f(x, y) = e^{-(x^2+y^2)} = \frac{1}{e^{x^2+y^2}}$ is maximum when $e^{x^2+y^2}$ is minimum, that is, when $x^2 + y^2$ is minimum. The minimum value of $x^2 + y^2$ on the given set is zero, obtained at $x = 0$ and $y = 0$. We conclude that the maximum value of f on the given set is

$$f(0, 0) = e^{-0^2-0^2} = e^0 = 1$$

f is minimum when $x^2 + y^2$ is maximum, that is, when $x^2 + y^2 = 1$. Thus, the minimum value of f on the given disk is obtained on the boundary of the disk, and it is $e^{-1} = \frac{1}{e}$.

31. Let $\mathcal{D} = \{(x, y) : x > 0, y > 0\}$. Show that $\mathcal{D}$ is not closed. Find a continuous function that does not have a global minimum value on $\mathcal{D}$.

SOLUTION The boundary of $\mathcal{D}$ consists of two rays: the positive x and y axes.

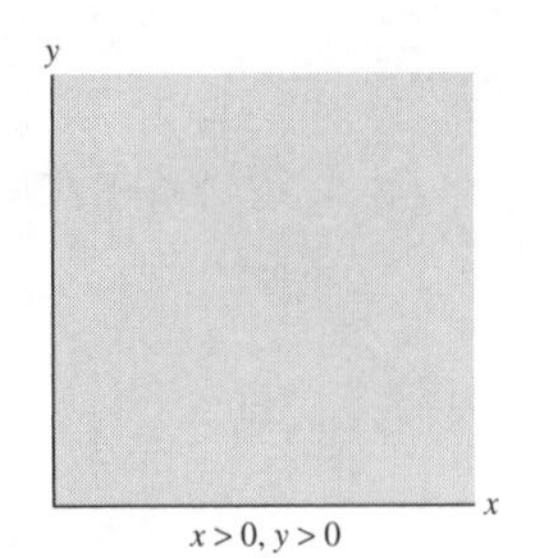

These rays are not contained in D, therefore D is not closed. The function $f(x, y) = -\frac{1}{x+y}$ is continuous on D (since $x + y \neq 0$ in D). However, f does not have a global minimum value on D, since $\lim_{x\to 0+}\left(-\frac{1}{x+x}\right) = -\infty$. Thus, when x approaches zero along the ray $\{(x, x) : x > 0\}$, the values of f are decreasing without bound.

33. Let $f(x, y) = (x + y)\ln(x^2 + y^2)$, defined for $(x, y) \neq (0, 0)$.

(a) Show that if (x, y) is a critical point of $f(x, y)$, then either $x = y$ or $x = -y$. Determine the critical points (four in all).

(b) CAS Plot a graph of $f(x, y)$ and use it to determine whether these critical points are maxima, minima, or neither.

SOLUTION

(a) This is best done by converting to polar coordinates. We see that $f(r, \theta) = (r\cos\theta + r\sin\theta)\ln r^2$, or in other words, $f(r, \theta) = (\cos\theta + \sin\theta)2r\ln r$. We take derivatives and find that

$$\frac{\partial f}{\partial r} = (\cos\theta + \sin\theta)(2\ln r + 2), \qquad \frac{\partial f}{\partial \theta} = (-\sin\theta + \cos\theta)(2r\ln r)$$

Since a critical point makes both equations equal to zero, and since $2\ln r + 2$ and $2r\ln r$ can't both be zero at the same r, we get that either $\cos\theta + \sin\theta = 0$ or $-\sin\theta + \cos\theta = 0$, which means either $x = y$ or $x = -y$. These solutions lead to $r = 1$ and $r = 1/e$, respectively, giving us our four critical points of $(1/e\sqrt{2}, 1/e\sqrt{2})$, $(-1/e\sqrt{2}, -1/e\sqrt{2})$, $(1/\sqrt{2}, -1/\sqrt{2})$, and $(-1/\sqrt{2}, 1/\sqrt{2})$.

(b) Looking at a graph, we see that $(1/\sqrt{2}, -1/\sqrt{2})$ and $(-1/\sqrt{2}, 1/\sqrt{2})$ are saddle points, while $(1/e\sqrt{2}, 1/e\sqrt{2})$ is a minimum and $(-1/e\sqrt{2}, -1/e\sqrt{2})$ is a maximum.

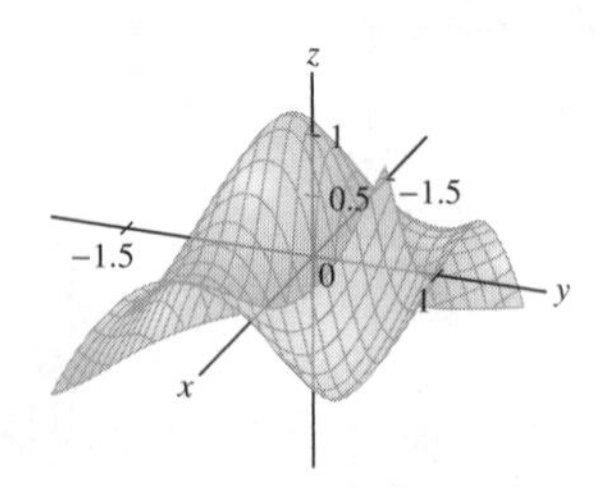

In Exercises 35–41, determine the global extreme values of the function on the given domain.

35. $f(x, y) = x^3 - 2y, \quad 0 \le x, y \le 1$

SOLUTION We use the following steps.

Step 1. Find the critical points. We set the first derivative equal to zero and solve:

$$f_x(x, y) = 3x^2 = 0, \quad f_y(x, y) = -2$$

The two equations have no solutions, hence there are no critical points.

Step 2. Check the boundary. The extreme values occur either at the critical points or at a point on the boundary of the domain. Since there are no critical points, the extreme values occur at boundary points. We consider each edge of the square $0 \le x, y \le 1$ separately.

The segment $\overline{OA}$: On this segment $y = 0$, $0 \le x \le 1$, and f takes the values $f(x, 0) = x^3$. The minimum value is $f(0, 0) = 0$ and the maximum value is $f(1, 0) = 1$.

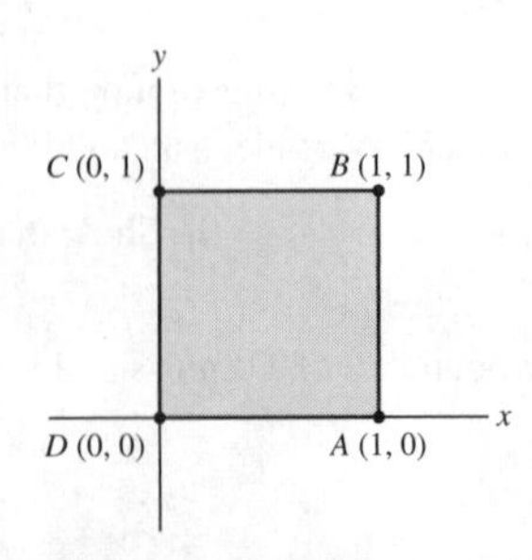

The segment $\overline{AB}$: On this segment $x = 1, 0 \le y \le 1$, and f takes the values $f(1, y) = 1 - 2y$. The minimum value is $f(1, 1) = 1 - 2 \cdot 1 = -1$ and the maximum value is $f(1, 0) = 1 - 2 \cdot 0 = 1$.
The segment $\overline{BC}$: On this segment $y = 1, 0 \le x \le 1$, and f takes the values $f(x, 1) = x^3 - 2$. The minimum value is $f(0, 1) = 0^3 - 2 = -2$ and the maximum value is $f(1, 1) = 1^3 - 2 = -1$.
The segment $\overline{OC}$: On this segment $x = 0, 0 \le y \le 1$, and f takes the values $f(0, y) = -2y$. The minimum value is $f(0, 1) = -2 \cdot 1 = -2$ and the maximum value is $f(0, 0) = -2 \cdot 0 = 0$.

Step 3. Conclusions. The values obtained in the previous steps are

$$f(0,0) = 0, \quad f(1,0) = 1, \quad f(1,1) = -1, \quad f(0,1) = -2$$

The smallest value is $f(0, 1) = -2$ and it is the global minimum of f on the square. The global maximum is the largest value $f(1, 0) = 1$.

37. $f(x, y) = x^2 + 2y^2, \quad 0 \le x, y \le 1$

SOLUTION The sum $x^2 + 2y^2$ is maximum at the point $(1, 1)$, where x^2 and y^2 are maximum. It is minimum if $x = y = 0$, that is, at the point $(0, 0)$. Hence,

$$\text{Global maximum} = f(1,1) = 1^2 + 2 \cdot 1^2 = 3$$

$$\text{Global minimum} = f(0,0) = 0^2 + 2 \cdot 0^2 = 0$$

39. $f(x, y) = x^3 + x^2y + 2y^2, \quad x, y \ge 0, x + y \le 1$

SOLUTION We use the following steps.

Step 1. Examine the critical points. We find the critical points of $f(x, y) = x^3 + x^2y + 2y^2$ in the interior of the domain (the standard region in the figure).

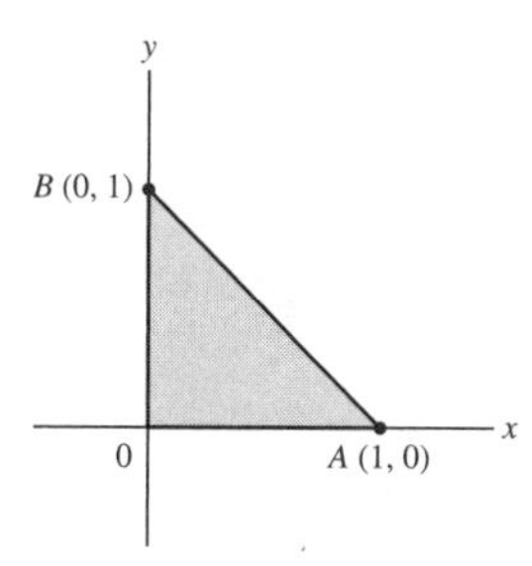

We set the partial derivatives of f equal to zero and solve:

$$f_x(x, y) = 3x^2 + 2xy = x(3x + 2y) = 0$$

$$f_y(x, y) = x^2 + 4y = 0$$

The first equation gives $x = 0$ or $y = -\frac{3}{2}x$. Substituting $x = 0$ in the second equation gives $4y = 0$ or $y = 0$. We obtain the critical point $(0, 0)$. We now substitute $y = -\frac{3}{2}x$ in the second equation and solve for x:

$$x^2 + 4 \cdot \left(-\frac{3}{2}x\right) = x^2 - 6x = x(x - 6) = 0 \quad \Rightarrow \quad x = 0, \quad x = 6$$

We get the critical points $(0, 0)$ and $(6, -9)$. None of the critical points $(0, 0)$ and $(6, -9)$ is in the interior of the domain.
Step 2. Check the boundary. The boundary consists of the three segments $\overline{OA}$, $\overline{OB}$, and $\overline{AB}$ shown in the figure. We consider each part of the boundary separately.

The segment $\overline{OA}$: On this segment $y = 0, 0 \le x \le 1$, and $f(x, y) = f(x, 0) = x^3$. The minimum value is $f(0, 0) = 0^3 = 0$ and the maximum value is $f(1, 0) = 1^3 = 1$.
The segment $\overline{OB}$: On this segment $x = 0, 0 \le y \le 1$, and $f(x, y) = f(0, y) = 2y^2$. The minimum value is $f(0, 0) = 2 \cdot 0^2 = 0$ and the maximum value is $f(0, 1) = 2 \cdot 1^2 = 2$.
The segment $\overline{AB}$: On this segment $y = 1 - x, 0 \le x \le 1$, and

$$f(x, y) = x^3 + x^2(1 - x) + 2(1 - x)^2 = x^3 + x^2 - x^3 + 2\left(1 - 2x + x^2\right) = 3x^2 - 4x + 2$$

We find the extreme values of $g(x) = 3x^2 - 4x + 2$ in the interval $0 \le x \le 1$. With the aid of the graph of $g(x)$, and with setting the derivative g' equal to 0, we find that the minimum value is

$$g\left(\frac{2}{3}\right) = f\left(\frac{2}{3}, \frac{1}{3}\right) = 3 \cdot \left(\frac{2}{3}\right)^2 - 4 \cdot \frac{2}{3} + 2 = \frac{2}{3}$$

and the maximum value is

$$g(0) = f(0, 1) = 3 \cdot 0^2 - 4 \cdot 0 + 2 = 2$$

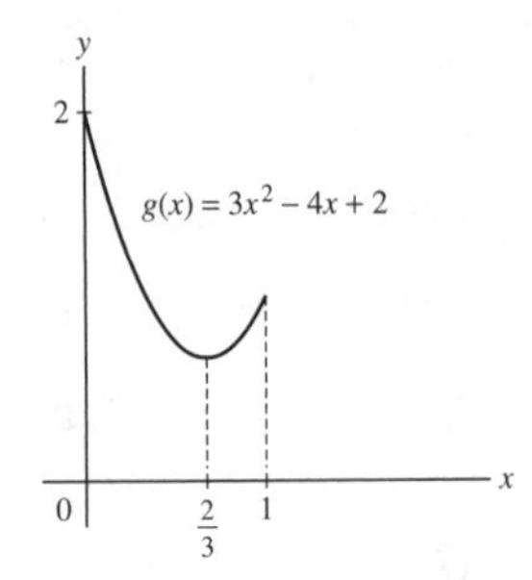

Step 3. Conclusions. We compare the values of $f(x, y)$ at the points obtained in step (2), and determine the global extrema of $f(x, y)$. This gives

$$f(0, 0) = 0, \quad f(1, 0) = 1, \quad f(0, 1) = 2, \quad f\left(\frac{2}{3}, \frac{1}{3}\right) = \frac{2}{3}$$

We conclude that the global minimum of f in the given domain is $f(0, 0) = 0$ and the global maximum is $f(0, 1) = 2$.

41. CAS $f(x, y) = x^3 y^5$, the set bounded by $x = 0$, $y = 0$, and $y = 1 - \sqrt{x}$. *Hint:* Use a computer algebra system to find the minimum along the boundary curve $y = 1 - \sqrt{x}$, which is parametrized by $(t, 1 - \sqrt{t})$ for $0 \le t \le 1$.

SOLUTION

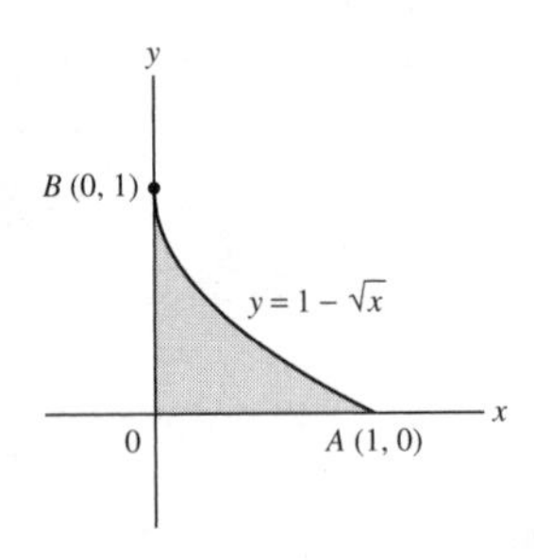

We use the following steps.

Step 1. Examine the critical points. We find the critical points of $f(x, y) = x^3 y^5$ in the interior of the domain. Setting the partial derivatives equal to zero and solving gives

$$f_x(x, y) = 3x^2 y^5 = 0, \quad f_y(x, y) = 5y^4 x^3 = 0$$

The solutions are all the points with at least one zero coordinate. These points are not in the interior of the region, hence there are no critical points in the interior of the region.

Step 2. Check the boundary. We consider each part of the boundary separately.

The segment $\overline{OA}$: On this segment $y = 0$ and $f(x, 0) = 0$, hence f has the constant value 0 on this part of the boundary.

The segment $\overline{OB}$: On this segment $x = 0$, hence $f(0, y) = 0^3 \cdot y^5 = 0$. f attains the constant value 0 on this part of the boundary.

The curve $y = 1 - \sqrt{x}$, $0 \le x \le 1$: On this curve we have $f(x, y) = g(x) = x^3 \left(1 - \sqrt{x}\right)^5$.

We find the points on the interval $0 \le x \le 1$ where $g(x) = x^3 \left(1 - \sqrt{x}\right)^5$ has maximum and minimum values. Differentiating g gives

$$g'(x) = 3x^2 \left(1 - \sqrt{x}\right)^5 + x^3 \cdot 5\left(1 - \sqrt{x}\right)^4 \cdot \left(\frac{-1}{2\sqrt{x}}\right) = 3x^2 \left(1 - \sqrt{x}\right)^5 - \frac{5}{2} x^{2.5} \left(1 - \sqrt{x}\right)^4$$

$$= x^2 \left(1 - \sqrt{x}\right)^4 \left(3 - 3\sqrt{x} - \frac{5}{2}\sqrt{x}\right) = x^2 \left(1 - \sqrt{x}\right)^4 \left(3 - \frac{11}{2}\sqrt{x}\right)$$

We solve $g'(x) = 0$ in the interval $0 < x < 1$:

$$g'(x) = x^2 \left(1 - \sqrt{x}\right)^4 \left(3 - \frac{11}{2}\sqrt{x}\right) = 0 \quad \Rightarrow \quad x = 0, \quad 1 - \sqrt{x} = 0, \quad 3 - \frac{11}{2}\sqrt{x} = 0$$

The solutions are $x = 0$, $x = 1$, and $x = \frac{36}{121}$. The critical point in the interval $0 < x < 1$ is $x = \frac{36}{121}$. We compute $g(x) = x^3\left(1 - \sqrt{x}\right)^5$ at the critical point and at the endpoints $x = 0$, $x = 1$:

$$g(0) = g(1) = 0, \quad g\left(\frac{36}{121}\right) = \left(\frac{36}{121}\right)^3\left(1 - \frac{6}{11}\right)^5 \approx 0.0005$$

The points where g has extreme values in the interval $0 \le x \le 1$ are $x = 0$, $x = 1$, and $x = \frac{36}{121}$. We find the y-coordinates of these points from $y = 1 - \sqrt{x}$:

$$x = 0: \quad y = 1 - \sqrt{0} = 1$$

$$x = 1: \quad y = 1 - \sqrt{1} = 0$$

$$x = \frac{36}{121}: \quad y = 1 - \sqrt{\frac{36}{121}} = \frac{5}{11}$$

We conclude that the global extrema of f on the curve $y = 1 - \sqrt{x}$, $0 \le x \le 1$ are obtained at the points $(0, 1)$, $(1, 0)$, and $\left(\frac{36}{121}, \frac{5}{11}\right)$.

Step 3. Conclusions. We examine the values of $f(x, y) = x^3y^5$ at the points obtained in the previous parts. The candidates for global extrema are $f = 0$, the values of f on the segments $\overline{OA}$ and $\overline{OB}$:

$$f(0, 1) = f(1, 0) = 0$$

$$f\left(\frac{36}{121}, \frac{5}{11}\right) = \left(\frac{36}{121}\right)^3\left(\frac{5}{11}\right)^5 = 0.0005$$

We conclude that the minimum value of $f(x, y)$ in the given domain is 0 and the maximum value is $f\left(\frac{36}{121}, \frac{5}{11}\right) \approx 0.0005$.

43. Consider a rectangular box B with a bottom and sides but no top such that B has minimal surface area among all boxes with fixed volume V.

(a) Do you think B is a cube as in the solution to Exercise 42? If not, how would you make its shape differ from a cube?

(b) Find the dimensions of B and compare with your response to (a).

SOLUTION

(a) Each of the variables x and y is the length of a side of three faces (for example, x is the length of the front, back, and bottom sides), whereas z is the length of a side of four faces.

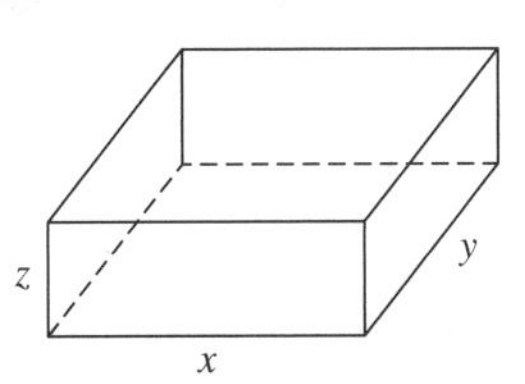

Therefore, the variables x, y, and z do not have equal influence on the surface area. We expect that in the box B with minimal surface area, z is smaller than $\sqrt[3]{V}$, which is the side of a cube with volume V.

(b) We must find the dimensions of the box B, with fixed volume V and with smallest possible surface area, when the top is not included.

Step 1. Find a function to be minimized. The surface area of the box with sides lengths x, y, z when the top is not included is

$$S = 2xz + 2yz + xy \tag{1}$$

To express the surface in terms of x and y only, we use the formula for the volume of the box, $V = xyz$, giving $z = \frac{V}{xy}$. We substitute in (1) to obtain

$$S = 2x \cdot \frac{V}{xy} + 2y \cdot \frac{V}{xy} + xy = \frac{2V}{y} + \frac{2V}{x} + xy$$

That is,

$$S = \frac{2V}{y} + \frac{2V}{x} + xy.$$

Step 2. Determine the domain. The variables x, y denote lengths, hence they must be nonnegative. Moreover, S is not defined for $x = 0$ or $y = 0$. Since there are no other limitations on the variables, the domain is

$$D = \{(x, y) : x > 0, y > 0\}$$

We must find the minimum value of S on D. Because this domain is neither closed nor bounded, we are not sure that a minimum value exists. However, it can be proved (in like manner as in Exercise 42) that S does have a minimum value on D. This value occurs at a critical point in D, hence we set the partial derivatives equal to zero and solve. This gives

$$S_x(x, y) = -\frac{2V}{x^2} + y = 0$$

$$S_y(x, y) = -\frac{2V}{y^2} + x = 0$$

The first equation gives $y = \frac{2V}{x^2}$. Substituting in the second equation yields

$$x - \frac{2V}{\frac{4V^2}{x^4}} = x - \frac{x^4}{2V} = x\left(1 - \frac{x^3}{2V}\right) = 0$$

The solutions are $x = 0$ and $x = (2V)^{1/3}$. The solution $x = 0$ is not included in D, so the only solution is $x = (2V)^{1/3}$.
We find the value of y using $y = \frac{2V}{x^2}$:

$$y = \frac{2V}{(2V)^{2/3}} = (2V)^{1/3}$$

We conclude that the critical point, which is the point where the minimum value of S in D occurs, is $\left((2V)^{1/3}, (2V)^{1/3}\right)$.
We find the corresponding value of z using $z = \frac{V}{xy}$. We get

$$z = \frac{V}{(2V)^{1/3}(2V)^{1/3}} = \frac{V}{2^{2/3}V^{2/3}} = \frac{V^{1/3}}{2^{2/3}} = \left(\frac{V}{4}\right)^{1/3}$$

We conclude that the sizes of the box with minimum surface area are

width: $x = (2V)^{1/3}$;
length: $y = (2V)^{1/3}$;
height: $z = \left(\frac{V}{4}\right)^{1/3}$.

We see that z is smaller than x and y as predicted.

Further Insights and Challenges

45. The power (in microwatts) of a laser is measured as a function of current (in milliamps). Find the linear least-squares fit (Exercise 44) for the data points.

Current (mA)	1.0	1.1	1.2	1.3	1.4	1.5
Laser power (μW)	0.52	0.56	0.82	0.78	1.23	1.50

SOLUTION By Exercise 44, the coefficients of the linear least-square fit $f(x) = mx + b$ are determined by the following equations:

$$m\sum_{j=1}^{n} x_j + bn = \sum_{j=1}^{n} y_j$$

$$m\sum_{j=1}^{n} x_j^2 + b\sum_{j=1}^{n} x_j = \sum_{j=1}^{n} x_j \cdot y_j \qquad (1)$$

In our case there are $n = 6$ data points:

$$(x_1, y_1) = (1, 0.52), (x_2, y_2) = (1.1, 0.56),$$

$$(x_3, y_3) = (1.2, 0.82), (x_4, y_4) = (1.3, 0.78),$$
$$(x_5, y_5) = (1.4, 1.23), (x_6, y_6) = (1.5, 1.50).$$

We compute the sums in (1):

$$\sum_{j=1}^{6} x_j = 1 + 1.1 + 1.2 + 1.3 + 1.4 + 1.5 = 7.5$$

$$\sum_{j=1}^{6} y_j = 0.52 + 0.56 + 0.82 + 0.78 + 1.23 + 1.50 = 5.41$$

$$\sum_{j=1}^{6} x_j^2 = 1^2 + 1.1^2 + 1.2^2 + 1.3^2 + 1.4^2 + 1.5^2 = 9.55$$

$$\sum_{j=1}^{6} x_j \cdot y_j = 1 \cdot 0.52 + 1.1 \cdot 0.56 + 1.2 \cdot 0.82 + 1.3 \cdot 0.78 + 1.4 \cdot 1.23 + 1.5 \cdot 1.50 = 7.106$$

Substituting in (1) gives the following equations:

$$\begin{aligned} 7.5m + 6b &= 5.41 \\ 9.55m + 7.5b &= 7.106 \end{aligned} \tag{2}$$

We multiply the first equation by 9.55 and the second by (-7.5), then add the resulting equations. This gives

$$\begin{array}{r} 71.625m + 57.3b = 51.6655 \\ + \ -71.625m - 56.25b = -53.295 \\ \hline 1.05b = -1.6295 \end{array} \quad \Rightarrow \quad b = -1.5519$$

We now substitute $b = -1.5519$ in the first equation in (2) and solve for m:

$$\begin{array}{r} 7.5m + 6 \cdot (-1.5519) = 5.41 \\ 7.5m = 14.7214 \end{array} \quad \Rightarrow \quad m = 1.9629$$

The linear least squares fit $f(x) = mx + b$ is thus

$$f(x) = 1.9629x - 1.5519.$$

47. The following problem was proposed by Fermat as a challenge to the Italian scientist Evangelista Torricelli (1608–1647), a student of Galileo and inventor of the barometer. Given three points $A = (a_1, a_2)$, $B = (b_1, b_2)$, and $C = (c_1, c_2)$ in the plane, find the point $P = (x, y)$ that minimizes the sum of the distances

$$f(x, y) = AP + BP + CP$$

(a) Write out $f(x, y)$ as a function of x and y, and show that $f(x, y)$ is differentiable except at the points A, B, C.

(b) Define the unit vectors

$$\mathbf{e} = \frac{\overrightarrow{AP}}{\|\overrightarrow{AP}\|}, \qquad \mathbf{f} = \frac{\overrightarrow{BP}}{\|\overrightarrow{BP}\|}, \qquad \mathbf{g} = \frac{\overrightarrow{CP}}{\|\overrightarrow{CP}\|}$$

Show that the condition $\nabla f = 0$ is equivalent to

$$\mathbf{e} + \mathbf{f} + \mathbf{g} = 0 \tag{3}$$

Prove that Eq. (3) holds if and only if the mutual angles between the unit vectors are all 120°.

(c) Define the Fermat point to be the point P such that angles between the segments $\overline{AP}, \overline{BP}, \overline{CP}$ are all 120°. Conclude that the **Fermat point** solves the minimization problem (Figure 23).

(d) Show that the Fermat point does not exist if one of the angles in $\triangle ABC$ is $> 120°$. Where does the minimum occur in this case? *Hint:* The minimum must occur at a point where $f(x, y)$ is not differentiable.

A B e f P g C

(A) P is the Fermat point (the angles between **e**, **f**, and **g** are all 120°).

B 140° A C

(B) Fermat point does not exist.

FIGURE 23

SOLUTION

(a)

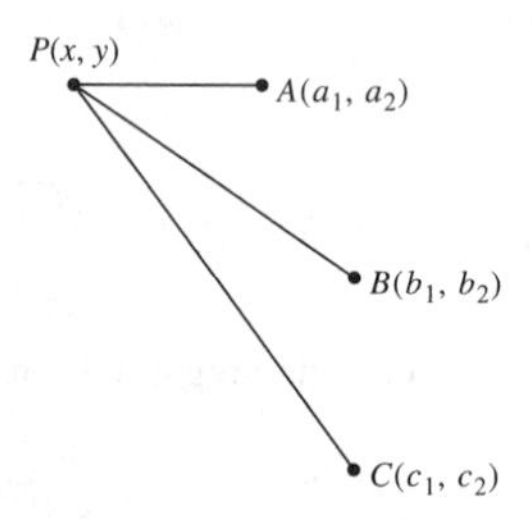

Using the formula for the length of a segment we obtain

$$f(x, y) = \sqrt{(x-a_1)^2 + (y-a_2)^2} + \sqrt{(x-b_1)^2 + (y-b_2)^2} + \sqrt{(x-c_1)^2 + (y-c_2)^2}$$

We compute the partial derivatives of f:

$$f_x(x, y) = \frac{x-a_1}{\sqrt{(x-a_1)^2 + (y-a_2)^2}} + \frac{x-b_1}{\sqrt{(x-b_1)^2 + (y-b_2)^2}} + \frac{x-c_1}{\sqrt{(x-c_1)^2 + (y-c_2)^2}} \quad (1)$$

$$f_y(x, y) = \frac{y-a_2}{\sqrt{(x-a_1)^2 + (y-a_2)^2}} + \frac{y-b_2}{\sqrt{(x-b_1)^2 + (y-b_2)^2}} + \frac{y-c_2}{\sqrt{(x-c_1)^2 + (y-c_2)^2}} \quad (2)$$

For all (x, y) other then (a_1, a_2), (b_1, b_2), (c_1, c_2) the partial derivatives are continuous, therefore the Criterion for Differentiability implies that f is differentiable at all points other than A, B, and C.

(b)

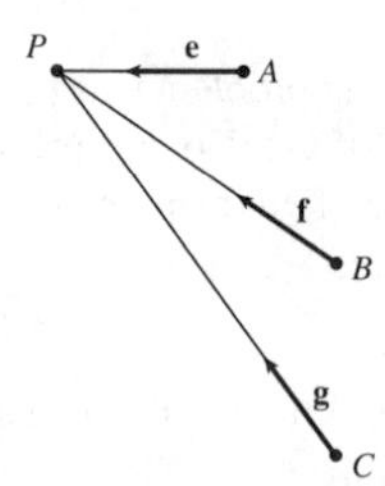

We compute the unit vectors **e**, **f**, and **g**:

$$\mathbf{e} = \frac{\langle x-a_1, y-a_2\rangle}{\sqrt{(x-a_1)^2 + (y-a_2)^2}}$$

$$\mathbf{f} = \frac{\langle x-b_1, y-b_2\rangle}{\sqrt{(x-b_1)^2 + (y-b_2)^2}}$$

$$\mathbf{g} = \frac{\langle x-c_1, y-c_2\rangle}{\sqrt{(x-c_1)^2 + (y-c_2)^2}}$$

We write the condition $\mathbf{e} + \mathbf{f} + \mathbf{g} = \mathbf{0}$:

$$\mathbf{e} + \mathbf{f} + \mathbf{g} = \frac{\langle x-a_1, y-a_2\rangle}{\sqrt{(x-a_1)^2 + (y-a_2)^2}} + \frac{\langle x-b_1, y-b_2\rangle}{\sqrt{(x-b_1)^2 + (y-b_2)^2}} + \frac{\langle x-c_1, y-c_2\rangle}{\sqrt{(x-c_1)^2 + (y-c_2)^2}}$$

$$= \left\langle \frac{x-a_1}{\sqrt{(x-a_1)^2+(y-a_2)^2}} + \frac{x-b_1}{\sqrt{(x-b_1)^2+(y-b_2)^2}} + \frac{x-c_1}{\sqrt{(x-c_1)^2+(y-c_2)^2}}, \right.$$

$$\left. \frac{y-a_2}{\sqrt{(x-a_1)^2+(y-a_2)^2}} + \frac{y-b_2}{\sqrt{(x-b_1)^2+(y-b_2)^2}} + \frac{y-c_2}{\sqrt{(x-c_1)^2+(y-c_2)^2}} \right\rangle$$

Combining with (1) and (2) we get

$$\mathbf{e}+\mathbf{f}+\mathbf{g} = \langle f_x(x,y), f_y(x,y) \rangle = \nabla f$$

Therefore, the condition $\nabla f = \mathbf{0}$ is equivalent to $\mathbf{e}+\mathbf{f}+\mathbf{g} = \mathbf{0}$. We now show that Eq. (3) holds if and only if the mutual angles between the unit vectors are all 120°. We place the axes so that the positive x-axis is in the direction of e.

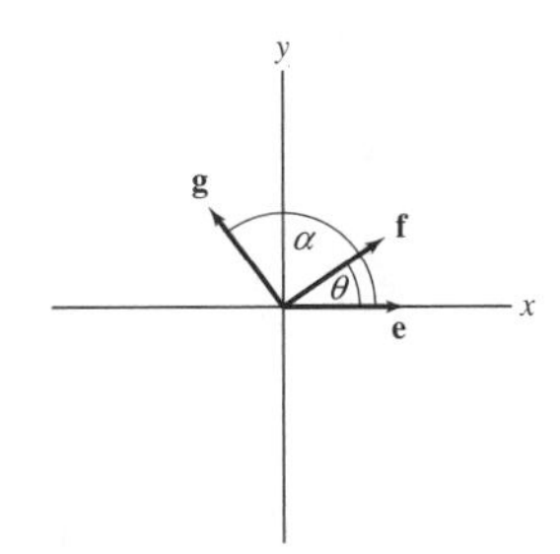

Let θ and α be the angles that **f** and **g** make with **e**, respectively. Hence,

$$\mathbf{e} = \langle 1, 0 \rangle, \mathbf{f} = \langle \cos\theta, \sin\theta \rangle, \mathbf{g} = \langle \cos\alpha, \sin\alpha \rangle$$

Substituting in $\mathbf{e}+\mathbf{f}+\mathbf{g} = \mathbf{0}$ we have

$$\langle \cos\theta + \cos\alpha + 1, \sin\theta + \sin\alpha \rangle = \langle 0, 0 \rangle$$

or

$$\cos\theta + \cos\alpha + 1 = 0$$
$$\sin\theta + \sin\alpha = 0$$

The second equation implies that

$$\sin\theta = -\sin\alpha = \sin(180+\alpha)$$

The solutions for $0 \le \alpha, \theta \le 360$ are

$$\theta = 180+\alpha, \quad \theta = 360-\alpha$$

We substitute each solution in the first equation and solve for α. This gives

$\theta = 180+\alpha$	$\theta = 360° - \alpha$
$\cos(180+\alpha)+\cos\alpha+1=0$	$\cos(360°-\alpha)+\cos\alpha+1=0$
$-\cos\alpha+\cos\alpha+1=0$	$\cos\alpha+\cos\alpha+1=0$
$1=0$	$2\cos\alpha=-1$
	$\cos\alpha = -\frac{1}{2}$

$$\Rightarrow \quad \begin{array}{ll} \alpha = 120° & \alpha = 240° \\ \theta = 360° - \alpha = 240° & \theta = 360° - \alpha = 120° \end{array}$$

We obtain the following vectors:

$$\mathbf{e} = \langle 1, 0 \rangle, \quad \mathbf{f} = \langle \cos 240°, \sin 240° \rangle, \quad \mathbf{g} = \langle \cos 120°, \sin 120° \rangle$$

or

$$\mathbf{e} = \langle 1, 0 \rangle, \quad \mathbf{f} = \langle \cos 120°, \sin 120° \rangle, \quad \mathbf{g} = \langle \cos 240°, \sin 240° \rangle$$

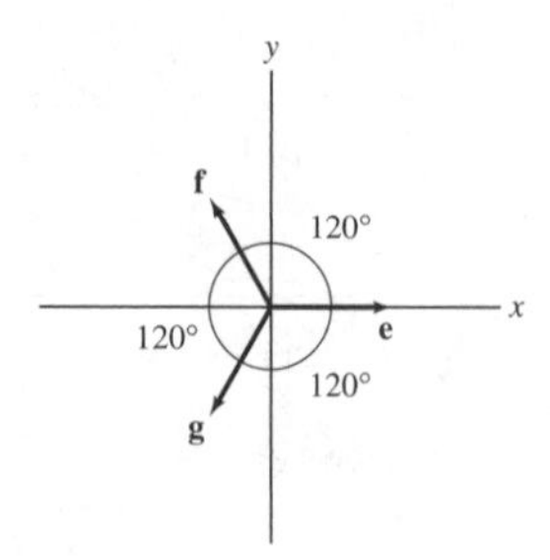

or

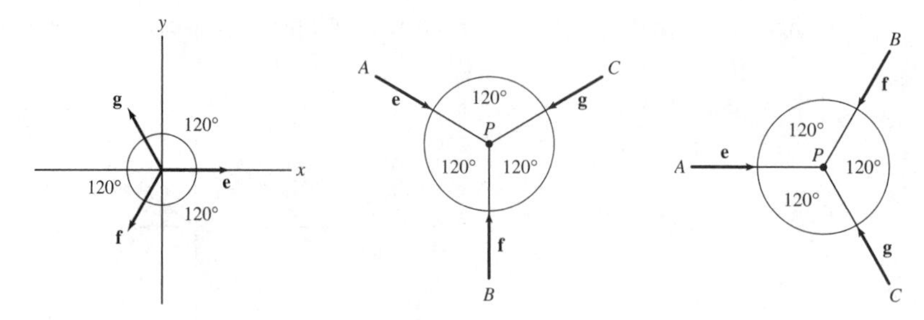

In either case the angles between the vectors are 120°.

(c) $f(x, y)$ has the minimum value at a critical point. The critical points are the points where f_x and f_y are 0 or do not exist, that is, the points A, B, C and the point where $\nabla f = \mathbf{0}$, which according to part (b) is the Fermat point. We now show that if the Fermat point P exists, then $f(P) \leq f(A), f(B), f(C)$.

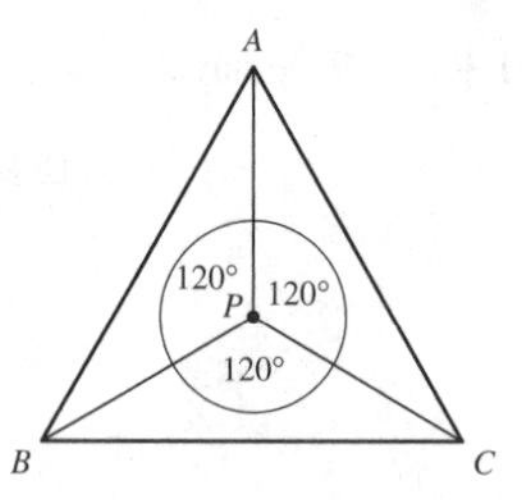

Suppose that the Fermat point P exists. The values of f at the critical points are

$$f(A) = \overline{AB} + \overline{AC}$$

$$f(B) = \overline{AB} + \overline{BC}$$

$$f(C) = \overline{AC} + \overline{BC}$$

$$f(P) = \overline{AP} + \overline{BP} + \overline{PC}$$

We show that $f(P) < f(A)$. Similarly it can be shown that also $f(P) < f(B)$ and $f(P) < f(C)$. By the Cosine Theorem for the triangles ABP and ACP we have

$$\overline{AB} = \sqrt{\overline{AP}^2 + \overline{BP}^2 - 2\overline{AP} \cdot \overline{BP} \cos 120^\circ} = \sqrt{\overline{AP}^2 + \overline{BP}^2 + \overline{AP} \cdot \overline{BP}}$$

$$\overline{AC} = \sqrt{\overline{AP}^2 + \overline{CP}^2 - 2\overline{AP} \cdot \overline{PC} \cos 120^\circ} = \sqrt{\overline{AP}^2 + \overline{CP}^2 + \overline{AP} \cdot \overline{PC}}$$

Hence

$$f(A) = \overline{AB} + \overline{AC} = \sqrt{\overline{AP}^2 + \overline{BP}^2 + \overline{AP} \cdot \overline{BP}} + \sqrt{\overline{AP}^2 + \overline{CP}^2 + \overline{AP} \cdot \overline{PC}}$$
$$\geq \overline{AP} + \overline{BP} + \overline{PC} = f(P)$$

The last inequality can be verified by squaring and transfering sides. It's best to use a computer to help with the algebra; it's a daunting task to do by hand.

(d) We show that if one of the angles of ΔABC is $\geq 120^\circ$, then the Fermat point does not exist. Notice that the Fermat point (if it exists) must fall inside the triangle ABC.

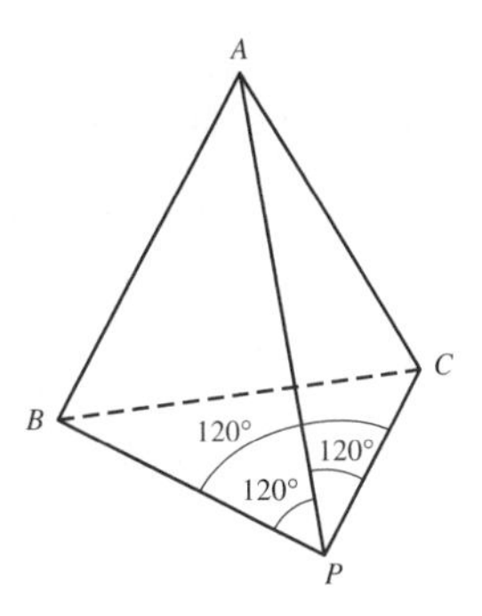

P cannot lie outside ΔABC

Suppose the Fermat point P exists.

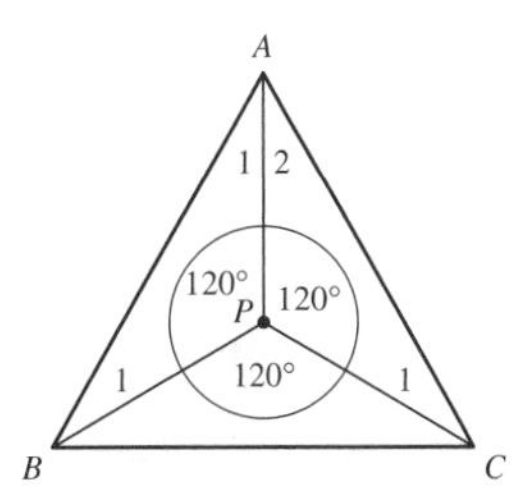

We sum the angles in the triangles ABP and ACP, obtaining

$$\sphericalangle A_1 + \sphericalangle B_1 + 120° = 180° \quad \Rightarrow \quad \sphericalangle A_1 = 60° - \sphericalangle B_1$$

$$\sphericalangle A_2 + \sphericalangle C_1 + 120° = 180° \quad \Rightarrow \quad \sphericalangle A_2 = 60° - \sphericalangle C_1$$

Therefore,

$$\sphericalangle A = \sphericalangle A_1 + \sphericalangle A_2 = \left(60° - \sphericalangle B_1\right) + \left(60° - \sphericalangle C_1\right) = 120° - (\sphericalangle B_1 + \sphericalangle C_1) < 120°$$

We thus showed that if the Fermat point exists, then $\sphericalangle A < 120°$. Similarly, one shows also that $\sphericalangle B$ and $\sphericalangle C$ must be smaller than 120°. We conclude that if one of the angles in ΔABC is equal or greater than 120°, then the Fermat point does not exist. In that case, the minimum value of $f(x, y)$ occurs at a point where f_x or f_y do not exist, that is, at one of the points A, B, or C.

15.8 Lagrange Multipliers: Optimizing with a Constraint (ET Section 14.8)

Preliminary Questions

1. Suppose that the maximum of $f(x, y)$ subject to the constraint $g(x, y) = 0$ occurs at a point $P = (a, b)$ such that $\nabla f_P \neq 0$. Which of the following are true?

(a) ∇f_P is tangent to $g(x, y) = 0$ at P.

(b) ∇f_P is orthogonal to $g(x, y) = 0$ at P.

SOLUTION

(a) Since the maximum of f subject to the constraint occurs at P, it follows by Theorem 1 that ∇f_P and ∇g_P are parallel vectors. The gradient ∇g_P is orthogonal to $g(x, y) = 0$ at P, hence ∇f_P is also orthogonal to this curve at P. We conclude that statement (b) is false (yet the statement can be true if $\nabla f_P = (0, 0)$).

(b) This statement is true by the reasoning given in the previous part.

2. Figure 8 shows a constraint $g(x, y) = 0$ and the level curves of a function f. In each case, determine whether f has a local minimum, local maximum, or neither at the labeled point.

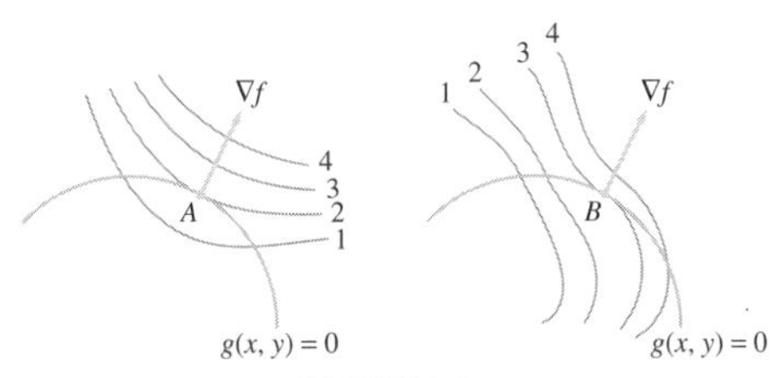

FIGURE 8

SOLUTION The level curve $f(x, y) = 2$ is tangent to the constraint curve at the point A. A close level curve that intersects the constraint curve is $f(x, y) = 1$, hence we may assume that f has a local maximum 2 under the constraint at A. The level curve $f(x, y) = 3$ is tangent to the constraint curve. However, in approaching B under the constraint, from one side f is increasing and from the other side f is decreasing. Therefore, $f(B)$ is neither local minimum nor local maximum of f under the constraint.

3. On the contour map in Figure 9:

(a) Identify the points where $\nabla f = \lambda \nabla g$ for some scalar λ.

(b) Identify the minimum and maximum values of $f(x, y)$ subject to $g(x, y) = 0$.

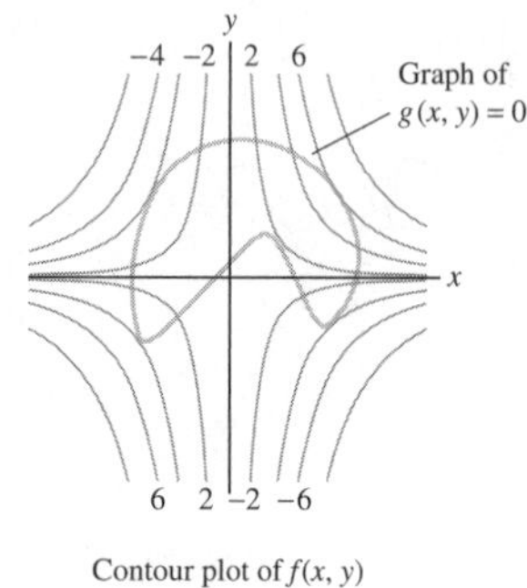

FIGURE 9 Contour map of $f(x, y)$; contour interval 2.

SOLUTION

(a) The gradient ∇g is orthogonal to the constraint curve $g(x, y) = 0$, and ∇f is orthogonal to the level curves of f. These two vectors are parallel at the points where the level curve of f is tangent to the constraint curve. These are the points A, B, C, D, E in the figure:

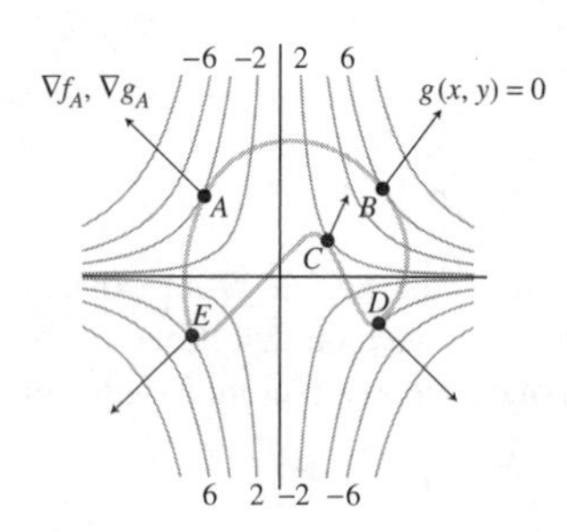

(b) The minimum and maximum occur where the level curve of f is tangent to the constraint curve. The level curves tangent to the constraint curve are

$$f(A) = -4, \quad f(C) = 2, \quad f(B) = 6, \quad f(D) = -4, \quad f(E) = 4$$

Therefore the global minimum of f under the constraint is -4 and the global maximum is 6.

Exercises

1. Use Lagrange multipliers to find the extreme values of the function $f(x, y) = 2x + 4y$ subject to the constraint $g(x, y) = x^2 + y^2 - 5 = 0$.

(a) Show that the Lagrange equation $\nabla f = \lambda \nabla g$ gives $\lambda x = 1$ and $\lambda y = 2$.

(b) Show that these equations imply $\lambda \neq 0$ and $y = 2x$.

(c) Use the constraint equation to determine the possible critical points (x, y).

(d) Evaluate $f(x, y)$ at the critical points and determine the minimum and maximum values.

SOLUTION

(a) The Lagrange equations are determined by the equality $\nabla f = \lambda \nabla g$. We find them:

$$\nabla f = \langle f_x, f_y \rangle = \langle 2, 4 \rangle, \quad \nabla g = \langle g_x, g_y \rangle = \langle 2x, 2y \rangle$$

Hence,

$$\langle 2, 4 \rangle = \lambda \langle 2x, 2y \rangle$$

or

$$\begin{aligned} \lambda(2x) &= 2 \\ \lambda(2y) &= 4 \end{aligned} \quad \Rightarrow \quad \begin{aligned} \lambda x &= 1 \\ \lambda y &= 2 \end{aligned}$$

(b) The Lagrange equations in part (a) imply that $\lambda \neq 0$. The first equation implies that $x = \frac{1}{\lambda}$ and the second equation gives $y = \frac{2}{\lambda}$. Therefore $y = 2x$.

(c) We substitute $y = 2x$ in the constraint equation $x^2 + y^2 - 5 = 0$ and solve for x and y. This gives

$$\begin{aligned} x^2 + (2x)^2 - 5 &= 0 \\ 5x^2 &= 5 \\ x^2 &= 1 \quad \Rightarrow \quad x_1 = -1, \quad x_2 = 1 \end{aligned}$$

Since $y = 2x$, we have $y_1 = 2x_1 = -2$, $y_2 = 2x_2 = 2$. The critical points are thus

$$(-1, -2) \quad \text{and} \quad (1, 2).$$

Extreme values can also occur at the points where $\nabla g = \langle 2x, 2y \rangle = \langle 0, 0 \rangle$. However, $(0, 0)$ is not on the constraint.

(d) We evaluate $f(x, y) = 2x + 4y$ at the critical points, obtaining

$$\begin{aligned} f(-1, -2) &= 2 \cdot (-1) + 4 \cdot (-2) = -10 \\ f(1, 2) &= 2 \cdot 1 + 4 \cdot 2 = 10 \end{aligned}$$

Since f is continuous and the graph of $g = 0$ is closed and bounded, global minimum and maximum points exist. So according to Theorem 1, we conclude that the maximum of $f(x, y)$ on the constraint is 10 and the minimum is -10.

3. Apply the method of Lagrange multipliers to the function $f(x, y) = (x^2 + 1)y$ subject to the constraint $x^2 + y^2 = 5$. *Hint:* First show that $y \neq 0$; then treat the cases $x = 0$ and $x \neq 0$ separately.

SOLUTION We first write out the Lagrange Equations. We have $\nabla f = \left\langle 2xy, x^2 + 1 \right\rangle$ and $\nabla g = \langle 2x, 2y \rangle$. Hence, the Lagrange Condition for $\nabla g \neq 0$ is

$$\begin{aligned} \nabla f &= \lambda \nabla g \\ \left\langle 2xy, x^2 + 1 \right\rangle &= \lambda \langle 2x, 2y \rangle \end{aligned}$$

We obtain the following equations:

$$\begin{aligned} 2xy &= \lambda(2x) \\ x^2 + 1 &= \lambda(2y) \end{aligned} \quad \Rightarrow \quad \begin{aligned} 2x(y - \lambda) &= 0 \\ x^2 + 1 &= 2\lambda y \end{aligned} \tag{1}$$

The second equation implies that $y \neq 0$, since there is no real value of x such that $x^2 + 1 = 0$. Likewise, $\lambda \neq 0$. The solutions of the first equation are $x = 0$ and $y = \lambda$.

Case 1: $x = 0$. Substituting $x = 0$ in the second equation gives $2\lambda y = 1$, or $y = \frac{1}{2\lambda}$. We substitute $x = 0$, $y = \frac{1}{2\lambda}$ (recall that $\lambda \neq 0$) in the constraint to obtain

$$0^2 + \frac{1}{4\lambda^2} = 5 \quad \Rightarrow \quad 4\lambda^2 = \frac{1}{5} \quad \Rightarrow \quad \lambda = \pm\frac{1}{\sqrt{20}} = \pm\frac{1}{2\sqrt{5}}$$

The corresponding values of y are

$$y = \frac{1}{2 \cdot \frac{1}{2\sqrt{5}}} = \sqrt{5} \quad \text{and} \quad y = \frac{1}{2 \cdot \left(-\frac{1}{2\sqrt{5}}\right)} = -\sqrt{5}$$

We obtain the critical points:

$$\left(0, \sqrt{5}\right) \quad \text{and} \quad \left(0, -\sqrt{5}\right)$$

Case 2: $x \neq 0$. Then the first equation in (1) implies $y = \lambda$. Substituting in the second equation gives

$$x^2 + 1 = 2\lambda^2 \quad \Rightarrow \quad x^2 = 2\lambda^2 - 1$$

We now substitute $y = \lambda$ and $x^2 = 2\lambda^2 - 1$ in the constraint $x^2 + y^2 = 5$ to obtain

$$\begin{aligned} 2\lambda^2 - 1 + \lambda^2 &= 5 \\ 3\lambda^2 &= 6 \\ \lambda^2 &= 2 \quad \Rightarrow \quad \lambda = \pm\sqrt{2} \end{aligned}$$

The solution (x, y) are thus

$$\lambda = \sqrt{2}: \quad y = \sqrt{2}, \quad x = \pm\sqrt{2 \cdot 2 - 1} = \pm\sqrt{3}$$

$$\lambda = -\sqrt{2}: \quad y = -\sqrt{2}, \quad x = \pm\sqrt{2 \cdot 2 - 1} = \pm\sqrt{3}$$

We obtain the critical points:

$$\left(\sqrt{3}, \sqrt{2}\right), \quad \left(-\sqrt{3}, \sqrt{2}\right), \quad \left(\sqrt{3}, -\sqrt{2}\right), \quad \left(-\sqrt{3}, -\sqrt{2}\right)$$

We conclude that the critical points are

$$\left(0, \sqrt{5}\right), \quad \left(0, -\sqrt{5}\right), \quad \left(\sqrt{3}, \sqrt{2}\right), \quad \left(-\sqrt{3}, \sqrt{2}\right), \quad \left(\sqrt{3}, -\sqrt{2}\right), \quad \left(-\sqrt{3}, -\sqrt{2}\right).$$

We now calculate $f(x, y) = \left(x^2 + 1\right) y$ at the critical points:

$$f\left(0, \sqrt{5}\right) = \sqrt{5} \approx 2.24$$
$$f\left(0, -\sqrt{5}\right) = -\sqrt{5} \approx -2.24$$
$$f\left(\sqrt{3}, \sqrt{2}\right) = f\left(-\sqrt{3}, \sqrt{2}\right) = 4\sqrt{2} \approx 5.66$$
$$f\left(\sqrt{3}, -\sqrt{2}\right) = f\left(-\sqrt{3}, -\sqrt{2}\right) = -4\sqrt{2} \approx -5.66$$

Since the constraint gives a closed and bounded curve, f achieves a minimum and a maximum under it. We conclude that the maximum of $f(x, y)$ on the constraint is $4\sqrt{2}$ and the minimum is $-4\sqrt{2}$.

In Exercises 4–13, find the minimum and maximum values of the function subject to the given constraint.

5. $f(x, y) = x^2 + y^2, \quad 2x + 3y = 6$

SOLUTION We find the extreme values of $f(x, y) = x^2 + y^2$ under the constraint $g(x, y) = 2x + 3y - 6 = 0$.

Step 1. Write out the Lagrange Equations. The gradients of f and g are $\nabla f = \langle 2x, 2y\rangle$ and $\nabla g = \langle 2, 3\rangle$. The Lagrange Condition is

$$\nabla f = \lambda \nabla g$$
$$\langle 2x, 2y\rangle = \lambda \langle 2, 3\rangle$$

We obtain the following equations:

$$2x = \lambda \cdot 2$$
$$2y = \lambda \cdot 3$$

Step 2. Solve for λ in terms of x and y. Notice that if $x = 0$, then the first equation gives $\lambda = 0$, therefore by the second equation also $y = 0$. The point $(0, 0)$ does not satisfy the constraint. Similarly, if $y = 0$ also $x = 0$. We therefore may assume that $x \neq 0$ and $y \neq 0$ and obtain by the two equations:

$$\lambda = x \quad \text{and} \quad \lambda = \frac{2}{3}y.$$

Step 3. Solve for x and y using the constraint. Equating the two expressions for λ gives

$$x = \frac{2}{3}y \quad \Rightarrow \quad y = \frac{3}{2}x$$

We substitute $y = \frac{3}{2}x$ in the constraint $2x + 3y = 6$ and solve for x and y:

$$2x + 3 \cdot \frac{3}{2}x = 6$$
$$13x = 12 \quad \Rightarrow \quad x = \frac{12}{13}, \quad y = \frac{3}{2} \cdot \frac{12}{13} = \frac{18}{13}$$

We obtain the critical point $\left(\frac{12}{13}, \frac{18}{13}\right)$.

Step 4. Calculate f at the critical point. We evaluate $f(x, y) = x^2 + y^2$ at the critical point:

$$f\left(\frac{12}{13}, \frac{18}{13}\right) = \left(\frac{12}{13}\right)^2 + \left(\frac{18}{13}\right)^2 = \frac{468}{169} \approx 2.77$$

Rewriting the constraint as $y = -\frac{2}{3}x + 2$, we see that as $|x| \to +\infty$ then so does $|y|$, and hence $x^2 + y^2$ is increasing without bound on the constraint as $|x| \to \infty$. We conclude that the value $468/169$ is the minimum value of f under the constraint, rather than the maximum value.

7. $f(x, y) = xy, \quad 4x^2 + 9y^2 = 32$

SOLUTION We find the extreme values of $f(x, y) = xy$ under the constraint $g(x, y) = 4x^2 + 9y^2 - 32 = 0$.

Step 1. Write out the Lagrange Equation. The gradient vectors are $\nabla f = \langle y, x\rangle$ and $\nabla g = \langle 8x, 18y\rangle$, hence the Lagrange Condition is

$$\nabla f = \lambda \nabla g$$
$$\langle y, x\rangle = \lambda \langle 8x, 18y\rangle$$

We obtain the following equations:

$$y = \lambda(8x)$$
$$x = \lambda(18y)$$

Step 2. Solve for λ in terms of x and y. If $x = 0$, then the Lagrange equations also imply that $y = 0$ and vice versa. Since the point $(0, 0)$ does not satisfy the equation of the constraint, we may assume that $x \neq 0$ and $y \neq 0$. The two equations give

$$\lambda = \frac{y}{8x} \quad \text{and} \quad \lambda = \frac{x}{18y}$$

Step 3. Solve for x and y using the constraint. We equate the two expressions for λ to obtain

$$\frac{y}{8x} = \frac{x}{18y} \quad \Rightarrow \quad 18y^2 = 8x^2 \quad \Rightarrow \quad y = \pm\frac{2}{3}x$$

We now substitute $y = \pm\frac{2}{3}x$ in the equation of the constraint and solve for x and y:

$$4x^2 + 9\cdot\left(\pm\frac{2}{3}x\right)^2 = 32$$
$$4x^2 + 9\cdot\frac{4x^2}{9} = 32$$
$$8x^2 = 32 \quad \Rightarrow \quad x = -2, \quad x = 2$$

We find y by the relation $y = \pm\frac{2}{3}x$:

$$y = \frac{2}{3}\cdot(-2) = -\frac{4}{3}, \quad y = -\frac{2}{3}\cdot(-2) = \frac{4}{3}, \quad y = \frac{2}{3}\cdot 2 = \frac{4}{3}, \quad y = -\frac{2}{3}\cdot 2 = -\frac{4}{3}$$

We obtain the following critical points:

$$\left(-2, -\frac{4}{3}\right), \quad \left(-2, \frac{4}{3}\right), \quad \left(2, \frac{4}{3}\right), \quad \left(2, -\frac{4}{3}\right)$$

Extreme values can also occur at the point where $\nabla g = \langle 8x, 18y\rangle = \langle 0, 0\rangle$, that is, at the point $(0, 0)$. However, the point does not lie on the constraint.

Step 4. Calculate f at the critical points. We evaluate $f(x, y) = xy$ at the critical points:

$$f\left(-2, -\frac{4}{3}\right) = f\left(2, \frac{4}{3}\right) = \frac{8}{3}$$
$$f\left(-2, \frac{4}{3}\right) = f\left(2, -\frac{4}{3}\right) = -\frac{8}{3}$$

Since f is continuous and the constraint is a closed and bounded set in R^2 (an ellipse), f attains global extrema on the constraint. We conclude that $\frac{8}{3}$ is the maximum value and $-\frac{8}{3}$ is the minimum value.

9. $f(x, y) = x^2 + y^2, \quad x^4 + y^4 = 1$

SOLUTION We find the extreme values of $f(x, y) = x^2 + y^2$ under the constraint $g(x, y) = x^4 + y^4 - 1 = 0$.

Step 1. Write out the Lagrange Equations. We have $\nabla f = \langle 2x, 2y\rangle$ and $\nabla g = \left\langle 4x^3, 4y^3\right\rangle$, hence the Lagrange Condition $\nabla f = \lambda \nabla g$ gives

$$\langle 2x, 2y\rangle = \lambda\left\langle 4x^3, 4y^3\right\rangle$$

or

$$\begin{aligned} 2x &= \lambda\left(4x^3\right) \\ 2y &= \lambda\left(4y^3\right) \end{aligned} \quad \Rightarrow \quad \begin{aligned} x &= 2\lambda x^3 \\ y &= 2\lambda y^3 \end{aligned} \tag{1}$$

Step 2. Solve for λ in terms of x and y. We first assume that $x \neq 0$ and $y \neq 0$. Then the Lagrange equations give

$$\lambda = \frac{1}{2x^2} \quad \text{and} \quad \lambda = \frac{1}{2y^2}$$

Step 3. Solve for x and y using the constraint. Equating the two expressions for λ gives

$$\frac{1}{2x^2} = \frac{1}{2y^2} \quad \Rightarrow \quad y^2 = x^2 \quad \Rightarrow \quad y = \pm x$$

We now substitute $y = \pm x$ in the equation of the constraint $x^4 + y^4 = 1$ and solve for x and y:

$$\begin{aligned} x^4 + (\pm x)^4 &= 1 \\ 2x^4 &= 1 \\ x^4 &= \frac{1}{2} \quad \Rightarrow \quad x = \frac{1}{2^{1/4}}, \quad x = -\frac{1}{2^{1/4}} \end{aligned}$$

The corresponding values of y are obtained by the relation $y = \pm x$. The critical points are thus

$$\left(\frac{1}{2^{1/4}}, \frac{1}{2^{1/4}}\right), \quad \left(\frac{1}{2^{1/4}}, -\frac{1}{2^{1/4}}\right), \quad \left(-\frac{1}{2^{1/4}}, \frac{1}{2^{1/4}}\right), \quad \left(-\frac{1}{2^{1/4}}, -\frac{1}{2^{1/4}}\right) \tag{2}$$

We examine the case $x = 0$ or $y = 0$. Notice that the point $(0, 0)$ does not satisfy the equation of the constraint, hence either $x = 0$ or $y = 0$ can hold, but not both at the same time.

Case 1: $x = 0$. Substituting $x = 0$ in the constraint $x^4 + y^4 = 1$ gives $y = \pm 1$. We thus obtain the critical points

$$(0, -1), \quad (0, 1) \tag{3}$$

Case 2: $y = 0$. We may interchange x and y in the discussion in case 1, and obtain the critical points:

$$(-1, 0), \quad (1, 0) \tag{4}$$

Combining (2), (3), and (4) we conclude that the critical points are

$$A_1 = \left(\frac{1}{2^{1/4}}, \frac{1}{2^{1/4}}\right), \quad A_2 = \left(\frac{1}{2^{1/4}}, -\frac{1}{2^{1/4}}\right), \quad A_3 = \left(-\frac{1}{2^{1/4}}, \frac{1}{2^{1/4}}\right),$$

$$A_4 = \left(-\frac{1}{2^{1/4}}, -\frac{1}{2^{1/4}}\right), \quad A_5 = (0, -1), \quad A_6 = (0, 1), \quad A_7 = (-1, 0), \quad A_8 = (1, 0)$$

The point where $\nabla g = \langle 4x^3, 4y^3 \rangle = \langle 0, 0 \rangle$, that is, $(0, 0)$, does not lie on the constraint.

Step 4. Compute f at the critical points. We evaluate $f(x, y) = x^2 + y^2$ at the critical points:

$$f(A_1) = f(A_2) = f(A_3) = f(A_4) = \left(\frac{1}{2^{1/4}}\right)^2 + \left(\frac{1}{2^{1/4}}\right)^2 = \frac{2}{2^{1/2}} = \sqrt{2}$$

$$f(A_5) = f(A_6) = f(A_7) = f(A_8) = 1$$

The constraint $x^4 + y^4 = 1$ is a closed and bounded set in R^2 and f is continuous on this set, hence f has global extrema on the constraint. We conclude that $\sqrt{2}$ is the maximum value and 1 is the minimum value.

11. $f(x, y, z) = 3x + 2y + 4z, \quad x^2 + 2y^2 + 6z^2 = 1$

SOLUTION We find the extreme values of $f(x, y, z) = 3x + 2y + 4z$ under the constraint $g(x, y, z) = x^2 + 2y^2 + 6z^2 - 1 = 0$.

Step 1. Write out the Lagrange Equations. The gradient vectors are $\nabla f = \langle 3, 2, 4 \rangle$ and $\nabla g = \langle 2x, 4y, 12z \rangle$, therefore the Lagrange Condition $\nabla f = \lambda \nabla g$ is:

$$\langle 3, 2, 4 \rangle = \lambda \langle 2x, 4y, 12z \rangle$$

The Lagrange equations are, thus:

$$\begin{aligned} 3 &= \lambda(2x) & & \frac{3}{2} = \lambda x \\ 2 &= \lambda(4y) & \Rightarrow \quad & \frac{1}{2} = \lambda y \\ 4 &= \lambda(12z) & & \frac{1}{3} = \lambda z \end{aligned}$$

Step 2. Solve for λ in terms of x, y, and z. The Lagrange equations imply that $x \neq 0$, $y \neq 0$, and $z \neq 0$. Solving for λ we get

$$\lambda = \frac{3}{2x}, \quad \lambda = \frac{1}{2y}, \quad \lambda = \frac{1}{3z}$$

Step 3. Solve for x, y, and z using the constraint. Equating the expressions for λ gives

$$\frac{3}{2x} = \frac{1}{2y} = \frac{1}{3z} \quad \Rightarrow \quad x = \frac{9}{2}z, \quad y = \frac{3}{2}z$$

Substituting $x = \frac{9}{2}z$ and $y = \frac{3}{2}z$ in the equation of the constraint $x^2 + 2y^2 + 6z^2 = 1$ and solving for z we get

$$\left(\frac{9}{2}z\right)^2 + 2\left(\frac{3}{2}z\right)^2 + 6z^2 = 1$$

$$\frac{123}{4}z^2 = 1 \quad \Rightarrow \quad z_1 = \frac{2}{\sqrt{123}}, z_2 = -\frac{2}{\sqrt{123}}$$

Using the relations $x = \frac{9}{2}z$, $y = \frac{3}{2}z$ we get

$$x_1 = \frac{9}{2} \cdot \frac{2}{\sqrt{123}} = \frac{9}{\sqrt{123}}, \quad y_1 = \frac{3}{2} \cdot \frac{2}{\sqrt{123}} = \frac{3}{\sqrt{123}}, \quad z_1 = \frac{2}{\sqrt{123}}$$

$$x_2 = \frac{9}{2} \cdot \frac{-2}{\sqrt{123}} = -\frac{9}{\sqrt{123}}, \quad y_2 = \frac{3}{2} \cdot \frac{-2}{\sqrt{123}} = -\frac{3}{\sqrt{123}}, \quad z_2 = -\frac{2}{\sqrt{123}}$$

We obtain the following critical points:

$$p_1 = \left(\frac{9}{\sqrt{123}}, \frac{3}{\sqrt{123}}, \frac{2}{\sqrt{123}}\right) \quad \text{and} \quad p_2 = \left(-\frac{9}{\sqrt{123}}, -\frac{3}{\sqrt{123}}, -\frac{2}{\sqrt{123}}\right)$$

Critical points are also the points on the constraint where $\nabla g = 0$. However, $\nabla g = \langle 2x, 4y, 12z \rangle = \langle 0, 0, 0 \rangle$ only at the origin, and this point does not lie on the constraint.

Step 4. Computing f at the critical points. We evaluate $f(x, y, z) = 3x + 2y + 4z$ at the critical points:

$$f(p_1) = \frac{27}{\sqrt{123}} + \frac{6}{\sqrt{123}} + \frac{8}{\sqrt{123}} = \frac{41}{\sqrt{123}} = \sqrt{\frac{41}{3}} \approx 3.7$$

$$f(p_2) = -\frac{27}{\sqrt{123}} - \frac{6}{\sqrt{123}} - \frac{8}{\sqrt{123}} = -\frac{41}{\sqrt{123}} = -\sqrt{\frac{41}{3}} \approx -3.7$$

Since f is continuous and the constraint is closed and bounded in R^3, f has global extrema under the constraint. We conclude that the minimum value of f under the constraint is about -3.7 and the maximum value is about 3.7.

13. $f(x, y, z) = xy + 3xz + 2yz, \quad 5x + 9y + z = 10$

SOLUTION We show that $f(x, y, z) = xy + 3xz + 2yz$ does not have minimum and maximum values subject to the constraint $g(x, y, z) = 5x + 9y + z - 10 = 0$. First notice that the curve $c_1 : (x, x, 10 - 14x)$ lies on the surface of the constraint since it satisfies the equation of the constraint. On c_1 we have,

$$f(x, y, z) = f(x, x, 10 - 14x) = x^2 + 3x(10 - 14x) + 2x(10 - 14x) = -69x^2 + 50x$$

Since $\lim_{x\to\infty}\left(-69x^2 + 50x\right) = -\infty$, f does not have minimum value on the constraint. Notice that the curve c_2 : $(x, -x, 10 + 4x)$ also lies on the surface of the constraint. The values of f on c_2 are

$$f(x, y, z) = f(x, -x, 10 + 4x) = -x^2 + 3x(10 + 4x) - 2x(10 + 4x) = 3x^2 + 10x$$

The limit $\lim_{x\to\infty}(3x^2 + 10x) = \infty$ implies that f does not have a maximum value subject to the constraint.

15. Use Lagrange multipliers to find the point (a, b) on the graph of $y = e^x$, where the value ab is as small as possible.

SOLUTION We must find the point where $f(x, y) = xy$ has a minimum value subject to the constraint $g(x, y) = e^x - y = 0$.

Step 1. Write out the Lagrange Equations. Since $\nabla f = \langle y, x \rangle$ and $\nabla g = \langle e^x, -1 \rangle$, the Lagrange Condition $\nabla f = \lambda \nabla g$ is

$$\langle y, x \rangle = \lambda \langle e^x, -1 \rangle$$

The Lagrange equations are thus

$$y = \lambda e^x$$

$$x = -\lambda$$

Step 2. Solve for λ in terms of x and y. The Lagrange equations imply that

$$\lambda = ye^{-x} \quad \text{and} \quad \lambda = -x$$

Step 3. Solve for x and y using the constraint. We equate the two expressions for λ to obtain

$$ye^{-x} = -x \quad \Rightarrow \quad y = -xe^{x}$$

We now substitute $y = -xe^x$ in the equation of the constraint and solve for x:

$$e^x - (-xe^x) = 0$$
$$e^x(1 + x) = 0$$

Since $e^x \neq 0$ for all x, we have $x = -1$. The corresponding value of y is determined by the relation $y = -xe^x$. That is,

$$y = -(-1)e^{-1} = e^{-1}$$

We obtain the critical point

$$(-1, e^{-1})$$

Step 4. Calculate f at the critical point. We evaluate $f(x, y) = xy$ at the critical point.

$$f(-1, e^{-1}) = (-1) \cdot e^{-1} = -e^{-1}$$

We conclude (see Remark) that the minimum value of xy on the graph of $y = e^x$ is $-e^{-1}$, and it is obtained for $x = -1$ and $y = e^{-1}$.

Remark: Since the constraint is not bounded, we need to justify the existence of a minimum value. The values $f(x, y) = xy$ on the constraint $y = e^x$ are $f(x, e^x) = h(x) = xe^x$. Since $h(x) > 0$ for $x > 0$, the minimum value (if it exists) occurs at a point $x < 0$. Since

$$\lim_{x\to-\infty} xe^x = \lim_{x\to-\infty} \frac{x}{e^{-x}} = \lim_{x\to-\infty} \frac{1}{-e^{-x}} = \lim_{x\to-\infty} -e^x = 0,$$

then for $x <$ some negative number $-R$, we have $|f(x) - 0| < 0.1$, say. Thus, on the bounded region $-R \le x \le 0$, f has a minimum value of $-e^{-1} \approx -0.37$, and this is thus a global minimum (for all x).

17. The surface area of a right-circular cone of radius r and height h is $S = \pi r\sqrt{r^2 + h^2}$, and its volume is $V = \frac{1}{3}\pi r^2 h$.

(a) Determine the ratio h/r for the cone with given surface area S and maximal volume V.

(b) What is the ratio h/r for a cone with given volume V and minimal surface area S?

(c) Does a cone with given volume V and maximal surface area exist?

SOLUTION

(a) Let S_0 denote a given surface area. We must find the ratio $\frac{h}{r}$ for which the function $V(r, h) = \frac{1}{3}\pi r^2 h$ has maximum value under the constraint $S(r, h) = \pi r\sqrt{r^2 + h^2} = \pi\sqrt{r^4 + h^2 r^2} = S_0$.

Step 1. Write out the Lagrange Equation. We have

$$\nabla V = \pi\left\langle \frac{2rh}{3}, \frac{r^2}{3} \right\rangle \quad \text{and} \quad \nabla S = \pi\left\langle \frac{2r^3 + h^2 r}{\sqrt{r^4 + h^2 r^2}}, \frac{hr^2}{\sqrt{r^4 + h^2 r^2}} \right\rangle$$

The Lagrange Condition $\nabla V = \lambda \nabla S$ gives the following equations:

$$\frac{2rh}{3} = \frac{2r^3 + h^2 r}{\sqrt{r^4 + h^2 r^2}}\lambda \quad \Rightarrow \quad \frac{2h}{3} = \frac{2r^2 + h^2}{\sqrt{r^4 + h^2 r^2}}\lambda$$

$$\frac{r^2}{3} = \frac{hr^2}{\sqrt{r^4 + h^2 r^2}}\lambda \quad \Rightarrow \quad \frac{1}{3} = \frac{h}{\sqrt{r^4 + h^2 r^2}}\lambda$$

Step 2. Solve for λ in terms of r and h. These equations yield two expressions for λ that must be equal:

$$\lambda = \frac{2h}{3}\frac{\sqrt{r^4 + h^2 r^2}}{2r^2 + h^2} = \frac{1}{3h}\sqrt{r^4 + h^2 r^2}$$

Step 3. Solve for r and h using the constraint. We have

$$\frac{2h}{3}\frac{\sqrt{r^4 + h^2 r^2}}{2r^2 + h^2} = \frac{1}{3h}\sqrt{r^4 + h^2 r^2}$$

$$2h\frac{1}{2r^2+h^2} = \frac{1}{h}$$

$$2h^2 = 2r^2 + h^2 \quad \Rightarrow \quad h^2 = 2r^2 \quad \Rightarrow \quad \frac{h}{r} = \sqrt{2}$$

We substitute $h^2 = 2r^2$ in the constraint $\pi r\sqrt{r^2+h^2} = S_0$ and solve for r. This gives

$$\pi r\sqrt{r^2+2r^2} = S_0$$

$$\pi r\sqrt{3r^2} = S_0$$

$$\sqrt{3}\pi r^2 = S_0 \quad \Rightarrow \quad r^2 = \frac{S_0}{\sqrt{3}\pi}, \quad h^2 = 2r^2 = \frac{2S_0}{\sqrt{3}\pi}$$

Extreme values can occur also at points on the constraint where $\nabla S = \left\langle \frac{2r^2+h^2r}{\sqrt{r^4+h^2r^2}}, \frac{hr^2}{\sqrt{r^4+h^2r^2}} \right\rangle = \langle 0, 0\rangle$, that is, at $(r,h) = (0,h)$, $h \neq 0$. However, since the radius of the cone is positive ($r > 0$), these points are irrelevant. We conclude that for the cone with surface area S_0 and maximum volume, the following holds:

$$\frac{h}{r} = \sqrt{2}, \quad h = \sqrt{\frac{2S_0}{\sqrt{3}\pi}}, \quad r = \sqrt{\frac{S_0}{\sqrt{3}\pi}}$$

For the surface area $S_0 = 1$ we get

$$h = \sqrt{\frac{2}{\sqrt{3}\pi}} \approx 0.6, \quad r = \sqrt{\frac{1}{\sqrt{3}\pi}} = 0.43$$

(b) We now must find the ratio $\frac{h}{r}$ that minimizes the function $S(r,h) = \pi r\sqrt{r^2+h^2}$ under the constraint

$$V(r,h) = \frac{1}{3}\pi r^2 h = V_0$$

Using the gradients computed in part (a), the Lagrange Condition $\nabla S = \lambda \nabla V$ gives the following equations:

$$\begin{aligned} \frac{2r^3+h^2r}{\sqrt{r^4+h^2r^2}} &= \lambda\frac{2rh}{3} \\ \frac{hr^2}{\sqrt{r^4+h^2r^2}} &= \lambda\frac{r^2}{3} \end{aligned} \quad \Rightarrow \quad \begin{aligned} \frac{2r^2+h^2}{\sqrt{r^4+h^2r^2}} &= \lambda\frac{2h}{3} \\ \frac{h}{\sqrt{r^4+h^2r^2}} &= \frac{\lambda}{3} \end{aligned}$$

These equations give

$$\frac{\lambda}{3} = \frac{1}{2h}\frac{2r^2+h^2}{\sqrt{r^4+h^2r^2}} = \frac{h}{\sqrt{r^4+h^2r^2}}$$

We simplify and solve for $\frac{h}{r}$:

$$\frac{2r^2+h^2}{2h} = h$$

$$2r^2 + h^2 = 2h^2$$

$$2r^2 = h^2 \quad \Rightarrow \quad \frac{h}{r} = \sqrt{2}$$

We conclude that the ratio $\frac{h}{r}$ for a cone with a given volume and minimal surface area is

$$\frac{h}{r} = \sqrt{2}$$

(c) The constant $V = 1$ gives $\frac{1}{3}\pi r^2 h = 1$ or $h = \frac{3}{\pi r^2}$. As $r \to \infty$, we have $h \to 0$, therefore

$$\lim_{\substack{r\to\infty \\ h\to 0}} S(r,h) = \lim_{\substack{r\to\infty \\ h\to 0}} \pi r\sqrt{r^2+h^2} = \infty$$

That is, S does not have maximum value on the constraint, hence there is no cone of volume 1 and maximal surface area.

19. Use Lagrange multipliers to find the maximum area of a rectangle inscribed in the ellipse (Figure 12):

$$\frac{x^2}{a^2} + \frac{y^2}{b^2} = 1$$

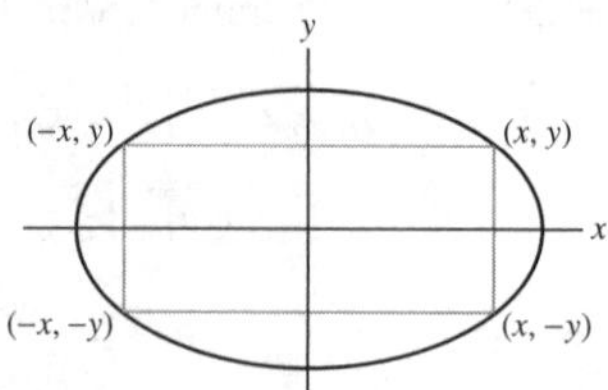

FIGURE 12 Rectangle inscribed in the ellipse $\frac{x^2}{a^2} + \frac{y^2}{b^2} = 1$.

SOLUTION Since (x, y) is in the first quadrant, $x > 0$ and $y > 0$. The area of the rectangle is $2x \cdot 2y = 4xy$. The vertices lie on the ellipse, hence their coordinates $(\pm x, \pm y)$ must satisfy the equation of the ellipse. Therefore, we must find the maximum value of the function $f(x, y) = 4xy$ under the constraint

$$g(x, y) = \frac{x^2}{a^2} + \frac{y^2}{b^2} = 1, \quad x > 0, \quad y > 0.$$

Step 1. Write out the Lagrange Equations. The gradient vectors are $\nabla f = \langle 4y, 4x \rangle$ and $\nabla g = \left\langle \frac{2x}{a^2}, \frac{2y}{b^2} \right\rangle$, hence the Lagrange Condition $\nabla f = \lambda \nabla g$ gives

$$\langle 4y, 4x \rangle = \lambda \left\langle \frac{2x}{a^2}, \frac{2y}{b^2} \right\rangle$$

or

$$\begin{aligned} 4y &= \lambda \left(\frac{2x}{a^2} \right) & & 2y = \lambda \frac{x}{a^2} \\ & & \Rightarrow & \\ 4x &= \lambda \left(\frac{2y}{b^2} \right) & & 2x = \lambda \frac{y}{b^2} \end{aligned}$$

Step 2. Solve for λ in terms of x and y. The Lagrange equations give the following two expressions for λ:

$$\lambda = \frac{2ya^2}{x}, \quad \lambda = \frac{2xb^2}{y}$$

Equating the two equations we get

$$\frac{2ya^2}{x} = \frac{2xb^2}{y}$$

Step 3. Solve for x and y using the constraint. We solve the equation in step 2 for y in terms of x:

$$\begin{aligned} \frac{2ya^2}{x} &= \frac{2xb^2}{y} \\ 2y^2a^2 &= 2x^2b^2 \\ y^2 &= \frac{x^2b^2}{a^2} \quad \Rightarrow \quad y = \frac{b}{a}x \end{aligned}$$

We now substitute $y = \frac{b}{a}x$ in the equation of the constraint $\frac{x^2}{a^2} + \frac{y^2}{b^2} = 1$ and solve for x:

$$\begin{aligned} \frac{x^2}{a^2} + \frac{\left(\frac{b}{a}x\right)^2}{b^2} &= 1 \\ \frac{x^2}{a^2} + \frac{x^2}{a^2} &= 1 \\ \frac{2x^2}{a^2} &= 1 \\ x^2 = \frac{a^2}{2} \quad &\Rightarrow \quad x = \frac{a}{\sqrt{2}} \end{aligned}$$

The corresponding value of y is obtained by the relation $y = \frac{b}{a}x$:

$$y = \frac{b}{a} \cdot \frac{a}{\sqrt{2}} = \frac{b}{\sqrt{2}}$$

We obtain the critical point $\left(\frac{a}{\sqrt{2}}, \frac{b}{\sqrt{2}}\right)$. Extreme values can also occur at points on the constraint where $\nabla g = \left\langle \frac{2x}{a^2}, \frac{2y}{b^2} \right\rangle = \langle 0, 0 \rangle$. However, the point $(0, 0)$ is not on the constraint. We conclude that if $f(x, y) = 4xy$ has a maximum value on the ellipse $\frac{x^2}{a^2} + \frac{y^2}{b^2} = 1$ with $x > 0$, $y > 0$, then it occurs at the point $\left(\frac{a}{\sqrt{2}}, \frac{b}{\sqrt{2}}\right)$ and the maximum value is

$$f\left(\frac{a}{\sqrt{2}}, \frac{b}{\sqrt{2}}\right) = 4 \cdot \frac{a}{\sqrt{2}} \cdot \frac{b}{\sqrt{2}} = 2ab$$

We now justify why the maximum value exists. We consider the problem of finding the extreme values of $f(x, y) = 4xy$ on the quarter ellipse $\frac{x^2}{a^2} + \frac{y^2}{b^2} = 1$ in the first quadrant. Since the constraint curve is bounded and $f(x, y)$ is continuous, f has a minimum and maximum values on the ellipse. The minimum volume occurs at the end points:

$$x = 0, \quad y = b \quad \Rightarrow \quad 4xy = 0 \qquad \text{or} \qquad x = a, \quad y = 0 \quad \Rightarrow \quad 4xy = 0$$

So the critical point $\left(\frac{a}{\sqrt{2}}, \frac{b}{\sqrt{2}}\right)$ must be a maximum.

21. Find the maximum value of $f(x, y) = x^a y^b$ for $x, y \ge 0$ on the unit circle, where $a, b > 0$ are constants.

SOLUTION We must find the maximum value of $f(x, y) = x^a y^b$ $(a, b > 0)$ subject to the constraint $g(x, y) = x^2 + y^2 = 1$.

Step 1. Write out the Lagrange Equations. We have $\nabla f = \left\langle ax^{a-1}y^b, bx^a y^{b-1} \right\rangle$ and $\nabla g = \langle 2x, 2y \rangle$. Therefore the Lagrange Condition $\nabla f = \lambda \nabla g$ is

$$\left\langle ax^{a-1}y^b, bx^a y^{b-1} \right\rangle = \lambda \langle 2x, 2y \rangle$$

or

$$\begin{aligned} ax^{a-1}y^b &= 2\lambda x \\ bx^a y^{b-1} &= 2\lambda y \end{aligned} \tag{1}$$

Step 2. Solve for λ in terms of x and y. If $x = 0$ or $y = 0$, f has the minimum value 0. We thus may assume that $x > 0$ and $y > 0$. The equations (1) imply that

$$\lambda = \frac{ax^{a-2}y^b}{2}, \quad \lambda = \frac{bx^a y^{b-2}}{2}$$

Step 3. Solve for x and y using the constraint. Equating the two expressions for λ and solving for y in terms of x gives

$$\begin{aligned} \frac{ax^{a-2}y^b}{2} &= \frac{bx^a y^{b-2}}{2} \\ ax^{a-2}y^b &= bx^a y^{b-2} \\ ay^2 &= bx^2 \\ y^2 &= \frac{b}{a}x^2 \quad \Rightarrow \quad y = \sqrt{\frac{b}{a}}x \end{aligned}$$

We now substitute $y = \sqrt{\frac{b}{a}}x$ in the constraint $x^2 + y^2 = 1$ and solve for $x > 0$. We obtain

$$\begin{aligned} x^2 + \frac{b}{a}x^2 &= 1 \\ (a+b)x^2 &= a \\ x^2 &= \frac{a}{a+b} \quad \Rightarrow \quad x = \sqrt{\frac{a}{a+b}} \end{aligned}$$

We find y using the relation $y = \sqrt{\frac{b}{a}}x$:

$$y = \sqrt{\frac{b}{a}}\sqrt{\frac{a}{a+b}} = \sqrt{\frac{ab}{a(a+b)}} = \sqrt{\frac{b}{a+b}}$$

We obtain the critical point:

$$\left(\sqrt{\frac{a}{a+b}}, \sqrt{\frac{b}{a+b}}\right)$$

Extreme points can also occur where $\nabla g = \mathbf{0}$, that is, $\langle 2x, 2y \rangle = \langle 0, 0 \rangle$ or $(x, y) = (0, 0)$. However, this point is not on the constraint.

Step 4. Conclusions. We compute $f(x, y) = x^a y^b$ at the critical point:

$$f\left(\sqrt{\frac{a}{a+b}}, \sqrt{\frac{b}{a+b}}\right) = \left(\frac{a}{a+b}\right)^{a/2}\left(\frac{b}{a+b}\right)^{b/2} = \frac{a^{a/2}b^{b/2}}{(a+b)^{(a+b)/2}} = \sqrt{\frac{a^a b^b}{(a+b)^{a+b}}}$$

The function $f(x, y) = x^a y^b$ is continuous on the set $x^2 + y^2 = 1$, $x \geq 0$, $y \geq 0$, which is a closed and bounded set in R^2, hence f has minimum and maximum values on the set. The minimum value is 0 (obtained at $(0, 1)$ and $(1, 0)$), hence the critical point that we found corresponds to the maximum value. We conclude that the maximum value of $x^a y^b$ on $x^2 + y^2 = 1$, $x > 0$, $y > 0$ is

$$\sqrt{\frac{a^a b^b}{(a+b)^{a+b}}}.$$

23. Find the maximum value of $f(x, y, z) = x^a y^b z^c$ for $x, y, z \geq 0$ on the unit sphere, where $a, b, c > 0$ are constants.

SOLUTION We must find the maximum value of $f(x, y, z) = x^a y^b z^c$ subject to the constraint $g(x, y, z) = x^2 + y^2 + z^2 - 1 = 0$, $x \geq 0$, $y \geq 0$, $z \geq 0$.

Step 1. Write the Lagrange Equations. The gradient vectors are $\nabla f = \left\langle ax^{a-1}y^b z^c, by^{b-1}x^a z^c, cz^{c-1}x^a y^b \right\rangle$ and $\nabla g = \langle 2x, 2y, 2z \rangle$, hence the Lagrange Condition $\nabla f = \lambda \nabla g$ gives the following equations:

$$\begin{aligned} ax^{a-1}y^b z^c &= \lambda(2x) \\ by^{b-1}x^a z^c &= \lambda(2y) \\ cz^{c-1}x^a y^b &= \lambda(2z) \end{aligned} \qquad (1)$$

Step 2. Solve for λ in terms of x, y, and z. If $x = 0$, $y = 0$, or $z = 0$, f attains the minimum value 0, therefore we may assume that $x \neq 0$, $y \neq 0$, and $z \neq 0$. The Lagrange equations (1) give

$$\lambda = \frac{ax^{a-2}y^b z^c}{2}, \quad \lambda = \frac{by^{b-2}x^a z^c}{2}, \quad \lambda = \frac{cz^{c-2}x^a y^b}{2}$$

Step 3. Solve for x, y, and z using the constraint. Equating the expressions for λ, we obtain the following equations:

$$\begin{aligned} ax^{a-2}y^b z^c &= by^{b-2}x^a z^c \\ ax^{a-2}y^b z^c &= cz^{c-2}x^a y^b \end{aligned} \qquad (2)$$

We solve for x and y in terms of z. We first divide the first equation by the second equation to obtain

$$\begin{aligned} 1 &= \frac{by^{b-2}x^a z^c}{cz^{c-2}x^a y^b} = \frac{b}{c}\frac{z^2}{y^2} \\ y^2 &= \frac{b}{c}z^2 \quad \Rightarrow \quad y = \sqrt{\frac{b}{c}}z \end{aligned} \qquad (3)$$

Both equations (2) imply that

$$\begin{aligned} by^{b-2}x^a z^c &= ax^{a-2}y^b z^c \\ by^{b-2}x^a z^c &= cz^{c-2}x^a y^b \end{aligned}$$

Dividing the first equation by the second equation gives

$$\begin{aligned} 1 &= \frac{ax^{a-2}y^b z^c}{cz^{c-2}x^a y^b} = \frac{a}{c}\frac{z^2}{x^2} \\ x^2 &= \frac{a}{c}z^2 \quad \Rightarrow \quad x = \sqrt{\frac{a}{c}}z \end{aligned} \qquad (4)$$

We now substitute x and y from (3) and (4) in the constraint $x^2 + y^2 + z^2 = 1$ and solve for z. This gives

$$\left(\sqrt{\frac{a}{c}}z\right)^2 + \left(\sqrt{\frac{b}{c}}z\right)^2 + z^2 = 1$$

$$\left(\frac{a}{c} + \frac{b}{c} + 1\right)z^2 = 1$$

$$\frac{a+b+c}{c}z^2 = 1 \quad \Rightarrow \quad z = \sqrt{\frac{c}{a+b+c}}$$

We find x and y using (4) and (3):

$$x = \sqrt{\frac{a}{c}}\sqrt{\frac{c}{a+b+c}} = \sqrt{\frac{ac}{c(a+b+c)}} = \sqrt{\frac{a}{a+b+c}}$$

$$y = \sqrt{\frac{b}{c}}\sqrt{\frac{c}{a+b+c}} = \sqrt{\frac{bc}{c(a+b+c)}} = \sqrt{\frac{b}{a+b+c}}$$

We obtain the critical point:

$$P = \left(\sqrt{\frac{a}{a+b+c}}, \sqrt{\frac{b}{a+b+c}}, \sqrt{\frac{c}{a+b+c}}\right)$$

We examine the point where $\nabla g = \langle 2x, 2y, 2z\rangle = \langle 0, 0, 0\rangle$, that is, $(0, 0, 0)$: This point does not lie on the constraint, hence it is not a critical point.

Step 4. Conclusions. We compute $f(x, y, z) = x^a y^b z^c$ at the critical point:

$$f(P) = \left(\sqrt{\frac{a}{a+b+c}}\right)^a \left(\sqrt{\frac{b}{a+b+c}}\right)^b \left(\sqrt{\frac{c}{a+b+c}}\right)^c = \sqrt{\frac{a^a b^b c^c}{(a+b+c)^{a+b+c}}}$$

Now, $f(x, y, z) = x^a y^b z^c$ is continuous on the set $x^2 + y^2 + z^2 = 1$, $x \ge 0$, $y \ge 0$, $z \ge 0$, which is closed and bounded in R^3. The minimum value is 0 (obtained at the points with at least one zero coordinate), therefore the critical point that we found corresponds to the maximum value. We conclude that the maximum value of $x^a y^b z^c$ subject to the constraint $x^2 + y^2 + z^2 = 1$, $x \ge 0$, $y \ge 0$, $z \ge 0$ is

$$\sqrt{\frac{a^a b^b c^c}{(a+b+c)^{a+b+c}}}$$

25. Let $f(x, y, z) = y + z - x^2$.

(a) Find the solutions to the Lagrange equations for f subject to the constraint $g(x, y, z) = x^2 - y^2 + z^3 = 0$. *Hint:* Show that at a critical point, λ, y, z must be nonzero and $x = 0$.

(b) Show that f has no minimum or maximum subject to the constraint. *Hint:* Consider the values of f at the points $(0, y^3, y^2)$, which satisfy the constraint.

(c) Does (b) contradict Theorem 1?

SOLUTION

(a) The gradient vectors are $\nabla f = \langle -2x, 1, 1\rangle$ and $\nabla g = \left\langle 2x, -2y, 3z^2\right\rangle$, hence the Lagrange Condition $\nabla f = \lambda \nabla g$ yields the following equations:

$$-2x = \lambda(2x) \qquad -2x = 2\lambda x \qquad (1)$$

$$1 = \lambda(-2y) \quad \Rightarrow \quad 1 = -2\lambda y \qquad (2)$$

$$1 = \lambda\left(3z^2\right) \qquad 1 = 3\lambda z^2 \qquad (3)$$

We solve the Lagrange equations. Equations (2) and (3) imply that $y \ne 0$, $z \ne 0$, $\lambda \ne 0$, and that

$$\lambda = -\frac{1}{2y} \quad \text{and} \quad \lambda = \frac{1}{3z^2}$$

Equating and solving for y in terms of z gives

$$-\frac{1}{2y} = \frac{1}{3z^2} \quad \Rightarrow \quad y = -\frac{3}{2}z^2$$

Equation (1) gives

$$(2\lambda + 2)x = 0 \quad \rightarrow \quad \lambda = -1 \quad \text{or} \quad x = 0$$

By equation (3), λ must be positive, hence the solution $\lambda = -1$ is not appropriate. We check the solution $x = 0$. Substituting $x = 0$ and $y = -\frac{3}{2}z^2$ in the equation of the constraint $x^2 - y^2 + z^3 = 0$ gives

$$0^2 - \left(-\frac{3}{2}z^2\right)^2 + z^3 = 0$$

$$-\frac{9}{4}z^4 + z^3 = 0$$

Since $z \neq 0$, we can divide by z^3 to obtain

$$-\frac{9}{4}z + 1 = 0 \quad \Rightarrow \quad z = \frac{4}{9}$$

We find y by the relation $y = -\frac{3}{2}z^2$:

$$y = -\frac{3}{2} \cdot \left(\frac{4}{9}\right)^2 = -\frac{3 \cdot 16}{2 \cdot 81} = -\frac{8}{27}$$

We obtain the following solution of the Lagrange equations:

$$\left(0, -\frac{8}{27}, \frac{4}{9}\right)$$

Extreme values may also occur at points on the constraint where $\nabla g = \left\langle 2x, -2y, 3z^2 \right\rangle = \langle 0, 0, 0 \rangle$.

(b) The points $(0, y^3, y^2)$ satisfy the constraint since $0^2 - (y^3)^2 + (y^2)^3 = -y^6 + y^6 = 0$. We examine the values of $f(x, y, z) = y + z - x^2$ at these points:

$$f(0, y^3, y^2) = y^3 + y^2 - 0^2 = y^3 + y^2$$

Since $\lim_{y \to -\infty} (y^3 + y^2) = -\infty$ and $\lim_{y \to \infty} (y^3 + y^2) = \infty$, f has no absolute minimum or absolute maximum subject to the constraint.

(c) Theorem 1 states that the solutions of the Lagrange equations are the only points where local extrema may occur subject to the constraint. As shown above, our function has no global extrema subject to the constraint, but this does not contradict Theorem 1. In fact, it is not hard to show that $f(x, y)$ does not have a local extremum at either $(0, 0, 0)$ or $(0, -8/27, 4/9)$. There is no contradiction.

27. Let Q be the point on an ellipse closest to a given point P outside the ellipse. It was known to the Greek mathematician Apollonius (third century BCE) that $\overline{PQ}$ is perpendicular to the tangent at Q (Figure 13). Explain in words why this conclusion is a consequence of the method of Lagrange multipliers. *Hint:* The circles centered at P are level curves of the function to be minimized.

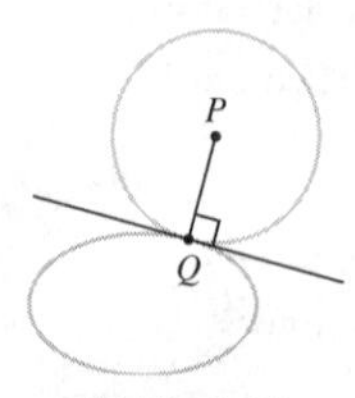

FIGURE 13

SOLUTION Let $P = (x_0, y_0)$. The distance d between the point P and a point $Q = (x, y)$ on the ellipse is minimum where the square d^2 is minimum (since the square function u^2 is increasing for $u \geq 0$). Therefore, we want to minimize the function

$$f(x, y, z) = (x - x_0)^2 + (y - y_0)^2 + (z - z_0)^2$$

subject to the constraint

$$g(x, y) = \frac{x^2}{a^2} + \frac{y^2}{b^2} = 1$$

The method of Lagrange indicates that the solution Q is the point on the ellipse where $\nabla f = \lambda \nabla g$, that is, the point on the ellipse where the gradients ∇f and ∇g are parallel. Since the gradient is orthogonal to the level curves of the function, ∇g is orthogonal to the ellipse $g(x, y) = 1$, and ∇f is orthogonal to the level curve of f passing through Q. But this level curve is a circle through Q centered at P, hence the parallel vectors ∇g and ∇f are orthogonal to the ellipse and to the circle centered at P respectively. We conclude that the point Q is the point at which the tangent to the ellipse is also the tangent to the circle through Q centered at P. That is, the tangent to the ellipse at Q is perpendicular to the radius $\overline{PQ}$ of the circle.

29. Find the maximum value of $f(x, y, z) = xy + xz + yz - 4xyz$ subject to the constraints $x + y + z = 1$ and $x, y, z \geq 0$.

SOLUTION

Step 1. Write out the Lagrange Equations. We have $\nabla f = \langle y + z - 4yz, x + z - 4xz, x + y - 4xy \rangle$ and $\nabla g = \langle 1, 1, 1 \rangle$, hence the Lagrange Condition $\nabla f = \lambda \nabla g$ yields the following equations:

$$y + z - 4yz = \lambda$$
$$x + z - 4xz = \lambda$$
$$x + y - 4xy = \lambda$$

Step 2. Solve for x, y, and z using the constraint. The Lagrange equations imply that

$$\begin{aligned} x + z - 4xz &= y + z - 4yz \\ x + y - 4xy &= y + z - 4yz \end{aligned} \quad \Rightarrow \quad \begin{aligned} x - 4xz &= y - 4yz \\ x - 4xy &= z - 4yz \end{aligned} \tag{1}$$

We solve for x and y in terms of z. The first equation gives

$$\begin{aligned} x - y + 4yz - 4xz &= 0 \\ x - y - 4z(x - y) &= 0 \\ (x - y)(1 - 4z) &= 0 \quad \Rightarrow \quad x = y \quad \text{or} \quad z = \frac{1}{4} \end{aligned} \tag{2}$$

The second equation in (1) gives:

$$\begin{aligned} x - z + 4yz - 4xy &= 0 \\ x - z - 4y(x - z) &= 0 \\ (x - z)(1 - 4y) &= 0 \quad \Rightarrow \quad x = z \quad \text{or} \quad y = \frac{1}{4} \end{aligned} \tag{3}$$

We examine the possible solutions.

1. $x = y$, $x = z$. Substituting $x = y = z$ in the equation of the constraint $x + y + z = 1$ gives $3z = 1$ or $z = \frac{1}{3}$. We obtain the solution

$$\left(\frac{1}{3}, \frac{1}{3}, \frac{1}{3}\right)$$

2. $x = y$, $y = \frac{1}{4}$. Substituting $x = y = \frac{1}{4}$ in the constraint $x + y + z = 1$ gives

$$\frac{1}{4} + \frac{1}{4} + z = 1 \quad \Rightarrow \quad z = \frac{1}{2}$$

We obtain the solution

$$\left(\frac{1}{4}, \frac{1}{4}, \frac{1}{2}\right)$$

3. $z = \frac{1}{4}$, $x = z$. Substituting $z = \frac{1}{4}$, $x = \frac{1}{4}$ in the constraint gives

$$\frac{1}{4} + y + \frac{1}{4} = 1 \quad \Rightarrow \quad y = \frac{1}{2}$$

We get the point

$$\left(\frac{1}{4}, \frac{1}{2}, \frac{1}{4}\right)$$

4. $z = \frac{1}{4}$, $y = \frac{1}{4}$. Substituting in the constraint gives $x + \frac{1}{4} + \frac{1}{4} = 1$ or $x = \frac{1}{2}$. We obtain the point

$$\left(\frac{1}{2}, \frac{1}{4}, \frac{1}{4}\right)$$

We conclude that the critical points are

$$P_1 = \left(\frac{1}{3}, \frac{1}{3}, \frac{1}{3}\right), \quad P_2 = \left(\frac{1}{4}, \frac{1}{4}, \frac{1}{2}\right)$$
$$P_3 = \left(\frac{1}{4}, \frac{1}{2}, \frac{1}{4}\right), \quad P_4 = \left(\frac{1}{2}, \frac{1}{4}, \frac{1}{4}\right) \tag{4}$$

Step 3. Conclusions. The constraint $x + y + z = 1, x \geq 0, y \geq 0, z \geq 0$ is the part of the plane $x + y + z = 1$ in the first octant. This is a closed and bounded set in R^3, hence f (which is a continuous function) has minimum and maximum value subject to the constraint. The extreme values occur at points from (4). We evaluate $f(x, y, z) = xy + xz + yz - 4xyz$ at these points:

$$f(P_1) = \frac{1}{3} \cdot \frac{1}{3} + \frac{1}{3} \cdot \frac{1}{3} + \frac{1}{3} \cdot \frac{1}{3} - 4 \cdot \frac{1}{3} \cdot \frac{1}{3} \cdot \frac{1}{3} = \frac{3}{9} - \frac{4}{27} = \frac{5}{27}$$
$$f(P_2) = f(P_3) = f(P_4) = \frac{1}{4} \cdot \frac{1}{4} + \frac{1}{4} \cdot \frac{1}{2} + \frac{1}{4} \cdot \frac{1}{2} - 4 \cdot \frac{1}{4} \cdot \frac{1}{4} \cdot \frac{1}{2} = \frac{3}{16}$$

We conclude that the maximum value of f subject to the constraint is

$$f(P_2) = f(P_3) = f(P_4) = \frac{3}{16}.$$

31. With the same set-up as in the previous problem, find the plane that minimizes V if the plane is constrained to pass through a point $P = (\alpha, \beta, \gamma)$ with $\alpha, \beta, \gamma > 0$.

SOLUTION The plane $\frac{x}{a} + \frac{y}{b} + \frac{z}{c} = 1$ must pass through the point $P(\alpha, \beta, \gamma)$, hence

$$\frac{\alpha}{a} + \frac{\beta}{b} + \frac{\gamma}{c} = 1$$

We thus must minimize the function $V(a, b, c) = \frac{1}{6}abc$ subject to the constant $g(a, b, c) = \frac{\alpha}{a} + \frac{\beta}{b} + \frac{\gamma}{c} = 1, a > 0, b > 0, c > 0$.

Step 1. Write out the Lagrange Equations. We have $\nabla V = \left\langle \frac{1}{6}bc, \frac{1}{6}ac, \frac{1}{6}ab \right\rangle$ and $\nabla g = \left\langle -\frac{\alpha}{a^2}, -\frac{\beta}{b^2}, -\frac{\gamma}{c^2} \right\rangle$, hence the Lagrange Condition $\nabla V = \lambda \nabla g$ yields the following equations:

$$\begin{aligned} \frac{1}{6}bc &= -\frac{\alpha}{a^2}\lambda & & \lambda = -\frac{a^2bc}{6\alpha} \\ \frac{1}{6}ac &= -\frac{\beta}{b^2}\lambda & \Rightarrow \quad & \lambda = -\frac{b^2ac}{6\beta} \\ \frac{1}{6}ab &= -\frac{\gamma}{c^2}\lambda & & \lambda = -\frac{c^2ab}{6\gamma} \end{aligned}$$

Step 2. Solve for a, b, c using the constraint. The Lagrange equations imply the following equations:

$$\frac{a^2bc}{\alpha} = \frac{c^2ab}{\gamma}$$
$$\frac{b^2ac}{\beta} = \frac{c^2ab}{\gamma}$$

We simplify the two equations to obtain

$$abc(\gamma a - \alpha c) = 0$$
$$abc(\gamma b - \beta c) = 0$$

Since $abc \neq 0$, these equations imply that

$$\begin{aligned} \gamma a - \alpha c = 0 \quad &\Rightarrow \quad a = \frac{\alpha}{\gamma}c \\ \gamma b - \beta c = 0 \quad &\Rightarrow \quad b = \frac{\beta}{\gamma}c \end{aligned} \tag{1}$$

We now substitute in the constraint $\frac{\alpha}{a} + \frac{\beta}{b} + \frac{\gamma}{c} = 1$ and solve for c. This gives

$$\frac{\alpha}{\frac{\alpha}{\gamma}c} + \frac{\beta}{\frac{\beta}{\gamma}c} + \frac{\gamma}{c} = 1$$

$$\frac{\gamma}{c}+\frac{\gamma}{c}+\frac{\gamma}{c}=1$$

$$\frac{3\gamma}{c}=1 \quad\Rightarrow\quad c=3\gamma$$

We find a and b using (1):

$$a=\frac{\alpha}{\gamma}\cdot 3\gamma=3\alpha,\quad b=\frac{\beta}{\gamma}\cdot 3\gamma=3\beta$$

We obtain the solution

$$P=(3\alpha,3\beta,3\gamma)$$

Step 3. Conclusions. Since V has a minimum value subject to the constraint, it occurs at the critical point. We substitute $a=3\alpha$, $b=3\beta$, and $c=3\gamma$ in the equation of the plane $\frac{x}{a}+\frac{y}{b}+\frac{z}{c}=1$ to obtain the following plane, which minimizes V:

$$\frac{x}{3\alpha}+\frac{y}{3\beta}+\frac{z}{3\gamma}=1 \quad\text{or}\quad \frac{x}{\alpha}+\frac{y}{\beta}+\frac{z}{\gamma}=3$$

33. Let L be the minimum length of a ladder that can reach over a fence of height h to a wall located a distance b behind the wall.

(a) Use Lagrange multipliers to show that $L=(h^{2/3}+b^{2/3})^{3/2}$ (Figure 16). *Hint:* Show that the problem amounts to minimizing $f(x,y)=(x+b)^2+(y+h)^2$ subject to $y/b=h/x$ or $xy=bh$.

(b) Show that the value of L is also equal to the radius of the circle with center $(-b,-h)$ that is tangent to the graph of $xy=bh$.

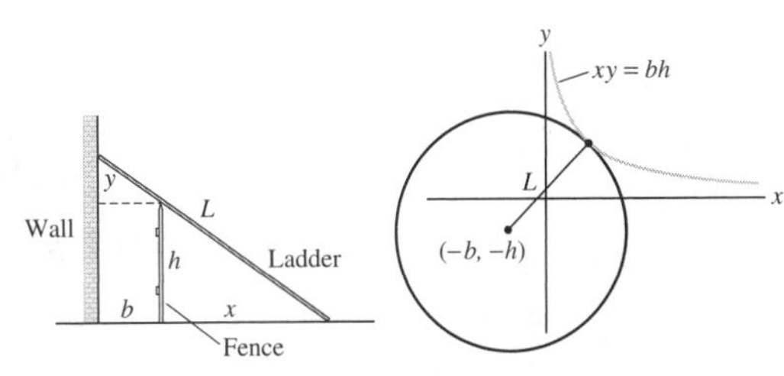

FIGURE 16

SOLUTION

(a) We denote by x and y the lengths shown in the figure, and express the length l of the ladder in terms of x and y.

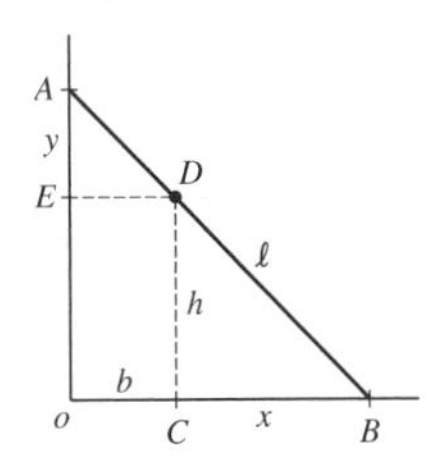

Using the Pythagorean Theorem, we have

$$l=\sqrt{\overline{OA}^2+\overline{OB}^2}=\sqrt{(y+h)^2+(x+b)^2} \tag{1}$$

Since the function u^2 is increasing for $u\geq 0$, l and l^2 have their minimum values at the same point. Therefore, we may minimize the function $f(x,y)=l^2(x,y)$, which is

$$f(x,y)=(x+b)^2+(y+h)^2$$

We now identify the constraint on the variables x and y. (Notice that h, b are constants while x and y are free to change). Using proportional lengths in the similar triangles ΔAED and ΔDCB, we have

$$\frac{\overline{AE}}{\overline{DC}}=\frac{\overline{ED}}{\overline{CB}}$$

That is,

$$\frac{y}{h}=\frac{b}{x} \quad\Rightarrow\quad xy=bh$$

We thus must minimize $f(x,y)=(x+b)^2+(y+h)^2$ subject to the constraint $g(x,y)=xy=bh$, $x>0$, $y>0$.

Step 1. Write out the Lagrange Equations. We have $\nabla f = \langle 2(x+b), 2(y+h)\rangle$ and $\nabla g = \langle y, x\rangle$, hence the Lagrange Condition $\nabla f = \lambda \nabla g$ gives the following equations:

$$2(x+b) = \lambda y$$
$$2(y+h) = \lambda x$$

Step 2. Solve for λ in terms of x and y. The equation of the constraint implies that $y \neq 0$ and $x \neq 0$. Therefore, the Lagrange equations yield

$$\lambda = \frac{2(x+b)}{y}, \quad \lambda = \frac{2(y+h)}{x}$$

Step 3. Solve for x and y using the constraint. Equating the two expressions for λ gives

$$\frac{2(x+b)}{y} = \frac{2(y+h)}{x}$$

We simplify:

$$x(x+b) = y(y+h)$$
$$x^2 + xb = y^2 + yh$$

The equation of the constraint implies that $y = \frac{bh}{x}$. We substitute and solve for $x > 0$. This gives

$$x^2 + xb = \left(\frac{bh}{x}\right)^2 + \frac{bh}{x} \cdot h$$
$$x^2 + xb = \frac{b^2h^2}{x^2} + \frac{bh^2}{x}$$
$$x^4 + x^3 b = b^2h^2 + bh^2 x$$
$$x^4 + bx^3 - bh^2 x - b^2h^2 = 0$$
$$x^3(x+b) - bh^2(x+b) = 0$$
$$\left(x^3 - bh^2\right)(x+b) = 0$$

Since $x > 0$ and $b > 0$, also $x + b > 0$ and the solution is

$$x^3 - bh^2 = 0 \quad \Rightarrow \quad x = \left(bh^2\right)^{1/3}$$

We compute y. Using the relation $y = \frac{bh}{x}$,

$$y = \frac{bh}{\left(bh^2\right)^{1/3}} = \frac{bh}{b^{1/3}h^{2/3}} = b^{2/3}h^{1/3} = \left(b^2h\right)^{1/3}$$

We obtain the solution

$$x = \left(bh^2\right)^{1/3}, \quad y = \left(b^2h\right)^{1/3} \tag{2}$$

Extreme values may also occur at the point on the constraint where $\nabla g = \mathbf{0}$. However, $\nabla g = \langle y, x\rangle = \langle 0, 0\rangle$ only at the point $(0, 0)$, which is not on the constraint.

Step 4. Conclusions. Notice that on the constraint $y = \frac{bh}{x}$ or $x = \frac{bh}{y}$, as $x \to 0+$ then $y \to \infty$, and as $x \to \infty$, then $y \to 0+$. Also, as $y \to 0+$, $x \to \infty$ and as $y \to \infty$, $x \to 0+$. In either case, $f(x, y)$ is increasing without bound. Using this property and the theorem on the existence of extreme values for a continuous function on a closed and bounded set (for a certain part of the constraint), one can show that f has a minimum value on the constraint. This minimum value occurs at the point (2). We substitute this point in (1) to obtain the following minimum length L:

$$L = \sqrt{\left(\left(b^2h\right)^{1/3} + h\right)^2 + \left(\left(bh^2\right)^{1/3} + b\right)^2}$$
$$= \sqrt{\left(b^2h\right)^{2/3} + 2h\left(b^2h\right)^{1/3} + h^2 + \left(bh^2\right)^{2/3} + 2b\left(bh^2\right)^{1/3} + b^2}$$
$$= \sqrt{b^{\frac{4}{3}}h^{2/3} + 2h^{\frac{4}{3}}b^{2/3} + h^2 + b^{2/3}h^{\frac{4}{3}} + 2b^{\frac{4}{3}}h^{2/3} + b^2}$$
$$= \sqrt{3b^{\frac{4}{3}}h^{2/3} + 3h^{\frac{4}{3}}b^{2/3} + h^2 + b^2}$$

$$= \sqrt{\left(h^{2/3}\right)^3 + 3\left(h^{2/3}\right)^2 b^{2/3} + 3h^{2/3}\left(b^{2/3}\right)^2 + \left(b^{2/3}\right)^3}$$

Using the identity $(\alpha + \beta)^3 = \alpha^3 + 3\alpha^2\beta + 3\alpha\beta^2 + \beta^3$, we conclude that

$$L = \sqrt{\left(h^{2/3} + b^{2/3}\right)^3} = \left(h^{2/3} + b^{2/3}\right)^{3/2}.$$

(b) The Lagrange Condition states that the gradient vectors ∇f_P and ∇g_P are parallel (where P is the minimizing point). The gradient ∇f_P is orthogonal to the level curve of f passing through P, which is a circle through P centered at $(-b, -h)$. ∇g_P is orthogonal to the level curve of g passing through P, which is the curve of the constraint $xy = bh$. We conclude that the circle and the curve $xy = bh$, both being perpendicular to parallel vectors, are tangent at P. The radius of the circle is the minimum value L, of $f(x, y)$.

35. Find the minimum value of $f(x, y, z) = x^2 + y^2 + z^2$ subject to two constraints, $x + 2y + z = 3$ and $x - y = 4$.

SOLUTION The constraint equations are

$$g(x, y, z) = x + 2y + z - 3 = 0, \quad h(x, y) = x - y - 4 = 0$$

Step 1. Write out the Lagrange Equations. We have $\nabla f = \langle 2x, 2y, 2z\rangle$, $\nabla g = \langle 1, 2, 1\rangle$, and $\nabla h = \langle 1, -1, 0\rangle$, hence the Lagrange Condition is

$$\begin{aligned} \nabla f &= \lambda \nabla g + \mu \nabla h \\ \langle 2x, 2y, 2z\rangle &= \lambda \langle 1, 2, 1\rangle + \mu \langle 1, -1, 0\rangle \\ &= \langle \lambda + \mu, 2\lambda - \mu, \lambda\rangle \end{aligned}$$

We obtain the following equations:

$$\begin{aligned} 2x &= \lambda + \mu \\ 2y &= 2\lambda - \mu \\ 2z &= \lambda \end{aligned}$$

Step 2. Solve for λ and μ. The first equation gives $\lambda = 2x - \mu$. Combining with the third equation we get

$$2z = 2x - \mu \qquad (1)$$

The second equation gives $\mu = 2\lambda - 2y$, combining with the third equation we get $\mu = 4z - 2y$. Substituting in (1) we obtain

$$\begin{aligned} 2z &= 2x - (4z - 2y) = 2x - 4z + 2y \\ 6z &= 2x + 2y \quad \Rightarrow \quad z = \frac{x + y}{3} \end{aligned} \qquad (2)$$

Step 3. Solve for x, y, and z using the constraints. The constraints give x and y as functions of z:

$$\begin{aligned} x - y = 4 \quad &\Rightarrow \quad y = x - 4 \\ x + 2y + z = 3 \quad &\Rightarrow \quad y = \frac{3 - x - z}{2} \end{aligned}$$

Combining the two equations we get

$$\begin{aligned} x - 4 &= \frac{3 - x - z}{2} \\ 2x - 8 &= 3 - x - z \\ 3x &= 11 - z \quad \Rightarrow \quad x = \frac{11 - z}{3} \end{aligned}$$

We find y using $y = x - 4$:

$$y = \frac{11 - z}{3} - 4 = \frac{-1 - z}{3}$$

We substitute x and y in (2) and solve for z:

$$\begin{aligned} z &= \frac{\frac{11-z}{3} + \frac{-1-z}{3}}{3} = \frac{11 - z - 1 - z}{9} = \frac{10 - 2z}{9} \\ 9z &= 10 - 2z \end{aligned}$$

$$9z = 10 - 2z$$
$$11z = 10 \quad \Rightarrow \quad z = \frac{10}{11}$$

We find x and y:

$$y = \frac{-1-z}{3} = \frac{-1-\frac{10}{11}}{3} = -\frac{21}{33} = -\frac{7}{11}$$
$$x = \frac{11-z}{3} = \frac{11-\frac{10}{11}}{3} = \frac{111}{33} = \frac{37}{11}$$

We obtain the solution

$$P = \left(\frac{37}{11}, -\frac{7}{11}, \frac{10}{11}\right)$$

Step 4. Calculate the critical values. We compute $f(x, y, z) = z^2 + y^2 + z^2$ at the critical point:

$$f(P) = \left(\frac{37}{11}\right)^2 + \left(-\frac{7}{11}\right)^2 + \left(\frac{10}{11}\right)^2 = \frac{1518}{121} = \frac{138}{11} \approx 12.545$$

As x tends to infinity, so also does $f(x, y, z)$ tend to ∞. Therefore f has no maximum value and the given critical point P must produce a minimum. We conclude that the minimum value of f subject to the two constraints is $f(P) = \frac{138}{11} \approx 12.545$.

Further Insights and Challenges

37. Assumptions Matter CAS Consider the problem of minimizing $f(x, y) = x$ subject to $g(x, y) = (x-1)^3 - y^2 = 0$.

(a) Show, without using calculus, that the minimum occurs at $P = (1, 0)$.

(b) Show that the Lagrange condition $\nabla f_P = \lambda \nabla g_P$ is not satisfied for any value of λ.

(c) Does this contradict Theorem 1?

SOLUTION

(a) The equation of the constraint can be rewritten as

$$(x-1)^3 = y^2 \quad \text{or} \quad x = y^{2/3} + 1$$

Therefore, at the points under the constraint, $x \geq 1$, hence $f(x, y) \geq 1$. Also at the point $P = (1, 0)$ we have $f(1, 0) = 1$, hence $f(1, 0) = 1$ is the minimum value of f under the constraint.

(b) We have $\nabla f = \langle 1, 0 \rangle$ and $\nabla g = \left\langle 3(x-1)^2, -2y \right\rangle$, hence the Lagrange Condition $\nabla f = \lambda \nabla g$ yields the following equations:

$$1 = \lambda \cdot 3(x-1)^2$$
$$0 = -2\lambda y$$

The first equation implies that $\lambda \neq 0$ and $x - 1 = \pm\frac{1}{\sqrt{3\lambda}}$. The second equation gives $y = 0$. Substituting in the equation of the constraint gives

$$(x-1)^3 - y^2 = \left(\frac{\pm 1}{\sqrt{3\lambda}}\right)^3 - 0^2 = \frac{\pm 1}{(3\lambda)^{3/2}} \neq 0$$

We conclude that the Lagrange Condition is not satisfied by any point under the constraint.

(c) Theorem 1 is not violated since at the point $P = (1, 0)$, $\nabla g = \mathbf{0}$, whereas the Theorem is valid for points where $\nabla g_P \neq \mathbf{0}$.

39. Consider the utility function $U(x_1, x_2) = x_1 x_2$ with budget constraint $p_1 x_1 + p_2 x_2 = c$.

(a) Show that the maximum of $U(x_1, x_2)$ subject to the budget constraint is equal to $c^2/(4p_1 p_2)$.

(b) Calculate the value of the Lagrange multiplier λ occurring in (a).

(c) Prove the following interpretation: λ is the rate of increase in utility per unit increase of total budget c.

SOLUTION

(a) By the earlier exercise, the utility is maximized at a point where the following equality holds:

$$\frac{U_{x_1}}{U_{x_2}} = \frac{p_1}{p_2}$$

Since $U_{x_1} = x_2$ and $U_{x_2} = x_1$, we get

$$\frac{x_2}{x_1} = \frac{p_1}{p_2} \quad \Rightarrow \quad x_2 = \frac{p_1}{p_2}x_1$$

We now substitute x_2 in terms of x_1 in the constraint $p_1x_1 + p_2x_2 = c$ and solve for x_1. This gives

$$p_1x_1 + p_2 \cdot \frac{p_1}{p_2}x_1 = c$$

$$2p_1x_1 = c \quad \Rightarrow \quad x_1 = \frac{c}{2p_1}$$

The corresponding value of x_2 is computed by $x_2 = \frac{p_1}{p_2}x_1$:

$$x_2 = \frac{p_1}{p_2} \cdot \frac{c}{2p_1} = \frac{c}{2p_2}$$

That is, $U(x_1, x_2)$ is maximized at the consumption level $x_1 = \frac{c}{2p_1}$, $x_2 = \frac{c}{2p_2}$. The maximum value is

$$U\left(\frac{c}{2p_1}, \frac{c}{2p_2}\right) = \frac{c}{2p_1} \cdot \frac{c}{2p_2} = \frac{c^2}{4p_1p_2}$$

(b) The Lagrange Condition $\nabla U = \lambda \nabla g$ for $U(x_1, x_2) = x_1x_2$ and $g(x_1, x_2) = p_1x_1 + p_2x_2 - c = 0$ is

$$\langle x_2, x_1 \rangle = \lambda \langle p_1, p_2 \rangle \qquad (1)$$

or

$$\begin{aligned} x_2 &= \lambda p_1 \\ x_1 &= \lambda p_2 \end{aligned} \quad \Rightarrow \quad \lambda = \frac{x_2}{p_1} = \frac{x_1}{p_2}$$

In part (a) we showed that at the maximizing point $x_1 = \frac{c}{2p_1}$, therefore the value of λ is

$$\lambda = \frac{x_1}{p_2} = \frac{c}{2p_1p_2}$$

(c) We compute $\frac{dU}{dc}$ using the Chain Rule:

$$\frac{dU}{dc} = \frac{\partial U}{\partial x_1}x_1'(c) + \frac{\partial U}{\partial x_2}x_2'(c) = x_2x_1'(c) + x_1x_2'(c) = \langle x_2, x_1 \rangle \cdot \langle x_1'(c), x_2'(c) \rangle$$

Substituting in (1) we get

$$\frac{dU}{dc} = \lambda \langle p_1, p_2 \rangle \cdot \langle x_1'(c), x_2'(c) \rangle = \lambda \left(p_1x_1'(c) + p_2x_2'(c)\right) \qquad (2)$$

We now use the Chain Rule to differentiate the equation of the constraint $p_1x_1 + p_2x_2 = c$ with respect to c:

$$p_1x_1'(c) + p_2x_2'(c) = 1$$

Substituting in (2), we get

$$\frac{dU}{dc} = \lambda \cdot 1 = \lambda$$

Using the approximation $\Delta U \approx \frac{dU}{dc}\Delta c$, we conclude that λ is the rate of increase in utility per unit increase of total budget L.

41. Let $B > 0$. Show that the maximum of $f(x_1, \dots, x_n) = x_1x_2 \cdots x_n$ subject to the constraints $x_1 + \cdots + x_n = B$ and $x_j \geq 0$ for $j = 1, \dots, n$ occurs for $x_1 = \cdots = x_n = B/n$. Use this to conclude that

$$(a_1a_2 \cdots a_n)^{1/n} \leq \frac{a_1 + \cdots + a_n}{n}$$

for all positive numbers $a_1, \dots, a_n$.

SOLUTION We first notice that the constraints $x_1 + \cdots + x_n = B$ and $x_j \geq 0$ for $j = 1, \ldots, n$ define a closed and bounded set in the nth dimensional space, hence f (continuous, as a polynomial) has extreme values on this set. The minimum value zero occurs where one of the coordinates is zero (for example, for $n = 2$ the constraint $x_1 + x_2 = B$, $x_1 \geq 0$, $x_2 \geq 0$ is a triangle in the first quadrant). We need to maximize the function $f(x_1, \ldots, x_n) = x_1 x_2 \cdots x_n$ subject to the constraints $g(x_1, \ldots, x_n) = x_1 + \cdots + x_n - B = 0$, $x_j \geq 0$, $j = 1, \ldots, n$.

Step 1. Write out the Lagrange Equations. The gradient vectors are

$$\nabla f = \left\langle x_2 x_3 \cdots x_n, x_1 x_3 \cdots x_n, \ldots, x_1 x_2 \cdots x_{n-1} \right\rangle$$
$$\nabla g = \langle 1, 1, \ldots, 1 \rangle$$

The Lagrange Condition $\nabla f = \lambda \nabla g$ yields the following equations:

$$x_2 x_3 \cdots x_n = \lambda$$
$$x_1 x_3 \cdots x_n = \lambda$$
$$x_1 x_2 \cdots x_{n-1} = \lambda$$

Step 2. Solving for $x_1, x_2, \ldots, x_n$ using the constraint. The Lagrange equations imply the following equations:

$$\begin{aligned} x_2 x_3 \cdots x_n &= x_1 x_2 \cdots x_{n-1} \\ x_1 x_3 \cdots x_n &= x_1 x_2 \cdots x_{n-1} \\ x_1 x_2 x_4 \cdots x_n &= x_1 x_2 \cdots x_{n-1} \\ &\vdots \\ x_1 x_2 \cdots x_{n-2} x_n &= x_1 x_2 \cdots x_{n-1} \end{aligned}$$

We may assume that $x_j \neq 0$ for $j = 1, \ldots, n$, since if one of the coordinates is zero, f has the minimum value zero. We divide each equation by its right-hand side to obtain

$$\begin{aligned} \frac{x_n}{x_1} &= 1 \\ \frac{x_n}{x_2} &= 1 \\ \frac{x_n}{x_3} &= 1 \\ &\vdots \\ \frac{x_n}{x_{n-1}} &= 1 \end{aligned} \quad \Rightarrow \quad \begin{aligned} x_1 &= x_n \\ x_2 &= x_n \\ x_3 &= x_n \\ &\vdots \\ x_{n-1} &= x_n \end{aligned}$$

Substituting in the constraint $x_1 + \cdots + x_n = B$ and solving for x_n gives

$$\underbrace{x_n + x_n + \cdots + x_n}_{n} = B$$

$$n x_n = B \quad \Rightarrow \quad x_n = \frac{B}{n}$$

Hence $x_1 = \cdots = x_n = \frac{B}{n}$.

Step 3. Conclusions. The maximum value of $f(x_1, \ldots, x_n) = x_1 x_2 \cdots x_n$ on the constraint $x_1 + \cdots + x_n = B, x_j \geq 0$, $j = 1, \ldots, n$ occurs at the point at which all coordinates are equal to $\frac{B}{n}$. The value of f at this point is

$$f\left(\frac{B}{n}, \frac{B}{n}, \ldots, \frac{B}{n}\right) = \left(\frac{B}{n}\right)^n$$

It follows that for any point $(x_1, \ldots, x_n)$ on the constraint, that is, for any point satisfying $x_1 + \cdots + x_n = B$ with x_j positive, the following holds:

$$f(x_1, \ldots, x_n) \leq \left(\frac{B}{n}\right)^n$$

That is,

$$x_1 \cdots x_n \leq \left(\frac{x_1 + \cdots + x_n}{n}\right)^n$$

or

$$(x_1 \cdots x_n)^{1/n} \le \frac{x_1 + \cdots + x_n}{n}.$$

43. Given constants E, E_1, E_2, E_3, consider the maximum of

$$S(x_1, x_2, x_3) = x_1 \ln x_1 + x_2 \ln x_2 + x_3 \ln x_3$$

subject to two constraints:

$$x_1 + x_2 + x_3 = N, \quad E_1 x_1 + E_2 x_2 + E_3 x_3 = E$$

Show that there is a constant μ such that $x_i = A^{-1} e^{\mu E_i}$ for $i = 1, 2, 3$, where $A = N^{-1}(e^{\mu E_1} + e^{\mu E_2} + e^{\mu E_3})$.

SOLUTION The constraints equations are

$$g(x_1, x_2, x_3) = x_1 + x_2 + x_3 - N = 0$$
$$h(x_1, x_2, x_3) = E_1 x_1 + E_2 x_2 + E_3 x_3 - E = 0$$

We first find the Lagrange equations. The gradient vectors are

$$\nabla S = \left\langle \ln x_1 + x_1 \cdot \frac{1}{x_1}, \ln x_2 + x_2 \cdot \frac{1}{x_2}, \ln x_3 + x_3 \cdot \frac{1}{x_3} \right\rangle = \langle 1 + \ln x_1, 1 + \ln x_2, 1 + \ln x_3 \rangle$$
$$\nabla g = \langle 1, 1, 1 \rangle, \quad \nabla h = \langle E_1, E_2, E_3 \rangle$$

The Lagrange Condition $\nabla f = \lambda \nabla g + \mu \nabla h$ gives the following equation:

$$\langle 1 + \ln x_1, 1 + \ln x_2, 1 + \ln x_3 \rangle = \lambda \langle 1, 1, 1 \rangle + \mu \langle E_1, E_2, E_3 \rangle = \langle \lambda + \mu E_1, \lambda + \mu E_2, \lambda + \mu E_3 \rangle$$

We obtain the Lagrange equations:

$$1 + \ln x_1 = \lambda + \mu E_1$$
$$1 + \ln x_2 = \lambda + \mu E_2$$
$$1 + \ln x_3 = \lambda + \mu E_3$$

We subtract the third equation from the other equations to obtain

$$\ln x_1 - \ln x_3 = \mu (E_1 - E_3)$$
$$\ln x_2 - \ln x_3 = \mu (E_2 - E_3)$$

or

$$\begin{aligned} \ln \frac{x_1}{x_3} &= \mu (E_1 - E_3) \\ \ln \frac{x_2}{x_3} &= \mu (E_2 - E_3) \end{aligned} \quad \Rightarrow \quad \begin{aligned} x_1 &= x_3 e^{\mu(E_1 - E_3)} \\ x_2 &= x_3 e^{\mu(E_2 - E_3)} \end{aligned} \tag{1}$$

Substituting x_1 and x_2 in the equation of the constraint $g(x_1, x_2, x_3) = 0$ and solving for x_3 gives

$$x_3 e^{\mu(E_1 - E_3)} + x_3 e^{\mu(E_2 - E_3)} + x_3 = N$$

We multiply by $e^{\mu E_3}$:

$$x_3 \left(e^{\mu E_1} + e^{\mu E_2} + e^{\mu E_3}\right) = N e^{\mu E_3}$$
$$x_3 = \frac{N e^{\mu E_3}}{e^{\mu E_1} + e^{\mu E_2} + e^{\mu E_3}}$$

Substituting in (1) we get

$$x_1 = \frac{N e^{\mu E_3}}{e^{\mu E_1} + e^{\mu E_2} + e^{\mu E_3}} \cdot e^{\mu(E_1 - E_3)} = \frac{N e^{\mu E_1}}{e^{\mu E_1} + e^{\mu E_2} + e^{\mu E_3}}$$
$$x_2 = \frac{N e^{\mu E_3}}{e^{\mu E_1} + e^{\mu E_2} + e^{\mu E_3}} \cdot e^{\mu(E_2 - E_3)} = \frac{N e^{\mu E_2}}{e^{\mu E_1} + e^{\mu E_2} + e^{\mu E_3}}$$

Letting $A = \frac{e^{\mu E_1} + e^{\mu E_2} + e^{\mu E_3}}{N}$, we obtain

$$x_1 = A^{-1} e^{\mu E_1}, \quad x_2 = A^{-1} e^{\mu E_2}, \quad x_3 = A^{-1} e^{\mu E_3}$$

The value of μ is determined by the second constraint $h(x_1, x_2, x_3) = 0$.

CHAPTER REVIEW EXERCISES

1. Given $f(x, y) = \dfrac{\sqrt{x^2 - y^2}}{x + 3}$,
(a) Sketch the domain of f.
(b) Calculate $f(3, 1)$ and $f(-5, -3)$.
(c) Find a point satisfying $f(x, y) = 1$.

SOLUTION

(a) f is defined where $x^2 - y^2 \geq 0$ and $x + 3 \neq 0$. We solve these two inequalities:

$$x^2 - y^2 \geq 0 \quad \Rightarrow \quad x^2 \geq y^2 \quad \Rightarrow \quad |x| \geq |y|$$
$$x + 3 \neq 0 \quad \Rightarrow \quad x \neq -3$$

Therefore, the domain of f is the following set:

$$D = \{(x, y) : |x| \geq |y|, x \neq -3\}$$

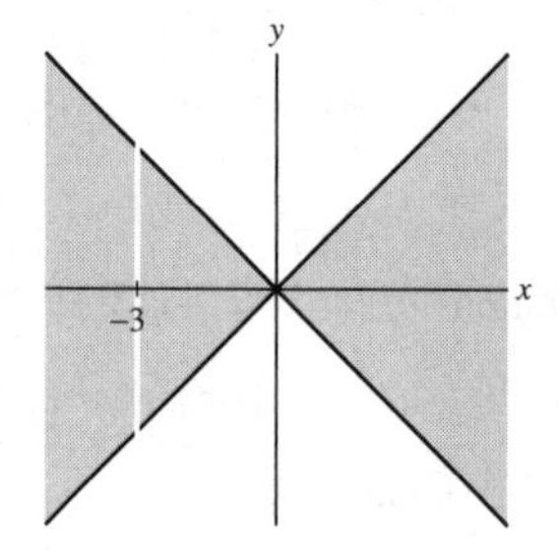

(b) To find $f(3, 1)$ we substitute $x = 3$, $y = 1$ in $f(x, y)$. We get

$$f(3, 1) = \frac{\sqrt{3^2 - 1^2}}{3 + 3} = \frac{\sqrt{8}}{6} = \frac{\sqrt{2}}{3}$$

Similarly, setting $x = -5$, $y = -3$, we get

$$f(-5, -3) = \frac{\sqrt{(-5)^2 - (-3)^2}}{-5 + 3} = \frac{\sqrt{16}}{-2} = -2.$$

(c) We must find a point (x, y) such that

$$f(x, y) = \frac{\sqrt{x^2 - y^2}}{x + 3} = 1$$

We choose, for instance, $y = 1$, substitute and solve for x. This gives

$$\frac{\sqrt{x^2 - 1^2}}{x + 3} = 1$$
$$\sqrt{x^2 - 1} = x + 3$$
$$x^2 - 1 = (x + 3)^2 = x^2 + 6x + 9$$
$$6x = -10 \quad \Rightarrow \quad x = -\frac{5}{3}$$

Thus, the point $\left(-\frac{5}{3}, 1\right)$ satisfies $f\left(-\frac{5}{3}, 1\right) = 1$.

3. Sketch the graph $f(x, y) = x^2 - y + 1$ and describe its vertical and horizontal traces.

SOLUTION The graph is shown in the following figure.

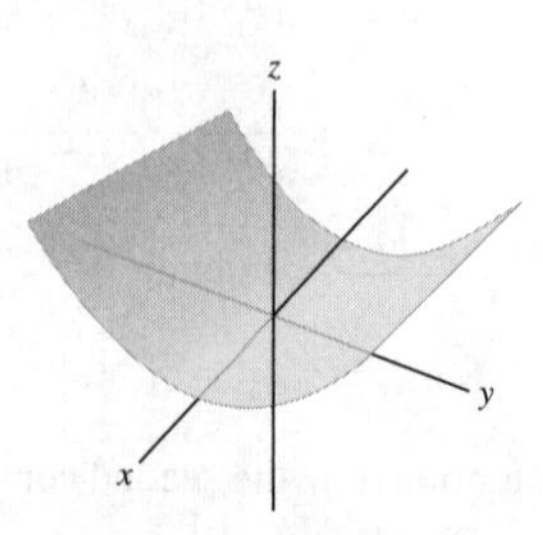

The trace obtained by setting $x = c$ is the line $z = c^2 - y + 1$ or $z = (c^2 + 1) - y$ in the plane $x = c$. The trace obtained by setting $y = c$ is the parabola $z = x^2 - c + 1$ in the plane $y = c$. The trace obtained by setting $z = c$ is the parabola $y = x^2 + 1 - c$ in the plane $z = c$.

5. Match the functions (a)–(d) with their graphs in Figure 1.

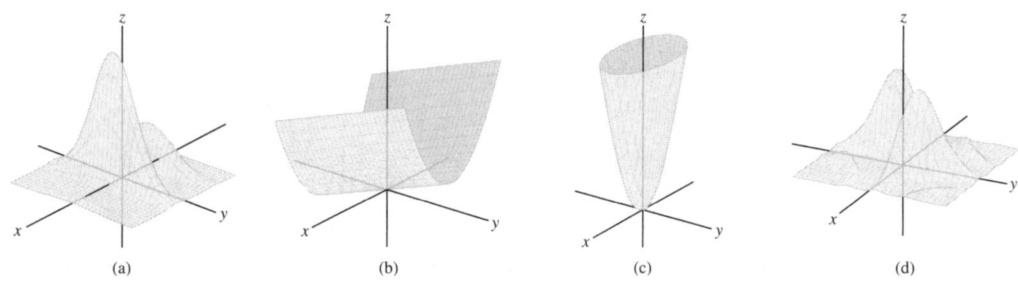

FIGURE 1

(a) $f(x, y) = x^2 + y$

(b) $f(x, y) = x^2 + 4y^2$

(c) $f(x, y) = \sin(4xy)e^{-x^2-y^2}$

(d) $f(x, y) = \sin(4x)e^{-x^2-y^2}$

SOLUTION The function $f = x^2 + y$ matches picture (b), as can be seen by taking the $x = 0$ slice. The function $f = x^2 + 4y^2$ matches picture (c), as can be seen by taking $z = c$ slices (giving ellipses). Since $\sin(4xy)e^{-x^2-y^2}$ is symmetric with respect to x and y, and so also is picture (d), we match $\sin(4xy)e^{-x^2-y^2}$ with (d). That leaves the third function, $\sin(4x)e^{-x^2-y^2}$, to match with picture (a).

7. Describe the level curves of:

(a) $f(x, y) = e^{4x-y}$

(b) $f(x, y) = \ln(4x - y)$

(c) $f(x, y) = 3x^2 - 4y^2$

(d) $f(x, y) = x + y^2$

SOLUTION

(a) The level curves of $f(x, y) = e^{4x-y}$ are the curves $e^{4x-y} = c$ in the xy-plane, where $c > 0$. Taking ln from both sides we get $4x - y = \ln c$. Therefore, the level curves are the parallel lines of slope 4, $4x - y = \ln c$, $c > 0$, in the xy-plane.

(b) The level curves of $f(x, y) = \ln(4x - y)$ are the curves $\ln(4x - y) = c$ in the xy-plane. We rewrite it as $4x - y = e^c$ to obtain the parallel lines of slope 4, with negative y-intercepts.

(c) The level curves of $f(x, y) = 3x^2 - 4y^2$ are the hyperbolas $3x^2 - 4y^2 = c$ in the xy plane.

(d) The level curves of $f(x, y) = x + y^2$ are the curves $x + y^2 = c$ or $x = c - y^2$ in the xy-plane. These are parabolas whose axis is the x-axis.

In Exercises 9–14, evaluate the limit or state that it does not exist.

9. $\displaystyle\lim_{(x,y)\to(1,-3)} (xy + y^2)$

SOLUTION The function $f(x, y) = xy + y^2$ is continuous everywhere because it is a polynomial, therefore we evaluate the limit using substitution:

$$\lim_{(x,y)\to(1,-3)} \left(xy + y^2\right) = 1 \cdot (-3) + (-3)^2 = 6$$

11. $\displaystyle\lim_{(x,y)\to(0,0)} \frac{xy + xy^2}{x^2 + y^2}$

SOLUTION We evaluate the limits as (x, y) approaches the origin along the lines $y = x$ and $y = 2x$:

$$\lim_{\substack{(x,y)\to(0,0)\\ \text{along } y=x}} \frac{xy + xy^2}{x^2 + y^2} = \lim_{x\to 0} \frac{x \cdot x + x \cdot x^2}{x^2 + x^2} = \lim_{x\to 0} \frac{x^2 + x^3}{2x^2} = \lim_{x\to 0} \frac{1 + x}{2} = \frac{1}{2}$$

$$\lim_{\substack{x\to(0,0)\\ \text{along } y=2x}} \frac{xy + xy^2}{x^2 + y^2} = \lim_{x\to 0} \frac{x \cdot 2x + x \cdot (2x)^2}{x^2 + (2x)^2} = \lim_{x\to 0} \frac{2x^2 + 4x^3}{5x^2} = \lim_{x\to 0} \frac{2 + 4x}{5} = \frac{2}{5}$$

Since the two limits are different, $f(x, y)$ does not approach one limit as $(x, y) \to (0, 0)$, therefore the limit does not exist.

13. $\lim_{(x,y)\to(1,-3)} (2x+y)e^{-x+y}$

SOLUTION The function $f(x, y) = (2x + y)e^{-x+y}$ is continuous, hence we evaluate the limit using substitution:

$$\lim_{(x,y)\to(1,-3)} (2x+y)e^{-x+y} = (2\cdot 1 - 3)e^{-1-3} = -e^{-4}$$

15. Let

$$f(x,y) = \begin{cases} \dfrac{(xy)^p}{x^4+y^4} & (x,y) \neq (0,0) \\ 0 & (x,y) = (0,0) \end{cases}$$

Use polar coordinates to show that $f(x, y)$ is continuous at all (x, y) if $p > 2$, but discontinuous at $(0, 0)$ if $p \le 2$.

SOLUTION We show using the polar coordinates $x = r\cos\theta$, $y = r\sin\theta$, that the limit of $f(x, y)$ as $(x, y) \to (0, 0)$ is zero for $p > 2$. This will prove that f is continuous at the origin. Since f is a rational function with nonzero denominator for $(x, y) \neq (0, 0)$, f is continuous there. We have

$$\lim_{(x,y)\to(0,0)} f(x,y) = \lim_{r\to 0+} \frac{(r\cos\theta)^p(r\sin\theta)^p}{(r\cos\theta)^4+(r\sin\theta)^4} = \lim_{r\to 0+} \frac{r^{2p}(\cos\theta\sin\theta)^p}{r^4\left(\cos^4\theta+\sin^4\theta\right)} \quad (1)$$

$$= \lim_{r\to 0+} \frac{r^{2(p-2)}(\cos\theta\sin\theta)^p}{\cos^4\theta+\sin^4\theta}$$

We use the following inequalities:

$$\left|\cos^4\theta\sin^4\theta\right| \le 1$$

$$\cos^4\theta+\sin^4\theta = \left(\cos^2\theta+\sin^2\theta\right)^2 - 2\cos^2\theta\sin^2\theta = 1 - \frac{1}{2}\cdot(2\cos\theta\sin\theta)^2$$

$$= 1 - \frac{1}{2}\sin^2 2\theta \ge 1 - \frac{1}{2} = \frac{1}{2}$$

Therefore,

$$0 \le \left|\frac{r^{2(p-2)}(\cos\theta\sin\theta)^p}{\cos^4\theta+\sin^4\theta}\right| \le \frac{r^{2(p-2)}\cdot 1}{\frac{1}{2}} = 2r^{2(p-2)}$$

Since $p - 2 > 0$, $\lim_{r\to 0+} 2r^{2(p-2)} = 0$, hence by the Squeeze Theorem the limit in (1) is also zero. We conclude that f is continuous for $p > 2$.

We now show that for $p < 2$ the limit of $f(x, y)$ as $(x, y) \to (0, 0)$ does not exist. We compute the limit as (x, y) approaches the origin along the line $y = x$.

$$\lim_{\substack{(x,y)\to(0,0)\\ \text{along } y=x}} f(x,y) = \lim_{x\to 0} \frac{(x^2)^p}{x^4+x^4} = \lim_{x\to 0}\frac{x^{2p}}{2x^4} = \lim_{x\to 0}\frac{x^{2(p-2)}}{2} = \infty$$

Therefore the limit of $f(x, y)$ as $(x, y) \to (0, 0)$ does not exist for $p < 2$. We now show that the limit $\lim_{(x,y)\to(0,0)} \frac{x^2y^2}{x^4+y^4}$ does not exist for $p = 2$ as well. We compute the limits along the line $y = 0$ and $y = x$:

$$\lim_{\substack{(x,y)\to(0,0)\\ \text{along } y=0}} \frac{x^2y^2}{x^4+y^4} = \lim_{x\to 0}\frac{x^2\cdot 0^2}{x^4+0^4} = \lim_{x\to 0}\frac{0}{x^4} = 0$$

$$\lim_{\substack{(x,y)\to(0,0)\\ \text{along } y=x}} \frac{x^2y^2}{x^4+y^4} = \lim_{x\to 0}\frac{x^2\cdot x^2}{x^4+x^4} = \lim_{x\to 0}\frac{x^4}{2x^4} = \frac{1}{2}$$

Since the limits along two paths are different, $f(x, y)$ does not approach one limit as $(x, y) \to (0, 0)$. We thus showed that if $p \le 2$, the limit $\lim_{(x,y)\to(0,0)} f(x, y)$ does not exist, and f is not continuous at the origin for $p \le 2$.

In Exercises 17–20, compute f_x and f_y.

17. $f(x, y) = 2x + y^2$

SOLUTION To find f_x we treat y as a constant, and to find f_y we treat x as a constant. We get

$$f_x = \frac{\partial}{\partial x}\left(2x + y^2\right) = \frac{\partial}{\partial x}(2x) + \frac{\partial}{\partial x}\left(y^2\right) = 2 + 0 = 2$$

$$f_y = \frac{\partial}{\partial y}\left(2x + y^2\right) = \frac{\partial}{\partial y}(2x) + \frac{\partial}{\partial y}\left(y^2\right) = 0 + 2y = 2y$$

19. $f(x, y) = \sin(xy)e^{-x-y}$

SOLUTION We compute f_x, treating y as a constant and using the Product Rule and the Chain Rule. We get

$$\begin{aligned} f_x &= \frac{\partial}{\partial x}\left(\sin(xy)e^{-x-y}\right) = \frac{\partial}{\partial x}\left(\sin(xy)\right)e^{-x-y} + \sin(xy)\frac{\partial}{\partial x}e^{-x-y} \\ &= \cos(xy)\cdot ye^{-x-y} + \sin(xy)\cdot(-1)e^{-x-y} = e^{-x-y}\left(y\cos(xy) - \sin(xy)\right) \end{aligned}$$

We compute f_y similarly, treating x as a constant. Notice that since $f(y, x) = f(x, y)$, the partial derivative f_y can be obtained from f_x by interchanging x and y. That is,

$$f_y = e^{-x-y}\left(x\cos(yx) - \sin(yx)\right).$$

21. Calculate f_{xxyz} for $f(x, y, z) = y\sin(x + z)$.

SOLUTION We differentiate f twice with respect to x, once with respect to y, and finally with respect to z. This gives

$$f_x = \frac{\partial}{\partial x}\left(y\sin(x+z)\right) = y\cos(x+z)$$

$$f_{xx} = \frac{\partial}{\partial x}\left(y\cos(x+z)\right) = -y\sin(x+z)$$

$$f_{xxy} = \frac{\partial}{\partial y}\left(-y\sin(x+z)\right) = -\sin(x+z)$$

$$f_{xxyz} = \frac{\partial}{\partial z}\left(-\sin(x+z)\right) = -\cos(x+z)$$

23. Find an equation of the tangent to the graph of $f(x, y) = xy^2 - xy + 3x^3y$ at $P = (1, 3)$.

SOLUTION The tangent plane has the equation

$$z = f(1,3) + f_x(1,3)(x-1) + f_y(1,3)(y-3) \tag{1}$$

We compute the partial derivatives of $f(x, y) = xy^2 - xy + 3x^3y$:

$$\begin{aligned} f_x(x,y) &= y^2 - y + 9x^2y \\ f_y(x,y) &= 2xy - x + 3x^3 \end{aligned} \quad\Rightarrow\quad \begin{aligned} f_x(1,3) &= 3^2 - 3 + 9\cdot 1^2\cdot 3 = 33 \\ f_y(1,3) &= 2\cdot 1\cdot 3 - 1 + 3\cdot 1^3 = 8 \end{aligned}$$

Also, $f(1, 3) = 1\cdot 3^2 - 1\cdot 3 + 3\cdot 1^3\cdot 3 = 15$. Substituting these values in (1), we obtain the following equation:

$$z = 15 + 33(x-1) + 8(y-3)$$

or

$$z = 33x + 8y - 42$$

25. Estimate $\sqrt{7.1^2 + 4.9^2 + 70.1}$ using the linear approximation. Compare with a calculator value.

SOLUTION The function whose value we want to approximate is

$$f(x, y, z) = \sqrt{x^2 + y^2 + z}$$

We use the linear approximation at the point $(7, 5, 70)$, hence $h = 7.1 - 7 = 0.1$, $k = 4.9 - 5 = -0.1$, and $l = 70.1 - 70 = 0.1$. We get

$$f(7.1, 4.9, 70.1) \approx f(7,5,70) + 0.1f_x(7,5,70) - 0.1f_y(7,5,70) + 0.1f_z(7,5,70) \tag{1}$$

We compute the partial derivatives of f:

$$f_x(x,y,z) = \frac{2x}{2\sqrt{x^2+y^2+z}} = \frac{x}{\sqrt{x^2+y^2+z}} \quad\Rightarrow\quad f_x(7,5,70) = \frac{7}{\sqrt{7^2+5^2+70}} = \frac{7}{12}$$

$$f_y(x,y,z) = \frac{2y}{2\sqrt{x^2+y^2+z}} = \frac{y}{\sqrt{x^2+y^2+z}} \quad \Rightarrow \quad f_y(7,5,70) = \frac{5}{\sqrt{7^2+5^2+70}} = \frac{5}{12}$$

$$f_z(x,y,z) = \frac{1}{2\sqrt{x^2+y^2+z}} \quad \Rightarrow \quad f_z(7,5,70) = \frac{1}{2\sqrt{7^2+5^2+70}} = \frac{1}{24}$$

Also, $f(7,5,70) = \sqrt{7^2+5^2+70} = 12$. Substituting the values in (1) we obtain the following approximation:

$$\sqrt{7.1^2+4.9^2+70.1} \approx 12 + 0.1\cdot\frac{7}{12} - 0.1\cdot\frac{5}{12} + 0.1\cdot\frac{1}{24} = 12\frac{1}{48} \approx 12.020833$$

That is,

$$\sqrt{7.1^2+4.9^2+70.1} \approx 12.020833$$

The value obtained using a calculator is 12.021647.

27. Jason earns $S(h,c) = 20h\left(1+\frac{c}{100}\right)^{1.5}$ dollars per month at a used car lot, where h is the number of hours worked and c is the number of cars sold. He has already worked 160 hours and sold 69 cars. Right now Jason wants to go home but wonders how much more he might earn if he stays another 10 minutes with a customer who is considering buying a car. Use the linear approximation to estimate how much extra money Jason will earn if he sells his 70th car during these 10 minutes.

SOLUTION We estimate the money earned in staying for $\frac{1}{6}$ hour more and selling one more car, using the linear approximation

$$\Delta S \approx S_h(a,b)\Delta h + S_c(a,b)\Delta c \tag{1}$$

By the given information, $a = 160$, $b = 69$, $\Delta h = \frac{1}{6}$, and $\Delta c = 1$. We compute the partial derivative of the function:

$$S(h,c) = 20h\left(1+\frac{c}{100}\right)^{1.5}$$

$$S_h(h,c) = 20\left(1+\frac{c}{100}\right)^{1.5} \quad \Rightarrow \quad S_h(160,69) = 43.94$$

$$S_c(h,c) = 20h\cdot 1.5\left(1+\frac{c}{100}\right)^{0.5}\cdot\frac{1}{100} = 0.3h\left(1+\frac{c}{100}\right)^{0.5} \quad \Rightarrow \quad S_c(160,69) = 62.4$$

Substituting the values in (1), we get the following approximation:

$$\Delta S = S_h(160,69)\cdot\frac{1}{6} + S_c(160,69)\cdot 1 = 43.94\cdot\frac{1}{6} + 62.4 \approx \$69.72$$

We see that John will make approximately \$69.72 more if he sells his 70th car during 10 min.

In Exercises 28–31, compute $\frac{d}{dt}f(\mathbf{c}(t))$ at the given value of t.

29. $f(x,y,z) = xz - y^2$, $\mathbf{c}(t) = (t, t^3, 1-t)$

SOLUTION We use the Chain Rule for Paths:

$$\frac{d}{dt}f(\mathbf{c}(t)) = \nabla f_{\mathbf{c}(t)}\cdot\mathbf{c}'(t) \tag{1}$$

We compute the gradient of f:

$$\nabla f = \left\langle \frac{\partial f}{\partial x}, \frac{\partial f}{\partial y}, \frac{\partial f}{\partial z}\right\rangle = \langle z, -2y, x\rangle$$

On the path, $x = t$, $y = t^3$, and $z = 1-t$. Therefore,

$$\nabla f_{\mathbf{c}(t)} = \left\langle 1-t, -2t^3, t\right\rangle$$

Also, $\mathbf{c}'(t) = \left\langle 1, 3t^2, -1\right\rangle$, hence by (1) we obtain

$$\frac{d}{dt}f(\mathbf{c}(t)) = \left\langle 1-t, -2t^3, t\right\rangle\cdot\left\langle 1, 3t^2, -1\right\rangle = 1 - t + 3t^2\left(-2t^3\right) - t = -6t^5 - 2t + 1$$

31. $f(x,y) = \tan^{-1}\frac{y}{x}$, $\mathbf{c}(t) = (\cos t, \sin t)$, $t = \frac{\pi}{3}$

SOLUTION We use the Chain Rule for Paths. We have

$$\nabla f = \left\langle \frac{\partial f}{\partial x}, \frac{\partial f}{\partial y} \right\rangle = \left\langle \frac{-\frac{y}{x^2}}{1+\left(\frac{y}{x}\right)^2}, \frac{\frac{1}{x}}{1+\left(\frac{y}{x}\right)^2} \right\rangle = \left\langle \frac{-y}{x^2+y^2}, \frac{x}{x^2+y^2} \right\rangle$$

On the path, $x = \cos t$ and $y = \sin t$. Therefore,

$$\nabla f_{\mathbf{c}(t)} = \left\langle -\frac{\sin t}{\cos^2 t + \sin^2 t}, \frac{\cos t}{\cos^2 t + \sin^2 t} \right\rangle = \langle -\sin t, \cos t \rangle$$

$$\mathbf{c}'(t) = \langle -\sin t, \cos t \rangle$$

At the point $t = \frac{\pi}{3}$ we have

$$\nabla f_{\mathbf{c}(\frac{\pi}{3})} = \left\langle -\sin\frac{\pi}{3}, \cos\frac{\pi}{3} \right\rangle = \left\langle -\frac{\sqrt{3}}{2}, \frac{1}{2} \right\rangle \quad \text{and} \quad \mathbf{c}'\left(\frac{\pi}{3}\right) = \left(-\sin\frac{\pi}{3}, \cos\frac{\pi}{3}\right) = \left\langle -\frac{\sqrt{3}}{2}, \frac{1}{2} \right\rangle$$

Therefore,

$$\left. \frac{d}{dt} f(\mathbf{c}(t)) \right|_{t=\frac{\pi}{3}} = \nabla f_{\mathbf{c}(\frac{\pi}{3})} \cdot \mathbf{c}'\left(\frac{\pi}{3}\right) = \left\langle -\frac{\sqrt{3}}{2}, \frac{1}{2} \right\rangle \cdot \left\langle -\frac{\sqrt{3}}{2}, \frac{1}{2} \right\rangle = \frac{3}{4} + \frac{1}{4} = 1$$

In Exercises 32–35, compute the directional derivative at P in the direction of **v**.

33. $f(x, y, z) = zx - xy^2$, $P = (1, 1, 1)$, $\mathbf{v} = \langle 2, -1, 2 \rangle$

SOLUTION We first normalize **v** to obtain a unit vector **u** in the direction of **v**:

$$\mathbf{u} = \frac{\langle 2, -1, 2 \rangle}{\sqrt{2^2 + (-1)^2 + 2^2}} = \left\langle \frac{2}{3}, -\frac{1}{3}, \frac{2}{3} \right\rangle$$

We compute the directional derivative using the following equality:

$$D_{\mathbf{u}} f(1, 1, 1) = \nabla f_{(1,1,1)} \cdot \mathbf{u}$$

The gradient vector at the point $(1, 1, 1)$ is the following vector:

$$\nabla f = \langle f_x, f_y, f_z \rangle = \langle z - y^2, -2xy, x \rangle \quad \Rightarrow \quad \nabla f_{(1,1,1)} = \langle 0, -2, 1 \rangle$$

Hence,

$$D_{\mathbf{u}} f(1, 1, 1) = \langle 0, -2, 1 \rangle \cdot \left\langle \frac{2}{3}, -\frac{1}{3}, \frac{2}{3} \right\rangle = 0 + \frac{2}{3} + \frac{2}{3} = \frac{4}{3}$$

35. $f(x, y, z) = \sin(xy + z)$, $P = (0, 0, 0)$, $\mathbf{v} = \mathbf{j} + \mathbf{k}$

SOLUTION We normalize **v** to obtain a vector **u** in the direction of **v**:

$$\mathbf{u} = \frac{1}{\sqrt{0^2 + 1^2 + 1^2}} \cdot \langle 0, 1, 1 \rangle = \frac{1}{\sqrt{2}} \langle 0, 1, 1 \rangle$$

By the Theorem on Evaluating Directional Derivatives,

$$D_{\mathbf{v}} f(P) = \nabla f_P \cdot \mathbf{u} \tag{1}$$

We compute the gradient vector:

$$\nabla f = \left\langle \frac{\partial f}{\partial x}, \frac{\partial f}{\partial y}, \frac{\partial f}{\partial z} \right\rangle = \langle y\cos(xy+z), x\cos(xy+z), \cos(xy+z) \rangle$$

Hence,

$$\nabla f_P = \langle 0, 0, 1 \rangle .$$

By (1) we conclude that

$$D_{\mathbf{v}} f(P) = \nabla f_P \cdot \mathbf{u} = \langle 0, 0, 1 \rangle \cdot \frac{1}{\sqrt{2}} \langle 0, 1, 1 \rangle = \frac{1}{\sqrt{2}}.$$

37. Find an equation of the tangent plane at $P = (0, 3, -1)$ to the surface with equation

$$ze^x + e^{z+1} = xy + y - 3$$

SOLUTION The surface is defined implicitly by the equation

$$F(x, y, z) = ze^x + e^{z+1} - xy - y + 3$$

The tangent plane to the surface at the point $(0, 3, -1)$ has the following equation:

$$0 = F_x(0, 3, -1)x + F_y(0, 3, -1)(y - 3) + F_z(0, 3, -1)(z + 1) \quad (1)$$

We compute the partial derivatives at the given point:

$$F_x(x, y, z) = ze^x - y \quad \Rightarrow \quad F_x(0, 3, -1) = -1e^0 - 3 = -4$$

$$F_y(x, y, z) = -x - 1 \quad \Rightarrow \quad F_y(0, 3, -1) = -0 - 1 = -1$$

$$F_z(x, y, z) = e^x + e^{z+1} \quad \Rightarrow \quad F_z(0, 3, -1) = e^0 + e^{-1+1} = 2$$

Substituting in (1) we obtain the following equation:

$$-4x - (y - 3) + 2(z + 1) = 0$$

$$-4x - y + 2z + 5 = 0$$

$$2z = 4x + y - 5 \quad \Rightarrow \quad z = 2x + 0.5y - 2.5$$

39. Let $f(x, y) = (x - y)e^x$. Use the Chain Rule to calculate $\frac{\partial f}{\partial u}$ and $\frac{\partial f}{\partial v}$, where $x = u - v$ and $y = u + v$.

SOLUTION First we calculate the Primary Derivatives:

$$\frac{\partial f}{\partial x} = e^x(x - y) + e^x = e^x(x - y + 1), \quad \frac{\partial f}{\partial y} = -e^x$$

Since $\frac{\partial x}{\partial u} = 1$, $\frac{\partial y}{\partial u} = 1$, $\frac{\partial x}{\partial v} = -1$, and $\frac{\partial y}{\partial v} = 1$, the Chain Rule gives

$$\frac{\partial f}{\partial u} = \frac{\partial f}{\partial x}\frac{\partial x}{\partial u} + \frac{\partial f}{\partial y}\frac{\partial y}{\partial u} = e^x(x - y + 1) \cdot 1 - e^x \cdot 1 = e^x(x - y + 1 - 1) = e^x(x - y)$$

$$\frac{\partial f}{\partial v} = \frac{\partial f}{\partial x}\frac{\partial x}{\partial v} + \frac{\partial f}{\partial y}\frac{\partial y}{\partial v} = e^x(x - y + 1) \cdot (-1) - e^x \cdot 1 = e^x(y - x - 2)$$

We now substitute $x = u - v$ and $y = u + v$ to express the partial derivatives in terms of u and v. We get

$$\frac{\partial f}{\partial u} = e^{u-v}(u - v - u - v) = -2ve^{u-v}$$

$$\frac{\partial f}{\partial v} = e^{u-v}(u + v - u + v - 2) = 2e^{u-v}(v - 1)$$

41. Express the partial derivatives $\frac{\partial f}{\partial r}$ and $\frac{\partial f}{\partial \theta}$ of a function $f(x, y, z)$ in terms of $\frac{\partial f}{\partial x}$, $\frac{\partial f}{\partial y}$, and $\frac{\partial f}{\partial z}$, where (r, θ, z) are cylindrical coordinates.

SOLUTION The cylinderical coordinates are

$$x = r\cos\theta, \quad y = r\sin\theta, \quad z = z \quad (1)$$

By the Chain Rule, we have

$$\frac{\partial f}{\partial r} = \frac{\partial f}{\partial x}\frac{\partial x}{\partial r} + \frac{\partial f}{\partial y}\frac{\partial y}{\partial r} + \frac{\partial f}{\partial z}\frac{\partial z}{\partial r}$$

$$\frac{\partial f}{\partial \theta} = \frac{\partial f}{\partial x}\frac{\partial x}{\partial \theta} + \frac{\partial f}{\partial y}\frac{\partial y}{\partial \theta} + \frac{\partial f}{\partial z}\frac{\partial z}{\partial \theta}$$

We use the relation in (1) to compute the following partial derivatives:

$$\frac{\partial x}{\partial r} = \cos\theta, \quad \frac{\partial y}{\partial r} = \sin\theta, \quad \frac{\partial z}{\partial r} = 0$$

$$\frac{\partial x}{\partial \theta} = -r\sin\theta, \quad \frac{\partial y}{\partial \theta} = r\cos\theta, \quad \frac{\partial z}{\partial \theta} = 0$$

$$\frac{\partial x}{\partial z} = 0, \quad \frac{\partial y}{\partial z} = 0, \quad \frac{\partial z}{\partial z} = 1$$

Substituting these derivatives in the Chain Rule gives

$$\frac{\partial f}{\partial r} = \frac{\partial f}{\partial x}\cos\theta + \frac{\partial f}{\partial y}\sin\theta + \frac{\partial f}{\partial z}\cdot 0 = \frac{\partial f}{\partial x}\cos\theta + \frac{\partial f}{\partial y}\sin\theta$$

$$\frac{\partial f}{\partial \theta} = \frac{\partial f}{\partial x}(-r\sin\theta) + \frac{\partial f}{\partial y}r\cos\theta + \frac{\partial f}{\partial z}\cdot 0 = -\frac{\partial f}{\partial x}r\sin\theta + \frac{\partial f}{\partial y}r\cos\theta$$

43. Let $g(u, v) = f(u^3 - v^3, v^3 - u^3)$. Prove that

$$v^2\frac{\partial g}{\partial u} - u^2\frac{\partial g}{\partial v} = 0$$

SOLUTION We are given the function $f(x, y)$, where $x = u^3 - v^3$ and $y = v^3 - u^3$. Using the Chain Rule we have the following derivatives:

$$\frac{\partial g}{\partial u} = \frac{\partial f}{\partial x}\frac{\partial x}{\partial u} + \frac{\partial f}{\partial y}\frac{\partial y}{\partial u}$$

$$\frac{\partial g}{\partial v} = \frac{\partial f}{\partial x}\frac{\partial x}{\partial v} + \frac{\partial f}{\partial y}\frac{\partial y}{\partial v} \qquad (1)$$

We compute the following partial derivatives:

$$\frac{\partial x}{\partial u} = 3u^2, \quad \frac{\partial y}{\partial u} = -3u^2$$

$$\frac{\partial x}{\partial v} = -3v^2, \quad \frac{\partial y}{\partial v} = 3v^2$$

Substituting in (1) we obtain

$$\frac{\partial g}{\partial u} = \frac{\partial f}{\partial x}\cdot 3u^2 + \frac{\partial f}{\partial y}\left(-3u^2\right) = 3u^2\left(\frac{\partial f}{\partial x} - \frac{\partial f}{\partial y}\right)$$

$$\frac{\partial g}{\partial v} = \frac{\partial f}{\partial x}\left(-3v^2\right) + \frac{\partial f}{\partial y}\left(3v^2\right) = -3v^2\left(\frac{\partial f}{\partial x} - \frac{\partial f}{\partial y}\right)$$

Therefore,

$$v^2\frac{\partial g}{\partial u} + u^2\frac{\partial g}{\partial v} = 3u^2v^2\left(\frac{\partial f}{\partial x} - \frac{\partial f}{\partial y}\right) - 3u^2v^2\left(\frac{\partial f}{\partial x} - \frac{\partial f}{\partial y}\right) = 0$$

45. Calculate $\frac{\partial z}{\partial x}$, where $xe^z + ze^y = x + y$.

SOLUTION The function $F(x, y, z) = xe^z + ze^y - x - y = 0$ defines z implicitly as a function of x and y. Using implicit differentiation, the partial derivative of z with respect to x is

$$\frac{\partial z}{\partial x} = -\frac{F_x}{F_z} \qquad (1)$$

We compute the partial derivatives F_x and F_z:

$$F_x = e^z - 1$$

$$F_z = xe^z + e^y$$

Substituting in (1) gives

$$\frac{\partial z}{\partial x} = -\frac{e^z - 1}{xe^z + e^y}.$$

In Exercises 47–50, find the critical points of the function and analyze them using the Second Derivative Test.

47. $f(x, y) = x^2 + 2y^2 - 4xy + 6x$

SOLUTION To find the critical points of the function $f(x, y) = x^2 + 2y^2 - 4xy + 6x$, we set the partial derivatives equal to zero and solve. This gives

$$f_x(x, y) = 2x - 4y + 6 = 0$$

$$f_y(x, y) = 4y - 4x = 0$$

By the second equation $y = x$. Substituting in the first equation gives

$$2x - 4x + 6 = 0$$

$$-2x = -6 \quad \Rightarrow \quad x = 3$$

There is one critical point $P = (3, 3)$. We now apply the Second Derivative Test to examine the critical point. We first find the second-order partials:

$$f_{xx}(x, y) = 2, \quad f_{yy}(x, y) = 4, \quad f_{xy}(x, y) = -4$$

Since $D = f_{xx} f_{yy} - f_{xy}^2 = 2 \cdot 4 - 16 = -8 < 0$, the point $(3, 3)$ is a saddle point.

49. $f(x, y) = e^{x+y} - xe^{2y}$

SOLUTION We find the critical point by setting the partial derivatives of $f(x, y) = e^{x+y} - xe^{2y}$ equal to zero and solve. This gives

$$f_x(x, y) = e^{x+y} - e^{2y} = 0$$

$$f_y(x, y) = e^{x+y} - 2xe^{2y} = 0$$

The first equation gives $e^{x+y} = e^{2y}$ and the second equation gives $e^{x+y} = 2xe^{2y}$. Equating the two expressions, dividing by the nonzero function e^{2y}, and solving for x, we obtain

$$e^{2y} = 2xe^{2y} \quad \Rightarrow \quad 1 = 2x \quad \Rightarrow \quad x = \frac{1}{2}$$

We now substitute $x = \frac{1}{2}$ in the first equation and solve for y, to obtain

$$e^{\frac{1}{2}+y} - e^{2y} = 0 \quad \Rightarrow \quad e^{\frac{1}{2}+y} = e^{2y} \quad \Rightarrow \quad \frac{1}{2} + y = 2y \quad \Rightarrow \quad y = \frac{1}{2}$$

There is one critical point, $\left(\frac{1}{2}, \frac{1}{2}\right)$. We examine the critical point using the Second Derivative Test. We compute the second derivatives at this point:

$$f_{xx}(x, y) = e^{x+y} \quad \Rightarrow \quad f_{xx}\left(\frac{1}{2}, \frac{1}{2}\right) = e^{\frac{1}{2}+\frac{1}{2}} = e$$

$$f_{yy}(x, y) = e^{x+y} - 4xe^{2y} \quad \Rightarrow \quad f_{yy}\left(\frac{1}{2}, \frac{1}{2}\right) = e^{\frac{1}{2}+\frac{1}{2}} - 4 \cdot \frac{1}{2} e^{2 \cdot \frac{1}{2}} = -e$$

$$f_{xy}(x, y) = e^{x+y} - 2e^{2y} \quad \Rightarrow \quad f_{xy}\left(\frac{1}{2}, \frac{1}{2}\right) = e^{\frac{1}{2}+\frac{1}{2}} - 2e^{2 \cdot \frac{1}{2}} = -e$$

Therefore the discriminant at the critical point is

$$D\left(\frac{1}{2}, \frac{1}{2}\right) = f_{xx} f_{yy} - f_{xy}^2 = e \cdot (-e) - (-e)^2 = -2e^2 < 0$$

We conclude that $\left(\frac{1}{2}, \frac{1}{2}\right)$ is a saddle point.

51. Prove that $f(x, y) = (x + 2y)e^{xy}$ has no critical points.

SOLUTION We find the critical points by setting the partial derivatives of $f(x, y) = (x + 2y)e^{xy}$ equal to zero and solving. We get

$$f_x(x, y) = e^{xy} + (x + 2y)ye^{xy} = e^{xy}\left(1 + xy + 2y^2\right) = 0$$

$$f_y(x, y) = 2e^{xy} + (x + 2y)xe^{xy} = e^{xy}\left(2 + x^2 + 2xy\right) = 0$$

We divide the two equations by the nonzero expression e^{xy} to obtain the following equations:

$$1 + xy + 2y^2 = 0$$

$$2 + 2xy + x^2 = 0$$

The first equation implies that $xy = -1 - 2y^2$. Substituting in the second equation gives

$$2 + 2\left(-1 - 2y^2\right) + x^2 = 0$$
$$2 - 2 - 4y^2 + x^2 = 0$$
$$x^2 = 4y^2 \quad \Rightarrow \quad x = 2y \quad \text{or} \quad x = -2y$$

We substitute in the first equation and solve for y:

$x = 2y$	$x = -2y$
$1 + 2y^2 + 2y^2 = 0$	$1 - 2y^2 + 2y^2 = 0$
$1 + 4y^2 = 0$	$1 = 0$
$y^2 = -\frac{1}{4}$	

In both cases there is no solution. We conclude that there are no solutions for $f_x = 0$ and $f_y = 0$, that is, there are no critical points.

53. Find the global extrema of $f(x, y) = 2xy - x - y$ on the domain $\{y \le 4, y \ge x^2\}$.

SOLUTION The region is shown in the figure.

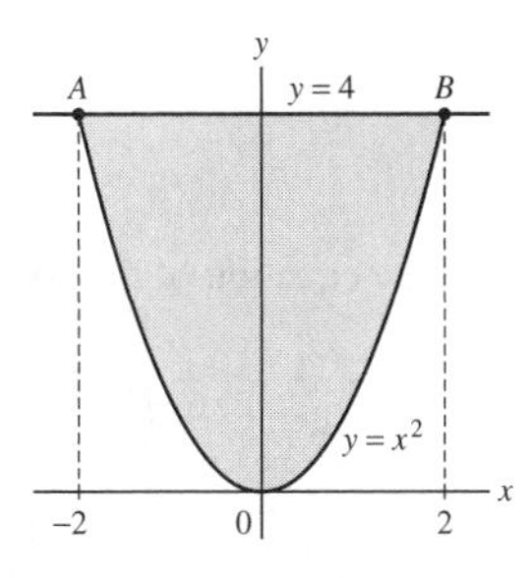

Step 1. Finding the critical points. We find the critical points in the interior of the domain by setting the partial derivatives equal to zero and solving. We get

$$f_x = 2y - 1 = 0$$
$$f_y = 2x - 1 = 0 \quad \Rightarrow \quad x = \frac{1}{2}, \quad y = \frac{1}{2}$$

The critical point is $\left(\frac{1}{2}, \frac{1}{2}\right)$. (It lies in the interior of the domain since $\frac{1}{2} < 4$ and $\frac{1}{2} > \left(\frac{1}{2}\right)^2$).

Step 2. Finding the global extrema on the boundary. We consider the two parts of the boundary separately.

The parabola $y = x^2$, $-2 \le x \le 2$:

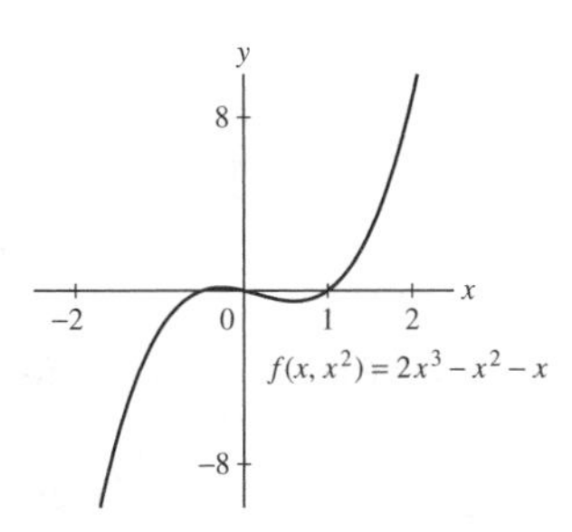

On this curve, $f(x, x^2) = 2 \cdot x \cdot x^2 - x - x^2 = 2x^3 - x^2 - x$. Using calculus in one variable or the graph of the function, we see that the minimum of $f(x, x^2)$ on the interval occurs at $x = -2$ and the maximum at $x = 2$. The corresponding points are $(-2, 4)$ and $(2, 4)$.

The segment $\overline{AB}$: On this segment $y = 4$, $-2 \le x \le 2$, hence $f(x, 4) = 2 \cdot x \cdot 4 - x - 4 = 7x - 4$. The maximum value occurs at $x = 2$ and the minimum value at $x = -2$. The corresponding points on the segment $\overline{AB}$ are $(-2, 4)$ and $(2, 4)$

Step 3. Conclusions. Since the global extrema occur either at critical points in the interior of the domain or on the boundary of the domain, the candidates for global extrema are the following points:

$$\left(\frac{1}{2}, \frac{1}{2}\right), \quad (-2, 4), \quad (2, 4)$$

We compute the values of $f = 2xy - x - y$ at these points:

$$f\left(\frac{1}{2}, \frac{1}{2}\right) = 2 \cdot \frac{1}{2} \cdot \frac{1}{2} - \frac{1}{2} - \frac{1}{2} = -\frac{1}{2}$$
$$f(-2, 4) = 2 \cdot (-2) \cdot 4 + 2 - 4 = -18$$
$$f(2, 4) = 2 \cdot 2 \cdot 4 - 2 - 4 = 10$$

We conclude that the global maximum is $f(2, 4) = 10$ and the global minimum is $f(-2, 4) = -18$.

55. Use Lagrange multipliers to find the minimum and maximum value of $f(x, y) = 3x - 2y$ on the circle $x^2 + y^2 = 4$.

SOLUTION

Step 1. Write out the Lagrange Equations. The constraint curve is $g(x, y) = x^2 + y^2 - 4 = 0$, hence $\nabla g = \langle 2x, 2y\rangle$ and $\nabla f = \langle 3, -2\rangle$. The Lagrange Condition $\nabla f = \lambda \nabla g$ is thus $\langle 3, -2\rangle = \lambda \langle 2x, 2y\rangle$. That is,

$$3 = \lambda \cdot 2x$$
$$-2 = \lambda \cdot 2y$$

Note that $\lambda \neq 0$.

Step 2. Solve for x and y using the constraint. The Lagrange equations gives

$$\begin{aligned} 3 &= \lambda \cdot 2x \\ -2 &= \lambda \cdot 2y \end{aligned} \quad \Rightarrow \quad \begin{aligned} x &= \frac{3}{2\lambda} \\ y &= -\frac{1}{\lambda} \end{aligned} \tag{1}$$

We substitute x and y in the equation of the constraint and solve for λ. We get

$$\left(\frac{3}{2\lambda}\right)^2 + \left(-\frac{1}{\lambda}\right)^2 = 4$$
$$\frac{9}{4\lambda^2} + \frac{1}{\lambda^2} = 4$$
$$\frac{1}{\lambda^2} \cdot \frac{13}{4} = 4 \quad \Rightarrow \quad \lambda = \frac{\sqrt{13}}{4} \quad \text{or} \quad \lambda = -\frac{\sqrt{13}}{4}$$

Substituting in (1), we obtain the points

$$x = \frac{6}{\sqrt{13}}, \quad y = -\frac{4}{\sqrt{13}}$$
$$x = -\frac{6}{\sqrt{13}}, \quad y = \frac{4}{\sqrt{13}}$$

The critical points are thus

$$P_1 = \left(\frac{6}{\sqrt{13}}, -\frac{4}{\sqrt{13}}\right)$$
$$P_2 = \left(-\frac{6}{\sqrt{13}}, \frac{4}{\sqrt{13}}\right)$$

Step 3. Calculate the value at the critical points. We find the value of $f(x, y) = 3x - 2y$ at the critical points:

$$f(P_1) = 3 \cdot \frac{6}{\sqrt{13}} - 2 \cdot \frac{-4}{\sqrt{13}} = \frac{26}{\sqrt{13}}$$
$$f(P_2) = 3 \cdot \frac{-6}{\sqrt{13}} - 2 \cdot \frac{4}{\sqrt{13}} = \frac{-26}{\sqrt{13}}$$

Thus, the maximum value of f on the circle is $\frac{26}{\sqrt{13}}$, and the minimum is $-\frac{26}{\sqrt{13}}$.

57. Find the minimum and maximum values of $f(x, y) = x^2y$ on the ellipse $4x^2 + 9y^2 = 36$.

SOLUTION We must find the minimum and maximum values of $f(x, y) = x^2y$ subject to the constraint $g(x, y) = 4x^2 + 9y^2 - 36 = 0$.

Step 1. Write out the Lagrange Equations. The gradient vectors are $\nabla f = \left\langle 2xy, x^2 \right\rangle$ and $\nabla g = \langle 8x, 18y\rangle$, hence the Lagrange Condition $\nabla f = \lambda \nabla g$ gives

$$\left\langle 2xy, x^2 \right\rangle = \lambda \langle 8x, 18y\rangle = \langle 8\lambda x, 18\lambda y\rangle$$

We obtain the following Lagrange Equations:

$$2xy = 8\lambda x$$
$$x^2 = 18\lambda y$$

Step 2. Solve for λ in terms of x and y. If $x = 0$, the equation of the constraint implies that $y = \pm 2$. The points $(0, 2)$ and $(0, -2)$ satisfy the Lagrange Equations for $\lambda = 0$. If $x \neq 0$, the second Lagrange Equation implies that $y \neq 0$. Therefore the Lagrange Equations give

$$2xy = 8\lambda x \quad \Rightarrow \quad \lambda = \frac{y}{4}$$

$$x^2 = 18\lambda y \quad \Rightarrow \quad \lambda = \frac{x^2}{18y}$$

Step 3. Solve for x and y using the constraint. We equate the two expressions for λ to obtain

$$\frac{y}{4} = \frac{x^2}{18y}$$
$$18y^2 = 4x^2$$

We now substitute $4x^2 = 18y^2$ in the equation of the constraint $4x^2 + 9y^2 = 36$ and solve for y. This gives

$$\begin{aligned} 18y^2 + 9y^2 &= 36 \\ 27y^2 &= 36 \end{aligned} \quad \Rightarrow \quad y^2 = \frac{36}{27} \quad \Rightarrow \quad y_1 = \frac{2}{\sqrt{3}}, \quad y_2 = -\frac{2}{\sqrt{3}}$$

We find the x-coordinates using $x^2 = \frac{9y^2}{2}$:

$$x^2 = \frac{9y^2}{2}$$

$$x^2 = \frac{9}{2} \cdot \frac{4}{3} = 6 \quad \Rightarrow \quad x_1 = \sqrt{6}, \quad x_2 = -\sqrt{6}$$

We obtain the following critical points:

$$P_1 = (0, 2), \quad P_2 = (0, -2), \quad P_3 = \left(\sqrt{6}, \frac{2}{\sqrt{3}}\right)$$

$$P_4 = \left(\sqrt{6}, -\frac{2}{\sqrt{3}}\right), \quad P_5 = \left(-\sqrt{6}, \frac{2}{\sqrt{3}}\right), \quad P_6 = \left(-\sqrt{6}, -\frac{2}{\sqrt{3}}\right)$$

Step 4. Conclusions. We evaluate the function $f(x, y) = x^2 y$ at the critical points:

$$f(P_1) = 0^2 \cdot 2 = 0$$

$$f(P_2) = 0^2 \cdot (-2) = 0$$

$$f(P_3) = f(P_5) = 6 \cdot \frac{2}{\sqrt{3}} = \frac{12}{\sqrt{3}}$$

$$f(P_4) = f(P_5) = 6 \cdot \left(-\frac{2}{\sqrt{3}}\right) = -\frac{12}{\sqrt{3}}$$

Since the min and max of f occur on the ellipse, it must occur at critical points. Thus, we conclude that the maximum and minimum of f subject to the constraint are $\frac{12}{\sqrt{3}}$ and $-\frac{12}{\sqrt{3}}$ respectively.

59. Find the extreme values of $f(x, y, z) = x + 2y + 3z$ subject to the two constraints $x + y + z = 1$ and $x^2 + y^2 + z^2 = 1$.

SOLUTION We must find the extreme values of $f(x, y, z) = x + 2y + 3z$ subject to the constraints $g(x, y, z) = x + y + z - 1 = 0$ and $h(x, y, z) = x^2 + y^2 + z^2 - 1 = 0$.

Step 1. Write out the Lagrange Equations. We have $\nabla f =< 1, 2, 3 >$, $\nabla g =< 1, 1, 1 >$, $\nabla h =< 2x, 2y, 2z >$, hence the Lagrange condition $\nabla f = \lambda \nabla g + \mu \nabla h$ gives

$$< 1, 2, 3 > = \lambda < 1, 1, 1 > +\mu < 2x, 2y, 2z > =< \lambda + 2\mu x, \lambda + 2\mu y, \lambda + 2\mu z >$$

or

$$\begin{aligned} 1 &= \lambda + 2\mu x \\ 2 &= \lambda + 2\mu y \\ 3 &= \lambda + 2\mu z \end{aligned}$$

Step 2. Solve for λ and μ. The Lagrange Equations give

$$\begin{aligned} 1 &= \lambda + 2\mu x & & \lambda=1 - 2\mu x \\ 2 &= \lambda + 2\mu y & \Rightarrow \quad & \lambda=2 - 2\mu y \\ 3 &= \lambda + 2\mu z & & \lambda=3 - 2\mu z \end{aligned}$$

Equating the three expressions for λ, we get the following equations:

$$\begin{aligned} 1 - 2\mu x &= 2 - 2\mu y & & 2\mu(y - x) = 1 \\ & & \Rightarrow & \\ 1 - 2\mu x &= 3 - 2\mu z & & \mu(z - x) = 2 \end{aligned}$$

The first equation implies that $\mu = \frac{1}{2(y-x)}$, and the second implies that $\mu = \frac{2}{z-x}$. Equating the two expressions for μ, we get

$$\frac{1}{2(y-x)} = \frac{2}{z-x}$$

$$z - x = 4y - 4x \quad \Rightarrow \quad z = 4y - 3x$$

Step 3. Solve for x, y, and z using the constraints. We substitute $z = 4y - 3x$ in the equations of the constraints and solve to find x and y. This gives

$$\begin{aligned} x + y + (4y - 3x) &= 1 & & y = \frac{1 + 2x}{5} \\ & & \Rightarrow & \\ x^2 + y^2 + (4y - 3x)^2 &= 1 & & 10x^2 + 17y^2 - 24xy = 1 \end{aligned}$$

Substituting in the second equation and solving for x, we get

$$y = \frac{1 + 2x}{5}$$

$$10x^2 + 17\left(\frac{1 + 2x}{5}\right)^2 - 24x \cdot \frac{1 + 2x}{5} = 1$$

$$250x^2 + 17(1 + 2x)^2 - 120x\,(1 + 2x) = 25$$

$$39x^2 - 26x - 4 = 0$$

$$x_{1,2} = \frac{26 \pm \sqrt{1300}}{78}$$

$$\Rightarrow x_1 = \frac{1}{3} + \frac{5\sqrt{13}}{39} \approx 0.8, \quad x_2 = \frac{1}{3} - \frac{5\sqrt{13}}{39} \approx -0.13$$

We find the y-coordinates using $y = \frac{1+2x}{5}$.

$$y_1 = \frac{1 + 2 \cdot 0.8}{5} = 0.52, \quad y_2 = \frac{1 - 2 \cdot 0.13}{5} = 0.15$$

Finally, we find the z-coordinate using $z = 4y - 3x$:

$$z_1 = 4 \cdot 0.52 - 3 \cdot 0.8 = -0.32, \quad z_2 = 4 \cdot 0.15 + 3 \cdot 0.13 = 0.99$$

We obtain the critical points:

$$P_1 = (0.8, 0.52, -0.32), \quad P_2 = (-0.13, 0.15, 0.99)$$

.

Step 4. Conclusions. We evaluate the function $f(x, y, z) = x + 2y + 3z$ at the critical points:

$$f(P_1) = 0.8 + 2 \cdot 0.52 - 3 \cdot 0.32 = 0.88$$
$$f(P_2) = -0.13 + 2 \cdot 0.15 + 3 \cdot 0.99 = 3.14 \qquad (1)$$

The two constraints determine the common points of the unit sphere $x^2 + y^2 + z^2 = 1$ and the plane $x + y + z = 1$. This set is a circle that is a closed and bounded set in R^3. Therefore, f has a minimum and maximum values on this set. These extrema are given in (1).

61. Find the dimensions of the box of maximum volume with its sides parallel to the coordinate planes that can be inscribed in the ellipsoid (Figure 4)

$$\left(\frac{x}{a}\right)^2 + \left(\frac{y}{b}\right)^2 + \left(\frac{z}{c}\right)^2 = 1$$

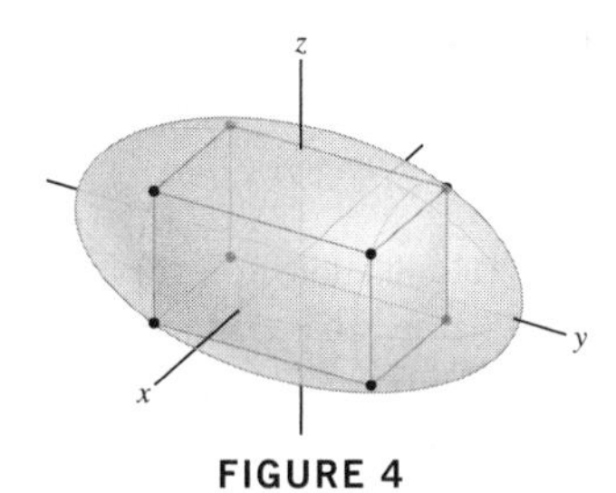

FIGURE 4

SOLUTION We denote the vertices of the box by $(\pm x, \pm y, \pm z)$, where $x \geq 0$, $y \geq 0$, $z \geq 0$. The volume of the box is

$$V(x, y, z) = 8xyz$$

The vertices of the box must satisfy the equation of the ellipsoid, hence,

$$g(x, y, z) = \frac{x^2}{a^2} + \frac{y^2}{b^2} + \frac{z^2}{c^2} - 1 = 0, \quad x \geq 0, \quad y \geq 0, \quad z \geq 0.$$

We need to maximize V due to the constraint: $g(x, y, z) = 0$, $x \geq 0$, $y \geq 0$, $z \geq 0$.

Step 1. Write out the Lagrange Equations. We have $\nabla V = 8 \langle yz, xz, xy \rangle$ and $\nabla g = \left\langle \frac{2x}{a^2}, \frac{2y}{b^2}, \frac{2z}{c^2} \right\rangle$, hence the Lagrange Condition $\nabla V = \lambda \nabla g$ gives the following equations:

$$yz = \lambda \frac{2x}{a^2}$$
$$xz = \lambda \frac{2y}{b^2}$$
$$xy = \lambda \frac{2z}{c^2}$$

Step 2. Solve for λ in terms of x, y, and z. If $x = 0$, $y = 0$, or $z = 0$, the volume of the box has the minimum value zero. We thus may assume that $x \neq 0$, $y \neq 0$, and $z \neq 0$. The Lagrange equations give

$$\lambda = \frac{a^2 yz}{2x}, \quad \lambda = \frac{b^2 xz}{2y}, \quad \lambda = \frac{c^2 xy}{2z}$$

Step 3. Solve for x, y, and z using the constraint. Equating the three expressions for λ yields the following equations:

$$\begin{aligned} \frac{a^2}{2}\frac{yz}{x} &= \frac{c^2}{2}\frac{xy}{z} \\ \frac{b^2}{2}\frac{xz}{y} &= \frac{c^2}{2}\frac{xy}{z} \end{aligned} \quad \Rightarrow \quad \begin{aligned} y\left(c^2x^2 - a^2z^2\right) &= 0 \\ x\left(c^2y^2 - b^2z^2\right) &= 0 \end{aligned}$$

Since $x > 0$ and $y > 0$, these equations imply that

$$\begin{aligned} c^2x^2 - a^2z^2 &= 0 \\ c^2y^2 - b^2z^2 &= 0 \end{aligned} \quad \Rightarrow \quad \begin{aligned} x &= \frac{az}{c} \\ y &= \frac{bz}{c} \end{aligned} \qquad (1)$$

We now substitute x and y in the equation of the constraint and solve for z. This gives

$$\frac{\left(\frac{az}{c}\right)^2}{a^2} + \frac{\left(\frac{bz}{c}\right)^2}{b^2} + \frac{z^2}{c^2} = 1$$

$$\frac{z^2}{c^2} + \frac{z^2}{c^2} + \frac{z^2}{c^2} = 1$$

$$\frac{3z^2}{c^2} = 1 \quad \Rightarrow \quad z = \frac{c}{\sqrt{3}}$$

We find x and y using (1):

$$x = \frac{a}{c}\frac{c}{\sqrt{3}} = \frac{a}{\sqrt{3}}, \quad y = \frac{b}{c}\frac{c}{\sqrt{3}} = \frac{b}{\sqrt{3}}$$

We obtain the critical point:

$$P = \left(\frac{a}{\sqrt{3}}, \frac{b}{\sqrt{3}}, \frac{c}{\sqrt{3}}\right)$$

Step 4. Conclusions. The function $V = 8xyz$ is a polynomial, hence it is continuous. The constraint defines a closed and compact set in R^3, hence f has extreme values on the constraint. The maximum value is obtained at the critical point P. We find it:

$$V(P) = 8\frac{a}{\sqrt{3}} \cdot \frac{b}{\sqrt{3}} \cdot \frac{c}{\sqrt{3}} = 8\frac{abc}{3\sqrt{3}}$$

We conclude that the dimensions of the box of maximum volume with sides parallel to the coordinate planes, which can be inscribed in the ellipsoid, are

$$x = \frac{a}{\sqrt{3}}, \quad y = \frac{b}{\sqrt{3}}, \quad z = \frac{c}{\sqrt{3}}.$$

63. A bead hangs on a string of length ℓ whose ends are fixed by thumbtacks located at points $(0, 0)$ and (a, b) on a bulletin board (Figure 5). The bead rests in the position that minimizes its height y. Use Lagrange multipliers to show that the two sides of the string make equal angles with the horizontal, that is, show that $\theta_1 = \theta_2$.

As an aside, note that the locus of the bead when pulled taut is an ellipse with foci at O and P, and the tangent line at the lowest point is horizontal. Therefore, this exercise provides another proof of the reflective property of the ellipse (a light ray emanating from one focus and bouncing off the ellipse is reflected to the other focus).

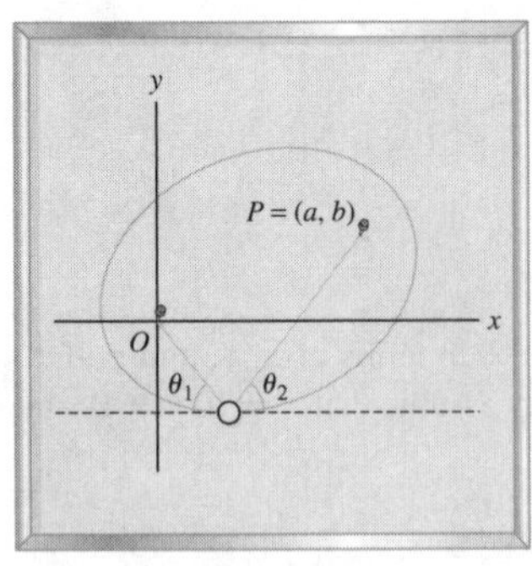

FIGURE 5

SOLUTION We denote by l_1 and l_2 the lengths shown in the figure.

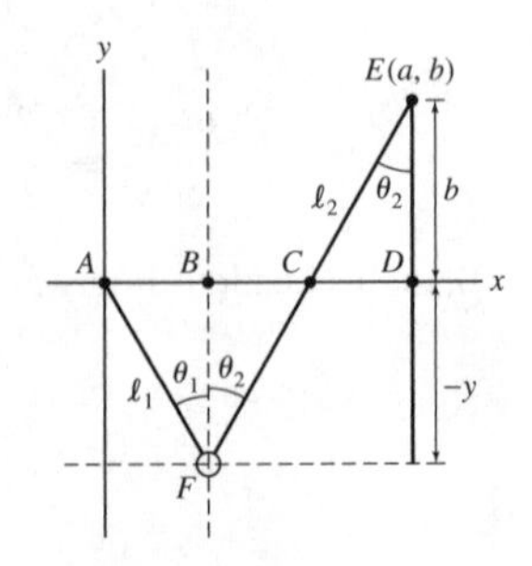

Hence,

$$l = l_1 + l_2 \tag{1}$$

We find l_1 and l_2 in terms of y, θ_1, and θ_2. Since $y < 0$, we have

$$\frac{-y}{l_1} = \cos\theta_1 \quad \Rightarrow \quad l_1 = -\frac{y}{\cos\theta_1}$$

$$\frac{b-y}{l_2} = \cos\theta_2 \quad \Rightarrow \quad l_2 = \frac{b-y}{\cos\theta_2}$$

Substituting in (1), we obtain

$$l = \frac{-y}{\cos\theta_1} + \frac{b-y}{\cos\theta_2} \tag{2}$$

Rewriting the above equation, we obtain our first constraint equation:

$$l = -y\sec\theta_1 + (b-y)\sec\theta_2 \tag{3}$$

For our second constraint, we note that $\tan\theta_1 = AB/(-y)$ and $\tan\theta_2 = BD/(b-y)$. Since $AB + BD = a$, we have

$$a = -y\tan\theta_1 + (b-y)\tan\theta_2 \tag{4}$$

Our objective is to minimize $f(y, \theta_1, \theta_2) = -y$ with the two constraint equations

$$g_1(y, \theta_1, \theta_2) = -y\sec\theta_1 + (b-y)\sec\theta_2$$

$$g_2(y, \theta_1, \theta_2) = -y\tan\theta_1 + (b-y)\tan\theta_2$$

We use the equation $\nabla f(y, \theta_1, \theta_2) = \lambda_1 \nabla g_1(y, \theta_1, \theta_2) + \lambda_2 \nabla g_2(y, \theta_1, \theta_2)$, which becomes

$$\langle 1, 0, 0\rangle = \lambda_1\langle -\sec\theta_1 - \sec\theta_2, -y\sec\theta_1\tan\theta_1, (b-y)\sec\theta_2\tan\theta_2\rangle$$
$$+ \lambda_2\langle -\tan\theta_1 - \tan\theta_2, -y\sec^2\theta_1, (b-y)\sec^2\theta_2\rangle$$

Note that λ_1 and λ_2 can't both be zero. If $\lambda_1 = 0$, then we get $0 = \lambda_2(-y\sec^2\theta_2)$, which doesn't happen. A similar argument holds for λ_2. Thus, we can assume that both λ_1 and λ_2 are nonzero. Looking at the second coordinate of the above equation, we get $\lambda_1(y\sec\theta_1\tan\theta_1) = \lambda_2(-y\sec^2\theta_1)$, which means $\lambda_1/\lambda_2 = -\sec\theta_1/\tan\theta_1 = -1/\sin\theta_1$. A similar analysis of the third coordinate gives $\lambda_1/\lambda_2 = -\sec\theta_2/\tan\theta_2 = -1/\sin\theta_2$. We conclude that $\sin\theta_1 = \sin\theta_2$, and hence $\theta_1 = \theta_2$, as desired.

16 MULTIPLE INTEGRATION

16.1 Integration in Several Variables (ET Section 15.1)

Preliminary Questions

1. In the Riemann sum $S_{8,4}$ for a double integral over $\mathcal{R} = [1, 5] \times [2, 10]$, what is the area of each subrectangle and how many subrectangles are there?

SOLUTION Each subrectangle has sides of length

$$\Delta x = \frac{5-1}{8} = \frac{1}{2}, \quad \Delta y = \frac{10-2}{4} = 2$$

Therefore the area of each subrectangle is $\Delta A = \Delta x \Delta y = \frac{1}{2} \cdot 2 = 1$, and the number of subrectangles is $8 \cdot 4 = 32$.

2. Estimate the double integral of a continuous function f over the small rectangle $\mathcal{R} = [0.9, 1.1] \times [1.9, 2.1]$ if $f(1, 2) = 4$.

SOLUTION Since we are given the value of f at one point in $\mathcal{R}$ only, we can only use the approximation S_{11} for the integral of f over $\mathcal{R}$. For S_{11} we have one rectangle with sides

$$\Delta x = 1.1 - 0.9 = 0.2, \quad \Delta y = 2.1 - 1.9 = 0.2$$

Hence, the area of the rectangle is $\Delta A = \Delta x \Delta y = 0.2 \cdot 0.2 = 0.04$. We obtain the following approximation:

$$\iint_{\mathcal{R}} f\,dA \approx S_{1,1} = f(1, 2)\Delta A = 4 \cdot 0.04 = 0.16$$

3. What is the integral of the constant function $f(x, y) = 5$ over the rectangle $[-2, 3] \times [2, 4]$?

SOLUTION The integral of f over the unit square $\mathcal{R} = [-2, 3] \times [2, 4]$ is the volume of the box of base $\mathcal{R}$ and height 5. That is,

$$\iint_{\mathcal{R}} 5\,dA = 5 \cdot \text{Area}(\mathcal{R}) = 5 \cdot 5 \cdot 2 = 50$$

4. What is the interpretation of $\iint_{\mathcal{R}} f(x, y)\,dA$ if $f(x, y)$ takes on both positive and negative values on $\mathcal{R}$?

SOLUTION The double integral $\iint_{\mathcal{R}} f(x, y)\,dA$ is the signed volume between the graph $z = f(x, y)$ for $(x, y) \in \mathcal{R}$, and the xy-plane. The region below the xy-plane is treated as negative volume.

5. Which of (a) or (b) is equal to $\int_1^2 \int_4^5 f(x, y)\,dy\,dx$?

(a) $\int_1^2 \int_4^5 f(x, y)\,dx\,dy$

(b) $\int_4^5 \int_1^2 f(x, y)\,dx\,dy$

SOLUTION The integral $\int_1^2 \int_4^5 f(x, y)\,dy\,dx$ is written with dy preceding dx, therefore the integration is first with respect to y over the interval $4 \le y \le 5$, and then with respect to x over the interval $1 \le x \le 2$. By Fubini's Theorem, we may replace the order of integration over the corresponding intervals. Therefore the given integral is equal to (b) rather than to (a).

6. For which of the following functions is the double integral over the rectangle in Figure 16 equal to zero? Explain your reasoning.

(a) $f(x, y) = x^2 y$

(b) $f(x, y) = xy^2$

(c) $f(x, y) = \sin x$

(d) $f(x, y) = e^x$

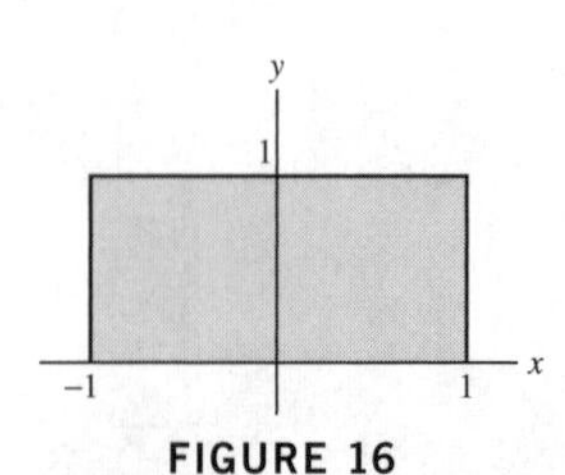

FIGURE 16

SOLUTION The double integral is the signed volume of the region between the graph of $f(x, y)$ and the xy-plane over $\mathcal{R}$. In (c) and (d) the function satisfies $f(-x, y) = -f(x, y)$, hence the region below the xy-plane, where $-1 \le x \le 0$ cancels with the region above the xy-plane, where $0 \le x \le 1$. Therefore, the double integral is zero. In (a) and (b), the function $f(x, y)$ is always positive on the rectangle, so the double integral is greater than zero.

Exercises

1. Compute the Riemann sum $S_{4,3}$ to estimate the double integral of $f(x, y) = xy$ over $\mathcal{R} = [1, 3] \times [1, 2.5]$. Use the regular partition and upper-right vertices of the subrectangles as sample points.

SOLUTION The rectangle $\mathcal{R}$ and the subrectangles are shown in the following figure:

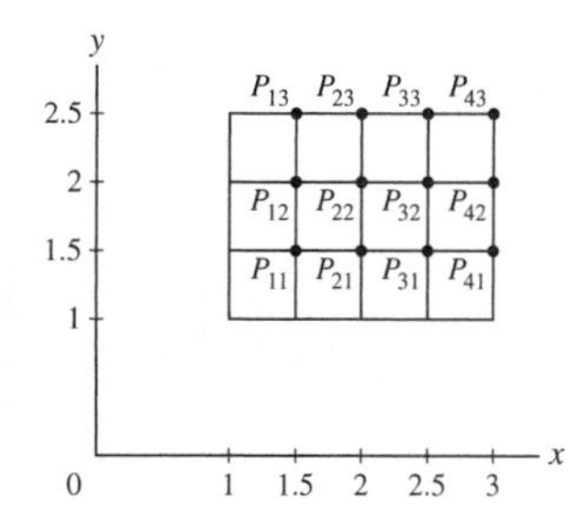

The subrectangles have sides of length

$$\Delta x = \frac{3-1}{4} = 0.5, \quad \Delta y = \frac{2.5-1}{3} = 0.5 \quad \Rightarrow \quad \Delta A = 0.5 \cdot 0.5 = 0.25$$

The upper right vertices are the following points:

$$\begin{array}{llll} P_{11} = (1.5, 1.5) & P_{21} = (2, 1.5) & P_{31} = (2.5, 1.5) & P_{41} = (3, 1.5) \\ P_{12} = (1.5, 2) & P_{22} = (2, 2) & P_{32} = (2.5, 2) & P_{42} = (3, 2) \\ P_{13} = (1.5, 2.5) & P_{23} = (2, 2.5) & P_{33} = (2.5, 2.5) & P_{43} = (3, 2.5) \end{array}$$

We compute $f(x, y) = xy$ at these points:

$$\begin{array}{lll} f(P_{11}) = 1.5 \cdot 1.5 = 2.25 & f(P_{12}) = 1.5 \cdot 2 = 3 & f(P_{13}) = 3.75 \\ f(P_{21}) = 2 \cdot 1.5 = 3 & f(P_{22}) = 2 \cdot 2 = 4 & f(P_{23}) = 5 \\ f(P_{31}) = 2.5 \cdot 1.5 = 3.75 & f(P_{32}) = 2.5 \cdot 2 = 5 & f(P_{33}) = 6.25 \\ f(P_{41}) = 3 \cdot 1.5 = 4.5 & f(P_{42}) = 3 \cdot 2 = 6 & f(P_{43}) = 7.5 \end{array}$$

Hence, $S_{4,3}$ is the following sum:

$$S_{4,3} = \sum_{i=1}^{4} \sum_{j=1}^{3} f(P_{ij}) \Delta A = 0.25(2.25 + 3 + 3.75 + 4.5 + 3 + 4 + 5 + 6 + 3.75 + 5 + 6.25 + 7.5) = 13.5$$

In Exercises 3–6, compute the Riemann sums for the double integral $\iint_{\mathcal{R}} f(x, y)\, dA$, *where* $\mathcal{R} = [1, 4] \times [1, 3]$, *for the grid and two choices of sample points shown in Figure 17.*

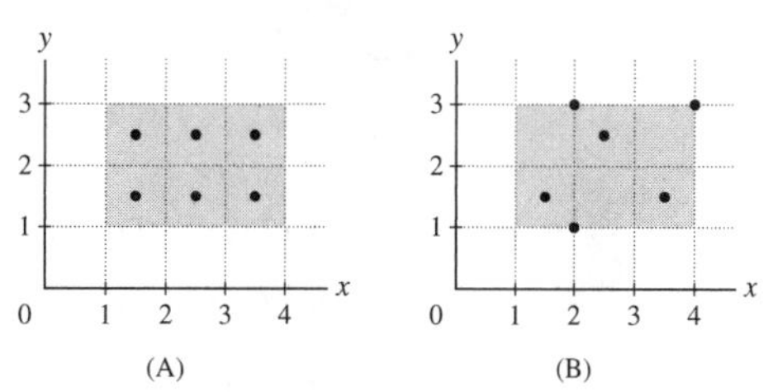

FIGURE 17

3. $f(x, y) = 2x + y$

SOLUTION The subrectangles have sides of length $\Delta x = \frac{4-1}{3} = 1$ and $\Delta y = \frac{3-1}{2} = 1$, and area $\Delta A = \Delta x \Delta y = 1$. We find the sample points in (A) and (B):

(A)

$$P_{11} = (1.5, 1.5)\ P_{21} = (2.5, 1.5)\ P_{31} = (3.5, 1.5)$$

$$P_{12} = (1.5, 2.5)\ P_{22} = (2.5, 2.5)\ P_{32} = (3.5, 2.5)$$

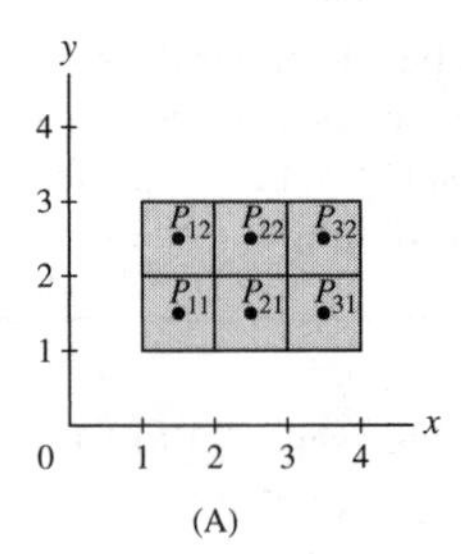

(A)

(B)

$$P_{11} = (1.5, 1.5)\ P_{21} = (2, 1)\ P_{31} = (3.5, 1.5)$$

$$P_{21} = (2, 3)\ P_{22} = (2.5, 2.5)\ P_{23} = (4, 3)$$

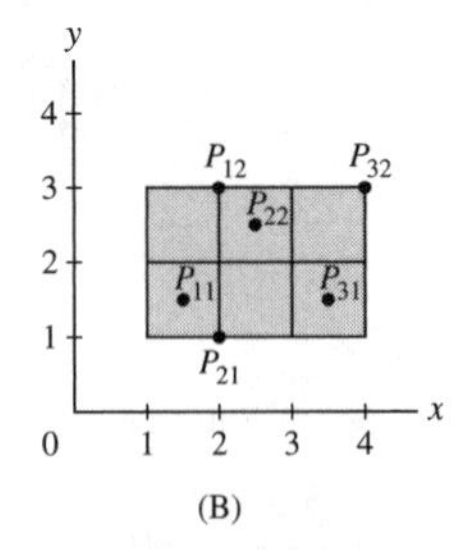

(B)

The Riemann Sum S_{32} is the following estimation of the double integral:

$$\iint_{\mathcal{R}} f(x, y)\, dA \approx S_{32} = \sum_{i=1}^{3} \sum_{j=1}^{2} f\left(P_{ij}\right) \Delta A = \sum_{i=1}^{3} \sum_{j=1}^{2} f\left(P_{ij}\right)$$

We compute S_{32} for the two choices of sample points (A) and (B), and the following function:

$$f(x, y) = 2x + y$$

We compute $f\left(P_{ij}\right)$ for the sample points computed above:

(A)

$$f\left(P_{11}\right) = f(1.5, 1.5) = 2 \cdot 1.5 + 1.5 = 4.5$$

$$f\left(P_{21}\right) = f(2.5, 1.5) = 2 \cdot 2.5 + 1.5 = 6.5$$

$$f\left(P_{31}\right) = f(3.5, 1.5) = 2 \cdot 3.5 + 1.5 = 8.5$$

$$f\left(P_{12}\right) = f(1.5, 2.5) = 2 \cdot 1.5 + 2.5 = 5.5$$

$$f\left(P_{22}\right) = f(2.5, 2.5) = 2 \cdot 2.5 + 2.5 = 7.5$$

$$f\left(P_{32}\right) = f(3.5, 2.5) = 2 \cdot 3.5 + 2.5 = 9.5$$

Hence,

$$S_{32} = \sum_{i=1}^{3} \sum_{j=1}^{2} f\left(P_{ij}\right) \Delta A = 4.5 + 6.5 + 8.5 + 5.5 + 7.5 + 9.5 = 42$$

(B)

$$f\left(P_{11}\right) = f(1.5, 1.5) = 2 \cdot 1.5 + 1.5 = 4.5$$

$$f\left(P_{21}\right) = f(2, 1) = 2 \cdot 2 + 1 = 5$$

$$f\left(P_{31}\right) = f(3.5, 1.5) = 2 \cdot 3.5 + 1.5 = 8.5$$

$$f\left(P_{21}\right) = f(2, 3) = 2 \cdot 2 + 3 = 7$$

$$f\left(P_{22}\right) = f(2.5, 2.5) = 2 \cdot 2.5 + 2.5 = 7.5$$

$$f\left(P_{23}\right) = f(4, 3) = 2 \cdot 4 + 3 = 11$$

Hence,

$$S_{32} = \sum_{i=1}^{3} \sum_{j=1}^{2} f\left(P_{ij}\right) \Delta A = 4.5 + 5 + 8.5 + 7 + 7.5 + 11 = 43.5$$

5. $f(x, y) = 4x$

SOLUTION We compute the values of f at the sample points:

(A)

$$f(P_{11}) = f(1.5, 1.5) = 4 \cdot 1.5 = 6$$
$$f(P_{21}) = f(2.5, 1.5) = 4 \cdot 2.5 = 10$$
$$f(P_{31}) = f(3.5, 1.5) = 4 \cdot 3.5 = 14$$
$$f(P_{12}) = f(1.5, 2.5) = 4 \cdot 1.5 = 6$$
$$f(P_{22}) = f(2.5, 2.5) = 4 \cdot 2.5 = 10$$
$$f(P_{32}) = f(3.5, 2.5) = 4 \cdot 3.5 = 14$$
$$\Delta x = \frac{4-1}{3} = 1, \quad \Delta y = \frac{3-1}{2} = 1$$

Hence $\Delta A = \Delta x \cdot \Delta y = 1$ and we get

$$S_{32} = \sum_{i=1}^{3} \sum_{j=1}^{2} f(P_{ij})\, \Delta A = 6 + 10 + 14 + 6 + 10 + 14 = 60$$

(B)

$$f(P_{11}) = f(1.5, 1.5) = 4 \cdot 1.5 = 6$$
$$f(P_{21}) = f(2, 1) = 4 \cdot 2 = 8$$
$$f(P_{31}) = f(3.5, 1.5) = 4 \cdot 3.5 = 14$$
$$f(P_{12}) = f(2, 3) = 4 \cdot 2 = 8$$
$$f(P_{22}) = f(2.5, 2.5) = 4 \cdot 2.5 = 10$$
$$f(P_{32}) = f(4, 3) = 4 \cdot 4 = 16$$

$\Delta A = 1$. Hence,

$$S_{32} = \sum_{i=1}^{3} \sum_{j=1}^{2} f(P_{ij})\, \Delta A = 6 + 8 + 14 + 8 + 10 + 16 = 62$$

7. Let $\mathcal{R} = [0, 1] \times [0, 1]$. Estimate $\iint_{\mathcal{R}} (x + y)\, dA$ by computing two different Riemann sums, each with at least six rectangles.

SOLUTION We define the following subrectangles and sample points:

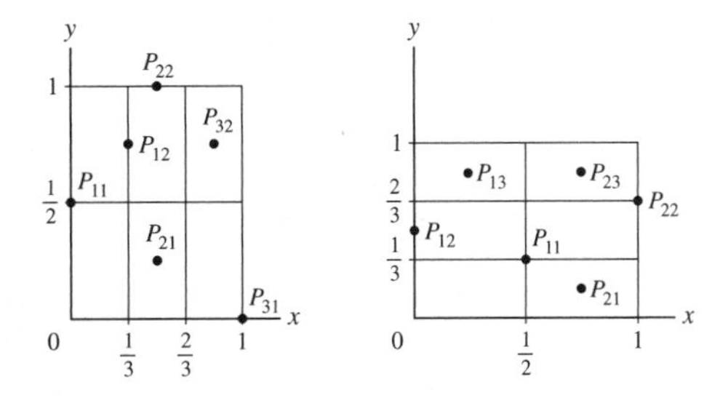

The sample points defined in the two figures are:

(A)

$$P_{11} = \left(0, \tfrac{1}{2}\right) \quad P_{21} = \left(\tfrac{1}{2}, \tfrac{1}{4}\right) \quad P_{31} = (1, 0)$$
$$P_{12} = \left(\tfrac{1}{3}, \tfrac{3}{4}\right) \quad P_{22} = \left(\tfrac{1}{2}, 1\right) \quad P_{32} = \left(\tfrac{5}{6}, \tfrac{3}{4}\right)$$

(B)

$$P_{11} = \left(\tfrac{1}{2}, \tfrac{1}{3}\right) \quad P_{21} = \left(\tfrac{3}{4}, \tfrac{1}{6}\right) \quad P_{12} = \left(0, \tfrac{1}{2}\right)$$
$$P_{22} = \left(1, \tfrac{2}{3}\right) \quad P_{13} = \left(\tfrac{1}{4}, \tfrac{5}{6}\right) \quad P_{23} = \left(\tfrac{3}{4}, \tfrac{5}{6}\right)$$

We compute the values of $f(x, y) = x + y$ at the sample points:

(A)

$$f(P_{11}) = f\left(0, \frac{1}{2}\right) = 0 + \frac{1}{2} = \frac{1}{2}$$

$$f(P_{21}) = f\left(\frac{1}{2}, \frac{1}{4}\right) = \frac{1}{2} + \frac{1}{4} = \frac{3}{4}$$

$$f(P_{31}) = f(1, 0) = 1 + 0 = 1$$

$$f(P_{12}) = f\left(\frac{1}{3}, \frac{3}{4}\right) = \frac{1}{3} + \frac{3}{4} = \frac{13}{12}$$

$$f(P_{22}) = f\left(\frac{1}{2}, 1\right) = \frac{1}{2} + 1 = \frac{3}{2}$$

$$f(P_{32}) = f\left(\frac{5}{6}, \frac{3}{4}\right) = \frac{5}{6} + \frac{3}{4} = \frac{19}{12}$$

Each subrectangle has sides of length $\Delta x = \frac{1}{3}$, $\Delta y = \frac{1}{2}$ and area $\Delta A = \Delta x \Delta y = \frac{1}{3} \cdot \frac{1}{2} = \frac{1}{6}$. We obtain the following Riemann sum:

$$S_{32} = \sum_{i=1}^{3} \sum_{j=1}^{2} f(P_{ij})\,\Delta A = \frac{1}{6}\left(\frac{1}{2} + \frac{3}{4} + 1 + \frac{13}{12} + \frac{3}{2} + \frac{19}{12}\right) = \frac{77}{72} \approx 1.069$$

(B)

$$f(P_{11}) = f\left(\frac{1}{2}, \frac{1}{3}\right) = \frac{1}{2} + \frac{1}{3} = \frac{5}{6}$$

$$f(P_{21}) = f\left(\frac{3}{4}, \frac{1}{6}\right) = \frac{3}{4} + \frac{1}{6} = \frac{11}{12}$$

$$f(P_{12}) = f\left(0, \frac{1}{2}\right) = 0 + \frac{1}{2} = \frac{1}{2}$$

$$f(P_{22}) = f\left(1, \frac{2}{3}\right) = 1 + \frac{2}{3} = \frac{5}{3}$$

$$f(P_{13}) = f\left(\frac{1}{4}, \frac{5}{6}\right) = \frac{1}{4} + \frac{5}{6} = \frac{13}{12}$$

$$f(P_{23}) = f\left(\frac{3}{4}, \frac{5}{6}\right) = \frac{3}{4} + \frac{5}{6} = \frac{19}{12}$$

Each subrectangle has sides of length $\Delta x = \frac{1}{2}$, $\Delta y = \frac{1}{3}$ and area $\Delta A = \Delta x \Delta y = \frac{1}{2} \cdot \frac{1}{3} = \frac{1}{6}$. We obtain the following Riemann sum:

$$S_{23} = \sum_{i=1}^{3} \sum_{j=1}^{2} f(P_{ij})\,\Delta A = \frac{1}{6}\left(\frac{5}{6} + \frac{11}{12} + \frac{1}{2} + \frac{5}{3} + \frac{13}{12} + \frac{19}{12}\right) = \frac{79}{72} \approx 1.097$$

9. Evaluate $\iint_{\mathcal{R}} (-5)\,dA$, where $\mathcal{R} = [2, 5] \times [4, 7]$.

SOLUTION The double integral is the signed volume of the box of base $\mathcal{R}$ and height -5. That is,

$$\iint_{\mathcal{R}} (-5)\,dA = -5 \cdot \text{Area}(\mathcal{R}) = -5 \cdot (5 - 2) \cdot (7 - 4) = -5 \cdot 9 = -45$$

11. The following table gives the approximate height at 1-ft intervals of a 5×4-ft mound of gravel. Estimate the volume of the mound by computing the average of the four Riemann sums $S_{5,4}$ with lower left, lower right, upper left, and upper right vertices of the subrectangles as sample points.

y \ x	0	1	2	3	4	5
4	0.4	0.4	0.6	0.8	0.6	0.6
3	0.8	1.6	1.8	2.5	2.1	0.8
2	0.5	1.5	3.2	3.5	2.1	0.6
1	0.4	0.8	1.3	1.5	1.4	0.6
0	0.3	0.3	0.5	0.8	0.5	0.4

SOLUTION The function $f(x, y)$ on the rectangle $\mathcal{R} = [0, 5] \times [0, 4]$ gives the height of the mound at each point $(x, y) \in \mathcal{R}$. We are given the values of f at some points. The volume is the double integral $\iint_{\mathcal{R}} f(x, y)\, dA$, which we estimate by an average of four Riemann sums. The computations are fairly tedious and we will not show them here. The Riemann sum using the lower-left corner of each rectange is 27.4, for lower-right it is 27.8, for upper-right we get 28.3, and for upper-left we get 27.8 The average is 27.825.

13. CAS Let $S_{N,N}$ be the Riemann sum for $\int_0^1 \int_0^1 e^{x^3 - y^3}\, dy\, dx$ using the regular partition and the lower left-hand vertex of each subrectangle as sample points. Use a computer algebra system to calculate $S_{N,N}$ for $N = 25, 50, 100$.

SOLUTION Using a computer algebra system, we compute $S_{N,N}$ to be 1.0731, 1.0783, and 1.0809.

In Exercises 15–16, use symmetry to evaluate the double integral.

15. $\iint_{\mathcal{R}} \sin x\, dA, \quad \mathcal{R} = [0, 2\pi] \times [0, 2\pi]$

SOLUTION Since $\sin(\pi + x) = -\sin x$, the region between the graph and the xy-plane where $\pi \le x \le 2\pi$, is below the xy-plane, and it cancels with the region above the xy-plane where $0 \le x \le \pi$. Hence,

$$\iint_{\mathcal{R}} \sin x\, dA = 0$$

In Exercises 17–32, evaluate the iterated integral.

17. $\int_1^3 \int_0^2 x^3 y\, dy\, dx$

SOLUTION We first compute the inner integral, treating x as a constant, then integrate the result with respect to x:

$$\int_1^3 \int_0^2 x^3 y\, dy\, dx = \int_1^3 x^3 \frac{y^2}{2}\bigg|_{y=0}^{2} dx = \int_1^3 x^3 \left(\frac{2^2}{2} - 0\right) dx = \int_1^3 2x^3\, dx = \frac{x^4}{2}\bigg|_1^3 = 40$$

19. $\int_0^2 \int_1^3 x^3 y\, dy\, dx$

SOLUTION We first evaluate the inner integral, treating x as a constant, then integrate the result with respect to x. We obtain

$$\int_0^2 \int_1^3 x^3 y\, dy\, dx = \int_0^2 x^3 \frac{y^2}{2}\bigg|_{y=1}^{3} dx = \int_0^2 x^3 \left(\frac{3^2}{2} - \frac{1^2}{2}\right) dx = \int_0^2 4x^3\, dx = x^4\bigg|_0^2 = 16$$

21. $\int_2^6 \int_1^4 x^2\, dx\, dy$

SOLUTION We use Iterated Integral of a Product Function to compute the integral as follows:

$$\int_2^6 \int_1^4 x^2\, dx\, dy = \int_2^6 \int_1^4 x^2 \cdot 1\, dx\, dy = \left(\int_1^4 x^2\, dx\right)\left(\int_2^6 1\, dy\right) = \left(\frac{x^3}{3}\bigg|_1^4\right)\left(y\bigg|_2^6\right)$$

$$= \left(\frac{4^3}{3} - \frac{1^3}{3}\right)(6 - 2) = 21 \cdot 4 = 84$$

23. $\int_0^1 \int_0^2 (x + 4y^3)\, dx\, dy$

SOLUTION We use additivity of the double integral to write

$$\int_0^1 \int_0^2 \left(x + 4y^3\right) dx\, dy = \int_0^1 \int_0^2 x\, dx\, dy + \int_0^1 \int_0^2 4y^3\, dx\, dy \qquad (1)$$

We now compute each of the double integrals using product of iterated integrals:

$$\int_0^1 \int_0^2 x\, dx\, dy = \left(\int_0^2 x\, dx\right)\left(\int_0^1 1\, dy\right) = \left(\frac{1}{2}x^2\bigg|_0^2\right)\left(y\bigg|_0^1\right) = 2 \cdot 1 = 2$$

$$\int_0^1 \int_0^2 4y^3\,dx\,dy = \left(\int_0^1 4y^3\,dy\right)\left(\int_0^2 1\,dx\right) = \left(y^4\Big|_0^1\right)\left(x\Big|_0^2\right) = 1\cdot 2 = 2$$

Substituting in (1) gives

$$\int_0^1 \int_0^2 (x + 4y^3)\,dx\,dy = 2 + 2 = 4.$$

25. $\displaystyle\int_0^1 \int_2^3 \sqrt{x+4y}\,dx\,dy$

SOLUTION We compute the inner integral, treating y as a constant. Then we evaluate the resulting integral with respect to y:

$$\int_0^1 \int_2^3 \sqrt{x+4y}\,dx\,dy = \int_0^1 \frac{2}{3}(x+4y)^{3/2}\Big|_{x=2}^3 dy = \int_0^1 \frac{2}{3}\left((3+4y)^{3/2} - (2+4y)^{3/2}\right)dy$$
$$= \frac{2}{3}\left(\frac{2}{5\cdot 4}(3+4y)^{5/2} - \frac{2}{5\cdot 4}(2+4y)^{5/2}\right)\Big|_0^1 = \frac{1}{15}\left(\left(7^{5/2} - 6^{5/2}\right) - \left(3^{5/2} - 2^{5/2}\right)\right)$$
$$= \frac{1}{15}\left(7^{5/2} - 6^{5/2} - 3^{5/2} + 2^{5/2}\right) \approx 2.102$$

27. $\displaystyle\int_1^2 \int_0^4 \frac{dy\,dx}{x+y}$

SOLUTION The inner integral is an iterated integral with respect to y. We evaluate it first and then compute the resulting integral with respect to x. This gives

$$\int_1^2 \int_0^4 \frac{dy\,dx}{x+y} = \int_1^2 \left(\int_0^4 \frac{dy}{x+y}\right)dx = \int_1^2 \ln(x+y)\Big|_{y=0}^4 dx = \int_1^2 (\ln(x+4) - \ln x)\,dx$$

We use the integral formula:

$$\int \ln(x+a)\,dx = (x+a)\left(\ln(x+a) - 1\right) + C$$

We get

$$\int_1^2 \int_0^4 \frac{dy\,dx}{x+y} = (x+4)\left(\ln(x+4) - 1\right) - x(\ln x - 1)\Big|_1^2 = 6(\ln 6 - 1) - 2(\ln 2 - 1) - (5(\ln 5 - 1) - (\ln 1 - 1))$$
$$= 6\ln 6 - 2\ln 2 - 5\ln 5 \approx 1.31$$

29. $\displaystyle\int_0^1 \int_2^3 \frac{1}{(x+4y)^3}\,dx\,dy$

SOLUTION We calculate the inner integral with respect to x, then we compute the resulting integral with respect to y. This gives

$$\int_0^1 \int_2^3 \frac{1}{(x+4y)^3}\,dx\,dy = \int_0^1 \left(\int_2^3 (x+4y)^{-3}\,dx\right)dy = \int_0^1 -\frac{1}{2}(x+4y)^{-2}\Big|_{x=2}^3 dy$$
$$= -\frac{1}{2}\int_0^1 \left((3+4y)^{-2} - (2+4y)^{-2}\right)dy = -\frac{1}{2\cdot 4}\left(-(3+4y)^{-1} + (2+4y)^{-1}\right)\Big|_0^1$$
$$= -\frac{1}{8}\left(\frac{1}{2+4y} - \frac{1}{3+4y}\right)\Big|_0^1 = -\frac{1}{8}\left(\left(\frac{1}{6} - \frac{1}{7}\right) - \left(\frac{1}{2} - \frac{1}{3}\right)\right)$$
$$= \frac{1}{8}\left(\frac{1}{7} - \frac{1}{6} + \frac{1}{2} - \frac{1}{3}\right) = \frac{1}{56}$$

31. $\displaystyle\int_1^2 \int_1^2 \ln(xy)\,dy\,dx$

SOLUTION We use $\ln xy = \ln x + \ln y$, additivity of the double integral and the double integral as a product of single integrals, to write

$$\int_1^2 \int_1^2 \ln(xy)\,dy\,dx = \int_1^2 \int_1^2 (\ln x + \ln y)\,dy\,dx = \int_1^2 \int_1^2 \ln x\,dy\,dx + \int_1^2 \int_1^2 \ln y\,dy\,dx$$
$$= \left(\int_1^2 \ln x\,dx\right)\left(\int_1^2 1\,dy\right) + \left(\int_1^2 \ln y\,dy\right)\left(\int_1^2 1\,dx\right) = 2\left(\int_1^2 \ln x\,dx\right)\left(\int_1^2 1\,dy\right)$$

We compute each integral:

$$\int_1^2 1\,dy = y\Big|_1^2 = 2 - 1 = 1$$
$$\int_1^2 \ln x\,dx = x(\ln x - 1)\Big|_1^2 = 2\cdot(\ln 2 - 1) - 1\cdot(\ln 1 - 1) = 2\ln 2 - 2 + 1 = 2\ln 2 - 1$$

Substituting in (1) we get

$$\int_1^2 \int_1^2 \ln(xy)\,dy\,dx = 2\cdot(2\ln 2 - 1)\cdot 1 = 4\ln 2 - 2 \approx 0.773$$

33. Let $f(x, y) = mxy^2$, where m is a constant. Find a value of m such that $\iint_{\mathcal{R}} f(x, y)\,dx\,dy = 1$, where $\mathcal{R} = [0, 1] \times [0, 2]$.

SOLUTION Since $f(x, y) = mxy^2$ is a product of a function of x and a function of y, we may compute the double integral as the product of two single integrals. That is,

$$\int_0^2 \int_0^1 mxy^2\,dx\,dy = m\left(\int_0^1 x\,dx\right)\left(\int_0^2 y^2\,dy\right) \tag{1}$$

We compute each integral:

$$\int_0^1 x\,dx = \frac{1}{2}x^2\Big|_0^1 = \frac{1}{2}\left(1^2 - 0^2\right) = \frac{1}{2}$$
$$\int_0^2 y^2\,dy = \frac{1}{3}y^3\Big|_0^2 = \frac{1}{3}\left(2^3 - 0^3\right) = \frac{8}{3}$$

We substitute in (1), equate to 1 and solve the resulting equation for m. This gives

$$m\cdot\frac{1}{2}\cdot\frac{8}{3} = 1 \Rightarrow m = \frac{3}{4}$$

In Exercises 34–37, evaluate the double integral of the function over the rectangle.

35. $\iint_{\mathcal{R}} x^2y\,dA, \quad \mathcal{R} = [-1, 1] \times [0, 2]$

SOLUTION We compute the double integral as the product of two single integrals:

$$\iint_{\mathcal{R}} x^2y\,dA = \int_0^2 \int_{-1}^1 x^2y\,dx\,dy = \left(\int_{-1}^1 x^2\,dx\right)\left(\int_0^2 y\,dy\right) = \left(\frac{1}{3}x^3\Big|_{-1}^1\right)\left(\frac{1}{2}y^2\Big|_0^2\right) = \frac{2}{3}\cdot\frac{1}{2}\cdot 4 = \frac{4}{3}$$

37. $\iint_{\mathcal{R}} \frac{y}{x+1}\,dA, \quad \mathcal{R} = [0, 2] \times [0, 4]$

SOLUTION We evaluate the integral as the product of two single integrals. This can be done since the function has the form $f(x, y) = g(x)h(y)$.

$$\iint_{\mathcal{R}} \frac{y}{x+1}\,dA = \int_0^4 \int_0^2 \frac{y}{x+1}\,dx\,dy = \left(\int_0^2 \frac{dx}{x+1}\right)\left(\int_0^4 y\,dy\right)$$
$$= \left(\ln(x+1)\Big|_0^2\right)\left(\frac{y^2}{2}\Big|_0^4\right) = (\ln 3 - \ln 1)\left(\frac{4^2}{2} - \frac{0^2}{2}\right) = 8\ln 3 \approx 8.79$$

In Exercises 38–41, use Eq. (3) to evaluate the integral.

39. $\iint_{\mathcal{R}} e^{2x+3y}\,dA, \quad \mathcal{R} = [0,1] \times [0,2]$

SOLUTION We use the property $e^{2x+3y} = e^{2x} \cdot e^{3y}$ and Eq. (2) to compute the double integral as the product of two single integrals. This gives

$$\iint_{\mathcal{R}} e^{2x+3y}\,dA = \int_0^2 \int_0^1 e^{2x}e^{3y}\,dx\,dy = \left(\int_0^1 e^{2x}\,dx\right)\left(\int_0^2 e^{3y}\,dy\right) = \left(\frac{1}{2}e^{2x}\bigg|_0^1\right)\left(\frac{1}{3}e^{3y}\bigg|_0^2\right)$$

$$= \frac{1}{2}\left(e^2 - e^0\right) \cdot \frac{1}{3}\left(e^6 - e^0\right) = \frac{1}{6}\left(e^2 - 1\right)\left(e^6 - 1\right) \approx 428.52$$

41. $\int_1^2 \int_0^2 \frac{x}{y}\,dx\,dy$

SOLUTION Using Eq. (2) we get

$$\int_1^2 \int_0^2 \frac{x}{y}\,dx\,dy = \left(\int_0^2 x\,dx\right)\left(\int_1^2 \frac{dy}{y}\right) = \left(\frac{x^2}{2}\bigg|_0^2\right)\left(\ln y\bigg|_1^2\right) = 2(\ln 2 - \ln 1) = 2\ln 2 \approx 1.39$$

43. CAS Use the inequality $0 \le \sin x \le x$ for $x \ge 0$ to show that

$$\int_0^1 \int_0^1 \sin(xy)\,dx\,dy \le \frac{1}{4}$$

Then use a computer algebra system to evaluate the double integral to three decimal places.

SOLUTION Since $\sin(xy) \le xy$, we get that

$$\int_0^1 \int_0^1 \sin(xy)\,dx\,dy \le \int_0^1 \int_0^1 xy\,dx\,dy = \int_0^1 \frac{y}{2}\,dy = \frac{1}{4}$$

Using a CAS, we find that the double integral is approximately 0.240.

45. Calculate a Riemann sum $S_{3,3}$ on the square $\mathcal{R} = [0,3] \times [0,3]$ for the function $f(x,y)$ whose contour plot is shown in Figure 19. Choose sample points and use the plot to find the values of $f(x,y)$ at these points.

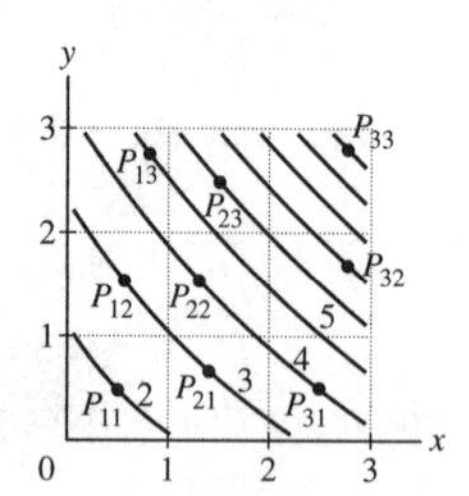

FIGURE 19 Contour plot of $f(x,y)$.

SOLUTION Each subrectangle is a square of side length 1, hence its area is $\Delta A = 1^2 = 1$. We choose the sample points shown in the figure. The contour plot shows the following values of f at the sample points:

$$f(P_{11}) = 2 \quad f(P_{21}) = 3 \quad f(P_{31}) = 4$$
$$f(P_{12}) = 3 \quad f(P_{22}) = 4 \quad f(P_{32}) = 7$$
$$f(P_{13}) = 5 \quad f(P_{23}) = 6 \quad f(P_{33}) = 10$$

The Riemann sum S_{33} is thus

$$S_{33} = \sum_{i=1}^{3}\sum_{j=1}^{3} f(P_{ij})\,\Delta A = \sum_{i=1}^{3}\sum_{j=1}^{3} f(P_{ij}) \cdot 1 = 2+3+4+3+4+7+5+6+10 = 44$$

Further Insights and Challenges

47. Prove the following extension of the FTC to two variables: If $\frac{\partial^2 F}{\partial x\,\partial y} = f(x, y)$, then

$$\iint_{\mathcal{R}} f(x, y)\,dA = F(b, d) - F(a, d) - F(b, c) + F(a, c)$$

where $\mathcal{R} = [a, b] \times [c, d]$.

SOLUTION By Fubini's Theorem we get

$$\iint_{\mathcal{R}} f(x, y)\,dx = \int_c^d \int_a^b \frac{\partial^2 F}{\partial x\,\partial y}\,dx\,dy = \int_c^d \left(\int_a^b \frac{\partial}{\partial x}\left(\frac{\partial F}{\partial y}\right) dx \right) dy \tag{1}$$

In the inner integral, y is considered as constant. Therefore, by the Fundamental Theorem of calculus (part I) for the variable x, we have

$$\int_a^b \frac{\partial}{\partial x}\left(\frac{\partial F}{\partial y}\right) dx = \left.\frac{\partial F}{\partial y}\right|_{x=b} - \left.\frac{\partial F}{\partial y}\right|_{x=a} = \frac{\partial F}{\partial y}(b, y) - \frac{\partial F}{\partial y}(a, y)$$

We substitute in (1), use additivity of the single integral and use again the Fundamental Theorem, this time for the variable y. This gives

$$\begin{aligned}
\iint_{\mathcal{R}} f(x, y)\,dx &= \int_c^d \left(\frac{\partial F}{\partial y}(b, y) - \frac{\partial F}{\partial y}(a, y) \right) dy = \int_c^d \frac{\partial F}{\partial y}(b, y)\,dy - \int_c^d \frac{\partial F}{\partial y}(a, y)\,dy \\
&= \left(F(b, y)\Big|_{y=d} - F(b, y)\Big|_{y=c} \right) - \left(F(a, y)\Big|_{y=d} - F(a, y)\Big|_{y=c} \right) \\
&= F(b, d) - F(b, c) - F(a, d) + F(a, c)
\end{aligned}$$

49. Find a function $F(x, y)$ satisfying $\frac{\partial^2 F}{\partial x\,\partial y} = 6x^2 y$ and use the result of Exercise 47 to evaluate $\iint_{\mathcal{R}} 6x^2 y\,dA$ for the rectangle $\mathcal{R} = [0, 1] \times [0, 4]$.

SOLUTION We integrate $\frac{\partial^2 F}{\partial x \partial y}$ with respect to x, taking zero as the constant of integration. We get $\frac{\partial F}{\partial y} = 6y\frac{x^3}{3} = 2yx^3$. We now integrate with respect to y, obtaining

$$F(x, y) = 2 \cdot \frac{y^2}{2} x^3 = y^2 x^3$$

In Exercise 47 we showed that

$$\int_c^d \int_a^b \frac{\partial^2 F}{\partial x\,\partial y}\,dx\,dy = F(b, d) - F(b, c) - F(a, d) + F(a, c)$$

For $F(x, y) = y^2 x^3$ we get

$$\iint_{\mathcal{R}} 6x^2 y\,dA = \int_0^4 \int_0^1 6x^2 y\,dx\,dy = F(1, 4) - F(1, 0) - F(0, 4) + F(0, 0) = 4^2 \cdot 1^3 - 0 - 0 + 0 = 16$$

16.2 Double Integrals over More General Regions (ET Section 15.2)

Preliminary Questions

1. Which of the following expressions do not make sense?

(a) $\int_0^1 \int_1^y f(x, y)\,dy\,dx$

(b) $\int_0^1 \int_1^x f(x, y)\,dy\,dx$

(c) $\int_1^y \int_0^1 f(x, y)\,dy\,dx$

(d) $\int_0^1 \int_x^1 f(x, y)\,dy\,dx$

SOLUTION

(a) This integral is the same as

$$\int_0^1 \int_1^y f(x,y)\,dy\,dx = \int_0^1 \left(\int_1^y f(x,y)\,dy \right) dx$$

The inner integral is an integral with respect to y, over the interval $[1, y]$. This does not make sense.
(b) This integral is the following iterated integral:

$$\int_0^1 \int_1^x f(x,y)\,dy\,dx = \int_0^1 \left(\int_1^x f(x,y)\,dy \right) dx$$

The inner integral is a function of x and it is integrated with respect to x over the interval $0 \le x \le 1$. The result is a number. This integral makes sense.
(c) This integral is the following iterated integral:

$$\int_1^y \left(\int_0^1 f(x,y)\,dy \right) dx$$

The inner integral is a function of x and it is integrated with respect to x over the interval $1 \le x \le y$. The result is a function of y. This expression makes sense but the value of the integral is a function of y rather than a fixed number.
(d) This integral is the following iterated integral:

$$\int_0^1 \left(\int_x^1 f(x,y)\,dy \right) dx$$

The inner integral is a function of x and it is integrated with respect to x. This makes sense.

2. Draw a domain in the plane that is neither vertically nor horizontally simple.

SOLUTION The following region cannot be described in the form $\{a \le x \le b, \alpha(x) \le y \le \beta(x)\}$ nor in the form $\{c \le y \le d, \alpha(y) \le x \le \beta(y)\}$, hence it is neither vertically nor horizontally simple.

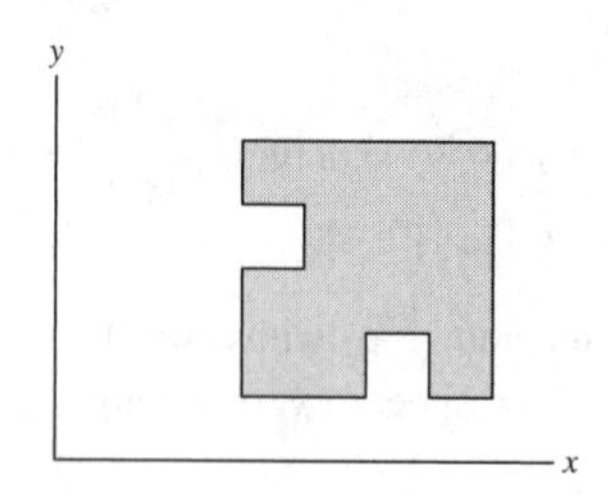

3. Which of the four regions in Figure 16 is the domain of integration for $\int_{-\sqrt{2}/2}^{0} \int_{x}^{\sqrt{1-x^2}} f(x,y)\,dy\,dx$?

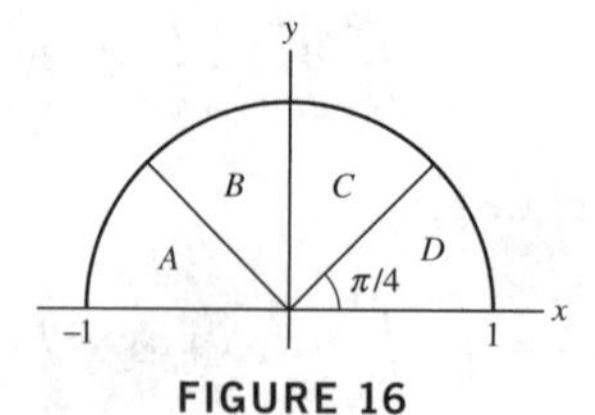

FIGURE 16

SOLUTION The region B is defined by the inequalities

$$-x \le y \le \sqrt{1-x^2}, \qquad -\frac{\sqrt{2}}{2} \le x \le 0$$

To compute $\int_{-\sqrt{2}/2}^{0} \int_{-x}^{\sqrt{1-x^2}} f(x,y)\,dy\,dx$, we first integrate with respect to y over the interval $-x \le y \le \sqrt{1-x^2}$, and then with respect to x over $-\frac{\sqrt{2}}{2} \le x \le 0$. That is, the domain of integration is B.

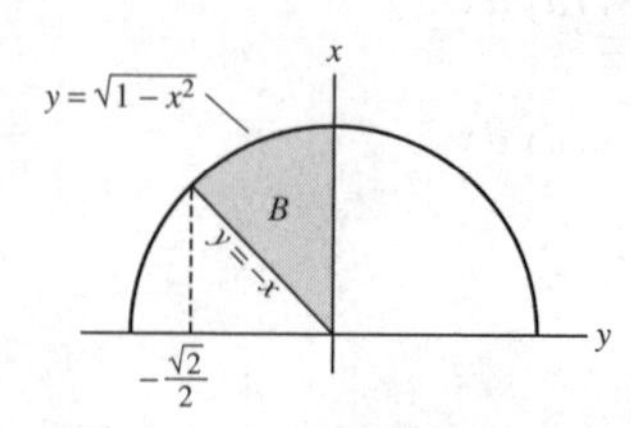

4. Let $\mathcal{D}$ be the unit circle. If the maximum value of $f(x, y)$ on $\mathcal{D}$ is 4, then the largest possible value of $\iint_{\mathcal{D}} f(x, y)\, dx\, dy$ is (choose the correct answer):

(a) 4 **(b)** 4π **(c)** $\frac{4}{\pi}$

SOLUTION The area of the unit circle is π and the maximum value of $f(x, y)$ on this region is $M = 4$, therefore we have,

$$\iint_{\mathcal{D}} f(x, y)\, dx\, dy \le 4\pi$$

The correct answer is (b).

Exercises

1. Calculate the Riemann sum for $f(x, y) = x - y$ and domain $\mathcal{D}$ in Figure 17 with two choices of sample points, • and ∘. Which do you think is a better approximation to the integral of f over $\mathcal{D}$? Why?

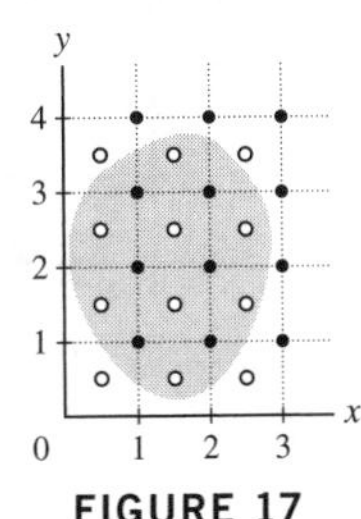

FIGURE 17

SOLUTION The subrectangles in Figure 17 have sides of length $\Delta x = \Delta y = 1$ and area $\Delta A = 1 \cdot 1 = 1$.

(a) Sample points •. There are six sample points that lie in the domain $\mathcal{D}$. We compute the values of $f(x, y) = x - y$ at these points:

$$f(1, 1) = 0, \quad f(1, 2) = -1, \quad f(1, 3) = -2,$$
$$f(2, 1) = 1, \quad f(2, 2) = 0, \quad f(2, 3) = -1$$

The Riemann sum is

$$S_{3,4} = (0 - 1 - 2 + 1 + 0 - 1) \cdot 1 = -3$$

(b) Sample points ∘. We compute the values of $f(x, y) = x - y$ at the eight sample points that lie in $\mathcal{D}$:

$$f(1.5, 0.5) = 1, \quad f(0.5, 1.5) = -1, \quad f(0.5, 2.5) = -2,$$
$$f(1.5, 3.5) = -2, \quad f(1.5, 1.5) = 0, \quad f(1.5, 2.5) = -1,$$
$$f(2.5, 1.5) = 1, \quad f(2.5, 2.5) = 0.$$

The corresponding Riemann sum is thus

$$S_{34} = (1 - 1 - 2 + 0 - 1 - 2 + 1 + 0) \cdot 1 = -4.$$

3. Let $\mathcal{D}$ be the domain defined by $0 \le x \le 1$, $x^2 \le y \le 4 - x^2$. Sketch $\mathcal{D}$ and express $\iint_{\mathcal{D}} y\, dA$ as an iterated integral and evaluate the result.

SOLUTION The domain $\mathcal{D}$ is shown in the following figure:

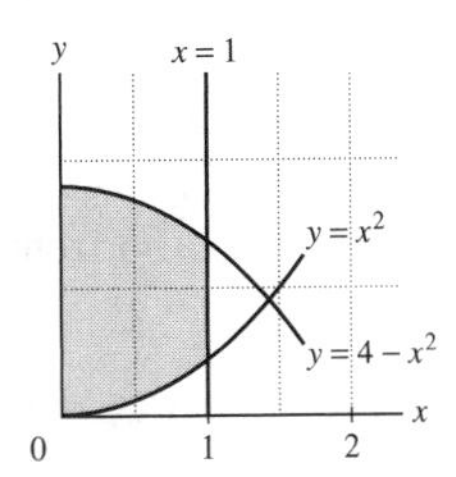

$\mathcal{D}$ is a vertically simple region and the limits of integration are

$$\underbrace{0 \le x \le 1}_{\text{limits of outer integral}} \qquad \underbrace{x^2 \le y \le 4 - x^2}_{\text{limits of inner integral}}$$

We follow three steps.

Step 1. Set up the iterated integral. The iterated integral is

$$\iint_{\mathcal{D}} y\,dA = \int_0^1 \int_{x^2}^{4-x^2} y\,dy\,dx.$$

Step 2. Evaluate the inner integral. We evaluate the inner integral with respect to y:

$$\int_{x^2}^{4-x^2} y\,dy = \frac{y^2}{2}\bigg|_{y=x^2}^{y=4-x^2} = \frac{1}{2}\left((4-x^2)^2 - (x^2)^2\right) = \frac{1}{2}(16 - 8x^2 + x^4 - x^4) = 8 - 4x^2$$

Step 3. Complete the computation. We integrate the resulting function with respect to x:

$$\iint_{\mathcal{D}} y\,dy = \int_0^1 (8 - 4x^2)\,dx = 8x - \frac{4}{3}x^3\bigg|_0^1 = 8 - \frac{4}{3} = \frac{20}{3} \approx 6.67$$

In Exercises 5–7, compute the double integral of $f(x, y) = x^2 y$ over the given shaded domain in Figure 20.

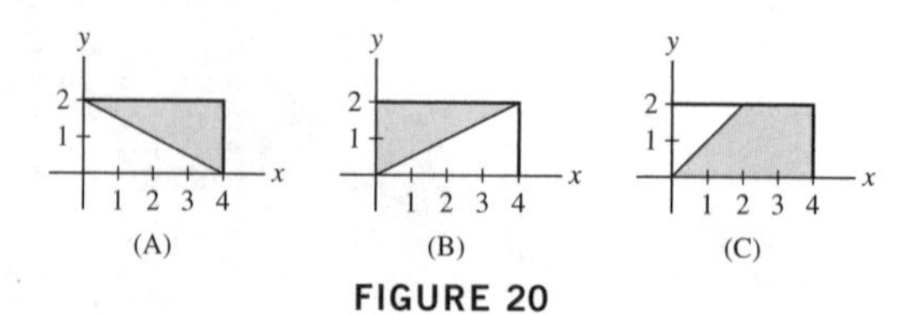

FIGURE 20

5. (A)

SOLUTION

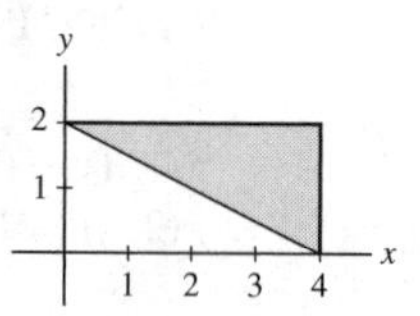

We describe the domain $\mathcal{D}$ as a vertically simple region. We find the equation of the line connecting the points $(0, 2)$ and $(4, 0)$.

$$y - 0 = \frac{2-0}{0-4}(x-4) \quad \Rightarrow \quad y = -\frac{1}{2}x + 2$$

Therefore the domain is described as a vertically simple region by the inequalities

$$0 \le x \le 4, \quad -\frac{1}{2}x + 2 \le y \le 2$$

We use Theorem 2 to evaluate the double integral:

$$\iint_{\mathcal{D}} x^2 y\,dA = \int_0^4 \int_{-\frac{x}{2}+2}^{2} x^2 y\,dy\,dx = \int_0^4 \frac{x^2 y^2}{2}\bigg|_{y=-\frac{x}{2}+2}^{2} dx = \int_0^4 \frac{x^2}{2}\left(2^2 - \left(-\frac{x}{2}+2\right)^2\right) dx$$

$$= \int_0^4 \left(x^3 - \frac{x^4}{8}\right) dx = \frac{x^4}{4} - \frac{x^5}{40}\bigg|_0^4 = \frac{4^4}{4} - \frac{4^5}{40} = \frac{192}{5} = 38.4$$

7. (C)

SOLUTION The domain in (C) is a horizontally simple region, described by the inequalities

$$0 \le y \le 2, \quad y \le x \le 4$$

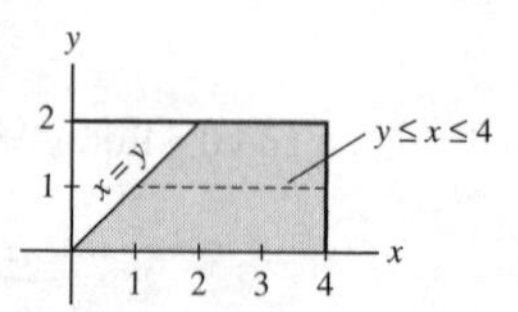

Using Theorem 2 we obtain the following integral:

$$\iint_{\mathcal{D}} x^2 y\, dA = \int_0^2 \int_y^4 x^2 y\, dx\, dy = \int_0^2 \frac{x^3 y}{3}\bigg|_{x=y}^{x=4} dy = \int_0^2 \frac{y}{3}\left(4^3 - y^3\right) dy = \int_0^2 \left(\frac{64y}{3} - \frac{y^4}{3}\right) dy$$

$$= \frac{32}{3}y^2 - \frac{y^5}{15}\bigg|_0^2 = \frac{32 \cdot 2^2}{3} - \frac{2^5}{15} = \frac{608}{15} \approx 40.53$$

9. Integrate $f(x, y) = x$ over the region bounded by $y = x^2$ and $y = x + 2$.

SOLUTION The domain of integration is shown in the following figure:

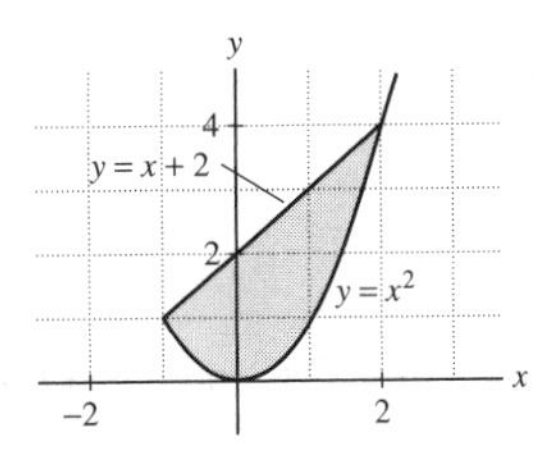

To find the inequalities defining the domain as a vertically simple region we first must find the x-coordinates of the two points where the line $y = x + 2$ and the parabola $y = x^2$ intersect. That is,

$$x + 2 = x^2 \quad \Rightarrow \quad x^2 - x - 2 = (x - 2)(x + 1) = 0$$

$$\Rightarrow \quad x_1 = -1, \quad x_2 = 2$$

We describe the domain by the following inequalities:

$$-1 \le x \le 2, \quad x^2 \le y \le x + 2$$

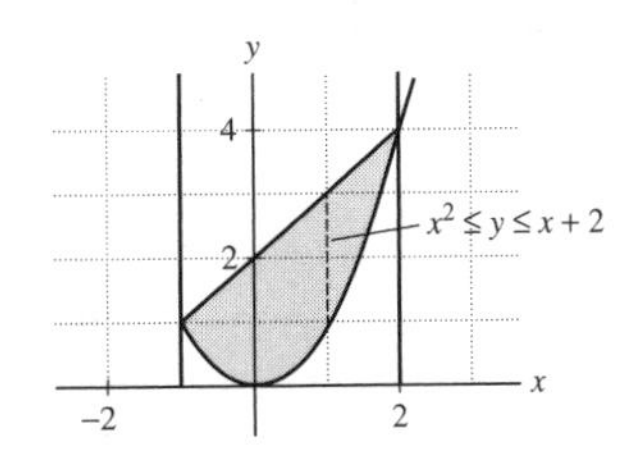

We now evaluate the integral of $f(x, y) = x$ over the vertically simple region $\mathcal{D}$:

$$\iint_{\mathcal{D}} x\, dA = \int_{-1}^2 \int_{x^2}^{x+2} x\, dy\, dx = \int_{-1}^2 xy\bigg|_{y=x^2}^{x+2} dx = \int_{-1}^2 x\left(x + 2 - x^2\right) dx$$

$$= \int_{-1}^2 \left(x^2 + 2x - x^3\right) dx = \frac{x^3}{3} + x^2 - \frac{x^4}{4}\bigg|_{-1}^2 = \left(\frac{8}{3} + 4 - 4\right) - \left(-\frac{1}{3} + 1 - \frac{1}{4}\right) = 2\frac{1}{4}$$

11. Sketch the region $\mathcal{D}$ between $y = x^2$ and $y = x(1 - x)$. Express $\mathcal{D}$ as an elementary region and calculate the integral of $f(x, y) = 2y$ over $\mathcal{D}$.

SOLUTION The region $\mathcal{D}$ between $y = x^2$ and $y = x(1 - x)$ is shown in the following figure:

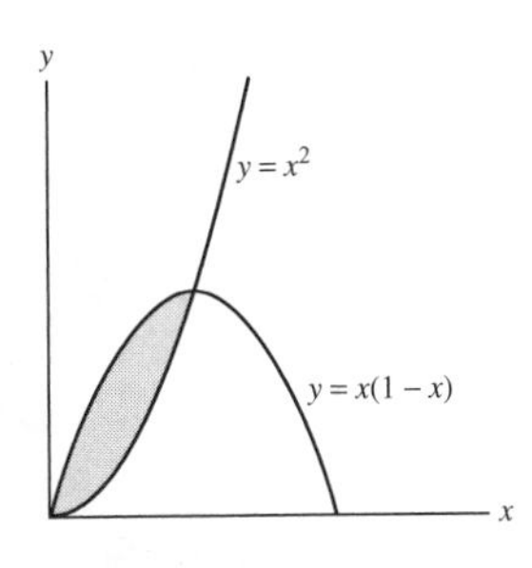

To find the inequalities for the vertically simple region $\mathcal{D}$, we first compute the x-coordinate of the point where the curves $y = x^2$ and $y = x(1 - x)$ intersect.

$$x^2 = x(1 - x) \quad \Rightarrow \quad x^2 = x - x^2$$

$$\Rightarrow \quad 2x^2 - x = x(2x-1) = 0$$

$$\Rightarrow \quad x_1 = 0, \quad x_2 = \frac{1}{2}$$

The region $\mathcal{D}$ is defined by the following inequalities:

$$0 \le x \le \frac{1}{2}, \quad x^2 \le y \le x(1-x)$$

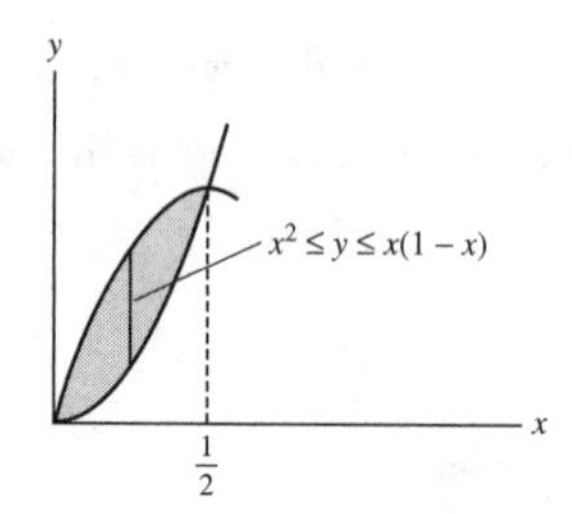

The double integral of f over $\mathcal{D}$ is computed using Theorem 2. That is,

$$\iint_{\mathcal{D}} 2y\,dA = \int_0^{1/2} \int_{x^2}^{x(1-x)} 2y\,dy\,dx = \int_0^{1/2} y^2 \Big|_{y=x^2}^{y=x(1-x)} dx = \int_0^{1/2} x^2(1-x)^2 - x^4\,dx$$

$$= \int_0^{1/2} x^2 - 2x^3 + x^4 - x^4\,dx = \int_0^{1/2} x^2 - 2x^3\,dx = \frac{1}{96} \approx 0.010$$

13. Calculate the integral of $f(x, y) = x$ over the region $\mathcal{D}$ bounded above by $y = x(2-x)$ and below by $x = y(2-y)$. *Hint:* Apply the quadratic formula to the lower boundary curve to solve for y as a function of x.

SOLUTION The two graphs are symmetric with respect to the line $y = x$, thus their point of intersection is $(1, 1)$. The region $\mathcal{D}$ is shown in the following figure:

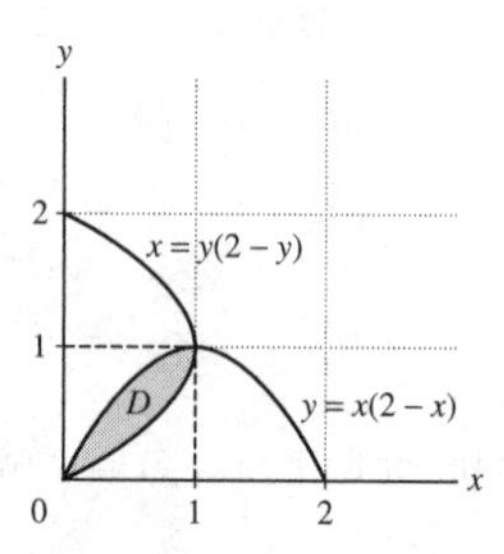

To find the inequalities defining the region $\mathcal{D}$ as a vertically simple region, we first must solve the lower boundary curve for y in terms of x. We get

$$x = y(2-y) = 2y - y^2$$

$$y^2 - 2y + x = 0$$

We solve the quadratic equation in y:

$$y = 1 \pm \sqrt{1-x}$$

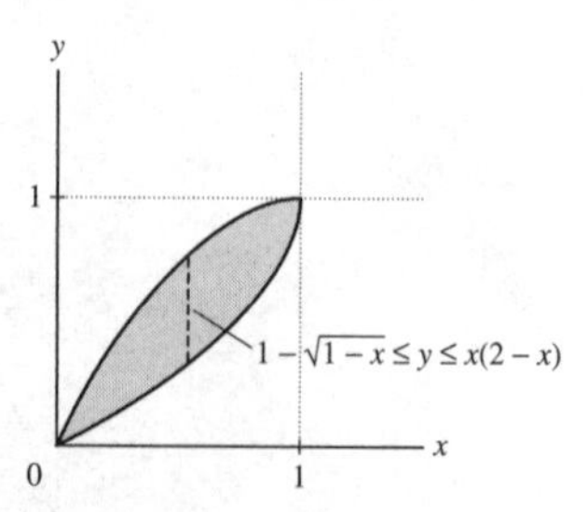

The domain $\mathcal{D}$ lies below the line $y = 1$, hence the appropriate solution is $y = 1 - \sqrt{1-x}$. We obtain the following inequalities for $\mathcal{D}$:

$$0 \le x \le 1, \quad 1 - \sqrt{1-x} \le y \le x(2-x)$$

We now evaluate the double integral of $f(x, y) = x$ over $\mathcal{D}$:

$$\iint_{\mathcal{D}} x\,dA = \int_0^1 \int_{1-\sqrt{1-x}}^{x(2-x)} x\,dy\,dx = \int_0^1 xy\Big|_{y=1-\sqrt{1-x}}^{x(2-x)} dx = \int_0^1 \left(x^2(2-x) - \left(x - x\sqrt{1-x}\right)\right) dx$$

$$= \int_0^1 \left(2x^2 - x^3 - x + x\sqrt{1-x}\right) dx = \frac{2x^3}{3} - \frac{x^4}{4} - \frac{x^2}{2}\Big|_0^1 + \int_0^1 x\sqrt{1-x}\,dx$$

$$= -\frac{1}{12} + \int_0^1 x\sqrt{1-x}\,dx$$

Using the substitution $u = \sqrt{1-x}$ it can be shown that $\int_0^1 x\sqrt{1-x}\,dx = \frac{4}{15}$. Therefore we get

$$\iint_{\mathcal{D}} x\,dA = -\frac{1}{12} + \frac{4}{15} = \frac{11}{60}$$

15. Integrate $f(x, y) = (x + y + 1)^{-2}$ over the triangle with vertices (0, 0), (4, 0), and (0, 8).

SOLUTION

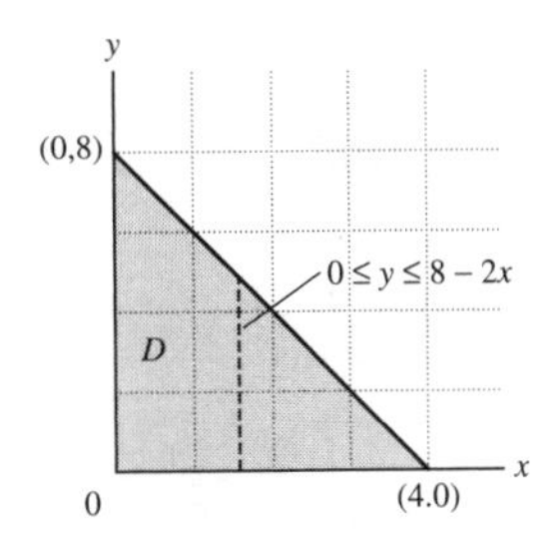

We describe the region $\mathcal{D}$ as a vertically simple region, but first we need to find the equation of the line joining the points (4, 0) and (0, 8):

$$y - 8 = \frac{8-0}{0-4}(x - 0) \quad \Rightarrow \quad y - 8 = -2x$$

$$\Rightarrow \quad y = 8 - 2x$$

We obtain the following inequalities for $\mathcal{D}$:

$$0 \le x \le 4, \quad 0 \le y \le 8 - 2x$$

We now evaluate the double integral of $f(x, y) = (x + y + 1)^{-2}$ over the triangle $\mathcal{D}$, by the following iterated integral:

$$\iint_{\mathcal{D}} f(x, y)\,dA = \int_0^4 \int_0^{8-2x} (x + y + 1)^{-2}\,dy\,dx = \int_0^4 -(x + y + 1)^{-1}\Big|_{y=0}^{8-2x} dx$$

$$= \int_0^4 \left(-(x + 8 - 2x + 1)^{-1} + (x + 0 + 1)^{-1}\right) dx = \int_0^4 \left(\frac{1}{x+1} - \frac{1}{9-x}\right) dx$$

$$= \ln(x+1) + \ln(9-x)\Big|_0^4 = \ln 5 + \ln 5 - (\ln 1 + \ln 9) = 2\ln 5 - \ln 9 = \ln\frac{25}{9} \approx 1.02$$

In Exercises 17–30, compute the integral of the function over the given region. Sketch the region.

17. $f(x, y) = x^3, \quad 0 \le x \le 2, \quad x^2 \le y \le 4$

SOLUTION The region $\mathcal{D}$ is shown in the following figure.

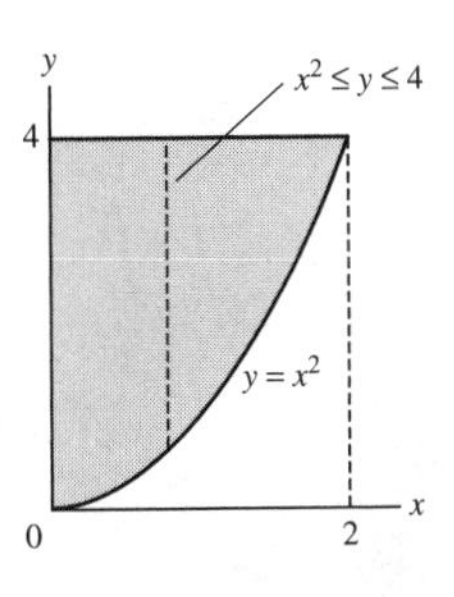

The given inequalities describe $\mathcal{D}$ as a vertically simple region, with given limits of integration. We obtain:

$$\iint_{\mathcal{D}} x^3\,dA = \int_0^2 \int_{x^2}^4 x^3\,dy\,dx = \int_0^2 x^3 y\Big|_{y=x^2}^4 dx = \int_0^2 x^3(4-x^2)\,dx = \int_0^2 (4x^3 - x^5)\,dx$$

$$= x^4 - \frac{x^6}{6}\Big|_0^2 = 2^4 - \frac{2^6}{6} = \frac{16}{3} \approx 5.33$$

19. $f(x, y) = x^2 y, \quad 1 \le x \le 3, \quad x \le y \le 2x + 1$

SOLUTION These inequalities describe $\mathcal{D}$ as a vertically simple region.

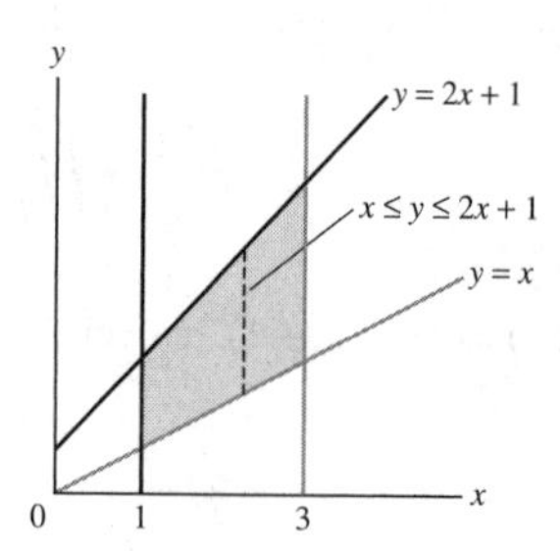

We compute the double integral of $f(x, y) = x^2 y$ on $\mathcal{D}$ by the following iterated integral:

$$\iint_{\mathcal{D}} x^2 y\,dA = \int_1^3 \int_x^{2x+1} x^2 y\,dy\,dx = \int_1^3 \frac{x^2 y^2}{2}\Big|_{y=x}^{2x+1} dx = \int_1^3 \frac{x^2}{2}\left((2x+1)^2 - x^2\right) dx$$

$$= \int_1^3 \left(\frac{3}{2}x^4 + 2x^3 + \frac{x^2}{2}\right) dx = \frac{3}{10}x^5 + \frac{x^4}{2} + \frac{x^3}{6}\Big|_1^3$$

$$= \frac{3 \cdot 3^5}{10} + \frac{3^4}{2} + \frac{3^3}{6} - \left(\frac{3}{10} + \frac{1}{2} + \frac{1}{6}\right) = \frac{1754}{15} \approx 116.93$$

21. $f(x, y) = 1, \quad 0 \le x \le 1, \quad 1 \le y \le e^x$

SOLUTION The domain $\mathcal{D}$ is a vertically simple region.

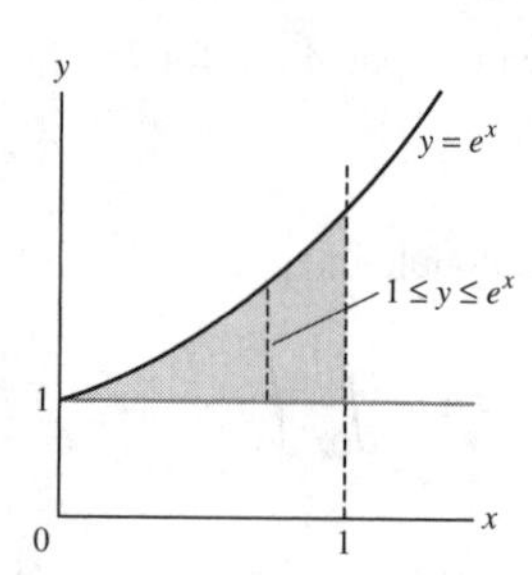

We compute the double integral of $f(x, y) = 1$ over $\mathcal{D}$ as the following iterated integral:

$$\iint_{\mathcal{D}} 1\,dA = \int_0^1 \int_1^{e^x} 1\,dy\,dx = \int_0^1 y\Big|_{y=1}^{e^x} dx = \int_0^1 (e^x - 1)\,dx = e^x - x\Big|_0^1$$

$$= (e^1 - 1) - (e^0 - 0) = e^1 - 1 - 1 = e - 2 \approx 0.718$$

23. $f(x, y) = x, \quad 0 \le x \le 1, \quad 1 \le y \le e^{x^2}$

SOLUTION We compute the double integral of $f(x, y) = x$ over the vertically simple region $\mathcal{D}$, as the following iterated integral:

$$\iint_{\mathcal{D}} x\,dA = \int_0^1 \int_1^{e^{x^2}} x\,dy\,dx = \int_0^1 xy\Big|_{y=1}^{e^{x^2}} dx = \int_0^1 \left(xe^{x^2} - x \cdot 1\right) dx$$

$$= \int_0^1 xe^{x^2}\,dx - \int_0^1 x\,dx = \int_0^1 xe^{x^2}\,dx - \frac{x^2}{2}\Big|_0^1 = \int_0^1 xe^{x^2}\,dx - \frac{1}{2} \tag{1}$$

The resulting integral can be computed using the substitution $u = x^2$. The value of this integral is

$$\int_0^1 xe^{x^2}\,dx = \frac{e-1}{2}$$

Combining with (1) we get

$$\iint_{\mathcal{D}} x\,dA = \frac{e-1}{2} - \frac{1}{2} = \frac{e-2}{2} \approx 0.359$$

25. $f(x, y) = \cos(2x + y), \quad \frac{1}{2} \le x \le \frac{\pi}{2}, \quad 1 \le y \le 2x$

SOLUTION The vertically simple region $\mathcal{D}$ defined by the given inequalities is shown in the figure:

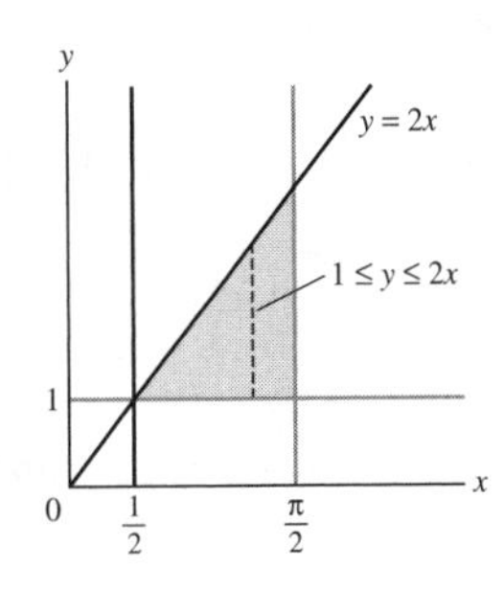

We compute the double integral of $f(x, y) = \cos(2x + y)$ over $\mathcal{D}$ as an iterated integral, as stated in Theorem 2. This gives

$$\begin{aligned}
\iint_{\mathcal{D}} \cos(2x+y)\,dA &= \int_{1/2}^{\pi/2}\int_1^{2x} \cos(2x+y)\,dy\,dx = \int_{1/2}^{\pi/2} \sin(2x+y)\Big|_{y=1}^{2x}\,dx \\
&= \int_{1/2}^{\pi/2} (\sin(2x+2x) - \sin(2x+1))\,dx = \int_{1/2}^{\pi/2} (\sin(4x) - \sin(2x+1))\,dx \\
&= -\frac{\cos 4x}{4} + \frac{\cos(2x+1)}{2}\Big|_{x=1/2}^{\pi/2} = -\frac{\cos\frac{4\pi}{2}}{4} + \frac{\cos\left(\frac{2\pi}{2}+1\right)}{2} - \left(-\frac{\cos 2}{4} + \frac{\cos 2}{2}\right) \\
&= -\frac{1}{4} + \frac{\cos(\pi+1)}{2} - \frac{\cos 2}{4} = -0.416
\end{aligned}$$

27. $f(x, y) = 2xy, \quad 0 \le y \le 1, \quad y^2 \le x \le y$

SOLUTION The intersection points of the graphs $x = y$ and $x = y^2$ are $(0, 0)$ $(1, 1)$. The horizontally simple region $\mathcal{D}$ is shown in the figure:

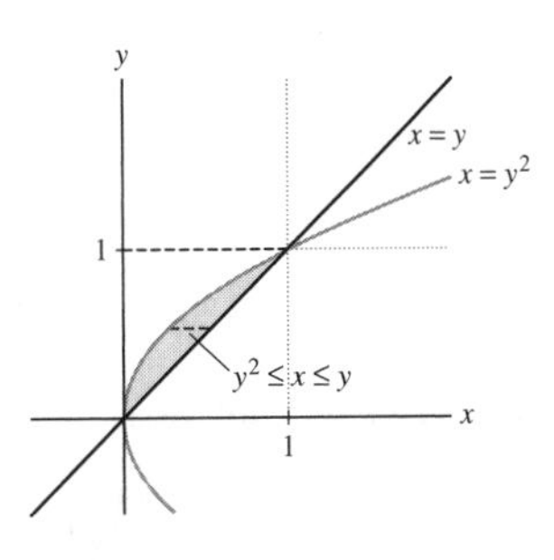

We compute the double integral of $f(x, y) = 2xy$ over $\mathcal{D}$, using Theorem 2. The limits of integration are determined by the inequalities:

$$0 \le y \le 1, \quad y^2 \le x \le y.$$

Defining $\mathcal{D}$, we get

$$\iint_{\mathcal{D}} 2xy\,dA = \int_0^1 \int_{y^2}^{y} 2xy\,dx\,dy = \int_0^1 x^2 y\Big|_{x=y^2}^{y} dy = \int_0^1 (y^2 \cdot y - y^4 \cdot y)\,dy$$

$$= \int_0^1 (y^3 - y^5)\,dy = \frac{y^4}{4} - \frac{y^6}{6}\Big|_0^1 = \frac{1}{4} - \frac{1}{6} = \frac{1}{12}$$

29. $f(x, y) = (x + y)^{-1}$, $1 \le y \le e$, $0 \le x \le y$

SOLUTION

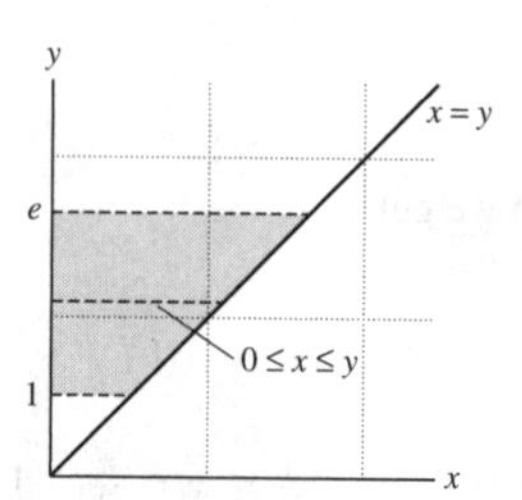

The double integral $f(x, y) = (x + y)^{-1}$ over the horizontally simple region $\mathcal{D}$ (shown in the figure) is computed, using Theorem 2, by the following iterated integral:

$$\iint_{\mathcal{D}} f(x, y)\,dA = \int_1^e \int_0^y (x + y)^{-1}\,dx\,dy = \int_1^e \ln(x + y)\Big|_{x=0}^{y} dy = \int_1^e (\ln(y + y) - \ln(0 + y))\,dy$$

$$= \int_1^e (\ln(2y) - \ln y)\,dy = \int_1^e \ln\frac{2y}{y}\,dy = \int_1^e \ln 2\,dy = (e - 1) \cdot \ln 2 \approx 1.19$$

In Exercises 31–34, sketch the domain of integration and express the iterated integral in the opposite order.

31. $\displaystyle\int_0^4 \int_x^4 f(x, y)\,dy\,dx$

SOLUTION The limits of integration correspond to the inequalities describing the following domain $\mathcal{D}$:

$$0 \le x \le 4, \quad x \le y \le 4$$

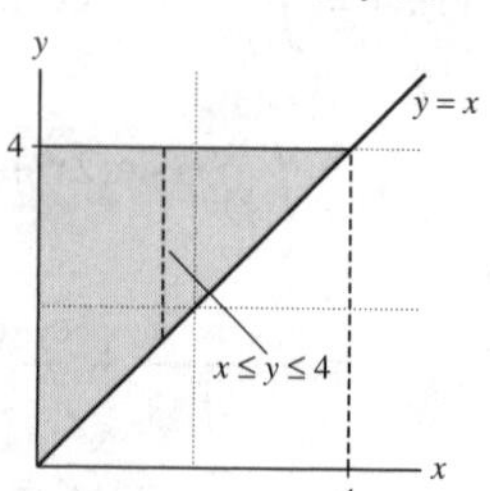

From the sketch of $\mathcal{D}$ we see that $\mathcal{D}$ can also be expressed as a horizontally simple region as follows:

$$0 \le y \le 4, \quad 0 \le x \le y$$

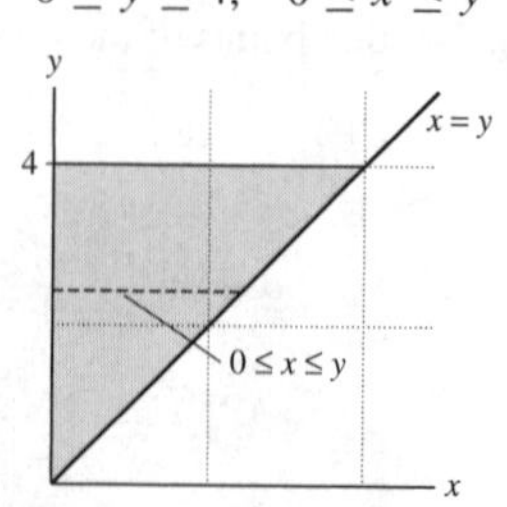

Therefore we can reverse the order of integration as follows:

$$\int_0^4 \int_x^4 f(x, y)\,dy\,dx = \int_0^4 \int_0^y f(x, y)\,dx\,dy.$$

33. $\displaystyle\int_4^9 \int_2^{\sqrt{y}} f(x, y)\,dx\,dy$

SOLUTION The limits of integration correspond to the following inequalities defining the horizontally simple region $\mathcal{D}$:

$$4 \le y \le 9, \quad 2 \le x \le \sqrt{y}$$

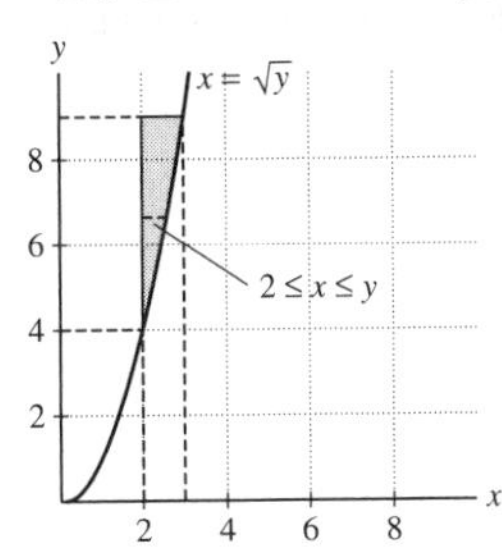

The region $\mathcal{D}$ can also be expressed as a vertically simple region. We first need to write the equation of the curve $x = \sqrt{y}$ in the form $y = x^2$. The corresponding inequalities are

$$2 \le x \le 3, \quad x^2 \le y \le 9$$

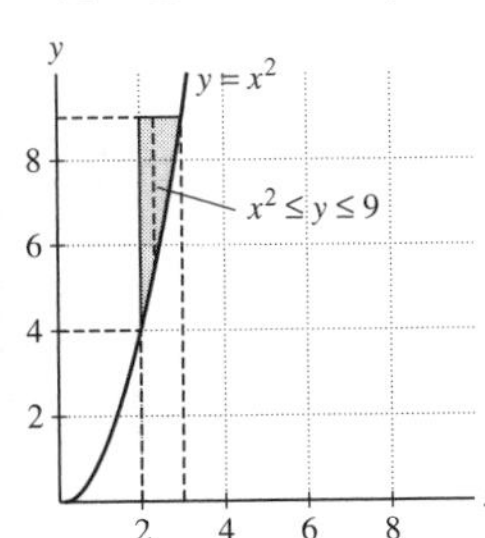

We now can write the iterated integral with reversed order of integration:

$$\int_4^9 \int_2^{\sqrt{y}} f(x, y)\, dx\, dy = \int_2^3 \int_{x^2}^9 f(x, y)\, dy\, dx.$$

35. According to Eq. (3), the area of a domain $\mathcal{D}$ is equal to $\iint_{\mathcal{D}} 1\, dA$. Prove that if $\mathcal{D}$ is the region between two curves $y = \alpha(x)$ and $y = \beta(x)$ for $a \le x \le b$, then

$$\iint_{\mathcal{D}} 1\, dA = \int_a^b (\beta(x) - \alpha(x))\, dx$$

The integral on the right is the area of $\mathcal{D}$ as defined in Chapter 6.

SOLUTION The region $\mathcal{D}$ is defined by the inequalities

$$a \le x \le b, \quad \alpha(x) \le y \le \beta(x)$$

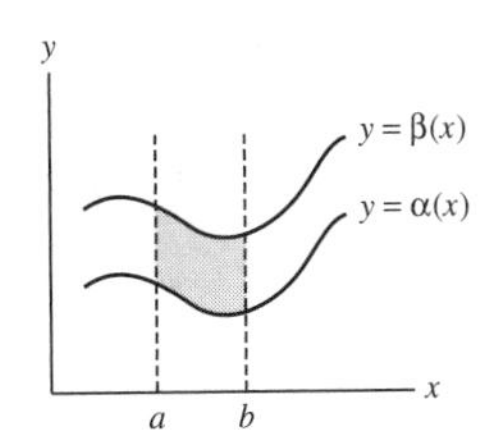

We compute the double integral of $f(x, y) = 1$ on $\mathcal{D}$, using Theorem 2, by evaluating the following iterated integral:

$$\int_{\mathcal{D}} 1\, dA = \int_a^b \int_{\alpha(x)}^{\beta(x)} 1\, dy\, dx = \int_a^b \left(\int_{\alpha(x)}^{\beta(x)} 1\, dy \right) dx = \int_a^b y \Big|_{y=\alpha(x)}^{\beta(x)} dx = \int_a^b \big(\beta(x) - \alpha(x)\big)\, dx$$

In Exercises 37–44, for each double integral, sketch the region. Then change the order of integration and evaluate. Explain the simplification achieved by interchanging the order.

37. $\displaystyle \int_0^1 \int_y^1 \frac{\sin x}{x}\, dx\, dy$

SOLUTION The limits of integration correspond to the following inequalities:

$$0 \le y \le 1, \quad y \le x \le 1$$

The horizontally simple region $\mathcal{D}$ is shown in the figure.

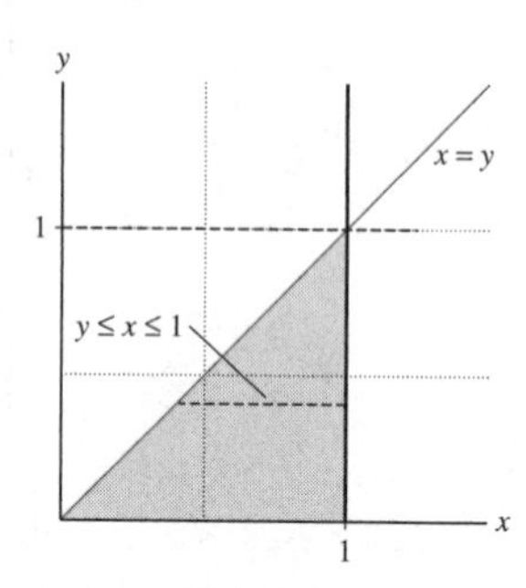

We see that $\mathcal{D}$ can also be described as a vertically simple region, by the following inequalities:

$$0 \le x \le 1, \quad 0 \le y \le x$$

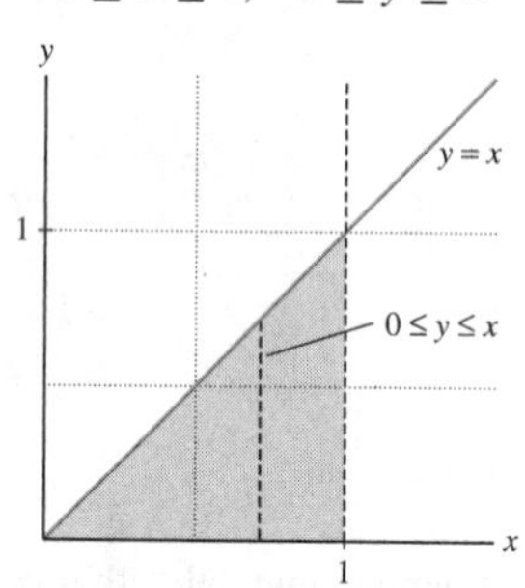

We evaluate the corresponding iterated integral:

$$\int_0^1 \int_0^x \frac{\sin x}{x}\, dy\, dx = \int_0^1 \frac{\sin x}{x} y \Big|_{y=0}^{x} dx = \int_0^1 \frac{\sin x}{x}(x-0)\, dx = \int_0^1 \sin x\, dx = -\cos x \Big|_0^1 = 1 - \cos 1 \approx 0.46$$

Trying to integrate in reversed order we obtain a complicated integral in the inner integral. That is,

$$\int_0^1 \int_y^1 \frac{\sin x}{x}\, dx\, dy = \int_0^1 \left(\int_y^1 \frac{\sin x}{x}\, dx \right) dy$$

Remark: $f(x, y) = \frac{\sin x}{x}$ is not continuous at the point $(0, 0)$ in $\mathcal{D}$. To make it continuous we need to define $f(0, 0) = 1$.

39. $\displaystyle\int_0^1 \int_{y=x}^1 x e^{y^3}\, dy\, dx$

SOLUTION The limits of integration define a vertically simple region $\mathcal{D}$ by the following inequalities:

$$0 \le x \le 1, \quad x \le y \le 1$$

This region can also be described as a horizontally simple region by the following inequalities (see figure):

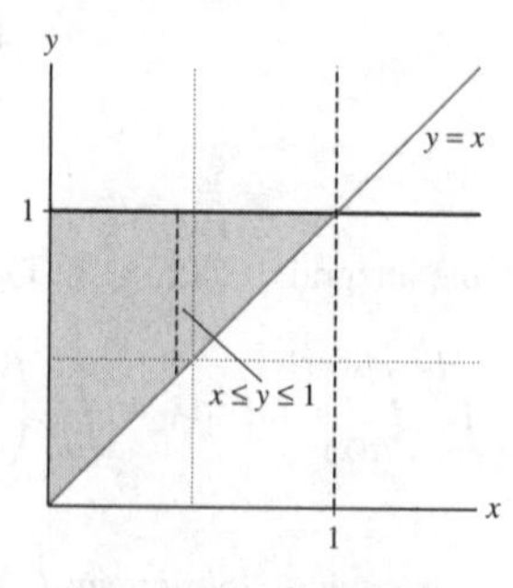

$$0 \le y \le 1, \quad 0 \le x \le y$$

We thus can rewrite the given integral in reversed order of integration as follows:

$$\int_0^1 \int_0^y x e^{y^3}\, dx\, dy = \int_0^1 \frac{x^2}{2} e^{y^3} \Big|_{x=0}^{y} dy = \int_0^1 e^{y^3} \left(\frac{y^2}{2} - 0 \right) dy = \int_0^1 \frac{1}{2} e^{y^3} y^2\, dy$$

We compute this integral using the substitution $u = y^3$, $du = 3y^2\,dy$. This gives

$$\int_0^1 \int_0^y xe^{y^3}\,dx\,dy = \int_0^1 \frac{1}{2}e^{y^3}y^2\,dy = \int_0^1 e^u \cdot \frac{1}{6}\,du = \frac{e^u}{6}\bigg|_0^1 = \frac{e-1}{6} \approx 0.286$$

Trying to evaluate the double integral in the original order of integration, we find that the inner integral is impossible to compute:

$$\int_0^1 \int_x^1 xe^{y^3}\,dy\,dx = \int_0^1 \left(\int_x^1 xe^{y^3}\,dy\right)dx$$

41. $\displaystyle\int_0^1 \int_0^{\pi/2} x\cos(xy)\,dx\,dy$

SOLUTION The domain of integration is the rectangle defined by the following inequalities:

$$0 \le y \le 1, \quad 0 \le x \le \frac{\pi}{2}$$

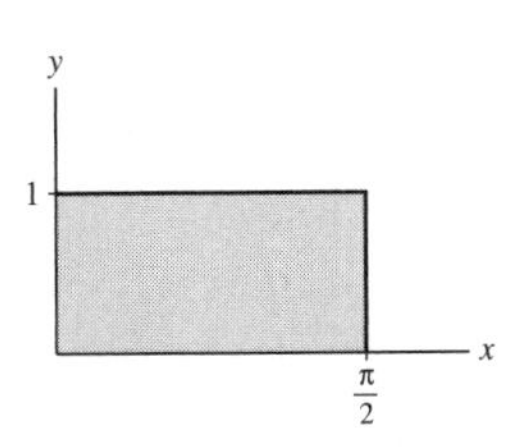

By Fubini's Theorem, the double integral of $f(x, y) = x\cos(xy)$ over the rectangle is equal to the iterated integral in either order. Hence,

$$\int_0^1 \int_0^{\pi/2} x\cos(xy)\,dx\,dy = \int_0^{\pi/2} \int_0^1 x\cos(xy)\,dy\,dx = \int_0^{\pi/2} x \cdot \frac{1}{x}(\sin xy)\bigg|_{y=0}^1 dx = \int_0^{\pi/2} \sin(xy)\bigg|_{y=0}^1 dx$$

$$= \int_0^{\pi/2} (\sin x - \sin 0)\,dx = \int_0^{\pi/2} \sin x\,dx = -\cos x\bigg|_0^{\pi/2}$$

$$= -\left(\cos\frac{\pi}{2} - \cos 0\right) = -(0-1) = 1$$

Trying to integrate in the original order of integration, we obtain

$$\int_0^1 \int_0^{\pi/2} x\cos(xy)\,dx\,dy = \int_0^1 \left(\int_0^{\pi/2} x\cos(xy)\,dx\right)dy \tag{1}$$

To compute the inner integral we would have to use integration by parts, whereas the integrals involved in computing the integral first with respect to x and then with respect to y were quite easy to compute.

43. $\displaystyle\int_0^9 \int_0^{\sqrt{y}} \frac{x^3\,dx\,dy}{(3x^2+y)^{1/2}}$

SOLUTION The limits of integration correspond to the inequalities describing the domain $\mathcal{D}$. That is,

$$0 \le y \le 9, \quad 0 \le x \le \sqrt{y}$$

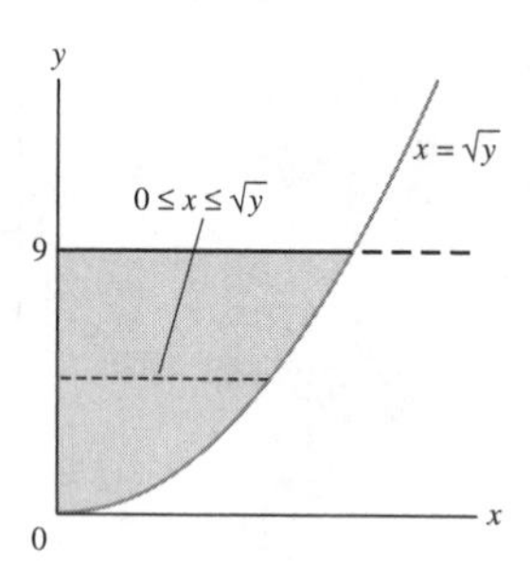

The horizontally simple region $\mathcal{D}$ (shown in the figure) can also be expressed as a vertically simple region. We rewrite the equation of the curve $x = \sqrt{y}$ as $y = x^2$ and describe $\mathcal{D}$ by the following inequalities:

$$0 \le x \le 3, \quad x^2 \le y \le 9$$

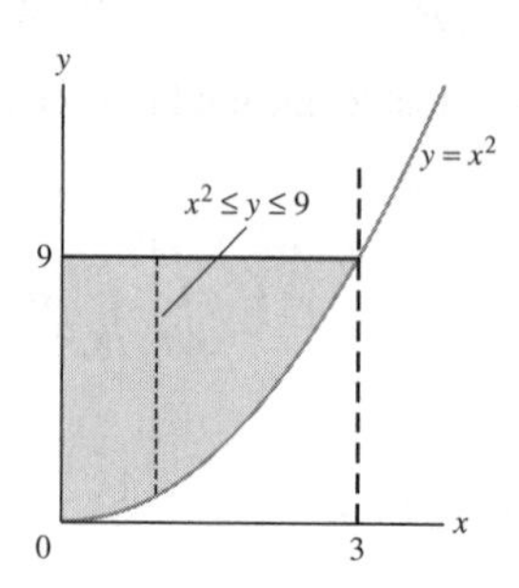

Using Theorem 2 we obtain the following integral:

$$\int_0^3 \int_{x^2}^9 \frac{x^3\,dy\,dx}{(3x^2+y)^{1/2}} = \int_0^3 \left(\int_{x^2}^9 \frac{x^3}{(3x^2+y)^{1/2}}\,dy \right) dx = \int_0^3 2x^3(3x^2+y)^{1/2}\Big|_{y=x^2}^9 \,dx$$

$$= \int_0^3 2x^3\left((3x^2+9)^{1/2} - (3x^2+x^2)^{1/2}\right) dx = \int_0^3 2x^3\left((3x^2+9)^{1/2} - 2x\right) dx$$

$$= \int_0^3 2x^3\sqrt{3x^2+9}\,dx - \int_0^3 4x^4\,dx = \int_0^3 2x^3\sqrt{3x^2+9}\,dx - \frac{4}{5}x^5\Big|_0^3$$

$$= \int_0^3 2x^3\sqrt{3x^2+9}\,dx - 194.4 \tag{1}$$

We compute the integral using the substitution $u = \sqrt{3x^2+9}$. Then,

$$du = \frac{6x}{2u}dx = \frac{3x}{u}dx, \quad \text{that is,} \quad x\,dx = \frac{u}{3}\,du.$$

Also $u^2 = 3x^2 + 9$ or $x^2 = \frac{u^2}{3} - 3$. Therefore,

$$\int_0^3 2x^3\sqrt{3x^2+9}\,dx = \int_0^3 2x^2\sqrt{3x^2+9}\cdot x\,dx = \int_3^6 \left(\frac{2u^2}{3} - 6\right) u \cdot \frac{u}{3}\,du$$

$$= \int_3^6 \left(\frac{2}{9}u^4 - 2u^2\right) du = \frac{2}{45}u^5 - \frac{2}{3}u^3\Big|_3^6$$

$$= \frac{2\cdot 6^5}{45} - \frac{2\cdot 6^3}{3} - \left(\frac{2\cdot 3^5}{45} - \frac{2\cdot 3^3}{3}\right) = \frac{9396}{45} = 208.8$$

Combining with (1) we get

$$\int_0^3 \int_{x^2}^9 \frac{x^3\,dy\,dx}{(3x^2+y)^{1/2}} = \int_0^3 2x^3\sqrt{3x^2+9}\,dx - 194.4 = 208.8 - 194.4 = 14.4$$

Trying to evaluate the iterated integral in the original order of integration gives:

$$\int_0^9 \int_0^{\sqrt{y}} \frac{x^3\,dx\,dy}{(3x^2+y)^{1/2}} = \int_0^9 \left(\int_0^{\sqrt{y}} \frac{x^3}{(3x^2+y)^{1/2}}\,dx \right) dy$$

The inner integral is much more difficult to compute than the integrals involved in the previous computation.

45. Sketch the region $\mathcal{D}$ bounded by the curves $y = e^x$, $y = e^{\sqrt{x}}$, $y = 1$, and $y = 2$, and compute $\iint_{\mathcal{D}} (\ln y)^{-1}\, dA$ by writing $\mathcal{D}$ as a horizontally simple region.

SOLUTION The region $\mathcal{D}$ bounded by the curves $y = e^x$, $y = e^{\sqrt{x}}$, $y = 1$, and $y = 2$ is shown in the following figure:

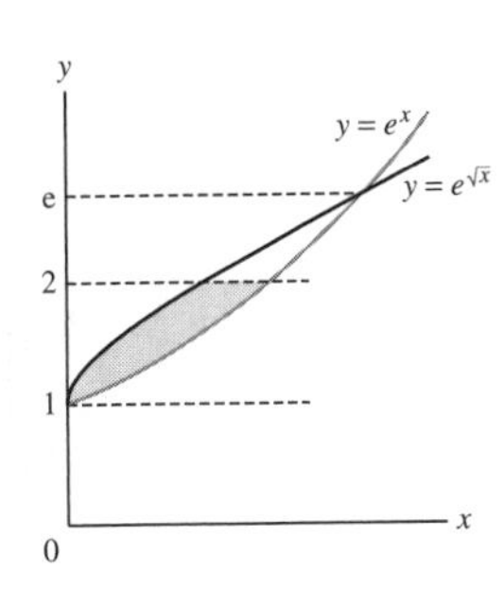

To express $\mathcal{D}$ as a horizontally simple region, we first must rewrite the equations of the curves $y = e^x$ and $y = e^{\sqrt{x}}$ with x as a function of y. That is,

$$y = e^x \quad \Rightarrow \quad x = \ln y$$
$$y = e^{\sqrt{x}} \quad \Rightarrow \quad \sqrt{x} = \ln y \quad \Rightarrow \quad x = \ln^2 y$$

We obtain the following inequalities:

$$1 \le y \le 2, \quad \ln^2 y \le x \le \ln y$$

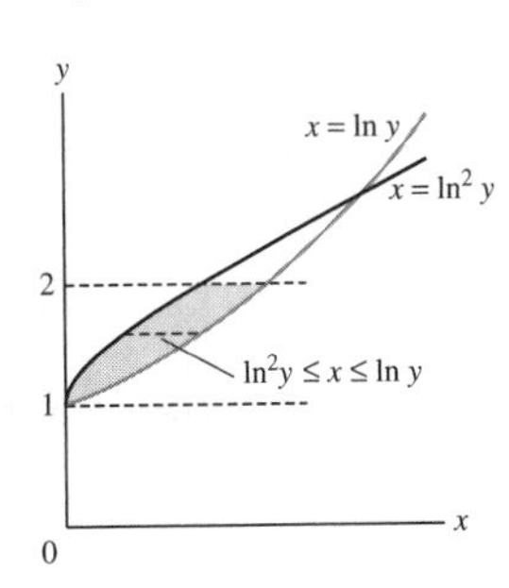

Using Theorem 2, we compute the double integral of $f(x, y) = (\ln y)^{-1}$ over $\mathcal{D}$ as the following iterated integral:

$$\iint_{\mathcal{D}} (\ln y)^{-1}\, dA = \int_1^2 \int_{\ln^2 y}^{\ln y} (\ln y)^{-1}\, dx\, dy = \int_1^2 (\ln y)^{-1} x \Big|_{x=\ln^2 y}^{\ln y} dy = \int_1^2 (\ln y)^{-1} \left(\ln y - \ln^2 y\right) dy$$
$$= \int_1^2 (1 - \ln y)\, dy = \int_1^2 1\, dy - \int_1^2 \ln y\, dy = y\Big|_1^2 - y(\ln y - 1)\Big|_1^2$$
$$= 1 - (2(\ln 2 - 1) - 1(\ln 1 - 1)) = 2 - 2\ln 2 \approx 0.614$$

In Exercises 47–50, calculate the double integral of $f(x, y)$ over the triangle indicated in Figure 22.

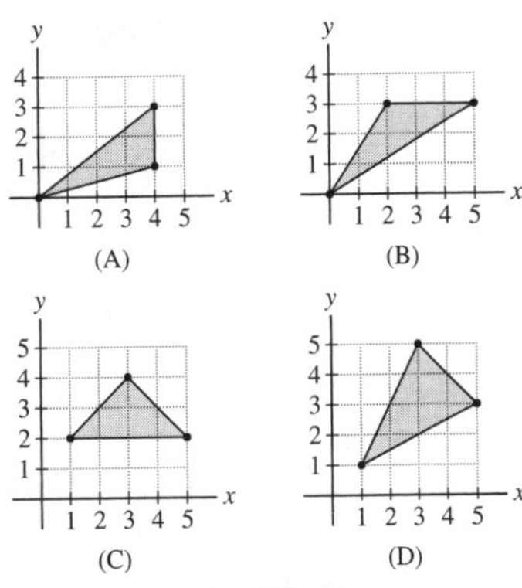

FIGURE 22

47. $f(x, y) = ye^x$, (A)

SOLUTION The equations of the lines OA and OB are $y = \frac{3}{4}x$ and $y = \frac{1}{4}x$, respectively. Therefore, the triangle may be expressed as a vertically simple region by the following inequalities:

$$0 \le x \le 4, \quad \frac{x}{4} \le y \le \frac{3x}{4}$$

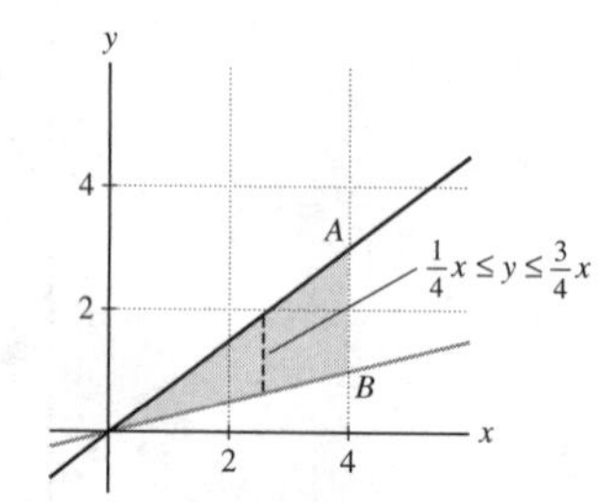

The double integral of $f(x, y) = ye^x$ over the triangle is the following iterated integral:

$$\int_0^4 \int_{x/4}^{3x/4} ye^x \, dy \, dx = \int_0^4 \frac{1}{2}y^2 e^x \Big|_{y=x/4}^{3x/4} dx = \int_0^4 \frac{1}{2}e^x \left(\frac{9x^2}{16} - \frac{x^2}{16}\right) dx = \int_0^4 \frac{x^2}{4} e^x \, dx$$

We use the integration formula

$$\int x^2 e^x \, dx = x^2 e^x - 2xe^x + 2e^x + C$$

This gives

$$\int_0^4 \int_{x/4}^{3x/4} ye^x \, dy \, dx = \frac{1}{4}\int_0^4 x^2 e^x \, dx = \frac{x^2 e^x}{4} - \frac{xe^x}{2} + \frac{e^x}{2}\Big|_0^4 = \frac{16e^4}{4} - \frac{4e^4}{2} + \frac{e^4}{2} - \frac{1}{2} = \frac{5e^4}{2} - \frac{1}{2} \approx 136$$

49. $f(x, y) = \frac{x}{y^2}$, (C)

SOLUTION To find the inequalities defining the triangle as a horizontally simple region, we first find the inequalities of the lines AB and BC:

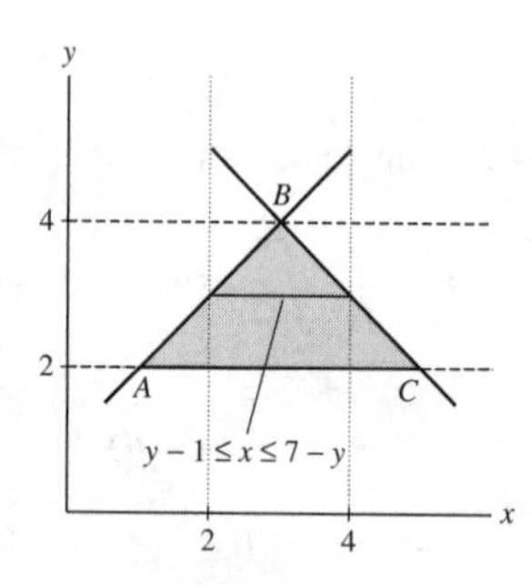

$$AB: \quad y - 2 = \frac{4-2}{3-1}(x - 1) \quad \Rightarrow \quad y - 2 = x - 1 \quad \Rightarrow \quad x = y - 1$$

$$BC: \quad y - 2 = \frac{4-2}{3-5}(x - 5) \quad \Rightarrow \quad y - 2 = 5 - x \quad \Rightarrow \quad x = 7 - y$$

We obtain the following inequalities for the triangle:

$$2 \le y \le 4, \quad y - 1 \le x \le 7 - y$$

The double integral of $f(x, y) = \frac{x}{y^2}$ over the triangle is the following iterated integral:

$$\int_2^4 \int_{y-1}^{7-y} \frac{x}{y^2} \, dx \, dy = \int_2^4 \frac{x^2}{2y^2}\Big|_{x=y-1}^{7-y} dy = \int_2^4 \frac{(7-y)^2 - (y-1)^2}{2y^2} \, dy = \int_2^4 \left(\frac{24}{y^2} - \frac{6}{y}\right) dy$$

$$= -\frac{24}{y} - 6\ln y \Big|_2^4 = -\frac{24}{4} - 6\ln 4 - \left(-\frac{24}{2} - 6\ln 2\right) = 6 - 6\ln 2 = 1.84$$

51. Calculate the double integral $f(x, y) = \dfrac{\sin y}{y}$ over the region $\mathcal{D}$ in Figure 23.

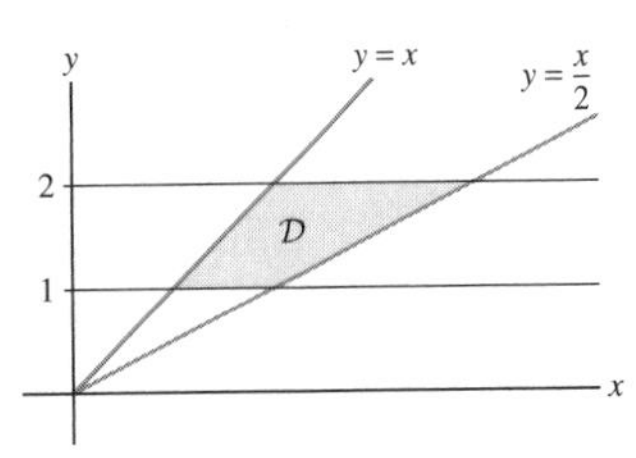

FIGURE 23

SOLUTION To describe $\mathcal{D}$ as a horizontally simple region, we rewrite the equations of the lines with x as a function of y, that is, $x = y$ and $x = 2y$. The inequalities for $\mathcal{D}$ are

$$1 \le y \le 2, \quad y \le x \le 2y$$

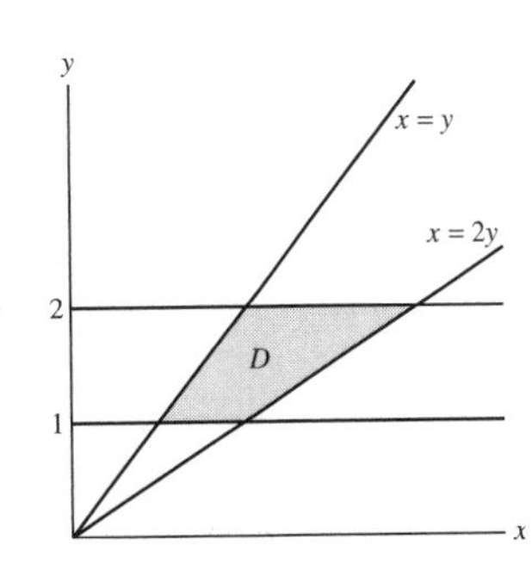

We now compute the double integral of $f(x, y) = \frac{\sin y}{y}$ over $\mathcal{D}$ by the following iterated integral:

$$\iint_{\mathcal{D}} \frac{\sin y}{y}\, dA = \int_1^2 \int_y^{2y} \frac{\sin y}{y}\, dx\, dy = \int_1^2 \frac{\sin y}{y} x \Big|_{x=y}^{2y} dy = \int_1^2 \frac{\sin y}{y}(2y - y)\, dy$$

$$= \int_1^2 \frac{\sin y}{y} \cdot y\, dy = \int_1^2 \sin y\, dy = -\cos y \Big|_1^2 = \cos 1 - \cos 2 \approx 0.956$$

53. Find the volume of the region in Figure 25 bounded by the paraboloids $z = x^2 + y^2$ and $z = 8 - x^2 - y^2$.

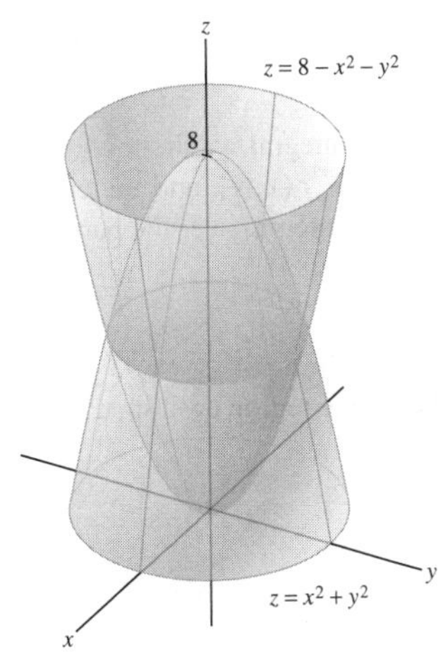

FIGURE 25

SOLUTION

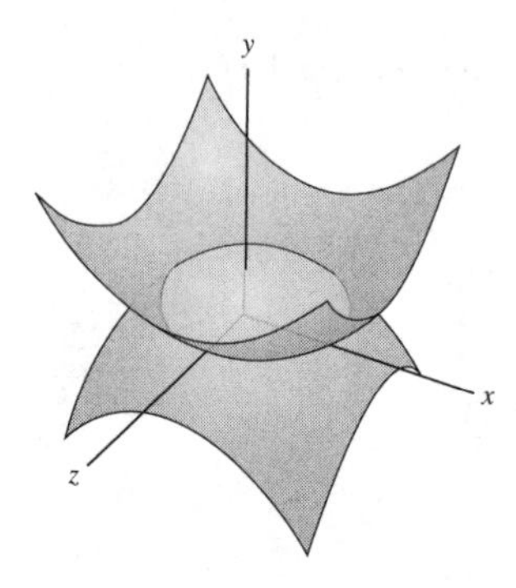

We first must identify the domain of integration in the xy-plane. To do this we equate the equations of the two surfaces. That is,

$$x^2 + y^2 = 8 - x^2 - y^2 \quad \Rightarrow \quad 2(x^2 + y^2) = 8 \quad \Rightarrow \quad x^2 + y^2 = 4$$

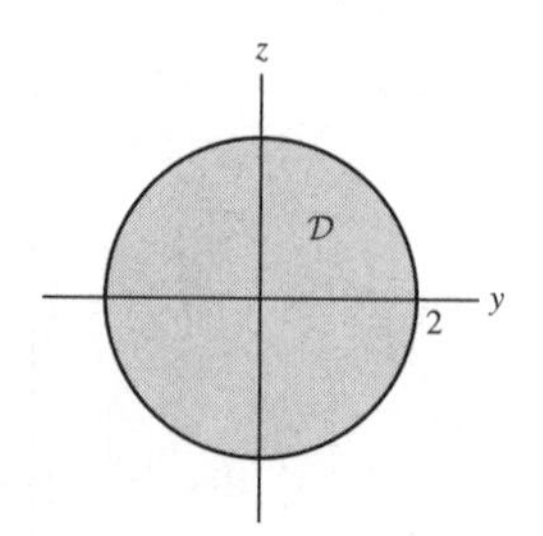

Therefore, the domain of integration is

$$\mathcal{D} = \left\{(x, y); x^2 + y^2 \le 4\right\}$$

The double integrals $\iint_{\mathcal{D}} (x^2 + y^2)\,dA$ and $\iint_{\mathcal{D}} (8 - x^2 - y^2)\,dA$ give the volumes of the regions bounded by the graphs of $z = x^2 + y^2$ and $z = 8 - x^2 - y^2$, respectively, and the xy-plane over the region $\mathcal{D}$. Therefore, the volume of the required region is the difference between these two integrals (see figure). That is,

$$V = \iint_{\mathcal{D}} (8 - x^2 - y^2)\,dA - \iint_{\mathcal{D}} (x^2 + y^2)\,dA = \iint_{\mathcal{D}} \left((8 - x^2 - y^2) - (x^2 + y^2)\right)\,dA$$

or

$$V = \iint_{\mathcal{D}} (8 - 2x^2 - 2y^2)\,dA$$

To compute the double integral we first notice that, since the function $f(x, y) = 8 - 2x^2 - 2y^2$ satisfies $f(x, -y) = f(x, y)$ and $f(-x, y) = f(x, y)$, and the circle $\mathcal{D}$ is symmetric with respect to the x and y axes, the double integral over $\mathcal{D}$ is four times the integral over the part $\mathcal{D}_1$ of the circle in the first quadrant. That is,

$$V = 4\iint_{\mathcal{D}_1} (8 - 2x^2 - 2y^2)\,dA$$

$\mathcal{D}_1$ is the vertically simple region described by the inequalities

$$0 \le x \le 2, \quad 0 \le y \le \sqrt{4 - x^2}$$

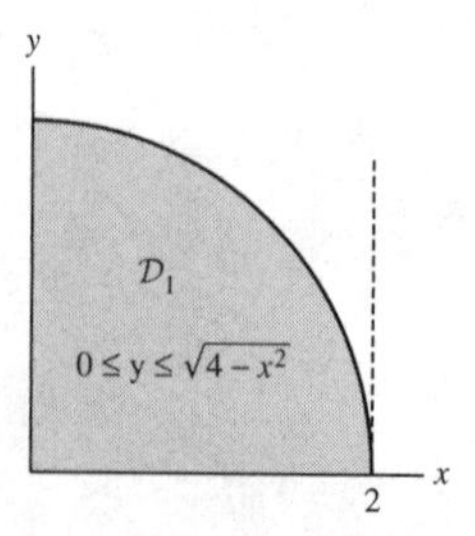

Therefore,

$$V = 4\iint_{\mathcal{D}_1} (8 - 2x^2 - 2y^2)\,dA = 4\int_0^2 \int_0^{\sqrt{4-x^2}} (8 - 2x^2 - 2y^2)\,dy\,dx$$

$$= 4\int_0^2 8y - 2x^2y - \frac{2}{3}y^3\Big|_{y=0}^{\sqrt{4-x^2}} dx = 4\int_0^2 8\sqrt{4-x^2} - 2x^2\sqrt{4-x^2} - \frac{2}{3}(4-x^2)^{2/3}\,dx$$

$$= 4\int_0^2 2\sqrt{4-x^2}(4-x^2) - \frac{2}{3}(4-x^2)^{3/2}\,dx = 4\int_0^2 \frac{4}{3}(4-x^2)^{3/2}\,dx = \frac{16}{3}\int_0^2 (4-x^2)^{3/2}\,dx$$

$$= \frac{16}{3}\cdot 4\int_0^2 \sqrt{4-x^2}\,dx - \frac{16}{3}\int_0^2 x^2\sqrt{4-x^2}\,dx \tag{1}$$

If we let $x = 2\sin\theta$, then $dx = 2\cos\theta\,d\theta$, and so

$$V = \frac{16}{3}\int_0^{\pi/2} (4 - 4\sin^2\theta)^{3/2}\cdot\cos\theta\,d\theta = \frac{16}{3}\int_0^{\pi/2} 8\cos^3\theta\cdot\cos\theta\,d\theta$$

$$= \frac{16}{3}\int_0^{\pi/2} 8\cos^4\theta\,d\theta = \frac{128}{3}\int_0^{\pi/2}\cos^4\theta\,d\theta$$

Using trig formulas and integration formulas, we find that $\int_0^{\pi/2}\cos^4\theta\,d\theta = 3\pi/8$, and so $V = 16\pi$.

55. Find the volume of the region enclosed by $z = 1 - y^2$ and $z = y^2 - 1$ for $0 \le x \le 2$.

SOLUTION

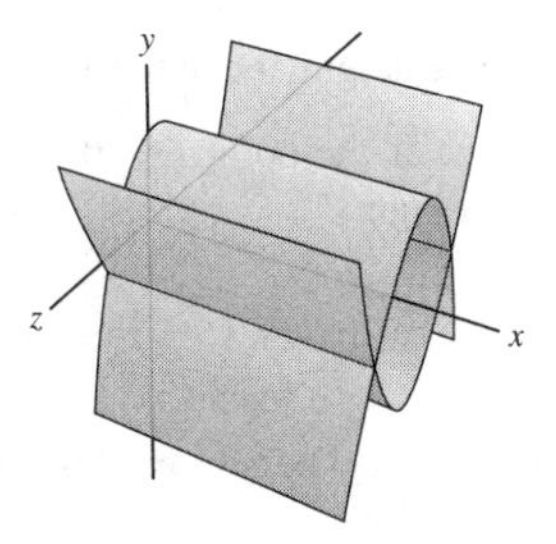

The volume of the region is the double integral of $f(y, z) = 2$ over the domain $\mathcal{D}$ in the yz-plane between the curves $z = 1 - y^2$ and $z = y^2 - 1$.

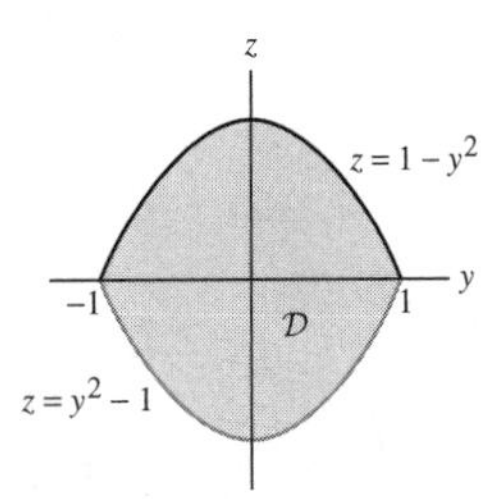

This domain is the vertically simple region described by the inequalities

$$-1 \le y \le 1, \quad y^2 - 1 \le z \le 1 - y^2$$

We compute the double integral as the following iterated integral:

$$V = \iint_{\mathcal{D}} 2\,dA = \int_{-1}^{1}\int_{y^2-1}^{1-y^2} 2\,dz\,dy = \int_{-1}^{1} 2z\Big|_{z=y^2-1}^{1-y^2} dy = \int_{-1}^{1} 2\left((1-y^2) - (y^2-1)\right)dy$$

$$= \int_{-1}^{1}(4 - 4y^2)\,dy = \int_0^1 (8 - 8y^2)\,dy = 8y - \frac{8}{3}y^3\Big|_0^1 = 8 - \frac{8}{3} = \frac{16}{3} \approx 5.33$$

57. A plate in the shape of the region bounded by $y = x^{-1}$ and $y = 0$ for $1 \le x \le 4$ has mass density $\rho(x, y) = y/x$. Calculate the total mass of the plate.

SOLUTION The total mass M of the plate is obtained by computing the double integral of mass density $\rho(x, y) = \frac{y}{x}$ over the region $\mathcal{D}$ shown in the figure.

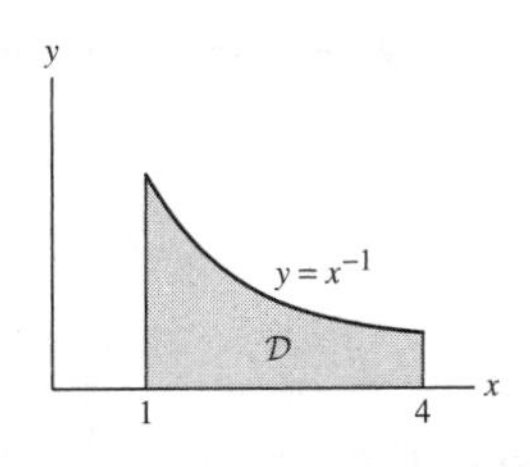

$\mathcal{D}$ is a vertically simple region defined by the following inequalities:

$$1 \le x \le 4, \quad 0 \le y \le x^{-1}$$

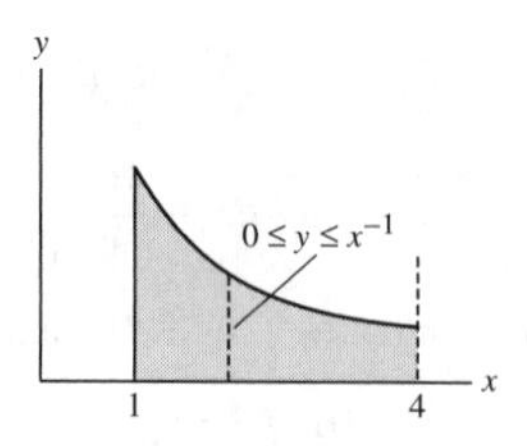

We compute the double integral using Theorem 2. That is,

$$M = \iint_{\mathcal{D}} \rho(x, y)\, dA = \int_1^4 \int_0^{x^{-1}} \frac{y}{x}\, dy\, dx = \int_1^4 \frac{y^2}{2x}\bigg|_{y=0}^{x^{-1}} dx = \int_1^4 \frac{\left(x^{-1}\right)^2 - 0^2}{2x}\, dx = \int_1^4 \frac{1}{2}x^{-3}\, dx$$

$$= -\frac{1}{4}x^{-2}\bigg|_{x=1}^{4} = -\frac{1}{4}(4^{-2} - 1^{-2}) = \frac{15}{64}$$

59. Calculate the average height above the x-axis of a point in the region $0 \le x \le 1, 0 \le y \le x^2$.

SOLUTION The height of the point (x, y) in the region $\mathcal{D}$ above the x-axis is $f(x, y) = y$. Therefore, the average height is the following value:

$$\overline{H} = \frac{1}{\text{Area}(\mathcal{D})} \iint_{\mathcal{D}} y\, dA \tag{1}$$

We first compute the integral. The region $\mathcal{D}$ is a vertically simple region defined by the inequalities

$$0 \le x \le 1, \quad 0 \le y \le x^2$$

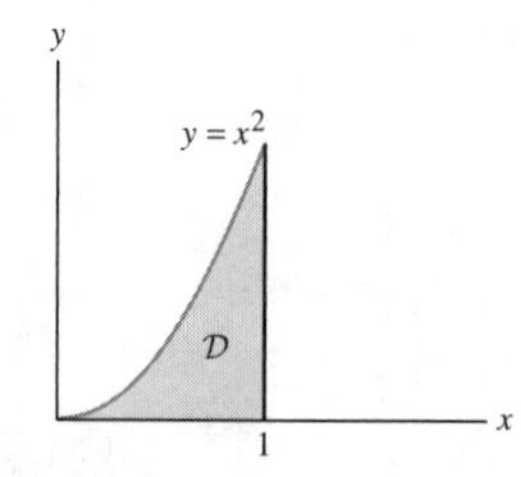

Therefore,

$$\iint_{\mathcal{D}} y\, dA = \int_0^1 \int_0^{x^2} y\, dy\, dx = \int_0^1 \frac{1}{2}y^2\bigg|_{y=0}^{x^2} dx = \int_0^1 \frac{1}{2}\left(x^4 - 0\right) dx = \int_0^1 \frac{1}{2}x^4\, dx = \frac{x^5}{10}\bigg|_0^1 = \frac{1}{10} \tag{2}$$

We compute the area of $\mathcal{D}$ using the formula for the area as a double integral:

$$\text{Area}(\mathcal{D}) = \iint_{\mathcal{D}} 1\, dA = \int_0^1 \int_0^{x^2} dy\, dx = \int_0^1 y\bigg|_{y=0}^{x^2} dx = \int_0^1 x^2\, dx = \frac{x^3}{3}\bigg|_0^1 = \frac{1}{3} \tag{3}$$

Substituting (2) and (3) in (1) we obtain

$$\overline{H} = \frac{1}{\frac{1}{3}} \cdot \frac{1}{10} = \frac{3}{10}$$

61. Calculate the average value of the x-coordinate of a point on the semicircle $x^2 + y^2 \le R^2$, $x \ge 0$. What is the average value of the y-coordinate?

SOLUTION The average value of the x-coordinates of a point on the semicircle $\mathcal{D}$ is

$$\overline{x} = \frac{1}{\text{Area}(\mathcal{D})} \iint_{\mathcal{D}} x\, dA$$

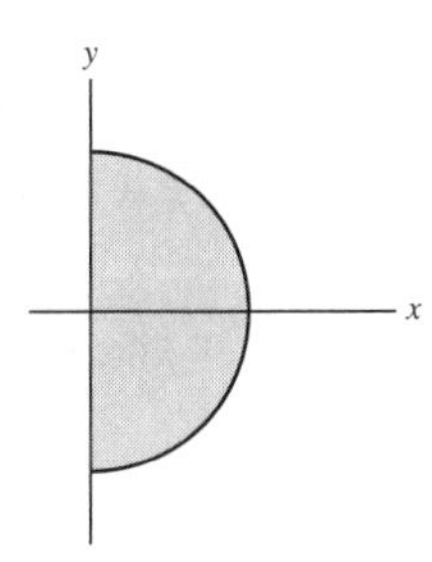

The area of the semicircle is $\frac{\pi R^2}{2}$. To compute the double integral, we identify the inequalities defining $\mathcal{D}$ as a horizontally simple region:

$$-R \le y \le R, \quad 0 \le x \le \sqrt{R^2 - y^2}$$

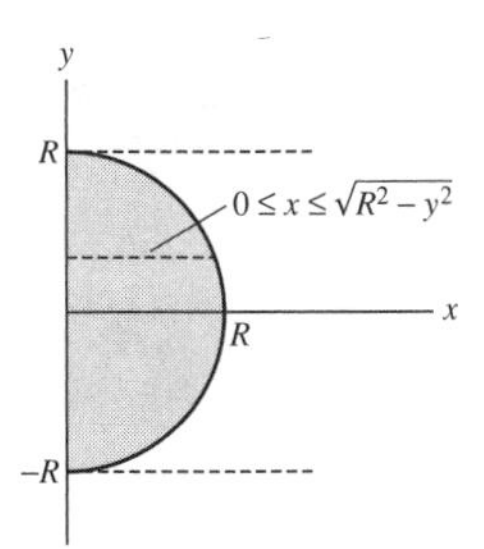

Therefore,

$$\overline{x} = \frac{1}{\frac{\pi R^2}{2}} \int_{-R}^{R} \int_{0}^{\sqrt{R^2-y^2}} x\,dx\,dy = \frac{2}{\pi R^2} \int_{-R}^{R} \frac{x^2}{2}\bigg|_{x=0}^{\sqrt{R^2-y^2}} dy = \frac{2}{\pi R^2} \int_{-R}^{R} \frac{\left(\sqrt{R^2-y^2}\right)^2 - 0^2}{2}\,dy$$

$$= \frac{1}{\pi R^2} \int_{-R}^{R} \left(R^2 - y^2\right) dy = \frac{2}{\pi R^2} \int_{0}^{R} \left(R^2 - y^2\right) dy = \frac{2}{\pi R^2} \left(R^2 y - \frac{y^3}{3}\right)\bigg|_{y=0}^{R}$$

$$= \frac{2}{\pi R^2} \left(R^3 - \frac{R^3}{3}\right) = \frac{2}{\pi R^2} \cdot \frac{2R^3}{3} = \frac{4R}{3\pi}$$

The average value of the x-coordinate is $\overline{x} = \frac{4R}{3\pi}$. The average value of the y-coordinate is

$$\overline{y} = \frac{1}{\text{Area}(\mathcal{D})} \iint_{\mathcal{D}} y\,dA = \frac{1}{\frac{\pi R^2}{2}} \int_{-R}^{R} \int_{0}^{\sqrt{R^2-y^2}} y\,dx\,dy = \frac{2}{\pi R^2} \int_{-R}^{R} yx\bigg|_{x=0}^{\sqrt{R^2-y^2}} dy$$

$$= \frac{2}{\pi R^2} \int_{-R}^{R} y\left(\sqrt{R^2-y^2} - 0\right) dy = \frac{2}{\pi R^2} \int_{-R}^{R} y\sqrt{R^2-y^2}\,dy = 0$$

(The integral of an odd function over a symmetric interval with respect to the x-axis is zero). The average value of the y-coordinate is $\overline{y} = 0$.

Remark: Since the region is symmetric with respect to the x-axis, we expect the average value of y to be zero.

63. What is the average value of the linear function $f(x, y) = mx + ny + p$ on the ellipse $\left(\frac{x}{a}\right)^2 + \left(\frac{y}{b}\right)^2 \le 1$? Argue by symmetry rather than calculation.

SOLUTION The average value of the linear function $f(x, y) = mx + ny + p$ over the ellipse $\mathcal{D}$ is

$$\overline{f} = \frac{1}{\text{Area}(\mathcal{D})} \iint_{\mathcal{D}} f(x, y)\,dA = \frac{1}{\text{Area}(\mathcal{D})} \iint_{\mathcal{D}} (mx + ny + p)\,dA$$

$$= m \cdot \underbrace{\frac{1}{\text{Area}(\mathcal{D})} \iint_{\mathcal{D}} x\,dA}_{I_1} + n \cdot \underbrace{\frac{1}{\text{Area}(\mathcal{D})} \iint_{\mathcal{D}} y\,dA}_{I_2} + \frac{1}{\text{Area}(\mathcal{D})} \iint_{\mathcal{D}} p\,dA \qquad (1)$$

I_1 and I_2 are the average values of the x and y coordinates of a point in the region enclosed by the ellipse. This region is symmetric with respect to the y-axis, hence $I_1 = 0$. It is also symmetric with respect to the x-axis, hence $I_2 = 0$. We use the formula

$$\iint_{\mathcal{D}} p\,dA = p \cdot \text{Area}(\mathcal{D})$$

to conclude by (1) that

$$\overline{f} = m \cdot 0 + n \cdot 0 + \frac{1}{\text{Area}(\mathcal{D})} \cdot p \cdot \text{Area}(\mathcal{D}) = p$$

65. Show that the average value of $f(x, y) = y$ (the y-coordinate of the centroid) of the sector in Figure 28 is equal to $\overline{y} = \left(\frac{2R}{3}\right)\left(\frac{\sin\theta}{\theta}\right)$.

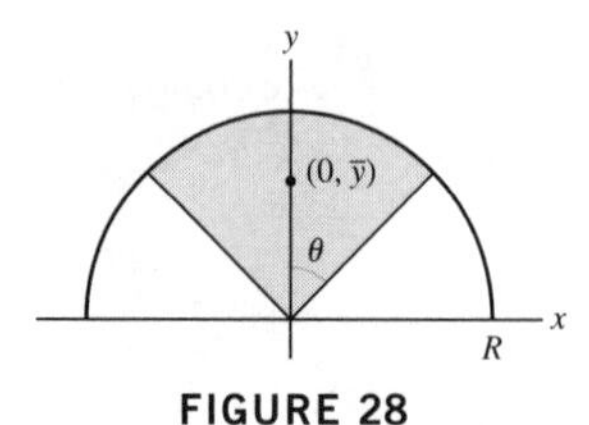

FIGURE 28

SOLUTION

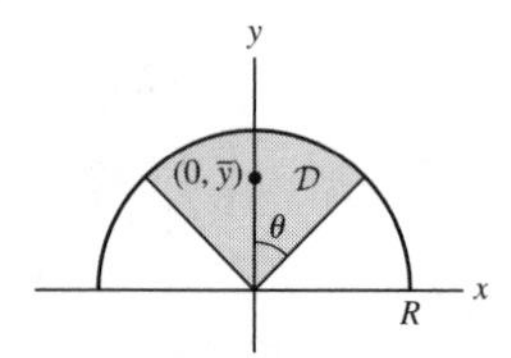

The y-coordinate of the centroid of the sector $\mathcal{D}$ is the average value of the y-coordinate of a point in the sector. That is,

$$\overline{y} = \frac{1}{\text{Area}(\mathcal{D})} \iint_{\mathcal{D}} y \, dA$$

The sector is symmetric with respect to the y-axis, and the integrand $f(x, y) = y$ satisfies $f(-x, y) = f(x, y)$, hence the double integral over $\mathcal{D}$ is twice the integral over the right half $\mathcal{D}_1$ of the sector. Also $\text{Area}(\mathcal{D}) = 2\text{Area}(\mathcal{D}_1)$, therefore

$$\overline{y} = \frac{1}{\text{Area}(\mathcal{D}_1)} \iint_{\mathcal{D}_1} y \, dA \tag{1}$$

We now find the inequalities describing $\mathcal{D}_1$ as a vertically simple region.

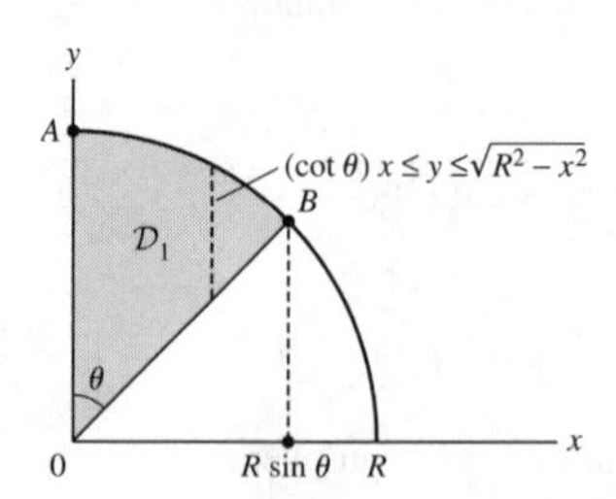

The circular bounding has the equation $y = \sqrt{\mathcal{R}^2 - x^2}$ and the line OB has the equation $y = (\cot\theta)x$. We obtain the following inequalities for $\mathcal{D}_1$:

$$0 \le x \le \mathcal{R}\sin\theta, \quad (\cot\theta)x \le y \le \sqrt{\mathcal{R}^2 - x^2}$$

Hence,

$$\iint_{\mathcal{D}_1} y \, dA = \int_0^{\mathcal{R}\sin\theta} \int_{(\cot\theta)x}^{\sqrt{\mathcal{R}^2 - x^2}} y \, dy \, dx = \int_0^{\mathcal{R}\sin\theta} \frac{y^2}{2}\bigg|_{y=(\cot\theta)x}^{\sqrt{\mathcal{R}^2 - x^2}} dx = \int_0^{\mathcal{R}\sin\theta} \frac{\mathcal{R}^2 - x^2 - (\cot\theta)^2 x^2}{2} \, dx$$

$$= \int_0^{\mathcal{R}\sin\theta} \left(\frac{\mathcal{R}^2}{2} - \frac{x^2}{2\sin^2\theta}\right) dx = \frac{\mathcal{R}^2}{2}x - \frac{x^3}{6\sin^2\theta}\bigg|_{x=0}^{\mathcal{R}\sin\theta} = \frac{\mathcal{R}^3\sin\theta}{2} - \frac{\mathcal{R}^3\sin^3\theta}{6\sin^2\theta} = \frac{\mathcal{R}^3\sin\theta}{3} \tag{2}$$

The area of the sector $\mathcal{D}_1$ is

$$\text{Area}(\mathcal{D}_1) = \frac{\mathcal{R}^2\theta}{2} \tag{3}$$

Substituting (2) and (3) in (1), we obtain the following solution:

$$\overline{y} = \frac{1}{\frac{\mathcal{R}^2\theta}{2}} \cdot \frac{\mathcal{R}^3 \sin\theta}{3} = \frac{2\mathcal{R}^3 \sin\theta}{3\mathcal{R}^2\theta} = \left(\frac{2\mathcal{R}}{3}\right)\left(\frac{\sin\theta}{\theta}\right)$$

67. Find a point P in $\mathcal{D} = [0, 1] \times [0, 4]$ such that $f(P) = \overline{f}$, where $\overline{f}$ is the average of $f(x, y) = xy^2$ on $\mathcal{D}$ (the existence of such a point is guaranteed by the MVT for Double Integrals).

SOLUTION We first compute the average $\overline{f}$ of $f(x, y) = xy^2$ on $\mathcal{D}$.

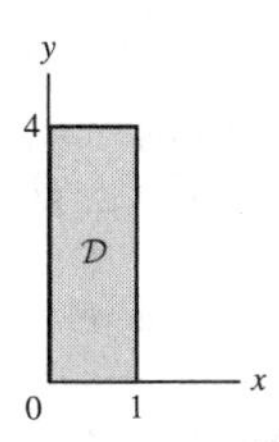

$\overline{f}$ is

$$\overline{f} = \frac{1}{\text{Area}(\mathcal{D})} \iint_{\mathcal{D}} xy^2\, dA = \frac{1}{4 \cdot 1} \int_0^1 \int_0^4 xy^2\, dy\, dx = \frac{1}{4} \int_0^1 \left. \frac{xy^3}{3} \right|_{y=0}^{4} dx$$

$$= \frac{1}{4} \int_0^1 \left(\frac{x \cdot 4^3}{3} - \frac{x \cdot 0^3}{3} \right) dx = \int_0^1 \frac{16x}{3}\, dx = \left. \frac{8x^2}{3} \right|_0^1 = \frac{8}{3}$$

We now must find a point $P = (a, b)$ in $\mathcal{D}$ such that

$$f(P) = ab^2 = \frac{8}{3}$$

We choose $b = 2$, obtaining

$$a \cdot 2^2 = \frac{8}{3} \quad \Rightarrow \quad a = \frac{2}{3}$$

The point $P = \left(\frac{2}{3}, 2\right)$ in the rectangle $\mathcal{D}$ satisfies

$$f(P) = \overline{f} = \frac{8}{3}$$

In Exercises 69–70, use Eq. (12) to estimate the double integral.

69. The following table lists the areas of the subdomains $\mathcal{D}_j$ of the domain $\mathcal{D}$ in Figure 29 and the values of a function $f(x, y)$ at sample points $P_j \in \mathcal{D}_j$. Estimate $\iint_{\mathcal{D}} f(x, y)\, dA$.

j	1	2	3	4	5	6
Area($\mathcal{D}_j$)	1.2	1.1	1.4	0.6	1.2	0.8
$f(P_j)$	9	9.1	9.3	9.1	8.9	8.8

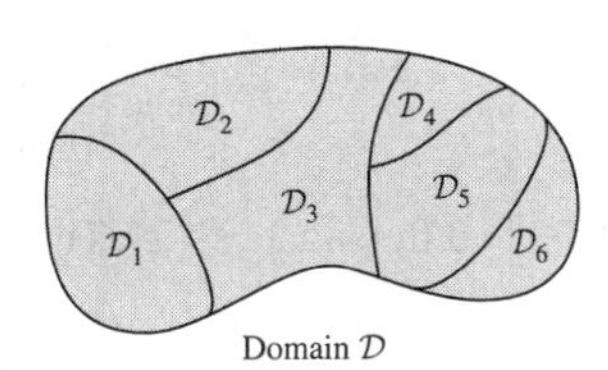

FIGURE 29

SOLUTION By Eq. (12) we have

$$\iint_{\mathcal{D}} f(x, y)\, dA \approx \sum_{j=1}^{6} f\left(P_j\right) \text{Area}\left(\mathcal{D}_j\right)$$

Substituting the data given in the table, we obtain

$$\iint_{\mathcal{D}} f(x, y)\, dA \approx 9 \cdot 1.2 + 9.1 \cdot 1.1 + 9.3 \cdot 1.4 + 9.1 \cdot 0.6 + 8.9 \cdot 1.2 + 8.8 \cdot 0.8 = 57.01$$

Thus,

$$\iint_{\mathcal{D}} f(x, y)\, dA \approx 57.01$$

Further Insights and Challenges

71. Suppose that $f(x, y)$ is a continuous function on a closed connected domain $\mathcal{D}$. The Intermediate Value Theorem (IVT) states that if $f(P) = a$ and $f(Q) = b$, where $P, Q \in \mathcal{D}$, then $f(x, y)$ takes on every value between a and b at some point in $\mathcal{D}$.

(a) Show, by constructing a counterexample, that the IVT is false if $\mathcal{D}$ is not connected.

(b) Prove the IVT in two variables as follows: Let $\mathbf{c}(t)$ be a path such that $\mathbf{c}(0) = P$ and $\mathbf{c}(1) = Q$ (such a path exists because $\mathcal{D}$ is connected). Apply the IVT in one variable to the composite function $f(\mathbf{c}(t))$.

SOLUTION

(a) Let $\mathcal{D}$ be the union of the disc $\mathcal{D}_1$ of radius $\frac{1}{2}$ centered at the origin, and the disc $\mathcal{D}_2$ of radius $\frac{1}{2}$ centered at $(1, 1)$.

Obviously $\mathcal{D}$ is not connected. We define a function $f(x, y)$ on $\mathcal{D}$ as follows:

$$f(x, y) = \begin{cases} 1 & (x, y) \in \mathcal{D}_1 \\ 2 & (x, y) \in \mathcal{D}_2 \end{cases}$$

f is continuous on $\mathcal{D}$, but it does not take on any value between 1 and 2.

(b) Let $\mathcal{D}$ be a closed connected domain and $f(x, y)$ a continuous function on $\mathcal{D}$. Suppose that $f(P) = a$ and $f(Q) = b$ where $P, Q \in \mathcal{D}$, and $a < c < b$. We show that f takes on the value c at a point in $\mathcal{D}$. Since $\mathcal{D}$ is connected, there is a curve $\gamma(t) = (x(t), y(t))$ lying entirely in $\mathcal{D}$, such that $\gamma(0) = P$ and $\gamma(1) = Q$. We consider the function

$$g(t) = f(x(t), y(t)), \quad 0 \le t \le 1$$

The composition $g(t)$ is continuous on the segment $0 \le t \le 1$, and c is an intermediate value of g on this segment (since $g(0) = a < c < b = g(1)$). Therefore, by the IVT, there exists $t_0 \in (0, 1)$ such that $g(t_0) = c$. The curve $\gamma(t)$ lies in $\mathcal{D}$, hence the point $\mathcal{R} = \gamma(t_0) = (x(t_0), y(t_0))$ is in $\mathcal{D}$ and the following holds:

$$f(\mathcal{R}) = f(x(t_0), y(t_0)) = g(t_0) = c.$$

73. Let $f(y)$ be a function of y alone and set $G(t) = \int_0^t \int_0^x f(y)\, dy\, dx$.

(a) Use the FTC to prove that $G''(t) = f(t)$.

(b) Show, by changing the order in the double integral, that $G(t) = \int_0^t (t - y) f(y)\, dy$. This shows that the "second antiderivative" of $f(y)$ can be expressed as a single integral.

SOLUTION

(a) Let $H(x) = \int_0^x f(y)\, dy$. Then $G(t) = \int_0^t H(x)\, dx$, and by the FTC we have

$$G'(t) = \frac{d}{dt} \int_0^t H(x)\, dx = H(t) = \int_0^t f(y)\, dy$$

We again use the FTC to differentiate $G'(t)$. This gives

$$G''(t) = \frac{d}{dt} \int_0^t f(y)\, dy = f(t)$$

(b) For a fixed t, the domain of integration is described by the following inequalities:

$$0 \le x \le t, \quad 0 \le y \le x$$

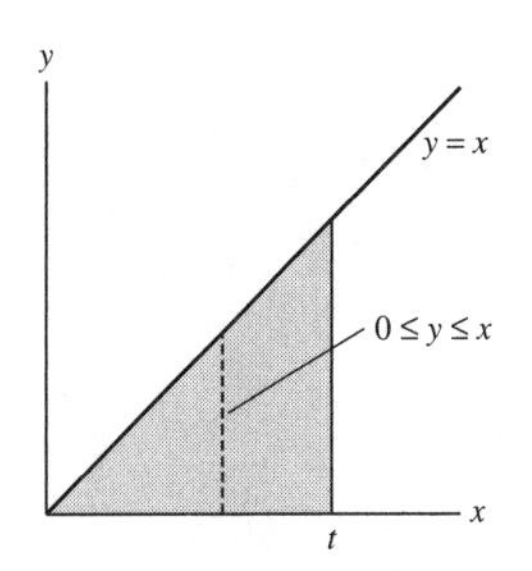

We describe the domain as a horizontally simple region by the inequalities

$$0 \le y \le t, \quad y \le x \le t$$

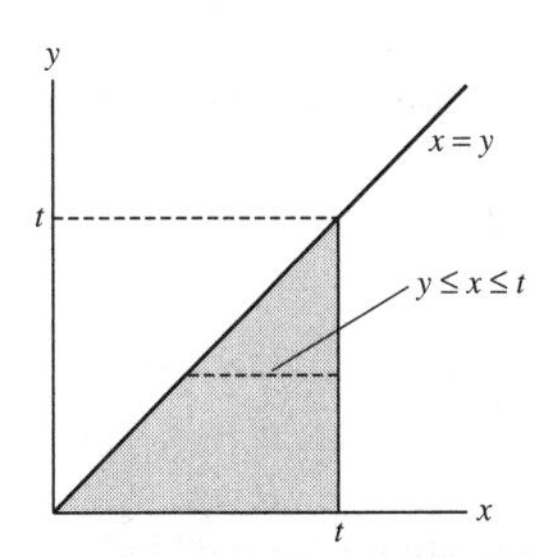

Then, the iterated integral for $G(t)$ can be computed in reverse order of integration as follows:

$$G(t) = \int_0^t \int_y^t f(y)\,dx\,dy = \int_0^t f(y)x\Big|_{x=y}^{t} dy = \int_0^t f(y)(t-y)\,dy = \int_0^t (t-y)f(y)\,dy$$

We, thus showed that

$$G(t) = \int_0^t (t-y)f(y)\,dy$$

16.3 Triple Integrals (ET Section 15.3)

Preliminary Questions

1. Which of (a)–(c) is not equal to $\int_0^1 \int_3^4 \int_6^7 f(x,y,z)\,dz\,dy\,dx$?

(a) $\int_6^7 \int_0^1 \int_3^4 f(x,y,z)\,dy\,dx\,dz$

(b) $\int_3^4 \int_0^1 \int_6^7 f(x,y,z)\,dz\,dx\,dy$

(c) $\int_3^4 \int_3^4 \int_6^7 f(x,y,z)\,dx\,dz\,dy$

SOLUTION The given integral, I, is a triple integral of f over the box $B = [0,1] \times [3,4] \times [6,7]$. In (a) the limits of integration are $0 \le x \le 1$, $3 \le y \le 4$, $6 \le z \le 7$, hence this integral is equal to I. In (b) the limits of integration are $0 \le x \le 1$, $3 \le y \le 4$, $6 \le z \le 7$, hence it is also equal to I. In (c) the limits of integration are $6 \le x \le 7$, $3 \le y \le 4$, $3 \le z \le 4$. This is the triple integral of f over the box $[6,7] \times [3,4] \times [3,4]$, which is different from $\mathcal{B}$. Therefore, the triple integral is usually unequal to I.

2. Which of the following does not represent a meaningful triple integral?

(a) $\int_0^1 \int_0^x \int_{x+y}^{2x+y} e^{x+y+z}\,dz\,dy\,dx$

(b) $\int_0^1 \int_0^z \int_{x+y}^{2x+y} e^{x+y+z}\,dz\,dy\,dx$

SOLUTION

(a) The limits of integration determine the following inequalities:

$$0 \le x \le 1, \quad 0 \le y \le x, \quad x+y \le z \le 2x+y$$

The integration is over the simple region $\mathcal{W}$, which lies between the planes $z = x+y$ and $z = 2x+y$ over the domain $\mathcal{D}_1 = \{(x,y) : 0 \le x \le 1,\ 0 \le y \le x\}$ in the xy-plane.

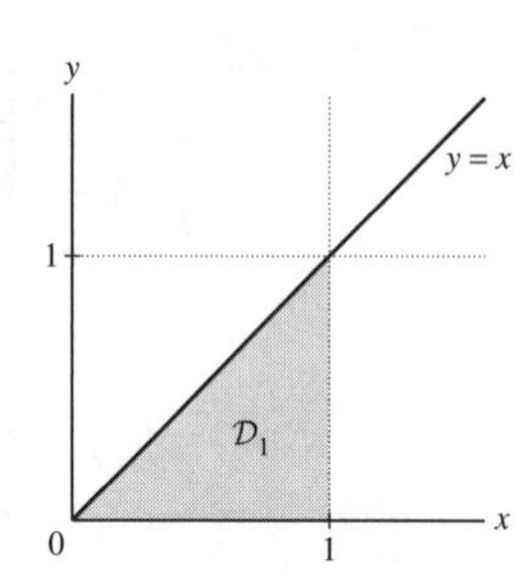

Thus, the integral represents a meaningful triple integral.

(b) Note that the inner integral is with respect to z, but then the middle integral has limits from 0 to z! This makes no sense.

3. Describe the projection of the region of integration $\mathcal{W}$ onto the xy-plane:

(a) $\displaystyle\int_0^1 \int_0^x \int_0^{x^2+y^2} f(x, y, z)\, dz\, dy\, dx$

(b) $\displaystyle\int_0^1 \int_0^{\sqrt{1-x^2}} \int_2^4 f(x, y, z)\, dz\, dy\, dx$

SOLUTION

(a) The region of integration is defined by the limits of integration, yielding the following inequalities:

$$0 \le x \le 1, \quad 0 \le y \le x, \quad 0 \le z \le x^2 + y^2$$

$\mathcal{W}$ is the region between the paraboloid $z = x^2 + y^2$ and the xy-plane which is above the triangle $\mathcal{D} = \{(x, y) : 0 \le x \le 1, 0 \le y \le x\}$ in the xy-plane. This triangle is the projection of $\mathcal{W}$ onto the xy-plane.

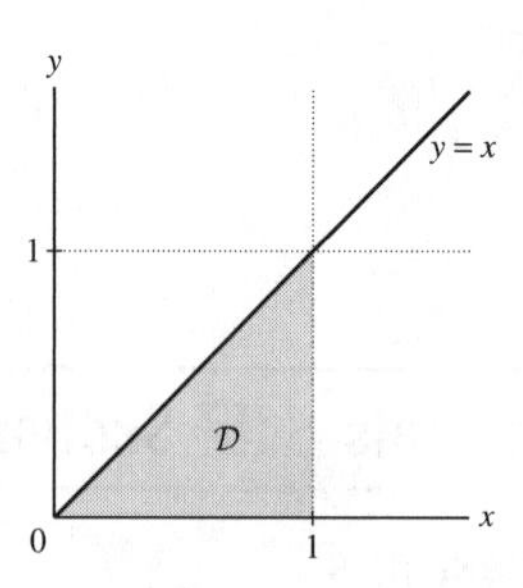

(b) The inequalities determined by the limits of integration are

$$0 \le x \le 1, \quad 0 \le y \le \sqrt{1 - x^2}, \quad 2 \le z \le 4$$

This is the region between the planes $z = 2$ and $z = 4$, which is above the region $\mathcal{D} = \left\{(x, y) : 0 \le x \le 1,\ 0 \le y \le \sqrt{1 - x^2}\right\}$ in the xy-plane. The projection $\mathcal{D}$ of $\mathcal{W}$ onto the xy-plane is the part of the unit disk in the first quadrant.

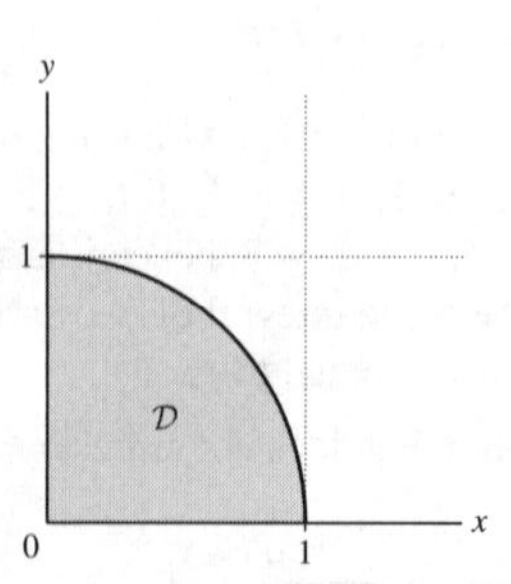

Exercises

In Exercises 1–10, evaluate $\iiint_{\mathcal{B}} f(x, y, z)\, dV$ *for the specified function* f *and box* $\mathcal{B}$.

1. $f(x, y, z) = z^4$; $\quad 2 \le x \le 8, 0 \le y \le 5, 0 \le z \le 1$

SOLUTION We write the triple integral as an iterated integral and compute it to obtain

$$\iiint_{\mathcal{B}} z^4\,dV = \int_2^8\int_0^5\int_0^1 z^4\,dz\,dy\,dx = \int_2^8\int_0^5\left(\int_0^1 z^4\,dz\right)dy\,dx = \int_2^8\int_0^5 \frac{1}{5}z^5\bigg|_{z=0}^{1}\,dy\,dx$$

$$= \int_2^8\int_0^5 \frac{1}{5}\,dy\,dx = \frac{1}{5}\int_2^8\int_0^5 dy\,dx = \frac{1}{5}\cdot 6\cdot 5 = 6$$

3. $f(x, y, z) = xz^2$; $[0, 2] \times [1, 6] \times [3, 4]$

SOLUTION The box $[0, 2] \times [1, 6] \times [3, 4]$ corresponds to the inequalities $0 \le x \le 2$, $1 \le y \le 6$, $3 \le z \le 4$. We write the integral as an iterated integral in any order we choose, and evaluate the inner, middle, and outer integral one after the other. This gives

$$\iint_{\mathcal{B}} xz^2\,dV = \int_0^2\int_1^6\int_3^4 xz^2\,dz\,dy\,dx = \int_0^2\int_1^6\left(\int_3^4 xz^2\,dz\right)dy\,dx = \int_0^2\int_1^6 \frac{xz^3}{3}\bigg|_{z=3}^{4}\,dy\,dx$$

$$= \int_0^2\int_1^6 \frac{x(4^3-3^3)}{3}\,dy\,dx = \int_0^2\left(\int_1^6 \frac{37}{3}x\,dy\right)dx = \int_0^2 \frac{37x}{3}y\bigg|_{y=1}^{6}\,dx$$

$$= \int_0^2 \frac{5\cdot 37x}{3}\,dx = \frac{185}{3}\cdot\frac{x^2}{2}\bigg|_0^2 = \frac{370}{3} = 123\frac{1}{3}$$

Alternatively, we can use the form $f(x, y, z) = xz^2 = h(x)g(y)l(z)$ to compute the triple integral as the product:

$$\iint_{\mathcal{B}} xz^2\,dV = \int_0^2\int_1^6\int_3^4 xz^2\,dz\,dy\,dx = \left(\int_0^2 x\,dx\right)\left(\int_1^6 1\,dy\right)\left(\int_3^4 z^2\,dz\right)$$

$$= \left(\frac{x^2}{2}\bigg|_0^2\right)\left(y\bigg|_1^6\right)\left(\frac{z^3}{3}\bigg|_3^4\right) = 2\cdot 5\cdot\frac{\left(4^3-3^3\right)}{3} = 123\frac{1}{3}$$

5. $f(x, y, z) = xe^{y-2z}$; $0 \le x \le 2, 0 \le y, z \le 1$

SOLUTION We write the triple integral as an iterated integral. Since $f(x, y, z) = xe^y \cdot e^{-2z}$, we may evaluate the iterated integral as the product of three single integrals. We get

$$\iiint_{\mathcal{B}} xe^{y-2z}\,dV = \int_0^2\int_0^1\int_0^1 xe^{y-2z}\,dz\,dy\,dx = \left(\int_0^2 x\,dx\right)\left(\int_0^1 e^y\,dy\right)\left(\int_0^1 e^{-2z}\,dz\right)$$

$$= \left(\frac{1}{2}x^2\bigg|_0^2\right)\left(e^y\bigg|_0^1\right)\left(-\frac{1}{2}e^{-2z}\bigg|_0^1\right) = 2(e-1)\cdot -\frac{1}{2}(e^{-2}-1) = (e-1)(1-e^{-2})$$

7. $f(x, y, z) = \dfrac{z}{x}$; $1 \le x \le 3, 0 \le y \le 2, 0 \le z \le 4$

SOLUTION We write the triple integral as an iterated integral and evaluate it using iterated integral of a product function. We get

$$\iiint_{\mathcal{B}} f(x, y, z)\,dV = \int_1^3\int_0^2\int_0^4 \frac{z}{x}\,dz\,dy\,dx = \left(\int_0^4 z\,dz\right)\left(\int_0^2 1\,dy\right)\left(\int_1^3 \frac{1}{x}\,dx\right)$$

$$= \left(\frac{1}{2}z^2\bigg|_0^4\right)\left(y\bigg|_0^2\right)\left(\ln x\bigg|_1^3\right) = 8\cdot 2\cdot(\ln 3 - \ln 1) = 16\ln 3$$

9. $f(x, y, z) = (x+z)^3$; $[0, a] \times [0, b] \times [0, c]$

SOLUTION We write the triple integral as an iterated integral and evaluate it to obtain

$$\iiint_{\mathcal{B}} f(x, y, z)\,dV = \int_0^a\int_0^b\int_0^c (x+z)^3\,dz\,dy\,dx = \int_0^a\int_0^b \frac{(x+z)^4}{4}\bigg|_{z=0}^{c}\,dy\,dx$$

$$= \int_0^a\int_0^b\left(\frac{(x+c)^4}{4} - \frac{x^4}{4}\right)dy\,dx = \int_0^a \frac{(x+c)^4 - x^4}{4}y\bigg|_{y=0}^{b}\,dx$$

$$= \int_0^a \frac{b}{4}\left[(x+c)^4 - x^4\right] dx = \frac{b}{4}\left[\frac{(x+c)^5}{5} - \frac{x^5}{5}\right]\Big|_{x=0}^{a}$$

$$= \frac{b}{4}\frac{(a+c)^5 - a^5 - c^5}{5} = \frac{b}{20}\left[(a+c)^5 - a^5 - c^5\right]$$

In Exercises 11–14, evaluate $\iiint_{\mathcal{W}} f(x, y, z)\, dV$ *for the function* f *and region* $\mathcal{W}$ *specified.*

11. $f(x, y, z) = x + y$; $\mathcal{W}$: $y \le z \le x, 0 \le y \le x, 0 \le x \le 1$

SOLUTION $\mathcal{W}$ is the region between the planes $z = y$ and $z = x$ lying over the triangle $\mathcal{D}$ in the xy-plane defined by the inequalities $0 \le y \le x, 0 \le x \le 1$.

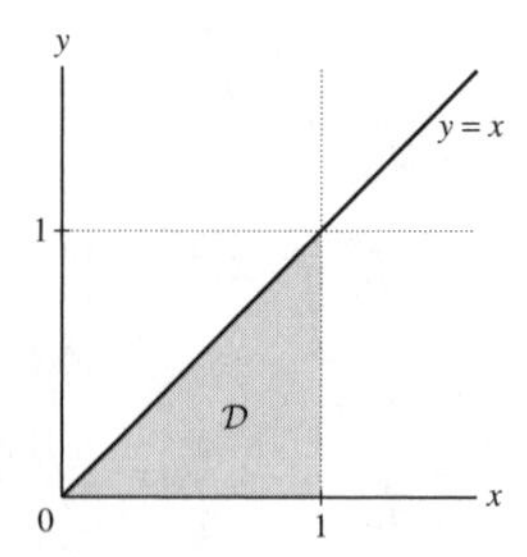

We compute the integral, using Theorem 2, by evaluating the following iterated integral:

$$\iiint_{\mathcal{W}} (x+y)\, dV = \iint_{\mathcal{D}} \left(\int_y^x (x+y)\, dz\right) dA = \iint_{\mathcal{D}} (x+y)z\Big|_{z=y}^{x} dA = \iint_{\mathcal{D}} (x+y)(x-y)\, dA$$

$$= \iint_{\mathcal{D}} \left(x^2 - y^2\right) dA = \int_0^1 \int_0^x \left(x^2 - y^2\right) dy\, dx = \int_0^1 \left(\int_0^x \left(x^2 - y^2\right) dy\right) dx$$

$$= \int_0^1 x^2 y - \frac{y^3}{3}\Big|_{y=0}^{x} dx = \int_0^1 \frac{2x^3}{3}\, dx = \frac{2}{12}x^4\Big|_0^1 = \frac{1}{6}$$

13. $f(x, y, z) = xyz$; $\mathcal{W}$: $0 \le z \le 1, 0 \le y \le \sqrt{1-x^2}, 0 \le x \le 1$

SOLUTION $\mathcal{W}$ is the region between the planes $z = 0$ and $z = 1$, lying over the part $\mathcal{D}$ of the disk in the first quadrant.

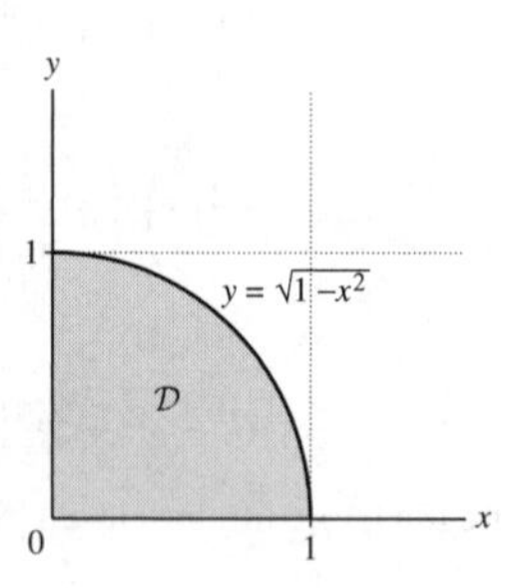

Using Theorem 2, we compute the triple integral as the following iterated integral:

$$\iiint_{\mathcal{W}} xyz\, dV = \iint_{\mathcal{D}} \left(\int_0^1 xyz\, dz\right) dA = \iint_{\mathcal{D}} \frac{xyz^2}{2}\Big|_{z=0}^{1} dA = \iint_{\mathcal{D}} \frac{xy}{2}\, dA$$

$$= \int_0^1 \left(\int_0^{\sqrt{1-x^2}} \frac{xy}{2}\, dy\right) dx = \int_0^1 \frac{xy^2}{4}\Big|_{y=0}^{\sqrt{1-x^2}} dx = \int_0^1 \frac{x(1-x^2)}{4}\, dx$$

$$= \int_0^1 \frac{x - x^3}{4}\, dx = \frac{x^2}{8} - \frac{x^4}{16}\Big|_0^1 = \frac{1}{8} - \frac{1}{16} = \frac{1}{16}$$

15. Calculate the integral of $f(x, y, z) = z$ over the region $\mathcal{W}$ below the upper hemisphere of the sphere $x^2 + y^2 + z^2 = 9$ lying over the unit square $0 \le x, y \le 1$ in Figure 12.

SOLUTION The upper surface is the sphere $z = \sqrt{9 - x^2 - y^2}$ and the lower surface is the plane $z = 0$.

The projection $\mathcal{D}$ of $\mathcal{W}$ onto the xy-plane is the unit square.

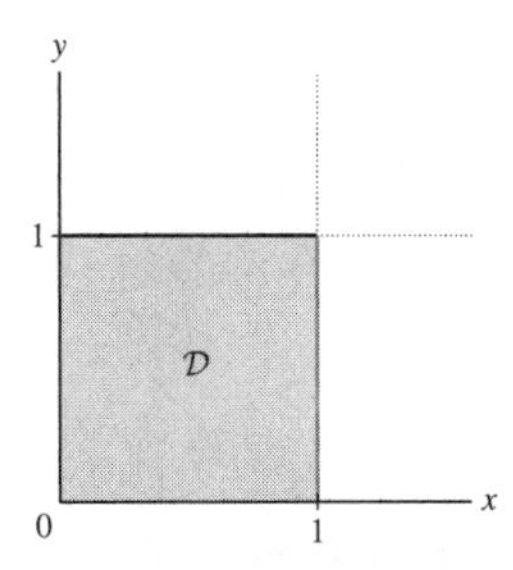

The triple integral of f over $\mathcal{W}$ is equal to the following iterated integral:

$$\iiint_{\mathcal{W}} z\,dV = \iint_{\mathcal{D}} \left(\int_0^{\sqrt{9-x^2-y^2}} z\,dz \right) dA = \iint_{\mathcal{D}} \frac{1}{2} z^2 \Big|_{z=0}^{\sqrt{9-x^2-y^2}} dA = \iint_{\mathcal{D}} \frac{9-x^2-y^2}{2}\,dA$$

$$= \int_0^1 \left(\int_0^1 \frac{9-x^2-y^2}{2}\,dx \right) dy = \int_0^1 \frac{9x}{2} - \frac{x^3}{6} - \frac{y^2 x}{2} \Big|_{x=0}^{1} dy = \int_0^1 \left(\frac{9}{2} - \frac{1}{6} - \frac{y^2}{2} - 0 \right) dy$$

$$= \int_0^1 \left(\frac{13}{3} - \frac{y^2}{2} \right) dy = \frac{13}{3} y - \frac{y^3}{6} \Big|_0^1 = \frac{13}{3} - \frac{1}{6} = \frac{25}{6} = 4\frac{1}{6}$$

17. Calculate the integral of $f(x, y, z) = e^{x+y+z}$ over the tetrahedron $\mathcal{W}$ in Figure 13.

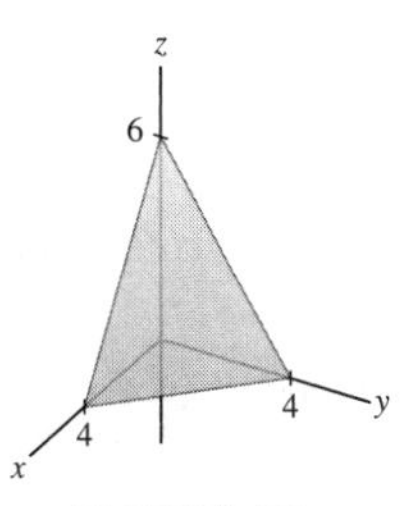

FIGURE 13

SOLUTION We first must find the equation of the upper surface, which is the plane through the points $A = (4, 0, 0)$, $B = (0, 4, 0)$, and $C = (0, 0, 6)$.

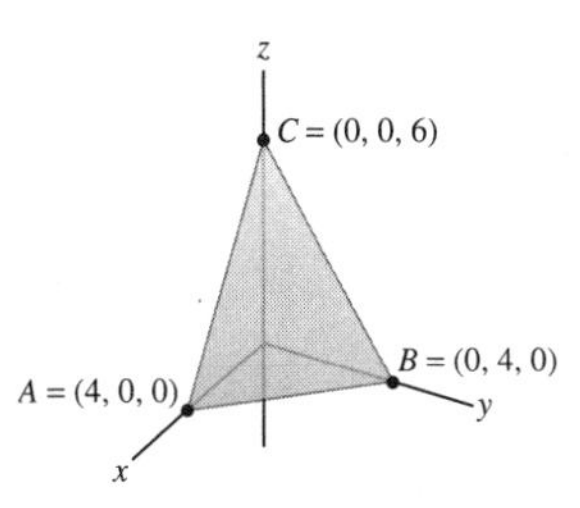

A vector normal to the plane is the following cross product:

$$\mathbf{n} = \overrightarrow{AB} \times \overrightarrow{AC} = \langle -4, 4, 0 \rangle \times \langle -4, 0, 6 \rangle = \begin{vmatrix} \mathbf{i} & \mathbf{j} & \mathbf{k} \\ -4 & 4 & 0 \\ -4 & 0 & 6 \end{vmatrix} = 24\mathbf{i} + 24\mathbf{j} + 16\mathbf{k} = 8\,(3\mathbf{i} + 3\mathbf{j} + 2\mathbf{k})$$

The equation of the plane with normal $3\mathbf{i} + 3\mathbf{j} + 2\mathbf{k}$ passing through $A = (4, 0, 0)$ is

$$3(x-4) + 3y + 2z = 0$$

$$3x + 3y + 2z = 12 \quad \Rightarrow \quad z = 6 - \frac{3}{2}x - \frac{3}{2}y$$

Thus, the tetrahedron $\mathcal{V}$ lies between the planes $z = 6 - \frac{3x}{2} - \frac{3y}{2}$ and $z = 0$. The projection $\mathcal{D}$ of $\mathcal{V}$ onto the xy-plane is the triangle enclosed by the line AB and the positive axes. We compute the equation of the line AB:

$$y - 0 = \frac{4-0}{0-4}(x-4) \quad \Rightarrow \quad y = -x + 4$$

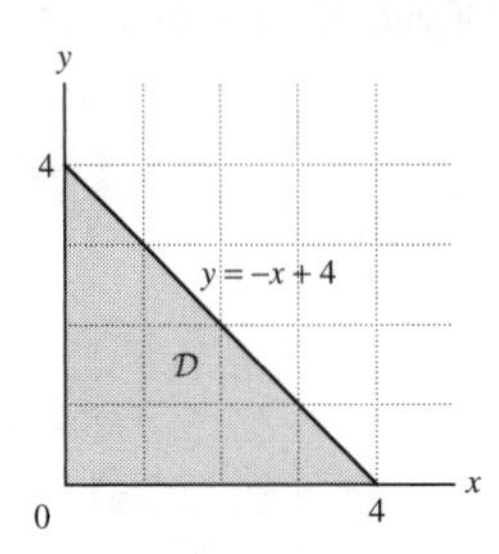

We now compute the triple integral of f over $\mathcal{V}$ as the following iterated integral:

$$\begin{aligned}
\iiint_{\mathcal{V}} e^{x+y+z}\,dV &= \iint_{\mathcal{D}} \left(\int_0^{6-\frac{3x}{2}-\frac{3y}{2}} e^{x+y+z}\,dz \right) dA = \iint_{\mathcal{D}} e^{x+y+z} \Big|_{z=0}^{6-\frac{3x}{2}-\frac{3y}{2}} dA \\
&= \iint_{\mathcal{D}} e^{x+y+6-\frac{3x}{2}-\frac{3y}{2}} - e^{x+y}\,dA = \iint_{\mathcal{D}} e^{6-\frac{1}{2}x-\frac{1}{2}y} - e^{x+y}\,dA \\
&= \int_0^4 \int_0^{4-x} e^{6-\frac{1}{2}x-\frac{1}{2}y} - e^{x+y}\,dy\,dx = \int_0^4 -2e^{6-\frac{1}{2}x-\frac{1}{2}y} - e^{x+y} \Big|_{y=0}^{4-x} dx \\
&= \int_0^4 -2e^{6-\frac{1}{2}x-\frac{1}{2}(4-x)} - e^{x+4-x} - \left(-2e^{6-\frac{1}{2}x} - e^x\right) dx \\
&= \int_0^4 -2e^4 - e^4 + 2e^{6-\frac{1}{2}x} + e^x\,dx = -3e^4x - 4e^{6-\frac{1}{2}x} + e^x \Big|_{x=0}^4 \\
&= -12e^4 - 4e^4 + e^4 + 4e^6 - 1 = -15e^4 + 4e^6 - 1 \approx 793.74
\end{aligned}$$

19. Integrate $f(x, y, z) = xz$ over the region in the first octant $(x, y, z \geq 0)$ above the parabolic cylinder $z = y^2$ and below the paraboloid $z = 8 - 2x^2 - y^2$.

SOLUTION We first find the projection of the region $\mathcal{W}$ onto the xy-plane. We find the curve of intersection between the upper and lower surfaces, by solving the following equation for $x, y \geq 0$:

$$8 - 2x^2 - y^2 = y^2 \quad \Rightarrow \quad y^2 = 4 - x^2 \quad \Rightarrow \quad y = \sqrt{4-x^2},\ x \geq 0$$

The projection $\mathcal{D}$ of $\mathcal{W}$ onto the xy-plane is the region bounded by the circle $x^2 + y^2 = 4$ and the positive axes.

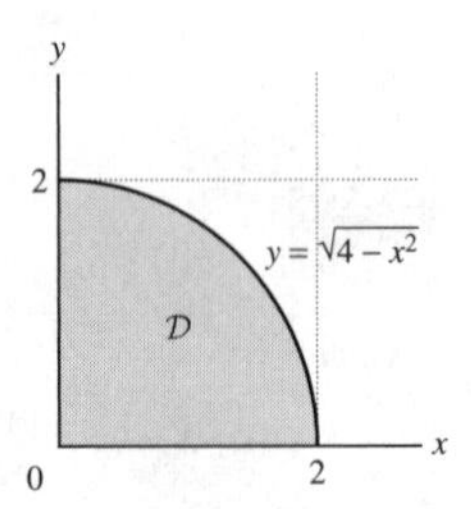

We now compute the triple integral over $\mathcal{W}$ by evaluating the following iterated integral:

$$\begin{aligned}
\iiint_{\mathcal{W}} xz\,dV &= \iint_{\mathcal{D}} \left(\int_{y^2}^{8-2x^2-y^2} xz\,dz \right) dA = \iint_{\mathcal{D}} \frac{xz^2}{2} \Big|_{z=y^2}^{8-2x^2-y^2} dA = \iint_{\mathcal{D}} \frac{x}{2} \left(\left(8 - 2x^2 - y^2\right)^2 - y^4 \right) dA \\
&= \iint_{\mathcal{D}} \frac{x}{2} \left(8 - 2x^2 - 2y^2\right)\left(8 - 2x^2\right) dA = \iint_{\mathcal{D}} 2x\left(4 - x^2\right)\left(4 - x^2 - y^2\right) dA
\end{aligned}$$

$$= \int_0^2 \left(2x\left(4-x^2\right) \int_0^{\sqrt{4-x^2}} \left(4-x^2-y^2\right) dy \right) dx$$

$$= \int_0^2 2x\left(4-x^2\right)\left(\left(4-x^2\right)y - \frac{y^3}{3}\right)\Bigg|_{y=0}^{\sqrt{4-x^2}} dx$$

$$= \int_0^2 2x\left(4-x^2\right)\left(\left(4-x^2\right)^{3/2} - \frac{\left(4-x^2\right)^{3/2}}{3}\right) dx = \int_0^2 \frac{4}{3}\left(4-x^2\right)^{5/2} x\, dx$$

We compute the integral using the substitution $u = 4 - x^2$, $du = -2x\, dx$. This gives

$$\iiint_{\mathcal{W}} xz\, dV = \int_0^2 \frac{4}{3}\left(4-x^2\right)^{5/2} x\, dx = \int_4^0 \frac{4}{3}u^{5/2}\left(-\frac{du}{2}\right) = \int_0^4 \frac{2}{3}u^{5/2}\, du = \frac{2}{3}\cdot\frac{2}{7}u^{7/2}\Big|_0^4$$

$$= \frac{4}{21}\cdot 4^{7/2} = \frac{2^9}{21} = \frac{512}{21} = 24\frac{8}{21}$$

21. Find the triple integral of the function z over the ramp in Figure 14. Here, z is the height above the ground.

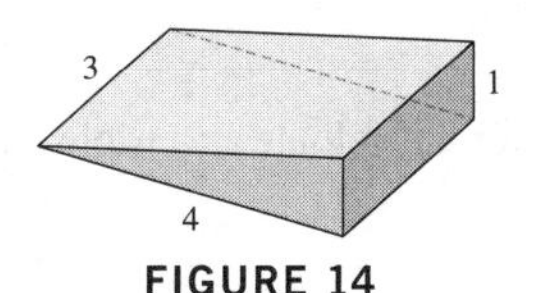

FIGURE 14

SOLUTION We place the coordinate axes as shown in the figure:

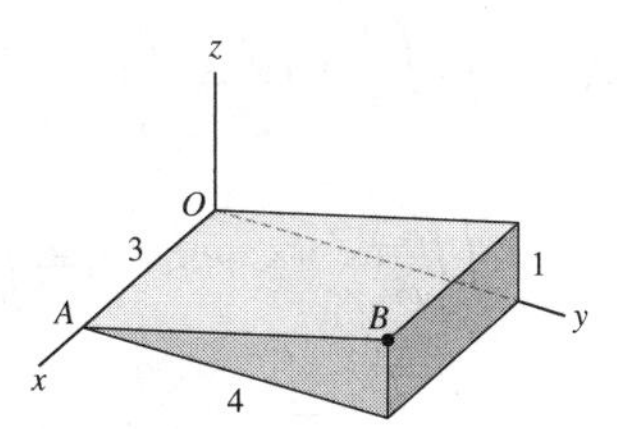

The upper surface is the plane passing through the points $O = (0,0,0)$, $A = (3,0,0)$, and $B = (3,4,1)$. We find a normal to this plane and then determine the equation of the plane. We get

$$\overrightarrow{OA} \times \overrightarrow{AB} = \langle 3,0,0\rangle \times \langle 0,4,1\rangle = \begin{vmatrix} \mathbf{i} & \mathbf{j} & \mathbf{k} \\ 3 & 0 & 0 \\ 0 & 4 & 1 \end{vmatrix} = -3\mathbf{j} + 12\mathbf{k} = 3\left(-\mathbf{j} + 4\mathbf{k}\right)$$

The plane is orthogonal to the vector $\langle 0,-1,4\rangle$ and passes through the origin, hence the equation of the plane is

$$0\cdot x - y + 4z = 0 \quad\Rightarrow\quad z = \frac{y}{4}$$

The projection of the region of integration $\mathcal{W}$ onto the xy-plane is the rectangle $\mathcal{D}$ defined by

$$0 \le x \le 3, \quad 0 \le y \le 4.$$

We now compute the triple integral of f over $\mathcal{W}$, as the following iterated integral:

$$\iiint_{\mathcal{W}} z\, dV = \iint_{\mathcal{D}} \left(\int_0^{y/4} z\, dz\right) dA = \iint_{\mathcal{D}} \frac{z^2}{2}\Big|_{z=0}^{y/4} dA = \iint_{\mathcal{D}} \frac{y^2}{32}\, dA = \int_0^4 \left(\int_0^3 \frac{y^2}{32}\, dx\right) dy$$

$$= \int_0^4 \frac{y^2 x}{32}\Big|_{x=0}^{3} dy = \int_0^4 \frac{3y^2}{32}\, dy = \frac{y^3}{32}\Big|_0^4 = \frac{4^3}{32} = 2$$

23. Find the volume of the solid in $\mathbf{R}^3$ bounded by $y = x^2$, $x = y^2$, $z = x + y + 5$, and $z = 0$.

SOLUTION The solid $\mathcal{W}$ is shown in the following figure:

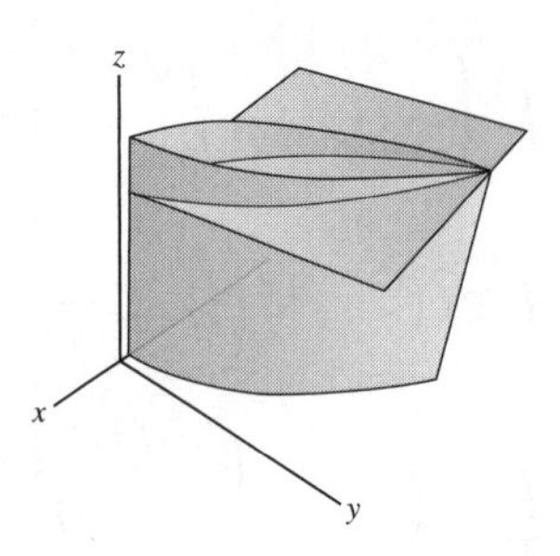

The upper surface is the plane $z = x + y + 5$ and the lower surface is the plane $z = 0$. The projection of $\mathcal{W}$ onto the xy-plane is the region in the first quadrant enclosed by the curves $y = x^2$ and $x = y^2$.

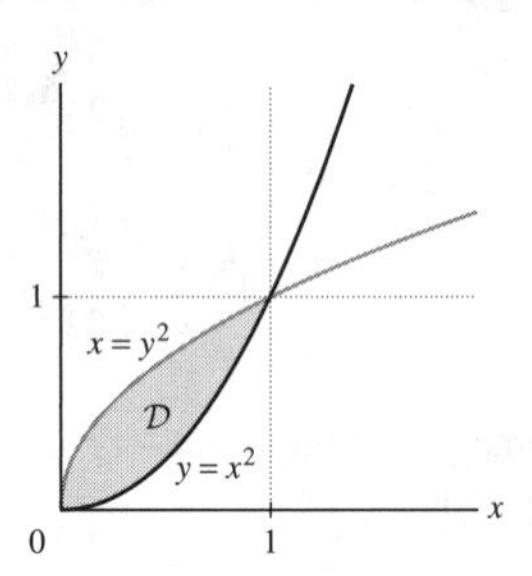

We use the formula for the volume as a triple integral to write

$$\text{Volume}(\mathcal{W}) = \iiint_{\mathcal{W}} 1\,dV$$

The triple integral is equal to the following iterated integral:

$$\text{Volume}(\mathcal{W}) = \iiint_{\mathcal{W}} 1\,dV = \iint_{\mathcal{D}} \left(\int_0^{x+y+5} 1\,dz \right) dA = \iint_{\mathcal{D}} z \Big|_{z=0}^{x+y+5} dA$$

$$= \iint_{\mathcal{D}} (x + y + 5)\,dA = \int_0^1 \left(\int_{x^2}^{\sqrt{x}} (x + y + 5)\,dy \right) dx = \int_0^1 xy + \frac{y^2}{2} + 5y \Big|_{y=x^2}^{\sqrt{x}} dx$$

$$= \int_0^1 \left(x\sqrt{x} + \frac{x}{2} + 5\sqrt{x} - \left(x^3 + \frac{x^4}{2} + 5x^2 \right) \right) dx$$

$$= \int_0^1 \left(-\frac{x^4}{2} - x^3 - 5x^2 + x^{3/2} + \frac{x}{2} + 5x^{1/2} \right) dx$$

$$= -\frac{x^5}{10} - \frac{x^4}{4} - \frac{5x^3}{3} + \frac{2}{5}x^{5/2} + \frac{x^2}{4} + \frac{10}{3}x^{3/2} \Big|_0^1$$

$$= -\frac{1}{10} - \frac{1}{4} - \frac{5}{3} + \frac{2}{5} + \frac{1}{4} + \frac{10}{3} = \frac{59}{30} = 1\frac{29}{30}$$

25. Calculate $\iiint_{\mathcal{W}} y\,dV$, where $\mathcal{W}$ is the region above $z = x^2 + y^2$ and below $z = 5$, and bounded by $y = 0$ and $y = 1$.

SOLUTION The region $\mathcal{W}$ is shown in the figure:

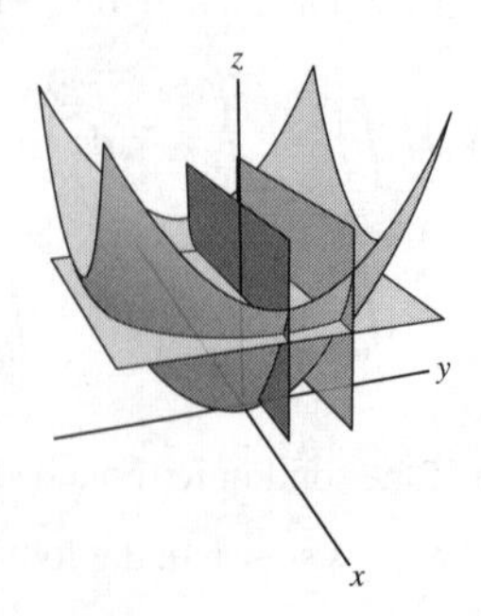

The upper surface is the plane $z = 5$ and the lower surface is the paraboloid $z = x^2 + y^2$. The projection of $\mathcal{W}$ onto the xy-plane is the part of the disk $x^2 + y^2 \le 5$ between the lines $y = 0$ and $y = 1$.

The triple integral of $f(x, y, z) = y$ over $\mathcal{W}$ is equal to the following iterated integral:

$$\iiint_{\mathcal{W}} y\,dV = \iint_{\mathcal{D}} \left(\int_{x^2+y^2}^{5} y\,dz \right) dA = \iint_{\mathcal{D}} yz\Big|_{z=x^2+y^2}^{5} dA = \iint_{\mathcal{D}} y\left(5 - x^2 - y^2\right) dA$$

$$= \int_0^1 \left(\int_{-\sqrt{5-y^2}}^{\sqrt{5-y^2}} y\left(5 - x^2 - y^2\right) dx \right) dy = 2\int_0^1 y\left(5x - \frac{x^3}{3} - y^2x\right)\Big|_{x=0}^{\sqrt{5-y^2}} dy$$

$$= 2\int_0^1 y\left(\left(5 - y^2\right)x - \frac{x^3}{3}\right)\Big|_{x=0}^{\sqrt{5-y^2}} dy = 2\int_0^1 y\left(\left(5 - y^2\right)^{3/2} - \frac{1}{3}\left(5 - y^2\right)^{3/2}\right) dy$$

$$= \int_0^1 \frac{4}{3}\left(5 - y^2\right)^{3/2} y\,dy \qquad (1)$$

We compute the integral using the substitution $u = 5 - y^2$, $du = -2y\,dy$:

$$\iiint_{\mathcal{W}} y\,dV = \int_0^1 \frac{4}{3}\left(5 - y^2\right)^{3/2} y\,dy = \int_5^4 -\frac{2}{3}u^{3/2}\,du = \int_4^5 \frac{2}{3}u^{3/2}\,du = \frac{4}{15}u^{5/2}\Big|_4^5$$

$$= \frac{4}{15}\left(5^{5/2} - 4^{5/2}\right) \approx 6.37$$

27. The solution to Example 4 expresses a triple integral in the three orders $dz\,dy\,dx$, $dx\,dz\,dy$, and $dy\,dz\,dx$. Write the integral in the three other orders: $dz\,dx\,dy$, $dx\,dy\,dz$, and $dy\,dx\,dz$.

SOLUTION In Example 4 we considered the triple integral $\iiint_{\mathcal{W}} xyz^2\,dV$, where $\mathcal{W}$ is the region bounded by

$$z = 4 - y^2, \quad z = 0, \quad y = 2x, \quad x = 0.$$

We now write the triple integral as an iterated integral in the orders $dz\,dx\,dy$, $dx\,dy\,dz$, and $dy\,dx\,dz$.

- $dz\,dx\,dy$: The upper surface $z = 4 - y^2$ projects onto the xy-plane on the triangle defined by the lines $y = 2$, $y = 2x$, and $x = 0$.

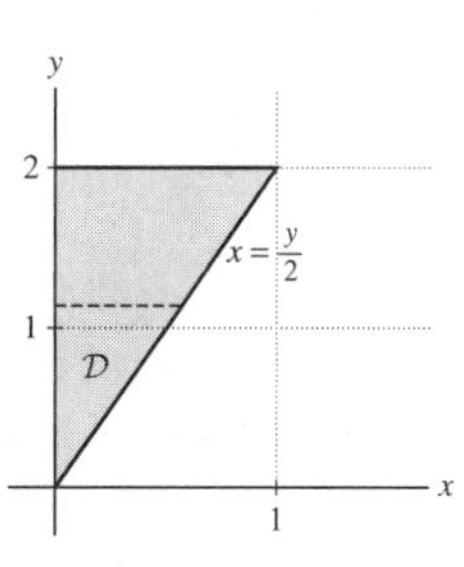

We express the line $y = 2x$ as $x = \frac{y}{2}$ and write the triple integral as

$$\iiint_{\mathcal{W}} xyz^2\,dV = \iint_{\mathcal{D}} \left(\int_0^{4-y^2} xyz^2\,dz \right) dA = \int_0^2 \int_0^{y/2} \int_0^{4-y^2} xyz^2\,dz\,dx\,dy$$

- $dx\,dy\,dz$: The projection of $\mathcal{W}$ onto the yz-plane is the domain T (see Example 4) defined by the inequalities

$$T : 0 \le y \le 2,\ 0 \le z \le 4 - y^2$$

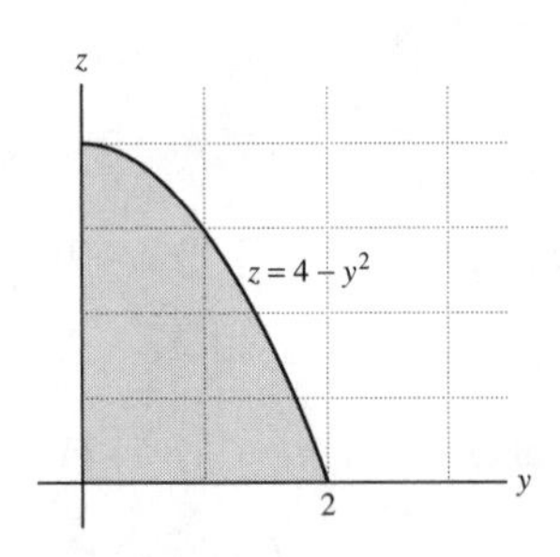

This region can also be expressed as

$$0 \le z \le 4, \quad 0 \le y \le \sqrt{4-z}$$

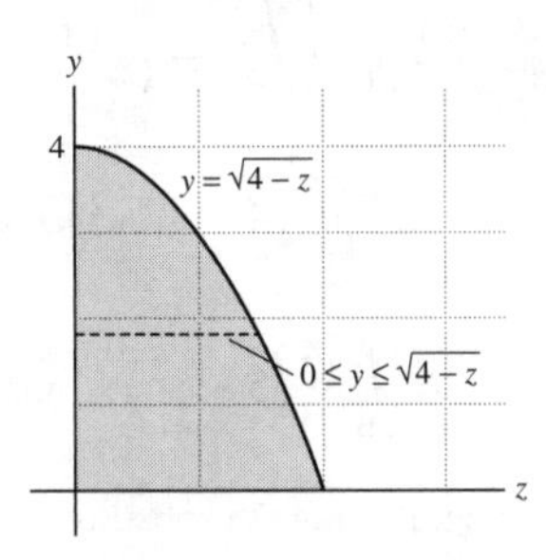

As explained in Example 4, the region $\mathcal{W}$ consists of all points lying between T and the "left-face" $y = 2x$, or $x = \frac{y}{2}$. Therefore, we obtain the following inequalities for $\mathcal{W}$:

$$0 \le z \le 4, \quad 0 \le y \le \sqrt{4-z}, \quad 0 \le x \le \frac{1}{2}y$$

This yields the following iterated integral:

$$\iiint_{\mathcal{W}} xyz^2\,dV = \iint_T \left(\int_0^{y/2} xyz^2\,dx\right) dA = \int_0^4 \int_0^{\sqrt{4-z}} \int_0^{y/2} xyz^2\,dx\,dy\,dz$$

- $dy\,dx\,dz$: As explained in Example 4, the projection of $\mathcal{W}$ onto the xz-plane is determined by the inequalities

$$S : 0 \le x \le 1,\ 0 \le z \le 4 - 4x^2$$

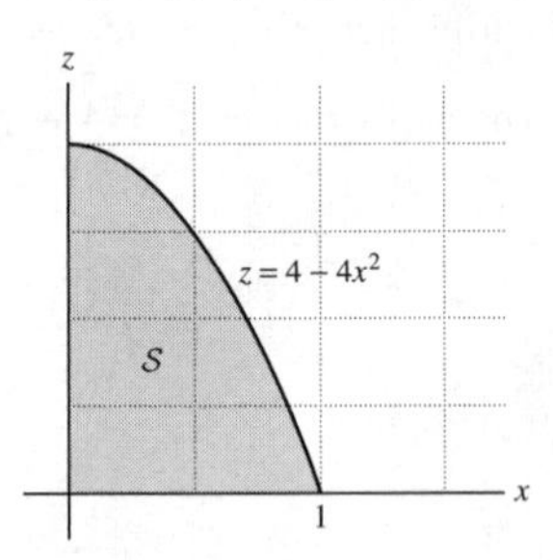

This region can also be described if we write x as a function of z:

$$z = 4 - 4x^2 \quad \Rightarrow \quad 4x^2 = 4 - z \quad \Rightarrow \quad x = \sqrt{1 - \frac{z}{4}}$$

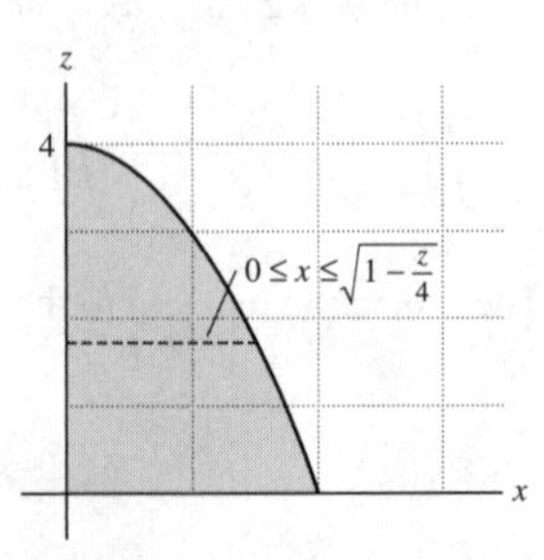

This gives the following inequalities of S:

$$S : 0 \le z \le 4,\ 0 \le x \le \sqrt{1 - \frac{z}{4}}$$

The upper surface $z = 4 - y^2$ can be described by $y = \sqrt{4-z}$, hence the limits of y are $2x \le y \le \sqrt{4-z}$. We obtain the following iterated integral:

$$\iiint_{\mathcal{W}} xyz^2\,dV = \iint_S \left(\int_{2x}^{\sqrt{4-z}} xyz^2\,dy\right) dA = \int_0^4 \int_0^{\sqrt{1-\frac{z}{4}}} \int_{2x}^{\sqrt{4-z}} xyz^2\,dy\,dx\,dz$$

29. Express the triple integral $\iiint_{\mathcal{W}} f(x, y, z)\,dV$ as an iterated integral in the two orders $dz\,dy\,dx$ and $dx\,dy\,dz$, where (Figure 17)

$$\mathcal{W} = \left\{(x, y, z) : \sqrt{x^2 + y^2} \le z \le 1\right\}$$

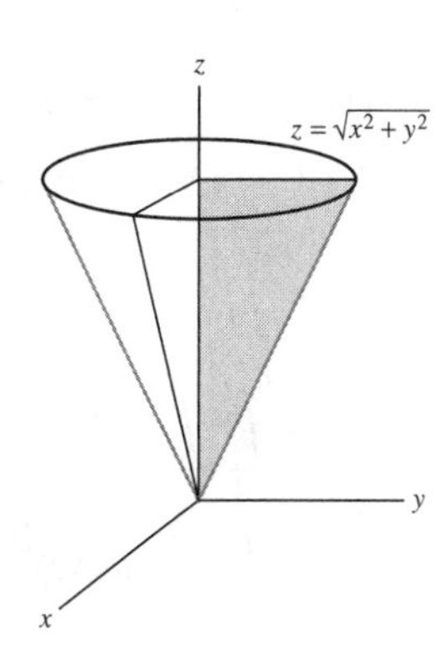

FIGURE 17

SOLUTION To express the triple integral as an iterated integral in order $dz\,dy\,dx$, we must find the projection of $\mathcal{W}$ onto the xy-plane. The upper circle is $\sqrt{x^2 + y^2} = 1$ or $x^2 + y^2 = 1$, hence the projection of $\mathcal{W}$ onto the xy plane is the disk

$$\mathcal{D} : x^2 + y^2 \le 1$$

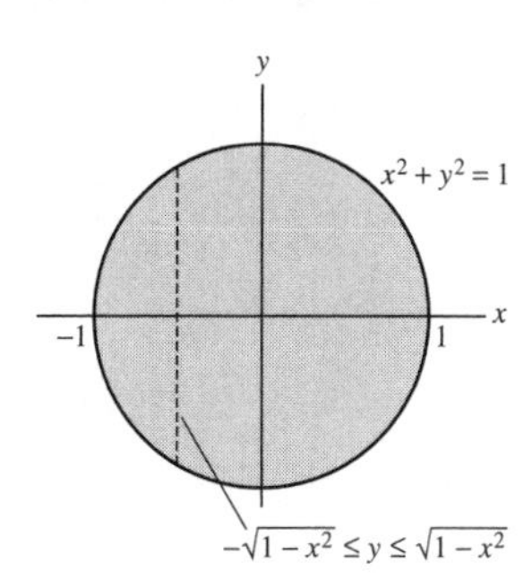

The upper surface is the plane $z = 1$ and the lower surface is $z = \sqrt{x^2 + y^2}$, therefore the triple integral over $\mathcal{W}$ is equal to the following iterated integral:

$$\iiint_{\mathcal{W}} f(x, y, z)\,dV = \iint_{\mathcal{D}} \left(\int_{\sqrt{x^2+y^2}}^{1} f(x, y, z)\,dz\right) dA = \int_{-1}^{1} \int_{-\sqrt{1-x^2}}^{\sqrt{1-x^2}} \int_{\sqrt{x^2+y^2}}^{1} f(x, y, z)\,dz\,dy\,dx$$

To express the triple integral as an iterated integral in order $dx\,dy\,dz$, we must find the projection of $\mathcal{W}$ onto the yz-plane. The points on the surface $z = \sqrt{x^2 + y^2}$ are $\left(x, y, \sqrt{x^2 + y^2}\right)$, hence the projection on the yz-plane consists of the points $(0, y, |y|)$, that is, $z = |y|$.

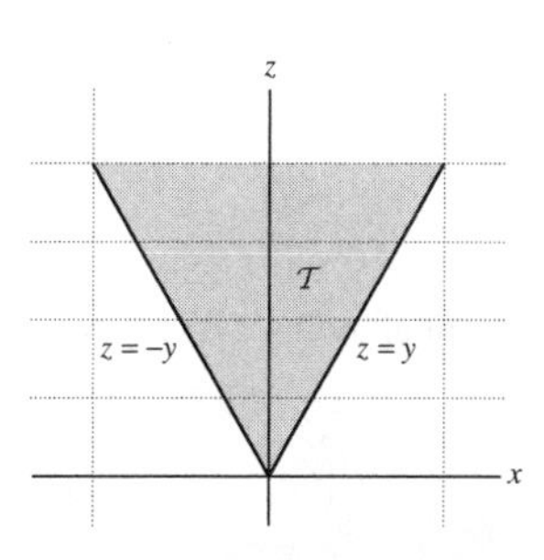

Therefore, the projection of $\mathcal{W}$ onto the yz-plane is the triangle defined by the lines $z = |y|$ and $z = 1$. The region $\mathcal{W}$ consists of all points lying between T and the surface $z = \sqrt{x^2 + y^2}$, or $x = \pm\sqrt{z^2 - y^2}$. $\mathcal{W}$ can be described by the inequalities

$$0 \le z \le 1$$
$$-z \le y \le z$$
$$-\sqrt{z^2 - y^2} \le x \le \sqrt{z^2 - y^2}$$

The triple integral is equal to the following iterated integral:

$$\iiint_{\mathcal{W}} f(x, y, z)\, dV = \int_0^1 \int_{-z}^{z} \int_{-\sqrt{z^2-y^2}}^{\sqrt{z^2-y^2}} f(x, y, z)\, dx\, dy\, dz.$$

31. Find the average of $f(x, y, z) = xy\sin(\pi z)$ over the cube $0 \le x, y, z \le 1$.

SOLUTION The volume of the cube is $V = 1$, hence the average of f over the cube is the following value:

$$\begin{aligned}
\overline{f} &= \iiint_{\mathcal{W}} xy\sin(\pi z)\, dV = \int_0^1 \int_0^1 \int_0^1 xy\sin(\pi z)\, dx\, dy\, dz \\
&= \int_0^1 \int_0^1 \frac{1}{2}x^2 y \sin(\pi z)\Big|_{x=0}^{1} dy\, dz = \int_0^1 \int_0^1 \frac{1}{2} y \sin(\pi z)\, dy\, dz \\
&= \int_0^1 \frac{y^2}{4}\sin(\pi z)\Big|_{y=0}^{1} dz = \int_0^1 \frac{1}{4}\sin(\pi z)\, dz = -\frac{1}{4\pi}\cos(\pi z)\Big|_0^1 \\
&= -\frac{1}{4\pi}(\cos\pi - \cos 0) = -\frac{1}{4\pi}(-1-1) = \frac{1}{2\pi}
\end{aligned}$$

33. Find the average of $f(x, y, z) = x^2 + y^2 + z^2$ over the region bounded by the planes $2y + z = 1$, $x = 0$, $x = 1$, $z = 0$, and $y = 0$.

SOLUTION The prism $\mathcal{W}$ bounded by the planes $2y + z = 1$, $x = 0$, $x = 1$, $z = 0$, and $y = 0$ is shown in the figure:

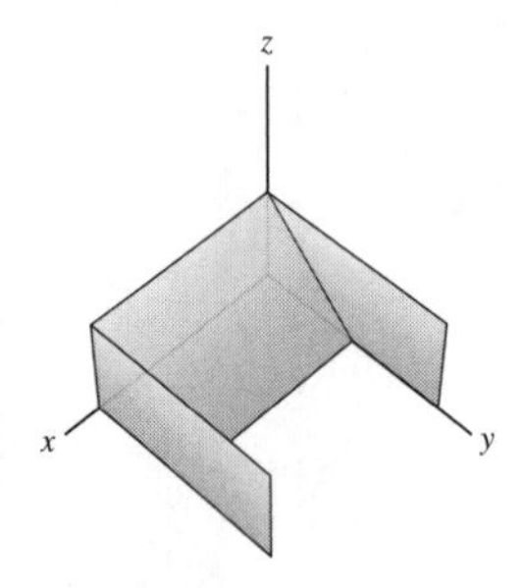

The average of f over $\mathcal{W}$ is

$$\overline{f} = \frac{1}{V}\iiint_{\mathcal{W}} f(x, y, z)\, dV \tag{1}$$

where $V = \text{Volume}(\mathcal{W})$. The projection of $\mathcal{W}$ onto the xy-plane is the rectangle

$$\mathcal{D}: 0 \le x \le 1,\ 0 \le y \le \frac{1}{2}$$

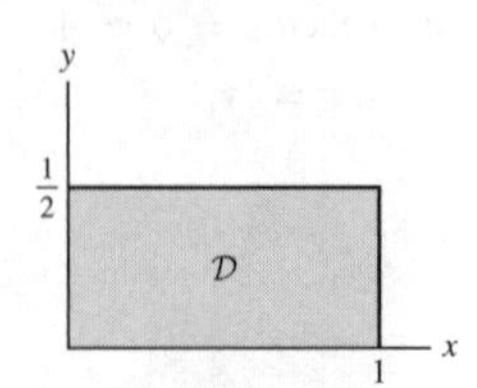

The region $\mathcal{W}$ consists of all points lying between the upper plane $z = 1 - 2y$ and the lower plane $z = 0$. The volume of the prism is

$$V = \frac{\frac{1}{2}\cdot 1}{2}\cdot 1 = \frac{1}{4}.$$

We compute the triple integral (1) using an iterated integral. That is,

$$\overline{f} = \frac{1}{\frac{1}{4}} \int_0^1 \int_0^{1/2} \int_0^{1-2y} \left(x^2 + y^2 + z^2\right) dz\,dy\,dx = 4 \int_0^1 \int_0^{1/2} \left(x^2 + y^2\right) z + \frac{z^3}{3} \Big|_{z=0}^{1-2y} dy\,dx$$

$$= 4 \int_0^1 \int_0^{1/2} \left(\left(x^2 + y^2\right)(1 - 2y) + \frac{(1-2y)^3}{3}\right) dy\,dx$$

$$= 4 \int_0^1 \int_0^{1/2} \left(x^2 - 2x^2y + y^2 - 2y^3 + \frac{(1-2y)^3}{3}\right) dy\,dx$$

$$= 4 \int_0^1 x^2y - x^2y^2 + \frac{y^3}{3} - \frac{y^4}{2} - \frac{(1-2y)^4}{24} \Big|_{y=0}^{1/2} dx$$

$$= 4 \int_0^1 \left(\frac{x^2}{4} + \frac{5}{96}\right) dx = 4 \left(\frac{x^3}{12} + \frac{5x}{96}\Big|_0^1\right) = \frac{13}{24}$$

The average value is $\overline{f} = \frac{13}{24}$.

35. Calculate the centroid of the tetrahedron in Figure 18.

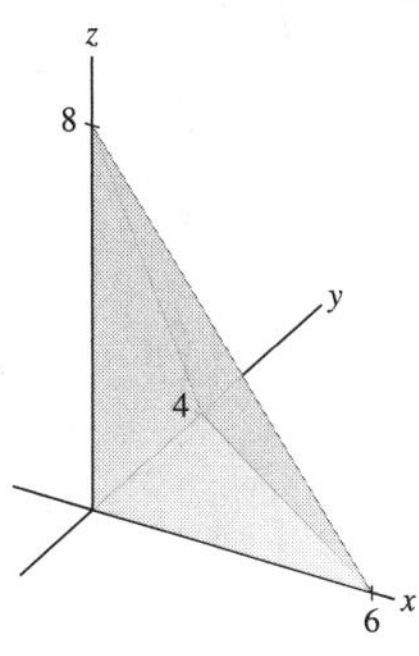

FIGURE 18

SOLUTION

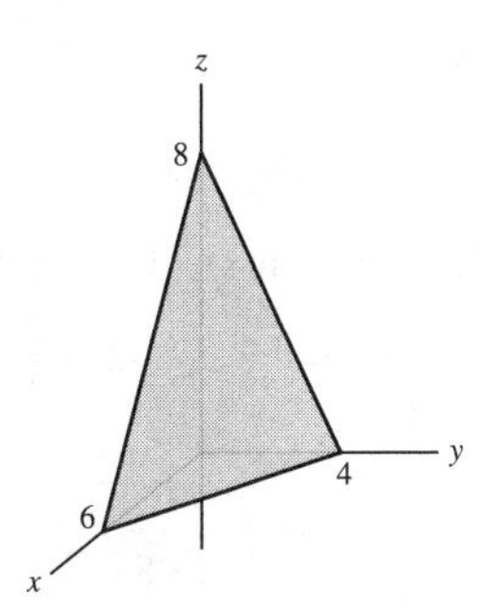

The centroid $P = (\overline{x}, \overline{y}, \overline{z})$ of the tetrahedron $\mathcal{W}$ is defined by

$$\overline{x} = \frac{1}{V} \iiint_{\mathcal{W}} x\,dv, \quad \overline{y} = \frac{1}{V} \iiint_{\mathcal{W}} y\,dv, \quad \overline{z} = \frac{1}{V} \iiint_{\mathcal{W}} z\,dV \tag{1}$$

The projection of $\mathcal{W}$ onto the xy-plane is the triangle defined by the line $AB : y = 4 - \frac{2}{3}x$ and the positive x and y axes.

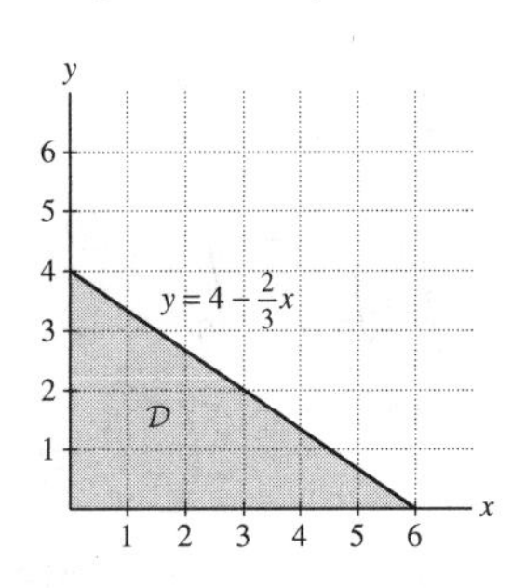

The upper surface is the plane passing through the points $A = (6, 0, 0)$, $B = (0, 4, 0)$, and $C = (0, 0, 8)$, and the lower surface is the plane $z = 0$. We find the equation of the upper plane. A vector normal to the plane is

$$\overrightarrow{AB} \times \overrightarrow{AC} = \langle -6, 4, 0\rangle \times \langle -6, 0, 8\rangle = \begin{vmatrix} \mathbf{i} & \mathbf{j} & \mathbf{k} \\ -6 & 4 & 0 \\ -6 & 0 & 8 \end{vmatrix}$$

$$= 32\mathbf{i} + 48\mathbf{j} + 24\mathbf{k} = 8(4\mathbf{i} + 6\mathbf{j} + 3\mathbf{k}) = 8\langle 4, 6, 3\rangle$$

We obtain the following equation of the upper plane:

$$4(x-6) + 6(y-0) + 3(z-0) = 0$$

$$4x + 6y + 3z = 24 \quad \Rightarrow \quad z = 8 - 2y - \frac{4}{3}x$$

We now can write the inequalities of the region $\mathcal{W}$:

$$0 \le x \le 6, \quad 0 \le y \le 4 - \frac{2x}{3}, \quad 0 \le z \le 8 - 2y - \frac{4x}{3} \tag{2}$$

First, we compute the volume of $\mathcal{W}$:

$$V = \text{Volume}(\mathcal{W}) = \iiint_{\mathcal{W}} 1\, dV = \int_0^6 \int_0^{4-\frac{2x}{3}} \int_0^{8-2y-\frac{4x}{3}} dz\, dy\, dx = \int_0^6 \int_0^{4-\frac{2x}{3}} z\Big|_{z=0}^{8-2y-\frac{4x}{3}} dy\, dx$$

$$= \int_0^6 \int_0^{4-\frac{2x}{3}} \left(8 - 2y - \frac{4x}{3}\right) dy\, dx = \int_0^6 8y - y^2 - \frac{4xy}{3}\Big|_{y=0}^{4-\frac{2x}{3}} dx$$

$$= \int_0^6 \left(\left(8 - \frac{4x}{3}\right)\left(4 - \frac{2x}{3}\right) - \left(4 - \frac{2x}{3}\right)^2\right) dx$$

$$= \int_0^6 \left(2\left(4 - \frac{2x}{3}\right)^2 - \left(4 - \frac{2x}{3}\right)^2\right) dx = \int_0^6 \left(4 - \frac{2x}{3}\right)^2 dx$$

$$= \frac{\left(4 - \frac{2}{3}x\right)^3}{3 \cdot \left(-\frac{2}{3}\right)}\Bigg|_0^6 = -\frac{1}{2}\left(4 - \frac{2}{3}x\right)^3\Bigg|_0^6 = 32$$

Thus,

$$V = 32 \tag{3}$$

We compute the triple integrals in (1). Using (2) we get

$$\iiint_{\mathcal{W}} x\, dV = \int_0^6 \int_0^{4-\frac{2x}{3}} \int_0^{8-2y-\frac{4x}{3}} x\, dz\, dy\, dx = \int_0^6 \int_0^{4-\frac{2x}{3}} xz\Big|_{z=0}^{8-2y-\frac{4x}{3}} dy\, dx$$

$$= \int_0^6 \int_0^{4-\frac{2x}{3}} x\left(8 - 2y - \frac{4x}{3}\right) dy\, dx = \int_0^6 \int_0^{4-\frac{2x}{3}} \left(x\left(8 - \frac{4x}{3}\right) - 2xy\right) dy\, dx$$

$$= \int_0^6 x\left(8 - \frac{4x}{3}\right)y - xy^2\Big|_{y=0}^{4-\frac{2x}{3}} dx = \int_0^6 \left(x\left(8 - \frac{4x}{3}\right)\left(4 - \frac{2x}{3}\right) - x\left(4 - \frac{2x}{3}\right)^2\right) dx$$

$$= \int_0^6 \left(2x\left(4 - \frac{2x}{3}\right)^2 - x\left(4 - \frac{2x}{3}\right)^2\right) dx = \int_0^6 x\left(4 - \frac{2x}{3}\right)^2 dx$$

$$= \int_0^6 \left(\frac{4x^3}{9} - \frac{16x^2}{3} + 16x\right) dx = \left(\frac{x^4}{9} - \frac{16x^3}{9} + 8x^2\right)\Bigg|_0^6 = 48$$

$$\iiint_{\mathcal{W}} y\, dV = \int_0^6 \int_0^{4-\frac{2x}{3}} \int_0^{8-2y-\frac{4x}{3}} y\, dz\, dy\, dx = \int_0^6 \int_0^{4-\frac{2x}{3}} yz\Big|_{z=0}^{8-2y-\frac{4x}{3}} dy\, dx$$

$$= \int_0^6 \int_0^{4-\frac{2x}{3}} y\left(8 - 2y - \frac{4x}{3}\right) dy\, dx = \int_0^6 \int_0^{4-\frac{2x}{3}} \left(\left(8 - \frac{4x}{3}\right)y - 2y^2\right) dy\, dx$$

$$= \int_0^6 \left(4 - \frac{2x}{3}\right)y^2 - \frac{2y^3}{3}\Big|_{y=0}^{4-\frac{2x}{3}} dx = \int_0^6 \left(\left(4 - \frac{2x}{3}\right)\left(4 - \frac{2x}{3}\right)^2 - \frac{2}{3}\left(4 - \frac{2x}{3}\right)^3\right) dx$$

$$= \int_0^6 \frac{1}{3}\left(4 - \frac{2x}{3}\right)^3 dx = \frac{\left(4 - \frac{2x}{3}\right)^4}{3 \cdot 4 \cdot \left(-\frac{2}{3}\right)}\Bigg|_0^6 = -\frac{1}{8}\left(4 - \frac{2x}{3}\right)^4\Bigg|_0^6 = 32$$

$$\iiint_{\mathcal{W}} z\,dV = \int_0^6 \int_0^{4-\frac{2x}{3}} \int_0^{8-2y-\frac{4x}{3}} z\,dz\,dy\,dx = \int_0^6 \int_0^{4-\frac{2x}{3}} \frac{z^2}{2}\Bigg|_{z=0}^{8-2y-\frac{4x}{3}} dy\,dx$$

$$= \frac{1}{2}\int_0^6 \int_0^{4-\frac{2x}{3}} \left(8 - 2y - \frac{4x}{3}\right)^2 dy\,dx = \frac{1}{2}\int_0^6 \frac{\left(8 - 2y - \frac{4x}{3}\right)^3}{3 \cdot (-2)}\Bigg|_{y=0}^{4-\frac{2x}{3}} dx$$

$$= -\frac{1}{12}\int_0^6 \left(\left(8 - 2\left(4 - \frac{2x}{3}\right) - \frac{4x}{3}\right)^3 - \left(8 - \frac{4x}{3}\right)^3\right) dx = \frac{1}{12}\int_0^6 \left(8 - \frac{4x}{3}\right)^3 dx$$

$$= \frac{1}{12}\frac{\left(8 - \frac{4x}{3}\right)^4}{4 \cdot \left(-\frac{4}{3}\right)}\Bigg|_0^6 = -\frac{1}{64}\left(8 - \frac{4x}{3}\right)^4\Bigg|_0^6 = \frac{8^4}{64} = 64$$

Using (1) and (3) gives

$$\overline{x} = \frac{48}{32} = 1.5, \quad \overline{y} = \frac{32}{32} = 1, \quad \overline{z} = \frac{64}{32} = 2$$

The centroid of the tetrahedron is thus $P = (1.5, 1, 2)$.

37. Find the center mass of the solid bounded by planes $x + y + z = 1$, $x = 0$, $y = 0$, and $z = 0$, assuming a mass density of $\rho(x, y, z) = \sqrt{z}$.

SOLUTION The coordinates of the center of mass are defined by

$$x_{CM} = \frac{\iiint_{\mathcal{W}} x\sqrt{z}\,dV}{\iiint_{\mathcal{W}} \sqrt{z}\,dV}$$
$$y_{CM} = \frac{\iiint_{\mathcal{W}} y\sqrt{z}\,dV}{\iiint_{\mathcal{W}} \sqrt{z}\,dV} \qquad (1)$$
$$z_{CM} = \frac{\iiint_{\mathcal{W}} z\sqrt{z}\,dV}{\iiint_{\mathcal{W}} \sqrt{z}\,dV}$$

$\mathcal{W}$ is the tetrahedron shown in the figure:

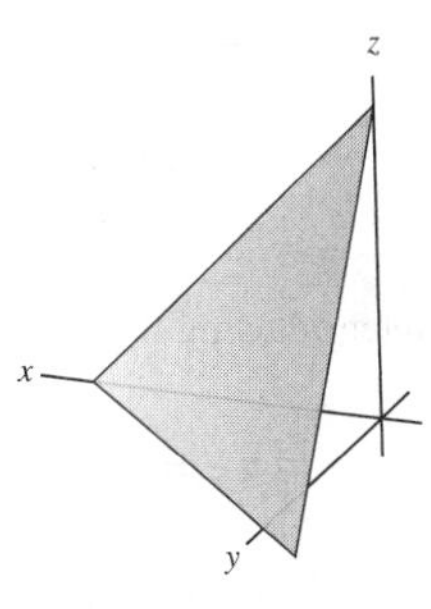

The projection $\mathcal{D}$ of $\mathcal{W}$ onto the xy-plane is the triangle determined by the line $AB : y = 1 - x$ and the positive x and y axes. The upper surface is the plane $z = 1 - x - y$ and the lower surface is $z = 0$. Therefore, we describe $\mathcal{W}$ by the following inequalities:

$$0 \le x \le 1, \quad 0 \le y \le 1 - x, \quad 0 \le z \le 1 - x - y \qquad (2)$$

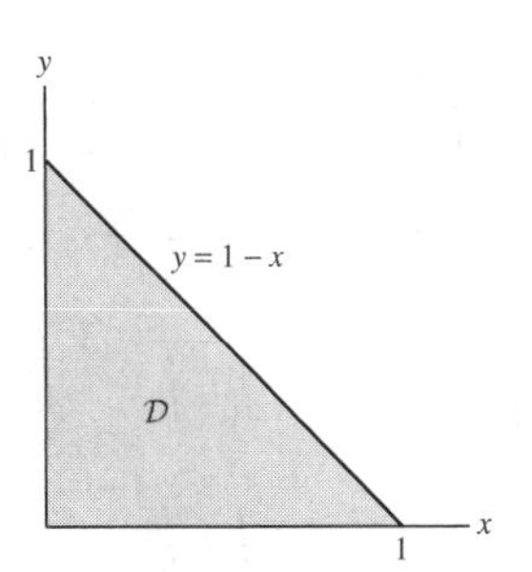

We compute the mass of the solid $\mathcal{W}$:

$$\iiint_{\mathcal{W}} \sqrt{z}\,dV = \int_0^1 \int_0^{1-x} \int_0^{1-x-y} \sqrt{z}\,dz\,dy\,dx = \int_0^1 \int_0^{1-x} \frac{2}{3} z^{3/2} \Big|_{z=0}^{1-x-y} dy\,dx$$
$$= \int_0^1 \int_0^{1-x} \frac{2}{3}(1-x-y)^{3/2} dy\,dx = \int_0^1 -\frac{2}{3}\cdot\frac{2}{5}(1-x-y)^{5/2}\Big|_{y=0}^{1-x} dx$$
$$= \int_0^1 -\frac{4}{15}\left((1-x-(1-x))^{5/2} - (1-x)^{5/2}\right) dx = \int_0^1 \frac{4}{15}(1-x)^{5/2}\,dx$$
$$= -\frac{4}{15}\cdot\frac{2}{7}(1-x)^{7/2}\Big|_0^1 = \frac{4}{15}\cdot\frac{2}{7} = \frac{8}{105}$$

That is,

$$\iiint_{\mathcal{W}} \sqrt{z}\,dV = \frac{8}{105} \tag{3}$$

We now compute the triple integrals in the numerators of (1). Using (2) we have

$$\iiint_{\mathcal{W}} x\sqrt{z}\,dV = \int_0^1 \int_0^{1-x} \int_0^{1-x-y} x\sqrt{z}\,dz\,dy\,dx = \int_0^1 \int_0^{1-x} \frac{2x}{3} z^{3/2}\Big|_{z=0}^{1-x-y} dy\,dx$$
$$= \int_0^1 \int_0^{1-x} \frac{2x}{3}(1-x-y)^{3/2} dy\,dx = \int_0^1 \frac{2x}{3}\cdot\left(-\frac{2}{5}\right)(1-x-y)^{5/2}\Big|_{y=0}^{1-x} dx$$
$$= \int_0^1 \frac{-4x}{15}\left((1-x-(1-x))^{5/2} - (1-x)^{5/2}\right) dx = \int_0^1 \frac{4x}{15}(1-x)^{5/2}\,dx$$

We compute the integral using the substitution $u = 1-x$, $du = -dx$. We get

$$\iiint_{\mathcal{W}} x\sqrt{z}\,dV = \int_1^0 \frac{4}{15}(1-u)u^{5/2}(-du) = \int_0^1 \frac{4}{15}\left(u^{5/2} - u^{7/2}\right) du$$
$$= \frac{4}{15}\left(\frac{2}{7}u^{7/2} - \frac{2}{9}u^{9/2}\right)\Big|_{u=0}^{1} = \frac{4}{15}\left(\frac{2}{7} - \frac{2}{9}\right) = \frac{16}{945} \tag{4}$$

We compute the numerator of y_{CM} in (1), using (2). This gives

$$\iiint_{\mathcal{W}} y\sqrt{z}\,dV = \int_0^1 \int_0^{1-x} \int_0^{1-x-y} y\sqrt{z}\,dz\,dy\,dx = \int_0^1 \int_0^{1-x} \frac{2y}{3} z^{3/2}\Big|_{z=0}^{1-x-y} dy\,dx$$
$$= \int_0^1 \int_0^{1-x} \frac{2y}{3}(1-x-y)^{3/2} dy\,dx \tag{5}$$

We compute the inner integral using the substitution $u = 1-x-y$, $du = -dy$. We get

$$\int_0^{1-x} \frac{2y}{3}(1-x-y)^{3/2}\,dy = \int_{1-x}^0 -\frac{2}{3}(1-x-u)u^{3/2}\,du = \int_0^{1-x} \frac{2}{3}\left((1-x)u^{3/2} - u^{5/2}\right) du$$
$$= \frac{2}{3}\left(\frac{2(1-x)}{5}u^{5/2} - \frac{2}{7}u^{7/2}\right)\Big|_{u=0}^{1-x} = \frac{4}{15}(1-x)(1-x)^{5/2} - \frac{4}{21}(1-x)^{7/2}$$
$$= \frac{8}{105}(1-x)^{7/2}$$

Combining with (5) we get

$$\iiint_{\mathcal{W}} y\sqrt{z}\,dV = \int_0^1 \frac{8}{105}(1-x)^{7/2}\,dx = -\frac{8}{105}\cdot\frac{2}{9}(1-x)^{9/2}\Big|_0^1 = \frac{16}{945} \tag{6}$$

Finally, we compute the denominator of z_{CM} in (1), using (2):

$$\iiint_{\mathcal{W}} z\sqrt{z}\,dV = \int_0^1 \int_0^{1-x} \int_0^{1-x-y} z^{3/2}dz\,dy\,dx = \int_0^1 \int_0^{1-x} \frac{2}{5} z^{5/2}\Big|_{z=0}^{1-x-y} dy\,dx$$
$$= \int_0^1 \int_0^{1-x} \frac{2}{5}(1-x-y)^{5/2} dy\,dx = \int_0^1 -\frac{2}{5}\cdot\frac{2}{7}(1-x-y)^{7/2}\Big|_{y=0}^{1-x} dx$$

$$= \int_0^1 \frac{-4}{35}\left((1-x-(1-x))^{7/2} - (1-x)^{7/2}\right) dx$$

$$= \int_0^1 \frac{4}{35}(1-x)^{7/2}\, dx = \frac{-4}{35}\cdot\frac{2}{9}(1-x)^{9/2}\Big|_0^1 = \frac{8}{315} \tag{7}$$

Substituting (3), (4), (6) and (7) in (1) we obtain the following center of mass:

$$(x_{\text{CM}}, y_{\text{CM}}, z_{\text{CM}}) = \left(\frac{\frac{16}{945}}{\frac{8}{105}}, \frac{\frac{16}{945}}{\frac{8}{105}}, \frac{\frac{8}{315}}{\frac{8}{105}}\right) = \left(\frac{2}{9}, \frac{2}{9}, \frac{1}{3}\right)$$

In Exercises 39–40, let $I = \int_0^1 \int_0^1 \int_0^1 f(x, y, z)\, dV$ *and let* $S_{N,N,N}$ *be the Riemann sum approximation*

$$S_{N,N,N} = \frac{1}{N^3} \sum_{i=1}^{N} \sum_{j=1}^{N} \sum_{k=1}^{N} f\left(\frac{i}{N}, \frac{j}{N}, \frac{k}{N}\right)$$

39. CAS Calculate $S_{N,N,N}$ for $f(x, y, z) = e^{x^2-y-z}$ for $N = 10, 20, 30$. Then evaluate I and find an N such that $S_{N,N,N}$ approximates I to two decimal places.

SOLUTION Using a CAS, we get $S_{N,N,N} \approx 0.561, 0.572$, and 0.576 for $N = 10, 20$, and 30, respectively. We get $I \approx 0.584$, and using $N = 100$ we get $S_{N,N,N} \approx 0.582$, accurate to two decimal places.

Further Insights and Challenges

41. Use Integration by Parts to verify Eq. (6).

SOLUTION If $C_n = \int_{-\pi/2}^{\pi/2} \cos^n \theta\, d\theta$, we use integration by parts to show that

$$C_n = \left(\frac{n-1}{n}\right) C_{n-2}.$$

We use integration by parts with $u = \cos^{n-1}\theta$ and $V' = \cos\theta$. Hence, $u' = (n-1)\cos^{n-2}\theta(-\sin\theta)$ and $v = \sin\theta$. Thus,

$$C_n = \int_{-\pi/2}^{\pi/2} \cos^n \theta\, d\theta = \int_{-\pi/2}^{\pi/2} \cos^{n-1}\theta \cos\theta\, d\theta = \cos^{n-1}\theta \sin\theta\Big|_{\theta=-\pi/2}^{\pi/2} + \int_{-\pi/2}^{\pi/2} (n-1)\cos^{n-2}\theta \sin^2\theta\, d\theta$$

$$= \cos^{n-1}\frac{\pi}{2}\sin\frac{\pi}{2} - \cos^{n+1}\left(-\frac{\pi}{2}\right)\sin\left(-\frac{\pi}{2}\right) + (n-1)\int_{-\pi/2}^{\pi/2} \cos^{n-2}\theta \sin^2\theta\, d\theta$$

$$= 0 + (n-1)\int_{-\pi/2}^{\pi/2} \cos^{n-2}\theta\left(1-\cos^2\theta\right) d\theta = (n-1)\int_{-\pi/2}^{\pi/2} \cos^{n-2}\theta\, d\theta - (n-1)\int_{-\pi/2}^{\pi/2} \cos^n\theta\, d\theta$$

$$= (n-1)C_{n-2} - (n-1)C_n$$

We obtain the following equality:

$$C_n = (n-1)C_{n-2} - (n-1)C_n$$

or

$$C_n + (n-1)C_n = (n-1)C_{n-2}$$

$$nC_n = (n-1)C_{n-2}$$

$$C_n = \frac{n-1}{n} C_{n-2}$$

16.4 Integration in Polar, Cylindrical, and Spherical Coordinates (ET Section 15.4)

Preliminary Questions

1. Which of the following represent the integral of $f(x, y) = x^2 + y^2$ over the unit circle?

(a) $\int_0^1 \int_0^{2\pi} r^2\, dr\, d\theta$

(b) $\int_0^{2\pi} \int_0^1 r^2\, dr\, d\theta$

(c) $\int_0^1 \int_0^{2\pi} r^3 \, dr \, d\theta$ **(d)** $\int_0^{2\pi} \int_0^1 r^3 \, dr \, d\theta$

SOLUTION The unit circle is described in polar coordinates by the inequalities

$$0 \le \theta \le 2\pi, \quad 0 \le r \le 1$$

Using double integral in polar coordinates, we have

$$\iint_{\mathcal{D}} f(x, y) \, dA = \int_0^{2\pi} \int_0^1 \left((r\cos\theta)^2 + (r\sin\theta)^2 \right) r \, dr \, d\theta = \int_0^{2\pi} \int_0^1 r^2 \left(\cos^2\theta + \sin^2\theta \right) r \, dr \, d\theta$$
$$= \int_0^{2\pi} \int_0^1 r^3 \, dr \, d\theta$$

Therefore (d) is the correct answer.

2. What are the limits of integration in $\iiint f(r, \theta, z) r \, dr \, d\theta \, dz$ if the integration extends over the following region?

(a) $x^2 + y^2 \le 4$, $-1 \le z \le 2$

(b) lower hemisphere of the sphere of radius 2, center at origin

SOLUTION

(a) This is a cylinder of radius 2. In the given region the z coordinate is changing between the values -1 and 2, and the angle θ is changing between the values $\theta = 0$ and 2π. Therefore the region is described by the inequalities

$$-1 \le z \le 2, \quad 0 \le \theta < 2\pi, \quad 0 \le r \le 2$$

Using triple integral in cylindrical coordinates gives

$$\int_{-1}^{2} \int_0^{2\pi} \int_0^2 f(P) \, r \, dr \, d\theta \, dz$$

(b) The sphere of radius 2 is $x^2 + y^2 + z^2 = r^2 + z^2 = 4$, or $r = \sqrt{4 - z^2}$.

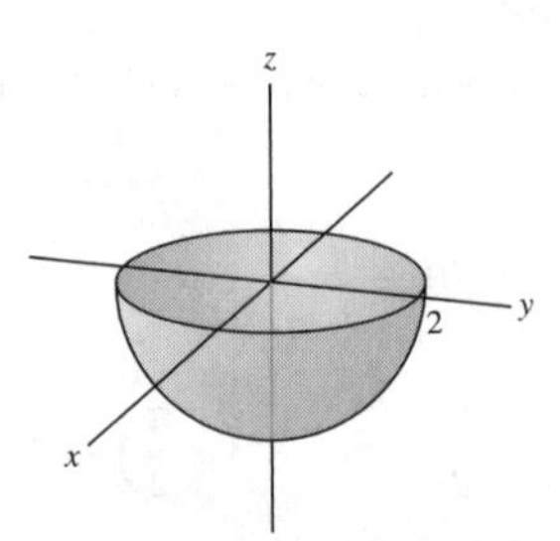

In the lower hemisphere we have $-2 \le z \le 0$ and $0 \le \theta < 2\pi$. Therefore, it has the description

$$-2 \le z \le 0, \quad 0 \le \theta < 2\pi, \quad 0 \le r \le \sqrt{4 - z^2}$$

We obtain the following integral in cylindrical coordinates:

$$\int_{-2}^{0} \int_0^{2\pi} \int_0^{\sqrt{4-z^2}} r \, dr \, d\theta \, dz$$

3. What are the limits of integration in

$$\iiint f(\rho, \phi, \theta) \rho^2 \sin\phi \, d\rho \, d\phi \, d\theta$$

if the integration extends over the following spherical regions centered at the origin?

(a) Sphere of radius 4

(b) Region between the spheres of radii 4 and 5

(c) Lower hemisphere of the sphere of radius 2

SOLUTION

(a) In the sphere of radius 4, θ varies from 0 to 2π, ϕ varies from 0 to π, and ρ varies from 0 to 4. Using triple integral in spherical coordinates, we obtain the following integral:

$$\int_0^{2\pi} \int_0^{\pi} \int_0^4 f(P) \rho^2 \sin\phi \, d\rho \, d\phi \, d\theta$$

(b) In the region between the spheres of radii 4 and 5, ρ varies from 4 to 5, ϕ varies from 0 to π, and θ varies from 0 to 2π. We obtain the following integral:

$$\int_0^{2\pi} \int_0^{\pi} \int_4^5 f(P)\rho^2 \sin\phi \, d\rho \, d\phi \, d\theta$$

(c) The inequalities in spherical coordinates for the lower hemisphere of radius 2 are

$$0 \le \theta \le 2\pi, \quad \frac{\pi}{2} \le \phi \le \pi, \quad 0 \le \rho \le 2$$

Therefore we obtain the following integral:

$$\int_0^{2\pi} \int_{\pi/2}^{\pi} \int_0^2 f(P)\rho^2 \sin\phi \, d\rho \, d\phi \, d\theta.$$

4. An ordinary rectangle of sides Δx and Δy has area $\Delta x \, \Delta y$, no matter where it is located in the plane. However, the area of a polar rectangle of sides Δr and $\Delta\theta$ depends on its distance from the origin. How is this difference reflected in the Change of Variables Formula for polar coordinates?

SOLUTION The area ΔA of a small polar rectangle is

$$\Delta A = \frac{1}{2}(r + \Delta r)^2 \Delta\theta - \frac{1}{2}r^2 \Delta\theta = r\,(\Delta r \Delta\theta) + \frac{1}{2}(\Delta r)^2 \Delta\theta \approx r\,(\Delta r \Delta\theta)$$

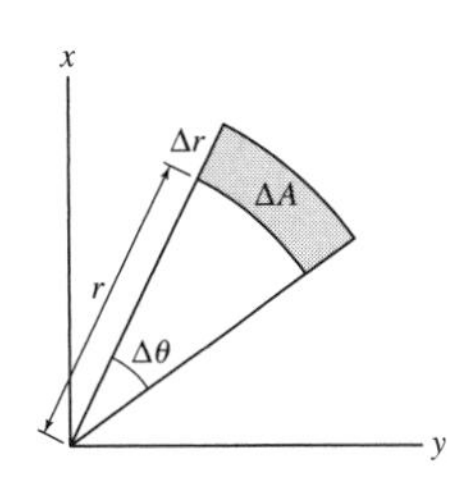

The factor r, due to the distance of the polar rectangle from the origin, appears in $dA = r\,dr\,d\theta$, in the Change of Variables formula.

Exercises

In Exercises 1–6, sketch the $\mathcal{D}$ indicated and integrate $f(x, y)$ over $\mathcal{D}$ using polar coordinates.

1. $f(x, y) = \sqrt{x^2 + y^2}, \quad x^2 + y^2 \le 2$

SOLUTION The domain $\mathcal{D}$ is the disk of radius $\sqrt{2}$ shown in the figure:

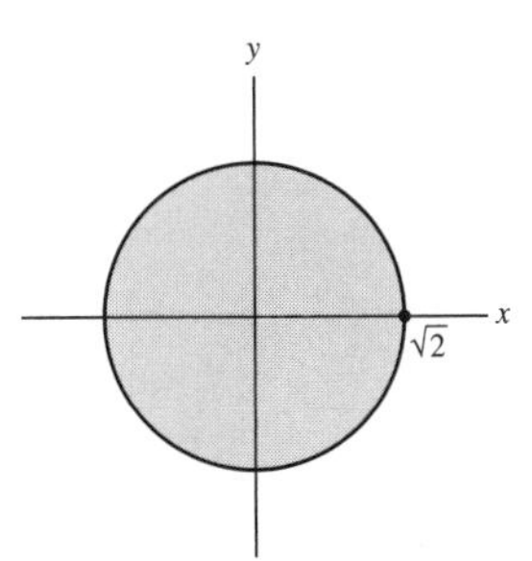

The inequalities defining $\mathcal{D}$ in polar coordinates are

$$0 \le \theta \le 2\pi, \quad 0 \le r \le \sqrt{2}$$

We describe $f(x, y) = \sqrt{x^2 + y^2}$ in polar coordinates:

$$f(x, y) = \sqrt{x^2 + y^2} = \sqrt{r^2} = r$$

Using change of variables in polar coordinates, we get

$$\iint_{\mathcal{D}} \sqrt{x^2 + y^2} \, dA = \int_0^{2\pi} \int_0^{\sqrt{2}} r \cdot r \, dr \, d\theta = \int_0^{2\pi} \int_0^{\sqrt{2}} r^2 \, dr \, d\theta = \int_0^{2\pi} \left. \frac{r^3}{3} \right|_{r=0}^{\sqrt{2}} d\theta$$

$$= \int_0^{2\pi} \frac{\left(\sqrt{2}\right)^3}{3} \, d\theta = \left. \frac{2\sqrt{2}}{3}\theta \right|_0^{2\pi} = \frac{4\sqrt{2}\pi}{3}$$

3. $f(x, y) = xy; \quad x \geq 0, \quad y \geq 0, \quad x^2 + y^2 \leq 4$

SOLUTION The domain $\mathcal{D}$ is the quarter circle of radius 2 in the first quadrant.

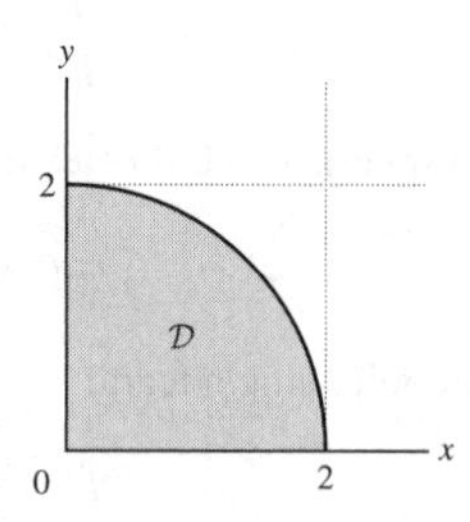

It is described by the inequalities

$$0 \leq \theta \leq \frac{\pi}{2}, \quad 0 \leq r \leq 2$$

We write f in polar coordinates:

$$f(x, y) = xy = (r\cos\theta)(r\sin\theta) = r^2\cos\theta\sin\theta = \frac{1}{2}r^2\sin 2\theta$$

Using change of variables in polar coordinates gives

$$\iint_{\mathcal{D}} xy\,dA = \int_0^{\pi/2}\int_0^2 \left(\frac{1}{2}r^2\sin 2\theta\right) r\,dr\,d\theta = \int_0^{\pi/2}\int_0^2 \frac{1}{2}r^3\sin 2\theta\,dr\,d\theta = \int_0^{\pi/2} \frac{1}{2}\cdot\frac{r^4}{4}\sin 2\theta\bigg|_{r=0}^{2} d\theta$$

$$= \int_0^{\pi/2} 2\sin 2\theta\,d\theta = -\cos 2\theta\bigg|_0^{\pi/2} = -(\cos\pi - \cos 0) = 2$$

5. $f(x, y) = y(x^2 + y^2)^{-1}; \quad y \geq \frac{1}{2}, \quad x^2 + y^2 \leq 1$

SOLUTION The region $\mathcal{D}$ is the part of the unit circle lying above the line $y = \frac{1}{2}$.

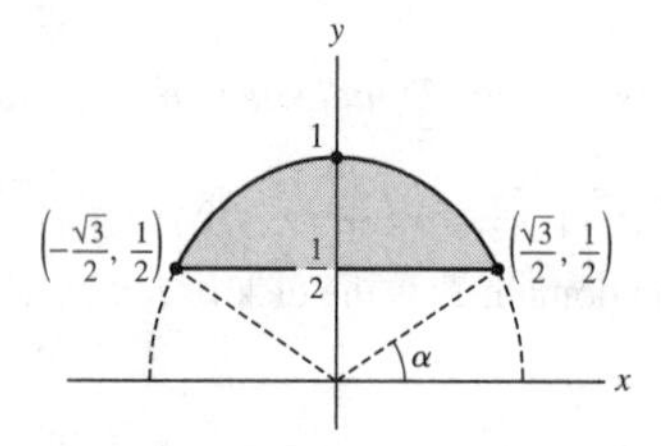

The angle α in the figure is

$$\alpha = \tan^{-1}\frac{\frac{1}{2}}{\frac{\sqrt{3}}{2}} = \tan^{-1}\frac{1}{\sqrt{3}} = \frac{\pi}{6}$$

Therefore, θ varies between $\frac{\pi}{6}$ and $\pi - \frac{\pi}{6} = \frac{5\pi}{6}$. The horizontal line $y = \frac{1}{2}$ has polar equation $r\sin\theta = \frac{1}{2}$ or $r = \frac{1}{2}\csc\theta$. The circle of radius 1 centered at the origin has polar equation $r = 1$. Therefore, r varies between $\frac{1}{2}\csc\theta$ and 1. The inequalities describing $\mathcal{D}$ in polar coordinates are thus

$$\frac{\pi}{6} \leq \theta \leq \frac{5\pi}{6}, \quad \frac{1}{2}\csc\theta \leq r \leq 1$$

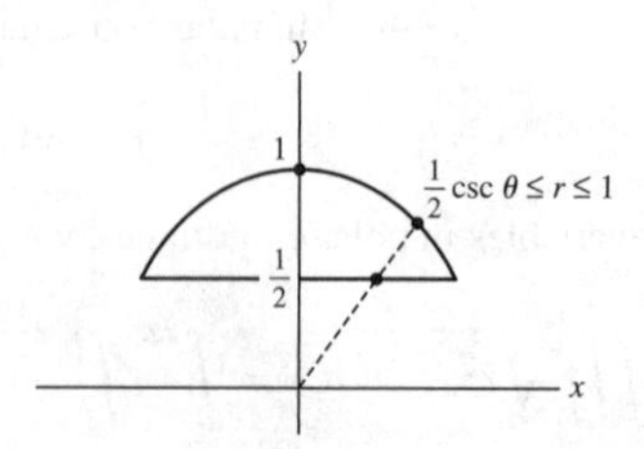

We write f in polar coordinates:

$$f(x, y) = y\left(x^2 + y^2\right)^{-1} = (r\sin\theta)\left(r^2\right)^{-1} = r^{-1}\sin\theta$$

Using change of variables in polar coordinates, we obtain

$$\iint_{\mathcal{D}} y(x^2+y^2)^{-1}\,dA = \int_{\pi/6}^{5\pi/6}\int_{\frac{1}{2}\csc\theta}^{1} r^{-1}\sin\theta\, r\,dr\,d\theta = \int_{\pi/6}^{5\pi/6}\int_{\frac{1}{2}\csc\theta}^{1}\sin\theta\,dr\,d\theta$$

$$= \int_{\pi/6}^{5\pi/6} r\sin\theta\Big|_{r=\frac{1}{2}\csc\theta}^{1}\,d\theta = \int_{\pi/6}^{5\pi/6}\left(\sin\theta - \frac{1}{2}\sin\theta\csc\theta\right)d\theta$$

$$= \int_{\pi/6}^{5\pi/6}\left(\sin\theta - \frac{1}{2}\right)d\theta = -\cos\theta - \frac{\theta}{2}\Big|_{\pi/6}^{5\pi/6} = -\cos\frac{5\pi}{6} - \frac{5\pi}{12} - \left(-\cos\frac{\pi}{6} - \frac{\pi}{12}\right)$$

$$= \frac{\sqrt{3}}{2} - \frac{\pi}{3} + \frac{\sqrt{3}}{2} = \sqrt{3} - \frac{\pi}{3} \approx 0.685$$

7. Find the volume of the wedge-shaped region (Figure 18) contained in the cylinder $x^2+y^2=9$ and bounded above by the plane $z=x$ and below by the xy-plane.

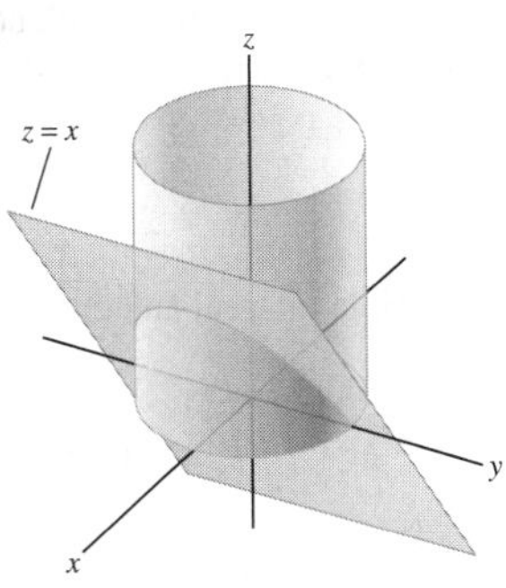

FIGURE 18

SOLUTION

Step 1. Express $\mathcal{W}$ in cylindrical coordinates. $\mathcal{W}$ is bounded above by the plane $z=x$ and below by $z=0$, therefore $0\le z\le x$, in particular $x\ge 0$. Hence, $\mathcal{W}$ projects onto the semicircle $\mathcal{D}$ in the xy-plane of radius 3, where $x\ge 0$.

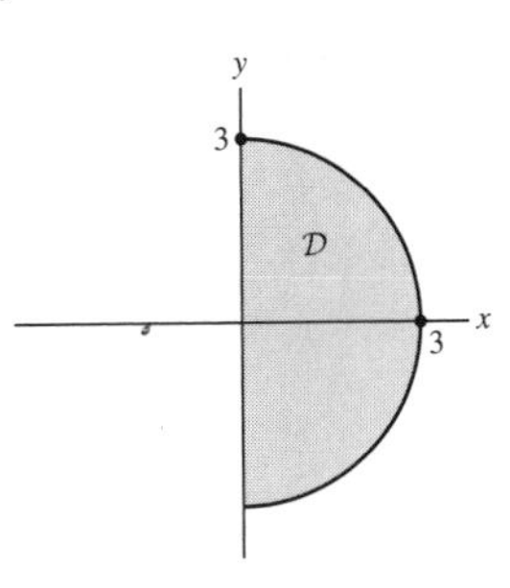

In polar coordinates,

$$\mathcal{D}: -\frac{\pi}{2}\le\theta\le\frac{\pi}{2},\ 0\le r\le 3$$

The upper surface is $z=x=r\cos\theta$ and the lower surface is $z=0$. Therefore,

$$\mathcal{W}: -\frac{\pi}{2}\le\theta\le\frac{\pi}{2},\ 0\le r\le 3,\ 0\le z\le r\cos\theta$$

Step 2. Set up an integral in cylindrical coordinates and evaluate. The volume of $\mathcal{W}$ is the triple integral $\iiint_{\mathcal{W}} 1\,dV$. Using change of variables in cylindrical coordinates gives

$$\iiint_{\mathcal{W}} 1\,dV = \int_{-\pi/2}^{\pi/2}\int_0^3\int_0^{r\cos\theta} r\,dz\,dr\,d\theta = \int_{-\pi/2}^{\pi/2}\int_0^3 rz\Big|_{z=0}^{r\cos\theta}\,dr\,d\theta = \int_{-\pi/2}^{\pi/2}\int_0^3 r^2\cos\theta\,dr\,d\theta$$

$$= \int_{-\pi/2}^{\pi/2}\frac{r^3}{3}\cos\theta\Big|_{r=0}^{3}\,d\theta = \int_{-\pi/2}^{\pi/2} 9\cos\theta\,d\theta = 9\sin\theta\Big|_{-\pi/2}^{\pi/2} = 9\left(\sin\frac{\pi}{2} - \sin\left(-\frac{\pi}{2}\right)\right) = 18$$

9. Use polar coordinates to compute the volume of the region defined by $4-x^2-y^2\le z\le 10-4x^2-4y^2$.

SOLUTION

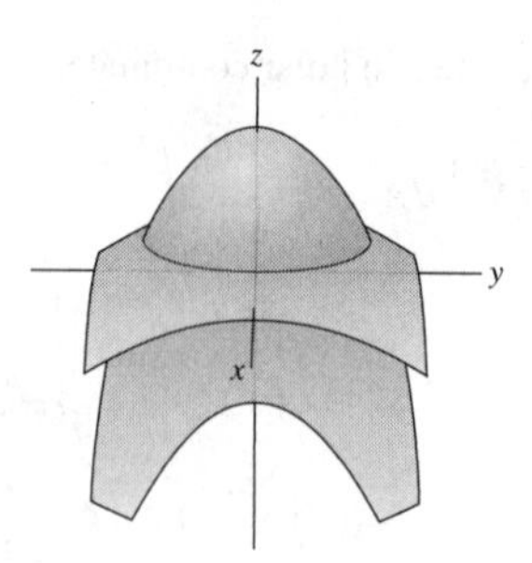

We first find the projection of $\mathcal{W}$ onto the xy-plane. The intersection curve of the upper and lower boundaries of $\mathcal{W}$ is obtained by solving

$$10 - 4x^2 - 4y^2 = 4 - x^2 - y^2$$
$$6 = 3(x^2 + y^2) \quad \Rightarrow \quad x^2 + y^2 = 2$$

Therefore, the projection of $\mathcal{W}$ onto the xy-plane is the circle $x^2 + y^2 \le 2$. The upper surface is $z = 10 - 4(x^2 + y^2)$ or $z = 10 - 4r^2$ and the lower surface is $z = 4 - (x^2 + y^2) = 4 - r^2$. Therefore, the inequalities for $\mathcal{W}$ in cylindrical coordinates are

$$0 \le \theta \le 2\pi, \quad 0 \le r \le \sqrt{2}, \quad 4 - r^2 \le z \le 10 - 4r^2$$

We use the volume as a triple integral and change of variables in cylindrical coordinates to write

$$V = \text{Volume}(\mathcal{W}) = \iiint_{\mathcal{W}} 1\, dV = \int_0^{2\pi} \int_0^{\sqrt{2}} \int_{4-r^2}^{10-4r^2} r\, dz\, dr\, d\theta = \int_0^{2\pi} \int_0^{\sqrt{2}} rz \Big|_{z=4-r^2}^{10-4r^2} dr\, d\theta$$
$$= \int_0^{2\pi} \int_0^{\sqrt{2}} r\left(10 - 4r^2 - \left(4 - r^2\right)\right) dr\, d\theta = \int_0^{2\pi} \int_0^{\sqrt{2}} \left(6r - 3r^3\right) dr\, d\theta$$
$$= \int_0^{2\pi} 3r^2 - \frac{3}{4}r^4 \Big|_{r=0}^{\sqrt{2}} d\theta = \int_0^{2\pi} (6 - 3)\, d\theta = 6\pi$$

11. Use polar coordinates to find the centroid of a quarter circle of radius R.

SOLUTION

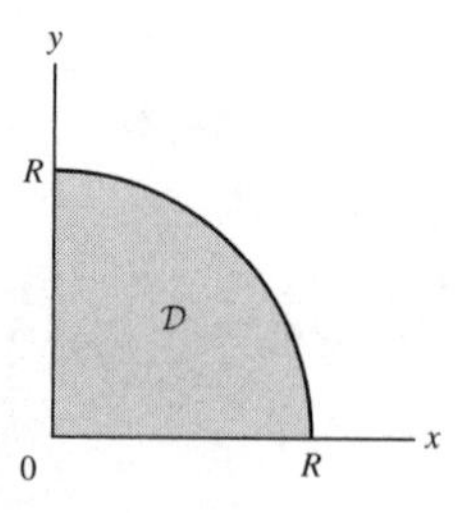

The centroid $P = (\overline{x}, \overline{y})$ has the following coordinates:

$$\overline{x} = \frac{1}{\text{Area}(\mathcal{D})} \iint_{\mathcal{D}} x\, dA = \frac{4}{\pi R^2} \iint_{\mathcal{D}} x\, dA$$
$$\overline{y} = \frac{1}{\text{Area}(\mathcal{D})} \iint_{\mathcal{D}} y\, dA = \frac{4}{\pi R^2} \iint_{\mathcal{D}} y\, dA$$

We compute the integrals using polar coordinates. The domain $\mathcal{D}$ is described in polar coordinates by the inequalities

$$\mathcal{D}: 0 \le \theta \le \frac{\pi}{2}, \quad 0 \le r \le R$$

The functions are $x = r\cos\theta$ and $y = r\sin\theta$, respectively. Using the Change of Variables Formula gives

$$\overline{x} = \frac{4}{\pi R^2} \int_0^{\pi/2} \int_0^R r\cos\theta \cdot r\, dr\, d\theta = \frac{4}{\pi R^2} \int_0^{\pi/2} \int_0^R r^2 \cos\theta\, dr\, d\theta = \frac{4}{\pi R^2} \int_0^{\pi/2} \frac{r^3 \cos\theta}{3} \Big|_{r=0}^{R} d\theta$$
$$= \frac{4}{\pi R^2} \int_0^{\pi/2} \frac{R^3 \cos\theta}{3}\, d\theta = \frac{4R}{3\pi} \sin\theta \Big|_0^{\pi/2} = \frac{4R}{3\pi}\left(\sin\frac{\pi}{2} - \sin 0\right) = \frac{4R}{3\pi}$$

And,

$$\overline{y} = \frac{4}{\pi R^2} \int_0^{\pi/2} \int_0^R r \sin\theta \cdot r\, dr\, d\theta = \frac{4}{\pi R^2} \int_0^{\pi/2} \int_0^R r^2 \sin\theta\, dr\, d\theta = \frac{4}{\pi R^2} \int_0^{\pi/2} \frac{r^3 \sin\theta}{3}\bigg|_{r=0}^{R} d\theta$$

$$= \frac{4}{\pi R^2} \int_0^{\pi/2} \frac{R^3 \sin\theta}{3}\, d\theta = \frac{4R}{3\pi}(-\cos\theta)\bigg|_0^{\pi/2} = \frac{4R}{3\pi}\left(-\cos\frac{\pi}{2} + \cos 0\right) = \frac{4R}{3\pi}$$

Notice that we can use the symmetry of $\mathcal{D}$ with respect to x and y to conclude that $\overline{y} = \overline{x}$, and save the computation of $\overline{y}$. We obtain the centroid $P = \left(\frac{4R}{3\pi}, \frac{4R}{3\pi}\right)$.

In Exercises 13–20, sketch the region of integration and evaluate by changing to polar coordinates.

13. $\displaystyle\int_{-2}^{2} \int_0^{\sqrt{4-x^2}} (x^2 + y^2)\, dy\, dx$

SOLUTION The domain $\mathcal{D}$ is described by the inequalities

$$\mathcal{D}: -2 \le x \le 2,\ 0 \le y \le \sqrt{4 - x^2}$$

That is, $\mathcal{D}$ is the semicircle $x^2 + y^2 \le 4$, $0 \le y$.

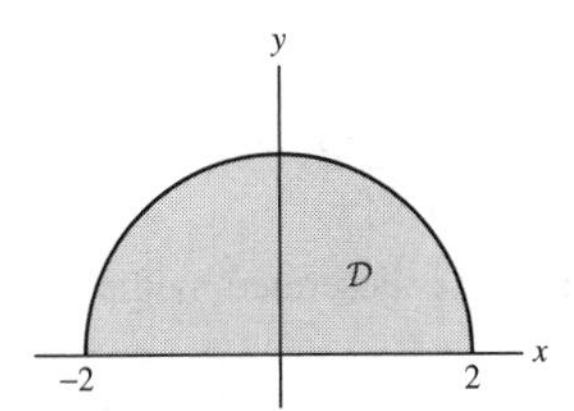

We describe $\mathcal{D}$ in polar coordinates:

$$\mathcal{D}: 0 \le \theta \le \pi,\ 0 \le r \le 2$$

The function f in polar coordinates is $f(x, y) = x^2 + y^2 = r^2$. We use the Change of Variables Formula to write

$$\int_{-2}^{2} \int_0^{\sqrt{4-x^2}} \left(x^2 + y^2\right) dy\, dx = \int_0^{\pi} \int_0^2 r^2 \cdot r\, dr\, d\theta = \int_0^{\pi} \int_0^2 r^3\, dr\, d\theta = \int_0^{\pi} \frac{r^4}{4}\bigg|_{r=0}^{2} d\theta = \int_0^{\pi} \frac{2^4}{4}\, d\theta = 4\pi$$

15. $\displaystyle\int_0^{1/2} \int_{\sqrt{3}x}^{\sqrt{1-x^2}} x\, dy\, dx$

SOLUTION The region of integration is described by the inequalities

$$0 \le x \le \frac{1}{2},\quad \sqrt{3}x \le y \le \sqrt{1 - x^2}$$

$\mathcal{D}$ is the circular sector shown in the figure.

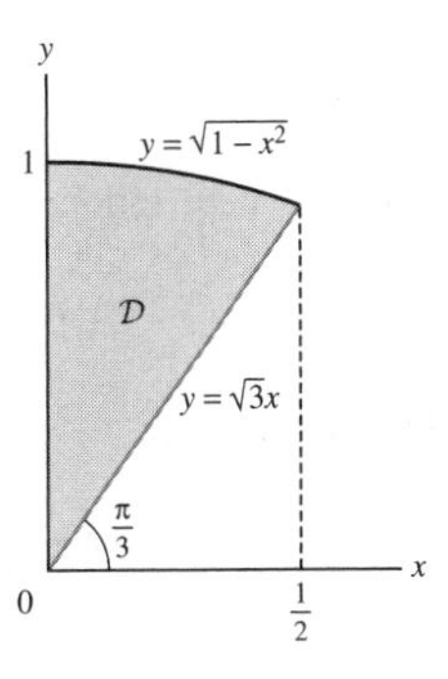

The ray $y = \sqrt{3}x$ in the first quadrant has the polar equation

$$r \sin\theta = \sqrt{3} r \cos\theta \quad\Rightarrow\quad \tan\theta = \sqrt{3} \quad\Rightarrow\quad \theta = \frac{\pi}{3}$$

Therefore, $\mathcal{D}$ lies in the angular sector $\frac{\pi}{3} \le \theta \le \frac{\pi}{2}$. Also, the circle $y = \sqrt{1 - x^2}$ has the polar equation $r = 1$, hence $\mathcal{D}$ can be described by the inequalities

$$\frac{\pi}{3} \le \theta \le \frac{\pi}{2},\quad 0 \le r \le 1$$

We use change of variables to obtain

$$\int_0^{1/2}\int_{\sqrt{3}x}^{\sqrt{1-x^2}} x\,dy\,dx = \int_{\pi/3}^{\pi/2}\int_0^1 r(\cos\theta)r\,dr\,d\theta = \int_{\pi/3}^{\pi/2}\int_0^1 r^2\cos\theta\,dr\,d\theta = \int_{\pi/3}^{\pi/2} \frac{r^3\cos\theta}{3}\bigg|_{r=0}^{1} d\theta$$

$$= \int_{\pi/3}^{\pi/2} \frac{\cos\theta}{3}\,d\theta = \frac{\sin\theta}{3}\bigg|_{\pi/3}^{\pi/2} = \frac{1}{3}\left(\sin\frac{\pi}{2} - \sin\frac{\pi}{3}\right) = \frac{1}{3}\left(1 - \frac{\sqrt{3}}{2}\right) \approx 0.045$$

17. $\displaystyle\int_0^4\int_0^{\sqrt{16-x^2}} \tan^{-1}\frac{y}{x}\,dy\,dx$

SOLUTION We note that this is an integral over the quarter circle of radius 4 in the first quadrant. Using the standard polar coordinates, we get:

$$\int_0^4\int_0^{\sqrt{16-x^2}} \tan^{-1}\frac{y}{x}\,dy\,dx = \int_0^{\pi/2}\int_0^4 \tan^{-1}\frac{r\sin\theta}{r\cos\theta}\,r\,dr\,d\theta = \int_0^{\pi/2}\int_0^4 \tan^{-1}\tan\theta\,r\,dr\,d\theta$$

$$= \int_0^{\pi/2}\int_0^4 \theta\,r\,dr\,d\theta = \frac{1}{2}r^2\bigg|_0^4 \cdot \frac{1}{2}\theta^2\bigg|_0^{\pi/2} = \pi^2$$

19. $\displaystyle\int_1^2\int_0^{\sqrt{2x-x^2}} \frac{1}{\sqrt{x^2+y^2}}\,dy\,dx$

SOLUTION The region is described by the inequalities

$$1 \le x \le 2, \quad 0 \le y \le \sqrt{2x - x^2}$$

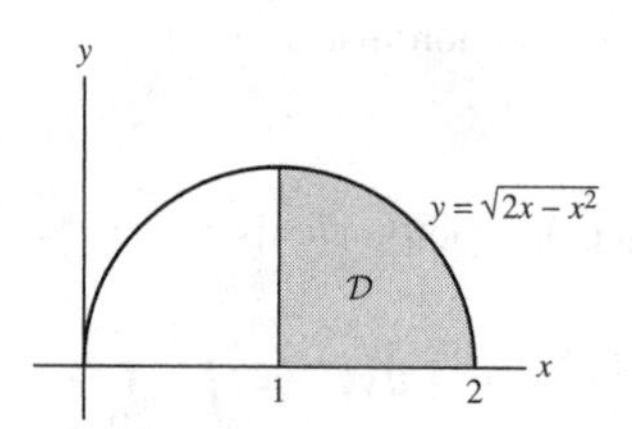

We first describe $\mathcal{D}$ in polar coordinates. The region lies in the angular sector $0 \le \theta \le \frac{\pi}{4}$. The circle $y = \sqrt{2x - x^2}$ or $(x-1)^2 + y^2 = 1$, $y \ge 0$ (obtained by completing the square) is the circle of radius 1 and center $(1, 0)$. Its polar equation is $r = 2\cos\theta$. The polar equation of the line $x = 1$ is $r\cos\theta = 1$ or $r = \sec\theta$.

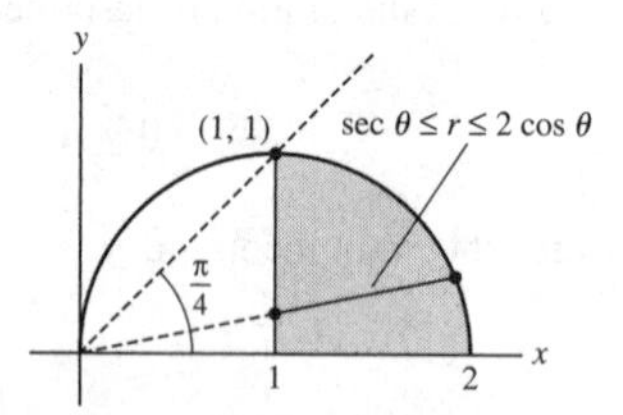

Therefore, $\mathcal{D}$ has the following description:

$$0 \le \theta \le \frac{\pi}{4}, \quad \sec\theta \le r \le 2\cos\theta$$

Using change of variables we get

$$\int_1^2\int_0^{\sqrt{2x-x^2}} \frac{dy\,dx}{\sqrt{x^2+y^2}} = \int_0^{\pi/4}\int_{\sec\theta}^{2\cos\theta} \frac{1}{r}\cdot r\,dr\,d\theta = \int_0^{\pi/4}\int_{\sec\theta}^{2\cos\theta} dr\,d\theta = \int_0^{\pi/4} r\bigg|_{r=\sec\theta}^{2\cos\theta} d\theta$$

$$= \int_0^{\pi/4} (2\cos\theta - \sec\theta)\,d\theta = 2\sin\theta\bigg|_0^{\pi/4} - \ln(\sec\theta + \tan\theta)\bigg|_0^{\pi/4}$$

$$= 2\sin\frac{\pi}{4} - \left(\ln\left(\sec\frac{\pi}{4} + \tan\frac{\pi}{4}\right) - \ln 1\right) = 2\cdot\frac{\sqrt{2}}{2} - \ln\left(\sqrt{2}+1\right)$$

$$= \sqrt{2} - \ln\left(1 + \sqrt{2}\right) \approx 0.533$$

In Exercises 21–26, calculate the integral over the given region by changing to polar coordinates.

21. $f(x, y) = (x^2 + y^2)^{-2}; \quad x^2 + y^2 \le 2, \quad x \ge 1$

SOLUTION The region $\mathcal{D}$ lies in the angular sector

$$-\frac{\pi}{4} \le \theta \le \frac{\pi}{4}$$

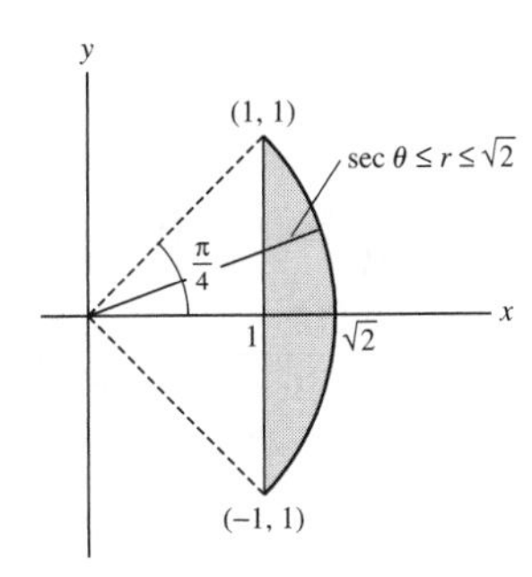

The vertical line $x = 1$ has polar equation $r\cos\theta = 1$ or $r = \sec\theta$. The circle $x^2 + y^2 = 2$ has polar equation $r = \sqrt{2}$. Therefore, $\mathcal{D}$ has the following description:

$$-\frac{\pi}{4} \le \theta \le \frac{\pi}{4}, \quad \sec\theta \le r \le \sqrt{2}$$

The function in polar coordinates is

$$f(x, y) = \left(x^2 + y^2\right)^{-2} = \left(r^2\right)^{-2} = r^{-4}.$$

Using change of variables we obtain

$$\iint_{\mathcal{D}} \left(x^2 + y^2\right)^{-2} dA = \int_{-\pi/4}^{\pi/4} \int_{\sec\theta}^{\sqrt{2}} r^{-4} r\,dr\,d\theta = \int_{-\pi/4}^{\pi/4} \int_{\sec\theta}^{\sqrt{2}} r^{-3}\,dr\,d\theta = \int_{-\pi/4}^{\pi/4} \frac{r^{-2}}{-2}\bigg|_{\sec\theta}^{\sqrt{2}} d\theta$$

$$= \int_{-\pi/4}^{\pi/4} \left(\frac{1}{2\sec^2\theta} - \frac{1}{4}\right) d\theta = 2\int_0^{\pi/4} \left(\frac{1}{2}\cos^2\theta - \frac{1}{4}\right) d\theta$$

$$= \left(\frac{\theta}{2} + \frac{\sin 2\theta}{4}\right)\bigg|_0^{\pi/4} - \frac{\theta}{2}\bigg|_0^{\pi/4} = \frac{\pi}{8} + \frac{1}{4} - \frac{\pi}{8} = \frac{1}{4}$$

23. $f(x, y) = |xy|, \quad x^2 + y^2 \le 1$

SOLUTION The unit disk is described in polar coordinates by

$$\mathcal{D}: 0 \le \theta \le 2\pi, \; 0 \le r \le 1$$

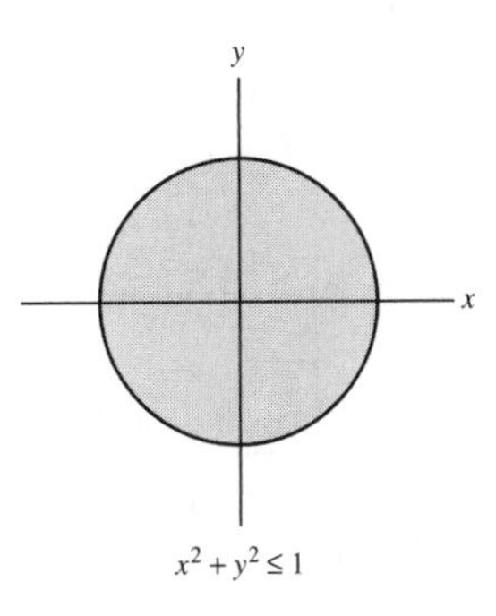

The function is $f(x, y) = |xy| = |r\cos\theta \cdot r\sin\theta| = \frac{1}{2}r^2|\sin 2\theta|$. Using change of variables we obtain

$$\iint_{\mathcal{D}} |xy|\,dA = \int_0^{2\pi} \int_0^1 \frac{1}{2}r^2|\sin 2\theta| \cdot r\,dr\,d\theta = \int_0^{2\pi} \int_0^1 \frac{1}{2}r^3|\sin 2\theta|\,dr\,d\theta$$

$$= \int_0^{2\pi} \frac{r^4}{8}|\sin 2\theta|\bigg|_{r=0}^1 d\theta = \int_0^{2\pi} \frac{1}{8}|\sin 2\theta|\,d\theta \tag{1}$$

The signs of $\sin 2\theta$ in the interval of integration are

For $0 \le \theta \le \frac{\pi}{2}$ or $\pi \le \theta \le \frac{3\pi}{2}$, $\sin 2\theta \ge 0$, hence $|\sin 2\theta| = \sin 2\theta$.

For $\frac{\pi}{2} \le \theta \le \pi$ or $\frac{3\pi}{2} \le \theta \le 2\pi$, $\sin 2\theta \le 0$, hence $|\sin 2\theta| = -\sin 2\theta$.

Therefore, by (1) we get

$$\iint_{\mathcal{D}} |xy|\, dA = \int_0^{\pi/2} \frac{1}{8}\sin 2\theta\, d\theta - \int_{\pi/2}^{\pi} \frac{1}{8}\sin 2\theta\, d\theta + \int_{\pi}^{3\pi/2} \frac{1}{8}\sin 2\theta\, d\theta - \int_{3\pi/2}^{2\pi} \frac{1}{8}\sin 2\theta\, d\theta$$

$$= -\frac{1}{16}\cos 2\theta\Big|_0^{\pi/2} + \frac{1}{16}\cos 2\theta\Big|_{\pi/2}^{\pi} - \frac{1}{16}\cos 2\theta\Big|_{\pi}^{3\pi/2} + \frac{1}{16}\cos 2\theta\Big|_{3\pi/2}^{2\pi}$$

$$= -\frac{1}{16}(\cos\pi - 1) + \frac{1}{16}(\cos 2\pi - \cos\pi) - \frac{1}{16}(\cos 3\pi - \cos 2\pi) + \frac{1}{16}(\cos 4\pi - \cos 3\pi)$$

$$= \frac{2}{16} + \frac{2}{16} + \frac{2}{16} + \frac{2}{16} = \frac{1}{2}$$

That is,

$$\iint_{\mathcal{D}} |xy|\, dA = \frac{1}{2}$$

25. $f(x, y) = x - y$; $x^2 + y^2 \le 1$, $x + y \ge 1$

SOLUTION As shown in Exercise 24, the region $\mathcal{D}$ is described by the following inequalities:

$$\mathcal{D}: 0 \le \theta \le \frac{\pi}{2}, \quad \frac{1}{\cos\theta + \sin\theta} \le r \le 1$$

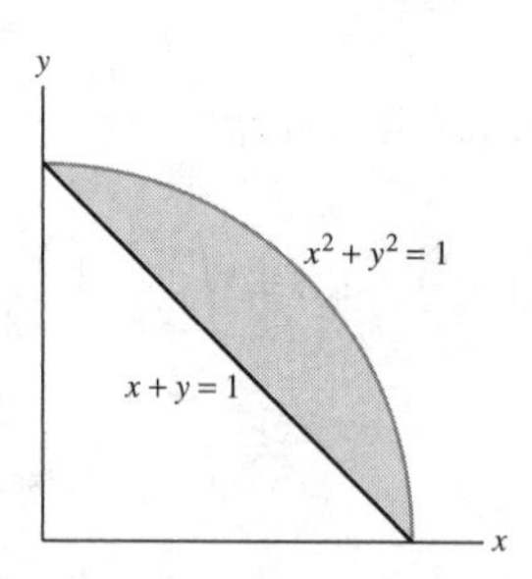

An easy solution is to note that $\iint_{\mathcal{D}} x - y\, dA = \iint_{\mathcal{D}} x\, dA - \iint_{\mathcal{D}} y\, dA$, the difference between the x and y coordinate of the centroid. But by symmetry these must be equal, so the entire double integral must be zero. However, let's go ahead and do this out the long way.

The function in polar coordinates is

$$f(x, y) = x - y = r\cos\theta - r\sin\theta = r(\cos\theta - \sin\theta)$$

Using the Change of Variables Formula we get

$$\iint_{\mathcal{D}} (x - y)\, dA = \int_0^{\pi/2} \int_{\frac{1}{\cos\theta+\sin\theta}}^{1} r(\cos\theta - \sin\theta) r\, dr\, d\theta = \int_0^{\pi/2} \int_{\frac{1}{\cos\theta+\sin\theta}}^{1} r^2(\cos\theta - \sin\theta)\, dr\, d\theta$$

$$= \int_0^{\pi/2} \frac{r^3(\cos\theta - \sin\theta)}{3}\Bigg|_{r=\frac{1}{\cos\theta+\sin\theta}}^{1} d\theta = \int_0^{\pi/2} \frac{\cos\theta - \sin\theta}{3}\left(1 - \frac{1}{(\cos\theta + \sin\theta)^3}\right) d\theta$$

$$= \int_0^{\pi/2} \frac{\cos\theta - \sin\theta}{3}\, d\theta - \frac{1}{3}\int_0^{\pi/2} \frac{\cos\theta - \sin\theta}{(\cos\theta + \sin\theta)^3}\, d\theta \qquad (1)$$

We compute the two integrals:

$$\int_0^{\pi/2} \frac{\cos\theta - \sin\theta}{3}\, d\theta = \frac{\sin\theta + \cos\theta}{3}\Bigg|_0^{\pi/2} = \frac{(1+0) - (0+1)}{3} = 0 \qquad (2)$$

We show that the second integral is zero, by showing that

$$\int_0^{\pi/2} \frac{\cos\theta}{(\cos\theta + \sin\theta)^3}\, d\theta = \int_0^{\pi/2} \frac{\sin\theta}{(\cos\theta + \sin\theta)^3}\, d\theta$$

We substitute $\theta = \frac{\pi}{2} - \alpha$, $d\theta = -d\alpha$ in the integral on the left-hand side:

$$\int_0^{\pi/2} \frac{\cos\theta}{(\cos\theta + \sin\theta)^3}\,d\theta = \int_{\pi/2}^{0} \frac{\cos\left(\frac{\pi}{2} - \alpha\right)(-d\alpha)}{\left(\cos\left(\frac{\pi}{2} - \alpha\right) + \sin\left(\frac{\pi}{2} - \alpha\right)\right)^{3/2}}$$

$$= \int_0^{\pi/2} \frac{\sin\alpha\,d\alpha}{(\sin\alpha + \cos\alpha)^{3/2}} = \int_0^{\pi/2} \frac{\sin\theta\,d\theta}{(\cos\alpha + \sin\alpha)^{3/2}}$$

We conclude that

$$\int_0^{\pi/2} \frac{\cos\theta - \sin\theta}{(\cos\theta + \sin\theta)^3}\,d\theta = 0 \tag{3}$$

Combining (1), (2), and (3) we conclude that

$$\iint_{\mathcal{D}} (x - y)\,dA = 0$$

27. Evaluate $\iint_{\mathcal{D}} \sqrt{x^2 + y^2}\,dA$, where $\mathcal{D}$ is the domain in Figure 21. *Hint:* Find the equation of the inner circle in polar coordinates and treat the right and left parts of the region separately.

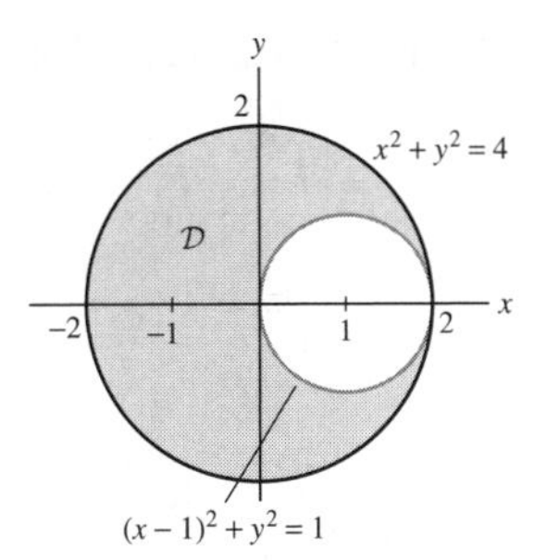

FIGURE 21

SOLUTION We denote by $\mathcal{D}_1$ and $\mathcal{D}_2$ the regions enclosed by the circles $x^2 + y^2 = 4$ and $(x-1)^2 + y^2 = 1$. Therefore,

$$\iint_{\mathcal{D}} \sqrt{x^2 + y^2}\,dx\,dy = \iint_{\mathcal{D}_1} \sqrt{x^2 + y^2}\,dx\,dy - \iint_{\mathcal{D}_2} \sqrt{x^2 + y^2}\,dx\,dy \tag{1}$$

We compute the integrals on the right hand-side.

$\mathcal{D}_1$:

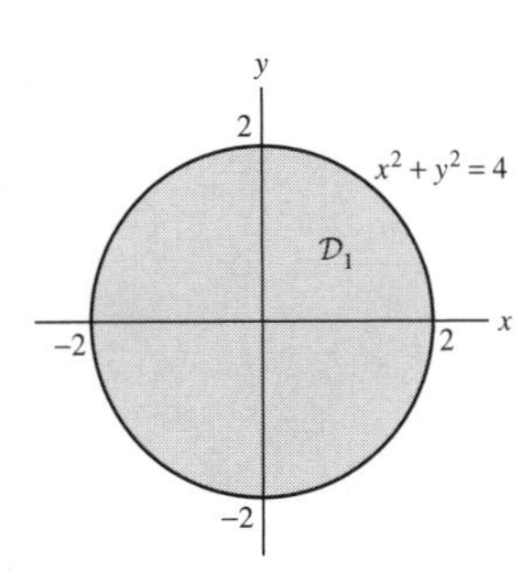

The circle $x^2 + y^2 = 4$ has polar equation $r = 2$, therefore $\mathcal{D}_1$ is determined by the following inequalities:

$$\mathcal{D}_1 : 0 \le \theta \le 2\pi,\ 0 \le r \le 2$$

The function in polar coordinates is $f(x, y) = \sqrt{x^2 + y^2} = r$. Using change of variables in the integral gives

$$\iint_{\mathcal{D}_1} \sqrt{x^2 + y^2}\,dx\,dy = \int_0^{2\pi}\int_0^2 r \cdot r\,dr\,d\theta = \int_0^{2\pi}\int_0^2 r^2\,dr\,d\theta = \int_0^{2\pi} \frac{r^3}{3}\Big|_{r=0}^{2}\,d\theta = \int_0^{2\pi} \frac{8}{3}\,d\theta = \frac{16\pi}{3} \tag{2}$$

$\mathcal{D}_2$:

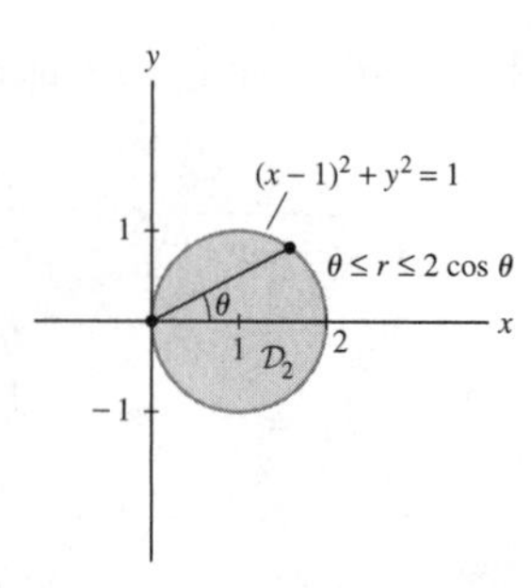

$\mathcal{D}_2$ lies in the angular sector $-\frac{\pi}{2} \le \theta \le \frac{\pi}{2}$. We find the polar equation of the circle $(x-1)^2+y^2=1$:

$$(x-1)^2+y^2 = x^2-2x+1+y^2 = x^2+y^2-2x+1 = 1 \quad \Rightarrow \quad x^2+y^2 = 2x$$
$$\Rightarrow \quad r^2 = 2r\cos\theta$$
$$\Rightarrow \quad r = 2\cos\theta$$

Thus, the domain $\mathcal{D}_2$ is defined by the following inequalities:

$$\mathcal{D}_2: -\frac{\pi}{2} \le \theta \le \frac{\pi}{2}, \quad 0 \le r \le 2\cos\theta$$

We use the change of variables in the integral and integration table to obtain

$$\iint_{\mathcal{D}_2} \sqrt{x^2+y^2}\,dx\,dy = \int_{-\pi/2}^{\pi/2}\int_0^{2\cos\theta} r\cdot r\,dr\,d\theta = \int_{-\pi/2}^{\pi/2}\int_0^{2\cos\theta} r^2\,dr\,d\theta = \int_{-\pi/2}^{\pi/2}\int_0^{2\cos\theta} \frac{r^3}{3}\bigg|_{r=0}^{2\cos\theta} d\theta$$
$$= \int_{-\pi/2}^{\pi/2} \frac{8\cos^3\theta}{3}\,d\theta = 2\int_0^{\pi/2}\frac{8\cos^3\theta}{3}\,d\theta = \frac{16}{3}\left(\frac{\cos^2\theta\sin\theta}{3} + \frac{2}{3}\sin\theta\right)\bigg|_{\theta=0}^{\pi/2}$$
$$= \frac{16}{3}\cdot\frac{2}{3}\sin\frac{\pi}{2} = \frac{32}{9} \tag{3}$$

Substituting (2) and (3) in (1), we obtain the following solution:

$$\iint_{\mathcal{D}} \sqrt{x^2+y^2}\,dx\,dy = \frac{16\pi}{3} - \frac{32}{9} = \frac{48\pi-32}{9} \approx 13.2.$$

Remark: The integral can also be evaluated using the hint as the sum of

$$\int_{\mathcal{D}^*}\iint \sqrt{x^2+y^2}\,dA \quad \text{and} \quad \int_{\mathcal{D}^{**}}\iint \sqrt{x^2+y^2}\,dA$$

where $\mathcal{D}^*$ is the left semicircle $x^2+y^2=4$ and $\mathcal{D}^*$ is the right part of $\mathcal{D}$. Since

$$\mathcal{D}^*: \frac{\pi}{2} \le \theta \le \frac{3\pi}{2}, \quad 0 \le r \le 2$$
$$\mathcal{D}^{**}: -\frac{\pi}{2} \le \theta \le \frac{\pi}{2}, \quad 2\cos\theta \le r \le 2$$

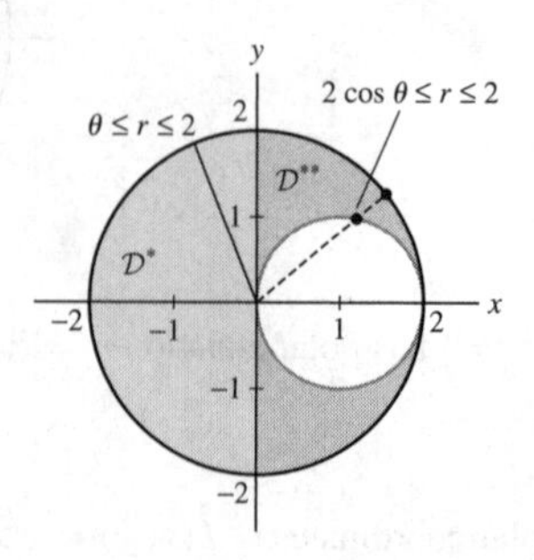

we get

$$\iint_{\mathcal{D}} \sqrt{x^2+y^2}\,dA = \int_{\pi/2}^{3\pi/2}\int_0^2 r^2\,dr\,d\theta + \int_{-\pi/2}^{\pi/2}\int_{2\cos\theta}^2 r^2\,dr\,d\theta$$

Obviously, computing the integrals leads to the same result.

29. Let $\mathcal{W}$ be the region between the paraboloids $z = x^2+y^2$ and $z = 8-x^2-y^2$.

(a) Describe $\mathcal{W}$ in cylindrical coordinates.

(b) Use cylindrical coordinates to compute the volume of $\mathcal{W}$.

SOLUTION

(a)

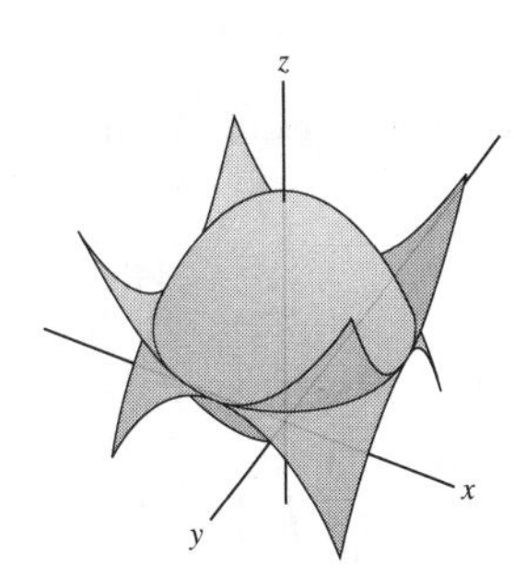

The paraboloids $z = x^2 + y^2$ and $z = 8 - (x^2 + y^2)$ have the polar equations $z = r^2$ and $z = 8 - r^2$, respectively. We find the curve of intersection by solving

$$8 - r^2 = r^2 \quad \Rightarrow \quad 2r^2 = 8 \quad \Rightarrow \quad r = 2$$

Therefore, the projection $\mathcal{D}$ of $\mathcal{W}$ onto the xy-plane is the region enclosed by the circle $r = 2$, and $\mathcal{D}$ has the following description:

$$\mathcal{D}: 0 \le \theta \le 2\pi,\ 0 \le r \le 2$$

The upper and lower boundaries of $\mathcal{W}$ are $z = 8 - r^2$ and $z = r^2$, respectively. Hence,

$$\mathcal{W}: 0 \le \theta \le 2\pi,\ 0 \le r \le 2,\ r^2 \le z \le 8 - r^2$$

(b) Using change of variables in cylindrical coordinates, we get

$$\text{Volume}(\mathcal{W}) = \iiint_{\mathcal{W}} 1\,dV = \int_0^{2\pi}\int_0^2\int_{r^2}^{8-r^2} r\,dz\,dr\,d\theta = \int_0^{2\pi}\int_0^2 rz\Big|_{z=r^2}^{8-r^2} dr\,d\theta = \int_0^{2\pi}\int_0^2 r\left(8 - 2r^2\right) dr\,d\theta$$

$$= \int_0^{2\pi}\int_0^2 \left(8r - 2r^3\right) dr\,d\theta = \int_0^{2\pi} 4r^2 - \frac{r^4}{2}\Big|_{r=0}^{2} d\theta = \int_0^{2\pi} 8\,d\theta = 16\pi$$

In Exercises 31–36, use cylindrical coordinates to calculate $\iiint_{\mathcal{W}} f(x, y, z)\,dV$ *for the given function and region.*

31. $f(x, y, z) = x^2 + y^2; \quad x^2 + y^2 \le 9, \quad 0 \le z \le 5$

SOLUTION The projection of $\mathcal{W}$ onto the xy-plane is the region inside the circle $x^2 + y^2 = 9$. In polar coordinates,

$$\mathcal{D}: 0 \le \theta \le 2\pi,\ 0 \le r \le 3$$

The upper and lower boundaries are the planes $z = 5$ and $z = 0$, respectively. Therefore, $\mathcal{W}$ has the following description in cylindrical coordinates:

$$\mathcal{W}: 0 \le \theta \le 2\pi,\ 0 \le r \le 3,\ 0 \le z \le 5$$

The integral in cylindrical coordinates is thus

$$\iiint_{\mathcal{W}} (x^2 + y^2)\,dV = \int_0^{2\pi}\int_0^3\int_0^5 r^2 \cdot r\,dz\,dr\,d\theta = \int_0^{2\pi}\int_0^3\int_0^5 r^3\,dz\,dr\,d\theta$$

$$= \left(\int_0^{2\pi} 1\,d\theta\right)\left(\int_0^3 r^3\,dr\right)\left(\int_0^5 1\,dz\right) = 2\pi \cdot 5 \cdot \frac{r^4}{4}\Big|_0^3 = \frac{5 \cdot 3^4\pi}{2} \approx 636.17$$

33. $f(x, y, z) = y; \quad x^2 + y^2 \le 1, \quad x \ge 0, \quad y \ge 0, \quad 0 \le z \le 2$

SOLUTION

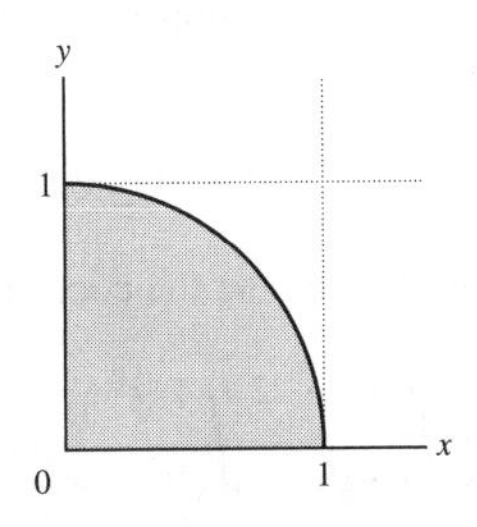

The projection of $\mathcal{W}$ onto the xy-plane is the quarter of the unit circle in the first quadrant. It is defined by the following polar equations:

$$\mathcal{D}: 0 \le \theta \le \frac{\pi}{2},\ 0 \le r \le 1$$

The upper and lower boundaries of $\mathcal{W}$ are the planes $z = 2$ and $z = 0$, respectively; hence, $\mathcal{W}$ has the following definition:

$$\mathcal{W}: 0 \le \theta \le \frac{\pi}{2},\ 0 \le r \le 1,\ 0 \le z \le 2$$

The function is $f(x, y, z) = y = r \sin\theta$. The integral in cylindrical coordinates is thus

$$\iiint_{\mathcal{W}} y\,dV = \int_0^{\pi/2}\int_0^1\int_0^2 (r\sin\theta)r\,dz\,dr\,d\theta = \int_0^{\pi/2}\int_0^1\int_0^2 r^2\sin\theta\,dz\,dr\,d\theta$$
$$= \left(\int_0^{\pi/2}\sin\theta\,d\theta\right)\left(\int_0^1 r^2\,dr\right)\left(\int_0^2 1\,dz\right) = \left(-\cos\theta\Big|_0^{\pi/2}\right)\left(\frac{r^3}{3}\Big|_0^1\right)\left(z\Big|_0^2\right) = 1\cdot\frac{1}{3}\cdot 2 = \frac{2}{3}$$

35. $f(x, y, z) = z,\quad x^2 + y^2 \le z \le 9$

SOLUTION

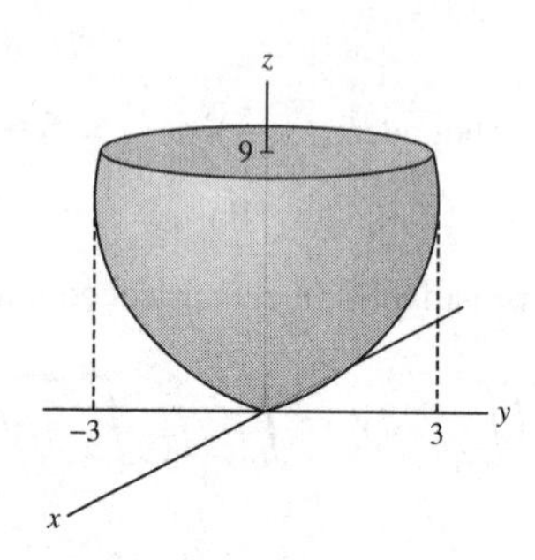

The upper boundary of $\mathcal{W}$ is the plane $z = 9$, and the lower boundary is $z = x^2 + y^2 = r^2$. Therefore, $r^2 \le z \le 9$. The projection $\mathcal{D}$ onto the xy-plane is the circle $x^2 + y^2 = 9$ or $r = 3$. That is,

$$\mathcal{D}: 0 \le \theta \le 2\pi,\ 0 \le r \le 3$$

The inequalities defining $\mathcal{W}$ in cylindrical coordinates are thus

$$\mathcal{W}: 0 \le \theta \le 2\pi,\ 0 \le r \le 3,\ r^2 \le z \le 9$$

Therefore, we obtain the following integral:

$$\iiint_{\mathcal{W}} z\,dV = \int_0^{2\pi}\int_0^3\int_{r^2}^9 zr\,dz\,dr\,d\theta = \int_0^{2\pi}\int_0^3 \frac{z^2r}{2}\Big|_{z=r^2}^9 dr\,d\theta = \int_0^{2\pi}\int_0^3 \frac{r(81-r^4)}{2}\,dr\,d\theta$$
$$= \int_0^{2\pi}\int_0^3 \frac{81r - r^5}{2}\,dr\,d\theta = \int_0^{2\pi} \frac{81r^2}{4} - \frac{r^6}{12}\Big|_0^3 d\theta = \int_0^{2\pi} 121.5\,d\theta = 243\pi$$

37. Find the height of the centroid (average value of the y-coordinate) for the region $\mathcal{W}$ in Figure 19, lying above the unit sphere $x^2 + y^2 + z^2 = 6$ and below the paraboloid $z = 4 - x^2 - y^2$.

SOLUTION The centroid is the point P with the following coordinates:

$$\overline{x} = \frac{1}{V}\iiint_{\mathcal{W}} x\,dV,\quad \overline{y} = \frac{1}{V}\iiint_{\mathcal{W}} y\,dV,\quad \overline{z} = \frac{1}{V}\iiint_{\mathcal{W}} z\,dV$$

In Exercise 8 we showed that the volume of the region is $V = 1.54\pi$. We also showed that $\mathcal{D}$ has the following definition in cylindrical coordinates:

$$0 \le \theta \le 2\pi,\quad 0 \le r \le \sqrt{2},\quad \sqrt{6-r^2} \le z \le 4 - r^2$$

Using this information we compute the coordinates of the centroid by the following integrals:

$$\overline{x} = \frac{1}{1.54\pi}\int_0^{2\pi}\int_0^{\sqrt{2}}\int_{\sqrt{6-r^2}}^{4-r^2}(r\cos\theta)r\,dz\,dr\,d\theta = \frac{1}{1.54\pi}\int_0^{2\pi}\int_0^{\sqrt{2}} r^2\cos\theta z\Big|_{z=\sqrt{6-r^2}}^{4-r^2} dr\,d\theta$$

$$= \frac{1}{1.54\pi} \int_0^{2\pi} \int_0^{\sqrt{2}} r^2 \cos\theta \left(4 - r^2 - \sqrt{6 - r^2}\right) dr\, d\theta$$

$$= \frac{1}{1.54\pi} \int_0^{2\pi} \cos\theta \int_0^{\sqrt{2}} \left(4r^2 - r^4 - r^2\sqrt{6 - r^2}\right) dr\, d\theta \qquad (1)$$

We denote the inner integral by a and compute the second integral to obtain

$$\overline{x} = \frac{1}{1.54\pi} \int_0^{2\pi} \cos\theta \cdot a\, d\theta = \frac{1}{1.54\pi} a \sin\theta \bigg|_0^{2\pi} = 0$$

The value $\overline{x} = 0$ is the result of the symmetry of $\mathcal{W}$ with respect to the yz-plane. Similarly, since $\mathcal{W}$ is symmetric with respect to the xz-plane, the average value of the y-coordinate is zero.

$\overline{y} = 0$:

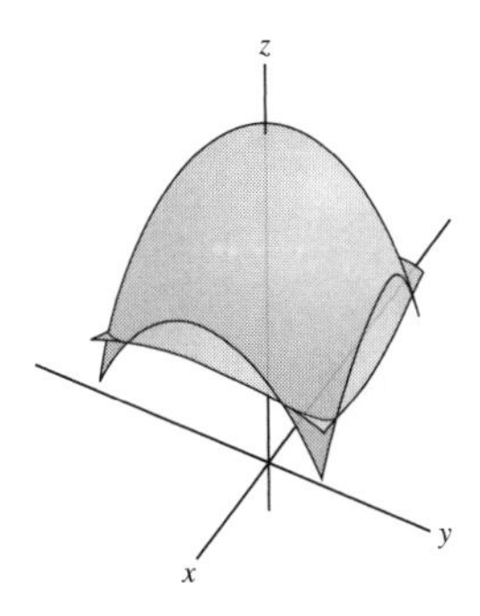

We compute the z-coordinate of the centroid:

$$\overline{z} = \frac{1}{1.54\pi} \int_0^{2\pi} \int_0^{\sqrt{2}} \int_{\sqrt{6-r^2}}^{4-r^2} zr\, dz\, dr\, d\theta = \frac{1}{1.54\pi} \int_0^{2\pi} \int_0^{\sqrt{2}} \frac{z^2 r}{2} \bigg|_{z=\sqrt{6-r^2}}^{4-r^2} dr\, d\theta$$

$$= \frac{1}{1.54\pi} \int_0^{2\pi} \int_0^{\sqrt{2}} \frac{r}{2} \left((4 - r^2)^2 - \left(\sqrt{6 - r^2}\right)^2 \right) dr\, d\theta$$

$$= \frac{1}{2 \cdot 1.54\pi} \int_0^{2\pi} \int_0^{\sqrt{2}} (r^5 - 7r^3 + 10r)\, dr\, d\theta$$

$$= \frac{1}{3.08\pi} \int_0^{2\pi} \frac{r^6}{6} - \frac{7r^4}{4} + 5r^2 \bigg|_{r=0}^{\sqrt{2}} d\theta = \frac{1}{3.08\pi} \cdot \frac{13}{3} \cdot 2\pi \approx 2.81$$

Therefore the centroid of $\mathcal{W}$ is

$$P = (0, 0, 2.81).$$

In Exercises 38–41, express the triple integral in cylindrical coordinates.

39. $\displaystyle\int_0^1 \int_{y=-\sqrt{1-x^2}}^{y=\sqrt{1-x^2}} \int_{z=0}^{4} f(x, y, z)\, dz\, dy\, dx$

SOLUTION The region of integration is determined by the limits of integration. That is,

$$\mathcal{W} : 0 \le x \le 1,\ -\sqrt{1 - x^2} \le y \le \sqrt{1 - x^2}\,,\ 0 \le z \le 4$$

Thus, the projection $\mathcal{D}$ of $\mathcal{W}$ onto the xy-plane is the semicircle $x^2 + y^2 = 1$, where $0 \le x \le 1$. This region is described by the polar inequalities

$$\mathcal{D} : -\frac{\pi}{2} \le \theta \le \frac{\pi}{2},\ 0 \le r \le 1$$

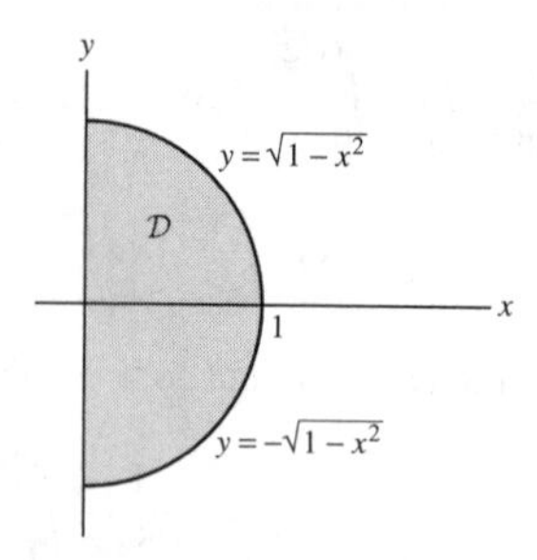

The upper and lower boundaries of $\mathcal{W}$ are the planes $z = 4$ and $z = 0$, respectively. Hence,

$$\mathcal{W}: -\frac{\pi}{2} \le \theta \le \frac{\pi}{2}, \ 0 \le r \le 1, \ 0 \le z \le 4$$

The integral in cylindrical coordinates is thus

$$\int_{-\pi/2}^{\pi/2} \int_0^1 \int_0^4 f(r\cos\theta, r\sin\theta, z) r \, dz \, dr \, d\theta$$

41. $\displaystyle\int_0^2 \int_{y=0}^{y=\sqrt{2x-x^2}} \int_{z=0}^{\sqrt{x^2+y^2}} f(x, y, z) \, dz \, dy \, dx$

SOLUTION The inequalities defining the region of integration are

$$\mathcal{W}: 0 \le x \le 2, \ 0 \le y \le \sqrt{2x - x^2}, \ 0 \le z \le \sqrt{x^2 + y^2}$$

The curve $y = \sqrt{2x - x^2}$ is the semicircle of radius 1 centered at $(1, 0)$, where $y \ge 0$. We find the polar equation of the semicircle:

$$r\sin\theta = \sqrt{2r\cos\theta - r^2\cos^2\theta}$$
$$r^2\sin^2\theta = 2r\cos\theta - r^2\cos^2\theta$$
$$r^2(\sin^2\theta + \cos^2\theta) = 2r\cos\theta$$
$$r^2 = 2r\cos\theta \quad \Rightarrow \quad r = 2\cos\theta \quad \text{and} \quad 0 \le \theta \le \frac{\pi}{2}$$

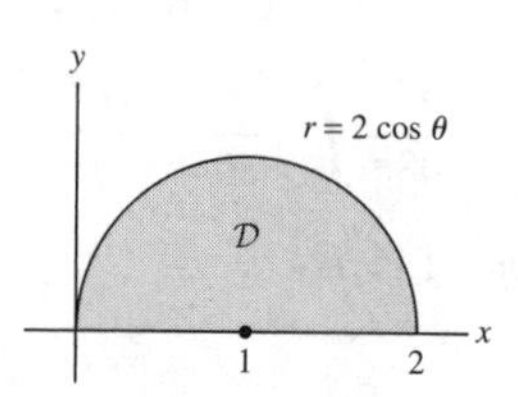

Therefore the projection of $\mathcal{W}$ onto the xy-plane is defined by the polar inequalities

$$\mathcal{D}: 0 \le \theta \le \frac{\pi}{2}, \ 0 \le r \le 2\cos\theta$$

The lower boundary of $\mathcal{W}$ is the plane $z = 0$ and the upper boundary is $z = \sqrt{x^2 + y^2} = r$, hence $\mathcal{W}$ has the following cylindrical definition:

$$\mathcal{W}: 0 \le \theta \le \frac{\pi}{2}, \ 0 \le r \le 2\cos\theta, \ 0 \le z \le r$$

The integral in cylindrical coordinates is

$$\int_0^{\pi/2} \int_0^{2\cos\theta} \int_0^r f(r\cos\theta, r\sin\theta, z) r \, dz \, dr \, d\theta$$

43. Find the equation of the right-circular cone in Figure 23 in cylindrical coordinates and compute its volume.

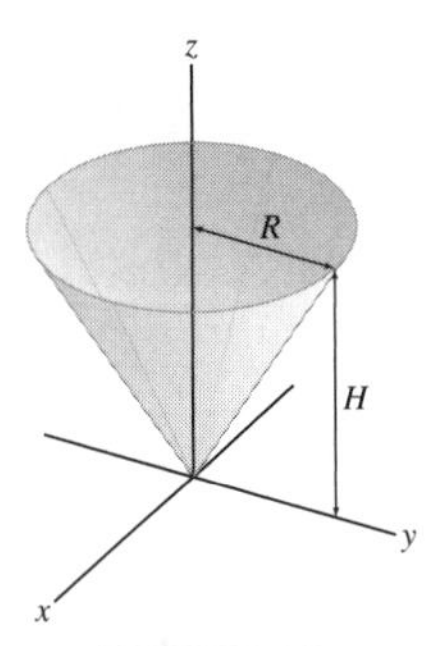

FIGURE 23

SOLUTION To find the equation of the surface we use proportion in similar triangles.

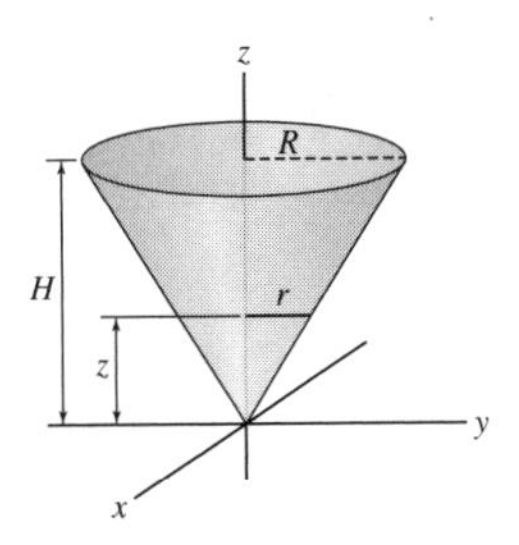

This gives

$$\frac{z}{H} = \frac{r}{R} \quad \Rightarrow \quad z = \frac{H}{R}r$$

The volume of the right circular cone is

$$V = \iiint_{\mathcal{W}} 1\,dV$$

The projection of $\mathcal{W}$ onto the xy-plane is the region $x^2 + y^2 \le R^2$, or in polar coordinates,

$$\mathcal{D} : 0 \le \theta \le 2\pi,\ 0 \le r \le R$$

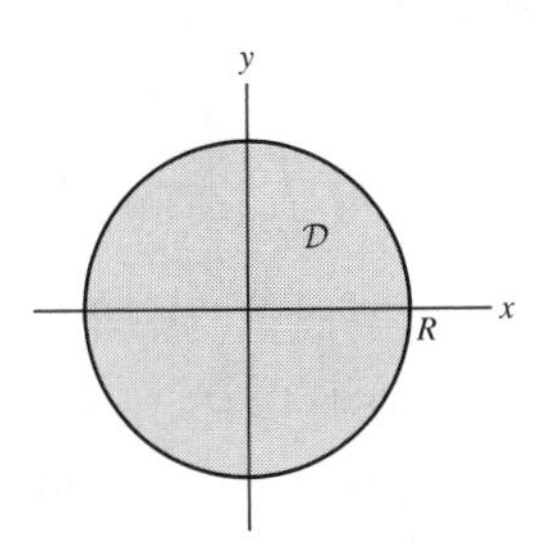

The upper and lower boundaries are the surfaces $z = H$ and $z = \frac{H}{R}r$, respectively. Hence $\mathcal{W}$ is determined by the following cylindrical inequalities:

$$\mathcal{W} : 0 \le \theta \le 2\pi,\ 0 \le r \le R,\ \frac{H}{R}r \le z \le H$$

We compute the volume using the following integral:

$$V = \iiint_{\mathcal{W}} 1\,dv = \int_0^{2\pi} \int_0^R \int_{\frac{H}{R}r}^{H} r\,dz\,dr\,d\theta = \int_0^{2\pi} \int_0^R rz\Big|_{z=\frac{Hr}{R}}^{H} dr\,d\theta = \int_0^{2\pi} \int_0^R r\left(H - \frac{Hr}{R}\right) dr\,d\theta$$

$$= \int_0^{2\pi} \int_0^R \left(rH - \frac{r^2 H}{R}\right) dr\,d\theta = \int_0^{2\pi} \frac{r^2 H}{2} - \frac{r^3 H}{3R}\Big|_{r=0}^{R} d\theta = \int_0^{2\pi} \frac{R^2 H}{6}\,d\theta = \frac{R^2 H}{6} \cdot 2\pi = \frac{\pi R^2 H}{3}$$

45. Use cylindrical coordinates to integrate $f(x, y, z) = z$ over the intersection of the sphere $x^2 + y^2 + z^2 = 4$ and the cylinder $x^2 + y^2 = 1$.

SOLUTION

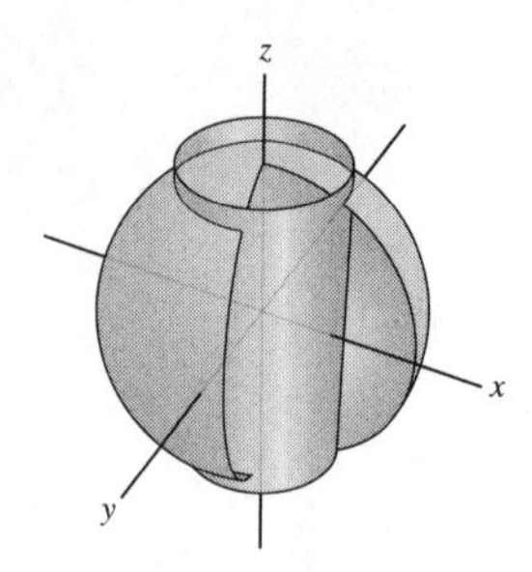

The region of integration projects onto the circle $\mathcal{D}$ of radius 1 in the xy-plane.

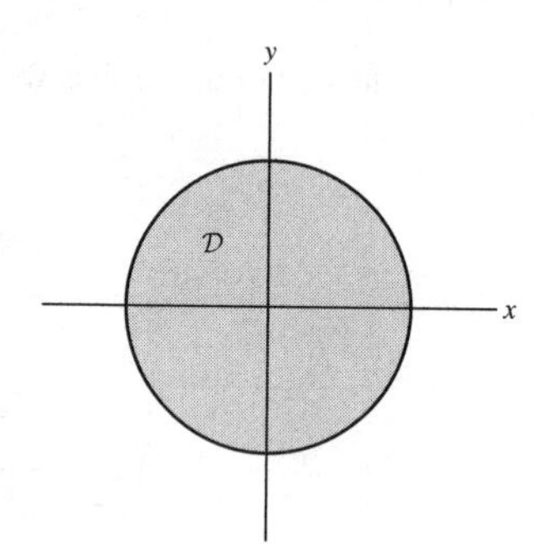

In polar coordinates,

$$\mathcal{D}: 0 \le \theta \le 2\pi,\ 0 \le r \le 1$$

The upper boundary of $\mathcal{W}$ is the sphere

$$x^2 + y^2 + z^2 = 4 \quad \text{or} \quad z = \sqrt{4 - (x^2 + y^2)} = \sqrt{4 - r^2}$$

and the lower boundary is $z = -\sqrt{4 - r^2}$. Therefore,

$$\mathcal{W}: 0 \le \theta \le 2\pi,\ 0 \le r \le 1,\ -\sqrt{4 - r^2} \le z \le \sqrt{4 - r^2}$$

We obtain the following integral in cylindrical coordinates:

$$\iiint_{\mathcal{W}} z\,dV = \int_0^{2\pi}\int_0^1\int_{-\sqrt{4-r^2}}^{\sqrt{4-r^2}} zr\,dz\,dr\,d\theta = \int_0^{2\pi}\int_0^1 \frac{z^2 r}{2}\bigg|_{z=-\sqrt{4-r^2}}^{\sqrt{4-r^2}} dr\,d\theta = \int_0^{2\pi}\int_0^1 0\,dr\,d\theta = 0$$

This should not surprise us since $\mathcal{W}$ is symmetric with respect to $z = 0$ and f is antisymmetric.

47. Evaluate the triple integral of $f(x, y, z) = x^2$ for region (11).

SOLUTION The function in spherical coordinates is $f(x, y, z) = x^2 = \rho^2 \cos^2\theta \sin^2\phi$. Using triple integrals in spherical coordinates and integration formulas gives

$$\begin{aligned}
\iiint_{\mathcal{W}} x^2\,dV &= \int_0^{\pi/3}\int_0^{\pi/2}\int_1^2 \left(\rho^2\cos^2\theta\,\sin^2\phi\right)\rho^2\sin\phi\,d\rho\,d\phi\,d\theta \\
&= \int_0^{\pi/3}\int_0^{\pi/2}\int_1^2 \rho^4\cos^2\theta\,\sin^3\phi\,d\rho\,d\phi\,d\theta \\
&= \left(\int_0^{\pi/3}\cos^2\theta\,d\theta\right)\left(\int_0^{\pi/2}\sin^3\phi\,d\phi\right)\left(\int_1^2 \rho^4 d\rho\right) \\
&= \left(\frac{\theta}{2} + \frac{\sin 2\theta}{4}\bigg|_0^{\pi/3}\right)\left(-\frac{\sin^2\phi\cos\phi}{3} - \frac{2}{3}\cos\phi\bigg|_0^{\pi/2}\right)\left(\frac{\rho^5}{5}\bigg|_1^2\right) \\
&= \left(\frac{\pi}{6} + \frac{\sqrt{3}}{8}\right)\left(\frac{2}{3}\right)\left(\frac{31}{5}\right) = \frac{62}{15}\left(\frac{\pi}{6} + \frac{\sqrt{3}}{8}\right) \approx 3.06
\end{aligned}$$

49. Find the volume of the region lying above the cone $\phi = \phi_0$ and below the sphere $\rho = R$.

SOLUTION

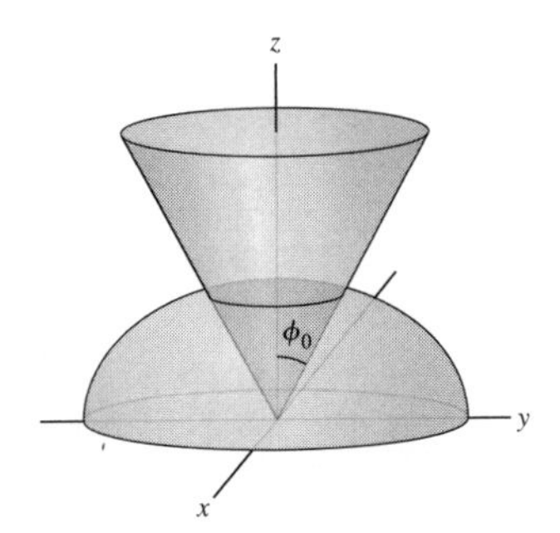

The region is described by the following inequalities in spherical coordinates:

$$\mathcal{W}: 0 \le \theta \le 2\pi,\ 0 \le \phi \le \phi_0,\ 0 \le \rho \le R$$

We compute the volume V of $\mathcal{W}$ using triple integrals in spherical coordinates:

$$V = \iiint_{\mathcal{W}} 1\, dV = \int_0^{2\pi} \int_0^{\phi_0} \int_0^R \rho^2 \sin\phi\, d\rho\, d\phi\, d\theta = \left(\int_0^{2\pi} 1\, d\theta\right)\left(\int_0^{\phi_0} \sin\phi\, d\phi\right)\left(\int_0^R \rho^2 d\rho\right)$$

$$= \left(\theta\Big|_0^{2\pi}\right)\left(-\cos\phi\Big|_0^{\phi_0}\right)\left(\frac{\rho^3}{3}\Big|_0^R\right) = 2\pi\left(1-\cos\phi_0\right)\cdot\frac{R^3}{3} = \frac{2\pi R^3\left(1-\cos\phi_0\right)}{3}$$

In Exercises 51–56, use spherical coordinates to calculate the triple integral of $f(x, y, z)$ over the given region.

51. $f(x, y, z) = y$; $x^2 + y^2 + z^2 \le 1$, $x, y, z \le 0$

SOLUTION

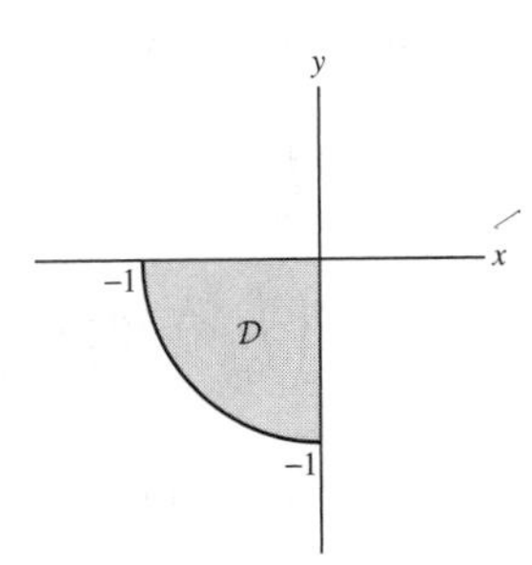

The region inside the unit sphere in the octant $x, y, z \le 0$ is defined by the inequalities

$$\mathcal{W}: \pi \le \theta \le \frac{3\pi}{2},\ \frac{\pi}{2} \le \phi \le \pi,\ 0 \le \rho \le 1$$

The function in spherical coordinates is $f(x, y, z) = y = \rho \sin\theta \sin\phi$. Using a triple integral in spherical coordinates, we obtain

$$\iiint_{\mathcal{W}} y\, dV = \int_{\pi}^{3\pi/2} \int_{\pi/2}^{\pi} \int_0^1 (\rho\sin\theta\sin\phi)\rho^2 \sin\phi\, d\rho\, d\phi\, d\theta = \int_{\pi}^{3\pi/2} \int_{\pi/2}^{\pi} \int_0^1 \rho^3 \sin\theta \sin^2\phi\, d\rho\, d\phi\, d\theta$$

$$= \left(\int_{\pi}^{3\pi/2} \sin\theta\, d\theta\right)\left(\int_{\pi/2}^{\pi} \sin^2\phi\, d\phi\right)\left(\int_0^1 \rho^3 d\rho\right) = \left(-\cos\theta\Big|_{\pi}^{3\pi/2}\right)\left(\frac{\theta}{2} - \frac{\sin 2\theta}{4}\Big|_{\pi/2}^{\pi}\right)\left(\frac{\rho^4}{4}\Big|_0^1\right)$$

$$= (-1)\cdot\left(\frac{\pi}{2} - \frac{\pi}{4}\right)\cdot\frac{1}{4} = -\frac{\pi}{16}$$

53. $f(x, y, z) = \rho^{-3}$, $2 \le x^2 + y^2 + z^2 \le 4$

SOLUTION

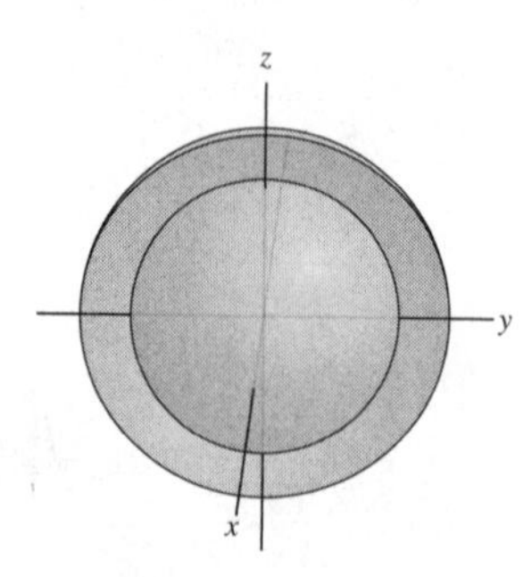

The lower and upper boundaries of $\mathcal{W}$ are the spheres $x^2 + y^2 + z^2 = 2$ and $x^2 + y^2 + z^2 = 4$. Therefore, ρ varies from $\sqrt{2}$ to 2, ϕ varies from 0 to π, and θ varies from 0 to 2π.

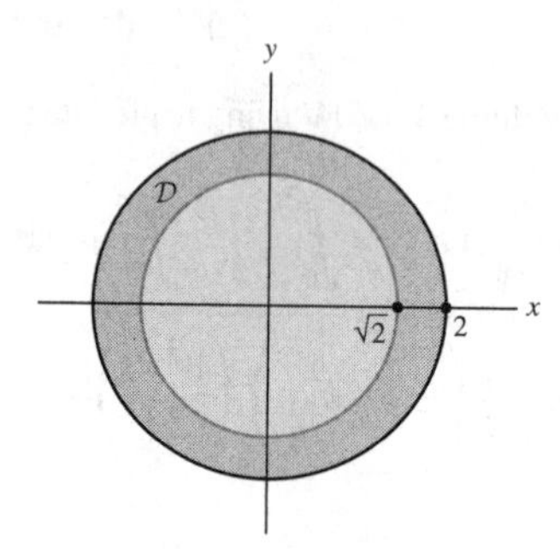

That is, $\mathcal{W}$ is defined by the inequalities

$$\mathcal{W} : 0 \le \theta \le 2\pi,\ 0 \le \phi \le \pi,\ \sqrt{2} \le \rho \le 2$$

The triple integral in spherical coordinates is thus

$$\iiint_{\mathcal{W}} f(x, y, z)\, dV = \int_0^{2\pi} \int_0^{\pi} \int_{\sqrt{2}}^{2} \rho^{-3} \rho^2 \sin\phi\, d\rho\, d\phi\, d\theta = \int_0^{2\pi} \int_0^{\pi} \int_{\sqrt{2}}^{2} \rho^{-1} \sin\phi\, d\rho\, d\phi\, d\theta$$

$$= \left(\int_0^{2\pi} 1\, d\theta\right)\left(\int_0^{\pi} \sin\phi\, d\phi\right)\left(\int_{\sqrt{2}}^{2} \rho^{-1}\, d\rho\right) = \left(\theta\Big|_0^{2\pi}\right)\left(-\cos\phi\Big|_0^{\pi}\right)\left(\ln\rho\Big|_{\sqrt{2}}^{2}\right)$$

$$= 2\pi \cdot 2 \cdot \left(\ln 2 - \frac{1}{2}\ln 2\right) = 2\pi \ln 2$$

55. $f(x, y, z) = 1; \quad x^2 + y^2 + z^2 \le 4z, \quad z \ge \sqrt{x^2 + y^2}$

SOLUTION The inequality $x^2 + y^2 + z^2 \le 4z$ can be rewritten as

$$x^2 + y^2 + z^2 - 4z \le 0 \quad \Rightarrow \quad x^2 + y^2 + (z - 2)^2 \le 4$$

This inequality defines the region inside the sphere of radius 2 centered at $(0, 0, 2)$. Therefore, $\mathcal{W}$ is the region inside the sphere, above the cone $z = \sqrt{x^2 + y^2}$.

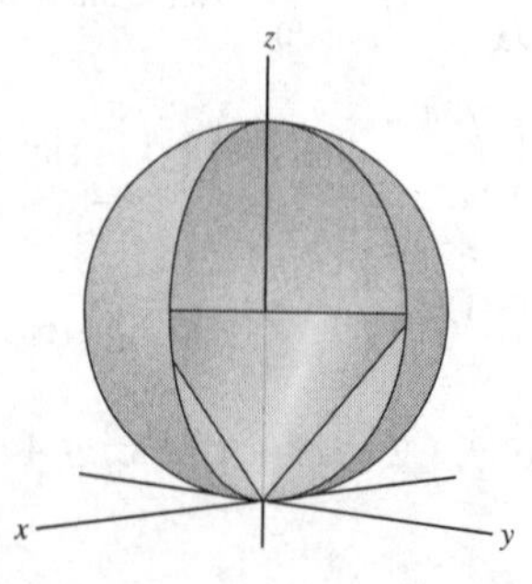

We write the equation of the sphere $x^2 + y^2 + z^2 = 4z$ in spherical coordinates:

$$(\rho\cos\theta\sin\phi)^2 + (\rho\sin\theta\sin\phi)^2 + (\rho\cos\phi)^2 = 4\rho\cos\phi$$

$$\rho^2\sin^2\phi\left(\cos^2\theta + \sin^2\theta\right) + \rho^2\cos^2\phi = 4\rho\cos\phi$$

$$\rho^2\sin^2\phi + \rho^2\cos^2\phi = 4\rho\cos\phi$$

$$\rho^2 = 4\rho\cos\phi$$
$$\rho = 4\cos\phi$$

We write the equation of the cone $z = \sqrt{x^2 + y^2}$ in spherical coordinates:

$$\rho\cos\phi = \sqrt{(\rho\cos\theta\sin\phi)^2 + (\rho\sin\theta\sin\phi)^2} = \sqrt{\rho^2\sin^2\phi} = \rho\sin\phi$$

or

$$\tan\phi = 1 \quad\Rightarrow\quad \phi = \frac{\pi}{4}$$

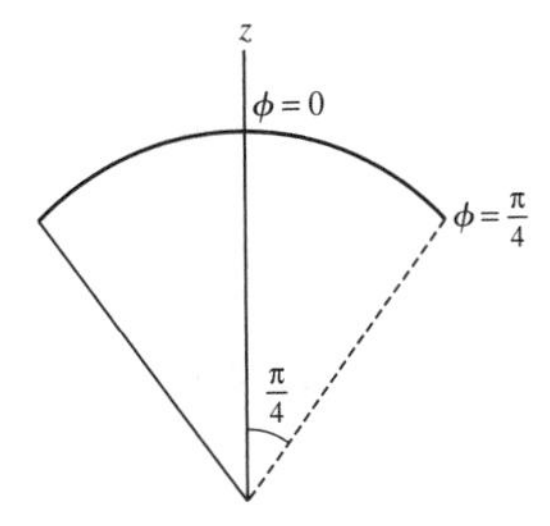

The spherical inequalities for $\mathcal{W}$ are thus

$$\mathcal{W}: 0 \le \theta \le 2\pi,\ 0 \le \phi \le \frac{\pi}{4},\ 0 \le \rho \le 4\cos\phi$$

We obtain the following integral:

$$\iiint_{\mathcal{W}} 1\,dV = \int_0^{2\pi}\int_0^{\pi/4}\int_0^{4\cos\phi} \rho^2\sin\phi\,d\rho\,d\phi\,d\theta = \int_0^{2\pi}\int_0^{\pi/4} \frac{\rho^3\sin\phi}{3}\bigg|_{\rho=0}^{4\cos\phi} d\phi\,d\theta$$
$$= \int_0^{2\pi}\int_0^{\pi/4} \frac{64\cos^3\phi\sin\phi}{3}\,d\phi\,d\theta = \left(\int_0^{2\pi}\frac{64}{3}\,d\theta\right)\left(\int_0^{\pi/4}\cos^3\phi\sin\phi\,d\phi\right)$$
$$= \frac{128\pi}{3}\int_0^{\pi/4}\cos^3\phi\sin\phi\,d\phi$$

We compute the integral using the substitution $u = \cos\phi$, $du = -\sin\phi\,d\phi$:

$$\iiint_{\mathcal{W}} 1\,dV = \frac{128\pi}{3}\int_1^{1/\sqrt{2}} u^3(-du) = \frac{128\pi}{3}\int_{1/\sqrt{2}}^1 u^3\,du = \frac{128\pi}{3}\cdot\frac{u^4}{4}\bigg|_{1/\sqrt{2}}^1 = 8\pi$$

57. Find the centroid of the region $\mathcal{W}$ bounded by the cone $\phi = \phi_0$ and the sphere $\rho = R$ as a function of ϕ_0 and R.

SOLUTION The centroid is the point $P = (\overline{x}, \overline{y}, \overline{z})$, where

$$\overline{x} = \frac{1}{V}\iiint_{\mathcal{W}} x\,dV,\quad \overline{y} = \frac{1}{V}\iiint_{\mathcal{W}} y\,dV,\quad \overline{z} = \frac{1}{V}\iiint_{\mathcal{W}} z\,dV \tag{1}$$

We first compute the volume V of $\mathcal{W}$. The region $\mathcal{W}$ has the following definition in spherical coordinates:

$$\mathcal{W}: 0 \le \theta \le 2\pi,\ 0 \le \phi \le \phi_0,\ 0 \le \rho \le R$$

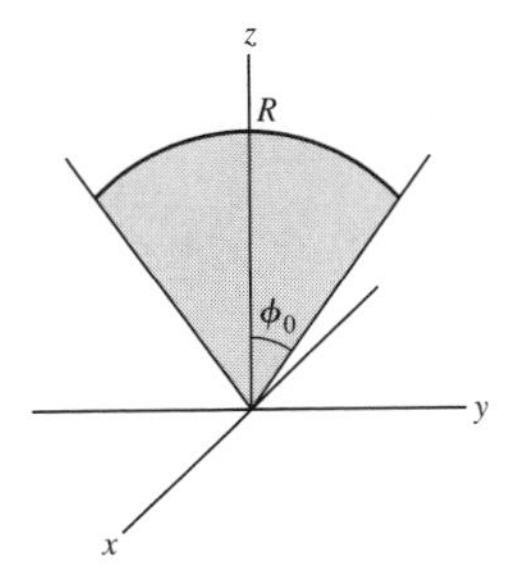

Hence,

$$V = \iiint_{\mathcal{W}} 1\,dV = \int_0^{2\pi}\int_0^{\phi_0}\int_0^R \rho^2\sin\phi\,d\rho\,d\phi\,d\theta = \left(\int_0^{2\pi} 1\,d\theta\right)\left(\int_0^{\phi_0}\sin\phi\,d\phi\right)\left(\int_0^R \rho^2 d\rho\right)$$

$$= 2\pi \cdot \left(-\cos\phi \Big|_0^{\phi_0} \right) \left(\frac{\rho^3}{3} \Big|_0^R \right) = 2\pi \cdot (1 - \cos\phi_0) \cdot \frac{R^3}{3} = \frac{2\pi R^3(1 - \cos\phi_0)}{3} \tag{2}$$

Since the region $\mathcal{D}$ is symmetric with respect to the z-axis, its centroid lies on the z-axis. That is, $\overline{x} = \overline{y} = 0$. To compute $\overline{z}$, we evaluate the following integral:

$$\iiint_{\mathcal{W}} z\,dV = \int_0^{2\pi} \int_0^{\phi_0} \int_0^R (\rho\cos\phi)\rho^2 \sin\phi\, d\rho\, d\phi\, d\theta = \int_0^{2\pi} \int_0^{\phi_0} \int_0^R \rho^3 \cdot \frac{\sin 2\phi}{2}\, d\rho\, d\phi\, d\theta$$

$$= \left(\int_0^{2\pi} 1\,d\theta \right) \left(\int_0^{\phi_0} \frac{\sin 2\phi}{2}\, d\phi \right) \left(\int_0^R \rho^3\, d\rho \right) = 2\pi \cdot \left(-\frac{\cos 2\phi}{4} \Big|_0^{\phi_0} \right) \left(\frac{\rho^4}{4} \Big|_0^R \right)$$

$$= 2\pi \cdot \frac{1 - \cos 2\phi}{4} \cdot \frac{R^4}{4} = \frac{R^4(1 - \cos 2\phi_0)\pi}{8}$$

Combining with (1) and (2) gives

$$\overline{z} = \frac{3}{2\pi R^3(1 - \cos\phi_0)} \cdot \frac{R^4(1 - \cos 2\phi_0)\pi}{8} = \frac{3R(1 - \cos 2\phi_0)}{16(1 - \cos\phi_0)}$$

The centroid is thus

$$P = \left(0, 0, \frac{3R(1 - \cos 2\phi_0)}{16(1 - \cos\phi_0)} \right).$$

59. Evaluate the triple integral of $f(x, y, z) = z(x^2 + y^2 + z^2)^{-3/2}$ over the part of the ball $x^2 + y^2 + z^2 \le 16$ defined by $z \ge 2$.

SOLUTION

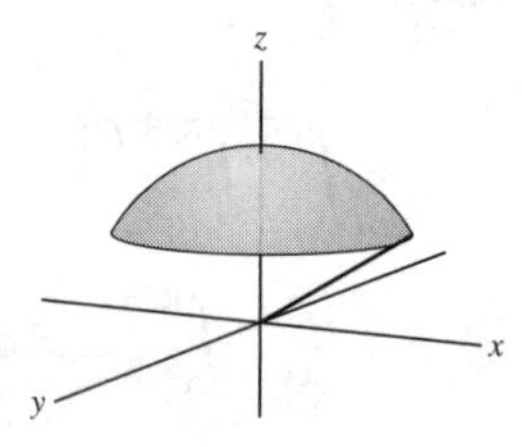

The equation of the sphere in spherical coordinates is $\rho^2 = 16$ or $\rho = 4$.

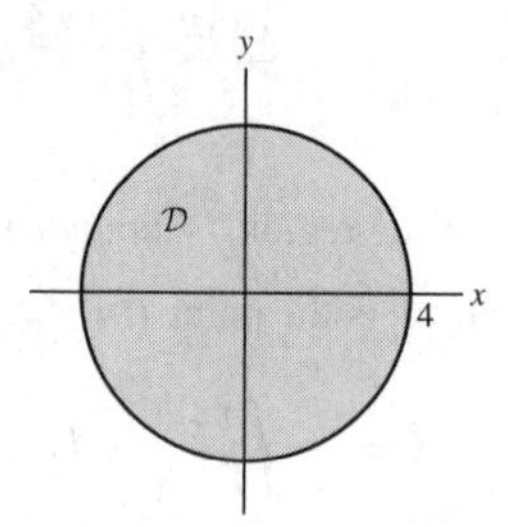

We write the equation of the plane $z = 2$ in spherical coordinates:

$$\rho\cos\phi = 2 \quad \Rightarrow \quad \rho = \frac{2}{\cos\phi}$$

To compute the interval of ϕ, we must find the value of ϕ corresponding to $\rho = 4$ on the plane $z = 2$. We get

$$4 = \frac{2}{\cos\phi} \quad \Rightarrow \quad \cos\phi = \frac{1}{2} \quad \Rightarrow \quad \phi = \frac{\pi}{3}$$

Therefore, ϕ is changing from 0 to $\frac{\pi}{3}$, θ is changing from 0 to 2π, and ρ is changing from $\frac{2}{\cos\phi}$ to 4. We obtain the following description for $\mathcal{W}$:

$$\mathcal{W}: 0 \le \theta \le 2\pi,\ 0 \le \phi \le \frac{\pi}{3},\ \frac{2}{\cos\phi} \le \rho \le 4$$

The function is

$$f(x, y, z) = z(x^2 + y^2 + z^2)^{-3/2} = \rho\cos\phi \cdot (\rho^2)^{-3/2} = \rho^{-2}\cos\phi$$

We use triple integrals in spherical coordinates to write

$$\iiint_{\mathcal{W}} f(x, y, z)\, dV = \int_0^{2\pi} \int_0^{\pi/3} \int_{2/\cos\phi}^{4} (\rho^{-2}\cos\phi)\rho^2 \sin\phi \, d\rho\, d\phi\, d\theta = \int_0^{2\pi} \int_0^{\pi/3} \int_{2/\cos\phi}^{4} \frac{\sin 2\phi}{2}\, d\rho\, d\phi\, d\theta$$

$$= \int_0^{2\pi} \int_0^{\pi/3} \frac{\sin 2\phi}{2}\rho \Big|_{\rho=\frac{2}{\cos\phi}}^{4} d\phi\, d\theta = \int_0^{2\pi} \int_0^{\pi/3} \left(2\sin 2\phi - \frac{\sin 2\phi}{2} \cdot \frac{2}{\cos\phi}\right) d\phi\, d\theta$$

$$= \int_0^{2\pi} \int_0^{\pi/3} (2\sin 2\phi - 2\sin\phi)\, d\phi\, d\theta = 2\pi \cdot \left(-\cos 2\phi + 2\cos\phi \Big|_{\phi=0}^{\pi/3}\right)$$

$$= 2\pi \cdot \left(-\cos\frac{2\pi}{3} + 2\cos\frac{\pi}{3} + 1 - 2\right) = \pi$$

61. Find the center of mass of a cylinder of radius 2 and height 4 and mass density e^{-z}, where z is the height above the base.

SOLUTION

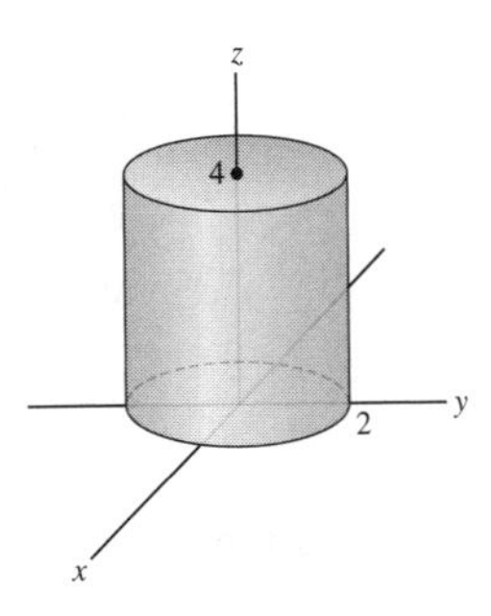

The center of mass is the point with the following coordinates:

$$x_{CM} = \frac{1}{M}\iiint_{\mathcal{W}} xe^{-z}\, dV, \quad y_{CM} = \frac{1}{M}\iiint_{\mathcal{W}} ye^{-z}\, dV, \quad z_{CM} = \frac{1}{M}\iiint_{\mathcal{W}} ze^{-z}\, dV$$

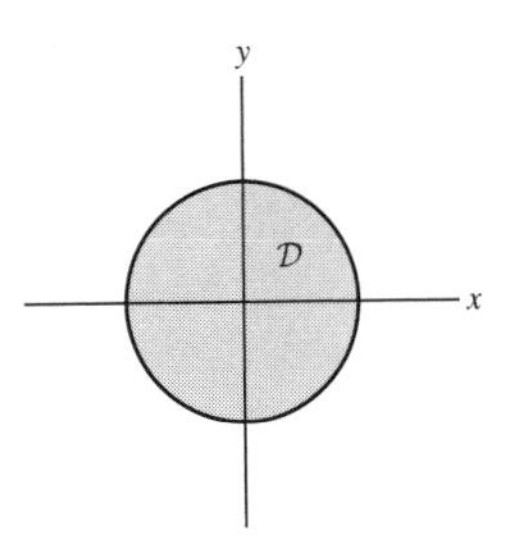

Since $\mathcal{W}$ is symmetric with respect to the z axis, and the functions xe^{-z} and ye^{-z} are odd with respect to the variables x and y, respectively, we have

$$x_{CM} = y_{CM} = 0$$

We need only to find z_{CM}. We first compute the mass M by the triple integral

$$M = \iiint_{\mathcal{W}} e^{-z}\, dV$$

To evaluate the integral we use cylindrical coordinates. The region $\mathcal{W}$ is described by the inequalities

$$\mathcal{W}: 0 \le \theta \le 2\pi,\ 0 \le r \le 2,\ 0 \le z \le 4$$

Using triple integrals in cylindrical coordinates, we obtain

$$M = \int_0^{2\pi} \int_0^{2} \int_0^{4} e^{-z} r\, dz\, dr\, d\theta = \left(\int_0^{2\pi} 1\, d\theta\right)\left(\int_0^{2} r\, dr\right)\left(\int_0^{4} e^{-z}\, dz\right)$$

$$= 2\pi \cdot \left(\frac{r^2}{2}\Big|_0^2\right)\left(-e^{-z}\Big|_0^4\right) = 2\pi \cdot 2 \cdot (1 - e^{-4}) = 4\pi(1 - e^{-4}) \approx 12.34$$

We compute z_{CM}:

$$z_{CM} = \frac{1}{12.34}\int_0^{2\pi}\int_0^2\int_0^4 ze^{-z}r\,dz\,dr\,d\theta = 0.08\left(\int_0^{2\pi} d\theta\right)\left(\int_0^2 r\,dr\right)\left(\int_0^4 ze^{-z}\,dz\right)$$

$$= 0.08\cdot 2\pi\cdot 2\int_0^4 ze^{-z}\,dz = 0.32\pi\int_0^4 ze^{-z}\,dz$$

We compute the integral using integration by parts:

$$z_{CM} = 0.32\pi\left(-ze^{-z}\Big|_0^4 - e^{-z}\Big|_0^4\right) = -0.32\pi e^{-z}(1+z)\Big|_0^4 = 0.32\pi(1-5e^{-4}) = 0.91$$

The center of mass is the point $(0, 0, 0.91)$.

In Exercises 63–65, compute the centroid of the shapes in Figure 24.

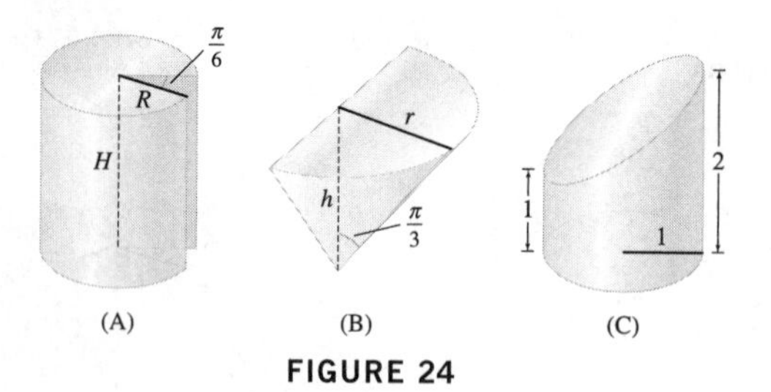

FIGURE 24

63. (A)

SOLUTION Since the mass distribution is uniform, we may assume that $\rho(x, y, z) = 1$, hence the center of mass is

$$x_{CM} = \frac{1}{V}\iiint_{\mathcal{W}} x\,dV, \quad y_{CM} = \frac{1}{V}\iiint_{\mathcal{W}} y\,dV, \quad z_{CM} = \frac{1}{V}\iiint_{\mathcal{W}} z\,dV$$

The inequalities describing $\mathcal{W}$ in cylindrical coordinates are

$$\mathcal{W}: 0 \le \theta \le \frac{\pi}{6},\ 0 \le r \le R,\ 0 \le z \le H$$

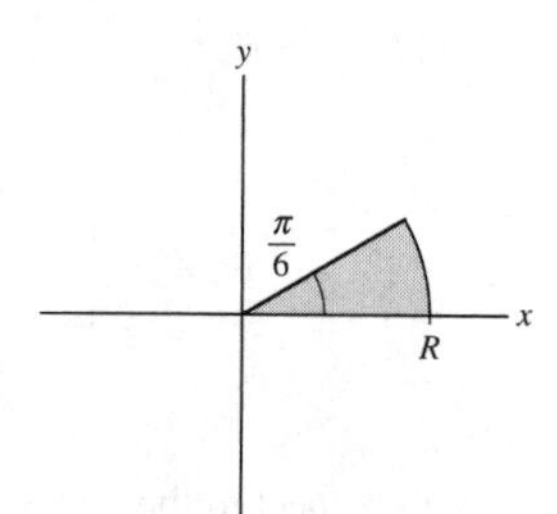

The entire cylinder has a total volume $\pi R^2 H$. The region $\mathcal{W}$ has the fraction $(2\pi - \frac{\pi}{6})/(2\pi)$ of this total volume, so

$$V = \frac{(2\pi - \frac{\pi}{6})}{2\pi}(\pi R^2 H) = \frac{11\pi R^2 H}{12}$$

We use cylindrical coordinates to compute the triple integrals:

$$x_{CM} = \frac{1}{V}\int_{\pi/6}^{2\pi}\int_0^R\int_0^H (r\cos\theta)r\,dz\,dr\,d\theta = \frac{12}{11\pi R^2 H}\left(\int_{\pi/6}^{2\pi}\cos\theta\,d\theta\right)\left(\int_0^R r^2\,dr\right)\left(\int_0^H dz\right)$$

$$= \frac{12}{11\pi R^2 H}\left(-\frac{1}{2}\right)\left(\frac{R^3}{3}\right)(H) = -\frac{2R}{11\pi}$$

$$y_{CM} = \frac{1}{V}\int_{\pi/6}^{2\pi}\int_0^R\int_0^H (r\sin\theta)r\,dzdrd\theta = \frac{12}{11\pi R^2 H}\left(\int_{\pi/6}^{2\pi}\sin\theta\,d\theta\right)\left(\int_0^R r^2\,dr\right)\left(\int_0^H dz\right)$$

$$= \frac{12}{11\pi R^2 H}\left(\frac{-2+\sqrt{3}}{2}\right)\left(\frac{R^3}{3}\right)(H) = \frac{2R}{11\pi}(\sqrt{3}-2)$$

$$z_{CM} = \frac{1}{V}\int_{\pi/6}^{2\pi}\int_0^R\int_0^H zr\,dz\,dr\,d\theta = \frac{12}{11\pi R^2 H}\left(\int_{\pi/6}^{2\pi} d\theta\right)\left(\int_0^R r\,dr\right)\left(\int_0^H z\,dz\right)$$

$$= \frac{12}{11\pi R^2 H}\left(\frac{11\pi}{6}\right)\left(\frac{R^2}{2}\right)\left(\frac{H^2}{2}\right) = \frac{H}{2}$$

Therefore, the center of mass is the following point:

$$\left(-\frac{2R}{11\pi}, \frac{2R}{11\pi}(\sqrt{3}-2), \frac{H}{2}\right)$$

65. (C)

SOLUTION Notice that our picture here is slightly different from the one in the book; we've arranged our picture so that the top slopes down in the positive y direction. Since the mass distribution is uniform, we may assume that $\rho(x, y, z) = 1$, hence the center of mass is

$$x_{\text{CM}} = \frac{1}{V}\iiint_{\mathcal{W}} x\,dV, \quad y_{\text{CM}} = \frac{1}{V}\iiint_{\mathcal{W}} y\,dV, \quad z_{\text{CM}} = \frac{1}{V}\iiint_{\mathcal{W}} z\,dV$$

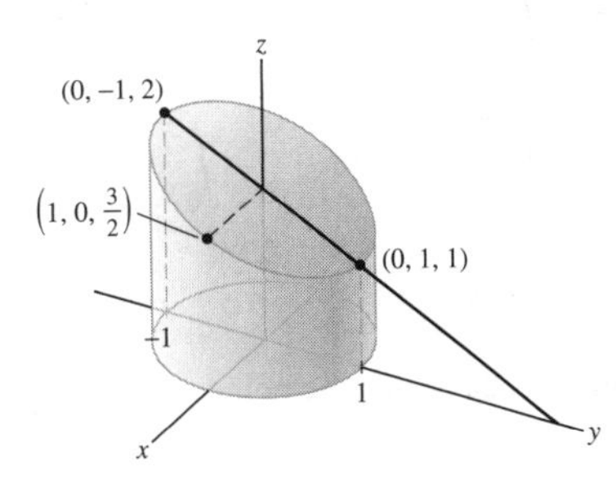

We first must find the equation of the upper plane. This plane is passing through the points $(0, -1, 2)$, $(0, 1, 1)$, and $\left(1, 0, \frac{3}{2}\right)$, hence it has the equation $y + 2z = 3$ or $z = \frac{3-y}{2} = \frac{3-r\sin\theta}{2}$.

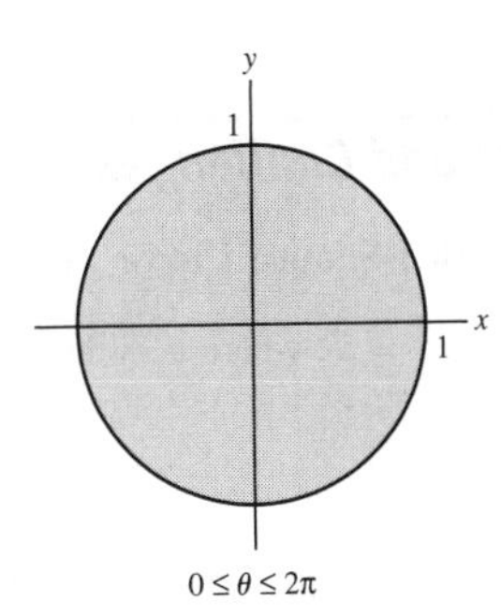

The region $\mathcal{W}$ has the following definition in cylindrical coordinates:

$$\mathcal{W}: 0 \le \theta \le 2\pi,\ 0 \le r \le 1,\ 0 \le z \le \frac{3 - r\sin\theta}{2}$$

We first find the volume of $\mathcal{W}$:

$$V = \iiint_{\mathcal{W}} 1\,dV = \int_0^{2\pi}\int_0^1\int_0^{(3-r\sin\theta)/2} r\,dz\,dr\,d\theta = \int_0^{2\pi}\int_0^1 rz\Big|_0^{(3-r\sin\theta)/2} dr\,d\theta$$

$$= \int_0^{2\pi}\int_0^1 r\frac{3-r\sin\theta}{2}\,dr\,d\theta = \int_0^{2\pi}\int_0^1\left(\frac{3r}{2} - \frac{r^2}{2}\sin\theta\right)dr\,d\theta = \int_0^{2\pi} \frac{3r^2}{4} - \frac{r^3\sin\theta}{6}\Big|_{r=0}^1 d\theta$$

$$= \int_0^{2\pi}\left(\frac{3}{4} - \frac{\sin\theta}{6}\right)d\theta = \frac{3}{4}\theta + \frac{\cos\theta}{6}\Big|_0^{2\pi} = \frac{3\pi}{2} + \frac{1}{6} - \frac{1}{6} = \frac{3\pi}{2}$$

$V = \frac{3\pi}{2}$: Since the region is symmetric with respect to the yz-plane, we have

$$x_{\text{CM}} = \frac{1}{V}\iiint_{\mathcal{W}} x\,dV = 0$$

We compute y_{CM}:

$$y_{\text{CM}} = \frac{2}{3\pi}\int_0^{2\pi}\int_0^1\int_0^{(3-r\sin\theta)/2}(r\sin\theta)r\,dz\,dr\,d\theta = \frac{2}{3\pi}\int_0^{2\pi}\int_0^1\int_0^{(3-r\sin\theta)/2} r^2\sin\theta\,dz\,dr\,d\theta$$

$$= \frac{2}{3\pi}\int_0^{2\pi}\int_0^1 r^2 \sin\theta\left(\frac{3 - r\sin\theta}{2}\right) dr\,d\theta = \frac{2}{3\pi}\int_0^{2\pi}\int_0^1 \left(\frac{3r^2\sin\theta}{2} - \frac{r^3\sin^2\theta}{2}\right) dr\,d\theta$$

$$= \frac{2}{3\pi}\int_0^{2\pi}\left(\frac{3\sin\theta}{2}\frac{r^3}{3} - \frac{r^4\sin^2\theta}{8}\bigg|_{r=0}^{1}\right) d\theta = \frac{2}{3\pi}\int_0^{2\pi}\left(\frac{\sin\theta}{2} - \frac{\sin^2\theta}{8}\right) d\theta$$

$$= \frac{1}{3\pi}\int_0^{2\pi}\sin\theta\,d\theta - \frac{1}{12\pi}\int_0^{2\pi}\sin^2\theta\,d\theta = -\frac{1}{12\pi}\left(\frac{\theta}{2} - \frac{\sin 2\theta}{4}\bigg|_{\theta=0}^{2\pi}\right) = -\frac{1}{12\pi}\cdot\pi = -\frac{1}{12}$$

Finally we find z_{CM}:

$$z_{CM} = \frac{2}{3\pi}\int_0^{2\pi}\int_0^1\int_0^{(3-r\sin\theta)/2} zr\,dz\,dr\,d\theta = \frac{2}{3\pi}\int_0^{2\pi}\int_0^1 \frac{z^2 r}{2}\bigg|_{z=0}^{(3-r\sin\theta)/2} dr\,d\theta$$

$$= \frac{2}{3\pi}\int_0^{2\pi}\int_0^1 \frac{r}{2}\left(\frac{3 - r\sin\theta}{2}\right)^2 dr\,d\theta = \frac{2}{8\cdot 3\pi}\int_0^{2\pi}\int_0^1\left(r^3\sin^2\theta - 6r^2\sin\theta + 9r\right) dr\,d\theta$$

$$= \frac{1}{12\pi}\int_0^{2\pi}\frac{r^4\sin^2\theta}{4} - 2r^3\sin\theta + \frac{9}{2}r^2\bigg|_0^1 d\theta = \frac{1}{12\pi}\int_0^{2\pi}\left(\frac{\sin^2\theta}{4} - 2\sin\theta + \frac{9}{2}\right) d\theta$$

$$= \frac{1}{48\pi}\int_0^{2\pi}\sin^2\theta\,d\theta - \frac{1}{6\pi}\int_0^{2\pi}\sin\theta\,d\theta + \frac{9}{24\pi}\int_0^{2\pi}d\theta = \frac{1}{48\pi}\left(\frac{\theta}{2} - \frac{\sin 2\theta}{4}\bigg|_0^{2\pi}\right) - 0 + \frac{9}{24\pi}\cdot 2\pi$$

$$= \frac{1}{48} + \frac{3}{4} = \frac{37}{48}$$

The center of mass is the following point:

$$\left(0, -\frac{1}{12}, \frac{37}{48}\right)$$

If you had arranged the axes differently, you could have computed the answer as $\left(0, \frac{1}{12}, \frac{37}{48}\right)$ (depending on orientation).

Further Insights and Challenges

67. Area Under the Bell-Shaped Curve Let $I = \int_{-\infty}^{\infty} e^{-x^2}\,dx$ be the area under the bell-shaped curve (Figure 26). By Fubini's Theorem, $I^2 = J$, where J is the improper double integral:

$$J = \int_{-\infty}^{\infty}\int_{-\infty}^{\infty} e^{-x^2-y^2}\,dx\,dy$$

Write J in polar coordinates and evaluate, showing that $J = \pi$ and $I = \sqrt{\pi}$.

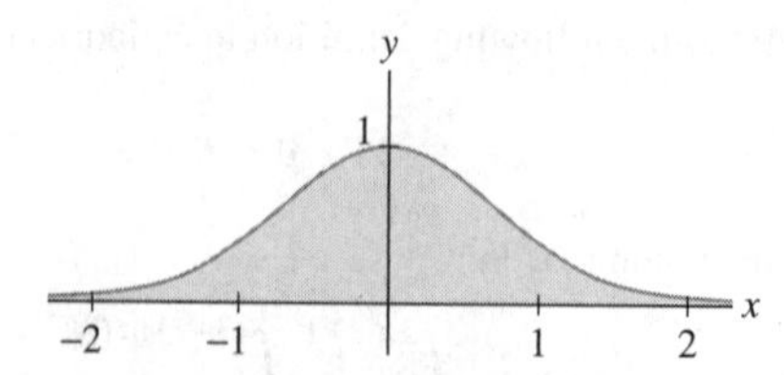

FIGURE 26 The bell-shaped curve $y = e^{-x^2}$.

SOLUTION The improper integral over the xy-plane can be computed as the limit as $\mathcal{R} \to \infty$ of the double integrals over the disk. $\mathcal{D}_{\mathcal{R}}$ is defined by

$$\mathcal{D}_{\mathcal{R}} : 0 \le \theta \le 2\pi,\ 0 \le r \le \mathcal{R}$$

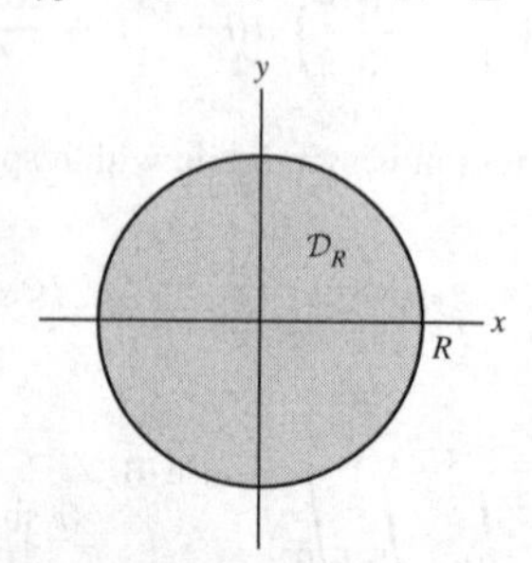

That is,

$$J = \lim_{\mathcal{R}\to\infty} \iint_{\mathcal{D}_{\mathcal{R}}} e^{-(x^2+y^2)}\, dx\, dy \tag{1}$$

We compute the double integral using polar coordinates. The function is $f(x, y) = e^{-(x^2+y^2)} = e^{-r^2}$, hence

$$\iint_{\mathcal{D}_{\mathcal{R}}} e^{-(x^2+y^2)}\, dx\, dy = \int_0^{2\pi}\int_0^{\mathcal{R}} e^{-r^2} r\, dr\, d\theta = \left(\int_0^{2\pi} 1\, d\theta\right)\left(\int_0^{\mathcal{R}} e^{-r^2} r\, dr\right) = 2\pi \int_0^{\mathcal{R}} e^{-r^2} r\, dr$$

We compute the integral using the substitution $u = r^2$, $du = 2r\, dr$. We get

$$\iint_{\mathcal{D}_{\mathcal{R}}} e^{-(x^2+y^2)}\, dx\, dy = 2\pi \int_0^{\mathcal{R}^2} e^{-u}\frac{du}{2} = \pi \int_0^{\mathcal{R}^2} e^{-u}\, du = \pi(-e^{-u})\Big|_0^{R^2} = \pi(1 - e^{-R^2}) \tag{2}$$

Combining (1) and (2), we get

$$J = \lim_{\mathcal{R}\to\infty}\left(\pi\left(1 - e^{-R^2}\right)\right) = \pi$$

On the other hand, using the Iterated Integral of a Product Function, we get

$$\pi = J = \int_{-\infty}^{\infty}\int_{-\infty}^{\infty} e^{-x^2-y^2}\, dx\, dy = \int_{-\infty}^{\infty}\int_{-\infty}^{\infty} e^{-x^2}\cdot e^{-y^2}\, dx\, dy = \left(\int_{-\infty}^{\infty} e^{-x^2}\, dx\right)\left(\int_{-\infty}^{\infty} e^{-y^2}\, dy\right) = I^2$$

That is,

$$I^2 = \pi \quad \Rightarrow \quad I = \sqrt{\pi}$$

69. Calculate the integral of $\cos z$ over the ball $x^2 + y^2 + z^2 \le 1$. *Hint:* Use cylindrical coordinates and integrate in the order $d\theta\, dz\, dr$.

SOLUTION The ball $x^2 + y^2 + z^2 \le 1$ is defined by the inequalities

$$\mathcal{W}: 0 \le \theta \le 2\pi,\ 0 \le r \le 1,\ -\sqrt{1 - r^2} \le z \le \sqrt{1 - r^2}$$

The integral in cylindrical coordinates is

$$\iiint_{\mathcal{W}} \cos z\, dV = \int_0^1 \int_{-\sqrt{1-r^2}}^{\sqrt{1-r^2}} \int_0^{2\pi} (\cos z) r\, d\theta\, dz\, dr = \int_0^1 \int_{-\sqrt{1-r^2}}^{\sqrt{1-r^2}} r(\cos z)\theta\Big|_{\theta=0}^{2\pi} dz\, dr$$

$$= \int_0^1 \int_{-\sqrt{1-r^2}}^{\sqrt{1-r^2}} 2\pi r \cos z\, dz\, dr = \int_0^1 2\pi r \sin z\Big|_{z=-\sqrt{1-r^2}}^{\sqrt{1-r^2}} dr = 4\pi \int_0^1 \sin\left(\sqrt{1 - r^2}\right) r\, dr$$

We compute the integral using the substitution $u = \sqrt{1 - r^2}$, $du = -\frac{r}{u}\, dr$. This gives

$$\int_{\mathcal{W}} \cos z\, dV = 4\pi \int_1^0 (\sin u)(-u\, du) = 4\pi \int_0^1 u \sin u\, du$$

Using Integration by Parts we obtain

$$\iiint_{\mathcal{W}} \cos z\, dV = 4\pi\left(-u\cos u + \sin u\Big|_0^1\right) = 4\pi(-\cos 1 + \sin 1) = 4\pi(\sin 1 - \cos 1)$$

71. Prove the formula

$$\iint_{\mathcal{D}} \ln r\, dA = -\frac{\pi}{2}$$

where $r = \sqrt{x^2 + y^2}$ and $\mathcal{D}$ is the unit disk $x^2 + y^2 \le 1$. This is an improper integral since $\ln r$ is not defined at $(0, 0)$, so integrate first over the annulus $a \le r \le 1$ and let $a \to 0$.

SOLUTION

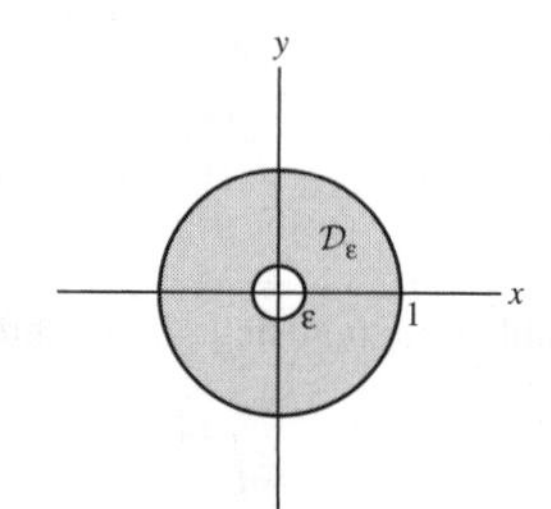

The improper integral I is computed by the limit as $a \to 0^+$ of the integrals over the annulus $\mathcal{D}_a$ defined by

$$\mathcal{D}_a : 0 \le \theta \le 2\pi,\ a \le r \le 1$$

Using double integrals in polar coordinates and integration by parts, we get

$$I_a = \int_0^{2\pi} \int_a^1 (\ln r) \cdot r\, dr\, d\theta = 2\pi \int_a^1 r \ln r\, dr = 2\pi \left(\frac{r^2 \ln r}{2} - \frac{r^2}{4} \bigg|_a^1 \right)$$

$$= 2\pi \left(\frac{\ln 1}{2} - \frac{1}{4} - \frac{a^2 \ln a}{2} + \frac{a^2}{4} \right) = \frac{\pi}{2} \left(a^2 - 2a^2 \ln a - 1 \right)$$

We now compute the limit of I_a as $a \to 0^+$. We use L'Hôpital's rule to obtain

$$I = \lim_{a\to 0+} \frac{\pi}{2}(a^2 - 2a^2 \ln a - 1) = -\frac{\pi}{2} - \pi \lim_{a\to 0+} a^2 \ln a = -\frac{\pi}{2} - \pi \lim_{a\to 0+} \frac{\ln a}{a^{-2}}$$

$$= -\frac{\pi}{2} - \pi \lim_{a\to 0+} \frac{a^{-1}}{-2a^{-3}} = -\frac{\pi}{2} + \frac{\pi}{2} \lim_{a\to 0+} a^2 = -\frac{\pi}{2}$$

16.5 Change of Variables (ET Section 15.5)

Preliminary Questions

1. Which of the following maps is linear?

(a) (uv, v) **(b)** $(u + v, u)$ **(c)** $(3, e^u)$

SOLUTION

(a) This map is not linear since it does not satisfy the linearity property:

$$\Phi(2u, 2v) = (2u \cdot 2v, 2v) = (4uv, 2v) = 2(2uv, v)$$

$$2\Phi(u, v) = 2(uv, v) \quad \Rightarrow \quad \Phi(2u, 2v) \ne 2\Phi(u, v)$$

(b) This map is linear since it has the form $\Phi(u, v) = (Au + Cv, Bu + Dv)$ where $A = C = 1$, $B = 1$, $D = 0$.

(c) This map is not linear since it does not satisfy the linearity properties. For example,

$$\left.\begin{aligned} \Phi(2u, 2v) &= (3, e^{2u}) \\ 2\Phi(u, v) &= 2(3, e^u) \end{aligned}\right\} \Rightarrow \quad \Phi(2u, 2v) \ne 2\Phi(u, v)$$

2. Suppose that Φ is a linear map such that $\Phi(2, 0) = (4, 0)$ and $\Phi(0, 3) = (-3, 9)$. Find the images of:

(a) $\Phi(1, 0)$ **(b)** $\Phi(1, 1)$ **(c)** $\Phi(2, 1)$

SOLUTION We denote the linear map by $\Phi(u, v) = (Au + Cv, Bu + Dv)$. By the given information we have

$$\Phi(2, 0) = (A \cdot 2 + C \cdot 0, B \cdot 2 + D \cdot 0) = (2A, 2B) = (4, 0)$$

$$\Phi(0, 3) = (A \cdot 0 + C \cdot 3, B \cdot 0 + D \cdot 3) = (3C, 3D) = (-3, 9)$$

Therefore,

$$\left.\begin{aligned} 2A &= 4 \\ 2B &= 0 \\ 3C &= -3 \\ 3D &= 9 \end{aligned}\right\} \Rightarrow \quad A = 2, \quad B = 0, \quad C = -1, \quad D = 3$$

The linear map is thus

$$\Phi(u, v) = (2u - v, 3v)$$

We now compute the images:

(a) $\Phi(1, 0) = (2 \cdot 1 - 0, 3 \cdot 0) = (2, 0)$

(b) $\Phi(1, 1) = (2 \cdot 1 - 1, 3 \cdot 1) = (1, 3)$

(c) $\Phi(2, 1) = (2 \cdot 2 - 1, 3 \cdot 1) = (3, 3)$

3. What is the area of $\Phi(\mathcal{R})$ if $\mathcal{R}$ is a rectangle of area 9 and Φ is a mapping whose Jacobian has constant value 4?

SOLUTION

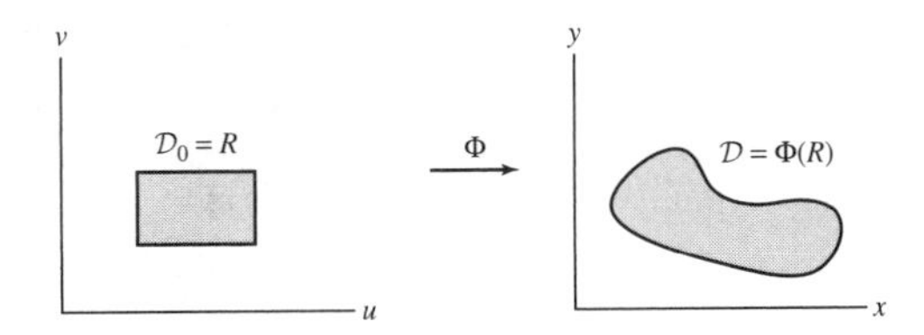

The areas of $\mathcal{D}_0 = \Phi(\mathcal{R})$ and $\mathcal{D} = \mathcal{R}$ are the following integrals:

$$\text{Area}(\mathcal{R}) = 9 = \iint_{\mathcal{D}_0} 1\, du\, dv$$

$$\text{Area}(\Phi(\mathcal{R})) = \iint_{\mathcal{D}} 1\, dx\, dy$$

Using the Change of Variables Formula, we have

$$\text{Area}\,(\Phi(\mathcal{R})) = \iint_{\mathcal{D}} 1\, dx\, dy = \iint_{\mathcal{D}_0} 1|\text{Jac}\Phi|\, du\, dv = \iint_{\mathcal{D}_0} 4\, du\, dv = 4\iint_{\mathcal{D}_0} 1\, du\, dv = 4 \cdot 9 = 36$$

The area of $\Phi(\mathcal{R})$ is 36.

4. Estimate the area of $\Phi(\mathcal{R})$, where $\mathcal{R} = [1, 1.2] \times [3, 3.1]$ and Φ is a mapping such that $\text{Jac}(\Phi)(1, 3) = 3$.

SOLUTION

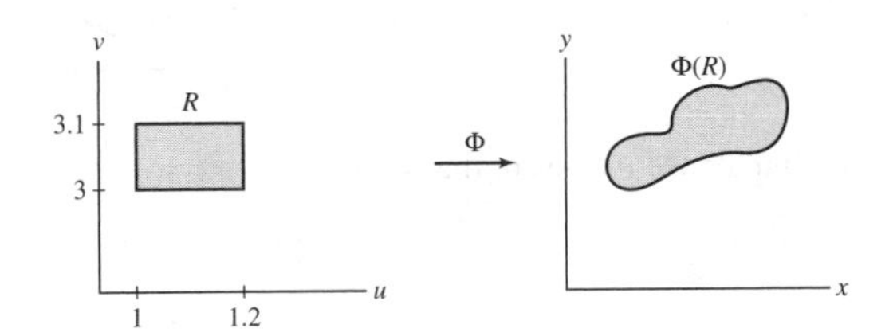

We use the following estimation:

$$\text{Area}\,(\Phi(\mathcal{R})) \approx |\text{Jac}\,(\Phi)\,(P)|\text{Area}(\mathcal{R})$$

The area of the rectangle $\mathcal{R}$ is

$$\text{Area}(\mathcal{R}) = 0.2 \cdot 0.1 = 0.02$$

We choose the sample point $P = (1, 3)$ in $\mathcal{R}$ to obtain the following estimation:

$$\text{Area}\,(\Phi(\mathcal{R})) \approx |\text{Jac}\,(\Phi)\,(1, 3)|\text{Area}(\mathcal{R}) = 3 \cdot 0.02 = 0.06$$

Exercises

1. Determine the image under $\Phi(u, v) = (2u, u + v)$ of the following sets:

(a) The u- and v-axes

(b) The rectangle $\mathcal{R} = [0, 5] \times [0, 7]$

(c) The line segment joining $(1, 2)$ and $(5, 3)$

(d) The triangle with vertices $(0, 1)$, $(1, 0)$, and $(1, 1)$

SOLUTION

(a) The image of the u-axis is obtained by substituting $v = 0$ in $\Phi(u, v) = (2u, u + v)$. That is,

$$\Phi(u, 0) = (2u, u + 0) = (2u, u).$$

The image of the u-axis is the set of points $(x, y) = (2u, u)$, which is the line $y = \frac{1}{2}x$ in the xy-plane. The image of the v-axis is obtained by substituting $u = 0$ in $\Phi(u, v) = (2u, u + v)$. That is,

$$\Phi(0, v) = (0, 0 + v) = (0, v).$$

Therefore, the image of the v-axis is the set $(x, y) = (0, v)$, which is the vertical line $x = 0$ (the y-axis).

(b) Since Φ is a linear map, the segment through points P and Q is mapped to the segment through $\Phi(P)$ and $\Phi(Q)$. We thus must find the images of the vertices of $\mathcal{R}$:

$$\Phi(0, 0) = (2 \cdot 0, 0 + 0) = (0, 0)$$

$$\Phi(5, 0) = (2 \cdot 5, 5 + 0) = (10, 5)$$

$$\Phi(5, 7) = (2 \cdot 5, 5 + 7) = (10, 12)$$

$$\Phi(0, 7) = (2 \cdot 0, 0 + 7) = (0, 7)$$

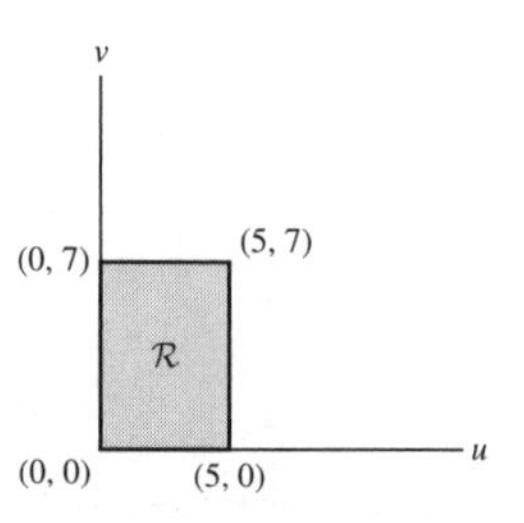

The image of $\mathcal{R}$ is the parallelogram with vertices (0, 0), (10, 5), (10, 12), and (0, 7) in the xy-plane.

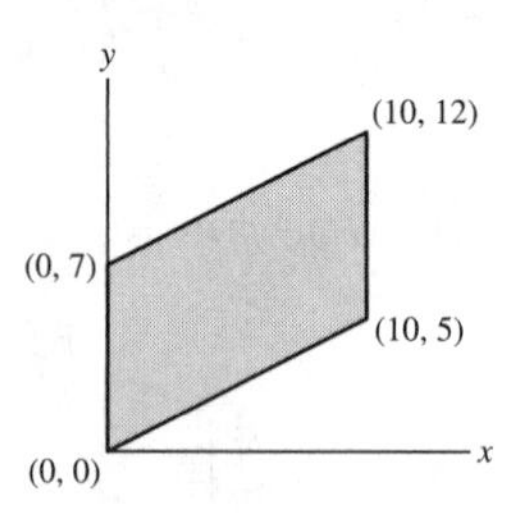

(c) We compute the images of the endpoints of the segment:

$$\Phi(1, 2) = (2 \cdot 1, 1 + 2) = (2, 3)$$

$$\Phi(5, 3) = (2 \cdot 5, 5 + 3) = (10, 8)$$

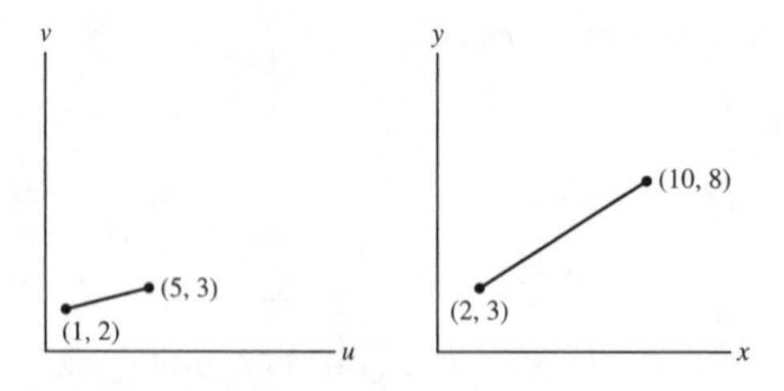

The image is the segment in the xy-plane joining the points (2, 3) and (10, 8).

(d) Since Φ is linear, the image of the triangle is the triangle whose vertices are the images of the vertices of the triangle. We compute these images:

$$\Phi(0, 1) = (2 \cdot 0, 0 + 1) = (0, 1)$$

$$\Phi(1, 0) = (2 \cdot 1, 1 + 0) = (2, 1)$$

$$\Phi(1, 1) = (2 \cdot 1, 1 + 1) = (2, 2)$$

Therefore the image is the triangle in the xy-plane whose vertices are at the points (0, 1), (2, 1), and (2, 2).

3. Let $\Phi(u, v) = (u^2, v)$. Is Φ one-to-one? If not, determine a domain on which Φ is one-to-one. Find the image under Φ of:

(a) The u- and v-axes
(b) The rectangle $\mathcal{R} = [-1, 1] \times [-1, 1]$
(c) The line segment joining $(0, 0)$ and $(1, 1)$
(d) The triangle with vertices $(0, 0)$, $(0, 1)$, and $(1, 1)$

SOLUTION Φ is not one-to-one since for any $u \neq 0$, (u, v) and $(-u, v)$ are two different points with the same image. However, Φ is one-to-one on the domain $\{(u, v) : u \geq 0\}$ and on the domain $\{(u, v) : u \leq 0\}$.

(a) The image of the u-axis is the set of the points

$$(x, y) = \Phi(u, 0) = (u^2, 0) \quad \Rightarrow \quad x = u^2, \quad y = 0$$

That is, the positive x-axis, including the origin. The image of the v-axis is the set of the following points:

$$(x, y) = \Phi(0, v) = (0^2, v) = (0, v) \quad \Rightarrow \quad x = 0, \quad y = v$$

That is, the line $x = 0$, which is the y-axis.

(b) The rectangle $\mathcal{R}$ is defined by

$$|u| \leq 1, \quad |v| \leq 1$$

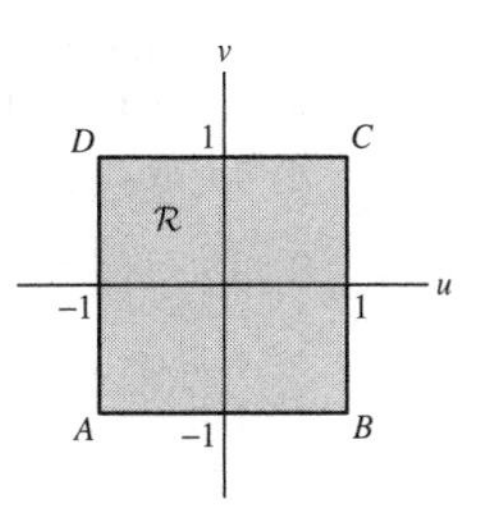

Since $x = u^2$ and $y = v$, we have $u = \pm\sqrt{x}$ and $v = y$ (depending on our choice of domain). Therefore, the inequalities for x and y are

$$|\pm\sqrt{x}| \leq 1, \quad |y| \leq 1$$

or

$$0 \leq x \leq 1 \quad \text{and} \quad -1 \leq y \leq 1.$$

We conclude that the image of $\mathcal{R}$ in the xy-plane is the rectangle $[0, 1] \times [-1, 1]$.

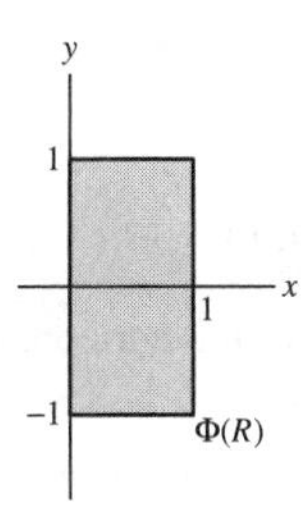

(c) The line segment joining the points $(0, 0)$ and $(1, 1)$ in the uv-plane is defined by

$$0 \leq u \leq 1, \quad v = u.$$

Substituting $u = \sqrt{x}$ and $v = y$, we get

$$0 \leq \sqrt{x} \leq 1, \quad y = \sqrt{x}$$

or

$$0 \leq x \leq 1, \quad y = \sqrt{x}$$

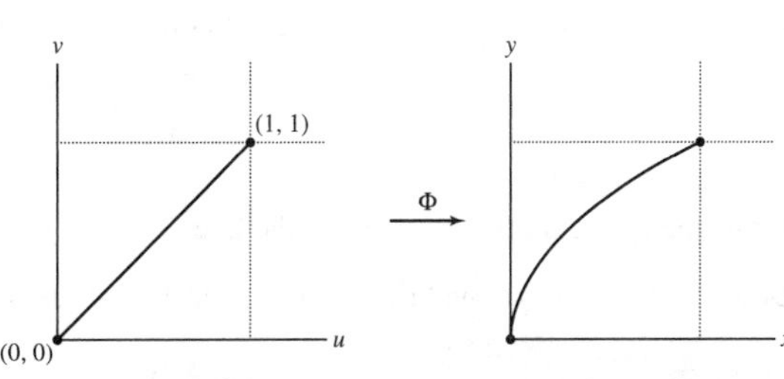

The image is the curve $y = \sqrt{x}$ for $0 \le x \le 1$.

(d) We identify the image of the sides of the triangle OAB.

The image of $\overline{OA}$: This segment is defined by $u = 0$ and $0 \le v \le 1$. That is,

$$\pm\sqrt{x} = 0 \quad \text{and} \quad 0 \le y \le 1$$

or

$$x = 0, \quad 0 \le y \le 1.$$

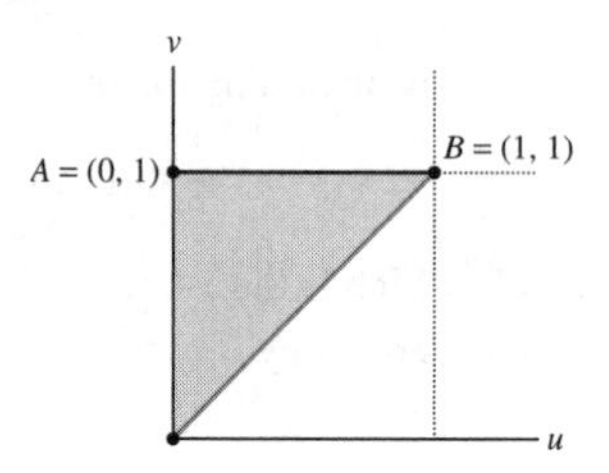

This is the segment joining the points (0, 0) and (0, 1) in the xy-plane.

The image of $\overline{AB}$: This segment is defined by $0 \le u \le 1$ and $v = 1$. That is,

$$0 \le \sqrt{x} \le 1, \quad y = 1$$

or

$$0 \le x \le 1, \quad y = 1.$$

This is the segment joining the points (0, 1) and (1, 1) in the xy-plane.

The image of $\overline{OB}$: In part (c) we showed that the image of the segment is the curve $y = \sqrt{x}$, $0 \le x \le 1$.

Therefore, the image of the triangle is the region shown in the figure:

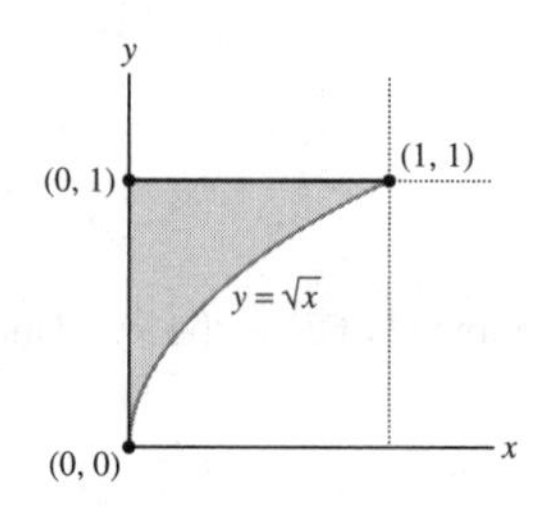

In Exercises 5–12, let $\Phi(u, v) = (2u + v, 5u + 3v)$ be a map from the uv-plane to the xy-plane.

5. Show that the image of the horizontal line $v = c$ is the line $y = \frac{5}{2}x + \frac{1}{2}c$. What is the image (in slope-intercept form) of the vertical line $u = c$?

SOLUTION The image of the vertical line $u = c$ is the set of the following points:

$$(x, y) = \Phi(c, v) = (2c + v, 5c + 3v) \quad \Rightarrow \quad x = 2c + v, \quad y = 5c + 3v$$

By the first equation, $v = x - 2c$. Substituting in the second equation gives

$$y = 5c + 3(x - 2c) = 5c + 3x - 6c = 3x - c$$

Therefore, the image of the line $u = c$ is the line $y = 3x - c$ in the xy-plane. The image of the horizontal line $v = c$ is the set of the following points:

$$(x, y) = \Phi(u, c) = (2u + c, 5u + 3c) \quad \Rightarrow \quad x = 2u + c, \quad y = 5u + 3c$$

The first equation implies $u = \frac{x-c}{2}$. Substituting in the second equation gives

$$y = 5\frac{(x - c)}{2} + 3c = \frac{5x}{2} + \frac{c}{2}$$

Therefore, the image of the line $v = c$ is the line $y = \frac{5x}{2} + \frac{c}{2}$ in the xy-plane.

7. Describe the image of the line $v = 4u$ under Φ in slope-intercept form.

SOLUTION We choose any two points on the line $v = 4u$, for example $(u, v) = (1, 4)$ and $(u, v) = (0, 0)$. By a property of linear maps, the image of the line $v = 4u$ under the linear map $\Phi(u, v) = (2u + v, 5u + 3v)$ is the line in the xy-plane through the points $\Phi(1, 4)$ and $\Phi(0, 0)$. We find these points:

$$\Phi(0, 0) = (2 \cdot 0 + 0, 5 \cdot 0 + 3 \cdot 0) = (0, 0)$$

$$\Phi(1, 4) = (2 \cdot 1 + 4, 5 \cdot 1 + 3 \cdot 4) = (6, 17)$$

We now find the slope-intercept equation of the line in the xy-plane through the points $(0, 0)$ and $(6, 17)$:

$$y - 0 = \frac{17 - 0}{6 - 0}(x - 0) \quad \Rightarrow \quad y = \frac{17}{6}x$$

9. Show that the inverse of Φ is

$$\Phi^{-1}(x, y) = (3x - y, -5x + 2y)$$

Hint: Show that $\Phi(\Phi^{-1}(x, y)) = (x, y)$ and $\Phi^{-1}(\Phi(u, v)) = (u, v)$.

SOLUTION By the definition of the inverse map, we must show that the given maps $\Phi^{-1}(x, y) = (3x - y, -5x + 2y)$ and $\Phi(u, v) = (2u + v, 5u + 3v)$ satisfy $\Phi\left(\Phi^{-1}(x, y)\right) = (x, y)$ and $\Phi^{-1}\left(\Phi(u, v)\right) = (u, v)$. We have

$$\Phi\left(\Phi^{-1}(x, y)\right) = \Phi(3x - y, -5x + 2y) = (2(3x - y) + (-5x + 2y), 5(3x - y) + 3(-5x + 2y)) = (x, y)$$

$$\Phi^{-1}\left(\Phi(u, v)\right) = \Phi^{-1}(2u + v, 5u + 3v) = (3(2u + v) - (5u + 3v), -5(2u + v) + 2(5u + 3v)) = (u, v)$$

We conclude that Φ^{-1} is the inverse of Φ.

11. Calculate $\text{Jac}(\Phi) = \dfrac{\partial(x, y)}{\partial(u, v)}$.

SOLUTION The Jacobian of the linear mapping $\Phi(u, v) = (2u + v, 5u + 3v)$ is the following determinant:

$$\text{Jac}(\Phi) = \frac{\partial(x, y)}{\partial(u, v)} = \begin{vmatrix} 2 & 1 \\ 5 & 3 \end{vmatrix} = 2 \cdot 3 - 5 \cdot 1 = 1$$

In Exercises 13–18, compute the Jacobian (at the point, if indicated).

13. $\Phi(u, v) = (3u + 4v, u - 2v)$

SOLUTION Using the Jacobian of linear mappings we get

$$\text{Jac}(\Phi) = \frac{\partial(x, y)}{\partial(u, v)} = \begin{vmatrix} 3 & 4 \\ 1 & -2 \end{vmatrix} = 3 \cdot (-2) - 1 \cdot 4 = -10$$

15. $\Phi(u, v) = (ue^v, ve^{3u}), \quad (u, v) = (1, 2)$

SOLUTION We have $x = ue^v$ and $y = ve^{3u}$. Thus,

$$\text{Jac}(\Phi) = \begin{vmatrix} \dfrac{\partial x}{\partial u} & \dfrac{\partial x}{\partial v} \\ \dfrac{\partial y}{\partial u} & \dfrac{\partial y}{\partial v} \end{vmatrix} = \begin{vmatrix} e^v & ue^v \\ 3ve^{3u} & e^{3u} \end{vmatrix} = e^v \cdot e^{3u} - ue^v \cdot 3ve^{3u} = e^{v+3u} - 3uve^{v+3u} = (1 - 3uv)e^{v+3u}$$

Hence, $\text{Jac}(\Phi)(1, 2) = (1 - 3 \cdot 1 \cdot 2)e^{2+3\cdot 1} = -5e^5$.

17. $\Phi(r, \theta) = (r \cos\theta, r \sin\theta)$

SOLUTION Since $x = r\cos\theta$ and $y = r\sin\theta$, the Jacobian of Φ is the following determinant:

$$\text{Jac}(\Phi) = \frac{\partial(x, y)}{\partial(r, \theta)} = \begin{vmatrix} \dfrac{\partial x}{\partial r} & \dfrac{\partial x}{\partial \theta} \\ \dfrac{\partial y}{\partial r} & \dfrac{\partial y}{\partial \theta} \end{vmatrix} = \begin{vmatrix} \cos\theta & -r\sin\theta \\ \sin\theta & r\cos\theta \end{vmatrix} = r\cos^2\theta + r\sin^2\theta = r(\cos^2\theta + \sin^2\theta) = r \cdot 1 = r$$

19. Find a linear mapping Φ that maps $[0, 1] \times [0, 1]$ to the parallelogram in the xy-plane spanned by the vectors $\langle 2, 3\rangle$ and $\langle 4, 1\rangle$.

SOLUTION

We denote the linear map by

$$\Phi(u, v) = (Au + Cv, Bu + Dv) \tag{1}$$

The image of the unit square $\mathcal{R} = [0, 1] \times [0, 1]$ under the linear map is the parallelogram whose vertices are the images of the vertices of $\mathcal{R}$. Two of vertices of the given parallelogram are $(2, 3)$ and $(4, 1)$. To find A, B, C, and D it suffices to determine four equations. Therefore, we ask that (notice that for linear maps $\Phi(0, 0) = (0, 0)$)

$$\Phi(0, 1) = (2, 3), \quad \Phi(1, 0) = (4, 1)$$

We substitute in (1) and solve for A, B, C, and D:

$$\begin{aligned} (A \cdot 0 + C \cdot 1, B \cdot 0 + D \cdot 1) &= (C, D) = (2, 3) \\ (A \cdot 1 + C \cdot 0, B \cdot 1 + D \cdot 0) &= (A, B) = (4, 1) \end{aligned} \quad \Rightarrow \quad \begin{aligned} C = 2, \quad D = 3 \\ A = 4, \quad B = 1 \end{aligned}$$

Substituting in (1) we obtain the following map:

$$\Phi(u, v) = (4u + 2v, u + 3v).$$

21. Let $\mathcal{D}$ be the parallelogram in Figure 14. Apply the Change of Variables Formula to the $\Phi(u, v) = (5u + 3v, u + 4v)$ to evaluate $\iint_{\mathcal{D}} xy\, dA$ as an integral over $\mathcal{D}_0 = [0, 1] \times [0, 1]$.

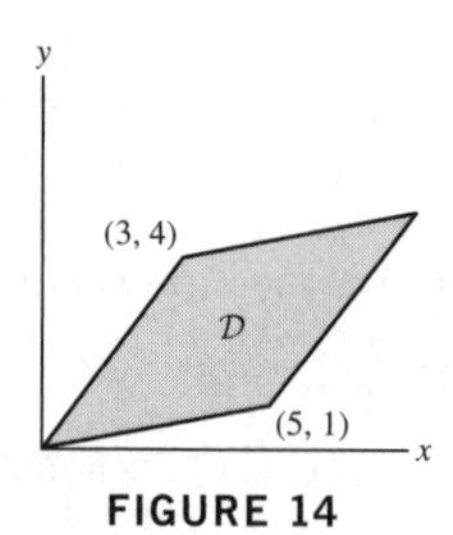

FIGURE 14

SOLUTION

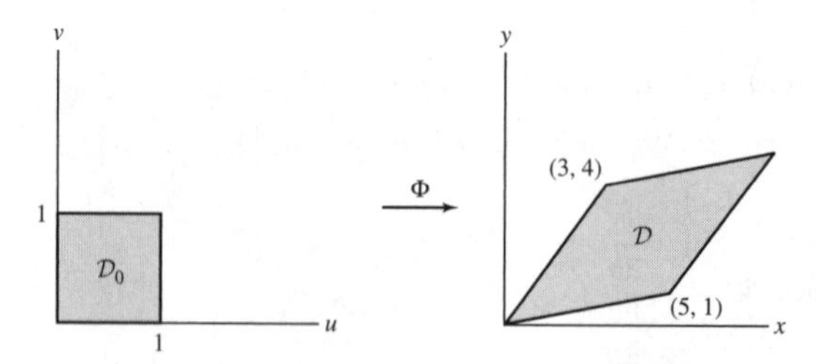

We express $f(x, y) = xy$ in terms of u and v. Since $x = 5u + 3v$ and $y = u + 4v$, we have

$$f(x, y) = xy = (5u + 3v)(u + 4v) = 5u^2 + 12v^2 + 23uv$$

The Jacobian of the linear map $\Phi(u, v) = (5u + 3v, u + 4v)$ is

$$\text{Jac}(\Phi) = \frac{\partial(x, y)}{\partial(u, v)} = \begin{vmatrix} 5 & 3 \\ 1 & 4 \end{vmatrix} = 20 - 3 = 17$$

Applying the Change of Variables Formula we get

$$\iint_{\mathcal{D}} xy\, dA = \iint_{\mathcal{D}_0} f(x, y) \left| \frac{\partial(x, y)}{\partial(u, v)} \right| du\, dv = \int_0^1 \int_0^1 (5u^2 + 12v^2 + 23uv) \cdot 17\, du\, dv$$

$$= 17 \int_0^1 \frac{5u^3}{3} + 12v^2 u + \frac{23u^2 v}{2} \Bigg|_{u=0}^{1} dv = 17 \int_0^1 \left(\frac{5}{3} + 12v^2 + \frac{23v}{2} \right) dv$$

$$= 17\left(\frac{5v}{3} + 4v^3 + \frac{23v^2}{4}\Big|_0^1\right) = 17\left(\frac{5}{3} + 4 + \frac{23}{4}\right) = \frac{2329}{12} \approx 194.08$$

23. Let $\Phi(u, v) = (3u + v, u - 2v)$. Use the Jacobian to determine the area of $\Phi(\mathcal{R})$ for:
(a) $\mathcal{R} = [0, 3] \times [0, 5]$ **(b)** $\mathcal{R} = [2, 5] \times [1, 7]$

SOLUTION The Jacobian of the linear map $\Phi(u, v) = (3u + v, u - 2v)$ is the following determinant:

$$\text{Jac}\Phi = \frac{\partial(x, y)}{\partial(u, v)} = \begin{vmatrix} 3 & 1 \\ 1 & -2 \end{vmatrix} = -6 - 1 = -7$$

By properties of linear maps, we have

$$\text{Area}\,(\Phi(\mathcal{R})) = |\text{Jac}\Phi|\text{Area}(\mathcal{R}) = 7 \cdot \text{Area}(\mathcal{R})$$

(a) The area of the rectangle $R = [0, 3] \times [0, 5]$ is $3 \cdot 5 = 15$, therefore the area of $\Phi(\mathcal{R})$ is

$$\text{Area}\,(\Phi(\mathcal{R})) = 7 \cdot 15 = 105$$

(b) The area of the rectangle $\mathcal{R} = [2, 5] \times [1, 7]$ is $3 \cdot 6 = 18$ hence the area of $\Phi(\mathcal{R})$ is

$$\text{Area}\,(\Phi(\mathcal{R})) = 7 \cdot 18 = 126.$$

25. With Φ as in Example 3, use the Change of Variables Formula to compute the area of the image of $[1, 4] \times [1, 4]$.

SOLUTION Let $\mathcal{R}$ represent the rectangle $[1, 4] \times [1, 4]$. We proceed as follows. Jac(Φ) is easily calculated as

$$\text{Jac}(T) = \frac{\partial(x, y)}{\partial(u, v)} = \begin{vmatrix} 1/v & -u/v^2 \\ v & u \end{vmatrix} = 2u/v$$

Now, the area is given by the Change of Variables Formula as

$$\iint_{\Phi(\mathcal{R})} 1\,dA = \iint_{\mathcal{R}} 1|\text{Jac}(\Phi)|\,du\,dv = \iint_{\mathcal{R}} 1|2u/v|\,du\,dv = \int_1^4 \int_1^4 2u/v\,du\,dv$$

$$= \int_1^4 2u\,du \cdot \int_1^4 \frac{1}{v}\,dv = (16 - 1)(\ln 4 - \ln 1) = 15 \ln 4$$

In Exercises 26–28, let $\mathcal{R}_0 = [0, 1] \times [0, 1]$ be the unit square. The translate of a map $\Phi_0(u, v) = (\phi(u, v), \psi(u, v))$ is a map

$$\Phi(u, v) = (a + \phi(u, v), b + \psi(u, v))$$

where a, b are constants. Observe that the map Φ_0 in Figure 16 maps $\mathcal{R}_0$ to the parallelogram $\mathcal{P}_0$ and the translate

$$\Phi_1(u, v) = (2 + 4u + 2v, 1 + u + 3v)$$

maps $\mathcal{R}_0$ to $\mathcal{P}_1$.

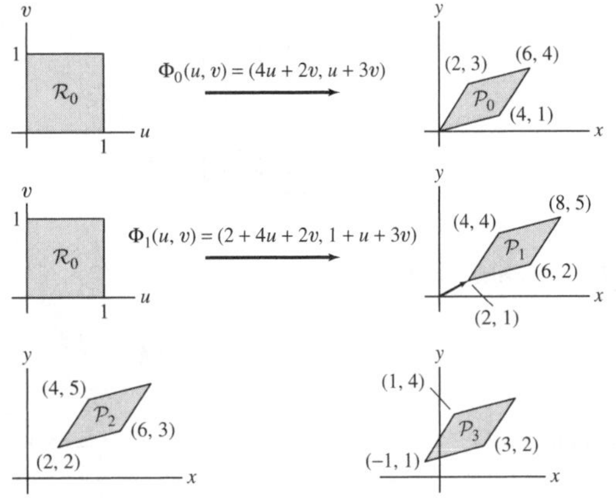

FIGURE 16

27. Sketch the parallelogram $\mathcal{P}$ with vertices (1, 1), (2, 4), (3, 6), (4, 9) and find the translate of a linear mapping that maps $\mathcal{R}_0$ to $\mathcal{P}$.

SOLUTION The parallelogram $\mathcal{P}$ is shown in the figure:

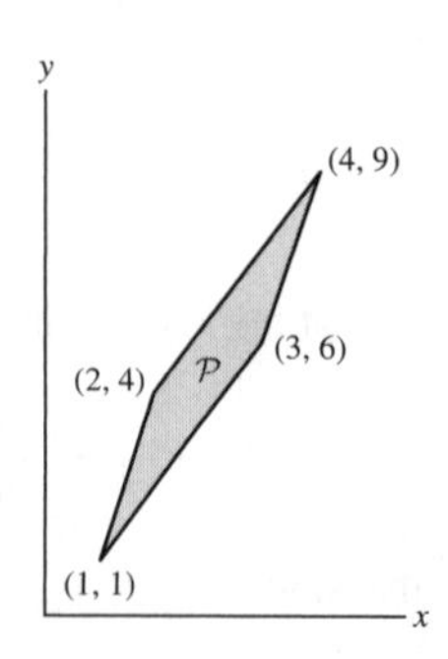

We first translate the parallelogram $\mathcal{P}$ one unit to the left and one unit downward to obtain a parallelogram $\mathcal{P}_0$ with a vertex at the origin.

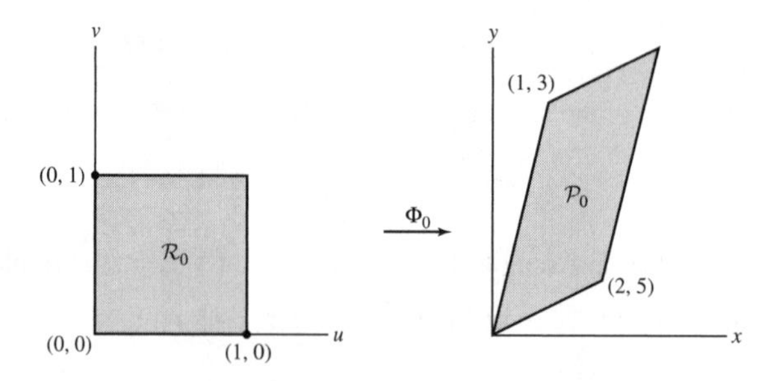

We find a linear map $\Phi_0(u, v) = (Au + Cv, Bu + Dv)$ that maps $\mathcal{R}_0$ to $\mathcal{P}_0$:

$$\Phi_0(0, 1) = (1, 3) \quad \Rightarrow \quad (C, D) = (1, 3) \quad \Rightarrow \quad C = 1, \quad D = 3$$

$$\Phi_0(1, 0) = (2, 5) \quad \Rightarrow \quad (A, B) = (2, 5) \quad \Rightarrow \quad A = 2, \quad B = 5$$

Therefore,

$$\Phi_0(u, v) = (2u + v, 5u + 3v)$$

Now we can determine the translate Φ of Φ_0 that maps $\mathcal{R}_0$ to $\mathcal{P}$. Since $\mathcal{P}$ is obtained by translating $\mathcal{P}_0$ one unit upward and one unit to the right, the map Φ is the following translate of Φ_0:

$$\Phi(u, v) = (1 + 2u + v, 1 + 5u + 3v)$$

29. Let $\mathcal{D} = \Phi(\mathcal{R})$, where $\Phi(u, v) = (u^2, u + v)$ and $\mathcal{R} = [1, 2] \times [0, 6]$. Calculate $\iint_{\mathcal{D}} y\,dA$. *Note:* It is not necessary to describe $\mathcal{D}$.

SOLUTION

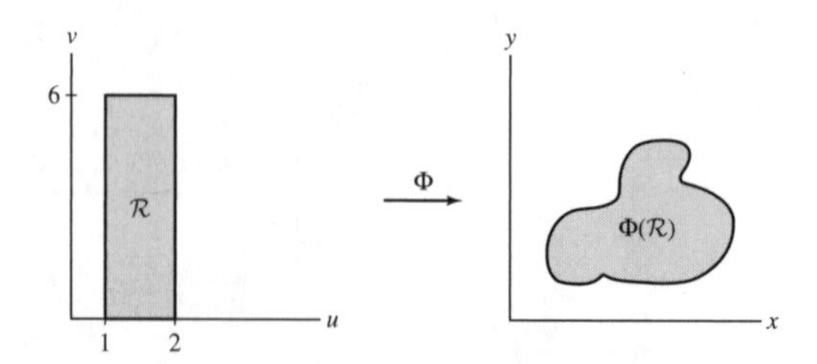

Changing variables, we have

$$\iint_{\mathcal{D}} y\,dA = \iint_{\mathcal{R}} (u + v) \left| \frac{\partial(x, y)}{\partial(u, v)} \right| du\,dv \tag{1}$$

We compute the Jacobian of Φ. Since $x = u^2$ and $y = u + v$, we have

$$\frac{\partial(x, y)}{\partial(u, v)} = \begin{vmatrix} \frac{\partial x}{\partial u} & \frac{\partial x}{\partial v} \\ \frac{\partial y}{\partial u} & \frac{\partial y}{\partial v} \end{vmatrix} = \begin{vmatrix} 2u & 0 \\ 1 & 1 \end{vmatrix} = 2u$$

We substitute in (1) and compute the resulting integral:

$$\iint_{\mathcal{D}} y\,dA = \int_0^6 \int_1^2 (u+v)\cdot 2u\,du\,dv = \int_0^6 \int_1^2 (2u^2 + 2uv)\,du\,dv = \int_0^6 \frac{2u^3}{3} + u^2 v \Big|_{u=1}^{2} dv$$

$$= \int_0^6 \left(\left(\frac{16}{3} + 4v \right) - \left(\frac{2}{3} + v \right) \right) dv = \int_0^6 \left(3v + \frac{14}{3} \right) dv = \frac{3}{2}v^2 + \frac{14}{3}v \Big|_0^6 = 82$$

31. Compute $\iint_{\mathcal{D}} (x+3y)\,dx\,dy$, where $\mathcal{D}$ is the shaded region in Figure 17. *Hint:* Use the map $\Phi(u, v) = (u - 2v, v)$.

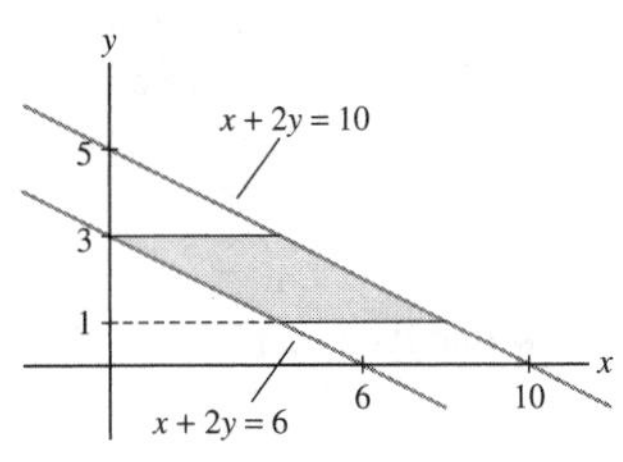

FIGURE 17

SOLUTION The boundary of $\mathcal{D}$ is defined by the lines $x + 2y = 6$, $x + 2y = 10$, $y = 1$, and $y = 3$.

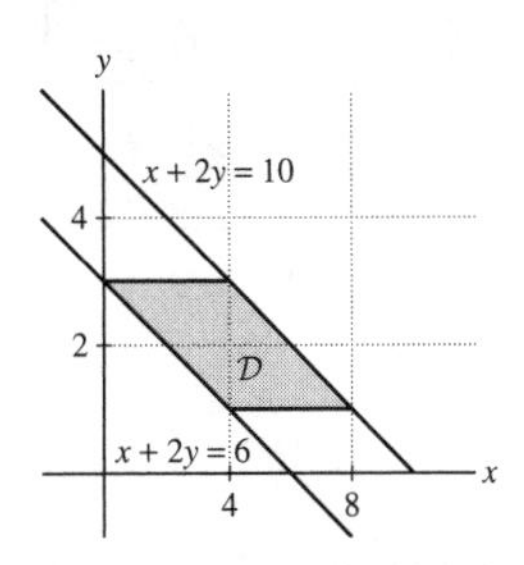

Therefore, $\mathcal{D}$ is mapped to a rectangle $\mathcal{D}_0$ in the uv-plane under the map

$$u = x + 2y, \quad v = y \tag{1}$$

or

$$(u, v) = \Phi^{-1}(x, y) = (x + 2y, y)$$

Since $\mathcal{D}$ is defined by the inequalities $6 \le x + 2y \le 10$ and $1 \le y \le 3$, the corresponding domain in the uv-plane is the rectangle

$$\mathcal{D}_0 : 6 \le u \le 10, \ 1 \le v \le 3$$

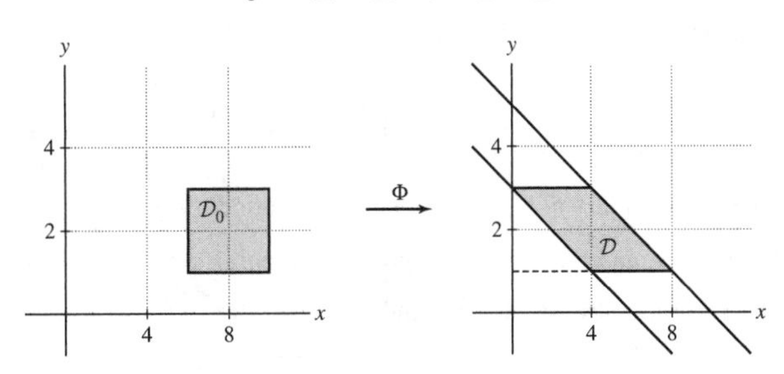

To find $\Phi(u, v)$ we must solve the equations (1) for x and y in terms of u and v. We obtain

$$\begin{matrix} u = x + 2y \\ v = y \end{matrix} \quad \Rightarrow \quad \begin{matrix} x = u - 2v \\ y = v \end{matrix} \quad \Rightarrow \quad \Phi(u, v) = (u - 2v, v)$$

We compute the Jacobian of the linear mapping Φ:

$$\text{Jac}(\Phi) = \frac{\partial(x, y)}{\partial(u, v)} = \begin{vmatrix} 1 & -2 \\ 0 & 1 \end{vmatrix} = 1 \cdot 1 + 2 \cdot 0 = 1$$

The function $f(x, y) = x + 3y$ expressed in terms of the new variables u and v is

$$f(x, y) = u - 2v + 3v = u + v$$

We now use the Change of Variables Formula to compute the required integral. We get

$$\iint_{\mathcal{D}} f(x, y)\,dx\,dy = \iint_{\mathcal{D}_0} (u+v)\left|\frac{\partial(x, y)}{\partial(u, v)}\right| du\,dv = \int_1^3 \int_6^{10} (u+v)\cdot 1\,du\,dv$$

$$= \int_1^3 \frac{u^2}{2} + vu\bigg|_{u=6}^{10} dv = \int_1^3 ((50+10v)-(18+6v))\,dv$$

$$= \int_1^3 (32+4v)\,dv = 32v + 2v^2\bigg|_1^3 = (96+18)-(32+2) = 80$$

33. Show that $T(u, v) = (u^2 - v^2, 2uv)$ maps the triangle $\mathcal{D}_0 = \{(u, v) : 0 \le v \le u \le 1\}$ to the domain $\mathcal{D}$ bounded by $x = 0$, $y = 0$, and $y^2 = 4 - 4x$. Use T to evaluate

$$\iint_{\mathcal{D}} \sqrt{x^2 + y^2}\,dx\,dy$$

SOLUTION We show that the boundary of $\mathcal{D}_0$ is mapped to the boundary of $\mathcal{D}$.

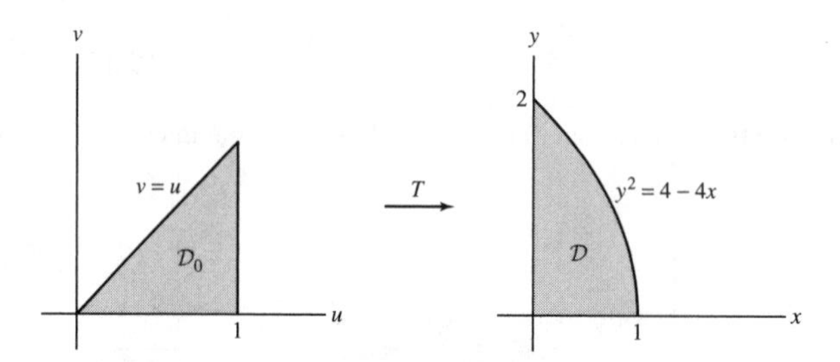

We have

$$x = u^2 - v^2 \quad \text{and} \quad y = 2uv$$

The line $v = u$ is mapped to the following set:

$$(x, y) = (u^2 - u^2, 2u^2) = (0, 2u^2) \quad \Rightarrow \quad x = 0, \quad y \ge 0$$

That is, the image of the line $u = v$ is the positive y-axis. The line $v = 0$ is mapped to the following set:

$$(x, y) = (u^2, 0) \quad \Rightarrow \quad x = u^2, \quad y = 0 \quad \Rightarrow \quad y = 0, \quad x \ge 0$$

Thus, the line $v = 0$ is mapped to the positive x-axis. We now show that the vertical line $u = 1$ is mapped to the curve $y^2 + 4x = 4$. The image of the line $u = 1$ is the following set:

$$(x, y) = (1 - v^2, 2v) \quad \Rightarrow \quad x = 1 - v^2, \quad y = 2v$$

We substitute $v = \frac{y}{2}$ in the equation $x = 1 - v^2$ to obtain

$$x = 1 - \left(\frac{y}{2}\right)^2 = 1 - \frac{y^2}{4} \quad \Rightarrow \quad 4x = 4 - y^2 \quad \Rightarrow \quad y^2 + 4x = 4$$

Since the boundary of $\mathcal{D}_0$ is mapped to the boundary of $\mathcal{D}$, we conclude that the domain $\mathcal{D}_0$ is mapped by T to the domain $\mathcal{D}$ in the xy-plane. We now compute the integral $\iint_{\mathcal{D}} \sqrt{x^2 + y^2}\,dx\,dy$. We express the function $f(x, y) = \sqrt{x^2 + y^2}$ in terms of the new variables u and v:

$$f(x, y) = \sqrt{(u^2 - v^2)^2 + (2uv)^2} = \sqrt{u^4 - 2u^2v^2 + v^4 + 4u^2v^2}$$

$$= \sqrt{u^4 + 2u^2v^2 + v^4} = \sqrt{(u^2 + v^2)^2} = u^2 + v^2$$

We compute the Jacobian of T:

$$\text{Jac}(T) = \frac{\partial(x, y)}{\partial(u, v)} = \begin{vmatrix} \frac{\partial x}{\partial u} & \frac{\partial x}{\partial v} \\ \frac{\partial y}{\partial u} & \frac{\partial y}{\partial v} \end{vmatrix} = \begin{vmatrix} 2u & -2v \\ 2v & 2u \end{vmatrix} = 4u^2 + 4v^2 = 4(u^2 + v^2)$$

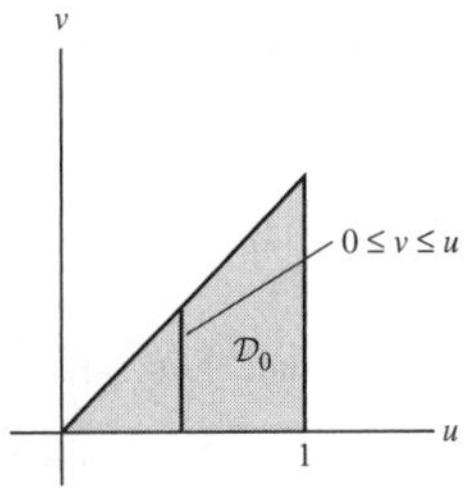

Using the Change of Variables Formula gives

$$\iint_{\mathcal{D}} \sqrt{x^2+y^2}\, dx\, dy = \iint_{\mathcal{D}_0} (u^2+v^2)\cdot 4(u^2+v^2)\, du\, dv = 4\int_0^1 \int_0^u (u^4+2u^2v^2+v^4)\, dv\, du$$

$$= 4\int_0^1 u^4 v + \frac{2}{3}u^2 v^3 + \frac{v^5}{5}\bigg|_{v=0}^{u} du = 4\int_0^1 \left(u^5 + \frac{2}{3}u^5 + \frac{u^5}{5}\right) du$$

$$= 4\int_0^1 \frac{28}{15}u^5\, du = \frac{112}{15}\cdot \frac{u^6}{6}\bigg|_0^1 = \frac{56}{45}$$

35. Calculate $\iint_{\mathcal{D}} e^{9x^2+4y^2}\, dA$, where $\mathcal{D}$ is the interior of the ellipse $\left(\frac{x}{2}\right)^2 + \left(\frac{y}{3}\right)^2 \le 1$.

SOLUTION We define a map that maps the unit disk $u^2+v^2 \le 1$ onto the interior of the ellipse. That is,

$$x = 2u, \quad y = 3v \quad \Rightarrow \quad \Phi(u,v) = (2u, 3v)$$

Since $\left(\frac{x}{2}\right)^2 + \left(\frac{y}{3}\right)^2 \le 1$ if and only if $u^2+v^2 \le 1$, Φ is the map we need.

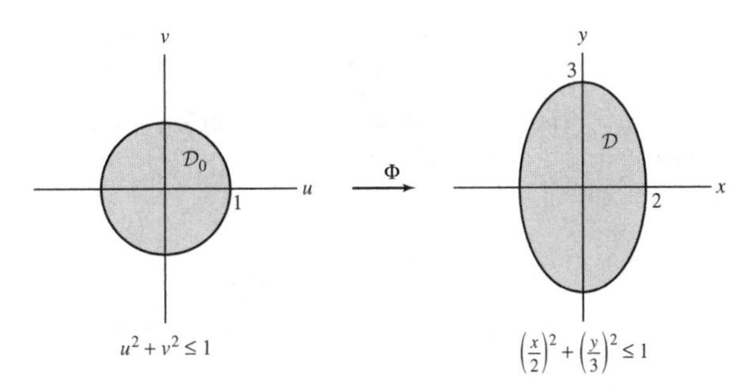

We express the function $f(x,y) = e^{9x^2+4y^2}$ in terms of u and v:

$$f(x,y) = e^{9(2u)^2+4(3v)^2} = e^{36u^2+36v^2} = e^{36(u^2+v^2)}$$

We compute the Jacobian of Φ:

$$\text{Jac}(\Phi) = \begin{vmatrix} \frac{\partial x}{\partial u} & \frac{\partial x}{\partial v} \\ \frac{\partial y}{\partial u} & \frac{\partial y}{\partial v} \end{vmatrix} = \begin{vmatrix} 2 & 0 \\ 0 & 3 \end{vmatrix} = 6$$

Using the Change of Variables Formula gives

$$\iint_{\mathcal{D}} e^{9x^2+4y^2}\, dA = \iint_{\mathcal{D}_0} e^{36(u^2+v^2)} \cdot 6\, du\, dv$$

We compute the integral using polar coordinates $u = r\cos\theta$, $v = r\sin\theta$:

$$\iint_{\mathcal{D}} e^{9x^2+4y^2}\, dA = \int_0^{2\pi}\int_0^1 6e^{36r^2}\cdot r\, dr\, d\theta = \left(6\int_0^{2\pi} d\theta\right)\left(\int_0^1 e^{36r^2} r\, dr\right)$$

$$= 12\pi \frac{e^{36r^2}}{72}\bigg|_{r=0}^{1} = \frac{12\pi(e^{36}-1)}{72} = \frac{\pi(e^{36}-1)}{6}$$

37. Sketch the domain $\mathcal{D}$ bounded by $y = x^2$, $y = \frac{1}{2}x^2$, and $y = x$. Use a change of variables with the map $x = uv$, $y = u^2$ to calculate

$$\iint_{\mathcal{D}} y^{-1}\,dx\,dy$$

This is an improper integral since $f(x, y) = y^{-1}$ is undefined at $(0, 0)$, but it becomes proper after changing variables.

SOLUTION The domain $\mathcal{D}$ is shown in the figure.

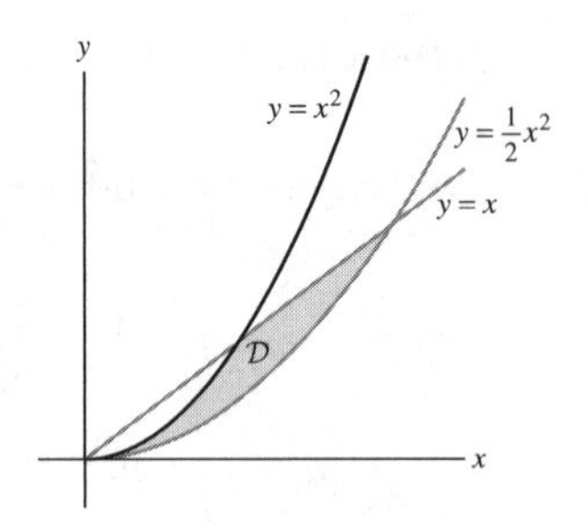

We must identify the domain $\mathcal{D}_0$ in the uv-plane. Notice that Φ is one-to-one, where $u \ge 0$ (or $u \le 0$), since in $\mathcal{D}$, $x \ge 0$, so it also follows by $x = uv$ that $v \ge 0$. Therefore, we search the domain $\mathcal{D}_0$ in the first quadrant of the uv-plane. To do this, we examine the curves that are mapped to the curves defining the boundary of $\mathcal{D}$. We examine each curve separately.

$y = x^2$: Since $x = uv$ and $y = u^2$ we get

$$u^2 = (uv)^2 \quad \Rightarrow \quad 1 = v^2 \quad \Rightarrow \quad v = 1$$

$y = \frac{1}{2}x^2$:

$$u^2 = \frac{1}{2}(uv)^2 \quad \Rightarrow \quad 1 = \frac{1}{2}v^2 \quad \Rightarrow \quad v^2 = 2 \quad \Rightarrow \quad v = \sqrt{2}$$

$y = x$: $u^2 = uv \quad \Rightarrow \quad v = u$. The region $\mathcal{D}_0$ is the region in the first quadrant of the uv-plane enclosed by the curves $v = 1$, $v = \sqrt{2}$, and $v = u$.

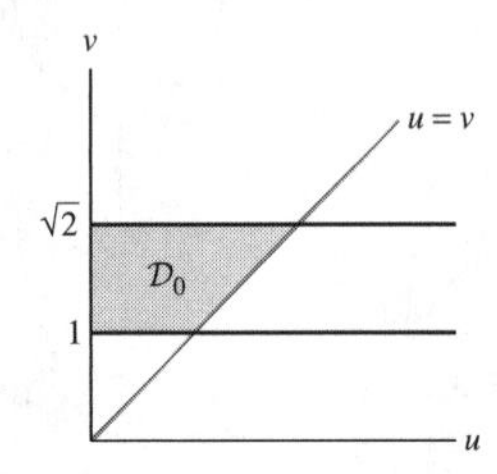

We now use change of variables to compute the integral $\iint_{\mathcal{D}} y^{-1}\,dx\,dy$. The function in terms of the new variables is $f(x, y) = u^{-2}$. We compute the Jacobian of $\Phi(u, v) = (x, y) = (uv, u^2)$:

$$\text{Jac}(\Phi) = \begin{vmatrix} \frac{\partial x}{\partial u} & \frac{\partial x}{\partial v} \\ \frac{\partial y}{\partial u} & \frac{\partial y}{\partial v} \end{vmatrix} = \begin{vmatrix} v & u \\ 2u & 0 \end{vmatrix} = -2u^2$$

Using the Change of Variables Formula gives

$$\iint_{\mathcal{D}} y^{-1}\,dx\,dy = \iint_{\mathcal{D}_0} u^{-2} \cdot 2u^2\,du\,dv = \int_1^{\sqrt{2}} \int_0^v 2\,du\,dv = \int_1^{\sqrt{2}} 2u\Big|_{u=0}^{v}\,dv = \int_1^{\sqrt{2}} 2v\,dv = v^2\Big|_1^{\sqrt{2}} = 2 - 1 = 1$$

39. Let Φ be the inverse of the map $T(x, y) = (xy, x^2y)$ from the xy-plane to the uv-plane. Let $\mathcal{D}$ be the domain in Figure 19. Show, by applying the Change of Variables Formula to the inverse $\Phi = T^{-1}$, that

$$\iint_{\mathcal{D}} e^{xy}\,dA = \int_{10}^{20} \int_{20}^{40} e^u v^{-1}\,dv\,du$$

and evaluate this result. *Hint:* Use Eq. (14) to compute $\text{Jac}(\Phi)$. It is not necessary to determine Φ explicitly.

FIGURE 19

SOLUTION The domain $\mathcal{D}$ is defined by the inequalities

$$\mathcal{D}: \ 10 \le xy \le 20, \ 20 \le x^2y \le 40$$

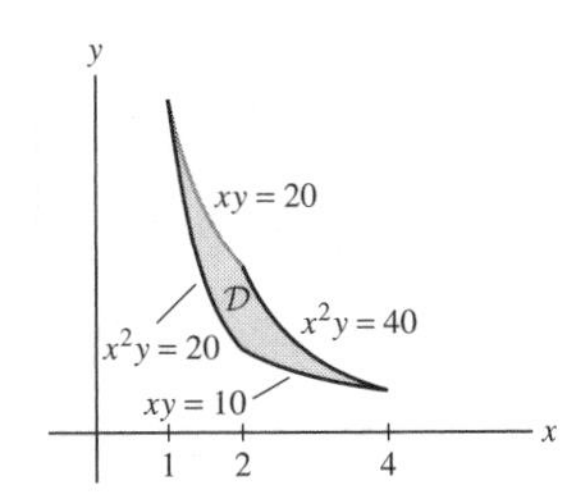

Since $u = xy$ and $v = x^2y$, the image $\mathcal{D}_0$ of $\mathcal{D}$ (in the uv-plane) under T is the rectangle

$$\mathcal{D}_0: \ 10 \le u \le 20, \ 20 \le v \le 40$$

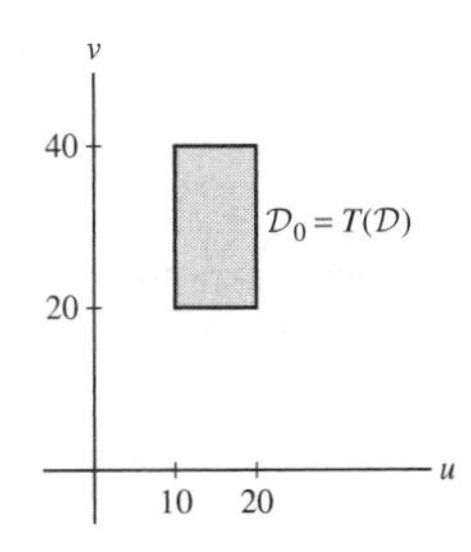

The function expressed in the new variables is

$$f(x, y) = e^{xy} = e^u$$

To find the Jacobian of the inverse Φ of T, we use the formula for the Jacobian of the inverse mapping. That is,

$$\frac{\partial(x, y)}{\partial(u, v)} = \left(\frac{\partial(u, v)}{\partial(x, y)}\right)^{-1}$$

We find the Jacobian of T. Since $u = xy$ and $v = x^2y$, we have

$$\text{Jac}(T) = \begin{vmatrix} \frac{\partial u}{\partial x} & \frac{\partial u}{\partial y} \\ \frac{\partial v}{\partial x} & \frac{\partial v}{\partial y} \end{vmatrix} = \begin{vmatrix} y & x \\ 2xy & x^2 \end{vmatrix} = yx^2 - 2x^2y = -x^2y$$

Hence,

$$\text{Jac}(\Phi) = -\frac{1}{x^2y}$$

We now compute the double integral $\iint_{\mathcal{D}} e^{xy}\, dA$ using the Change of Variables Formula. Since $y > 0$ in $\mathcal{D}$, we have $|\text{Jac}(\Phi)| = |-\frac{1}{x^2y}| = \frac{1}{x^2y} = v^{-1}$. Therefore,

$$\iint_{\mathcal{D}} e^{xy}\, dA = \iint_{\mathcal{D}_0} e^u v^{-1}\, dv\, du = \int_{10}^{20} \int_{20}^{40} e^u v^{-1}\, dv\, du = \left(\int_{10}^{20} e^u\, du\right)\left(\int_{20}^{40} v^{-1}\, dv\right)$$

$$= e^u \Big|_{10}^{20} \cdot \ln v \Big|_{20}^{40} = (e^{20} - e^{10})(\ln(40) - \ln(20)) = (e^{20} - e^{10}) \ln 2$$

41. Let $I = \iint_{\mathcal{D}} (x^2 - y^2)\,dx\,dy$, where

$$\mathcal{D} = \{(x, y) : 2 \le xy \le 4, 0 \le x - y \le 3, x \ge 0, y \ge 0\}$$

(a) Show that the mapping $u = xy$, $v = x - y$ maps $\mathcal{D}$ to the rectangle $\mathcal{R} = [2, 4] \times [0, 3]$.
(b) Compute $\partial(x, y)/\partial(u, v)$ by first computing $\partial(u, v)/\partial(x, y)$.
(c) Use the Change of Variables Formula to show that I is equal to the integral of $f(u, v) = v$ over $\mathcal{R}$ and evaluate.

SOLUTION

(a) The domain $\mathcal{D}$ is defined by the inequalities

$$\mathcal{D}: 2 \le xy \le 4,\ 0 \le x - y \le 3,\ x \ge 0,\ y \ge 0$$

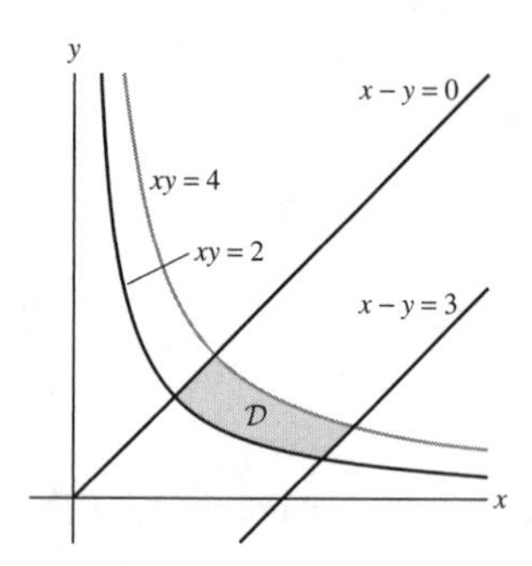

Since $u = xy$ and $v = x - y$, the image of $\mathcal{D}$ under this mapping is the rectangle defined by

$$\mathcal{D}_0 : 2 \le u \le 4,\ 0 \le v \le 3$$

That is, $\mathcal{D}_0 = [2, 4] \times [0, 3]$.

(b) We compute the Jacobian $\frac{\partial(u,v)}{\partial(x,y)}$ and then use the formula for the Jacobian of the inverse mapping to compute $\frac{\partial(x,y)}{\partial(u,v)}$. Since $u = xy$ and $v = x - y$, we have

$$\frac{\partial(u, v)}{\partial(x, y)} = \begin{vmatrix} \frac{\partial u}{\partial x} & \frac{\partial u}{\partial y} \\ \frac{\partial v}{\partial x} & \frac{\partial v}{\partial y} \end{vmatrix} = \begin{vmatrix} y & x \\ 1 & -1 \end{vmatrix} = -y - x = -(x + y)$$

Therefore,

$$\frac{\partial(x, y)}{\partial(u, v)} = \left(\frac{\partial(u, v)}{\partial(x, y)}\right)^{-1} = -\frac{1}{x + y}$$

(c) In $\mathcal{D}$, $x \ge 0$ and $y \ge 0$, hence $\left|\frac{\partial(x,y)}{\partial(u,v)}\right| = \frac{1}{x+y}$. Using the change of variable formula gives:

$$I = \iint_{\mathcal{D}_0} (x^2 - y^2) \cdot \frac{1}{x + y}\,du\,dv = \iint_{\mathcal{D}_0} (x - y)\,du\,dv = \int_0^3 \int_2^4 v\,du\,dv$$
$$= \left(\int_0^3 v\,dv\right)\left(\int_2^4 du\right) = \left(\frac{v^2}{2}\bigg|_0^3\right)\left(u\bigg|_2^4\right) = \frac{9}{2} \cdot 2 = 9$$

43. Derive formula (10) in Section 16.4 for integration in spherical coordinates from the general Change of Variables Formula.

SOLUTION The spherical coordinates are

$$x = \rho\cos\theta\sin\phi, \quad y = \rho\sin\theta\sin\phi, \quad z = \rho\cos\phi$$

Suppose that a region $\mathcal{W}$ in the (x, y, z)-plane is the image of a region $\mathcal{W}_0$ in the (θ, ϕ, ρ)-space defined by:

$$\mathcal{W}_0 : \theta_1 \le \theta \le \theta_2,\ \phi_1 \le \phi \le \phi_2,\quad \rho_1(\theta, \phi) \le \rho \le \rho_2(\theta, \phi) \tag{1}$$

Then, by the Change of Variables Formula, we have

$$\iiint_{\mathcal{W}} f(x, y, z) = \iiint_{\mathcal{W}_0} f(\rho\cos\theta\sin\phi, \rho\sin\theta\sin\phi, \rho\cos\phi) = \left|\frac{\partial(x, y, z)}{\partial(\theta, \phi, \rho)}\right| d\rho\,d\phi\,d\theta \tag{2}$$

We compute the Jacobian:

$$\frac{\partial(x,y,z)}{\partial(\theta,\phi,\rho)} = \begin{vmatrix} \frac{\partial x}{\partial \theta} & \frac{\partial x}{\partial \phi} & \frac{\partial x}{\partial \rho} \\ \frac{\partial y}{\partial \theta} & \frac{\partial y}{\partial \phi} & \frac{\partial y}{\partial \rho} \\ \frac{\partial z}{\partial \theta} & \frac{\partial z}{\partial \phi} & \frac{\partial z}{\partial \rho} \end{vmatrix} = \begin{vmatrix} -\rho\sin\theta\sin\phi & \rho\cos\theta\cos\phi & \cos\theta\sin\phi \\ \rho\cos\theta\sin\phi & \rho\sin\theta\cos\phi & \sin\theta\sin\phi \\ 0 & -\rho\sin\phi & \cos\phi \end{vmatrix}$$

$$= -\rho\sin\theta\sin\phi \begin{vmatrix} \rho\sin\theta\cos\phi & \sin\theta\sin\phi \\ -\rho\sin\phi & \cos\phi \end{vmatrix} - \rho\cos\theta\cos\phi \begin{vmatrix} \rho\cos\theta\sin\phi & \sin\theta\sin\phi \\ 0 & \cos\phi \end{vmatrix}$$

$$+ \cos\theta\sin\phi \begin{vmatrix} \rho\cos\theta\sin\phi & \rho\sin\theta\cos\phi \\ 0 & -\rho\sin\phi \end{vmatrix}$$

$$= -\rho\sin\theta\sin\phi(\rho\sin\theta\cos^2\phi + \rho\sin\theta\sin^2\phi) - \rho\cos\theta\cos\phi(\rho\cos\theta\cos\phi\sin\phi - 0)$$

$$+ \cos\theta\sin\phi(-\rho^2\cos\theta\sin^2\phi - 0)$$

$$= -\rho^2\sin^2\theta\sin\phi(\cos^2\phi + \sin^2\phi) - \rho^2\cos^2\theta\cos^2\phi\sin\phi - \rho^2\cos^2\theta\sin^3\phi$$

$$= -\rho^2\sin^2\theta\sin\phi - \rho^2\cos^2\theta\cos^2\phi\sin\phi - \rho^2\cos^2\theta\sin^3\phi$$

$$= -\rho^2\sin\phi(\sin^2\theta + \cos^2\theta\cos^2\phi + \cos^2\theta\sin^2\phi)$$

$$= -\rho^2\sin\phi\left(\sin^2\theta + \cos^2\theta(\cos^2\phi + \sin^2\phi)\right)$$

$$= -\rho^2\sin\phi(\sin^2\theta + \cos^2\theta) = -\rho^2\sin\phi$$

Since $0 \le \phi \le \pi$, we have $\sin\phi \ge 0$. Therefore,

$$\left|\frac{\partial(x,y,z)}{\partial(\theta,\phi,\rho)}\right| = \rho^2\sin\phi \tag{3}$$

Combining (1), (2), and (3) gives

$$\iiint_{\mathcal{W}} f(x,y,z)\,dv = \int_{\theta_1}^{\theta_2}\int_{\phi_1}^{\phi_2}\int_{\rho_1(\theta,\phi)}^{\rho_2(\theta,\phi)} f(\rho\cos\theta\sin\phi, \rho\sin\theta\sin\phi, \rho\cos\phi)\rho^2\sin\phi\,d\rho\,d\phi\,d\theta$$

Further Insights and Challenges

45. Use the map

$$x = \frac{\sin u}{\cos v}, \qquad y = \frac{\sin v}{\cos u}$$

to evaluate the integral

$$\int_0^1\int_0^1 \frac{dx\,dy}{1 - x^2y^2}$$

This is an improper integral since the integrand is infinite if $x = \pm 1$, $y = \pm 1$, but the Change of Variables shows that the result is finite.

SOLUTION We express the function $f(x,y) = \frac{1}{1-x^2y^2}$ in terms of the new variables u and v:

$$1 - x^2y^2 = 1 - \frac{\sin^2 u\,\sin^2 v}{\cos^2 v\,\cos^2 u} = 1 - \left(\frac{\sin u}{\cos u}\right)^2 \cdot \left(\frac{\sin v}{\cos v}\right)^2 = 1 - \tan^2 u\tan^2 v$$

Hence,

$$f(x,y) = \frac{1}{1 - \tan^2 u\tan^2 v}$$

We compute the Jacobian of the mapping:

$$\frac{\partial(x,y)}{\partial(u,v)} = \begin{vmatrix} \frac{\partial x}{\partial u} & \frac{\partial x}{\partial v} \\ \frac{\partial y}{\partial u} & \frac{\partial y}{\partial v} \end{vmatrix} = \begin{vmatrix} \frac{\cos u}{\cos v} & \frac{\sin u\sin v}{\cos^2 v} \\ \frac{\sin v\sin u}{\cos^2 u} & \frac{\cos v}{\cos u} \end{vmatrix} = \frac{\cos u}{\cos v}\cdot\frac{\cos v}{\cos u} - \frac{\sin u\sin v}{\cos^2 v}\cdot\frac{\sin v\sin u}{\cos^2 u}$$

$$= 1 - \frac{\sin^2 u}{\cos^2 u} \cdot \frac{\sin^2 v}{\cos^2 v} = 1 - \tan^2 u \tan^2 v$$

Now, since $0 \le x \le 1$ and $0 \le y \le 1$, we have $0 \le \frac{\sin u}{\cos v} \cdot \frac{\sin v}{\cos u} \le 1$ or $0 \le \tan u \tan v \le 1$. Therefore, $0 \le \tan^2 u \tan^2 v \le 1$, hence

$$\left| \frac{\partial(x, y)}{\partial(u, v)} \right| = 1 - \tan^2 u \tan^2 v$$

We now identify a domain $\mathcal{D}_0$ in the uv-plane that is mapped by Φ onto $\mathcal{D}$ and Φ is one-to-one on $\mathcal{D}_0$.

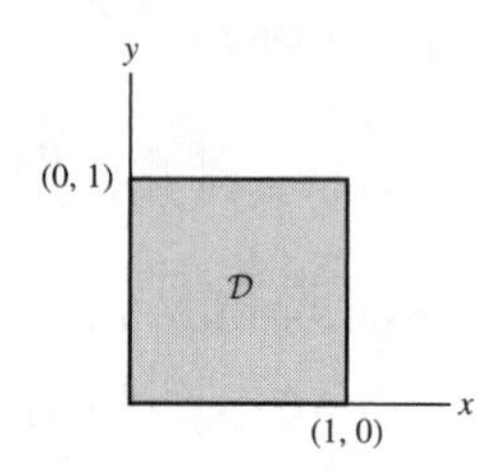

We examine each segment on the boundary of $\mathcal{D}$ separately.

$y = 0$:

$$\frac{\sin v}{\cos u} = 0 \quad \Rightarrow \quad \sin v = 0 \quad \Rightarrow \quad v = \pi k$$

$x = 0$:

$$\frac{\sin u}{\cos v} = 0 \quad \Rightarrow \quad \sin u = 0 \quad \Rightarrow \quad u = \pi k$$

$y = 1$:

$$\frac{\sin v}{\cos u} = 1 \quad \Rightarrow \quad \sin v = \cos u \quad \Rightarrow \quad v + u = \frac{\pi}{2} + 2\pi k \quad \text{or} \quad v - u = \frac{\pi}{2} + 2\pi k \tag{1}$$

$x = 1$:

$$\frac{\sin u}{\cos v} = 1 \quad \Rightarrow \quad \sin u = \cos v \quad \Rightarrow \quad v + u = \frac{\pi}{2} + 2\pi k \quad \text{or} \quad u - v = \frac{\pi}{2} + 2\pi k \tag{2}$$

One of the possible regions $\mathcal{D}_0$ is obtained by choosing $k = 0$ in all solutions. We get

$$v = 0, \quad u = 0, \quad \left(v + u = \frac{\pi}{2} \text{ or } v - u = \frac{\pi}{2} \right), \quad \left(u + v = \frac{\pi}{2} \text{ or } u - v = \frac{\pi}{2} \right)$$

The corresponding regions are:

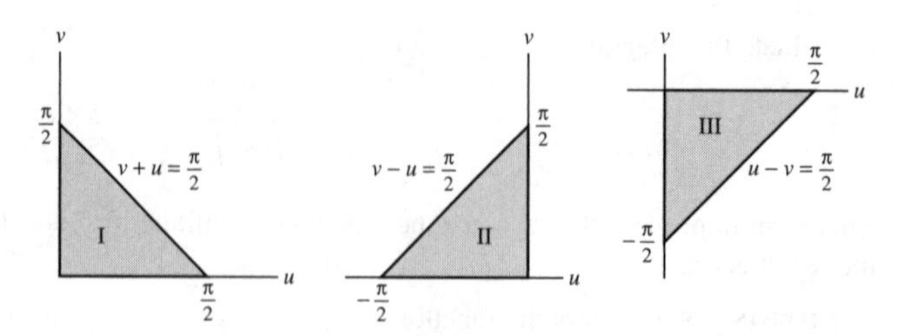

In II, $x = \frac{\sin u}{\cos v} < 0$ and in III $y = \frac{\sin v}{\cos u} < 0$, therefore these regions are not mapped to the unit square in the xy-plane. The appropriate region is I.

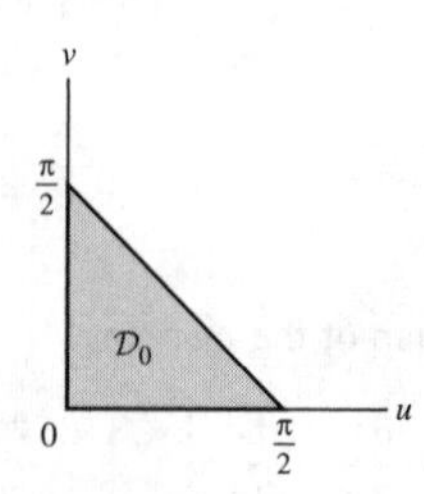

We now use the Change of Variables Formula and the result obtained previously to obtain the following integral:

$$\int_0^1 \int_0^1 \frac{dx\,dy}{1 - x^2y^2} = \iint_{\mathcal{D}_0} \frac{1}{1 - \tan^2 u \tan^2 v} \cdot (1 - \tan^2 u \tan^2 v)\,du\,dv$$

$$= \iint_{\mathcal{D}_0} 1\,du\,dv = \text{Area}(\mathcal{D}_0) = \frac{\frac{\pi}{2} \cdot \frac{\pi}{2}}{2} = \frac{\pi^2}{8}$$

47. Let Φ be a linear map. Prove Eq. (4) in the following steps.

(a) For any set $\mathcal{D}$ in the uv-plane and any vector $\mathbf{u}$, let $\mathcal{D} + \mathbf{u}$ be the set obtained by translating all points in $\mathcal{D}$ by $\mathbf{u}$. By linearity, Φ maps $\mathcal{D} + \mathbf{u}$ to the translate $\Phi(\mathcal{D}) + \Phi(\mathbf{u})$ [Figure 20(C)]. Therefore, if Eq. (4) holds for $\mathcal{D}$, it also holds for $\mathcal{D} + \mathbf{u}$.

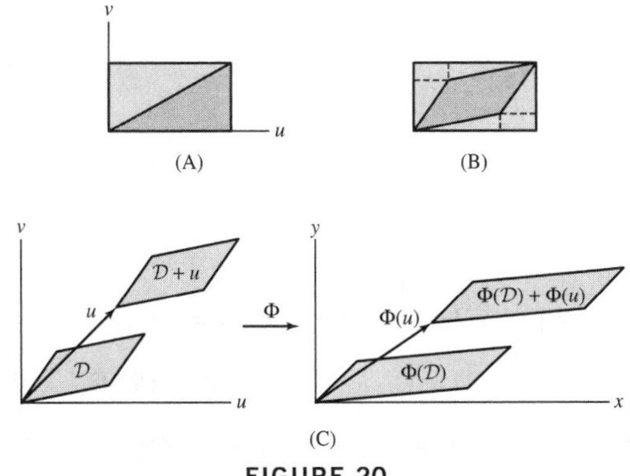

FIGURE 20

(b) In the text, we verified Eq. (4) for the unit rectangle. Use linearity to show that Eq. (4) also holds for all rectangles with vertex at the origin and sides parallel to the axes. Then argue that it also holds for each triangular half of such a rectangle, as in Figure 20(A).

(c) Figure 20(B) shows that the area of a parallelogram is a difference of the areas of rectangles and triangles covered by steps (a) and (b). Use this to prove Eq. (4) for arbitrary parallelograms.

SOLUTION We must show that if Φ is a linear map, then

$$\text{Area}\,(\Phi(\mathcal{D})) = |\text{Jac}\,(\Phi)\,|\text{Area}(\mathcal{D}) \tag{1}$$

(a) For any vector $\mathbf{v} \in \mathcal{D}$, $\mathbf{v} + \mathbf{u}$ is in $\mathcal{D} + \mathbf{u}$. We show that $\Phi(\mathbf{v} + \mathbf{u})$ is in $\Phi(\mathcal{D}) + \Phi(\mathbf{u})$. By linearity, we have

$$\Phi(\mathbf{v} + \mathbf{u}) = \Phi(\mathbf{v}) + \Phi(\mathbf{u})$$

Since $\mathbf{v} \in \mathcal{D}$, $\Phi(\mathbf{v}) \in \Phi(\mathcal{D})$, hence $\Phi(\mathbf{v}) + \Phi(\mathbf{u}) \in \Phi(\mathcal{D}) + \Phi(\mathbf{u})$. Therefore, Φ maps $\mathcal{D} + \mathbf{u}$ to $\Phi(\mathcal{D}) + \Phi(\mathbf{u})$.

(b) Let $\mathcal{D}$ be the rectangle $\mathcal{D} = [0, a] \times [0, b]$ in the uv-plane, and $\Phi(\mathcal{D})$ be the image of $\mathcal{D}$ under the linear mapping Φ:

$$\Phi(u, v) = (Au + Cv, Bu + Dv)$$

Suppose that $a > 0$, $b > 0$, and $AD > BC$. For other cases, the proof is similar. To determine $\Phi(\mathcal{D})$, we compute the images of the vertices $(0, b)$ and $(a, 0)$ of $\mathcal{D}$:

$$\Phi(0, b) = (A \cdot 0 + Cb, B \cdot 0 + Db) = (Cb, Db)$$

$$\Phi(a, 0) = (Aa + C \cdot 0, B \cdot a + D \cdot 0) = (Aa, Ba)$$

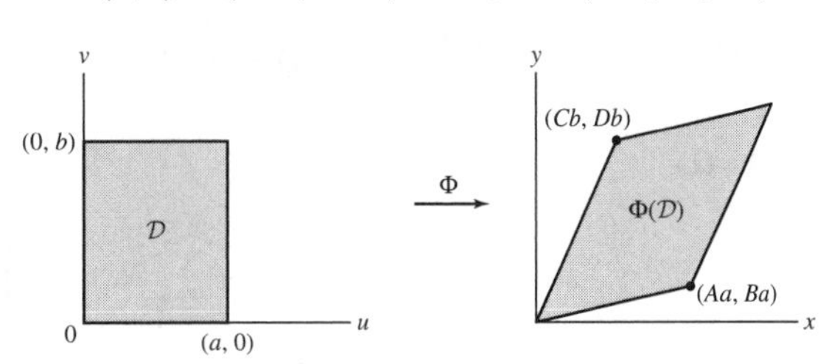

Therefore, $\Phi(\mathcal{D})$ is the parallelogram spanned by the vectors $\langle Cb, Db \rangle$ and $\langle Aa, Ba \rangle$.

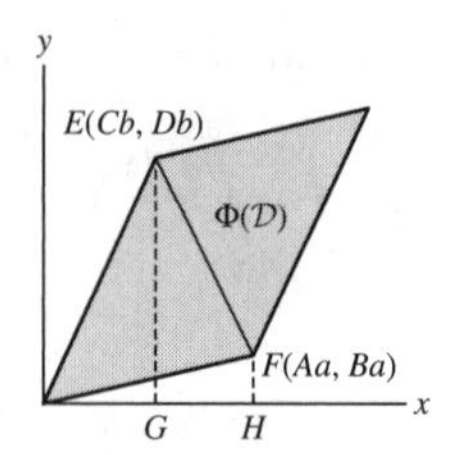

The area of the parallelogram is

$$2\left(S_{\text{tri}(OEG)} + S_{\text{tra}(GEFH)} - S_{\text{tri}(OFH)}\right) = 2\left(\frac{EG \cdot OG}{2} + \frac{(EG + FH)GH}{2} - \frac{FH \cdot OH}{2}\right)$$

$$= 2\left(\frac{Db \cdot Cb}{2} + \frac{(Db + Ba)(Aa - Cb)}{2} - \frac{Ba \cdot Aa}{2}\right)$$

$$= DCb^2 + ADab - DCb^2 + ABa^2 - BCab - ABa^2$$

$$= ab(AD - BC)$$

That is,

$$\text{Area}\,(\Phi(\mathcal{D})) = (AD - BC)ab$$

Since $|\text{Jac}(\Phi)| = AD - BC$ and $\text{Area}(\mathcal{D}) = ab$, we get

$$\text{Area}\,(\Phi(\mathcal{D})) = |\text{Jac}\,(\Phi)\,|\text{Area}(\mathcal{D}).$$

Now, let $\mathcal{D}_1$ be a triangular half of the parallelogram $\mathcal{D}$. Then, $\Phi(\mathcal{D}_1)$ is a triangular half of the parallelogram $\Phi(\mathcal{D})$. Using the result above, we have

$$\text{Area}\Phi(\mathcal{D}_1) = \frac{1}{2}\text{Area}\Phi(\mathcal{D}) = \frac{1}{2} \cdot |\text{Jac}(\Phi)|\text{Area}(\mathcal{D}) = |\text{Jac}(\Phi)|\frac{\text{Area}(\mathcal{D})}{2} = |\text{Jac}(\Phi)|\text{Area}(\mathcal{D}_1)$$

(c) By part (a), if we show that Eq. (4) holds for parallelograms with vertex at the origin, it holds for all other parallelograms (since each parallelogram can be translated to a parallelogram with a vertex at the origin). We consider a parallelogram with a vertex at the origin, and inscribe it in a rectangle $\mathcal{D}^*$; as shown in the figure:

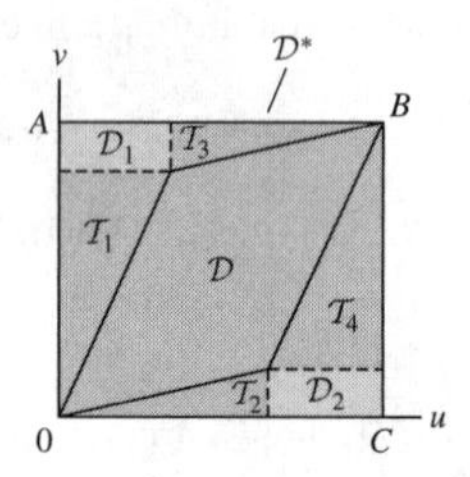

We denote the triangles and rectangles as shown in the figure. By parts (a) and (b), it follows that Eq. (4) holds for each one of the rectangles and triangles. That is,

$$\begin{aligned} \text{Area}\Phi(\mathcal{D}^*) &= |\text{Jac}(\Phi)|\text{Area}(\mathcal{D}^*) \\ \text{Area}\Phi(\mathcal{D}_i) &= |\text{Jac}(\Phi)|\text{Area}(\mathcal{D}_i), \quad i = 1, 2 \\ \text{Area}\Phi(T_i) &= |\text{Jac}(\Phi)|\text{Area}(T_i), \quad i = 1, 2, 3, 4 \end{aligned} \tag{2}$$

Since

$$\sum_{i=1}^{2} \text{Area}\Phi(\mathcal{D}_i) + \sum_{i=1}^{4} \text{Area}\Phi(T_i) + \text{Area}\Phi(\mathcal{D}) = \text{Area}\Phi(\mathcal{D}^*)$$

we have by (1),

$$|\text{Jac}(\Phi)|\left(\sum_{i=1}^{2} \text{Area}\,(\mathcal{D}_i) + \sum_{i=1}^{4} \text{Area}(T_i)\right) + \text{Area}\Phi(\mathcal{D}) = |\text{Jac}\Phi|\text{Area}(\mathcal{D}^*)$$

Translating sides gives

$$|\text{Jac}(\Phi)|\left(\text{Area}(\mathcal{D}^*) - \sum_{i=1}^{2} \text{Area}(\mathcal{D}_i) - \sum_{i=1}^{4} \text{Area}(T_i)\right) = \text{Area}\Phi(\mathcal{D})$$

But the difference in the brackets on the left-hand side is the area of the parallelogram $\mathcal{D}$. Therefore, we get

$$|\text{Jac}(\Phi)|\text{Area}(\mathcal{D}) = \text{Area}\Phi(\mathcal{D})$$

We thus showed that Eq. (4) holds for $\mathcal{D}$.

49. Let $\Phi_1 : \mathcal{D}_1 \to \mathcal{D}_2$ and $\Phi_2 : \mathcal{D}_2 \to \mathcal{D}_3$ be C^1 maps, and let $\Phi_2 \circ \Phi_1 : \mathcal{D}_1 \to \mathcal{D}_3$ be the composite map. Use the Multivariable Chain Rule and Exercise 48 to show that

$$\text{Jac}(\Phi_2 \circ \Phi_1) = \text{Jac}(\Phi_2)\text{Jac}(\Phi_1)$$

SOLUTION

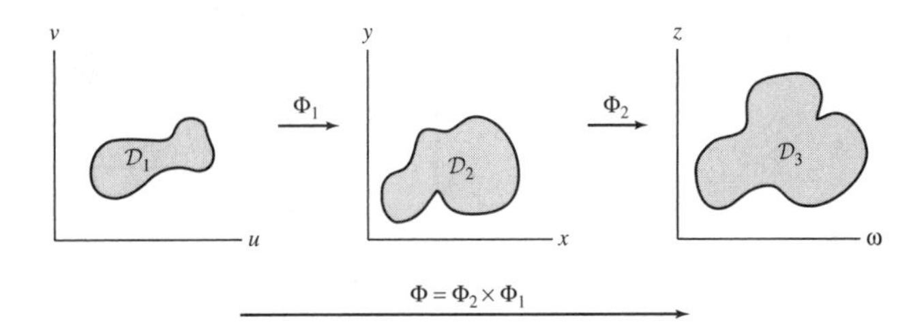

Let $\Phi = \Phi_2 \circ \Phi_1$. We have

$$\text{Jac}(\Phi_1) = \begin{vmatrix} \frac{\partial x}{\partial u} & \frac{\partial x}{\partial v} \\ \frac{\partial y}{\partial u} & \frac{\partial y}{\partial v} \end{vmatrix} \tag{1}$$

$$\text{Jac}(\Phi_2) = \begin{vmatrix} \frac{\partial \omega}{\partial x} & \frac{\partial \omega}{\partial y} \\ \frac{\partial z}{\partial x} & \frac{\partial z}{\partial y} \end{vmatrix} \tag{2}$$

$$\text{Jac}(\Phi) = \begin{vmatrix} \frac{\partial \omega}{\partial u} & \frac{\partial \omega}{\partial v} \\ \frac{\partial z}{\partial u} & \frac{\partial z}{\partial v} \end{vmatrix} \tag{3}$$

We use the multivariable Chain Rule to write

$$\frac{\partial \omega}{\partial u} = \frac{\partial \omega}{\partial x}\frac{\partial x}{\partial u} + \frac{\partial \omega}{\partial y}\frac{\partial y}{\partial u}, \qquad \frac{\partial z}{\partial u} = \frac{\partial z}{\partial x}\frac{\partial x}{\partial u} + \frac{\partial z}{\partial y}\frac{\partial y}{\partial u}$$

$$\frac{\partial \omega}{\partial v} = \frac{\partial \omega}{\partial x}\frac{\partial x}{\partial v} + \frac{\partial \omega}{\partial y}\frac{\partial y}{\partial v}, \qquad \frac{\partial z}{\partial v} = \frac{\partial z}{\partial x}\frac{\partial x}{\partial v} + \frac{\partial z}{\partial y}\frac{\partial y}{\partial v}$$

Substituting in (3) we obtain

$$\text{Jac}(\Phi) = \begin{vmatrix} \frac{\partial \omega}{\partial x}\frac{\partial x}{\partial u} + \frac{\partial \omega}{\partial y}\frac{\partial y}{\partial u} & \frac{\partial \omega}{\partial x}\frac{\partial x}{\partial v} + \frac{\partial \omega}{\partial y}\frac{\partial y}{\partial v} \\ \frac{\partial z}{\partial x}\frac{\partial x}{\partial u} + \frac{\partial z}{\partial y}\frac{\partial y}{\partial u} & \frac{\partial z}{\partial x}\frac{\partial x}{\partial v} + \frac{\partial z}{\partial y}\frac{\partial y}{\partial v} \end{vmatrix}$$

We now use the definition of the product of two matrices, given in Exercise 48, equalities (1) and (2), and the equality proved in Exercise 48, to write

$$\text{Jac}(\Phi) = \left| \begin{pmatrix} \frac{\partial \omega}{\partial x} & \frac{\partial \omega}{\partial y} \\ \frac{\partial z}{\partial x} & \frac{\partial z}{\partial y} \end{pmatrix} \begin{pmatrix} \frac{\partial x}{\partial u} & \frac{\partial x}{\partial v} \\ \frac{\partial y}{\partial u} & \frac{\partial y}{\partial v} \end{pmatrix} \right| = \begin{vmatrix} \frac{\partial \omega}{\partial x} & \frac{\partial \omega}{\partial y} \\ \frac{\partial z}{\partial x} & \frac{\partial z}{\partial y} \end{vmatrix} \begin{vmatrix} \frac{\partial x}{\partial u} & \frac{\partial x}{\partial v} \\ \frac{\partial y}{\partial u} & \frac{\partial y}{\partial v} \end{vmatrix} = \text{Jac}(\Phi_2)\text{Jac}(\Phi_1)$$

That is,

$$\text{Jac}(\Phi_2 \circ \Phi_1) = \text{Jac}(\Phi_2)\text{Jac}(\Phi_1)$$

51. Let $(\overline{x}, \overline{y})$ be the centroid of a domain $\mathcal{D}$. For $\lambda > 0$, let $\lambda\mathcal{D}$ be the **dilate** of $\mathcal{D}$, defined by

$$\lambda\mathcal{D} = \{(\lambda x, \lambda y) : (x, y) \in \mathcal{D}\}$$

Use the Change of Variables Formula to prove that the centroid of $\lambda\mathcal{D}$ is $(\lambda\overline{x}, \lambda\overline{y})$.

SOLUTION The centroid of $\mathcal{D}$ has the following coordinates, where $S = \text{Area}(\mathcal{D})$:

$$\overline{x} = \frac{1}{S}\iint_{\mathcal{D}} x\,dx\,dy, \quad \overline{y} = \frac{1}{S}\iint_{\mathcal{D}} y\,dx\,dy \tag{1}$$

The centroid of $\lambda\mathcal{D}$ is the following point:

$$\overline{u} = \frac{1}{\text{Area}(\lambda\mathcal{D})}\iint_{\lambda\mathcal{D}} u\,du\,dv, \quad \overline{v} = \frac{1}{\text{Area}(\lambda\mathcal{D})}\iint_{\lambda\mathcal{D}} v\,du\,dv \tag{2}$$

We compute the double integrals in (2) using change of variables with the following mapping:

$$u = \lambda x, \quad v = \lambda y$$

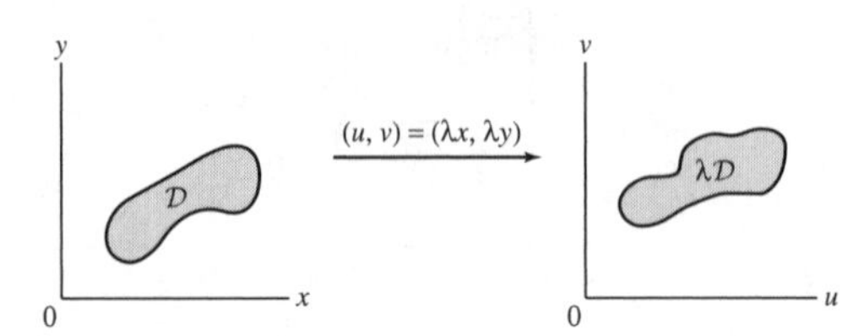

Therefore $(u, v) \in \lambda\mathcal{D}$ if and only if $(x, y) \in \mathcal{D}$, hence the image of $\lambda\mathcal{D}$ under this mapping is the domain $\mathcal{D}$ in the xy-plane. The Jacobian of the linear mapping $(u, v) = (\lambda x + 0y, 0x + \lambda y)$ is

$$\frac{\partial(u, v)}{\partial(x, y)} = \begin{vmatrix} \lambda & 0 \\ 0 & \lambda \end{vmatrix} = \lambda^2$$

We compute the integrals:

$$\iint_{\lambda\mathcal{D}} u\,du\,dv = \iint_{\mathcal{D}} \lambda x \cdot \lambda^2\,dx\,dy = \lambda^3 \iint_{\mathcal{D}} x\,dx\,dy = \lambda^3 S\overline{x}$$

$$\iint_{\lambda\mathcal{D}} v\,du\,dv = \iint_{\mathcal{D}} \lambda y \cdot \lambda^2\,dx\,dy = \lambda^3 \iint_{\mathcal{D}} y\,dx\,dy = \lambda^3 S\overline{y}$$

Substituting in (2) gives

$$\overline{u} = \frac{\lambda^3 S}{\text{Area}(\lambda\mathcal{D})}\overline{x}, \quad \overline{v} = \frac{\lambda^3 S}{\text{Area}(\lambda\mathcal{D})}\overline{y} \tag{3}$$

We now compute the area of $\lambda\mathcal{D}$ using the same mapping:

$$\text{Area}(\lambda\mathcal{D}) = \iint_{\lambda\mathcal{D}} 1\,du\,dv = \iint_{\mathcal{D}} \lambda^2\,dx\,dy = \lambda^2 \iint_{\mathcal{D}} dx\,dy = \lambda^2\text{Area}(\mathcal{D}) = \lambda^2 S$$

Substituting in (3), we obtain the following centroid of $\lambda\mathcal{D}$:

$$\overline{u} = \frac{\lambda^3 S\overline{x}}{\lambda^2 S} = \lambda\overline{x}, \quad \overline{v} = \frac{\lambda^3 S\overline{y}}{\lambda^2 S} = \lambda\overline{y}$$

The centroid of $\lambda\mathcal{D}$ is $(\lambda\overline{x}, \lambda\overline{y})$.

CHAPTER REVIEW EXERCISES

1. Calculate the Riemann sum $S_{3,4}$ for $\int_1^2 \int_2^3 x^2 y\,dx\,dy$ using two choices of sample points:

(a) Lower-left vertex

(b) Midpoint of rectangle

Then calculate the exact value of the double integral.

SOLUTION

(a) The rectangle $[2, 3] \times [1, 2]$ is divided into 3×4 subrectangles as shown in the figure:

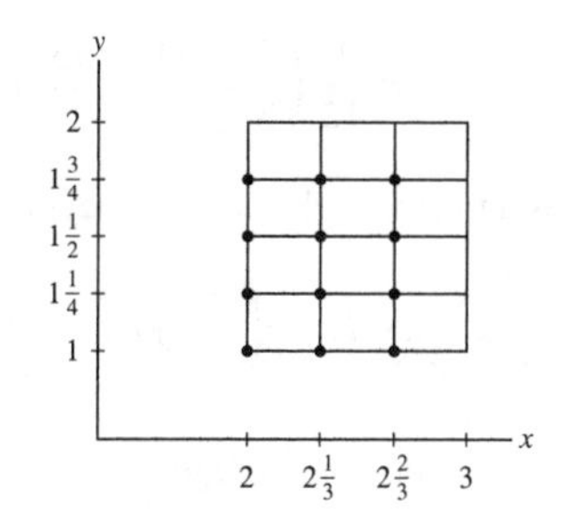

The lower-left vertices of the subrectangles are

$$P_{11} = (2, 1)\ P_{21} = \left(2\frac{1}{3}, 1\right)\ P_{31} = \left(2\frac{2}{3}, 1\right)$$

$$P_{12} = \left(2, 1\frac{1}{4}\right)\ P_{22} = \left(2\frac{1}{3}, 1\frac{1}{4}\right)\ P_{32} = \left(2\frac{2}{3}, 1\frac{1}{4}\right)$$

$$P_{13} = \left(2, 1\frac{1}{2}\right)\ P_{23} = \left(2\frac{1}{3}, 1\frac{1}{2}\right)\ P_{33} = \left(2\frac{2}{3}, 1\frac{1}{2}\right)$$

$$P_{14} = \left(2, 1\frac{3}{4}\right)\ P_{24} = \left(2\frac{1}{3}, 1\frac{3}{4}\right)\ P_{34} = \left(2\frac{2}{3}, 1\frac{3}{4}\right)$$

Also $\Delta x = \frac{3-2}{3} = \frac{1}{3}$, $\Delta y = \frac{2-1}{4} = \frac{1}{4}$, hence $\Delta A = \frac{1}{3} \cdot \frac{1}{4} = \frac{1}{12}$. The Riemann sum $S_{3,4}$ is the following sum:

$$S_{3,4} = \frac{1}{12}\left(2^2 \cdot 1 + \frac{49}{9} \cdot 1 + \frac{64}{9} \cdot 1 + 2^2 \cdot \frac{5}{4} + \frac{49}{9} \cdot \frac{5}{4} + \frac{64}{9} \cdot \frac{5}{4}\right.$$

$$\left. +2^2 \cdot \frac{3}{2} + \frac{49}{9} \cdot \frac{3}{2} + \frac{64}{9} \cdot \frac{3}{2} + 2^2 \cdot \frac{7}{4} + \frac{49}{9} \cdot \frac{7}{4} + \frac{64}{9} \cdot \frac{7}{4}\right)$$

$$= \frac{1}{12}\left(\left(1 + \frac{5}{4} + \frac{3}{2} + \frac{7}{4}\right)\left(4 + \frac{49}{9} + \frac{64}{9}\right)\right) = \frac{1639}{216} \approx 7.588$$

(b) The midpoints of the subrectangles are

$$P_{11} = \left(2\frac{1}{6}, 1\frac{1}{8}\right)\ P_{21} = \left(2\frac{1}{2}, 1\frac{1}{8}\right)\ P_{31} = \left(2\frac{5}{6}, 1\frac{1}{8}\right)$$

$$P_{12} = \left(2\frac{1}{6}, 1\frac{3}{8}\right)\ P_{22} = \left(2\frac{1}{2}, 1\frac{3}{8}\right)\ P_{32} = \left(2\frac{5}{6}, 1\frac{3}{8}\right)$$

$$P_{13} = \left(2\frac{1}{6}, 1\frac{5}{8}\right)\ P_{23} = \left(2\frac{1}{2}, 1\frac{5}{8}\right)\ P_{33} = \left(2\frac{5}{6}, 1\frac{5}{8}\right)$$

$$P_{14} = \left(2\frac{1}{6}, 1\frac{7}{8}\right)\ P_{24} = \left(2\frac{1}{2}, 1\frac{7}{8}\right)\ P_{34} = \left(2\frac{5}{6}, 1\frac{7}{8}\right)$$

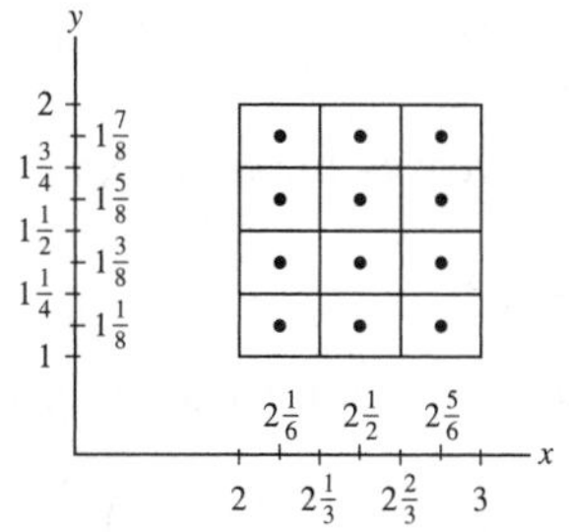

Also $\Delta x = \frac{1}{3}$, $\Delta y = \frac{1}{4}$, hence $\Delta A = \frac{1}{3} \cdot \frac{1}{4} = \frac{1}{12}$. The Riemann sum $S_{3,4}$ is

$$S_{3,4} = \frac{1}{12}\left(\left(\frac{13}{6}\right)^2 \cdot \frac{9}{8} + \left(\frac{13}{6}\right)^2 \cdot \frac{11}{8} + \left(\frac{13}{6}\right)^2 \cdot \frac{13}{8} + \left(\frac{13}{6}\right)^2 \cdot \frac{15}{8}\right.$$

$$+\left(\frac{5}{2}\right)^2 \cdot \frac{9}{8} + \left(\frac{5}{2}\right)^2 \cdot \frac{11}{8} + \left(\frac{5}{2}\right)^2 \cdot \frac{13}{8} + \left(\frac{5}{2}\right)^2 \cdot \frac{15}{8} + \left(\frac{17}{6}\right)^2 \cdot \frac{9}{8}$$

$$\left. +\left(\frac{17}{6}\right)^2 \cdot \frac{11}{8} + \left(\frac{17}{6}\right)^2 \cdot \frac{13}{8} + \left(\frac{17}{6}\right)^2 \cdot \frac{15}{8}\right)$$

$$= \frac{1}{12}\left(\frac{9}{8}+\frac{11}{8}+\frac{13}{8}+\frac{15}{8}\right)\left(\left(\frac{13}{6}\right)^2+\left(\frac{5}{2}\right)^2+\left(\frac{17}{6}\right)^2\right) \approx 9.486$$

We compute the exact value of the double integral, using an iterated integral of a product function. We get

$$\int_1^2\int_2^3 x^2y\,dx\,dy = \left(\int_1^2 y\,dy\right)\left(\int_2^3 x^2\,dx\right) = \left(\frac{y^2}{2}\Big|_1^2\right)\left(\frac{x^3}{3}\Big|_2^3\right) = \frac{4-1}{2}\cdot\frac{27-8}{3} = 9.5$$

3. Let $\mathcal{D}$ be the shaded domain in Figure 1.

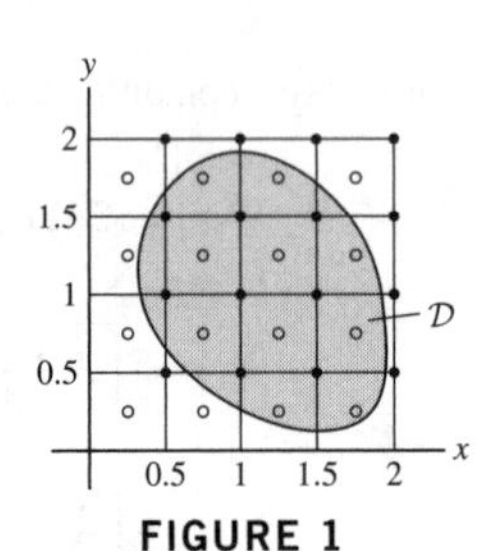

FIGURE 1

Estimate $\iint_{\mathcal{D}} xy\,dA$ by the Riemann sum whose sample points are the midpoints of the squares in the grid.

SOLUTION The subrectangles have sides of length $\Delta x = \Delta y = 0.5$ and area $\Delta A = 0.5^2 = 0.25$. Of sixteen sample points only ten lie in $\mathcal{D}$. The sample points that lie in $\mathcal{D}$ are

$$(0.75, 0.75),\ (0.75, 1.25),\ (0.75, 1.75),\ (1.25, 0.25),\ (1.25, 0.75),$$
$$(1.25, 1.25),\ (1.25, 1.75),\ (1.75, 0.25),\ (1.75, 0.75),\ (1.75, 1.25)$$

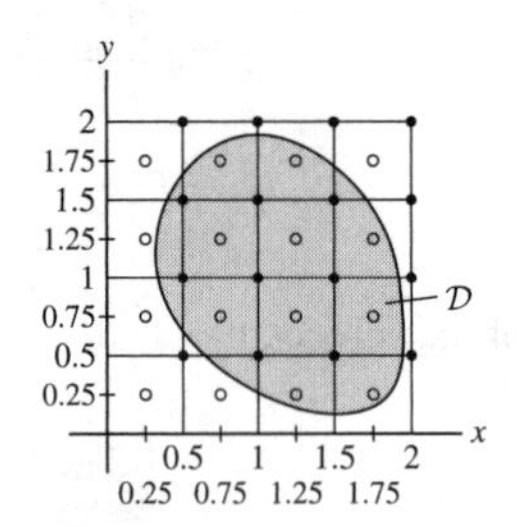

The Riemann sum S_{44} is thus

$$\begin{aligned} S_{44} &= 0.25\,(f(0.75, 0.75) + f(0.75, 1.25) + f(0.75, 1.75) + f(1.25, 0.25) + f(1.25, 0.75)\\ &\quad + f(1.25, 1.25) + f(1.25, 1.75) + f(1.75, 0.25) + f(1.75, 0.75) + f(1.75, 1.25))\\ &= 0.25\left(0.75^2 + 0.75\cdot 1.25 + 0.75\cdot 1.75 + 1.25\cdot 0.25 + 1.25\cdot 0.75 + 1.25^2\right.\\ &\quad \left. + 1.25\cdot 1.75 + 1.75\cdot 0.25 + 1.75\cdot 0.75 + 1.75\cdot 1.25\right)\\ &= 0.25\cdot 11.75 = 2.9375 \end{aligned}$$

In Exercises 5–8, evaluate the iterated integral.

5. $\displaystyle\int_0^2\int_3^5 y(x-y)\,dx\,dy$

SOLUTION First we evaluate the inner integral treating y as a constant:

$$\int_3^5 y(x-y)\,dx = y\left(\frac{x^2}{2} - yx\right)\Big|_{x=3}^5 = y\left(\left(\frac{25}{2}-5y\right)-\left(\frac{9}{2}-3y\right)\right) = y(8-2y) = 8y - 2y^2$$

Now we integrate this result with respect to y:

$$\int_0^2 (8y - 2y^2)\,dy = 4y^2 - \frac{2}{3}y^3\Big|_0^2 = 16 - \frac{16}{3} = \frac{32}{3}$$

Therefore,

$$\int_0^2\int_3^5 y(x-y)\,dx\,dy = \frac{32}{3}.$$

7. $\displaystyle\int_0^{\pi/3}\int_0^{\pi/6} \sin(x+y)\,dx\,dy$

SOLUTION We compute the inner integral treating y as a constant:

$$\int_0^{\pi/6} \sin(x+y)\,dx = -\cos(x+y)\Big|_{x=0}^{\pi/6} = -\cos\left(\frac{\pi}{6}+y\right)+\cos y = \cos y - \cos\left(y+\frac{\pi}{6}\right)$$

We now integrate the result with respect to y:

$$\int_0^{\pi/3}\int_0^{\pi/6} \sin(x+y)\,dx\,dy = \int_0^{\pi/3}\left(\cos y - \cos\left(y+\frac{\pi}{6}\right)\right)dy = \sin y - \sin\left(y+\frac{\pi}{6}\right)\Big|_0^{\pi/3}$$

$$= \sin\frac{\pi}{3} - \sin\left(\frac{\pi}{3}+\frac{\pi}{6}\right) - \left(\sin 0 - \sin\frac{\pi}{6}\right) = \frac{\sqrt{3}}{2} - 1 + \frac{1}{2} = \frac{\sqrt{3}-1}{2}$$

In Exercises 9–14, sketch the domain $\mathcal{D}$ and calculate $\iint_{\mathcal{D}} f(x,y)\,dA$.

9. $\mathcal{D} = \{0 \le x \le 4,\ 0 \le y \le x\}$, $\quad f(x,y) = \cos y$

SOLUTION The domain $\mathcal{D}$ is shown in the figure:

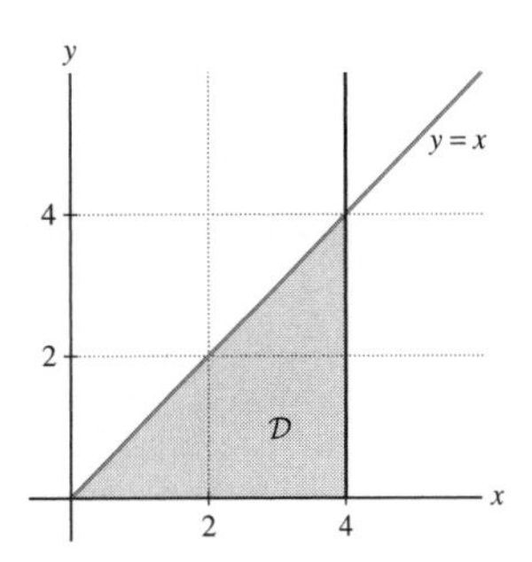

We compute the double integral, considering $\mathcal{D}$ as a vertically simple region. We describe $\mathcal{D}$ by the inequalities

$$0 \le x \le 4, \quad 0 \le y \le x.$$

We now write the double integral as an iterated integral and compute:

$$\iint_{\mathcal{D}} \cos y\,dA = \int_0^4\int_0^x \cos y\,dy\,dx = \int_0^4 \sin y\Big|_{y=0}^{x} dx$$

$$= \int_0^4 (\sin x - \sin 0)dx = \int_0^4 \sin x\,dx = -\cos x\Big|_0^4 = 1 - \cos 4$$

11. $\mathcal{D} = \{0 \le x \le 1,\ 1-x \le y \le 2-x\}$ $\quad f(x,y) = e^{x+2y}$

SOLUTION

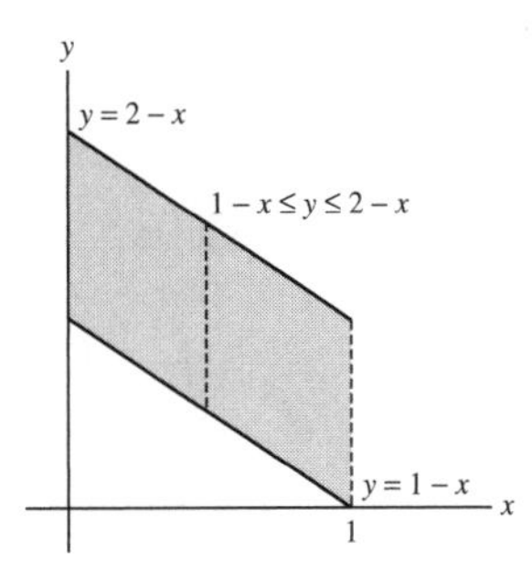

$\mathcal{D}$ is a vertically simple region, hence the double integral over $\mathcal{D}$ is the following iterated integral:

$$\iint_{\mathcal{D}} e^{x+2y}\,dA = \int_0^1\int_{1-x}^{2-x} e^{x+2y}\,dy\,dx = \int_0^1 \frac{1}{2}e^{x+2y}\Big|_{y=1-x}^{2-x} dx = \int_0^1\left(\frac{1}{2}e^{x+2(2-x)} - \frac{1}{2}e^{x+2(1-x)}\right)dx$$

$$= \int_0^1\left(\frac{1}{2}e^{4-x} - \frac{1}{2}e^{2-x}\right)dx = -\frac{1}{2}e^{4-x} + \frac{1}{2}e^{2-x}\Big|_0^1 = -\frac{1}{2}e^3 + \frac{1}{2}e + \frac{1}{2}e^4 - \frac{1}{2}e^2$$

$$= \frac{1}{2}e(e^3 - e^2 - e + 1) = \frac{1}{2}e(e+1)(e-1)^2$$

13. $\mathcal{D} = \{0 \le y \le 1,\ 0.5y^2 \le x \le y^2\} \quad f(x, y) = ye^{1+x}$

SOLUTION

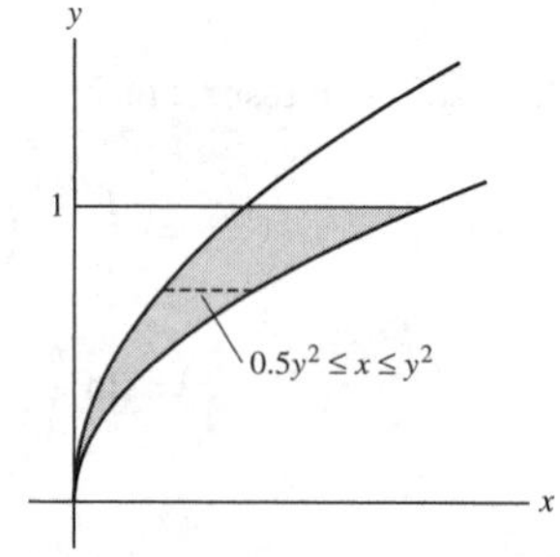

The region is horizontally simple, hence the double integral is equal to the following iterated integral:

$$\iint_{\mathcal{D}} ye^{1+x}\, dA = \int_0^1 \int_{0.5y^2}^{y^2} ye^{1+x}\, dx\, dy = \int_0^1 ye^{1+x}\Big|_{x=0.5y^2}^{y^2} dy$$

$$= \int_0^1 y\left(e^{1+y^2} - e^{1+0.5y^2}\right) dy = \int_0^1 ye^{1+y^2}\, dy - \int_0^1 ye^{1+0.5y^2}\, dy$$

We compute the integrals using the substitutions $u = 1 + y^2$, $du = 2y\, dy$, and $v = 1 + 0.5y^2$, $dv = y\, dy$, respectively. We get

$$\iint_{\mathcal{D}} ye^{1+x}\, dA = \frac{1}{2}\int_1^2 e^u\, du - \int_1^{1.5} e^v dv = \frac{1}{2}e^u\Big|_1^2 - e^v\Big|_1^{1.5} = \frac{1}{2}(e^2 - e) - (e^{3/2} - e)$$

$$= \frac{1}{2}e^2 + \frac{1}{2}e - e^{3/2} = 0.5(e^2 - 2e^{1.5} + e)$$

15. Express $\displaystyle\int_{-3}^{3}\int_0^{9-x^2} f(x, y)\, dy\, dx$ as an iterated integral in the order $dx\, dy$.

SOLUTION The limits of integration correspond to the inequalities describing the domain $\mathcal{D}$:

$$-3 \le x \le 3, \quad 0 \le y \le 9 - x^2.$$

A quick sketch verifies that this is the region under the upper part of the parabola $y = 9 - x^2$, that is, the part that is above the x-axis. Therefore, the double integral can be rewritten as the following sum:

$$\int_{-3}^{3}\int_0^{9-x^2} f(x, y)\, dy\, dx = \int_0^9 \int_{-\sqrt{9-y}}^{\sqrt{9-y}} f(x, y)\, dx\, dy$$

17. Verify directly that

$$\int_2^3 \int_0^2 \frac{dy\, dx}{1 + x - y} = \int_0^2 \int_2^3 \frac{dx\, dy}{1 + x - y}$$

SOLUTION We compute the two iterated integrals:

$$I_1 = \int_2^3 \int_0^2 \frac{dy\, dx}{1 + x - y} = \int_2^3 \left(\int_0^2 \frac{dy}{1 + x - y}\right) dx = \int_2^3 -\ln(1 + x - y)\Big|_{y=0}^{2} dx$$

$$= \int_2^3 (-\ln(1 + x - 2) + \ln(1 + x - 0))\, dx = \int_2^3 (\ln(1 + x) - \ln(x - 1))\, dx$$

$$= (1 + x)(\ln(1 + x) - 1) - (x - 1)(\ln(x - 1) - 1)\Big|_2^3$$

$$= 4(\ln 4 - 1) - 2(\ln 2 - 1) - (3(\ln 3 - 1) - (\ln 1 - 1)) = 6\ln 2 - 3\ln 3$$

$$I_2 = \int_0^2 \int_2^3 \frac{dx\, dy}{1 + x - y} = \int_0^2 \left(\int_2^3 \frac{dx}{1 + x - y}\right) dy = \int_0^2 \ln(1 + x - y)\Big|_{x=2}^{3} dy$$

$$= \int_0^2 (\ln(1+3-y) - \ln(1+2-y))\,dy = \int_0^2 (\ln(4-y) - \ln(3-y))\,dy$$

$$= \int_{-2}^{0} (\ln(4+u) - \ln(3+u))\,du = (4+u)\,(\ln(4+u)-1) - (3+u)\,(\ln(3+u)-1)\Big|_{u=-2}^{0}$$

$$= 4(\ln 4 - 1) - 3(\ln 3 - 1) - (2(\ln 2 - 1) - (\ln 1 - 1)) = 4\ln 4 - 3\ln 3 - 2\ln 2 = 6\ln 2 - 3\ln 3$$

The two integrals are equal.

19. Rewrite $\displaystyle\int_0^1 \int_{-\sqrt{1-y^2}}^{\sqrt{1-y^2}} \frac{y\,dx\,dy}{(1+x^2+y^2)^2}$ by interchanging the order of integration and evaluate.

SOLUTION This integral gets simpler if we change the order of integration. We first identify the region $\mathcal{D}$ by the limits of integration. That is,

$$\mathcal{D}: 0 \le y \le 1,\ -\sqrt{1-y^2} \le x \le \sqrt{1-y^2}$$

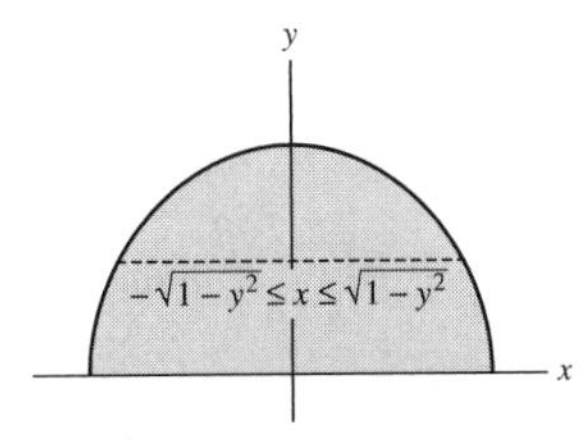

The semicircle $x^2 + y^2 = 1$, $y \ge 0$ can be rewritten as $y = \sqrt{1-x^2}$. The inequalities describing $\mathcal{D}$ as a vertically simple region are thus

$$\mathcal{D}: -1 \le x \le 1,\ 0 \le y \le \sqrt{1-x^2}$$

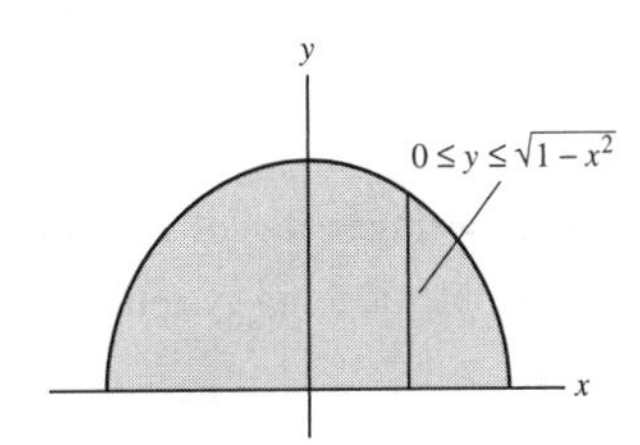

We obtain the following iterated integral:

$$\iint_{\mathcal{D}} \frac{y}{(1+x^2+y^2)^2}\,dx\,dy = \int_{-1}^{1} \int_0^{\sqrt{1-x^2}} \frac{y}{(1+x^2+y^2)^2}\,dy\,dx \tag{1}$$

We compute the inner integral with respect to y, using the substitution $u = 1 + x^2 + y^2$, $du = 2y\,dy$ (x is considered as a constant). This gives

$$\int_0^{\sqrt{1-x^2}} \frac{y\,dy}{(1+x^2+y^2)^2} = \int_{1+x^2}^{2} \frac{\frac{1}{2}\,du}{u^2} = -\frac{1}{2u}\Big|_{1+x^2}^{2} = -\frac{1}{4} + \frac{1}{2(1+x^2)}$$

We compute the outer integral in (1):

$$\iint_{\mathcal{D}} \frac{y}{(1+x^2+y^2)}\,dx\,dy = \int_{-1}^{1} \left(-\frac{1}{4} + \frac{1}{2(1+x^2)}\right) dx = \int_0^1 \left(-\frac{1}{2} + \frac{1}{1+x^2}\right) dx$$

$$= -\frac{x}{2} + \tan^{-1} x\Big|_0^1 = \left(-\frac{1}{2} + \tan^{-1} 1\right) - 0 = -\frac{1}{2} + \frac{\pi}{4} = \frac{\pi}{4} - \frac{1}{2}$$

21. Find the center of mass of the sector of central angle 2θ (symmetric with respect to the y-axis) in Figure 3, assuming that the mass density is $\rho(x, y) = x^2$.

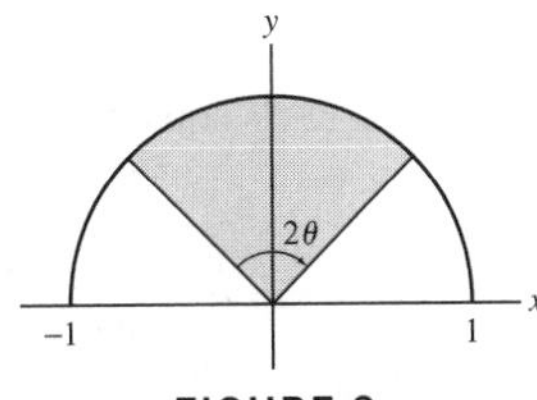

FIGURE 3

SOLUTION Since the region is symmetric with respect to the y-axis, the x-coordinate of the center of mass is zero, hence the center of mass is located on the y-axis. We first compute the mass:

$$M = \iint \rho\, dA = \int_0^1 \int_{\pi/2-\theta}^{\pi/2+\theta} r^2 \cos^2\alpha \cdot r\, dr\, d\alpha = \frac{1}{4}r^4\Big|_0^1 \cdot \left[\frac{\alpha}{2} + \frac{\sin 2\alpha}{4}\right]\Big|_{\pi/2-\theta}^{\pi/2+\theta}$$

$$= \frac{1}{4}\left[\theta + \frac{1}{4}(\sin(\pi+2\theta) - \sin(\pi-2\theta))\right] = \frac{1}{4}\left[\theta + \frac{1}{4}(-\sin(2\theta) - \sin(2\theta))\right] = \frac{1}{4}\left[\theta - \frac{1}{2}\sin(2\theta)\right]$$

We now compute y_{CM}.

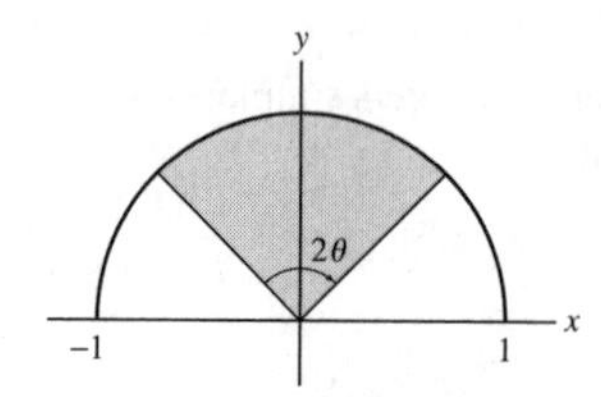

We obtain the integral

$$y_{CM} = \frac{1}{M}\int_0^1 \int_{\frac{\pi}{2}-\theta}^{\pi/2+\theta} r\sin\alpha \cdot r^2\cos^2\alpha \cdot r\, d\alpha\, dr = \frac{1}{M}\frac{1}{5}r^5\Big|_0^1 \cdot \left[\frac{-1}{3}\cos^3\alpha\right]\Big|_{\pi/2-\theta}^{\pi/2+\theta}$$

$$= \frac{1}{15M}\left[\cos^3(\pi/2-\theta) - \cos^3(\pi/2+\theta)\right] = \frac{1}{15M}\left[\sin^3(\theta) + \sin^3(\theta)\right] = \frac{2\sin^3(\theta)}{15M}$$

Substituting in the previous value for the mass M, we obtain

$$y_{CM} = \frac{8\sin^3\theta}{15(\theta - \frac{1}{2}\sin 2\theta)}$$

23. CAS Express the average value of $f(x,y) = e^{xy}$ over the ellipse $\frac{x^2}{2} + y^2 = 1$ as an iterated integral and evaluate numerically using a computer algebra system.

SOLUTION

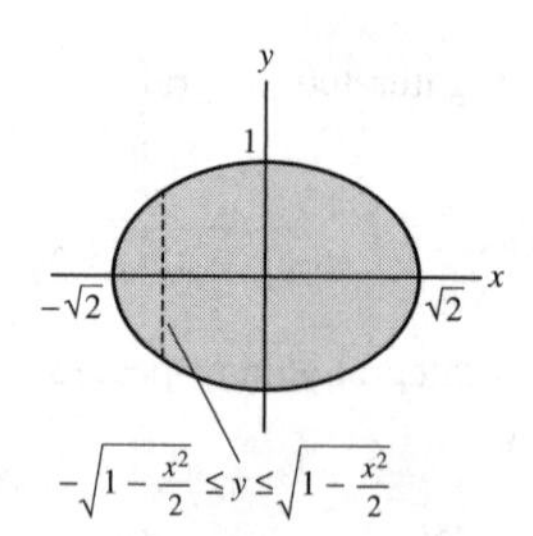

The area of the ellipse $\frac{x^2}{a^2} + \frac{y^2}{b^2} = 1$ is πab, hence the area of the given ellipse is

$$\text{Area}(\mathcal{D}) = \pi \cdot \sqrt{2} \cdot 1 = \pi\sqrt{2}$$

The average value of $f(x,y) = e^{xy}$ inside the ellipse is

$$\overline{f} = \frac{1}{\pi\sqrt{2}} \iint_{\mathcal{D}} e^{xy}\, dA \qquad (1)$$

$\mathcal{D}$ is described as a vertically simple region by the inequalities

$$\mathcal{D}: -\sqrt{2} \le x \le \sqrt{2},\ -\sqrt{1-\frac{x^2}{2}} \le y \le \sqrt{1-\frac{x^2}{2}}$$

Therefore, the double integral (1) is equal to the following iterated integral, which we compute using a CAS:

$$\overline{f} = \frac{1}{\pi\sqrt{2}} \int_{-\sqrt{2}}^{\sqrt{2}} \int_{-\sqrt{1-\frac{x^2}{2}}}^{\sqrt{1-\frac{x^2}{2}}} e^{xy}\, dy\, dx \approx 1.0421$$

25. Calculate $\iiint_{\mathcal{B}} (xy+z)\,dV$, where

$$\mathcal{B} = \{0 \le x \le 2,\ 0 \le y \le 1,\ 1 \le z \le 3\}$$

as an iterated integral in two different ways.

SOLUTION The triple integral over the box may be evaluated in any order. For instance,

$$\iiint_{\mathcal{B}} (xy+z)\,dV = \int_0^2\int_0^1\int_1^3 (xy+z)\,dz\,dy\,dx = \int_0^1\int_0^2\int_1^3 (xy+z)\,dz\,dx\,dy = \int_1^3\int_0^2\int_0^1 (xy+z)\,dy\,dx\,dz$$

We compute the integral in two of the possible orders:

$$\iiint_{\mathcal{B}} (xy+z)\,dV = \int_0^2\int_0^1\int_1^3 (xy+z)\,dz\,dy\,dx = \int_0^2\int_0^1 xyz + \frac{z^2}{2}\bigg|_{z=1}^{3}\,dy\,dx$$

$$= \int_0^2\int_0^1 \left(\left(3xy+\frac{9}{2}\right) - \left(xy+\frac{1}{2}\right)\right)dy\,dx = \int_0^2\int_0^1 (2xy+4)\,dy\,dx$$

$$= \int_0^2 xy^2 + 4y\bigg|_{y=0}^{1}\,dx = \int_0^2 (x+4)\,dx = \frac{x^2}{2} + 4x\bigg|_0^2 = \frac{4}{2} + 8 = 10$$

$$\iiint_{\mathcal{B}} (xy+z)\,dV = \int_0^1\int_0^2\int_1^3 (xy+z)\,dz\,dx\,dy = \int_0^1\int_0^2 xyz + \frac{z^2}{2}\bigg|_{z=1}^{3}\,dx\,dy$$

$$= \int_0^1\int_0^2 \left(\left(3xy+\frac{9}{2}\right) - \left(xy+\frac{1}{2}\right)\right)dx\,dy = \int_0^1\int_0^2 (2xy+4)\,dx\,dy$$

$$= \int_0^1 x^2y + 4x\bigg|_{x=0}^{2}\,dy = \int_0^1 (4y+8)\,dy = 2y^2 + 8y\bigg|_0^1 = 2 + 8 = 10$$

27. Evaluate $I = \int_{-1}^{1}\int_0^{\sqrt{1-x^2}}\int_0^1 (x+y+z)\,dz\,dy\,dx$. Then rewrite I as an iterated integral in the order $dy\,dz\,dx$ and evaluate a second time.

SOLUTION We compute the triple integral:

$$I_1 = \int_{-1}^{1}\int_0^{\sqrt{1-x^2}}\int_0^1 (x+y+z)\,dz\,dy\,dx = \int_{-1}^{1}\int_0^{\sqrt{1-x^2}} (x+y)z + \frac{z^2}{2}\bigg|_{y=0}^{1}\,dy\,dx$$

$$= \int_{-1}^{1}\int_0^{\sqrt{1-x^2}} \left(x+y+\frac{1}{2}\right)dy\,dx = \int_{-1}^{1} \left(x+\frac{1}{2}\right)y + \frac{y^2}{2}\bigg|_{y=0}^{\sqrt{1-x^2}}\,dx$$

$$= \int_{-1}^{1} \left(x+\frac{1}{2}\right)\sqrt{1-x^2} + \frac{1-x^2}{2}\,dx = \int_{-1}^{1} x\sqrt{1-x^2}\,dx + \int_{-1}^{1}\frac{1}{2}\sqrt{1-x^2}\,dx + \int_{-1}^{1}\frac{1-x^2}{2}\,dx \qquad (1)$$

The first integral is zero since the integrand is an odd function. Therefore, using Integration Formulas we get

$$I_1 = \int_0^1 \sqrt{1-x^2}\,dx + \int_0^1 (1-x^2)\,dx = \frac{x}{2}\sqrt{1-x^2} + \frac{1}{2}\sin^{-1}x\bigg|_0^1 + \left(x - \frac{x^3}{3}\right)\bigg|_0^1$$

$$= \frac{1}{2}\sin^{-1}1 + \frac{2}{3} = \frac{\pi}{4} + \frac{2}{3}$$

To determine the limits of integration in the order $dy\,dz\,dx$, we must identify the projection of $\mathcal{W}$ onto the xz-plane. The region $\mathcal{W}$ is the region bounded by the planes $z=0$ and $z=1$ over the semicircle $x^2+y^2=1$, $y \ge 0$ in the xy-plane. That is, $\mathcal{W}$ is the half cylinder $x^2+y^2=1$, $y \ge 0$ bounded by the planes $z=0$ and $z=1$. The projection of $\mathcal{W}$ onto the xz-plane is the rectangle S defined by

$$S:\ -1 \le x \le 1,\ 0 \le z \le 1$$

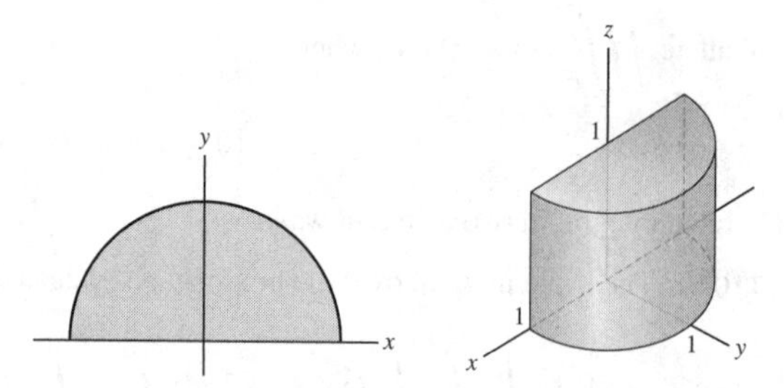

The region $\mathcal{W}$ consists of all points lying between S and cylinder $y = \sqrt{1-x^2}$. That is,

$$\mathcal{W}: -1 \le x \le 1,\ 0 \le z \le 1,\ 0 \le y \le \sqrt{1-x^2}$$

These inequalities determine the limits of integration in the order $dy\,dz\,dx$. We have

$$\int_{-1}^{1}\int_{0}^{\sqrt{1-x^2}}\int_{0}^{1}(x+y+z)\,dz\,dy\,dx = \int_{-1}^{1}\int_{0}^{1}\int_{0}^{\sqrt{1-x^2}}(x+y+z)\,dy\,dz\,dx$$

We compute the second triple integral:

$$I_2 = \int_{-1}^{1}\int_{0}^{1}\int_{0}^{\sqrt{1-x^2}}(x+y+z)\,dy\,dz\,dx = \int_{-1}^{1}\int_{0}^{1}(x+z)y + \frac{y^2}{2}\bigg|_{y=0}^{\sqrt{1-x^2}}dz\,dx$$

$$= \int_{-1}^{1}\int_{0}^{1}\left((x+z)\sqrt{1-x^2} + \frac{1-x^2}{2}\right)dz\,dx = \int_{-1}^{1}\int_{0}^{1}\left(x\sqrt{1-x^2} + \frac{1-x^2}{2} + z\sqrt{1-x^2}\right)dz\,dx$$

$$= \int_{-1}^{1}\left(x\sqrt{1-x^2} + \frac{1-x^2}{2}\right)z + \frac{z^2\sqrt{1-x^2}}{2}\bigg|_{z=0}^{1}dx = \int_{-1}^{1}\left(x\sqrt{1-x^2} + \frac{1-x^2}{2} + \frac{\sqrt{1-x^2}}{2}\right)dx$$

$$= \int_{-1}^{1}x\sqrt{1-x^2}\,dx + \int_{0}^{1}(1-x^2)\,dx + \int_{0}^{1}\sqrt{1-x^2}\,dx$$

The first integral is zero since the function is odd. Using integration formulas we obtain

$$I_2 = \int_{0}^{1}(1-x^2)\,dx + \int_{0}^{1}\sqrt{1-x^2}\,dx = x - \frac{x^3}{3}\bigg|_{0}^{1} + \left(\frac{x}{2}\sqrt{1-x^2} + \frac{1}{2}\sin^{-1}x\right)\bigg|_{0}^{1}$$

$$= \left(1-\frac{1}{3}\right) + \frac{1}{2}\sin^{-1}1 = \frac{2}{3} + \frac{\pi}{4}$$

The two integrals are equal as expected.

29. Find the center of mass of the first octant of the ball $x^2+y^2+z^2=1$, assuming a mass density of $\rho(x,y,z)=x$.

SOLUTION

(a) The solid is the part of the unit sphere in the first octant. The inequalities defining the projection of the solid onto the xy-plane are

$$\mathcal{D}: 0 \le y \le 1,\ 0 \le x \le \sqrt{1-y^2}$$

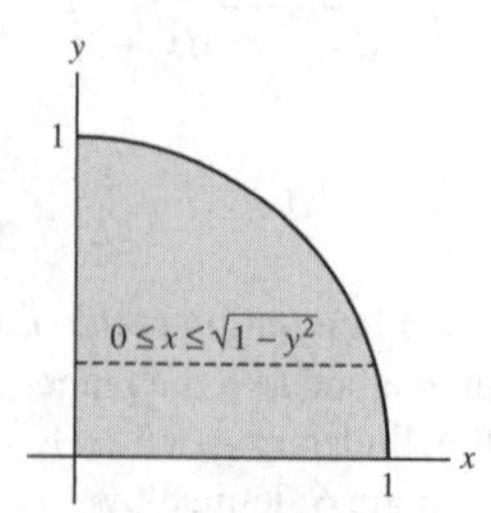

$\mathcal{W}$ is the region bounded by $\mathcal{D}$ and the sphere $z=\sqrt{1-x^2-y^2}$, hence $\mathcal{W}$ is defined by the inequalities

$$\mathcal{W}: 0 \le y \le 1,\ 0 \le x \le \sqrt{1-y^2},\ 0 \le z \le \sqrt{1-x^2-y^2} \tag{1}$$

We first must compute the mass of the solid. Using the mass as a triple integral, we have

$$M = \int_{0}^{1}\int_{0}^{\sqrt{1-y^2}}\int_{0}^{\sqrt{1-x^2-y^2}} x\,dz\,dx\,dy = \int_{0}^{1}\int_{0}^{\sqrt{1-y^2}} xz\bigg|_{z=0}^{\sqrt{1-x^2-y^2}}dx\,dy$$

$$= \int_0^1 \int_0^{\sqrt{1-y^2}} x\sqrt{1-x^2-y^2}\,dx\,dy$$

We compute the inner integral using the substitution $u = \sqrt{1-x^2-y^2}$, $du = -\frac{x}{u}\,dx$, or $x\,dx = -u\,du$. We get

$$\int_0^{\sqrt{1-y^2}} x\sqrt{1-x^2-y^2}\,dx = \int_{\sqrt{1-y^2}}^0 u(-u\,du) = \int_0^{\sqrt{1-y^2}} u^2\,du = \frac{u^3}{3}\bigg|_0^{\sqrt{1-y^2}} = \frac{(1-y^2)^{3/2}}{3} \tag{2}$$

We substitute in (2) and compute the resulting integral substituting $y = \sin t$, $dy = \cos t\,dt$:

$$M = \int_0^1 \frac{(1-y^2)^{3/2}}{3}\,dy = \frac{1}{3}\int_0^{\pi/2} (1-\sin^2 t)^{3/2}\cos t\,dt = \frac{1}{3}\int_0^{\pi/2} \cos^4 t\,dt$$

$$= \frac{1}{3}\left(\frac{\cos^3 t\sin t}{4} + \frac{3}{4}\left(\frac{t}{2} + \frac{\sin 2t}{4}\right)\bigg|_0^{\pi/2}\right) = \frac{1}{4}\cdot\frac{\pi}{4} = \frac{\pi}{16}$$

That is, $M = \frac{\pi}{16}$. We now find the coordinates of the center of mass. To compute x_{CM} we use the definition of $\mathcal{D}$ as a vertically simple region to obtain a simpler integral. That is, we write the inequalities for $\mathcal{W}$ as

$$\mathcal{W} : 0 \le x \le 1,\ 0 \le y \le \sqrt{1-x^2},\ 0 \le z \le \sqrt{1-x^2-y^2} \tag{3}$$

Thus,

$$x_{\text{CM}} = \frac{1}{M}\iiint_{\mathcal{W}} x\rho\,dV = \frac{16}{\pi}\int_0^1\int_0^{\sqrt{1-x^2}}\int_0^{\sqrt{1-x^2-y^2}} x^2\,dz\,dy\,dx = \frac{16}{\pi}\int_0^1\int_0^{\sqrt{1-x^2}} x^2 z\bigg|_{z=0}^{\sqrt{1-x^2-y^2}} dy\,dx$$

$$= \frac{16}{\pi}\int_0^1\int_0^{\sqrt{1-x^2}} x^2\sqrt{1-x^2-y^2}\,dy\,dx = \frac{16}{\pi}\int_0^1 x^2\left(\int_0^{\sqrt{1-x^2}}\sqrt{1-x^2-y^2}\,dy\right)dx \tag{4}$$

We compute the inner integral using Integration Formulas:

$$\int_0^{\sqrt{1-x^2}}\sqrt{1-x^2-y^2}\,dy = \frac{y}{2}\sqrt{1-x^2-y^2} + \frac{1-x^2}{2}\sin^{-1}\frac{y}{\sqrt{1-x^2}}\bigg|_{y=0}^{\sqrt{1-x^2}}$$

$$= \frac{1-x^2}{2}\sin^{-1} 1 = \frac{1-x^2}{2}\cdot\frac{\pi}{2} = \frac{\pi}{4}(1-x^2)$$

Substituting in (4) gives

$$x_{\text{CM}} = \frac{16}{\pi}\int_0^1 x^2\cdot\frac{\pi}{4}(1-x^2)\,dx = 4\int_0^1 (x^2-x^4)\,dx = 4\left(\frac{x^3}{3} - \frac{x^5}{5}\right)\bigg|_0^1 = 4\left(\frac{1}{3} - \frac{1}{5}\right) = \frac{8}{15}$$

(b) We compute the y-coordinate of the center of mass, using (1):

$$y_{\text{CM}} = \frac{1}{M}\iiint_{\mathcal{W}} y\rho\,dV = \frac{16}{\pi}\int_0^1\int_0^{\sqrt{1-y^2}}\int_0^{\sqrt{1-x^2-y^2}} yx\,dz\,dx\,dy = \frac{16}{\pi}\int_0^1\int_0^{\sqrt{1-y^2}} yxz\bigg|_{z=0}^{\sqrt{1-x^2-y^2}} dx\,dy$$

$$= \frac{16}{\pi}\int_0^1\int_0^{\sqrt{1-y^2}} yx\sqrt{1-x^2-y^2}\,dx\,dy = \frac{16}{\pi}\int_0^1 y\left(\int_0^{\sqrt{1-y^2}} x\sqrt{1-x^2-y^2}\,dx\right)dy$$

The inner integral was computed in (2), therefore,

$$y_{\text{CM}} = \frac{16}{\pi}\int_0^1 y\cdot\frac{(1-y^2)^{3/2}}{3}\,dy = \frac{16}{3\pi}\int_0^1 y(1-y^2)^{3/2}\,dy$$

We compute the integral using the substitution $u = 1-y^2$, $du = -2y\,dy$. We get

$$y_{\text{CM}} = \frac{16}{3\pi}\int_1^0 u^{3/2}\cdot\left(-\frac{du}{2}\right) = \frac{8}{3\pi}\int_0^1 u^{3/2}\,du = \frac{8}{3\pi}\cdot\frac{2}{5}\cdot u^{5/2}\bigg|_0^1 = \frac{16}{15\pi}$$

Finally we find the z-coordinate of the center of mass, using (1):

$$z_{\text{CM}} = \frac{1}{M}\iiint_{\mathcal{W}} z\rho\,dV = \frac{16}{\pi}\int_0^1\int_0^{\sqrt{1-y^2}}\int_0^{\sqrt{1-x^2-y^2}} zx\,dz\,dx\,dy = \frac{16}{\pi}\int_0^1\int_0^{\sqrt{1-y^2}} \frac{z^2 x}{2}\bigg|_{z=0}^{\sqrt{1-x^2-y^2}} dx\,dy$$

$$= \frac{16}{\pi}\int_0^1\int_0^{\sqrt{1-y^2}} \frac{x}{2}(1-x^2-y^2)\,dx\,dy = \frac{8}{\pi}\int_0^1\int_0^{\sqrt{1-y^2}} (x-x^3-xy^2)\,dx\,dy$$

$$= \frac{8}{\pi}\int_0^1 \frac{x^2}{2}-\frac{x^4}{4}-\frac{x^2y^2}{2}\Big|_{x=0}^{\sqrt{1-y^2}} dy = \frac{8}{\pi}\int_0^1 \left(\frac{1-y^2}{2}-\frac{(1-y^2)^2}{4}-\frac{(1-y^2)y^2}{2}\right) dy$$

$$= \frac{2}{\pi}\int_0^1 (y^4-2y^2+1)\,dy = \frac{2}{\pi}\left(\frac{y^5}{5}-\frac{2y^3}{3}+y\right)\Big|_{y=0}^{1} = \frac{2}{\pi}\left(\frac{1}{5}-\frac{2}{3}+1\right) = \frac{16}{15\pi}$$

The center mass is the following point:

$$P = \left(\frac{8}{15}, \frac{16}{15\pi}, \frac{16}{15\pi}\right).$$

31. Use polar coordinates to calculate $\iint_{\mathcal{D}} \sqrt{x^2+y^2}\,dA$, where $\mathcal{D}$ is the region in the first quadrant bounded by the spiral $r=\theta$, the circle $r=1$, and the x-axis.

SOLUTION The region of integration, shown in the figure, has the following description in polar coordinates:

$$\mathcal{D}: 0 \le \theta \le 1,\ \theta \le r \le 1$$

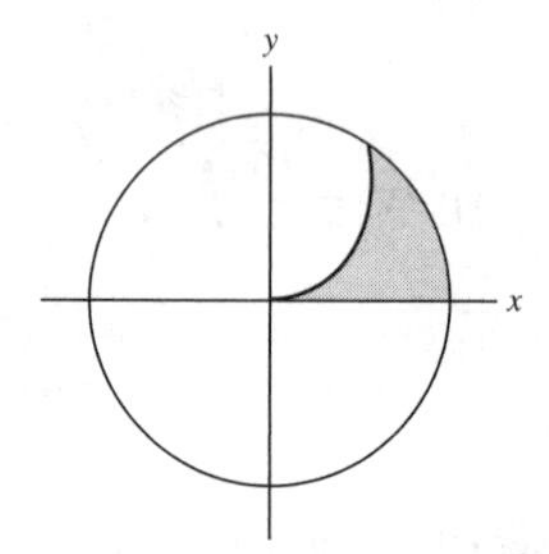

The function is $f(x, y) = \sqrt{x^2+y^2} = r$. We convert the double integral to polar coordinates and compute to obtain

$$\iint_{\mathcal{D}} \sqrt{x^2+y^2}\,dA = \int_0^1\int_\theta^1 r\cdot r\,dr\,d\theta = \int_0^1\int_\theta^1 r^2\,dr\,d\theta = \int_0^1 \frac{r^3}{3}\Big|_{r=\theta}^{1} d\theta$$

$$= \int_0^1 \left(\frac{1}{3}-\frac{\theta^3}{3}\right) d\theta = \frac{\theta}{3}-\frac{\theta^4}{12}\Big|_0^1 = \frac{1}{3}-\frac{1}{12} = \frac{1}{4}$$

33. Use cylindrical coordinates to find the total mass of the solid bounded by $z = 8-x^2-y^2$ and $z = x^2+y^2$, assuming a mass density of $f(x, y, z) = (x^2+y^2)^{1/2}$.

SOLUTION The mass of the solid $\mathcal{W}$ is the following integral:

$$M = \iiint_{\mathcal{W}} (x^2+y^2)^{1/2}\,dV$$

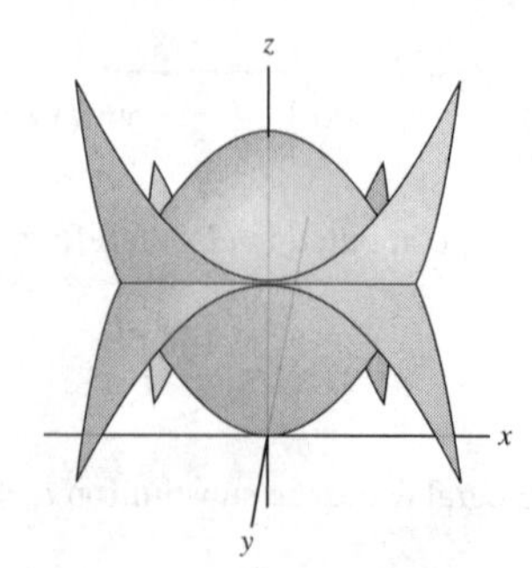

The projection of $\mathcal{W}$ on the xy-plane is obtained by equating the equations of the two surfaces:

$$\begin{aligned} 8-x^2-y^2 &= x^2+y^2 \\ 2(x^2+y^2) &= 8 \end{aligned} \quad \Rightarrow \quad x^2+y^2 = 4$$

We conclude that the projection is the disk $\mathcal{D}: x^2+y^2 \le 4$.

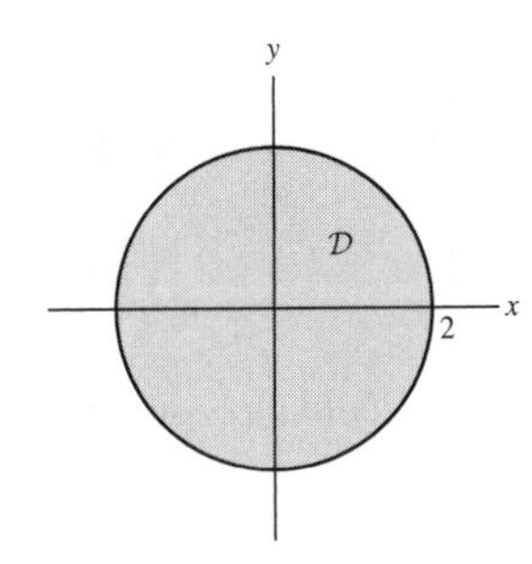

Therefore, $\mathcal{W}$ is described by

$$\mathcal{W}: x^2 + y^2 \le z \le 8 - (x^2 + y^2),\ (x, y) \in \mathcal{D}$$

Thus,

$$M = \iint_{\mathcal{D}} \int_{x^2+y^2}^{8-(x^2+y^2)} (x^2 + y^2)^{1/2}\, dz\, dx\, dy$$

We convert the integral to cylindrical coordinates. The inequalities for $\mathcal{W}$ are

$$0 \le r \le 2, \quad 0 \le \theta \le 2\pi, \quad r^2 \le z \le 8 - r^2.$$

Also, $(x^2 + y^2)^{1/2} = r$, hence we obtain the following integral:

$$M = \int_0^2 \int_0^{2\pi} \int_{r^2}^{8-r^2} r \cdot r\, dz\, d\theta\, dr = \int_0^2 \int_0^{2\pi} \int_{r^2}^{8-r^2} r^2\, dz\, d\theta\, dr = \int_0^2 \int_0^{2\pi} r^2 z \Big|_{z=r^2}^{8-r^2} d\theta\, dr$$

$$= \int_0^2 \int_0^{2\pi} r^2(8 - r^2 - r^2)\, d\theta\, dr = \int_0^2 \int_0^{2\pi} (8r^2 - 2r^4)\, d\theta\, dr = \left(\int_0^{2\pi} 1\, d\theta\right)\left(\int_0^2 (8r^2 - 2r^4)\, dr\right)$$

$$= 2\pi\left(\frac{8r^3}{3} - \frac{2}{5}r^5 \Big|_0^2\right) = \frac{256}{15}\pi \approx 53.62$$

35. Find the volume of the solid contained in the cylinder $x^2 + y^2 = 1$ below the curve $z = (x + y)^2$ and above the curve $z = -(x - y)^2$.

SOLUTION

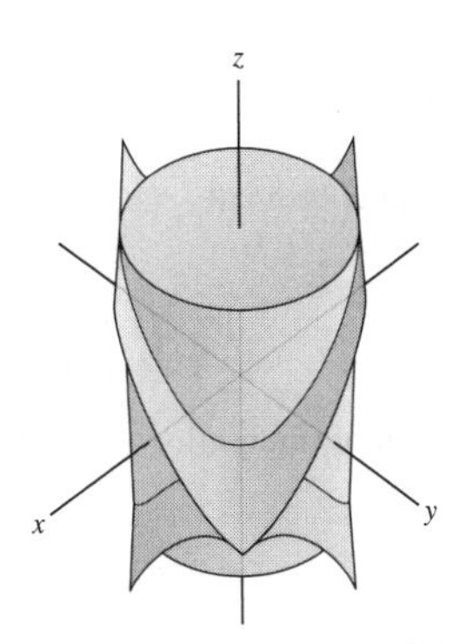

We rewrite the equations of the surfaces using cylindrical coordinates:

$$z = (x + y)^2 = x^2 + y^2 + 2xy = r^2 + 2(r\cos\theta)(r\sin\theta) = r^2(1 + \sin 2\theta)$$
$$z = -(x - y)^2 = -(x^2 + y^2 - 2xy) = -(r^2 - 2r^2\cos\theta\sin\theta) = -r^2(1 - \sin 2\theta)$$

The projection of the solid onto the xy-plane is the unit disk. Therefore, the solid is described by the following inequalities:

$$\mathcal{W}: 0 \le \theta \le 2\pi,\ 0 \le r \le 1,\ -r^2(1 - \sin 2\theta) \le z \le r^2(1 + \sin 2\theta)$$

Expressing the volume as a triple integral and converting the triple integral to cylindrical coordinates, we get

$$V = \text{Volume}(\mathcal{W}) = \iiint_{\mathcal{W}} 1\, dv = \int_0^{2\pi} \int_0^1 \int_{-r^2(1-\sin 2\theta)}^{r^2(1+\sin 2\theta)} r\, dz\, dr\, d\theta$$

$$= \int_0^{2\pi}\int_0^1 rz\Big|_{z=-r^2(1-\sin 2\theta)}^{r^2(1+\sin 2\theta)} dr\,d\theta = \int_0^{2\pi}\int_0^1 r\left(r^2(1+\sin 2\theta)+r^2(1-\sin 2\theta)\right) dr\,d\theta$$

$$= \int_0^{2\pi}\int_0^1 r^3\cdot 2\,dr\,d\theta = \left(\int_0^{2\pi} 2d\theta\right)\left(\int_0^1 r^3\,dr\right) = 4\pi\cdot\frac{r^4}{4}\Big|_0^1 = \pi$$

37. Express in cylindrical coordinates and evaluate:

$$\int_0^1\int_0^{\sqrt{1-x^2}}\int_0^{\sqrt{x^2+y^2}} z\,dz\,dy\,dx$$

SOLUTION We evaluate the integral by converting it to cylindrical coordinates. The projection of the region of integration onto the xy-plane, as defined by the limits of integration, is

$$\mathcal{D}: 0\le x\le 1,\ 0\le y\le\sqrt{1-x^2}$$

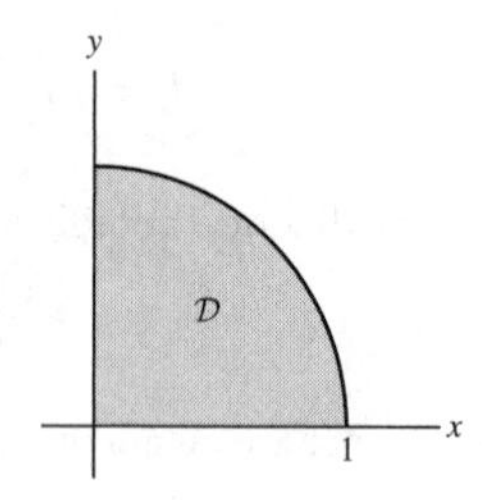

That is, $\mathcal{D}$ is the part of the disk $x^2+y^2\le 1$ in the first quadrant. The inequalities defining $\mathcal{D}$ in polar coordinates are

$$\mathcal{D}: 0\le\theta\le\frac{\pi}{2},\ 0\le r\le 1$$

The upper surface is $z=\sqrt{x^2+y^2}=r$ and the lower surface is $z=0$. Therefore, the inequalities defining the region of integration in cylindrical coordinates are

$$\mathcal{W}: 0\le\theta\le\frac{\pi}{2},\ 0\le r\le 1,\ 0\le z\le r$$

Converting the double integral to cylindrical coordinates gives

$$I = \int_0^{\pi/2}\int_0^1\int_0^r zr\,dz\,dr\,d\theta = \int_0^{\pi/2}\int_0^1 \frac{z^2r}{2}\Big|_{z=0}^r dr\,d\theta = \int_0^{\pi/2}\int_0^1 \frac{r^3}{2}\,dr\,d\theta$$

$$= \left(\int_0^{\pi/2} d\theta\right)\left(\int_0^1 \frac{r^3}{2}\,dr\right) = \frac{\pi}{2}\cdot\frac{r^4}{8}\Big|_0^1 = \frac{\pi}{16}$$

39. Convert to spherical coordinates and evaluate:

$$\int_{-2}^2\int_{-\sqrt{4-x^2}}^{\sqrt{4-x^2}}\int_0^{\sqrt{4-x^2-y^2}} e^{-(x^2+y^2+z^2)^{3/2}}\,dz\,dy\,dx$$

SOLUTION The region of integration as defined by the limits of integration is

$$\mathcal{W}: -2\le x\le 2,\ -\sqrt{4-x^2}\le y\le\sqrt{4-x^2},\ 0\le z\le\sqrt{4-x^2-y^2}$$

That is, $\mathcal{W}$ is the region enclosed by the sphere $x^2+y^2+z^2=4$ and the xy-plane. We see that the region of integration is the upper half-ball $x^2+y^2+z^2\le 4$, hence the inequalities defining $\mathcal{W}$ in spherical coordinates are

$$\mathcal{W}: 0\le\theta\le 2\pi,\ 0\le\phi\le\frac{\pi}{2},\ 0\le\rho\le 2$$

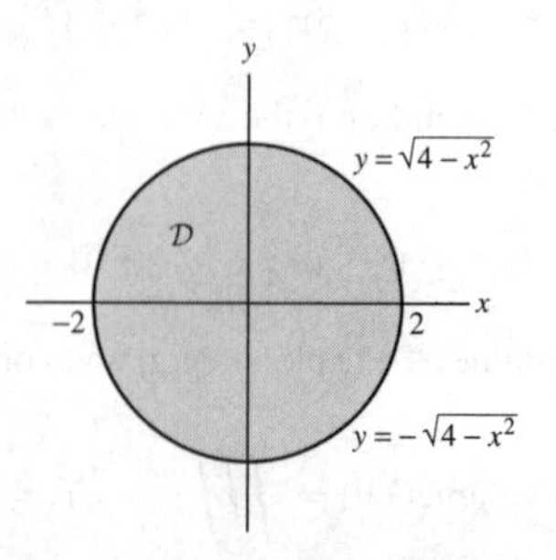

The function is $f(x,y,z)=e^{-(x^2+y^2+z^2)^{3/2}}=e^{-(\rho^2)^{3/2}}=e^{-\rho^3}$, therefore the integral in spherical coordinates is

$$I=\int_0^{2\pi}\int_0^{\pi/2}\int_0^2 e^{-\rho^3}\rho^2\sin\phi\,d\rho\,d\phi\,d\theta=\left(\int_0^{2\pi}d\theta\right)\left(\int_0^{\pi/2}\sin\phi\,d\phi\right)\left(\int_0^2 e^{-\rho^3}\rho^2 d\rho\right)$$

$$=2\pi\left(-\cos\phi\Big|_0^{\pi/2}\right)\int_0^2 e^{-\rho^3}\rho^2 d\rho=2\pi\int_0^2 e^{-\rho^3}\rho^2 d\rho$$

We compute the integral using the substitution $u=\rho^3$, $du=3\rho^2 d\rho$. We get

$$I=2\pi\int_0^8 e^{-u}\frac{du}{3}=\frac{2\pi}{3}(-e^{-u})\Big|_0^8=\frac{2\pi}{3}(-e^{-8}+1)=\frac{2\pi\left(-1+e^8\right)}{3e^8}$$

41. Compute the Jacobian of the map

$$\Phi(r,s)=\left(e^r\cosh(s),e^r\sinh(s)\right)$$

SOLUTION We have $x=e^r\cosh(s)$ and $y=e^r\sinh(s)$. Therefore,

$$\mathrm{Jac}(\Phi)=\frac{\partial(x,y)}{\partial(r,s)}=\begin{vmatrix}\frac{\partial x}{\partial r} & \frac{\partial x}{\partial s}\\ \frac{\partial y}{\partial r} & \frac{\partial y}{\partial s}\end{vmatrix}=\begin{vmatrix}e^r\cosh(s) & e^r\sinh(s)\\ e^r\sinh(s) & e^r\cosh(s)\end{vmatrix}$$

$$=e^{2r}\cosh^2(s)-e^{2r}\sinh^2(s)=e^{2r}(\cosh^2(s)-\sinh^2(s))=e^{2r}$$

43. Use the map

$$\Phi(u,v)=\left(\frac{u+v}{2},\frac{u-v}{2}\right)$$

to compute $\iint_{\mathcal{R}}((x-y)\sin(x+y))^2\,dx\,dy$, where $\mathcal{R}$ is the square with vertices $(\pi,0)$, $(2\pi,\pi)$, $(\pi,2\pi)$, and $(0,\pi)$.

SOLUTION We express $f(x,y)=((x-y)\sin(x+y))^2$ in terms of u and v. Since $x=\frac{u+v}{2}$ and $y=\frac{u-v}{2}$, we have $x-y=v$ and $x+y=u$. Hence, $f(x,y)=v^2\sin^2 u$. We find the Jacobian of the linear transformation:

$$\mathrm{Jac}(\Phi)=\begin{vmatrix}\frac{1}{2} & \frac{1}{2}\\ \frac{1}{2} & -\frac{1}{2}\end{vmatrix}=-\frac{1}{4}-\frac{1}{4}=-\frac{1}{2}$$

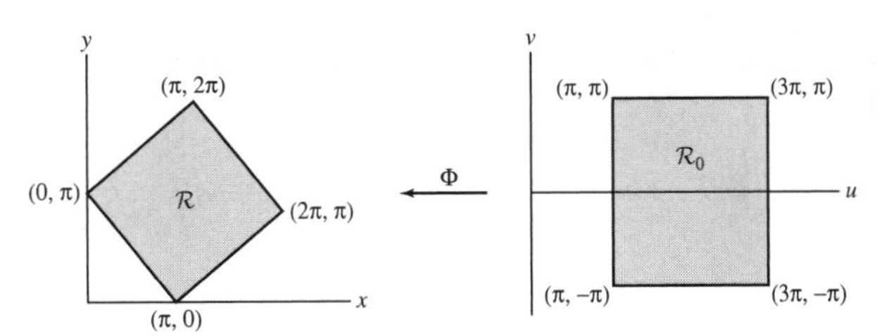

To compute the vertices of the quadrangle $\mathcal{R}$ mapped by Φ onto $\mathcal{R}$, we first find the inverse of Φ by solving the following equations for u, v in terms of x and y:

$$\begin{array}{l}x=\frac{u+v}{2}\\ y=\frac{u-v}{2}\end{array}\quad\Rightarrow\quad\begin{array}{l}u+v=2x\\ u-v=2y\end{array}\quad\Rightarrow\quad u=x+y,\quad v=x-y$$

Hence,

$$\Phi^{-1}(x,y)=(x+y,x-y)$$

We now compute the vertices of P as the following images:

$$\Phi^{-1}(\pi,0)=(\pi,\pi)$$
$$\Phi^{-1}(2\pi,\pi)=(3\pi,\pi)$$
$$\Phi^{-1}(\pi,2\pi)=(3\pi,-\pi)$$

$$\Phi^{-1}(0, \pi) = (\pi, -\pi)$$

Finally, we apply the change of variable formula to compute the integral:

$$\iint_{\mathcal{R}} (x-y)^2 \sin^2(x+y)\, dx\, dy = \iint_{R_0} v^2 \sin^2 u |\text{Jac}(\Phi)|\, du\, dv = \frac{1}{2}\int_{\pi}^{3\pi}\int_{-\pi}^{\pi} v^2 \sin^2 u\, dv\, du$$

$$= \frac{1}{2}\left(\int_{\pi}^{3\pi} (\sin^2 u)\, du\right)\left(\int_{-\pi}^{\pi} v^2\, dv\right) = \left(\frac{u}{4} - \frac{\sin 2u}{8}\Big|_{\pi}^{3\pi}\right)\left(\frac{v^3}{3}\Big|_{-\pi}^{\pi}\right)$$

$$= \left(\frac{3\pi}{4} - \frac{\pi}{4}\right)\cdot\frac{2\pi^3}{3} = \frac{\pi^4}{3}$$

45. Calculate the integral of $f(x, y) = e^{3x-2y}$ over the parallelogram in Figure 6.

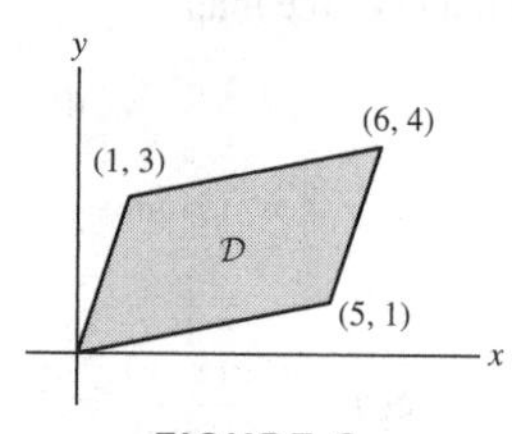

FIGURE 6

SOLUTION The equation of the boundary lines are $y = \frac{1}{5}x$, $y = \frac{1}{5}x + \frac{14}{5}$, $y = 3x$, and $y = 3x - 14$. These equations may be written as

$$x - 5y = 0, \quad x - 5y = -14, \quad 3x - y = 0, \quad 3x - y = 14$$

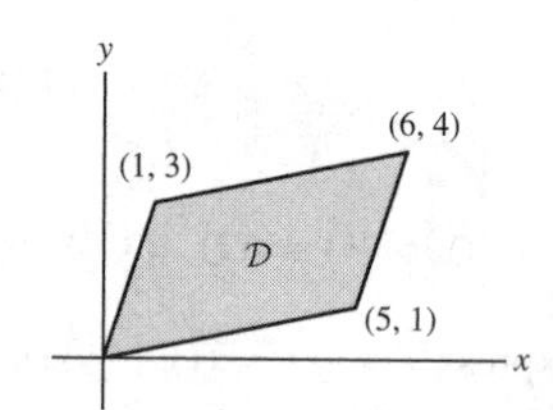

We define the following map:

$$\Phi^{-1}(x, y) = (u, v) = (x - 5y, 3x - y)$$

Φ^{-1} maps the boundary lines to the lines $u = 0$, $u = -14$, $v = 0$, and $v = 14$ in the (u, v)-plane. Therefore the image of $\mathcal{D}$ under Φ^{-1} is the rectangle $\mathcal{R} = [-14, 0] \times [0, 14]$ in the (u, v)-plane. Using the Change of variables Formula, we have

$$\iint_{\mathcal{D}} f(x, y)\, dx\, dy = \iint_{\mathcal{R}} f\,(x(u, v), y(u, v))\, |\text{Jac}(\Phi)|\, du\, dv \tag{1}$$

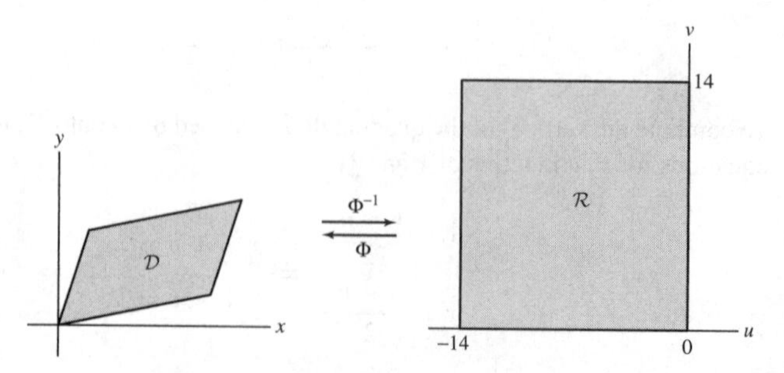

We compute the inverse Φ of Φ^{-1} by solving the equations $u = x - 5y$, $v = 3x - y$ for x, y in terms of u, v. We get

$$\begin{array}{l} u = x - 5y \\ v = 3x - y \end{array} \quad\Rightarrow\quad x = \frac{-u + 5v}{14}, \quad y = \frac{-3u + v}{14}$$

Therefore the function $f(x, y) = e^{3x-2y}$ in terms of u and v is

$$f\,(x(u, v), y(u, v)) = e^{3\left(\frac{-u+5v}{14}\right) - 2\left(\frac{-3u+v}{14}\right)} = e^{\frac{3u+13v}{14}} \tag{2}$$

The Jacobian of $\Phi(u,v) = (x,y) = \left(\frac{-u+5v}{14}, \frac{-3u+v}{14}\right)$ is

$$\mathrm{Jac}(\Phi) = \begin{vmatrix} \frac{\partial x}{\partial u} & \frac{\partial x}{\partial v} \\ \frac{\partial y}{\partial u} & \frac{\partial y}{\partial v} \end{vmatrix} = \begin{vmatrix} -\frac{1}{14} & \frac{5}{14} \\ -\frac{3}{14} & \frac{1}{14} \end{vmatrix} = \frac{1}{14} \tag{3}$$

Substituting (2) and (3) in (1) gives

$$\iint_{\mathcal{D}} e^{3x-2y}\,dx\,dy = \int_0^{14}\int_{-14}^{0} e^{\frac{3u+13v}{14}} \cdot \frac{1}{14}\,du\,dv = \frac{1}{14}\int_0^{14}\int_{-14}^{0} e^{\frac{3u}{14}} \cdot e^{\frac{13v}{14}}\,du\,dv$$

$$= \frac{1}{14}\left(\int_{-14}^{0} e^{\frac{3u}{14}}\,du\right)\left(\int_0^{14} e^{\frac{13v}{14}}\,dv\right) = \frac{1}{14}\left(\frac{14}{3}e^{\frac{3u}{14}}\Big|_{u=-14}^{0}\right)\left(\frac{14}{13}e^{\frac{13v}{14}}\Big|_{v=0}^{14}\right)$$

$$= \frac{1}{3}(1-e^{-3}) \cdot \frac{14}{13}(e^{13}-1) = \frac{14}{39}(1-e^{-3})(e^{13}-1)$$

17 LINE AND SURFACE INTEGRALS

17.1 Vector Fields (ET Section 16.1)

Preliminary Questions

1. Which of the following is a unit vector field in the plane?

(a) $\mathbf{F} = \langle y, x \rangle$

(b) $\mathbf{F} = \left\langle \frac{y}{\sqrt{x^2+y^2}}, \frac{x}{\sqrt{x^2+y^2}} \right\rangle$

(c) $\mathbf{F} = \left\langle \frac{y}{x^2+y^2}, \frac{x}{x^2+y^2} \right\rangle$

SOLUTION

(a) The length of the vector $\langle y, x \rangle$ is

$$\|\langle y, x \rangle\| = \sqrt{y^2 + x^2}$$

This value is not 1 for all points, hence it is not a unit vector field.

(b) We have

$$\left\| \left\langle \frac{y}{\sqrt{x^2+y^2}}, \frac{x}{\sqrt{x^2+y^2}} \right\rangle \right\| = \sqrt{\left(\frac{y}{\sqrt{x^2+y^2}}\right)^2 + \left(\frac{x}{\sqrt{x^2+y^2}}\right)^2}$$

$$= \sqrt{\frac{y^2}{x^2+y^2} + \frac{x^2}{x^2+y^2}} = \sqrt{\frac{y^2+x^2}{x^2+y^2}} = 1$$

Hence the field is a unit vector field, for $(x, y) \neq (0, 0)$.

(c) We compute the length of the vector:

$$\left\| \left\langle \frac{y}{x^2+y^2}, \frac{x}{x^2+y^2} \right\rangle \right\| = \sqrt{\left(\frac{y}{x^2+y^2}\right)^2 + \left(\frac{x}{x^2+y^2}\right)^2} = \sqrt{\frac{y^2+x^2}{(x^2+y^2)^2}} = \sqrt{\frac{1}{x^2+y^2}}$$

Since the length is not identically 1, the field is not a unit vector field.

2. Sketch an example of a nonconstant vector field in the plane in which each vector is parallel to $\langle 1, 1 \rangle$.

SOLUTION The non-constant vector $\langle x, x \rangle$ is parallel to the vector $\langle 1, 1 \rangle$.

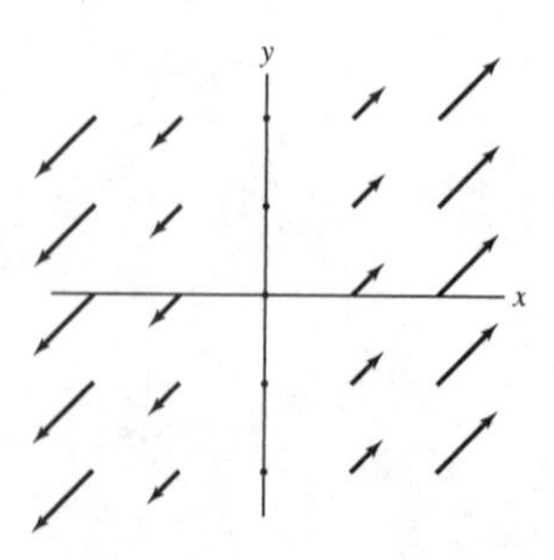

3. Show that the vector field $\mathbf{F} = \langle -z, 0, x \rangle$ is orthogonal to the position vector $\overrightarrow{OP}$ at each point P. Give an example of another vector field with this property.

SOLUTION The position vector at $P = (x, y, z)$ is $\langle x, y, z \rangle$. We must show that the following dot product is zero:

$$\langle x, y, z \rangle \cdot \langle -z, 0, x \rangle = x \cdot (-z) + y \cdot 0 + z \cdot x = 0$$

Therefore, the vector field $\mathbf{F} = \langle -z, 0, x\rangle$ is orthogonal to the position vector. Another vector field with this property is $\mathbf{F} = \langle 0, -z, y\rangle$, since

$$\langle 0, -z, y\rangle \cdot \langle x, y, z\rangle = 0 \cdot x + (-z) \cdot y + y \cdot z = 0$$

4. Give an example of a potential function for $\langle yz, xz, xy\rangle$ other than $\varphi(x, y, z) = xyz$.

SOLUTION Since any two potential functions of a gradient vector field differ by a constant, a potential function for the given field other than $\phi(x, y, z) = xyz$ is, for instance, $\phi_1(x, y, z) = xyz + 1$.

Exercises

1. Compute and sketch the vector assigned to the points $P = (1, 2)$ and $Q = (-1, -1)$ by the vector field $\mathbf{F} = \langle x^2, x\rangle$.

SOLUTION The vector assigned to $P = (1, 2)$ is obtained by substituting $x = 1$ in $\mathbf{F}$, that is,

$$\mathbf{F}(1, 2) = \langle 1^2, 1\rangle = \langle 1, 1\rangle$$

Similarly,

$$\mathbf{F}(-1, -1) = \langle (-1)^2, -1\rangle = \langle 1, -1\rangle$$

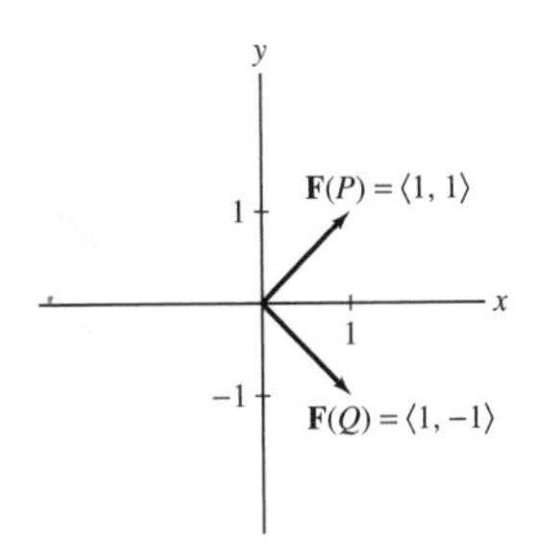

3. Compute and sketch the vector assigned to the points $P = (0, 1, 1)$ and $Q = (2, 1, 0)$ by the vector field $\mathbf{F} = \langle xy, z^2, x\rangle$.

SOLUTION To find the vector assigned to the point $P = (0, 1, 1)$, we substitute $x = 0$, $y = 1$, $z = 1$ in $\mathbf{F} = \langle xy, z^2, x\rangle$. We get

$$\mathbf{F}(P) = \langle 0 \cdot 1, 1^2, 0\rangle = \langle 0, 1, 0\rangle$$

Similarly, $\mathbf{F}(Q)$ is obtained by substituting $x = 2$, $y = 1$, $z = 0$ in $\mathbf{F}$. That is,

$$\mathbf{F}(Q) = \langle 2 \cdot 1, 0^2, 2\rangle = \langle 2, 0, 2\rangle$$

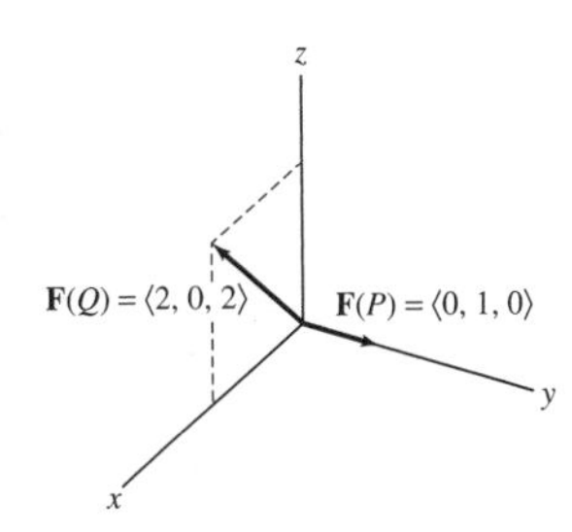

In Exercises 5–13, sketch the following planar vector fields by drawing the vectors attached to points with integer coordinates in the rectangle $-3 \le x, y \le 3$. Instead of drawing the vectors with their true lengths, scale them if necessary to avoid overlap.

5. $\mathbf{F} = \langle 1, 0\rangle$

SOLUTION The constant vector field $\langle 1, 0\rangle$ is shown in the figure:

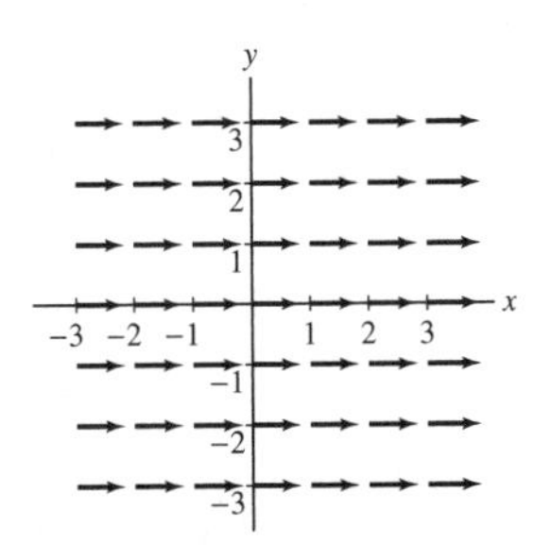

7. $\mathbf{F} = \langle 0, x \rangle$

SOLUTION We sketch the vector field $\mathbf{F}(x, y) = \langle 0, x \rangle$:

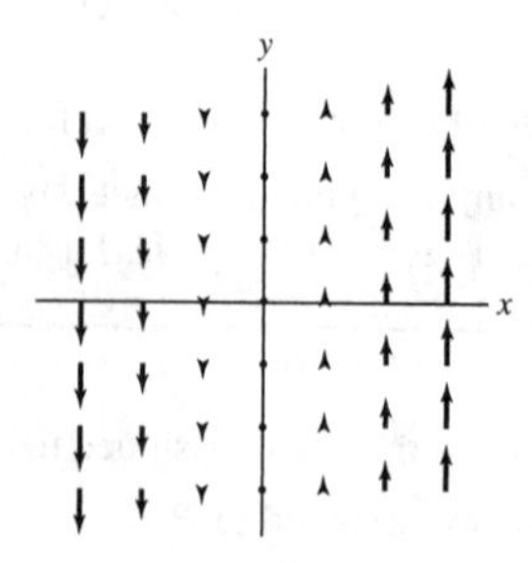

9. $\mathbf{F} = \langle 1, 1 \rangle$

SOLUTION We sketch the graph of the constant vector field $\mathbf{F}(x, y) = \langle 1, 1 \rangle$:

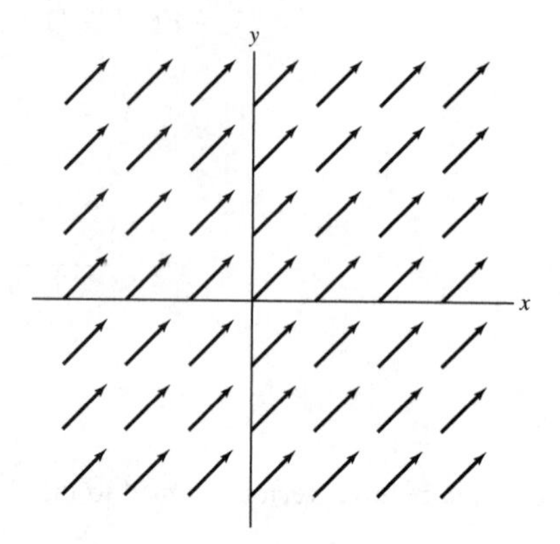

11. $\mathbf{F} = \left\langle \frac{x}{x^2 + y^2}, \frac{y}{x^2 + y^2} \right\rangle$

SOLUTION

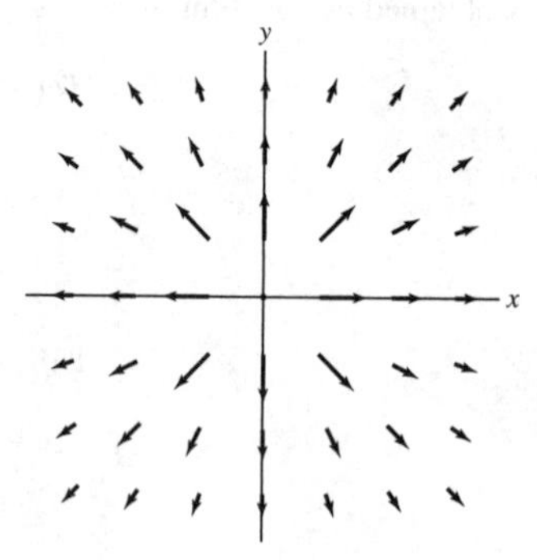

13. $\mathbf{F} = \left\langle \frac{y}{x^2 + y^2}, \frac{-x}{x^2 + y^2} \right\rangle$

SOLUTION

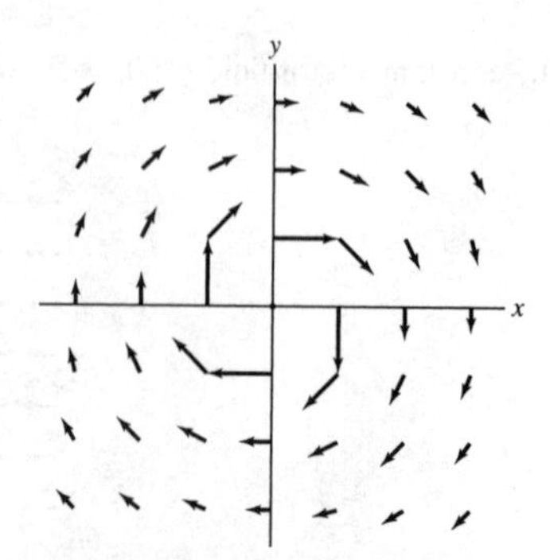

In Exercises 14–17, match the planar vector field with the corresponding plot in Figure 7.

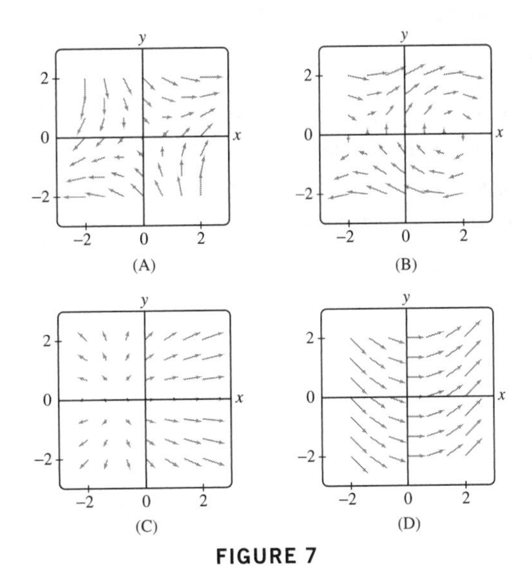

FIGURE 7

15. $\mathbf{F} = \langle 2x + 2, y \rangle$

SOLUTION We compute the images of the point $(0, 2)$, for instance, and identify the corresponding graph accordingly:

$$\mathbf{F}(x, y) = \langle 2x + 2, y \rangle \quad \Rightarrow \quad \mathbf{F}(0, 2) = \langle 2, 2 \rangle \quad \Rightarrow \quad \text{Plot(C)}$$

17. $\mathbf{F} = \langle x + y, x - y \rangle$

SOLUTION We compute the images of the point $(0, 2)$, for instance, and identify the corresponding graph accordingly:

$$\mathbf{F}(x, y) = \langle x + y, x - y \rangle \quad \Rightarrow \quad \mathbf{F}(0, 2) = \langle 2, -2 \rangle \quad \Rightarrow \quad \text{Plot(A)}$$

In Exercises 18–21, match the three-dimensional vector field with the corresponding plot in Figure 8.

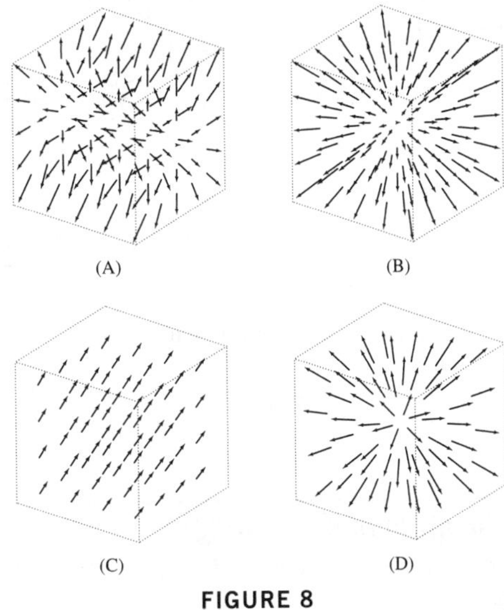

FIGURE 8

19. $\mathbf{F} = \langle x, 0, z \rangle$

SOLUTION This vector field is shown in (A) (by process of elimination).

21. $\mathbf{F} = \langle 1, 1, 1 \rangle$

SOLUTION The constant vector field $\langle 1, 1, 1 \rangle$ is shown in plot (C).

In Exercises 22–25, find a potential function for the vector field $\mathbf{F}$ *by inspection.*

23. $\mathbf{F} = \left\langle ye^{xy}, xe^{xy} \right\rangle$

SOLUTION The function $\varphi(x, y) = e^{xy}$ satisfies $\frac{\partial \varphi}{\partial x} = ye^{xy}$ and $\frac{\partial \varphi}{\partial y} = xe^{xy}$, hence φ is a potential function for the given vector field.

25. $\mathbf{F} = \langle 2xze^{x^2}, 0, e^{x^2} \rangle$

SOLUTION The function $\varphi(x, y, z) = ze^{x^2}$ satisfies $\frac{\partial \varphi}{\partial y} = 0$, $\frac{\partial \varphi}{\partial x} = 2xze^{x^2}$ and $\frac{\partial \varphi}{\partial z} = e^{x^2}$, hence φ is a potential function for the given vector field.

27. Which of (A) or (B) in Figure 9 is the contour plot of a potential function for the vector field $\mathbf{F}$? Recall that the gradient vectors are perpendicular to the level curves.

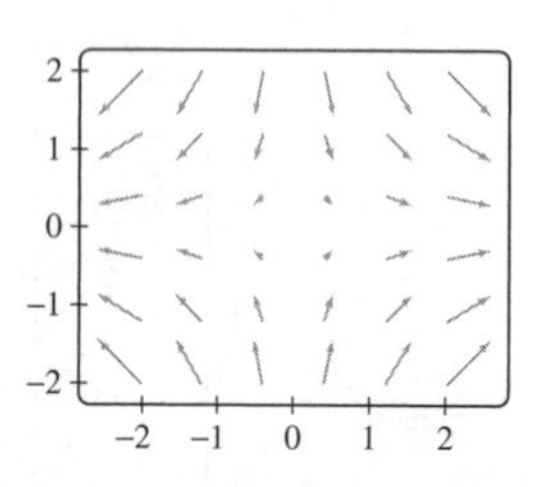

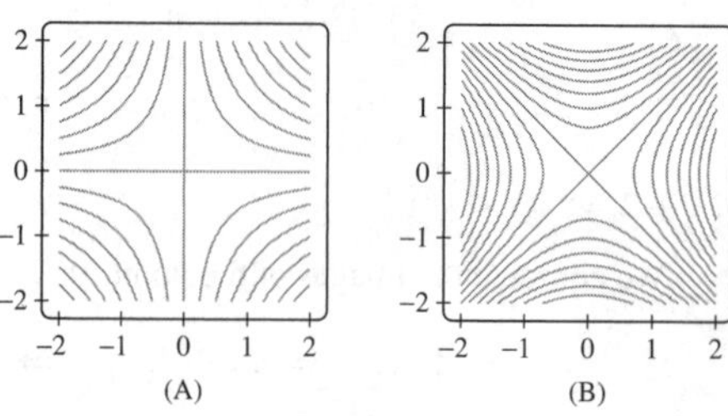

FIGURE 9

SOLUTION By the equality $\nabla\varphi = \mathbf{F}$ and since the gradient vectors are perpendicular to the level curves, it follows that the vectors $\mathbf{F}$ are perpendicular to the corresponding level curves of φ. This property is satisfied in (B) and not satisfied in (A). Therefore (B) is the contour plot of φ.

29. Let $\varphi = \ln r$, where $r = \sqrt{x^2 + y^2}$. Express $\nabla\varphi$ in terms of the unit radial vector $\mathbf{e}_r$ in $\mathbf{R}^2$.

SOLUTION Since $r = (x^2 + y^2 + z^2)^{1/2}$, we have $\varphi = \ln (x^2 + y^2 + z^2)^{1/2} = \frac{1}{2}\ln(x^2 + y^2 + z^2)$. We compute the partial derivatives:

$$\frac{\partial \varphi}{\partial x} = \frac{1}{2}\frac{2x}{x^2 + y^2 + z^2} = \frac{x}{r^2}$$

$$\frac{\partial \varphi}{\partial y} = \frac{1}{2}\frac{2y}{x^2 + y^2 + z^2} = \frac{y}{r^2}$$

$$\frac{\partial \varphi}{\partial z} = \frac{1}{2}\frac{2z}{x^2 + y^2 + z^2} = \frac{z}{r^2}$$

Therefore, the gradient of φ is the following vector:

$$\nabla\varphi = \left\langle \frac{\partial \varphi}{\partial x}, \frac{\partial \varphi}{\partial y}, \frac{\partial \varphi}{\partial z} \right\rangle = \left\langle \frac{x}{r^2}, \frac{y}{r^2}, \frac{z}{r^2} \right\rangle = \frac{1}{r}\left\langle \frac{x}{r}, \frac{y}{r}, \frac{z}{r} \right\rangle = \frac{\mathbf{e}_r}{r}$$

Further Insights and Challenges

31. Show that any vector field of the form $\mathbf{F} = \langle f(x), g(y), h(z) \rangle$ has a potential function. Assume that f, g, and h are continuous.

SOLUTION Let $F(x)$, $G(y)$, and $H(z)$ be antiderivatives of $f(x)$, $g(y)$, and $h(z)$, respectively. That is, $F'(x) = f(x)$, $G'(y) = g(y)$, and $H'(y) = h(z)$. We define the function

$$\varphi(x, y, z) = F(x) + G(y) + H(z)$$

Then,

$$\frac{\partial \varphi}{\partial x} = F'(x) = f(x), \quad \frac{\partial \varphi}{\partial x} = G'(y) = g(y), \quad \frac{\partial \varphi}{\partial z} = H'(z) = h(z)$$

Therefore, $\nabla\varphi = \mathbf{F}$, or φ is a potential function for $\mathbf{F}$.

33. Show that if $\nabla\varphi(x, y) = \mathbf{0}$ for all (x, y) in a disk $\mathcal{D}$ in $\mathbf{R}^2$, then φ is constant on $\mathcal{D}$. *Hint:* Given points $P = (a, b)$ and $Q = (c, d)$ in $\mathcal{D}$, let $R = (c, b)$ (Figure 13). Use single-variable calculus to show that φ is constant along the segments $\overline{PR}$ and $\overline{RQ}$ and conclude that $\varphi(P) = \varphi(R)$.

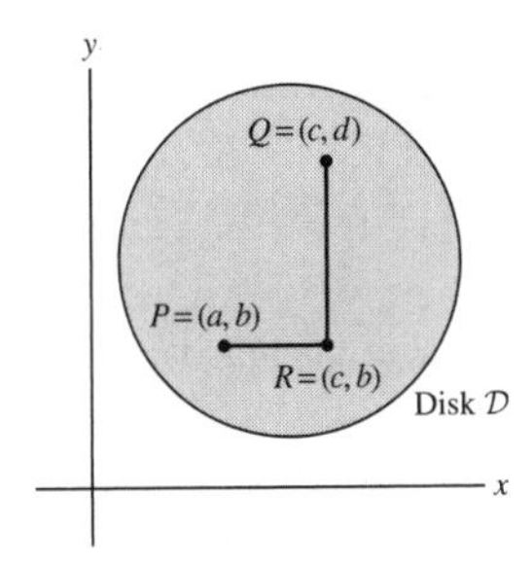

FIGURE 13

SOLUTION Given any two points $P = (a, b)$ and $Q = (c, d)$ in $\mathcal{D}$, we must show that

$$\varphi(P) = \varphi(Q)$$

We consider the point $R = (c, b)$ and the segments $\overline{PR}$ and $\overrightarrow{RQ}$. (We assume that (c, b) is in $\mathcal{D}$; if not, just use $R' = (a, d)$.)

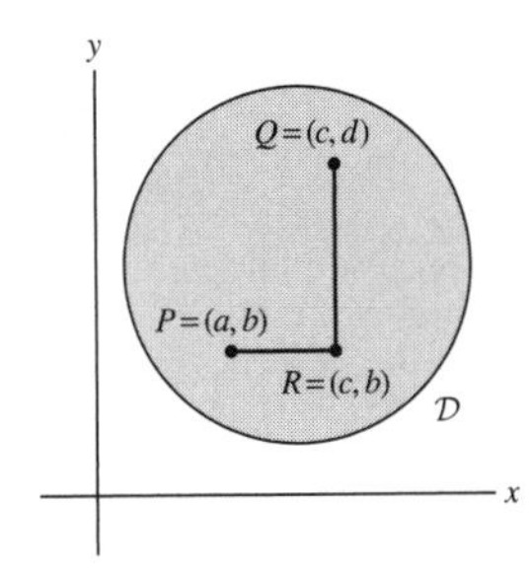

Since $\frac{\partial\varphi}{\partial x}(x, y) = 0$ in $\mathcal{D}$, in particular $\frac{\partial\varphi}{\partial x}(x, b) = 0$ for $a \le x \le c$. Therefore, for $a \le x \le c$ we have

$$\varphi(x, b) = \int_a^x \frac{\partial\varphi}{\partial u}(u, b)\, du + \varphi(a, b) = \int_a^x 0\, du + \varphi(a, b) = k + \varphi(a, b)$$

Substituting $x = a$ determines $k = 0$. Hence,

$$\varphi(x, b) = \varphi(a, b) \quad \text{for} \quad a \le x \le c$$

In particular,

$$\varphi(c, b) = \varphi(a, b) \quad \Rightarrow \quad \varphi(R) = \varphi(P) \tag{1}$$

Similarly, since $\frac{\partial\varphi}{\partial y}(x, y) = 0$ in $\mathcal{D}$, we have $\frac{\partial\varphi}{\partial y}(c, y) = 0$ for $b \le y \le d$. Therefore for $b \le y \le d$ we have

$$\varphi(c, y) = \int_b^y \frac{\partial\varphi}{\partial v}(c, v)\, dv + \varphi(c, b) = \int_b^y 0\, dv + \varphi(c, b) = k + \varphi(c, b)$$

Substituting $y = b$ gives $\varphi(c, b) = k + \varphi(c, b)$ or $k = 0$. Therefore,

$$\varphi(c, y) = \varphi(c, b) \quad \text{for} \quad b \le y \le d$$

In particular,

$$\varphi(c, d) = \varphi(c, b) \quad \Rightarrow \quad \varphi(Q) = \varphi(R) \tag{2}$$

Combining (1) and (2) we obtain the desired equality $\varphi(P) = \varphi(Q)$. Since P and Q are any two points in $\mathcal{D}$, we conclude that φ is constant on $\mathcal{D}$.

17.2 Line Integrals (ET Section 16.2)

Preliminary Questions

1. What is the line integral of the constant function $f(x, y, z) = 10$ over a curve $\mathcal{C}$ of length 5?

SOLUTION Since the length of C is the line integral $\int_C 1\,ds = 5$, we have

$$\int_C 10\,ds = 10\int_C 1\,ds = 10 \cdot 5 = 50$$

2. Which of the following have a zero line integral over the vertical segment from (0, 0) to (0, 1)?

(a) $f(x, y) = x$

(b) $f(x, y) = y$

(c) $\mathbf{F} = \langle x, 0\rangle$

(d) $\mathbf{F} = \langle y, 0\rangle$

(e) $\mathbf{F} = \langle 0, x\rangle$

(f) $\mathbf{F} = \langle 0, y\rangle$

SOLUTION The vertical segment from (0, 0) to (0, 1) has the parametrization

$$\mathbf{c}(t) = (0, t), \quad 0 \le t \le 1$$

Therefore, $\mathbf{c}'(t) = \langle 0, 1\rangle$ and $\|\mathbf{c}'(t)\| = 1$. The line integrals are thus computed by

$$\int_C f(x, y)\,ds = \int_0^1 f(\mathbf{c}(t))\,\|\mathbf{c}'(t)\|\,dt \tag{1}$$

$$\int_C \mathbf{F} \cdot d\mathbf{s} = \int_0^1 \mathbf{F}(\mathbf{c}(t)) \cdot \mathbf{c}'(t)\,dt \tag{2}$$

(a) We have $f(\mathbf{c}(t)) = x = 0$. Therefore by (1) the line integral is zero.

(b) By (1), the line integral is

$$\int_C f(x, y)\,ds = \int_0^1 t \cdot 1\,dt = \frac{1}{2}t^2\Big|_0^1 = \frac{1}{2} \ne 0$$

(c) This vector line integral is computed using (2). Since $\mathbf{F}(\mathbf{c}(t)) = \langle x, 0\rangle = \langle 0, 0\rangle$, the vector line integral is zero.

(d) By (2) we have

$$\int_C \mathbf{F} \cdot d\mathbf{s} = \int_0^1 \langle t, 0\rangle \cdot \langle 0, 1\rangle\,dt = \int_0^1 0\,dt = 0$$

(e) The vector integral is computed using (2). Since $\mathbf{F}(\mathbf{c}(t)) = \langle 0, x\rangle = \langle 0, 0\rangle$, the line integral is zero.

(f) For this vector field we have

$$\int_C \mathbf{F} \cdot d\mathbf{s} = \int_0^1 \mathbf{F}(\mathbf{c}(t)) \cdot \mathbf{c}'(t)\,dt = \int_0^1 \langle 0, t\rangle \cdot \langle 0, 1\rangle\,dt = \int_0^1 t\,dt = \frac{t^2}{2}\Big|_0^1 = \frac{1}{2} \ne 0$$

So, we conclude that (a), (c), (d), and (e) have an integral of zero.

3. State whether true or false. If false, give the correct statement.

(a) The scalar line integral does not depend on how you parametrize the curve.

(b) If you reverse the orientation of the curve, neither the vector nor the scalar line integral changes sign.

SOLUTION

(a) True: It can be shown that any two parametrizations of the curve yield the same value for the scalar line integral, hence the statement is true.

(b) False: For the definition of the scalar line integral, there is no need to specify a direction along the path, hence reversing the orientation of the curve does not change the sign of the integral. However, reversing the orientation of the curve changes the sign of the vector line integral.

4. Let C be a curve of length 5. What is the value of $\displaystyle\int_C \mathbf{F} \cdot d\mathbf{s}$ if

(a) $\mathbf{F}(P)$ is normal to C at all points P on C?

(b) $\mathbf{F}(P) = \mathbf{T}(P)$ at all points P on C, where $\mathbf{T}(P)$ is the unit tangent vector pointing in the forward direction along the curve?

SOLUTION

(a) The vector line integral is the integral of the tangential component of the vector field along the curve. Since $\mathbf{F}(P)$ is normal to C at all points P on C, the tangential component is zero, hence the line integral $\int_C \mathbf{F} \cdot d\mathbf{s}$ is zero.

(b) In this case we have

$$\mathbf{F}(P) \cdot \mathbf{T}(P) = \mathbf{T}(P) \cdot \mathbf{T}(P) = \|\mathbf{T}(P)\|^2 = 1$$

Therefore,

$$\int_C \mathbf{F} \cdot d\mathbf{s} = \int_C (\mathbf{F} \cdot \mathbf{T})\,ds = \int_C 1\,ds = \text{Length of } C = 5.$$

Exercises

1. Let $f(x, y, z) = x + yz$ and let C be the line segment from $P = (0, 0, 0)$ to $(6, 2, 2)$.

(a) Calculate $f(\mathbf{c}(t))$ and $ds = \|\mathbf{c}'(t)\|\,dt$ for the parametrization $\mathbf{c}(t) = (6t, 2t, 2t)$ for $0 \le t \le 1$.

(b) Evaluate $\int_C f(x, y, z)\,ds$.

SOLUTION

(a) We substitute $x = 6t$, $y = 2t$, $z = 2t$ in the function $f(x, y, z) = x + yz$ to find $f(\mathbf{c}(t))$:

$$f(\mathbf{c}(t)) = 6t + (2t)(2t) = 6t + 4t^2$$

We differentiate the vector $c(t)$ and compute the length of the derivative vector:

$$\mathbf{c}'(t) = \frac{d}{dt}\langle 6t, 2t, 2t\rangle = \langle 6, 2, 2\rangle \quad \Rightarrow \quad \mathbf{c}'(t) = \sqrt{6^2 + 2^2 + 2^2} = \sqrt{44} = 2\sqrt{11}$$

Hence,

$$ds = \|\mathbf{c}'(t)\|\,dt = 2\sqrt{11}\,dt$$

(b) Computing the scalar line integral, we obtain

$$\int_C f(x, y, z)\,ds = \int_0^1 f(\mathbf{c}(t))\,\|\mathbf{c}'(t)\|\,dt = \int_0^1 (6t + 4t^2)\cdot 2\sqrt{11}\,dt$$

$$= 2\sqrt{11}\left(3t^2 + \frac{4}{3}t^3\right)\Big|_0^1 = 2\sqrt{11}\left(3 + \frac{4}{3}\right) = \frac{26\sqrt{11}}{3}$$

3. Let $\mathbf{F} = \langle y^2, x^2\rangle$ and let C be the $y = x^{-1}$ for $1 \le x \le 2$, oriented from left to right.

(a) Calculate $\mathbf{F}(\mathbf{c}(t))$ and $d\mathbf{s} = \mathbf{c}'(t)\,dt$ for the parametrization $\mathbf{c}(t) = (t, t^{-1})$.

(b) Calculate the dot product $\mathbf{F}(\mathbf{c}(t))\cdot\mathbf{c}'(t)\,dt$ and evaluate $\int_C \mathbf{F}\cdot d\mathbf{s}$.

SOLUTION

(a) We calculate $\mathbf{F}(\mathbf{c}(t))$ by substituting $x = t$, $y = t^{-1}$ in $\mathbf{F} = \left\langle y^2, x^2\right\rangle$. We get

$$\mathbf{F}(\mathbf{c}(t)) = \left\langle (t^{-1})^2, t^2\right\rangle = \left\langle t^{-2}, t^2\right\rangle$$

We compute $\mathbf{c}'(t)$:

$$\mathbf{c}'(t) = \frac{d}{dt}\left\langle t, t^{-1}\right\rangle = \left\langle 1, -t^{-2}\right\rangle \quad \Rightarrow \quad d\mathbf{s} = \left\langle 1, -t^{-2}\right\rangle dt$$

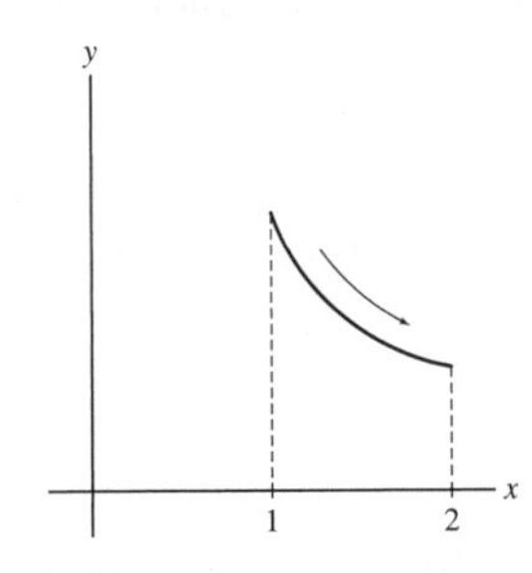

(b) We compute the dot product:

$$\mathbf{F}(\mathbf{c}(t))\cdot\mathbf{c}'(t) = \left\langle t^{-2}, t^2\right\rangle\cdot\left\langle 1, -t^{-2}\right\rangle = t^{-2}\cdot 1 + t^2\cdot(-t^{-2}) = t^{-2} - 1$$

Computing the vector line integral, we obtain

$$\int_C \mathbf{F}\cdot d\mathbf{s} = \int_1^2 \mathbf{F}(\mathbf{c}(t))\cdot\mathbf{c}'(t)\,dt = \int_1^2 (t^{-2} - 1)\,dt = -t^{-1} - t\Big|_1^2 = \left(-\frac{1}{2} - 2\right) - (-1 - 1) = -\frac{1}{2}$$

In Exercises 5–8, calculate the integral of the given scalar function or vector field over the curve $\mathbf{c}(t) = (\cos t, \sin t, t)$ for $0 \le t \le \pi$.

5. $f(x, y, z) = x^2 + y^2 + z^2$

SOLUTION

Step 1. Compute $\|\mathbf{c}'(t)\|$. We differentiate $\mathbf{c}(t)$:

$$\mathbf{c}'(t) = \frac{d}{dt}\langle \cos t, \sin t, t\rangle = \langle -\sin t, \cos t, 1\rangle$$

Hence,

$$\|\mathbf{c}'(t)\| = \sqrt{(-\sin t)^2 + \cos^2 t + 1^2} = \sqrt{\sin^2 t + \cos^2 t + 1} = \sqrt{2}$$

$$ds = \|\mathbf{c}'(t)\|\, dt = \sqrt{2}\, dt$$

Step 2. Write out $f(\mathbf{c}(t))$. We substitute $x = \cos t$, $y = \sin t$, $z = t$ in $f(x, y, z) = x^2 + y^2 + z^2$ to obtain

$$f(\mathbf{c}(t)) = \cos^2 t + \sin^2 t + t^2 = 1 + t^2$$

Step 3. Compute the line integral. Using the Theorem on Scalar Line Integrals we obtain

$$\int_C (x^2 + y^2 + z^2)\, ds = \int_0^{\pi} f(\mathbf{c}(t))\, \|\mathbf{c}'(t)\|\, dt = \int_0^{\pi} (1 + t^2)\sqrt{2}\, dt = \sqrt{2}\left(t + \frac{t^3}{3}\right)\Big|_0^{\pi} = \sqrt{2}\left(\pi + \frac{\pi^3}{3}\right)$$

7. $\mathbf{F} = \langle x, y, z\rangle$

SOLUTION

Step 1. Calculate the integrand. We write out the vectors:

$$\mathbf{c}(t) = (\cos t, \sin t, t)$$
$$\mathbf{F}(\mathbf{c}(t)) = \langle x, y, z\rangle = \langle \cos t, \sin t, t\rangle$$
$$\mathbf{c}'(t) = \langle -\sin t, \cos t, 1\rangle$$

The integrand is the dot product:

$$\mathbf{F}(\mathbf{c}(t)) \cdot \mathbf{c}'(t) = \langle \cos t, \sin t, t\rangle \cdot \langle -\sin t, \cos t, 1\rangle = -\cos t \sin t + \sin t \cos t + t = t$$

Step 2. Evaluate the integral. We use the Theorem on Vector Line Integrals to evaluate the integral:

$$\int_C \mathbf{F}\, d\mathbf{s} = \int_0^{\pi} \mathbf{F}(\mathbf{c}(t)) \cdot \mathbf{c}'(t)\, dt = \int_0^{\pi} t\, dt = \frac{1}{2}t^2\Big|_0^{\pi} = \frac{\pi^2}{2}$$

9. Calculate the total mass of a circular piece of wire of radius 4 cm centered at the origin whose mass density is $\rho(x, y) = x^2$ g/cm.

SOLUTION The total mass is the following integral:

$$M = \int_C x^2\, ds$$

We use the following parametrization of the wire:

$$\mathbf{c}(t) = (4\cos t, 4\sin t), \quad 0 \le t \le 2\pi$$

Hence,

$$\mathbf{c}'(t) = \langle -4\sin t, 4\cos t\rangle \quad \Rightarrow \quad \|\mathbf{c}'(t)\| = \sqrt{(-4\sin t)^2 + (4\cos t)^2} = 4$$

We compute the line integral using the Theorem on Scalar Line Integrals. We get

$$M = \int_0^{2\pi} \rho(\mathbf{c}(t))\, \|\mathbf{c}'(t)\|\, dt = \int_0^{2\pi} (4\cos t)^2 \cdot 4\, dt$$
$$= 64\int_0^{2\pi} \cos^2 t\, dt = 64\left(\frac{t}{2} + \frac{\sin 2t}{4}\right)\Big|_0^{2\pi} = 64 \cdot \frac{2\pi}{2} = 64\pi \text{g}$$

11. The values of a function $f(x, y, z)$ and vector field $\mathbf{F}(x, y, z)$ are given at six sample points along the path ABC in Figure 11. Estimate the line integrals of f and $\mathbf{F}$ along ABC.

Point	$f(x, y, z)$	$\mathbf{F}(x, y, z)$
$(1, \frac{1}{6}, 0)$	3	$\langle 1, 0, 2\rangle$
$(1, \frac{1}{2}, 0)$	3.3	$\langle 1, 1, 3\rangle$
$(1, \frac{5}{6}, 0)$	3.6	$\langle 2, 1, 5\rangle$
$(1, 1, \frac{1}{6})$	4.2	$\langle 3, 2, 4\rangle$
$(1, 1, \frac{1}{2})$	4.5	$\langle 3, 3, 3\rangle$
$(1, 1, \frac{5}{6})$	4.2	$\langle 5, 3, 3\rangle$

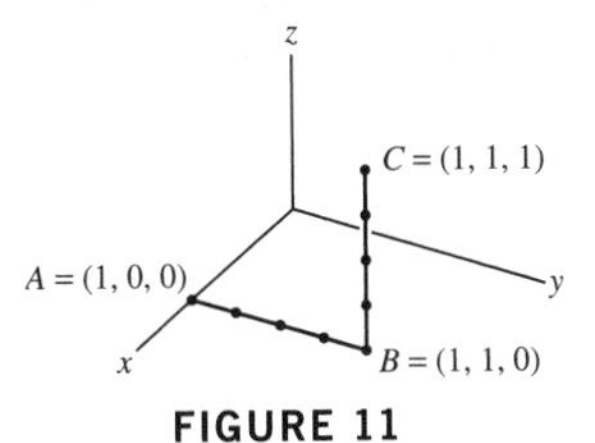

FIGURE 11

SOLUTION

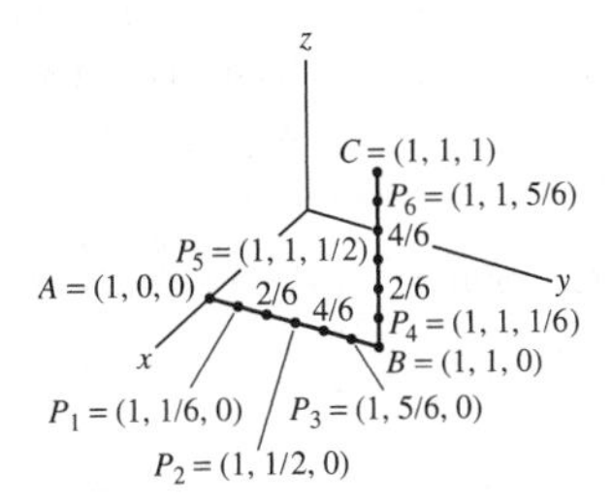

We write the integrals as sum of integrals and estimate each integral by a Riemann Sum. That is,

$$\int_{ABC} f(x, y, z)\,ds = \int_{AB} f(x, y, z)\,ds + \int_{BC} f(x, y, z)\,d\mathbf{s} \approx \sum_{i=1}^{3} f(P_i)\,\Delta s_i + \sum_{i=4}^{6} f(P_i)\Delta s_i \tag{1}$$

$$\int_{ABC} \mathbf{F}\cdot d\mathbf{s} = \int_{AB} \mathbf{F}\cdot d\mathbf{s} + \int_{BC} \mathbf{F}\cdot d\mathbf{s} = \int_{AB} (\mathbf{F}\cdot\mathbf{T})ds + \int_{BC} (\mathbf{F}\cdot\mathbf{T})ds$$

On AB, the unit tangent vector is $\mathbf{T} = \langle 1, 0, 0\rangle$, hence $\mathbf{F}\cdot\mathbf{T} = F_1$. On BC, the unit tangent vector is $\mathbf{T} = \langle 0, 0, 1\rangle$, hence $\mathbf{F}\cdot\mathbf{T} = F_3$. Therefore,

$$\int_{ABC} \mathbf{F}\,ds = \int_{AB} F_1\,ds + \int_{BC} F_3\,ds \approx \sum_{i=1}^{3} F_1(P_i)\,\Delta s_i + \sum_{i=4}^{6} F_3(P_i)\,\Delta s_i \tag{2}$$

We consider the partitions of AB and BC to three subarcs with equal length $\Delta s_i = \frac{1}{3}$, therefore (1) and (2) give

$$\int_{ABC} f(x, y, z)\,ds \approx \frac{1}{3}\left(f(P_1) + f(P_2) + f(P_3) + f(P_4) + f(P_5) + f(P_6)\right)$$

$$\int_{ABC} \mathbf{F}d\mathbf{s} = \frac{1}{3}\left(F_1(P_1) + F_1(P_2) + F_1(P_3) + F_3(P_4) + F_3(P_5) + F_3(P_6)\right)$$

We now substitute the values of the functions at the sample points to obtain the following approximations:

$$\int_{ABC} f(x, y, z)\,ds \approx \frac{1}{3}(3 + 3.3 + 3.6 + 4.2 + 4.5 + 4.2) = 7.6$$

$$\int_{ABC} \mathbf{F}\cdot d\mathbf{s} \approx \frac{1}{3}(1 + 1 + 2 + 4 + 3 + 3) = \frac{14}{3} = 4\frac{2}{3}$$

13. Figure 13 shows three vector fields. In each case, determine whether the line integral around the circle (oriented counterclockwise) is positive, negative, or zero.

(A)

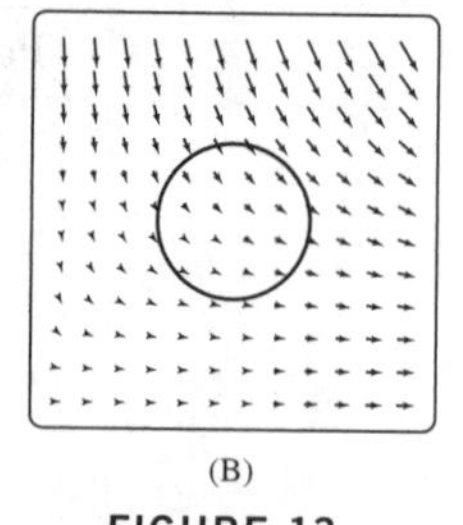

(B)

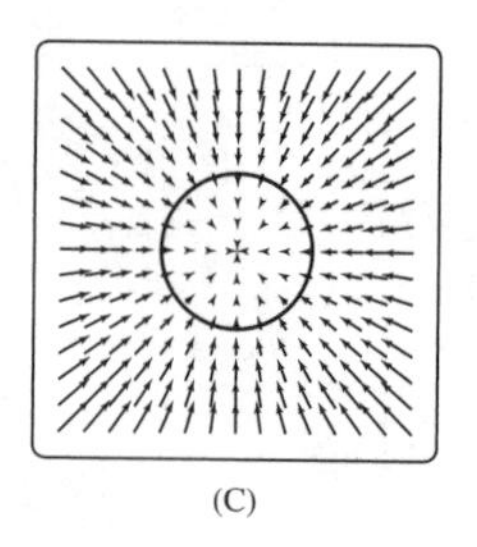

(C)

FIGURE 13

SOLUTION The vector line integral of F is the integral of the tangential component of F along the curve. The positive direction of a curve is counterclockwise.

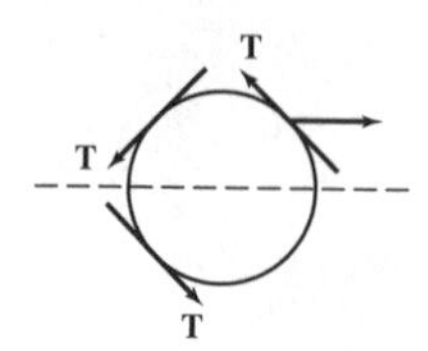

For the vector field in (A), the line integral around the circle is zero because the contribution of the negative tangential components from the upper part of the circle is the same as the contribution of the positive tangential components from the lower part. For the vector in (B) the contribution of the negative tangential component appear to dominate over the positive contribution, hence the line integral is negative. In (C), the vector field is orthogonal to the unit tangent vector at each point, hence the line integral is zero.

In Exercises 15–22, compute the line integral of the scalar function over the curve.

15. $f(x, y, z) = z^2$, $\mathbf{c}(t) = (2t, 3t, 4t)$ for $0 \le t \le 2$

SOLUTION

Step 1. Compute $\|\mathbf{c}'(t)\|$ We have

$$\mathbf{c}'(t) = \frac{d}{dt}\langle 2t, 3t, 4t\rangle = \langle 2, 3, 4\rangle \quad \Rightarrow \quad \|\mathbf{c}'(t)\| = \sqrt{2^2 + 3^2 + 4^2} = \sqrt{29}$$

Step 2. Write out $f(\mathbf{c}(t))$ We substitute $z = 4t$ in $f(x, y, z) = z^2$ to obtain:

$$f(\mathbf{c}(t)) = 16t^2$$

Step 3. Compute the line integral. By the Theorem on Scalar Line Integrals we have

$$\int_C f(x, y, z)\, ds = \int_0^2 f(\mathbf{c}(t))\, \|\mathbf{c}'(t)\|\, dt = \int_0^2 16t^2 \cdot \sqrt{29}\, dt = \sqrt{29} \cdot \frac{16}{3}t^3 \Big|_0^2 = \frac{128\sqrt{29}}{3} \approx 229.8$$

17. $f(x, y) = \sqrt{1 + 9xy}$, $y = x^3$ for $0 \le x \le 1$

SOLUTION The curve is parametrized by $\mathbf{c}(t) = \left(t, t^3\right)$ for $0 \le t \le 1$

Step 1. Compute $\|\mathbf{c}'(t)\|$. We have

$$\mathbf{c}'(t) = \frac{d}{dt}\left\langle t, t^3\right\rangle = \left\langle 1, 3t^2\right\rangle \quad \Rightarrow \quad \|\mathbf{c}'(t)\| = \sqrt{1 + 9t^4}$$

Step 2. Write out $f(\mathbf{c}(t))$. We substitute $x = t$, $y = t^3$ in $f(x, y) = \sqrt{1 + 9xy}$ to obtain

$$f(\mathbf{c}(t)) = \sqrt{1 + 9t \cdot t^3} = \sqrt{1 + 9t^4}$$

Step 3. Compute the line integral. We use the Theorem on Scalar Line Integrals to write

$$\int_C f(x, y)\, ds = \int_0^1 f(\mathbf{c}(t))\, \|\mathbf{c}'(t)\|\, dt = \int_0^1 \sqrt{1 + 9t^4}\sqrt{1 + 9t^4}\, dt = \int_0^1 \left(1 + 9t^4\right) dt$$

$$= t + \frac{9t^5}{5}\Big|_0^1 = \frac{14}{5} = 2.8$$

19. $f(x, y, z) = xe^{z^2}$, piecewise linear path from $(0, 0, 1)$ to $(0, 2, 0)$ to $(1, 1, 1)$.

SOLUTION Let $\mathcal{C}_1$ be the segment joining the points $(0, 0, 1)$ and $(0, 2, 0)$ and $\mathcal{C}_2$ be the segment joining the points $(0, 2, 0)$ and $(1, 1, 1)$. We parametrize $\mathcal{C}_1$ and $\mathcal{C}_2$ by the following parametrization:

$$\mathcal{C}_1 : \mathbf{c}_1(t) = (0, 2t, 1-t),\ 0 \le t \le 1$$
$$\mathcal{C}_2 : \mathbf{c}_2(t) = (t, 2-t, t),\ 0 \le t \le 1$$

For $\mathcal{C} = \mathcal{C}_1 + \mathcal{C}_1$ we have

$$\int_{\mathcal{C}} f(x, y, z)\, ds = \int_{\mathcal{C}_1} f(x, y, z)\, ds + \int_{\mathcal{C}_2} f(x, y, z)\, ds \tag{1}$$

We compute the integrals on the right hand side.

- The integral over $\mathcal{C}_1$: We have

$$\mathbf{c}_1'(t) = \frac{d}{dt}\langle 0, 2t, 1-t\rangle = \langle 0, 2, -1\rangle \quad \Rightarrow \quad \|\mathbf{c}_1'(t)\| = \sqrt{0+4+1} = \sqrt{5}$$
$$f(\mathbf{c}(t)) = xe^{z^2} = 0 \cdot e^{(1-t)^2} = 0$$

Hence,

$$\int_{\mathcal{C}_1} f(x, y, z)\, ds = \int_0^1 f(\mathbf{c}_1(t))\, \|\mathbf{c}_1'(t)\|\, dt = \int_0^1 0\, dt = 0 \tag{2}$$

- The integral over $\mathcal{C}_2$: We have

$$\mathbf{c}_2'(t) = \frac{d}{dt}\langle t, 2-t, t\rangle = \langle 1, -1, 1\rangle \quad \Rightarrow \quad \|\mathbf{c}_2'(t)\| = \sqrt{1+1+1} = \sqrt{3}$$
$$f(\mathbf{c}_2(t)) = xe^{z^2} = te^{t^2}$$

Hence,

$$\int_{\mathcal{C}_2} f(x, y, z)\, ds = \int_0^1 te^{t^2}\sqrt{3}\, dt \tag{3}$$

Using the substitution $u = t^2$ we find that

$$\int_{\mathcal{C}_2} f(x, y, z)\, ds = \int_0^1 \frac{\sqrt{3}}{2} e^u\, du = \frac{\sqrt{3}}{2}(e-1) \approx 1.488$$

Hence,

$$\int_{\mathcal{C}} f(x, y, z)\, ds \approx 1.488$$

21. $f(x, y, z) = 2x^2 + 8z, \quad \mathbf{c}(t) = (e^t, t^2, t), \quad 0 \le t \le 1$

SOLUTION

Step 1. Compute $\|\mathbf{c}'(t)\|$.

$$\mathbf{c}'(t) = \frac{d}{dt}\left\langle e^t, t^2, t\right\rangle = \left\langle e^t, 2t, 1\right\rangle \quad \Rightarrow \quad \|\mathbf{c}'(t)\| = \sqrt{e^{2t} + 4t^2 + 1}$$

Step 2. Write out $f(\mathbf{c}(t))$. We substitute $x = e^t$, $y = t^2$, $z = t$ in $f(x, y, z) = 2x^2 + 8z$ to obtain:

$$f(\mathbf{c}(t)) = 2e^{2t} + 8t$$

Step 3. Compute the line integral. We have

$$\int_{\mathcal{C}} f(x, y, z)\, ds = \int_0^1 f(\mathbf{c}(t))\, \|\mathbf{c}'(t)\|\, dt = \int_0^1 (2e^{2t} + 8t)\sqrt{e^{2t} + 4t^2 + 1}\, dt$$

We compute the integral using the substitution $u = e^{2t} + 4t^2 + 1$, $du = 2e^{2t} + 8t\, dt$. We get:

$$\int_{\mathcal{C}} f(x, y, z)\, ds = \int_2^{e^2+5} u^{1/2}\, du = \frac{2}{3}u^{3/2}\Big|_2^{e^2+5} = \frac{2}{3}\left((e^2+5)^{3/2} - 2^{3/2}\right)$$

In Exercises 23–35, compute the line integral of the vector field over the oriented curve.

23. $\mathbf{F} = \langle x^2, xy \rangle$, line segment from (0, 0) to (2, 2)

SOLUTION The oriented line segment is parametrized by

$$\mathbf{c}(t) = (t, t), \quad t \text{ varies from 0 to 2.}$$

Therefore,

$$\mathbf{F}(\mathbf{c}(t)) = \left\langle x^2, xy \right\rangle = \left\langle t^2, t \cdot t \right\rangle = \left\langle t^2, t^2 \right\rangle$$

$$\mathbf{c}'(t) = \frac{d}{dt}\langle t, t \rangle = \langle 1, 1 \rangle$$

The integrand is the dot product:

$$\mathbf{F}(\mathbf{c}(t)) \cdot \mathbf{c}'(t) = \left\langle t^2, t^2 \right\rangle \cdot \langle 1, 1 \rangle = t^2 + t^2 = 2t^2$$

We now use the Theorem on vector line integral to compute $\int_C \mathbf{F} \cdot d\mathbf{s}$:

$$\int_C \mathbf{F} \cdot d\mathbf{s} = \int_0^2 \mathbf{F}(\mathbf{c}(t)) \cdot \mathbf{c}'(t)\, dt = \int_0^2 2t^2\, dt = \left.\frac{2t^3}{3}\right|_0^2 = \frac{16}{3}$$

25. $\mathbf{F} = \langle x^2, xy \rangle$, circle $x^2 + y^2 = 9$ oriented clockwise

SOLUTION

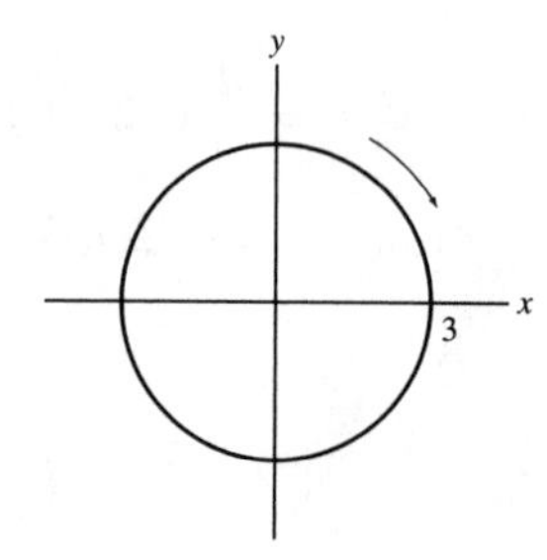

The oriented path is parametrized by

$$\mathbf{c}(t) = (3\cos t, 3\sin t); \quad t \text{ is changing from } 2\pi \text{ to } 0.$$

We compute the integrand:

$$\mathbf{F}(\mathbf{c}(t)) = \left\langle x^2, xy \right\rangle = \left\langle 9\cos^2 t, 9\cos t \sin t \right\rangle$$

$$\mathbf{c}'(t) = \langle -3\sin t, 3\cos t \rangle$$

$$\mathbf{F}(\mathbf{c}(t)) \cdot \mathbf{c}'(t) = \left\langle 9\cos^2 t, 9\cos t \sin t \right\rangle \cdot \langle -3\sin t, 3\cos t \rangle = -27\cos^2 t \sin t + 27\cos^2 t \sin t = 0$$

Hence,

$$\int_C \mathbf{F} \cdot d\mathbf{s} = \int_{2\pi}^0 \mathbf{F}(\mathbf{c}(t)) \cdot \mathbf{c}'(t)\, dt = \int_{2\pi}^0 0\, dt = 0$$

27. $\mathbf{F} = \langle xy, x + y \rangle$, $\mathbf{c}(t) = (1 + t^{-1}, t^2)$ for $1 \le t \le 2$

SOLUTION

Step 1. Calculate the integrand. We write the vectors and compute the integrand:

$$\mathbf{c}(t) = \left(1 + t^{-1}, t^2\right)$$

$$\mathbf{F}(\mathbf{c}(t)) = \langle xy, x + y \rangle = \left\langle \left(1 + t^{-1}\right) t^2, 1 + t^{-1} + t^2 \right\rangle = \left\langle t^2 + t, 1 + t^{-1} + t^2 \right\rangle$$

$$\mathbf{c}'(t) = \frac{d}{dt}\left\langle 1 + t^{-1}, t^2 \right\rangle = \left\langle -t^{-2}, 2t \right\rangle$$

The integrand is the dot product:

$$\mathbf{F}(\mathbf{c}(t)) \cdot \mathbf{c}'(t) = \left\langle t^2 + t, 1 + t^{-1} + t^2 \right\rangle \cdot \left\langle -t^{-2}, 2t \right\rangle = \left(t^2 + t\right)\left(-t^{-2}\right) + \left(1 + t^{-1} + t^2\right) \cdot 2t$$

$$= -1 - t^{-1} + 2t + 2 + 2t^3 = 2t^3 + 2t + 1 - t^{-1}$$

Step 2. Evaluate the integral. The vector line integral is

$$\int_C \mathbf{F}\cdot d\mathbf{s} = \int_1^2 \mathbf{F}(\mathbf{c}(t))\cdot \mathbf{c}'(t)\,dt = \int_1^2 \left(2t^3+2t+1-t^{-1}\right)dt = \frac{t^4}{2}+t^2+t-\ln t\Big|_1^2$$
$$= \left(\frac{16}{2}+4+2-\ln 2\right)-\left(\frac{1}{2}+1+1-\ln 1\right) = 11.5-\ln 2$$

29. $\mathbf{F} = \langle 3zy^{-1}, 4x, -y\rangle$, $\mathbf{c}(t) = (e^t, e^t, t)$ for $-1 \le t \le 1$

SOLUTION

Step 1. Calculate the integrand. We write out the vectors and compute the integrand:

$$\mathbf{c}(t) = \left(e^t, e^t, t\right)$$
$$\mathbf{F}(\mathbf{c}(t)) = \left\langle 3zy^{-1}, 4x, -y\right\rangle = \left\langle 3te^{-t}, 4e^t, -e^t\right\rangle$$
$$\mathbf{c}'(t) = \left\langle e^t, e^t, 1\right\rangle$$

The integrand is the dot product:

$$\mathbf{F}(\mathbf{c}(t))\cdot\mathbf{c}'(t) = \left\langle 3te^{-t}, 4e^t, -e^t\right\rangle\cdot\left\langle e^t, e^t, 1\right\rangle = 3te^{-t}\cdot e^t + 4e^t\cdot e^t - e^t\cdot 1 = 3t+4e^{2t}-e^t$$

Step 2. Evaluate the integral. The vector line integral is:

$$\int_C \mathbf{F}\cdot d\mathbf{s} = \int_{-1}^1 \mathbf{F}(\mathbf{c}(t))\cdot\mathbf{c}'(t)\,dt = \int_{-1}^1\left(3t+4e^{2t}-e^t\right)dt = 0+\int_{-1}^1\left(4e^{2t}-e^t\right)dt = 2e^{2t}-e^t\Big|_{-1}^1$$
$$= \left(2e^2-e\right)-\left(2e^{-2}-e^{-1}\right) = 2\left(e^2-e^{-2}\right)-\left(e-e^{-1}\right)\approx 12.157$$

31. $\mathbf{F} = \left\langle \dfrac{-y}{(x^2+y^2)^2}, \dfrac{x}{(x^2+y^2)^2}\right\rangle$, circle of radius R with center at the origin oriented counterclockwise

SOLUTION

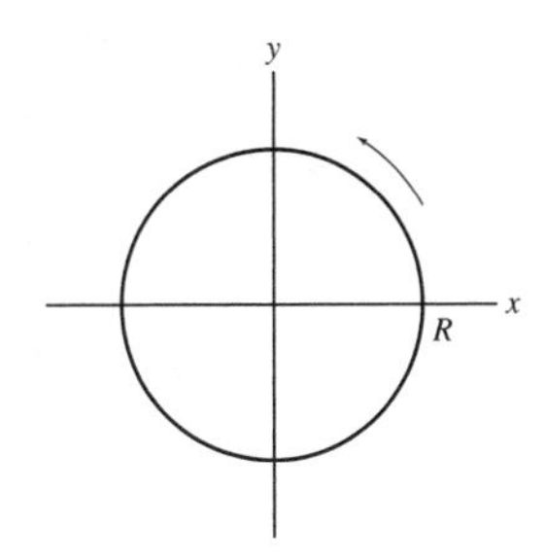

The path has the following parametrization:

$$\mathbf{c}(t) = \langle R\cos t, R\sin t\rangle, \quad 0\le t\le 2\pi$$

Step 1. Calculate the integrand. Since $x^2+y^2=R^2$ on the circle, we have

$$\mathbf{F}(\mathbf{c}(t)) = \left\langle \frac{-y}{(x^2+y^2)^2}, \frac{x}{(x^2+y^2)^2}\right\rangle = \left\langle \frac{-R\sin t}{R^4}, \frac{R\cos t}{R^4}\right\rangle = \frac{1}{R^3}\langle -\sin t, \cos t\rangle$$
$$\mathbf{c}'(t) = \frac{d}{dt}\langle R\cos t, R\sin t\rangle = R\langle -\sin t, \cos t\rangle$$

The integrand is the dot product:

$$\mathbf{F}(\mathbf{c}(t))\cdot\mathbf{c}'(t) = \frac{1}{R^3}\langle -\sin t, \cos t\rangle\cdot R\langle -\sin t, \cos t\rangle = \frac{1}{R^2}(\cos^2 t+\sin^2 t) = \frac{1}{R^2}$$

Step 2. Evaluate the integral. We obtain the following integral:

$$\int_C \mathbf{F}\cdot d\mathbf{s} = \int_0^{2\pi}\mathbf{F}(\mathbf{c}(t))\cdot\mathbf{c}'(t)\,dt = \int_0^{2\pi}\frac{dt}{R^2} = \frac{2\pi}{R^2}$$

33. $\mathbf{F} = \left\langle z^2, x, y\right\rangle$, $\mathbf{c}(t) = (\cos t, \tan t, t)$ for $0\le t\le \frac{\pi}{4}$

SOLUTION

Step 1. Calculate the integrand. The vectors are:

$$\mathbf{c}(t) = (\cos t, \tan t, t)$$
$$\mathbf{F}(\mathbf{c}(t)) = \left\langle z^2, x, y \right\rangle = \left\langle t^2, \cos t, \tan t \right\rangle$$
$$\mathbf{c}'(t) = \frac{d}{dt}\langle \cos t, \tan t, t\rangle = \left\langle -\sin t, \frac{1}{\cos^2 t}, 1 \right\rangle$$

The integrand is the dot product

$$\mathbf{F}(\mathbf{c}(t)) \cdot \mathbf{c}'(t) = \left\langle t^2, \cos t, \tan t \right\rangle \cdot \left\langle -\sin t, \frac{1}{\cos^2 t}, 1 \right\rangle = -t^2 \sin t + \frac{1}{\cos t} + \tan t$$

Step 2. Evaluate the integral. The Theorem on vector line integrals gives the following integral:

$$\int_C \mathbf{F} \cdot d\mathbf{s} = \int_0^{\pi/4} \mathbf{F}(\mathbf{c}(t)) \cdot \mathbf{c}'(t)\, dt = \int_0^{\pi/4} \left(-t^2 \sin t + \frac{1}{\cos t} + \tan t \right) dt$$
$$= t^2 \cos t - 2t \sin t - 2\cos t \Big|_0^{\pi/4} + \ln\left(\frac{1}{\cos t} + \tan t \right) \Big|_0^{\pi/4} - \ln(\cos t) \Big|_0^{\pi/4}$$
$$= \left(\left(\frac{\pi^2}{16} \cdot \frac{1}{\sqrt{2}} - 2 \cdot \frac{\pi}{4} \cdot \frac{1}{\sqrt{2}} - 2 \cdot \frac{1}{\sqrt{2}} \right) - (-2) \right) + \left(\ln\left(\sqrt{2}+1\right) - \ln 1 \right) - \left(\ln \frac{1}{\sqrt{2}} - \ln 1 \right)$$
$$= \frac{\pi^2}{16\sqrt{2}} - \frac{\pi}{2\sqrt{2}} - \sqrt{2} + 2 + \ln\left(1+\sqrt{2}\right) + \ln\sqrt{2} = \frac{\pi(\pi-8)}{16\sqrt{2}} + \ln\left(2+\sqrt{2}\right) + 2 - \sqrt{2} \approx 1.139$$

35. $\mathbf{F} = \left\langle z^3, yz, x \right\rangle$, circle of radius 2 in the yz-plane with center at the origin oriented clockwise when viewed from the positive x-axis

SOLUTION

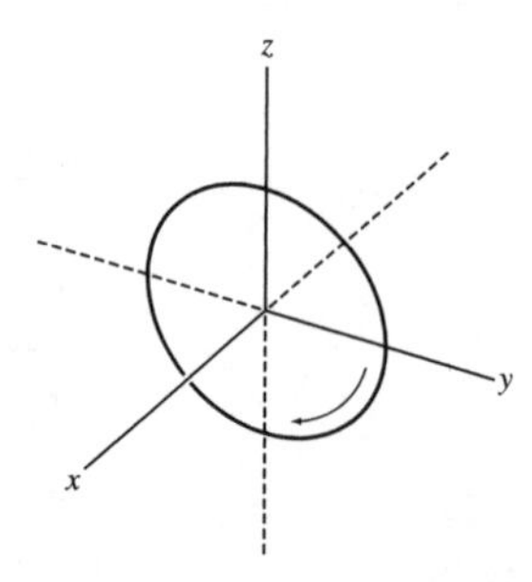

The oriented path has the following parametrization:

$$\mathbf{c}(t) = (0, 2\cos t, 2\sin t)$$

t is changing from 2π to 0.

Step 1. Calculate the integrand. We write out the vectors and compute the integrand:

$$\mathbf{c}(t) = (0, 2\cos t, 2\sin t)$$
$$\mathbf{F}(\mathbf{c}(t)) = \left\langle z^3, yz, x \right\rangle = \left\langle 8\sin^3 t, 4\cos t \sin t, 0 \right\rangle$$
$$\mathbf{c}'(t) = \langle 0, -2\sin t, 2\cos t\rangle$$

The integrand is the dot product:

$$\mathbf{F}(\mathbf{c}(t)) \cdot \mathbf{c}'(t) = \left\langle 8\sin^3 t, 4\cos t \sin t, 0 \right\rangle \cdot \langle 0, -2\sin t, 2\cos t\rangle = -8\cos t \sin^2 t$$

Step 2. Evaluate the integral. We obtain the following vector line integral:

$$\int_C \mathbf{F} \cdot d\mathbf{s} = \int_{2\pi}^{0} \mathbf{F}(\mathbf{c}(t)) \cdot \mathbf{c}'(t)\, dt = \int_{2\pi}^{0} -8\cos t \sin^2 t\, dt = \int_0^{2\pi} 8\sin^2 t \cos t\, dt = 8\left(\frac{\sin^3 t}{3} \Big|_0^{2\pi} \right) = 0$$

37. CAS Use a CAS to calculate $\int_C \langle e^{x-y}, e^{x+y}\rangle \cdot d\mathbf{s}$ to four decimal places, where C is the curve $y = \sin x$ for $0 \le x \le \pi$, oriented from left to right.

SOLUTION Using the parameterization $\mathbf{c}(t) = \langle t, \sin t\rangle$, our integral becomes $\int_0^{\pi} \left\langle e^{t-\sin t}, e^{t+\sin t}\right\rangle \cdot \langle 1, \cos t\rangle\, dt$, which is calculated to be -4.5088.

In Exercises 38–39, calculate the line integral of $\mathbf{F} = \langle e^z, e^{x-y}, e^y\rangle$ *over the given path.*

39. The path ABC in Figure 16

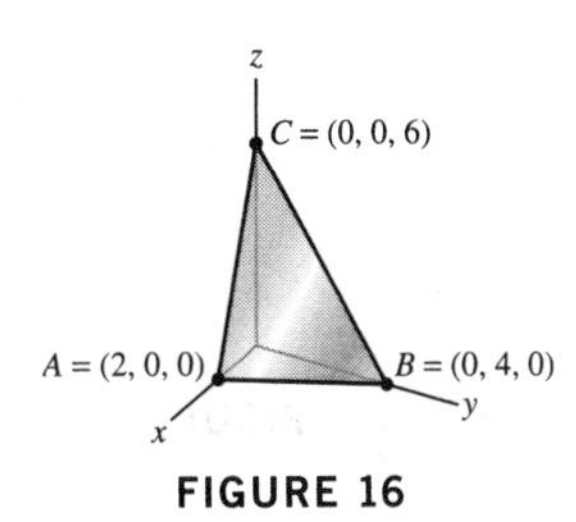

FIGURE 16

SOLUTION

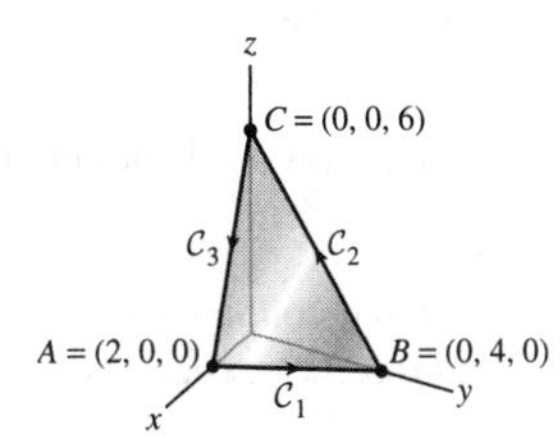

We denote by C_1, C_2, C_3 the oriented segments from A to B, from B to C and from C to A. We parametrize these paths by,

$$\begin{aligned}
C_1&: \mathbf{c}_1(t) = (1-t)(2,0,0) + t(0,4,0) = (2-2t, 4t, 0),\ 0 \le t \le 1 & \quad & \mathbf{c}_1'(t) = \langle -2, 4, 0\rangle \\
C_2&: \mathbf{c}_2(t) = (1-t)(0,4,0) + t(0,0,6) = (0, 4-4t, 6t),\ 0 \le t \le 1 & \Rightarrow \quad & \mathbf{c}_2'(t) = \langle 0, -4, 6\rangle \\
C_3&: \mathbf{c}_3(t) = (1-t)(0,0,6) + t(2,0,0) = (2t, 0, 6-6t),\ 0 \le t \le 1 & \quad & \mathbf{c}_3'(t) = \langle 2, 0, -6\rangle
\end{aligned}$$

Since $C = C_1 + C_2 + C_3$ we have,

$$\int_C \mathbf{F} \cdot d\mathbf{s} = \sum_{i=1}^{3} \int_{C_i} \mathbf{F} \cdot d\mathbf{s} \tag{1}$$

We compute the integrals on the right-hand side:

$$\int_{C_1} \mathbf{F} \cdot d\mathbf{s} = \int \left\langle e^0, e^{2-6t}, e^{4t}\right\rangle \cdot \langle -2, 4, 0\rangle\, dt = \int_0^1 \left\langle 1, e^{2-6t}, e^{4t}\right\rangle \cdot \langle -2, 4, 0\rangle\, dt$$

$$= \int_0^1 \left(-2 + 4e^{2-6t}\right) dt = -2t - \frac{2}{3}e^{2-6t}\Big|_0^1 = \frac{2}{3}e^2 - \frac{2}{3}e^{-4} - 2$$

$$\int_{C_2} \mathbf{F} \cdot d\mathbf{s} = \int_0^1 \left\langle e^{6t}, e^{-4+4t}, e^{4-4t}\right\rangle \cdot \langle 0, -4, 6\rangle\, dt = \int_0^1 \left(-4e^{-4+4t} + 6e^{4-4t}\right) dt$$

$$= -e^{-4+4t} - \frac{3}{2}e^{4-4t}\Big|_0^1 = \frac{3}{2}e^4 + e^{-4} - \frac{5}{2}$$

$$\int_{C_3} \mathbf{F} \cdot d\mathbf{s} = \int_0^1 \left\langle e^{6-6t}, e^{2t}, e^0\right\rangle \cdot \langle 2, 0, -6\rangle\, dt = \int_0^1 \left(2e^{6-6t} - 6\right) dt$$

$$= -\frac{1}{3}e^{6-6t} - 6t\Big|_0^1 = \frac{1}{3}e^6 - \frac{19}{3}$$

We substitute these values in (1) to obtain the solution:

$$\int_C \mathbf{F} \cdot d\mathbf{s} = \left(\frac{2}{3}e^2 - \frac{2}{3}e^{-4} - 2\right) + \left(\frac{3}{2}e^4 + e^{-4} - \frac{5}{2}\right) + \left(\frac{1}{3}e^6 - \frac{19}{3}\right)$$

$$= \frac{1}{3}e^6 + \frac{3}{2}e^4 + \frac{2}{3}e^2 - \frac{65}{6} + \frac{1}{3}e^{-4}$$

In Exercises 41–44, let **F** *be the* **vortex vector field** *(so-called because it swirls around the origin as shown in Figure 18)*

$$\mathbf{F} = \left\langle \frac{-y}{x^2+y^2}, \frac{x}{x^2+y^2} \right\rangle$$

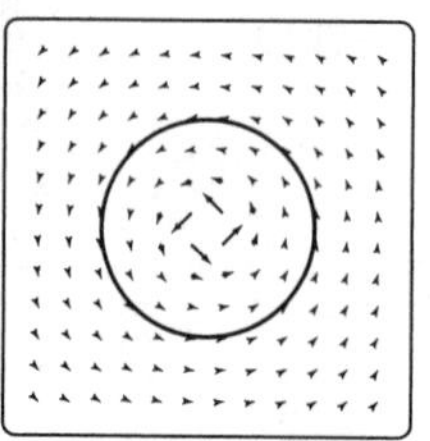

FIGURE 18 Vector field $\mathbf{F} = \left\langle \frac{-y}{x^2+y^2}, \frac{x}{x^2+y^2} \right\rangle$.

41. Let $I = \int_C \mathbf{F} \cdot d\mathbf{s}$, where C is the circle of radius 2 centered at the origin oriented counterclockwise (Figure 18).

(a) Do you expect I to be positive, negative, or zero?

(b) Evaluate I.

(c) Verify that I changes sign when C is oriented in the clockwise direction.

SOLUTION

(a) When the circle is oriented counterclockwise, the dot product of **F** with the unit tangent vector at each point along the circle is positive. Therefore, we expect the vector line integral I to be positive.

(b)

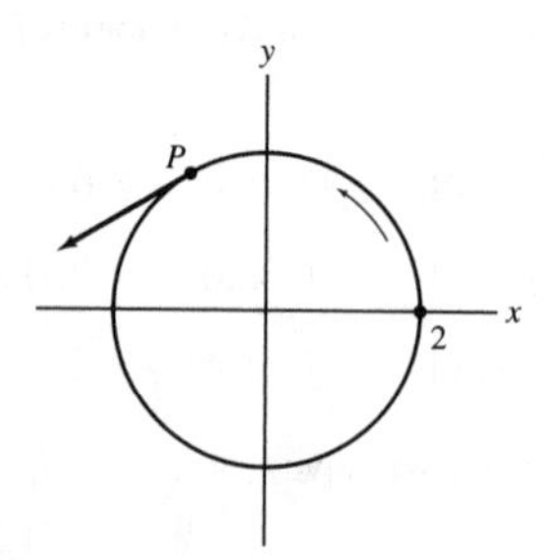

The circle of radius 2 oriented counterclockwise has the parametrization:

$$\mathbf{c}(t) = (2\cos t, 2\sin t), \quad 0 \le t \le 2\pi$$

Hence,

$$\mathbf{F}(\mathbf{c}(t)) = \left\langle \frac{-2\sin t}{4\cos^2 t + 4\sin^2 t}, \frac{2\cos t}{4\cos^2 t + 4\sin^2 t} \right\rangle = \frac{1}{2}\langle -\sin t, \cos t\rangle$$

$$\mathbf{c}'(t) = \langle -2\sin t, 2\cos t\rangle$$

Therefore, the integrand is the dot product,

$$\mathbf{F}(\mathbf{c}(t)) \cdot \mathbf{c}'(t) = \frac{1}{2}\langle -\sin t, \cos t\rangle \cdot \langle -2\sin t, 2\cos t\rangle = \sin^2 t + \cos^2 t = 1$$

We obtain the following integral:

$$\int_C \mathbf{F} \cdot d\mathbf{s} = \int_0^{2\pi} \mathbf{F}(\mathbf{c}(t)) \cdot \mathbf{c}'(t)\, dt = \int_0^{2\pi} 1\, dt = 2\pi$$

(c) When C is oriented in the clockwise direction, the parameter t is changing from 2π to 0, therefore, the line integral is,

$$\int_C \mathbf{F} \cdot d\mathbf{s} = \int_{2\pi}^{0} \mathbf{F}(\mathbf{c}(t)) \cdot \mathbf{c}'(t)\, dt = -\int_0^{2\pi} 1\, dt = -2\pi$$

43. Calculate $\int_{\mathcal{A}} \mathbf{F} \cdot d\mathbf{s}$, where $\mathcal{A}$ is the arc of angle θ_0 on the circle of radius R centered at the origin oriented counterclockwise. *Note:* $\mathcal{A}$ begins at $(R, 0)$ and ends at $(R\cos\theta_0, R\sin\theta_0)$.

SOLUTION We use the following parametrization for $\mathcal{A}$:

$$\mathbf{c}(\theta) = (R\cos\theta, R\sin\theta), \quad 0 \le \theta \le \theta_0$$

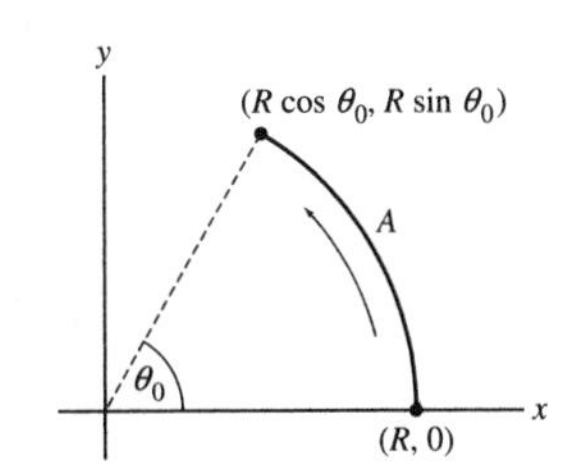

Hence,

$$\mathbf{c}'(\theta) = \langle -R\sin\theta, R\cos\theta \rangle$$

$$\mathbf{F}(\mathbf{c}(\theta)) = \left\langle \frac{-y}{x^2+y^2}, \frac{x}{x^2+y^2} \right\rangle = \left\langle \frac{-R\sin\theta}{R^2\cos^2\theta + R^2\sin^2\theta}, \frac{R\cos\theta}{R^2\cos^2\theta + R^2\sin^2\theta} \right\rangle$$

$$= \left\langle \frac{-R\sin\theta}{R^2}, \frac{R\cos\theta}{R^2} \right\rangle = \left\langle \frac{-\sin\theta}{R}, \frac{\cos\theta}{R} \right\rangle$$

The integrand is the dot product:

$$\mathbf{F}(\mathbf{c}(\theta)) \cdot \mathbf{c}'(\theta) = \left\langle \frac{-\sin\theta}{R}, \frac{\cos\theta}{R} \right\rangle \cdot \langle -R\sin\theta, R\cos\theta \rangle = \sin^2\theta + \cos^2\theta = 1$$

We now compute the line integral using the Theorem on vector line integrals. We get:

$$\int_{\mathcal{A}} \mathbf{F} \cdot d\mathbf{s} = \int_0^{\theta_0} \mathbf{F}(\mathbf{c}(\theta)) \cdot \mathbf{c}'(\theta)\, d\theta = \int_0^{\theta_0} 1\, d\theta = \theta\Big|_0^{\theta_0} = \theta_0$$

45. Calculate the line integral of the constant vector field $\mathbf{F} = \langle 2, -1, 4 \rangle$ along the segment $\overline{PQ}$, where:

(a) $P = (0, 0, 0)$, $Q = (1, 0, 0)$

(b) $P = (0, 0, 0)$, $Q = (4, 3, 5)$

(c) $P = (3, 2, 3)$, $Q = (4, 8, 12)$

SOLUTION

(a) The segment $\overline{PQ}$, where $P = (0, 0, 0)$ and $Q = (1, 0, 0)$ is parametrized by

$$\mathbf{c}(t) = (t, 0, 0), \quad 0 \le t \le 1$$

Therefore, $\mathbf{c}'(t) = \langle 1, 0, 0 \rangle$ and we obtain the integral,

$$\int_C \mathbf{F} \cdot d\mathbf{s} = \int_0^1 \mathbf{F}(\mathbf{c}(t)) \cdot \mathbf{c}'(t)\, dt = \int_0^1 \langle 2, -1, 4 \rangle \cdot \langle 1, 0, 0 \rangle\, dt = \int_0^1 2\, dt = 2$$

(b) The segment $\overline{PQ}$, where $P = (0, 0, 0)$ and $Q = (4, 3, 5)$ has the parametrization,

$$\mathbf{c}(t) = (4t, 3t, 5t), \quad 0 \le t \le 1$$

Therefore, $\mathbf{c}'(t) = \langle 4, 3, 5 \rangle$ and we obtain the following integral:

$$\int_C \mathbf{F} \cdot d\mathbf{s} = \int_0^1 \mathbf{F}(\mathbf{c}(t)) \cdot \mathbf{c}'(t)\, dt = \int_0^1 \langle 2, -1, 4 \rangle \cdot \langle 4, 3, 5 \rangle\, dt = \int_0^1 25\, dt = 25$$

(c) The segment $\overline{PQ}$, where $P = (3, 2, 3)$ and $Q = (4, 8, 12)$ has the parametrization,

$$\mathbf{c}(t) = (3 + t, 2 + 6t, 3 + 9t), \quad 0 \le t \le 1$$

Therefore, $\mathbf{c}'(t) = \langle 1, 6, 9 \rangle$ and we obtain the line integral

$$\int_C \mathbf{F} \cdot d\mathbf{s} = \int_0^1 \mathbf{F}(\mathbf{c}(t)) \cdot \mathbf{c}'(t)\, dt = \int_0^1 \langle 2, -1, 4 \rangle \cdot \langle 1, 6, 9 \rangle\, dt = \int_0^1 32\, dt = 32$$

47. Figure 19 shows the vector field $\mathbf{F}(x, y) = \langle x, x \rangle$.

(a) Are $\int_{\overline{AB}} \mathbf{F} \cdot d\mathbf{s}$ and $\int_{\overline{DC}} \mathbf{F} \cdot d\mathbf{s}$ equal? If not, which is larger?

(b) Which is smaller: the line integral of $\mathbf{F}$ over the path ADC or ABC?

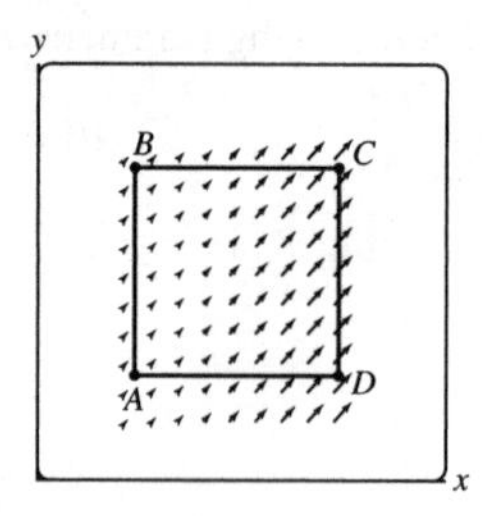

FIGURE 19

SOLUTION

(a) Since x is constant on $\overline{AB}$ and $\overline{DC}$, $\mathbf{F}(x, y) = \langle x, x \rangle$ is also constant on these segments.

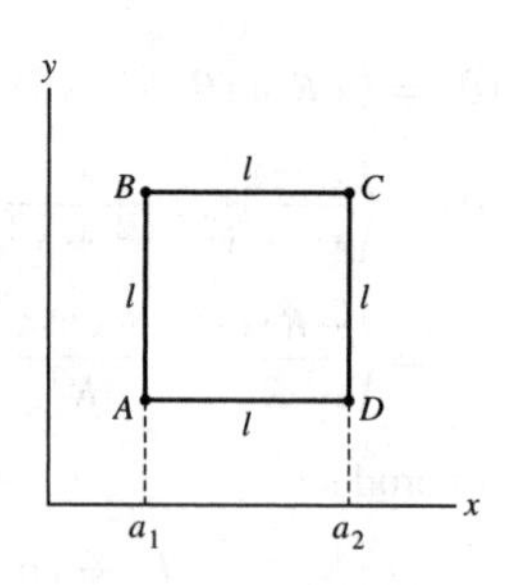

Let a_1 and a_2 denote the constant values of x on the segments $\overline{AB}$ and $\overline{DC}$ respectively, and l denote the lengths of these segments. By Exercise 46 we have

$$\int_{AB} \mathbf{F} \cdot d\mathbf{s} = \langle a_1, a_1 \rangle \cdot \langle 0, l \rangle = a_1 \cdot 0 + a_1 \cdot l = a_1 l$$

$$\int_{DC} \mathbf{F} \cdot d\mathbf{s} = \langle a_2, a_2 \rangle \cdot \langle 0, l \rangle = a_2 \cdot 0 + a_2 \cdot l = a_2 l$$

Since $a_1 < a_2$ we have $\int_{AB} \mathbf{F} \cdot d\mathbf{s} < \int_{DC} \mathbf{F} \cdot d\mathbf{s}$.

(b) We compute the integral over BC. This segment is parametrized by:

$$\mathbf{c}(t) = (a_1 + lt, b),\ 0 \le t \le 1.$$

Hence,

$$\mathbf{F}(\mathbf{c}(t)) = \langle x, x \rangle = \langle a_1 + lt, a_1 + lt \rangle,\ \mathbf{c}'(t) = \langle l, 0 \rangle$$

$$\mathbf{F}(\mathbf{c}(t)) \cdot \mathbf{c}'(t) = \langle a_1 + lt, a_1 + lt \rangle \cdot \langle l, 0 \rangle = a_1 l + l^2 t$$

Thus,

$$\int_{BC} \mathbf{F} \cdot d\mathbf{s} = \int_0^1 \left(a_1 l + l^2 t \right) dt = a_1 l t + \frac{l^2 t^2}{2} \Bigg|_{t=0}^{1} = a_1 l + \frac{l^2}{2}$$

We see that the line integral does not depend on b, therefore,

$$\int_{AD} \mathbf{F} \cdot d\mathbf{s} = \int_{BC} \mathbf{F} \cdot d\mathbf{s} \tag{1}$$

In part (a) we showed that:

$$\int_{AB} \mathbf{F} \cdot d\mathbf{s} < \int_{DC} \mathbf{F} \cdot d\mathbf{s} \tag{2}$$

Combining (1) and (2) gives:

$$\int_{ABC} \mathbf{F} \cdot d\mathbf{s} = \int_{AB} \mathbf{F} \cdot d\mathbf{s} + \int_{BC} \mathbf{F} \cdot d\mathbf{s} < \int_{DC} \mathbf{F} \cdot d\mathbf{s} + \int_{AD} \mathbf{F} \cdot d\mathbf{s} = \int_{ADC} \mathbf{F} \cdot d\mathbf{s}$$

49. Calculate the work done by the force field $\mathbf{F} = \langle x, y, z\rangle$ along the path $(\cos t, \sin t, t)$ for $0 \le t \le 3\pi$.

SOLUTION The work done by the force field $\mathbf{F}$ is the line integral:

$$W = \int_C \mathbf{F} \cdot d\mathbf{s}$$

We compute the integrand:

$$\begin{aligned}
\mathbf{F}(\mathbf{c}(t)) &= \langle x, y, z\rangle = \langle \cos t, \sin t, t\rangle \\
\mathbf{c}'(t) &= \frac{d}{dt}\langle \cos t, \sin t, t\rangle = \langle -\sin t, \cos t, 1\rangle \\
\mathbf{F}(\mathbf{c}(t)) \cdot \mathbf{c}'(t) &= \langle \cos t, \sin t, t\rangle \cdot \langle -\sin t, \cos t, 1\rangle = -\cos t \sin t + \sin t \cos t + t = t
\end{aligned}$$

We obtain the following integral:

$$W = \int_0^{3\pi} \mathbf{F}(\mathbf{c}(t)) \cdot \mathbf{c}'(t)\, dt = \int_0^{3\pi} t\, dt = \frac{t^2}{2}\Big|_0^{3\pi} = \frac{9\pi^2}{2}$$

Further Insights and Challenges

51. As observed in the text, the value of a scalar line integral does not depend on the choice of parametrization. Prove this directly. Namely, suppose that $\mathbf{c}_1(t)$ and $\mathbf{c}(t)$ are two parametrizations of C and that $\mathbf{c}_1(t) = \mathbf{c}(\varphi(t))$, where $\varphi(t)$ is an increasing function. Use the Change of Variables Formula to verify that

$$\int_c^d f(\mathbf{c}_1(t))\|\mathbf{c}_1'(t)\|\, dt = \int_a^b f(\mathbf{c}(t))\|\mathbf{c}'(t)\|\, dt$$

where $a = \varphi(c)$ and $b = \varphi(d)$.

SOLUTION We compute the integral $\int_a^b f(\mathbf{c}(t))\,\|\mathbf{c}'(t)\|\, dt$ using the substitution $t = \varphi(u)$, $a = \varphi(c)$, $b = \varphi(d)$. We get:

$$\int_a^b f(\mathbf{c}_1(t))\,\|\mathbf{c}'(t)\|\, dt = \int_{\varphi^{-1}(a)}^{\varphi^{-1}(b)} f\left(\mathbf{c}\left(\varphi(t)\right)\right)\|\mathbf{c}'\left(\varphi(t)\right)\|\varphi'(u)\, du \tag{1}$$

Since φ is an increasing function, $\varphi'(u) > 0$ for all u, therefore:

$$\left\|\mathbf{c}'\left(\varphi(u)\right)\right\| \varphi'(u) = \left\|\mathbf{c}'\left(\varphi(u)\right)\varphi'(u)\right\| \tag{2}$$

By the Chain Rule for vector valued functions, we have,

$$\frac{d}{du}\mathbf{c}\left(\varphi(u)\right) = \varphi'(u)\mathbf{c}'\left(\varphi(u)\right) \tag{3}$$

Combining (2) and (3) gives:

$$\left\|\mathbf{c}'\left(\varphi(u)\right)\right\| \varphi'(u) = \left\|\frac{d}{du}\mathbf{c}\left(\varphi(u)\right)\right\| = \left\|\frac{d}{du}\mathbf{c}_1(u)\right\| = \|\mathbf{c}_1'(u)\| \tag{4}$$

We substitute (4) in (1) to obtain:

$$\int_a^b f(\mathbf{c}(t))\,\|\mathbf{c}'(t)\|\, dt = \int_c^d f(\mathbf{c}_1(u))\,\|\mathbf{c}_1'(u)\|\, du = \int_c^d f(\mathbf{c}_1(t))\,\|\mathbf{c}_1'(t)\|\, dt$$

The last step is simply replacing the dummy variable of integration u by t.

53. Use Eq. (8) to calculate the average value of $f(x, y) = x - y$ along the segment from $P = (2, 1)$ to $Q = (5, 5)$.

SOLUTION We can parametrize this line segment by

$$\mathbf{c}(t) = (2 + 3t, 1 + 4t), \quad 0 \le t \le 1$$

Therefore,

$$\mathbf{c}'(t) = \langle 3, 4 \rangle \quad \Rightarrow \quad \|\mathbf{c}'(t)\| = \sqrt{9 + 16} = 5$$

We compute the length of the curve,

$$L = \int_0^1 \|\mathbf{c}'(t)\|\, dt = \int_0^1 \sqrt{5}\, dt = \sqrt{5}$$

Thus, using our values for x and y given above, we find that

$$\text{Av}(f) = \frac{1}{L} \int_C x - y\, dt = \frac{1}{\sqrt{5}} \int_0^1 (2 + 3t) - (1 + 4t)\, dt = \frac{1}{\sqrt{5}} \int_0^1 1 - t\, dt = \frac{1}{2\sqrt{5}}$$

55. The temperature (in degrees centigrade) at a point P on a circular wire of radius 2 cm centered at the origin is equal to the square of the distance from P to $P_0 = (2, 0)$. Compute the average temperature along the wire.

SOLUTION

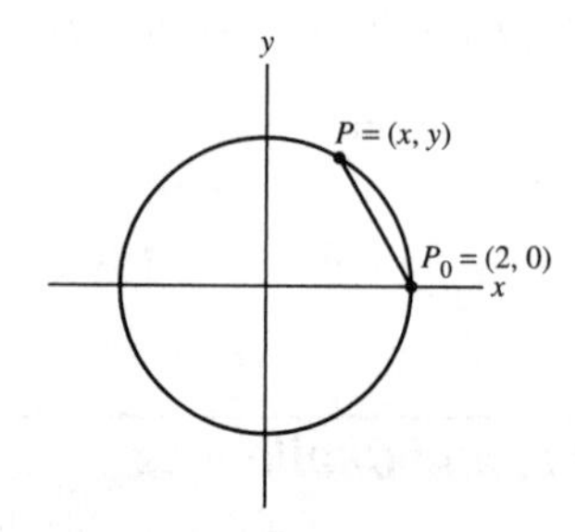

The temperature at a point $P(x, y)$ on the wire is given by the function,

$$T(x, y) = (x - 2)^2 + y^2$$

The length of the wire is the length of the circle of radius 2, $L = 2\pi \cdot 2 = 4\pi$. Therefore, the average temperature along the wire is,

$$\text{Av}(T) = \frac{1}{L} \int_C T\, ds = \frac{1}{4\pi} \int_C \left((x - 2)^2 + y^2\right) ds$$

To compute the line integral, we parametrize the circle by:

$$\mathbf{c}(t) = (2\cos t, 2\sin t), \quad 0 \le t \le 2\pi.$$

Then,

$$\mathbf{c}(t) = \langle -2\sin t, 2\cos t \rangle \quad \Rightarrow \quad \|\mathbf{c}'(t)\| = \sqrt{4\sin^2 t + 4\cos^2 t} = 2$$

We express T in terms of the parameter:

$$\begin{aligned} T(\mathbf{c}(t)) &= (x - 2)^2 + y^2 = (2\cos t - 2)^2 + (2\sin t)^2 = 4\cos^2 t - 8\cos t + 4 + 4\sin^2 t \\ &= 4\left(\cos^2 + \sin^2 t\right) + 4 - 8\cos t = 8(1 - \cos t) \end{aligned}$$

We obtain the integral,

$$\text{Av}(T) = \frac{1}{4\pi} \int_0^{2\pi} T(\mathbf{c}(t)) \|\mathbf{c}'(t)\|\, dt = \frac{1}{4\pi} \int_0^{2\pi} 16(1 - \cos t)\, dt = \frac{4}{\pi} \left(t - \sin t \Big|_0^{2\pi}\right) = \frac{4 \cdot 2\pi}{\pi} = 8$$

57. Let $\mathbf{F} = \langle y, x \rangle$. Prove that if C is any path from (a, b) to (c, d), then

$$\int_C \mathbf{F} \cdot d\mathbf{s} = cd - ab$$

SOLUTION

We denote a parametrization of the path by:

$$\mathbf{c}(t) = (x(t), y(t)), \quad t_0 \le t \le t_1$$
$$\mathbf{c}(t_0) = (a, b), \quad \mathbf{c}(t_1) = (c, d)$$

By the Theorem on vector line integrals we have

$$\int_{\mathbf{c}} \mathbf{F} \cdot d\mathbf{s} = \int_{t_0}^{t_1} \mathbf{F}(\mathbf{c}(t)) \cdot \mathbf{c}'(t)\, dt = \int_{t_0}^{t_1} \langle y(t), x(t) \rangle \cdot \langle x'(t), y'(t) \rangle\, dt$$
$$= \int_{t_0}^{t_1} \left(y(t)x'(t) + x(t)y'(t) \right) dt = \int_{t_0}^{t_1} \frac{d}{dt}(x(t)y(t))\, dt$$

The last equality follows from the Product Rule for differentiation. We now use the Fundamental Theorem of Calculus to obtain:

$$\int_{\mathbf{c}} \mathbf{F} \cdot d\mathbf{s} = x(t)y(t)\Big|_{t=t_0}^{t_1} = x(t_1)\, y(t_1) - x(t_0)\, y(t_0) = cd - ab$$

17.3 Conservative Vector Fields (ET Section 16.3)

Preliminary Questions

1. The following statement is false. *If* **F** *is a gradient vector field, then the line integral of* **F** *along every curve is zero.* Which single word must be added to make it true?

SOLUTION The missing word is "closed" (curve). The line integral of a gradient vector field along every closed curve is zero.

2. Which of the following statements are true for all vector fields, which are true only for conservative vector fields?

(a) The line integral along a path from P to Q does not depend on which path is chosen.

(b) The line integral over an oriented curve $\mathcal{C}$ does not depend on how the $\mathcal{C}$ is parametrized.

(c) The line integral around a closed curve is zero.

(d) The line integral changes sign if the orientation is reversed.

(e) The line integral is equal to the difference of a potential function at the two endpoints.

(f) The line integral is equal to the integral of the tangential component along the curve.

(g) The cross partials of the components are equal.

SOLUTION

(a) This statement is true only for conservative vector fields.

(b) This statement is true for all vector fields.

(c) This statement holds only for conservative vector fields.

(d) This is a property of all vector fields.

(e) Only conservative vector fields have a potential function, and the line integral is computed by using the potential function as stated.

(f) All vector fields' line integrals share this property.

(g) The cross-partials of the components of a conservative field are equal. For other fields, the cross-partials of the components may or may not equal.

3. Let **F** be a vector field on an open, connected domain $\mathcal{D}$. Which of the following statements are always true, and which are true under additional hypotheses on $\mathcal{D}$?

(a) If **F** has a potential function, then **F** is conservative.

(b) If **F** is conservative, then the cross partials of **F** are equal.

(c) If the cross partials of **F** are equal, then **F** is conservative.

SOLUTION

(a) This statement is always true, since every gradient vector field is conservative.

(b) If **F** is conservative on a connected domain $\mathcal{D}$, then **F** has a potential function $\mathcal{D}$ and consequently the cross partials of **F** are equal in $\mathcal{D}$.

(c) If the cross partials of **F** are equal in a simply-connected region $\mathcal{D}$, then **F** is a gradient vector field in $\mathcal{D}$.

4. Let $\mathcal{C}$, $\mathcal{D}$, and $\mathcal{E}$ be the oriented curves in Figure 15 and let $\mathbf{F} = \nabla\varphi$ be a gradient vector field such that $\int_{\mathcal{C}} \mathbf{F} \cdot d\mathbf{s} = 4$. What are the values of the following integrals?

(a) $\int_{\mathcal{D}} \mathbf{F} \cdot d\mathbf{s}$ **(b)** $\int_{\mathcal{E}} \mathbf{F} \cdot d\mathbf{s}$

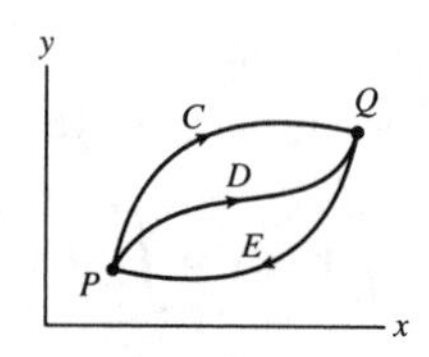

FIGURE 15

SOLUTION Since **F** is a gradient vector field the integrals over closed paths are zero. Therefore, by the equivalent conditions for path independence we have:

(a) $\int_D \mathbf{F} \cdot d\mathbf{s} = \int_C \mathbf{F} \cdot d\mathbf{s} = 4$

(b) $\int_E \mathbf{F} \cdot d\mathbf{s} = \int_{-C} \mathbf{F} \cdot d\mathbf{s} = -\int_C \mathbf{F} \cdot d\mathbf{s} = -4$

Exercises

1. Let $\varphi(x, y, z) = xy\sin(yz)$. Evaluate $\int_{\mathbf{c}} \nabla\varphi \cdot d\mathbf{s}$, where **c** is any path from $(0, 0, 0)$ to $(1, 1, \pi)$.

SOLUTION By the Fundamental Theorem for Gradient Vector Fields, we have:

$$\int_{\mathbf{c}} \nabla\varphi \cdot d\mathbf{s} = \varphi(1, 1, \pi) - \varphi(0, 0, 0) = 1 \cdot 1 \sin\pi - 0 = 0$$

In Exercises 3–8, verify that $\mathbf{F} = \nabla\varphi$ *and evaluate the line integral of* **F** *over the given path.*

3. $\mathbf{F} = \langle 3, 6y \rangle$, $\varphi(x, y, z) = 3x + 3y^2$; $\mathbf{c}(t) = (t, 2t^{-1})$ for $1 \le t \le 4$

SOLUTION The gradient of $\varphi = 3x + 3y^2$ is:

$$\nabla\varphi = \left\langle \frac{\partial\varphi}{\partial x}, \frac{\partial\varphi}{\partial y} \right\rangle = \langle 3, 6y \rangle = \mathbf{F}$$

Using the Fundamental Theorem for Gradient Vector Fields, we have:

$$\int_{\mathbf{c}} \mathbf{F} \cdot d\mathbf{s} = \varphi(\mathbf{c}(4)) - \varphi(\mathbf{c}(1)) = \varphi\left(4, \frac{1}{2}\right) - \varphi(1, 2) = \left(3 \cdot 4 + 3 \cdot \frac{1}{4}\right) - (3 \cdot 1 + 3 \cdot 4) = -\frac{9}{4}$$

5. $\mathbf{F} = \langle xy^2, x^2y \rangle$, $\varphi(x, y) = \frac{1}{2}x^2y^2$; upper half of the unit circle centered at the origin oriented counterclockwise

SOLUTION We compute the gradient of $\varphi(x, y) = \frac{1}{2}x^2y^2$:

$$\nabla\varphi = \left\langle \frac{\partial\varphi}{\partial x}, \frac{\partial\varphi}{\partial y} \right\rangle = \langle xy^2, x^2y \rangle = \mathbf{F}$$

We now use the Fundamental Theorem of Gradient Vector Fields. The terminal point is $Q = (-1, 0)$ and the initial point is $P = (1, 0)$, therefore:

$$\int_{\mathbf{c}} \mathbf{F} \cdot d\mathbf{s} = \varphi(Q) - \varphi(P) = \varphi(-1, 0) - \varphi(1, 0) = \frac{1}{2} \cdot (-1)^2 \cdot 0^2 - \frac{1}{2} \cdot 1^2 \cdot 0^2 = 0$$

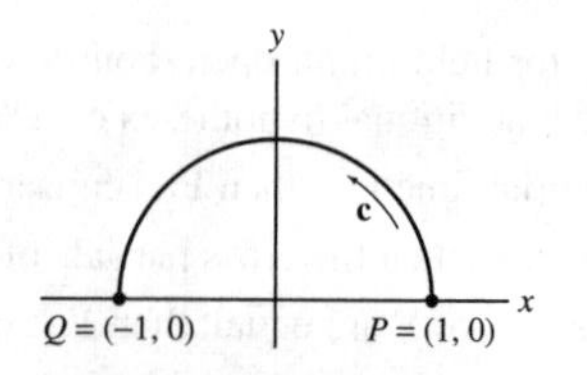

7. $\mathbf{F} = \langle ye^z, xe^z, xye^z \rangle$, $\varphi(x, y, z) = xye^z$; $\mathbf{c}(t) = (t^2, t^3, t-1)$ for $1 \le t \le 2$

SOLUTION We verify that $\mathbf{F}$ is the gradient of φ:

$$\nabla\varphi = \left\langle \frac{\partial\varphi}{\partial x}, \frac{\partial\varphi}{\partial y}, \frac{\partial\varphi}{\partial z} \right\rangle = \langle ye^z, xe^z, xye^z \rangle = \mathbf{F}$$

We use the Fundamental Theorem for Gradient Vectors with the initial point $\mathbf{c}(1) = (1, 1, 0)$ and terminal point $\mathbf{c}(2) = (4, 8, 1)$, to obtain:

$$\int_{\mathbf{c}} \mathbf{F} \cdot d\mathbf{s} = \varphi(4, 8, 1) - \varphi(1, 1, 0) = 32e - 1$$

9. Find a potential function for $\mathbf{F} = \langle 2xy + 5, x^2 - 4z, -4y \rangle$ and evaluate

$$\int_{\mathbf{c}} \mathbf{F} \cdot d\mathbf{s}$$

where $\mathbf{c}(t) = (t^2, \sin(\pi t), e^{t^2 - 2t})$ for $0 \le t \le 2$.

SOLUTION We find a potential function $\varphi(x, y, z)$ for $\mathbf{F}$, using the following steps.

Step 1. Use the condition $\frac{\partial\varphi}{\partial x} = \mathbf{F}_1$. φ is an antiderivative of $\mathbf{F}_1 = 2xy + 5$ when y and z are fixed, therefore,

$$\varphi(x, y, z) = \int (2xy + 5)\, dx = x^2 y + 5x + g(y, z) \tag{1}$$

Step 2. Use the condition $\frac{\partial\varphi}{\partial y} = \mathbf{F}_2$. We have,

$$\frac{\partial}{\partial y}\left(x^2 y + 5x + g(y, z)\right) = x^2 - 4z$$

$$x^2 + g_y(y, z) = x^2 - 4z \quad \Rightarrow \quad g_y(y, z) = -4z$$

We integrate with respect to y, holding z fixed:

$$g(y, z) = \int -4z\, dy = -4zy + h(z)$$

Combining with (1) gives:

$$\varphi(x, y, z) = x^2 y + 5x - 4zy + h(z) \tag{2}$$

Step 3. Use the condition $\frac{\partial\varphi}{\partial z} = \mathbf{F}_3$. We have,

$$\frac{\partial}{\partial z}\left(x^2 y + 5x - 4zy + h(z)\right) = -4y$$

$$-4y + h'(z) = -4y$$

$$h'(z) = 0 \quad \Rightarrow \quad h(z) = c$$

Substituting in (2) we obtain the general potential function:

$$\varphi(x, y, z) = x^2 y + 5x - 4zy + c$$

To compute the line integral we need one of the potential functions. We choose $c = 0$ to obtain the function,

$$\varphi(x, y, z) = x^2 y + 5x - 4zy$$

We now use the Fundamental Theorem for Gradient Vector Fields to evaluate the line integral:

$$\int_{\mathbf{c}} \mathbf{F} \cdot d\mathbf{s} = \varphi\left(\gamma(2) - \varphi\left(\gamma(0)\right)\right) = \varphi(4, 0, 1) - \varphi(0, 0, 1) = 20 - 0 = 20$$

11. Find the line integral of $\mathbf{F} = \langle 2xyz, x^2 z, x^2 y \rangle$ over any path from $(0, 0, 0)$ to $(3, 2, 1)$.

SOLUTION We first show that the cross partials condition is satisfied:

$$\left.\begin{aligned} \frac{\partial F_1}{\partial y} &= \frac{\partial}{\partial y}(2xyz) = 2xz \\ \frac{\partial F_2}{\partial x} &= \frac{\partial}{\partial x}\left(x^2 z\right) = 2xz \end{aligned}\right\} \quad \Rightarrow \quad \frac{\partial F_1}{\partial y} = \frac{\partial F_2}{\partial x}$$

$$\left.\begin{aligned}\frac{\partial F_2}{\partial z} &= \frac{\partial}{\partial z}\left(x^2 z\right) = x^2 \\ \frac{\partial F_3}{\partial y} &= \frac{\partial}{\partial y}\left(x^2 y\right) = x^2\end{aligned}\right\} \Rightarrow \quad \frac{\partial F_2}{\partial z} = \frac{\partial F_3}{\partial y}$$

$$\left.\begin{aligned}\frac{\partial F_3}{\partial x} &= \frac{\partial}{\partial x}\left(x^2 y\right) = 2xy \\ \frac{\partial F_1}{\partial z} &= \frac{\partial}{\partial z}(2xyz) = 2xy\end{aligned}\right\} \Rightarrow \quad \frac{\partial F_3}{\partial x} = \frac{\partial F_1}{\partial z}$$

Since the cross partials condition is satisfied at all points, **F** is conservative. We find a potential function for **F**.

Step 1. Use the condition $\frac{\partial \varphi}{\partial x} = F_1$. φ is an antiderivative of $F_1 = 2xyz$ when y and z are fixed. Therefore:

$$\varphi(x, y, z) = \int 2xyz\,dx = x^2 yz + g(y, z) \tag{1}$$

Step 2. Use the condition $\frac{\partial \varphi}{\partial y} = F_2$. By (1) we have:

$$\frac{\partial \varphi}{\partial y} = \left(x^2 yz + g(y, z)\right) = x^2 z$$

$$x^2 z + g_y(y, z) = x^2 z \quad \Rightarrow \quad g_y(y, z) = 0$$

It follows that $g(y, z) = h(z)$. Substituting in (1) gives

$$\varphi(x, y, z) = x^2 yz + h(z) \tag{2}$$

Step 3. Use the condition $\frac{\partial \varphi}{\partial z} = F_3$. This condition along with (2) gives:

$$\frac{\partial}{\partial z} = \left(x^2 yz + h(z)\right) = x^2 y$$

$$x^2 y + h'(z) = x^2 z$$

$$h'(z) = 0 \quad \Rightarrow \quad h(z) = C$$

Substituting in (2) we get

$$\varphi(x, y, z) = x^2 yz + C$$

Since only one potential function is needed, we choose the one corresponding to $C = 0$. That is,

$$\varphi(x, y, z) = x^2 yz$$

Using the Fundamental Theorem for Gradient Vectors we obtain:

$$\int_{\mathbf{c}} \mathbf{F} \cdot d\mathbf{s} = \varphi(3, 2, 1) - \varphi(0, 0, 0) = 3^2 \cdot 2 \cdot 1 - 0 = 18$$

In Exercises 12–17, determine whether the vector field is conservative and, if so, find a potential function.

13. $\mathbf{F} = \langle 0, x, y \rangle$

SOLUTION Since $\frac{\partial F_1}{\partial y} = \frac{\partial}{\partial y}(0) = 0$ and $\frac{\partial F_2}{\partial x} = \frac{\partial}{\partial x}(x) = 1$, we have $\frac{\partial F_1}{\partial y} \neq \frac{\partial F_2}{\partial x}$. Therefore **F** does not satisfy the cross-partial condition, hence **F** is not conservative.

15. $\mathbf{F} = \left\langle y, x, z^3 \right\rangle$

SOLUTION We examine whether the field $\mathbf{F} = \left\langle y, x, z^3 \right\rangle$ satisfies the cross partials condition.

$$\left.\begin{aligned}\frac{\partial F_1}{\partial y} &= \frac{\partial}{\partial y}(y) = 1 \\ \frac{\partial F_2}{\partial x} &= \frac{\partial}{\partial x}(x) = 1\end{aligned}\right\} \Rightarrow \quad \frac{\partial F_1}{\partial y} = \frac{\partial F_2}{\partial x}$$

$$\left.\begin{aligned}\frac{\partial F_2}{\partial z} &= \frac{\partial}{\partial z}(x) = 0 \\ \frac{\partial F_3}{\partial y} &= \frac{\partial}{\partial y}\left(z^3\right) = 0\end{aligned}\right\} \Rightarrow \quad \frac{\partial F_2}{\partial z} = \frac{\partial F_3}{\partial y}$$

$$\left.\begin{aligned}\frac{\partial F_3}{\partial x} &= \frac{\partial}{\partial x}\left(z^3\right) = 0\\ \frac{\partial F_1}{\partial z} &= \frac{\partial}{\partial z}(y) = 0\end{aligned}\right. \quad \Rightarrow \quad \frac{\partial F_3}{\partial x} = \frac{\partial F_1}{\partial z}$$

Since **F** satisfies the cross partials condition everywhere, **F** is conservative. We find a potential function for **F**.

Step 1. Use the condition $\frac{\partial \varphi}{\partial x} = F_1$. φ is an antiderivative of $F_1 = y$ when y and z are fixed. Therefore:

$$\varphi(x, y, z) = \int y\,dx = yx + g(y, z) \tag{1}$$

Step 2. Use the condition $\frac{\partial \varphi}{\partial y} = F_2$. By (1) we have:

$$\frac{\partial}{\partial y}\left(yx + g(y, z)\right) = x$$

$$x + g_y(y, z) = x \quad \Rightarrow \quad g_y(y, z) = 0$$

Therefore, $g(y, z) = g(z)$. Substituting in (1) gives:

$$\varphi(x, y, z) = yx + g(z) \tag{2}$$

Step 3. Use the condition $\frac{\partial \varphi}{\partial z} = F_3$. Using (2) we have:

$$\frac{\partial}{\partial z}\left(yx + g(z)\right) = z^3$$

$$g'(z) = z^3 \quad \Rightarrow \quad g(z) = \frac{1}{4}z^4 + c$$

Substituting in (2) gives the following general potential function:

$$\varphi(x, y, z) = yx + \frac{1}{4}z^4 + c$$

Choosing $c = 0$ we obtain the potential:

$$\varphi(x, y, z) = yx + \frac{z^4}{4}.$$

17. $\mathbf{F} = \langle \cos z, 2y, -x \sin z\rangle$

SOLUTION We examine whether **F** satisfies the cross partials condition:

$$\left.\begin{aligned}\frac{\partial F_1}{\partial y} &= \frac{\partial}{\partial y}(\cos z) = 0\\ \frac{\partial F_2}{\partial x} &= \frac{\partial}{\partial x}(2y) = 0\end{aligned}\right. \quad \Rightarrow \quad \frac{\partial F_1}{\partial y} = \frac{\partial F_2}{\partial x}$$

$$\left.\begin{aligned}\frac{\partial F_2}{\partial z} &= \frac{\partial}{\partial z}(2y) = 0\\ \frac{\partial F_3}{\partial y} &= \frac{\partial}{\partial y}(-x \sin z) = 0\end{aligned}\right. \quad \Rightarrow \quad \frac{\partial F_2}{\partial z} = \frac{\partial F_3}{\partial y}$$

$$\left.\begin{aligned}\frac{\partial F_3}{\partial x} &= \frac{\partial}{\partial x}(-x \sin z) = -\sin z\\ \frac{\partial F_1}{\partial z} &= \frac{\partial}{\partial z}(\cos z) = -\sin z\end{aligned}\right. \quad \Rightarrow \quad \frac{\partial F_3}{\partial x} = \frac{\partial F_1}{\partial z}$$

We see that the conditions are satisfied, therefore **F** is conservative. We find a potential function for **F**.

Step 1. Use the condition $\frac{\partial \varphi}{\partial x} = F_1$. $\varphi(x, y, z)$ is an antiderivative of $F_1 = \cos z$ when y and z are fixed, therefore:

$$\varphi(x, y, z) = \int \cos z\,dx = x \cos z + g(y, z) \tag{1}$$

Step 2. Use the condition $\frac{\partial \varphi}{\partial y} = F_2$. Using (1) we get:

$$\frac{\partial}{\partial y}\left(x \cos z + g(y, z)\right) = 2y$$

$$g_y(y, z) = 2y$$

We integrate with respect to y, holding z fixed:

$$g(y, z) = \int 2y\, dy = y^2 + g(z)$$

Substituting in (1) gives

$$\varphi(x, y, z) = x \cos z + y^2 + g(z) \tag{2}$$

Step 3. Use the condition $\frac{\partial \varphi}{\partial z} = F_3$. By (2) we have

$$\frac{\partial}{\partial z}\left(x \cos z + y^2 + g(z)\right) = -x \sin z$$

$$-x \sin z + g'(z) = -x \sin z$$

$$g'(z) = 0 \quad \Rightarrow \quad g(z) = c$$

Substituting in (2) we obtain the general potential function:

$$\varphi(x, y, z) = x \cos z + y^2 + c$$

Choosing $c = 0$ gives the potential function:

$$\varphi(x, y, z) = x \cos z + y^2.$$

19. Let $\mathbf{F} = \left\langle \frac{1}{x}, \frac{-1}{y} \right\rangle$. Calculate the work against F required to move an object from $(1, 1)$ to $(3, 4)$ along any path in the first quadrant.

SOLUTION $\mathbf{F}$ is a conservative force, since $\mathbf{F} = -\nabla \varphi$ with potential energy $\varphi(x, y) = \ln y - \ln x$. The work required to move an object from $(1, 1)$ to $(3, 4)$ along any path $\mathcal{C}$ is equal to the change in potential energy:

$$\text{Work against } \mathbf{F} = -\int_{\mathcal{C}} \mathbf{F} \cdot d\mathbf{s} = \varphi(3, 4) - \varphi(1, 1) = (\ln 4 - \ln 3) - (\ln 1 - \ln 1) = \ln 4 - \ln 3$$

21. The vector field $\mathbf{F}$ in Figure 17 is horizontal and appears to depend on only the x-coordinate. Suppose that $\mathbf{F} = \langle g(x), 0 \rangle$. Prove that

$$\int_{\overline{PR}} \mathbf{F} \cdot d\mathbf{s} = \int_{\overline{PQ}} \mathbf{F} \cdot d\mathbf{s}$$

by showing that both integrals are equal to $\int_a^b g(x)\, dx$.

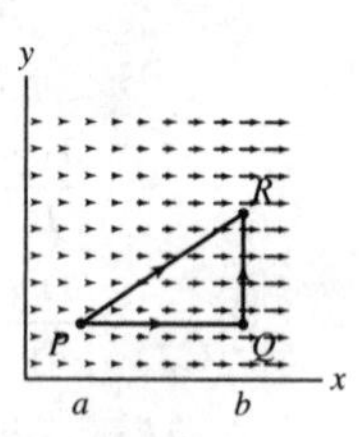

FIGURE 17

SOLUTION The vector field $\mathbf{F}$ has the form:

$$\mathbf{F} = \langle g(x), 0 \rangle$$

Since $\frac{\partial F_1}{\partial y} = \frac{\partial}{\partial y}(g(x)) = 0$ and $\frac{\partial F_2}{\partial x} = 0$, we have $\frac{\partial F_1}{\partial y} = \frac{\partial F_2}{\partial x}$ consequently $\mathbf{F}$ is conservative. Therefore:

$$\int_{\overline{PR}} \mathbf{F} \cdot d\mathbf{s} = \int_{\overline{PQ}} \mathbf{F} \cdot d\mathbf{s} + \int_{\overline{QR}} \mathbf{F} \cdot d\mathbf{s} \tag{1}$$

The vector field $\mathbf{F}$ is orthogonal to $\overline{QR}$, therefore the tangential component of $\mathbf{F}$ at each point along $\overline{QR}$ is zero. We conclude that:

$$\int_{\overline{QR}} \mathbf{F} \cdot d\mathbf{s} = 0$$

Substituting in (1) we get:

$$\int_{\overline{PR}} \mathbf{F} \cdot d\mathbf{s} = \int_{\overline{PQ}} \mathbf{F} \cdot d\mathbf{s}$$

Let's try it again, this time using the hint. We compute the integral along $\overline{PQ}$, using the parametrization

$$\overline{PQ} : c(t) = (t, y_0), \quad a \le t \le b$$

we get:

$$\int_{\overline{PQ}} \mathbf{F} \cdot d\mathbf{s} = \int_a^b \mathbf{F}(c(t)) \cdot c'(t)\, dt = \int_a^b \langle g(t), 0 \rangle \cdot \langle 1, 0 \rangle\, dt = \int_a^b g(t)\, dt = \int_a^b g(x)\, dx$$

If we now compute the integral along $\overline{PR}$ with the parameterization

$$\overline{PR} : c(t) = (t, y(t)), \quad a \le t \le b,$$

we get the same thing:

$$\int_{\overline{PR}} \mathbf{F} \cdot d\mathbf{s} = \int_a^b \mathbf{F}(c(t)) \cdot c'(t)\, dt = \int_a^b \langle g(t), 0 \rangle \cdot \langle 1, y' \rangle\, dt = \int_a^b g(t)\, dt = \int_a^b g(x)\, dx$$

We see, again, that these are equal.

23. How much energy (in joules) does it take to carry a 2-kg object from sea level along any path to the top of a hill that is 1,000 m high? Assume that the force of gravity $\mathbf{F}$ is constant $-mg$ in the vertical direction, where $g = 9.8\ \text{m/s}^2$. *Hint:* Find a potential function for $\mathbf{F}$.

SOLUTION The force of gravity is $\mathbf{F} = \langle 0, 0, -mg \rangle$, therefore $\mathbf{F} = -\nabla \varphi$ for $\varphi(x, y, z) = mgz$. The work performed by the gravitational field is the line integral of $\mathbf{F}$ over the path. Since $\mathbf{F}$ is conservative, the energy is independent of the path connecting the two points. Using the Fundamental Theorem for Gradient Vector Fields we have:

$$W = -\int_{\mathbf{c}} \mathbf{F} \cdot d\mathbf{s} = \varphi(z = 1000) - \varphi(z = 0) = mg \cdot 1000 = 2 \cdot 9.8 \cdot 10^3 = 19{,}600 \text{ joules}$$

Further Insights and Challenges

25. The vector field $\mathbf{F} = \left\langle \frac{x}{x^2 + y^2}, \frac{y}{x^2 + y^2} \right\rangle$ is defined on the domain $\mathcal{D} = \{(x, y) \neq (0, 0)\}$.

(a) Show that $\mathbf{F}$ satisfies the cross-partials condition on $\mathcal{D}$.
(b) Show that $\varphi(x, y) = \frac{1}{2} \ln(x^2 + y^2)$ is a potential function for $\mathbf{F}$.
(c) Is $\mathcal{D}$ simply connected?
(d) Do these results contradict Theorem 4?

SOLUTION

(a) We compute the partials of $\mathbf{F}$:

$$\frac{\partial F_2}{\partial x} = \frac{\partial}{\partial x}\left(\frac{y}{x^2 + y^2}\right) = \frac{-2xy}{(x^2 + y^2)^2}$$

$$\frac{\partial F_1}{\partial y} = \frac{\partial}{\partial y}\left(\frac{x}{x^2 + y^2}\right) = \frac{-2yx}{(x^2 + y^2)^2}$$

The cross partials are equal in $\mathcal{D}$.

(b) We compute the gradient of $\varphi(x, y) = \frac{1}{2} \ln\left(x^2 + y^2\right)$:

$$\nabla \varphi = \left\langle \frac{\partial \varphi}{\partial x}, \frac{\partial \varphi}{\partial x} \right\rangle = \frac{1}{2}\left\langle \frac{2x}{x^2 + y^2}, \frac{2y}{x^2 + y^2} \right\rangle = \left\langle \frac{x}{x^2 + y^2}, \frac{y}{x^2 + y^2} \right\rangle = \mathbf{F}$$

(c) $\mathcal{D}$ is not simply-connected since it has a "hole" at the origin.

(d) The requirement in Theorem 4 (that the domain be simply connected) is a sufficient condition for a vector field with equal cross-partials to have a potential function. It is not necessary, since as in our example, even if the domain is not simply-connected the field may have a gradient function. Moreover, for any closed curve in $\mathcal{D}$, φ have the same value after completing one round along c. This is perhaps best seen by noting that $\varphi = \log(r)$ in polar coordinates, which will be independent of θ. Therefore,

$$\int_c \mathbf{F} \cdot d\mathbf{s} = 0$$

Hence, $\mathbf{F}$ is conservative.

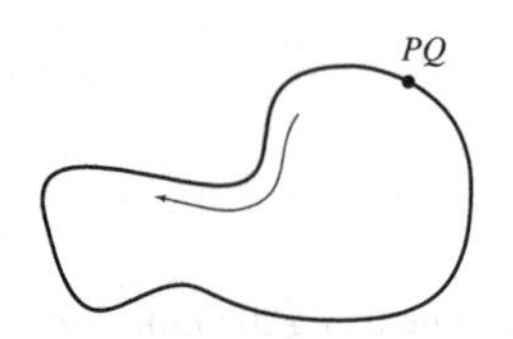

17.4 Parametrized Surfaces and Surface Integrals (ET Section 16.4)

Preliminary Questions

1. What is the surface integral of the function $f(x, y, z) = 10$ over a surface of total area 5?

SOLUTION Using Surface Integral and Surface Area we have:

$$\iint_S f(x, y, z)\, dS = \iint_{\mathcal{D}} f(\Phi(u, v))\, \|\mathbf{n}(u, v)\|\, du\, dv = \iint_{\mathcal{D}} 10\|\mathbf{n}(u, v)\|\, du\, dv$$

$$= 10 \iint_{\mathcal{D}} \|\mathbf{n}(u, v)\|\, du\, dv = 10\,\text{Area}(S) = 10 \cdot 5 = 50$$

2. What interpretation can we give to the length $\|\mathbf{n}\|$ of the normal vector for a parametrization $\Phi(u, v)$?

SOLUTION The approximation:

$$\text{Area}\left(S_{ij}\right) \approx \|\mathbf{n}\left(u_{ij}, v_{ij}\right)\| \text{Area}\left(R_{ij}\right)$$

tells that $\|\mathbf{n}\|$ is a distortion factor that indicates how much the area of a small rectangle R_{ij} is altered under the map ϕ.

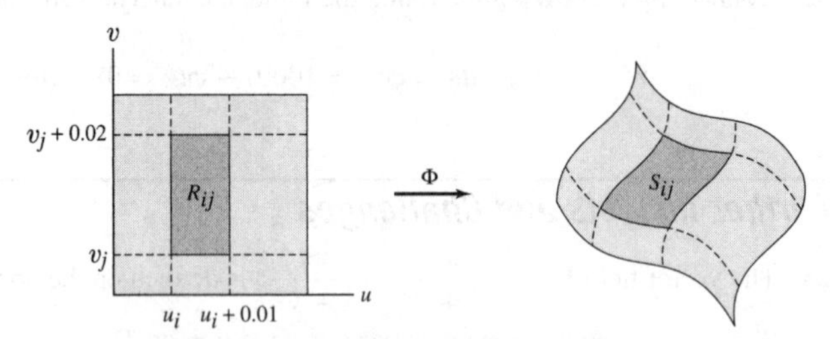

3. A parametrization maps a rectangle of size 0.01×0.02 in the uv-plane onto a small patch $\mathcal{S}$ of a surface. Estimate Area($\mathcal{S}$) if $\mathbf{T}_u \times \mathbf{T}_v = \langle 1, 2, 2\rangle$ at a sample point in the rectangle.

SOLUTION We use the estimation

$$\text{Area}(S) \approx \|\mathbf{n}(u, v)\| \text{Area}(R)$$

where $\mathbf{n}(u, v) = \mathbf{T}_u \times \mathbf{T}_v$ at a sample point in R. We get:

$$\text{Area}(S) \approx \| \langle 1, 2, 2\rangle \| \cdot 0.01 \cdot 0.02 = \sqrt{1^2 + 2^2 + 2^2} \cdot 0.0002 = 0.0006$$

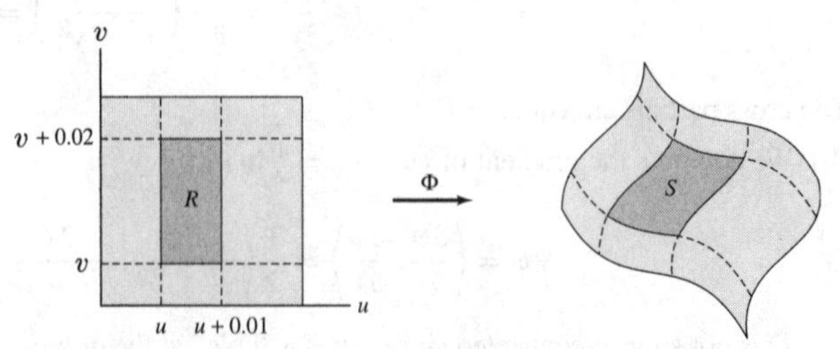

4. A small surface $\mathcal{S}$ is divided into three small pieces, each of area 0.2. Estimate $\iint_{\mathcal{S}} f(x, y, z)\, dS$ if $f(x, y, z)$ takes the values 0.9, 1, and 1.1 at sample points in these three pieces.

SOLUTION We use the approximation obtained by the Riemann Sum:

$$\iint_S f(x, y, z)\, dS \approx \sum_{ij} f\left(P_{ij}\right) \text{Area}\left(S_{ij}\right) = 0.9 \cdot 0.2 + 1 \cdot 0.2 + 1.1 \cdot 0.2 = 0.6$$

5. A surface $\mathcal{S}$ has a parametrization whose domain is the square $0 \le u, v \le 2$ such that $\|\mathbf{n}(u,v)\| = 5$ for all (u,v). What is Area($\mathcal{S}$)?

SOLUTION Writing the surface area as a surface integral where $\mathcal{D}$ is the square $[0,2]\times[0,2]$ in the uv-plane, we have:

$$\text{Area}(S) = \iint_{\mathcal{D}} \|\mathbf{n}(u,v)\|\,du\,dv = \iint_{\mathcal{D}} 5\,du\,dv = 5\iint_{\mathcal{D}} 1\,du\,dv = 5\text{Area}(\mathcal{D}) = 5\cdot 2^2 = 20$$

6. What is the outward-pointing unit normal to the sphere of radius 3 centered at the origin at $P = (2,2,1)$?

SOLUTION The outward-pointing normal to the sphere of radius $R = 3$ centered at the origin is the following vector:

$$\langle \cos\theta\sin\phi, \sin\theta\sin\phi, \cos\phi\rangle \tag{1}$$

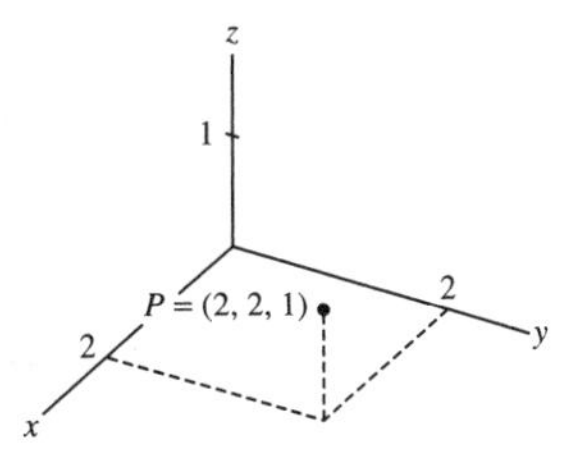

We compute the values in (1) corresponding to $P = (2,2,1)$: $x = y = 2$, $z = 1$ hence $0 \le \theta \le \frac{\pi}{2}$ and $0 < \phi < \frac{\pi}{2}$. We get:

$$\cos\phi = \frac{z}{\rho} = \frac{1}{3} \quad\Rightarrow\quad \sin\phi = \sqrt{1-\left(\frac{1}{3}\right)^2} = \frac{2\sqrt{2}}{3}$$

$$\cos\theta = \frac{x}{\rho\sin\phi} = \frac{2}{3\cdot\frac{2\sqrt{2}}{3}} = \frac{1}{\sqrt{2}} \quad\Rightarrow\quad \sin\theta = \sqrt{1-\frac{1}{2}} = \frac{1}{\sqrt{2}}$$

Substituting in (1) we get the following unit normal:

$$\left\langle \frac{1}{\sqrt{2}}\cdot\frac{2\sqrt{2}}{3}, \frac{1}{\sqrt{2}}\cdot\frac{2\sqrt{2}}{3}, \frac{1}{3}\right\rangle = \left\langle \frac{2}{3}, \frac{2}{3}, \frac{1}{3}\right\rangle$$

Exercises

1. Match the parametrization with the surface in Figure 16.

(a) $(u, \cos v, \sin v)$

(b) $(u, u+v, v)$

(c) (u, v^3, v)

(d) $(\cos u\sin v, 3\cos u\sin v, \cos v)$

(e) $(u, u(2+\cos v), u(2+\sin v))$

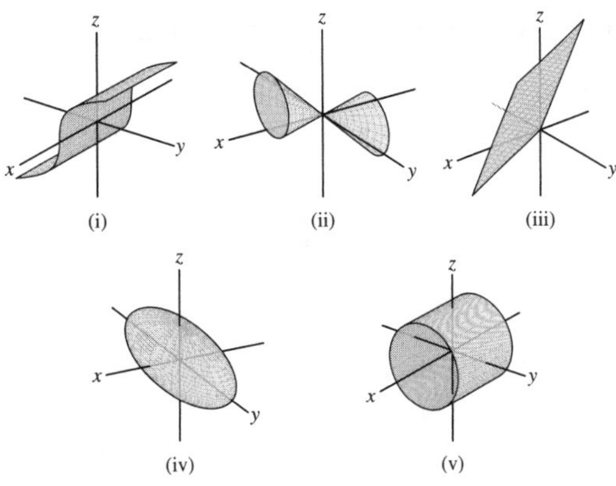

FIGURE 16

SOLUTION (a) = (v), because the y and z coordinates describe a circle with fixed radius.

(b) = (iii), because the coordinates are all linear in u and v.

(c) = (i), because the parametrization gives $y = z^3$.

(d) = (iv), an ellipsoid.

(e) = (ii), because the y and z coordinates describe a circle with varying radius.

3. Show that $\Phi(u, v) = (2u + 1, u - v, 3u + v)$ parametrizes the plane $2x - y - z = 2$. Then:

(a) Calculate $\mathbf{T}_u$, $\mathbf{T}_v$, and $\mathbf{n}(u, v)$.

(b) Find the area of $S = \Phi(\mathcal{D})$, where $\mathcal{D} = \{(u, v) : 0 \le u \le 2, 0 \le v \le 1\}$.

(c) Express $f(x, y, z) = yz$ in terms of u and v and evaluate $\iint_S f(x, y, z)\, dS$.

SOLUTION We show that $x = 2u + 1$, $y = u - v$, and $z = 3u + v$ satisfy the equation of the plane,

$$2x - y - z = 2(2u + 1) - (u - v) - (3u + v) = 4u + 2 - u + v - 3u - v = 2$$

Moreover, for any x, y, z satisfying $2x - y - z = z$, there are values of u and v such that $x = 2u + 1$, $y = u - v$, and $z = 3u + v$, since the following equations can be solved for u and v:

$$\begin{array}{c} x = 2u + 1 \\ y = u - v \\ z = 3u + v \\ 2x - y - z = 2 \end{array} \quad \Rightarrow \quad u = \frac{x-1}{2}, \quad v = \frac{x-1}{2} - y$$

We conclude that $\Phi(u, v)$ parametrizes the whole plane $2x - y - z = 2$.

(a) The tangent vectors $\mathbf{T}_u$ and $\mathbf{T}_v$ are:

$$\mathbf{T}_u = \frac{\partial \phi}{\partial u} = \frac{\partial}{\partial u}(2u + 1, u - v, 3u + v) = \langle 2, 1, 3 \rangle$$

$$\mathbf{T}_v = \frac{\partial \phi}{\partial v} = \frac{\partial}{\partial v}(2u + 1, u - v, 3u + v) = \langle 0, -1, 1 \rangle$$

The normal vector is the following cross product:

$$\mathbf{n}(u, v) = \mathbf{T}_u \times \mathbf{T}_v = \begin{vmatrix} \mathbf{i} & \mathbf{j} & \mathbf{k} \\ 2 & 1 & 3 \\ 0 & -1 & 1 \end{vmatrix} = \begin{vmatrix} 1 & 3 \\ -1 & 1 \end{vmatrix}\mathbf{i} - \begin{vmatrix} 2 & 3 \\ 0 & 1 \end{vmatrix}\mathbf{j} + \begin{vmatrix} 2 & 1 \\ 0 & -1 \end{vmatrix}\mathbf{k}$$

$$= 4\mathbf{i} - 2\mathbf{j} - 2\mathbf{k} = \langle 4, -2, -2 \rangle$$

(b) That area of $S = \Phi(\mathcal{D})$ is the following surface integral:

$$\text{Area}(S) = \iint_{\mathcal{D}} \|\mathbf{n}(u, v)\|\, du\, dv = \iint_{\mathcal{D}} \|\langle 4, -2, -2 \rangle\|\, du\, dv = \sqrt{24} \iint_{\mathcal{D}} 1\, du\, dv$$

$$= \sqrt{24}\,\text{Area}(\mathcal{D}) = \sqrt{24} \cdot 2 \cdot 1 = 4\sqrt{6}$$

(c) We express $f(x, y, z) = yz$ in terms of the parameters u and v:

$$f(\phi(u, v)) = (u - v)(3u + v) = 3u^2 - 2uv - v^2$$

Using the Theorem on Surface Integrals we have:

$$\iint_S f(x, y, z)\, dS = \iint_{\mathcal{D}} f(\phi(u, v))\, \|\mathbf{n}(u, v)\|\, du\, dv = \iint_{\mathcal{D}} \left(3u^2 - 2uv - v^2\right) \|\langle 4, -2, -2 \rangle\|\, du\, dv$$

$$= \sqrt{24} \int_0^1 \int_0^2 \left(3u^2 - 2uv - v^2\right) du\, dv = \sqrt{24} \int_0^1 \left(u^3 - u^2 v - v^2 u\right)\Big|_{u=0}^{2} dv$$

$$= \sqrt{24} \int_0^1 \left(8 - 4v - 2v^2\right) dv = \sqrt{24}\left(8v - 2v^2 - \frac{2}{3}v^3\right)\Big|_0^1 = \frac{32\sqrt{6}}{3}$$

5. Let $\Phi(x, z) = (x, y, xy)$.

(a) Calculate $\mathbf{T}_x$, $\mathbf{T}_y$, and $\mathbf{n}(x, y)$.

(b) Let S be the part of the surface with parameter domain $\mathcal{D} = \{(x, y) : x^2 + y^2 \le 1, x \ge 0, y \ge 0\}$. Verify the following formula and evaluate using polar coordinates:

$$\iint_S 1\, dS = \iint_{\mathcal{D}} \sqrt{1 + x^2 + y^2}\, dx\, dy$$

(c) Verify the following formula and evaluate:

$$\iint_S z\, dS = \int_0^{\pi/2} \int_0^1 (\sin\theta\cos\theta) r^3 \sqrt{1 + r^2}\, dr\, d\theta$$

SOLUTION

(a) The tangent vectors are:

$$\mathbf{T}_x = \frac{\partial \phi}{\partial x} = \frac{\partial}{\partial x}(x, y, xy) = \langle 1, 0, y \rangle$$

$$\mathbf{T}_y = \frac{\partial \phi}{\partial y} = \frac{\partial}{\partial y}(x, y, xy) = \langle 0, 1, x \rangle$$

The normal vector is the cross product:

$$\mathbf{n}(x, y) = \mathbf{T}_x \times \mathbf{T}_y = \begin{vmatrix} \mathbf{i} & \mathbf{j} & \mathbf{k} \\ 1 & 0 & y \\ 0 & 1 & x \end{vmatrix} = \begin{vmatrix} 0 & y \\ 1 & x \end{vmatrix}\mathbf{i} - \begin{vmatrix} 1 & y \\ 0 & x \end{vmatrix}\mathbf{j} + \begin{vmatrix} 1 & 0 \\ 0 & 1 \end{vmatrix}\mathbf{k}$$

$$= -y\mathbf{i} - x\mathbf{j} + \mathbf{k} = \langle -y, -x, 1 \rangle$$

(b) Using the Theorem on evaluating surface integrals we have:

$$\iint_S 1\, dS = \iint_{\mathcal{D}} \|\mathbf{n}(x, y)\|\, dx\, dy = \iint_{\mathcal{D}} \|\langle -y, -x, 1 \rangle\|\, dx\, dy = \iint_{\mathcal{D}} \sqrt{y^2 + x^2 + 1}\, dx\, dy$$

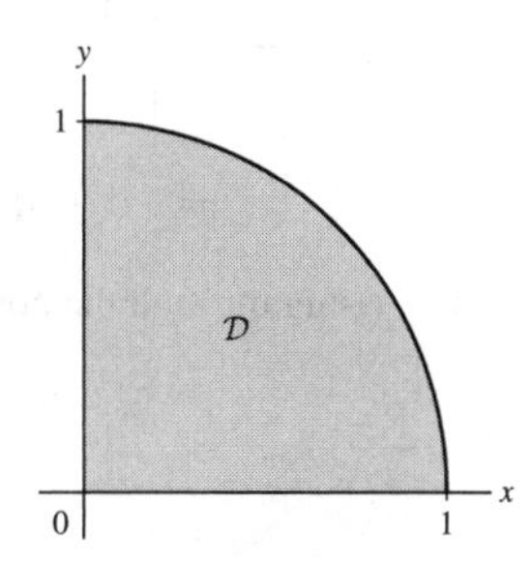

We convert the integral to polar coordinates $x = r\cos\theta$, $y = r\sin\theta$. The new region of integration is:

$$0 \le r \le 1, \quad 0 \le \theta \le \frac{\pi}{2}.$$

We get:

$$\iint_S 1\, dS = \int_0^{\pi/2} \int_0^1 \sqrt{r^2 + 1} \cdot r\, dr\, d\theta = \int_0^{\pi/2} \left(\int_0^1 \sqrt{r^2 + 1} \cdot r\, dr \right) d\theta$$

$$= \int_0^{\pi/2} \left(\int_1^2 \frac{\sqrt{u}}{2}\, du \right) d\theta = \int_0^{\pi/2} \frac{2\sqrt{2} - 1}{3}\, d\theta = \frac{\left(2\sqrt{2} - 1\right)\pi}{6}$$

(c) The function z expressed in terms of the parameters x, y is $f(\Phi(x, y)) = xy$. Therefore,

$$\iint_S z\, dS = \iint_{\mathcal{D}} xy \cdot \|\mathbf{n}(x, y)\|\, dx\, dy = \iint_{\mathcal{D}} xy\sqrt{1 + x^2 + y^2}\, dx\, dy$$

We compute the double integral by converting it to polar coordinates. We get:

$$\iint_S z\, dS = \int_0^{\pi/2} \int_0^1 (r\cos\theta)(r\sin\theta)\sqrt{1 + r^2} \cdot r\, dr\, d\theta = \int_0^{\pi/2} \int_0^1 (\sin\theta\cos\theta) r^3 \sqrt{1 + r^2}\, dr\, d\theta$$

$$= \left(\int_0^{\pi/2} (\sin\theta\cos\theta)\, d\theta \right) \left(\int_0^1 r^3 \sqrt{1 + r^2}\, dr \right) \tag{1}$$

We compute each integral in (1). Using the substitution $u = 1 + r^2$, $du = 2r\, dr$ we get:

$$\int_0^1 r^3\sqrt{1 + r^2}\, dr = \int_0^1 r^2\sqrt{1 + r^2} \cdot r\, dr = \int_1^2 \left(u^{3/2} - u^{1/2}\right) \frac{du}{2} = \frac{u^{5/2}}{5} - \frac{u^{3/2}}{3}\bigg|_1^2 = \frac{2\left(\sqrt{2} + 1\right)}{15}$$

Also,

$$\int_0^{\pi/2} \sin\theta\cos\theta\, d\theta = \int_0^{\pi/2} \frac{\sin 2\theta}{2}\, d\theta = -\frac{\cos 2\theta}{4}\bigg|_0^{\pi/2} = \frac{1}{2}$$

We substitute the integrals in (1) to obtain the following solution:

$$\iint_S z\,dS = \frac{1}{2}\cdot\frac{2\left(\sqrt{2}+1\right)}{15} = \frac{\sqrt{2}+1}{15}$$

In Exercises 7–10, calculate $\mathbf{T}_u$, $\mathbf{T}_v$, and $\mathbf{n}(u,v)$ for the parametrized surface at the given point. Then find the equation of the tangent plane to the surface at that point.

7. $\Phi(u,v) = (2u+v, u-4v, 3u);\quad u=1,\quad v=4$

SOLUTION The tangent vectors are the following vectors,

$$\mathbf{T}_u = \frac{\partial\Phi}{\partial u} = \frac{\partial}{\partial u}(2u+v, u-4v, 3u) = \langle 2, 1, 3\rangle$$

$$\mathbf{T}_v = \frac{\partial\Phi}{\partial v} = \frac{\partial}{\partial v}(2u+v, u-4v, 3u) = \langle 1, -4, 0\rangle$$

The normal is the cross product:

$$\mathbf{n}(u,v) = \mathbf{T}_u \times \mathbf{T}_v = \begin{vmatrix} \mathbf{i} & \mathbf{j} & \mathbf{k} \\ 2 & 1 & 3 \\ 1 & -4 & 0 \end{vmatrix} = \begin{vmatrix} 1 & 3 \\ -4 & 0 \end{vmatrix}\mathbf{i} - \begin{vmatrix} 2 & 3 \\ 1 & 0 \end{vmatrix}\mathbf{j} + \begin{vmatrix} 2 & 1 \\ 1 & -4 \end{vmatrix}\mathbf{k}$$

$$= 12\mathbf{i} + 3\mathbf{j} - 9\mathbf{k} = 3\langle 4, 1, -3\rangle$$

The equation of the plane passing through the point $P : \Phi(1,4) = (6, -15, 3)$ with the normal vector $\langle 4, 1, -3\rangle$ is:

$$\langle x-6, y+15, z-3\rangle \cdot \langle 4, 1, -3\rangle = 0$$

or

$$4(x-6) + y + 15 - 3(z-3) = 0$$

$$4x + y - 3z = 0$$

9. $\Phi(\theta,\phi) = (\cos\theta\sin\phi, \sin\theta\sin\phi, \cos\phi);\quad \theta = \frac{\pi}{2},\quad \phi = \frac{\pi}{4}$

SOLUTION We compute the tangent vectors:

$$\mathbf{T}_\theta = \frac{\partial\Phi}{\partial\theta} = \frac{\partial}{\partial\theta}(\cos\theta\sin\phi, \sin\theta\sin\phi, \cos\phi) = \left\langle -\sin\theta\sin\phi, \cos\theta\sin\phi, 0\right\rangle$$

$$\mathbf{T}_\phi = \frac{\partial\Phi}{\partial\phi} = \frac{\partial}{\partial\phi}(\cos\theta\sin\phi, \sin\theta\sin\phi, \cos\phi) = \left\langle \cos\theta\cos\phi, \sin\theta\cos\phi, -\sin\phi\right\rangle$$

The normal vector is the cross product:

$$\mathbf{n}(\theta,\phi) = \mathbf{T}_\theta \times \mathbf{T}_\phi = \begin{vmatrix} \mathbf{i} & \mathbf{j} & \mathbf{k} \\ -\sin\theta\sin\phi & \cos\theta\sin\phi & 0 \\ \cos\theta\cos\phi & \sin\theta\cos\phi & -\sin\phi \end{vmatrix}$$

$$= \left(-\cos\theta\sin^2\phi\right)\mathbf{i} - \left(\sin\theta\sin^2\phi\right)\mathbf{j} + \left(-\sin^2\theta\sin\phi\cos\phi - \cos^2\theta\cos\phi\sin\phi\right)\mathbf{k}$$

$$= -\left(\cos\theta\sin^2\phi\right)\mathbf{i} - \left(\sin\theta\sin^2\phi\right)\mathbf{j} - (\sin\phi\cos\phi)\mathbf{k}$$

The tangency point and the normal at this point are,

$$P = \Phi\left(\frac{\pi}{2}, \frac{\pi}{4}\right) = \left(\cos\frac{\pi}{2}\sin\frac{\pi}{4}, \sin\frac{\pi}{2}\sin\frac{\pi}{4}, \cos\frac{\pi}{4}\right) = \left(0, \frac{\sqrt{2}}{2}, \frac{\sqrt{2}}{2}\right)$$

$$\mathbf{n}\left(\frac{\pi}{2}, \frac{\pi}{4}\right) = -\frac{1}{2}\mathbf{j} - \frac{1}{2}\mathbf{k} = -\frac{1}{2}(\mathbf{j}+\mathbf{k}) = -\frac{1}{2}\langle 0, 1, 1\rangle$$

The equation of the plane orthogonal to the vector $\langle 0, 1, 1\rangle$ and passing through $P = \left(0, \frac{\sqrt{2}}{2}, \frac{\sqrt{2}}{2}\right)$ is:

$$\left\langle x, y - \frac{\sqrt{2}}{2}, z - \frac{\sqrt{2}}{2}\right\rangle \cdot \langle 0, 1, 1\rangle = 0$$

or

$$y - \frac{\sqrt{2}}{2} + z - \frac{\sqrt{2}}{2} = 0$$
$$y + z = \sqrt{2}$$

11. Use the normal vector computed in Exercise 8 to estimate the area of the small patch of the surface $\Phi(u, v) = (u^2 - v^2, u + v, u - v)$ defined by

$$2 \le u \le 2.1, \qquad 3 \le v \le 3.2$$

SOLUTION We denote the rectangle $\mathcal{D} = \{(u, v) : 2 \le u \le 2.1,\ 3 \le v \le 3.2\}$. Using the sample point corresponding to $u = 2$, $v = 3$ we obtain the following estimation for the area of $S = \Phi(\mathcal{D})$:

$$\text{Area}(S) \approx \|\mathbf{n}(2, 3)\|\text{Area}(\mathcal{D}) = \|\mathbf{n}(2, 3)\| \cdot 0.1 \cdot 0.2 = 0.02\|\mathbf{n}(2, 3)\| \tag{1}$$

In Exercise 8 we found that $\mathbf{n}(2, 3) = 2\langle -1, -1, 5\rangle$. Therefore,

$$\|\mathbf{n}(2, 3)\| = 2\sqrt{1^2 + 1^2 + 5^2} = 2\sqrt{27}$$

Substituting in (1) gives the following estimation:

$$\text{Area}(S) \approx 0.02 \cdot 2 \cdot \sqrt{27} \approx 0.2078.$$

13. A surface $\mathcal{S}$ has a parametrization $\Phi(u, v)$ with domain $0 \le u \le 2$, $0 \le v \le 4$ such that the following partial derivatives are constant:

$$\frac{\partial \Phi}{\partial u} = \langle 2, 0, 1\rangle, \qquad \frac{\partial \Phi}{\partial v} = \langle 4, 0, 3\rangle$$

What is the surface area of $\mathcal{S}$?

SOLUTION Since the partial derivatives are constant, the normal vector is also constant. We find it by computing the cross product:

$$\mathbf{n} = \mathbf{T}_u \times \mathbf{T}_v = \frac{\partial \Phi}{\partial u} \times \frac{\partial \Phi}{\partial v} = \begin{vmatrix} \mathbf{i} & \mathbf{j} & \mathbf{k} \\ 2 & 0 & 1 \\ 4 & 0 & 3 \end{vmatrix} = -2\mathbf{j} = \langle 0, -2, 0\rangle \Rightarrow \|\mathbf{n}\| = 2$$

We denote the rectangle $\mathcal{D} = \{(u, v) : 0 \le u \le 2,\ 0 \le v \le 4\}$, and use the surface area to compute the area of $S = \Phi(\mathcal{D})$. We obtain:

$$\text{Area}(S) = \iint_{\mathcal{D}} \|\mathbf{n}\|\, du\, dv = \iint_{\mathcal{D}} 2\, du\, dv = 2\iint_{\mathcal{D}} 1\, du\, dv = 2 \cdot \text{Area}(\mathcal{D}) = 2 \cdot 2 \cdot 4 = 16$$

15. CAS Let $\mathcal{S}$ be the surface with parametrization

$$\Phi(u, v) = \big((3 + \sin v)\cos u, (3 + \sin v)\sin u, v\big)$$

for $0 \le u, v \le 2\pi$. Using a computer algebra system:

(a) Plot $\mathcal{S}$ from several different viewpoints. Is $\mathcal{S}$ best described as a "vase that holds water" or a "bottomless vase"?

(b) Calculate the normal vector $\mathbf{n}(u, v)$.

(c) Calculate the surface area of $\mathcal{S}$ to four decimal places.

SOLUTION

(a) We show the graph of $\mathcal{S}$ here.

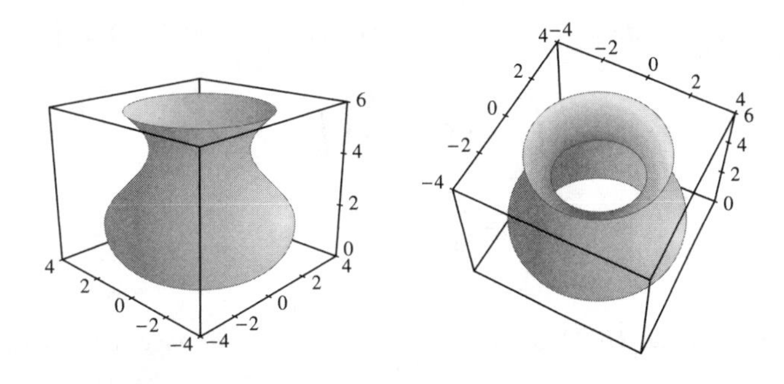

Note that it is best described as a "bottomless vase".

(b) We compute the tangent and normal vectors:

$$\mathbf{T}_u = \frac{\partial \Phi}{\partial u} = \langle (3+\sin v)(-\sin u), (3+\sin v)(\cos u), 0\rangle$$

$$\mathbf{T}_v = \frac{\partial \Phi}{\partial v} = \langle \cos v \cos u, \cos v \sin u, 1\rangle$$

The normal vector is the cross product:

$$\mathbf{n}(u,v) = \mathbf{T}_u \times \mathbf{T}_v = \begin{vmatrix} \mathbf{i} & \mathbf{j} & \mathbf{k} \\ (3+\sin v)(-\sin u) & (3+\sin v)(\cos u) & 0 \\ \cos v \cos u & \cos v \sin u & 1 \end{vmatrix}$$

$$= ((3+\sin v)\cos u)\mathbf{i} + ((3+\sin v)\sin u)\mathbf{j} - ((3+\sin v)\cos v)\mathbf{k}$$

Hence,

$$\|\mathbf{n}(u,v)\| = (3+\sin v)\sqrt{1+\cos^2 v}$$

We obtain the following area:

$$\text{Area}(S) = \iint_{\mathcal{D}} \|\mathbf{n}\|\,du\,dv = \int_0^{2\pi}\int_0^{2\pi} (3+\sin v)\sqrt{1+\cos^2 v}\,du\,dv \approx 144.0181$$

17. Use spherical coordinates to compute the surface area of a sphere of radius R.

SOLUTION The sphere of radius R centered at the origin has the following parametrization in spherical coordinates:

$$\Phi(\theta,\phi) = (R\cos\theta\sin\phi, R\sin\theta\sin\phi, R\cos\phi),\ 0 \le \theta \le 2\pi,\ 0 \le \phi \le \pi$$

The length of the normal vector is:

$$\|\mathbf{n}\| = R^2 \sin\phi$$

Using the integral for surface area gives:

$$\text{Area}(S) = \iint_{\mathcal{D}} \|\mathbf{n}\|\,d\theta\,d\phi = \int_0^{2\pi}\int_0^{\pi} R^2 \sin\phi\,d\phi\,d\theta = \left(\int_0^{2\pi} R^2\,d\theta\right)\left(\int_0^{\pi}\sin\phi\,d\phi\right)$$

$$= 2\pi R^2 \cdot \left(-\cos\phi\Big|_0^{\pi}\right) = 2\pi R^2 \cdot 2 = 4\pi R^2$$

19. Compute the integral of x^2 over the octant of the unit sphere centered at the origin, where $x, y, z \ge 0$.

SOLUTION The octant of the unit sphere centered at the origin, where $x, y, z \ge 0$ has the following parametrization in spherical coordinates:

$$\Phi(\theta,\phi) = (\cos\theta\sin\phi, \sin\theta\sin\phi, \cos\phi), \quad 0 \le \theta \le \frac{\pi}{2}, \quad 0 \le \phi \le \frac{\pi}{2}$$

The length of the normal vector is:

$$\|\mathbf{n}\| = \sin\phi$$

The function x^2 expressed in terms of the parameters is $\cos^2\theta\sin^2\phi$. Using the theorem on computing surface integrals we obtain,

$$\iint_S x^2\,dS = \int_0^{\pi/2}\int_0^{\pi/2}\left(\cos^2\theta\sin^2\phi\right)(\sin\phi)\,d\phi\,d\theta = \int_0^{\pi/2}\int_0^{\pi/2}\cos^2\theta\sin^3\phi\,d\phi\,d\theta$$

$$= \left(\int_0^{\pi/2}\cos^2\theta\,d\theta\right)\left(\int_0^{\pi/2}\sin^3\phi\,d\phi\right) = \left(\frac{\theta}{2}+\frac{\sin 2\theta}{4}\right)\Bigg|_{\theta=0}^{\pi/2} \cdot \left(-\frac{\sin^2\phi\cos\phi}{3} - \frac{2}{3}\cos\phi\right)\Bigg|_{\phi=0}^{\pi/2}$$

$$= \frac{\pi}{4}\cdot\frac{2}{3} = \frac{\pi}{6}$$

In Exercises 21–32, calculate $\iint_S f(x,y,z)\,dS$ *for the given surface and function.*

21. $\Phi(u, v) = (u\cos v, u\sin v, u), \quad 0 \le u, v \le 1; \qquad f(x, y, z) = z(x^2 + y^2)$

SOLUTION

Step 1. Compute the tangent and normal vectors. We have:

$$\mathbf{T}_u = \frac{\partial\Phi}{\partial u} = \frac{\partial}{\partial u}(u\cos v, u\sin v, u) = \langle \cos v, \sin v, 1\rangle$$

$$\mathbf{T}_v = \frac{\partial\Phi}{\partial v} = \frac{\partial}{\partial v}(u\cos v, u\sin v, u) = \langle -u\sin v, u\cos v, 0\rangle$$

The normal vector is the cross product:

$$\mathbf{n} = \mathbf{T}_u \times \mathbf{T}_v = \begin{vmatrix} \mathbf{i} & \mathbf{j} & \mathbf{k} \\ \cos v & \sin v & 1 \\ -u\sin v & u\cos v & 0 \end{vmatrix}$$

$$= (-u\cos v)\mathbf{i} - (u\sin v)\mathbf{j} + \left(u\cos^2 v + u\sin^2 v\right)\mathbf{k}$$

$$= (-u\cos v)\mathbf{i} - (u\sin v)\mathbf{j} + u\mathbf{k} = \langle -u\cos v, -u\sin v, u\rangle$$

We compute the length of $\mathbf{n}$:

$$\|\mathbf{n}\| = \sqrt{(-u\cos v)^2 + (-u\sin v)^2 + u^2} = \sqrt{u^2\left(\cos^2 v + \sin^2 v + 1\right)} = \sqrt{u^2\cdot 2} = \sqrt{2}|u| = \sqrt{2}u$$

Notice that in the region of integration $u \ge 0$, therefore $|u| = u$.

Step 2. Calculate the surface integral. We express the function $f(x, y, z) = z\left(x^2 + y^2\right)$ in terms of the parameters u, v:

$$f(\Phi, (u, v)) = u\left(u^2\cos^2 v + u^2\sin^2 v\right) = u\cdot u^2 = u^3$$

We obtain the following integral:

$$\iint_S f(x, y, z)\,dS = \int_0^1\int_0^1 f(\Phi, (u, v))\,\|\mathbf{n}\|\,du\,dv = \int_0^1\int_0^1 u^3\cdot\sqrt{2}u\,du\,dv$$

$$= \left(\int_0^1 \sqrt{2}\,dv\right)\left(\int_0^1 u^4\,du\right) = \sqrt{2}\cdot\frac{u^5}{5}\bigg|_0^1 = \frac{\sqrt{2}}{5}$$

23. $x^2 + y^2 = 4, \quad 0 \le z \le 4; \qquad f(x, y, z) = e^{-z}$

SOLUTION The cylinder has the following parametrization in cylindrical coordinates:

$$\Phi(\theta, z) = (2\cos\theta, 2\sin\theta, z), \ 0 \le \theta \le 2\pi, \ 0 \le z \le 4$$

Step 1. Compute the tangent and normal vectors. The tangent vectors are the partial derivatives:

$$\mathbf{T}_\theta = \frac{\partial\Phi}{\partial\theta} = \frac{\partial}{\partial\theta}(2\cos\theta, 2\sin\theta, z) = \langle -2\sin\theta, 2\cos\theta, 0\rangle$$

$$\mathbf{T}_z = \frac{\partial}{\partial z}(2\cos\theta, 2\sin\theta, z) = \langle 0, 0, 1\rangle$$

The normal vector is their cross product:

$$\mathbf{n}(\theta, z) = \mathbf{T}_\theta \times \mathbf{T}_z = \begin{vmatrix} \mathbf{i} & \mathbf{j} & \mathbf{k} \\ -2\sin\theta & 2\cos\theta & 0 \\ 0 & 0 & 1 \end{vmatrix} = (2\cos\theta)\mathbf{i} + (2\sin\theta)\mathbf{j} = \langle 2\cos\theta, 2\sin\theta, 0\rangle$$

The length of the normal vector is thus

$$\|\mathbf{n}(\theta, z)\| = \sqrt{(2\cos\theta)^2 + (2\sin\theta)^2 + 0} = \sqrt{4\left(\cos^2\theta + \sin^2\theta\right)} = \sqrt{4} = 2$$

Step 2. Calculate the surface integral. The surface integral equals the following double integral:

$$\iint_S f(x, y, z)\,dS = \iint_{\mathcal{D}} f(\Phi(\theta, z))\,\|\mathbf{n}\|\,d\theta\,dz = \int_0^{2\pi}\int_0^4 e^{-z}\cdot 2\,d\theta\,dz$$

$$= \left(\int_0^{2\pi} 2\,d\theta\right)\left(\int_0^4 e^{-z}\,dz\right) = 4\pi\cdot\left(-e^{-z}\right)\bigg|_0^4 = 4\pi\left(1 - e^{-4}\right)$$

25. $z = 4 - x^2 - y^2, \quad z \geq 0; \qquad f(x, y, z) = z(x^2 + y^2)$

SOLUTION We use the formula for the surface integral over a graph:

$$\iint_S f(x, y, z)\, dS = \iint_{\mathcal{D}} f(x, y, g(x, y)) \sqrt{1 + g_x^2 + g_y^2}\, dx\, dy \tag{1}$$

Since $z = g(x, y) = 4 - x^2 - y^2$, we have:

$$f(x, y, g(x, y)) = \left(4 - x^2 - y^2\right)\left(x^2 + y^2\right) = \left(4 - \left(x^2 + y^2\right)\right)\left(x^2 + y^2\right)$$

$$\sqrt{1 + g_x^2 + g_y^2} = \sqrt{1 + (-2x)^2 + (-2y)^2} = \sqrt{1 + 4\left(x^2 + y^2\right)}$$

The domain of integration $\mathcal{D}$ is determined by the inequality:

$$\mathcal{D}: z = 4 - x^2 - y^2 \geq 0 \Rightarrow x^2 + y^2 \leq 4$$

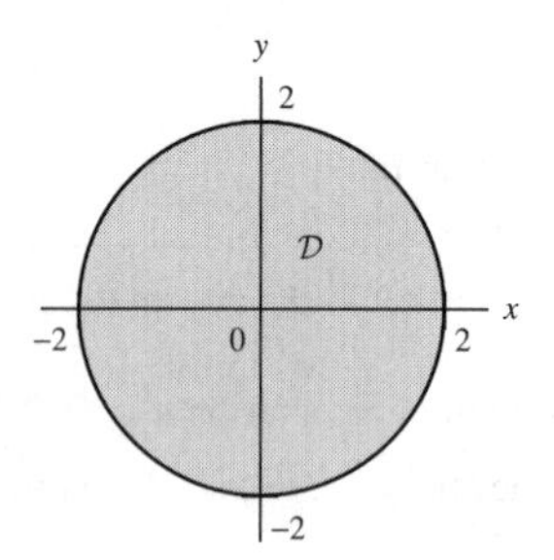

By (1) we obtain:

$$\iint_S f(x, y, z)\, dS = \iint_{\mathcal{D}} \left(4 - (x^2 + y^2)\right) \cdot (x^2 + y^2)\sqrt{1 + 4(x^2 + y^2)}\, dx\, dy$$

We convert the integral to polar coordinates $x = r\cos\theta$, $y = r\sin\theta$ to obtain:

$$\iint_S f(x, y, z)\, dS = \int_0^{2\pi} \int_0^2 \left(4 - r^2\right) r^2 \sqrt{1 + 4r^2} \cdot r\, dr\, d\theta$$

$$= 2\pi \left(\int_0^2 4r^3 \sqrt{1 + 4r^2}\, dr - \int_0^2 r^5 \sqrt{1 + 4r^2}\, dr \right) \tag{2}$$

We compute the integrals using the (somewhat unusual) substitution $u = \sqrt{1 + 4r^2}$, $du = \frac{4r}{u} dr$. This gives:

$$\int_0^2 4r^3 \sqrt{1 + 4r^2}\, dr = \int_0^2 r^2 \sqrt{1 + 4r^2} \cdot 4r\, dr = \int_1^{\sqrt{17}} \frac{u^2 - 1}{4} u^2\, du$$

$$= \int_1^{\sqrt{17}} \frac{u^4 - u^2}{4}\, du = \frac{1}{4}\left(\frac{u^5}{5} - \frac{u^3}{3}\right)\Bigg|_1^{\sqrt{17}} = \frac{391\sqrt{17} + 1}{30}$$

$$\int_0^2 r^5 \sqrt{1 + 4r^2}\, dr = \int_0^2 r^4 \sqrt{1 + 4r^2} \cdot r\, dr = \int_1^{\sqrt{17}} \frac{\left(u^2 - 1\right)^2}{16} \cdot \frac{u^2}{4}\, du = \frac{1}{64} \int_1^{\sqrt{17}} \left(u^6 - 2u^4 + u^2\right) du$$

$$= \frac{1}{64}\left(\frac{u^7}{7} - \frac{2u^5}{5} + \frac{u^3}{3} \Bigg|_1^{\sqrt{17}} \right) = \frac{7769\sqrt{17} - 1}{840}$$

Substituting the integrals in (2) gives:

$$\iint_S f(x, y, z)\, dS = 2\pi \left(\frac{391\sqrt{17} + 1}{30} - \frac{7769\sqrt{17} - 1}{840} \right) = \frac{\left(3179\sqrt{17} + 29\right)\pi}{420} \approx 98.26$$

27. $y = 9 - z^2, \quad 0 \leq x, z \leq 3; \qquad f(x, y, z) = z$

SOLUTION We use the formula for the surface integral over a graph $y = g(x, z)$:

$$\iint_S f(x, y, z)\, dS = \iint_{\mathcal{D}} f(x, g(x, z), z) \sqrt{1 + g_x^2 + g_z^2}\, dx\, dz \tag{1}$$

Since $y = g(x, z) = 9 - z^2$, we have $g_x = 0$, $g_z = -2z$, hence:

$$\sqrt{1 + g_x^2 + g_z^2} = \sqrt{1 + 4z^2}$$
$$f(x, g(x, z), z) = z$$

The domain of integration is the square $[0, 3] \times [0, 3]$ in the xz-plane. By (1) we get:

$$\iint_S f(x, y, z)\, dS = \int_0^3 \int_0^3 z\sqrt{1 + 4z^2}\, dz\, dx = \left(\int_0^3 1\, dx\right)\left(\int_0^3 z\sqrt{1 + 4z^2}\, dz\right) = 3\int_0^3 z\sqrt{1 + 4z^2}\, dz$$

We use the substitution $u = 1 + 4z^2$, $du = 8z\, dz$ to compute the integral. This gives:

$$\iint_S f(x, y, z)\, dS = 3\int_0^3 z\sqrt{1 + 4z^2}\, dz = 3\int_1^{37} \frac{u^{1/2}}{8}\, du = \frac{37\sqrt{37} - 1}{4} \approx 56$$

29. Part of the plane $x + y + z = 1$, where $x, y, z \geq 0$; $\quad f(x, y, z) = z$

SOLUTION We let $z = g(x, y) = 1 - x - y$ and use the formula for the surface integral over the graph of $z = g(x, y)$, where $\mathcal{D}$ is the parameter domain in the xy-plane. That is:

$$\iint_S f(x, y, z)\, dS = \iint_{\mathcal{D}} f(x, y, g(x, y))\sqrt{1 + g_x^2 + g_y^2}\, dx\, dy \tag{1}$$

We have, $g_x = -1$ and $g_y = -1$ therefore:

$$\sqrt{1 + g_x^2 + g_y^2} = \sqrt{1 + (-1)^2 + (-1)^2} = \sqrt{3}$$

We express the function $f(x, y, z) = z$ in terms of the parameters x and y:

$$f(x, y, g(x, y)) = z = 1 - x - y$$

The domain of integration is the triangle $\mathcal{D}$ in the xy-plane shown in the figure.

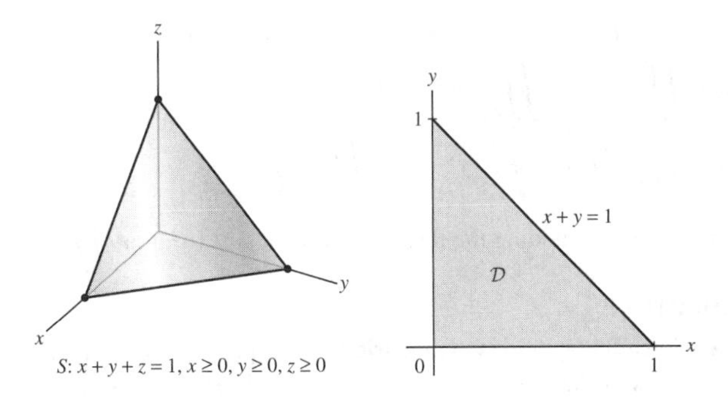

By (1) we get:

$$\iint_S f(x, y, z)\, dS = \int_0^1 \int_0^{1-y} (1 - x - y)\sqrt{3}\, dx\, dy = \sqrt{3}\int_0^1 x - \frac{x^2}{2} - yx\Big|_{x=0}^{1-y} dy$$
$$= \sqrt{3}\int_0^1 \left((1 - y)^2 - \frac{(1 - y)^2}{2}\right) dy = \frac{\sqrt{3}}{2}\int_0^1 \left(1 - 2y + y^2\right) dy$$
$$= \frac{\sqrt{3}}{2}\left(y - y^2 + \frac{y^3}{3}\right)\Big|_0^1 = \frac{\sqrt{3}}{6}$$

31. Part of the surface $x = z^3$, where $0 \leq x, y \leq 1$; $\quad f(x, y, z) = x$

SOLUTION We let $z = g(x, y) = x^{1/3}$ and use the formula for the surface integral over a graph:

$$\iint_S f(x, y, z)\, dS = \iint_{\mathcal{D}} f(x, y, g(x, y))\sqrt{1 + g_x^2 + g_y^2}\, dx\, dy \tag{1}$$

where $\mathcal{D}$ is the square $[0, 1] \times [0, 1]$ in the xy-plane. We compute the integrand in (1):

$$g_x = \frac{1}{3}x^{-2/3}, \quad g_y = 0 \quad \Rightarrow \quad \sqrt{1 + g_x^2 + g_y^2} = \sqrt{1 + \frac{1}{9}x^{-4/3}}$$

$$f(x, y, g(x, y)) = x$$

Substituting in (1) we get:

$$\iint_S f(x, y, z)\, dS = \int_0^1 \int_0^1 x\sqrt{1 + \frac{1}{9}x^{-4/3}}\, dx\, dy = \int_0^1 x\sqrt{1 + \frac{1}{9}x^{-4/3}}\, dx$$

We compute the integral using the substitution $\frac{1}{3}x^{-2/3} = \tan\theta$. Then:

$$\sqrt{1 + \frac{1}{9}x^{-4/3}} = \sqrt{1 + \tan^2\theta} = \frac{1}{\cos\theta}$$

$$-\frac{2}{9}x^{-5/3}\, dx = \frac{1}{\cos^2\theta}\, d\theta \quad \Rightarrow \quad x\, dx = -\frac{9}{2}\cdot\frac{1}{\cos^2\theta}x^{8/3}\, d\theta = -\frac{1}{18}\frac{\cos^2\theta}{\sin^4\theta}\, d\theta$$

Hence:

$$\sqrt{1 + \frac{1}{9}x^{-4/3}}\cdot x\, dx = -\frac{1}{18}\frac{\cos\theta}{\sin^4\theta}\, d\theta$$

We obtain the following integral, which we compute by substituting $u = \sin\theta$, $du = \cos\theta\, d\theta$:

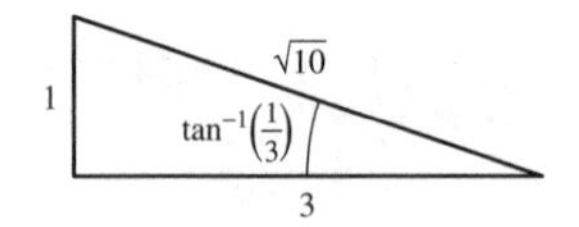

$$\iint_S f(x, y, z)\, dS = \int_{\pi/2}^{\tan^{-1}\frac{1}{3}} -\frac{1}{18}\frac{\cos\theta}{\sin^4\theta}\, d\theta = -\frac{1}{18}\int_1^{\sin\left(\tan^{-1}\frac{1}{3}\right) = \frac{1}{\sqrt{10}}} \frac{1}{u^4}\, du$$

$$= \frac{1}{54}\cdot\frac{1}{u^3}\Big|_1^{1/(\sqrt{10})} = \frac{1}{54}\left(10\sqrt{10} - 1\right) \approx 0.567$$

33. Let S be the sphere of radius R centered at the origin. Explain the following equalities using symmetry:

(a) $\iint_S x\, dS = \iint_S y\, dS = \iint_S z\, dS = 0$

(b) $\iint_S x^2\, dS = \iint_S y^2\, dS = \iint_S z^2\, dS$

Then show, by adding the three integrals in part (b), that $\iint_S x^2\, dS = \frac{4}{3}\pi R^4$.

SOLUTION

(a) Since the sphere is symmetric with respect to the yz-plane, the surface integrals of x over the hemispheres on the two sides of the plane cancel each other and the result is zero. The two other integrals are zero due to the symmetry of the sphere with respect to the xz and xy-planes.

(b) Since the sphere is symmetric with respect to the xy, xz and yz-planes, interchanging x and y in the integral for $\iint_S x^2\, dS$ does not change the value of the integral and the result is $\iint_S y^2\, dS$. The equality for $\iint_S z^2\, dS$ is explained similarly.

On the sphere, we have $x^2 + y^2 + z^2 = R^2$ so, using properties of integrals, the integral for surface area, and the surface area of the sphere of radius R we obtain:

$$\iint_S x^2\, dS + \iint_S y^2\, dS + \iint_S z^2\, dS = \iint_S \left(x^2 + y^2 + z^2\right) dS = R^2\iint_S 1\, dS$$

$$= R^2\cdot \text{Area}(S) = R^2\cdot 4\pi R^2 = 4\pi R^4$$

Combining with (b) we conclude that the value of each of the integrals is $\frac{4}{3}\pi R^4$. That is:

$$\iint_S x^2\, dS = \iint_S y^2\, dS = \iint_S z^2\, dS = \frac{4}{3}\pi R^4.$$

35. Find the area of the portion of the plane $2x + 3y + 4z = 28$ lying above the rectangle $1 \le x \le 3$, $2 \le y \le 5$ in the xy-plane.

SOLUTION We rewrite the equation of the plane as:

$$z = g(x, y) = -\frac{x}{2} - \frac{3}{4}y + 7 \tag{1}$$

The domain of the parameters is the rectangle $\mathcal{D} = [1, 3] \times [2, 5]$ in the xy-plane. Using the integral for surface area and the surface integral over a graph we have:

$$\text{Area}(S) = \iint_S 1\, dS = \iint_{\mathcal{D}} \sqrt{1 + g_x^2 + g_z^2}\, dx\, dy \tag{2}$$

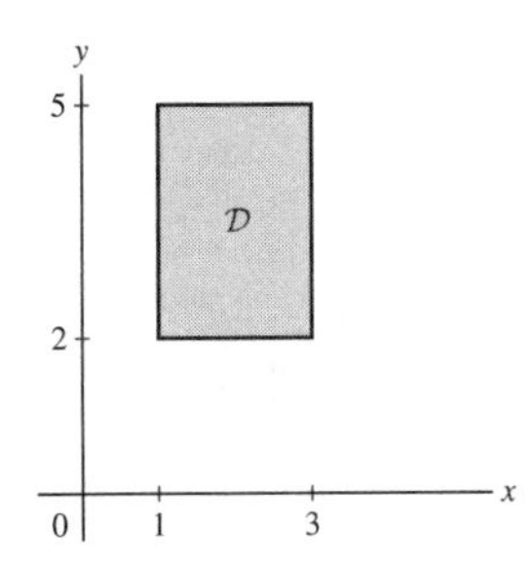

By (1) we have:

$$g_x = -\frac{1}{2},\ g_y = -\frac{3}{4} \Rightarrow \sqrt{1 + g_x^2 + g_z^2} = \sqrt{1 + \frac{1}{4} + \frac{9}{16}} = \frac{\sqrt{29}}{4}$$

We substitute in (2) to obtain:

$$\text{Area}(S) = \iint_{\mathcal{D}} \frac{\sqrt{29}}{4}\, dx\, dy = \frac{\sqrt{29}}{4} \iint_{\mathcal{D}} 1\, dx\, dy = \frac{\sqrt{29}}{4}\text{Area}(\mathcal{D}) = \frac{\sqrt{29}}{4} \cdot 3 \cdot 2 = \frac{3\sqrt{29}}{2}$$

37. Compute the integral of $f(x, y, z) = z^2(x^2 + y^2 + z^2)^{-1}$ over the cap of the sphere $x^2 + y^2 + z^2 = 4$ defined by $z \geq 1$.

SOLUTION We use spherical coordinates to parametrize the cap S.

$$\Phi(\theta, \phi) = (2\cos\theta\sin\phi, 2\sin\theta\sin\phi, 2\cos\phi)$$

$$\mathcal{D}\ :\ 0 \leq \theta \leq 2\pi,\ 0 \leq \phi \leq \phi_0$$

The angle ϕ_0 is determined by $\cos\phi_0 = \frac{1}{2}$, that is, $\phi_0 = \frac{\pi}{3}$. The length of the normal vector in spherical coordinates is:

$$\|\mathbf{n}\| = R^2 \sin\phi = 4\sin\phi$$

We express the function $f(x, y, z) = z^2\left(x^2 + y^2 + z^2\right)^{-1}$ in terms of the parameters:

$$f\left(\Phi(\theta, \phi)\right) = (2\cos\phi)^2 4^{-1} = \cos^2\phi$$

Using the theorem on computing the surface integral we get:

$$\iint_S f(x, y, z)\, dS = \iint_{\mathcal{D}} f\left(\Phi(\theta, \phi)\right) \|\mathbf{n}\|\, d\phi\, d\theta = \int_0^{2\pi} \int_0^{\pi/3} \left(\cos^2\phi\right) \cdot 4\sin\phi\, d\phi\, d\theta$$

$$= \left(\int_0^{2\pi} 4\, d\theta\right)\left(\int_0^{\pi/3} \cos^2\phi\sin\phi\, d\phi\right) = 8\pi\left(-\frac{\cos\phi}{3}\right)\Big|_0^{\pi/3}$$

$$= 8\pi\left(\left(\frac{1}{8} - \frac{1}{6}\right) - \left(-\frac{1}{3}\right)\right) = 8\pi \cdot \frac{7}{24} = \frac{7\pi}{3}$$

39. Let S be the portion of the sphere $x^2 + y^2 + z^2 = 9$, where $1 \leq x^2 + y^2 \leq 4$ and $z \geq 0$ (Figure 20). Find a parametrization of S in polar coordinates and use it to compute:

(a) The area of S

(b) $\iint_S z^{-1}\, dS$

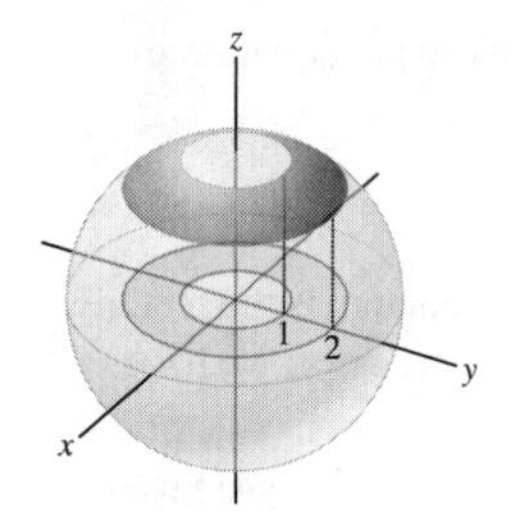

FIGURE 20

SOLUTION

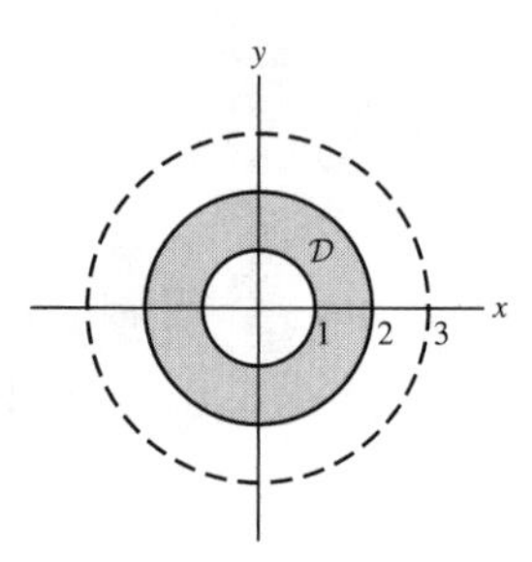

We parametrize S by spherical coordinates as follows:

$$\Phi(\theta, \phi) = (3\cos\theta\sin\phi, 3\sin\theta\sin\phi, 3\cos\phi)$$

$$\mathcal{D}: 0 \le \theta \le 2\pi,\ \phi_0 \le \phi \le \phi_1$$

The angles ϕ_0 and ϕ_1 are determined by,

$$\sin\phi_0 = \frac{1}{3} \quad \Rightarrow \quad \phi_0 = \sin^{-1}\frac{1}{3}$$

$$\sin\phi_1 = \frac{2}{3} \quad \Rightarrow \quad \phi_1 = \sin^{-1}\frac{2}{3}$$

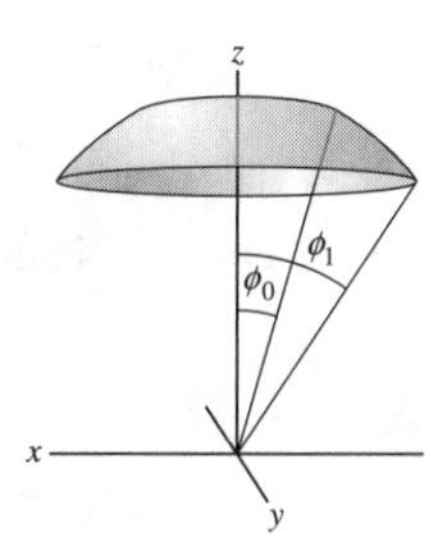

The length of the normal is:

$$\|\mathbf{n}\| = R^2\sin\phi = 9\sin\phi$$

(a) Using the integral for the surface area we have,

$$\text{Area}(S) = \iint_{\mathcal{D}} \|\mathbf{n}\|\,d\phi\,d\theta = \int_0^{2\pi}\int_{\sin^{-1}(1/3)}^{\sin^{-1}(2/3)} 9\sin\phi\,d\phi\,d\theta = \left(\int_0^{2\pi} 9\,d\phi\right)\left(\int_{\sin^{-1}(1/3)}^{\sin^{-1}(2/3)} \sin\phi\,d\phi\right)$$

$$= 18\pi\left(-\cos\phi\Big|_{\phi_0=\sin^{-1}(1/3)}^{\phi_1=\sin^{-1}(2/3)}\right) = 18\pi\left(-\frac{\sqrt{5}}{3} + \frac{\sqrt{8}}{3}\right) = 6\pi\left(\sqrt{8} - \sqrt{5}\right) \approx 11.166$$

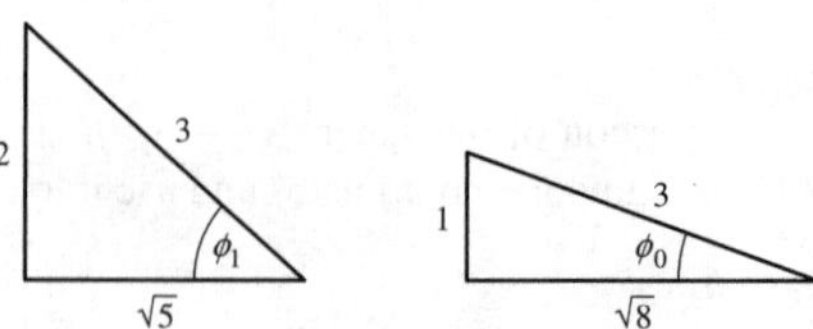

(b) We express the function $f(x, y, z) = z^{-1}$ in the terms of the parameters:

$$f\left(\Phi(\theta, \phi)\right) = (3\cos\phi)^{-1} = \frac{\sec\phi}{3}$$

Using the surface integral as a double integral we obtain:

$$\iint_S z^{-1}\,dS = \iint_{\mathcal{D}} f\left(\Phi(\theta,\phi)\right)\|\mathbf{n}\|\,d\phi\,d\theta = \int_0^{2\pi}\int_{\phi_0}^{\phi_1}\frac{\sec\phi}{3}\cdot 9\sin\phi\,d\phi\,d\theta = \int_0^{2\pi}\int_{\phi_0}^{\phi_1} 3\tan\phi\,d\phi\,d\theta$$

$$= \left(\int_0^{2\pi} 3\,d\theta\right)\left(\int_{\phi_0}^{\phi_1}\tan\phi\,d\phi\right) = 6\pi\left(\ln(\sec\phi)\Big|_{\phi_0=\sin^{-1}(1/3)}^{\phi_1=\sin^{-1}(2/3)}\right)$$

$$= 6\pi\left(\ln\frac{3}{\sqrt{5}} - \ln\frac{3}{\sqrt{8}}\right) = 6\pi\ln\sqrt{\frac{8}{5}} = 3\pi\ln 1.6 \approx 4.43$$

41. Find the surface area of the portion S of the cone $z^2 = x^2 + y^2$, where $z \geq 0$, contained within the cylinder $y^2 + z^2 \leq 1$.

SOLUTION We rewrite the equation of the cone as $x = \pm\sqrt{z^2 - y^2}$. The projection of the cone onto the yz-plane is obtained by setting $x = 0$ in the equation of the cone, that is,

$$x = 0 = \sqrt{z^2 - y^2} \quad \Rightarrow \quad z = \pm y$$

Since on S, $z \geq 0$, we get $z = |y|$. We conclude that the projection of the upper part of the cone $x^2 + y^2 = z^2$ onto the yz-plane is the region between the lines $z = y$ and $z = -y$ on the upper part of the yz-plane. Therefore, the projection $\mathcal{D}$ of S onto the yz-plane is the region shown in the figure:

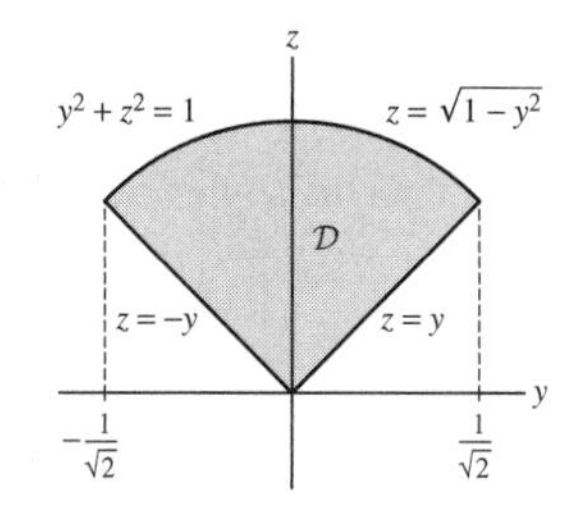

The area of S is the surface integral:

$$\text{Area}(S) = \iint_S dS$$

We compute the integral using a surface integral over a graph. Since $x = g(y, z) = \pm\sqrt{z^2 - y^2}$ we have,

$$g_z = \pm\frac{z}{\sqrt{z^2 - y^2}}, \quad g_y = \pm\frac{y}{\sqrt{z^2 - y^2}}$$

Hence, (notice that $z \geq 0$ on S):

$$\sqrt{1 + g_y^2 + g_z^2} = \sqrt{1 + \frac{z^2}{z^2 - y^2} + \frac{y^2}{z^2 - y^2}} = \sqrt{\frac{2z^2}{z^2 - y^2}} = \frac{z\sqrt{2}}{\sqrt{z^2 - y^2}}$$

We obtain the following integral:

$$\text{Area}(S) = \iint_{\mathcal{D}} \sqrt{1 + g_y^2 + g_z^2}\,dy\,dz = \iint_{\mathcal{D}} \frac{z\sqrt{2}}{\sqrt{z^2 - y^2}}\,dz\,dy$$

Using symmetry gives:

$$\text{Area}(S) = 2\int_0^{1/(\sqrt{2})}\int_y^{\sqrt{1-y^2}} \frac{z\sqrt{2}}{\sqrt{z^2 - y^2}}\,dz\,dy = 2\sqrt{2}\int_0^{1/(\sqrt{2})}\left(\int_y^{\sqrt{1-y^2}}\frac{z\,dz}{\sqrt{z^2 - y^2}}\right)dy \qquad (1)$$

We compute the inner integral using the substitution $u = \sqrt{z^2 - y^2}$, $du = \frac{z}{u}\,dz$. We get:

$$\int_y^{\sqrt{1-y^2}} \frac{z\,dz}{\sqrt{z^2 - y^2}} = \int_0^{\sqrt{1-y^2}}\frac{u\,du}{u} = \int_0^{\sqrt{1-2y^2}} du = \sqrt{1 - 2y^2}$$

We substitute in (1) and compute the resulting integral using the substitution $t = \sqrt{2}y$. We get:

$$\text{Area}(S) = \int_0^{1/(\sqrt{2})}\sqrt{1 - 2y^2}\,dy = 2\sqrt{2}\int_0^1\sqrt{1 - t^2}\frac{dt}{\sqrt{2}} = 2\int_0^1\sqrt{1 - t^2}\,dt = 2\cdot\frac{\pi}{4} = \frac{\pi}{2}$$

43. Prove a famous result of Archimedes: The surface area of the portion of the sphere of radius r between two horizontal planes $z = a$ and $z = b$ is equal to the surface area of the corresponding portion of the circumscribed cylinder (Figure 22).

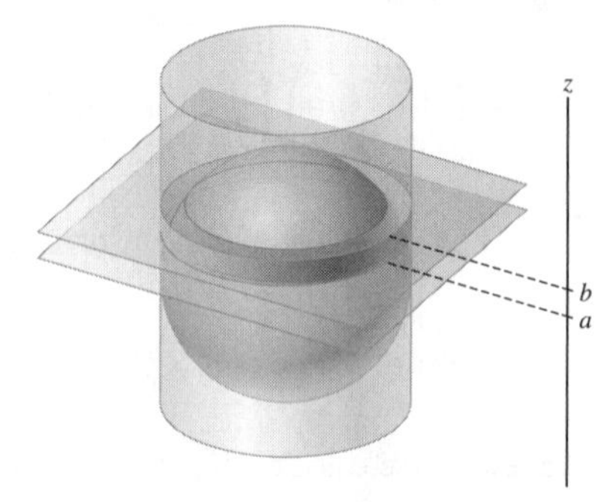

FIGURE 22

SOLUTION We compute the area of the portion of the sphere between the planes a and b. The portion S_1 of the sphere has the parametrization,

$$\Phi(\theta, \phi) = (r\cos\theta\sin\phi, r\sin\theta\sin\phi, r\cos\phi)$$

where,

$$\mathcal{D}_1 : 0 \le \theta \le 2\pi,\ \phi_0 \le \phi \le \phi_1$$

If we assume $0 < a < b$, then the angles ϕ_0 and ϕ_1 are determined by,

$$\cos\phi_0 = \frac{b}{r} \quad \Rightarrow \quad \phi_0 = \cos^{-1}\frac{b}{r}$$
$$\cos\phi_1 = \frac{a}{r} \quad \Rightarrow \quad \phi_1 = \cos^{-1}\frac{a}{r}$$

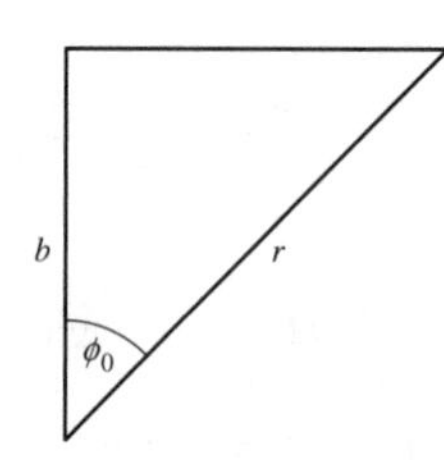

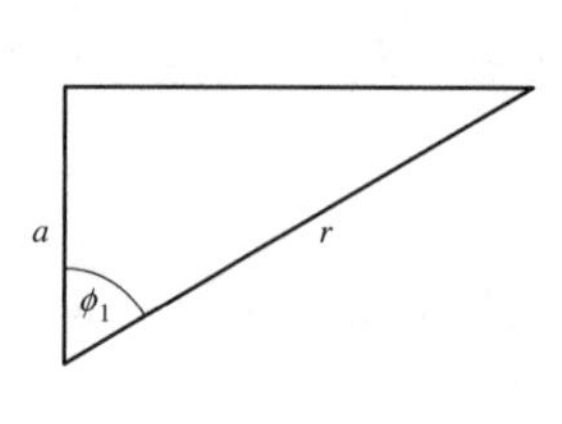

The length of the normal vector is $\|\mathbf{n}\| = r^2\sin\phi$. We obtain the following integral:

$$\text{Area}\,(S_1) = \iint_{\mathcal{D}_1} \|\mathbf{n}\|\,d\phi\,d\theta = \int_0^{2\pi}\int_{\phi_0}^{\phi_1} r^2\sin\phi\,d\phi\,d\theta = \left(\int_0^{2\pi} r^2 d\phi\right)\left(\int_{\phi_1}^{\phi_2} \sin\phi\,d\phi\right)$$

$$= 2\pi r^2\left(-\cos\phi\Big|_{\phi=\cos^{-1}\frac{b}{r}}^{\cos^{-1}\frac{a}{r}}\right) = 2\pi r^2\left(-\frac{a}{r} + \frac{b}{r}\right) = 2\pi r(b-a)$$

The area of the part S_2 of the cylinder of radius r between the planes $z = a$ and $z = b$ is:

$$\text{Area}\,(S_2) = 2\pi r \cdot (b-a)$$

We see that the two areas are equal:

$$\text{Area}\,(S_1) = \text{Area}\,(S_2)$$

Further Insights and Challenges

45. Use Eq. (13) to compute the surface area of $z = 4 - y^2$ for $0 \le y \le 2$ rotated about the z-axis.

SOLUTION Since $g(y) = 4 - y^2$, we have $g'(y) = -2y$. By Eq. (13) we obtain the following integral,

$$\text{Area}(S) = 2\pi\int_0^2 |y|\sqrt{1 + (-2y)^2}\,dy = 2\pi\int_0^2 y \cdot \sqrt{1 + 4y^2}\,dy$$

We compute the integral using the substitution $u = 1 + 4y^2$, $du = 8y\,dy$. We get:

$$\text{Area}(S) = 2\pi \int_1^{17} u^{1/2} \cdot \frac{du}{8} = 2\pi \frac{2}{3} \cdot \frac{u^{3/2}}{8}\bigg|_1^{17} = \frac{\pi}{6}\left(17\sqrt{17} - 1\right) \approx 36.18$$

47. Area of a Torus Let T be the torus obtained by rotating the circle in the yz-plane of radius a centered at $(0, b, 0)$ about the z-axis (Figure 24). We assume that $b > a > 0$.

(a) Use Eq. (13) to show that

$$\text{Area(T)} = 4\pi \int_{b-a}^{b+a} \frac{ay}{\sqrt{a^2 - (b-y)^2}}\,dy$$

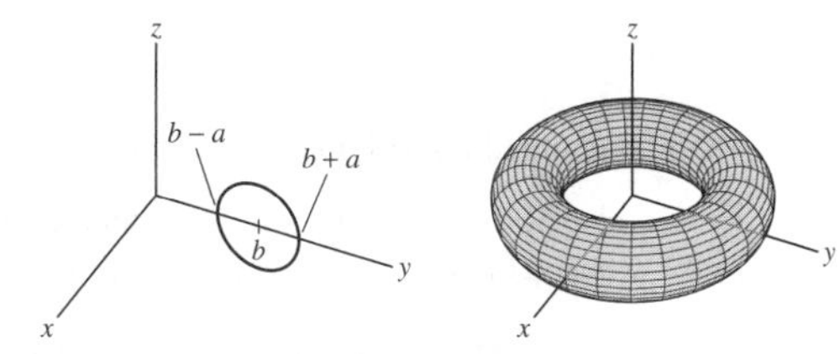

FIGURE 24 The torus obtained by rotating a circle of radius a.

(b) Show that $\text{Area(T)} = 4\pi^2 ab$. *Hint:* Rewrite the integral using substitution.

SOLUTION

(a) Using symmetry, the area of the surface obtained by rotating the upper part of the circle is half the area of the torus.

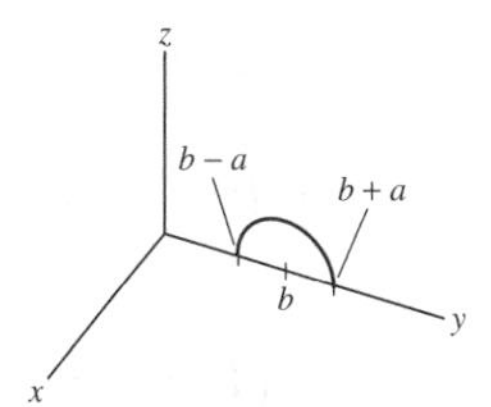

The rotated graph is $z = g(y) = \sqrt{a^2 - (y-b)^2}$, $b - a \le y \le b + a$. So, we have,

$$g'(y) = \frac{-2(y-b)}{2\sqrt{a^2 - (y-b)^2}} = -\frac{y-b}{\sqrt{a^2 - (y-b)^2}}$$

$$\sqrt{1 + g'(y)^2} = \sqrt{1 + \frac{(y-b)^2}{a^2 - (y-b)^2}} = \sqrt{\frac{a^2 - (y-b)^2 + (y-b)^2}{a^2 - (y-b)^2}} = \frac{a}{\sqrt{a^2 - (y-b)^2}}$$

We now use symmetry and Eq. (13) to obtain the following area of the torus (we assume that $b - a > 0$, hence $y > 0$):

$$\text{Area }(\mathbf{T}) = 2 \cdot 2\pi \int_{b-a}^{b+a} |y|\sqrt{1 + g'(y)^2}\,dy = 4\pi \int_{b-a}^{b+a} \frac{ay}{\sqrt{a^2 - (y-b)^2}}\,dy \tag{1}$$

(b) We compute the integral using the substitution $u = \frac{y-b}{a}$, $du = \frac{1}{a}\,dy$. We get:

$$\int_{b-a}^{b+a} \frac{ay}{\sqrt{a^2 - (y-b)^2}}\,dy = \int_{-1}^{1} \frac{a^2u + ab}{\sqrt{a^2 - a^2u^2}}a\,du = \int_{-1}^{1} \frac{a^2u + ab}{\sqrt{1 - u^2}}\,du = \int_{-1}^{1} \frac{a^2u}{\sqrt{1-u^2}}\,du + \int_{-1}^{1} \frac{ab}{\sqrt{1-u^2}}\,du$$

The first integral is zero since the integrand is an odd function. We get:

$$\int_{b-a}^{b+a} \frac{ay}{\sqrt{a^2 - (y-b)^2}}\,dy = 2\int_0^1 \frac{ab}{\sqrt{1-u^2}}\,du = 2ab \sin^{-1} u\bigg|_0^1 = 2ab\left(\frac{\pi}{2} - 0\right) = \pi ab$$

Substituting in (1) gives the following area:

$$\text{Area }(\mathbf{T}) = 4\pi \cdot \pi ab = 4\pi^2 ab$$

49. Compute the surface area of the torus in Exercise 47 using Pappus's Theorem.

SOLUTION The generating curve is the circle of radius a in the (y, z)-plane centered at the point $(0, b, 0)$. The length of the generating curve is $L = \pi a$.

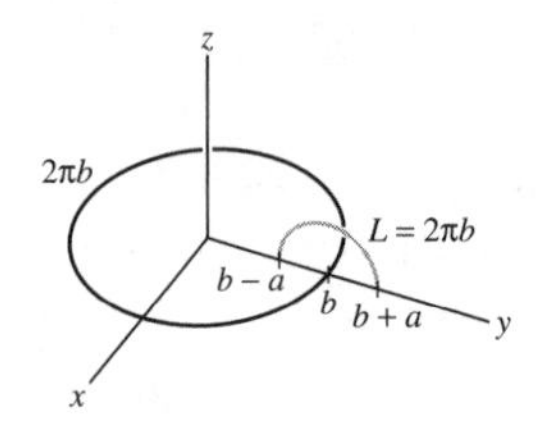

The center of mass of the circle is at the center $(\overline{y}, \overline{z}) = (b, 0)$, and it traverses a circle of radius b centered at the origin. Therefore, the center of mass makes a distance of $2\pi b$. Using Pappus' Theorem, the area of the torus is:

$$L \cdot 2\pi a = 2\pi a \cdot 2\pi b = 4\pi^2 ab.$$

51. Calculate the gravitational potential φ for a hemisphere of radius R with uniform mass distribution.

SOLUTION In Exercise 50(b) we expressed the potential φ for a sphere of radius R. To find the potential for a hemisphere of radius R, we need only to modify the limits of the angle ϕ to $0 \le \phi \le \frac{\pi}{2}$. This gives the following integral:

$$\varphi(0, 0, r) = \varphi(r) = -\frac{Gm}{4\pi}\int_0^{\pi/2}\int_0^{2\pi}\frac{\sin\phi\, d\theta\, d\phi}{\sqrt{R^2 + r^2 - 2Rr\cos\phi}} = -\frac{Gm}{4\pi}\cdot 2\pi\int_0^{\pi/2}\frac{\sin\phi\, d\phi}{\sqrt{R^2 + r^2 - 2Rr\cos\phi}}$$

$$= -\frac{Gm}{4\pi}\int_0^{\pi/2}\frac{\sin\phi\, d\phi}{\sqrt{R^2 + r^2 - 2Rr\cos\phi}}$$

We compute the integral using the substitution $u = R^2 + r^2 - 2Rr\cos\phi, \quad du = 2Rr\sin\phi\, d\phi$. We obtain:

$$\varphi(r) = -\frac{Gm}{2}\int_{(R-r)^2}^{R^2+r^2}\frac{\frac{du}{2Rr}}{\sqrt{u}} = -\frac{Gm}{4Rr}\int_{(R-r)^2}^{R^2+r^2}u^{-1/2}\,du = -\frac{Gm}{4Rr}\cdot 2u^{1/2}\Big|_{u=(R-r)^2}^{R^2+r^2}$$

$$= -\frac{Gm}{2Rr}\left(\left(R^2 + r^2\right)^{1/2} - \left((R-r)^2\right)^{1/2}\right) = -\frac{Gm}{2Rr}\left(\sqrt{R^2 + r^2} - |R - r|\right)$$

53. Let S be the part of the graph $z = g(x, y)$ lying over a domain $\mathcal{D}$ in the xy-plane. Let $\phi = \phi(x, y)$ be the angle between the normal to S and the vertical. Prove the formula

$$\text{Area}(S) = \iint_{\mathcal{D}}\frac{dA}{|\cos\phi|}$$

SOLUTION

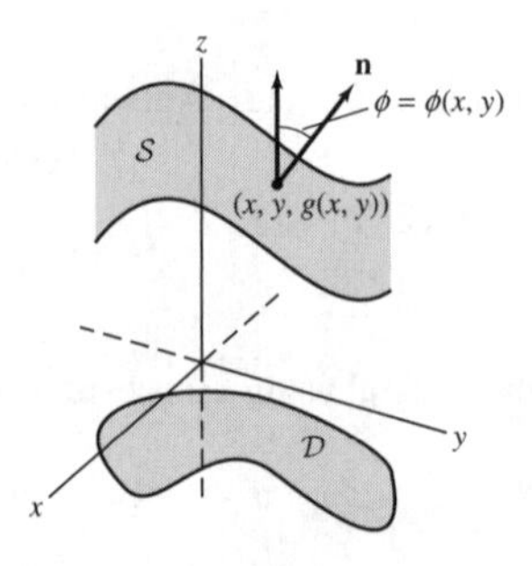

Using the Surface Integral over a Graph we have:

$$\text{Area}(S) = \iint_S 1\, dS = \iint_{\mathcal{D}}\sqrt{1 + g_x^2 + g_y^2}\, dA \qquad (1)$$

In parametrizing the surface by $\phi(x, y) = (x, y, g(x, y)), (x, y) = \mathcal{D}$, we have:

$$\mathbf{T}_x = \frac{\partial\Phi}{\partial x} = \langle 1, 0, g_x\rangle$$

$$\mathbf{T}_y = \frac{\partial\Phi}{\partial y} = \langle 0, 1, g_y\rangle$$

Hence,

$$\mathbf{n} = \mathbf{T}_x \times \mathbf{T}_y = \begin{vmatrix} \mathbf{i} & \mathbf{j} & \mathbf{k} \\ 1 & 0 & g_x \\ 0 & 1 & g_y \end{vmatrix} = -g_x\mathbf{i} - g_y\mathbf{j} + \mathbf{k} = \langle -g_x, -g_y, 1 \rangle$$

$$\|\mathbf{n}\| = \sqrt{g_x^2 + g_y^2 + 1}$$

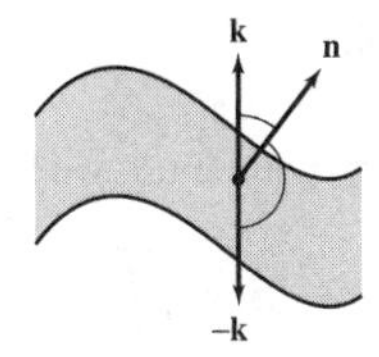

There are two adjacent angles between the normal **n** and the vertical, and the cosines of these angles are opposite numbers. Therefore we take the absolute value of $\cos\phi$ to obtain a positive value for Area(S). Using the Formula for the cosine of the angle between two vectors we get:

$$|\cos\phi| = \frac{|\mathbf{n}\cdot\mathbf{k}|}{\|\mathbf{n}\|\,\|\mathbf{k}\|} = \frac{|\langle -g_x, -g_y, 1\rangle \cdot \langle 0, 0, 1\rangle|}{\sqrt{1 + g_x^2 + g_y^2}\cdot 1} = \frac{1}{\sqrt{1 + g_x^2 + g_y^2}}$$

Substituting in (1) we get:

$$\text{Area}(S) = \iint_{\mathcal{D}} \frac{dA}{|\cos\phi|}$$

17.5 Surface Integrals of Vector Fields (ET Section 16.5)

Preliminary Questions

1. Let **F** be a vector field and $\Phi(u, v)$ a parametrization of a surface $\mathcal{S}$, and set $\mathbf{n} = \mathbf{T}_u \times \mathbf{T}_v$. Which of the following is the normal component of **F**?

(a) $\mathbf{F}\cdot\mathbf{n}$ **(b)** $\mathbf{F}\cdot\mathbf{e_n}$

SOLUTION The normal component of **F** is $\mathbf{F}\cdot\mathbf{e}_n$ rather than $\mathbf{F}\cdot\mathbf{n}$.

2. The vector surface integral $\iint_{\mathcal{S}} \mathbf{F}\cdot d\mathbf{S}$ is equal to the scalar surface integral of the function (choose the correct answer):

(a) $\|\mathbf{F}\|$

(b) $\mathbf{F}\cdot\mathbf{n}$, where **n** is a normal vector

(c) $\mathbf{F}\cdot\mathbf{e_n}$, where $\mathbf{e_n}$ is the unit normal vector

SOLUTION The vector surface integral $\iint_S \mathbf{F}\cdot d\mathbf{S}$ is defined as the scalar surface integral of the normal component of **F** on the oriented surface. That is, $\iint_S \mathbf{F}\cdot d\mathbf{S} = \iint_S (\mathbf{F}\cdot\mathbf{e}_n)\, dS$ as stated in (c).

3. $\iint_{\mathcal{S}} \mathbf{F}\cdot d\mathbf{S}$ is zero if (choose the correct answer):

(a) **F** is tangent to $\mathcal{S}$ at every point.

(b) **F** is perpendicular to $\mathcal{S}$ at every point.

SOLUTION Since $\iint_S \mathbf{F}\cdot d\mathbf{S}$ is equal to the scalar surface integral of the normal component of **F** on S, this integral is zero when the normal component is zero at every point, that is, when **F** is tangent to S at every point as stated in (a).

4. If $\mathbf{F}(P) = \mathbf{e_n}(P)$ at each point on $\mathcal{S}$, then $\iint_{\mathcal{S}} \mathbf{F}\cdot d\mathbf{S}$ is equal to (choose the correct answer):

(a) Zero **(b)** Area($\mathcal{S}$) **(c)** Neither

SOLUTION If $\mathbf{F}(P) = \mathbf{e}_n(P)$ at each point on S, then,:

$$\iint_S \mathbf{F}\cdot d\mathbf{S} = \iint_S (\mathbf{e}_n\cdot\mathbf{e}_n)\, dS = \iint_S \|\mathbf{e}_n\|^2\, dS = \iint_S 1\, dS = \text{Area}(S)$$

Therefore, (b) is the correct answer.

5. Let $\mathcal{S}$ be the disk $x^2 + y^2 \le 1$ in the xy-plane oriented with normal in the positive z-direction. Determine $\iint_{\mathcal{S}} \mathbf{F} \cdot d\mathbf{S}$ for each of the following vector constant fields:

(a) $\mathbf{F} = \langle 1, 0, 0 \rangle$

(b) $\mathbf{F} = \langle 0, 0, 1 \rangle$

(c) $\mathbf{F} = \langle 1, 1, 1 \rangle$

SOLUTION The unit normal vector to the oriented disk is $\mathbf{e}_n = \langle 0, 0, 1 \rangle$.

(a) Since $\mathbf{F} \cdot \mathbf{e}_n = \langle 1, 0, 0 \rangle \cdot \langle 0, 0, 1 \rangle = 0$, $\mathbf{F}$ is perpendicular to the unit normal vector at every point on S, therefore $\iint_S \mathbf{F} \cdot d\mathbf{S} = 0$.

(b) Since $\mathbf{F} = \mathbf{e}_n$ at every point on S, we have:

$$\iint_S \mathbf{F} \cdot d\mathbf{S} = \iint_S (\mathbf{e}_n \cdot \mathbf{e}_n)\, dS = \iint_S \|\mathbf{e}_n\|^2\, dS = \iint_S 1\, dS = \text{Area}(S) = \pi$$

(c) For $\mathbf{F} = \langle 1, 1, 1 \rangle$ we have:

$$\iint_S \mathbf{F} \cdot d\mathbf{S} = \iint_S (\mathbf{F} \cdot \mathbf{e}_n)\, dS = \iint_S \langle 1, 1, 1 \rangle \cdot \langle 0, 0, 1 \rangle\, dS = \iint_S 1\, dS = \text{Area}(S) = \pi$$

6. Estimate $\iint_{\mathcal{S}} \mathbf{F} \cdot d\mathbf{S}$, where $\mathcal{S}$ is a tiny oriented surface of area 0.05 and the value of $\mathbf{F}$ at a sample point in $\mathcal{S}$ is a vector of length 2 making an angle $\frac{\pi}{4}$ with the normal to the surface.

SOLUTION

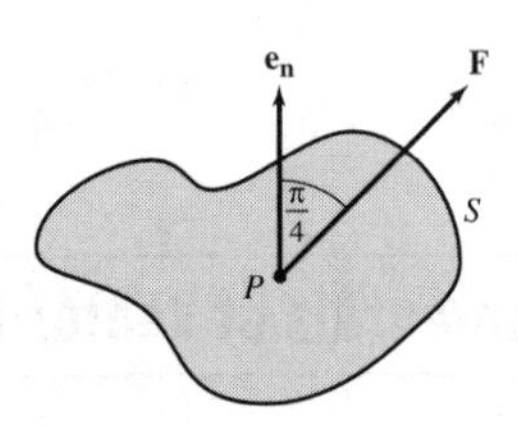

Since S is a tiny surface, we may assume that the dot product $\mathbf{F} \cdot \mathbf{e}_n$ on S is equal to the dot product at the sample point. This gives the following approximation:

$$\iint_S \mathbf{F} \cdot d\mathbf{S} = \iint_S (\mathbf{F} \cdot \mathbf{e}_n)\, d\mathbf{S} \approx \iint_S (\mathbf{F}(P) \cdot \mathbf{e}_n(P))\, d\mathbf{S} = \mathbf{F}(P) \cdot \mathbf{e}_n(P) \iint_S 1 d\mathbf{S} = \mathbf{F}(P) \cdot \mathbf{e}_n \text{Area}(S)$$

That is,

$$\iint_S \mathbf{F} \cdot d\mathbf{S} \approx \mathbf{F}(P) \cdot \mathbf{e}_n(P)\text{Area}(S) \tag{1}$$

We are given that $\text{Area}(S) = 0.05$. We compute the dot product:

$$\mathbf{F}(P) \cdot \mathbf{e}_n(P) = \|\mathbf{F}(P)\| \|\mathbf{e}_n(P)\| \cos\frac{\pi}{4} = 2 \cdot 1 \cdot \frac{1}{\sqrt{2}} = \sqrt{2}$$

Combining with (1) gives the following estimation:

$$\iint_S \mathbf{F} \cdot d\mathbf{S} \approx 0.05\sqrt{2} \approx 0.0707.$$

7. A small surface $\mathcal{S}$ is divided into three pieces of area 0.2. Estimate $\iint_{\mathcal{S}} \mathbf{F} \cdot d\mathbf{S}$ if $\mathbf{F}$ is a unit vector field making angles of 85, 90, and 95° with the normal at sample points in these three pieces.

SOLUTION

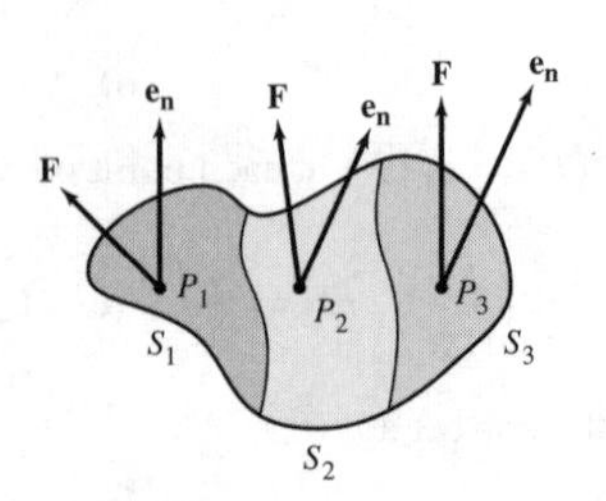

We estimate the vector surface integral by the following sum:

$$\iint_S \mathbf{F}\cdot d\mathbf{S} = \mathbf{F}(P_1)\cdot \mathbf{e}_n(P_1)\,\text{Area}(S_1) + \mathbf{F}(P_2)\cdot \mathbf{e}_n(P_2)\,\text{Area}(S_2) + \mathbf{F}(P_3)\cdot \mathbf{e}_n(P_3)\,\text{Area}(S_3)$$

$$= 0.2\left(\mathbf{F}(P_1)\cdot \mathbf{e}_n(P_1) + \mathbf{F}(P_2)\cdot \mathbf{e}_n(P_2) + \mathbf{F}(P_3)\cdot \mathbf{e}_n(P_3)\right)$$

We compute the dot product. Since $\mathbf{F}$ and $\mathbf{e}_n$ are unit vectors, we have:

$$\mathbf{F}(P_1)\cdot \mathbf{e}_n(P_1) = \cos 85^\circ \approx 0.0872$$

$$\mathbf{F}(P_2)\cdot \mathbf{e}_n(P_2) = \cos 90^\circ = 0$$

$$\mathbf{F}(P_3)\cdot \mathbf{e}_n(P_3) = \cos 95^\circ \approx -0.0872$$

Substituting gives the following estimation:

$$\iint_S \mathbf{F}\cdot d\mathbf{S} \approx 0.2(0.0872 + 0 - 0.0872) = 0.$$

Exercises

1. Let $\mathbf{F} = \langle y, z, x\rangle$ and let $\mathcal{S}$ be the oriented surface parametrized by $\Phi(u, v) = (u^2 - v, u + v, v^2)$ for $0 \le u \le 2$, $-1 \le v \le 1$. Calculate:

(a) $\mathbf{n}$ and $\mathbf{F}\cdot\mathbf{n}$ as functions of u and v

(b) The normal component of $\mathbf{F}$ to the surface at $P = (3, 3, 1) = \Phi(2, 1)$

(c) $\iint_{\mathcal{S}} \mathbf{F}\cdot d\mathbf{S}$

SOLUTION

(a) The tangent vectors are,

$$\mathbf{T}_u = \frac{\partial\Phi}{\partial u} = \frac{\partial}{\partial u}\left(u^2 - v, u + v, v^2\right) = \langle 2u, 1, 0\rangle$$

$$\mathbf{T}_v = \frac{\partial\Phi}{\partial v} = \frac{\partial}{\partial v}\left(u^2 - v, u + v, v^2\right) = \langle -1, 1, 2v\rangle$$

The normal vector is their cross product:

$$\mathbf{n} = \mathbf{T}_u \times \mathbf{T}_v = \begin{vmatrix} \mathbf{i} & \mathbf{j} & \mathbf{k} \\ 2u & 1 & 0 \\ -1 & 1 & 2v \end{vmatrix} = (2v)\mathbf{i} - (4uv)\mathbf{j} + (2u+1)\mathbf{k} = \langle 2v, -4uv, 2u+1\rangle$$

We write $\mathbf{F} = \langle y, z, x\rangle$ in terms of the parameters $x = u^2 - v$, $y = u + v$, $z = v^2$ and then compute $\mathbf{F}\cdot\mathbf{n}$:

$$\mathbf{F}(\Phi(u, v)) = \langle y, z, x\rangle = \left\langle u + v, v^2, u^2 - v\right\rangle$$

$$\mathbf{F}(\Phi(u, v))\cdot \mathbf{n}(u, v) = \left\langle u + v, v^2, u^2 - v\right\rangle \cdot \langle 2v, -4uv, 2u+1\rangle$$

$$= 2v(u + v) - 4uv\cdot v^2 + (2u+1)\left(u^2 - v\right)$$

$$= 2u^3 - 4uv^3 + 2v^2 + u^2 - v$$

(b) At the point $P = (3, 3, 1) = \Phi(2, 1)$ we have:

$$\mathbf{F}(P) = \langle 3, 1, 3\rangle$$

$$\mathbf{n}(P) = \langle 2\cdot 1, -4\cdot 2\cdot 1, 2\cdot 2 + 1\rangle = \langle 2, -8, 5\rangle$$

$$\mathbf{e}_n(P) = \frac{\mathbf{n}(P)}{\|\mathbf{n}(P)\|} = \frac{\langle 2, -8, 5\rangle}{\sqrt{4 + 64 + 25}} = \frac{1}{\sqrt{93}}\langle 2, -8, 5\rangle$$

Hence, the normal component of $\mathbf{F}$ to the surface at P is the dot product:

$$\mathbf{F}(P)\cdot \mathbf{e}_n(P) = \langle 3, 1, 3\rangle \cdot \frac{1}{\sqrt{93}}\langle 2, -8, 5\rangle = \frac{1}{\sqrt{93}}(6 - 8 + 15) = \frac{13}{\sqrt{93}}$$

(c) Using the definition of the vector surface integral and the dot product in part (a), we have:

$$\iint_S \mathbf{F} \cdot d\mathbf{S} = \iint_{\mathcal{D}} \mathbf{F}(\phi(u,v)) \cdot \mathbf{n}(u,v)\, du\, dv = \int_0^2 \int_{-1}^1 \left(2u^3 - 4uv^3 + 2v^2 + u^2 - v\right) dv\, du$$

$$= \int_0^2 2u^3 v - uv^4 + \frac{2}{3}v^3 + u^2 v - \frac{1}{2}v^2 \Big|_{v=-1}^{1} du$$

$$= \int_0^2 \left(2u^3 - u + \frac{2}{3} + u^2 - \frac{1}{2}\right) - \left(-2u^3 - u - \frac{2}{3} - u^2 - \frac{1}{2}\right) du$$

$$= \int_0^2 \left(4u^3 + 2u^2 + \frac{4}{3}\right) du = u^4 + \frac{2}{3}u^3 + \frac{4u}{3}\Big|_0^2 = 24$$

3. Let $\mathcal{S}$ be the square in the xy-plane shown in Figure 13, oriented with the normal pointing in the positive z-direction. Estimate

$$\iint_S \mathbf{F} \cdot d\mathbf{S}$$

where $\mathbf{F}$ is a vector field whose values at the labeled points are

$$\mathbf{F}(A) = \langle 2, 6, 4 \rangle, \qquad \mathbf{F}(B) = \langle 1, 1, 7 \rangle$$

$$\mathbf{F}(C) = \langle 3, 3, -3 \rangle, \qquad \mathbf{F}(D) = \langle 0, 1, 8 \rangle$$

SOLUTION The unit normal vector to S is $\mathbf{e}_n = \langle 0, 0, 1 \rangle$. We estimate the vector surface integral $\iint_S \mathbf{F} \cdot d\mathbf{S}$ using the division and sample points given in Figure 12.

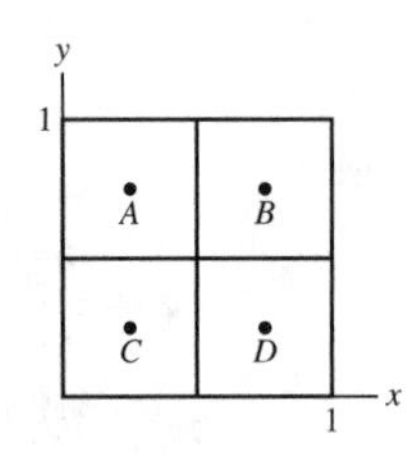

Each subsquare has area $\frac{1}{4}$, therefore we obtain the following estimation:

$$\iint_S \mathbf{F} \cdot d\mathbf{S} \approx (\mathbf{F}(A) \cdot \mathbf{e}_n + \mathbf{F}(B) \cdot \mathbf{e}_n + \mathbf{F}(C) \cdot \mathbf{e}_n + \mathbf{F}(D) \cdot \mathbf{e}_n) \cdot \frac{1}{4}$$

$$= (\langle 2,6,4 \rangle \cdot \langle 0,0,1 \rangle + \langle 1,1,7 \rangle \cdot \langle 0,0,1 \rangle + \langle 3,3,-3 \rangle \cdot \langle 0,0,1 \rangle + \langle 0,1,8 \rangle \cdot \langle 0,0,1 \rangle) \cdot \frac{1}{4}$$

$$= (4 + 7 - 3 + 8) \cdot \frac{1}{4} = 4$$

In Exercises 5–17, compute the surface integral over the given oriented surface.

5. $\mathbf{F} = \langle y, z, x \rangle$, plane $3x - 4y + z = 1$,
$0 \le x, y \le 1$, upward-pointing normal

SOLUTION We rewrite the equation of the plane as $z = 1 - 3x + 4y$, and parametrize the plane by:

$$\Phi(x,y) = (x, y, 1 - 3x + 4y)$$

Here, the parameter domain is the square $\mathcal{D} = \{(x,y) : 0 \le x,\ y \le 1\}$ in the xy-plane.

Step 1. Compute the tangent and normal vectors.

$$\mathbf{T}_x = \frac{\partial \Phi}{\partial x} = \frac{\partial}{\partial x}(x, y, 1 - 3x + 4y) = \langle 1, 0, -3 \rangle$$

$$\mathbf{T}_y = \frac{\partial \Phi}{\partial y} = \frac{\partial}{\partial y}(x, y, 1 - 3x + 4y) = \langle 0, 1, 4 \rangle$$

$$\mathbf{T}_x \times \mathbf{T}_y = \begin{vmatrix} \mathbf{i} & \mathbf{j} & \mathbf{k} \\ 1 & 0 & -3 \\ 0 & 1 & 4 \end{vmatrix} = 3\mathbf{i} - 4\mathbf{j} + \mathbf{k} = \langle 3, -4, 1 \rangle$$

Since the plane is oriented with upward pointing normal, the normal vector $\mathbf{n}$ is:

$$\mathbf{n} = \langle 3, -4, 1 \rangle$$

Step 2. Evaluate the dot product $\mathbf{F}\cdot\mathbf{n}$. We write $\mathbf{F}$ in terms of the parameters:

$$\mathbf{F}(\Phi(x,y)) = \langle y, z, x\rangle = \langle y, 1-3x+4y, x\rangle$$

The dot product $\mathbf{F}\cdot\mathbf{n}$ is thus

$$\mathbf{F}(\Phi(x,y))\cdot\mathbf{n} = \langle y, 1-3x+4y, x\rangle\cdot\langle 3,-4,1\rangle = 3y-4(1-3x+4y)+x = 13x-13y-4$$

Step 3. Evaluate the surface integral. The surface integral is equal to the following double integral:

$$\iint_S \mathbf{F}\cdot d\mathbf{S} = \iint_{\mathcal{D}} \mathbf{F}(\Phi(x,y))\cdot\mathbf{n}(x,y)\,dx\,dy = \int_0^1\int_0^1 (13x-13y-4)\,dx\,dy$$
$$= \int_0^1 \frac{13x^2}{2} - 13yx - 4x\Big|_{x=0}^{1} dy = \int_0^1\left(\frac{13}{2}-13y-4\right)dy = \frac{5y}{2}-\frac{13y^2}{2}\Big|_0^1 = -4$$

7. $\mathbf{F} = \langle 0, 3, x^2\rangle$, hemisphere $x^2+y^2+z^2=9$, $z\ge 0$, outward-pointing normal

SOLUTION We parametrize the hemisphere S by:

$$\Phi(\theta,\phi) = (3\cos\theta\sin\phi, 3\sin\theta\sin\phi, 3\cos\phi),\ 0\le\theta\le 2\pi,\ 0\le\phi\le\frac{\pi}{2}$$

Step 1. Compute the normal vector. As seen in the text, the normal vector that points to the outside of the hemisphere is:

$$\mathbf{n} = \mathbf{T}_\phi\times\mathbf{T}_\theta = \sin\phi\langle\cos\theta\sin\phi, \sin\theta\sin\phi, \cos\phi\rangle$$

For $0\le\phi\le\frac{\pi}{2}$ we have $\sin\phi\cos\phi\ge 0$, therefore $\mathbf{n}$ points to the outside of the hemisphere.

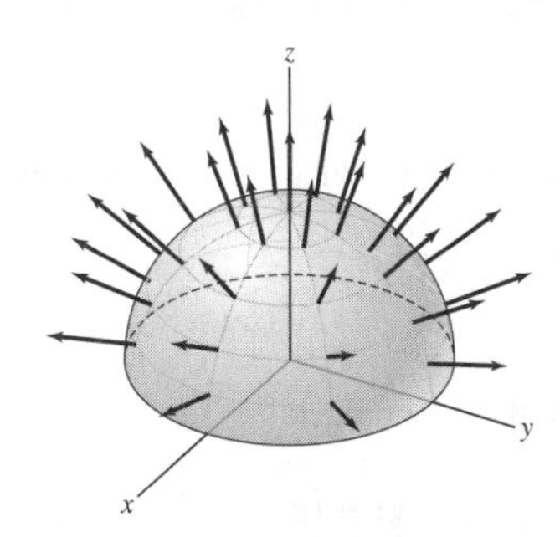

Step 2. Evaluate the dot product $\mathbf{F}\cdot\mathbf{n}$. We express the vector field in terms of the parameters:

$$\mathbf{F}(\Phi(\theta,\phi)) = \left\langle 0,3,x^2\right\rangle = \left\langle 0,3,9\cos^2\theta\sin^2\phi\right\rangle$$

Hence:

$$\mathbf{F}(\Phi(\theta,\phi))\cdot\mathbf{n}(\theta,\phi) = \left\langle 0,3,9\cos^2\theta\sin^2\phi\right\rangle\cdot\sin\phi\langle\cos\theta\sin\phi,\sin\theta\sin\phi,\cos\phi\rangle$$
$$= \sin\phi\left(3\sin\theta\sin\phi + 9\cos^2\theta\sin^2\phi\cos\phi\right)$$
$$= 3\sin\theta\sin^2\phi + 9\cos^2\theta\sin^3\phi\cos\phi$$

Step 3. Evaluate the surface integral. The surface integral is equal to the following double integral:

$$\iint_S \mathbf{F}\cdot d\mathbf{S} = \iint_{\mathcal{D}} \mathbf{F}(\Phi(\theta,\phi))\cdot\mathbf{n}(\theta,\phi)\,d\theta\,d\phi$$
$$= \int_0^{\pi/2}\int_0^{2\pi}\left(3\sin\theta\sin^2\phi + 9\cos^2\theta\sin^3\phi\cos\phi\right)d\theta\,d\phi$$
$$= \int_0^{\pi/2}\int_0^{2\pi} 3\sin\theta\sin^2\phi\,d\theta\,d\phi + \int_0^{\pi/2}\int_0^{2\pi} 9\cos^2\theta\sin^3\phi\cos\phi\,d\theta\,d\phi$$
$$= \left(\int_0^{\pi/2}\sin^2\phi\,d\phi\right)\left(\int_0^{2\pi}3\sin\theta\,d\theta\right) + \left(\int_0^{\pi/2}9\sin^3\phi\cos\phi\,d\phi\right)\left(\int_0^{2\pi}\cos^2\theta\,d\theta\right)$$
$$= \left(\frac{\phi}{2}-\frac{\sin 2\phi}{4}\Big|_{\phi=0}^{\pi/2}\right)\left(-3\cos\theta|_{\theta=0}^{2\pi}\right) + \left(\frac{9\sin^4\phi}{4}\Big|_{\phi=0}^{\pi/2}\right)\left(\frac{\theta}{2}+\frac{\sin 2\theta}{4}\Big|_{\theta=0}^{2\pi}\right)$$

$$= 0 + \frac{9}{4} \cdot \pi = \frac{9\pi}{4}$$

9. $\mathbf{F} = \langle e^z, z, x \rangle$, $z = 9 - x^2 - y^2, z \geq 0$, upward-pointing normal

SOLUTION

Step 1. Find a parametrization. We use x and y as parameters and parametrize the surface by:

$$\Phi(x, y) = \left(x, y, 9 - x^2 - y^2\right)$$

The parameter domain $\mathcal{D}$ is determined by the condition $z = 9 - x^2 - y^2 \geq 0$, or $x^2 + y^2 \leq 9$. That is:

$$\mathcal{D} = \left\{(x, y) : x^2 + y^2 \leq 9\right\}$$

Step 2. Compute the tangent and normal vectors. We have:

$$\mathbf{T}_x = \frac{\partial \Phi}{\partial x} = \frac{\partial}{\partial x}\left(x, y, 9 - x^2 - y^2\right) = \langle 1, 0, -2x \rangle$$

$$\mathbf{T}_y = \frac{\partial \Phi}{\partial y} = \frac{\partial}{\partial y}\left(x, y, 9 - x^2 - y^2\right) = \langle 0, 1, -2y \rangle$$

We compute the cross product of the tangent vectors:

$$\mathbf{T}_x \times \mathbf{T}_y = \begin{vmatrix} \mathbf{i} & \mathbf{j} & \mathbf{k} \\ 1 & 0 & -2x \\ 0 & 1 & -2y \end{vmatrix} = (2x)\mathbf{i} + (2y)\mathbf{j} + \mathbf{k} = \langle 2x, 2y, 1 \rangle$$

Since the z-component is positive, the vector points upward, and we have:

$$\mathbf{n} = \langle 2x, 2y, 1 \rangle$$

Step 3. Evaluate the dot product $\mathbf{F} \cdot \mathbf{n}$. We first express the vector field in terms of the parameters x and y, by setting $z = 9 - x^2 - y^2$. We get:

$$\mathbf{F}(\Phi(x, y)) = \langle e^z, z, x \rangle = \left\langle e^{9-x^2-y^2}, 9 - x^2 - y^2, x \right\rangle$$

We now compute the dot product:

$$\mathbf{F}(\Phi(x, y)) \cdot \mathbf{n}(x, y) = \left\langle e^{9-x^2-y^2}, 9 - x^2 - y^2, x \right\rangle \cdot \langle 2x, 2y, 1 \rangle = 2xe^{9-x^2-y^2} + 2y(9 - x^2 - y^2) + x$$

Step 4. Evaluate the surface integral.

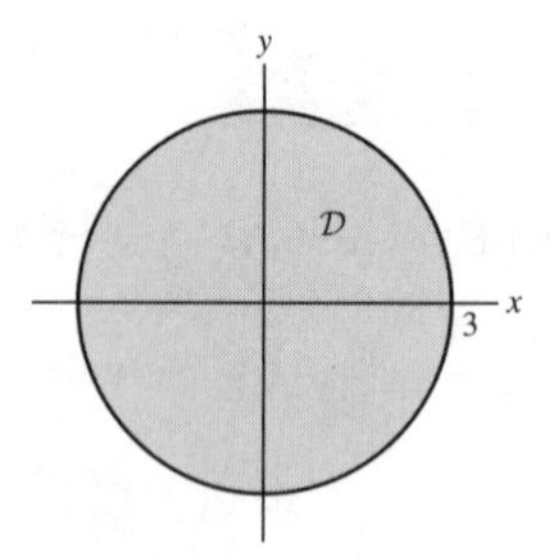

The surface integral is equal to the following double integral:

$$\iint_S \mathbf{F} \cdot d\mathbf{S} = \iint_{\mathcal{D}} \mathbf{F}(\Phi(x, y)) \cdot \mathbf{n}(x, y)\, dx\, dy = \iint_{\mathcal{D}} \left(2xe^{9-(x^2+y^2)} + 2y\left(9 - (x^2 + y^2)\right) + x\right) dx\, dy$$

We convert the integral to polar coordinates to obtain:

$$\iint_S \mathbf{F} \cdot d\mathbf{S} = \int_0^3 \int_0^{2\pi} \left(2r\cos\theta e^{9-r^2} + 2r\sin\theta(9 - r^2) + r\cos\theta\right) r\, d\theta\, dr$$

$$= \int_0^3 \left(2r^2 e^{9-r^2} + r^2\right)\sin\theta - 2r^2(9 - r^2)\cos\theta \Big|_{\theta=0}^{2\pi} dr = \int_0^3 0\, dr = 0$$

11. $\mathbf{F} = y^2\mathbf{i} + 2\mathbf{j} - x\mathbf{k}$, portion of the plane $x + y + z = 1$ in the octant $x, y, z \geq 0$, upward-pointing normal

SOLUTION

We parametrize the surface by:

$$\Phi(x, y) = (x, y, 1 - x - y),$$

using the parameter domain $\mathcal{D}$ shown in the figure.

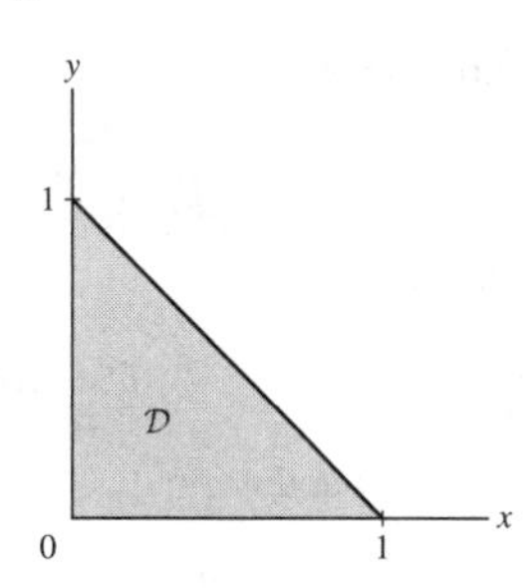

Step 1. Compute the tangent and normal vectors. We have:

$$\mathbf{T}_x = \frac{\partial \Phi}{\partial x} = \frac{\partial}{\partial x}(x, y, 1 - x - y) = \langle 1, 0, -1 \rangle$$

$$\mathbf{T}_y = \frac{\partial \Phi}{\partial y} = \frac{\partial}{\partial y}(x, y, 1 - y) = \langle 0, 1, -1 \rangle$$

$$\mathbf{T}_x \times \mathbf{T}_y = \begin{vmatrix} \mathbf{i} & \mathbf{j} & \mathbf{k} \\ 1 & 0 & -1 \\ 0 & 1 & -1 \end{vmatrix} = \mathbf{i} + \mathbf{j} + \mathbf{k} = \langle 1, 1, 1 \rangle$$

Since the normal points downward, the z-component must be negative, hence:

$$\mathbf{n} = \langle -1, -1, -1 \rangle$$

Step 2. Evaluate the dot product $\mathbf{F} \cdot \mathbf{n}$.

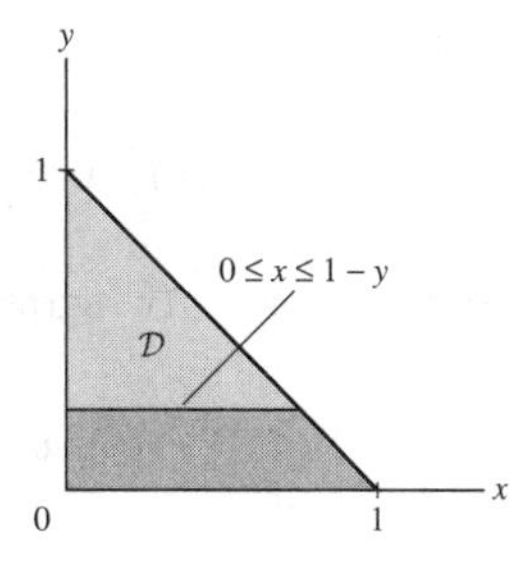

We compute the dot product:

$$\mathbf{F}(\Phi(x, y)) \cdot \mathbf{n} = \left\langle y^2, 2, -x \right\rangle \cdot \langle -1, -1, -1 \rangle = -y^2 - 2 + x$$

Step 3. Evaluate the surface integral. The surface integral is equal to the following double integral:

$$\iint_S \mathbf{F} \cdot d\mathbf{S} = \iint_{\mathcal{D}} \mathbf{F}(\Phi(x, y)) \cdot \mathbf{n}\, dx\, dy = \int_0^1 \int_0^{1-y} \left(-y^2 - 2 + x\right) dx\, dy = \int_0^1 -y^2 x - 2x + \frac{x^2}{2}\bigg|_{x=0}^{1-y} dy$$

$$= \int_0^1 \left(-y^2(1 - y) - 2(1 - y) + \frac{(1 - y)^2}{2}\right) dy = \int_0^1 \left(y^3 - y^2 + 2(y - 1) + \frac{(y - 1)^2}{2}\right) dy$$

$$= \frac{y^4}{4} - \frac{y^3}{3} + (y - 1)^2 + \frac{(y - 1)^3}{6}\bigg|_0^1 = \left(\frac{1}{4} - \frac{1}{3}\right) - \left(1 - \frac{1}{6}\right) = -\frac{11}{12}$$

13. $\mathbf{F} = \langle xz, yz, z^{-1} \rangle$, disk of radius 3 at height 4 parallel to the xy-plane, upward-pointing normal

SOLUTION

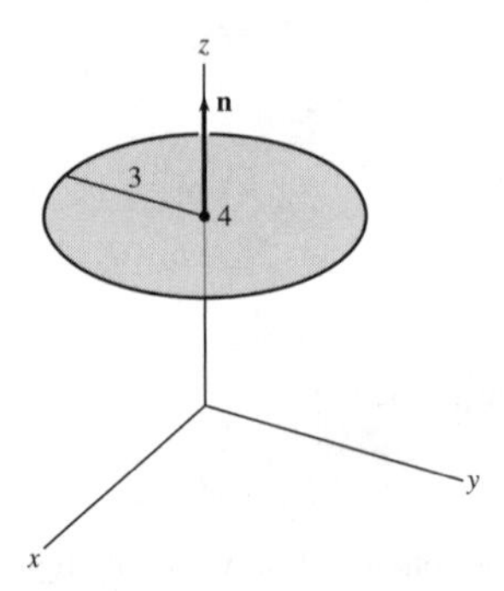

We parametrize the surface S by:

$$\Phi(\theta, r) = (r\cos\theta, r\sin\theta, 4)$$

with the parameter domain:

$$\mathcal{D} = \{(\theta, r) : 0 \le \theta \le 2\pi, 0 \le r \le 3\}$$

Step 1. Compute the tangent and normal vectors. We have:

$$\mathbf{T}_\theta = \frac{\partial \Phi}{\partial \theta} = \frac{\partial}{\partial \theta}(r\cos\theta, r\sin\theta, 4) = \langle -r\sin\theta, r\cos\theta, 0 \rangle$$

$$\mathbf{T}_r = \frac{\partial \Phi}{\partial r} = \frac{\partial}{\partial r}(r\cos\theta, r\sin\theta, 4) = \langle \cos\theta, \sin\theta, 0 \rangle$$

$$\mathbf{T}_\theta \times \mathbf{T}_r = \begin{vmatrix} \mathbf{i} & \mathbf{j} & \mathbf{k} \\ -r\sin\theta & r\cos\theta & 0 \\ \cos\theta & \sin\theta & 0 \end{vmatrix} = \left(-r\sin^2\theta - r\cos^2\theta\right)\mathbf{k} = -r\mathbf{k} = \langle 0, 0, -r \rangle$$

Since the orientation of S is with an upward pointing normal, the z-coordinate of $\mathbf{n}$ must be positive. Hence:

$$\mathbf{n} = \langle 0, 0, r \rangle$$

Step 2. Evaluate the dot product $\mathbf{F} \cdot \mathbf{n}$. We first express $\mathbf{F}$ in terms of the parameters:

$$\mathbf{F}(\Phi(\theta, r)) = \left\langle xz, yz, z^{-1} \right\rangle = \left\langle r\cos\theta \cdot 4, r\sin\theta \cdot 4, 4^{-1} \right\rangle = \left\langle 4r\cos\theta, 4r\sin\theta, \frac{1}{4} \right\rangle$$

We now compute the dot product:

$$\mathbf{F}(\Phi(\theta, r)) \cdot \mathbf{n}(\theta, r) = \left\langle 4r\cos\theta, 4r\sin\theta, \frac{1}{4} \right\rangle \cdot \langle 0, 0, r \rangle = \frac{r}{4}$$

Step 3. Evaluate the surface integral. The surface integral is equal to the following double integral:

$$\iint_S \mathbf{F} \cdot d\mathbf{S} = \iint_{\mathcal{D}} \mathbf{F}(\Phi(\theta, r)) \cdot \mathbf{n}(\theta, r)\, dr\, d\theta = \int_0^{2\pi} \int_0^3 \frac{r}{4}\, dr\, d\theta = 2\pi \int_0^3 \frac{r}{4}\, dr = 2\pi \cdot \left.\frac{r^2}{8}\right|_0^3 = \frac{9\pi}{4}$$

15. $\mathbf{F} = \langle 0, 0, e^{y+z} \rangle$, boundary of unit cube $0 \le x, y, z \le 1$, outward-pointing normal

SOLUTION

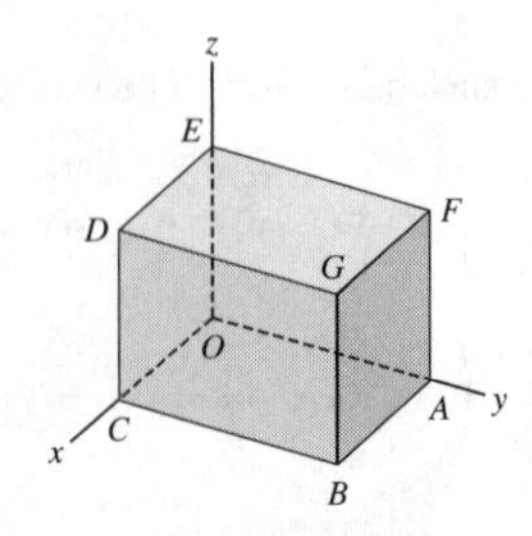

We denote the faces of the cube by:

$$S_1 = \text{Face } OABC \quad S_2 = \text{Face } DGEF \quad S_3 = \text{Face } ABGF$$
$$S_4 = \text{Face } OCDE \quad S_5 = \text{Face } BCDG \quad S_6 = \text{Face } OAFE$$

- On S_1

$$\Phi_1(x, y) = (x, y, 0)$$

and $\mathbf{n}_1 = \langle 0, 0, -1 \rangle$. Thus,

$$\mathbf{F}(\Phi_1(x, y)) \cdot \mathbf{n}_1 = \left\langle 0, 0, e^y \right\rangle \cdot \langle 0, 0, -1 \rangle = -e^y$$

- On S_2

$$\Phi_2(x, y) = (x, y, 1)$$

and $\mathbf{n}_2 = \langle 0, 0, 1 \rangle$. Thus,

$$\mathbf{F}(\Phi_2(x, y)) \cdot \mathbf{n}_2 = \left\langle 0, 0, e^{y+1} \right\rangle \cdot \langle 0, 0, 1 \rangle = e^{y+1}$$

- On any other surface S_i, $3 \le i \le 6$, we have

$$\mathbf{F}(\Phi_1(x, y)) \cdot \mathbf{n}_i = 0,$$

because the z-component of $\mathbf{n}_i = 0$ and the x, y components of $\mathbf{F}$ equal 0. Thus,

$$\begin{aligned}
\iint_S \mathbf{F} \cdot d\mathbf{S} &= \iint_{S_1} \mathbf{F} \cdot d\mathbf{S} + \iint_{S_2} \mathbf{F} \cdot d\mathbf{S} = \int_0^1 \int_0^1 -e^y \, dx \, dy + \int_0^1 \int_0^1 e^{y+1} \, dx \, dy \\
&= \int_0^1 \int_0^1 \left(e^{y+1} - e^y \right) dx \, dy = \int_0^1 \left(e^{y+1} - e^y \right) dy \\
&= \int_0^1 e^y (e-1) \, dy = (e-1) e^y \Big|_0^1 = (e-1)^2
\end{aligned}$$

17. $\mathbf{F} = \langle y, z, 0 \rangle$, $\quad \Phi(u, v) = (u^3 - v, u + v, v^2)$, $0 \le u \le 2$, $0 \le v \le 3$, downward-pointing normal

SOLUTION

Step 1. Compute the tangent and normal vectors. We have,

$$\begin{aligned}
\mathbf{T}_u &= \frac{\partial \Phi}{\partial u} = \frac{\partial}{\partial u} \left(u^3 - v, u + v, v^2 \right) = \left\langle 3u^2, 1, 0 \right\rangle \\
\mathbf{T}_v &= \frac{\partial \Phi}{\partial v} = \frac{\partial}{\partial v} \left(u^3 - v, u + v, v^2 \right) = \langle -1, 1, 2v \rangle
\end{aligned}$$

$$\mathbf{T}_u \times \mathbf{T}_v = \begin{vmatrix} \mathbf{i} & \mathbf{j} & \mathbf{k} \\ 3u^2 & 1 & 0 \\ -1 & 1 & 2v \end{vmatrix} = (2v)\mathbf{i} - \left(6u^2 v \right) \mathbf{j} + \left(3u^2 + 1 \right) \mathbf{k} = \left\langle 2v, -6u^2 v, 3u^2 + 1 \right\rangle$$

Since the normal is pointing downward, the z-coordinate is negative, hence,

$$\mathbf{n} = \left\langle -2v, 6u^2 v, -3u^2 - 1 \right\rangle$$

Step 2. Evaluate the dot product $\mathbf{F} \cdot \mathbf{n}$. We first express $\mathbf{F}$ in terms of the parameters:

$$\mathbf{F}(\Phi(u, v)) = \langle y, z, 0 \rangle = \left\langle u + v, v^2, 0 \right\rangle$$

We compute the dot product:

$$\begin{aligned}
\mathbf{F}(\Phi(u, v)) \cdot \mathbf{n}(u, v) &= \left\langle u + v, v^2, 0 \right\rangle \cdot \left\langle -2v, 6u^2 v, -3u^2 - 1 \right\rangle \\
&= -2v(u + v) + 6u^2 v \cdot v^2 + 0 = -2vu - 2v^2 + 6u^2 v^3
\end{aligned}$$

Step 3. Evaluate the surface integral. The surface integral is equal to the following double integral:

$$\begin{aligned}
\iint_S \mathbf{F} \cdot d\mathbf{S} &= \iint_{\mathcal{D}} \mathbf{F}(\Phi(u, v)) \cdot \mathbf{n}(u, v) \, du \, dv = \int_0^3 \int_0^2 \left(-2uv - 2v^2 + 6u^2 v^3 \right) du \, dv \\
&= \int_0^3 -u^2 v - 2v^2 u + 2u^3 v^3 \Big|_{u=0}^2 dv = \int_0^3 \left(16v^3 - 4v^2 - 4v \right) dv = 4v^4 - \frac{4}{3} v^3 - 2v^2 \Big|_0^3 = 270
\end{aligned}$$

19. Let $\mathbf{e_r}$ be the unit radial vector and $r = \sqrt{x^2+y^2+z^2}$. Calculate the integral of $\mathbf{F} = e^{-r}\mathbf{e_r}$ over:

(a) The upper-hemisphere of $x^2+y^2+z^2 = 9$, outward-pointing normal

(b) The octant $x, y, z \ge 0$ of the unit sphere centered at the origin

SOLUTION

(a) We parametrize the upper-hemisphere by,

$$\Phi : x = 3\cos\theta\sin\phi,\ y = 3\sin\theta\sin\phi,\ z = 3\cos\phi$$

with the parameter domain:

$$\mathcal{D} = \left\{(\theta,\phi) : 0 \le \theta < 2\pi, 0 \le \phi < \frac{\pi}{2}\right\}$$

The outward pointing normal is (see Eq. (4) in sec. 17.4):

$$\mathbf{n} = 9\sin\phi\mathbf{e}_r$$

We compute the dot product $\mathbf{F}\cdot\mathbf{n}$ on the sphere. On the sphere $r = 3$, hence,

$$\mathbf{F}\cdot\mathbf{n} = e^{-r}\,\mathbf{e}_r\cdot\mathbf{n} = e^{-3}\mathbf{e}_r\cdot 9\sin\phi\,\mathbf{e}_r = 9e^{-3}\sin\phi\,\mathbf{e}_r\cdot\mathbf{e}_r = 9\,e^{-3}\sin\phi$$

We obtain the following integral:

$$\iint_S \mathbf{F}\cdot d\mathbf{S} = \iint_{\mathcal{D}} (\mathbf{F}\cdot\mathbf{n})\,d\phi\,d\theta = \int_0^{2\pi}\int_0^{\pi/2} 9e^{-3}\sin\phi\,d\phi\,d\theta$$

$$= 18\pi e^{-3}\int_0^{\pi/2}\sin\phi\,d\phi = 18\pi e^{-3}\left(-\cos\phi\Big|_0^{\pi/2}\right) = 18\pi e^{-3}$$

(b) We parametrize the first octant of the sphere by,

$$\Phi : x = \cos\theta\sin\phi,\ y = \sin\theta\sin\phi,\ z = \cos\phi$$

with the parameter domain:

$$\mathcal{D} = \left\{(\theta,\phi) : 0 \le \theta < \frac{\pi}{2}, 0 \le \phi < \frac{\pi}{2}\right\}$$

The outward pointing normal is (as seen above):

$$\mathbf{n} = 1\sin\phi\mathbf{e}_r$$

We compute the dot product $\mathbf{F}\cdot\mathbf{n}$ on the sphere. On the sphere $r = 1$, hence,

$$\mathbf{F}\cdot\mathbf{n} = e^{-r}\,\mathbf{e}_r\cdot\mathbf{n} = e^{-1}\mathbf{e}_r\cdot\sin\phi\,\mathbf{e}_r = e^{-1}\sin\phi\,\mathbf{e}_r\cdot\mathbf{e}_r = e^{-1}\sin\phi$$

We obtain the following integral:

$$\iint_S \mathbf{F}\cdot d\mathbf{S} = \iint_{\mathcal{D}} (\mathbf{F}\cdot\mathbf{n})\,d\phi\,d\theta = \int_0^{\pi/2}\int_0^{\pi/2} e^{-1}\sin\phi\,d\phi\,d\theta$$

$$= \frac{\pi}{2}e^{-1}\int_0^{\pi/2}\sin\phi\,d\phi = \frac{\pi}{2}e^{-1}\left(-\cos\phi\Big|_0^{\pi/2}\right) = \frac{\pi}{2}e^{-1}$$

21. The electric field due to a point charge located at the origin is $\mathbf{E} = k\frac{\mathbf{e}_r}{r^2}$, where k is a constant. Calculate the flux of $\mathbf{E}$ through the disk D of radius 2 parallel to the xy-plane with center $(0, 0, 3)$.

SOLUTION Let $r = \sqrt{x^2+y^2+z^2}$ and $\hat{r} = \sqrt{x^2+y^2}$. We parametrize the disc by:

$$\Phi(\hat{r},\theta) = (\hat{r}\cos\theta, \hat{r}\sin\theta, 3)$$

$$\mathbf{T}_{\hat{r}} = \frac{\partial\Phi}{\partial\hat{r}} = \langle\cos\theta, \sin\theta, 0\rangle$$

$$\mathbf{T}_\theta = \frac{\partial\Phi}{\partial\theta} = \langle-\hat{r}\sin\theta, \hat{r}\cos\theta, 0\rangle$$

$$\mathbf{n} = \mathbf{T}_{\hat{r}}\times\mathbf{T}_\theta = \begin{vmatrix} \mathbf{i} & \mathbf{j} & \mathbf{k} \\ \cos\theta & \sin\theta & 0 \\ -\hat{r}\sin\theta & \hat{r}\cos\theta & 0 \end{vmatrix} = \langle 0, 0, \hat{r}\rangle$$

Now,

$$\mathbf{E}\cdot\mathbf{n} = k\frac{\mathbf{e}_r}{r^2}\cdot\left\langle 0,0,\hat{r}\right\rangle = \frac{k\hat{r}}{r^3}\langle x,y,z\rangle\cdot\langle 0,0,1\rangle = \frac{zk\hat{r}}{r^3}$$

Since on the disk $z = 3$, we get:

$$\mathbf{E}\cdot\mathbf{n} = 3k\frac{\hat{r}}{r^3} \text{ and } r = \sqrt{\hat{r}^2+9}$$

so $\mathbf{E}\cdot\mathbf{n} = 3k\frac{\hat{r}}{\left(\sqrt{\hat{r}^2+9}\right)^3}$.

$$\iint_{\mathcal{D}} \mathbf{E}\cdot d\mathbf{S} = \int_0^{2\pi}\int_0^2 \frac{3k\hat{r}}{(\hat{r}^2+9)^{3/2}}\,d\hat{r}\,d\theta = 6\pi k\int_0^2 \frac{\hat{r}}{(\hat{r}^2+9)^{3/2}}\,d\hat{r}$$

Substituting $u = \hat{r}^2+9$ and $\frac{1}{2}\,du = \hat{r}\,d\hat{r}$, we get:

$$\iint_{\mathcal{D}} \mathbf{E}\cdot d\mathbf{S} = 3\pi k\int_9^{13}\frac{du}{u^{3/2}} = -6\pi k u^{-1/2}\Big|_9^{13} = \left(2-\frac{6}{\sqrt{13}}\right)\pi k$$

23. Let $\mathbf{v} = \left\langle x,0,z\right\rangle$ be the velocity field (in ft/s) of a fluid in $\mathbf{R}^3$. Calculate the flow rate (in ft^3/s) through the upper hemisphere of the sphere $x^2+y^2+z^2 = 1$ $(z \ge 0)$.

SOLUTION We use the spherical coordinates:

$$x = \cos\theta\sin\phi,\quad y = \sin\theta\sin\phi,\quad z = \cos\phi$$

with the parameter domain

$$0 \le \theta < 2\pi,\quad 0 \le \phi \le \frac{\pi}{2}$$

The normal vector is (see Eq. (4) in Section 17.4):

$$\mathbf{n} = \mathbf{T}_\phi \times \mathbf{T}_\theta = \sin\phi\left\langle\cos\theta\sin\phi,\sin\theta\sin\phi,\cos\phi\right\rangle$$

We express the function in terms of the parameters:

$$\mathbf{v} = \langle x,0,z\rangle = \left\langle\cos\theta\sin\phi,0,\cos\phi\right\rangle$$

Hence,

$$\mathbf{v}\cdot\mathbf{n} = \left\langle\cos\theta\sin\phi,0,\cos\phi\right\rangle\cdot\sin\phi\left\langle\cos\theta\sin\phi,\sin\theta\sin\phi,\cos\phi\right\rangle = \sin\phi\left(\cos^2\theta\sin^2\phi+\cos^2\phi\right)$$

The flow rate of the fluid through the upper hemisphere S is equal to the flux of the velocity vector through S. That is,

$$\begin{aligned}\iint_S \mathbf{v}\cdot d\mathbf{S} &= \int_0^{\pi/2}\int_0^{2\pi}\left(\cos^2\theta\sin^3\phi+\sin\phi\cos^2\phi\right)d\theta\,d\phi\\ &= \left(\int_0^{2\pi}\cos^2\theta\,d\theta\right)\left(\int_0^{\pi/2}\sin^3\phi\,d\phi\right)+2\pi\int_0^{\pi/2}\sin\phi\cos^2\phi\,d\phi\\ &= \pi\cdot 0+2\pi\left(\frac{1}{3}\right) = \frac{2\pi}{3}\text{ft}^3/\text{s}\end{aligned}$$

25. Calculate the flow rate of a fluid with velocity field $\mathbf{v} = \left\langle x,y,x^2y\right\rangle$ (in ft/s) through the portion of the ellipse $\left(\frac{x}{2}\right)^2+\left(\frac{y}{3}\right)^2 = 1$ in the xy-plane, where $x, y \ge 0$, oriented with the normal in the positive z-direction.

SOLUTION We use the following parametrization for the surface (see remark at the end of the solution):

$$\Phi: x = 2r\cos\theta,\ y = 3r\sin\theta,\ z = 0$$

$$0 \le \theta \le \frac{\pi}{2},\quad 0 \le r \le 1 \tag{1}$$

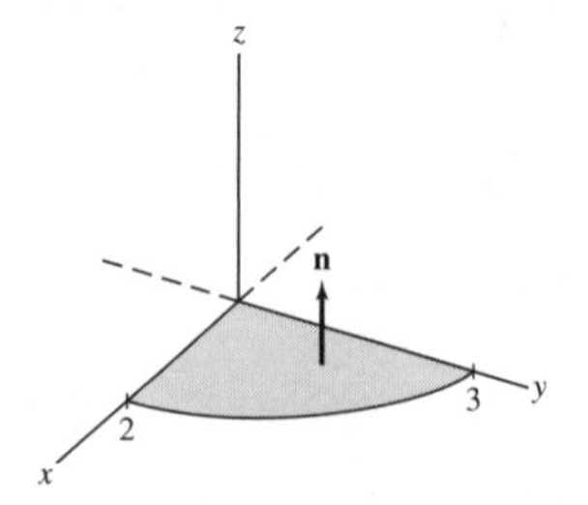

Step 1. Compute the tangent and normal vectors. We have,

$$\mathbf{T}_r = \frac{\partial \Phi}{\partial r} = \frac{\partial}{\partial r}(2r\cos\theta, 3r\sin\theta, 0) = \langle 2\cos\theta, 3\sin\theta, 0\rangle$$

$$\mathbf{T}_\theta = \frac{\partial \Phi}{\partial \theta} = \frac{\partial}{\partial \theta}(2r\cos\theta, 3r\sin\theta, 0) = \langle -2r\sin\theta, 3r\cos\theta, 0\rangle$$

$$\mathbf{T}_r \times \mathbf{T}_\theta = \begin{vmatrix} \mathbf{i} & \mathbf{j} & \mathbf{k} \\ 2\cos\theta & 3\sin\theta & 0 \\ -2r\sin\theta & 3r\cos\theta & 0 \end{vmatrix} = \left(6r\cos^2\theta + 6r\sin^2\theta\right)\mathbf{k} = 6r\mathbf{k}$$

Since the normal points to the positive z-direction, the normal vector is,

$$\mathbf{n} = 6r\mathbf{k} = \langle 0, 0, 6r\rangle$$

Step 2. Compute the dot product $\mathbf{v} \cdot \mathbf{n}$. We write the velocity vector in terms of the parameters:

$$\mathbf{v} = \left\langle x, y, x^2 y\right\rangle = \left\langle 2r\cos\theta, 3r\sin\theta, 4r^2\cos^2\theta \cdot 3r\sin\theta\right\rangle$$

$$= \left\langle 2r\cos\theta, 3r\sin\theta, 12r^3\cos^2\theta\sin\theta\right\rangle$$

Hence,

$$\mathbf{v} \cdot \mathbf{n} = 12r^3\cos^2\theta\sin\theta \cdot 6r = 72r^4\cos^2\theta\sin\theta$$

Step 3. Compute the flux. The flow rate of the fluid is the flux of the velocity vector through S. That is,

$$\iint_S \mathbf{v} \cdot d\mathbf{S} = \int_0^{\pi/2}\int_0^1 72r^4\cos^2\theta\sin\theta\, dr\, d\theta = \left(\int_0^1 72r^4\, dr\right)\left(\int_0^{\pi/2}\cos^2\theta\sin\theta\, d\theta\right)$$

$$= \left(\frac{72}{5}r^5\Big|_0^1\right)\left(-\frac{\cos^3\theta}{3}\Big|_{\theta=0}^{\pi/2}\right) = \frac{72}{5}\cdot\left(0 + \frac{1}{3}\right) = \frac{24}{5} = 4.8\ \text{ft}^3/\text{s}$$

Remark: We explain why (1) parametrizes the given portion of the ellipse. At any point (x, y) which satisfies (1) we have,

$$\left(\frac{x}{2}\right)^2 + \left(\frac{y}{3}\right)^3 = r^2\cos^2\theta + r^2\sin^2\theta = r^2 \le 1$$

Therefore (x, y) is inside the ellipse $\left(\frac{x}{2}\right)^2 + \left(\frac{y}{2}\right)^2 = 1$. The limits of θ determine the part of the region inside the ellipse in the first quadrant.

In Exercises 26–27, let $\mathcal{T}$ be the triangular region with vertices $(1, 0, 0)$, $(0, 1, 0)$, *and* $(0, 0, 1)$ *oriented with upward-pointing normal vector (Figure 15). Assume distances are in meters.*

27. Calculate the flow rate through $\mathcal{T}$ if $\mathbf{v} = -\mathbf{j}$ m/s.

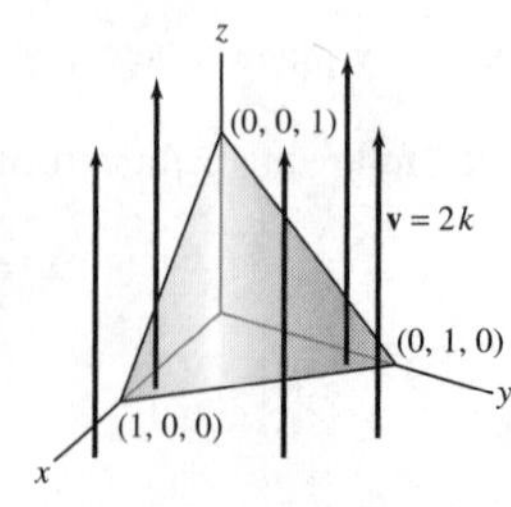

FIGURE 15

SOLUTION We compute the flow rate through $\mathcal{T}$. Since the unit normal vector is $\mathbf{e_n} = \left\langle \frac{1}{\sqrt{3}}, \frac{1}{\sqrt{3}}, \frac{1}{\sqrt{3}} \right\rangle$ we have,

$$\mathbf{v} \cdot \mathbf{e_n} = \langle 0, -1, 0 \rangle \cdot \left\langle \frac{1}{\sqrt{3}}, \frac{1}{\sqrt{3}}, \frac{1}{\sqrt{3}} \right\rangle = \frac{-1}{\sqrt{3}}$$

Therefore, the flow rate through $\mathcal{T}$ is the following flux:

$$\iint_S \mathbf{v} \cdot d\mathbf{S} = \iint_S (\mathbf{v} \cdot \mathbf{e_n})\, dS = \iint_S \frac{-1}{\sqrt{3}}\, dS = -\text{Area}(S)/\sqrt{3} = -\frac{\sqrt{3}}{2}\frac{1}{\sqrt{3}} = \frac{-1}{2}$$

The upward pointing normal to the projection $\mathcal{D}$ of $\mathcal{T}$ onto the xy-plane is $\mathbf{n} = \langle 0, 0, 1 \rangle$. Since $\mathbf{v} = \langle 0, -1, 0 \rangle$ is orthogonal to $\mathbf{n}$, the flux of $\mathbf{v}$ through $\mathcal{D}$ is zero.

In Exercises 29–30, a varying current $i(t)$ flows through a long straight wire in the xy-plane as in Example 5. The current produces a magnetic field **B** *whose magnitude at a distance r from the wire is $\|\mathbf{B}\| = \frac{\mu_0 i}{2\pi r}$ T, where $\mu_0 = 4\pi \cdot 10^{-7}$ T-m/A. Furthermore,* **B** *points into the page at points P in the xy-plane.*

29. Assume that $i(t) = t(12 - t)$ A (t in seconds). Calculate the flux $\Phi(t)$, at time t, of **B** through a rectangle of dimensions $L \times H = 3 \times 2$ m, whose top and bottom edges are parallel to the wire and whose bottom edge is located $d = 0.5$ m above the wire (similar to Figure 11). Then use Faraday's Law to determine the voltage drop around the rectangular loop (the boundary of the rectangle) at time t.

SOLUTION

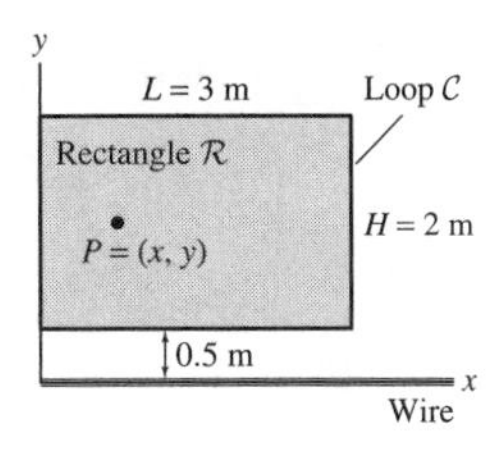

We choose the coordinate system as shown in the figure. Therefore the rectangle $\mathcal{R}$ is the region:

$$\mathcal{R} = \{(x, y) : 0 \le x \le 3, 0.5 \le y \le 2.5\}$$

Since the magnetic field points into the page and $\mathcal{R}$ is oriented with normal vector pointing out of the page (as in Example 5) we have $\mathbf{B} = -\|\mathbf{B}\|\mathbf{k}$ and $\mathbf{n} = \mathbf{e_n} = \mathbf{k}$. Hence:

$$\mathbf{B} \cdot \mathbf{n} = \|\mathbf{B}\|\,(-\mathbf{k}) \cdot \mathbf{k} = -\|\mathbf{B}\| = -\frac{\mu_0 i}{2\pi r}$$

The distance from $P = (x, y)$ in $\mathcal{R}$ to the wire is $r = y$, hence, $\mathbf{B} \cdot \mathbf{n} = -\frac{\mu_0 i}{2\pi y}$. We now compute the flux $\Phi(t)$ of **B** through the rectangle $\mathcal{R}$, by evaluating the following double integral:

$$\Phi(t) = \iint_{\mathcal{R}} \mathbf{B} \cdot d\mathbf{S} = \iint_{\mathcal{R}} \mathbf{B} \cdot \mathbf{n}\, dy\, dx = \int_0^3 \int_{0.5}^{2.5} -\frac{\mu_0 i}{2\pi y}\, dy\, dx = -\frac{\mu_0 i}{2\pi} \int_0^3 \int_{0.5}^{2.5} \frac{1}{y}\, dy\, dx$$

$$= -\frac{3\mu_0 i}{2\pi} \int_{0.5}^{2.5} \frac{dy}{y} = -\frac{3\mu_0 i}{2\pi}(\ln 2.5 - \ln 0.5) = -\frac{3\mu_0 i}{2\pi} \ln \frac{2.5}{0.5}$$

$$= \frac{-3 \cdot 4\pi \cdot 10^{-7} \ln 5}{2\pi} t(12 - t) = -9.65 \times 10^{-7} t(12 - t)\ \text{T/m}^2$$

We now use Faraday's Law to determine the voltage drop around the boundary $\mathcal{C}$ of the rectangle. By Faraday's Law, the voltage drop around $\mathcal{C}$, when $\mathcal{C}$ is oriented according to the orientation of $\mathcal{R}$ and the Right Hand Rule (that is, counterclockwise) is,

$$\int_{\mathcal{C}} \mathbf{E} \cdot d\mathbf{S} = -\frac{d\Phi}{dt} = -\frac{d}{dt}\left(-9.65 \cdot 10^{-7} t(12 - t)\right) = 9.65 \cdot 10^{-7} \cdot 2(6 - t) = 1.93 \cdot 10^{-6}(6 - t)\ \text{volts}$$

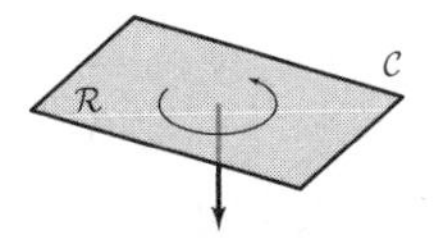

Further Insights and Challenges

31. A point mass m is located at the origin. Let Q be the flux of the gravitational field $\mathbf{F} = -Gm\frac{\mathbf{e}_r}{r^2}$ through the cylinder $x^2 + y^2 = R^2$ for $a \le z \le b$, including the top and bottom (Figure 17). Show that $Q = -4\pi Gm$ if $a < 0 < b$ (m lies inside the cylinder) and $Q = 0$ if $0 < a < b$ (m lies outside the cylinder).

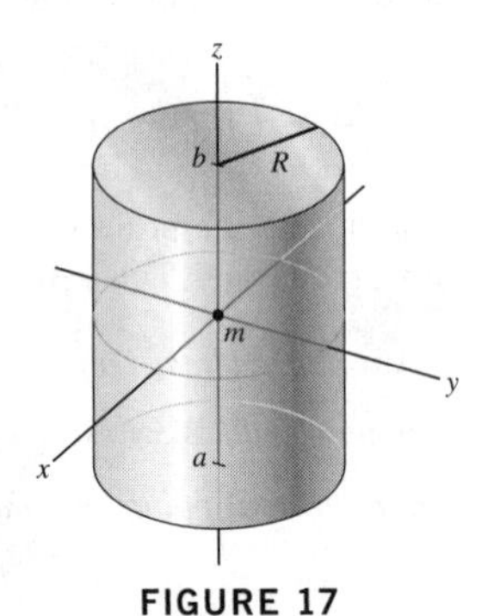

FIGURE 17

SOLUTION Let the surface be oriented with normal vector pointing outward.

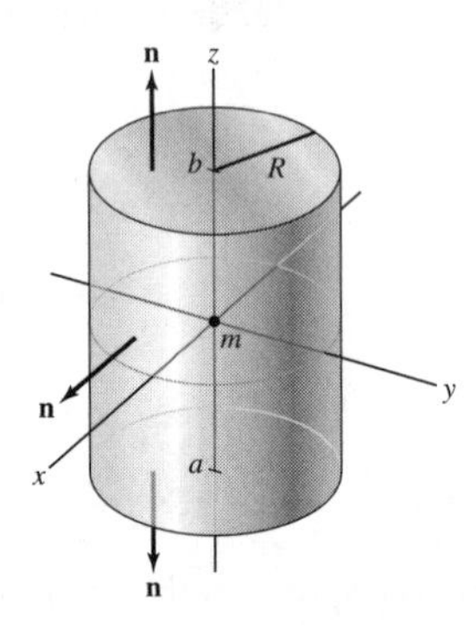

We denote by S_1, S_2 and S_3 the cylinder, the top and the bottom respectively. These surfaces are parametrized by:

- S_1:

$$\Phi_1(\theta, z) = (R\cos\theta, R\sin\theta, z),\ 0 \le \theta < 2\pi,\ a \le z \le b,\ \mathbf{n} = R\langle \cos\theta, \sin\theta, 0\rangle$$

- S_2:

$$\Phi_2(\theta, r) = (r\cos\theta, r\sin\theta, b),\ 0 \le \theta < 2\pi,\ 0 \le r \le R,\ \mathbf{n} = \langle 0, 0, r\rangle$$

- S_3:

$$\Phi_3(\theta, r) = (r\cos\theta, r\sin\theta, a),\ 0 \le \theta < 2\pi,\ 0 \le r \le R,\ \mathbf{n} = \langle 0, 0, -r\rangle$$

Using properties of integrals we have,

$$Q = \iint_S \mathbf{F}\cdot d\mathbf{S} = \iint_{S_1} \mathbf{F}\cdot d\mathbf{S} + \iint_{S_2} \mathbf{F}\cdot d\mathbf{S} + \iint_{S_3} \mathbf{F}\cdot d\mathbf{S} \tag{1}$$

Let us assume that $a < 0$. We compute the integrals over each part of the surface S separately.

- S_1: On S_1, we have:

$$\mathbf{F}(\Phi_1(\theta, z)) = -Gm\frac{\mathbf{e}_r}{r^2} = -\frac{Gm}{\left(R^2+z^2\right)^{3/2}}\langle R\cos\theta, R\sin\theta, z\rangle$$

Hence,

$$\mathbf{F}(\Phi_1(\theta, z))\cdot \mathbf{n}(\theta, z) = -\frac{Gm}{\left(R^2+z^2\right)^{3/2}}\langle R\cos\theta, R\sin\theta, z\rangle \cdot R\langle \cos\theta, \sin\theta, 0\rangle = -\frac{GmR^2}{\left(R^2+z^2\right)^{3/2}}$$

We obtain the following integral:

$$\iint_{S_1} \mathbf{F}\cdot d\mathbf{S} = \int_0^{2\pi}\int_a^b -\frac{GmR^2}{\left(R^2+z^2\right)^{3/2}}\,dz\,d\theta = -2\pi GmR^2\int_a^b \frac{dz}{\left(R^2+z^2\right)^{3/2}}$$

We compute the integral using the substitution $z = R\tan t$. This gives:

$$\iint_{S_1} \mathbf{F}\cdot d\mathbf{S} = -2\pi GmR^2 \int_{\tan^{-1}\frac{a}{R}}^{\tan^{-1}\frac{b}{R}} \frac{\cos t}{R^2}\,dt = -2\pi Gm \sin t\Big|_{t=\tan^{-1}\frac{a}{R}}^{\tan^{-1}\frac{b}{R}}$$

$$= -2\pi Gm\left(\frac{b}{\sqrt{b^2+R^2}} - \frac{a}{\sqrt{a^2+R^2}}\right) \qquad (2)$$

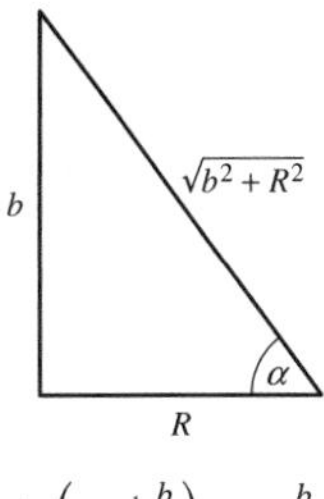

$\sin\left(\tan^{-1}\frac{b}{R}\right) = \frac{b}{\sqrt{b^2+R^2}}$

- S_2:

$$\mathbf{F}(\Phi_2(\theta, r)) = -Gm\frac{\mathbf{e}_r}{r^2} = -\frac{Gm}{(r^2+b^2)^{3/2}}\langle r\cos\theta, r\sin\theta, b\rangle$$

Hence,

$$\mathbf{F}(\Phi_2(\theta, r))\cdot\mathbf{n}(\theta, r) = -\frac{Gm}{(r^2+b^2)^{3/2}}\langle r\cos\theta, r\sin\theta, b\rangle\cdot\langle 0,0,r\rangle = -\frac{Gmbr}{(r^2+b^2)^{3/2}}$$

We obtain the following integral:

$$\iint_{S_2} \mathbf{F}\cdot d\mathbf{S} = \int_0^{2\pi}\int_0^R -\frac{Gmbr}{(r^2+b^2)^{3/2}}\,dr\,d\theta = -2\pi Gmb\int_0^R \frac{r\,dr}{(r^2+b^2)^{3/2}}$$

We compute the integral using the substitution $t = r^2+b^2$, $dt = 2r\,dr$, and we get:

$$\iint_{S_2} \mathbf{F}\cdot d\mathbf{S} = -\pi Gmb\int_{b^2}^{R^2+b^2}\frac{dt}{t^{3/2}} = 2\pi Gmb\frac{1}{\sqrt{t}}\Big|_{t=b^2}^{R^2+b^2} = 2\pi Gmb\left(\frac{1}{\sqrt{b^2+R^2}} - \frac{1}{b}\right) \qquad (3)$$

- S_3:

$$\mathbf{F}(\Phi_3(\theta, r)) = -Gm\frac{\mathbf{e}_r}{r^2} = -\frac{Gm}{(r^2+a^2)^{3/2}}\langle r\cos\theta, r\sin\theta, a\rangle$$

$$\mathbf{F}(\Phi_3(\theta, r))\cdot\mathbf{n}(\theta, r) = -\frac{Gm}{(r^2+a^2)^{3/2}}\langle r\cos\theta, r\sin\theta, a\rangle\cdot\langle 0,0,-r\rangle = \frac{Gmar}{(r^2+a^2)^{3/2}}$$

Hence, by the same computation as for S_2 we get (notice that since $a < 0$, we have $\sqrt{a^2} = -a$):

$$\iint_{S_3} \mathbf{F}\cdot d\mathbf{S} = \int_0^{2\pi}\int_0^R \frac{Gmar}{(r^2+a^2)^{3/2}}\,dr\,d\theta = -2\pi Gma\frac{1}{\sqrt{t}}\Big|_{t=a^2}^{R^2+a^2}$$

$$= -2\pi Gma\left(\frac{1}{\sqrt{R^2+a^2}} - \frac{1}{\sqrt{a^2}}\right) = -2\pi Gma\left(\frac{1}{\sqrt{R^2+a^2}} + \frac{1}{a}\right) \qquad (4)$$

Substituting (2), (3), and (4) in (1) we get:

$$Q = -2\pi Gm\left(\frac{b}{\sqrt{b^2+R^2}} - \frac{a}{\sqrt{a^2+R^2}}\right) + 2\pi Gmb\left(\frac{1}{\sqrt{b^2+R^2}} - \frac{1}{b}\right) - 2\pi Gma\left(\frac{1}{\sqrt{R^2+a^2}} + \frac{1}{a}\right)$$

$$= -2\pi Gm - 2\pi Gm = -4\pi Gm$$

If $0 < a < b$ the only difference is in the integral in (4). In this case $\sqrt{a} = a$ therefore,

$$\iint_{S_3} \mathbf{F}\cdot d\mathbf{S} = -2\pi Gma\left(\frac{1}{\sqrt{R^2+a^2}} - \frac{1}{\sqrt{a^2}}\right) = -2\pi Gma\left(\frac{1}{\sqrt{R^2+a^2}} - \frac{1}{a}\right).$$

Therefore, adding the integrals gives:

$$Q = -2\pi Gm\left(\frac{b}{\sqrt{b^2+R^2}} - \frac{a}{\sqrt{a^2+R^2}}\right) + 2\pi Gmb\left(\frac{1}{\sqrt{b^2+R^2}} - \frac{1}{b}\right) - 2\pi Gma\left(\frac{1}{\sqrt{R^2+a^2}} - \frac{1}{a}\right)$$
$$= -2\pi Gm + 2\pi Gm = 0$$

In Exercises 32–33, let $\mathcal{S}$ be the surface with parametrization

$$\Phi(u, v) = \left(\left(1 + v\cos\frac{u}{2}\right)\cos u,\ \left(1 + v\cos\frac{u}{2}\right)\sin u,\ v\sin\frac{u}{2}\right)$$

for $0 \le u \le 2\pi$, $-\frac{1}{2} \le v \le \frac{1}{2}$.

33. CAS It is not possible to integrate a vector field over $\mathcal{S}$ because $\mathcal{S}$ is not orientable. However, it is possible to integrate functions over $\mathcal{S}$. Using a computer algebra system:

(a) Verify that

$$\|\mathbf{n}(u, v)\|^2 = 1 + \frac{3}{4}v^2 + 2v\cos\frac{u}{2} + \frac{1}{2}v^2\cos u$$

(b) Compute the surface area of $\mathcal{S}$ to four decimal places.

(c) Compute $\iint_{\mathcal{S}} (x^2 + y^2 + z^2)\, dS$ to four decimal places.

SOLUTION

(a) Using a CAS, we discover that

$$\mathbf{n}(u, v) = \frac{\partial \mathbf{n}}{\partial u} \times \frac{\partial \mathbf{n}}{\partial v} = \left\langle \frac{1}{2}\left(-v\cos\left(\frac{u}{2}\right) + 2\cos u + v\cos\left(\frac{3u}{2}\right)\right)\sin\left(\frac{u}{2}\right),\right.$$
$$\frac{1}{4}\left(v + 2\cos(u/2) + 2v\cos(u) - 2\cos\left(\frac{3u}{2}\right) - v\cos(2u)\right),$$
$$\left. -\cos\left(\frac{u}{2}\right)\left(1 + v\cos\left(\frac{u}{2}\right)\right)\right\rangle$$

and after taking the norm of this, we find that

$$\|\mathbf{n}(u, v)\|^2 = 1 + \frac{3}{4}v^2 + 2v\cos\frac{u}{2} + \frac{1}{2}v^2\cos u$$

(b) We calculate the area of $\mathcal{S}$ as follows:

$$A(\mathcal{S}) = \iint \|\mathbf{n}(u, v)\|\, du\, dv = \int_{-1/2}^{1/2}\int_0^{2\pi} \sqrt{1 + \frac{3}{4}v^2 + 2v\cos\frac{u}{2} + \frac{1}{2}v^2\cos u}\, du\, dv \approx 6.3533$$

(c) We proceed as follows. Since

$$x^2 + y^2 + z^2 = \left(\left(1 + v\cos\frac{u}{2}\right)\cos u\right)^2 + \left(\left(1 + v\cos\frac{u}{2}\right)\sin u\right)^2 + \left(v\sin\frac{u}{2}\right)^2$$

and

$$\|\mathbf{n}(u, v)\| = \sqrt{1 + \frac{3}{4}v^2 + 2v\cos\frac{u}{2} + \frac{1}{2}v^2\cos u}$$

then, substituting these expressions into the double integral $\iint_{\mathcal{S}}(x^2 + y^2 + z^2)\, dS = \iint_{\mathcal{S}}(x^2 + y^2 + z^2)\|\mathbf{n}(u, v)\|\, dudv$, and integrating over $0 \le u \le 2\pi$, $-\frac{1}{2} \le v \le \frac{1}{2}$, we find that

$$\iint_{\mathcal{S}} (x^2 + y^2 + z^2)\, dS \approx 7.4003$$

CHAPTER REVIEW EXERCISES

1. Compute the vector assigned to the point $P = (-3, 5)$ by the vector field:

(a) $\mathbf{F} = \langle xy, y - x\rangle$

(b) $\mathbf{F} = \langle 4, 8\rangle$

(c) $\mathbf{F} = \langle 3^{x+y}, \log_2(x+y) \rangle$

SOLUTION

(a) Substituting $x = -3$, $y = 5$ in $\mathbf{F} = \langle xy, y - x \rangle$ we obtain:

$$\mathbf{F} = \langle -3 \cdot 5, 5 - (-3) \rangle = \langle -15, 8 \rangle$$

(b) The constant vector field $\mathbf{F} = \langle 4, 8 \rangle$ assigns the vector $\langle 4, 8 \rangle$ to all the vectors. Thus:

$$\mathbf{F}(-3, 5) = \langle 4, 8 \rangle$$

(c) Substituting $x = -3$, $y = 5$ in $\mathbf{F} = \langle 3^{x+y}, \log_2(x+y) \rangle$ we obtain

$$\mathbf{F} = \left\langle 3^{-3+5}, \log_2(-3+5) \right\rangle = \left\langle 3^2, \log_2(2) \right\rangle = \langle 9, 1 \rangle$$

In Exercises 3–6, sketch the vector field.

3. $\mathbf{F}(x, y) = \langle y, 1 \rangle$

SOLUTION Notice that the vector field is constant along horizontal lines.

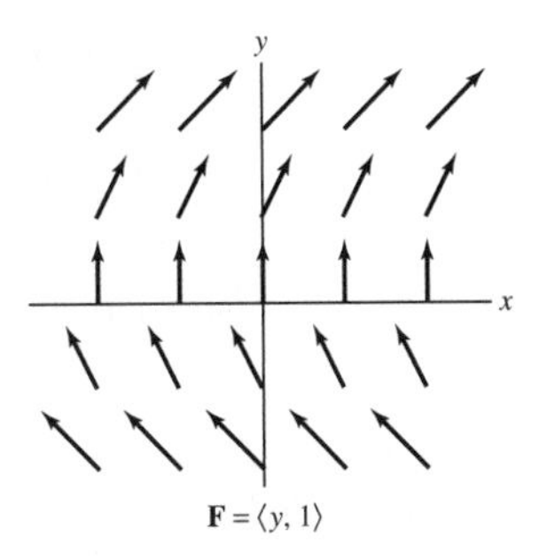

$\mathbf{F} = \langle y, 1 \rangle$

5. $\nabla\varphi$, where $\varphi(x, y) = x^2 - y$

SOLUTION The gradient of $\varphi(x, y) = x^2 - y$ is the following vector:

$$\mathbf{F}(x, y) = \left\langle \frac{\partial\varphi}{\partial x}, \frac{\partial\varphi}{\partial y} \right\rangle = \langle 2x, -1 \rangle$$

This vector is sketched in the following figure:

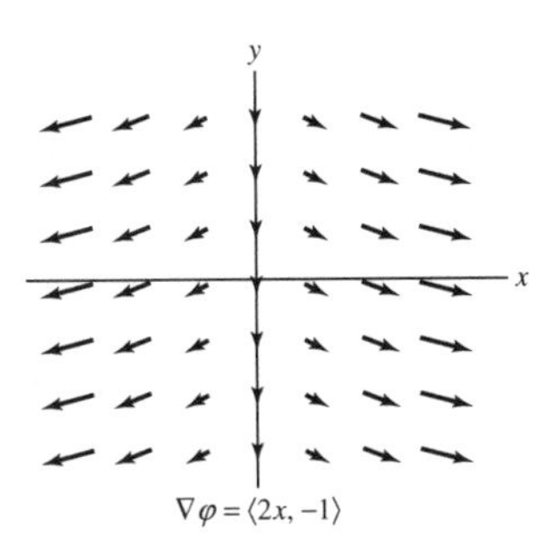

$\nabla\varphi = \langle 2x, -1 \rangle$

In Exercises 7–14, determine whether or not the vector field is conservative and, if so, find a potential function.

7. $\mathbf{F}(x, y, z) = \langle \sin x, e^y, z \rangle$

SOLUTION We examine the cross partials of $\mathbf{F}$. Since $\mathbf{F}_1 = \sin x$, $\mathbf{F}_2 = e^y$, $\mathbf{F}_3 = z$ we have:

$$\begin{array}{lll} \dfrac{\partial \mathbf{F}_1}{\partial y} = 0 & \dfrac{\partial \mathbf{F}_2}{\partial z} = 0 & \dfrac{\partial \mathbf{F}_3}{\partial x} = 0 \\ \dfrac{\partial \mathbf{F}_2}{\partial x} = 0 & \dfrac{\partial \mathbf{F}_3}{\partial y} = 0 & \dfrac{\partial \mathbf{F}_1}{\partial z} = 0 \end{array} \quad \Rightarrow \quad \frac{\partial \mathbf{F}_1}{\partial y} = \frac{\partial \mathbf{F}_2}{\partial x}, \quad \frac{\partial \mathbf{F}_2}{\partial z} = \frac{\partial \mathbf{F}_3}{\partial y}, \quad \frac{\partial \mathbf{F}_3}{\partial x} = \frac{\partial \mathbf{F}_1}{\partial z}$$

Since the cross partials are equal, $\mathbf{F}$ is conservative. We denote the potential field by $\varphi(x, y, z)$. So we have:

$$\varphi_x = \sin x \quad \varphi_y = e^y \quad \varphi_z = z$$

By integrating we get:

$$\varphi(x, y, z) = \int \sin x \, dx = -\cos x + C(y, z)$$

$$\varphi_y = C_y = e^y \quad \Rightarrow \quad C(y, z) = e^y + D(z)$$

$$\varphi(x, y, z) = -\cos x + e^y + D(z)$$

$$\varphi_z = D_z = z \quad \Rightarrow \quad D(z) = \frac{z^2}{2}$$

We conclude that $\varphi(x, y, z) = -\cos x + e^y + \frac{z^2}{2}$. Indeed:

$$\nabla\varphi = \left\langle \frac{\partial\varphi}{\partial x}, \frac{\partial\varphi}{\partial y}, \frac{\partial\varphi}{\partial z} \right\rangle = \left\langle \sin x, e^y, z \right\rangle = \mathbf{F}$$

9. $\mathbf{F}(x, y, z) = \left\langle xyz, \frac{1}{2}x^2 z, 2z^2 y \right\rangle$

SOLUTION No. We show that the cross partials for x and z are not equal. Since the equality of the cross partials is a necessary condition for a field to be a gradient vector field, we conclude that $\mathbf{F}$ is not a gradient field. We have:

$$\left.\begin{aligned} \frac{\partial F_1}{\partial z} &= \frac{\partial}{\partial z}(xyz) = xy \\ \frac{\partial F_3}{\partial x} &= \frac{\partial}{\partial x}(2z^2 y) = 0 \end{aligned}\right\} \quad \Rightarrow \quad \frac{\partial F_1}{\partial z} \neq \frac{\partial F_3}{\partial x}$$

Therefore the cross partials condition is not satisfied, hence $\mathbf{F}$ is not a gradient vector field.

11. $\mathbf{F}(x, y, z) = \left\langle \frac{y}{1+x^2}, \tan^{-1} x, 2z \right\rangle$

SOLUTION We examine the cross partials of $\mathbf{F}$. Since $\mathbf{F}_1 = \frac{y}{1+x^2}$, $\mathbf{F}_2 = \tan^{-1} x$, $\mathbf{F}_3 = 2z$ we have:

$$\left.\begin{aligned} \frac{\partial \mathbf{F}_1}{\partial y} &= \frac{1}{1+x^2} \\ \frac{\partial \mathbf{F}_2}{\partial x} &= \frac{1}{1+x^2} \end{aligned}\right\} \quad \Rightarrow \quad \frac{\partial \mathbf{F}_1}{\partial y} = \frac{\partial \mathbf{F}_2}{\partial x}$$

$$\left.\begin{aligned} \frac{\partial \mathbf{F}_2}{\partial z} &= 0 \\ \frac{\partial \mathbf{F}_3}{\partial y} &= 0 \end{aligned}\right\} \quad \Rightarrow \quad \frac{\partial \mathbf{F}_2}{\partial z} = \frac{\partial \mathbf{F}_3}{\partial y}$$

$$\left.\begin{aligned} \frac{\partial \mathbf{F}_3}{\partial x} &= 0 \\ \frac{\partial \mathbf{F}_1}{\partial z} &= 0 \end{aligned}\right\} \quad \Rightarrow \quad \frac{\partial \mathbf{F}_3}{\partial x} = \frac{\partial \mathbf{F}_1}{\partial z}$$

Since the cross partials are equal, $\mathbf{F}$ is conservative. We denote the potential field by $\varphi(x, y, z)$. We have:

$$\varphi_x = \frac{y}{1+x^2}, \quad \varphi_y = \tan^{-1}(x), \quad \varphi_z = 2z$$

By integrating we get:

$$\varphi(x, y, z) = \int \frac{y}{1+x^2} \, dx = y \tan^{-1}(x) + c(y, z)$$

$$\varphi_y = \tan^{-1}(x) + c_y(y, z) = \tan^{-1}(x) \quad \Rightarrow \quad c_y(y, z) = 0 \quad \Rightarrow \quad c(y, z) = c(z)$$

Hence $\varphi(x, y, z) = y \tan^{-1}(x) + c(z)$. $\varphi_z = c'(z) = 2z \Rightarrow c(z) = z^2$. We conclude that $\varphi(x, y, z) = y \tan^{-1}(x) + z^2$. Indeed:

$$\nabla\varphi = \left\langle \frac{\partial\varphi}{\partial x}, \frac{\partial\varphi}{\partial y}, \frac{\partial\varphi}{\partial z} \right\rangle = \left\langle \frac{y}{1+x^2}, \tan^{-1} x, 2z \right\rangle = \mathbf{F}$$

13. $\mathbf{F}(x, y, z) = \left\langle xe^{2x}, ye^{2z}, ze^{2y} \right\rangle$

SOLUTION We have:

$$\frac{\partial F_3}{\partial y} = \frac{\partial}{\partial y}\left(ze^{2y}\right) = 2ze^{2y}$$

$$\frac{\partial F_2}{\partial z} = \frac{\partial}{\partial z}\left(ye^{2z}\right) = 2ye^{2y}$$

Since $\frac{\partial F_3}{\partial y} \neq \frac{\partial F_2}{\partial z}$, the cross-partials condition is not satisfied , hence **F** is not conservative.

15. Calculate $\int_{\mathbf{c}} \nabla\varphi \cdot d\mathbf{s}$, where $\varphi(x, y, z) = x^4y^3z^2$ and $\mathbf{c}(t) = (t^2, 1+t, t^{-1})$ for $1 \le t \le 3$.

SOLUTION The initial point P and the terminal point Q are the following points:

$$P = \mathbf{c}(1) = \left(1^2, 1+1, 1^{-1}\right) = (1, 2, 1)$$

$$Q = \mathbf{c}(3) = \left(3^2, 1+3, 3^{-1}\right) = \left(9, 4, \frac{1}{3}\right)$$

Using the Fundamental Theorem for Gradient Vector Fields we obtain:

$$\int_c \nabla\varphi \cdot d\mathbf{s} = \varphi(Q) - \varphi(P) = 9^4 \cdot 4^3 \cdot \left(\frac{1}{3}\right)^2 - 1^4 \cdot 2^3 \cdot 1^2 = 46{,}648$$

In Exercises 17–20, compute the line integral $\int_C f(x, y)\, ds$ *for the given function and path or curve.*

17. $f(x, y) = xy$, the path $\mathbf{c}(t) = (t, 2t-1)$ for $0 \le t \le 1$

SOLUTION

Step 1. Compute $ds = \|\mathbf{c}'(t)\|\, dt$. We differentiate $\mathbf{c}(t) = (t, 2t-1)$ and compute the length of the derivative vector:

$$\mathbf{c}'(t) = \langle 1, 2\rangle \Rightarrow \|\mathbf{c}'(t)\| = \sqrt{1^2 + 2^2} = \sqrt{5}$$

Hence,

$$ds = \|\mathbf{c}'(t)\|\, dt = \sqrt{5}\, dt$$

Step 2. Write out $f(\mathbf{c}(t))$ and evaluate the line integral. We have:

$$f(\mathbf{c}(t)) = xy = t(2t-1) = 2t^2 - t$$

Using the Theorem on Scalar Line Integral we have:

$$\int_C f(x, y)\, ds = \int_0^1 f(\mathbf{c}(t))\, \|\mathbf{c}'(t)\|\, dt = \int_0^1 \left(2t^2 - t\right)\sqrt{5}\, dt = \sqrt{5}\left(\frac{2}{3}t^3 - \frac{1}{2}t^2\right)\Big|_0^1 = \sqrt{5}\left(\frac{2}{3} - \frac{1}{2}\right) = \frac{\sqrt{5}}{6}$$

19. $f(x, y, z) = e^x - \frac{y}{2\sqrt{2}z}$, the path $\mathbf{c}(t) = \left(\ln t, \sqrt{2}t, \frac{1}{2}t^2\right)$ for $1 \le t \le 2$

SOLUTION

Step 1. Compute $ds = \|\mathbf{c}'(t)\|\, dt$. We have:

$$\mathbf{c}'(t) = \frac{d}{dt}\left\langle \ln t, \sqrt{2}t, \frac{1}{2}t^2\right\rangle = \left\langle \frac{1}{t}, \sqrt{2}, t\right\rangle$$

$$\|\mathbf{c}'(t)\| = \sqrt{\left(\frac{1}{t}\right)^2 + \left(\sqrt{2}\right)^2 + t^2} = \sqrt{\frac{1}{t^2} + 2 + t^2} = \sqrt{\left(\frac{1}{t} + t\right)^2} = \frac{1}{t} + t$$

Hence:

$$ds = \|\mathbf{c}'(t)\|\, dt = \left(t + \frac{1}{t}\right) dt$$

Step 2. Write out $f(\mathbf{c}(t))$ and evaluate the integral.

$$f(\mathbf{c}(t)) = e^x - \frac{y}{2\sqrt{2}z} = e^{\ln t} - \frac{\sqrt{2}t}{2\sqrt{2}\cdot\frac{1}{2}t^2} = t - \frac{1}{t}$$

We use the Theorem on Scalar Line Integrals to compute the line integral:

$$\int_C f(x,y)\,ds = \int_1^2 f(\mathbf{c}(t))\,\|\mathbf{c}'(t)\|\,dt = \int_1^2 \left(t - \frac{1}{t}\right)\left(t + \frac{1}{t}\right)dt$$

$$= \int_1^2 \left(t^2 - \frac{1}{t^2}\right)dt = \frac{t^3}{3} + \frac{1}{t}\Big|_1^2 = \left(\frac{8}{3} + \frac{1}{2}\right) - \left(\frac{1}{3} + 1\right) = \frac{11}{6}$$

21. Find the total mass of an L-shaped rod consisting of the segments $(2t, 2)$ and $(2, 2-2t)$ for $0 \le t \le 1$ (length in centimeters) with mass density $\rho(x,y) = x^2y$ g/cm.

SOLUTION

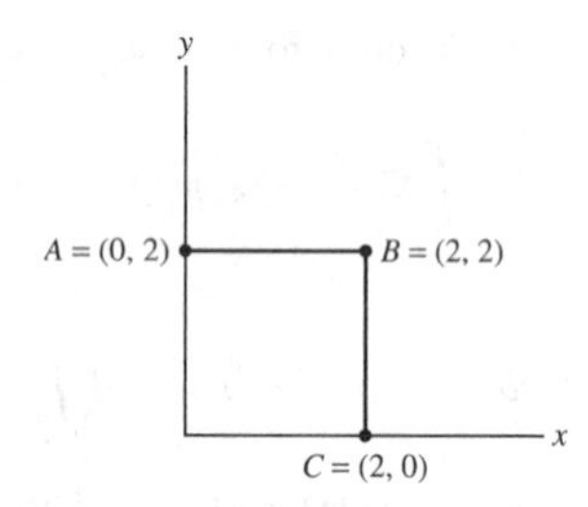

The total mass of the rod is the following sum:

$$M = \int_{\overline{AB}} x^2y\,ds + \int_{\overline{BC}} x^2y\,ds \tag{1}$$

The segment $\overline{AB}$ is parametrized by $\mathbf{c}_1(t) = (2t, 2)$, $0 \le t \le 1$. Hence

$$\mathbf{c}_1'(t) = \langle 2, 0\rangle, \quad \|\mathbf{c}_1'(t)\| = 2$$

and

$$f(\mathbf{c}_1(t)) = x^2y = (2t)^2\cdot 2 = 8t^2.$$

The segment $\overline{BC}$ is parametrized by $\mathbf{c}_2(t) = (2, 2-2t)$, $0 \le t \le 1$. Hence

$$\mathbf{c}_2'(t) = \langle 0, -2\rangle, \quad \|\mathbf{c}_2'(t)\| = 2$$

and

$$f(\mathbf{c}_2(t)) = x^2y = 2^2(2-2t) = 8 - 8t.$$

Using these values, the Theorem on Scalar Line Integrals and (1) we get:

$$M = \int_0^1 8t^2\cdot 2\,dt + \int_0^1 (8-8t)\cdot 2\,dt = \frac{16t^3}{3}\Big|_0^1 + 16t - 8t^2\Big|_0^1 = \frac{40}{3} = 13\frac{1}{3}$$

23. Calculate $\int_{C_1} y^3\,dx + x^2y\,dy$, where C_1 is the oriented curve in Figure 1(A).

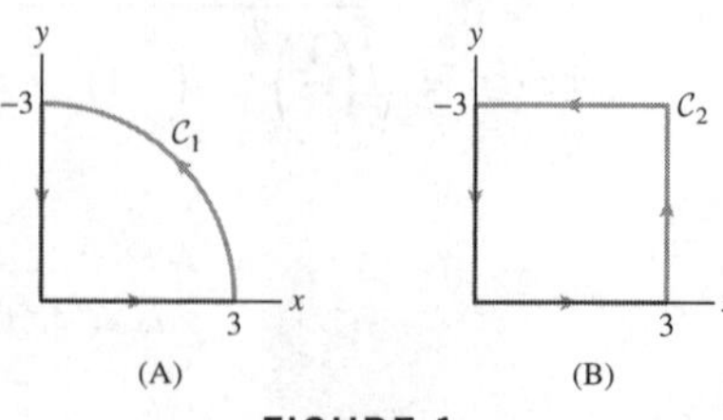

FIGURE 1

SOLUTION We compute the line integral as the sum of the line integrals over the segments $\overline{AO}$, $\overline{OB}$ and the circular arc BA.

The vector field is $\mathbf{F} = \left\langle y^3, x^2 y \right\rangle$. We have:

$$\int_{C_1} \mathbf{F} \cdot d\mathbf{s} = \int_{\overline{AO}} \mathbf{F} \cdot d\mathbf{s} + \int_{\overline{OB}} \mathbf{F} \cdot d\mathbf{s} + \int_{\text{arc } BA} \mathbf{F} \cdot d\mathbf{s} \tag{1}$$

We compute each integral separately.

- The line integral over $\overline{AO}$. The segment $\overline{AO}$ is parametrized by $\mathbf{c}(t) = (0, -t), \quad -3 \le t \le 0$. Hence:

$$\mathbf{F}(\mathbf{c}(t)) = \left\langle y^3, x^2 y \right\rangle = \left\langle -t^3, 0 \right\rangle$$

$$\mathbf{c}'(t) = \langle 0, -1 \rangle$$

$$\mathbf{F}(\mathbf{c}(t)) \cdot \mathbf{c}'(t) = \left\langle -t^3, 0 \right\rangle \cdot \langle 0, -1 \rangle = 0$$

Therefore:

$$\int_{\overline{AO}} \mathbf{F} \cdot d\mathbf{s} = \int_{-3}^{0} \mathbf{F}(\mathbf{c}(t)) \cdot \mathbf{c}'(t)\, dt = 0 \tag{2}$$

- The line integral over $\overline{OB}$. We parametrize the segment $\overline{OB}$ by $\mathbf{c}(t) = (t, 0), \quad 0 \le t \le 3$. Hence:

$$\mathbf{F}(\mathbf{c}(t)) = \left\langle y^3, x^2 y \right\rangle = \langle 0, 0 \rangle$$

$$\mathbf{c}'(t) = \langle 1, 0 \rangle$$

$$\mathbf{F}(\mathbf{c}(t)) \cdot \mathbf{c}'(t) = 0$$

Therefore:

$$\int_{\overline{OB}} \mathbf{F} \cdot d\mathbf{s} = \int_{0}^{3} \mathbf{F}(\mathbf{c}(t)) \cdot \mathbf{c}'(t)\, dt = 0 \tag{3}$$

- The line integral over the circular arc BA. We parametrize the circular arc by $\mathbf{c}(t) = (3\cos t, 3\sin t), 0 \le t \le \frac{\pi}{2}$. Then $\mathbf{c}'(t) = \langle -3\sin t, 3\cos t \rangle$ and $\mathbf{F}(\mathbf{c}(t)) = \left\langle y^3, x^2 y \right\rangle = \left\langle 27\sin^3 t, 27\cos^2 t \sin t \right\rangle$. We compute the dot product:

$$\mathbf{F}(\mathbf{c}(t)) \cdot \mathbf{c}'(t) = 27\left\langle \sin^3 t, \cos^2 t \sin t \right\rangle \cdot \langle -3\sin t, 3\cos t \rangle = 81\left(-\sin^{-4} t + \cos^3 t \sin t \right)$$

We obtain the integral:

$$\int_{\text{arc } BA} \mathbf{F} \cdot d\mathbf{s} = \int_0^{\pi/2} 81\left(-\sin^4 t + \cos^3 t \sin t \right) dt$$

$$= 81\left(\frac{\sin^3 t \cos t}{4} - \frac{3}{4}\left(\frac{t}{2} - \frac{\sin 2t}{4} \right) - \frac{\cos^4 t}{4} \right)\Bigg|_{t=0}^{\pi/2}$$

$$= 81\left(-\frac{3\pi}{4 \cdot 4} + \frac{1}{4} \right) = -\frac{243}{16}\pi + 20\frac{1}{4} \tag{4}$$

Combining (1), (2), (3), and (4) gives:

$$\int_{C_1} \mathbf{F} \cdot d\mathbf{s} = 0 + 0 - \frac{243}{16}\pi + 20.25 \approx -27.463$$

In Exercises 25–28, compute the line integral $\int_{\mathbf{c}} \mathbf{F} \cdot d\mathbf{s}$*. for the given vector field and path.*

25. $\mathbf{F}(x, y) = \left\langle \dfrac{2y}{x^2 + 4y^2}, \dfrac{x}{x^2 + 4y^2} \right\rangle,$

the path $\mathbf{c}(t) = \left(\cos t, \frac{1}{2}\sin t \right)$ for $0 \le t \le 2\pi$

SOLUTION

Step 1. Calculate the integral $\mathbf{F}(\mathbf{c}(t)) \cdot \mathbf{c}'(t)$.

$$\mathbf{c}(t) = \left(\cos t, \frac{1}{2}\sin t\right)$$

$$\mathbf{F}(\mathbf{c}(t)) = \left\langle \frac{2y}{x^2+4y^2}, \frac{x}{x^2+4y^2} \right\rangle = \left\langle \frac{2\cdot\frac{1}{2}\cdot\sin t}{\cos^2 t + 4\cdot\frac{1}{4}\sin^2 t}, \frac{\cos t}{\cos^2 t + 4\cdot\frac{1}{4}\sin^2 t} \right\rangle$$

$$= \left\langle \frac{\sin t}{\cos^2 t+\sin^2 t}, \frac{\cos t}{\cos^2 t+\sin^2 t} \right\rangle = \langle \sin t, \cos t\rangle$$

$$\mathbf{c}'(t) = \left\langle -\sin t, \frac{1}{2}\cos t \right\rangle$$

The integral is the dot product:

$$\mathbf{F}(\mathbf{c}(t)) \cdot \mathbf{c}'(t) = \langle \sin t, \cos t\rangle \cdot \left\langle -\sin t, \frac{1}{2}\cos t\right\rangle = -\sin^2 t + \frac{1}{2}\cos^2 t = \frac{1}{2}\cos 2t - \frac{1}{2}\sin^2 t$$

Step 2. Evaluate the line integral.

$$\int_C \mathbf{F}\cdot d\mathbf{s} = \int_0^{2\pi} \mathbf{F}(\mathbf{c}(t)) \cdot \mathbf{c}'(t)\, dt = \int_0^{2\pi} \left(\frac{1}{2}\cos 2t - \frac{1}{2}\sin^2 t\right) dt = \frac{\sin 2t}{4} - \frac{t}{4} + \frac{\sin 2t}{8}\Big|_0^{2\pi} = -\frac{\pi}{2}$$

27. $\mathbf{F}(x, y) = \langle x^2 y, y^2 z, z^2 x\rangle$, the path $\mathbf{c}(t) = \left(e^{-t}, e^{-2t}, e^{-3t}\right)$ for $0 \le t < \infty$

SOLUTION

Step 1. Calculate the integrand $\mathbf{F}(\mathbf{c}(t)) \cdot \mathbf{c}'(t)$.

$$\mathbf{c}(t) = \left(e^{-t}, e^{-2t}, e^{-3t}\right)$$

$$\mathbf{c}'(t) = \left\langle e^{-t}, -2e^{-2t}, -3e^{-3t}\right\rangle$$

$$\mathbf{F}(\mathbf{c}(t)) = \left\langle x^2 y, y^2 z, z^2 x\right\rangle = \left\langle e^{-2t}\cdot e^{-2t}, e^{-4t}\cdot e^{-3t}, e^{-6t}\cdot e^{-t}\right\rangle = \left\langle e^{-4t}, e^{-7t}, e^{-7t}\right\rangle$$

The integrand is the dot product:

$$\mathbf{F}(\mathbf{c}(t)) \cdot \mathbf{c}'(t) = \left\langle e^{-4t}, e^{-7t}, e^{-7t}\right\rangle \cdot \left\langle e^{-t}, -2e^{-2t}, -3e^{-3t}\right\rangle = -e^{-5t} - 2e^{-9t} - 3e^{-10t}$$

Step 2. Evaluate the line integral.

$$\int_C \mathbf{F}\cdot d\mathbf{s} = \int_0^{\infty} \mathbf{F}(\mathbf{c}(t)) \cdot \mathbf{c}'(t)\, dt = \int_0^{\infty} \left(-e^{-5t} - 2e^{-9t} - 3e^{-10t}\right) dt$$

$$= \lim_{R\to\infty} \left(\frac{1}{5}e^{-5R} + \frac{2}{9}e^{-9R} + \frac{3}{10}e^{-10R}\right) - \left(\frac{1}{5} + \frac{2}{9} + \frac{3}{10}\right) = 0 - \frac{13}{18} = -\frac{13}{18}$$

29. Consider the line integrals $\int_{\mathbf{c}} \mathbf{F}\cdot d\mathbf{s}$ for the vector fields $\mathbf{F}$ and paths $\mathbf{c}$ in Figure 2. Which two of the line integrals appear to have a value of zero? Which of the other two is negative?

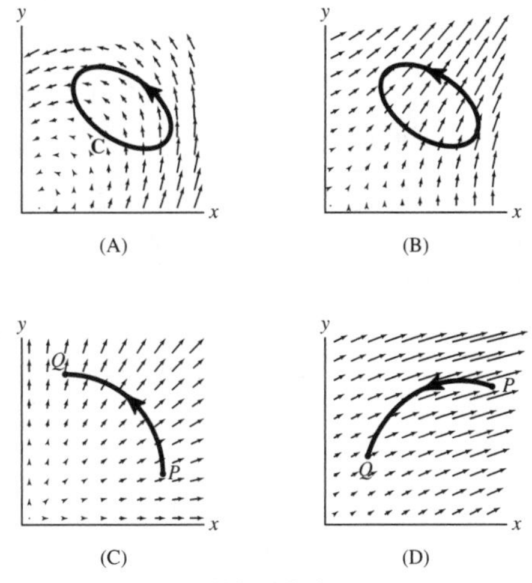

FIGURE 2

SOLUTION In (A), the line integral around the ellipse appears to be positive, because the negative tangential components from the lower part of the curve appears to be smaller than the positive contribution of the tangential components from the upper part.

In (B), the line integral around the ellipse appears to be zero, since **F** is orthogonal to the ellipse at all points except for two points where the tangential components of **F** cancel each other.

In (C), **F** is orthogonal to the path, hence the tangential component is zero at all points on the curve. Therefore the line integral $\int_C \mathbf{F} \cdot d\mathbf{s}$ is zero.

In (D), the direction of **F** is opposite to the direction of the curve. Therefore the dot product $\mathbf{F} \cdot \mathbf{T}$ is negative at each point along the curve, resulting in a negative line integral.

31. Find constants a, b, c such that

$$\Phi(u, v) = (u + av, bu + v, 2u - c)$$

parametrizes the plane $3x - 4y + z = 5$. Calculate $\mathbf{T}_u$, $\mathbf{T}_v$, and $\mathbf{n}(u, v)$.

SOLUTION We substitute $x = u + av$, $y = bu + v$ and $z = 2u - c$ in the equation of the plane $3x - 4y + z = 5$, to obtain:

$$5 = 3x - 4y + z = 3(u + av) - 4(bu + v) + 2u - c = (5 - 4b)u + (3a - 4)v - c$$

or

$$(5 - 4b)u + (3a - 4)v - (5 + c) = 0$$

This equation must be satisfied for all u and v, therefore the following must hold:

$$\begin{array}{rcl} 5 - 4b = 0 & & b = \dfrac{5}{4} \\ 3a - 4 = 0 & \Rightarrow & a = \dfrac{4}{3} \\ 5 + c = 0 & & c = -5 \end{array}$$

We obtain the following parametrization for the plane $3x - 4y + z = 5$:

$$\phi(u, v) = \left(u + \frac{4}{3}v, \frac{5}{4}u + v, 2u + 5\right)$$

We compute the tangent vectors $\mathbf{T}_u$ and $\mathbf{T}_v$:

$$\mathbf{T}_u = \frac{\partial \phi}{\partial u} = \left\langle 1, \frac{5}{4}, 2 \right\rangle; \mathbf{T}_v = \frac{\partial \phi}{\partial v} = \left\langle \frac{4}{3}, 1, 0 \right\rangle$$

The normal vector is their cross product:

$$\mathbf{n} = \mathbf{T}_u \times \mathbf{T}_v = \begin{vmatrix} \mathbf{i} & \mathbf{j} & \mathbf{k} \\ 1 & \frac{5}{4} & 2 \\ \frac{4}{3} & 1 & 0 \end{vmatrix} = \begin{vmatrix} \frac{5}{4} & 2 \\ 1 & 0 \end{vmatrix} \mathbf{i} - \begin{vmatrix} 1 & 2 \\ \frac{4}{3} & 0 \end{vmatrix} \mathbf{j} + \begin{vmatrix} 1 & \frac{5}{4} \\ \frac{4}{3} & 1 \end{vmatrix} \mathbf{k}$$

$$= -2\mathbf{i} + \frac{8}{3}\mathbf{j} + \left(1 - \frac{5}{3}\right)\mathbf{k} = \left\langle -2, \frac{8}{3}, -\frac{2}{3} \right\rangle$$

33. Let $\mathcal{S}$ be the surface parametrized by

$$\Phi(u, v) = \left(2u \sin \frac{v}{2}, 2u \cos \frac{v}{2}, 3v\right)$$

for $0 \le u \le 1$ and $0 \le v \le 2\pi$.

(a) Calculate the tangent vectors $\mathbf{T}_u$ and $\mathbf{T}_v$, and normal vector $\mathbf{n}(u, v)$ at $P = \Phi(1, \frac{\pi}{3})$.

(b) Find the equation of the tangent plane at P.

(c) Compute the surface area of $\mathcal{S}$.

SOLUTION

(a) The tangent vectors are the partial derivatives:

$$\mathbf{T}_u = \frac{\partial \varphi}{\partial u} = \frac{\partial}{\partial u}\left\langle 2u \sin \frac{v}{2}, 2u \cos \frac{v}{2}, 3v\right\rangle = \left\langle 2 \sin \frac{v}{2}, 2 \cos \frac{v}{2}, 0\right\rangle$$

$$\mathbf{T}_v = \frac{\partial \varphi}{\partial v} = \frac{\partial}{\partial v}\left\langle 2u \sin \frac{v}{2}, 2u \cos \frac{v}{2}, 3v\right\rangle = \left\langle u \cos \frac{v}{2}, -u \sin \frac{v}{2}, 3\right\rangle$$

The normal vector is their cross-product:

$$\mathbf{n} = \mathbf{T}_u \times \mathbf{T}_v = \begin{vmatrix} \mathbf{i} & \mathbf{j} & \mathbf{k} \\ 2\sin\frac{v}{2} & 2\cos\frac{v}{2} & 0 \\ u\cos\frac{v}{2} & -u\sin\frac{v}{2} & 3 \end{vmatrix} = \begin{vmatrix} 2\cos\frac{v}{2} & 0 \\ -u\sin\frac{v}{2} & 3 \end{vmatrix}\mathbf{i} - \begin{vmatrix} 2\sin\frac{v}{2} & 0 \\ u\cos\frac{v}{2} & 3 \end{vmatrix}\mathbf{j} + \begin{vmatrix} 2\sin\frac{v}{2} & 2\cos\frac{v}{2} \\ u\cos\frac{v}{2} & -u\sin\frac{v}{2} \end{vmatrix}\mathbf{k}$$

$$= \left(6 \cos \frac{v}{2}\right)\mathbf{i} - \left(6 \sin \frac{v}{2}\right)\mathbf{j} + \left(-2u \sin^2 \frac{v}{2} - 2u \cos^2 \frac{v}{2}\right)\mathbf{k}$$

$$= \left(6 \cos \frac{v}{2}\right)\mathbf{i} - \left(6 \sin \frac{v}{2}\right)\mathbf{j} - 2u\mathbf{k} = \left\langle 6 \cos \frac{v}{2}, -6 \sin \frac{v}{2}, -2u\right\rangle$$

At the point $P = \Phi\left(1, \frac{\pi}{3}\right)$, $u = 1$ and $v = \frac{\pi}{3}$. The tangents and the normal vector at this point are,

$$\mathbf{T}_u\left(1, \frac{\pi}{3}\right) = \left\langle 2 \sin \frac{\pi}{6}, 2 \cos \frac{\pi}{6}, 0\right\rangle = \left\langle 1, \sqrt{3}, 0\right\rangle$$

$$\mathbf{T}_v\left(1, \frac{\pi}{3}\right) = \left\langle 1 \cdot \cos \frac{\pi}{6}, -1 \cdot \sin \frac{\pi}{6}, 3\right\rangle = \left\langle \frac{\sqrt{3}}{2}, -\frac{1}{2}, 3\right\rangle$$

$$\mathbf{n}\left(1, \frac{\pi}{3}\right) = \left\langle 6 \cos \frac{\pi}{6}, -6 \sin \frac{\pi}{6}, -2 \cdot 1\right\rangle = \left\langle 3\sqrt{3}, -3, -2\right\rangle$$

(b) A normal to the plane is $\mathbf{n}\left(1, \frac{\pi}{3}\right) = \left\langle 3\sqrt{3}, -3, -2\right\rangle$ found in part (a). We find the tangency point:

$$P = \phi\left(1, \frac{\pi}{3}\right) = \left(2 \cdot 1 \sin \frac{\pi}{6}, 2 \cdot 1 \cos \frac{\pi}{6}, 3 \cdot \frac{\pi}{3}\right) = \left(1, \sqrt{3}, \pi\right)$$

The equation of the tangent plane is, thus,

$$\left\langle x - 1, y - \sqrt{3}, z - \pi\right\rangle \cdot \left\langle 3\sqrt{3}, -3, -2\right\rangle = 0$$

or

$$3\sqrt{3}(x - 1) - 3\left(y - \sqrt{3}\right) - 2(z - \pi) = 0$$

$$3\sqrt{3}x - 3y - 2z + 2\pi = 0$$

(c) In part (a) we found the normal vector:

$$\mathbf{n} = \left\langle 6 \cos \frac{v}{2}, -6 \sin \frac{v}{2}, -2u\right\rangle$$

We compute the length of $\mathbf{n}$:

$$\|\mathbf{n}\| = \sqrt{36 \cos^2 \frac{v}{2} + 36 \sin^2 \frac{v}{2} + 4u^2} = \sqrt{36 + 4u^2} = 2\sqrt{9 + u^2}$$

Using the Integral for the Surface Area we get:

$$\text{Area}(S) = \iint_{\mathcal{D}} \|n(u, v)\|\, du\, dv = \int_0^{2\pi} \int_0^1 2\sqrt{9 + u^2}\, du\, dv = 4\pi \int_0^1 \sqrt{9 + u^2}\, du$$

$$= 4\pi \left(\frac{u}{2}\sqrt{u^2 + 9} + \frac{9}{2} \ln\left(u + \sqrt{9 + u^2}\right)\Big|_{u=0}^{1}\right) = 4\pi\left(\frac{1}{2}\sqrt{10} + \frac{9}{2} \ln\left(1 + \sqrt{10}\right) - \frac{9}{2} \ln 3\right)$$

$$= 2\sqrt{10}\pi + 18\pi \ln\left(1+\sqrt{10}\right) - 18\pi \ln 3 = 2\sqrt{10}\pi + 18\pi \ln \frac{1+\sqrt{10}}{3} \approx 38.4$$

35. CAS Express the surface area of the surface $z = 10 - x^2 - y^2$, $-1 \le x \le 1$, $-3 \le y \le 3$ as a double integral. Evaluate the integral numerically using a CAS.

SOLUTION We use the Surface Integral over a graph. Let $g(x, y) = 10 - x^2 - y^2$. Then $g_x = -2x$, $g_y = -2y$ hence $\sqrt{1 + g_x{}^2 + g_y{}^2} = \sqrt{1 + 4x^2 + 4y^2}$. The area at the surface is the following integral which we compute using a CAS:

$$\text{Area}(S) = \iint_{\mathcal{D}} \sqrt{1 + g_x{}^2 + g_y{}^2}\, dx\, dy = \int_{-3}^{3} \int_{-1}^{1} \sqrt{1 + 4x^2 + 4y^2}\, dx\, dy \approx 41.8525$$

37. Calculate $\iint_{\mathcal{S}} \left(x^2 + y^2\right) e^{-z}\, dS$, where $\mathcal{S}$ is the cylinder with equation $x^2 + y^2 = 9$ for $0 \le z \le 10$.

SOLUTION We parametrize the cylinder $\mathcal{S}$ by,

$$\Phi(\theta, z) = (3\cos\theta, 3\sin\theta, z)$$

with the parameter domain:

$$0 \le \theta \le 2\pi, \quad 0 \le z \le 10.$$

We compute the tangent and normal vectors:

$$\mathbf{T}_\theta = \frac{\partial\phi}{\partial\theta} = \frac{\partial}{\partial\theta}\langle 3\cos\theta, 3\sin\theta, z\rangle = \langle -3\sin\theta, 3\cos\theta, 0\rangle$$

$$\mathbf{T}_z = \frac{\partial\phi}{\partial\theta} = \frac{\partial}{\partial\theta}\langle 3\cos\theta, 3\sin\theta, z\rangle = \langle 0, 0, 1\rangle$$

The normal vector is their cross product:

$$\mathbf{n} = \mathbf{T}_\theta \times \mathbf{T}_z = \begin{vmatrix} \mathbf{i} & \mathbf{j} & \mathbf{k} \\ -3\sin\theta & 3\cos\theta & 0 \\ 0 & 0 & 1 \end{vmatrix} = \begin{vmatrix} 3\cos\theta & 0 \\ 0 & 1 \end{vmatrix}\mathbf{i} - \begin{vmatrix} -3\sin\theta & 0 \\ 0 & 1 \end{vmatrix}\mathbf{j} + \begin{vmatrix} -3\sin\theta & 3\cos\theta \\ 0 & 0 \end{vmatrix}\mathbf{k}$$

$$= (3\cos\theta)\mathbf{i} + (3\sin\theta)\mathbf{j} = 3\langle\cos\theta, \sin\theta, 0\rangle$$

We compute the length of the normal vector:

$$\|\mathbf{n}\| = 3\sqrt{\cos^2\theta + \sin^2\theta + 0} = 3$$

We now express the function $f(x, y, z) = \left(x^2 + y^2\right) e^{-z}$ in terms of the parameters:

$$f(\phi(\theta, z)) = \left(x^2 + y^2\right) e^{-z} = \left(9\cos^2\theta + 9\sin^2\theta\right) e^{-z} = 9e^{-z}$$

Using the Theorem on Surface Integrals, we obtain:

$$\iint_{\mathcal{S}} \left(x^2 + y^2\right) e^{-z} dS = \int_0^{10} \int_0^{2\pi} 9e^{-z} 3\, d\theta\, dz = 27 \cdot 2\pi \int_0^{10} e^{-z} dz = 54\pi \left(-e^{-z}\right)\Big|_{z=0}^{10}$$

$$= 54\pi\left(-e^{-10} + 1\right) \approx 54\pi$$

39. Let $\mathcal{S}$ be a small patch of surface with a parametrization $\Phi(u, v)$ for $0 \le u, v \le 0.1$ such that the normal vector $\mathbf{n}(u, v)$ for $(u, v) = (0, 0)$ is $\mathbf{n} = \langle 2, -2, 4\rangle$. Use Eq. (1) in Section 17.4 to estimate the surface area of $\mathcal{S}$.

SOLUTION

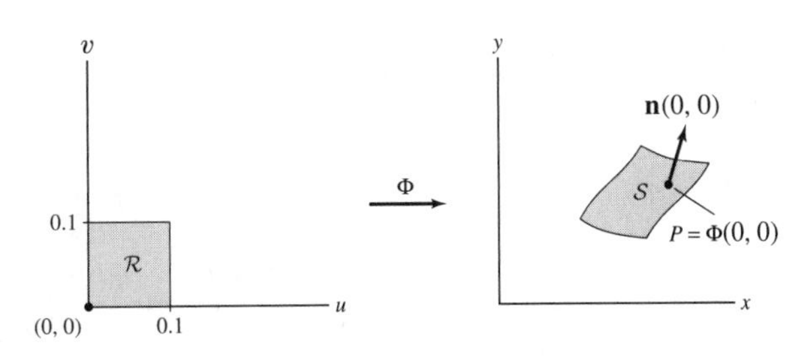

We use Eq. (1) in section 17.4 with $(u_{ij}, v_{ij}) = (0,0)$, $\mathcal{R}_{ij} = \mathcal{R} = [0, 0.1] \times [0, 0.1]$ in the (u, v)-plane and $\mathcal{S}_{ij} = \mathcal{S} = \Phi(\mathcal{R})$, in the (x, y)-plane to obtain the following estimation for the area of $\mathcal{S}$:

$$\text{Area}(\mathcal{S}) \approx \|\mathbf{n}(0,0)\| \text{Area}(\mathcal{R})$$

That is:

$$\text{Area}(\mathcal{S}) \approx \| \langle 2, -2, 4 \rangle \| 0.1^2 = \sqrt{2^2 + (-2)^2 + 4^2} \cdot (0.1)^2 = 0.02\sqrt{6} \approx 0.049$$

In Exercises 41–46, compute $\iint_{\mathcal{S}} \mathbf{F} \cdot d\mathbf{S}$ *for the given oriented surface or parametrized surface.*

41. $\mathbf{F}(x, y, z) = \langle y, x^2 y, e^{xz} \rangle$, $x^2 + y^2 = 9$, $-3 \le z \le 3$, outward-pointing normal

SOLUTION The part of the cylinder is parametrized by:

$$\Phi(\theta, z) = (3\cos\theta, 3\sin\theta, z), \quad 0 \le \theta \le 2\pi, \quad -3 \le z \le 3$$

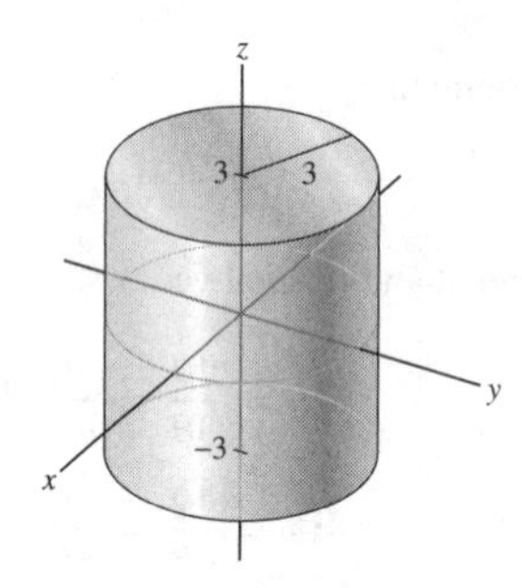

Step 1. Compute the tangent and normal vectors.

$$\mathbf{T}_\theta = \frac{\partial \Phi}{\partial \theta} = \frac{\partial}{\partial \theta} \langle 3\cos\theta, 3\sin\theta, z \rangle = \langle -3\sin\theta, 3\cos\theta, 0 \rangle$$

$$\mathbf{T}_z = \frac{\partial \Phi}{\partial z} = \frac{\partial}{\partial z} \langle 3\cos\theta, 3\sin\theta, z \rangle = \langle 0, 0, 1 \rangle$$

We compute the cross product:

$$\mathbf{T}_\theta \times \mathbf{T}_z = ((-3\sin\theta)\mathbf{i} + (3\cos\theta)\mathbf{j}) \times \mathbf{k} = (3\sin\theta)\mathbf{j} + (3\cos\theta)\mathbf{i} = \langle 3\cos\theta, 3\sin\theta, 0 \rangle$$

The outward pointing normal is (when $\theta = 0$, the x-component must be positive):

$$\mathbf{n} = \langle 3\cos\theta, 3\sin\theta, 0 \rangle$$

Step 2. Evaluate the dot product $\mathbf{F} \cdot \mathbf{n}$. We write $\mathbf{F}(x, y, z) = \langle y, x^2 y, e^{xz} \rangle$ in terms of the parameters by substituting $x = 3\cos\theta$, $y = 3\sin\theta$. We get:

$$\mathbf{F}(\Phi(\theta, z)) = \left\langle 3\sin\theta, 9\cos^2\theta \cdot 3\sin\theta, e^{3z\cos\theta} \right\rangle = \left\langle 3\sin\theta, 27\cos^2\theta\sin\theta, e^{3z\cos\theta} \right\rangle$$

Hence:

$$\begin{aligned} \mathbf{F}(\Phi(\theta, z)) \cdot \mathbf{n} &= \left\langle 3\sin\theta, 27\cos^2\theta\sin\theta, e^{3z\cos\theta} \right\rangle \cdot \langle 3\cos\theta, 3\sin\theta, 0 \rangle \\ &= 9\sin\theta\cos\theta + 81\cos^2\theta\sin^2\theta \end{aligned}$$

Step 3. Evaluate the surface integral. The surface integral is equal to the following double integral (we use the trigonometric identities $\sin\theta\cos\theta = \frac{\sin 2\theta}{2}$ and $\sin^2 2\theta = \frac{1}{2}(1 - \cos 4\theta)$):

$$\begin{aligned} \iint_{\mathcal{S}} \mathbf{F} \cdot d\mathbf{S} &= \int_0^{2\pi} \int_{-3}^{3} \mathbf{F}(\Phi(\theta, z)) \cdot \mathbf{n}(\theta, z)\, dz\, d\theta = \int_0^{2\pi} \int_{-3}^{3} \left(9\sin\theta\cos\theta + 81\cos^2\theta\sin^2\theta\right) d\theta \\ &= 6\int_0^{2\pi} \left(\frac{9}{2}\sin 2\theta + 81 \cdot \left(\frac{\sin 2\theta}{2}\right)^2\right) d\theta = \frac{6 \cdot 9}{2} \int_0^{2\pi} \sin 2\theta\, d\theta + \frac{6 \cdot 81}{8} \int_0^{2\pi} (1 - \cos 4\theta)\, d\theta \\ &= \frac{54}{2} - \left(\frac{\cos 2\theta}{2}\right)\Big|_0^{2\pi} + \frac{243}{4}\left(\theta - \frac{\sin 4\theta}{4}\right)\Big|_0^{2\pi} = \frac{243}{4} \cdot 2\pi = \frac{243}{2}\pi \approx 381.7 \end{aligned}$$

43. $\mathbf{F}(x, y, z) = \langle x^2, y^2, x^2 + y^2 \rangle$, $\quad x^2 + y^2 + z^2 = 4$, $z \ge 0$, outward-pointing normal

SOLUTION The upper hemisphere is parametrized by:

$$\Phi(\theta, \phi) = (2\cos\theta\sin\phi, 2\sin\theta\sin\phi, 2\cos\phi), \quad 0 \le \theta \le 2\pi, \quad 0 \le \phi \le \frac{\pi}{2}$$

As seen in section 17.4, since $0 \le \phi \le \frac{\pi}{2}$ then the outward-pointing normal is:

$$\mathbf{n} = 4\sin\phi \langle \cos\theta\sin\phi, \sin\theta\sin\phi, \cos\phi \rangle$$

We express $\mathbf{F}$ in terms of the parameters:

$$\begin{aligned}
\mathbf{F}(\Phi(\theta, \phi)) &= \left\langle x^2, y^2, x^2 + y^2 \right\rangle = \left\langle 4\cos^2\theta\sin^2\phi, 4\sin^2\theta\sin^2\phi, 4\sin^2\phi\left(\cos^2\theta + \sin^2\theta\right) \right\rangle \\
&= \left\langle 4\cos^2\theta\sin^2\phi, 4\sin^2\theta\sin^2\phi, 4\sin^2\phi \right\rangle
\end{aligned}$$

The dot product $\mathbf{F} \cdot \mathbf{n}$ is thus

$$\begin{aligned}
\mathbf{F}(\Phi(\theta, \phi)) \cdot \mathbf{n}(\theta, \phi) &= 4\sin\left(4\cos^3\theta\,\sin^3\phi + 4\sin^3\theta\,\sin^3\phi + 4\sin^2\phi\,\cos\phi\right) \\
&= 16\left(\cos^3\theta\,\sin^4\phi + \sin^3\theta\,\sin^4\phi + \sin^3\phi\,\cos\phi\right)
\end{aligned}$$

We obtain the following integral:

$$\begin{aligned}
\iint_S \mathbf{F} \cdot d\mathbf{s} &= \iint_{\mathcal{D}} \mathbf{F}(\Phi(\theta, \phi)) \cdot \mathbf{n}(\theta, \phi)\, d\theta\, d\phi \\
&= 16\int_0^{\pi/2}\int_0^{2\pi}\left(\cos^3\theta\,\sin^4\phi + \sin^3\theta\,\sin^4\phi + \sin^3\phi\,\cos\phi\right) d\theta\, d\phi \\
&= 16\left(\int_0^{2\pi}\cos^3\theta\, d\theta\right)\left(\int_0^{\pi/2}\sin^4\phi\, d\phi\right) \\
&\quad + 16\left(\int_0^{2\pi}\sin^3\theta\, d\theta\right)\left(\int_0^{\pi/2}\sin^4\phi\, d\phi\right) + 32\pi\int_0^{\pi/2}\sin^3\phi\cos v\, d\phi
\end{aligned}$$

Since $\int_0^{2\pi}\cos^3\theta\, d\theta = \int_0^{2\pi}\sin^3\theta\, d\theta = 0$, we get:

$$\iint_S \mathbf{F} \cdot d\mathbf{s} = 32\pi\int_0^{\pi/2}\sin^3\phi\cos\phi\, d\phi = 32\pi\left(\frac{\sin^4\phi}{4}\Big|_{\phi=0}^{\pi/2}\right) = 32\pi\left(\frac{1-0}{4}\right) = 8\pi$$

45. $\mathbf{F}(x, y, z) = \langle 0, 0, xze^{xy} \rangle$, $\quad z = xy$, $0 \le x, y \le 1$,
upward-pointing normal

SOLUTION We parametrize the surface by:

$$\Phi(x, y) = (x, y, xy)$$

Where the parameter domain is the square:

$$\mathcal{D} = \{(x, y) : 0 \le x \le 1, 0 \le y \le 1\}$$

Step 1. Compute the tangent and normal vectors.

$$\mathbf{T}_x = \frac{\partial\Phi}{\partial x} = \frac{\partial}{\partial x}\langle x, y, xy \rangle = \langle 1, 0, y \rangle$$

$$\mathbf{T}_y = \frac{\partial\Phi}{\partial y} = \frac{\partial}{\partial y}\langle x, y, xy \rangle = \langle 0, 1, x \rangle$$

$$\mathbf{T}_x \times \mathbf{T}_y = \begin{vmatrix} \mathbf{i} & \mathbf{j} & \mathbf{k} \\ 1 & 0 & y \\ 0 & 1 & x \end{vmatrix} = \begin{vmatrix} 0 & y \\ 1 & x \end{vmatrix}\mathbf{i} - \begin{vmatrix} 1 & y \\ 0 & x \end{vmatrix}\mathbf{j} + \begin{vmatrix} 1 & 0 \\ 0 & 1 \end{vmatrix}\mathbf{k} = -y\mathbf{i} - x\mathbf{j} + \mathbf{k} = \langle -y, -x, 1 \rangle$$

Since the normal points upwards, the z-coordinate is positive. Therefore the normal vector is:

$$\mathbf{n} = \langle -y, -x, 1 \rangle$$

Step 2. Evaluate the dot product $\mathbf{F} \cdot \mathbf{n}$. We express $\mathbf{F}$ in terms of x and y:

$$\mathbf{F}(\Phi(x, y)) = \langle 0, 0, xze^{xy} \rangle = \langle 0, 0, x(xy)e^{xy} \rangle = \left\langle 0, 0, x^2ye^{xy} \right\rangle$$

Hence:

$$\mathbf{F}(\Phi(x, y)) \cdot \mathbf{n}(x, y) = \left\langle 0, 0, x^2ye^{xy} \right\rangle \cdot \langle -y, -x, 1 \rangle = x^2ye^{xy}$$

Step 3. Evaluate the surface integral. The surface integral is equal to the following double integral:

$$\iint_S \mathbf{F} \cdot d\mathbf{s} = \iint_{\mathcal{D}} \mathbf{F}(\Phi(x, y)) \cdot \mathbf{n}(x, y)\, dx\, dy$$

$$= \int_0^1 \int_0^1 x^2ye^{xy}\, dy\, dx = \int_0^1 x^2 \left(\int_0^1 ye^{xy}\, dy \right) dx \tag{1}$$

We evaluate the inner integral using integration by parts:

$$\int_0^1 ye^{xy}\, dy = \frac{y}{x}e^{xy}\Big|_{y=0}^1 - \int_0^1 \frac{1}{x}e^{xy}\, dy = \frac{e^x}{x} - \frac{1}{x^2}e^{xy}\Big|_{y=0}^1 = \frac{e^x}{x} - \frac{1}{x^2}\left(e^x - 1\right)$$

Substituting this integral in (1) gives:

$$\iint_S \mathbf{F} \cdot d\mathbf{s} = \int_0^1 \left(xe^x - \left(e^x - 1\right)\right) dx = \int_0^1 xe^x\, dx - \int_0^1 \left(e^x - 1\right) dx$$

$$= \int_0^1 xe^x\, dx - \left(e^x - x\right)\Big|_0^1 = \int_0^1 xe^x\, dx - (e - 2)$$

Using integration by parts we have:

$$\iint_S F \cdot dS = xe^x - e^x \Big|_0^1 - (e - 2) = 1 - (e - 2) = 3 - e$$

47. Calculate the total charge on the cylinder

$$x^2 + y^2 = R^2, \qquad 0 \le z \le H$$

if the charge density in cylindrical coordinates is $\rho(\theta, z) = Kz^2 \cos^2 \theta$, where K is a constant.

SOLUTION The total change on the surface $\mathcal{S}$ is $\iint_{\mathcal{S}} \rho\, dS$. We parametrize the surface by,

$$\Phi(\theta, z) = (R\cos\theta, R\sin\theta, Hz)$$

with the parameter domain,

$$0 \le \theta \le 2\pi,\ 0 \le z \le 1.$$

We compute the tangent and normal vectors:

$$\mathbf{T}_\theta = \frac{\partial \Phi}{\partial \theta} = \frac{\partial}{\partial \theta}\langle R\cos\theta, R\sin\theta, Hz \rangle = \langle -R\sin\theta, R\cos\theta, 0 \rangle$$

$$\mathbf{T}_z = \frac{\partial \Phi}{\partial z} = \frac{\partial}{\partial z}\langle R\cos\theta, R\sin\theta, Hz \rangle = \langle 0, 0, H \rangle$$

The normal vector is their cross product:

$$\mathbf{n} = \mathbf{T}_\theta \times \mathbf{T}_z = \begin{vmatrix} \mathbf{i} & \mathbf{j} & \mathbf{k} \\ -R\sin\theta & R\cos\theta & 0 \\ 0 & 0 & H \end{vmatrix}$$

$$= \begin{vmatrix} R\cos\theta & 0 \\ 0 & H \end{vmatrix}\mathbf{i} - \begin{vmatrix} -R\sin\theta & 0 \\ 0 & H \end{vmatrix}\mathbf{j} + \begin{vmatrix} -R\sin\theta & R\cos\theta \\ 0 & 0 \end{vmatrix}\mathbf{k}$$

$$= (RH\cos\theta)\mathbf{i} + (RH\sin\theta)\mathbf{j} = RH\langle \cos\theta, \sin\theta, 0 \rangle$$

We find the length of $\mathbf{n}$:

$$\|\mathbf{n}\| = RH\sqrt{\cos^2\theta + \sin^2\theta} = RH$$

We compute $\rho(\Phi(\theta, z))$:

$$\rho(\Phi(\theta, z)) = K(Hz)^2 \cos^2\theta = KH^2z^2\cos^2\theta$$

Using the Theorem on Surface Integrals we obtain:

$$\iint_{\mathcal{S}} \rho \cdot dS = \iint_{\mathcal{D}} \rho(\Phi(\theta, z)) \cdot \|\mathbf{n}(\theta, z)\|\, dz\, d\theta = \int_0^{2\pi}\int_0^1 KH^2z^2\cos^2\theta \cdot HR\, dz\, d\theta$$

$$= \left(\int_0^1 KH^3Rz^2 dz\right)\left(\int_0^{2\pi} \cos^2\theta\, d\theta\right) = \left(\frac{KH^3Rz^3}{3}\Big|_0^1\right)\left(\frac{\theta}{2} + \frac{\sin 2\theta}{4}\Big|_0^{2\pi}\right)$$

$$= \frac{KH^3R}{3} \cdot \pi = \frac{\pi}{3}KH^3R$$

49. With $\mathbf{v}$ as in Exercise 48, calculate the flow rate across the part of the elliptic cylinder $\frac{x^2}{4} + y^2 = 1$ where $x, y \geq 0$, and $0 \leq z \leq 4$.

SOLUTION The flow rate of a fluid with velocity field $\mathbf{v} = \langle 2x, y, xy\rangle$ through the elliptic cylinder S is the surface integral:

$$\iint_{\mathcal{S}} \mathbf{v} \cdot d\mathbf{S} \qquad (1)$$

To compute this integral, we parametrize $\mathcal{S}$ by,

$$\Phi(\theta, z) = (2\cos\theta, \sin\theta, z), \quad 0 \leq \theta \leq \frac{\pi}{2}, \quad 0 \leq z \leq 4$$

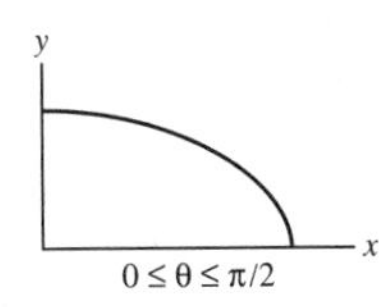

Step 1. Compute the tangent and normal vectors.

$$\mathbf{T}_\theta = \frac{\partial\Phi}{\partial\theta} = \frac{\partial}{\partial\theta}\langle 2\cos\theta, \sin\theta, z\rangle = \langle -2\sin\theta, \cos\theta, 0\rangle$$

$$\mathbf{T}_z = \frac{\partial\Phi}{\partial z} = \frac{\partial}{\partial z}\langle 2\cos\theta, \sin\theta, z\rangle = \langle 0, 0, 1\rangle$$

$$\mathbf{n} = \mathbf{T}_\theta \times \mathbf{T}_z = \begin{vmatrix} \mathbf{i} & \mathbf{j} & \mathbf{k} \\ -2\sin\theta & \cos\theta & 0 \\ 0 & 0 & 1 \end{vmatrix} = (\cos\theta)\mathbf{i} + (2\sin\theta)\mathbf{j} = \langle \cos\theta, 2\sin\theta, 0\rangle$$

Step 2. Compute the dot product $\mathbf{v} \cdot \mathbf{n}$

$$\mathbf{v}(\Phi(\theta, z)) \cdot \mathbf{n} = \langle 4\cos\theta, \sin\theta, 2\cos\theta\sin\theta\rangle \cdot \langle \cos\theta, 2\sin\theta, 0\rangle = 4\cos^2\theta + 2\sin^2\theta$$

$$= 2\cos^2\theta + 2\left(\cos^2\theta + \sin^2\theta\right) = 2\cos^2\theta + 2$$

Step 3. Evaluate the flux of $\mathbf{v}$. The flux of $\mathbf{v}$ in (1) is equal to the following double integral (we use the equality $2\cos^2\theta = 1 + \cos 2\theta$ in our calculation):

$$\iint_{\mathcal{S}} \mathbf{v} \cdot d\mathbf{S} = \iint_{\mathcal{D}} \mathbf{v}(\Phi(\theta, z)) \cdot \mathbf{n}\, d\theta\, dz = \int_0^4\int_0^{\pi/2} \left(2\cos^2\theta + 2\right) d\theta\, dz$$

$$= 4\int_0^{\pi/2} \left(2\cos^2\theta + 2\right) d\theta = 4\int_0^{\pi/2} (3 + \cos 2\theta)\, d\theta = 4\left(3\theta + \frac{\sin 2\theta}{2}\Big|_{\theta=0}^{\pi/2}\right) = 6\pi$$

18 FUNDAMENTAL THEOREMS OF VECTOR ANALYSIS

18.1 Green's Theorem (ET Section 17.1)

Preliminary Questions

1. Which vector field $\mathbf{F}$ is being integrated in the line integral $\oint x^2\,dy - e^y\,dx$?

SOLUTION The line integral can be rewritten as $\oint -e^y\,dx + x^2\,dy$. This is the line integral of $\mathbf{F} = \left\langle -e^y, x^2 \right\rangle$ along the curve.

2. Draw a domain in the shape of an ellipse and indicate with an arrow the boundary orientation of the boundary curve. Do the same for the annulus (the region between two concentric circles).

SOLUTION The orientation on $\mathcal{C}$ is counterclockwise, meaning that the region enclosed by $\mathcal{C}$ lies to the left in traversing $\mathcal{C}$.

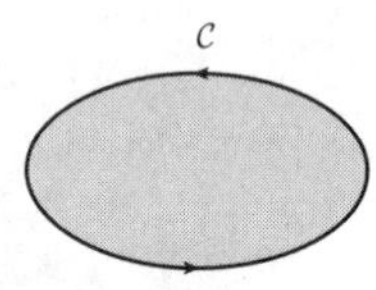

For the annulus, the inner boundary is oriented clockwise and the outer boundary is oriented counterclockwise. The region between the circles lies to the left while traversing each circle.

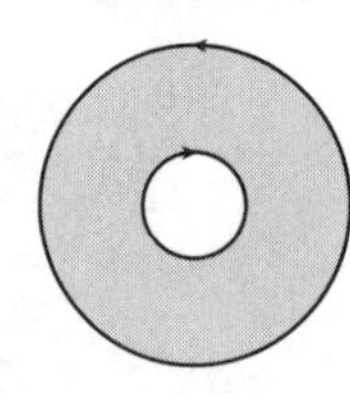

3. The circulation of a gradient vector field around a closed curve is zero. Is this fact consistent with Green's Theorem? Explain.

SOLUTION Green's Theorem asserts that

$$\int_{\mathcal{C}} \mathbf{F} \cdot d\mathbf{s} = \int_{\mathcal{C}} P\,dx + Q\,dy = \iint_{\mathcal{D}} \left(\frac{\partial Q}{\partial x} - \frac{\partial P}{\partial y} \right) dA \qquad (1)$$

If $\mathbf{F}$ is a gradient vector field, the cross partials are equal, that is,

$$\frac{\partial P}{\partial y} = \frac{\partial Q}{\partial x} \quad \Rightarrow \quad \frac{\partial Q}{\partial x} - \frac{\partial P}{\partial y} = 0 \qquad (2)$$

Combining (1) and (2) we obtain $\int_{\mathcal{C}} \mathbf{F} \cdot d\mathbf{s} = 0$. That is, Green's Theorem implies that the integral of a gradient vector field around a simple closed curve is zero.

4. Which of the following vector fields possess the following property: For every simple closed curve $\mathcal{C}$, $\int_{\mathcal{C}} \mathbf{F} \cdot d\mathbf{s}$ is equal to the area enclosed by $\mathcal{C}$?

(a) $\mathbf{F} = \langle -y, 0 \rangle$

(b) $\mathbf{F} = \langle x, y \rangle$

(c) $\mathbf{F} = \left\langle \sin(x^2), x + e^{y^2} \right\rangle$

SOLUTION By Green's Theorem,

$$\int_{\mathcal{C}} \mathbf{F} \cdot d\mathbf{s} = \iint_{\mathcal{D}} \left(\frac{\partial Q}{\partial x} - \frac{\partial P}{\partial y} \right) dx\,dy \qquad (1)$$

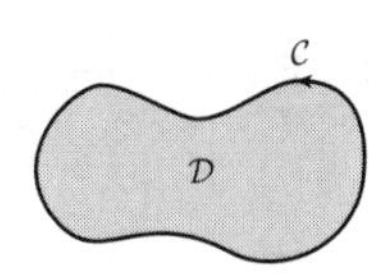

We compute the curl of each one of the given fields.

(a) Here, $P = -y$ and $Q = 0$, hence $\frac{\partial Q}{\partial x} - \frac{\partial P}{\partial y} = 0 - (-1) = 1$. Therefore, by (1),

$$\int_C \mathbf{F} \cdot d\mathbf{s} = \iint_{\mathcal{D}} 1\, dx\, dy = \text{Area}(\mathcal{D})$$

(b) We have $P = x$ and $Q = y$, therefore $\frac{\partial Q}{\partial x} - \frac{\partial P}{\partial y} = 0 - 0 = 0$. By (1) we get

$$\int_C \mathbf{F} \cdot d\mathbf{s} = \iint_{\mathcal{D}} 0\, dx\, dy = 0 \neq \text{Area}(\mathcal{D})$$

(c) In this vector field we have $P = \sin(x^2)$ and $Q = x + e^{y^2}$. Therefore,

$$\frac{\partial Q}{\partial x} - \frac{\partial P}{\partial y} = 1 - 0 = 1.$$

By (1) we obtain

$$\int_C \mathbf{F} \cdot d\mathbf{s} = \iint_{\mathcal{D}} 1\, dx\, dy = \text{Area}(\mathcal{D}).$$

Exercises

1. Verify Green's Theorem for the line integral $\oint_C xy\, dx + y\, dy$, where C is the unit circle, oriented counterclockwise.

SOLUTION

Step 1. Evaluate the line integral. We use the parametrization $\gamma(\theta) = \langle \cos\theta, \sin\theta \rangle$, $0 \le \theta \le 2\pi$ of the unit circle. Then

$$dx = -\sin\theta\, d\theta, \quad dy = \cos\theta\, d\theta$$

and

$$xy\, dx + y\, dy = \cos\theta \sin\theta(-\sin\theta\, d\theta) + \sin\theta\cos\theta\, d\theta = \left(-\cos\theta\sin^2\theta + \sin\theta\cos\theta\right) d\theta$$

The line integral is thus

$$\int_C xy\, dx + y\, dy = \int_0^{2\pi} \left(-\cos\theta\sin^2\theta + \sin\theta\cos\theta\right) d\theta$$

$$= \int_0^{2\pi} -\cos\theta\sin^2\theta\, d\theta + \int_0^{2\pi} \sin\theta\cos\theta\, d\theta = -\frac{\sin^3\theta}{3}\bigg|_0^{2\pi} - \frac{\cos 2\theta}{4}\bigg|_0^{2\pi} = 0 \qquad (1)$$

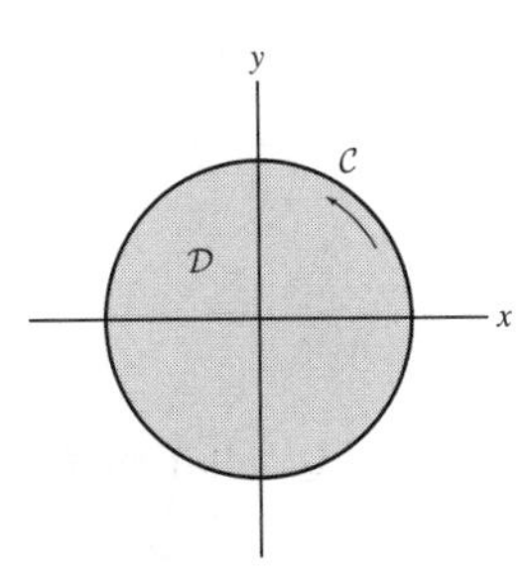

Step 2. Evaluate the double integral. Since $P = xy$ and $Q = y$, we have

$$\frac{\partial Q}{\partial x} - \frac{\partial P}{\partial y} = 0 - x = -x$$

We compute the double integral in Green's Theorem:

$$\iint_{\mathcal{D}} \left(\frac{\partial Q}{\partial x} - \frac{\partial P}{\partial y}\right) dx\, dy = \iint_{\mathcal{D}} -x\, dx\, dy = -\iint_{\mathcal{D}} x\, dx\, dy$$

The integral of x over the disk $\mathcal{D}$ is zero, since by symmetry the positive and negative values of x cancel each other. Therefore,

$$\iint_{\mathcal{D}} \left(\frac{\partial Q}{\partial x} - \frac{\partial P}{\partial y} \right) dx\, dy = 0 \tag{2}$$

Step 3. Compare. The line integral in (1) is equal to the double integral in (2), as stated in Green's Theorem.

In Exercises 3–11, use Green's Theorem to evaluate the line integral. Orient the curve counterclockwise unless otherwise indicated.

3. $\oint_C y^2\, dx + x^2\, dy$, where C is the boundary of the unit square $0 \le x \le 1, 0 \le y \le 1$

SOLUTION

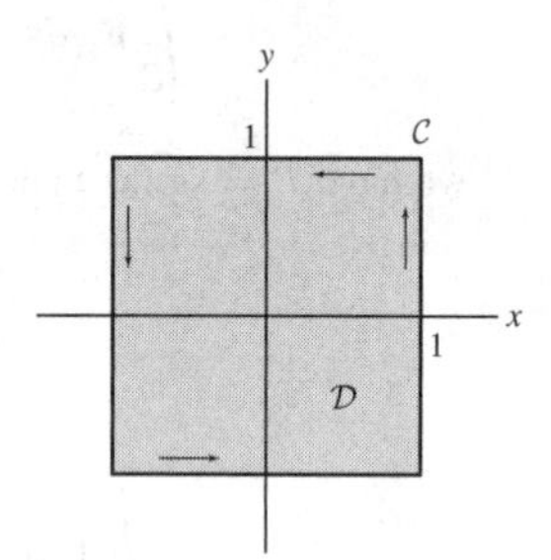

We have $P = y^2$ and $Q = x^2$, therefore

$$\frac{\partial Q}{\partial x} - \frac{\partial P}{\partial y} = 2x - 2y$$

Using Green's Theorem we obtain

$$\int_C y^2\, dx + x^2\, dy = \iint_{\mathcal{D}} \frac{\partial Q}{\partial x} - \frac{\partial P}{\partial y}\, dA = \iint_{\mathcal{D}} (2x - 2y)\, dx\, dy = 2\iint_{\mathcal{D}} x\, dx\, dy - 2\iint_{\mathcal{D}} y\, dx\, dy$$

By symmetry, the positive and negative values of x cancel each other in the first integral, so this integral is zero. The second double integral is zero by similar reasoning. Therefore,

$$\int_C y^2\, dx + x^2\, dy = 0 - 0 = 0$$

5. $\oint_C x^2 y\, dx$, where C is the unit circle centered at the origin

SOLUTION

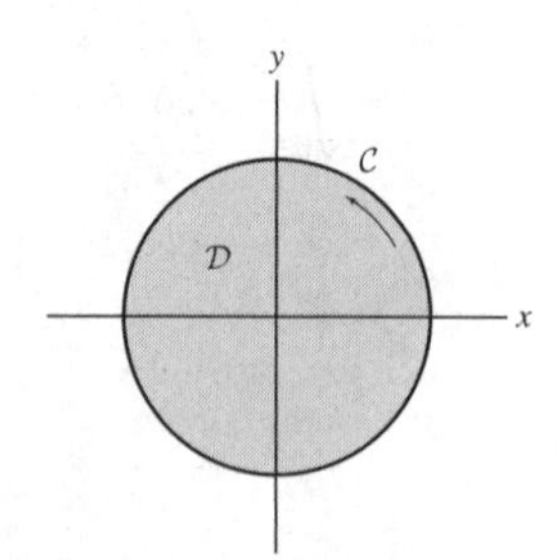

In this function $P = x^2 y$ and $Q = 0$. Therefore,

$$\frac{\partial Q}{\partial x} - \frac{\partial P}{\partial y} = 0 - x^2 = -x^2$$

We obtain the following integral:

$$I = \int_C x^2 y\, dx = \iint_{\mathcal{D}} \left(\frac{\partial Q}{\partial x} - \frac{\partial P}{\partial y} \right) dA = \iint_{\mathcal{D}} -x^2\, dA$$

We convert the integral to polar coordinates. This gives

$$I = \int_0^{2\pi} \int_0^1 -r^2 \cos^2\theta \cdot r\, dr\, d\theta = \int_0^{2\pi} \int_0^1 -r^3 \cos^2\theta\, dr\, d\theta$$

$$= \left(\int_0^{2\pi} \cos^2\theta\, d\theta\right)\left(\int_0^1 -r^3\, dr\right) = \left(\frac{\theta}{2} + \frac{\sin 2\theta}{4}\Big|_{\theta=0}^{2\pi}\right)\left(-\frac{r^4}{4}\Big|_{r=0}^{1}\right) = \pi \cdot \left(-\frac{1}{4}\right) = -\frac{\pi}{4}$$

7. $\oint_C \mathbf{F} \cdot d\mathbf{s}$, where $\mathbf{F} = \langle x^2, x^2 \rangle$ and C consists of the arcs $y = x^2$ and $y = x$ for $0 \le x \le 1$

SOLUTION By Green's Theorem,

$$I = \int_C \mathbf{F} \cdot d\mathbf{s} = \iint_{\mathcal{D}} \left(\frac{\partial Q}{\partial x} - \frac{\partial P}{\partial y}\right) dA$$

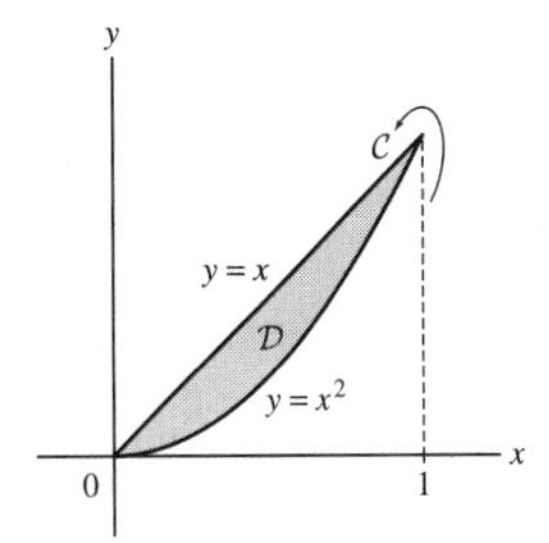

We have $P = Q = x^2$, therefore

$$\frac{\partial Q}{\partial x} - \frac{\partial P}{\partial y} = 2x - 0 = 2x$$

Hence,

$$I = \iint_{\mathcal{D}} 2x\, dA = \int_0^1 \int_{x^2}^{x} 2x\, dy\, dx = \int_0^1 2xy\Big|_{y=x^2}^{x} dx = \int_0^1 2x(x - x^2)\, dx = \int_0^1 (2x^2 - 2x^3)\, dx$$

$$= \frac{2}{3}x^3 - \frac{1}{2}x^4\Big|_0^1 = \frac{2}{3} - \frac{1}{2} = \frac{1}{6}$$

9. The line integral of $\mathbf{F} = \langle x^3, 4x \rangle$ around the boundary of the parallelogram in Figure 15 (note the orientation)

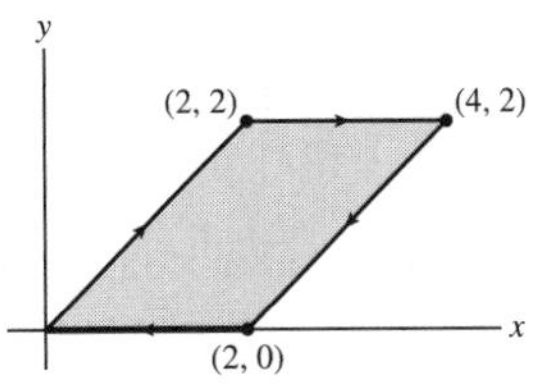

FIGURE 15

SOLUTION We have $P = x^3$ and $Q = 4x$, therefore

$$\frac{\partial Q}{\partial x} - \frac{\partial P}{\partial y} = 4 - 0 = 4$$

Hence, Green's Theorem implies

$$\int_C x^3 x + 4x\, dy = -\iint_{\mathcal{D}} \left(\frac{\partial Q}{\partial x} - \frac{\partial P}{\partial y}\right) dA = -\iint_{\mathcal{D}} 4\, dA = 4\,\text{Area}(\mathcal{D}) = -16$$

11. $\int_C xy\, dx + (x^2 + x)\, dy$, where C is the path in Figure 16

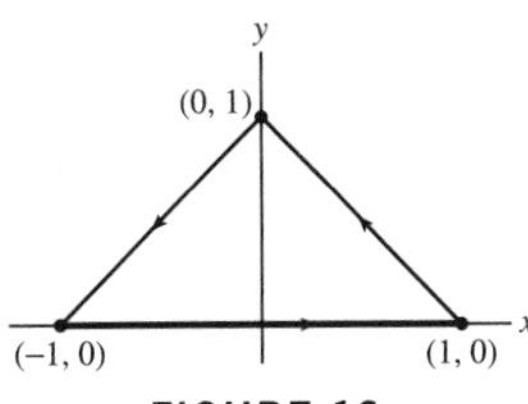

FIGURE 16

SOLUTION

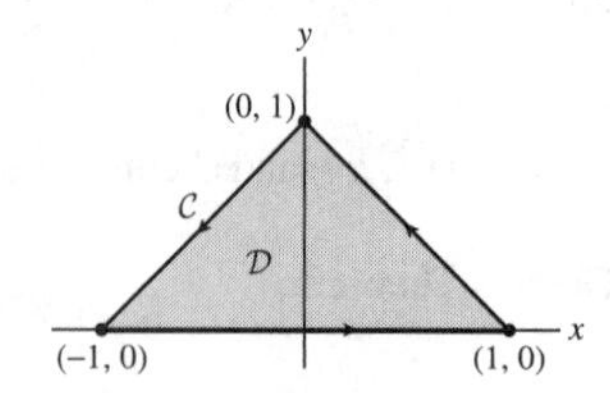

In the given function, $P = xy$ and $Q = x^2 + x$. Therefore,

$$\frac{\partial Q}{\partial x} - \frac{\partial P}{\partial y} = 2x + 1 - x = x + 1$$

By Green's Theorem we obtain the following integral:

$$\int_C xy\,dx + (x^2 + x)\,dy = \iint_{\mathcal{D}} \left(\frac{\partial Q}{\partial x} - \frac{\partial P}{\partial y}\right) dA = \iint_{\mathcal{D}} (x + 1)\,dA = \iint_{\mathcal{D}} x\,dA + \iint_{\mathcal{D}} 1\,dA$$

By symmetry, the positive and negative values of x cancel each other, causing the first integral to be zero. Thus,

$$\int_C xy\,dx + (x^2 + x)\,dy = 0 + \iint_{\mathcal{D}} dA = \text{Area}(\mathcal{D}) = \frac{2 \cdot 1}{2} = 1.$$

In Exercises 13–16, use Eq. (5) to calculate the area of the given region.

13. The circle of radius 3 centered at the origin

SOLUTION By Eq. (5), we have

$$A = \frac{1}{2} \int_C x\,dy - y\,dx$$

We parametrize the circle by $x = 3\cos\theta$, $y = 3\sin\theta$, hence,

$$x\,dy - y\,dx = 3\cos\theta \cdot 3\cos\theta\,d\theta - 3\sin\theta(-3\sin\theta)\,d\theta = (9\cos^2\theta + 9\sin^2\theta)\,d\theta = 9\,d\theta$$

Therefore,

$$A = \frac{1}{2} \int_C x\,dy - y\,dx = \frac{1}{2} \int_0^{2\pi} 9\,d\theta = \frac{9}{2} \cdot 2\pi = 9\pi.$$

15. The region between the x-axis and the cycloid parametrized by $\mathbf{c}(t) = (t - \sin t, 1 - \cos t)$ for $0 \le t \le 2\pi$ (Figure 17)

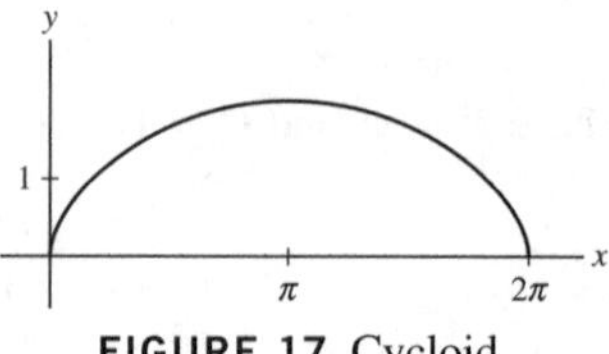

FIGURE 17 Cycloid.

SOLUTION By Eq. (5), the area is the following integral:

$$A = \frac{1}{2} \int_C x\,dy - y\,dx$$

where C is the closed curve determined by the segment OA and the cycloid Γ.

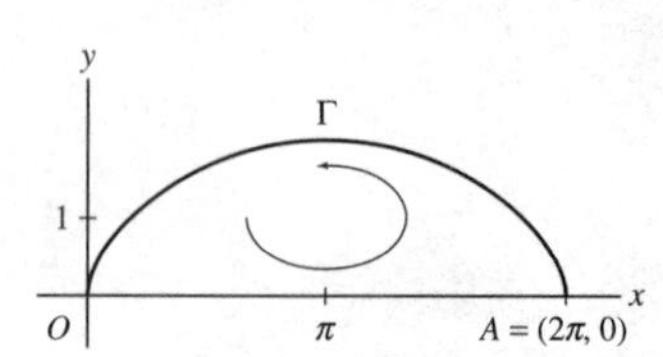

Therefore,

$$A = \frac{1}{2}\int_{OA} x\,dy - y\,dx + \frac{1}{2}\int_{\Gamma} x\,dy - y\,dx \tag{1}$$

We compute the two integrals. The segment OA is parametrized by $\langle t, 0\rangle$, $t = 0$ to $t = 2\pi$. Hence, $x = t$ and $y = 0$. Therefore,

$$x\,dy - y\,dx = t \cdot 0\,dt - 0 \cdot dt = 0$$

$$\int_{OA} x\,dy - y\,dx = 0 \tag{2}$$

On Γ we have $x = t - \sin t$ and $y = 1 - \cos t$, therefore

$$\begin{aligned} x\,dy - y\,dx &= (t - \sin t)\sin t\,dt - (1 - \cos t)(1 - \cos t)\,dt \\ &= (t\sin t - \sin^2 t - 1 + 2\cos t - \cos^2 t)\,dt = (t\sin t + 2\cos t - 2)\,dt \end{aligned}$$

Hence,

$$\begin{aligned} \int_{\Gamma} x\,dy - y\,dx &= \int_{2\pi}^{0} (t\sin t + 2\cos t - 2)\,dt = \int_0^{2\pi} (2 - 2\cos t - t\sin t)\,dt \\ &= 2t - 2\sin t + t\cos t - \sin t\Big|_0^{2\pi} = 2t - 3\sin t + t\cos t\Big|_0^{2\pi} = 6\pi \end{aligned} \tag{3}$$

Substituting (2) and (3) in (1) we get

$$A = \frac{1}{2} \cdot 0 + \frac{1}{2} \cdot 6\pi = 3\pi.$$

17. Let $x^3 + y^3 = 3xy$ be the **folium of Descartes** (Figure 18).

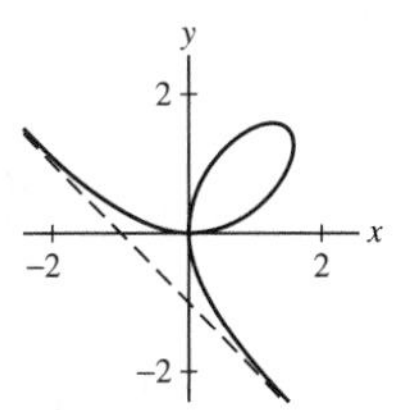

FIGURE 18 Folium of Descartes.

(a) Show that the folium has a parametrization in terms of $t = y/x$ given by

$$x = \frac{3t}{1+t^3}, \qquad y = \frac{3t^2}{1+t^3} \quad (-\infty < t < \infty) \quad (t \neq -1)$$

(b) Show that

$$x\,dy - y\,dx = \frac{9t^2}{(1+t^3)^2}\,dt$$

Hint: By the Quotient Rule,

$$x^2\,d\left(\frac{y}{x}\right) = x\,dy - y\,dx$$

(c) Find the area of the loop of the folium.

SOLUTION

(a) We show that $x = \frac{3t}{1+t^3}$, $y = \frac{3t^2}{1+t^3}$ satisfy the equation $x^3 + y^3 - 3xy = 0$ of the folium:

$$\begin{aligned} x^3 + y^3 - 3xy &= \left(\frac{3t}{1+t^3}\right)^3 + \left(\frac{3t^2}{1+t^3}\right)^3 - 3 \cdot \frac{3t}{1+t^3} \cdot \frac{3t^2}{1+t^3} \\ &= \frac{27t^3 + 27t^6}{(1+t^3)^3} - \frac{27t^3(1+t^3)}{(1+t^3)^3} = \frac{27t^3\left(1 + t^3 - (1+t^3)\right)}{(1+t^3)^3} = \frac{0}{(1+t^3)^3} = 0 \end{aligned}$$

This proves that the curve parametrized by $x = \frac{3t}{1+t^3}$, $y = \frac{3t^2}{1+t^3}$ lies on the folium of Descartes. This parametrization parametrizes the whole folium since the two equations can be solved for t in terms of x and y. That is,

$$\left.\begin{aligned} x &= \frac{3t}{1+t^3} \\ y &= \frac{3t^2}{1+t^3} \end{aligned}\right\} \quad \Rightarrow \quad t = \frac{y}{x}$$

A glance at the graph of the folium shows that any line $y = tx$, with slope t, intersects the folium exactly once. Thus, there is a one-to-one relationship between the values of t and the points on the graph.

(b) We differentiate the two sides of $t = \frac{y}{x}$ with respect to t. Using the Quotient Rule gives

$$1 = \frac{x\frac{dy}{dt} - y\frac{dx}{dt}}{x^2}$$

or

$$x\frac{dy}{dt} - y\frac{dx}{dt} = x^2 = \left(\frac{3t}{1+t^3}\right)^2$$

This equality can be written in the form

$$x\,dy - y\,dx = \frac{9t^2}{(1+t^3)^2}\,dt$$

(c) We use the formula for the area enclosed by a closed curve and the result of part (b) to find the required area. That is,

$$A = \frac{1}{2}\int_C x\,dy - y\,dx = \frac{1}{2}\int_0^\infty \frac{9t^2}{(1+t^3)^2}\,dt$$

From our earlier discussion on the parametrization of the folium, we see that the loop is traced when the parameter t is increasing along the interval $0 \le t < \infty$. We compute the improper integral using the substitution $u = 1 + t^3$, $du = 3t^2\,dt$. This gives

$$A = \frac{1}{2}\lim_{R\to\infty}\int_0^R \frac{9t^2}{(1+t^3)^2}\,dt = \frac{1}{2}\lim_{R\to\infty}\int_1^{1+R^3}\frac{3\,du}{u^2} = \frac{3}{2}\lim_{R\to\infty} -\frac{1}{u}\bigg|_{u=1}^{1+R^3}$$

$$= \frac{3}{2}\lim_{R\to\infty}\left(1 - \frac{1}{1+R^3}\right) = \frac{3}{2}(1-0) = \frac{3}{2}$$

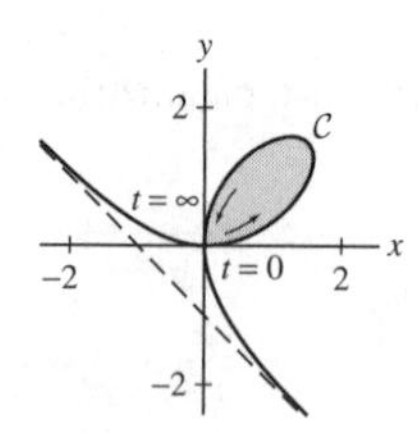

19. Show that if C is a simple closed curve, then

$$\oint_C -y\,dx = \oint_C x\,dy$$

and both integrals are equal to the area enclosed by C.

SOLUTION We show that $\int_C y\,dx + x\,dy = 0$ by showing that the vector field $\mathbf{F} = \langle y, x\rangle$ is conservative. Indeed, since $P = y$ and $Q = x$, we have $\frac{\partial Q}{\partial x} = 1$ and $\frac{\partial P}{\partial y} = 1$. Therefore, the cross partials are equal and therefore $\mathbf{F}$ is conservative. By the formula for the area enclosed by a simple closed curve, the area enclosed by C is

$$A = \frac{1}{2}\int_C x\,dy - y\,dx = \frac{1}{2}\int_C x\,dy + \frac{1}{2}\int_C -y\,dx$$

Using the equality obtained above, we have

$$A = \frac{1}{2}\int_C x\,dy + \frac{1}{2}\int_C x\,dy = \int_C x\,dy = \int_C -y\,dx.$$

21. Let $\mathbf{F} = \langle 2xe^y, x + x^2e^y \rangle$ and let C be the quarter-circle path from A to B in Figure 21. Evaluate $I = \oint_C \mathbf{F} \cdot d\mathbf{s}$ as follows:

(a) Find a function $\varphi(x, y)$ such that $\mathbf{F} = \mathbf{G} + \nabla\varphi$, where $\mathbf{G} = \langle 0, x \rangle$.

(b) Show that the line integrals of $\mathbf{G}$ along the segments $\overline{OA}$ and $\overline{OB}$ are zero.

(c) Use Green's Theorem to show that

$$I = \varphi(B) - \varphi(A) + 4\pi$$

and evaluate I.

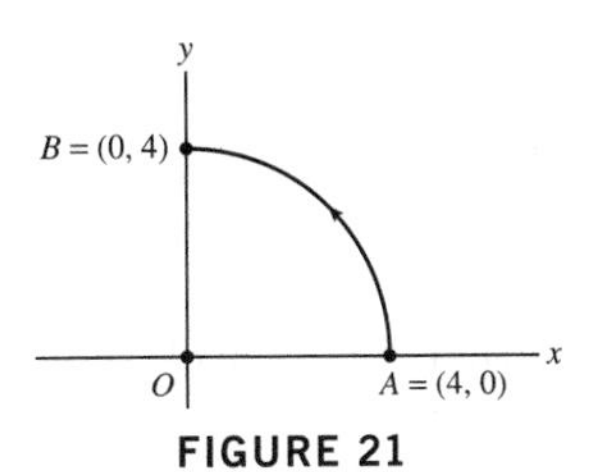

FIGURE 21

SOLUTION

(a) We need to find a potential function $\varphi(x, y)$ for the difference

$$\mathbf{F} - \mathbf{G} = \left\langle 2xe^y, x + x^2e^y \right\rangle - \langle 0, x \rangle = \left\langle 2xe^y, x^2e^y \right\rangle$$

We let $\varphi(x, y) = x^2e^y$.

(b) We use the parametrizations $\overline{AO} : \langle t, 0 \rangle$, $0 \le t \le 4$ and $\overline{OB} : \langle 0, t \rangle$, $0 \le t \le 4$ to evaluate the integrals of $\mathbf{G} = \langle 0, x \rangle$. We get

$$\int_{\overline{OA}} \mathbf{G} \cdot d\mathbf{s} = \int_0^4 \langle 0, t \rangle \cdot \langle 1, 0 \rangle \, dt = \int_0^4 0 \, dt = 0$$

$$\int_{\overline{OB}} \mathbf{G} \cdot d\mathbf{s} = \int_0^4 \langle 0, 0 \rangle \cdot \langle 0, 1 \rangle \, dt = \int_0^4 0 \, dt = 0$$

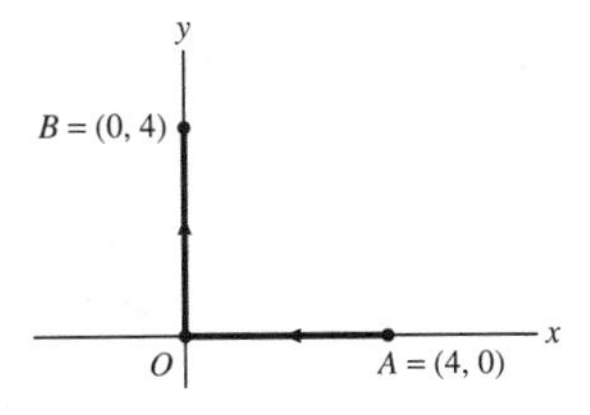

(c) Since $\mathbf{F} - \mathbf{G} = \nabla\varphi$, we have

$$\int_C (\mathbf{F} - \mathbf{G}) \cdot d\mathbf{s} = \varphi(B) - \varphi(A) = \int_C \mathbf{F} \cdot d\mathbf{s} - \int_C \mathbf{G} \cdot d\mathbf{s} = I - \int_C \mathbf{G} \cdot d\mathbf{s}$$

That is,

$$I = \varphi(B) - \varphi(A) + \int_C \mathbf{G} \cdot d\mathbf{s} \tag{1}$$

To compute the line integral on the right-hand side, we rewrite it as

$$\int_C \mathbf{G} \cdot d\mathbf{s} = \int_{\overline{BO} + \overline{OA} + C} \mathbf{G} \cdot d\mathbf{s} - \int_{\overline{BO}} \mathbf{G} \cdot d\mathbf{s} - \int_{\overline{OA}} \mathbf{G} \cdot d\mathbf{s}$$

Using part (b) we may write

$$\int_C \mathbf{G} \cdot d\mathbf{s} = \int_{\overline{BO} + \overline{OA} + C} \mathbf{G} \cdot d\mathbf{s} \tag{2}$$

We now use Green's Theorem. Since $\mathbf{G} = \langle 0, x \rangle$, we have $P = 0$ and $Q = x$, hence $\frac{\partial Q}{\partial x} - \frac{\partial P}{\partial y} = 1 - 0 = 1$. Thus,

$$\int_{\overline{BO} + \overline{OA} + C} \mathbf{G} \cdot d\mathbf{s} = \iint_{\mathcal{D}} 1 \, dA = \text{Area}(\mathcal{D}) = \frac{\pi \cdot 4^2}{4} = 4\pi \tag{3}$$

Combining (1), (2), and (3), we obtain

$$I = \varphi(B) - \varphi(A) + 4\pi$$

Since $\varphi(x, y) = x^2 e^y$, we conclude that

$$I = \varphi(0, 4) - \varphi(4, 0) + 4\pi = 0 - 4^2 e^0 + 4\pi = 4\pi - 16.$$

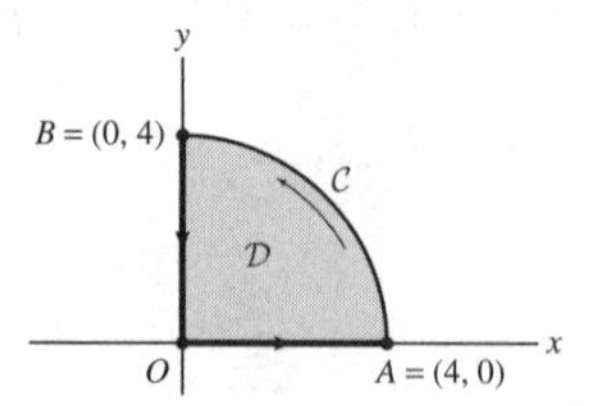

23. Evaluate $I = \int_{\mathcal{C}} (\sin x + y)\,dx + (3x + y)\,dy$ for the nonclosed path $ABCD$ in Figure 23. *Hint:* Use the method of Exercise 22.

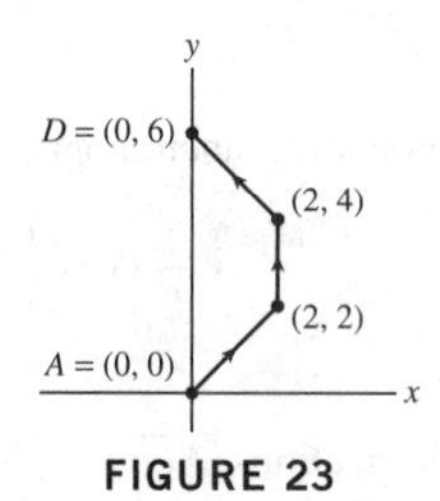

FIGURE 23

SOLUTION

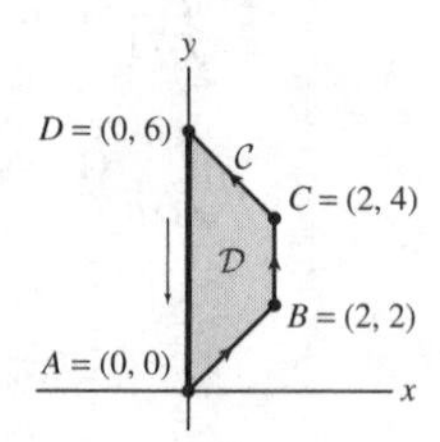

Let $\mathbf{F} = \langle \sin x + y, 3x + y \rangle$, hence $P = \sin x + y$ and $Q = 3x + y$. We denote by $\mathcal{C}_1$ the closed path determined by $\mathcal{C}$ and the segment $\overline{DA}$. Then by Green's Theorem,

$$\int_{\mathcal{C}_1} P\,dx + Q\,dy = \iint_{\mathcal{D}} \left(\frac{\partial Q}{\partial x} - \frac{\partial P}{\partial y} \right) dA = \iint_{\mathcal{D}} (3 - 1)\,dA = 2 \iint_{\mathcal{D}} dA = 2\,\text{Area}(\mathcal{D}) \tag{1}$$

The area of $\mathcal{D}$ is the area of the trapezoid $ABCD$, that is,

$$\text{Area}(\mathcal{D}) = \frac{\left(\overline{BC} + \overline{AD}\right) h}{2} = \frac{(2 + 6) \cdot 2}{2} = 8.$$

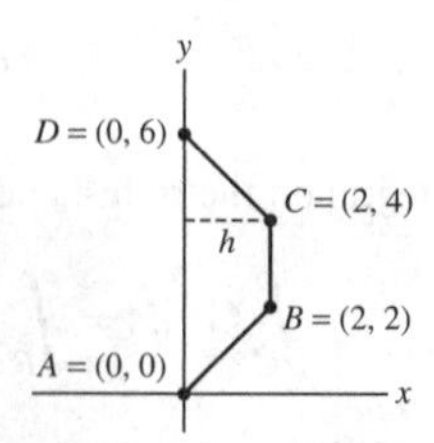

Combining with (1) we get

$$\int_{\mathcal{C}_1} P\,dx + Q\,dy = 2 \cdot 8 = 16$$

Using properties of line integrals, we have

$$\int_{\mathcal{C}} P\,dx + Q\,dy + \int_{\overline{DA}} P\,dx + Q\,dy = 16 \tag{2}$$

We compute the line integral over $\overline{DA}$, using the parametrization

$$\overline{DA}: x = 0, \ \ y = t, t \text{ varies from 6 to 0.}$$

We get

$$\int_{\overline{DA}} P\,dx + Q\,dy = \int_6^0 F(0,t) \cdot \frac{d}{dt}\langle 0, t\rangle \, dt = \int_6^0 \langle \sin 0 + t, 3\cdot 0 + t\rangle \cdot \langle 0, 1\rangle \, dt$$

$$= \int_6^0 \langle t, t\rangle \cdot \langle 0, 1\rangle \, dt = \int_6^0 t\,dt = \frac{t^2}{2}\bigg|_{t=6}^{0} = -18$$

We substitute in (2) and solve for the required integral:

$$\int_C P\,dx + Q\,dy - 18 = 16 \quad \text{or} \quad \int_C P\,dx + Q\,dy = 34.$$

25. Let $\mathbf{F}$ be the velocity field. Estimate the circulation of $\mathbf{F}$ around a circle of radius $R = 0.05$ with center P, assuming that $\text{curl}_z(\mathbf{F})(P) = -3$. In which direction would a small paddle placed at P rotate? How fast would it rotate (in radians per second) if $\mathbf{F}$ is expressed in meters per second?

SOLUTION We use the following estimation:

$$\int_C \mathbf{F} \cdot d\mathbf{s} \approx \text{curl}(\mathbf{F})(P)\text{Area}(\mathcal{D}) \tag{1}$$

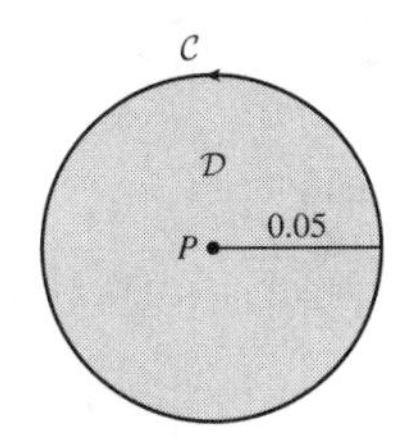

We are given that $\text{curl}(\mathbf{F})(P) = -3$. Also, the area of the disk of radius $R = 0.05$ is $\pi \cdot 0.05^2 = 0.0025\pi$. Therefore, we obtain the following estimation:

$$\int_C \mathbf{F} \cdot d\mathbf{s} \approx -3 \cdot 0.0025\pi \approx -0.024.$$

Since the curl is negative, the paddle would rotate in the clockwise direction. Using the formula $|\text{curl}(\mathbf{F})| = 2\omega$, we see that the angular speed is $\omega = 1.5$ radians per second.

27. Referring to Figure 25, suppose that

$$\oint_{C_2} \mathbf{F} \cdot d\mathbf{s} = 3\pi, \qquad \oint_{C_3} \mathbf{F} \cdot d\mathbf{s} = 4\pi$$

Use Green's Theorem to determine the circulation of $\mathbf{F}$ around C_1, assuming that $\text{curl}_z(\mathbf{F}) = 9$ on the shaded region.

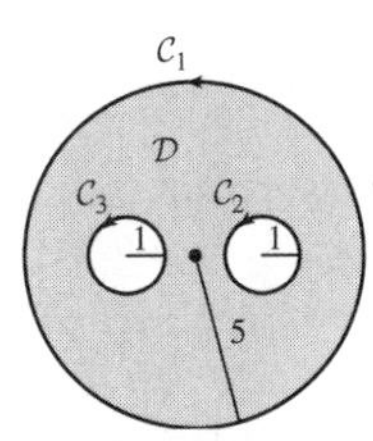

FIGURE 25

SOLUTION We must calculate $\int_{C_1} \mathbf{F} \cdot d\mathbf{s}$. We use Green's Theorem for the region $\mathcal{D}$ between the three circles C_1, C_2, and C_3. Because of orientation, the line integrals $\int_{-C_2} \mathbf{F} \cdot d\mathbf{s} = -\int_{C_2} \mathbf{F} \cdot d\mathbf{s}$ and $\int_{-C_3} \mathbf{F} \cdot d\mathbf{s} = -\int_{C_3} \mathbf{F} \cdot d\mathbf{s}$ must be used in applying Green's Theorem. That is,

$$\int_{C_1} \mathbf{F} \cdot d\mathbf{s} - \int_{C_2} \mathbf{F} \cdot d\mathbf{s} - \int_{C_3} \mathbf{F} \cdot d\mathbf{s} = \iint_{\mathcal{D}} \text{curl}(\mathbf{F})\,dA$$

We substitute the given information to obtain

$$\int_{C_1} \mathbf{F} \cdot d\mathbf{s} - 3\pi - 4\pi = \iint_{\mathcal{D}} 9\,dA = 9\iint_{\mathcal{D}} 1 \cdot dA = 9\,\text{Area}(\mathcal{D}) \tag{1}$$

The area of $\mathcal{D}$ is computed as the difference of areas of discs. That is,

$$\text{Area}(\mathcal{D}) = \pi \cdot 5^2 - \pi \cdot 1^2 - \pi \cdot 1^2 = 23\pi$$

We substitute in (1) and compute the desired circulation:

$$\int_{C_1} \mathbf{F} \cdot d\mathbf{s} - 7\pi = 9 \cdot 23\pi$$

or

$$\int_{C_1} \mathbf{F} \cdot d\mathbf{s} = 214\pi.$$

29. Use the result of Exercise 28 to compute the areas of the polygons in Figure 26. Check your result for the area of the triangle in (A) using geometry.

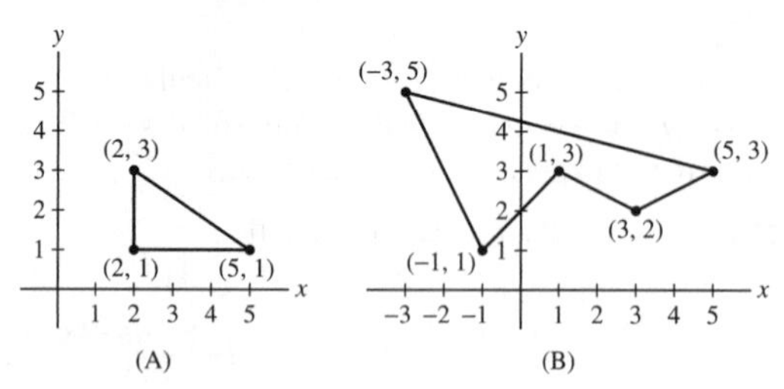

FIGURE 26

SOLUTION

(a) The vertices of the triangle are

$$(x_1, y_1) = (x_4, y_4) = (2, 1), \quad (x_2, y_2) = (5, 1), \quad (x_3, y_3) = (2, 3)$$

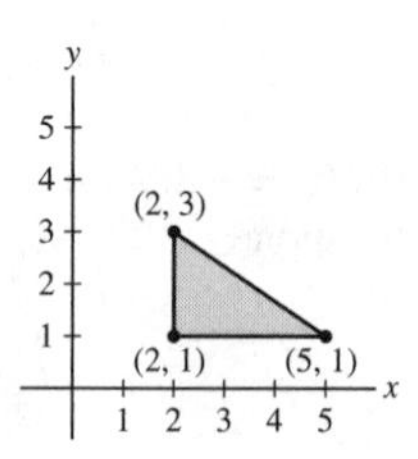

Using the formula obtained in Exercise 28, the area of the triangle is the following sum:

$$\begin{aligned} A &= \frac{1}{2}\big((x_1y_2 - x_2y_1) + (x_2y_3 - x_3y_2) + (x_3y_1 - x_1y_3)\big) \\ &= \frac{1}{2}\big((2 \cdot 1 - 5 \cdot 1) + (5 \cdot 3 - 2 \cdot 1) + (2 \cdot 1 - 2 \cdot 3)\big) = \frac{1}{2}(-3 + 13 - 4) = 3 \end{aligned}$$

We verify our result using the formula for the area of a triangle:

$$A = \frac{1}{2}bh = \frac{1}{2} \cdot (5 - 2) \cdot (3 - 1) = 3$$

(b) The vertices of the polygon are

$$(x_1, y_1) = (x_6, y_6) = (-1, 1)$$
$$(x_2, y_2) = (1, 3)$$
$$(x_3, y_3) = (3, 2)$$
$$(x_4, y_4) = (5, 3)$$
$$(x_5, y_5) = (-3, 5)$$

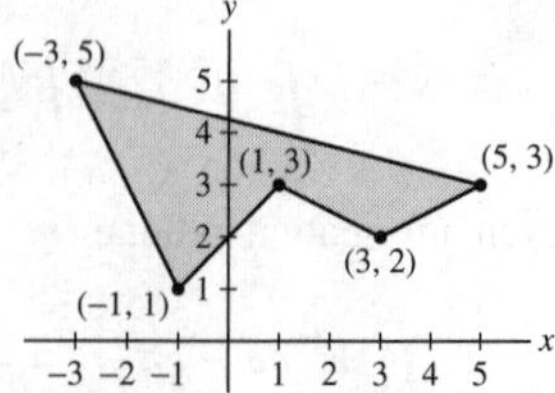

Using the formula in part (a), the area of the polygon is the following sum:

$$A = \frac{1}{2}\big((x_1y_2 - x_2y_1) + (x_2y_3 - x_3y_2) + (x_3y_4 - x_4y) + (x_4y_5 - x_5y_4) + (x_5y_1 - x_1y_5)\big)$$

$$= \frac{1}{2}\Big((-1\cdot 3 - 1\cdot 1) + (1\cdot 2 - 3\cdot 3) + (3\cdot 3 - 5\cdot 2) + \big(5\cdot 5 - (-3)\cdot 3\big) + \big(-3\cdot 1 - (-1)\cdot 5\big)\Big)$$

$$= \frac{1}{2}(-4 - 7 - 1 + 34 + 2) = 12$$

Further Insights and Challenges

In Exercises 30–31, let **F** *be the vortex vector field [defined for* $(x, y) \neq (0, 0)$*]:*

$$\mathbf{F} = \left\langle \frac{-y}{x^2 + y^2}, \frac{x}{x^2 + y^2} \right\rangle$$

31. Prove that if $\mathcal{C}$ is a simple closed curve whose interior contains the origin, then $\oint_{\mathcal{C}} \mathbf{F} \cdot d\mathbf{s} = 2\pi$ (Figure 27). *Hint:* Apply Green's Theorem to the domain between $\mathcal{C}$ and $\mathcal{C}_R$ where R is so small that $\mathcal{C}_R$ is contained in $\mathcal{C}$.

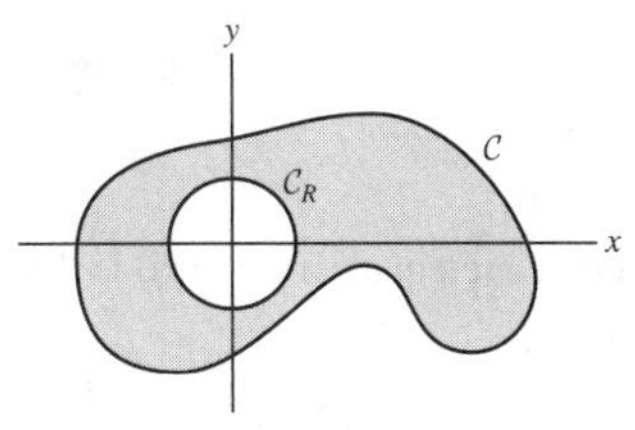

FIGURE 27

SOLUTION Let $R > 0$ be sufficiently small so that the circle $\mathcal{C}_R$ is contained in $\mathcal{C}$.

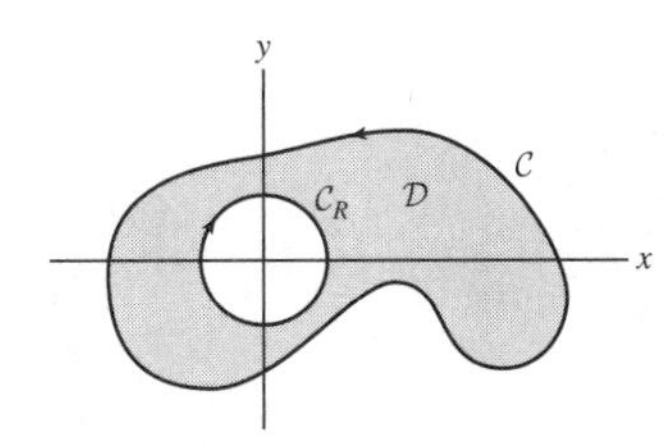

Let $\mathcal{D}$ denote the region between $\mathcal{C}_R$ and $\mathcal{C}$. We apply Green's Theorem to the region $\mathcal{D}$. The curve $\mathcal{C}$ is oriented counterclockwise and $\mathcal{C}_R$ is oriented clockwise. We have

$$\int_{\mathcal{C}} \mathbf{F} \cdot d\mathbf{s} + \int_{\mathcal{C}_R} \mathbf{F} \cdot d\mathbf{s} = \iint_{\mathcal{D}} \operatorname{curl}(\mathbf{F})\, dA \tag{1}$$

From Exercise 30(b) we know that $\int_{\mathcal{C}_R} \mathbf{F} \cdot d\mathbf{s} = -2\pi$. Since $\mathcal{D}$ does not contain the origin, we have by part (a) of Exercise 30, curl $(\mathbf{F}) = 0$ on $\mathcal{D}$. Substituting in (1) we obtain

$$\int_{\mathcal{C}} \mathbf{F} \cdot d\mathbf{s} - 2\pi = \iint_{\mathcal{D}} 0\, dA = 0$$

or

$$\int_{\mathcal{C}} \mathbf{F} \cdot d\mathbf{s} = 2\pi.$$

In Exercises 32–34, the conjugate *of a vector field* $\mathbf{F} = \langle P, Q \rangle$ *is the vector field* $\mathbf{F}^* = \langle -Q, P \rangle$.

33. The normal component of **F** at a point P on a simple closed $\mathcal{C}$ is the quantity $\mathbf{F}(P) \cdot \mathbf{n}(P)$, where $\mathbf{n}(P)$ is the outward-pointing unit normal vector. The **flux** of **F** across curve $\mathcal{C}$ is defined as the line integral of the normal component around $\mathcal{C}$ (Figure 28). Show that the flux across $\mathcal{C}$ is equal to $\oint_{\mathcal{C}} \mathbf{F}^* \cdot d\mathbf{s}$.

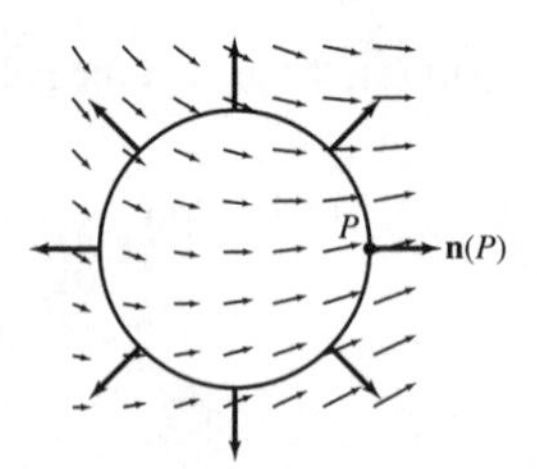

FIGURE 28 The flux of **F** is the integral of the normal component $\mathbf{F}\cdot\mathbf{n}$ around the curve.

SOLUTION We must show that if C is a simple closed curve, then

$$\int_C (\mathbf{F}\cdot\mathbf{n})ds = \int_C \mathbf{F}^*\cdot d\mathbf{s}$$

By the definition of the vector line integral, the line integral of the vector field $\mathbf{F}^*$ over C is

$$\int_C \mathbf{F}^*\cdot d\mathbf{s} = \int_C (\mathbf{F}^*\cdot\mathbf{T})\cdot ds \tag{1}$$

In Exercise 32 we showed that $\mathbf{F}^*$ is a rotation of $\mathbf{F}$ by $\frac{\pi}{2}$ counterclockwise. Therefore, the angle between $\mathbf{F}^*$ and the tangent $\mathbf{T}$ is equal to the angle between $\mathbf{F}$ and the normal $\mathbf{n}$. Also, $\|\mathbf{F}^*\| = \|\mathbf{F}\|$. Hence,

$$\mathbf{F}^*\cdot\mathbf{T} = \|\mathbf{F}^*\|\,\|\mathbf{T}\|\cos\theta = \|\mathbf{F}\|\cos\theta$$

$$\mathbf{F}\cdot\mathbf{n} = \|\mathbf{F}\|\,\|\mathbf{n}\|\cos\theta = \|\mathbf{F}\|\cos\theta$$

Since the dot products are equal, we conclude that

$$\int_C (\mathbf{F}^*\cdot\mathbf{T})\cdot ds = \int_C (\mathbf{F}\cdot\mathbf{n})ds \tag{2}$$

Combining (1) and (2) we obtain

$$\int_C \mathbf{F}^*\cdot d\mathbf{s} = \int_C (\mathbf{F}\cdot\mathbf{n})ds.$$

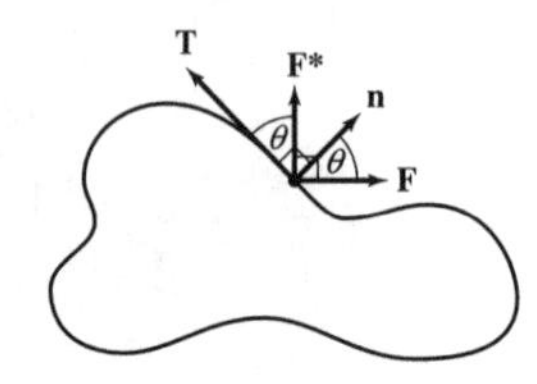

In Exercises 35–38, the **Laplace operator** Δ *is defined by*

$$\Delta\varphi = \frac{\partial^2\varphi}{\partial x^2} + \frac{\partial^2\varphi}{\partial y^2}$$

9

35. Let $\mathbf{F} = \nabla\varphi$. Show that $\text{curl}_z(\mathbf{F}^*) = \Delta\varphi$, where $\mathbf{F}^*$ is the conjugate vector field (defined in Exercises 32–34).

SOLUTION For a vector field $\mathbf{F} = \langle P, Q\rangle$, the conjugate vector field is $\mathbf{F}^* = \langle -Q, P\rangle$. By the given information,

$$\mathbf{F} = \nabla\varphi = \left\langle \frac{\partial\varphi}{\partial x}, \frac{\partial\varphi}{\partial y}\right\rangle \quad\Rightarrow\quad \mathbf{F}^* = \left\langle -\frac{\partial\varphi}{\partial y}, \frac{\partial\varphi}{\partial x}\right\rangle$$

We compute the curl of $\mathbf{F}^*$:

$$\text{curl}(\mathbf{F}^*) = \frac{\partial}{\partial x}\left(\frac{\partial\varphi}{\partial x}\right) - \frac{\partial}{\partial y}\left(\frac{-\partial\varphi}{\partial y}\right) = \frac{\partial^2\varphi}{\partial x^2} + \frac{\partial^2\varphi}{\partial y^2} = \Delta\varphi$$

37. Let $P = (a, b)$ and let $C(r)$ be the circle of radius r centered at P. The average value of a continuous function φ on $C(r)$ is defined as the integral

$$I_\varphi(r) = \frac{1}{2\pi}\int_0^{2\pi} \varphi(a + r\cos\theta, b + r\sin\theta)\,d\theta$$

(a) Show that

$$\frac{\partial\varphi}{\partial\mathbf{n}}(a + r\cos\theta, b + r\sin\theta) = \frac{\partial\varphi}{\partial r}(a + r\cos\theta, b + r\sin\theta)$$

(b) Use differentiation under the integral sign to prove that

$$\frac{d}{dr}I_\varphi(r) = \frac{1}{2\pi r}\int_{\mathcal{C}(r)} \frac{\partial\varphi}{\partial\mathbf{n}}\,ds$$

(c) Use Exercise 36 to conclude that

$$\frac{d}{dr}I_\varphi(r) = \frac{1}{2\pi r}\iint_{\mathcal{D}(r)} \Delta\varphi\,dA$$

where $\mathcal{D}(r)$ is the interior of $\mathcal{C}(r)$.

SOLUTION In this solution, $\varphi_r(a + r\cos\theta, b + r\sin\theta)$ denotes the partial derivative φ_r computed at $(a + r\cos\theta, b + r\sin\theta)$, whereas $\frac{\partial}{\partial r}\varphi(a + r\cos\theta, b + r\sin\theta)$ is the derivative of the composite function.

(a) Since $\frac{\partial\varphi}{\partial n} = \nabla\varphi\cdot\mathbf{n}$, we first express the gradient vector in terms of polar coordinates. We use the Chain Rule and the derivatives:

$$\theta_x = -\frac{\sin\theta}{r}, \quad \theta_y = \frac{\cos\theta}{r}, \quad r_x = \cos\theta, \quad r_y = \sin\theta$$

We get

$$\begin{aligned}
\varphi_x &= \varphi_r r_x + \varphi_\theta\theta_x = \varphi_r\cos\theta + \varphi_\theta\left(-\frac{\sin\theta}{r}\right)\\
\varphi_y &= \varphi_r r_y + \varphi_\theta\theta_y = \varphi_r\sin\theta + \varphi_\theta\left(\frac{\cos\theta}{r}\right)
\end{aligned} \tag{1}$$

Hence,

$$\nabla\varphi = \left\langle \varphi_r\cos\theta - \varphi_\theta\frac{\sin\theta}{r},\ \varphi_r\sin\theta + \varphi_\theta\frac{\cos\theta}{r}\right\rangle$$

We use the following parametrization for $\mathcal{C}(r)$:

$$\mathcal{C}(r) : \mathbf{c}(\theta) = \langle a + r\cos\theta,\ b + r\sin\theta\rangle, \quad 0 \le \theta \le 2\pi$$

The unit normal vector is

$$\mathbf{n} = \langle\cos\theta, \sin\theta\rangle.$$

We compute the dot product:

$$\begin{aligned}
\frac{\partial\varphi}{\partial\mathbf{n}} &= \nabla\varphi\cdot\mathbf{n} = \left\langle \varphi_r\cos\theta - \varphi_\theta\frac{\sin\theta}{r},\ \varphi_r\sin\theta + \varphi_\theta\frac{\cos\theta}{r}\right\rangle\cdot\langle\cos\theta, \sin\theta\rangle\\
&= \varphi_r\cos^2\theta - \varphi_\theta\frac{\sin\theta\cos\theta}{r} + \varphi_r\sin^2\theta + \varphi_\theta\frac{\cos\theta\sin\theta}{r} = \varphi_r\left(\cos^2\theta + \sin^2\theta\right) = \varphi_r
\end{aligned}$$

That is,

$$\frac{\partial\varphi}{\partial\mathbf{n}}(a + r\cos\theta, b + r\sin\theta) = \varphi_r(a + r\cos\theta, b + r\sin\theta) \tag{2}$$

(b) We compute the following derivative using the Chain Rule and (1):

$$\begin{aligned}
\frac{\partial}{\partial r}\varphi(a + r\cos\theta, b + r\sin\theta) &= \varphi_x\frac{\partial}{\partial r}(a + r\cos\theta) + \varphi_y\frac{\partial}{\partial r}(b + r\sin\theta)\\
&= \left(\varphi_r\cos\theta - \varphi_\theta\frac{\sin\theta}{r}\right)\cos\theta + \left(\varphi_r\sin\theta + \varphi_\theta\frac{\cos\theta}{r}\right)\sin\theta\\
&= \varphi_r\cos^2\theta - \varphi_\theta\frac{\sin\theta\cos\theta}{r} + \varphi_r\sin^2\theta + \varphi_\theta\frac{\cos\theta\sin\theta}{r}\\
&= \varphi_r(a + r\cos\theta, b + r\sin\theta)
\end{aligned} \tag{3}$$

We now differentiate $I_\varphi(r)$ under the integral sign, and use (3) and (2) to obtain

$$\begin{aligned}
\frac{d}{dr}I_\varphi(r) &= \frac{1}{2\pi}\int_0^{2\pi}\frac{\partial}{\partial r}\varphi(a + r\cos\theta, b + r\sin\theta)\,d\theta = \frac{1}{2\pi}\int_0^{2\pi}\varphi_r(a + r\cos\theta, b + r\sin\theta)\,d\theta\\
&= \frac{1}{2\pi}\int_0^{2\pi}\frac{\partial\varphi}{\partial\mathbf{n}}(a + r\cos\theta, b + r\sin\theta)\,d\theta
\end{aligned} \tag{4}$$

On the other hand, since $\mathbf{c}'(\theta) = \langle -r\sin\theta, r\cos\theta\rangle$, we have

$$\int_{\mathcal{C}(r)} \frac{\partial\varphi}{\partial\mathbf{n}}\,ds = \int_0^{2\pi} \frac{\partial\varphi}{\partial\mathbf{n}}(a + r\cos\theta, b + r\sin\theta)\,\|\mathbf{c}'(\theta)\|\,d\theta = \int_0^{2\pi} \frac{\partial\varphi}{\partial\mathbf{n}}(a + r\cos\theta, b + r\sin\theta) r\,d\theta$$

$$= r\int_0^{2\pi} \frac{\partial\varphi}{\partial\mathbf{n}}(a + r\cos\theta, b + r\sin\theta)\,d\theta \tag{5}$$

Combining (4) and (5) we get

$$\frac{d}{dr}I_\varphi(r) = \frac{1}{2\pi r}\int_{\mathcal{C}(r)} \frac{\partial\varphi}{\partial\mathbf{n}}\,ds.$$

(c) We combine the result of part (b) and Exercise 36 to conclude

$$\frac{d}{dr}I_\varphi(r) = \frac{1}{2\pi r}\int_{\mathcal{C}(r)} \frac{\partial\varphi}{\partial\mathbf{n}}\,ds = \frac{1}{2\pi r}\iint_{\mathcal{D}(r)} \Delta\varphi\,dA.$$

In Exercises 39–40, let $\mathcal{D}$ be the region bounded by a simple closed curve $\mathcal{C}$. A function $\varphi(x, y)$ on $\mathcal{D}$ (whose second-order partial derivatives exist and are continuous) is called **harmonic** *if $\Delta\varphi = 0$, where $\Delta\varphi$ is the Laplace operator defined in Eq. (9).*

39. Use the results of Exercises 37 and 38 to prove the **mean-value property** of harmonic functions: If φ is harmonic, then $I_\varphi(r) = \varphi(P)$ for all r.

SOLUTION In Exercise 37 we showed that

$$\frac{d}{dr}I_\varphi(r) = \frac{1}{2\pi r}\iint_{\mathcal{D}} \Delta\varphi\,dA$$

If φ is harmonic, $\Delta\varphi = 0$. Therefore the right-hand side of the equality is zero, and we get

$$\frac{d}{dr}I_\varphi(r) = 0$$

We conclude that $I_\varphi(r)$ is constant, that is, $I_\varphi(r)$ has the same value for all r. The constant value is determined by the limit $\lim_{r\to 0} I_\varphi(r) = \varphi(P)$ obtained in Exercise 38. That is, $I_\varphi(r) = \varphi(P)$ for all r.

18.2 Stokes' Theorem (ET Section 17.2)

Preliminary Questions

1. Indicate with an arrow the boundary orientation of the boundary curves of the surfaces in Figure 13, oriented by the outward-pointing normal vectors.

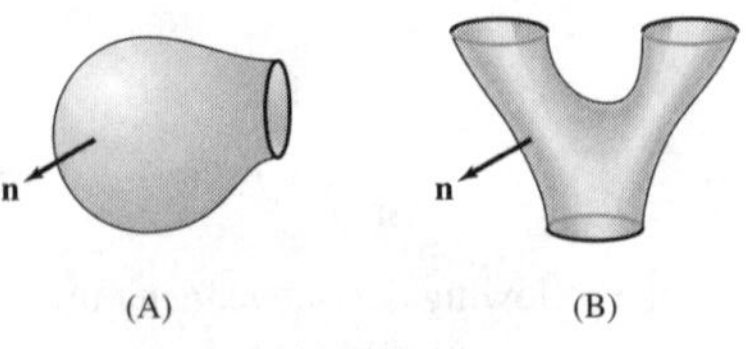

FIGURE 13

SOLUTION The indicated orientation is defined so that if the normal vector is moving along the boundary curve, the surface lies to the left. Since the surfaces are oriented by the outward-pointing normal vectors, the induced orientation is as shown in the figure:

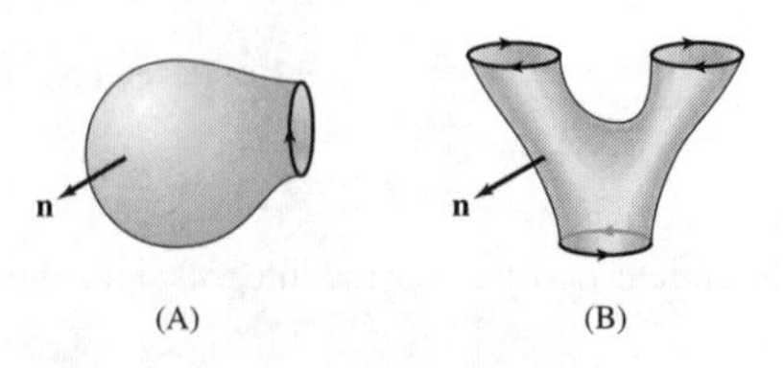

2. Let $\mathbf{F} = \text{curl}(\mathbf{A})$. Which of the following are related by Stokes' Theorem?

(a) The circulation of $\mathbf{A}$ and flux of $\mathbf{F}$.

(b) The circulation of $\mathbf{F}$ and flux of $\mathbf{A}$.

SOLUTION Stokes' Theorem states that the circulation of **A** is equal to the flux of **F**. The correct answer is (b).

3. What is the definition of a vector potential?

SOLUTION A vector field **A** such that $\mathbf{F} = \text{curl}(\mathbf{A})$ is a vector potential for **F**.

4. Which of the following statements is correct?

(a) The flux of curl(**F**) through every oriented surface is zero.

(b) The flux of curl(**F**) through every closed, oriented surface is zero.

SOLUTION Statement (b) is the correct statement. The flux of curl(**F**) through an oriented surface is not necessarily zero, unless the surface is closed.

5. Which condition on **F** guarantees that the flux through $\mathcal{S}_1$ is equal to the flux through $\mathcal{S}_2$ for any two oriented surfaces $\mathcal{S}_1$ and $\mathcal{S}_2$ with the same oriented boundary?

SOLUTION If **F** has a vector potential **A**, then by a corollary of Stokes' Theorem,

$$\iint_{\mathcal{S}} \mathbf{F} \cdot d\mathbf{s} = \int_{\mathcal{C}} \mathbf{A} \cdot d\mathbf{s}$$

Therefore, if two oriented surfaces $\mathcal{S}_1$ and $\mathcal{S}_2$ have the same oriented boundary curve, $\mathcal{C}$, then

$$\iint_{\mathcal{S}_1} \mathbf{F} \cdot d\mathbf{s} = \int_{\mathcal{C}} \mathbf{A} \cdot d\mathbf{s} \quad \text{and} \quad \iint_{\mathcal{S}_2} \mathbf{F} \cdot d\mathbf{s} = \int_{\mathcal{C}} \mathbf{A} \cdot d\mathbf{s}$$

Hence,

$$\iint_{\mathcal{S}_1} \mathbf{F} \cdot d\mathbf{s} = \iint_{\mathcal{S}_2} \mathbf{F} \cdot d\mathbf{s}$$

Exercises

In Exercises 1–4, calculate curl(**F**).

1. $\mathbf{F} = \langle z - y^2, x + z^3, y + x^2 \rangle$

SOLUTION We have

$$\text{curl}(\mathbf{F}) = \begin{vmatrix} \mathbf{i} & \mathbf{j} & \mathbf{k} \\ \frac{\partial}{\partial x} & \frac{\partial}{\partial y} & \frac{\partial}{\partial z} \\ z - y^2 & x + z^3 & y + x^2 \end{vmatrix} = (1 - 3z^2)\mathbf{i} - (2x - 1)\mathbf{j} + (1 + 2y)\mathbf{k} = \langle 1 - 3z^2, 1 - 2x, 1 + 2y \rangle$$

3. $\mathbf{F} = \langle e^y, \sin x, \cos x \rangle$

SOLUTION We have

$$\text{curl}(\mathbf{F}) = \begin{vmatrix} \mathbf{i} & \mathbf{j} & \mathbf{k} \\ \frac{\partial}{\partial x} & \frac{\partial}{\partial y} & \frac{\partial}{\partial z} \\ e^y & \sin x & \cos x \end{vmatrix} = 0\mathbf{i} - (-\sin x)\mathbf{j} + (\cos x - e^y)\mathbf{k} = \langle 0, \sin x, \cos x - e^y \rangle$$

In Exercises 5–8, verify Stokes' Theorem for the given vector field and surface, oriented with an upward-pointing normal.

5. $\mathbf{F} = \langle 2xy, x, y + z \rangle$, the surface $z = 1 - x^2 - y^2$ for $x^2 + y^2 \le 1$

SOLUTION We must show that

$$\int_{\mathcal{C}} \mathbf{F} \cdot d\mathbf{s} = \iint_{\mathcal{S}} \text{curl}(\mathbf{F}) \cdot d\mathbf{S}$$

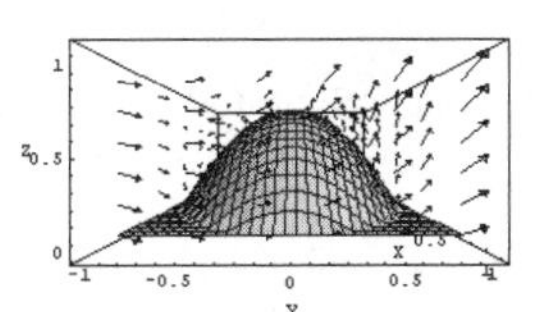

Step 1. Compute the line integral around the boundary curve. The boundary curve $\mathcal{C}$ is the unit circle oriented in the counterclockwise direction. We parametrize $\mathcal{C}$ by

$$\gamma(t) = (\cos t, \sin t, 0), \quad 0 \le t \le 2\pi$$

Then,

$$\mathbf{F}(\gamma(t)) = \langle 2\cos t \sin t, \cos t, \sin t\rangle$$
$$\gamma'(t) = \langle -\sin t, \cos t, 0\rangle$$
$$\mathbf{F}(\gamma(t)) \cdot \gamma'(t) = \langle 2\cos t \sin t, \cos t, \sin t\rangle \cdot \langle -\sin t, \cos t, 0\rangle = -2\cos t\ \sin^2 t + \cos^2 t$$

We obtain the following integral:

$$\int_{\mathcal{C}} \mathbf{F}\,ds = \int_0^{2\pi} \left(-2\cos t\ \sin^2 t + \cos^2 t\right) dt = -\frac{2\ \sin^3 t}{3} + \frac{t}{2} + \frac{\sin 2t}{4}\bigg|_0^{2\pi} = \pi \tag{1}$$

Step 2. Compute the flux of the curl through the surface. We parametrize the surface by

$$\Phi(\theta, t) = \left(t\cos\theta, t\sin\theta, 1 - t^2\right), \quad 0 \le t \le 1, \quad 0 \le \theta \le 2\pi$$

We compute the normal vector:

$$\mathbf{T}_\theta = \frac{\partial \Phi}{\partial \theta} = \langle -t\sin\theta, t\cos\theta, 0\rangle$$

$$\mathbf{T}_t = \frac{\partial \Phi}{\partial t} = \langle \cos\theta, \sin\theta, -2t\rangle$$

$$\mathbf{T}_\theta \times \mathbf{T}_t = \begin{vmatrix} \mathbf{i} & \mathbf{j} & \mathbf{k} \\ -t\sin\theta & t\cos\theta & 0 \\ \cos\theta & \sin\theta & -2t \end{vmatrix} = (-2t^2\cos\theta)\mathbf{i} - (2t^2\sin\theta)\mathbf{j} - t(\sin^2\theta + \cos^2\theta)\mathbf{k}$$
$$= (-2t^2\cos\theta)\mathbf{i} - (2t^2\sin\theta)\mathbf{j} - t\mathbf{k}$$

Since the normal is always supposed to be pointing upward, the z-coordinate of the normal vector must be positive. Therefore, the normal vector is

$$\mathbf{n} = \left\langle 2t^2\cos\theta, 2t^2\sin\theta, t\right\rangle$$

We compute the curl:

$$\text{curl}(\mathbf{F}) = \begin{vmatrix} \mathbf{i} & \mathbf{j} & \mathbf{k} \\ \frac{\partial}{\partial x} & \frac{\partial}{\partial y} & \frac{\partial}{\partial z} \\ 2xy & x & y+z \end{vmatrix} = \mathbf{i} + (1-2x)\mathbf{k} = \langle 1, 0, 1-2x\rangle$$

We compute the curl in terms of the parameters:

$$\text{curl}(\mathbf{F}) = \langle 1, 0, 1 - 2t\cos\theta\rangle$$

Hence,

$$\text{curl}(\mathbf{F}) \cdot \mathbf{n} = \langle 1, 0, 1 - 2t\cos\theta\rangle \cdot \left\langle 2t^2\cos\theta, 2t^2\sin\theta, t\right\rangle = 2t^2\cos\theta + t - 2t^2\cos\theta = t$$

The surface integral is thus

$$\iint_S \text{curl}(\mathbf{F}) \cdot d\mathbf{S} = \int_0^{2\pi}\int_0^1 t\,dt\,d\theta = 2\pi \int_0^1 t\,dt = 2\pi \cdot \frac{t^2}{2}\bigg|_0^1 = \pi \tag{2}$$

The values of the integrals in (1) and (2) are equal, as stated in Stokes' Theorem.

7. $\mathbf{F} = \left\langle e^{y-z}, 0, 0\right\rangle$, the square with vertices $(1, 0, 1)$, $(1, 1, 1)$, $(0, 1, 1)$, and $(0, 0, 1)$

SOLUTION

Step 1. Compute the integral around the boundary curve. The boundary consists of four segments $\mathcal{C}_1$, $\mathcal{C}_2$, $\mathcal{C}_3$, and $\mathcal{C}_4$ shown in the figure:

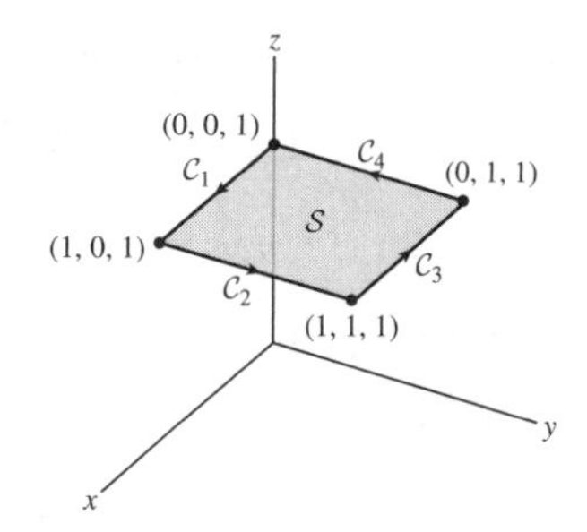

We parametrize the segments by

$$C_1 : \gamma_1(t) = (t, 0, 1), \quad 0 \le t \le 1$$
$$C_2 : \gamma_2(t) = (1, t, 1), \quad 0 \le t \le 1$$
$$C_3 : \gamma_3(t) = (1-t, 1, 1), \quad 0 \le t \le 1$$
$$C_4 : \gamma_4(t) = (0, 1-t, 1), \quad 0 \le t \le 1$$

We compute the following values:

$$\mathbf{F}(\gamma_1(t)) = \langle e^{y-z}, 0, 0\rangle = \left\langle e^{-1}, 0, 0\right\rangle$$
$$\mathbf{F}(\gamma_2(t)) = \langle e^{y-z}, 0, 0\rangle = \left\langle e^{t-1}, 0, 0\right\rangle$$
$$\mathbf{F}(\gamma_3(t)) = \langle e^{y-z}, 0, 0\rangle = \langle 1, 0, 0\rangle$$
$$\mathbf{F}(\gamma_4(t)) = \langle e^{y-z}, 0, 0\rangle = \left\langle e^{-t-1}, 0, 0\right\rangle$$

Hence,

$$\mathbf{F}(\gamma_1(t)) \cdot \gamma_1'(t) = \left\langle e^{-1}, 0, 0\right\rangle \cdot \langle 1, 0, 0\rangle = e^{-1}$$
$$\mathbf{F}(\gamma_2(t)) \cdot \gamma_2'(t) = \left\langle e^{t-1}, 0, 0\right\rangle \cdot \langle 0, 1, 0\rangle = 0$$
$$\mathbf{F}(\gamma_3(t)) \cdot \gamma_3'(t) = \langle 1, 0, 0\rangle \cdot \langle -1, 0, 0\rangle = -1$$
$$\mathbf{F}(\gamma_4(t)) \cdot \gamma_4'(t) = \left\langle e^{-t-1}, 0, 0\right\rangle \cdot \langle 0, -1, 0\rangle = 0$$

We obtain the following integral:

$$\int_{C} \mathbf{F} \cdot d\mathbf{s} = \sum_{i=1}^{4} \int_{C_i} \mathbf{F} \cdot d\mathbf{s} = \int_0^1 e^{-1}\,dt + 0 + \int_0^1 (-1)\,dt + 0 = e^{-1} - 1$$

Step 2. Compute the curl.

$$\text{curl}(\mathbf{F}) = \begin{vmatrix} \mathbf{i} & \mathbf{j} & \mathbf{k} \\ \frac{\partial}{\partial x} & \frac{\partial}{\partial y} & \frac{\partial}{\partial z} \\ e^{y-z} & 0 & 0 \end{vmatrix} = -e^{y-z}\ \mathbf{j}\ - e^{y-z}\ \mathbf{k}\ = \left\langle 0, -e^{y-z}, -e^{y-z}\right\rangle$$

Step 3. Compute the flux of the curl through the surface. We parametrize the surface by

$$\Phi(x, y) = (x, y, 1), \quad 0 \le x,\ y \le 1$$

The upward pointing normal is $\mathbf{n} = \langle 0, 0, 1\rangle$. We express curl($\mathbf{F}$) in terms of the parameters x and y:

$$\text{curl}(\mathbf{F})\,(\Phi(x, y)) = \left\langle 0, -e^{y-1}, -e^{y-1}\right\rangle$$

Hence,

$$\text{curl}(\mathbf{F}) \cdot \mathbf{n} = \left\langle 0, -e^{y-1}, -e^{y-1}\right\rangle \cdot \langle 0, 0, 1\rangle = -e^{y-1}$$

The surface integral is thus

$$\iint_{S} \text{curl}(\mathbf{F}) \cdot d\mathbf{S} = \iint_{\mathcal{D}} -e^{y-1}\,dA = \int_0^1 \int_0^1 -e^{y-1}\,dy\,dx = \int_0^1 -e^{y-1}\,dy = -e^{y-1}\Big|_0^1$$

$$= -1 + e^{-1} = e^{-1} - 1 \tag{1}$$

We see that the integrals in (1) and (2) are equal.

In Exercises 9–10, use Stokes' Theorem to compute the flux of curl(**F**) *through the given surface.*

9. $\mathbf{F} = \langle z, y, x \rangle$, the hemisphere $x^2 + y^2 + z^2 = 1, x \geq 0$

SOLUTION By Stokes' Theorem,

$$\int_C \mathbf{F} \cdot d\mathbf{s} = \iint_S \text{curl}(\mathbf{F}) \cdot d\mathbf{S}$$

We compute the line integral. The boundary curve, which is the unit circle in the (y, z) plane, is parametrized by

$$\gamma(t) = (0, \cos t, \sin t), \quad 0 \leq t \leq 2\pi$$

We compute the following values:

$$\mathbf{F}(\gamma(t)) = \langle z, y, x \rangle = \langle \sin t, \cos t, 0 \rangle$$
$$\gamma'(t) = \langle 0, -\sin t, \cos t \rangle$$
$$\mathbf{F}(\gamma(t)) \cdot \gamma'(t) = \langle \sin t, \cos t, 0 \rangle \cdot \langle 0, -\sin t, \cos t \rangle = -\sin t \cos t = -\frac{1}{2} \sin 2t$$

Hence,

$$\int_C \mathbf{F} \cdot d\mathbf{s} = \int_0^{2\pi} \mathbf{F}(\gamma(t)) \cdot \gamma'(t)\, dt = \int_0^{2\pi} -\frac{1}{2} \sin 2t\, dt = \left. \frac{\cos 2t}{4} \right|_0^{2\pi} = 0$$

Then also

$$\iint_S \text{curl}(\mathbf{F}) \cdot d\mathbf{S} = 0$$

11. Let $\mathcal{S}$ be the surface of the cylinder (not including the top and bottom) of radius 2 for $1 \leq z \leq 6$, oriented with outward-pointing normal (Figure 14).

(a) Indicate with an arrow the orientation of $\partial\mathcal{S}$ (the top and bottom circles).

(b) Verify Stokes' Theorem for $\mathcal{S}$ and $\mathbf{F} = \langle yz^2, 0, 0 \rangle$.

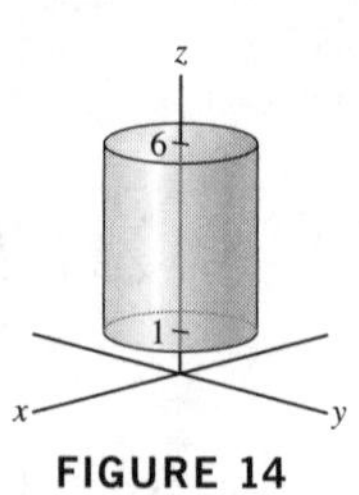

FIGURE 14

SOLUTION

(a) The induced orientation is defined so that as the normal vector travels along the boundary curve, the surface lies to its left. Therefore, the boundary circles on top and bottom have opposite orientations, which are shown in the figure.

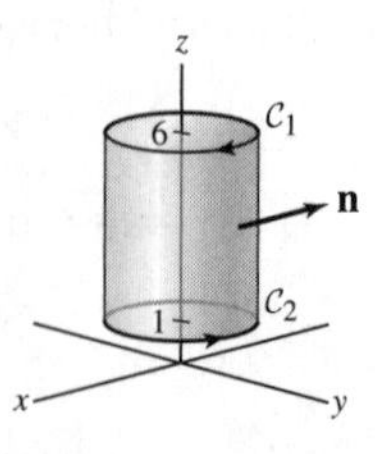

(b) We verify Stokes' Theorem for $\mathcal{S}$ and $\mathbf{F} = \langle yz^2, 0, 0 \rangle$.

Step 1. Compute the integral around the boundary circles. We use the following parametrizations:

$$\mathcal{C}_1 : \gamma_1(t) = (2\cos t, 2\sin t, 6), \quad t \text{ from } 2\pi \text{ to } 0$$
$$\mathcal{C}_2 : \gamma_2(t) = (2\cos t, 2\sin t, 1), \quad t \text{ from } 0 \text{ to } 2\pi$$

We compute the following values:

$$\mathbf{F}(\gamma_1(t)) = \left\langle yz^2, 0, 0\right\rangle = \langle 72\sin t, 0, 0\rangle,$$

$$\gamma_1'(t) = \langle -2\sin t, 2\cos t, 0\rangle$$

$$\mathbf{F}(\gamma_1(t)) \cdot \gamma_1'(t) = \langle 72\sin t, 0, 0\rangle \cdot \langle -2\sin t, 2\cos t, 0\rangle = -144\sin^2 t$$

$$\mathbf{F}(\gamma_2(t)) = \left\langle yz^2, 0, 0\right\rangle = \langle 2\sin t, 0, 0\rangle,$$

$$\gamma_2'(t) = \langle -2\sin t, 2\cos t, 0\rangle$$

$$\mathbf{F}(\gamma_2(t)) \cdot \gamma_2'(t) = \langle 2\sin t, 0, 0\rangle \cdot \langle -2\sin t, 2\cos t, 0\rangle = -4\sin^2 t$$

The line integral is thus

$$\int_C \mathbf{F}\cdot d\mathbf{s} = \int_{C_1} \mathbf{F}\cdot d\mathbf{s} + \int_{C_2} \mathbf{F}\cdot d\mathbf{s} = \int_{2\pi}^{0} (-144\sin^2 t)\, dt + \int_0^{2\pi} (-4\sin^2 t)\, dt$$

$$= \int_0^{2\pi} 140\sin^2 t\, dt = 140\int_0^{2\pi} \frac{1-\cos 2t}{2}\, dt = 70\cdot 2\pi - \left.\frac{70\sin 2t}{2}\right|_0^{2\pi} = 140\pi$$

Step 2. Compute the curl

$$\operatorname{curl}(\mathbf{F}) = \begin{vmatrix} \mathbf{i} & \mathbf{j} & \mathbf{k} \\ \frac{\partial}{\partial x} & \frac{\partial}{\partial y} & \frac{\partial}{\partial z} \\ yz^2 & 0 & 0 \end{vmatrix} = (2yz)\mathbf{j} - z^2\mathbf{k} = \left\langle 0, 2yz, -z^2\right\rangle$$

Step 3. Compute the flux of the curl through the surface. We parametrize $\mathcal{S}$ by

$$\Phi(\theta, z) = (2\cos\theta, 2\sin\theta, z), \quad 0 \le \theta \le 2\pi, \quad 1 \le z \le 6$$

In Example 2 of Chapter 17.4 it is shown that the outward pointing normal is

$$\mathbf{n} = \langle 2\cos\theta, 2\sin\theta, 0\rangle$$

We compute the dot product:

$$\operatorname{curl}(\mathbf{F})(\Phi(\theta, z)) \cdot \mathbf{n} = \left\langle 0, 4z\sin\theta, -z^2\right\rangle \cdot \langle 2\cos\theta, 2\sin\theta, 0\rangle = 8z\sin^2\theta$$

We obtain the following integral (and use the integral we computed before):

$$\iint_{\mathcal{S}} \operatorname{curl}(\mathbf{F}) \cdot d\mathbf{S} = \int_1^6 \int_0^{2\pi} 8z\sin^2\theta\, d\theta\, dz = \left(\int_1^6 8z\, dz\right)\left(\int_0^{2\pi} \sin^2\theta\, d\theta\right) = \left.4z^2\right|_1^6 \cdot \pi = 140\pi$$

The line integral and the flux have the same value. This verifies Stokes' Theorem.

13. Let I be the flux of $\mathbf{F} = \left\langle e^y, 2xe^{x^2}, z^2\right\rangle$ through the upper hemisphere $\mathcal{S}$ of the unit sphere.

(a) Let $\mathbf{G} = \left\langle e^y, 2xe^{x^2}, 0\right\rangle$. Find a vector field $\mathbf{A}$ such that $\operatorname{curl}(\mathbf{A}) = \mathbf{G}$.

(b) Use Stokes' Theorem to show that the flux of $\mathbf{G}$ through $\mathcal{S}$ is zero. *Hint:* Calculate the circulation of $\mathbf{A}$ around $\partial\mathcal{S}$.

(c) Calculate I. *Hint:* Use (b) to show that I is equal to the flux of $\left\langle 0, 0, z^2\right\rangle$ through $\mathcal{S}$.

SOLUTION

(a) We search for a vector field $\mathbf{A}$ so that $\mathbf{G} = \operatorname{curl}(\mathbf{A})$. That is,

$$\left\langle \frac{\partial \mathbf{A}_3}{\partial y} - \frac{\partial \mathbf{A}_2}{\partial z}, \frac{\partial \mathbf{A}_1}{\partial z} - \frac{\partial \mathbf{A}_3}{\partial x}, \frac{\partial \mathbf{A}_2}{\partial x} - \frac{\partial \mathbf{A}_1}{\partial y}\right\rangle = \left\langle e^y, 2xe^{x^2}, 0\right\rangle$$

We note that the third coordinate of this curl vector must be zero; this can be satisfied if $\mathbf{A}_1 = 0$ and $\mathbf{A}_2 = 0$. With this in mind, we let $\mathbf{A} = \left\langle 0, 0, e^y - e^{x^2}\right\rangle$. The vector field $\mathbf{A} = \left\langle 0, 0, e^y - e^{x^2}\right\rangle$ satisfies this equality. Indeed,

$$\frac{\partial \mathbf{A}_3}{\partial y} - \frac{\partial \mathbf{A}_2}{\partial z} = e^y, \quad \frac{\partial \mathbf{A}_1}{\partial z} - \frac{\partial \mathbf{A}_3}{\partial x} = 2xe^{x^2}, \quad \frac{\partial \mathbf{A}_2}{\partial x} - \frac{\partial \mathbf{A}_1}{\partial y} = 0$$

(b) We found that $\mathbf{G} = \text{curl}(\mathbf{A})$, where $\mathbf{A} = \left\langle 0, 0, e^y - e^{x^2} \right\rangle$. We compute the flux of $\mathbf{G}$ through $\mathcal{S}$. By Stokes' Theorem,

$$\iint_{\mathcal{S}} \mathbf{G} \cdot d\mathbf{S} = \iint_{\mathcal{S}} \text{curl}(\mathbf{A}) \cdot d\mathbf{S} = \int_{\mathcal{C}} \mathbf{A} \cdot d\mathbf{s}$$

The boundary $\mathcal{C}$ is the circle $x^2 + y^2 = 1$, parametrized by

$$\gamma(t) = (\cos t, \sin t, 0), \quad 0 \le t \le 2\pi$$

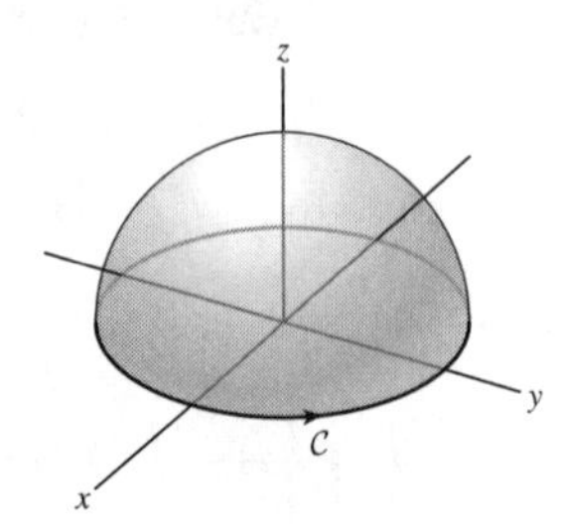

We compute the following values:

$$\mathbf{A}\left(\gamma(t)\right) = \left\langle 0, 0, e^y - e^{x^2} \right\rangle = \left\langle 0, 0, e^{\sin t} - e^{\cos^2 t} \right\rangle$$

$$\gamma'(t) = \langle -\sin t, \cos t, 0 \rangle$$

$$\mathbf{A}\left(\gamma(t)\right) \cdot \gamma'(t) = \left\langle 0, 0, e^{\sin t} - e^{\cos^2 t} \right\rangle \cdot \langle -\sin t, \cos t, 0 \rangle = 0$$

Therefore,

$$\int_{\mathcal{C}} \mathbf{A} \cdot d\mathbf{s} = \int_0^{2\pi} 0 \, dt = 0$$

(c) We rewrite the vector field $\mathbf{F} = \left\langle e^y, 2xe^{x^2}, z^2 \right\rangle$ as

$$\mathbf{F} = \left\langle e^y, 2xe^{x^2}, z^2 \right\rangle = \left\langle e^y, 2xe^{x^2}, 0 \right\rangle + \left\langle 0, 0, z^2 \right\rangle = \text{curl}(\mathbf{A}) + \left\langle 0, 0, z^2 \right\rangle$$

Therefore,

$$\iint_{\mathcal{S}} \mathbf{F} \cdot d\mathbf{S} = \iint_{\mathcal{S}} \text{curl}(\mathbf{A}) \cdot d\mathbf{S} + \iint_{\mathcal{S}} \left\langle 0, 0, z^2 \right\rangle \cdot d\mathbf{S} \tag{1}$$

In part (b) we showed that the first integral on the right-hand side is zero. Therefore,

$$\iint_{\mathcal{S}} \mathbf{F} \cdot d\mathbf{S} = \iint_{\mathcal{S}} \left\langle 0, 0, z^2 \right\rangle \cdot d\mathbf{S} \tag{2}$$

The upper hemisphere is parametrized by

$$\Phi(\theta, \phi) = (\cos\theta \sin\phi, \sin\theta \sin\phi, \cos\phi), \quad 0 \le \theta \le 2\pi, \quad 0 \le \phi \le \frac{\pi}{2}.$$

with the outward pointing normal

$$\mathbf{n} = \sin\phi \left\langle \cos\theta \sin\phi, \sin\theta \sin\phi, \cos\phi \right\rangle$$

See Example 4, Section 17.4. We have

$$\left\langle 0, 0, \cos^2\phi \right\rangle \cdot \mathbf{n} = \sin\phi \cos^3\phi$$

Therefore,

$$\iint_{\mathcal{S}} \left\langle 0, 0, z^2 \right\rangle \cdot d\mathbf{S} = \int_0^{2\pi} \int_0^{\pi/2} \sin\phi \cos^3\phi \, d\phi \, d\theta = 2\pi \int_0^{\pi/2} \sin\phi \cos^3\phi \, d\phi$$

$$= 2\pi \frac{-\cos^4\phi}{4} \Big|_0^{\pi/2} = -\frac{\pi}{2}(0 - 1) = \frac{\pi}{2}$$

Combining with (2) we obtain the solution

$$\iint_{\mathcal{S}} \mathbf{F} \cdot d\mathbf{S} = \frac{\pi}{2}.$$

15. Let **A** be the vector potential and **B** the magnetic field of the infinite solenoid of radius R in Example 5. Use Stokes' Theorem to compute:

(a) The flux of **B** through a circle in the xy-plane of radius $r < R$

(b) The circulation of **A** around the boundary $\mathcal{C}$ of a surface lying outside the solenoid

SOLUTION

(a) In Example 5 it is shown that $\mathbf{B} = \text{curl}(\mathbf{A})$, where

$$\mathbf{A} = \begin{cases} \frac{1}{2}R^2 B\left\langle -\frac{y}{r^2}, \frac{x}{r^2}, 0\right\rangle & \text{if} \quad r > R \\ \frac{1}{2}B\langle -y, x, 0\rangle & \text{if} \quad r < R \end{cases} \tag{1}$$

Therefore, using Stokes' Theorem, we have ($\mathcal{S}$ is the disk of radius r in the xy-plane)

$$\iint_{\mathcal{S}} \mathbf{B} \cdot d\mathbf{S} = \iint_{\mathcal{S}} \text{curl}(\mathbf{A}) \cdot d\mathbf{S} = \int_{\partial\mathcal{S}} \mathbf{A} \cdot d\mathbf{s} \tag{2}$$

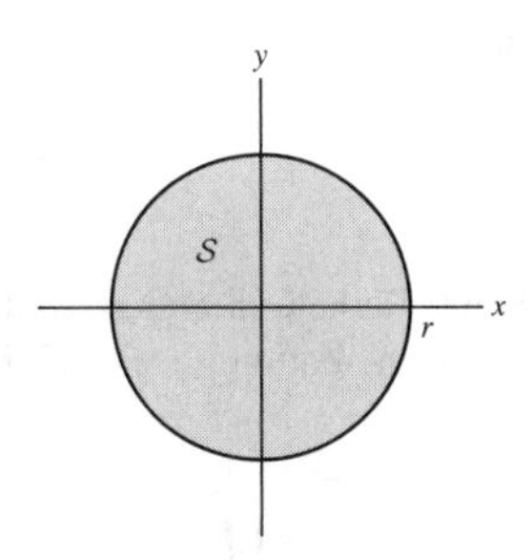

We parametrize the circle $\mathcal{C} = \partial\mathcal{S}$ by $\mathbf{c}(t) = \langle r\cos t, r\sin t, 0\rangle$, $0 \le t \le 2\pi$. Then

$$\mathbf{c}'(t) = \langle -r\sin t, r\cos t, 0\rangle$$

By (1) for $r < R$,

$$\mathbf{A}(\mathbf{c}(t)) = \frac{1}{2}B\langle -r\sin t, r\cos t, 0\rangle$$

Hence,

$$\mathbf{A}(\mathbf{c}(t)) \cdot \mathbf{c}'(t) = \frac{1}{2}B\langle -r\sin t, r\cos t, 0\rangle \cdot \langle -r\sin t, r\cos t, 0\rangle = \frac{1}{2}B\left(r^2\sin^2 t + r^2\cos^2 t\right) = \frac{1}{2}r^2 B$$

Now, by (2) we get

$$\iint_{\mathcal{S}} \mathbf{B} \cdot d\mathbf{S} = \int_{\partial\mathcal{S}} \mathbf{A} \cdot d\mathbf{S} = \int_0^{2\pi} \frac{1}{2}r^2\mathbf{B}\, dt = \frac{1}{2}r^2\mathbf{B}\int_0^{2\pi} dt = r^2\mathbf{B}\pi$$

(b) Outside the solenoid **B** is the zero field, hence $\mathbf{B} = \mathbf{0}$ on every domain lying outside the solenoid. Therefore, Stokes' Theorem implies that

$$\int_{\partial\mathcal{S}} \mathbf{A} \cdot d\mathbf{S} = \iint_{\mathcal{S}} \text{curl}(\mathbf{A}) \cdot d\mathbf{S} = \iint_{\mathcal{S}} \mathbf{B} \cdot d\mathbf{S} = \iint_{\mathcal{S}} \mathbf{0} \cdot d\mathbf{S} = 0.$$

17. A uniform magnetic field **B** has constant strength b teslas in the z-direction [i.e., $\mathbf{B} = \langle 0, 0, b\rangle$].

(a) Verify that $\mathbf{A} = \frac{1}{2}\mathbf{B} \times \mathbf{r}$ is a vector potential for **B**, where $\mathbf{r} = \langle x, y, 0\rangle$.

(b) Calculate the flux of **B** through the rectangle with vertices A, B, C, and D in Figure 17.

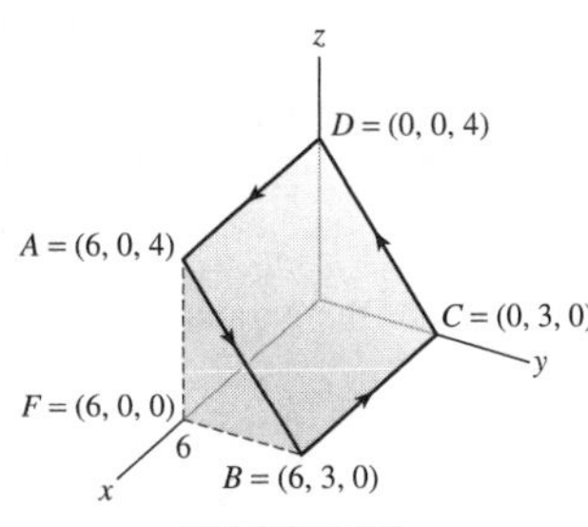

FIGURE 17

SOLUTION

(a) We compute the vector $\mathbf{A} = \frac{1}{2}\mathbf{B} \times \mathbf{r}$. Since $\mathbf{B} = b\mathbf{k}$ and $\mathbf{r} = x\mathbf{i} + y\mathbf{j}$, we have

$$\mathbf{A} = \frac{1}{2}\mathbf{B} \times \mathbf{r} = \frac{1}{2}b\mathbf{k} \times (x\mathbf{i} + y\mathbf{j}) = \frac{1}{2}b(x\mathbf{k} \times \mathbf{i} + y\mathbf{k} \times \mathbf{j}) = \frac{1}{2}b(x\mathbf{j} - y\mathbf{i}) = \left\langle -\frac{by}{2}, \frac{bx}{2}, 0 \right\rangle$$

We now show that $\text{curl}(\mathbf{A}) = \mathbf{B}$. We compute the curl of $\mathbf{A}$:

$$\text{curl}\,(\mathbf{A}) = \begin{vmatrix} \mathbf{i} & \mathbf{j} & \mathbf{k} \\ \frac{\partial}{\partial x} & \frac{\partial}{\partial y} & \frac{\partial}{\partial z} \\ -\frac{by}{2} & \frac{bx}{2} & 0 \end{vmatrix} = \left\langle 0, 0, \frac{b}{2} + \frac{b}{2} \right\rangle = \langle 0, 0, b \rangle = \mathbf{B}$$

Therefore, $\mathbf{A}$ is a vector potential for $\mathbf{B}$.

(b) Let $\mathcal{S}$ be the rectangle $\square ABCD$ and let $\mathcal{C}$ be the boundary of $\mathcal{S}$. Since $\mathbf{B} = \text{Curl}(\mathbf{A})$, we see that $\mathbf{B}$ has a vector potential. It follows, as explained in this section, that the flux of $\mathbf{B}$ through rectangle $\mathcal{S}$ is equal to the flux of $\mathbf{B}$ through any surface with the same boundary $\mathcal{C}$. Let $\mathcal{S}'$ be the wedge-shaped box with four sides and open top. Since the boundary of $\mathcal{S}'$ is also $\mathcal{C}$, we have

$$\iint_{\mathcal{S}} \mathbf{B} \cdot d\mathbf{S} = \iint_{\mathcal{S}'} \mathbf{B} \cdot d\mathbf{S}$$

The vector field $\mathbf{B}$ points in the $\mathbf{k}$ direction, so it has zero flux through the three vertical sides of $\mathcal{S}'$. On the other hand, the unit normal vector to the bottom face of $\mathcal{S}'$ is $\mathbf{k}$, so the normal component of $\mathbf{B}$ along the bottom face is equal to b. We obtain

$$\iint_{\mathcal{S}'} \mathbf{B} \cdot d\mathbf{S} = \iint_{\text{Bottom Face of } \mathcal{S}'} b\, dA$$

$$= b(\text{Area of Bottom Face of } \mathcal{S}') = 18b$$

19. Let $\mathbf{F} = \langle -z^2, 2zx, 4y - x^2 \rangle$ and let $\mathcal{C}$ be a simple closed curve in the plane $x + y + z = 4$ that encloses a region of area 16 (Figure 18). Calculate $\oint_{\mathcal{C}} \mathbf{F} \cdot d\mathbf{s}$, where $\mathcal{C}$ is oriented in the counterclockwise direction (when viewed from above the plane).

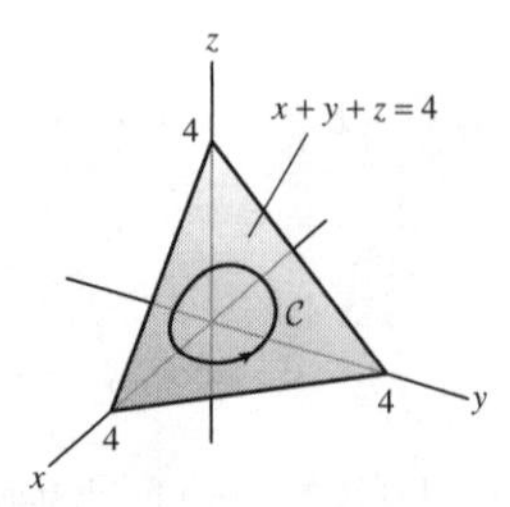

FIGURE 18

SOLUTION We denote by $\mathcal{S}$ the region enclosed by $\mathcal{C}$. Then by Stokes' Theorem,

$$\int_{\mathcal{C}} \mathbf{F} \cdot d\mathbf{s} = \iint_{\mathcal{S}} \text{curl}(\mathbf{F}) \cdot d\mathbf{s} \tag{1}$$

We compute the curl of $\mathbf{F} = \langle -z^2, 2zx, 4y - x^2 \rangle$:

$$\text{curl}(\mathbf{F}) = \begin{vmatrix} \mathbf{i} & \mathbf{j} & \mathbf{k} \\ \frac{\partial}{\partial x} & \frac{\partial}{\partial y} & \frac{\partial}{\partial z} \\ -z^2 & 2zx & 4y - x^2 \end{vmatrix} = \langle 4 - 2x, 2x - 2z, 2z \rangle$$

The plane $x + y + z = 4$ has the parametrization

$$\Phi(x, y) = \langle x, y, 4 - x - y \rangle$$

Hence,

$$\frac{\partial \Phi}{\partial x} \times \frac{\partial \Phi}{\partial y} = \langle 1, 0, -1 \rangle \times \langle 0, 1, -1 \rangle = (\mathbf{i} - \mathbf{k})(\mathbf{j} - \mathbf{k}) = \mathbf{k} + \mathbf{j} + \mathbf{i} = \langle 1, 1, 1 \rangle$$

The normal determined by the induced orientation is

$$\mathbf{n} = \langle 1, 1, 1 \rangle$$

Let $\mathcal{D}$ be the parameter domain in the parametrization $\Phi(x, y) = (x, y, 4 - x - y)$ of $\mathcal{S}$; that is, $\mathcal{D}$ will be the base triangle in the xy plane that lies underneath the pyramid in the picture. To compute the surface integral in (1) we compute the values

$$\text{curl}(\mathbf{F})\,(\Phi(x, y)) = \langle 4 - 2x, 2x - 2(4 - x - y), 2(4 - x - y) \rangle = \langle 4 - 2x, -8 + 4x + 2y, 8 - 2x - 2y \rangle$$

$$\text{curl}(\mathbf{F}) \cdot \mathbf{n} = \langle 4 - 2x, -8 + 4x + 2y, 8 - 2x - 2y \rangle \cdot \langle 1, 1, 1 \rangle = 4 - 2x - 8 + 4x + 2y + 8 - 2x - 2y = 4$$

Therefore, using (1) and (2) we obtain

$$\int_{\mathcal{C}} \mathbf{F} \cdot d\mathbf{s} = \iint_{\mathcal{S}} \text{curl}(\mathbf{F}) \cdot d\mathbf{S} = \iint_{\mathcal{D}} 4\, dA = 4\text{Area}(\mathcal{D}) = 4 \cdot \frac{16}{2} = 32$$

21. Let $\mathcal{C}$ be the triangular boundary of the portion of the plane $\frac{x}{a} + \frac{y}{b} + \frac{z}{c} = 1$ lying in the octant $x, y, z \geq 0$. Use Stokes' Theorem to find positive constants a, b, c such that the line integral of $\mathbf{F} = \langle y^2, 2z + x, 2y^2 \rangle$ around $\mathcal{C}$ is zero. *Hint:* Choose constants so that curl($\mathbf{F}$) is orthogonal to the normal vector.

SOLUTION

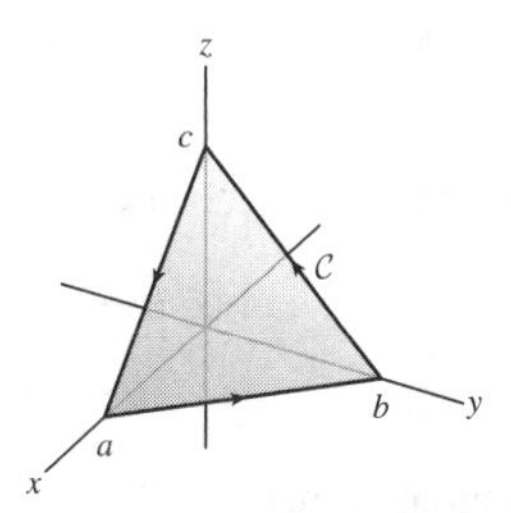

We compute the curl:

$$\text{curl}(\mathbf{F}) = \begin{vmatrix} \mathbf{i} & \mathbf{j} & \mathbf{k} \\ \frac{\partial}{\partial x} & \frac{\partial}{\partial y} & \frac{\partial}{\partial z} \\ y^2 & 2z + x & 2y^2 \end{vmatrix} = (4y - 2)\mathbf{i} + (1 - 2y)\mathbf{k} = \langle 4y - 2, 0, 1 - 2y \rangle$$

The outward-pointing normal to the plane is $\mathbf{n} = \left\langle \frac{1}{a}, \frac{1}{b}, \frac{1}{c} \right\rangle$. Hence,

$$\text{curl}(\mathbf{F}) \cdot \mathbf{n} = \langle 4y - 2, 0, 1 - 2y \rangle \cdot \left\langle \frac{1}{a}, \frac{1}{b}, \frac{1}{c} \right\rangle = \frac{4y - 2}{a} + \frac{1 - 2y}{c}$$

Note that if we choose $a = 2$ and $c = 1$, then curl($\mathbf{F}$) $\cdot\, \mathbf{n} = 0$ (here, b can be any positive number). Thus, using Stokes' Theorem, we have

$$\int_{\mathcal{C}} \mathbf{F} \cdot d\mathbf{s} = \iint_{\mathcal{D}} \text{curl}(\mathbf{F}) \cdot \mathbf{n}\, dA = 0$$

23. You know two things about a vector field $\mathbf{F}$:

(a) $\mathbf{F}$ has a vector potential $\mathbf{A}$ (but $\mathbf{A}$ is unknown).

(b) $\mathbf{F}(x, y, 0) = \langle 0, 0, 1 \rangle$ for all (x, y).

Determine the flux of $\mathbf{F}$ through the surface $\mathcal{S}$ in Figure 20.

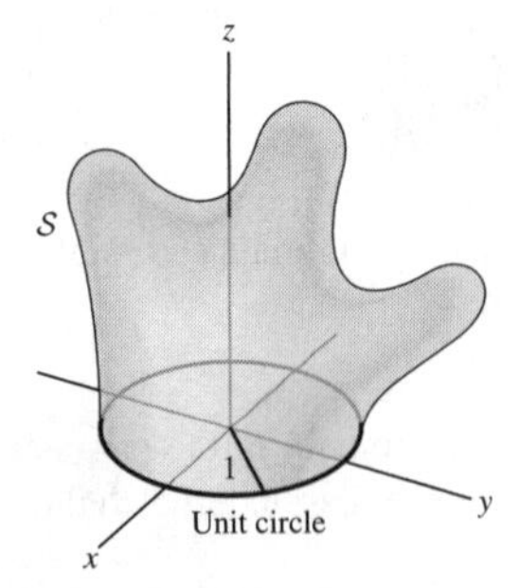

FIGURE 20

SOLUTION Since **F** has a vector potential—that is, **F** is the curl of a vector field—the flux of **F** through a surface depends only on the boundary curve $\mathcal{C}$. Now, the surface $\mathcal{S}$ and the unit disc $\mathcal{S}_1$ in the xy-plane share the same boundary $\mathcal{C}$. Therefore,

$$\iint_{\mathcal{S}} \mathbf{F} \cdot d\mathbf{S} = \iint_{\mathcal{S}_1} \mathbf{F} \cdot d\mathbf{S} \tag{1}$$

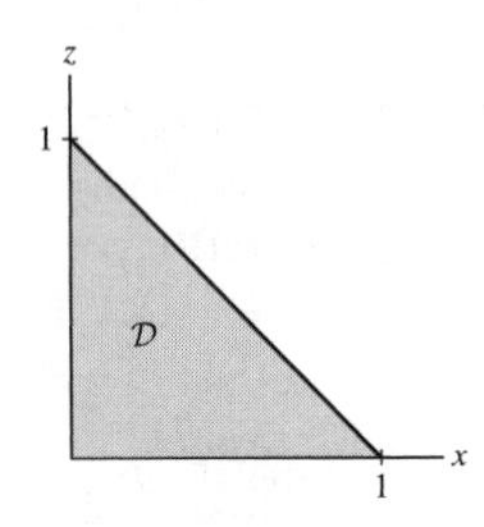

We compute the flux of **F** through $\mathcal{S}_1$, using the parametrization

$$\mathcal{S}_1 : \Phi(r, \theta) = (r\cos\theta, r\sin\theta, 0), \quad 0 \le r \le 1, \quad 0 \le \theta \le 2\pi$$

$$\mathbf{n} = \langle 0, 0, 1 \rangle$$

By the given information, we have

$$\mathbf{F}(\Phi(r, \theta)) = \mathbf{F}(r\cos\theta, r\sin\theta, 0) = \langle 0, 0, 1 \rangle$$

Hence,

$$\mathbf{F}(\Phi(r, \theta)) \cdot \mathbf{n} = \langle 0, 0, 1 \rangle \cdot \langle 0, 0, 1 \rangle = 1$$

We obtain the following integral:

$$\iint_{\mathcal{S}_1} \mathbf{F} \cdot d\mathbf{S} = \int_0^{2\pi} \int_0^1 \mathbf{F}(\Phi(r, \theta)) \cdot \mathbf{n}\, dr\, d\theta = \int_0^{2\pi} \int_0^1 1\, dr\, d\theta = 2\pi$$

Combining with (1) we obtain

$$\iint_{\mathcal{S}} \mathbf{F} \cdot d\mathbf{S} = 2\pi$$

25. Use Eq. (8) to prove that if **a** is a constant vector, then $\text{curl}(\varphi\mathbf{a}) = \nabla\varphi \times \mathbf{a}$.

SOLUTION By Eq. (8) we have

$$\text{curl}(\varphi\mathbf{a}) = \varphi\, \text{curl}(\mathbf{a}) + \nabla\varphi \times \mathbf{a}$$

Since **a** is a constant vector, all the partial derivatives of the components of **a** are zero, hence the curl of **a** is the zero vector:

$$\text{curl}(\mathbf{a}) = \mathbf{0}$$

Thus we obtain

$$\text{curl}(\varphi\mathbf{a}) = \nabla\varphi \times \mathbf{a}$$

27. Verify the identity

$$\boxed{\text{curl}(\nabla(\varphi)) = \mathbf{0}}$$

7

SOLUTION We have

$$\nabla\varphi = \left\langle \frac{\partial\varphi}{\partial x}, \frac{\partial\varphi}{\partial y}, \frac{\partial\varphi}{\partial z} \right\rangle$$

We compute each component of curl $(\nabla(\varphi))$. The first component is

$$\frac{\partial}{\partial y}\left(\frac{\partial\varphi}{\partial z}\right) - \frac{\partial}{\partial z}\left(\frac{\partial\varphi}{\partial y}\right) = \frac{\partial^2\varphi}{\partial y\partial z} - \frac{\partial^2\varphi}{\partial z\partial y} = 0$$

The second component of $\text{curl}(\nabla(\varphi))$ is

$$\frac{\partial}{\partial z}\left(\frac{\partial\varphi}{\partial x}\right) - \frac{\partial}{\partial x}\left(\frac{\partial\varphi}{\partial z}\right) = \frac{\partial^2\varphi}{\partial z\partial x} - \frac{\partial^2\varphi}{\partial x\partial z} = 0$$

And the third component is

$$\frac{\partial}{\partial x}\left(\frac{\partial\varphi}{\partial y}\right) - \frac{\partial}{\partial y}\left(\frac{\partial\varphi}{\partial x}\right) = \frac{\partial^2\varphi}{\partial x\partial y} - \frac{\partial^2\varphi}{\partial y\partial x} = 0$$

We conclude that

$$\text{curl}(\nabla(\varphi)) = \mathbf{0}$$

29. Assume that the second partial derivatives of φ and ψ exist and are continuous. Use(7) and (8) to prove that

$$\oint_{\partial\mathcal{S}} \varphi\nabla(\psi)\cdot d\mathbf{s} = \int_{\mathcal{S}} \nabla(\varphi)\times\nabla(\psi)\cdot d\mathbf{s}$$

where $\mathcal{S}$ is a smooth surface with boundary $\partial\mathcal{S}$.

SOLUTION By Stokes' Theorem, we have

$$\int_{\partial\mathcal{S}} \varphi\nabla(\psi)\cdot d\mathbf{s} = \iint_{\mathcal{S}} \text{curl}(\varphi\nabla\psi)\cdot d\mathbf{S}$$

We now use Eq.(8) to evaluate the curl of $\varphi\nabla\psi$. That is,

$$\begin{aligned}\int_{\partial\mathcal{S}} \varphi\nabla(\psi)\cdot d\mathbf{s} &= \iint_{\mathcal{S}} \left(\varphi\text{curl}(\nabla\psi) + \nabla\varphi\times\nabla\psi\right)\cdot d\mathbf{S}\\ &= \iint_{\mathcal{S}} \varphi\text{curl}(\nabla\psi)\cdot d\mathbf{S} + \iint_{\mathcal{S}} \nabla(\varphi)\times\nabla(\psi)\cdot d\mathbf{S} \qquad (1)\end{aligned}$$

Now, since the gradient field $\nabla\varphi$ is conservative, this field satisfies the cross-partials condition. In other words,

$$\text{curl}(\nabla\varphi) = \mathbf{0}$$

Combining with(1) we obtain

$$\int_{\partial\mathcal{S}} \varphi(\nabla\psi)\cdot d\mathbf{s} = \iint_{\mathcal{S}} \mathbf{0}\cdot d\mathbf{S} + \iint_{\mathcal{S}} \nabla(\varphi)\times\nabla(\psi)\cdot d\mathbf{S} = \iint_{\mathcal{S}} \nabla(\varphi)\times\nabla(\psi)\cdot d\mathbf{S}$$

Further Insights and Challenges

31. Complete the proof of Theorem 1 by proving the equality

$$\oint_{\mathcal{C}} F_3(x, y, z)\mathbf{k}\cdot d\mathbf{s} = \int_{\mathcal{S}} \text{curl}(F_3(x, y, z)\mathbf{k})\cdot d\mathbf{S}$$

where $\mathcal{S}$ is the graph of a function $z = f(x, y)$ over a domain $\mathcal{D}$ in the xy-plane whose boundary is a simple closed curve.

SOLUTION Let $(x(t), y(t))$, $a \le t \le b$ be a parametrization of the boundary curve $\mathcal{C}_0$ of the domain $\mathcal{D}$.

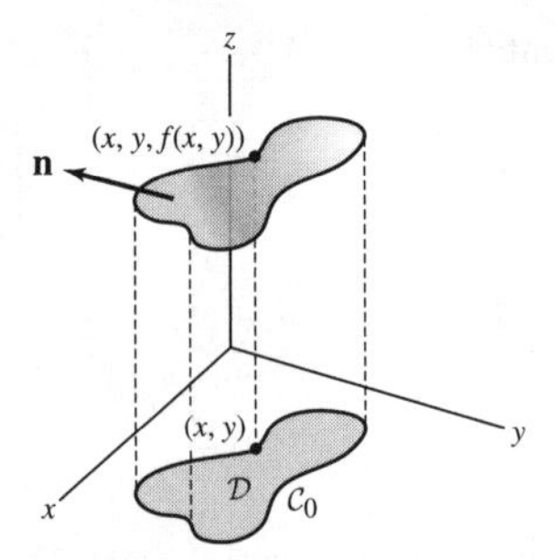

The boundary curve $\mathcal{C}$ of $\mathcal{S}$ projects on $\mathcal{C}_0$ and has the parametrization

$$\gamma(t) = (x(t), y(t), f(x(t), y(t))), \quad a \le t \le b$$

Let

$$\mathbf{F} = \langle 0, 0, F_3(x, y, z)\rangle$$

We must show that

$$\int_{\mathcal{C}} \mathbf{F} \cdot d\mathbf{s} = \iint_{\mathcal{S}} \operatorname{curl}(\mathbf{F}) \cdot d\mathbf{S} \tag{1}$$

We first compute the surface integral, using the parametrization

$$\mathcal{S} : \Phi(x, y) = (x, y, f(x, y))$$

The normal vector is

$$\begin{aligned}\mathbf{n} &= \frac{\partial \Phi}{\partial x} \times \frac{\partial \Phi}{\partial y} = \langle 1, 0, f_x(x, y)\rangle \times \langle 0, 1, f_y(x, y)\rangle = (\mathbf{i} + f_x(x, y)\mathbf{k}) \times (\mathbf{j} + f_y(x, y)\mathbf{k}) \\ &= -f_y(x, y)\mathbf{j} - f_x(x, y)\mathbf{i} + \mathbf{k} = \langle -f_x(x, y), -f_y(x, y), 1\rangle\end{aligned}$$

We compute the curl of $\mathbf{F}$:

$$\operatorname{curl}(\mathbf{F}) = \begin{vmatrix} \mathbf{i} & \mathbf{j} & \mathbf{k} \\ \frac{\partial}{\partial x} & \frac{\partial}{\partial y} & \frac{\partial}{\partial z} \\ 0 & 0 & F_3(x, y, z) \end{vmatrix} = \left\langle \frac{\partial F_3(x, y, z)}{\partial y}, -\frac{\partial F_3(x, y, z)}{\partial x}, 0 \right\rangle$$

Hence,

$$\begin{aligned}\operatorname{curl}(\mathbf{F})\,(\Phi(x, y)) \cdot \mathbf{n} &= \left\langle \frac{\partial F_3}{\partial y}(x, y, f(x, y)) - \frac{\partial F_3}{\partial x}(x, y, f(x, y)), 0 \right\rangle \cdot \langle -f_x(x, y), -f_y(x, y), 1\rangle \\ &= -\frac{\partial F_3(x, y, f(x, y))}{\partial y} f_x(x, y) + \frac{\partial F_3(x, y, f(x, y))}{\partial x} f_y(x, y)\end{aligned}$$

The surface integral is thus

$$\iint_{\mathcal{S}} \operatorname{curl}(\mathbf{F}) \cdot d\mathbf{S} = \iint_{\mathcal{D}} \left(-\frac{\partial F_3(x, y, f(x, y))}{\partial y} f_x(x, y) + \frac{\partial F_3(x, y, f(x, y))}{\partial x} f_y(x, y) \right) dx\,dy \tag{2}$$

We now evaluate the line integral in (1). We have

$$\begin{aligned}\mathbf{F}(\gamma(t)) \cdot \gamma'(t) &= \left\langle 0, 0, \mathbf{F}_3\big(x(t), y(t), f(x(t), y(t))\big)\right\rangle \cdot \left\langle x'(t), y'(t), \frac{d}{dt} f(x(t), y(t))\right\rangle \\ &= \mathbf{F}_3\big(x(t), y(t), f(x(t), y(t))\big) \frac{d}{dt} f(x(t), y(t))\end{aligned} \tag{3}$$

Using the Chain Rule gives

$$\frac{d}{dt} f(x(t), y(t)) = f_x(x(t), y(t))x'(t) + f_y(x(t), y(t))y'(t)$$

Substituting in (3), we conclude that the line integral is

$$\int_{\mathcal{C}} \mathbf{F} \cdot d\mathbf{s} = \int_a^b \Big(F_3\big(x(t), y(t), f(x(t), y(t))\big) \cdot \big(f_x(x(t), y(t))x'(t) + f_y(x(t), y(t))y'(t)\big)\Big)\, dt \tag{4}$$

We consider the following vector field:

$$\mathbf{G}(x,y) = \left\langle F_3\big(x,y,f(x,y)\big) f_x(x,y),\ F_3\big(x,y,f(x,y)\big) f_y(x,y)\right\rangle$$

Then the integral in (4) is the line integral of the planar vector field $\mathbf{G}$ over $\mathcal{C}_0$. That is,

$$\int_{\mathcal{C}} \mathbf{F}\cdot d\mathbf{s} = \int_{\mathcal{C}_0} \mathbf{G}\cdot d\mathbf{s}$$

Therefore, we may apply Green's Theorem and write

$$\int_{\mathcal{C}} \mathbf{F}\cdot d\mathbf{s} = \int_{\mathcal{C}_0} \mathbf{G}\cdot d\mathbf{s} = \iint_{\mathcal{D}} \left(\frac{\partial}{\partial x}\Big(F_3\big(x,y,f(x,y)\big) f_y(x,y)\Big) - \frac{\partial}{\partial y}\Big(F_3\big(x,y,f(x,y)\big) f_x(x,y)\Big)\right) dx\,dy \tag{5}$$

We use the Product Rule to evaluate the integrand:

$$\frac{\partial F_3}{\partial x}(x,y,f(x,y))\, f_y(x,y) + F_3\big(x,y,f(x,y)\big) f_{yx}(x,y) - \frac{\partial F_3}{\partial y}\big(x,y,f(x,y)\big) f_x(x,y) - F_3\big(x,y,f(x,y)\big) f_{xy}(x,y)$$

$$= \frac{\partial F_3}{\partial x}\big(x,y,f(x,y)\big) f_y(x,y) - \frac{\partial F_3}{\partial y}\big(x,y,f(x,y)\big) f_x(x,y)$$

Substituting in (5) gives

$$\int_{\mathcal{C}} \mathbf{F}\cdot d\mathbf{s} = \iint_{\mathcal{D}} \left(\frac{\partial F_3\,(x,y,f(x,y))}{\partial x} f_y(x,y) - \frac{\partial F_3\,(x,y,f(x,y))}{\partial y} f_x(x,y)\right) dx\,dy \tag{6}$$

Equations(2) and(6) give the same result, hence

$$\int_{\mathcal{C}} \mathbf{F}\cdot d\mathbf{s} = \iint_{\mathcal{S}} \operatorname{curl}(\mathbf{F})\cdot d\mathbf{s}$$

for

$$\mathbf{F} = \langle 0, 0, F_3(x,y,z)\rangle$$

18.3 Divergence Theorem (ET Section 17.3)

Preliminary Questions

1. What is the flux of $\mathbf{F} = \langle 1,0,0\rangle$ through a closed surface?

SOLUTION The divergence of $\mathbf{F} = \langle 1,0,0\rangle$ is $\operatorname{div}(\mathbf{F}) = \frac{\partial P}{\partial x} + \frac{\partial Q}{\partial y} + \frac{\partial R}{\partial z} = 0$, therefore the Divergence Theorem implies that the flux of $\mathbf{F}$ through a closed surface $\mathcal{S}$ is

$$\iint_{\mathcal{S}} \mathbf{F}\cdot d\mathbf{S} = \iiint_{\mathcal{W}} \operatorname{div}(\mathbf{F})\,dV = \iiint_{\mathcal{W}} 0\,dV = 0$$

2. Justify the following statement: The flux of $\mathbf{F} = \langle x^3, y^3, z^3\rangle$ through every closed surface is positive.

SOLUTION The divergence of $\mathbf{F} = \left\langle x^3, y^3, z^3\right\rangle$ is

$$\operatorname{div}(\mathbf{F}) = 3x^2 + 3y^2 + 3z^2$$

Therefore, by the Divergence Theorem, the flux of $\mathbf{F}$ through a closed surface $\mathcal{S}$ is

$$\iint_{\mathcal{S}} \mathbf{F}\cdot d\mathbf{S} = \iiint_{\mathcal{W}} (3x^2 + 3y^2 + 3z^2)\,dV$$

Since the integrand is positive for all $(x,y,z) \neq (0,0,0)$, the triple integral, hence also the flux, is positive.

3. Which of the following expressions are meaningful (where $\mathbf{F}$ is a vector field and φ is a function)? Of those that are meaningful, which are automatically zero?

(a) $\operatorname{div}(\nabla\varphi)$ **(b)** $\operatorname{curl}(\nabla\varphi)$ **(c)** $\nabla\operatorname{curl}(\varphi)$

(d) $\operatorname{div}(\operatorname{curl}(\mathbf{F}))$ **(e)** $\operatorname{curl}(\operatorname{div}(\mathbf{F}))$ **(f)** $\nabla(\operatorname{div}(\mathbf{F}))$

SOLUTION

(a) The divergence is defined on vector fields. The gradient is a vector field, hence $\text{div}(\nabla\varphi)$ is defined. It is not automatically zero since for $\varphi = x^2 + y^2 + z^2$ we have

$$\text{div}(\nabla\varphi) = \text{div}\,\langle 2x, 2y, 2z\rangle = 2 + 2 + 2 = 6 \neq 0$$

(b) The curl acts on vector valued functions, and $\nabla\varphi$ is such a function. Therefore, $\text{curl}(\nabla\varphi)$ is defined. Since the gradient field $\nabla\varphi$ is conservative, the cross partials of $\nabla\varphi$ are equal, or equivalently, $\text{curl}(\nabla\varphi)$ is the zero vector.

(c) The curl is defined on vector fields rather than on scalar functions. Therefore, $\text{curl}(\varphi)$ is undefined. Obviously, $\nabla\text{curl}(\varphi)$ is also undefined.

(d) The curl is defined on the vector field $\mathbf{F}$ and the divergence is defined on the vector field $\text{curl}(\mathbf{F})$. Therefore the expression $\text{div}\,(\text{curl}(\mathbf{F}))$ is meaningful. We show that this vector is automatically zero:

$$\begin{aligned}
\text{div}\,(\text{curl}\,(\mathbf{F})) &= \text{div}\left\langle \frac{\partial F_3}{\partial y} - \frac{\partial F_2}{\partial z}, \frac{\partial F_1}{\partial z} - \frac{\partial F_3}{\partial x}, \frac{\partial F_2}{\partial x} - \frac{\partial F_1}{\partial y}\right\rangle \\
&= \frac{\partial}{\partial x}\left(\frac{\partial F_3}{\partial y} - \frac{\partial F_2}{\partial z}\right) + \frac{\partial}{\partial y}\left(\frac{\partial F_1}{\partial z} - \frac{\partial F_3}{\partial x}\right) + \frac{\partial}{\partial z}\left(\frac{\partial F_2}{\partial x} - \frac{\partial F_1}{\partial y}\right) \\
&= \frac{\partial^2 F_3}{\partial x \partial y} - \frac{\partial^2 F_2}{\partial x \partial z} + \frac{\partial^2 F_1}{\partial y \partial z} - \frac{\partial^2 F_3}{\partial y \partial x} + \frac{\partial^2 F_2}{\partial z \partial x} - \frac{\partial^2 F_1}{\partial z \partial y} \\
&= \left(\frac{\partial^2 F_3}{\partial x \partial y} - \frac{\partial^2 F_3}{\partial y \partial x}\right) + \left(\frac{\partial^2 F_2}{\partial z \partial x} - \frac{\partial^2 F_2}{\partial x \partial z}\right) + \left(\frac{\partial^2 F_1}{\partial y \partial z} - \frac{\partial^2 F_1}{\partial z \partial y}\right) \\
&= 0 + 0 + 0 = 0
\end{aligned}$$

(e) The curl acts on vector valued functions, whereas $\text{div}(\mathbf{F})$ is a scalar function. Therefore the expression $\text{curl}\,(\text{div}(\mathbf{F}))$ has no meaning.

(f) $\text{div}(\mathbf{F})$ is a scalar function, hence $\nabla(\text{div}\mathbf{F})$ is meaningful. It is not necessarily the zero vector as shown in the following example:

$$\begin{aligned}
\mathbf{F} &= \left\langle x^2, y^2, z^2 \right\rangle \\
\text{div}\,(\mathbf{F}) &= 2x + 2y + 2z \\
\nabla(\text{div}\mathbf{F}) &= \langle 2, 2, 2\rangle \neq \langle 0, 0, 0\rangle
\end{aligned}$$

4. Which of the following statements is correct (where $\mathbf{F}$ is a continuously differentiable vector field defined everywhere)?

(a) The flux of $\text{curl}(\mathbf{F})$ through all surfaces is zero.

(b) If $\mathbf{F} = \nabla\varphi$, then the flux of $\mathbf{F}$ through all surfaces is zero.

(c) The flux of $\text{curl}(\mathbf{F})$ through all closed surfaces is zero.

SOLUTION

(a) This statement holds only for conservative fields. If $\mathbf{F}$ is not conservative, there exist closed curves such that $\int_C \mathbf{F} \cdot ds \neq 0$, hence by Stokes' Theorem $\iint_S \text{curl}(\mathbf{F}) \cdot d\mathbf{S} \neq 0$.

(b) This statement is false. Consider the unit sphere S in the three-dimensional space and the function $\varphi(x, y, z) = x^2 + y^2 + z^2$. Then $\mathbf{F} = \nabla\varphi = \langle 2x, 2y, 2z\rangle$ and $\text{div}\,(\mathbf{F}) = 2 + 2 + 2 = 6$. Using the Divergence Theorem, we have ($\mathcal{W}$ is the unit ball in R^3)

$$\iint_S \mathbf{F} \cdot d\mathbf{S} = \iiint_{\mathcal{W}} \text{div}(\mathbf{F})\, dV = \iiint_{\mathcal{W}} 6\, dV = 6 \iiint_{\mathcal{W}} dV = 6\,\text{Vol}(\mathcal{W})$$

(c) This statement is correct, as stated in the corollary of Stokes' Theorem in section 18.2.

5. How does the Divergence Theorem imply that the flux of $\mathbf{F} = \left\langle x^2, y - e^z, y - 2zx\right\rangle$ through a closed surface is equal to the enclosed volume?

SOLUTION By the Divergence Theorem, the flux is

$$\iint_S \mathbf{F} \cdot d\mathbf{S} = \iiint_{\mathcal{W}} \text{div}(\mathbf{F})\, dV = \iiint_{\mathcal{W}} (2x + 1 - 2x)\, dV = \iiint_{\mathcal{W}} 1\, dV = \text{Volume}(\mathcal{W})$$

Therefore the statement is true.

Exercises

In Exercises 1–4, compute the divergence of the vector field.

1. $\mathbf{F} = \langle xy, yz, y^2 - x^3 \rangle$

SOLUTION The divergence of $\mathbf{F}$ is

$$\text{div}(\mathbf{F}) = \frac{\partial}{\partial x}(xy) + \frac{\partial}{\partial y}(yz) + \frac{\partial}{\partial z}(y^2 - x^3) = y + z + 0 = y + z$$

3. $\mathbf{F} = \langle x - 2zx^2, z - xy, z^2x^2 \rangle$

SOLUTION

$$\text{div}(\mathbf{F}) = \frac{\partial}{\partial x}(x - 2zx^2) + \frac{\partial}{\partial y}(z - xy) + \frac{\partial}{\partial z}(z^2x^2) = (1 - 4zx) + (-x) + (2zx^2) = 1 - 4zx - x + 2zx^2$$

In Exercises 5–8, verify the Divergence Theorem for the vector field and region.

5. $\mathbf{F} = \langle z, x, y \rangle$ and the box $[0, 4] \times [0, 2] \times [0, 3]$

SOLUTION Let $\mathcal{S}$ be the surface of the box and $\mathcal{R}$ the region enclosed by $\mathcal{S}$.

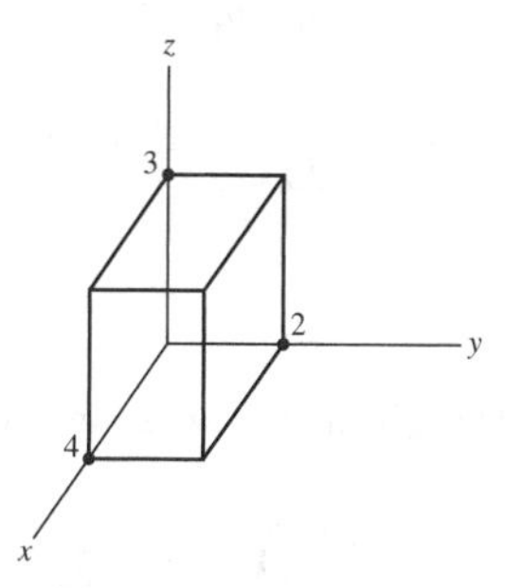

We first compute the surface integral in the Divergence Theorem:

$$\iint_{\mathcal{S}} \mathbf{F} \cdot d\mathbf{S} = \iiint_{\mathcal{R}} \text{div}(\mathbf{F})\, dV \tag{1}$$

We denote by $\mathcal{S}_i$, $i = 1, \ldots, 6$, the faces of the box, starting at the face on the xz-plane and moving counterclockwise, then moving to the bottom and the top. We use parametrizations

$$\mathcal{S}_1 : \Phi_1(x, z) = (x, 0, z), \quad 0 \le x \le 4, \quad 0 \le z \le 3$$
$$\mathbf{n} = \langle 0, -1, 0 \rangle$$
$$\mathcal{S}_2 : \Phi_2(y, z) = (0, y, z), \quad 0 \le y \le 2, \quad 0 \le z \le 3$$
$$\mathbf{n} = \langle -1, 0, 0 \rangle$$
$$\mathcal{S}_3 : \Phi_3(x, z) = (x, 2, z), \quad 0 \le x \le 4, \quad 0 \le z \le 3$$
$$\mathbf{n} = \langle 0, 1, 0 \rangle$$
$$\mathcal{S}_4 : \Phi_4(y, z) = (4, y, z), \quad 0 \le y \le 2, \quad 0 \le z \le 3$$
$$\mathbf{n} = \langle 1, 0, 0 \rangle$$
$$\mathcal{S}_5 : \Phi_5(x, y) = (x, y, 0), \quad 0 \le x \le 4, \quad 0 \le y \le 2$$
$$\mathbf{n} = \langle 0, 0, -1 \rangle$$
$$\mathcal{S}_6 : \Phi_6(x, y) = (x, y, 3), \quad 0 \le x \le 4, \quad 0 \le y \le 2$$
$$\mathbf{n} = \langle 0, 0, 1 \rangle$$

Then,

$$\iint_{\mathcal{S}_1} \mathbf{F} \cdot d\mathbf{S} = \int_0^3 \int_0^4 \mathbf{F}(\Phi_1(x, z)) \cdot \langle 0, -1, 0 \rangle\, dx\, dz = \int_0^3 \int_0^4 \langle z, x, 0 \rangle \cdot \langle 0, -1, 0 \rangle\, dx\, dz$$
$$= \int_0^3 \int_0^4 -x\, dx\, dz = 3 \frac{-x^2}{2}\Big|_0^4 = -24$$

$$\iint_{S_2} \mathbf{F}\cdot d\mathbf{S} = \int_0^3 \int_0^2 \mathbf{F}(\Phi_2(y,z)) \cdot \langle -1,0,0\rangle \, dy\, dz = \int_0^3 \int_0^2 \langle z,0,y\rangle \cdot \langle -1,0,0\rangle \, dy\, dz$$

$$= \int_0^3 \int_0^2 -z \, dy\, dz = 2 \cdot \frac{-z^2}{2}\bigg|_0^3 = -9$$

$$\iint_{S_3} \mathbf{F}\cdot d\mathbf{S} = \int_0^3 \int_0^4 \mathbf{F}(\Phi_3(x,z)) \cdot \langle 0,1,0\rangle \, dx\, dz = \int_0^3 \int_0^4 \langle z,x,2\rangle \cdot \langle 0,1,0\rangle \, dx\, dz$$

$$= \int_0^3 \int_0^4 x \, dx\, dz = 3 \cdot \frac{x^2}{2}\bigg|_0^4 = 24$$

$$\iint_{S_4} \mathbf{F}\cdot d\mathbf{S} = \int_0^3 \int_0^2 \mathbf{F}(\Phi_4(y,z)) \cdot \langle 1,0,0\rangle \, dy\, dz = \int_0^3 \int_0^2 \langle z,4,y\rangle \cdot \langle 1,0,0\rangle \, dy\, dz$$

$$= \int_0^3 \int_0^2 z \, dy\, dz = 2 \cdot \frac{z^2}{2}\bigg|_0^3 = 9$$

$$\iint_{S_5} \mathbf{F}\cdot d\mathbf{S} = \int_0^2 \int_0^4 \mathbf{F}(\Phi_5(x,y)) \cdot \langle 0,0,-1\rangle \, dx\, dy = \int_0^2 \int_0^4 \langle 0,x,y\rangle \cdot \langle 0,0,-1\rangle \, dx\, dy$$

$$= \int_0^2 \int_0^4 -y \, dx\, dy = 4 \cdot \frac{-y^2}{2}\bigg|_0^2 = -8$$

$$\iint_{S_6} \mathbf{F}\cdot d\mathbf{S} = \int_0^2 \int_0^4 \mathbf{F}(\Phi_6(x,y)) \cdot \mathbf{n} \, dx\, dy = \int_0^2 \int_0^4 \langle 3,x,y\rangle \cdot \langle 0,0,1\rangle \, dx\, dy$$

$$= \int_0^2 \int_0^4 y \, dx\, dy = 4 \cdot \frac{y^2}{2}\bigg|_0^2 = 8$$

We add the integrals to obtain the surface integral

$$\iint_{S} \mathbf{F}\cdot d\mathbf{S} = \sum_{i=1}^{6} \iint_{S_i} \mathbf{F}\cdot d\mathbf{S} = -24 - 9 + 24 + 9 - 8 + 8 = 0 \tag{2}$$

We now evaluate the triple integral in (1). We compute the divergence of $\mathbf{F} = \langle z, x, y\rangle$:

$$\operatorname{div}(\mathbf{F}) = \frac{\partial}{\partial x}(z) + \frac{\partial}{\partial y}(x) + \frac{\partial}{\partial z}(y) = 0$$

Hence,

$$\iiint_{\mathcal{R}} \operatorname{div}(\mathbf{F})\, dV = \iiint_{\mathcal{R}} 0\, dV = 0 \tag{3}$$

The equality of the integrals in (2) and (3) verifies the Divergence Theorem.

7. $\mathbf{F} = \langle 2x, 3z, 3y\rangle$ and the region $x^2 + y^2 \le 1, 0 \le z \le 2$

SOLUTION

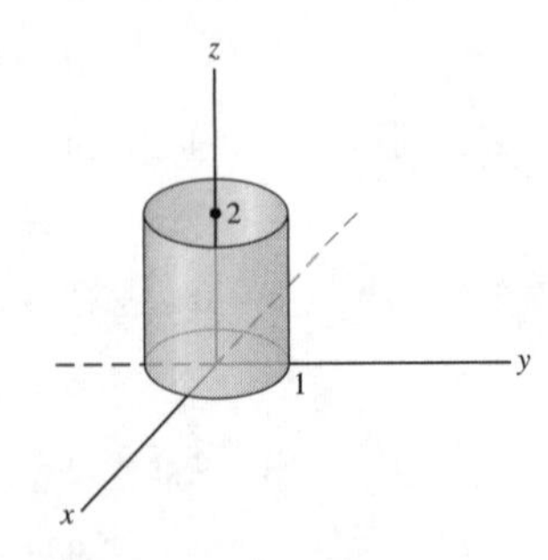

Let $\mathcal{S}$ be the surface of the cylinder and $\mathcal{R}$ the region enclosed by $\mathcal{S}$. We compute the two sides of the Divergence Theorem:

$$\iint_{\mathcal{S}} \mathbf{F}\cdot d\mathbf{S} = \iiint_{\mathcal{R}} \operatorname{div}(\mathbf{F})\, dV \tag{1}$$

We first calculate the surface integral.

Step 1. Integral over the side of the cylinder. The side of the cylinder is parametrized by

$$\Phi(\theta, z) = (\cos\theta, \sin\theta, z), \quad 0 \le \theta \le 2\pi, \quad 0 \le z \le 2$$

$$\mathbf{n} = \langle \cos\theta, \sin\theta, 0 \rangle$$

Then,

$$\mathbf{F}\left(\Phi(\theta, z)\right) \cdot \mathbf{n} = \langle 2\cos\theta, 3z, 3\sin\theta \rangle \cdot \langle \cos\theta, \sin\theta, 0 \rangle = 2\cos^2\theta + 3z\sin\theta$$

We obtain the integral

$$\iint_{\text{side}} \mathbf{F} \cdot d\mathbf{S} = \int_0^2 \int_0^{2\pi} \left(2\cos^2\theta + 3z\sin\theta\right) d\theta\, dz = 4\int_0^{2\pi} \cos^2\theta\, d\theta + \left(\int_0^2 3z\, dz\right)\left(\int_0^{2\pi} \sin\theta\, d\theta\right)$$

$$= 4 \cdot \left(\frac{\theta}{2} + \frac{\sin 2\theta}{4}\Big|_0^{2\pi}\right) + 0 = 4\pi$$

Step 2. Integral over the top of the cylinder. The top of the cylinder is parametrized by

$$\Phi(x, y) = (x, y, 2)$$

with parameter domain $\mathcal{D} = \left\{(x, y) : x^2 + y^2 \le 1\right\}$. The upward pointing normal is

$$\mathbf{n} = \mathbf{T}_x \times \mathbf{T}_y = \langle 1, 0, 0 \rangle \times \langle 0, 1, 0 \rangle = \mathbf{i} \times \mathbf{j} = \mathbf{k} = \langle 0, 0, 1 \rangle$$

Also,

$$\mathbf{F}\left(\Phi(x, y)\right) \cdot \mathbf{n} = \langle 2x, 6, 3y \rangle \cdot \langle 0, 0, 1 \rangle = 3y$$

Hence,

$$\iint_{\text{top}} \mathbf{F} \cdot d\mathbf{S} = \iint_{\mathcal{D}} 3y\, dA = 0$$

The last integral is zero due to symmetry.

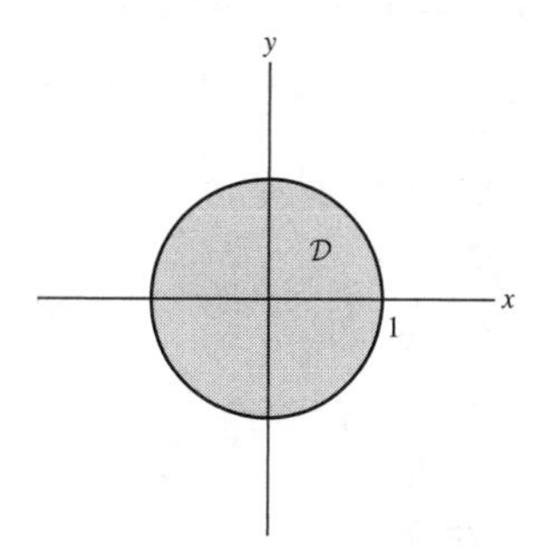

Step 3. Integral over the bottom of the cylinder. We parametrize the bottom by

$$\Phi(x, y) = (x, y, 0), \quad (x, y) \in \mathcal{D}$$

The downward pointing normal is $\mathbf{n} = \langle 0, 0, -1 \rangle$. Then

$$\mathbf{F}\left(\Phi(x, y)\right) \cdot \mathbf{n} = \langle 2x, 0, 3y \rangle \cdot \langle 0, 0, -1 \rangle = -3y$$

We obtain the following integral, which is zero due to symmetry:

$$\iint_{\text{bottom}} \mathbf{F} \cdot d\mathbf{S} = \iint_{\mathcal{D}} -3y\, dA = 0$$

Adding the integrals we get

$$\iint_{\mathcal{S}} \mathbf{F} \cdot d\mathbf{S} = \iint_{\text{side}} \mathbf{F} \cdot d\mathbf{S} + \iint_{\text{top}} \mathbf{F} \cdot d\mathbf{S} + \iint_{\text{bottom}} \mathbf{F} \cdot d\mathbf{S} = 4\pi + 0 + 0 = 4\pi \tag{2}$$

Step 4. Compare with integral of divergence.

$$\operatorname{div}(\mathbf{F}) = \operatorname{div}\langle 2x, 3z, 3y \rangle = \frac{\partial}{\partial x}(2x) + \frac{\partial}{\partial y}(3z) + \frac{\partial}{\partial z}(3y) = 2$$

$$\iiint_{\mathcal{R}} \operatorname{div}(\mathbf{F})\, dV = \iiint_{\mathcal{R}} 2\, dV = 2\iiint_{\mathcal{R}} dV = 2\,\mathrm{Vol}(\mathcal{R}) = 2\cdot\pi\cdot 2 = 4\pi \tag{3}$$

The equality of (2) and (3) verifies the Divergence Theorem.

In Exercises 9–16, use the Divergence Theorem to evaluate the surface integral $\iint_{\mathcal{S}} \mathbf{F}\cdot d\mathbf{S}$.

9. $\mathbf{F} = \langle x, y, z\rangle$, $\mathcal{S}$ is the sphere $x^2+y^2+z^2=1$.

SOLUTION We compute the divergence of $\mathbf{F} = \langle x, y, z\rangle$:

$$\operatorname{div}(\mathbf{F}) = \frac{\partial}{\partial x}(x) + \frac{\partial}{\partial y}(y) + \frac{\partial}{\partial z}(z) = 1+1+1 = 3$$

Using the Divergence Theorem and the volume of the sphere, we obtain

$$\iint_{\mathcal{S}} \mathbf{F}\cdot d\mathbf{S} = \iiint_{\mathcal{W}} \operatorname{div}(\mathbf{F})\, dV = \iiint_{\mathcal{W}} 3\, dV = 3\iiint_{\mathcal{W}} 1\, dV = 3\ \text{Volume}(\mathcal{W})$$
$$= 3\cdot\frac{4}{3}\pi\cdot 1^3 = 4\pi$$

11. $\mathbf{F} = \left\langle x^3, 0, z^3\right\rangle$, $\mathcal{S}$ is the sphere $x^2+y^2+z^2=4$.

SOLUTION We compute the divergence of $\mathbf{F} = \left\langle x^3, 0, z^3\right\rangle$:

$$\operatorname{div}(\mathbf{F}) = \frac{\partial}{\partial x}(x^3) + \frac{\partial}{\partial y}(0) + \frac{\partial}{\partial z}(z^3) = 3x^2+3z^2 = 3(x^2+z^2)$$

Using the Divergence Theorem we obtain ($\mathcal{W}$ is the region inside the sphere)

$$\iint_{\mathcal{S}} \mathbf{F}\cdot d\mathbf{S} = \iiint_{\mathcal{W}} \operatorname{div}(\mathbf{F})\, dV = \iiint_{\mathcal{W}} 3(x^2+z^2)\, dV$$

We convert the integral to spherical coordinates. We have

$$x^2+z^2 = \rho^2\cos^2\theta\sin^2\phi + \rho^2\cos^2\phi = \rho^2\cos^2\theta\sin^2\phi + \rho^2(1-\sin^2\phi)$$
$$= -\rho^2\sin^2\phi(1-\cos^2\theta) + \rho^2 = -\rho^2\sin^2\phi\sin^2\theta + \rho^2 = \rho^2(1-\sin^2\phi\sin^2\theta)$$

We obtain the following integral:

$$\iint_{\mathcal{S}} \mathbf{F}\cdot d\mathbf{S} = 3\int_0^{2\pi}\int_0^{\pi}\int_0^{2} \rho^2(1-\sin^2\phi\sin^2\theta)\cdot\rho^2\sin\phi\, d\rho\, d\phi\, d\theta$$
$$= 3\int_0^{2\pi}\int_0^{\pi}\int_0^{2} \rho^4(\sin\phi - \sin^3\phi\sin^2\theta) d\rho\, d\phi\, d\theta$$
$$= 3\int_0^{2\pi}\int_0^{\pi}\int_0^{2} \rho^4\sin\phi\, d\rho\, d\phi\, d\theta - 3\int_0^{2\pi}\int_0^{\pi}\int_0^{2} \rho^4\sin^3\phi\sin^2\theta\, d\rho\, d\phi\, d\theta$$
$$= 6\pi\left(\int_0^{\pi}\sin\phi\, d\phi\right)\left(\int_0^{2}\rho^4\, d\rho\right) - 3\left(\int_0^{2\pi}\sin^2\theta\, d\theta\right)\left(\int_0^{\pi}\sin^3\phi\, d\phi\right)\left(\int_0^{2}\rho^4\, d\rho\right)$$
$$= 6\pi\left(-\cos\phi\Big|_{\phi=0}^{\pi}\right)\left(\frac{\rho^5}{5}\Big|_{\rho=0}^{2}\right)\left(-3\frac{\theta}{2} - \frac{\sin 2\theta}{4}\Big|_{\theta=0}^{2\pi}\right)\cdot\left(-\frac{\sin^2\phi\cos\phi}{3} - \frac{2}{3}\cos\phi\Big|_{\phi=0}^{\pi}\right)\left(\frac{\rho^5}{5}\Big|_{\rho=0}^{2}\right)$$
$$= 12\pi\cdot\frac{32}{5} - 3\pi\cdot\frac{4}{3}\cdot\frac{32}{5} = \frac{256\pi}{5}$$

13. $\mathbf{F} = \left\langle x, y^2, z+y\right\rangle$, $\mathcal{S}$ is the boundary of the region contained in the cylinder $x^2+y^2=4$ between the planes $z=x$ and $z=8$.

SOLUTION Let $\mathcal{W}$ be the region enclosed by $\mathcal{S}$.

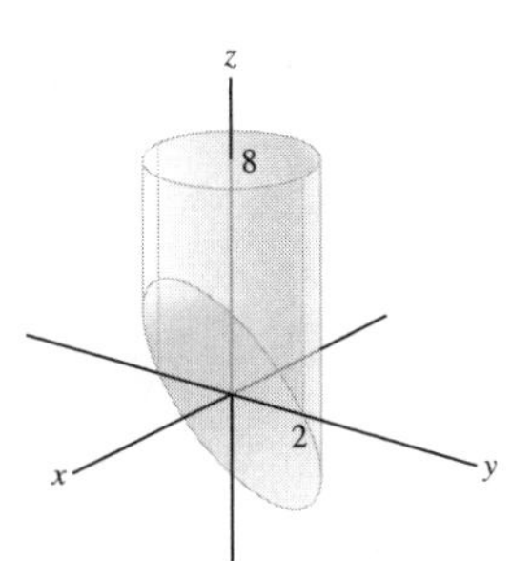

We compute the divergence of $\mathbf{F} = \left\langle x, y^2, z + y \right\rangle$:

$$\text{div}(\mathbf{F}) = \frac{\partial}{\partial x}(x) + \frac{\partial}{\partial y}(y^2) + \frac{\partial}{\partial z}(z + y) = 1 + 2y + 1 = 2 + 2y.$$

By the Divergence Theorem we have

$$\iint_{\mathcal{S}} \mathbf{F} \cdot d\mathbf{S} = \iiint_{\mathcal{W}} \text{div}(\mathbf{F})\, dV = \iiint_{\mathcal{W}} (2 + 2y)\, dV$$

We compute the triple integral. Denoting by $\mathcal{D}$ the disk $x^2 + y^2 \le 4$ in the xy-plane, we have

$$\iint_{\mathcal{S}} \mathbf{F} \cdot d\mathbf{S} = \iint_{\mathcal{D}} \int_x^8 (2 + 2y)\, dz\, dx\, dy = \iint_{\mathcal{D}} (2 + 2y)z \Big|_{z=x}^{8} dx\, dy = \iint_{\mathcal{D}} (2 + 2y)(8 - x)\, dx\, dy$$

We convert the integral to polar coordinates:

$$\begin{aligned}
\iint_{\mathcal{S}} \mathbf{F} \cdot d\mathbf{S} &= \int_0^{2\pi} \int_0^2 (2 + 2r\sin\theta)(8 - r\cos\theta) r\, dr\, d\theta \\
&= \int_0^{2\pi} \int_0^2 \left(16r + 2r^2(8\sin\theta - \cos\theta) - r^3 \sin 2\theta\right) dr\, d\theta \\
&= \int_0^{2\pi} 8r^2 + \frac{2}{3}r^3(8\sin\theta - \cos\theta) - \frac{r^4}{4}\sin 2\theta \Big|_{r=0}^{2} d\theta \\
&= \int_0^{2\pi} \left(32 + \frac{16}{3}(8\sin\theta - \cos\theta) - 4\sin 2\theta\right) d\theta \\
&= 64\pi + \frac{128}{3}\int_0^{2\pi} \sin\theta\, d\theta - \frac{16}{3}\int_0^{2\pi} \cos\theta\, d\theta - \int_0^{2\pi} 4\sin 2\theta\, d\theta = 64\pi
\end{aligned}$$

15. $\mathbf{F} = \langle x + y, z, z - x \rangle$, $\mathcal{S}$ is the boundary of the region between the paraboloid $z = 9 - x^2 - y^2$ and the xy-plane.

SOLUTION We compute the divergence of $\mathbf{F} = \langle x + y, z, z - x \rangle$,

$$\text{div}(\mathbf{F}) = \frac{\partial}{\partial x}(x + y) + \frac{\partial}{\partial y}(z) + \frac{\partial}{\partial z}(z - x) = 1 + 0 + 1 = 2.$$

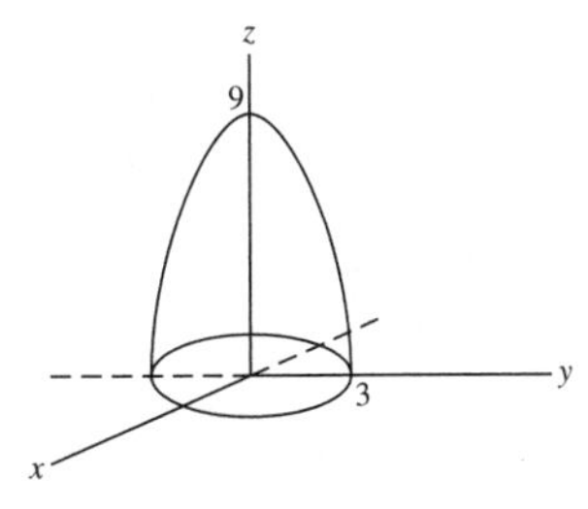

Using the Divergence Theorem we have

$$\iint_{\mathcal{S}} \mathbf{F} \cdot d\mathbf{S} = \iiint_{\mathcal{W}} \text{div}(\mathbf{F})\, dV = \iiint_{\mathcal{W}} 2\, dV$$

We compute the triple integral:

$$\iint_{\mathcal{S}} \mathbf{F} \cdot d\mathbf{S} = \iiint_{\mathcal{W}} 2\, dV = \iint_{\mathcal{D}} \int_0^{9-x^2-y^2} 2\, dz\, dx\, dy = \iint_{\mathcal{D}} 2z \Big|_0^{9-x^2-y^2} dx\, dy$$

$$= \iint_{\mathcal{W}} 2(9 - x^2 - y^2)\, dx\, dy$$

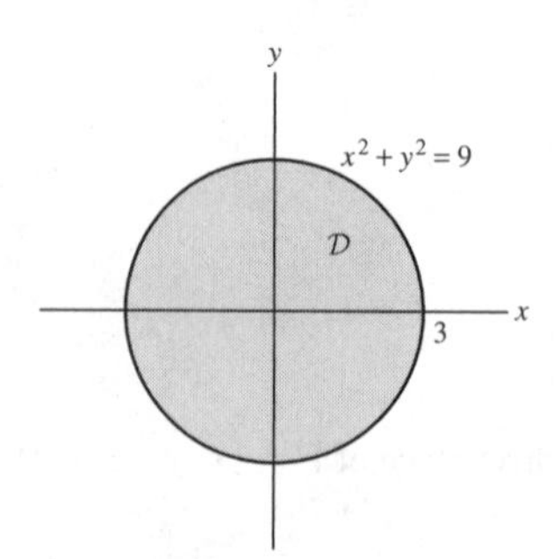

We convert the integral to polar coordinates:

$$x = r\cos\theta, \quad y = r\sin\theta, \quad 0 \le r \le 3, \quad 0 \le \theta \le 2\pi$$

$$\iint_{\mathcal{S}} \mathbf{F}\cdot d\mathbf{S} = \int_0^{2\pi}\int_0^3 2\left(9 - r^2\right) r\, dr\, d\theta = 4\pi \int_0^3 (9r - r^3)\, dr = 4\pi\left(\frac{9r^2}{2} - \frac{r^4}{4}\Big|_0^3\right) = 81\pi$$

17. Let $\mathcal{W}$ be the region in Figure 16 bounded by the cylinder $x^2 + y^2 = 9$, the plane $z = x + 1$, and the xy-plane. Use the Divergence Theorem to compute the flux of $\mathbf{F} = \langle z, x, y + 2z\rangle$ through the boundary of $\mathcal{W}$.

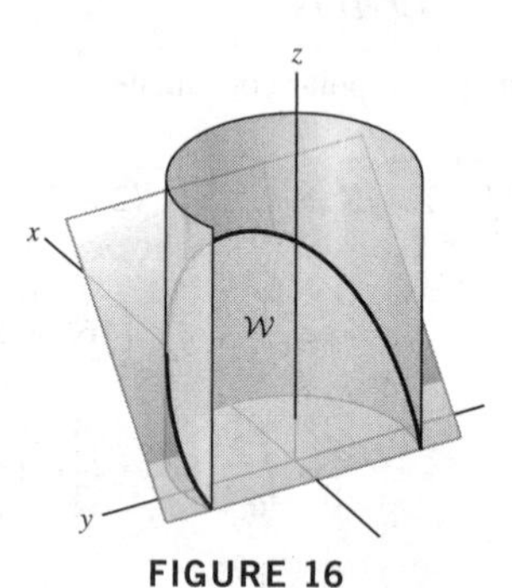

FIGURE 16

SOLUTION We compute the divergence of $\mathbf{F} = \langle z, x, y + 2z\rangle$:

$$\text{div}(\mathbf{F}) = \frac{\partial}{\partial x}(z) + \frac{\partial}{\partial y}(x) + \frac{\partial}{\partial z}(y + 2z) = 2$$

By the Divergence Theorem we have

$$\iint_{\mathcal{S}} \mathbf{F}\cdot d\mathbf{S} = \iiint_{\mathcal{W}} \text{div}(\mathbf{F})\, dV = \iiint_{\mathcal{W}} 2\, dV$$

To compute the triple integral, we identify the projection $\mathcal{D}$ of the region on the xy-plane. $\mathcal{D}$ is the region in the xy plane enclosed by the circle $x^2 + y^2 = 9$ and the line $0 = x + 1$ or $x = -1$. We obtain the following integral:

$$\iint_{\mathcal{S}} \mathbf{F}\cdot d\mathbf{S} = \iiint_{\mathcal{W}} 2\, dV = \iint_{\mathcal{D}}\int_0^{x+1} 2\, dz\, dx\, dy = \iint_{\mathcal{D}} 2z\Big|_{z=0}^{x+1} dx\, dy = \iint_{\mathcal{D}} (2x + 2)\, dx\, dy$$

We compute the double integral as the difference of two integrals: the integral over the disk $\mathcal{D}_2$ of radius 3, and the integral over the part $\mathcal{D}_1$ of the disk, shown in the figure.

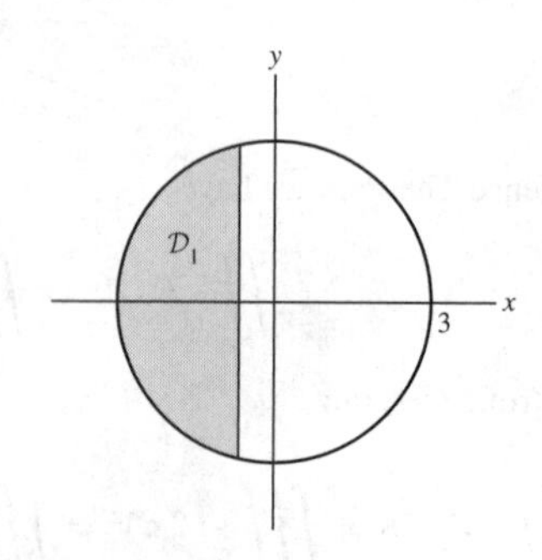

We obtain

$$\iint_S \mathbf{F}\cdot d\mathbf{S} = \iint_{\mathcal{D}_2} (2x+2)\,dx\,dy - \iint_{\mathcal{D}_1} (2x+2)\,dx\,dy$$

$$= \iint_{\mathcal{D}_2} 2x\,dx\,dy + \iint_{\mathcal{D}_2} 2\,dx\,dy - \iint_{\mathcal{D}_1} (2x+2)\,dx\,dy$$

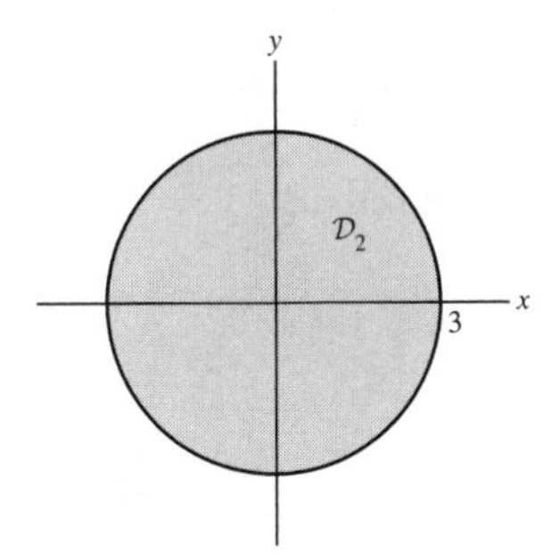

The first integral is zero due to symmetry. The second integral is twice the area of $\mathcal{D}_2$, that is, $2\cdot\pi\cdot 3^2 = 18\pi$. Therefore,

$$\iint_S \mathbf{F}\cdot d\mathbf{S} = 18\pi - \iint_{\mathcal{D}_1} (2x+2)\,dx\,dy$$

We compute the double integral over the upper part of $\mathcal{D}_1$. Due to symmetry, this integral is equal to half of the integral over $\mathcal{D}_1$.

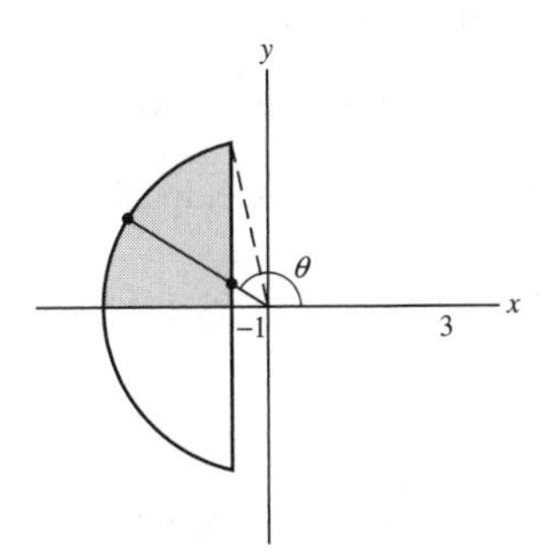

We describe the region in polar coordinates:

$$1.91 \le \theta \le \pi, \qquad \frac{-1}{\cos\theta} \le r \le 3$$

Then

$$\iint_{\mathcal{D}_1} (2x+2)\,dx\,dy = 2\int_{1.91}^{\pi}\int_{-1/\cos\theta}^{3} (2r\cos\theta + 2)r\,dr\,d\theta$$

$$= \int_{1.91}^{\pi}\int_{-1/\cos\theta}^{3} (4r^2\cos\theta + 4r)\,dr\,d\theta = \int_{1.91}^{\pi} \left.\frac{4r^3\cos\theta}{3} + 2r^2\right|_{r=\frac{-1}{\cos\theta}}^{3} d\theta$$

$$= \int_{1.91}^{\pi}\left(36\cos\theta + 18 + \frac{4}{3}\frac{\cos\theta}{\cos^3\theta} - \frac{2}{\cos^2\theta}\right)d\theta$$

$$= \int_{1.91}^{\pi}\left(36\cos\theta + 18 - \frac{2}{3}\cdot\frac{1}{\cos^2\theta}\right)d\theta = 36\sin\theta + 18\theta - \left.\frac{2}{3}\tan\theta\right|_{1.91}^{\pi}$$

$$= 18\pi - \left(36\sin 1.91 + 18\cdot 1.91 - \frac{2}{3}\tan 1.91\right) = -13.67$$

so we have

$$\iint_S \mathbf{F}\cdot d\mathbf{S} = 18\cdot\pi - \iint_{\mathcal{D}_1} (2x+2)\,dx\,dy = 70.22.$$

19. Volume as a Surface Integral Let $\mathbf{F} = \langle x, y, z\rangle$. Prove that if $\mathcal{W}$ is a region $\mathbf{R}^3$ with a smooth boundary $\mathcal{S}$, then

$$\text{Volume}(\mathcal{W}) = \frac{1}{3}\iint_S \mathbf{F}\cdot d\mathbf{S} \qquad \boxed{9}$$

SOLUTION Using the volume as a triple integral we have

$$\text{Volume}(\mathcal{W}) = \iiint_{\mathcal{W}} 1\,dV \tag{1}$$

We compute the surface integral of **F** over $\mathcal{S}$, using the Divergence Theorem. Since $\text{div}(\mathbf{F}) = \frac{\partial}{\partial x}(x) + \frac{\partial}{\partial y}(y) + \frac{\partial}{\partial z}(z) = 3$, we get

$$\iint_{\mathcal{S}} \mathbf{F}\cdot d\mathbf{S} = \iiint_{\mathcal{W}} \text{div}(\mathbf{F})\,dV = \iiint_{\mathcal{W}} 3\,dV = 3\iiint_{\mathcal{W}} 1\,dV \tag{2}$$

We combine (1) and (2) to obtain

$$\iint_{\mathcal{S}} \mathbf{F}\cdot d\mathbf{S} = 3\cdot \text{volume}(\mathcal{W})$$

or

$$\text{Volume}(\mathcal{W}) = \frac{1}{3}\iint_{\mathcal{S}} \mathbf{F}\cdot d\mathbf{S}$$

21. Show that $\Phi(a\cos\theta\sin\phi, b\sin\theta\sin\phi, c\cos\phi)$ is a parametrization of the ellipsoid

$$\left(\frac{x}{a}\right)^2 + \left(\frac{y}{b}\right)^2 + \left(\frac{z}{c}\right)^2 = 1$$

Then use Eq. (9) to calculate the volume of the ellipsoid as a surface integral over its boundary.

SOLUTION For the given parametrization,

$$x = a\cos\theta\sin\phi, \quad y = b\sin\theta\sin\phi, \quad z = c\cos\phi \tag{1}$$

We show that it satisfies the equation of the ellipsoid

$$\begin{aligned}
\left(\frac{x}{a}\right)^2 + \left(\frac{y}{b}\right)^2 + \left(\frac{z}{c}\right)^2 &= \left(\frac{a\cos\theta\sin\phi}{a}\right)^2 + \left(\frac{b\sin\theta\sin\phi}{b}\right)^2 + \left(\frac{c\cos\phi}{c}\right)^2 \\
&= \cos^2\theta\sin^2\phi + \sin^2\theta\sin^2\phi + \cos^2\phi \\
&= \sin^2\phi(\cos^2\theta + \sin^2\theta) + \cos^2\phi \\
&= \sin^2\phi + \cos^2\phi = 1
\end{aligned}$$

Conversely, for each (x, y, z) on the ellipsoid, there exists θ and ϕ so that (1) holds. Therefore $\Phi(\theta, \phi)$ parametrizes the whole ellipsoid. Let $\mathcal{W}$ be the interior of the ellipsoid $\mathcal{S}$. Then by Eq. (9):

$$\text{Volume}(\mathcal{W}) = \frac{1}{3}\iint_{\mathcal{S}} \mathbf{F}\cdot d\mathbf{S}, \quad \mathbf{F} = \langle x, y, z\rangle$$

We compute the surface integral, using the given parametrization. We first compute the normal vector:

$$\begin{aligned}
\frac{\partial\Phi}{\partial\theta} &= \langle -a\sin\theta\sin\phi, b\cos\theta\sin\phi, 0\rangle \\
\frac{\partial\Phi}{\partial\phi} &= \langle a\cos\theta\cos\phi, b\sin\theta\cos\phi, -c\sin\phi\rangle \\
\frac{\partial\Phi}{\partial\theta}\times\frac{\partial\Phi}{\partial\phi} &= -ab\sin^2\theta\sin\phi\cos\phi\mathbf{k} - ac\sin\theta\sin^2\phi\mathbf{j} - ab\cos^2\theta\sin\phi\cos\phi\mathbf{k} - bc\cos\theta\sin^2\phi\mathbf{i} \\
&= \left\langle -bc\cos\theta\sin^2\phi, -ac\sin\theta\sin^2\phi, -ab\sin\phi\cos\phi\right\rangle
\end{aligned}$$

Hence, the outward pointing normal is

$$\mathbf{n} = \left\langle bc\cos\theta\sin^2\phi, ac\sin\theta\sin^2\phi, ab\sin\phi\cos\phi\right\rangle$$

$$\begin{aligned}
\mathbf{F}\left(\Phi(\theta,\phi)\right)\cdot\mathbf{n} &= \langle a\cos\theta\sin\phi, b\sin\theta\sin\phi, c\cos\phi\rangle\cdot\left\langle bc\cos\theta\sin^2\phi, ac\sin\theta\sin^2\phi, ab\sin\phi\cos\phi\right\rangle \\
&= abc\cos^2\theta\sin^3\phi + abc\sin^2\theta\sin^3\phi + abc\sin\phi\cos^2\phi \\
&= abc\sin^3\phi(\cos^2\theta + \sin^2\theta) + abc\sin\phi\cos^2\phi \\
&= abc\sin^3\phi + abc\sin\phi\cos^2\phi = abc\sin^3\phi + abc\sin\phi(1 - \sin^2\phi)
\end{aligned}$$

$$= abc \sin \phi$$

We obtain the following integral:

$$\text{Volume}(\mathcal{W}) = \frac{1}{3}\iint_{\mathcal{S}} \mathbf{F} \cdot d\mathbf{S} = \frac{1}{3}\int_0^{2\pi}\int_0^{\pi} abc \sin \phi \, d\phi \, d\theta$$

$$= \frac{2\pi abc}{3}\int_0^{\pi} \sin \phi \, d\varphi = \frac{2\pi abc}{3}\left(-\cos \phi\Big|_0^{\pi}\right) = \frac{4\pi abc}{3}$$

23. Find and prove a Product Rule expressing $\text{div}(f\mathbf{F})$ in terms of $\text{div}(\mathbf{F})$ and ∇f.

SOLUTION Let $\mathbf{F} = \langle P, Q, R\rangle$. We compute $\text{div}(f\mathbf{F})$:

$$\text{div}(f\mathbf{F}) = \text{div}\,\langle fP, fQ, fR\rangle = \frac{\partial}{\partial x}(fP) + \frac{\partial}{\partial y}(fQ) + \frac{\partial}{\partial z}(fR)$$

Applying the product rule for scalar functions we obtain

$$\text{div}(f\mathbf{F}) = \left(f\frac{\partial P}{\partial x} + \frac{\partial f}{\partial x}P\right) + \left(f\frac{\partial Q}{\partial y} + \frac{\partial f}{\partial y}Q\right) + \left(f\frac{\partial R}{\partial z} + \frac{\partial f}{\partial z}R\right)$$

$$= f\left(\frac{\partial P}{\partial x} + \frac{\partial Q}{\partial y} + \frac{\partial R}{\partial z}\right) + \frac{\partial f}{\partial x}P + \frac{\partial f}{\partial y}Q + \frac{\partial f}{\partial z}R = f\text{div}(\mathbf{F}) + \mathbf{F} \cdot \nabla f$$

We thus proved the following identity:

$$\text{div}(f\mathbf{F}) = f\text{div}(\mathbf{F}) + \mathbf{F} \cdot \nabla f$$

25. Prove that $\text{div}(\nabla f \times \nabla g) = 0$.

SOLUTION We compute the cross product:

$$\nabla f \times \nabla g = \langle f_x, f_y, f_z\rangle \times \langle g_x, g_y, g_z\rangle = \begin{vmatrix} \mathbf{i} & \mathbf{j} & \mathbf{k} \\ f_x & f_y & f_z \\ g_x & g_y & g_z \end{vmatrix}$$

$$= \langle f_y g_z - f_z g_y, f_z g_x - f_x g_z, f_x g_y - f_y g_x\rangle$$

We now compute the divergence of this vector. Using the Product Rule for scalar functions and the equality of the mixed partials, we obtain

$$\text{div}(\nabla f \times \nabla g) = \frac{\partial}{\partial x}(f_y g_z - f_z g_y) + \frac{\partial}{\partial y}(f_z g_x - f_x g_z) + \frac{\partial}{\partial z}(f_x g_y - f_y g_x)$$

$$= f_{yx}g_z + f_y g_{zx} - f_{zx}g_y - f_z g_{yx} + f_{zy}g_x + f_z g_{xy} - f_{xy}g_z - f_x g_{zy} + f_{xz}g_y + f_x g_{yz}$$

$$- f_{yz}g_x - f_y g_{xz}$$

$$= (f_{yx} - f_{xy})g_z + (g_{zx} - g_{xz})f_y + (f_{xz} - f_{zx})g_y + (g_{xy} - g_{yx})f_z$$

$$+ (f_{zy} - f_{yz})g_x + (g_{yz} - g_{zy})f_x = 0$$

In Exercises 26–28, let Δ denote the Laplace operator defined by

$$\Delta\varphi = \frac{\partial\varphi^2}{\partial x^2} + \frac{\partial\varphi^2}{\partial y^2} + \frac{\partial\varphi^2}{\partial z^2}$$

27. A function φ satisfying $\Delta\varphi = 0$ is called **harmonic**.

(a) Show that $\Delta\varphi = \text{div}(\nabla\varphi)$ for any function φ.

(b) Show that φ is harmonic if and only if $\text{div}(\nabla\varphi) = 0$.

(c) Show that if $\mathbf{F}$ is the gradient of a harmonic function, then $\text{curl}(F) = 0$ and $\text{div}(F) = 0$.

(d) Show $\mathbf{F} = \left\langle xz, -yz, \frac{1}{2}(x^2 - y^2)\right\rangle$ is the gradient of a harmonic function. What is the flux of $\mathbf{F}$ through a closed surface?

SOLUTION

(a) We compute the divergence of $\nabla\varphi$:

$$\operatorname{div}(\nabla\varphi) = \operatorname{div}\left(\left\langle \frac{\partial\varphi}{\partial x}, \frac{\partial\varphi}{\partial y}, \frac{\partial\varphi}{\partial z}\right\rangle\right) = \frac{\partial^2\varphi}{\partial x^2} + \frac{\partial^2\varphi}{\partial y^2} + \frac{\partial^2\varphi}{\partial z^2} = \Delta\varphi$$

(b) In part (a) we showed that $\Delta\varphi = \operatorname{div}(\nabla\varphi)$. Therefore $\Delta\varphi = 0$ if and only if $\operatorname{div}(\nabla\varphi) = 0$. That is, φ is harmonic if and only if $\nabla\varphi$ is divergence free.

(c) We are given that $\mathbf{F} = \nabla\varphi$, where $\Delta\varphi = 0$. In part (b) we showed that

$$\operatorname{div}(\mathbf{F}) = \operatorname{div}(\nabla\varphi) = 0$$

We now show that $\operatorname{curl}(\mathbf{F}) = 0$. We have

$$\operatorname{curl}(\mathbf{F}) = \operatorname{curl}(\nabla\varphi) = \operatorname{curl}\langle \varphi_x, \varphi_y, \varphi_z\rangle = \begin{vmatrix} \mathbf{i} & \mathbf{j} & \mathbf{k} \\ \frac{\partial}{\partial x} & \frac{\partial}{\partial y} & \frac{\partial}{\partial z} \\ \varphi_x & \varphi_y & \varphi_z \end{vmatrix}$$

$$= \langle \varphi_{zy} - \varphi_{yz}, \varphi_{xz} - \varphi_{zx}, \varphi_{yx} - \varphi_{xy}\rangle = \langle 0, 0, 0\rangle = \mathbf{0}$$

The last equality is due to the equality of the mixed partials.

(d) We first show that $\mathbf{F} = \left\langle xz, -yz, \frac{x^2-y^2}{2}\right\rangle$ is the gradient of a harmonic function. We let $\varphi = \frac{x^2z}{2} - \frac{y^2z}{2}$ such that $\mathbf{F} = \nabla\varphi$. Indeed,

$$\nabla\varphi = \left\langle \frac{\partial\varphi}{\partial x}, \frac{\partial\varphi}{\partial y}, \frac{\partial\varphi}{\partial z}\right\rangle = \left\langle xz, -yz, \frac{x^2-y^2}{2}\right\rangle = \mathbf{F}$$

We show that φ is harmonic, that is, $\Delta\varphi = 0$. We compute the partial derivatives:

$$\frac{\partial\varphi}{\partial x} = xz \quad\Rightarrow\quad \frac{\partial^2\varphi}{\partial x^2} = z$$

$$\frac{\partial\varphi}{\partial y} = -yz \quad\Rightarrow\quad \frac{\partial^2\varphi}{\partial y^2} = -z$$

$$\frac{\partial\varphi}{\partial z} = \frac{x^2-y^2}{2} \quad\Rightarrow\quad \frac{\partial^2\varphi}{\partial z^2} = 0$$

Therefore,

$$\Delta\varphi = \frac{\partial^2\varphi}{\partial x^2} + \frac{\partial^2\varphi}{\partial y^2} + \frac{\partial^2\varphi}{\partial z^2} = z - z + 0 = 0$$

Since $\mathbf{F}$ is the gradient of a harmonic function, we know by part (c) that $\operatorname{div}(\mathbf{F}) = 0$. Therefore, by the Divergence Theorem, the flux of $\mathbf{F}$ through a closed surface is zero:

$$\iint_S \mathbf{F}\cdot d\mathbf{S} = \iiint_W \operatorname{div}(\mathbf{F})\,dV = \iiint_W 0\,dV = 0$$

29. The electric field due to a unit electric dipole oriented in the $\mathbf{k}$ direction is $\mathbf{E} = \nabla\left(\frac{z}{\rho^3}\right)$, where $\rho = (x^2+y^2+z^2)^{1/2}$ (Figure 17). Let $\mathbf{e}_r = \rho^{-1}\langle x, y, z\rangle$.

(a) Show that $\mathbf{E} = \rho^{-3}\mathbf{k} - 3z\rho^{-4}\mathbf{e}_r$.

(b) Calculate the flux of $\mathbf{E}$ through a sphere centered at the origin.

(c) Calculate $\operatorname{div}(\mathbf{E})$.

(d) Can we use the Divergence Theorem to compute the flux of $\mathbf{E}$ through a sphere centered at the origin?

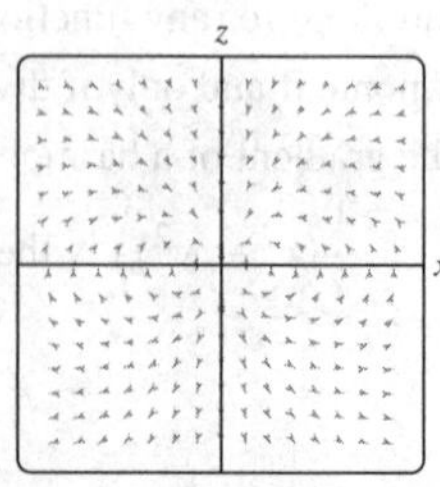

FIGURE 17 The dipole vector field restricted to the xz-plane.

SOLUTION

(a) We first compute the partial derivatives of ρ:

$$\frac{\partial \rho}{\partial x} = \frac{1}{2}(x^2+y^2+z^2)^{-1/2} \cdot 2x = \frac{x}{\rho}$$

$$\frac{\partial \rho}{\partial y} = \frac{1}{2}(x^2+y^2+z^2)^{-1/2} \cdot 2y = \frac{y}{\rho}$$

$$\frac{\partial \rho}{\partial z} = \frac{1}{2}(x^2+y^2+z^2)^{-1/2} \cdot 2z = \frac{z}{\rho} \qquad (1)$$

We compute the partial derivatives of $\frac{z}{\rho^3}$, using the Chain Rule and the partial derivatives in (1):

$$\frac{\partial}{\partial x}\left(\frac{z}{\rho^3}\right) = z\frac{\partial}{\partial x}(\rho^{-3}) = z \cdot (-3)\rho^{-4}\frac{\partial \rho}{\partial x} = -3z \cdot \rho^{-4}\frac{x}{\rho} = -\frac{3zx}{\rho^5} = -3z\rho^{-5}x$$

$$\frac{\partial}{\partial y}\left(\frac{z}{\rho^3}\right) = z\frac{\partial}{\partial y}(\rho^{-3}) = z \cdot (-3)\rho^{-4}\frac{\partial \rho}{\partial y} = -3z \cdot \rho^{-4}\frac{y}{\rho} = -3z\rho^{-5}y$$

$$\frac{\partial}{\partial z}\left(\frac{z}{\rho^3}\right) = \frac{\partial}{\partial z}(z \cdot \rho^{-3}) = 1 \cdot \rho^{-3} + z \cdot (-3)\rho^{-4}\frac{\partial \rho}{\partial z} = \rho^{-3} - 3z \cdot \rho^{-4} \cdot \frac{z}{\rho} = \rho^{-3} - 3z^2\rho^{-5}$$

Therefore,

$$\mathbf{E} = \nabla\left(\frac{z}{\rho^3}\right) = -3z\rho^{-5}x\mathbf{i} - 3z\rho^{-5}y\mathbf{j} + (\rho^{-3} - 3z^2\rho^{-5})\mathbf{k}$$

$$= \rho^{-3}\mathbf{k} - 3z\rho^{-4} \cdot \rho^{-1}(x\mathbf{i} + y\mathbf{j} + z\mathbf{k}) = \rho^{-3}\mathbf{k} - 3z\rho^{-4}\mathbf{e}_r$$

(b) To compute the flux $\iint_S \mathbf{E} \cdot d\mathbf{S}$ we use the parametrization $\Phi(\theta, \phi) = (R\cos\theta\sin\phi, R\sin\theta\sin\phi, R\cos\phi)$, $0 \le \theta \le 2\pi$, $0 \le \phi \le \pi$:

$$\mathbf{n} = R^2 \sin\phi \mathbf{e}_r$$

We compute $\mathbf{E}\left(\Phi(\theta, \phi)\right) \cdot \mathbf{n}$. Since $\rho = R$ on $\mathcal{S}$, we get

$$\mathbf{E}\left(\Phi(\theta,\phi)\right) \cdot \mathbf{n} = \left(R^{-3}\mathbf{k} - 3zR^{-4}\mathbf{e}_r\right) \cdot R^2 \sin\phi\mathbf{e}_r = R^{-1}\sin\phi\mathbf{k} \cdot \mathbf{e}_r - 3zR^{-2}\sin\phi$$

$$= R^{-1}\sin\phi\mathbf{k} \cdot R^{-1}(x\mathbf{i} + y\mathbf{j} + z\mathbf{k}) - 3zR^{-2}\sin\phi$$

$$= R^{-2}z\sin\phi - 3zR^{-2}\sin\phi = -2zR^{-2}\sin\phi$$

$$= -2R\cos\phi \cdot R^{-2}\sin\phi = -R^{-1}\sin 2\phi$$

Hence,

$$\iint_S \mathbf{E} \cdot d\mathbf{S} = \int_0^{2\pi}\int_0^{\pi} -R^{-1}\sin 2\phi \, d\phi \, d\theta = -\frac{2\pi}{R}\int_0^{\pi}\sin 2\phi d\phi = \frac{\pi}{R}\cos 2\phi\bigg|_{\phi=0}^{\pi} = 0$$

(c) We use part (a) to write the vector $\mathbf{E}$ componentwise:

$$\mathbf{E} = \rho^{-3}\mathbf{k} - 3z\rho^{-4}\mathbf{e}_r = \rho^{-3}\mathbf{k} - 3z\rho^{-4}\rho^{-1}\langle x, y, z\rangle = \left\langle -3z\rho^{-5}x, -3z\rho^{-5}y, -3z^2\rho^{-5} + \rho^{-3}\right\rangle$$

To find div($\mathbf{E}$) we compute the following derivatives, using (1) and the laws of differentiation. This gives

$$\frac{\partial}{\partial x}(-3z\rho^{-5}x) = -3z\frac{\partial}{\partial x}(\rho^{-5}x) = -3z\left(-5\rho^{-6}\frac{\partial \rho}{\partial x}x + \rho^{-5} \cdot 1\right)$$

$$= -3z\left(-5\rho^{-6}x\frac{x}{\rho} + \rho^{-5}\right) = 3z\rho^{-7}(5x^2 - \rho^2)$$

Similarly,

$$\frac{\partial}{\partial y}(-3z\rho^{-5}y) = 3z\rho^{-7}(5y^2 - \rho^2)$$

and

$$\frac{\partial}{\partial z}(-3z^2\rho^{-5} + \rho^{-3}) = -6z\rho^{-5} - 3z^2(-5)\rho^{-6}\frac{\partial \rho}{\partial z} - 3\rho^{-4}\frac{\partial \rho}{\partial z}$$

$$= -6z\rho^{-5} + 15z^2\rho^{-6}\frac{z}{\rho} - 3\rho^{-4}\frac{z}{\rho} = 3z\rho^{-7}(5z^2 - 3\rho^2)$$

Hence,

$$\text{div}(\mathbf{E}) = 3z\rho^{-7}(5x^2 - \rho^2 + 5y^2 - \rho^2 + 5z^2 - 3\rho^2) = 15z\rho^{-7}(x^2 + y^2 + z^2 - \rho^2)$$
$$= 15z\rho^{-7}(\rho^2 - \rho^2) = 0$$

(d) Since $\mathbf{E}$ is not defined at the origin, which is inside the ball $\mathcal{W}$, we cannot use the Divergence Theorem to compute the flux of $\mathbf{E}$ through the sphere.

31. Let $I = \iint_{\mathcal{S}} \mathbf{F} \cdot d\mathbf{S}$, where

$$\mathbf{F} = \left\langle \frac{2yz}{\rho^2}, -\frac{xz}{\rho^2}, -\frac{xy}{\rho^2} \right\rangle$$

($\rho = \sqrt{x^2 + y^2 + z^2}$) and $\mathcal{S}$ is the boundary of a region $\mathcal{W}$.

(a) Check that $\mathbf{F}$ is divergence-free.

(b) Show that $I = 0$ if $\mathcal{S}$ is a sphere centered at the origin. Note that the Divergence Theorem cannot be used. Why not?

(c) Give an argument showing that $I = 0$ for all $\mathcal{S}$.

SOLUTION

(a) To find div($\mathbf{F}$), we first compute the partial derivatives of $\rho = \sqrt{x^2 + y^2 + z^2}$:

$$\frac{\partial \rho}{\partial x} = \frac{2x}{2\sqrt{x^2 + y^2 + z^2}} = \frac{x}{\rho}, \quad \frac{\partial \rho}{\partial y} = \frac{2y}{2\sqrt{x^2 + y^2 + z^2}} = \frac{y}{\rho}, \quad \frac{\partial \rho}{\partial z} = \frac{2z}{2\sqrt{x^2 + y^2 + z^2}} = \frac{z}{\rho}$$

We compute the partial derivatives:

$$\frac{\partial}{\partial x}\left(\frac{2yz}{\rho^2}\right) = 2yz\frac{\partial}{\partial x}(\rho^{-2}) = 2yz \cdot (-2)\rho^{-3}\frac{\partial \rho}{\partial x} = -4yz \cdot \rho^{-3}\frac{x}{\rho} = -4xyz\rho^{-4}$$

$$\frac{\partial}{\partial y}\left(-\frac{xz}{\rho^2}\right) = -xz\frac{\partial}{\partial y}(\rho^{-2}) = -xz \cdot (-2)\rho^{-3}\frac{\partial \rho}{\partial y} = 2zx \cdot \rho^{-3}\frac{y}{\rho} = 2xyz\rho^{-4}$$

$$\frac{\partial}{\partial z}\left(-\frac{xy}{\rho^2}\right) = -xy\frac{\partial}{\partial z}(\rho^{-2}) = -xy \cdot (-2)\rho^{-3}\frac{\partial \rho}{\partial z} = 2xy\rho^{-3}\frac{z}{\rho} = 2xyz\rho^{-4}$$

The divergence of $\mathbf{F}$ is the sum of these partials. That is,

$$\text{div}(\mathbf{F}) = -4xyz\rho^{-4} + 2xyz\rho^{-4} + 2xyz\rho^{-4} = 0$$

We conclude that $\mathbf{F}$ is divergence-free.

(b) We compute the flux of $\mathbf{F}$ over $\mathcal{S}$, using the following parametrization:

$$\mathcal{S} : \Phi(\theta, \phi) = (R\cos\theta\sin\phi, R\sin\theta\sin\phi, R\cos\phi), \quad 0 \le \theta \le 2\pi, \quad 0 \le \phi \le \pi$$
$$\mathbf{n} = R^2 \sin\phi\, \mathbf{e}_r \quad \text{where } \mathbf{e}_r = \rho^{-1}\langle x, y, z\rangle$$

We compute the dot product:

$$\mathbf{F} \cdot \mathbf{n} = \left\langle \frac{2yz}{\rho^2}, -\frac{xz}{\rho^2}, -\frac{xy}{\rho^2} \right\rangle \cdot \langle x, y, z\rangle \rho^{-1} \cdot R^2 \sin\phi = (2xyz - xyz - xyz)\rho^{-3} \cdot R^2 \sin\phi = 0$$

Therefore, $\mathbf{F}(\Phi(\theta, \phi)) \cdot \mathbf{n} = 0$, so we have

$$\iint_{\mathcal{S}} \mathbf{F} \cdot d\mathbf{S} = \int_0^{2\pi}\int_0^{\pi} \mathbf{F}(\Phi(\theta, \phi)) \cdot \mathbf{n}\, d\phi\, d\theta = \int_0^{2\pi}\int_0^{\pi} 0\, d\phi\, d\theta = 0$$

The Divergence Theorem cannot be used since $\mathbf{F}$ is not defined at the origin, which is inside the ball with the boundary $\mathcal{S}$.

(c) In part (b) we showed that $\mathbf{F} \cdot \mathbf{n} = 0$ for all values of R. Therefore, $I = 0$ over all spheres centered at the origin.

Further Insights and Challenges

33. Assume that φ is harmonic. Show that $\text{div}(\varphi\nabla\varphi) = \|\nabla\varphi\|^2$ and conclude that

$$\iint_{\mathcal{S}} \varphi D_{\mathbf{e}_n}\varphi \, dS = \iiint_{\mathcal{W}} \|\nabla\varphi\|^2 \, dV$$

SOLUTION In Exercise 23 we proved the following Product Rule:

$$\text{div}(f\mathbf{F}) = \nabla f \cdot \mathbf{F} + f\text{div}(\mathbf{F})$$

We use this rule for $f = \varphi$ and $\mathbf{F} = \nabla\varphi$ to obtain

$$\text{div}(\varphi\nabla\varphi) = \nabla\varphi \cdot \nabla\varphi + \varphi\text{div}(\nabla\varphi) = \|\nabla\varphi\|^2 + \varphi\text{div}(\nabla\varphi) \tag{1}$$

By Exercise 27 part (a),

$$\text{div}(\nabla\varphi) = \Delta\varphi \tag{2}$$

Also, since φ is harmonic,

$$\Delta\varphi = 0 \tag{3}$$

Combining (1), (2), and (3), we obtain

$$\text{div}(\varphi\nabla\varphi) = \|\nabla\varphi\|^2 + \varphi \cdot 0 = \|\nabla\varphi\|^2 \tag{4}$$

Now, by the Theorem on evaluating directional derivatives,

$$D_{\mathbf{e}_n}\varphi = \nabla\varphi \cdot \mathbf{e}_n$$

Hence,

$$\iint_{\mathcal{S}} \varphi D_{\mathbf{e}_n}\varphi \, dS = \iint_{\mathcal{S}} (\varphi\nabla\varphi \cdot \mathbf{e}_n) \, dS \tag{5}$$

By the definition of the vector surface integral we have

$$\iint_{\mathcal{S}} \varphi\nabla\varphi \cdot d\mathbf{S} = \iint_{\mathcal{S}} (\varphi\nabla\varphi \cdot \mathbf{e}_n) \, dS \tag{6}$$

Combining (5) and (6) and using the Divergence Theorem and equality (4), we get

$$\iint_{\mathcal{S}} \varphi D_{\mathbf{e}_n}\varphi \, dS = \iint_{\mathcal{S}} \varphi\nabla\varphi \cdot d\mathbf{S} = \iiint_{\mathcal{W}} \text{div}(\varphi\nabla\varphi) \, dV = \iiint_{\mathcal{W}} \|\nabla\varphi\|^2 \, dV$$

35. Show that $\mathbf{F} = \langle 2y - 1, 3z^2, 2xy \rangle$ has a vector potential and find one.

SOLUTION Since $\text{div}(\mathbf{F}) = \frac{\partial}{\partial x}(2y-1) + \frac{\partial}{\partial y}(3z^2) + \frac{\partial}{\partial z}(2xy) = 0$, we know by Exercise 34 that $\mathbf{F}$ has a vector potential $\mathbf{A}$, which is

$$\mathbf{A} = \langle f, 0, g \rangle \tag{1}$$

$$f(x, y, z) = -\int_{y_0}^{y} R(x, t, z) \, dt + \int_{z_0}^{z} Q(x, y_0, t) \, dt$$

$$g(x, y, z) = \int_{y_0}^{y} P(x, t, z) \, dt$$

Hence, $P(x, y, z) = 2y - 1$, $Q(x, y, z) = 3z^2$, and $R(x, y, z) = 2xy$. We choose $z_0 = y_0 = 0$ and find f and g:

$$f(x, y, z) = -\int_0^y 2xt \, dt + \int_0^z 3t^2 \, dt = -xt^2\Big|_{t=0}^{y} + t^3\Big|_{t=0}^{z} = -xy^2 + z^3$$

$$g(x, y, z) = \int_0^y (2t - 1) \, dt = t^2 - t\Big|_{t=0}^{y} = y^2 - y$$

Substituting in (1) we obtain the vector potential

$$\mathbf{A} = \left\langle z^3 - xy^2, 0, y^2 - y \right\rangle$$

37. A vector field with a vector potential has zero flux through every closed surface in its domain. In the text, we observed that although the inverse-square radial vector field $\mathbf{F} = \frac{\mathbf{e}_r}{\rho^2}$ satisfies $\operatorname{div}(\mathbf{F}) = 0$, $\mathbf{F}$ cannot have a vector potential on its domain $\{(x, y, z) \neq (0, 0, 0)\}$ because the flux of $\mathbf{F}$ through a sphere containing the origin is nonzero.

(a) Show that the method of Exercise 34 produces a vector potential $\mathbf{A}$ such that $\mathbf{F} = \operatorname{curl}(\mathbf{A})$ on the restricted domain $\mathcal{D}$ consisting of $\mathbf{R}^3$ with the y-axis removed.

(b) Show that $\mathbf{F}$ also has a vector potential on the domains obtained by removing either the x-axis or the z-axis from $\mathbf{R}^3$.

(c) Does the existence of a vector potential on these restricted domains contradict the fact that the flux of $\mathbf{F}$ through a sphere containing the origin is nonzero?

SOLUTION

(a) We have $\mathbf{F}(x, y, z) = \frac{\mathbf{e}_r}{\rho^2} = \frac{\langle x,y,z\rangle}{(x^2+y^2+z^2)^{3/2}}$, hence

$$P(x, y, z) = \frac{x}{(x^2 + y^2 + z^2)^{3/2}}$$

$$Q(x, y, z) = \frac{y}{(x^2 + y^2 + z^2)^{3/2}}$$

$$R(x, y, z) = \frac{z}{(x^2 + y^2 + z^2)^{3/2}}$$

In Exercise 34, we defined the functions (taking $y_0 = z_0 = 0$)

$$f(x, y, z) = -\int_0^y \frac{z}{(x^2 + t^2 + z^2)^{3/2}}\,dt + \int_0^z Q(x, 0, t)\,dt = -\int_0^y \frac{z}{(x^2 + t^2 + z^2)^{3/2}}\,dt$$

$$g(x, y, z) = \int_0^y \frac{x}{(x^2 + t^2 + z^2)^{3/2}}\,dt$$

These functions are defined for $(x, z) \neq (0, 0)$, since the points with $x = 0$ and $z = 0$ are on the y-axis. (Notice that for any fixed $(x, z) \neq (0, 0)$ the interval of integration do not intersect the y-axis, therefore they are contained in the domain $\mathcal{D}$.) For $(x, z) \neq (0, 0)$ we have by the Fundamental Theorem of Calculus

$$\frac{\partial g}{\partial y} = \frac{\partial}{\partial y}\int_0^y \frac{x}{(x^2 + t^2 + z^2)^{3/2}}\,dt = \frac{x}{(x^2 + y^2 + z^2)^{3/2}} = P(x, y, z)$$

$$\begin{aligned}
\frac{\partial f}{\partial z} - \frac{\partial g}{\partial x} &= -\int_0^y \frac{(x^2 + t^2 + z^2)^{3/2} - z \cdot \frac{3}{2}(x^2 + t^2 + z^2)^{1/2} \cdot 2z}{(x^2 + t^2 + z^2)^3}\,dt \\
&\quad - \int_0^y \frac{(x^2 + t^2 + z^2)^{3/2} - x \cdot \frac{3}{2}(x^2 + t^2 + z^2)^{1/2} \cdot 2x}{(x^2 + t^2 + z^2)^3}\,dt \\
&= -\int_0^y \frac{(x^2 + t^2 + z^2)^{1/2}(x^2 + t^2 + z^2 - 3z^2)}{(x^2 + t^2 + z^2)^3}\,dt \\
&\quad - \int_0^y \frac{(x^2 + t^2 + z^2)^{1/2}(x^2 + t^2 + z^2 - 3x^2)}{(x^2 + t^2 + z^2)^3}\,dt \\
&= -\int_0^y \frac{x^2 + t^2 - 2z^2 + t^2 + z^2 - 2x^2}{(x^2 + t^2 + z^2)^{5/2}}\,dt \\
&= \int_0^y \frac{x^2 - 2t^2 + z^2}{(x^2 + t^2 + z^2)^{5/2}}\,dt = \frac{y}{(x^2 + y^2 + z^2)^{3/2}} = Q(x, y, z)
\end{aligned}$$

The last integral can be verified by showing that

$$\frac{\partial}{\partial y}\left(\frac{y}{(x^2 + y^2 + z^2)^{3/2}}\right) = \frac{x^2 - 2y^2 + z^2}{(x^2 + y^2 + z^2)^{5/2}}$$

and

$$\frac{\partial f}{\partial y} = -\frac{\partial}{\partial y}\int_0^y \frac{z}{(x^2 + t^2 + z^2)^{3/2}}\,dt = -\frac{z}{(x^2 + y^2 + z^2)^{3/2}} = -R(x, y, z)$$

We conclude that the vector $\mathbf{A} = \langle f, 0, g\rangle$ is a vector potential of $\mathbf{F}$ in $\mathcal{D}$, since

$$\text{curl}(\mathbf{A}) = \left\langle \frac{\partial g}{\partial y}, \frac{\partial f}{\partial z} - \frac{\partial g}{\partial x}, -\frac{\partial f}{\partial y} \right\rangle = \langle P, Q, R\rangle = \mathbf{F}.$$

(b) Suppose we remove the x-axis. In this case, we let

$$\mathbf{A} = \langle 0, f, g\rangle$$

$$g(x, y, z) = -\int_{x_0}^{x} Q(t, y, z)dt + \int_{y_0}^{y} P(x_0, t, z)\,dt$$

$$f(x, y, z) = \int_{x_0}^{x} R(t, y, z)\,dt$$

Using similar procedure to that in Exercise 34, one can show that

$$\mathbf{F} = \text{curl}(\mathbf{A}).$$

In removing the z-axis the proof is similar, with corresponding modifications of the functions in Exercise 34.
(c) The ball inside any sphere containing the origin must intersect the x, y, and z axes; therefore, $\mathbf{F}$ does not have a vector potential in the ball, and the flux of $\mathbf{F}$ through the sphere may differ from zero, as in our example.

CHAPTER REVIEW EXERCISES

1. Let $\mathbf{F}(x, y) = \langle x + y^2, x^2 - y\rangle$ and let $\mathcal{C}$ be the unit circle, oriented counterclockwise. Evaluate $\oint_{\mathcal{C}} \mathbf{F}\cdot d\mathbf{s}$ directly as a line integral and using Green's Theorem.

SOLUTION We parametrize the unit circle by $\mathbf{c}(t) = (\cos t, \sin t)$, $0 \le t \le 2\pi$. Then, $\mathbf{c}'(t) = \langle -\sin t, \cos t\rangle$ and $\mathbf{F}(\mathbf{c}(t)) = (\cos t + \sin^2 t, \cos^2 t - \sin t)$. We compute the dot product:

$$\begin{aligned}
\mathbf{F}(\mathbf{c}(t))\cdot\mathbf{c}'(t) &= \left\langle \cos t + \sin^2 t, \cos^2 t - \sin t\right\rangle\cdot\langle -\sin t, \cos t\rangle \\
&= (-\sin t)(\cos t + \sin^2 t) + \cos t(\cos^2 t - \sin t) \\
&= \cos^3 t - \sin^3 t - 2\sin t\cos t
\end{aligned}$$

The line integral is thus

$$\begin{aligned}
\int_{\mathcal{C}} \mathbf{F}(\mathbf{c}(t))\cdot\mathbf{c}'(t)\,dt &= \int_0^{2\pi} \left(\cos^3 t - \sin^3 t - 2\sin t\cos t\right) dt \\
&= \int_0^{2\pi} \cos^3 t\,dt - \int_0^{2\pi} \sin^3 t\,dt - \int_0^{2\pi} \sin 2t\,dt \\
&= \frac{\cos^2 t\sin t}{3} + \frac{2\sin t}{3}\Big|_0^{2\pi} + \left(\frac{\sin^2 t\cos t}{3} + \frac{2\cos t}{3}\right)\Big|_0^{2\pi} + \frac{\cos 2t}{2}\Big|_0^{2\pi} = 0
\end{aligned}$$

We now compute the integral using Green's Theorem. We compute the curl of $\mathbf{F}$. Since $P = x + y^2$ and $Q = x^2 - y$, we have

$$\frac{\partial Q}{\partial x} - \frac{\partial P}{\partial y} = 2x - 2y$$

Thus,

$$\int_{\mathcal{C}} \mathbf{F}\cdot d\mathbf{s} = \iint_{\mathcal{D}} (2x - 2y)\,dx\,dy$$

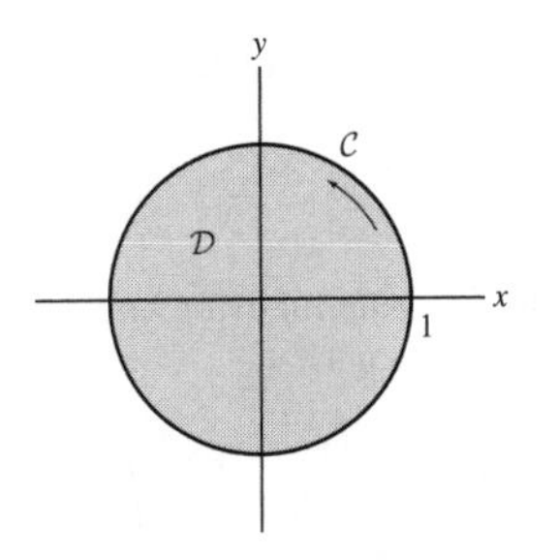

We compute the double integral by converting to polar coordinates. We get

$$\int_{C} \mathbf{F} \cdot d\mathbf{s} = \int_0^{2\pi} \int_0^1 (2r\cos\theta - 2r\sin\theta) r\, dr\, d\theta = \int_0^{2\pi} \int_0^1 2r^2(\cos\theta - \sin\theta)\, dr\, d\theta$$

$$= \left(\int_0^1 2r^2\, dr\right)\left(\int_0^{2\pi} (\cos\theta - \sin\theta)\, d\theta\right) = \left(\frac{2}{3}r^3\Big|_0^1\right)\left(\sin\theta + \cos\theta\Big|_0^{2\pi}\right) = \frac{2}{3}(1-1) = 0$$

In Exercises 3–6, use Green's Theorem to evaluate the line integral around the given closed curve.

3. $\oint_C xy^3\, dx + x^3 y\, dy$, where C is the rectangle $-1 \le x \le 2$, $-2 \le y \le 3$, oriented counterclockwise.

SOLUTION

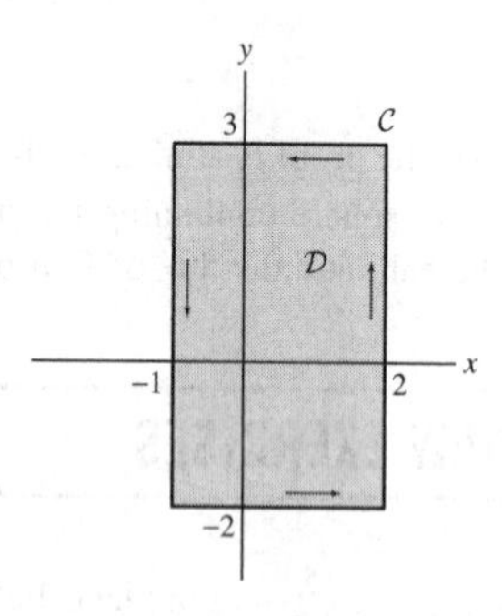

Since $P = xy^3$, $Q = x^3 y$ the curl of $\mathbf{F}$ is

$$\frac{\partial Q}{\partial x} - \frac{\partial P}{\partial y} = 3x^2 y - 3xy^2$$

By Green's Theorem we obtain

$$\int_C xy^3\, dx + x^3 y\, dy = \iint_{\mathcal{D}} (3x^2y - 3xy^2)\, dx\, dy = \int_{-2}^{3} \int_{-1}^{2} (3x^2y - 3xy^2)\, dx\, dy$$

$$= \int_{-2}^{3} x^3 y - \frac{3x^2y^2}{2}\Big|_{x=-1}^{2} dy = \int_{-2}^{3} \left((8y - 6y^2) - \left(-y - \frac{3y^2}{2}\right)\right) dy$$

$$= \int_{-2}^{3} \left(-\frac{9y^2}{2} + 9y\right) dy = -\frac{3y^3}{2} + \frac{9y^2}{2}\Big|_{-2}^{3} = \left(-\frac{81}{2} + \frac{81}{2}\right) - (12 + 18) = -30$$

5. $\oint_C y^2\, dx - x^2\, dy$, where C consists of the arcs $y = x^2$ and $y = \sqrt{x}$, $0 \le x \le 1$, oriented clockwise.

SOLUTION We compute the curl of $\mathbf{F}$.

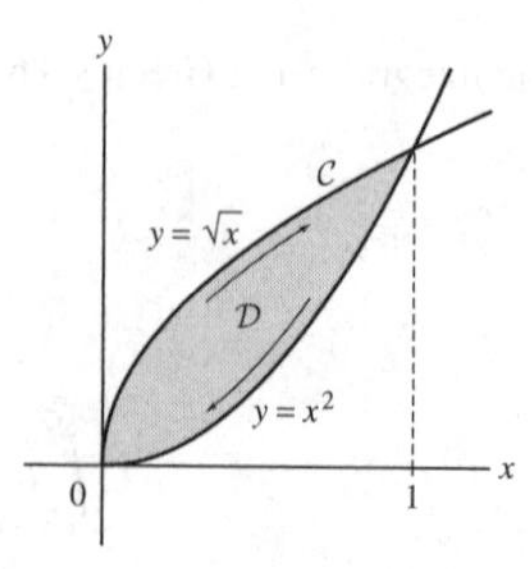

We have $P = y^2$ and $Q = -x^2$, hence

$$\frac{\partial Q}{\partial x} - \frac{\partial P}{\partial y} = -2x - 2y$$

We now compute the line integral using Green's Theorem. Since the curve is oriented clockwise, we consider the negative of the double integrals. We get

$$\int_C y^2\, dx - x^2\, dy = -\iint_{\mathcal{D}} (-2x - 2y)\, dA = -\int_0^1 \int_{x^2}^{\sqrt{x}} (-2x - 2y)\, dy\, dx$$

$$= \int_0^1 2xy + y^2 \Big|_{y=x^2}^{\sqrt{x}} dx = \int_0^1 \left((2x\sqrt{x} + x) - (2x \cdot x^2 + x^4) \right) dx$$

$$= \int_0^1 (-x^4 - 2x^3 + 2x^{3/2} + x)\, dx = -\frac{x^5}{5} - \frac{x^4}{2} + \frac{4x^{5/2}}{5} + \frac{x^2}{2} \Big|_0^1$$

$$= -\frac{1}{5} - \frac{1}{2} + \frac{4}{5} + \frac{1}{2} = \frac{3}{5}$$

7. Let $\mathbf{c}(t) = \left(t^2(1-t), t(t-1)^2\right)$.

(a) GU Plot the path $\mathbf{c}(t)$ for $0 \le t \le 1$.

(b) Calculate the area of the region enclosed by $\mathbf{c}(t)$ for $0 \le t \le 1$.

SOLUTION

(a) The path $\mathbf{c}(t)$ for $0 \le t \le 1$ is shown in the figure:

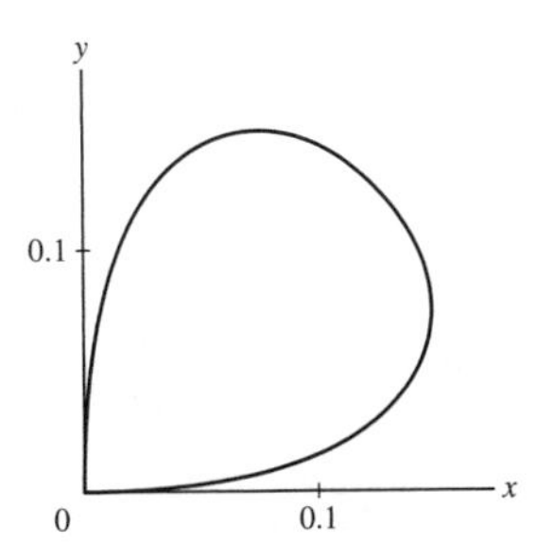

Note that the path is traced out clockwise as t goes from 0 to 1.

(b) We use the formula for the area enclosed by a closed curve,

$$A = \frac{1}{2} \int_C (x\, dy - y\, dx)$$

We compute the line integral. Since $x = t^2(1-t)$ and $y = t(t-1)^2$, we have

$$dx = \left(2t(1-t) - t^2\right) dt = \left(2t - 3t^2\right) dt$$

$$dy = (t-1)^2 + t \cdot 2(t-1) = (t-1)(3t-1)\, dt$$

Therefore,

$$x\, dy - y\, dx = t^2(1-t) \cdot (t-1)(3t-1)\, dt - t(t-1)^2 \cdot (2t - 3t^2)\, dt = t^2(t-1)^2\, dt$$

We obtain the following integral (note that the path must be counterclockwise):

$$A = \frac{1}{2} \int_1^0 -t^2(t-1)^2\, dt = \frac{1}{2} \int_0^1 (t^4 - 2t^3 + t^2)\, dt = \frac{1}{2} \left(\frac{t^5}{5} - \frac{t^4}{2} + \frac{t^3}{3} \Big|_0^1 \right) = \frac{1}{60}$$

In Exercises 9–12, calculate the curl and divergence of the vector field.

9. $\mathbf{F} = y\mathbf{i} - z\mathbf{k}$

SOLUTION We compute the curl of the vector field,

$$\text{curl}(\mathbf{F}) = \begin{vmatrix} \mathbf{i} & \mathbf{j} & \mathbf{k} \\ \frac{\partial}{\partial x} & \frac{\partial}{\partial y} & \frac{\partial}{\partial z} \\ y & 0 & -z \end{vmatrix}$$

$$= \left(\frac{\partial}{\partial y}(-z) - \frac{\partial}{\partial z}(0) \right) \mathbf{i} - \left(\frac{\partial}{\partial x}(-z) - \frac{\partial}{\partial z}(y) \right) \mathbf{j} + \left(\frac{\partial(0)}{\partial x} - \frac{\partial(y)}{\partial y} \right) \mathbf{k}$$

$$= 0\mathbf{i} + 0\mathbf{j} - 1\mathbf{k} = -\mathbf{k}$$

The divergence of $\mathbf{F}$ is

$$\text{div}(\mathbf{F}) = \frac{\partial}{\partial x}(y) + \frac{\partial}{\partial y}(0) + \frac{\partial}{\partial z}(-z) = 0 + 0 - 1 = -1.$$

11. $\mathbf{F} = \nabla(e^{-x^2-y^2-z^2})$

SOLUTION In Exercise 8 we proved the identity $\text{curl}(\nabla\varphi) = \mathbf{0}$. Here, $\varphi = e^{-x^2-y^2-z^2}$, and we have $\text{curl}\left(\nabla\left(e^{-x^2-y^2-z^2}\right)\right) = \mathbf{0}$. To compute $\text{div}\,\mathbf{F}$, we first write $\mathbf{F}$ explicitly:

$$\mathbf{F} = \nabla\left(e^{-x^2-y^2-z^2}\right) = \left\langle -2xe^{-x^2-y^2-z^2}, -2ye^{-x^2-y^2-z^2}, -2ze^{-x^2-y^2-z^2}\right\rangle = \langle P, Q, R\rangle$$

$$\begin{aligned}\text{div}(\mathbf{F}) &= \frac{\partial P}{\partial x} + \frac{\partial Q}{\partial y} + \frac{\partial R}{\partial z}\\ &= \left(-2e^{-x^2-y^2-z^2} + 4x^2e^{-x^2-y^2-z^2}\right) + \left(-2e^{-x^2-y^2-z^2} + 4y^2e^{-x^2-y^2-z^2}\right)\\ &\quad + \left(-2e^{-x^2-y^2-z^2} + 4z^2e^{-x^2-y^2-z^2}\right)\\ &= 2e^{-x^2-y^2-z^2}\left(2(x^2+y^2+z^2) - 3\right)\end{aligned}$$

13. Recall that if F_1, F_2, and F_3 are differentiable functions of one variable, then

$$\text{curl}\left(\langle F_1(x), F_2(y), F_3(z)\rangle\right) = \mathbf{0}$$

Use this to calculate the curl of

$$\mathbf{F} = \left\langle x^2 + y^2, \ln y + z^2, z^3\sin(z^2)e^{z^3}\right\rangle$$

SOLUTION We use the linearity of the curl and the property mentioned in the exercise to compute the curl of $\mathbf{F}$:

$$\begin{aligned}\text{curl}\,\mathbf{F} &= \text{curl}\left(\left\langle x^2+y^2, \ln y + z^2, z^3\sin\left(z^2\right)e^{z^3}\right\rangle\right) = \text{curl}\left(\left\langle x^2, \ln y, z^3\sin(z^2)e^{z^3}\right\rangle\right) + \text{curl}\left(\left\langle y^2, z^2, 0\right\rangle\right)\\ &= 0 + \text{curl}\left\langle y^2, z^2, 0\right\rangle = \left\langle \frac{\partial}{\partial y}(0) - \frac{\partial}{\partial z}z^2, \frac{\partial}{\partial z}y^2 - \frac{\partial}{\partial x}(0), \frac{\partial}{\partial x}z^2 - \frac{\partial}{\partial y}y^2\right\rangle = \langle -2z, 0, -2y\rangle\end{aligned}$$

15. Verify the identities of Exercises 22 and 24 in Section 18.3 for the vector fields $\mathbf{F} = \langle xz, ye^x, yz\rangle$ and $\mathbf{G} = \langle z^2, xy^3, x^2y\rangle$.

SOLUTION We first show $\text{div}(\text{curl}(\mathbf{F})) = 0$. Let $\mathbf{F} = \langle P, Q, R\rangle = \langle xz, ye^x, yz\rangle$. We compute the curl of $\mathbf{F}$:

$$\text{curl}\mathbf{F} = \begin{vmatrix} \mathbf{i} & \mathbf{j} & \mathbf{k}\\ \frac{\partial}{\partial x} & \frac{\partial}{\partial y} & \frac{\partial}{\partial z}\\ P & Q & R\end{vmatrix} = \left\langle \frac{\partial R}{\partial y} - \frac{\partial Q}{\partial z}, \frac{\partial P}{\partial z} - \frac{\partial R}{\partial x}, \frac{\partial Q}{\partial x} - \frac{\partial P}{\partial y}\right\rangle$$

Substituting in the appropriate values for P, Q, R and taking derivatives, we get

$$\text{curl}\mathbf{F} = \langle z - 0, x - 0, ye^x - 0\rangle$$

Thus,

$$\text{div}\,(\text{curl}(\mathbf{F})) = (z)_x + (x)_y + (ye^x)_z = 0 + 0 + 0 = 0.$$

Likewise, for $\mathbf{G} = \langle P, Q, R\rangle = \left\langle z^2, xy^3x^2y\right\rangle$, we compute the curl of $\mathbf{G}$:

$$\text{curl}\mathbf{G} = \begin{vmatrix} \mathbf{i} & \mathbf{j} & \mathbf{k}\\ \frac{\partial}{\partial x} & \frac{\partial}{\partial y} & \frac{\partial}{\partial z}\\ P & Q & R\end{vmatrix} = \left\langle \frac{\partial R}{\partial y} - \frac{\partial Q}{\partial z}, \frac{\partial P}{\partial z} - \frac{\partial R}{\partial x}, \frac{\partial Q}{\partial x} - \frac{\partial P}{\partial y}\right\rangle$$

Substituting in the appropriate values for P, Q, R and taking derivatives, we get

$$\text{curl}\mathbf{G} = \left\langle x^2 - 0, 2z - 2xy, y^3 - 0\right\rangle$$

Thus,

$$\text{div}\,(\text{curl}(\mathbf{G})) = (x^2)_x + (2z - 2xy)_y + (y^3)_z = 2x - 2x = 0.$$

We now work on the second identity. For $\mathbf{F} = \langle xz, ye^x, yz\rangle$ and $\mathbf{G} = \langle z^2, xy^3, x^2y\rangle$, it is easy to calculate

$$\mathbf{F}\times\mathbf{G} = \langle x^2y^2e^x - xy^4z, yz^3 - x^3yz, x^2y^3z - yz^2e^x\rangle$$

Thus,

$$\text{div}(\mathbf{F} \times \mathbf{G}) = (2xy^2e^x + x^2y^2e^x - y^4z) + (z^3 - x^3z) + (x^2y^3 - 2yze^x)$$

On the other hand, from our work above,

$$\text{curl}\mathbf{F} = \langle z, x, ye^x \rangle$$

$$\text{curl}\mathbf{G} = \left\langle x^2, 2z - 2xy, y^3 \right\rangle$$

So, we calculate

$$\begin{aligned}\mathbf{G} \cdot \text{curl}\,\mathbf{F} - \mathbf{F} \cdot \text{curl}\,\mathbf{G} &= z^2 \cdot z + xy^3 \cdot x + x^2y \cdot ye^x - xz \cdot x^2 - ye^x \cdot (2z - 2xy) - yz \cdot y^3 \\ &= z^3 + x^2y^3 + x^2y^2e^x + 2xy^2e^x - x^3z - 2yze^x - y^4z \\ &= (2xy^2e^x + x^2y^2e^x - y^4z) + (z^3 - x^3z) + (x^2y^3 - 2yze^x) = \text{div}(\mathbf{F} \times \mathbf{G})\end{aligned}$$

17. Prove that if **F** is a gradient vector field, then the flux of curl(**F**) through a smooth surface $\mathcal{S}$ (whether closed or not) is equal to zero.

SOLUTION If **F** is a gradient vector field, then **F** is conservative; therefore the line integral of **F** over any closed curve is zero. Combining with Stokes' Theorem yields

$$\iint_{\mathcal{S}} \text{curl}(\mathbf{F}) \cdot d\mathbf{S} = \int_{\partial\mathcal{S}} \mathbf{F} \cdot d\mathbf{s} = 0$$

In Exercises 19–20, let $\mathbf{F} = \langle z^2, x + z, y^2 \rangle$ *and let* $\mathcal{S}$ *be the upper half of the ellipsoid* $\frac{x^2}{4} + y^2 + z^2 = 1$*, oriented by outward-pointing normals.*

19. Use Stokes' Theorem to compute $\iint_{\mathcal{S}} \text{curl}(\mathbf{F}) \cdot d\mathbf{S}$.

SOLUTION We compute the curl of $\mathbf{F} = \left\langle z^2, x + z, y^2 \right\rangle$:

$$\text{curl}(\mathbf{F}) = \begin{vmatrix} \mathbf{i} & \mathbf{j} & \mathbf{k} \\ \frac{\partial}{\partial x} & \frac{\partial}{\partial y} & \frac{\partial}{\partial z} \\ z^2 & x + z & y^2 \end{vmatrix} = (2y - 1)\mathbf{i} - (0 - 2z)\mathbf{j} + (1 - 0)\mathbf{k} = \langle 2y - 1, 2z, 1 \rangle$$

Let $\mathcal{C}$ denote the boundary of $\mathcal{S}$, that is, the ellipse $\frac{x^2}{4} + y^2 = 1$ in the xy-plane, oriented counterclockwise. Then by Stoke's Theorem we have

$$\iint_{\mathcal{S}} \text{curl}(\mathbf{F}) \cdot d\mathbf{S} = \int_{\mathcal{C}} \mathbf{F} \cdot d\mathbf{s} \qquad (1)$$

We parametrize $\mathcal{C}$ by

$$\mathcal{C} : r(t) = (2\cos t, \sin t, 0), \quad 0 \le t \le 2\pi$$

Then

$$\mathbf{F}(r(t)) \cdot r'(t) = \left\langle 0, 2\cos t, \sin^2 t \right\rangle \cdot \langle -2\sin t, \cos t, 0 \rangle = 2\cos^2 t$$

Combining with (1) gives

$$\iint_{\mathcal{S}} \text{curl}(\mathbf{F}) \cdot d\mathbf{s} = \int_0^{2\pi} 2\cos^2 t\, dt = t + \frac{\sin 2t}{2}\bigg|_0^{2\pi} = 2\pi$$

21. Use Stokes' Theorem to evaluate $\oint_{\mathcal{C}} \langle y, z, x \rangle \cdot d\mathbf{s}$, where $\mathcal{C}$ is the curve in Figure 2.

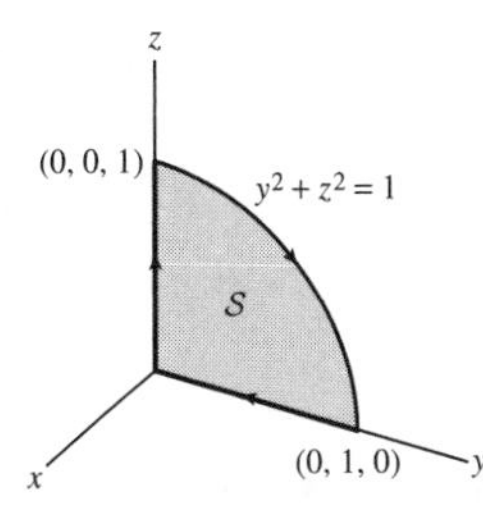

FIGURE 2

SOLUTION We compute the curl of $\mathbf{F} = \langle y, z, x \rangle$:

$$\text{curl}(\mathbf{F}) = \begin{vmatrix} \mathbf{i} & \mathbf{j} & \mathbf{k} \\ \frac{\partial}{\partial x} & \frac{\partial}{\partial y} & \frac{\partial}{\partial z} \\ y & z & x \end{vmatrix} = -\mathbf{i} - \mathbf{j} - \mathbf{k} = \langle -1, -1, -1 \rangle$$

By Stokes' Theorem, we have

$$\int_{\mathcal{C}} \langle y, z, x \rangle \cdot d\mathbf{s} = \iint_{\mathcal{S}} \text{curl}(\mathbf{F}) \cdot d\mathbf{S} = \iint_{\mathcal{S}} (\text{curl}(\mathbf{F}) \cdot \mathbf{e}_n)\, dS$$

Since the boundary $\mathcal{C}$ of the quarter circle $\mathcal{S}$ is oriented clockwise, the induced orientation on $\mathcal{S}$ is normal pointing in the negative x direction. Thus,

$$\mathbf{e}_n = \langle -1, 0, 0 \rangle .$$

Hence,

$$\text{curl}(\mathbf{F}) \cdot \mathbf{e}_n = \langle -1, -1, -1 \rangle \cdot \langle -1, 0, 0 \rangle = 1.$$

Combining with (1) we get

$$\int_{\mathcal{C}} \langle y, z, x \rangle \cdot d\mathbf{s} = \iint_{\mathcal{S}} 1\, ds = \text{Area}\,(\mathcal{S}) = \frac{\pi}{4}$$

In Exercises 23–26, use the Divergence Theorem to calculate $\iint_{\mathcal{S}} \mathbf{F} \cdot d\mathbf{S}$ *for the given vector field and surface.*

23. $\mathbf{F} = \langle xy, yz, x^2z + z^2 \rangle$, $\mathcal{S}$ is the boundary of the box $[0, 1] \times [2, 4] \times [1, 5]$.

SOLUTION

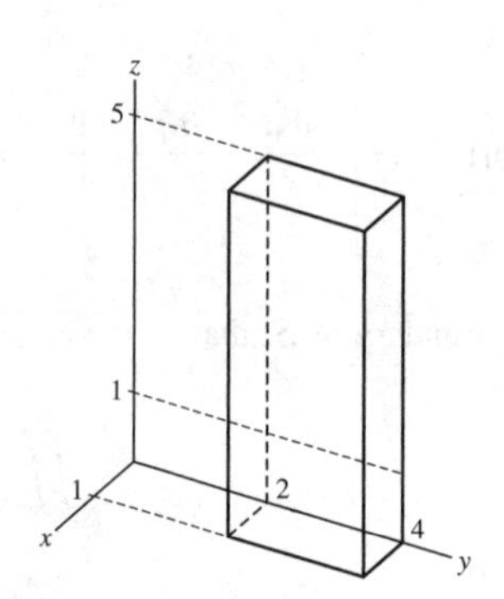

We compute the divergence of $\mathbf{F} = \langle xy, yz, x^2z + z^2 \rangle$:

$$\text{div}(\mathbf{F}) = \frac{\partial}{\partial x} xy + \frac{\partial}{\partial y} yz + \frac{\partial}{\partial z}(x^2z + z^2) = y + z + x^2 + 2z = x^2 + y + 3z$$

The Divergence Theorem gives

$$\begin{aligned}
\iint_{\mathcal{S}} \langle xy, yz, x^2z + z^2 \rangle \cdot d\mathbf{S} &= \int_1^5 \int_2^4 \int_0^1 (x^2 + y + 3z)\, dx\, dy\, dz = \int_1^5 \int_2^4 \frac{x^3}{3} + (y + 3z)x \bigg|_{x=0}^{1} dy\, dz \\
&= \int_1^5 \int_1^4 \left(\frac{1}{3} + y + 3z \right) dy\, dz = \int_1^5 \frac{1}{3}y + \frac{1}{2}y^2 + 3zy \bigg|_{y=1}^{4} dz \\
&= \int_1^5 \left(\left(\frac{4}{3} + \frac{16}{2} + 12z \right) - \left(\frac{1}{3} + \frac{1}{2} + 3z \right) \right) dz = \int_1^5 \left(\frac{17}{2} + 9z \right) dz \\
&= \frac{17z}{2} + \frac{9z^2}{2} \bigg|_1^5 = \frac{(17 \cdot 5 + 9 \cdot 25) - (17 + 9)}{2} = 142
\end{aligned}$$

25. $\mathbf{F} = \langle xyz + xy, \frac{1}{2}y^2(1 - z) + e^x, e^{x^2+y^2} \rangle$, $\mathcal{S}$ is the boundary of the solid bounded by the cylinder $x^2 + y^2 = 16$ and the planes $z = 0$ and $z = y - 4$.

SOLUTION We compute the divergence of $\mathbf{F}$:

$$\text{div}(\mathbf{F}) = \frac{\partial}{\partial x}(xyz + xy) + \frac{\partial}{\partial y}\left(\frac{y^2}{2}(1-z) + e^x\right) + \frac{\partial}{\partial z}(e^{x^2+y^2}) = yz + y + y(1-z) = 2y$$

Let $\mathcal{S}$ denote the surface of the solid $\mathcal{W}$. The Divergence Theorem gives

$$\iint_{\mathcal{S}} \mathbf{F} \cdot d\mathbf{S} = \iiint_{\mathcal{W}} \text{div}(\mathbf{F})\, dV = \iiint_{\mathcal{W}} 2y\, dV = \iint_{\mathcal{D}} \int_{y-4}^{0} 2y\, dz\, dx\, dy$$

$$= \iint_{\mathcal{D}} 2yz\Big|_{z=y-4}^{0} dx\, dy = \iint_{\mathcal{D}} 2y\,(0 - (y-4))\, dx\, dy = \iint_{\mathcal{D}} (8y - 2y^2)\, dx\, dy$$

We convert the integral to polar coordinates:

$$\iint_{\mathcal{S}} \mathbf{F} \cdot d\mathbf{S} = \int_0^{2\pi} \int_0^4 (8r\cos\theta - 2r^2\cos^2\theta) r\, dr\, d\theta$$

$$= 8\left(\int_0^4 r^2\, dr\right)\left(\int_0^{2\pi} \cos\theta\, d\theta\right) - \left(\int_0^4 r^3\, dr\right)\left(\int_0^{2\pi} 2\cos^2\theta\, d\theta\right)$$

$$= 0 - \left(\frac{r^4}{4}\Big|_0^4\right)\left(\theta + \frac{\sin 2\theta}{2}\Big|_0^{2\pi}\right) = -\frac{4^4}{4} \cdot 2\pi = -128\pi$$

27. Find the volume of a region $\mathcal{W}$ if

$$\iint_{\partial\mathcal{W}} \left\langle x + xy + z, x + 3y - \frac{1}{2}y^2, 4z \right\rangle \cdot d\mathbf{S} = 16$$

SOLUTION Let $\mathbf{F} = \left\langle x + xy + z, x + 3y - \frac{1}{2}y^2, 4z\right\rangle$. We compute the divergence of $\mathbf{F}$:

$$\text{div}(\mathbf{F}) = \frac{\partial}{\partial x}(x + xy + z) + \frac{\partial}{\partial y}\left(x + 3y - \frac{1}{2}y^2\right) + \frac{\partial}{\partial z}(4z) = 1 + y + 3 - y + 4 = 8$$

Using the Divergence Theorem and the given information, we obtain

$$16 = \iint_{\mathcal{S}} \mathbf{F} \cdot d\mathbf{S} = \iint_{\mathcal{W}} \text{div}(\mathbf{F})\, dV = \iint_{\mathcal{W}} 8\, dV = 8\iint_{\mathcal{W}} 1\, dV = 8\,\text{Volume}\,(\mathcal{W})$$

That is,

$$16 = 8\,\text{Volume}\,(\mathcal{W})$$

or

$$\text{Volume}\,(\mathcal{W}) = 2$$

In Exercises 29–32, let $\mathbf{F}$ *be a vector field whose curl and divergence at the origin are*

$$\text{curl}(\mathbf{F})(0,0,0) = \langle 2, -1, 4\rangle, \qquad \text{div}(\mathbf{F})(0,0,0) = -2$$

29. Estimate $\oint_C \mathbf{F} \cdot d\mathbf{s}$, where C is the circle of radius 0.03 in the xy-plane centered at the origin.

SOLUTION We use the estimation

$$\int_C \mathbf{F} \cdot d\mathbf{s} \approx (\text{curl}(\mathbf{F})(\mathbf{0}) \cdot \mathbf{e}_n)\,\text{Area}(\mathcal{R})$$

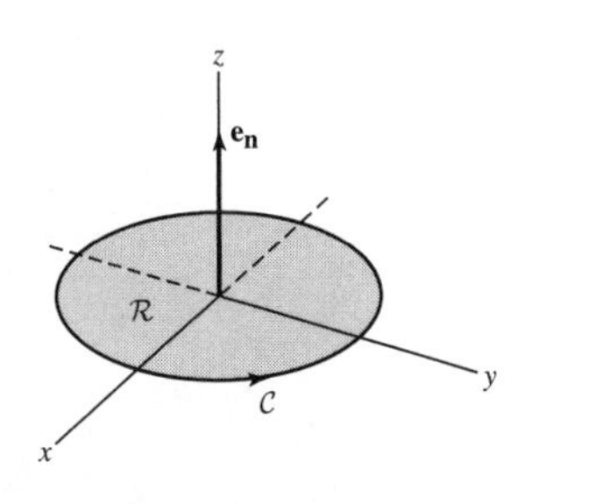

The unit normal vector to the disk $\mathcal{R}$ is $\mathbf{e}_n = \mathbf{k} = \langle 0, 0, 1 \rangle$. The area of the disk is

$$\text{Area}(\mathcal{R}) = \pi \cdot 0.03^2 = 0.0009\pi.$$

Using the given curl at the origin, we have

$$\int_C \mathbf{F} \cdot d\mathbf{s} \approx \langle 2, -1, 4 \rangle \cdot \langle 0, 0, 1 \rangle \cdot 0.0009\pi = 4 \cdot 0.0009\pi \approx 0.0113$$

31. Suppose that **F** is the velocity field of a fluid and imagine placing a small paddle wheel at the origin. Find the equation of the plane in which the paddle wheel should be placed to make it rotate as quickly as possible.

SOLUTION The paddle wheel has the maximum spin when the circulation of the velocity field **F** around the wheel is maximum. The maximum circulation occurs when $\mathbf{e}_n$, and the curl of **F** at the origin (i.e., the vector $\langle 2, -1, 4 \rangle$) point in the same direction. Therefore, the plane in which the paddle wheel should be placed is the plane through the origin with the normal $\langle 2, -1, 4 \rangle$. This plane has the equation, $2x - y + 4z = 0$.

33. The velocity field of a fluid (in meters per second) is

$$\mathbf{F} = \langle x^2 + y^2, 0, z^2 \rangle$$

Let $\mathcal{W}$ be the region between the hemisphere

$$\mathcal{S} = \left\{ (x, y, z) : x^2 + y^2 + z^2 = 1, \quad x, y, z \geq 0 \right\}$$

and the disk

$$\mathcal{D} = \left\{ (x, y, 0) : x^2 + y^2 \leq 1 \right\}$$

in the xy-plane.

(a) Show that no fluid flows across $\mathcal{D}$.

(b) Use (a) to show that the rate of fluid flow across $\mathcal{S}$ is equal to $\iiint_{\mathcal{W}} \operatorname{div}(\mathbf{F})\, dV$. Compute this triple integral using spherical coordinates.

SOLUTION

(a) To show that no fluid flows across $\mathcal{D}$, we show that the normal component of **F** at each point on $\mathcal{D}$ is zero. At each point $P = (x, y, 0)$ on the xy-plane,

$$\mathbf{F}(P) = \left\langle x^2 + y^2, 0, 0^2 \right\rangle = \left\langle x^2 + y^2, 0, 0 \right\rangle.$$

Moreover, the unit normal vector to the xy-plane is $\mathbf{e}_n = (0, 0, 1)$. Therefore,

$$\mathbf{F}(P) \cdot \mathbf{e}_n = \left\langle x^2 + y^2, 0, 0 \right\rangle \cdot \langle 0, 0, 1 \rangle = 0.$$

Since $\mathcal{D}$ is contained in the xy-plane, we conclude that the normal component of **F** at each point on $\mathcal{D}$ is zero. Therefore, no fluid flows across $\mathcal{D}$.

(b) By the Divergence Theorem and the linearity of the flux we have

$$\iint_{\mathcal{S}} \mathbf{F} \cdot d\mathbf{S} + \iint_{\mathcal{D}} \mathbf{F} \cdot d\mathbf{S} = \iiint_{\mathcal{W}} \operatorname{div}(\mathbf{F})\, dV$$

Since the flux through the disk $\mathcal{D}$ is zero, we have

$$\iint_{\mathcal{S}} \mathbf{F} \cdot d\mathbf{S} = \iiint_{\mathcal{W}} \operatorname{div}(\mathbf{F})\, dV \tag{1}$$

To compute the triple integral, we first compute $\operatorname{div}(\mathbf{F})$:

$$\operatorname{div}(\mathbf{F}) = \frac{\partial}{\partial x}(x^2 + y^2) + \frac{\partial}{\partial y}(0) + \frac{\partial}{\partial z}(z^2) = 2x + 2z = 2(x + z).$$

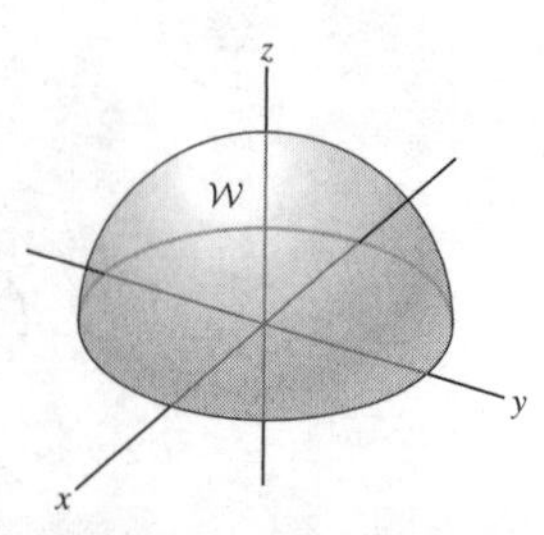

Using spherical coordinate we get

$$\iiint_{\mathcal{W}} \operatorname{div}(\mathbf{F})\,dV = 2\int_0^{\pi/2}\int_0^{2\pi}\int_0^1 (\rho\sin\phi\cos\theta + \rho\cos\phi)\rho^2\sin\phi\,d\rho\,d\phi$$

$$= 2\int_0^1 \rho^3 d\rho\left(\left(\int_0^{\pi/2}\sin^2\phi\,d\phi\right)\left(\int_0^{2\pi}\cos\theta\,d\theta\right) + 2\pi\int_0^{\pi/2}\cos\phi\sin\phi\,d\rho\right)$$

$$= \frac{1}{2}\left(0 + \pi\int_0^{\pi/2}\sin 2\phi\,d\phi\right) = \frac{\pi}{2}\left(-\frac{\cos 2\phi}{2}\right)\bigg|_0^{\pi/2} = -\frac{\pi}{4}(-1-1) = \frac{\pi}{2}$$

Combining with (1) we obtain the flux:

$$\iint_{\mathcal{S}} \mathbf{F}\cdot d\mathbf{S} = \frac{\pi}{2}$$

35. Let $\varphi(x, y) = x + \frac{x}{x^2+y^2}$. The vector field $\mathbf{F} = \nabla\varphi$ (Figure 5) provides a model in the plane of the velocity field of an incompressible, irrotational fluid flowing past a cylindrical obstacle (in this case, the obstacle is the unit circle $x^2 + y^2 = 1$).

(a) Verify that $\mathbf{F}$ is irrotational [by definition, $\mathbf{F}$ is irrotational if $\operatorname{curl}(\mathbf{F}) = \mathbf{0}$].

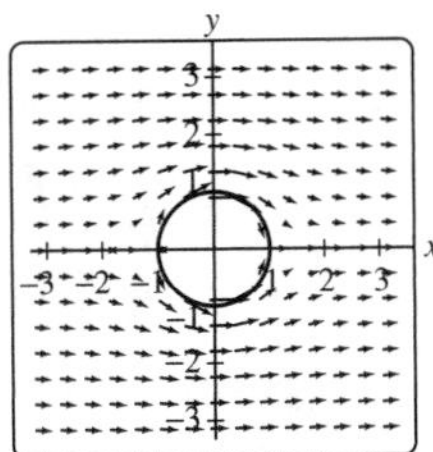

FIGURE 5 The vector field $\nabla\varphi$ for $\varphi(x, y) = x + \frac{x}{x^2+y^2}$.

(b) Verify that $\mathbf{F}$ is tangent to the unit circle at each point along the unit circle except $(1, 0)$ and $(-1, 0)$ (where $\mathbf{F} = \mathbf{0}$).

(c) What is the circulation of $\mathbf{F}$ around the unit circle?

(d) Calculate the line integral of $\mathbf{F}$ along the upper and lower halves of the unit circle separately.

SOLUTION

(a) In Exercise 8, we proved the identity $\operatorname{curl}(\nabla\varphi) = \mathbf{0}$. Since $\mathbf{F}$ is a gradient vector field, it is irrotational; that is, $\operatorname{curl}(\mathbf{F}) = \mathbf{0}$ for $(x, y) \neq (0, 0)$, where $\mathbf{F}$ is defined.

(b) We compute $\mathbf{F}$ explicitly:

$$\mathbf{F} = \nabla\varphi = \left\langle \frac{\partial\varphi}{\partial x}, \frac{\partial\varphi}{\partial y}\right\rangle = \left\langle 1 + \frac{y^2 - x^2}{(x^2+y^2)^2}, -\frac{2xy}{(x^2+y^2)^2}\right\rangle$$

Now, using $x = \cos t$ and $y = \sin t$ as a parametrization of the circle, we see that

$$\mathbf{F} = \left\langle 1 + \sin^2 t - \cos^2 t, -2\cos t\sin t\right\rangle = \left\langle 2\sin^2 t, -2\cos t\sin t\right\rangle,$$

and so

$$\mathbf{F} = 2\sin t\,\langle \sin t, -\cos t\rangle = 2\sin t\,\langle y, -x\rangle,$$

which is clearly perpendicular to the radial vector $\langle x, y\rangle$ for the circle.

(c) We use our expression of $\mathbf{F}$ from Part (b):

$$\mathbf{F} = \nabla\varphi = \left\langle 1 + \frac{y^2 - x^2}{(x^2+y^2)^2}, -\frac{2xy}{(x^2+y^2)^2}\right\rangle$$

Now, $\mathbf{F}$ is not defined at the origin and therefore we cannot use Green's Theorem to compute the line integral along the unit circle. We thus compute the integral directly, using the parametrization

$$\mathbf{c}(t) = (\cos t, \sin t), \quad 0 \leq t \leq 2\pi.$$

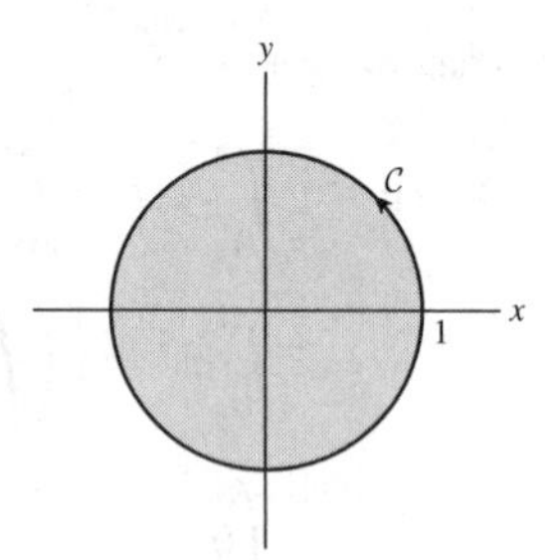

Then,

$$\mathbf{F}(\mathbf{c}(t)) \cdot c'(t) = \left\langle 1 + \frac{\sin^2 t - \cos^2 t}{(\cos^2 t + \sin^2 t)^2}, -\frac{2\cos t \sin t}{(\cos^2 t + \sin^2 t)^2} \right\rangle \cdot \langle -\sin t, \cos t \rangle$$

$$= \left\langle 1 + \sin^2 t - \cos^2 t, -2\cos t, \sin t \right\rangle \cdot \langle -\sin t, \cos t \rangle = \left\langle 2\sin^2 t, -2\cos t \sin t \right\rangle \cdot \langle -\sin t, \cos t \rangle$$

$$= -2\sin^3 t - 2\cos^2 t \sin t = -2\sin t(\sin^2 t + \cos^2 t) = -2\sin t$$

Hence,

$$\int_C \mathbf{F} \cdot d\mathbf{s} = \int_0^{2\pi} -2\sin t \, dt = 0$$

(d) We denote by $\mathcal{C}_1$ and $\mathcal{C}_2$ the upper and lower halves of the unit circle. Using part (c) we have

$$\int_{\mathcal{C}_1} \mathbf{F} \cdot d\mathbf{s} + \int_{\mathcal{C}_2} \mathbf{F} \cdot d\mathbf{s} = 0 \quad \Rightarrow \quad \int_{\mathcal{C}_2} \mathbf{F} \cdot d\mathbf{s} = -\int_{\mathcal{C}_1} \mathbf{F} \cdot d\mathbf{s} \tag{1}$$

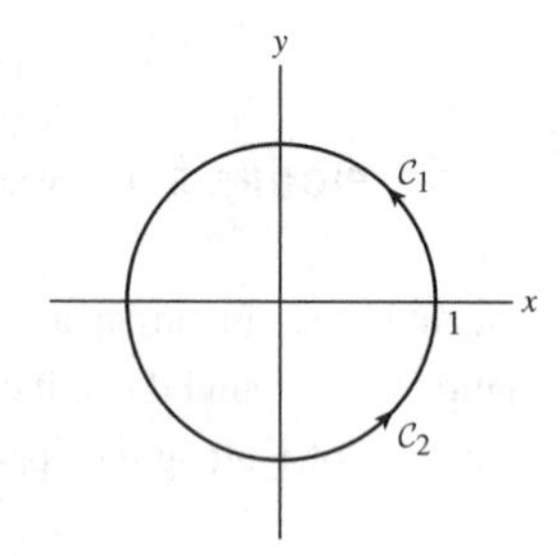

To compute the circulation along $\mathcal{C}_1$, we compute the integral as in part (c), only that the limits of integration are now $t = 0$ and $t = \pi$. Using the computations in part (c) we obtain

$$\int_{\mathcal{C}_1} \mathbf{F} \cdot d\mathbf{s} = \int_0^{\pi} -2\sin^2 t \, dt = -4$$

Therefore, by (1),

$$\int_{\mathcal{C}_2} \mathbf{F} \cdot d\mathbf{s} = 4.$$

37. In Section 18.1, we showed that if $\mathcal{C}$ is a simple closed curve, oriented counterclockwise, then the line integral is

$$\text{Area enclosed by } \mathcal{C} = \frac{1}{2} \oint_{\mathcal{C}} x\,dy - y\,dx \qquad \boxed{1}$$

Suppose that $\mathcal{C}$ is a path from P to Q that is not closed but has the property that every line through the origin intersects $\mathcal{C}$ in at most one point, as in Figure 7. Let $\mathcal{R}$ be the region enclosed by $\mathcal{C}$ and the two radial segments joining P and Q to the origin. Show that the line integral in Eq. (1) is equal to the area of $\mathcal{R}$. *Hint:* Show that the line integral of $\mathbf{F} = \langle -y, x \rangle$ along the two radial segments is zero and apply Green's Theorem.

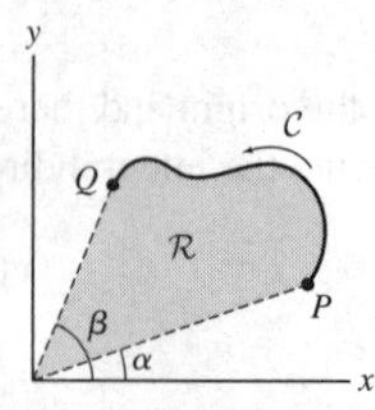

FIGURE 7

SOLUTION

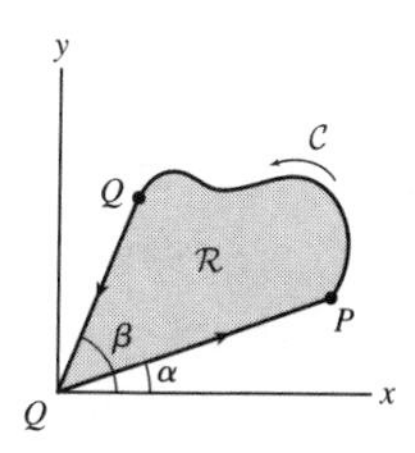

Let $\mathbf{F} = \langle -y, x\rangle$. Then $P = -y$ and $Q = x$, and $\frac{\partial Q}{\partial x} - \frac{\partial P}{\partial y} = 2$. By Green's Theorem, we have

$$\int_{\mathcal{C}} -y\,dx + x\,dy + \int_{\overline{QO}} -y\,dx + x\,dy + \int_{\overline{OP}} -y\,dx + x\,dy = \iint_{\mathcal{R}} 2\,dA = 2\iint_{\mathcal{R}} dA$$

Denoting by A the area of the region $\mathcal{R}$, we obtain

$$A = \frac{1}{2}\int_{\mathcal{C}} -y\,dx + x\,dy + \frac{1}{2}\int_{\overline{QO}} -y\,dx + x\,dy + \frac{1}{2}\int_{\overline{OP}} -y\,dx + x\,dy \tag{1}$$

We parametrize the two segments by

$$\begin{aligned} \overline{QO} &: \mathbf{c}(t) = (t, t\tan\beta) \\ \overline{OP} &: \mathbf{d}(t) = (t, t\tan\alpha) \end{aligned} \quad \Rightarrow \quad \begin{aligned} \mathbf{c}'(t) &= \langle 1, \tan\beta\rangle \\ \mathbf{d}'(t) &= \langle 1, \tan\alpha\rangle \end{aligned}$$

Then,

$$\mathbf{F}(\mathbf{c}(t)) \cdot \mathbf{c}'(t) = \langle -t\tan\beta, t\rangle \cdot \langle 1, \tan\beta\rangle = -t\tan\beta + t\tan\beta = 0$$
$$\mathbf{F}(\mathbf{d}(t)) \cdot \mathbf{d}'(t) = \langle -t\tan\alpha, t\rangle \cdot \langle 1, \tan\alpha\rangle = -t\tan\alpha + t\tan\alpha = 0$$

Therefore,

$$\int_{\overline{QO}} \mathbf{F}\cdot d\mathbf{s} = \int_{\overline{OP}} \mathbf{F}\cdot d\mathbf{s} = 0.$$

Combining with (1) gives

$$A = \frac{1}{2}\int_{\mathcal{C}} -y\,dx + x\,dy.$$

39. Prove the following generalization of Eq. (1). Let $\mathcal{C}$ be a simple closed curve in the plane (Figure 8)

$$\mathcal{S}: \quad ax + by + cz + d = 0$$

Then the area of the region R enclosed by $\mathcal{C}$ is equal to

$$\frac{1}{2\|\mathbf{n}\|}\oint_{\mathcal{C}} (bz - cy)\,dx + (cx - az)\,dy + (ay - bx)\,dz$$

where $\mathbf{n} = \langle a, b, c\rangle$ is the normal to $\mathcal{S}$ and $\mathcal{C}$ is oriented as the boundary of $\mathcal{R}$ (relative to the normal vector $\mathbf{n}$). *Hint:* Apply Stokes' Theorem to $\mathbf{F} = \langle bz - cy, cx - az, ay - bx\rangle$.

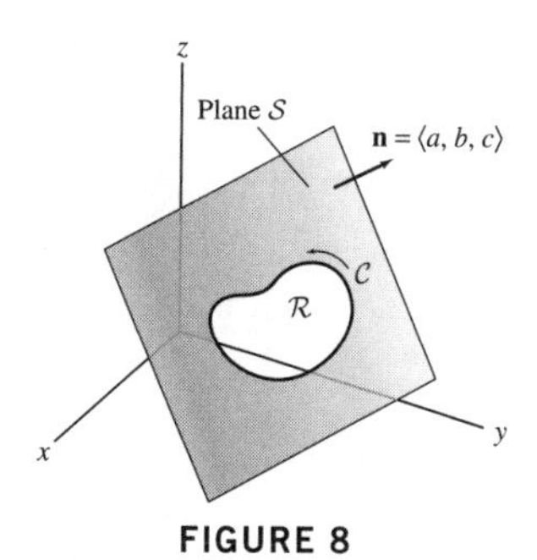

FIGURE 8

SOLUTION By Stokes' Theorem,

$$\iint_{\mathcal{S}} \text{curl}(\mathbf{F}) \cdot d\mathbf{S} = \iint_{\mathcal{S}} (\text{curl}(\mathbf{F}) \cdot \mathbf{e}_n)\,dS = \int_{\mathcal{C}} \mathbf{F}\cdot d\mathbf{s} \tag{1}$$

We compute the curl of $\mathbf{F}$:

$$\text{curl}(\mathbf{F}) = \begin{vmatrix} \mathbf{i} & \mathbf{j} & \mathbf{k} \\ \frac{\partial}{\partial x} & \frac{\partial}{\partial y} & \frac{\partial}{\partial z} \\ bz - cy & cx - az & ay - bx \end{vmatrix} = 2a\mathbf{i} + 2b\mathbf{j} + 2c\mathbf{k} = 2\langle a, b, c\rangle$$

The unit normal to the plane $ax + by + cz + d = 0$ is

$$\mathbf{e}_n = \frac{\langle a, b, c\rangle}{\sqrt{a^2 + b^2 + c^2}}$$

Therefore,

$$\begin{aligned} \text{curl}(\mathbf{F}) \cdot \mathbf{e}_n &= 2\langle a, b, c\rangle \cdot \frac{1}{\sqrt{a^2 + b^2 + c^2}} \langle a, b, c\rangle \\ &= \frac{2}{\sqrt{a^2 + b^2 + c^2}} (a^2 + b^2 + c^2) = 2\sqrt{a^2 + b^2 + c^2} \end{aligned}$$

Hence,

$$\iint_{\mathcal{S}} \text{curl}(\mathbf{F}) \cdot d\mathbf{S} = \iint_{\mathcal{S}} \text{curl}(\mathbf{F}) \cdot \mathbf{e}_n \, dS = \iint_{\mathcal{S}} 2\sqrt{a^2 + b^2 + c^2} \, dS = 2\sqrt{a^2 + b^2 + c^2} \iint_{\mathcal{S}} 1 \, dS \tag{2}$$

The sign of $\iint_{\mathcal{S}} 1 \, d\mathbf{S}$ is determined by the orientation of $\mathcal{S}$. Since the area is a positive value, we have

$$\left| \iint_{\mathcal{S}} 1 \, ds \right| = \text{Area}(\mathcal{S})$$

Therefore, (2) gives

$$\left| \iint_{\mathcal{S}} \text{curl}(\mathbf{F}) \cdot d\mathbf{S} \right| = 2\sqrt{a^2 + b^2 + c^2} \, \text{Area}(\mathcal{S})$$

Combining with (1) we obtain

$$2\sqrt{a^2 + b^2 + c^2} \, \text{Area}(\mathcal{S}) = \left| \int_{\mathcal{C}} \mathbf{F} \cdot d\mathbf{s} \right|$$

or

$$\text{Area}(\mathcal{S}) = \frac{1}{2\sqrt{a^2 + b^2 + c^2}} = \frac{1}{2\|\mathbf{n}\|} \cdot \left| \int_{\mathcal{C}} (bz - cy) \, dx + (cx - az) \, dy + (ay - bx) \, dz \right|$$